Lecture Notes in Computer Science 4641

Commenced Publication in 1973
Founding and Former Series Editors:
Gerhard Goos, Juris Hartmanis, and Jan van Leeuwen

Anne-Marie Kermarrec   Luc Bougé
Thierry Priol (Eds.)

Euro-Par 2007
Parallel Processing

13th International Euro-Par Conference
Rennes, France, August 28-31, 2007
Proceedings

 Springer

Volume Editors

Anne-Marie Kermarrec
IRISA/INRIA
Campus de Beaulieu
35042 Rennes Cedex, France
E-mail: Anne-Marie.Kermarrec@irisa.fr

Luc Bougé
IRISA/ENS Cachan
Campus de Beaulieu
35042 Rennes Cedex, France
E-mail: Luc.Bouge@bretagne.ens-cachan.fr

Thierry Priol
IRISA/INRIA
Campus de Beaulieu
35042 Rennes Cedex, France
E-mail: Thierry.Priol@irisa.fr

Library of Congress Control Number: 2007933330

CR Subject Classification (1998): C.1-4, D.1-4, F.1-3, G.1-2, H.2

LNCS Sublibrary: SL 1 – Theoretical Computer Science and General Issues

ISSN   0302-9743
ISBN-10   3-540-74465-7 Springer Berlin Heidelberg New York
ISBN-13   978-3-540-74465-8 Springer Berlin Heidelberg New York

Springer is a part of Springer Science+Business Media

springer.com

© Springer-Verlag Berlin Heidelberg 2007
Printed in Germany

Typesetting: Camera-ready by author, data conversion by Scientific Publishing Services, Chennai, India
Printed on acid-free paper   SPIN: 12112029   06/3180   5 4 3 2 1 0

Preface

Euro-Par is an annual series of international conferences dedicated to the promotion and advancement of all aspects of parallel and distributed computing. The major themes can be divided into the broad categories of hardware, software, algorithms and applications for parallel and distributed computing. The objective of Euro-Par is to provide a forum to promote the development of parallel computing both as an industrial technique and an academic discipline, extending the frontier of both the state of the art and the state of the practice. This is particularly important at a time when parallel and distributed computing is undergoing strong and sustained development and experiencing real industrial take-up. The main audience for, and participants in, Euro-Par are researchers in academic departments, government laboratories and industrial organizations. Euro-Par's objective is to be the primary choice of such professionals for the presentation of new results in their specific fields of expertise. Euro-Par is also interested in applications that demonstrate the effectiveness of the main Euro-Par themes.

Previous Euro-Par conferences took place in Stockholm, Lyon, Passau, Southampton, Toulouse, Munich, Manchester, Paderborn, Klagenfurt, Pisa, Lisbon and Dresden. The next conference will take place in Las Palmas de Gran Canaria (Canary Islands, Spain), organized by colleagues of the University Autonoma of Barcelona (UAB) and the University of Las Palmas de Gran Canaria (ULPGC).

The Euro-Par conference series is traditionally organized in cooperation with the International Federation for Information Processing (IFIP) and the Association for Computer Machinery (ACM). Euro-Par has a permanent Web site where its history and organization are described: http://www.europar.org/.

Euro-Par 2007, the 13th conference in the Euro-Par series, was organized by the Brittany Research Center of the French National Institute for Research in Computer Science and Control (INRIA), in the framework of the IRISA Research Laboratory, a joint lab between CNRS, INRIA, University of Rennes 1, INSA Rennes, and ENS Cachan, Brittany Campus.

CoreGRID is the European Research Network on Foundations, Software Infrastructures and Applications for large-scale distributed, GRID and Peer-to-Peer Technologies, a domain which lies at the heart of the Euro-Par interest. For the first time, the CoreGRID Symposium, a major event for this network of excellence, has been concomitantly organized and co-located with Euro-Par. For its first edition, the CoreGRID Symposium attracted more than 50 submissions from all over Europe. This will also be the occasion for a number of CoreGRID Working Groups to organize their regular meetings.

Also, four prominent workshops were organized with the conference:

GECON 2007, the 4th International Workshop on Grid Economics and Business Models

VHPC 2007, the Workshop on Virtualization in High-Performance Cluster and Grid Computing

HPPC 2007, the Workshop on Highly Parallel Processing on a Chip

UNICORE Summit 2007, the Annual User Group Meeting for the UNICORE environment

This co-location of workshops is intended to be the start of a general trend. Euro-Par is eager to develop and stabilize a set of co-located workshops in the coming years. As in 2006, Springer acknowledges this initiative by publishing the proceedings of these workshops in a special volume of the LNCS series.

Euro-Par 2007 was able to attract three most renowned invited speakers. Their talks highlighted some of the most recent trends in parallel and distributed computing:

Rachid Guerraoui (EPFL, Lausanne, Switzerland): *Sommersby: Transactions are back: but are they the same?*

Steve Hand (University of Cambridge, UK): *Virtualizing the Data Center with Xen*

André Seznec (IRISA, INRIA, Rennes, France): *15 mm × 15 mm: the new frontier of parallel computing*

Compared to the traditional conference format, Euro-Par 2007 tried to shrink the number of topics, and thereby get a larger number of submissions per topic. This was done on purpose, as we felt that it would improve the reliability of the reviewing and the selection process. Fourteen topics were defined and advertised, covering a large variety of aspects of parallel and distributed computing.

Each topic was initially supervised by a committee of four: a global chair, a local chair, and two vice-chairs. If more than 20 papers were submitted, this committee was extended with additional members to keep the reviewing load within reasonable bounds.

The call for papers attracted a total of 333 submissions. For all of the submitted papers, at least three and often four individual review reports were collected. A total of 89 papers were finally accepted for publication, of which 3 received the special honor of being nominated as *Best Papers*. The global acceptance rate was thus below 27 %. The authors of accepted papers come from 27 countries, with the four main contributing countries — USA, France, Spain and Germany, — accounting for more than 60 % of the authors of accepted papers.

Hosting Euro-Par 2007 in Rennes would not have been possible without the support and the help of different institutions and numerous people.

Although we are thankful to many more people, we are particularly grateful to Édith Blin, as she put a huge amount of work in the organization of the conference, always combining efficiency and enthusiasm, smoothing consistently the whole process of organizing the conference.

We would also like to thank Marin Bertier and Yann Busnel for their tremendous help; Marin was in charge of putting together the proceedings and was an extremely reliable helper on many other fronts. Yann Busnel successfully found his way around the submission software and ensured a smooth and efficient submission and review management process.

We are obviously most thankful to the main contributors of the conference, namely, the authors of all the submitted papers, the Program Committee members and the numerous reviewers. Their commitment greatly contributed to the success of the conference. We are also extremely grateful to our three invited speakers, Rachid Guerraoui, Steven Hand and André Seznec, for accepting our invitation. We would like to thank them for their most valuable contribution to the conference.

We also would like to address our special thanks to the Euro-Par Steering Committee and especially to Christian Lengauer for his support all along the preparation of the conference. We also strongly benefited from the help and the experience of the former organization team of Euro-Par 2006 of Dresden. José Cuhna was extremely helpful and we thank him for his support.

Euro-Par 2007 was hosted on the University Campus and we would like to thank the Department of Computer science (IFSIC) of the University of Rennes 1 for the support and infrastructure. We gratefully acknowledge the great financial and organizational support of INRIA and IRISA as well as the support of our institutional sponsors the University of Rennes 1, the Regional Council, Rennes Métropole, the local council, the Métivier Foundation, and the *Pôle de competitivité Images & Réseaux* and the city of Rennes.

Finally, we are grateful to Springer for agreeing to publish the proceedings. It has been a great pleasure to organize Euro-par 2007 and we hope the header will enjoy these proceedings.

June 2007   Anne-Marie Kermarrec
                                                                    Luc Bougé
                                                                  Thierry Priol

Organization

Euro-Par Steering Committee

Chair

Christian Lengauer   University of Passau, Germany

Vice-Chair

Luc Bougé   ENS Cachan, France

European Representatives

José Cunha	New University of Lisbon, Portugal
Marco Danelutto	University of Pisa, Italy
Rainer Feldmann	University of Paderborn, Germany
Christos Kaklamanis	Computer Technology Institute, Greece
Paul Kelly	Imperial College, UK
Harald Kosch	University of Passau, Germany
Thomas Ludwig	University of Heidelberg, Germany
Emilio Luque	Universitat Autònoma de Barcelona, Spain
Luc Moreau	University of Southampton, UK
Wolfgang E. Nagel	Technische Universität Dresden, Germany
Rizos Sakellariou	University of Manchester, UK

Non-European Representatives

Jack Dongarra	University of Tennessee at Knoxville, USA
Shinji Tomita	Kyoto University, Japan

Honorary Members

Ron Perrott	Queen's University Belfast, UK
Karl Dieter Reinartz	University of Erlangen-Nuremberg, Germany

Observers

Anne-Marie Kermarrec	IRISA/INRIA, Rennes, France
Domingo Benitez	University of Las Palmas, Gran Canaria, Spain

Euro-Par 2007 Local Organization

Euro-Par 2007 was organized by the IRISA/INRIA research laboratory in Rennes.

Conference Chairs

Anne-Marie Kermarrec   IRISA/INRIA
Luc Bougé   IRISA/ENS Cachan
Thierry Priol   IRISA/INRIA

General Organization

Édith Blin   IRISA/INRIA

Technical Support

Étienne Rivière, Yann Busnel

Publicity

Gabriel Antoniu

Proceedings

Marin Bertier

Secretariat

Patricia Houée-Barbedet, Violaine Tygréat

CoreGRID Coordination

Païvi Palosaari, Olivia Vasselin

Euro-Par 2006 Program Committee

Topic 1: Support Tools and Environments
Chair   Liviu Iftode (Rutgers University, Piscataway, USA)
Local Chair   Christine Morin (IRISA, INRIA, Rennes, France)
Vice-Chairs   Marios Dikaiakos (University of Cyprus, Nicosia, Cyprus)
                Erich Focht (NEC HPC Europe, Stuttgart, Germany)

Topic 2: Performance Prediction and Evaluation

Chair   Wolfgang Nagel (Zentrum für Informationsdienste und
                 Hochleistungsrechnen (ZIH), Dresden, Germany)
Local Chair   Bruno Gaujal (LIG, INRIA, Grenoble, France)
Vice-Chairs   Tugrul Dayar (Bilkent University, Ankara, Turkey)
                 Nihal Pekergin (PRISM, University of Versailles, France)

Topic 3: Scheduling and Load-Balancing

Chair   Henri Casanova (University of Hawai'i at Manoa, Honolulu,
                 USA)
Local Chair   Olivier Beaumont (LaBRI, ENSEIRB, Bordeaux, France)
Vice-Chairs   Uwe Schwiegelshohn (Institut for Roboterforschung, University
                 of Dortmund, Germany)
                 Marek Tudruj (Intitute of Computer Science, Polish Academy of
                 Science, Warsaw, Poland)

Topic 4: High-Performance Architectures and Compilers

Chair   Michael O'Boyle (Institute for Computing Systems
                 Architecture, Edinburgh, UK)
Local Chair   François Bodin (IRISA, University of Rennes, France)
Vice-Chairs   Jose Gonzalez (Intel Barcelona Research Center, Spain)
                 Lucian Vintan (University of Sibiu, Romania)

Topic 5: Parallel and Distributed Databases

Chair   Marta Patiño-Martinez (Universidad Politecnica de Madrid,
                 Spain)
Local Chair   Genoveva Vargas-Solar (LIG, CNRS, Grenoble, France)
Vice-Chairs   Elena Baralis (Politecnico di Torino, Italy)
                 Bettina Kemme (McGill University, Montreal, Canada)

Topic 6: Grid and Cluster Computing

Chair   Rosa M. Badia (Barcelona Supercomputing Center, Universitat
                 Politecnica de Catalunya, Spain)
Local Chair   Christian Pérez (IRISA, INRIA, Rennes, France)
Vice-Chairs   Artur Andrzejak (Zuse-Institute, Berlin, Germany)
                 Alvaro Arenas (CCLRC Rutherford Appleton Laboratory,
                 Chilton, Oxfordshire, UK)
Members   Franck Cappello (INRIA Futurs, Saclay, France)
                 Marco Danelutto (Univesity of Pisa, Italy)
                 Ramin Yahyapour (University of Dortmund, Germany)

Topic 7: Peer-to-Peer Computing

Chair   Alberto Montresor (University of Trento, Povo, Italy)
Local Chair   Fabrice Le Fessant (INRIA, Saclay, France)
Vice-Chairs   Dick Epema (Delft University of Technology, The Netherlands)
                 Spyros Voulgaris (ETH Zurich, Switzerland)

Topic 8: Distributed Systems and Algorithms

Chair   Luís Rodrigues (University of Lisbon, Portugal)
Local Chair   Achour Mostefaoui (IRISA, University of Rennes, France)
Vice-Chairs   Christof Fetzer (Dresden University of Technology, Germany)
                  Philippas Tsigas (Chalmers University of Technology, Göteborg,
                      Sweden)
Members   Filipe Araújo (University of Coimbra, Portugal)
                  Anders Gidenstam (Max-Planck-Institut für
                      Informatik, Saarbrücken, Germany)
                  Antonino Virgillito (Università di Roma, Italy)

Topic 9: Parallel and Distributed Programming

Chair   Luc Moreau (University of Southampton, UK)
Local Chair   Emmanuel Jeannot (LORIA, INRIA, Nancy, France)
Vice-Chairs   George Bosilca (University of Tennessee, Knoxville, USA)
                  Antonio J. Plaza (University of Extremadura, Càceres, Spain)
Members   Alexey Lastovetsky (University College, Dublin, Ireland)
                  Simon Miles (University of Southampton, UK)
                  Juri Papay (Electronics Computer Science, OMII, UK)

Topic 10: Parallel Numerical Algorithms

Chair   Iain Duff (CCLRC Rutherford Appleton Lab, UK)
Local Chair   Michel Daydé (IRIT, ENSEEIHT, Toulouse, France)
Vice-Chairs   Matthias Bollhoefer (Technical University of Braunschweig,
                      Germany)
                  Anne Trefethen (University of Oxford, UK)

Topic 11: Distributed and High-Performance Multimedia

Chair   Harald Kosch (University of Passau, Germany)
Local Chair   Laurent Amsaleg (IRISA, CNRS, Rennes, France)
Vice-Chairs   Eric Pauwels (CWI, Amsterdam, The Netherlands)
                  Björn Jónsson (Reykjavik University, Iceland)

Topic 12: Theory and Algorithms for Parallel Computation

Chair   Nir Shavit (Tel-Aviv University, Israel)
Local Chair   Nicolas Schabanel (UMI CNRS - Universidad de Chile, Santiago
                      de Chile, Chile)
Vice-Chairs   Pascal Felber (University of Neuchatel, Switzerland)
                  Christos Kaklamanis (University of Patras, Computer
                      Technology Institute, Greece)

Topic 13: High-Performance Networks

Chair   Thilo Kielmann (Vrije Universiteit, Amsterdam,
                      The Netherlands)

Local Chair   Pascale Primet (ENS Lyon, INRIA, École centrale de Lyon,
                France)
Vice-Chairs   Tomohiro Kudoh (AIST, Keio University, Japan)
                Bruce Lowekamp (College of William and Mary, Williamsburg,
                USA)

Topic 14: Mobile and Ubiquitous Computing

Chair   Nuno Preguiça (Universidade Nova de Lisboa, Portugal)
Local Chair   Éric Fleury (INSA de Lyon, INRIA, France)
Vice-Chairs   Holger Karl (University of Paderborn, Germany)
                Gerd Kortuem (Lancaster University, UK)
Members   Carlos Baquero (Universidade do Minho, Portugal)
                Marcelo Dias de Amorim (LIP6, University Paris 6, France)
                Pedro Marron (University of Stuttgart, Germany)
                Vasughi Sundramoorthy (Lancaster University,
                UK)
                Andreas Willig (Technical University of Berlin, Germany)
                Artur Ziviani (National Laboratory for Scientific Computing,
                Petrópolis, Brazil)

Euro-Par 2006 Referees

Jaume Abella	Rosa M. Badia
Manuel E. Acacio	Arati Baliga
Jean-Thomas Acquaviva	Euripides Bampis
Henoc Agbota	Carlos Baquero
German-Othon Aguilar-Tapia	Ranieri Baraglia
Toufik Ahmed	Elena Baralis
Marco Aldinucci	Manuel Barbosa
Alex Aleta	Olivier Beaumont
Guillaume Alléon	Khalid Belhajjame
Paulo Sérgio Almeida	Fehmi Ben Abdesslem
José Almeida	Richard Bennett
Patrick Amestoy	Jalel BenOthman
Laurent Amsaleg	Gregorio Bernabe
Matthias Andree	Carlo Bertolli
Artur Andrzejak	Marco Biazzini
Philippe d'Anfray	Urs Bischoff
Gabriel Antoniu	François Bodin
Daniele Apiletti	Aniruddha Bohra
Filipe Araujo	Matthias Bollhoefer
Alvaro Arenas	Silvia Bonomi
Stefano Arteconi	Thomas Bopp
Cevdet Aykanat	George Bosilca
Benjamin Aziz	Vincent Boudet

Aurélien Bouteiller
Céline Boutros Saab
Andrey Brito
Giulia Bruno

Qiong Cai
Franck Cappello
Ioannis Caragiannis
Damiano Carra
Nuno Carvalho
Henri Casanova
Roberto Cascella
António Casimiro
Hind Castel
Christophe Cérin
Tania Cerquitelli
Ali Cevahir
Yong Chen
Silvia Chiusano
Pierre-Nicolas Clauss
Thomas Claveirole
Josep M. Codina
Massimo Coppola
David Coquil
Julita Corbalan
Olivier Coulaud
Bruno Crispo
Victor Cuevas-Vicenttin

Georges DaCosta
Marco Danelutto
Christian Dannewitz
Anwitaman Datta
Tugrul Dayar
Michel Daydé
Yves Denneulin
Frédéric Desprez
Marcelo Dias de Amorim
Marios Dikaiakos
Mario Döller
Henrique Joao Domingos
Maciej Drozdowski
Dominique Dudkowski
Iain Duff
Pierre-François Dutot

Christos Efstratiou
Ahmed Eleuch

Nahid Emad
Dick Epema
Lionel Eyraud-Dubois

Weijian Fang
Tobias Farrell
Pascal Felber
Martin Feller
Manel Fernandez
Eric Fleury
Andreas Florides
Erich Focht
Victor Fonte
Jean-Michel Fourneau
Bjoern Franke
Stefan Freitag

Efstratios Gallopoulos
Anurag Garg
Paolo Garza
Matthias Gauger
Bruno Gaujal
Enric Gibert
Anders Gidenstam
Luc Giraud
Harald Gjermundrod
Antonio Gomes
Jose Gonzalez
Manuel Gonzalo
Anastasios Gounaris
Christian Grimme
Abdou Guermouche
Romaric Guillier
Ronan Guivarch
Jens Gustedt

Claire Hanen
Thomas Herault
Mikael Högqvist
Günther Hölbling
Guillaume Huard
Eduardo Huedo
Felix Hupfeld
Emmanuel Hyon

Liviu Iftode
Yiannis Ioannou

Katia Jaffrès-Runser
Mathieu Jan
Emmanuel Jeannot
Emmanuel Jeanvoine
Yvon Jégou
Mark Jelasity
Zbigniew Jerzak
Ernesto Jimenez
Ricardo Jimenez-Peris
Björn Jónsson

Christos Kaklamanis
Panagiotis Kanellopoulos
Holger Karl
Bettina Kemme
Thilo Kielmann
Ralf Klasing
Thomas Martin Knoll
Andreas Knüpfer
Björn Kolbeck
Derrick Kondo
Charalampos Konstantopoulos
Ibrahim Korpeoglu
Gerd Kortuem
Harald Kosch
Tomohiro Kudoh

Andreas Lachenmann
Amit Lakhani
Margit Lang
Tobias Langhammer
Alexey Lastovetsky
Fernando Latorre
Fabrice Le Fessant
Benedicte Le Grand
Adrien Lèbre
Jose Legatheaux Martins
Franck Legendre
Arnaud Legrand
Pierre Lemarinier
Joachim Lepping
Hermann Lichte
Yi Lin
Antonia Lopes
Rui Jorge Lopes
Nuno Lopes

Bruce Lowekamp
Thomas Ludwig

Carles Madriles
Grigorios Magklis
Loris Marchal
Pedro Marcuello
Dan Marinescu
Pedro Jose Marron
Brian Matthews
Pedro Medeiros
Jean-François Méhaut
Hein Meling
Alessia Milani
Simon Miles
Daniel Minder
Marine Minier
José Mocito
Lynda Mokdad
Ruben S. Montero
Alberto Montresor
Luc Moreau
Diana-Guadalupe Moreno-Garcia
Ricardo Morla
Monika Moser
Achour Mostefaoui
Luca Mottola
Grégory Mounié
Matthias Mueller
Ralph Mueller-Pfefferkorn
Anelise Munaretto

Wolfgang E. Nagel
Syed Naqvi

Michael O'Boyle
Sebastian Obermeier
Paulo Oliveira
Salvatore Orlando

Antoniadis Panayotis
Evi Papaioannou
Alexander Papaspyrou
Juri Papay
Marta Patiño-Martinez
Eric Pauwels
Nihal Pekergin
Christian Pérez

Francisco Perez-Sorrosal
Florence Perronnin
Kathrin Peter
Serge Petiton
Sylvain Peyronnet
Gert Pfeifer
Guillaume Pierre
José-Alejandro Pineiro
Jörg Platte
Antonio Plaza
Nuno Preguiça
Pascale Primet

Leonardo Querzoni
Martin Quinson

Bruno Raffin
Ravi Reddy
Olivier Richard
Torvald Riegel
Thomas Röblitz
Jean-Louis Roch
Luis Rodrigues
Thomas Ropars
Liliana Rosa
Enrico Rukzio
Vladimir Rychkov

Jorge Salas
Stefano Salvini
Jesus Sanchez
Uluc Saranli
Olga Saukh
Robert Sauter
Yanos Sazeides
Nicolas Schabanel
Stefan Schamberger
Florian Schintke
Lars Schley
Michael Schoettner
Thorsten Schütt
Uwe Schwiegelshohn
Sirio Scipioni
Ali Aydin Selcuk
Damian Serrano
Michael Sessinghaus
Nir Shavit

Gheorghe Silaghi
Peter Sloot
Steve Smaldone
Sebastien Soudan
Stella Stars
Jan Stender
Achim Streit
Jaspal Subhlok
Vasughi Sundramoorthy
Martin Süßkraut
Alan Sussman
Frédéric Suter

Hanh Tan
Cristian Tapus
Zahir Tari
Andrei Tchernykh
Sebastien Tixeuil
Nicola Tonellotto
Corinne Touati
Anne Trefethen
Philippas Tsigas
George Tsouloupas
Sara Tucci Piergiovanni
Marek Tudruj
Ata Turk

Bora Ucar

Karthik Vaidyanathan
Stefan Valentin
Geoffroy Vallée
Genoveva Vargas-Solar
Xavier Vera
Aline Viana
João Paulo Vilela
Jean-Marc Vincent
Lucian N. Vintan
Antonino Virgillito
Spyros Voulgaris

Frédéric Wagner
Bin Wang
Ute Wappler
Philipp Wieder
Christian Wietfeld
Andreas Willig
Dereje Woldegabrael

Polychronis Xekalakis
Wei Xing

Ramin Yahyapour
Erica Yang

Marcia Zangrilli
José Luis Zechinelli-Martini
Wolfgang Ziegler
Artur Ziviani

Table of Contents

Topic 3: Scheduling and Load-Balancing

Topic 4: High-Performance Architectures and Compilers

Topic 5: Parallel and Distributed Databases

Topic 6: Grid and Cluster Computing

Topic 7: Peer-to-Peer Computing

Topic 8: Distributed Systems and Algorithms

Topic 9: Parallel and Distributed Programming

Topic 10: Parallel Numerical Algorithms

Topic 11: Distributed and High-Performance Multimedia

Topic 12: Theory and Algorithms for Parallel Computation

Topic 13: High-Performance Networks

Topic 14: Mobile and Ubiquitous Computing

Topic 1
Support Tools and Environments

Liviu Iftode, Christine Morin, Marios Dikaiakos, and Erich Focht

Topic Chairs

Despite an impressive body of research, parallel and distributed computing remains a complex task prone to subtle software bugs, which can affect both the correctness and the performance of the computation. The increasing demand to distribute computing over large-scale distributed platforms, such as grids and large clusters, overlaps with an increasing pressure to make computing more dependable. To address these challenges, the parallel and distributed computing community continuously requires better tools and environments to design, program, debug, test, tune, and monitor programs. This topic aims to bring together tool designers, developers, and users to share their concerns, ideas, solutions, and products, covering a wide range of platforms.

This year, eighteen submitted papers were reviewed in Topic 1 and eight papers were accepted. The accepted papers can be grouped in three categories, reflecting the diversity of tools needed for parallel and distributed computing.

Three papers relate to profiling tools. The paper "A Profiling Tool for Detecting Cache Critical Data Structures" describes dprof, a tool to visualize cache misses in a multithreaded program. The dprof tool is able to correlate cache miss performance with the program code, providing levels of granularity related to the whole data structures, function calls or individual variables. The paper "On Using Incremental profiling for the Performance Analysis of Shared Memory Parallel Applications" presents an improvement of the ompP profiling tool for OpenMP adding a temporal dimension to profiling data. The paper "Automatic Structure Extraction from MPI Applications Tracefiles" addresses the important issue of analysing huge trace files obtained from large message-passing parallel applications executed on hundreds or thousands of processors in supercomputers or grids. It proposes an automatic reduction technique of trace files for extracting useful patterns.

Two papers describe tools for automatic tuning of parallel applications. The paper "Automatic Generation of Dynamic Tuning Techniques" presents in the framework of the MATE environment a performance problem specification language, which can be used to automatically generate a so-called tunlet to tune the running application for better performance. The paper "Fine Tuning Algorithmic Skeletons" describes a tool for developers to identify the performance bottlenecks through monitoring execution in a skeleton-driven parallel processing runtime (Calcium).

Three papers present scheduling frameworks. The paper "A Scheduling Toolkit for Multiprocessortask Programming with Dependencies" addresses the issue of scheduling M-task applications that exploit data and task parallelism at

A.-M. Kermarrec, L. Bougé, and T. Priol (Eds.): Euro-Par 2007, LNCS 4641, pp. 1–2, 2007.

the same time. The STK scheduling toolkit aims at closing the gap between the specification and the execution of M-task programs by automatically determining valid schedules for parallel target platforms. The paper "Building Portable Thread Schedulers for Hierarchical Multiprocessors: the BubbleSched Framework" presents the design and implementation the BubbleSched framework, whose goal is to ease the development and evaluation of customized user level thread schedulers for large shared-memory NUMA machines. The paper "Makefile::Parallel Dependency Specification Language" describes a tool to schedule and monitor the execution of parallel applications on clusters, which is based on a Makefile-like language to specify dependencies between processes. We would like to thank the panel of reviewers for their precious time and effort in the selection process.

Automatic Structure Extraction from MPI Applications Tracefiles*

Marc Casas, Rosa M. Badia, and Jesús Labarta

Barcelona Supercomputing Center (BSC),
Technical University of Catalonia (UPC),
Campus Nord, Modul C6, Jordi Girona, 1-3, 08034 Barcelona

Abstract. The process of obtaining useful message passing applications tracefiles for performance analysis in supercomputers is a large and tedious task. When using hundreds or thousands of processors, the tracefile size can grow up to 10 or 20 GB. It is clear that analyzing or even storing these large traces is a problem. The methodology we have developed and implemented performs an automatic analysis that can be applied to huge tracefiles, which obtains its internal structure and selects meaningful parts of the tracefile. The paper presents the methodology and results we have obtained from real applications.

1   Motivation and Goal

In the recent years, parallel platforms have amazingly increased in performance and in number of nodes and processors. Thus, the study of the execution of applications in these platforms has become a hard and tedious work. A complete timestamped sequence of events of an application, that is, a tracefile of the whole application, results in a huge file (10-20 GB). It is impossible to handle this amount of data with tools like Paraver [1]. Also, often, some parts of the trace are perturbed, and the analysis of these parts can be misleading. A third problem is the identification of the most representative regions of the tracefile.

To reduce tracefiles sizes, the process of application tracing must be carefully controlled, enabling the tracing in the interesting parts of the application and disabling otherwise. The number of events of the tracefile (hardware counters, instrumented routines, etc...) must be limited. This process is tedious and large and requires knowledge on the source code of the application.

For these reasons, several authors [8,9] believe that the development and utilization of trace based techniques is not useful. However, techniques based on tracefiles allow a very detailed study of the variations on space (set of processes) and time that could affect notably the performance of the application. Therefore, there is a need for developing techniques that allow to handle large event traces.

The goal of our approach is to start from very large tracefiles of the whole application, allowing simple tracing methodologies, and then analyzing them automatically. The underlying philosophy is to use resources that are generally

---

* Funded by project TIN2004-07739-CO2-01 and by a FPI grant from spanish gov.

A.-M. Kermarrec, L. Bougé, and T. Priol (Eds.): Euro-Par 2007, LNCS 4641, pp. 3–12, 2007.

available (Disk, CPU, ...) in order to avoid spending an expensive resource: analyst time. The tool we have implemented will, first, warn the analyst about those parts of the trace perturbed by an external factor not related to the application or to the machine itself. Second, the tool will give a description of the internal structure of the application and will identify and extract the most relevant parts of the trace.

There are other approaches to either avoid or handle large event traces. KO-JAK [2] is a tool for automatic detection of performance bottlenecks and inefficient behavior. Our methodology could be applied before KOJAK to reduce the size of the tracefile. VAMPIR Next Generation tool (VNG) [3,4] consists of two major components: A parallel analysis server and a visualization client, each of them executed on a different platform. An important VNG feature is the utilization of the data structure Complete Call Graph (CCG). It holds the full event stream including time information in a tree. It is also possible to compress the CCG into a compressed Call Graph (cCCG) in order to achieve a compressed representation of trace data. Compression errors can be maintained in a given range. Finally, the main goal of VNG is to make huge event traces accessible for interactive program analysis. Therefore, the VNG approach is different from ours since it does not perform an automatic analysis of the internal structure of the event traces. Related to VNG, there is another tool called DeWiz [5]. It is based, as VNG, on the event graph model. Two important characteristic of DeWiz are modularity, which enables it to be executed in distributed computing infrastructures, and automatic analysis, which enables it to detect significative information on the event graph. However, DeWiz is unable to find event based structure since it works using graph based methods. In that context, our work satisfies the need for an automatic performance analysis based on structural properties of the application. Furthermore, these structural properties of the application are event based and, for that reason, they have clear physical meaning. Other previous work in [6] presents a proposal for dynamic periodicity detection of iterations in parallel applications.

The paper is organized as follows: First, an explanation of the methodology we have developed and implemented is presented in Section 2. Next, a presentation of the results we have obtained using this methodology is described in section 3. Finally, conclusions and future work are shown in Section 4.

2   Methodology

The starting point is a Paraver tracefile generated with OMPItrace package [11]. This tracefile consists of a complete timestamped sequence of events of the whole execution of an application. The first problem we consider is the detection of perturbed regions of the tracefile. The second phase consists in a search for the internal structure of the trace, based on periodicities. In those two phases, we will use signals to characterize properties of the tracefile. Both phases will be handled using techniques of signal processing. Mainly, we will use non-linear filtering in both phases and spectral analysis in the second one.

2.1   Clean-Up

In this phase, our tool performs an analysis oriented to the identification of the perturbed regions of the tracefile. By perturbed regions we mean those regions with distorted relative timing behavior of the application being analyzed. In [7] is shown a detailed discussion about these distortions. Furthermore, all these perturbations have a common characteristic: They are neither caused by the application nor the architecture. They are caused by external factors such as tracing packages, unknown system activity, etc. Different phenomena or metrics can be identified as being the cause of a significant perturbation of the program behavior. An example of those phenomena is flushing, which is caused by the fact that tracing packages keep individual records in a buffer in memory during the tracing process. The problem is that when the buffer is full, these records will have to be flushed to disk. This flushing will take a significative time, will affect the execution and the statistics derived from the tracefile. Also, the flushing does not appear simultaneously in all the processes although it is typical that the flush of the different processes occur in bursts.

Identifying perturbed regions. The flushing phenomenon will be characterized by a signal indicating for each instant of time the number of processors flushing to disk. We derive that signal from a Paraver tracefile that contains flushing events, indicating when each process starts and finish the flushing to disk. Figure 1 shows an example of a flushing signal. In this kind of signals we frequently observe interleaved small bursts with flushing peaks and periods without flushing. The tracefile is perturbed not only during flushing but also in instants right after flushing peaks. Therefore, we want to consider the bursts of flushing as a single perturbed region.

With this objective, we will use a set of morphological filters, defined in the context of Mathematical Morphology. These filters are non-linear and are based on the minimum and maximum operations, aiming at the study of structural properties of the signal. The two basic morphological filters are **Erosion** and **Dilation**. The first has the property of eroding those regions of the signal with values different to zero. The second filter has the property of dilating the regions of the signal different to zero. Both operators have associated a width that has to be specified before the filter is applied. If we combine the two operators doing a Dilation followed by an Erosion we obtain an interesting result: First, the Dilation will merge the small regions with their larger or nearby neighbors. After that, the Erosion will allow us to return towards the initial signal, except in the cases that two different regions have been merged by the Dilation. With this combination, we obtain a new morphological operator called **Closing**. figure 2 shows the result of performing a Closing to the signal represented in figure 1. Note that the small regions that appear in figure 1 have been merged into larger region. The fundamental concepts and the formal definitions of Mathematical Morphology are described in [15].

In summary, following the methodology described in this section, we do three steps: First, we generate from the initial tracefile a signal, for example the number of processes flushing to disk. Second, we apply a Closing in order to merge

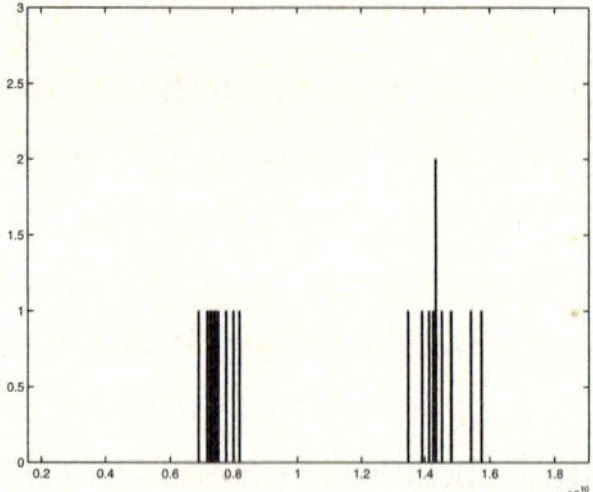

Fig. 1. Signal before closing

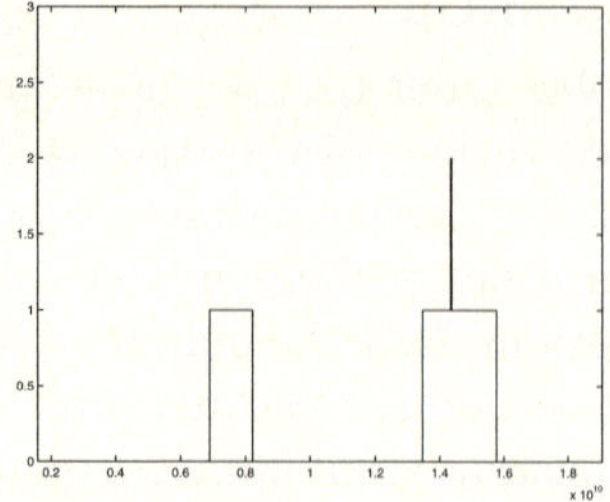

Fig. 2. Signal after closing

the pulses of the signal that are too close in an unified burst. The width associated to the Erosion and the Dilation are the same. Its choice is based on the minimum span of time we consider useful for the analysis. The pulses of the resulting signal indicate which are the perturbed (thus, useless) regions of the tracefile and the areas of the signal that equals zero indicate which are the non-perturbed regions. Finally, the process of identification of structure explained in the next section is applied to the non-perturbed regions of the trace.

2.2 Structure

There are two main characteristics about the structure that we will look for in a trace. First, the periodic structure is based on the identification of several different trace regions that are very similar. Furthermore, we will say that two trace regions are very similar if the signal that represents a metric is very similar in the two regions. Another approach based in this type of structure is presented in [17]. The second main characteristic we look for is the hierarchy of the periodicities. The periodic structure will be expressed in different levels: The first level is the structure within the original trace, the second level of periodicities is the structure within one period of the first level, and so on. In order to obtain that hierarchical structure, our algorithm will be recursive, i. e., when the internal structure of one level is detected, we will apply again the algorithm within one period of that level.

The information derived from this hierarchical structure based on periodicities is useful in, at least, two aspects. First, our tool will show the structure to the analyst as a first approach to the execution of the application under analysis. Second, the tool will provide the user with chunks of traces which are cut from the original trace. These small traces are representative parts of the original trace at different levels of the structure.

These chops of the original tracefile will allow an accurate analysis, but the tool is also reporting several metrics (percentage of time in MPI [10], ...) for each of the regions to give a first approach to the performance obtained by the application.

Metrics. There are many metrics that can be used to identify periodicities. Indeed, any user function correctly defined can be considered to generate a signal that describes several aspects of the execution. The limitation is that there are metrics that need some information not always included in the tracefile. The instantaneous MFLOPs ratio might be representative of the program structure, but needs hardware counters. If these counters are not included in the original tracefile, it will be impossible to generate the signal. Several examples of metrics that can be used without hardware counters and that can capture the global structure of the application are the following:

Number of MPI Point to Point Calls. This metric is represented by a signal which indicates how many MPI Point to Point Calls are being executed in a given moment.

Number of MPI Collective Calls. Very similar to the above metric, but considering MPI Collective Calls.

Specific MPI Call. This metric is represented by a signal which says how many calls to a specific MPI function are being executed in a given moment.

Autocorrelation. To find the internal structure of the application we apply the Autocorrelation function [14], to the signal generated from the tracefile:

$$A(k) = \sum_{i=0}^{N-1} (x_i - \mu)(x_{i+k} - \mu) \tag{1}$$

where μ is the arithmetic mean of the set $\{x_i\}$. This set is generated sampling the signal obtained from the tracefile. The higher values of the function $A(k)$ will be reached when k is equal to one of the main periods of $\{x_i\}$. However, for accuracy reasons [14], the numerical values of $A(k)$ are not obtained following (1). It is possible to obtain the value of $A(k)$ function performing, first, a Discrete Fourier Transform (DFT) and, after that, an Inverse Discrete Fourier Transform (IDFT) taking the square of the modulus of each spectral coefficient obtained with the DFT [14]. This method can we implemented using a FFT library. An important feature of FFT is its computational complexity, $O(n\ log(n))$, which allows us to calculate the values of $A(k)$ in reasonable time.

Periodicities. Once we have the values of the Autocorrelation function, the principal periodicities are selected [18]. We will select the maximum of the relatives maximums. In other words, the period we select, T, will satisfy the following:

$$A(T) = Max\{A(k)|A(k-1) < A(k) > A(k+1), k > 0\} \tag{2}$$

A remarkable point here is that it is possible that the signal does not have meaningful periods or that has 2 or 3 significant periods. Therefore, there is a need for a method to estimate the correctness of the period obtained. The approach taken is the following: assuming that T is the period identified and that M is the set of those k where $A(k)$ has a relative maximum.

$$\forall k, (k \in M \wedge k \neq T) \Rightarrow 0.9 > \frac{A(k)}{A(T)} \tag{3}$$

If the above formula holds, the 90 % of the value of $A(T)$ is higher than all the values in the rest of the maximum values. In that case, we will assume that T is a good approximation to the main period. We will check the logical formula (3) every time we perform an autocorrelation. If the formula is true, we will assume the correctness of the results. If not, we will perform a closing to the original signal in order to filter the small oscillations that can perturb the results and repeat the process. We use the closing filter in order to obtain a coarse-grained description of the signal. This replacement of a fine-grained description with a lower-resolution coarse-grained model will outline the global signal behavior.

Once a "good" period is identified, we select a region of the signal containing an iteration of this period and apply the methodology again to look for inner structure. At the same time, we cut the original tracefile in order to provide to the analyst one period on every periodic zone we found.

Finally, this methodology needs the execution of intense processes. In order to perform these executions and take advantage of several concurrent processes, we have implemented the methodology with GRID Superscalar [12], a grid programming environment developed at BSC.

3   Results

We have applied the methodology explained above to four real applications: Liso [19] with 74 processors, Idris [20] with 200 processors, Gadget [21] with 256 processors and Linpack[22] with 2048 processors. These have been executed and traced in MareNostrum. The structure found in these applications is based on the first of the metrics explained in section 2.2, the MPI Point to Point calls.

In figure 3 we show graphically a part of the structure of the Liso tracefile. This structure is shown with a Paraver visualization. In that visualization, the horizontal axis represents the time and the vertical axis the different processes. Black color means that a given process in a given instant of time is not executing any MPI Call. On the other hand, a light colored point (green when printing or visualizing in color) represents that an MPI call is being executed by the process. The picture first shows a visualization of the whole tracefile. The flushing regions are also outlined. In the second part of figure 3 we show the structure of the first region without flushing. In that case, we show, first, a region with a non-periodic structure that corresponds with the initialization phase of the application. The span of the initialization phase is 18029 ms. After that, there is periodic region with 5 iterations. The span is 47306 ms and the period shown is 9010 ms. Finally, in figure 3 we show one of the iterations of the periodic zone.

Table 5 shows that the automatic system has been able to detect the structure shown in figure 3. Furthermore, we can see the results of the automatic analysis for the whole Liso application. The first five rows correspond to the structure represented in figure 3.

Table 1. Idris. Table Representation.

Span/#it/T			Flushing
Level 0	Level 1	Level 2	
1091695/1/-	205740/1		
	885955/9/102870	550/5/110	
		590/1/-	
		540/7/80	
		101190/1/-	

Table 2. Idris. Tree Representation.

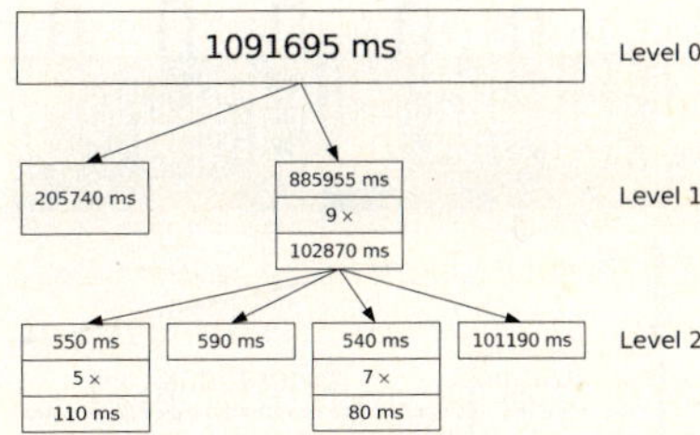

Table 3. Gadget

Span/#it/T		Flushing
Level 0	Level 1	
1097460/1/-	55000/1/-	
	811126/23/35095	
	286334/1/-	X

Table 4. Linpack

Span/#it/T		Flushing
Level 0	Level 1	
223168/1/-	82850/1/-	
	140318/130/1130	

Mainly, in table 5 we show two characteristics of the execution: First, from left to right we show the hierarchy. Second, from top to bottom we show the temporal sequence.

In the first column, we show the duration of the whole execution in milliseconds. Next, in the second column of table 5 there is a decomposition of the total elapsed time of execution. In this second column, we show a set of numbers in each cell. The first number is the total time span of the region, the second is the number of periods found in that region and, finally, the third is the duration of each period. In the regions where no periodicity has been found a dash line is written. This second column refers to the first level of hierarchical structure, i. e., is the structure over the original trace. For example, the MPI Point to Point calls distribution of the first (18020 ms) and second (47306 ms) regions of the tracefile shown in figure 3 are represented in the first and second cells of the second column.

The third column contains the second level of the structure, i.e, the structure that can be found in one of the periods of the first level. For example, the second, third, fourth and fifth cells of the third column are the decomposition of one of the periods of level 1. The MPI Point to Point distribution is shown on figure 3. Here it can be identified the second level of structure, with 6 periods (of 545 ms) of communication, a computation period (of 1005 ms), 3 periods more of communication (each of 550 ms) and a final computation period (of 3180 ms).

Finally, the fourth column shows the existence of flushing events in a given region of the tracefile.

The output of the tool is basically the information contained in this table plus the names of the files where can be found the chops of the original tracefile.

In table 1 we show the structure detected in Idris application. In table 2 we show another possible representation of the same information. Finally, in tables 3 and 4 the structure found in Gadget and Linpack applications is shown.

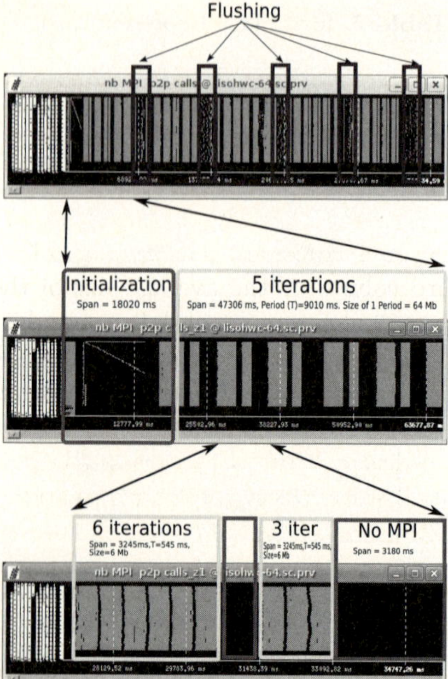

Fig. 3. On top, visualization of the whole Liso tracefile and the flushing zones. In the middle, the first region without flushing is shown. We highlight a region without periodic structure and a region with 5 iterations. At the bottom, we show one of these iterations.

Table 5. Liso application structure detected by our system. The time units are milliseconds. The first three columns show the hierarchical levels of periodicity. Level 0 column shows the total elapsed time, Level 1 column shows the different phases detected by the automatic system and Level 2 shows the internal structure of one of the periods of Level 1. In the first three columns, each cell contains three numbers: The total span of the region, the number of the periods found in that region and the duration of each period.

Span/#it/T			Flushing
Level 0	Level 1	Level 2	
356352/1/-	18020/1/-		
	47306/5/9010	3245/6/545	
		970/1/-	
		1615/3/550	
		3180/1/-	
	10273/1/-		X
	49166/5/9105	1490/3/545	
		1140/1/-	
		1585/3/550	
		3030/1/-	
		1860/3/535	
	11880/1/-		X
	60773/7/9100	3215/6/540	
		1004/1/-	
		1615/3/550	
		3266/1/-	
	11576/1/-		X
	49253/5/9145	1650/3/550	
		2825/1/-	
		3400/6/550	
		1270/1/-	
	12269/1/-		X
	48347/5/9045	3185/6/550	
		1015/1/-	
		1625/3/560	
		3220/1/-	
	13312/1/-		X
	24177/3/8905	1875/1/-	
		3425/6/550	
		1010/1/-	
		1635/3/555	
		960/1/-	

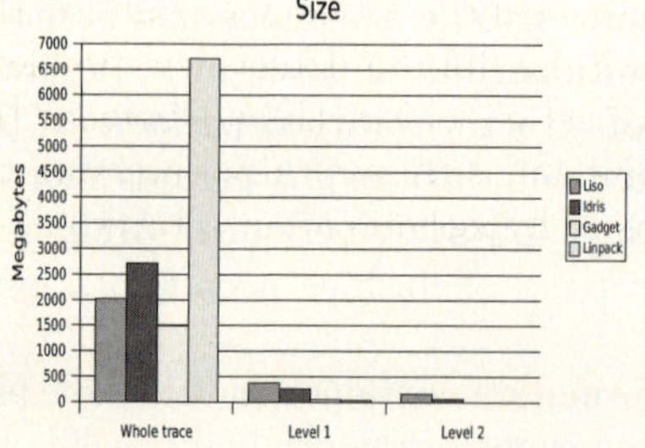

Fig. 4. Sum of the sizes of all the representative traces of each level

Table 6. Sizes of all the representative traces of each level

Application	Total Trace Size	Level 1 Size	Level 2 Size
Liso	2.02 Gb	64 Mb	6 Mb
Idris	2.7 Gb	250 Mb	25 Mb
Gadget	2.7 Gb	53 Mb	
Linpack	6.7 Gb	46 Mb	

Table 6 shows the average size of the chops of the original trace. As we have said in section 2.2, every time the system finds a periodic region it selects one of the periods of that region and cuts the tracefile to provide the analyst with representative chops of the original tracefile. The sizes shown in table 6 are, first, the size of the total tracefile and, second, the average size of the periods of the first level. For example, Liso tracefile has 6 periodic regions, one of these regions in every non-flushing zones. If we take one period of every periodic region and then we cut the tracefile, we will obtain 6 small tracefiles. The average of its sizes is the value we show. Finally, the third column is the average size of the second level periods. Note the large reduction in the amount of data to study.

Finally, in figure 4 we represent, first, the size of the whole trace. Next, we show the total sum of the sizes of the first and second level chops. Obviously, if there is only one periodic region, the value shown in figure 4 is the same as the value represented in table 6. The first level of Idris application is an example. The most important thing, however, is that the global behavior of the applications, with the exception of the flushing and initialization regions, is contained in Level 1 tracefiles. We have reduced notably the amount of data to be analyzed in order to study the performance of the applications.

4   Conclusions and Future Work

In this paper we have analyzed the possibility of automatically deriving the internal structure of a tracefile. This structure has two main properties: First, it is based on periodicities and, second, is hierarchical. We have shown that is possible to, first, detect the perturbed regions of the tracefile and, second, derive the internal structure of non-perturbed regions. It is useful in many aspects: It makes easier the process of tracing the application, it avoids the spending of time studying perturbed zones, it gives the internal structure of the tracefile and, finally, gives the most representative regions of it. In conclusion, we have reduced the problem of analyzing a huge tracefile (10 or 20 Gb) to the study of several hundreds of Mb.

In the future, this tool will perform an analysis of other parallelization problems such as Load Imbalance, efficiency, overhead, etc... Our objective is to automatize all the process of analysis and visualization of Paraver tracefiles with the intention to reduce the time required to analyze a tracefile. Finally, this tool will incorporate an expert system. It will be able to detect new problems and learn about it. What is more, the potential of tools such as Dimemas [13] will be used with the objective of predicting and automatically detecting the performance of message passing applications in hypothetic architectures.

References

1. Paraver: performance visualization and analysis, `http://www.cepba.upc.es/Paraver/`
2. KOJAK: Kit for Objective Judgment and Knowledge-based Detection of Performance Bottlenecks, `http://www.fz-juelich.de/zam/kojak/`

3. Knuepfer, A., Brunst, H., Nagel, W.E.: High Performance Event Trace Visualization. In: Proc. PDP 2005, pp. 258–263 (2005)
4. Brunst, H., Kranzlmuller, D., Nagel, W.E.: Tools for Scalable Parallel Program Analysis - Vampir VNG and DeWiz. In: DAPSYS2004, 93-102 (2004)
5. Kranzlmuller, D., Scarpa, M., Volkert, J.: DeWiz - A Modular tool Architecture for Parallel Program Analysis. In: Proc. Euro-Par 2003, pp. 74–80 (2003)
6. Freitag, F., Corbalán, J., Labarta, J.: A Dynamic Periodicity Detector: Application to Speedup Computation. In: IPDPS2001 (2001)
7. Mohr, B., Traff, J.L.: Initial Design of a Test Suite for Automatic Performance Analysis Tools. In: IPDPS (2003)
8. Nataraj, A., Malony, A., Shende, S., Morris, A.: Kernel-Level Measurement for Integrated Parallel Performance Views: the KTAU Project. In: IEEE International Conference on Cluster Computing (2006)
9. Vetter, J.S., Worley, P.H.: Asserting Performance Expectations. In: Supercomputing, ACM/IEEE2002, Conference (2002)
10. The Message Passing Interface (MPI) standard, `http://www-unix.mcs.anl.gov/mpi/`
11. OMPItrace manual, `http://www.cepba.upc.es/paraver/docs/OMPItrace.pdf`
12. Badia, R.M., Labarta, J., Sirvent, R., Perez, J.M., Cela, J.M., Grima, R.: Programming grid applications with GRID Superscalar. Journal of Grid Computing 1(2) (2003)
13. Dimemas: performance prediction for message passing applications 3rd ed. pp. 40–45, McGraw-Hill, New York, (1999), `http://www.cepba.upc.es/Dimemas/`
14. Press, W.H., Flannery, B.P., Teukolsky, S.A., Vetterling, W.T.: Correlation and Autocorrelation Using the FFT. In: 13.2 in Numerical Recipes in FORTRAN: The Art of Scientific Computing, 2nd edn., pp. 538–539. Cambridge University Press, Cambridge, England (1992)
15. Serra, J.: Image Analysis and Mathematical Morphology. Academic Press, London (1982)
16. Simon, B., Odom, J., DeRose, L., Ekanadham, K., Hollingsworth, J.K., Sbaraglia, S.: Using Dynamic Tracing Sampling to Measure Long Running Programs. In: Proceedings of the 2005 ACM/IEEE conference on Supercomputing (2005)
17. Sherwood, T., Perelman, E., Hamerly, G., Calder, B.: Automatically Characterizing Large Scale Program Behavior. In: 10th International Conference on Architectural Support for Programming Languages and Operating Systems (ASPLOS) (2002)
18. De Chevigne, A., Kawahara, H.: YIN, a fundamental frequency estimator for speech and music. Journal of Acoustical Society of America (2002)
19. Hoyas, S., Jimeńez, J.: Scaling of velocity fluctuations in turbulent channels up to Re=2003. Physics of fluids (2006)
20. Teysser, R.: Cosmological hydrodynamics with adaptive mesh refinement - A new high resolution code called RAMSES. Astronomy & Astrophysics (2002)
21. Springel, V., Yoshida, N., White, S.D.M.: Gadget: a code for collisionless and gas-dynamical cosmological simulations. New Astronomy, vol. 6 (2001)
22. Linpack benchmark: `http://www.netlib.org/linpack/`

Automatic Generation of Dynamic Tuning Techniques[*]

Paola Caymes-Scutari, Anna Morajko, Tomàs Margalef, and Emilio Luque

Departament d'Arquitectura de Computadors i Sistemes Operatius, E.T.S.E,
Universitat Autònoma de Barcelona, 08193-Bellaterra (Barcelona) Spain

Abstract. The use of parallel/distributed programming increases as it enables high performance computing. However, to cover the expectations of high performance, a high degree of expertise is required. Fortunately, in general, every parallel application follows a particular programming scheme, such as Master/Worker, Pipeline, etc. By studying the bottlenecks of these schemes, the performance problems they present can be mathematically modelled. In this paper we present a performance problem specification language to automate the development of tuning techniques, called "tunlets". Tunlets can be incorporated into MATE (Monitoring, Analysis and Tuning Environment) which dynamically adapts the applications to the current conditions of the execution environment. In summary, each tunlet provides an automatic way to monitor, analyze and tune the application according to its mathematical model.

1   Introduction

Nowadays, parallel/distributed applications are used in many science and engineering fields. They may be data intensive and may perform complex algorithms. Their main goal is to solve problems as fast as possible. Performance is a crucial issue on parallel/distributed programming. When a programmer develops an application, he/she expects to reach certain performance indexes. Therefore, it is necessary to carry out a performance analysis and tuning phase to fulfill the expectations. However, there are many applications that depend on the input data set or even can vary their behaviour during one particular execution according to the data evolution. In such cases, it is not worthy to carry out a postmortem analysis and tuning, since the conclusions based on one execution could be wrong for a new one. It is necessary to carry out a dynamic and automatic tuning of the application during its execution without stopping, recompiling nor rerunning it. In this context, the MATE environment was developed.

MATE (Monitoring, Analysis and Tuning Environment) [1,2] provides dynamic automatic tuning of parallel/distributed applications. The process followed by MATE to steer applications is showed in figure 1. During runtime, MATE automatically instruments a running application to gather information about its behaviour. The analysis phase receives events, which hold the collected information. Then, the performance functions are evaluated using that

---

[*] This work has been supported by the MCyT under contract TIN2004-03388.

A.-M. Kermarrec, L. Bougé, and T. Priol (Eds.): Euro-Par 2007, LNCS 4641, pp. 13–22, 2007.

information, in order to detect bottlenecks in the execution. According to the result of the evaluation, a solution is determined to improve the behaviour of the application. Finally, the application is dynamically tuned by applying the given solution. To modify the application execution on the fly, MATE uses the technique called *dynamic instrumentation* [3].

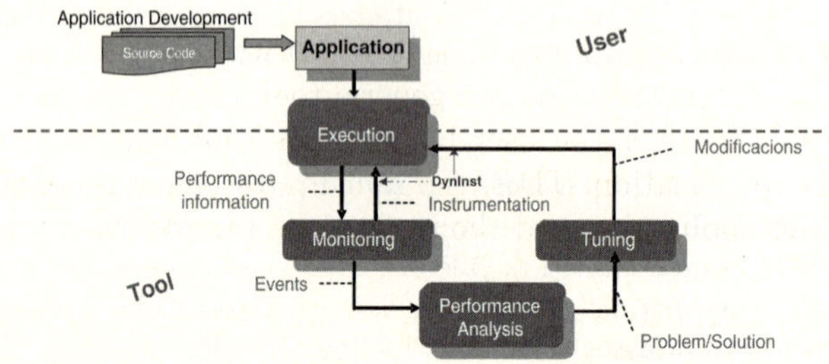

Fig. 1. Operation of MATE

All the information on how to solve a specific performance problem, is encapsulated as a piece of software called "**tunlet**". When MATE is executed, it loads a collection of tunlets in order to incorporate the knowledge to adapt the applications. Each tunlet implements a performance model and is used through the execution of the application to conduct the monitoring, analysis and tuning processes. Each tunlet condenses the description of a particular performance problem. The knowledge is represented by using the following terms: *measure points*, i.e. the locations in a process where the instrumentation must be inserted, such as a variable value or a function call; *performance functions*, i.e. activating conditions and/or formulas that model the application; *tuning points/actions*, i.e. the application components that must be changed to improve the performance.

In MATE, each performance problem should be separately tackled in a particular tunlet. This has as a consequence the need of implementing a tuning component for each performance problem. Until now, if users want to add a new tunlet to solve a problem not included previously, they should study the code of MATE to correctly implement their tunlet. Each tunlet is a C/C++ shared library that must be implemented using the Dynamic Tuning API provided by MATE (DTAPI). This added more complexity and effort to the programming and tuning tasks.

The goal of this work, is to develop a tool to automatically generate tunlets from specifications. The measure points, performance functions, tuning actions, and information about the application declared in the specification, will be used by the generator to automatically create the structures to allow the straightforward insertion of the new tunlet into the tuning environment.

So, in this paper we present a mechanism to automatically generate dynamic tuning techniques. In Section 2, we describe the performance problem specification language proposed to automatically create tunlets. In Section 3, we present a use case of the language: we provide an overview of the performance model to tune the number of workers in a Master/Worker application, and show the deduced specification needed to generate the tunlet. Section 4 shows some experimental results obtained by applying the automatically generated tunlet. Finally, Section 5 summarizes the conclusions of this work.

2 Automatic Generation of Tunlets

In this section we describe the whole tool which was designed to automatically generate tunlets. Notice that this is a general tool and can be used for any parallel application, whenever the user has enough knowledge about it to define the parts of the specification. Thus, the user has to define a set of abstractions in function of the application and the performance model to specify the tunlet. Such abstractions makes the user think in the tunlet as MATE does. In consequence, this tool makes easier and more transparent the usage of MATE. In the following, we firstly present the specification language, and secondly, the automatic generation process is described.

2.1 Tunlet Specification Language

When a language is defined, it is needed to analyze and consider what must be included, i.e. the elements, the relationships among them, its syntax and semantics [4,5]. In the particular case of specifying tunlets, it is needed to examine and consider the elements of the performance model and of the application, having in mind how the Analyzer represents and uses the knowledge, and the DTAPI that tunlets should use to correctly work in MATE.

From the **performance model** point of view, it is needed to consider the *measure points*, the *performance functions*, and the *tuning actions and points*, owing to they provide the metrics and the means to evaluate and adapt the behaviour of the application. In addition, it is necessary to determine the variables, functions, etc. in the specific application needed to interpret the model, i.e. to stablish a correlation between performance model parameters and entities in the application, to be able to collect the necessary information. Thus, from the point of view of the **application** we need to be aware of the *programming model* it follows, i.e. how the different kinds of processes or *actors* are involved in the scheme; the *variables or values* we can manipulate, both to get their value or to change them, and the functions whose execution we need to collect the information and send it as *events*.

Therefore, the specification of a tunlet is divided into three different sections, which we describe in the following paragraphs. The grammar of the tunlet specification language has been defined, but it is not included in this paper due to legibility and space reasons.

With regard to **measure points**, it constitutes the larger part of the specification. The user must define:

– The **actors** of the programming model: the types of processes co-existing in parallel, such as master and worker in the Master/Worker model, or each one of the phases in the Pipeline model. It is needed to declare the name

of the actor and the class in which is included or defined, and the name
of the executable file. These three elements are needed to obey the DTAPI
requirements. Some additional information is required:

- the minimal and maximal quantity of this type of actor could co-execute
  is needed to generate the structures to manage the behaviour information
  of each process along the succesives iterations.
- a completion condition to detect when the actor reached the end of its
  tasks along an iteration.
- the actor's attributes, i.e. the properties that should be registered in each
  iteration; for example for a worker, to catch the computation time along
  the iteration could be interesting.

- The **events** to capture, such as entries or exits of functions. Each event is
  defined by its name, the actor it is asociated with, the place in the source
  code and a certain key number to indicate if the event controls the beginning or the end of the iteration. Some attributes -that is some information
  measured when an event occurs- can be associated to a particular event. The
  quantity of bytes sent, can be an interesting metric caught when an event
  that indicates the exit of the sending function occurs. Each event has two
  default attributes: *tid* and *timestamp*, to indicate the task it was caught in
  and the specific instant in which it happened.
- The **variables**, i.e. the entities in the program whose value can vary. They
  can be instrumented or tuned in the application. For each one must be
  declared: name, data type, if in the program it is a variable, a parameter or
  a function output, and the actor who has visibility of it. In general, these
  entities are used in determining the value of the attributes of the remaining
  specification elements.
- The **iteration information**, includes an attribute to indicate the current
  iteration, and, according to the performance model, all the additional information necessary to describe the behaviour of each iteration, such as total
  communication or processing time in the iteration.
- The **model parameters** are the attributes of the performance model, whose
  value generally is calculated as a function of the attributes of actors or events.
- In general, all the elements in the specification with a set of attributes, must
  declare for each attribute its name, the data type, the initialization value and
  the way to update its value (see below). Finally, if the attribute depends on
  another attribute or event, the name of such entity should be expressed.

Regarding to the performance functions, they are the implementation in
C/C++ language of the performance expressions of the model. As in every element through the specification, the functions will depend on entities included
in the specification. The necessary mathematical libraries have to be declared in
the beginning of the specification.

Finally, for each tuning point it must be declared the name, the way in which
its value must be calculated, a condition to apply the tuning, and some information about synchronization: the appropriate place and instant to change the
value of the point.

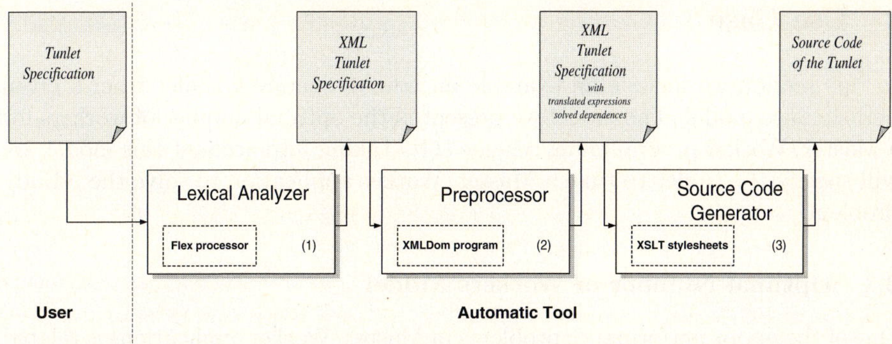

Fig. 2. Generation of a tunlet from a specification

To simplify the task of the users and reduce their involvement in implementation aspects, the expressions used to define the *initialization* (**init**) and *value* (**value**) attributes (for *attributes*, *iteration information* and *performance model parameters* sections)must be defined by using the user entities (variables, events, actors, etc.) included along the specification. Thus, to access each actor's data, we use a positional access, and selecting the right attribute such as in any data structure by using the dot (*actor[i].attribute*). Information asociated with events and iteration information are accessed in a similar way (*event.attribute, iter.attribute*).

2.2   Tunlets Generation

Once the user defined the specification, the translation process to obtain the source code of the tunlet follows several steps, shown in Figure 2:

1. *Lexical Analysis*: The input of the analyzer is the specification of the tunlet, written in a text file. The output is an equivalent specification but following the XML syntax [6]. This step exists for user-friendliness reassons, and consists in translating the specification into its equivalent XML specification. The lexical analyzer was implemented in Flex [7].
2. *Preprocess*: the input is the XML specification of the tunlet (obtained in the previous lexical analysis phase). In this phase, existing dependences among attributes and events through the specification are solved (i.e. ordered to avoid inconsistencies in the behaviour of the tunlet); in addition, the expressions to calculate the initialization and value of each entity are translated into internal structures of MATE. The output is the same XML specification but with the expressions translated and the dependences solved. The preprocessor is an XMLDom program [8]
3. *Source Code Generation*: the input is the XML specification obtained in preprocess phase. The output are a set of C/C++ files, with the source code of the tunlet. This last step in generating a tunlet, consists in extracting information from the different sections of the specification to conform the source code of the tunlet. The generator was implemented as several XSLT stylesheets [9,10].

3   Use Case

In this section we include an example on how to specify a tunlet from a given performance model. The model we present is the optimal number of workers for a Master/Worker programming scheme [11]. Taking into account this model, we will specify the tunlet to tune a Master/Worker application to solve the NBody problem.

3.1   Optimal Number of Workers Model

One of the major performance problems in Master/Worker applications is related to the quantity of workers used to process the tasks. The performance model we are using deals with homogeneous applications, and it is assumed there are only one process for each processor.

The following expression indicates how to calculate the number of workers suitable to improve the application performance:

$$Nopt = \sqrt{\frac{\lambda V + Tc}{tl}}$$

where $Nopt$ represents the number of workers needed to minimize the execution time. This expression depends on the following parameters, which should be captured as indicated:

- tl (network latency, in milliseconds) and λ (sending a byte cost -bandwidth inverse relation- in $\frac{ms}{byte}$). They must be calculated at the beginning of the execution and should be periodically updated to allow the adaptation of the system to the network load conditions.
- V is the total data volume($\sum(v_i + v_m)$) expressed in bytes, where:
  - v_i (size of tasks sent to each worker$_i$, in bytes) must be captured when master sends tasks to the workers.
  - v_m (size of the answer send back to the master for each worker, in bytes) must be captured when master receives answers from the workers.
- Tc is the total computing time ($\sum(tc_i)$), in ms, where:
  - tc_i(computing time of the worker i, in ms). Each worker computing time is needed to calculate the total computing time (Tc).

This expression is obtained by deriving the expression that models the execution time of an iteration, in order to minimize it. Such expression is defined in function of computing time and communication time, which is influenced by the latency and bandwidth (more details can be obtained from [11]).

3.2   Number of Workers Tunlet Specification

In this section we define the specification of a tunlet to tune the number of workers by considering the above-mentioned performance model. In the following, we analyze how to specify each one of the involved entities. Notice that given the

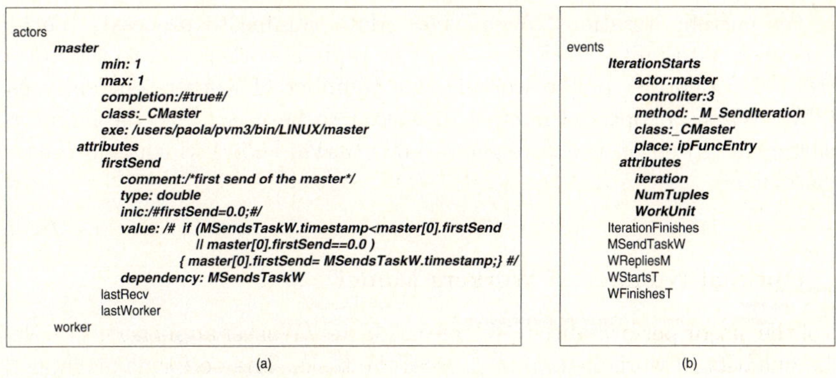

Fig. 3. (a) Actors in the application. (b) Events to catch during the execution.

length of the entire specification, in each subsection we only show one element in details (written in **_Bold and Italic_** fonts), even the rest of the elements are mentioned too. To start with the *measure points* section, we declare two kinds of actors: *master* and *worker*. As shows the figure 3*(a)*, the Master process is defined in the *_CMaster* and *_MyMaster* classes, and its executable is called "*master*". The *min* and *max* properties indicate that only one instance of the master can exist along the execution. The interesting attributes for the master are the time in which the first task in the iteration is sent (*firstSend*), the time in which the last answer is received (*lastRecv*) and the identifier of the worker that finished the processing at last (*lastWorker*). These three attributes are necessary to calculate communication time, needed in the expression used to obtain λ. In the case of the *firstSend* attribute, it is a *double* variable, whose initial value is 0.0. This value changes when an incoming *MSendsTaskW* event has a *timestamp* earlier than it, or that is the first *MSendsTaskW* event received. This fact clearly indicates that the value of *firstSend* depends on the reception of the MSendsTaskW events.

In the *events* section there are the events to determine the beginning and ending of an iteration (*IterationStarts* and *IterationFinishes*), the communication time (*MSendTaskW* and *WRepliesM*), and the computing time (*WStartsT* and *WFinishesT*). All of them are shown in the figure 3*(b)*. In particular, we describe the *IterationStarts* specification. This event has to be caught when *master* executes the *_M_SendIteration* method of *_CMaster* class, particularly at the entry of the method, which is indicated by the *place* property (*ipFuncEntry*). The attributes that the event should record, are *iteration*, *NumTuples* and *WorkUnit*, which are needed to register the current iteration, the data volume to being processed along an iteration, and the size of each data unit. These three attributes defined in the application, are specified in the *variables* section. The last two of them are useful to calculate the value of v_i.

The *variables* which could be "manipulated" in the application in order to dynamically obtain their value or change them, are *Iteration* (variable used to store the current iteration), *NumTuples* (data volume to process), *WorkUnit* (size of each data unit), *ResSize* (size of the reply a worker send to the master), *workerTID* (identifier of the worker), *nw* (number of workers currently used) and *NoptWorkers* (optimal number of worker to be used). See the figure 4*(a)*. In particular, *Iteration* is an integer variable (*asVarValue*) which is in the scope of the master.

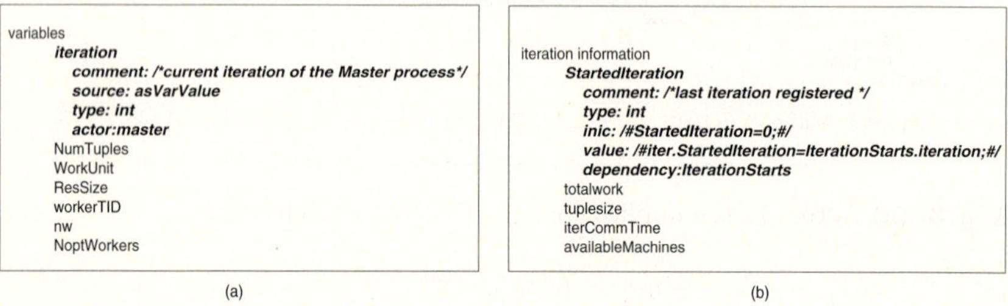

Fig. 4. (a) Susceptible of change or reading variables. (b) Iteration information.

In the case of the *iteration information* section, it includes the default attribute -*StartedIteration*- and some additional attributes needed to the particular performance model: *totalwork*, *tuplesize* and *iterCommTime*. As shows the figure 4*(b) StartedIteration* has to be initialized in 0, and its value should be updated to *iteration* each time an *IterationStarts* event happens.

This is all about the entities the users have to determine and define in order to obtain the values of each parameter of the model necessary to evaluate the performance functions. In figure 5*(a)* we enumerate the performance parameters of the model. In particular, v_i (or *vi*) is an integer whose value depends on *IterationStart* event, due to it carries the value of *NumTuples* and *WorkUnit*, needed to calculate the size of tasks sent to each worker, as explained in section 3.1. As can be seen in figure 5*(b)*, the specification includes the performance function to calculate the optimal number of workers ($pf()$), which is the implementation

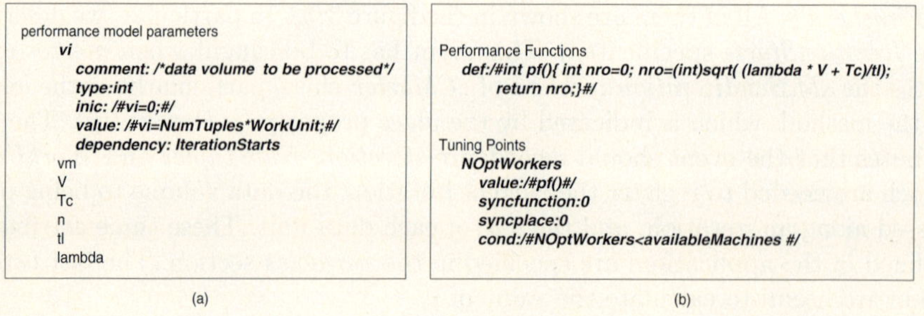

Fig. 5. (a) Performance parameters. (b) Performance function and Tuning points.

of the expression presented in Section 3.1. When *pf()* is evaluated, the result is assigned to the tuning point, *NoptWorkers*. Depending on the condition, in this case if the currently available resources are enough, the tuning is effected instantaneously, due to the *syncfunction* and *syncplace* do not present particular requirements.

4   Experiments

In this section we want to validate the usefulness of the tunlet automatically generated for the example presented in Section 3. We compare the execution time of the application when executed under the automatically generated tunlet, under the hand made tunlet and when executed by itself in different fixed number of workers (1, 2, 4, 8, 16, 19). To conduct the experiments, we selected a 2D N-Body created by using the Master/Worker framework [12]. Experiments were conducted on a cluster of homogeneous Pentium 4, 1.8 Ghz, (SuSE Linux 8.0) connected by 100Mb/sec network. We created certain load patterns, so that we can introduce and modify certain external loads to simulate the systems timesharing. Each experiment was performed many times and the average of the execution time for the application was calculated and is shown in Table 1. In the last two scenarios, the application started with one worker and then, during the execution, the number of workers has been dynamically changed according to the model described in Section 3.1. As can be seen, the automatic development of the tunlet from the specification by using our proposed tool, results in a tunlet capable of tuning the application in a time comparable to the hand made tunlet. This means that the performance of MATE is not degraded by the use of tunlets automatically generated.

Table 1. Execution time (in seconds) of NBody when executed in different scenarios

Fixed number of workers	1	2	4	8	16	19
Execution Time	64,49	34,61	18,09	10,37	11,83	15,49
NBody + handmade tunlet	Starting with 1 worker and then tuning					
Execution Time	10,92					
NBody + generated tunlet	Starting with 1 worker and then tuning					
Execution Time	10,89					

5   Conclusion

The increasing use of high performance computing and the necessity of improving the use of the resources, reflects the need of tools to assist the user in tuning his/her parallel applications. We presented a tunlet specification language. Each tunlet is defined from the performance model of the problem under consideration and information about the application and the parallel programming scheme it follows. Specifications are automatically translated into source code to be included in MATE. In this way, if a user knows a specific performance problem in the application and knows the mathematical model to overcome the problem, he or she can develop a specification of it which will automatically be translated into a tunlet to use in MATE in order to tune the program during runtime. This allows for extending MATE and facilitates its usage, due to the user does not need to know any MATE implementation details. In consequence, we make easier the use of dynamic and automatic tuning of parallel applications.

References

1. Morajko, A., Morajko, O., Jorba, J., Margalef, T., Luque, E.: Dynamic Performance Tuning of Distributed Programming Libraries. In: Sloot, P.M.A., Abramson, D., Bogdanov, A.V., Gorbachev, Y.E., Dongarra, J.J., Zomaya, A.Y. (eds.) ICCS 2003. LNCS, vol. 2660, pp. 191–200. Springer, Heidelberg (2003)
2. Morajko, A., Morajko, O., Margalef, T., Luque, E.: MATE: Dynamic Performance Tuning Environment. In: Danelutto, M., Vanneschi, M., Laforenza, D. (eds.) Euro-Par 2004. LNCS, vol. 3149, pp. 98–107. Springer, Heidelberg (2004)
3. Buck, B., Hollingsworth, J.K.: An API for Runtime Code Patching, University of Maryland, Computer Science Department. Journal of High Performance Computing Applications (2000)
4. Aho, A., Ullman, J.: The Theory of Parsing, Translation, and Compiling, vol. 1. Prentice Hall, Englewood Cliffs (1972)
5. Aho, A., Sethi, R., Ullman, J.: Compilers - Principles, Techniques, and Tools. Addison-Wesley Publishing Company, London, UK (1986)
6. Extensible Markup Language (XML), `http://www.w3.org/XML/`
7. Paxon, V.: Flex, a fast scanner generator (1995),
   `http://www.gnu.org/software/flex/manual/`
8. Document Object Model (DOM), `http://www.w3.org/DOM/`
9. XSL Transformations (XSLT) - Version 1.0, `http://www.w3.org/TR/xslt`
10. XQuery 1.0, XPath 2.0, and XSLT 2.0 Functions and Operators,
    `http://www.w3.org/2005/04/xpath-functions`
11. César, E., Mesa, J.G., Sorribes, J., Luque, E.: Modeling Master-Worker Applications in POETRIES. In: IEEE 9th International Workshop HIPS 2004, IPDPS, April, 2004, pp. 22–30 (2004)
12. Mesa, J.G.: Framework Master/Worker, Universitat Autònoma de Barcelona, Departament d'Informàtica. Master (2004)

A Scheduling Toolkit for Multiprocessor-Task Programming with Dependencies

Jörg Dümmler, Raphael Kunis*, and Gudula Rünger

Chemnitz University of Technology
Department of Computer Science
09107 Chemnitz, Germany
{djo,krap,ruenger}@cs.tu-chemnitz.de

Abstract. The performance of many scientific applications for distributed memory platforms can be increased by utilizing multiprocessor-task programming. To obtain the minimum parallel runtime an appropriate schedule that takes the computation and communication performance of the target platform into account is required. However, many tools and environments for multiprocessor-task programming lack the support for an integrated scheduler. This paper presents a scheduling toolkit, which provides this support and integrates popular scheduling algorithms. The implemented scheduling algorithms provide an infrastructure to automatically determine a schedule for multiprocessor-tasks with dependencies represented by a task graph.

1 Introduction

Large applications for distributed memory platforms often show an insufficient scalability for a high number of processors resulting from a large communication overhead especially caused by collective communication operations, like broadcast operations. The increasing popularity of large homogeneous cluster systems and hierarchical cluster-of-cluster systems necessitates a programming model that helps to reduce this overhead. Multiprocessor-task programming can address this problem by splitting an application into a set of multiprocessor-tasks (also called M-tasks or modules), which can be executed in parallel on disjoint processor groups if there are no preventing data dependencies.

The development of M-task applications, which exploit data and task parallelism at the same time, is usually more complex and error-prone compared to the development of pure data or task parallel applications. Therefore a variety of software tools, frameworks and language extensions have been proposed, see e.g. [1,2,3]. Most of these approaches belong to the class of data parallel languages with support for task parallelism or to the class of task parallel languages with the option to specify data parallelism.

An efficient execution of an M-task application requires a schedule, which defines the execution order and the processor groups for all M-tasks. The schedule

---

* Supported by Deutsche Forschungsgemeinschaft (DFG).

A.-M. Kermarrec, L. Bougé, and T. Priol (Eds.): Euro-Par 2007, LNCS 4641, pp. 23–32, 2007.

that leads to a minimum runtime strongly depends on the hardware characteristics of the target machine, e.g. the communication performance, and is therefore not portable. Scheduling algorithms that are often based on heuristics or approximation procedures can be used to calculate a schedule depending on the structure of the application and hardware specific parameters. Most of the existing tools used to develop M-task applications require the developer to manually specify a schedule. This increases the required implementation effort and leads to a poor portability.

In this paper we introduce the scheduling toolkit (STK) that offers support for scheduling M-task applications and may be used by other tools for M-task programming. STK takes a description of the structure of an M-task application in form of a hierarchical directed acyclic graph with annotated cost information and a description of the parallel target platform as an input and produces a feasible M-task schedule that is adopted to the target platform as an output. The cost information in the input may be measured runtimes or runtime predictions in form of symbolic runtime formulas. STK contains implementations of popular algorithms for M-task applications with precedence constraints. Furthermore we present a comparison of the implemented algorithms.

This paper is structured as follows. Section 2 explains the programming model used for STK. The structure and functionality of STK is described in Section 3. The algorithm library of STK, which includes all implemented scheduling algorithms, is presented in Section 4, which additionally includes a comparison of the performance of the scheduling algorithms. Section 5 explains the graphical user interface while Section 6 concludes this paper.

2   The M-Task Programming Model with Dependencies

An M-task programming model with dependencies was described e.g. in [4] or within the TwoL(Two Level)-system[5]. M-task applications can be modeled by a hierarchical annotated directed acyclic graph (dag) $G = (V, E)$. The node set V in STK consists of three different types of nodes: the unique *Start Node Q* that is a predecessor of all other nodes, the unique *Stop Node R* that succeeds all other nodes and a set of regular nodes M that correspond to the execution of an M-task. An M-task $m \in M$ is a parallel program part that can be executed on any group of processors $g_m \in \mathcal{P}(\{1, \ldots, P\}), g_m \neq \emptyset$ of a P-processor homogeneous target platform. The expected execution time is described by the runtime function $T_m : [1, \ldots, P] \to \mathbb{R}^+$ that is annotated at each M-task $m \in M$.

There are basic M-tasks, whose runtime functions are supplied by the application developer, and complex M-tasks, which consist of other M-tasks and are represented by an M-task dag. The runtime functions for basic M-tasks in STK can be specified by either giving measured runtimes or by using symbolic runtime formulas. Symbolic runtime formulas may be derived by adopting a cost model like BSP[6], LogP[7], or by fitting measured runtimes to a function prototype. Depending on the cost model the runtime functions T_m may require further parameters. The runtime functions for complex M-tasks are determined during the scheduling process by running a scheduling algorithm on the associated internal M-task dag.

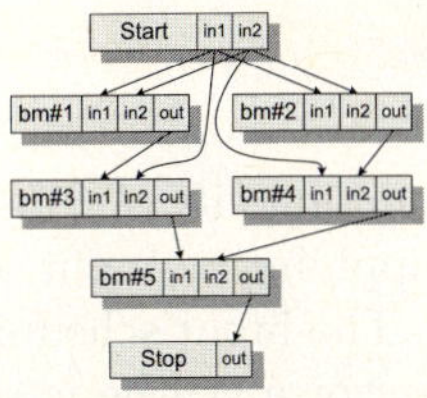
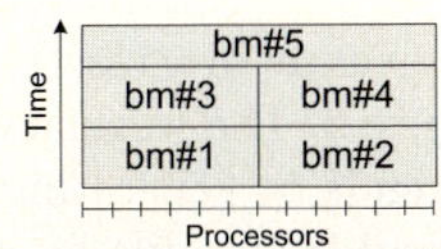

Fig. 1. Example of an M-task application represented by a directed acyclic graph (left) and an appropriate schedule (right). The figure shows the parameter dependencies resulting in a multigraph. The final M-task dag can be derived by combining multiple edges between a pair of nodes into a single edge.

Each node $v \in V$ of the M-task dag has a set of input parameters I_v and a set of output parameters O_v, which are data structures used for the communication between M-tasks. Each parameter $a \in \bigcup_{v \in V}(I_v \cup O_v)$ has a data type DT_a, which specifies the size and the memory layout, and a data distribution $DD_a(v)$, which determines which processor owns which part of the data structure. The data distribution may change during program execution and is therefore dependent on the node $v \in V$ that uses the parameter. There is a directed edge $e = (v_1, v_2) \in E$ in the M-task dag if an output parameter of v_1 is an input parameter of v_2 without being modified by another M-task. The edges may be obtained by a dataflow analysis. An edge $e = (v_1, v_2)$ leads to a data re-distribution if the processor group changes, i.e. $g_{v_1} \neq g_{v_2}$, or a parameter requires a different data distribution, i.e. $\exists a \in O_{v_1} \cap I_{v_2}$ with $DD_a(v_1) \neq DD_a(v_2)$. The costs for the re-distribution operation is denoted as $T_{re}(v_1, v_2)$.

There may be multiple implementations of each basic M-task, which accomplish the same task and require the same parameters with the same data types but may use different data distributions or have different runtime estimations. In case of multiple implementations of an M-task $m \in M$ the runtime function T_m and the data distributions $DD_a(m)$ for all $a \in I_m \cup O_m$ additionally depend on the implementation. An example M-task dag is shown in Figure 1(left).

A schedule S assigns each M-task $m \in M$ a processor group g_m and a starting time T_{S_m}, i.e. $S(m) = (g_m, T_{S_m})$. A feasible schedule has to guarantee that all predecessor M-tasks of m have finished their execution and necessary data re-distributions have been carried out before starting the execution of an M-task $m \in M$, i.e. $T_{S_n} + T_n(|g_n|) + T_{re}(n, m) \leq T_{S_m}$ for all $n \in V$ and $(n, m) \in E$. Additionally each processor of the target machine can process one task at any given time index at most, i.e. if $[T_{S_m}, T_{S_m} + T_m(|g_m|)] \cap [T_{S_n}, T_{S_n} + T_n(|g_n|)] \neq \emptyset$ then $g_m \cap g_n = \emptyset$ for all $m, n \in M$. The makespan of the schedule $C_{max}(S)$ is called the time where all output parameters are available at the Stop Node R, i.e.

$$C_{max}(S) = \max_{m \in M}(T_{S_m} + T_m(|g_m|) + T_{re}(m, R)).$$

The determination of a feasible schedule with a minimum makespan is called the scheduling problem. Figure 1(right) shows an example of a feasible schedule without data re-distributions.

3 Scheduling Toolkit

This section gives an overview of the structure and functionality of *STK*, which is shown in Figure 2. The workflow of *STK* starts by supplying an input scheduling problem and ends by generating an output schedule. The input scheduling problem can be read from a file by using the *Input Parser* or a synthetic scheduling problem can be created by using the *Problem Generator*.

The input used by *STK* is split into 3 XML-based files, the *SchedulingProblem*, the *ProblemDesc* and the *MachineDesc*. The *SchedulingProblem* input is the main part and describes the structure of an M-task application. This input file consists of a definition of data types and data distribution types and a set of M-task definitions, where each definition contains a parameter list with data types and a set of implementations. Each implementation fixes the data distribution for all parameters of the underlying M-task. Implementations of basic M-tasks furthermore contain a cost information in form of measured runtimes or a symbolic runtime formula. Implementations of complex M-tasks contain an M-task dag. The M-task dag is defined by specifying a set of nodes, where each node refers to an M-task definition in the input file, and a set of edges. Listing 3.1 shows an input description for the example application from Figure 1(left). The definition of data types and data distribution types has been omitted. The root M-task dag representing the whole application is referenced as *MainModule*. As a good schedule for a given application usually depends on the size of the input data or other problem specific values, this information is encapsulated in the *ProblemDesc* input part and may also be changed by using the graphical user interface. In this way, schedules for different problem instances may be created without altering the main input file. An example for such an input file is given in Listing 3.2. Finally all machine dependent information that may be required for symbolic runtime formulas is stored in the *MachineDesc* input part. There may be constants, e.g. the time needed to execute an arithmetical operation, and functions, e.g. the time required for a broadcast operation dependent on the size of the data to be transmitted and the number of processors. Listing 3.3 presents an example *MachineDesc* input.

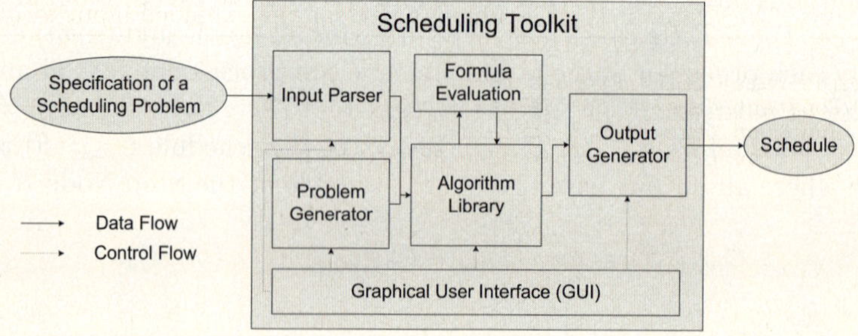

Fig. 2. Dataflow of the scheduling toolkit

Listing 3.1. Example of a main input file for *STK*

```
<SchedulingProblem Name="Example">
    <ProblemParam Name="n"/>
    <MachineParam Name="t_op"/><MachineParam Name="t_bc"/>
    <MainModule ModuleRef="2"/>

    <!-- Definition of data types (vector) and data distributions -->

    <Module Name="bm" Id="1">
        <Param Name="in1" Id="1" Type="vector"/>
        <Param Name="in2" Id="1" Type="vector"/>
        <Param Name="out" Id="2" Type="vector"/>
        <Implementation Name="bm_block" Id="1"><BasicModule>
            <Runtime Formula="T_par(p, n, t_op, t_bc)=n*t_op/p+t_bc(p, n)"/>
        </BasicModule></Implementation>
    </Module>

    <Module Name="main" Id="2">
        <Param Name="in1" Id="1"/><Param Name="in2" Id="2"/>
        <Param Name="out" Id="3"/>
        <Implementation Name="main-complex" Id="1"><ComplexModule>
            <!-- Specification of data distributions for all parameters -->
            <StartNode Name="Start" Id="1"/>
            <Node Name="bm#1" Id="2" ModuleRef="1"/>
            <Node Name="bm#2" Id="3" ModuleRef="1"/>
            <Node Name="bm#3" Id="4" ModuleRef="1"/>
            <Node Name="bm#4" Id="5" ModuleRef="1"/>
            <Node Name="bm#5" Id="6" ModuleRef="1"/>
            <StopNode Name="Stop" Id="7"/>
            <Edge Id="1" SourceNodeId="1" SourceParamId="1"
                TargetNodeId="2" TargetParamId="1"/>
            <Edge Id="2" SourceNodeId="1" SourceParamId="2"
                TargetNodeId="2" TargetParamId="2"/>
            <!-- further edge definitions -->
        </ComplexModule></Implementation>
    </Module>
</SchedulingProblem>
```

Listing 3.2. Example of a ProblemDesc-file for the extended input format

```
<ProblemDesc>
    <Constant Name="n" Value="1000"/>
</ProblemDesc>
```

Listing 3.3. Example of a MachineDesc-file for the extended input format

```
<MachineDesc>
    <Machine Name="CLiC" Processors="12">
        <Constant Name="t_op" Value="0.000000069"/>
        <Function Name="t_bc" Formula=
            "t_bc(p,n)=(0.0383+0.474e-6*log(p))*n"/>
    </Machine>
</MachineDesc>
```

The *Problem Generator* includes several algorithms for creating specific
M-task dags, e.g. SP-graphs, in-trees, out-trees, or general dags. SP-graphs [8]
are dags that are built according to a recursive definition. An SP-graph is a single
node or the series or parallel composition of two SP-graphs. In-trees are a special

Listing 3.4. Example of an output-file for the given input files

```
<GlobalSchedule Name="Example">
  <MainSchedule ScheduleRef="1"/>
     <Schedule Id="1" ModuleRef="4" ImplementationRef="1" Length="17.267858">
        <ProcessorMap>
           <Group FirstProc="0" NumProcs="16"/>
           <Machine Name="CLiC"/>
        </ProcessorMap>
        <ModuleCall Name="bm#1" Id="2" StartTime="0.0" EndTime="17.250608"
                              ModuleRef="1" ImplementationRef="1">
           <ProcessorGroup>
              <Group FirstProc="0" NumProcs="6"/>
           </ProcessorGroup>
        </ModuleCall>
        <!-- further module calls sorted by StartTime -->
     </Schedule>
  </MainSchedule>
</GlobalSchedule>
```

kind of tree where every node has exactly one outgoing edge meaning all edges are directed towards a single root node. Out-trees are the opposite where every node has exactly one incoming edge and all edges are directed away from a single root node. Furthermore several cost models for representing the computational costs of the nodes of the generated M-task dags are supported. The generated scheduling problems can be used to test the robustness and the performance of scheduling algorithms. The generation of the synthetic M-task dags can be influenced by a number of specifications. These specifications are: type of the dag (SP-graph, in-tree, out-tree or general dag), number of nodes in the dag, maximum degree of a node (sum of incoming edges and outgoing edges has to be smaller than this value for any node in the dag), a depth ratio to generate flat or deep graphs, a ratio of communication to computation costs, a ratio of the occurrence of more than one implementation for the nodes, and the runtime cost model.

The *Algorithm Library* is the core of *STK* and includes implementations of several scheduling algorithms for M-task programs with dependencies. The scheduling algorithms take an internal representation of a scheduling problem as input and produce an internal representation of a schedule, which can be saved to the output format by the *Output Generator*. The output contains a feasible schedule description for the whole application. This output file of a schedule contains the schedule for the *MainModule* of the input including the length of the schedule, the processor group of the overall scheduling problem, and a list of module calls ordered by their starting time. Each module call is defined by the processors the module is proposed to run on (processor group), the starting time and the finishing time of the module, and the chosen implementation. Parts of the output file for the example input problem are given in Listing 3.4.

STK supports symbolic runtime formulas to represent cost information. This is a very flexible approach as many different cost models can be used to derive these runtime formulas, e.g. BSP, LogP, Amdahl's law, or even unit runtimes. The *Formula Evaluation* subsystem encapsulates all functionality to handle the symbolic runtime formulas.

The *Graphical User Interface*(GUI) enables the user to control all steps in scheduling an M-task dag. Section 5 gives further insight into the features of the GUI. Additionally *STK* may be controlled via a command line interface.

4   Scheduling Algorithm Library

This section gives an overview of the implemented scheduling algorithms. Most of the implemented scheduling algorithms are defined for non-hierarchical M-task dags consisting only of basic M-tasks. Therefore hierarchical input scheduling problems have to be decomposed into a set of non-hierarchical problems first and the resulting schedules of the non-hierarchical schedulers have to be composed into a hierarchical schedule finally. Currently the following algorithms for non-hierarchical M-task dags are contained in our algorithm library:

Allocation-and-Scheduling-based algorithms try to solve the scheduling problem with precedence constraints in a two-step approach introduced in [4] consisting of an allocation step and a scheduling step. Implemented algorithms of this class are *Task-parallel, Data-parallel, TSAS (Two Step Allocation and Scheduling)*[4], *CPA (Critical Path and Area-based scheduling)*[9], *CPR (Critical Path Reduction)*[10] and a modified version of [11] that we call *MSAA (Modified SP-graph approximation algorithm)*.

Layer-based algorithms are scheduling algorithms, which are based on shrinking and decomposing an M-task dag into layers. These algorithms, which were introduced within the *(TwoL)* system with the *TwoL-Level*[12] and the *TwoL-Tree*[13] algorithm, consist of three phases, the *shrinking phase*, the *layering phase* and the *layer-scheduling phase*. In addition we extended the *Dual-3/2* approximation algorithm[14] for scheduling M-tasks without dependencies by a shrinking and a layering phase for M-tasks with dependencies.

In the following we present a comparison of the implemented scheduling algorithms based on the makespan of the produced schedules. For this purpose we consider the average makespan achieved for 100 different synthetic M-task dags belonging to the class of SP-graphs that were generated by the *Problem Generator* (see Section 3). The generated runtime formulas are based on Amdahl's law. We consider task graphs with 10 to 200 nodes and target platforms with 16 and 256 available processors. The obtained results are presented in Figure 3. Because the results of the *Data-parallel* scheduler are very slow and increase linearly we only show the first values so that the results of the other schedulers can be shown in more detail.

The *Data-parallel* and the *Task-parallel* schedulers generate the slowest schedules. We therefore focus on the specialized M-task scheduling algorithms, whose results lie closely together for 16 available processors. For 10 nodes *MSAA* constructs the schedules with the minimum average makespan. For 20 and more nodes *CPR* achieves the best results. The schedules delivered by *TSAS* for 16 available processors have a considerably higher average makespan compared to the best schedulers. Considering the average over all tested numbers of nodes the schedules of *TSAS* have a 22% higher makespan compared to *CPR*.

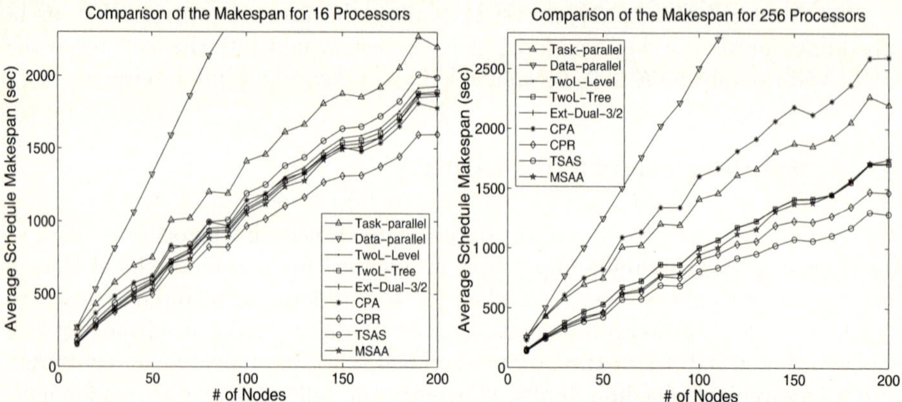

Fig. 3. Comparison of the average makespan of different scheduling algorithms for task graphs with 10 to 200 nodes and platforms with 16 (left) and 256 (right) processors

TSAS, whose performance was rather poor for 16 available processors, achieves the best average results for 256 available processors. For up to 60 nodes *MSAA* produces the second lowest makespans followed by *CPR*. Compared to *TSAS* the schedules constructed by *CPR* have a 12% higher average makespan followed by *MSAA* (18%). *CPA*, whose schedules require 95% more time compared to *TSAS*, is clearly outperformed by the other scheduling algorithms. *Ex-Dual-3/2*, *TwoL-Level* and *TwoL-Tree* exhibit a similar performance by constructing schedules that are about 14% slower compared to *CPR* for 16 processors and about 26% slower compared to *TSAS* for 256 processors.

Considering the obtained schedules for each synthetic M-task dag in isolation it can be seen that the schedule with the minimum makespan is never constructed by the *Task-parallel* or the *Data-parallel* scheduler. *CPA* builds a schedule with minimum makespan in very few cases, it is mostly dominated by *CPR*. *TwoL-Tree* is mostly better than *TwoL-Level* by a small amount. In almost all cases the best schedule is produced by *TSAS*, *TwoL-Level*, *CPR*, *Ext-Dual-3/2*, or *MSAA*. Altogether these results show that a mixed task and data parallel schedule outperforms a pure task or data parallel execution.

5   The Graphical User Interface

The graphical user interface (GUI) of *STK* is shown in Figures 4. The main window provides the functionality to load an M-task dag with annotated runtime information (*Load SchedulingProblem*), to load a specific problem instance (*Load ProblemDesc*), to load a description of a target machine (*Load MachineDesc*), to generate a synthetic M-task dag with synthetic runtime information and a synthetic target machine (*Generator*), to run a scheduling algorithm on the currently loaded problem (*Scheduler*), to validate the obtained schedule (*Validate Schedule*), to save the obtained schedule to an XML file in the output format (*Save Schedule*), and to save the currently loaded problem (*Save Problem*).

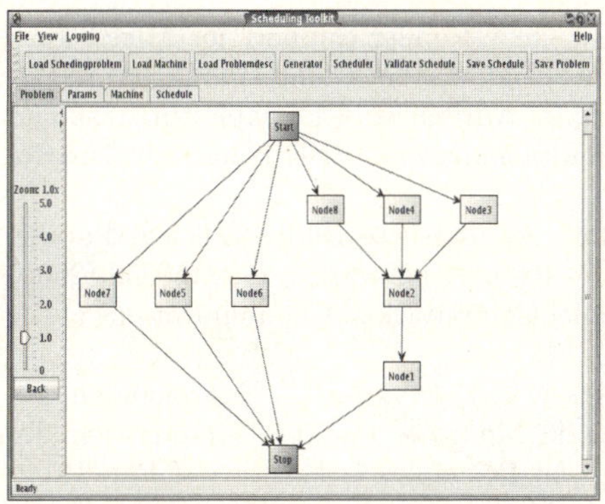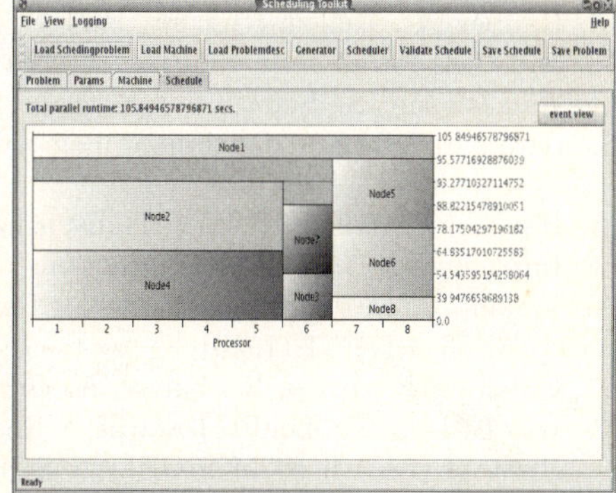

Fig. 4. The scheduling toolkit *STK* with a loaded scheduling problem (left) and the corresponding schedule (right)

The main part of the GUI allows the visualization of a loaded M-task dag presented in Figure 4(left), the editing of problem specific and machine dependent parameters and the visualization of schedules obtained by running one of the included scheduling algorithms shown in Figure 4(right). Furthermore, there are dialogs for the configuration of the scheduling algorithms and for the configuration of the generator for synthetic task graphs.

6   Conclusion and Future Work

In this paper we have introduced the scheduling toolkit *STK*, which offers a scheduling environment for M-task programming with dependencies. The aim of *STK* is to generate a schedule for a specified input scheduling problem, which can be used in isolation or for other tools for the development of M-Task applications. This is necessary because existing tools for developing M-task applications require a manual specification of the schedule. Our solution closes the gap between the specification and the execution of M-task programs by automatically determining good schedules. Through the possibility of including different scheduling algorithms it is possible to adaptively take the applications requirements and hardware details of the target machine into account.

Future steps in the development of *STK* include the support of a wider range of scheduling problems and heterogeneous target platforms. Currently we are working at the integration of *STK* into the TwoL Component System[3].

References

1. Bal, H., Haines, M.: Approaches for integrating task and data parallelism. IEEE Concurrency 6, 74–84 (1998)
2. Sips, H., van Reeuwijk, K.: An integrated annotation and compilation framework for task and data parallel programming in java. In: Proc. of 12th Int. Conf. on Par. Comp (ParCo'03) (2004)

3. Rauber, T., Reilein-Ruß, R., Rünger, G.: On Compiler Support for Mixed Task and Data Parallelism. In: Proc. of 12th Int. Conf. on Par. Comp (ParCo'03) (2004)
4. Ramaswamy, S., Sapatnekar, S., Banerjee, P.: A framework for exploiting task and data parallelism on distributed memory multicomputers. IEEE Trans. Par. Distrib. Syst. 8, 1098–1116 (1997)
5. Rauber, T., Rünger, G.: A Transformation Approach to Derive Efficient Parallel Implementations. IEEE Transactions on Software Engineering 26, 315–339 (2000)
6. Valiant, L.G.: A Bridging Model for Parallel Computation. Communications of the ACM 33, 103–111 (1990)
7. Culler, D.E., Karp, R., Sahay, A., Schauser, K.E., Santos, E., Subramonian, R., von Eicken, T.: LogP: Towards a Realistic Model of Parallel Computation. In: Proc. of the 4th ACM SIGPLAN Symp. on Principles & Practice of Par. Progr (PPOPP93), San Diego, CA, pp. 1–12. ACM Press, New York (1993)
8. Valdes, J., Tarjan, R.E., Lawler, E.L.: The recognition of series parallel digraphs. Technical report (1979)
9. Radulescu, A., van Gemund, A.: A Low-Cost Approach towards Mixed Task and Data Parallel Scheduling. In: Proc. of the 2001 Int. Conf. on Par. Processing, pp. 69–76. IEEE Computer Society, Los Alamitos (2001)
10. Radulescu, A., Nicolescu, C., van Gemund, A., Jonker, P.: CPR: Mixed Task and Data Parallel Scheduling for Distributed Systems. In: IPDPS '01: Proc. of the 15th Int. Par. & Distr. Processing Symp., p. 39. IEEE Computer Society, Los Alamitos (2001)
11. Lepere, R., Trystram, D., Woeginger, G.J.: Approximation algorithms for scheduling malleable tasks under precedence constraints. In: Meyer auf der Heide, F. (ed.) ESA 2001. LNCS, vol. 2161, Springer, Heidelberg (2001)
12. Rauber, T., Rünger, G.: Compiler support for task scheduling in hierarchical execution models. J. Syst. Archit. 45, 483–503 (1998)
13. Rauber, T., Rünger, G.: Scheduling of data parallel modules for scientific computing. In: Proc. of the 9th SIAM Conf. on Par. Processing for Scientific Computing (PPSC), San Antonio, Texas, USA (1999)
14. Mounie, G., Rapine, C., Trystram, D.: A $\frac{3}{2}$-Approximation Algorithm for Scheduling Independent Monotonic Malleable Tasks. SIAM Journal on Computing 37, 401–412 (2007)

Makefile::Parallel
Dependency Specification Language

Alberto Simões[*], Rúben Fonseca, and José João Almeida

Departamento de Informática
Universidade do Minho, Braga, Portugal
`{ambs,rubenfonseca,jj}@di.uminho.pt`

Abstract. Some processes are not easy to be programmed from scratch for parallel machines (clusters), but can be easily split on simple steps. Makefile::Parallel is a tool which lets users specify how processes depend on each other.

The language syntax resembles the well known Makefile[1] format, but instead of specifying files or targets dependencies, Makefile::Parallel specifies processes (or jobs) dependencies.

The scheduler reads the specification and submits jobs to the cluster scheduler (in our case, Rocks PBS) waiting them to end. When each process finishes, dependencies are calculated and direct dependent jobs are submitted.

Makefile::Parallel language includes features to specify parametric rules, used to split and join processes dependencies: some tasks can be split into smaller jobs working on different portions of files, and at the end, another process can be used to join results.

1 Introduction

More and more, researchers have access to multi-processors machines and, as well, clusters. The problem is that most researchers do not have time to learn how to program for parallel machines. Thus, they use their usual programs on clusters taking advantage just on the processor speeds and the big amount of available memory.

We propose a tool, Makefile::Parallel, to specify how small processes (or programs) depend on each other, to create parallelism at the level of the program, instead of the usual parallelism at the instruction level. Makefile::Parallel main goals are:

- use a compact language to specify dependencies: in our main case study we are dealing with more than one hundred jobs. To specify their dependencies manually is time consuming and error prone;
- reuse a well known language syntax that is being used in related tasks for years: the Makefile syntax.

[*] Partially supported by grant POSI/PLP/43931/2001 from Fundação para a Ciência e Tecnologia (Portugal), co-financed by POSI.

A.-M. Kermarrec, L. Bougé, and T. Priol (Eds.): Euro-Par 2007, LNCS 4641, pp. 33–41, 2007.

- embed other languages to reuse their expressiveness. In our pmakefiles — the name we gave to the text specification files — we can define actions both in Bash and Perl.
- support parametric rules: in some situations we want to instantiate rules with different values, accordingly with results from a previous job. Thus, these rules need to be instantiated in runtime.
- create information for profiling pmakefiles and create reports.

Next section will present the Makefile::Parallel Domain Specific Language, showing the language grammar and explaining its versatility. Follows a section on the scheduler implementation and how it interacts with the main cluster scheduler. Before the final remarks, there is a section detailing our case study.

2   Makefile::Parallel Language

Makefile::Parallel language syntax is heavily inspired on Makefile's syntax. The main difference is that instead of defining dependencies between files or targets, we define dependencies between jobs. The main idea can be seen as the formalization of a PERT[2][1] network.

Figure 1 shows a simplified BNF version of the Makefile::Parallel grammar. The grammar was implemented in YAPP[3], a Perl version of the well known `yacc` parser generator.

To explain the language we will use the example shown on figure 2: suppose we have a program we want to test with a different parameter, an integer ranging from 3 to 10. We need to create a directory to save the data, and at the end we want to remove some temporary files.

To solve the problem presented above we defined three rules:

- There is a job to *prepare* the output directory. This same rule defines a variable to be used on the next rule, a parametric one, named *run* (see section 2.2 for detailed explanation on these rules). The variable is defined as a set of values that the variable p can have.
- The parametric rule `run$p` will instantiate with values from 3 to 10, and run the program being tested with different parameters.
- Finally, the last rule `cleanup` depends on the *run* set of rules, and cleanups some temporary files.

Values between parenthesis indicate the expected walltime. This time is used by ROCKS[4] to schedule the job in an adequate queue, and to kill the job in case it is taking too long.

Following the walltime (a required parameter) there is an optional parameter in square brackets: the number of CPUs needed for the job. Because we are parallelizing processes, does not mean we can not have processes that, themselves, are parallel, thus needing more than one processor.

---

[1] PERT (Program Evaluation and Review Technique) charts were first developed in the 1950s by the Navy to help manage very large, complex projects with a high degree of intertask dependency.

$$
\begin{aligned}
\mathit{jobs} &\rightarrow \mathit{job\ jobs} \\
&\rightarrow \mathit{job} \\
\mathit{job} &\rightarrow \mathit{jobName}\ \text{`:'}\ \mathit{deps\ walltime\ nrCpus\ actions} \\
\mathit{walltime} &\rightarrow \text{`('}\ \mathbf{TIME}\ \text{`)'} \\
\mathit{nrCpus} &\rightarrow \text{`['}\ \mathbf{INT}\ \text{`]'} \\
&\rightarrow \epsilon \\
\mathit{jobName} &\rightarrow \mathbf{ID} \\
&\rightarrow \mathbf{ID\ VAR} \\
\mathit{deps} &\rightarrow \mathit{jobName\ deps} \\
&\rightarrow \epsilon \\
\mathit{actions} &\rightarrow \mathit{action} \\
&\rightarrow \mathit{actions\ action} \\
\mathit{action} &\rightarrow \mathit{shellCommand} \\
&\rightarrow \mathit{perlCommand} \\
&\rightarrow \mathit{setDefinition} \\
\mathit{shellCommand} &\rightarrow \mathbf{TAB\ SHELL} \\
\mathit{perlCommand} &\rightarrow \mathbf{TAB}\ \text{`sub\{'}\ \mathbf{PERL}\ \text{`\}'} \\
\mathit{setDefinition} &\rightarrow \mathbf{TAB\ VAR}\ \text{`<-'}\ \mathbf{SHELL} \\
&\rightarrow \mathbf{TAB\ VAR}\ \text{`<-'}\ \text{`sub\{'}\ \mathbf{PERL}\ \text{`\}'}
\end{aligned}
$$

Fig. 1. Makefile::Parallel DSL grammar

```
prepare:  (5:00)
    mkdir OutputData
    p <- sub{ print "$_\n" for (3..10) }

run$p: prepare (20:00:00) [2]
    runMyProgram -p $p InputData > OutputData/run.$p

cleanup: run$p (5:00)
    for a in @p; do rm -f OutputData/run.${a}.tmp; done
```

Fig. 2. Simple Makefile::Parallel example

2.1   Support for Both Shell and Perl Actions

Although most processes are binary programs, sometimes we need some specific small actions to put files in some special places and to change some file contents. With that in mind we support Bash (as makefile) and, because we work especially with Perl (and the scheduler is implemented in Perl), we also support actions written in Perl.

In the example above all actions are in Bash for simplicity.

2.2   Support for Parametric Rules

It is quite usual to test some specific programs with different algorithms or different parameters. In other cases, we need to split a big job on small chunks to be processed independently.

Both situations need the use of parametric rules: rules where variables get replaced by a set of values. For instance, on the example above, the variable $p on the *run* rule gets replaced by the values of the p set. This set is defined on the *prepare* rule with some Perl code. That Perl code returns a text file (in the *stdout*) with a value per line. These values are the p set elements.

So, p gets the values from 3 to 10: $p = \{3, 4, 5, 6, 7, 8, 9, 10\}$. Then, the rule *run$p* is replaced by:

```
run3: prepare(20:00:00) [2]
      runMyProgram -p 3 InputData > OutputData/run.3

run4: prepare(20:00:00) [2]
      runMyProgram -p 4 InputData > OutputData/run.4
...


run10: prepare(20:00:00) [2]
       runMyProgram -p 10 InputData > OutputData/run.4
```

In the *cleanup* rule, something similar happens. The dependency list is expanded with p values, and the special list variable @p in the rule action is also expanded to all values it can take. Note that @p is expanded correctly accordingly with its context (Bash or Perl). If we did not have parametric rules we would need to write:

```
cleanup: run3 run4 run5 run6 run7 run8 run9 run10 (5:00)
         for a in run3 run4 run5 run6 run7 run8 run9 run10; do
              rm -f OutputData/run.${a}.tmp; done
```

Note that the variable ${a} is a standard Bash variable that will be instantiated during run-time.

3   Makefile::Parallel Scheduler

The scheduler — pmake — is written in the Perl language. It takes a specification file and schedules jobs accordingly with their dependencies.

Since the beginning we had in mind to develop more than one scheduler subsystem to submit jobs. While most clusters use ROCKS, there are other scheduler systems. Also, a simple SSH scheduler could be created. Thus, the code was modularized: an abstract class to represent any scheduler, and a set of subclasses implementing all the methods for a specific system.

```
do {
    launch_rules_with_satisfied_dependencies()
    terminated_processes = gather_terminated_processes()
    for( process in terminated_processes ) {
        if ( defines_parameters( process ) ) {
            parameters = calculate_parameters( process )
            expand_dependency_graph(parameters)
        }
    }

    save_journal()
    sleep(10)
} while(! all_processes_executed())

generate_profiling_information()
print_report()
```

Fig. 3. Scheduler engine behavior algorithm

3.1 Scheduler Behavior

The basic scheduler behavior is specified on figure 3. The scheduler visits the dependency graph specification generated by the parser and, for every job with fulfilled dependencies, send them to run on the selected subsystem. The scheduler then waits for any of the running jobs to die or to end, by asking their state to the subsystem.

When a process ends, the scheduler gets information from the subsystem (return code, CPU time, memory used). If the return code does not indicate failure, all variables defined in the rule are instantiated (evaluating the Perl or Bash definition), and the dependency graph is modified on-the-fly to reflect the instantiated variables. Then, all processes not yet executed are browsed, and if any is found with all dependencies fulfilled, it is submitted to the system. The process continues all over again until no processes need to be executed.

From time to time, the scheduler saves some part of his internal state to permanent storage (hard disk), in the form of a journal. This is useful in case the specification is stopped by an error, or some kind of problem exists with the subsystem (as power shortages). Later, the user can simply pass an option to **pmake**, and the scheduler will bypass all processes ended correctly.

Two subsystem implementations were developed for now: a PBS designed for clusters running mainly ROCKS, and a Local, designed for desktop processing.

3.2 PBS Subsystem

High-performance clusters are the computing tool of choice for a wide range of scientific disciplines. Yet straightforward software installation, management, and monitoring for large-scale clusters have been consistent and nagging problems for non-cluster experts. The free ROCKS cluster distribution takes a fresh perspective on cluster installation and management to dramatically simplify version tracking, cluster management and integration[4]. The toolkit centers around a Linux distribution based on the Red Hat Enterprise line, and includes work from many popular cluster and grid specific projects.

One utility found in any cluster toolkit is a Portable Batch System (PBS). Basically, scheduling software let the cluster run like a batch system, allowing the allocation of cluster resources, such as CPU time and memory, on a job-by-job basis. Jobs are queued and run as resources become available, subject to the priorities established[5]. PBS is a powerful and versatile system.

When the scheduler emits a process to run on this subsystem, a PBS script is generated and sent to the cluster queue using the command `qsub`. To check if a process is still running Makefile::Parallel uses the `qstat` program with the appropriate parameters.

Eventually some processes finish. When the scheduler detects that (using the previous call to this subsystem), it asks for additional information about the dead process using the output of the `tracejob` program. This program returns detailed information, including (real) CPU time, return code, and memory usage of the process.

If the scheduler needs to stop a running process, the subsystem can call `qdel` to kill the process if it already running or to remove it from the queue otherwise.

While ROCKS supports dependencies between jobs (you can specify jobs dependency when submitting a job) it is not versatile enough for most users needs.

3.3   Local Scheduler Subsystem

Since most of modern computers are becoming multiprocessor by nature, a local scheduler is a good way of exploring parallelization on small to medium workflows.

With this in mind, a local scheduler subsystem was implemented. Running jobs on the development desktop machine allowed faster bug tracking and less time of coding. At the same time, one could not have access to a multicomputer cluster so this subsystem can be used in a variety of situations.

`pmake` local scheduler operation resembles the *GNU make* program on many aspects. First of all, if invoked with no parameters, `pmake` takes the specification and run it sequentially — the jobs are run as if a single pipeline exists. This could be optimal on a desktop machine.

However, as the level of multiprocessors rises the user may want to use the additional processor power available. For this reason, this subsystem was designed to accept a parameter that specifies how many parallel pipelines the scheduler must support on this execution.

This subsystem uses the *fork-exec-perror* paradigm and trust the operating system the correct map of the job to a free processor (both logical or physical).

4   Case Study

We have at our disposal a multicomputer cluster formed by approximately 140
CPUs and 50 nodes. The cluster runs Linux and ROCKS. This means that we
can use the TORQUE Resource Manager to schedule our jobs.

Although we are using Makefile::Parallel on two different research fields (Bio
Informatics and Natural Language Processing) we just present here the later,
because it is the most interesting and was the real motivation for this work.

The code shown on figure 4 is part of a bigger pmakefile, working in production for the word-alignment[6] of big parallel texts (bitexts: texts and respective
translations) and extraction of translation examples[7]. The word-alignment task
needs to create big sparse matrixes in memory and for big texts (with more than
300 MBytes of text files) this matrix does not fit on main memory. Thus the solution is to split the big text in smaller pieces, and process them independently.
At the end the processing result is merged up.

```
codify: (20:00:00)
    nat-codify -id=EurLex EurLex-PT EurLex-EN
    i <- sub{ $nr = `cat EurLex/nat.cnf |grep nr-chunks|cut -f 2 -d "="`;
                printf("%03d\n",$_) for (1..$nr); }

initmat$i: codify  (20:00:00)
    nat-initmat EurLex/source.$i.crp EurLex/target.$i.crp EurLex/mat.$i.in

ipfp$i: initmat$i (20:00:00)
    nat-ipfp 5 EurLex/source.$i.crp EurLex/target.$i.crp \
                EurLex/mat.$i.in EurLex/mat.$i.out
    rm -f EurLex/mat.$i.in

postipfp$i: ipfp$i (20:00:00)
    nat-mat2dic EurLex/mat.$i.out EurLex/dict.$i
    rm -f EurLex/mat.$i.out

postbin$i: postipfp$i (20:00:00)
    nat-postbin EurLex/dict.$i \
                EurLex/source.$i.crp.partials EurLex/target.$i.crp.partials \
                EurLex/source.lex EurLex/target.lex \
                EurLex/source-target.$i.bin EurLex/target-source.$i.bin
    rm -f EurLex/dict.$i

dicA: postbin$i (20:00:00)
    for a in @i; do \
            nat-dict add EurLex/source-target.bin EurLex/source-target.${a}.bin; \
        done
    for a in @i; do rm -f EurLex/source-target.${a}.bin; done

dicB: postbin$i (20:00:00)
    for a in @i; do \
            nat-dict add EurLex/target-source.bin EurLex/target-source.${a}.bin; \
        done
    for a in @i; do rm -f EurLex/target-source.${a}.bin; done

dump: dicA dicB (20:00:00)
    nat-dumpDicts -self EurLex
```

Fig. 4. Subset of NATools pmakefile

```
[...]
2006/12/12 10:49:22 The job "ipfp005" is ready to run. Launching
2006/12/12 10:49:22 Launched "ipfp005" (23996)
2006/12/12 10:49:52 Process 23996 (ipfp005) has terminated [30s]
2006/12/12 10:49:52 The job "postipfp005" is ready to run. Launching
2006/12/12 10:49:52 Launched "postipfp005" (23997)
2006/12/12 10:50:02 Process 23997 (postipfp005) has terminated [10s]
[...]
```

Fig. 5. Makefile::Parallel log output

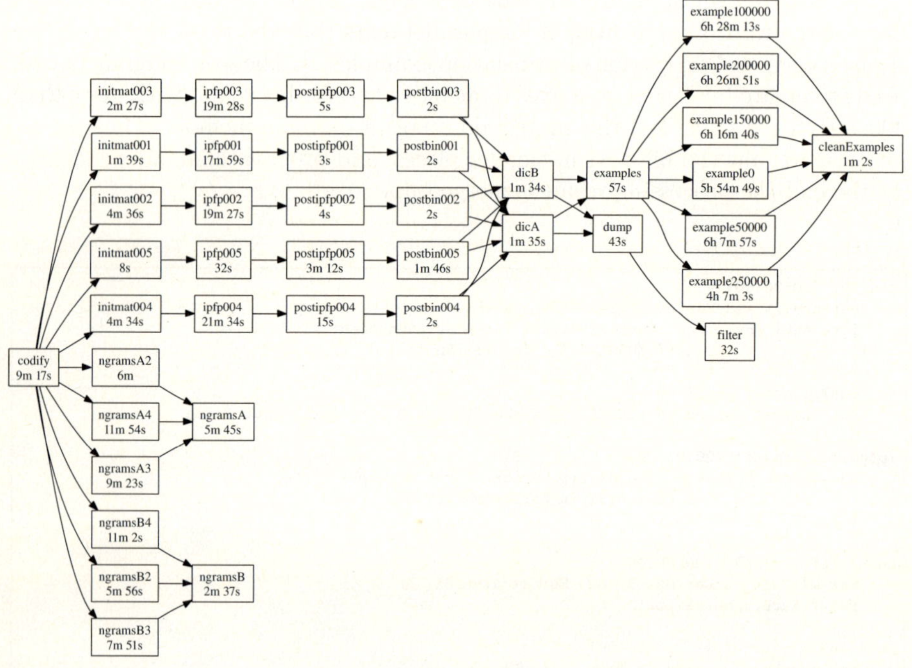

Fig. 6. Process dependency graph

```
ID          Start Time          End Time            Elapsed
codify      2006-12-12T10:41:10 2006-12-12T10:49:11 8m 1s
ngramsA     2006-12-12T10:49:11 2006-12-12T11:07:46 18m 34s
ngramsB     2006-12-12T10:49:11 2006-12-12T11:05:44 16m 33s
initmat001  2006-12-12T10:49:11 2006-12-12T10:50:12 1m
initmat002  2006-12-12T10:49:11 2006-12-12T10:50:43 1m 31s
initmat003  2006-12-12T10:49:11 2006-12-12T10:51:03 1m 51s
[...]
```

Fig. 7. Makefile::Parallel report output

Figure 6 shows graphically this process. Important to note that this graph is
generated by Graphviz[8] at run time with times for each task, and their dependencies. In fact we are getting 25% to 15% of the time needed in sequential mode
in our tasks (this value highly depends on the current cluster load, as expected).

During process `pmake` will output progress information to a log file as shown in figure 5. This log shows the processes being launched and finishing, as well as the time spent for each one. Together with the graph shown in figure 6, `pmake` generates also a file with timing information for easy profiling as can be seen on figure 7.

5   Conclusion

Makefile::Parallel proven to be a useful tool to specify jobs dependencies and to schedule them efficiently.

Being able to perform a concise description makes it easier to define dependencies correctly than defining them by hand.

The syntax is versatile enough to be applied in more than one research area. The parser is quite simple, and the fact of supporting the embed of external languages makes it yet more powerful.

Report facilities and graph generation are useful for week report generation about cluster usage and project evolution. The generated graphs are useful to explain how software works, how the different jobs interact together, and what are the critical jobs (and paths).

Worth saying that Makefile::Parallel has almost no running cost since it spends most of the time sleeping, waiting for events to happen.

References

1. Campbell, D., Grevstad, C.: A tutorial for make. In: ACM'85: Proceedings of the 1985 ACM annual conference on The range of computing: mid-80's perspective, New York, NY, USA, pp. 374–380. ACM Press, New York (1985)
2. Douglas, D.E.: Pert and simulation. In: WSC '78: Proceedings of the 10th conference on Winter simulation, Piscataway, NJ, USA, pp. 89–98. IEEE Press, Orlando, Florida, USA (1978)
3. Desarmenien, F.: Parse: Yapp — perl extension for generating and using lalr parsers. Perl module (2001), `http://search.cpan.org/dist/Parse-Yapp/`
4. Sacerdoti, F.D., Chandrai, S., Bhatia, K.: Grid systems deployment & management using rocks. In: IEEE International Conference on Cluster Computing, San Diego, IEEE Computer Society Press, Los Alamitos (2004)
5. Sloan, J.D.: High Performance Linux Clusters with OSCAR, Rocks, OpenMosix, and MPI. O'Reilly (2004)
6. Simões, A.M., Almeida, J.J.: Natools – a statistical word aligner workbench. SEPLN (September 2003)
7. Simões, A., Almeida, J.J.: Combinatory examples extraction for machine translation. In : Lonning, J.T., Oepen, S. (eds.) 11th Annual Conference of the European Association for Machine Translation, 19–20, June 2006, Oslo, Norway, (2006)
8. Gansner, E.R., North, S.C.: An open graph visualization system and its applications to software engineering. Software — Practice and Experience 30(11), 1203–1233 (2000)

Building Portable Thread Schedulers for Hierarchical Multiprocessors: The BubbleSched Framework

Samuel Thibault, Raymond Namyst, and Pierre-André Wacrenier

INRIA Futurs - LaBRI — 351 cours de la libération — 33405 Talence cedex, France
{thibault,namyst,wacrenier}@labri.fr

Abstract. Exploiting full computational power of current more and more hierarchical multiprocessor machines requires a very careful distribution of threads and data among the underlying non-uniform architecture. Unfortunately, most operating systems only provide a poor scheduling API that does not allow applications to transmit valuable *scheduling hints* to the system. In a previous paper [1], we showed that using a *bubble*-based thread scheduler can significantly improve applications' performance in a portable way. However, since multithreaded applications have various scheduling requirements, there is no universal scheduler that could meet all these needs. In this paper, we present a framework that allows scheduling experts to implement and experiment with customized thread schedulers. It provides a powerful API for dynamically distributing bubbles among the machine in a high-level, portable, and efficient way. Several examples show how experts can then develop, debug and tune their own portable *bubble schedulers*.

Keywords: *Threads, Scheduling, Bubbles, NUMA, SMP, Multi-Core, SMT.*

1 Introduction

Both AMD and INTEL now provide quad-core chips and are heading for 100-cores chips. This is, in the low-end market, the emerging part of a deep trend, in the scientific computation market, towards more and more complex machines (e.g. SUN WILDFIRE, SGI ALTIX, BULL NOVASCALE). Such large shared-memory machines are typically based on Non-Uniform Memory Architectures (NUMA). Recent technologies such as Simultaneous Multi-Threading (SMT) and multi-core chips make these architectures even more hierarchical.

Exploiting these machines efficiently is a real challenge, and a thread scheduler is faced with dilemmas when trying to take into account the memory hierarchy and the CPU utilization simultaneously. On NUMA machines for instance, threads should generally be scheduled as close to their data as possible, but bandwidth-consuming threads should rather be distributed over different chips.

A.-M. Kermarrec, L. Bougé, and T. Priol (Eds.): Euro-Par 2007, LNCS 4641, pp. 42–51, 2007.

The core scheduler of operating systems can often be influenced, but it misses the precise application behavior: for instance, adaptively-refined meshes entail very irregular and unpredictable behavior. A good solution would be to let application programmers take control of the scheduling, but writing a whole scheduler for hierarchical machines is a very difficult task.

In a previous paper [1], we introduced the *bubble* scheduling concept that helps to express the inherent parallel structure of multithreaded applications in a way that can be efficiently exploited by the underlying thread scheduler. *Bubbles* are abstractions to group threads which "work together" in a recursive way. The first *proof-of-concept* implementation of our bubble scheduler was featuring a generic hard-coded scheduler. However, applications may have different scheduling requirements and thus may attach different semantics to *bubbles*, enforcing memory affinity or emphasizing a high frequency of global synchronization operations for instance. Obviously, no generic scheduler can meet all these needs. In this paper, we present *BubbleSched*, a framework designed to ease the development and the evaluation of customized, high-level thread schedulers.

2   On the Design of Thread Schedulers

Designing a thread scheduler for hierarchical machines is complex because it means finding an application-specific compromise between lots of constraints: favoring affinities between threads and memory, taking advantage of all computational power, reducing synchronization cost, etc.

2.1   What Input Can a Scheduler Expect?

To make appropriate decisions at execution time, a thread scheduler can combine a number of parameters to evaluate the goodness of each potential scheduling action. These parameters can be collected from several places, at different times.

At runtime, some useful knowledge about the target machine can be discovered. The scheduler can not only get the number of processors but also the architecture hierarchy: how processors and memory banks are connected, how cache levels are shared between processors, etc. Moreover, indication about how well the threads are using the underlying processors can be fetched from performance counters. Some NUMA chips can also report the ratio of remote memory accesses. The scheduler can hence check whether threads and data are properly localized, or enable automatic migration policies [2].

The compiler can also provide information about the application behavior. Data access patterns may be analyzed [3] and used to choose runtime allocation policies. An OpenMP compiler can sometimes accurately estimate the amount of data shared by the threads involved in the same parallel section.

Last but not least, *programmers* can also provide relevant information about the application behavior: how threads will mostly access data, what threads are I/O-bounded, etc.) Such information can help the scheduler to find a good trade-off regarding the co-location of threads and data, and more generally can help to determine the combination of scheduling policies that will perform best.

To sum it up, a lot of information is at the scheduler's disposal or can be collected at run time. Programmers can sometimes even provide additional scheduling hints. All we need is a reasonable way to pass this information from the programmer to the scheduler.

2.2   On the Importance of Scheduling Guidance

Opportunistic policies based on Self-Scheduling [4] are the most natural approaches regarding thread scheduling. They use a centralized list of ready threads (FreeBSD 4, Linux 2.4, Windows 2K) or a distributed one (FreeBSD 5, Irix, Linux 2.6) associated to load balancing mechanisms. Such an approach scales and adapts to new workloads, but it does not use affinity information from the compiler or the programmer, and thus can not achieve best performance.

For very regular problems, a *predetermined approach* can be used. The idea is to compute *a priori* a good distribution of threads and data that will be enforced during execution. The machine being dedicated to the application, thread scheduling can be fully controlled by binding exactly one kernel thread to each processor. This approach, used in the PaStiX solver for sparse linear systems [5], gives excellent performance for regular problems, but as soon as the solving time depends on the data or intermediate results, behavior prediction fails and performance may actually get worse than by using an opportunistic approach.

An intermediate approach is based on *negotiation*. Some language extensions such as OpenMP, High Performance Fortran (HPF) or Unified Parallel C (UPC) let programmers write parallel applications by simply annotating the source code. The distribution and scheduling decisions then belong to the compiler. Such extensions get good performance by exploiting information from programmers, but the expressiveness is limited to "Fork-Join" decomposition schemes, and programmers can not express unbalanced parallelism for instance.

Negotiation yet appears to be a promising approach, because programmers just have to give suggestions and clues, sometimes indirectly, about the behavior of the threads. However, the runtime system has only little control over the operating system's thread scheduler. We believe that a good approach is to extend the scheduler interface so as to allow applications to transmit scheduling hints that will persist inside the scheduler during the lifetime of the threads.

2.3   Towards a Toolbox for Developing Thread Schedulers

In [1], we proposed a model that allows programmers to model the relationships between threads using nested sets called **bubbles**. Figure 1 illustrates this: four threads are grouped as pairs in bubbles (*e.g.* they work on the same data), which are themselves grouped along another thread in a larger bubble. This lets express relations like data sharing, collective operations, or more generally a particular scheduling policy need (serialization, gang scheduling, etc.) We also automatically model hierarchical machines with a hierarchy of runqueues. To each component of each hierarchical level of the machine is associated one runqueue: one per logical processor, core, chip, NUMA node, and one for the

Fig. 1. Expressing thread relationships: graphical and tree-based representations

whole machine. Our ground scheduler then uses a hierarchical Self-Scheduling algorithm. Idle processors scan all runqueues that span them, and executes the first thread that they find, from bottom to top. For instance, if the thread is on a runqueue that represents a chip, it may be run by any processor of this chip.

3   BubbleSched: A Framework for Building Portable Schedulers

To tackle the delicate issue of scheduling an application on hierarchical machines in an efficient, flexible and portable way, we propose a new platform for easily writing customized schedulers, based on our high-level *bubble* abstractions.

Our platform allows programmers of specialized scientific libraries or parallel programming environments to easily write thread schedulers for various application classes on these machines. It provides a high-level API for writing powerful and portable schedulers that manipulate threads grouped into bubbles, as well as tracing tools to help analyzing the dynamic behavior of these schedulers. Programmers can hence focus on algorithmic issues rather than on technical details.

3.1   An API for Writing Scheduling Strategies

By default, bubbles are placed on the machine runqueue upon start-up. In the previous example of Figure 1 running on a bi-dual-core machine, this leads to Figure 2(a): the bubble hierarchy is kept at the top of the hierarchy of runqueues. Such a distribution permits the use of all processors of the machine, since all of them have the opportunity to run any thread. This does not however take affinities between threads and processors into account, since no relation between them is used. On the contrary, Figure 2(b) shows how threads can be spatially distributed in a very affinity-aware way. However, if some threads sleep, the corresponding processors become idle, resulting to a partial CPU usage.

We designed a programming interface [6] for manipulating bubbles and threads among the runqueues so as to achieve compromises between distributions of Figures 2(a) and 2(b). Threads and bubbles are equally considered as **entities**, while bubbles and runqueues are equally considered as **scheduling holders**, so that we end up with entities (threads or bubbles) that we can schedule on holders (bubbles or runqueues). Primitives are then provided for manipulating entities in holders. Runqueues can be accessed through vectors, and can be walked thanks to "father" and "son" pointers. Some functions permit to gather statistics about

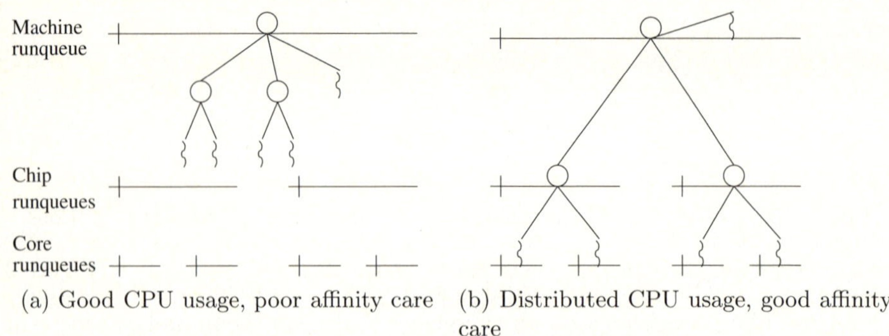

(a) Good CPU usage, poor affinity care   (b) Distributed CPU usage, good affinity care

Fig. 2. Possible distributions of threads and bubbles among the machine

bubbles so as to take appropriate decisions. This includes the total number of threads and the number of running threads, but also the accumulated expected and current CPU computation time or memory usage, or the cache miss rates. To handle concurrency, one can use some fine-grain locking functions, but since bubble scheduling decisions are generally rather medium-term oriented, one can also just use a coarse-grain function which locks a whole sub-part of the machine (runqueues and bubbles). Once locked, one can enumerate entities held by bubbles and runqueues, and redistribute them at will. This permits to elaborate complex manipulations without having to care about locking.

Writing a scheduler actually reduces to writing some hook functions. The `bubble_schedule()` hook is called when the ground Self-Scheduler encounters a bubble while searching for the next thread to execute. By default, it just finds a thread in the bubble (or its sub-bubbles) and switches to it. The `bubble_tick()` hook is called when a time-slice for a bubble expires, permiting periodic operations on bubbles. Of course, mere "daemon" threads can also be started to perform background operations. Programmers may hence manipulate threads with a high level of abstraction by deciding the placement of bubbles on runqueues.

3.2   Implementation Examples of Scheduling Algorithms

Our programming interface, though quite simple, permits to develop a wide range of powerful "bubble schedulers", a few examples are provided below.

Burst Scheduling. A first example is our previously-described "burst algorithm" [1]: a Self-Scheduling algorithm "pulls" towards processors the bubbles, which "burst" (*i.e.* release their content on a runqueue) in an opportunistic way when they reach hierarchy levels given by programmers. To implement this, the `bubble_schedule()` hook pulls the bubble a bit towards the current processor, and its content is released if needed. In a few iterations, bubbles and threads get distributed as on Figure 2(b). Moreover, bubbles are periodically "regenerated" through the `bubble_tick()` hook: their original content is put back into them, and they are put back on the machine runqueue for a new distribution. This permits to dynamically adapt the distribution to new workloads while keeping affinities in mind. This quite simple algorithm was tested with *heat conduction and advection* simulations. The results showed that this permits to get the same performance benefit as manual thread binding, but in a portable way.

Gang Scheduling. In the 1980's, OUSTERHOUT proposed to group data and threads into *gangs*, and schedule gangs on machines instead of threads on processors. To realize such *gang scheduler*, we express gangs thanks to bubbles, and a "daemon thread" performs the scheduling, see the dozen lines of code on Figure 3. The daemon uses its own `nosched_rq` where it puts all bubbles (i.e. gangs) aside, but leaves one on the main runqueue for some time during which the basic Self-Scheduler can schedule the threads of the bubble. The result is as expected: time slices are equally distributed between gangs, and then threads within gangs share their execution time within these time slices. Programmers can easily tinker with this (change the periodicity, etc.) without technical knowledge.

```
runqueue_t nosched_rq;
while(1) {
  /* Wait for next time slice */
  delay(timeslice);
  holder_lock(&main_rq);
  holder_lock(&nosched_rq);
  /* Put all entities aside */
  runqueue_for_each_entry(&main_rq, &e) {
    get_entity(e);
    put_entity(e, &nosched_rq);
  }
  /* Put one entity on main runqueue */
  if (!runqueue_empty(&nosched_rq)) {
    e = runqueue_entry(&nosched_rq);
    get_entity(e);
    put_entity(e, &main_rq);
  }
  holder_unlock(&main_rq);
  holder_unlock(&nosched_rq);
}
```

name	pr	cpu%	s	cpu
gang sched	42	0.0	I	0
0-0	43	26.4	R	3
0-1	43	26.6	R	0
0-2	43	26.8	R	2
0-3	43	25.6	R	1
0-4	43	26.6	R	3
1-0	43	21.9	R	2
1-1	43	22.2	R	0
1-2	43	22.5	R	3
1-3	43	22.1	R	2
1-4	43	21.9	R	0
1-5	43	22.6	R	1
2-0	43	19.2	R	3
2-1	43	18.9	R	1
2-2	43	19.3	R	0
2-3	43	19.8	R	2
2-4	43	19.1	R	3
2-5	43	19.3	R	1
2-6	43	19.2	R	2

Fig. 3. A gang scheduler: source code and statistics obtained with our `top`-like tool

An interesting use of this gang scheduler is to emulate a network of virtual machines on a single one in a fair way: each virtual machine is represented by a bubble which contains its thread. By running the gang scheduler, we get very fair execution: each machine gets time slices during which its threads can run on the machine. The result are shown on Figure 3, where three busy-looping gangs are sharing a four-processor machine (gang 0 has 5 threads `[0-4]`, etc).

Work-Stealing Scheduling. One of our currently in-progress algorithms is based on work stealing: the hierarchy of bubbles is first settled on the list of processor 0. Then, when `bubble_schedule()` is called on an idle processor, it will use our helper functions to look for work to steal locally (on the runqueue of the other processor of the same chip for instance), then more globally, until finding work to steal. The actual work "steal" is a non-trivial algorithmic problem:

only a part of the bubble hierarchy should be pulled toward the idle processor, and the structure of the hierarchy should be taken into account as much as possibleAll attributes and statistics attached to bubbles should be carefully taken into account in heuristics, so as to get a distribution suited to the application. These are only purely algorithmic issues though: no technical problems remain.

4   Implementation

Our BubbleSched platform is currently implemented as an extension of the MAR-CELportable two-level thread library with a low additional overhead ($\sim 5\%$ on context switches [1]). This library uses operating system functions to detect the machine architecture, bind one kernel-level thread to each processor and it then performs fast user-level context switches between user-level threads. Therefore, assuming no other application is running, it keeps complete control over thread scheduling on processors in userspace. Operations on bubbles can hence be also done in user space, and "Daemon threads" are just MARCEL threads.

Our BubbleSched platform also includes a debugger to help understanding the behavior of a scheduler. A lightweight trace of events (thread birth, sleep, wake up, bubble placement) is recorded during the execution of the application [7] and analyzed off-line. This trace can be converted into an animated movie that shows interactively the series of scheduling events and placement decisions that occurred during the execution. Thus, programmers can easily replay the scheduling decisions at will to find out their algorithmic flaws.

5   Evaluation

We show the usefulness of a programmable scheduling platform by introducing an example that illustrates well the dilemma between distributing threads over the machine or keeping related threads close together. It was run on a bi-dual-core Opteron system whose NUMA factor between the dual-cores is around 1.4.

We experimented with SuperLU_MT, a thread-parallel solver for large, sparse systems of linear equations (LU factorization). Figure 4 first shows how well SuperLU scales on the target machine. The speedup is quite close to perfect, up to the number of processors, but because of cache and synchronization affinities, using a number of threads greater than the number of processors (4) just makes performances really bad when using generic schedulers like NPTL (LINUX 2.6.17) or original Marcel with a single shared runqueue for all processors.

We consider a situation where a multi-scale scientific application needs to perform LU factorizations jobs "on demand". There are many ways to process these jobs, depending on how many threads to run per job and how to schedule them. Using a mere batch scheduler (running each job 4-way up to completion), or a completely distributed scheduler (running each job 1-way up to completion) may not be wise, in case other parallel parts of the application (running on other machines) need the job result sharply.

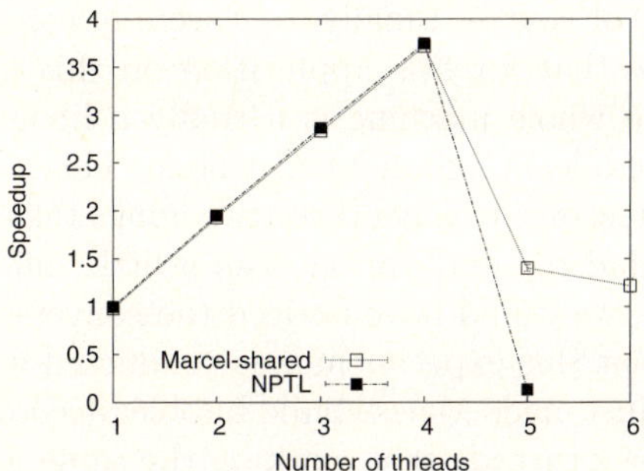

Fig. 4. Parallel speedup of a single job

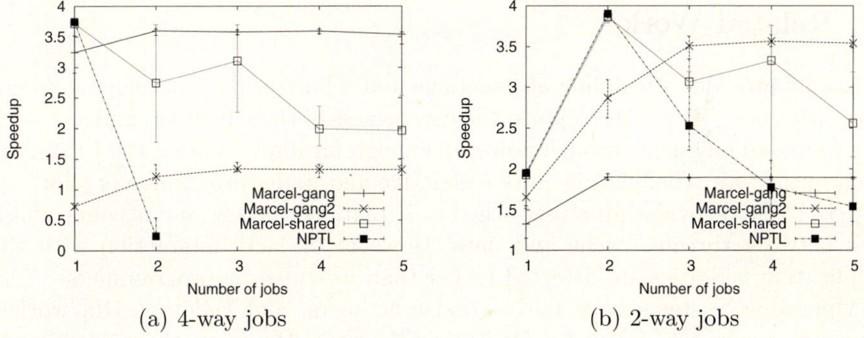

(a) 4-way jobs   (b) 2-way jobs

Fig. 5. Parallel speedup of mixed jobs, using generic or gang scheduling policies

Figures 5(a) and 5(b) show the results obtained using several approaches. On figure 5(a), all jobs were run by using four threads, while on figure 5(b), all jobs were performed by using two threads. Both figures clearly show that the generic schedulers of NPTL and original Marcel get very bad and erratic performances (min-max bars are very tall), because they equally process all threads of all jobs without taking affinities into account. We used our gang scheduler of section 3.2 for running the jobs with a time slice of 0.2ms (the solving time of one job with only one thread is around 10s) and a single gang scheduler over the whole 4-way machine ("Marcel-gang" curve), but we also tried executing two gang schedulers, one on each 2-way NUMA node of the machine ("Marcel-gang2" curve).

The bottom curve of Figure 5(b) shows that running only one gang scheduler for 2-way jobs obviously limits the speedup to a value of 2. The bottom curve of Figure 5(a) shows that running 4-way jobs on 2-way NUMA nodes also get a limited speedup (because there are more threads than processors). The top curve of Figure 5(a) shows that our gang scheduler performs quite well at running jobs on the machine: each job seems to achieve a speedup of approximately 3.5, whatever how many there are. The "Marcel-gang2" curve of Figure 5(b) shows that the two gang schedulers achieve this quite well too, provided that there are several jobs, of course. Finally, by carefully comparing these last two curves, one can notice that for this application on this machine, running one gang scheduler for the whole machine is actually a little better than running two separate gang schedulers on each NUMA node. This is probably because we chose, for repartition reasons, to have these two gangs take and put back jobs to a common job pool. Had the application been a little different (less cache- but more memory-bound), we would have noticed the converse.

It must be noted that this experiment was conducted without modifying the application at all. We just made Marcel build bubbles according to natural thread creation affiliation: since threads that work on the same job are created by the same thread, the resulting bubble hierarchy naturally maps to the job. It was then just a matter of starting gang schedulers with appropriate parameters.

6   Related Work

Bossa [8] provides scheduling abstractions and a language for developing schedulers. However, its goal is to *prove* the correctness of the scheduler, and as a such the proposed language, though powerful enough for implementing the LINUX 2.2 mono-processor scheduler, is quite restrictive and limits programmers a lot.

ELiTE [9] provides an a user-scheduling platform taking into account affinities between threads, cache and data. However, it lacks interaction with the application: affinities are detected rather than provided by programmers.

Operating Systems have fairly good schedulers, and Fedorova [10] worked on cache-aware schedulers for Operating Systems. However, these are targetted towards "blind" multi-application situations and hence can't benefit from knowledge provided by the programmer of a scientific application.

Several operating systems provide facilities for distributing kernel threads along the machine by grouping them into sets: liblgroup on Solaris, NSG on Tru64 and libnuma on Linux. These look very much like single level bubbles, but no possibility of nested sets is provided, which limits the affinity expressiveness. Moreover, none of them provides the degree of control that we provide: with *BubbleSched*, the application has hooks at the very heart of the scheduler to react to events like *thread wake up* or *processor idleness*.

7   Conclusion

In this paper, we present the BubbleSched platform, a tool for designing and prototyping specialized schedulers for specific application domains and libraries (e.g. adaptive mesh refinement algorithms, SPMD codes) for which threads behavior and memory affinities can be predicted to some extent. It provides programmers both with a way to express the application structure, and several high-level scheduling distribution primitives that let scheduling experts write *bubble algorithms* that tightly "drive" the thread scheduler by implementing some hooks. Examples of implementing scheduling strategies have shown how easy this is and how powerful it can be. Non-expert programmers may try different combinations of existing strategies to schedule threads, focusing on algorithmic issues rather than on gory details. Actually, part of this work was done in collaboration with researchers at the CEA (french Atomic Energy Commission) who have been developing huge HPC applications for a few decades and who are looking for a tool allowing them to transfer their expertise to the underlying runtime system. The conduction-advection application described in the paper is an example of such application where we have demonstrated the interest of our approach.

This work opens lots of future prospects. In the short term several algorithmic approaches will be tested and tuned to schedule real applications. On a longer run, a generic tunable scheduler could take into account as much information as possible from the hardware, the compiler and programmers. An integration to the LINUX kernel could even be considered, since a bubble hierarchy naturally exists through the notions of threads, processes, sessions and users.

References

1. Thibault, S.: A flexible thread scheduler for hierarchical multiprocessor machines. In: Second International Workshop on Operating Systems, Programming Environments and Management Tools for High-Performance Computing on Clusters (COSET-2), Cambridge / USA. ICS / ACM / IRISA (2005)
2. Marathe, J., Mueller, F.: Hardware Profile-guided Automatic Page Placement for ccNUMA Systems. In: Sixth Symposium on Principles and Practice of Parallel Programming (March 2006)
3. Shen, X., Gao, Y., Ding, C., Archambault, R.: Lightweight reference affinity analysis. In: 19th ACM International Conference on Supercomputing, Cambridge, MA, USA, pp. 131–140. ACM Press, New York (2005)
4. Durand, D., Montaut, T., Kervella, L., Jalby, W.: Impact of memory contention on dynamic scheduling on NUMA multiprocessors. In: Int. Conf. on Parallel and Distributed Systems, vol. 7. IEEE Computer Society Press, Los Alamitos (1996)
5. Hénon, P., Ramet, P., Roman, J.: PaStiX: A parallel sparse direct solver based on a static scheduling for mixed 1d/2d block distributions. In: Proceedings of the 15 IPDPS 2000 Workshops on Parallel and Distributed Processing (January 2000)
6. Thibault, S.: BubbleSched API, `http://runtime.futurs.inria.fr/marcel/doc/`
7. Danjean, V., Namyst, R.: An efficient multi-level trace toolkit for multi-threaded applications. In: EuroPar, Lisbonne, Portugal (September 2005)
8. Barreto, L.P., Muller, G.: Bossa: une approche langage á la conception d'ordonnanceurs de processus. In: Rencontres francophones en Parallélisme, Architecture, Systéme et Composant (RenPar 14), Hammamet, Tunisie (April 2002)
9. Steckermeier, M., Bellosa, F.: Using locality information in userlevel scheduling. Technical Report TR-95-14, University of Erlangen-Nürnberg (December 1995)
10. Fedorova, A.: Operating System Scheduling for Chip Multithreaded Processors. PhD thesis, Harvard University, Cambridge, Massachusetts (2006)

A Profiling Tool for Detecting Cache-Critical Data Structures

Jie Tao[1,2], Tobias Gaugler[3], and Wolfgang Karl[3]

[1] Department of Computer Science and Technology
Jilin University, P.R. China
[2] Institut für wissenschaftliches Rechnen
Forschungszentrum Karlsruhe GmbH, Germany
[3] Institut für Technische Informatik
Universität Karlsruhe (TH), Germany
jie.tao@iwr.fzk.de

Abstract. A poor cache behavior can significantly prohibit achieving high speedup and scalability of parallel applications. This means optimizing a program with respect to cache locality can potentially introduce considerable performance gain. As a consequence, programmers usually perform cache locality optimization for acquiring the expected performance of their applications.

Within this work, we developed a data profiling tool *dprof* with the goal of supporting the users in this task by allowing them to detect the optimization targets in their programs. In contrast to similar tools which mostly focus on code regions, we address data structures because they are the direct objects that programmers have to work with. Based on the Performance Monitoring Unit (PMU) provided by modern processors, *dprof* is capable of finding cache-critical variables, arrays, or even a segment of an array. It can also locate theses access hotspots to the most concrete position such as individual functions and code lines. This feature allows the user to apply *dprof* for efficient cache optimization.

1   Motivation

Amdahl's law shows the speedup and scalability of a parallel program is influenced by the parallelism: if no sufficient parallel tasks exist, an additional processor can not introduce a performance gain. Besides the parallelism, however, overhead of thread/process management and the memory behavior are also critical factors that decide the parallel performance [4]. The latter indicates both the cache behavior and the performance of the main memory in case of shared memory programs on multiprocessor machines with distributed memories. As the cache performance represents a common case, increasing programmers try to optimize their applications with respect to the cache efficiency.

With the growth of processor speed at exponential rate, the gap between memory and processor speed is continuously widening. Actually, cache is introduced

A.-M. Kermarrec, L. Bougé, and T. Priol (Eds.): Euro-Par 2007, LNCS 4641, pp. 52–61, 2007.

into the computer system for bridging this gap. However, general architectures are not tailed to applications; hence most programs, both sequential and parallel, show a significant number of cache misses when initially executed on a specific architecture. This scenario could be more critical for chip-multiprocessor (CMP) systems because on this kind of machines per-processor cache is smaller in comparison to the caches on uniprocessor or SMP machines. According to an existing research work [3], up to a 4-fold of cache miss rate has been measured on a 4-processor CMP as on a uniprocessor system.

The first challenge for cache locality optimization is to know the access bottlenecks, i.e. code regions or data structures which cause the most cache misses. To acquire this knowledge, cache miss events must be traced at the runtime during the execution of a program.

Actually, modern microprocessor architectures provide a hardware Performance Monitoring Unit (PMU) for collecting information about specific runtime events. However, theses PMUs only give global statistics on individual events, for example, the number of total misses at each cache location. The information in this form does not suffice for detecting bottlenecks.

In this case, we developed the profiling tool *dprof* in order to achieve high-level, user-understandable information capable of showing the optimization objects. *Dprof* is based on specific PMUs that are capable of delivering addresses of both the instruction and the data associated with a cache miss, like the PMU of the Intel Itanium 2 processor [8]. Using this special ability of such PMUs, it first generates a cache miss trace that records the captured runtime cache misses with address information included. In the next step, the virtual addresses in the trace file are mapped to code lines and data structures in the source program. For this, we rely on both the debugging information in the binary code for static data structures and a self-made library for dynamic ones. Based on this mapping information as well as the source code, miss events in the trace file are assigned to individual data structures and miss statistics on them are generated. This statistical information can be given at different granularity, covering e.g the whole problem, a single function, or a code line.

Additionally, *dprof* produces a histogram file, which records the number of cache misses with each individual data block of the whole working set of a program. This histogram is specifically formed in the way that it can be delivered to an existing visualization tool [12] to highlight the hot variables. These access hotspots in the form of graphical representations give the user a straightforward overview of the data structures with cache problems.

The first version of this profiler is implemented on Itanium 2, but can be ported to other processors, such as the Pentium 4, which also provide address information.

The rest of the paper is organized as follows. Section 2 describes the profiling tool in detail, including the data acquisition and the address mapping. This is followed by initial experimental results with several OpenMP applications in Section 3. Section 4 introduces several related work. The paper concludes in Section 5 with a short summary and some future directions.

2   Data Structure Profiling

Dprof aims at giving programmers possible support in their task of cache locality optimization, in the base of restricted information provided by available hardware resources. A large part of the development work is done with acquiring the needed knowledge for detecting hot data structures without involving the programmers. In this section, we first give a global overview of this profiling tool and then describe the implementation details.

***Dprof* Infrastructure.** Figure 1 illustrates our global concept of data profiling with performance counters. As shown on the right side of this figure, the performance registers are accessed during the execution of a program and the captured cache misses are recorded in a trace file. The access to performance counters is done through *pfmon* [6], an interface, similar to the PAPI [1] library, for establishing a connection between user-level software and the kernel-level hardware counters. Also at the running time of the program, as shown on the left side of the figure, dynamic data structures are instrumented with *libmalloc_hook*, a self-developed library. Static data structures, as shown in the middle of the figure, are achieved from the debugging information in the binary code of the program.

In the next step, the captured misses are processed, and associated to the data structures and code lines, according to the mapping information. Program source code is also required in this step, because it contains information for excluding those cache misses caused by other processes, e.g. a system process. The achieved result can be either in a text form for directly delivering to the users or in the form required by our visualization tool for graphical presentation.

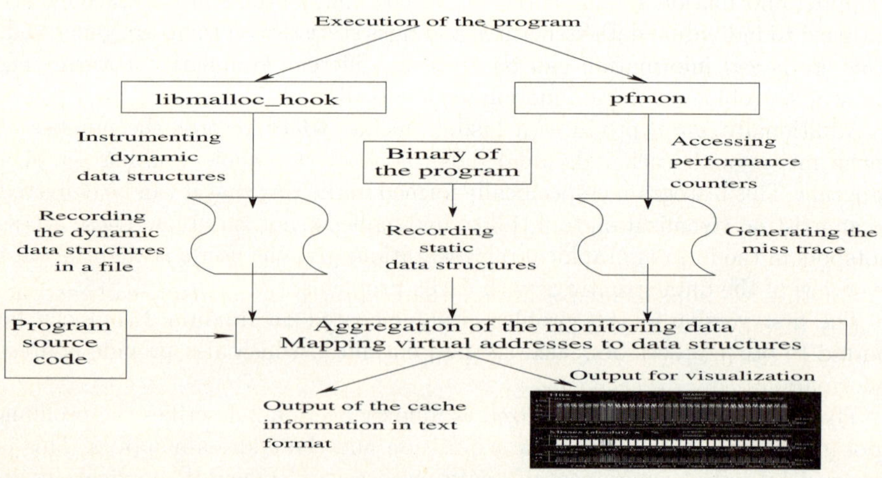

Fig. 1. *Dprof* overall infrastructure

Accessing the Performance Counters of Itanium 2. The Itanium 2 PMU has four 48-bits counters which enable the concurrent monitoring of up to four micro-architectural events. Among more than one hundred possible events, several can be used to measure cache misses and total references at each cache level. More specifically, the Itanium architectures provide specific registers, i.e. Event Address Register (EAR), which can be programmed to capture both the event and the instruction which caused the event.

When the data EAR is programmed to trace cache misses, the processor samples load misses and delivers for each sampled miss the data address, the instruction address, and the latency for accessing the data. The sampling follows an approach of interrupt, where a register counts cache misses and an interrupt is caused when the counter overflows. By giving the counter an initial value the sampling rate can be controlled.

The sampling rate is a tradeoff between the accuracy of gathered cache performance data and the intrusion influence on the program execution. A sampling rate of 1 enables to capture all cache misses, however, introduces high overhead. For this work a high accuracy is not necessary because we aim at finding access hotspots rather than measuring the exact number of cache misses. Hence, in order to reduce both the overhead and the trace size, we use an appropriate lower sampling frequency that can still present the correct hot data structures even only resulting in a lossy trace. This frequency is chosen based on the experiments with several applications. In the following section we demonstrate this result.

Overhead. *Dprof* introduces overhead in two scenarios: instrumenting *malloc*s and reading performance counters. According to our experimental results, the former is in most cases lower than 1% of the overall execution time. The overhead for accessing performance counters depends on the sampling rate. In our case of a low frequency, this penalty is under 4% of the total running time of the program.

Mapping Cache Misses to the Program Source. The addresses in the miss trace have to be mapped to the source code because they are not visible to programmers. As mentioned, the mapping information of static data structures can be extracted from the debugging information in the binary code. For dynamic data structures, however, the allocation is first performed after the execution of the program. Their mapping information can hence only be obtained at the runtime.

For this, we implemented a library that captures the *malloc()* calls and acquires thereby the start address and size of the allocated memory space. In addition, the instruction address of a *malloc()* call is also gathered. This information enables to order each memory allocation to the cache code line where a dynamic data structure is declared.

Based on the mapping information between virtual addresses and data structures, we could associate each captured cache miss to the responsible variable in the program. The instruction address in the miss record allows us further to locate the code line as well as the function, where the miss occurs.

Dprof is capable of delivering access hotspots at different granularities:

1. Data structure overview: statistics on the whole data structures within a program. Information includes number of misses at each cache level and the access latency for the misses.

2. Data structure per function. Miss statistics on data structures are individually calculated for each function in the program. In addition, time for both executing the function and accessing the data is provided and separately computed. This information shows how critical the penalty of memory operations is and this knowledge allows the user to determine whether the cache optimization is essential.

3. Data structure per code line. Number of misses of a data structure is individually calculated for each code line. This allows the programmers to locate cache problems to the most concrete position in the program source.

4. Data block: the finest granularity of miss statistics. In this case, a data structure is divided into blocks of cache line size and for each block the number of misses is computed. This information enables a separate processing of data segments, which is helpful when the optimization on the whole data structure is not efficient.

3   Initial Experimental Results

We applied *dprof* to examine several applications from the NAS OpenMP benchmark suite [2,10]. The measurement was conducted on a 4-way Itanium 2 machine with an L1 data cache of 16KB, an L2 cache of 256KB, and an L3 cache of 3MB. All applications were executed using four threads.

Accuracy and Overhead of Sampling. The first experiment aims at studying the impact of sampling rates. For this, we tested all applications using different sampling rates. As similar results have been acquired, we only show the MG code as an example.

As depicted in Table 1, we measured the sampling overhead and the number of cache misses with three main data structures in the code. In addition, we computed the miss distribution, which presents the percentage of the misses with a single data structure to the total miss. This metric highlights the variable with poor cache performance.

In Table 1, a sampling rate of 1/10 means that every 10th miss is recorded and correspondingly a rate of 1/100 enables to capture every 100th miss event. First, it can be seen that the absolute number of cache misses varies significantly over different sampling frequency, and a higher sampling rate results in the collection of more cache misses but also introduces higher overhead. Nevertheless, the miss distribution does not alter a lot over the sampling rates. Even with the lowest rate of 1/10000 we can clearly see that variable u and r are access bottlenecks. For the following experiments we use the middle rate of 1/1000.

Access Hotspots. *Dprof* gives the user a clear view of the access bottlenecks, from global to concrete. It first shows which data structure in the whole program

Table 1. Number of cache misses with different sampling frequency

		Sampling rate				
		1/10	1/100	1/1000	1/10000	1/100000
Overhead (%)		29.17	9.83	3.29	0.83	0.43
Number of cache misses	$u[l][k][j]$	116540	101629	39045	3950	390
	$v[k][j]$	21682	15686	5529	574	49
	$r[l][k][j]$	85841	72824	22647	2347	233
Miss distribution	$u[l][k][j]$	0.52	0.53	0.58	0.57	0.58
	$v[k][j]$	0.09	0.08	0.08	0.08	0.07
	$r[l][k][j]$	0.38	0.38	0.34	0.34	0.35

has cache problems (a global overview), then the function where most cache misses with this data structure occur, and finally the concrete code line. Again we use the MG program to demonstrate the experimental results.

Table 2 demonstrates the global overview of cache misses with the MG application. In this table, the first two columns give the source files and line number where the data structures are accessed and declared. This is followed by the name of the observed data structure. Misses at different cache level as well as the number of total miss are depicted in the last four columns. Note that the concrete values in this table could be different with the results in the previous table because with the sampling technique, each run gives a different result even using the same sampling rate.

Table 2. Miss overview of main data structures

srcFile	decla.	data structure	L1Miss	L2Miss	L3Miss	total miss
mg.c	206	$u[l][k]$	6	0	3	9
mg.c	208	$u[l][k][j]$	32247	442	586	33275
mg.c	214	$v[k]$	1	0	1	2
mg.c	216	$v[k][j]$	3039	0	191	3230
mg.c	223	$r[l][k]$	2	1	1	4
mg.c	225	$r[l][k][j]$	17028	218	209	17455

The MG program clearly shows that variable $u[l][k][j]$ and $r[l][k][j]$ are the access bottlenecks. The former introduces more than the half of cache misses, while the later causes about 37% of the total cache miss.

Now we further examine the miss behavior of variable $u[l][k][j]$. Table 3 presents all functions containing this variable. For each cache level we measured both the number of misses and the latency for accessing the missing data. Here, the miss statistics does not cover the total misses at a cache level, rather it is the misses found in the next level cache. The values in the last two columns are the sum of misses and latency of all caches. Hence, the total misses in the table are actually total L1 misses and the total latency is the time (in CPU cycles) for accessing the data missing in L1.

Table 3. Functions containing the hot data structure

Function	L1 cache		L2 cache		L3 cache		total	
	#miss	latency	#miss	latency	#miss	latency	#miss	latency
interp	2703	13515	16	224	108	28291	2827	42030
comm3	19	95	0	0	8	2035	27	2130
resid	21804	109020	246	3445	253	65940	22303	178405
psinv	1176	5880	0	0	81	21280	1257	27160

It can be clearly seen that *resid* is the hot function where optimization has to be performed. This function results in totally 22303 L1 misses, whereby 21804 ones can be tackled with L2 and another 246 with L3. However, still 65940 CPU cycles are lost for the data in the main memory which, according to our experimental results, is more than 20% of the time used to run this function.

Table 4. Hot data structures in a single function

Code line	L1 cache		L2 cache		L3 cache		total	
	#miss	latency	#miss	latency	#miss	latency	#miss	latency
531	2294	11470	113	1582	40	10480	2447	23532
532	2441	12205	0	0	0	0	2441	12205
533	2326	11630	131	1835	47	12009	2504	25474
534	2438	12190	1	14	166	43451	2605	55655
538	2454	12270	1	14	0	0	2455	12284

For optimization, *dprof* further gives the code line where the programmer has to do some transformations. As shown in Table 4, line 534 must be optimized because it shows a high memory access latency. Other lines has similar total misses, but most of the missing data can be found in other caches without accessing the main memory. Hence, the cache problem with them is not as critical as that with line 534.

Visualization of Bottlenecks. The profiling results can also be visualized using YACO [12], a visualization tool for displaying the runtime cache access behavior. Figure 2 and 3 are two sample views with the CG application.

Figure 2 is the Variable Miss Overview which shows the global access hotspots. Within this view, main data structures of a program are horizontally listed and separated with a bidirectional arrow. Corresponding to each data structure, the numbers of cache misses with all parallel threads at a selected cache level are illustrated in bars. For this example, four colored bars represent the four threads running the application. It can be seen that CG has an identical behavior with all threads, each showing a dominating misses with variable p and a. The visualization makes these bottlenecks more visible.

Figure 3 is the Data Structure Information view that exhibits the miss distribution within a single array, in this case array p. Within this diagram, the

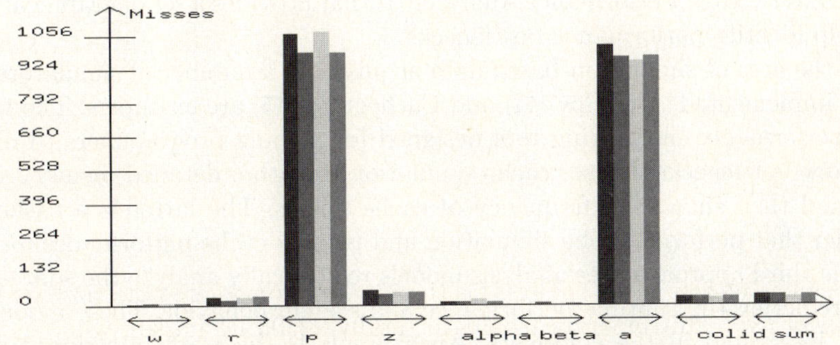

Fig. 2. Sample view: Variable Miss Overview

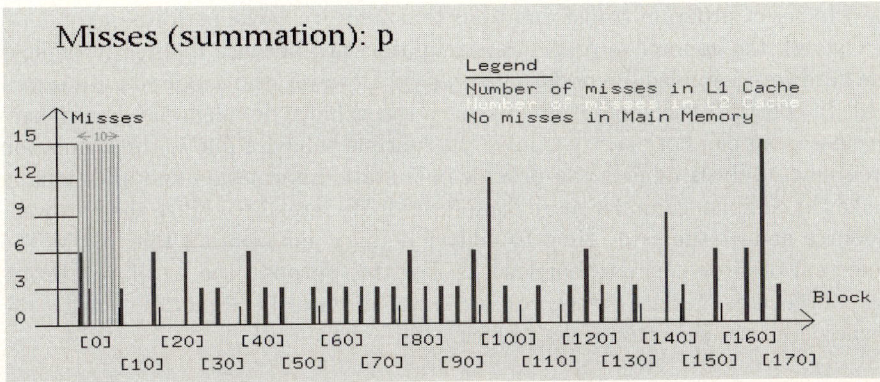

Fig. 3. Sample view: Data Structure Information

array is divided into data blocks in size of a cache line and the number of misses to each block is presented. As shown in Figure 3, this view allows the user to detect individual data blocks with excessive cache misses. For a focus of this hot block, a mask (the vertical lines on the most left side of the diagram) is provided and can be used to enclose the specific block for acquiring information about the corresponding array elements contained in it.

The block-based miss statistics can also be visualized in a per-function way using the 3D Phase Information view of YACO. This view contains several 2D diagrams, similar to that shown in Figure 3, each corresponding to a single function. This enables to highlight both the critical function and the data blocks.

4   Related Work

Currently, there are mainly three approaches for gathering performance data: hardware counter, simulation, and analysis models. The first approach is usually deployed by performance analyzer to acquire source data. A well-known example is the Intel VTune Performance Analyzer [7] that provides a set of graphical views to help identify performance bottlenecks.

In the area of simulation-based data acquisition, a number of simulators have been implemented. MemSpy [11] and Cachegrind [15] are examples. The former is a performance monitoring tool designed for helping programmers to discern memory bottlenecks. It uses cache simulation to gather detailed memory statistics and then shows the frequency of cache misses. The latter is a cache-miss profiler that performs cache simulation and records cache performance metrics.

The third approach uses analysis models to statically analyze the source code for predicting the runtime performance and system behavior. These models are often integrated in the compilers. An example is the work introduced in [5], where a framework is built in a compiler to automatically diagnose the cause of cache misses. This framework is based on the Cache Miss Equations (CMEs), an analytical representation of cache misses in a loop nest. The diagnosis is then used to select program transformations that improve cache performance.

Overall, the counter approach is straightforward because it directly deploys the hardware provided by modern processors. However, the acquired data is limited. In contrast, the simulation approach can achieve detailed and comprehensive data, but can not exactly exhibit the runtime behavior due to the simulation accuracy. Analysis models can provide only static information and additionally make the complex compiler more complicated. We intend to utilize the hardware resource and at the same time to deliver detailed information that allows the detection of more concrete bottlenecks. For this purpose, the developed profiling tool focuses on data structures and especially has the ability of showing the cache performance within a data block.

5   Conclusion

In this paper we introduce a profiling tool for acquiring cache performance data using hardware counters. Goal of this work is to provide the users possible support in their task of cache locality optimization, in the base of restricted information which can be gathered by available hardware resources. A specific feature of this profiler is the ability of mapping the counter information to data structures in the source program and showing the most concrete access bottlenecks, actually the optimization targets.

As the initial step, we have applied the profiling tool to analyze the cache problems of several applications. In the following, we will start with the optimization process and use traditional optimization strategies, such as data and code transformation, to correct the detected memory penalty.

References

1. Dongarra, J., London, K., Moore, S., Mucci, P., Terpstra, D.: Using PAPI For Hardware Performance Monitoring On Linux Systems. In: Linux Clusters: The HPC Revolution (June 2001)

2. Bailey, D., et al.: The NAS Parallel Benchmarks. Technical Report RNR-94-007, Department of Mathematics and Computer Science, Emory University (March 1994)
3. Fung, S.: Improving Cache Locality for Thread-Level Speculation. Master's thesis, University of Toronto (2005)
4. Fürlinger, K., Gerndt, M.: Analyzing Overheads and Scalability Characteristics of OpenMP Applications. In: Proceedings of the 7th International Meeting on High Performance Computing for Computational Science (July 2006)
5. Ghosh, S., Martonosi, M., Malik, S.: Automated Cache Optimizations using CME Driven Diagnosis. In: Proceedings of the 2000 International Conference on Supercomputing, pp. 316–326 (2000)
6. HP. Perfmon Project Web Site. Available at
   `http://www.hpl.hp.com/research/linux/perfmon/`
7. Intel Corporation. Intel VTune Performance Analyzer. Available at
   `http://www.cts.com.au/vt.html`
8. Intel Corporation. Intel Itanium Architecture Software Developer's Manual, vol. 1–3 (2002). Available at
   `http://developer.intel.com/design/itanium/manuals/iiasdmanual.htm`
9. Intel Corporation. IA-32 Intel Architecture Software Developer's Manual, vol. 1–3. Available at Intel's developer website (2004)
10. Jin, H., Frumkin, M., Yan, J.: The OpenMP Implementation of NAS Parallel Benchmarks and Its Performance. Technical Report NAS-99-011, NASA Ames Research Center (October 1999)
11. Martonosi, M., Gupta, A., Anderson, T.: Tuning Memory Performance of Sequential and Parallel Programs. Computer 28(4), 32–40 (1995)
12. Quaing, B., Tao, J., Karl, W.: YACO: A User Conducted Visualization Tool for Supporting Cache Optimization. In: Yang, L.T., Rana, O.F., Di Martino, B., Dongarra, J.J. (eds.) HPCC 2005. LNCS, vol. 3726, pp. 694–703. Springer, Heidelberg (2005)
13. Sun Microsystems. UltraSPARC IIi User's Manual October (1997), available at
   `http://www.sun.com/processors/documentation.html`
14. Welbon, E., et al.: The POWER2 Performance Monitor. IBM Journal of Research and Development 38(5) (1994)
15. WWW. Cachegrind: a cache-miss profiler. Available at
   `http://developer.kde.org/~sewardj/docs-2.2.0/cg_main.html#cg-top`

On Using Incremental Profiling for the Performance Analysis of Shared Memory Parallel Applications

Karl Fuerlinger[1], Michael Gerndt[2], and Jack Dongarra[1]

[1] Innovative Computing Laboratory,
Department of Computer Science,
University of Tennessee
`{karl,dongarra}@cs.utk.edu`
[2] Lehrstuhl für Rechnertechnik und Rechnerorganisation,
Institut für Informatik,
Technische Universität München
`gerndt@in.tum.de`

Abstract. Profiling is often the method of choice for performance analysis of parallel applications due to its low overhead and easily comprehensible results. However, a disadvantage of profiling is the loss of temporal information that makes it impossible to causally relate performance phenomena to events that happened prior or later during execution. We investigate techniques to add temporal dimension to profiling data by incrementally capturing profiles during the runtime of the application and discuss the insights that can be gained from this type of performance data. The context in which we explore these ideas is an existing profiling tool for OpenMP applications.

1   Introduction

Performance is an important concern for many developers of parallel applications and a large number of tools are available that support can be used for analyzing performance and to identify potential bottlenecks.

Most tools for parallel performance analysis capture data during the execution of an application to later analyze it offline. Depending on the kind of data recorded, *tracing* and *profiling* is commonly distinguished. Profiling denotes the reporting of performance data in a summarized style (such as fraction of total execution time spent in a certain subroutine call) and tracing refers to the recording of individual time-stamped program execution events.

Profiling is often preferred over tracing since it generates less performance data and greatly facilitates a manual interpretation. However, a detailed analysis of the interaction of processes or threads can sometimes require the level of insight only time-stamped event recoding offers. More generally, a significant drawback of profiling is the fact that the *temporal* dimension of performance data is completely lost. That is, with "one-shot" profiles is not possible to relate

A.-M. Kermarrec, L. Bougé, and T. Priol (Eds.): Euro-Par 2007, LNCS 4641, pp. 62–71, 2007.
© Springer-Verlag Berlin Heidelberg 2007

performance phenomena in profiling reports to the time when they occurred in the application and to explain the reason and consequences in terms of causal relations with other phenomena or events happening earlier or later.

In this paper we investigate the utility of incremental profiling (or "profiling over time") for performance analysis of parallel applications. We discuss general approaches to add a temporal component to profiling data and the kind of insights that can be derived from incremental profiling. The context in which we explore these ideas is our own profiling tool for OpenMP applications, called ompP [1]. Examples that show the utility of our proposed approach come from several applications from the SPEC OpenMP benchmark suite (SPEC OMPM2001) [2].

The rest of this paper is organized as follows: Sect. 2 briefly introduces our profiling tool and describes its existing capabilities. Sect. 3 then lists general options to add temporal dimension to performance data and describes the approach we have taken with ompP. Sect. 4 serves as an evaluation of our ideas: we describe the new types of performance data of incremental profiling (often graphical views) that are available to the user and show examples of their utility on application examples that come from the SPEC OpenMP benchmark suite. We describe related work in Sect. 5 and conclude and discuss further directions for our work in Sect. 6.

2   The OpenMP Profiler ompP

ompP is a profiling tool for OpenMP applications that relies on the Opari instrumenter [3] for source code instrumentation. ompP is a static library linked to the target application and delivers a text-based profiling report that is meant to be easily comprehensible by the user at program termination. As opposed to subroutine-based profiling tools like gprof, ompP is able to report timings and counts for various OpenMP constructs in terms of the user execution model [4]. As ompP is based on instrumentation-added measurement calls (and not on PC sampling, like gprof), it is *exact* in the sense that the reported values are not subject to sampling inaccuracy. An example for the profiling data delivered by ompP is shown in Fig. 1.

```
R00002 main.c (34-37) (default) CRITICAL
   TID      execT     execC      bodyT     enterT      exitT   PAPI_TOT_INS
     0       3.00         1       1.00       2.00       0.00           1595
     1       1.00         1       1.00       0.00       0.00           6347
     2       2.00         1       1.00       1.00       0.00           1595
     3       4.00         1       1.00       3.00       0.00           1595
   SUM      10.01         4       4.00       6.00       0.00          11132
```

Fig. 1. An example for ompP's profiling data for an OpenMP `critical` section. The body of this critical section contained a call to `sleep(1.0)` only.

The first line gives the source code location and type of construct. Different counts and timing categories (depending on the particular type of OpenMP construct) are listed as column headers and each line lists the accumulated values for a particular thread, while the last line sums over all threads. ompP also supports the measurement of hardware performance counter data for individual threads and OpenMP constructs as well as user-defined regions using PAPI [5]. The user selects a counter to measure via environment variables and the summed counter values then appear as additional columns in the profiles, as shown in Fig 1.

In addition to the flat region profile shown in Fig. 1, ompP performs an overhead analysis in which four well-defined overhead classes (synchronization, load imbalance, thread management, limited parallelism) are quantitatively evaluated. Furthermore ompP tries to detect common inefficiency situations, such as load imbalance in parallel loops, contention for locks and critical sections, etc. The profiling reports contains a list of the discovered instances of these – so called – *performance properties* [6] sorted by their severity (negative impact on performance).

3   Adding Temporal Dimension to Profiling Data

A straightforward way to add a temporal component to profiling-type performance data is to capture profiles at several points during the execution of the target application (and not just at the end) and to analyze how the profiles change between those capture points. Alternatively (and equivalently), the changes between capture points can be recorded incrementally and the overall state at capture time can later be recovered.

Several trigger events for the collection of profiling reports are conceivable:

- **Timer based, fixed:** Profiles are captured in regular fixed intervals during the execution of the target application. No prior knowledge or modification of the application is required. Since data samples are captured in uniform intervals, they are easy to visualize and comprehend. The overhead with respect to storage space for profiling reports is predicable as it depends on the duration of the program run and the capture interval only.
- **Timer based, adaptive:** This method dynamically adapts the duration between two capture points based on amount of change in profiling data that has been observed. Possible measures for this change are the number of different constructs executed and their invocation count. This option has the potential advantage of offering a finer-grained insight into phases of execution where a lot of change occurs while avoiding profiling reports for phases of largely uniform behavior.
- **Overflow based:** This is another method to correlate the profiling frequency with the activity of the program. With this technique, profiling reports are generated when a hardware counter overflows a pre-set threshold. For floating-point intensive applications, it may for example be beneficial to trigger profiling reports each n floating point operations have occurred.

Other potential triggers for profiling reports are the number of cache misses or the occurrence of page faults.

- **User added**: A method to trigger the generation of profiling reports can be exposed to the user. This technique is especially useful for phase-based programs, where program phases are iteratively executed.

In this paper we investigate the simplest form of incremental profiling described above, capturing profiles in regular, fixed-length intervals during the entire execution time of the application.

4   Evaluation of Incremental Profiling

For both profiling and tracing, the following dimensions of performance data can generally be distinguished:

- Kind of data: Describes which type of data is measured or reported to the user. Examples include timing data, execution counts, performance counter values, and so on.
- Source code location: Data can be collected globally for the whole program or for particular source code entities such as subroutines, OpenMP constructs, basic blocks, individual statements, etc.
- Threads or processes dimension: Measured data can either be reported for individual threads or processes or accumulated over the whole set (by summing or averaging, for example).
- Time dimension: Describes when a particular measurement was made (timestamp) or for which time duration values have been measured.

An appealing property of profiling data is its low dimensionality, i.e., it can often be comprehended textually (like gprof output) or it can be visualized as 1D or 2D graphs in a straightforward way. Adding a new (temporal) dimension jeopardizes this advantage and requires more sophisticated performance data displays. We came up with the following types of useful performance views that can be extracted from the incremental profiling reports delivered by `ompP`:[1]

Performance properties over time: Performance properties [6] offer a very compact way to represent performance analysis knowledge and their change over time can thus be easily visualized. A property like "Imbalance in parallel region `foo.f` `(23-42)` with severity of 4.5%" carries all relevant context information with it. The severity value denotes the percentage of total execution time improvement that can be expected if the cause for the inefficiency is completely removed. The threads dimension is collapsed in the specification of the property and the source code dimension is encoded as the properties context (`foo.f` `(23-42)` in the above example).

---

[1] All examples shown in this section come from an execution of the SPEC OpenMP medium size benchmark suite on a 32 CPU SGI Altix machine, based on Itanium-2 processors with 1.6 GHz and 6 MB L3 cache, used in batch mode.

Properties over time can be visualized as a 2D line-plot, where the x-axis is the time and the y-axis denotes severity values and a line is drawn for each property from the first time it was detected until program termination. Depending on the particular test application, valuable information can be deduced, depending on the behavior of the property graphs. For example, in the example graph shown in Fig. 2 it is evident that the severity of the properties appears to be continuously increasing as time proceeds, indicating that the imbalance situations in this code will become increasingly significant with longer runtime (e.g., larger data sets or more iterations).

Other applications from the SPEC OpenMP benchmark suite showed other interesting features such as initialization routines that generated high initial overheads which amortized over time (i.e., the severity decreased).

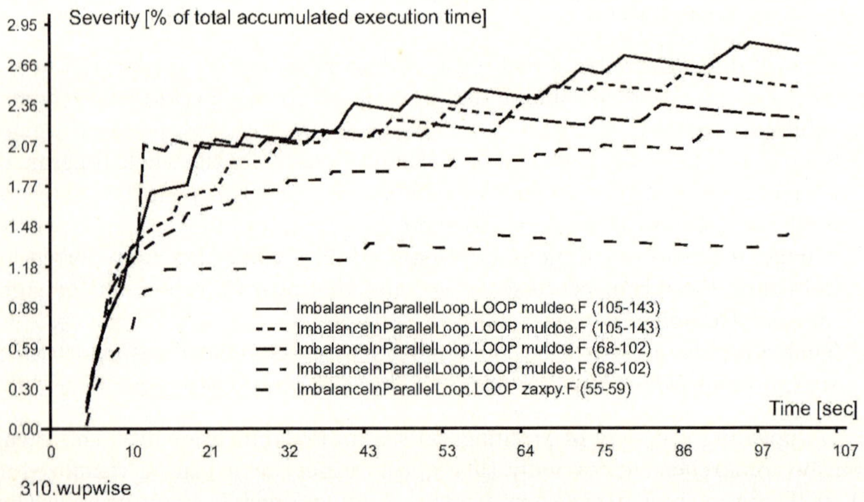

Fig. 2. An example for the "performance properties over time" display for the 310.wupwise application. Shown are the five most severe performance properties.

Region invocations over time: Depending on the size of the test application and the analyst's familiarity with the source code, it can be valuable to know when and how often a particular OpenMP construct, such as a parallel loop, was executed. The region-invocation over time displays offers this functionality.

As shown in Fig. 3 the graph gives the number of invocations of a particular region, this particular case shows the two most time-consuming parallel regions of the 328.fma3d application.

This view is most useful when aggregating (e.g., summing) over all threads, but in certain cases it can be valuable (for critical sections and locks, for example) to see which particular thread executed a construct at which time.

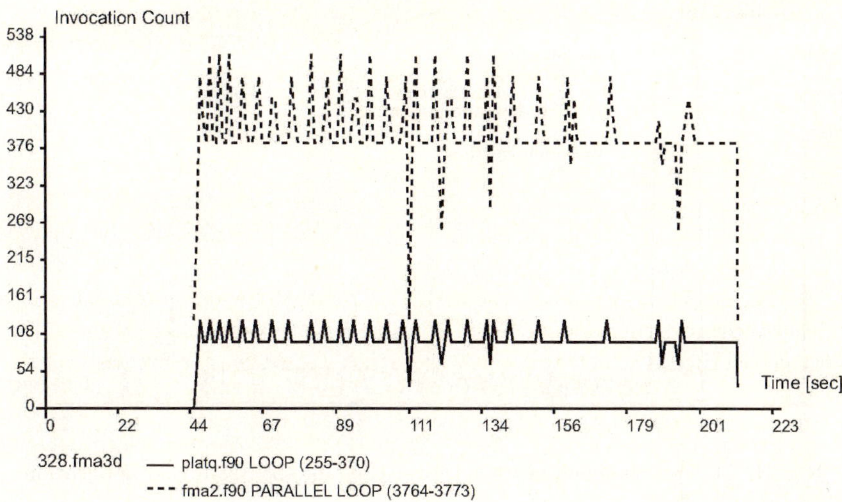

Fig. 3. This graph shows the number of region invocations over time for the 328.fma3d application

A surface plot can be used for visualization in this case or a heatmap display similar to the one used for visualizing performance counters (see below).

Region execution time over time: This display is similar to the region invocation over time display but shows the execution accumulated execution time between dump intervals instead of the invocation count. Again this display allows the user to see when particular portions of the code actually get executed.

Overheads over time: ompP evaluates four overhead classes based on the flat profiling data for individual parallel regions and for the program as a whole. For example, the time required to enter a critical section is attributed to the containing parallel region as synchronization time. A detailed discussion and motivation of this classification scheme can be found in [7].

The overheads over time display plots the incurred overheads (either for a particular parallel region or for the entire application) over the execution timeline of the application, as shown in Fig. 4. This graph gives the overheads as percentage of total aggregated CPU time. Hence, for an execution with 32 CPUs, a overhead percentage of 50 means that 16 CPUs are not doing useful work. The total overhead incurred over the entire program run is thus the integral of the overhead function (area under the graph) and the graph shows when a particular overhead was incurred. In the example in Fig. 4, the most noticeable overhead is synchronization overhead starting at about 30 seconds of execution and lasting for several seconds. A closer examination of the OpenMP profiling reports reveals that this overhead is caused by critical section contention. One thread after the other enters the critical section and performs a time-consuming initialization operation.

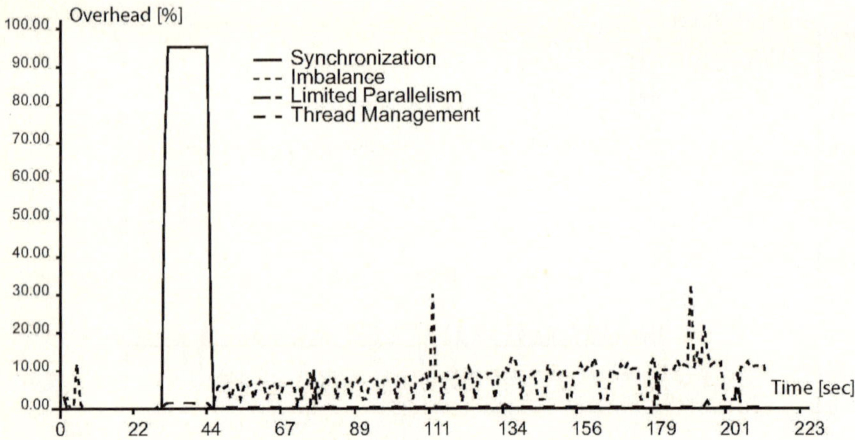

Fig. 4. This graph shows overheads over time for the 328.fma3d application

This effectively serializes the execution for more than 10 seconds and shows up as a overhead of $31/32 = 97\%$ in the overheads graph.

Performance counter heatmaps: This display is used to visualize hardware performance counter values over time and for several threads.

Fig. 5 shows examples of the performance counter heatmap display. The x-axis corresponds to the time (in seconds) while the y-axis corresponds to the thread ID. A color gradient (or gray-scale) coding is indicating high or low counter values. A tile is not filled if no data samples are available for that time period. This type of display is supported for both the entire program as well as for individual OpenMP regions. The example in Fig. 5a shows the DATA_EAR_CACHE_LAT1024 counter for the SPEC OpenMP application 318.galgel (note that the middle part of the timeline has been cut out to facilitate presentation). This counter measures the number of cache misses that took longer than 1024 cycles to be satisfied and thus roughly corresponds to remote memory accesses on the NUMA architecture of the SGI Alitx machine.

Depending on the selected hardware counters, this view offers very interesting insight into the behavior of the applications. Phenomena that we were able to identify with this kind of performance display include:

– The homogeneity or heterogeneity of threads. E.g., often threads 16, 8, and 24 would show markedly different behavior compared to other threads in a 32 thread run. Possible reasons for this difference in behavior might be in the application itself (related to the algorithm) but they could also come from the machine organization or system software layer (mapping of threads to processors and their arrangement in the machine an its interconnect).

As another example, Fig. 5c gives the number of retired floating point operations for the 324.apsi application and this graph shows a marked

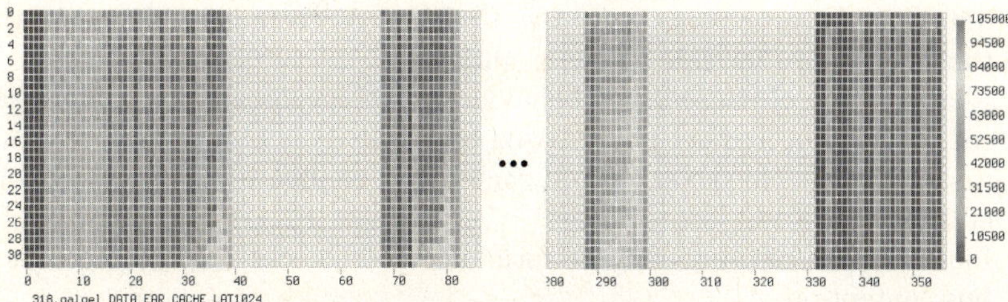

(a) Cache misses that took longer than 1024 cycles to be satisfied for the 318.galgel application

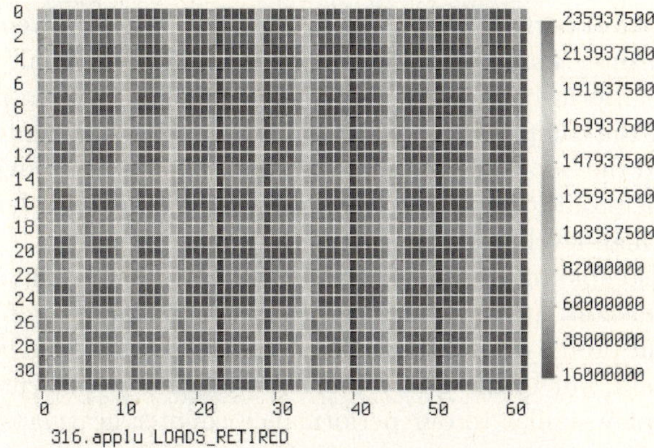

(b) Retired load instructions for the 316.applu application

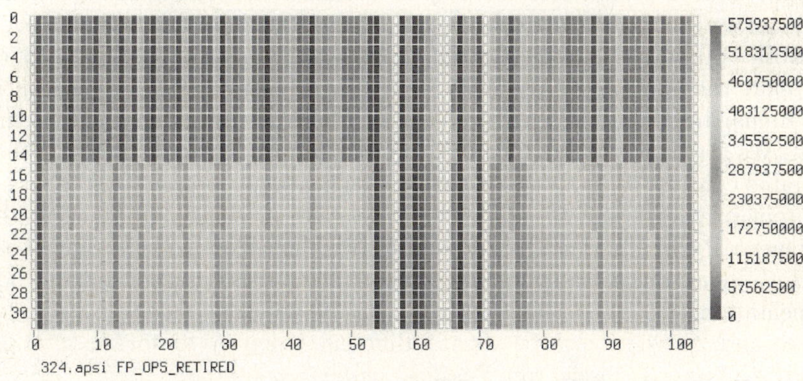

(c) Retired floating point operations for the 324.apsi application

Fig. 5. Example performance counter heatmaps. Time is displayed on the horizontal axis (in seconds), the vertical axis lists the threads (32 in this case). The middle part of 5a of the display has been cut out.

difference for processors 0 to 14 vs 15 to 31. We were not able to identify
the exact cause for this behavior until now.

- Identification of phase-based behavior, as in Figs. 5a and 5b, some applications show a marked phase-based behavior. It is also evident in many
  cases that the characteristics of each phase change from iteration to iteration.
- Identification of temporary performance bottlenecks such as short-term
  bus-contention.

5   Related Work

There are a number of performance analysis tools for OpenMP. Vendor-specific
tools such as the Intel Thread Profiler and Sun Studio are usually limited to
their respective platform but have the advantage of being able to make use of
internal details of the compiler's OpenMP implementation and runtime system.
Both the Intel and the Sun tool are based on sampling and can provide the user
with some timeline profile displays. Neither of those tools however has a concept
similar to `ompP`'s high-level abstraction of performance properties.

TAU [8,9] is also able to profile and trace OpenMP applications by utilizing
the Opari instrumenter. Its performance data visualizer Paraprof supports a
number of different profile displays and also supports interactive 3D exploration
of performance data, but to the best of our knowledge does not currently have
support for a display similar to our performance counter heatmaps. The TAU
toolset also contains a utility to convert TAU trace files to profiles which can
generate profile series and interval profiles.

OProfile and its predecessor the Digital Continuous Profiling Infrastructure
(DCPI) are system-wide statistical profilers based on hardware counter overflows. Both approaches rely on a profiling daemon running in the background
and support the dumping of profiling reports at any time. Data acquisition in
a style similar to our incremental profiling approach would thus be easy to implement. We are, however, not aware of any study using OProfile or DPCI that
investigated continuous profiling for parallel applications. In practice, the necessity of root privileges and the difficulty of relating profiling data back to the
user's OpenMP threading model are major stumbling blocks when using those
tools. Both issues are not a concern with `ompP` since it is based on source code
instrumentation.

6   Conclusion and Future Work

We have presented a study on the utility of incremental profiling for performance
analysis of shared memory parallel applications. Our results indicate that valuable information about the temporal behavior of applications can be discovered
by incremental profiling and that this technique strikes a good balance between
the level of detail offered by tracing and the simplicity and efficiency of profiling. Using incremental profiling we where able to acquire new insights into

the behavior of applications which can due to the lack of temporal data not be gained from pure profiling. The most interesting features are the revelation of iterative behavior, the identification of short-term contention for resources, and the temporal localization of overheads and execution patterns.

Future work is planned in several areas. Firstly, we plan to support other triggers for capturing profiles, most importantly user-added and overflow based. Secondly, be plan to integrate our profiling data with TAU's Paraprof viewer in order to interactively explore the incremental profiling data delivered by `ompP`. Thirdly, we plan to test our ideas in the context of MPI as well, a planned integrated MPI/OpenMP profiling tool based on `mpiP` [10] and `ompP` is the first step in this direction.

References

1. Fuerlinger, K., Gerndt, M.: ompP: A profiling tool for OpenMP. In: Proceedings of the First International Workshop on OpenMP (IWOMP2005), Eugene, Oregon, USA (May 2005) (Accepted for publication)
2. Aslot, V., Eigenmann, R.: Performance characteristics of the SPEC OMP2001 benchmarks. SIGARCH Comput. Archit. News 29(5), 31–40 (2001)
3. Mohr, B., Malony, A.D., Shende, S.S., Wolf, F.: Towards a performance tool interface for OpenMP: An approach based on directive rewriting. In: Proceedings of the Third Workshop on OpenMP (EWOMP'01) (September 2001)
4. Itzkowitz, M., Mazurov, O., Copty, N., Lin, Y.: An OpenMP runtime API for profiling Accepted by the OpenMP ARB as an official ARB White Paper, available online at `http://www.compunity.org/futures/omp-api.html`
5. Browne, S., Dongarra, J., Garner, N., Ho, G., Mucci, P.J.: A portable programming interface for performance evaluation on modern processors. Int. J. High Perform. Comput. Appl. 14(3), 189–204 (2000)
6. Gerndt, M., Fürlinger, K.: Specification and detection of performance problems with ASL. Concurrency and Computation: Practice and Experience (to appear 2006)
7. Fuerlinger, K., Gerndt, M.: Analyzing overheads and scalability characteristics of OpenMP applications. In: Dayde, R. (ed.) VECPAR 2006, Rio de Janeiro, Brasil, vol. 4395, pp. 39–51. Springer, Heidelberg (2006)
8. Shende, S.S., Malony, A.D.: The TAU parallel performance system. International Journal of High Performance Computing Applications (ACTS Collection Special Issue) (2005)
9. Malony, A.D., Shende, S.S., Bell, R., Li, K., Li, L., Trebon, N.: Advances in the TAU performance analysis system. In: Getov, V., Gerndt, M., Hoisie, A., Malony, A., Miller, B. (eds.) Performance Analysis and Grid Computing, pp. 129–144. Kluwer, Dordrecht (2003)
10. Vetter, J.S., Mueller, F.: Communication characteristics of large-scale scientific applications for contemporary cluster architectures. J. Parallel Distrib. Comput. 63(9), 853–865 (2003)

Fine Tuning Algorithmic Skeletons

Denis Caromel and Mario Leyton

INRIA Sophia-Antipolis, CNRS, I3S, UNSA. 2004, Route des Lucioles, BP 93,
F-06902 Sophia-Antipolis Cedex, France
`First.Last@sophia.inria.fr`

Abstract. Algorithmic skeletons correspond to a high-level programming model that takes advantage of nestable programming patterns to hide the complexity of parallel/distributed applications. Programmers have to: define the nested skeleton structure, and provide the muscle (sequential) portions of code which are specific to a problem.

An inadequate structure definition, or inefficient muscle code can lead to performance degradation of the application. Related research has focused on the problem of performing optimization to the skeleton structure. Nevertheless, to our knowledge, no focus has been done on how to aide the programmer to write performance efficient muscle code.

We present the Calcium skeleton framework as the environment in which to perform fine tuning of algorithmic skeletons. Calcium provides structured parallelism in Java using ProActive. ProAcitve is a grid middleware implementing the active object programming model, and providing a deployment framework.

Finally, using a skeleton solution of the NQueens counting problems in Calcium, we validate the fine tuning approach on a grid environment.

1   Introduction

Algorithmic skeletons correspond to a high level programming model which was introduced by Cole [1]. Skeletons take advantage of common programming patterns to hide the complexity of parallel and distributed applications. Starting from a basic set of patterns (skeletons), more complex patterns can be built by nesting the basic ones. All the non-functional aspects regarding parallelization and distribution are implicitly defined by the composed parallel structure. Once the structure has been defined, the programmer completes the program by providing the application's sequential functional aspects to the skeleton, which we refer to as *muscle* codes.

Skeletons are considered a high level programming paradigm because lower level details are hidden from the programmer. Achieving high performance for an application becomes the responsibility of the skeleton framework by performing optimizations on the skeleton structure [2,3], and adapting to the environment's dynamicity [4]. However, while these techniques are known to improve performance, by themselves they are not sufficient. The functional aspects of the application (ie the muscle), which is provided by the programmer, can be inefficient or generate performance degradations inside the framework.

A.-M. Kermarrec, L. Bougé, and T. Priol (Eds.): Euro-Par 2007, LNCS 4641, pp. 72–81, 2007.

Detecting the performance degradation, providing the programmer with an explanation, and suggesting how to solve the performance bugs are the main motivations of this work. The challenge arises because skeleton programming is a high level programming model. All the complex details of the parallelism and distribution are hidden from the programmer. Therefore, the programmer is unaware of how her muscle code will affect the performance of the application. Inversely, low level information of the framework has no meaning to the programmer to fine tune her muscle code.

In this paper we contribute by: (i) providing performance metrics that apply, not only to master-slave, but also to other common skeleton patterns; (ii) taking into consideration the nesting of task and data parallel skeletons as a producer/consumer problem; (iii) introducing the concept of muscle workout; and (iv) introducing a blaming phase that relates performance inefficiency causes with the actual muscle code. In this paper we also present a skeleton framework called *Calcium*. Calcium is written in Java and provides a library for nesting task and data parallel skeletons using dynamic task generation. Support for distributed programming in Calcium is provided by ProActive's [5] active object model [6], and deployment framework [7].

2   Calcium Skeleton Framework

The Calcium skeleton framework is greatly inspired on Lithium [8,9] and its successor Muskel [10]. It is written in Java [11] and is provided as a library. To achieve distributed computation Calcium uses ProActive. ProActive is a Grid middleware [12] providing, among others, a deployment framework [7], and a programming model based active objects with transparent first class futures [6].

Basic task and data parallel skeletons supported in Calcium can be combined and nested to solve more complex applications, in the following way:

$$\triangle ::= farm(\triangle) \mid pipe(\triangle_1, \triangle_2) \mid seq(f_e) \mid$$
$$if(f_b, \triangle_{true}, \triangle_{false}) \mid while(f_b, \triangle) \mid for(i, \triangle)$$
$$map(f_d, \triangle, f_c) \mid fork(f_d, \triangle_1, ..., \triangle_n, f_c) \mid d\&c(f_d, f_b, \triangle, f_c)$$

Where the task parallel skeletons are: *farm* for task replication; *pipe* for staged computation; *seq* for wrapping execution functions; *if* for conditional branching; and *while*/*for* for iteration. The data parallel skeletons are: *map* for single instruction multiple data; *fork* for multiple instruction multiple data; and *d&c* for divide and conquer.

The Calcium framework can be viewed as a producer/consumer problem. A central task pool stores and keeps track of tasks. Tasks are inputed into the task pool by the users who provide the initial configuration of the state parameter. Interpreters consume tasks from the task pool, compute the tasks according to the skeleton instructions, and return the computed tasks to the task pool. Additionally, new tasks can be dynamically produced by the interpreters when data parallelism is encountered, in a similar fashion as in [13]. Dynamically produced tasks are referred to as subtasks, while the task that created them is referred to as the parent task.

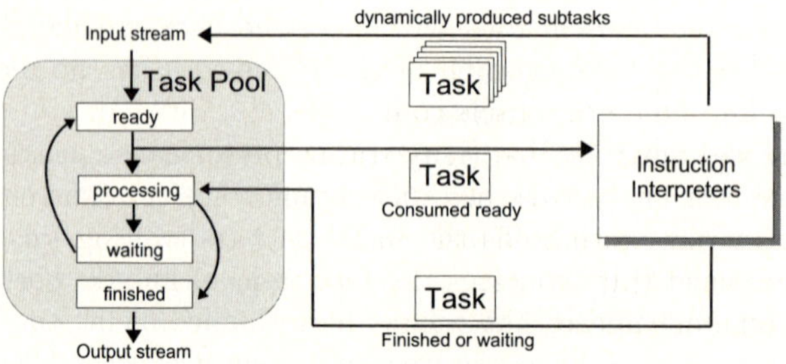

Fig. 1. Calcium Framework Task Flow

As shown in Figure 2(a), a (sub)task is mainly composed of: a skeleton instruction stack, and a state parameter. The instruction stack corresponds to the parsing and execution of the skeleton program, and each instruction is specified with the associated muscle codes. The state parameter is the glue between the skeleton instructions, since the result of one instruction is passed as parameter to the following one.

Since subtasks can also spawn their own subtasks, a task tree is generated as shown in Figure 2(b). The root of the tree corresponds to a task inputed into the framework by the user, while the rest of the nodes represent dynamically generated tasks. In the task tree, the tasks that represent the leaf nodes are

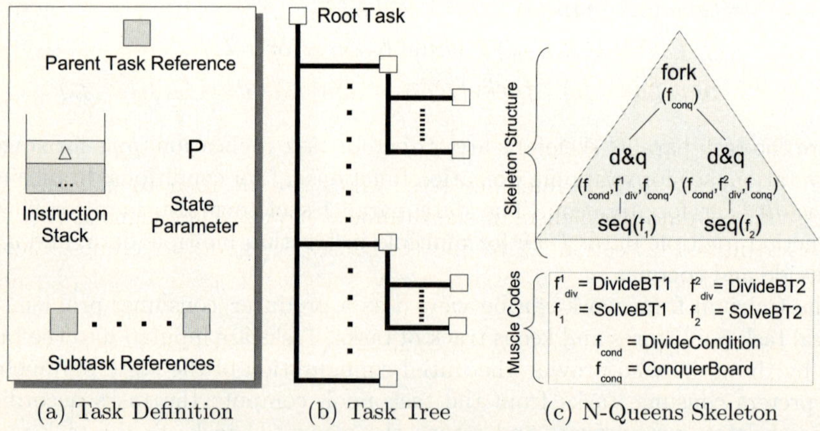

(a) Task Definition   (b) Task Tree   (c) N-Queens Skeleton

Fig. 2. Skeleton Programming Concepts

the ones that are ready for execution or being executed, while the inner nodes represent tasks that are waiting for their subtasks to finish.

When a task has no unfinished children, and no further skeleton instructions need to be executed, then the task is finished. When all brother tasks are finished, they are returned to the parent task for reduction. The parent may then continue with the execution of its own skeleton, and perhaps generate new subtasks. When a root task reaches the finished state it can be delivered to the user.

The task pool is therefore composed of tasks in 4 different states: ready, processing, waiting and finished. The ready state represents tasks that are currently ready for execution. The processing state keeps track of tasks currently being executed. The waiting state holds the tasks that are waiting for their children to finish. The finished state contains the root tasks that have been finished but have not yet been collected by the user.

3   Muscle Tuning of Algorithmic Skeletons

The global concepts that are presented in this section are depicted in Figure 3. After the execution of an application, the performance metrics are used to determine the causes of the performance inefficiency. Once the causes are identified, a blaming process takes places by considering the workout of the muscle code. The result of the blaming process yields the blamed muscle code. The programmer can then analyze the blamed code to fine tune her application.

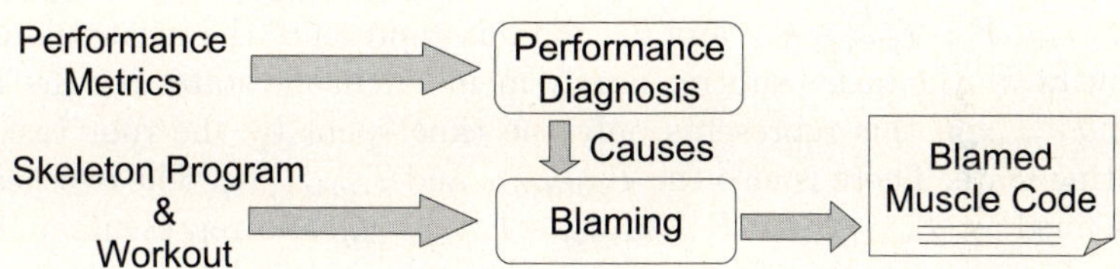

Fig. 3. Finding the tunable muscle code

3.1   Performance Diagnosis

Figure 4 shows a generic inference tree, which can be used to diagnose the performance of data parallelism in algorithmic skeletons. The diagnosis uses the metrics identified in section 3.2 to find the causes of performance bugs. Causes can correspond to two types. *External causes*, such as the framework deployment overhead, or framework overload; and *tunable causes*, which are related with the muscle code of the skeleton program. Since external causes are unrelated with the application's code, in this paper we focus on the tunable causes.

The inference tree has been constructed by considering the performance issues we have experienced with the skeleton framework. As such, the thresholds $\{\alpha, \beta, \ldots\}$ have been determined experimentally, and are provided to the users in three flavors: *weak, normal,* and *strong.*

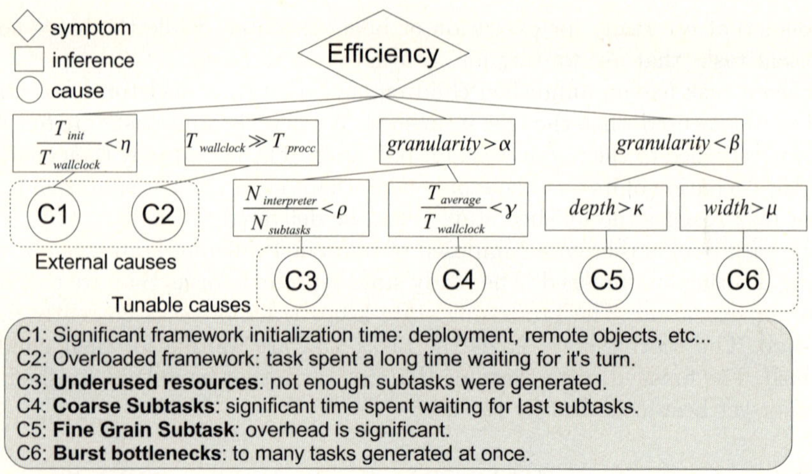

Fig. 4. Generic Cause Inference Tree for Data Parallelism

3.2 Performance Metrics

Task Pool: Number of tasks in each state: $N_{processing}$, N_{ready}, $N_{waiting}$, and $N_{finished}$.

Time: Time spent by a task in each state: $T_{processing}$, T_{ready}, $T_{waiting}$, and $T_{finished}$. For $T_{processing}$ and T_{ready} this represents the accumulated time spent by all the task's subtree family members in the state. For the $T_{waiting}$ and $T_{finished}$, this represents only the time spent by the root task in the waiting state. There is also the $T_{wallclock}$ and $T_{computing}$. The overhead time is defined as: $T_{overhead} = T_{processing} - T_{computing}$, and represents mainly the cost of communication time between the task pool and the interpreters.

Granularity: Provides a reference of the the task granularity achieved during the execution of data parallel skeletons by monitoring the task tree: size, span, and depth ($span^{depth} = size$), and the $granularity = \dfrac{T_{computing}}{T_{overhead}}$.

3.3 Muscle Workout

Let $m_0, ..., m_k$ be an indexed representation of all the muscle code inside a skeleton program $\triangle$. We will say that workout is a function that, given a skeleton program and a state parameter, after the application is executed, returns a list of all the executed muscle codes with the computation time for each instruction:

$$workout(\triangle, p) = [(m_i, t_0), ..., (m_j, t_n)]$$

The skeleton workout represents a trace of how the muscle codes were executed for this skeleton program. The same muscle code can appear more than once in the workout, having different T execution times.

3.4   Code Blaming

Since algorithmic skeletons abstract lower layers of the infrastructure from the programming of the application, low level causes have no meaning to the programmer. Therefore, we must link the causes with something that the programmer can relate to, and this corresponds to the muscle codes which have been implemented by the programmer. The process of relating lower level causes with the responsible code is what we refer to as *code blaming*.

A blaming mechanism must link each inferred cause with the relevant muscle codes. Thus, the blaming must consider: lower level causes (the result of the performance diagnosis), the skeleton program, and the muscle code's workout.

Let us recall that for any skeleton program, it's muscle codes must belong to one of the following types: $f_{execution}, f_{condition}, f_{division}, f_{conquer}$. Additionally, the semantics of the muscle code depend on the skeleton pattern where it is used. A simple implementation of a blaming algorithm is the following:

(C3) Underused resources: Blame the most invoked $f_{cond} \in \{d\&q\}$ and $f_{div} \in \{map, d\&q, fork\}$. Suggest incrementing the times f_{cond} returns *true*, and suggest that f_{div} divides into more parts.

(C4) Coarse subtasks: Blame the least invoked $f_{cond} \in \{d\&q\}$. Suggest incrementing the times f_{cond} returns *true*.

(C5) Fine grain subtasks: Blame the most invoked $f_{cond} \in \{d\&q\}$. Suggest reducing the number of times f_{cond} returns *true*.

(C6) Burst Bottlenecks: Blame the $f_{div} \in \{map, d\&q, fork\}$ which generated the most number of subtasks. Suggest modifying f_{div} to perform less divisions per invocation.

While more sophisticated blaming mechanisms can be introduced, as we shall see in the NQueens test case, even this simple blaming mechanism can provide valuable information to the programmer.

4   NQueens Test Case

The experiments were conducted using the *sophia* and *orsay* sites of Grid5000 [14], a french national grid infrastructure. The machines used AMD Opteron CPU at 2Ghz, and 1 GB RAM. The task pool was located on the sophia site, while the interpreters where located on the orsay site. The communication link between sophia and orsay was of $1[\frac{Mbit}{sec}]$ with $20[ms]$ latency.

As a test case, we implemented a solution of the NQueens counting problem: How many ways can n non attacking queens be placed on a chessboard of nxn? Our implementation is a skeleton approach of the Takaken algorithm [15], which takes advantage of symmetries to count the solutions. The skeleton program is shown in Figure 2(c), where two **d&c** are forked to be executed simultaneously. The first **d&c** uses the *backtrack1* algorithm which counts solutions with 8 symmetries, while the other **d&c** uses the *backtrack2* algorithm that counts solutions with 1, 2, and 4 symmetries.

W	Speedup	Efficiency	Granularity	Size	Width	Depth	T_avg [ms]	T_avg/T_wall
16	11,68	0,12	0,13	24892	11	4,2	0,59	0,05
17	97,27	0,97	5,33	2178	13	3	6,8	3,45
18	67,59	0,68	59,18	166	13	2	90,99	33,89

(a) Performance Metrics

Muscle Code	w = 16		w = 17		w = 18	
	Invoqued [#]	Time[ms]	Invoqued [#]	Time[ms]	Invoqued [#]	Time[ms]
DivideBT1	259	19	18	1	1	1
DivideBT2	1918	154	147	26	10	1
ForkDefaultDivide	1	2	1	2	1	2
SolveBT2	19872	13619957	1771	13730299	137	13910314
ConquerBoard	2178	175	166	15	12	1
SolveBT1	2842	1103186	241	1097325	17	1207444
DivideCondition	24891	523	2177	64	165	10

(b) Workout Summary

Fig. 5. Performance Metrics & Workout Summary for $n = 20$, $w \in \{16, 17, 18\}$

We have chosen the NQueens problem because the fine tuning of the application is centered in one parameter, and therefore represents a comprehensible test case. Task division is achieved by specifying the first $n - w$ queens on the board, where w is the number of queens that have to be backtracked. The problem is thus known to be $O(n^n)$, with exponential increase of the number of tasks as w decreases. Therefore, the programmer of the NQueens problem must tune a proper value for w.

Since the output of the blaming phase corresponds to the blamed muscle code, it is up to the user to identify the parameters that change the behavior of the muscle code. Therefore, the same approach can be used for applications that require tuning of multiple parameters.

We tested a problem instances for $n = 20$, $w \in \{16, 17, 18\}$, $nodes = 100$. Relevant metrics are shown in Figure 5(a), and a summary of the workout is shown in Figure 5(b). From the performance point of view, several guides can be obtained by simple inspection of the workout summary. For example, that the method `DivideCondition` must remain computationally lite because it is

```
//(n = 20, w = 16)
  Performance inefficiency found.
    Cause: Subtask are too fine grain, overhead is significant.
    Blamed Code: public boolean nqueens.DivideCondition.evalCondition(nqueens.Board)
    Suggested Action: This method should return true less often.


//(n = 20, w = 17)
  No inefficiency found.


//(n = 20, w = 18)
  Performance inefficiency found.
    Cause: Subtask are too coarse, significant time spent waiting for last subtasks.
    Blamed Code: public boolean nqueens.DivideCondition.evalCondition(nqueens.Board)
    Suggested Action: This method should return true more often.
```

Fig. 6. Fine Tuning Output for $n = 20, w \in \{16, 17, 18\}$

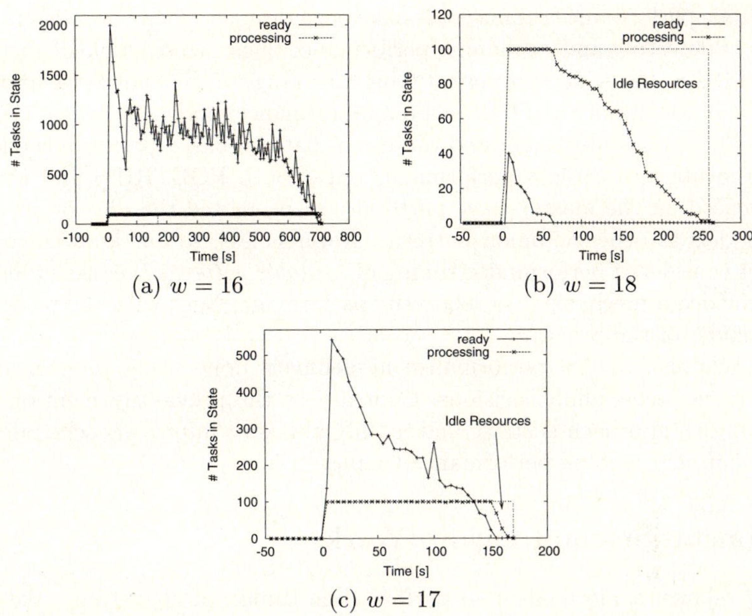

(a) $w = 16$

(b) $w = 18$

(c) $w = 17$

Fig. 7. Number of ready and processing subtasks for $n = 20$, with 100 nodes

called the most number of times, and that further optimizations of `SolveBT2` will provide the most performance gain.

The result of the blaming process is shown in Figure 6, where the $f_{condition}$ muscle code (`DivideCondition`) appears as the main tuning point of the application. For $w = 16$ the results suggest that the granularity of the subtasks is to fine, while for $w = 18$ the granularity is to coarse. To better understand why this takes place Figures 7(a), 7(c), 7(b) show the number of subtasks in ready and processing state during the computation. The figures concur with the fine tuning output. As we can see, the fine tuning of a skeleton program can have a big impact on the performance of the application. In the case of the NQueens test case, the performance improves up to 4 times when choosing $w = 17$ instead of $w \in \{16, 18\}$.

5   Related Work

Calcium is mainly inspired by Lithium [8,9] and Muskel [10] frameworks, where skeleton are provided to the programmer through a Java API. Research related to Lithium and Muskel has mainly focused on providing optimizations based on: skeleton rewriting techniques [8,2], task lookahead, and server-to-server lazy binding [3].

In Calcium we explore deeper into the internals of the skeleton task pool. Understanding it's design is vital to comprehend and expose adequate performance metrics concerning the operation of the skeleton framework.

Dynamic performance tuning tools have been designed to aid developers in the process of detecting and explaining performance bugs. An example of such tools is POETRIES [16], which proposes taking advantage of the knowledge about the structure of the application to develop a performance model. Hercules [17] is another tool that has also suggested the use of pattern based performance knowledge to locate and explain performance bugs. Both POETRIES and Hercules have focused on the master-slave pattern, and promoted the idea of extending their models to other common patterns. Nevertheless, to our knowledge, none have yet considered performance tuning of *nestable patterns* (i.e. skeletons), nor have provided a mechanism to relate the performance bugs with the responsible muscle codes of the program.

For skeletons, in [18] performance modeling is done using process algebra for improving scheduling decisions. Contrary to the previously mentioned approaches, this approach is static and mainly aimed at improving scheduling decisions, not at providing performance tuning.

6   Conclusions and Future Work

We have shown a mechanism to perform fine tuning of algorithmic skeletons' muscle code. The approach extends previous performance diagnosis techniques that take advantage on pattern knowledge by: taking into consideration netsable skeleton patterns, and relating the performance inefficiency causes with the skeleton's responsible muscle code. This is necessary because skeleton programming is a higher-level programming model, and as such, low level causes of performance inefficiencies have no meaning to the programmer.

The proposed approach can be applied to fine tune applications that are composed of nestable skeleton patterns. To program such applications we have presented the Calcium skeleton framework. The relation of inefficiency causes with the responsible muscle code is found by taking advantage of the skeleton structure, which implicitly informs the role of each muscle code.

We have validated the approach with a test case of the NQueens counting problem. The experiments where conducted on Grid5000 with up to a 100 nodes.

In the future we would like to improve the performance diagnosis inference tree, and the code blaming mechanism to consider, among others, stateful skeletons. Additionally, in the case where many causes are found, we would like to provide a prioritization scheme to be able to help the programmer decide which are the most significant and relevant muscle codes that must be fine tuned.

References

1. Cole, M.: Algorithmic skeletons: structured management of parallel computation. MIT Press, Cambridge, MA, USA (1991)
2. Aldinucci, M., Danelutto, M.: Stream parallel skeleton optimization. In: Proc. of PDCS: Intl. Conference on Parallel and Distributed Computing and Systems, Cambridge, Massachusetts, USA, November 1999, pp. 955–962. IASTED, ACTA press (1999)

3. Aldinucci, M., Danelutto, M., Dünnweber, J.: Optimization techniques for implementing parallel skeletons in grid environments. In: Gorlatch, S. (ed.) Proc. of CMPP: Intl. Workshop on Constructive Methods for Parallel Programming, Stirling, Scotland, UK, July 2004, pp. 35–47. Universität Münster, Germany (2004)

4. Danelutto, M.: Qos in parallel programming through application managers. In: PDP '05: Proceedings of the 13th Euromicro Conference on Parallel, Distributed and Network-Based Processing (PDP'05), Washington, DC, USA, pp. 282–289. IEEE Computer Society Press, Los Alamitos (2005)

5. ProActive: http://proactive.objectweb.org

6. Caromel, D.: Toward a method of object-oriented concurrent programming. Communications of the ACM 36(9), 90–102 (1993)

7. Baude, F., Caromel, D., Mestre, L., Huet, F., Vayssière, J.: Interactive and descriptor-based deployment of object-oriented grid applications. In: Proceedings of the 11th IEEE International Symposium on High Performance Distributed Computing, Edinburgh, Scotland, July 2002, pp. 93–102. IEEE Computer Society Press, Los Alamitos (2002)

8. Aldinucci, M., Danelutto, M., Teti, P.: An advanced environment supporting structured parallel programming in Java. Future Generation Computer Systems 19(5), 611–626 (2003)

9. Danelutto, M., Teti, P.: Lithium: A structured parallel programming enviroment in Java. In: Sloot, P.M.A., Tan, C.J.K., Dongarra, J.J., Hoekstra, A.G. (eds.) ICCS 2002. LNCS, vol. 2330, pp. 844–853. Springer, Heidelberg (2002)

10. Danelutto, M., Dazzi, P.: Joint structured/unstructured parallelism exploitation in Muskel (to appear). In: Alexandrov, V.N., van Albada, G.D., Sloot, P.M.A., Dongarra, J.J. (eds.) ICCS 2006. LNCS, vol. 3991, Springer, Heidelberg (2006)

11. Microsystems, S.: Java, http://java.sun.com

12. Caromel, D., Delbe, C., Costanzo, A., Leyton, M.: Proactive: an integrated platform for programming and running applications on grids and p2p systems. Computational Methods in Science and Technology 12 (2006)

13. Priebe, S.: Dynamic task generation and transformation within a nestable workpool skeleton. In: Nagel, W.E., Walter, W.V., Lehner, W. (eds.) Euro-Par 2006. LNCS, vol. 4128, pp. 615–624. Springer, Heidelberg (2006)

14. Grid5000: Official web site, http://www.grid5000.fr

15. Takaken: N queens problem, http://www.ic-net.or.jp/home/takaken/e/queen/

16. Cesar, E., Mesa, J.G., Sorribes, J., Luque, E.: Modeling master-worker applications in poetries. hips 00, 22–30 (2004)

17. Li, L., Malony, A.: Model-based performance diagnosis of master-worker parallel computations. In: Nagel, W.E., Walter, W.V., Lehner, W. (eds.) Euro-Par 2006. LNCS, vol. 4128, pp. 35–46. Springer, Heidelberg (2006)

18. Benoit, A., Cole, M., Gilmore, S., Hillston, J.: Evaluating the performance of skeleton-based high level parallel programs. In: Bubak, M., van Albada, G.D., Sloot, P.M.A., Dongarra, J.J. (eds.) ICCS 2004. LNCS, vol. 3036, pp. 299–306. Springer, Heidelberg (2004)

Topic 2
Performance Prediction and Evaluation

Wolfgang Nagel, Bruno Gaujal, Tugrul Dayar, and Nihal Pekergin

Topic Chairs

Parallel algorithms used to be evaluated using some version of the PRAM model where actual execution platforms are abstracted as ideal parallel machines. On the other hand the performance of hardware is often given in terms of individual pick performances which can be useless for actual applications. The real challenge for performance predictions and evaluations of parallel systems is to combine the top and low layers points of view, where congestions, control mechanisms, failures or even breakdowns do alter the behavior of a large distributed platform running a parallel application. In many different ways, this challenge is addressed in all the papers presented in the track, using either new modelling techniques, or sophisticated measures, or experimental approaches.

This track is made of eight papers.

In "Decision Trees and MPI Collective Algorithm Selection Problem", the authors study the applicability of decision trees to the MPI collective algorithm selection problem.

In "Profiling of Task-based Applications on Shared Memory Machines: Scalability and Bottlenecks", new measures are introduced to measure the scalability of task pools algorithms using a non-intrusive profiling method.

In "Search Strategies for Automatic Performance Analysis Tools", a set of agents distributed on the parallel machine are used to/ analyse the performance of parallel programs.

In "Experiences Understanding Performance in a Commercial Scale-Out Environment", the new performance challenges faced in clusters of loosely connected machines are studied using a systematic experimental approach.

In "TAUoverSupermon: Low-Overhead Online Parallel Performance Monitoring", two systems - Tuning and Analysis Utility (TAU) and Supermon are combined to perform online monitoring of large distributed applications with a very low overhead.

In "Practical Differential Profiling", the authors present eGprof, a tool that facilitate performance profile comparisons.

In "Detecting Application Load Imbalance on High End Massively Parallel Systems", a new set of indirect metrics are used to identify and measure application load imbalance. These metrics have been added to the Cray performance measurement and analysis infrastructure.

Finally, in "A First Step Towards Automatically Building Network Representations", the authors present new algorithms to design tools which automatically build a topological network model through some measurements and show that they accurately predict the running time of simple applications over actual platforms.

A.-M. Kermarrec, L. Bougé, and T. Priol (Eds.): Euro-Par 2007, LNCS 4641, p. 83, 2007.

TAUoverSupermon: Low-Overhead Online Parallel Performance Monitoring

Aroon Nataraj[1], Matthew Sottile[2], Alan Morris[1],
Allen D. Malony[1], and Sameer Shende[1]

[1] Department of Computer and Information Science
University of Oregon, Eugene, OR, USA
{anataraj,amorris,malony,sameer}@cs.uoregon.edu
[2] Los Alamos National Laboratory, Los Alamos, NM, USA
matt@lanl.gov

Abstract. Online application performance monitoring allows tracking performance characteristics during execution as opposed to doing so post-mortem. This opens up several possibilities otherwise unavailable such as real-time visualization and application performance steering that can be useful in the context of long-running applications. As HPC systems grow in size and complexity, the key challenge is to keep the online performance monitor scalable and low overhead while still providing a useful performance reporting capability. Two fundamental components that constitute such a performance monitor are the measurement and transport systems. We adapt and combine two existing, mature systems - TAU and Supermon - to address this problem. TAU performs the measurement while Supermon is used to collect the distributed measurement state. Our experiments show that this novel approach leads to very low-overhead application monitoring as well as other benefits unavailable from using a transport such as NFS.

Keywords: Online performance measurement, cluster monitoring.

1   Introduction

Online or real-time application performance monitoring tracks performance characteristics during execution and makes that information available to consumers at runtime. In contrast to post-mortem performance data analysis, developing monitoring capabilities opens up several possibilities otherwise unavailable, such as real-time visualization and application performance steering. One advantage of performance monitoring can be found in the tracking, adaptation, and control of long-running applications. The simple ability to detect problems early in a running application and terminating the execution has practical benefits to savings in computing resource usage. However, as HPC systems grow in size and complexity, building efficient, scalable parallel performance monitoring systems that minimize adverse impact on applications is a challenge.

Two fundamental components constitute an online application performance monitor: 1) the performance measurement system and 2) the transport system.

A.-M. Kermarrec, L. Bougé, and T. Priol (Eds.): Euro-Par 2007, LNCS 4641, pp. 85–96, 2007.

The infrastructure for performance measurement defines what performance metrics can be captured for events in individual execution contexts (e.g., processes and threads). Effectively, the measurement system is the performance data producer. The transport system enables querying the parallel/distributed performance state from the different contexts and delivers the data to monitoring consumers. The transport system acts as a bridge between where the data is generated to where it is consumed, but it can also be used to control the measurement system to control type and rate of performance data production.

While any performance measurement raises issues of overhead and perturbation, online monitoring introduces additional concerns. In static application performance measurement (with post-mortem data access and analysis), the measurement sub-systems of individual contexts are isolated as they perform local measurements and are impacted little by the scale of the parallel application. In contrast, the need to retrieve and integrate the global performance state of individual contexts means measurement sub-systems are no longer isolated and must support periodic interactions from the monitor, potentially affecting performance behavior and scalability. The challenge is to create an online measurement system that is scalable and very low overhead and can still provide useful performance analysis capabilities.

Our solution, *TAUoverSupermon* (*ToS*), adapts two existing mature systems – TAU [1] (as the measurement system) and Supermon [2] (as the transport system) – to address the problem of scalable, low-overhead online performance monitoring of parallel applications. We describe the ToS design and architecture in Section 3. The efficiency of our approach is demonstrated in Section 4 through experiments to evaluate *ToS* performance and scalability. An example to demonstrate application/system performance correlation is provided in Section 5. Section 6 examines related work. The paper concludes in Section 7 with final remarks and a brief discussion of future directions. We begin with a discussion of the rationale behind our approach in the next section, Section 2.

2   Rationale and Approach

There are several approaches to building a parallel application monitoring framework, combining existing performance measurement infrastructure with a system to transport performance data to monitoring consumers. The simplest is to use filesystem capabilities for data movement. Here, the performance measurement layer implements operations to write performance data to files. The filesystem aggregates the performance information which can be seen by consumers when they perform file read operations. Concurrent file I/O between the parallel application and performance data consumers must be possible for this alternative to be of any use. While NFS support provides for robust implementation of this approach, file I/O overheads can be high (both on the producer and consumer sides), the monitor control must also be implemented through file transactions, and NFS has known scalability issues. Furthermore, in the rare case that a globally visible filesystem is unavailable, a file-based transport system is simply not an option.

If a file system is not used, some alternative transport facility is necessary. A measurement system could be extended to include this support, but it would quickly lead to incompatibilities between monitors and significant programming investment required to build a scalable, low-overhead transport layer. Instead, where transport functionality has been developed in other system-level monitoring tools, an application-level performance monitor can leverage the functionality for performance data transport. Our approach couples the Supermon cluster monitor with the TAU application measurement system to provide the measurement and transport capabilities required for online parallel application performance monitoring. TAU performs the role of the performance producer (source) and is adapted to use Supermon as the transport from which consumers (sinks) query the distributed performance state.

The rationale behind TAUoverSupermon is based on the following criteria:

Reduced overhead. Using a traditional filesystem as monitor transport incurs high overhead. Currently, TAU allows runtime performance output through the filesystem, but suffers high overheads as discussed in Section 4.

Autonomous operation. Keeping the transport outside TAU makes available an online transport facility to any other system components that want to use it. Supermon can be used for several purposes, such as its default system monitoring role, independently of TAU.

Separation of concerns. Concerns such as portability, scalability, and robustness can be considered by the measurement and transport systems separately. Both TAU and Supermon are mature, standalone systems whose implementations are optimized for their respective purposes.

Performance correlation. This approach allows close correlation between system level information (such as from the OS and hardware sensors) with application performance (see Section 5). This facilitates determining the root cause of performance problems that may originate from outside of the application.

Light-weight control. Feedback to the measurement system is important for controlling monitoring system overhead. The control path should be light weight to avoid having its use be a significant contributing factor.

3   The *TAUoverSupermon* Architecture

The high-level architecture of ToS is shown in Figure 1. It is composed of the following interacting components:

TAU. The TAU measurement system (in blue) generates performance data for each thread of execution due to application instrumentation. TAU implements an API to output performance data during or at the end of execution.

Supermon. The Supermon transport, including the root and intermediate Supermon daemons are located on intermediate (or service) nodes. The mon daemons (shown in green), are located on each compute node where the application is running.

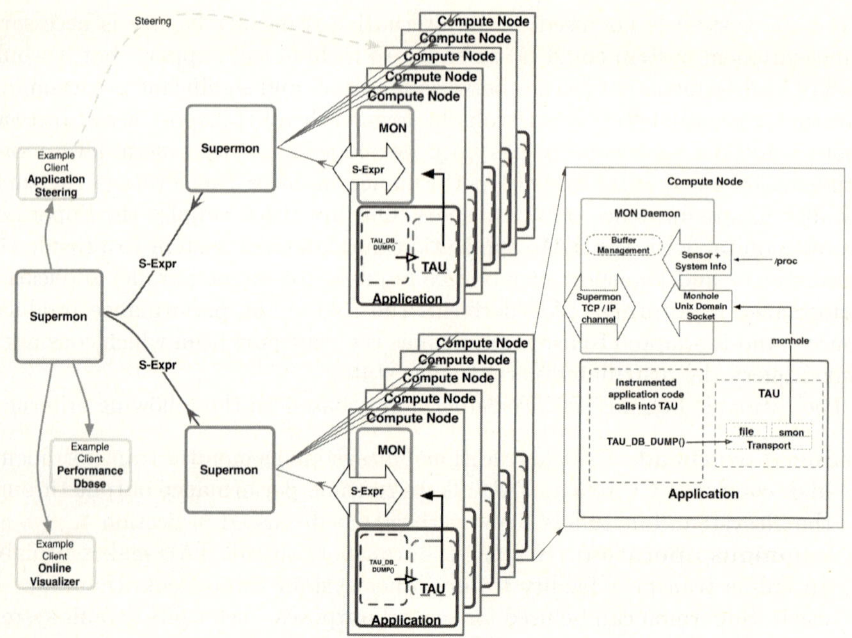

Fig. 1. *TAUoverSupermon* Architecture

Clients. Monitor clients (sinks) of the online performance data (shown in yellow) can be located anywhere on the network connected to the cluster.

In the following sections we discuss each of the components in detail, their implementation, and the changes we made to adapt them for *ToS*.

3.1   TAU Performance System

TAU [1] is an integrated toolkit for parallel performance instrumentation, measurement, analysis, and visualization of large-scale parallel applications. TAU provides robust support for multi-threaded, message passing, or mixed-mode programs. Multiple instrumentation options are available, including automatic source instrumentation. The TAU measurement system implements both parallel profiling and tracing. We are mainly concerned with profile-based performance data in this paper. TAU supports an API to read the performance profile data stored in its internal data structures at runtime, and to output the performance data to a monitoring system using the TAU DB DUMP() call. It is this output interface that will be extended to support Supermon.

3.2   Supermon Cluster Monitor

Supermon [2] is a scalable monitoring system for high performance computing systems. Its current implementation includes a set of socket-based servers that gather, organize and transfer monitoring data in a format based on LISP symbolic expressions (called s-exprs). Its architecture is hierarchical, whereby each node is treated as a leaf node in a tree, while a series of data concentrators gather data from the nodes and transport it to a root. The root then provides data to clients. Supermon's primary purpose has been in monitoring system-level performance such as those reported by hardware-sensors and OS performance data. The Supermon architecture builds on prior experiences with prior implementations based on SunRPC, followed by a non-hierarchical wire protocol over sockets, to achieve a low-overhead, highly extensible monitoring system in the current design. A Supermon system consists of the following components:

Kernel module. Within the compute node OS is a kernel module that exports system-level parameters locally as a file in the */proc* pseudo-filesystem formatted as s-exprs to avoid overhead from parsing the inconsistent formats of the various data sources.

***mon* daemon.** The *mon* daemon, on each compute node, reads the file under */proc* and makes available the system-level metrics as s-exprs over the network via TCP/IP.

monhole. *mon* also listens on a Unix Domain Socket (UDS) locally accepting data from any other sources outside the kernel. This interface to *mon* is called the *monhole*.

Supermon daemon. On the service node(s), there is the Supermon daemon. This talks to each of the *mon* daemons and queries and collects the data from them. This data includes the */proc* system-level parameters as well as data submitted through the *monhole* interface. Supermon then makes this data available as another s-expression to any interested clients. Using the same s-expr format as the *mon* daemons, Supermon daemons may act as clients to other Supermon daemons to create a scalable, hierarchical structure for transporting data from sources to sinks.

3.3   Coupling TAU and Supermon

Figure 1 depicts the interaction between the application instrumented with TAU and the *mon* daemon on the compute node through the *monhole* interface. Below we describe the changes made to Supermon and TAU to build the *ToS* system.

Adapting Supermon

The *mon* daemon provides the UDS-based *monhole* interface for external sources to inject data. We tested and updated *monhole* for TAU's use and made its buffering characteristics more flexible. The buffering policy name refers to how the existing data in the *mon* daemon buffer is managed due to a new write or a read. Some possible policies are:

***REPLACE-READ*:** Existing data is replaced (i.e., overwritten) on a write, and the buffer remains unaffected on a read.

***FILL-DRAIN*:** Writes append to buffer and reads empty the buffer.

REPLACE-DRAIN: Writes replace buffer data and reads empty the buffer.
FILL(K)-READ: Writes append data, but the buffer is unaffected by reads.
  Given a ring buffer of finite size K, repeated writes $(> K)$ will overwrite
  data.

The buffer policy is important as it determines how the buffer is affected by multiple concurrent clients and what data is retrieved. It also determines what guarantees regarding data loss can be made (e.g., when sampling rate does not match data generation rate), and what memory overhead on the data source is required to support maintaining data for clients to read.

Initially, the *monhole* only supported the simple and efficient *REPLACE-READ*. This policy has several advantages: i) slow sinks will not cause infinitely large buffers to be maintained at the *mon* daemon, and ii) multiple sinks can query the data simultaneously without race conditions. However, the policy suffers from potential data-loss when sink read rate (even transiently) is less than source generation rate, or bandwidth waste when sinks query too frequently and receive the same data. A small configurable buffer under *FILL(K)-READ* can alleviate the former, whereas a *REPLACE-DRAIN* strategy can remedy the latter when a single client is used. For these reasons, we implemented a runtime-configurable buffer strategy. The repair mechanism for hierarchical topologies was also fixed in Supermon.

Adapting TAU

Prior to our work with Supermon, TAU assumed the presence of a shared network filesystem for performance monitoring. Buffered file I/O routines were used in the TAU monitoring API. We first made the notion of *transport* a first-class entity by creating a generic transport class. To keep changes isolated to a small portion of the TAU code base, the generic transport class needed to expose interfaces exactly like the file I/O calls in the standard I/O library, *stdlib*. As shown in Figure 1, two implementations of this transport class were created: one for the default *stdlib* file I/O and the other for use with the *monhole* interface. The type and nature of the transport being used is kept hidden from the TAU API. The type of transport can be fixed statically at compile-time or can be communicated to the application via an environment variable at application startup. While read/write operations on the *monhole* are performed directly, other operations such as directory creation are not directly available and need to be forwarded to sinks (on control nodes). This framework allows easy extension by adding new custom transports to TAU in the future.

4   Investigating Performance and Scalability

To evaluate TAUoverSupermon we use the NAS Parallel LU application (Class C) benchmark [3] instrumented with TAU under different configurations. The choice of the benchmark was guided by the need for a representative parallel workload; one that triggers a sufficient number of events, so as to study the overhead generated as a function of number of profile measurements that take place.

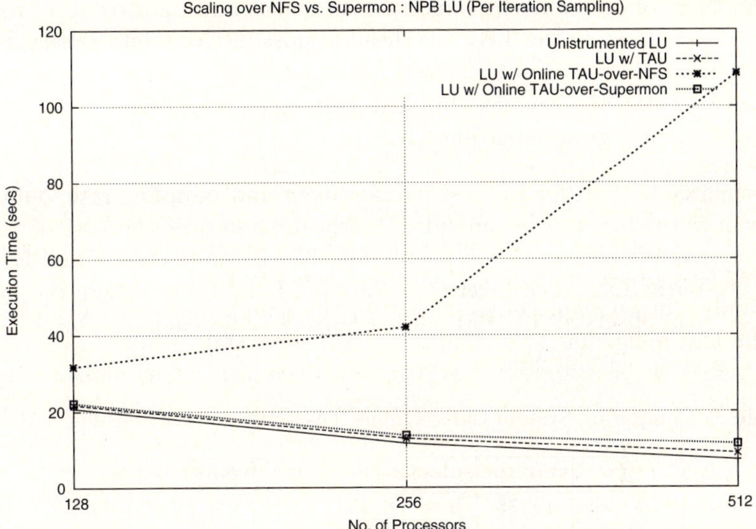

Fig. 2. Execution Time and Overhead

LU has a mix of routine and MPI events and an understood parallel algorithm that lets us relate overhead to scaling behavior. We compare the performance of NPB LU under the following configurations:

LU-none: Basic LU without instrumentation.
LU-PM: LU instrumented with TAU for post-mortem measurement data.
LU-NFS: LU instrumented with TAU for online measurement using NFS.
LU-ToS: LU instrumented with TAU for online measurement using Supermon.

Online measurement data retrieval of LU is performed at a frequency of *once per iteration*. We repeat each of the runs over 128, 256 and 512 nodes to examine scalability. The Atlas cluster from Lawrence Livermore National Lab, with quad dual-core Opteron Linux nodes running Infiniband, serves as our test environment. The metrics we use are the total runtime reported by LU and the overhead as % dilation, computed as the total runtime under some configuration divided by the total runtime of *LU-none* configuration.

In Figure 2 we plot the runtime of the LU benchmark under the different configurations as the processor count increases. The following observations are clear:

- TAU measurements (LU-PM) contributed 4.7% (N=128) to 24.6% (N=512) overhead. Re-runing the LU-PM (N=512) with TAU configured to use the light weight cycle counter (*rdtsc*) for timing brought the overhead down to just 2.5%.
- Overhead of online performance measurement and data-retrieval using NFS is at least 52.71% and grows super-linearly as the number of CPUs increase to a staggering 1402.6%.

- Overhead of online performance measurement and data-retrieval using Supermon is close to the TAU overhead of post-mortem data retrieval (as low as 6.83%).
- As LU scales, the savings obtained from using Supermon transport as opposed to NFS grow super-linearly.

It is remarkable that, for the test measurement and sampling rate, online measurement with *ToS* can be provided nearly for free over the cost of the *postmortem* run. We also ran experiments for the 128 node case (Class B) on the MCR cluster at Lawrence Livermore National Laboratory. There the following dilations were observed: LU-PM 8.5%, LU-NFS 72.6% and LU-ToS 9.1%.

Table 1. Comparing System Calls: Online TAU-NFS vs. Postmortem TAU (secs)

Type	rename	select	open	writev	read	close	write
Tau-NFS	**11.75**	9.46	**8.55**	4.02	3.22	2.50	0.63
Tau-PM	**0**	5.94	**0.03**	3.95	3.22	0	0.60

Why is there such a dramatic difference in performance between using the NFS transport and Supermon? To further investigate what aspects of the system contribute to the significant savings accrued, we use KTAU [4] to measure kernel-level events. Smaller LU-PM and LU-NFS experiments on 4 nodes (of Pentium III dual-CPU over Ethernet) are run, this time under a KTAU kernel. Table 1 compares the runtime of the largest system calls under both configurations, as measured by KTAU. Surprisingly the largest differences are seen in `sys_rename` and `sys_open` and not in the read/write calls. Why?

When files are used to transport performance data from TAU to a monitoring client, there is a problem of read consistency. If the client polls for new data, how does it know when and how much data is new? TAU uses a two-stage process: 1) write to a temporary file, then 2) rename the file to the filename being polled by the client. This approach employs the `rename` and `open` meta-data operations on every performance data dump. These meta-data operations are synchronous and blocking (between the client and the server), unlike the buffered read/write operations in NFS. The impact of these simultaneous meta-data operations grows significantly as node-count increases. In the case of the Supermon transport, these operations are not performed locally. Instead they are also made asynchronous (non-blocking) and performed by the sink (on the control/head node). Another aspect to note is the *'per iteration sampling frequency'* used (instead of, say, a fixed *1Hz sampling*). Because of the strong scaling nature of LU, as the number of nodes increase, the iterations become shorter and the overhead per unit time from data retrieval increases. When the dump operation is relatively costly, as in NFS, it results in the superlinear scaling behavior. In addition, the variability in the time taken by each NFS dump operation across the ranks leads to magnification of the overhead.

5   Online Application/System Performance Correlation

To give a sense of the power of online performance monitoring, we report results from a performance investigation of the Uintah Computational Framework [5] where the application performance is correlated with runtime system actions. Figure 3 shows the execution timeline of a single iteration of an Uintah application (*bigbar* using the Material Point Method) where performance data from 1 Hz monitoring is plotted. The performance data is coming from two sources: the application and the system-level monitoring. Both sources used Supermon for the transport and the data streams were available as separate s-expressions on the Supermon channel. The execution took place on a 32-processor Pentium Xeon Linux cluster in our lab. The cluster is served by two separate physical networks, one providing connectivity among back-end compute nodes (through interface eth0) and another providing NFS connectivity (through eth1).

What stands out in the figure are the phases of the application's computation and the correlated network demands on the two separate interfaces. The phases are numbered (and distinctly colored) so as to differentiate them. On the xaxis is the time elapsed since the start of the application (the iteration shown falls between 850 and 1200 seconds). The left y-axis plots the difference in task duration between consecutive samples. On the right y-axis are plotted differences in bytes transmitted between samples and this is overlayed on the application phases as two solid lines - magenta for interface eth0 and blue for interface eth1.

For each monitoring interval, the profile sample is drawn to show the performance data for the dominant events. These phases would not be apparent if the profile data was not sampled periodically by the monitor. In this way, application and system performance can be correlated to better understand runtime effects.

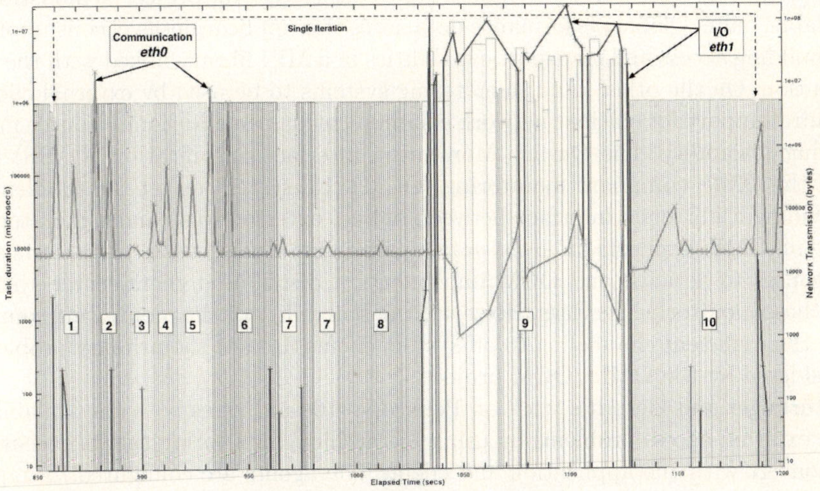

Fig. 3. Correlating Uintah Phases with System-level metrics

For instance, the impact of MPI and checkpoint operations on communication and I/O are clearly apparent. Tasks 1 through 8 mostly perform communication (seen from the eth0 curve), whereas task 9 (which is checkpointing) performs I/O to NFS (over eth1). Then Task 11 (MPI_Allreduce) ends the iteration. This correlation would be infeasible by direct measurement from within the application alone as it is unaware of system-level factors (e.g. the network topology and interfaces exercised).

6   Related Work

TAUoverSupermon owes its heritage to a long line of online performance monitoring projects. On-line automated computational steering frameworks like Falcon [6], Autopilot [7], Active Harmony [8], and MOSS [9] use a distributed system of sensors to collect data about an application's behavior and actuators to make modifications to application variables. These systems have built-in transport support and require the application to be modified to expose steerable parameters. In contrast, *ToS* couples two independent, standalone systems, and builds on a lower-level interface between TAU and Supermon which allows for more flexibility in its specific use. While we have not applied *ToS* to steering, we have demonstrated measurement control with Supermon using reverse channels supported in the *monhole*. It is conceivable that higher-level methods provided by these tools could also be layered on *ToS*.

It is important to distinguish between monitoring systems intended for *introspective* versus *extrospective* use. Scalability and low overhead for global performance access is important for introspective monitoring. Paradyn's Distributed Performance Consultant [10] supports introspective online performance diagnosis and uses a high-performance data transport and reduction system, MRNet [11], to address scalability issues [12]. Our TAUg [13] project demonstrated scalable, online global performance data access for application-level consumption by building access and transport capabilities in a MPI library linked with the application. On the other hand, monitoring systems to be used by external clients require support for efficient network communications, in addition to source monitoring scalability. The On-line Monitoring Interface Specification (OMIS) [14] and the OMIS compliant monitoring (OCM) [15] system target the problem of providing a universal interface between online, external tools and a monitoring system. OMIS supports an *event-action* paradigm to map events to requests and responses to actions, and OCM implements a distributed client-server system for these monitoring services. However, the scalability of the monitoring sources and their efficient channeling to off-system clients are not the primary problems considered by the OMIS/OCM project.

Fürlinger and Gerndt's work on Periscope [16] addresses both the scalability and external access problems by using hierarchical monitoring agents executing in concert with the application and client. The agents are configured to implement data reduction and evaluate performance properties, routing the results to interactive clients for use in performance diagnosis and steering. MRNet can also be used for extrospective monitoring. It is organized as a hierarchy of processes, created separately from the application processes, allowing it to connect to remote monitor sinks. Like MRNet-based tools, TAU can use Supermon in a flexible and scalable manner for both introspective and extrospective monitoring. The *ToS* work reported here demonstrates this performance monitoring functionality. It also shows how the *ToS* approach imposes few reengineering requirements on the monitoring sources and clients, allowing for a clean, light-weight implementation. It is interesting to note, that we could build a *TAUoverMRNet* monitoring system, and have plans in this regard.

7   Conclusions and Future Work

The desire to perform very-low overhead online application performance measurement led us to investigate alternatives to the traditional *'store performance data to shared-filesystem'* approach. We created a large-scale online application performance monitor by using Supermon as the underlying transport for the TAU measurement system. Experiments demonstrate that the *TAUoverSupermon* solution provides significantly lower overhead and greater scalability. Another demonstrated advantage to using an existing cluster-monitor as the transport is that it allows close correlation of application performance with system-level performance information. This facilitates separating performance effects that originate from within an application and those that are due to external effects outside the control of the application itself.

The scalability of a parallel performance monitoring system depends on several factors related to how it is designed and engineered as well as to how the system is used. Here we have demonstrated reduction in overheads for source data generation and transport. We are also experimenting with strategies to improve scalability further by reducing the number of nodes touched per query (e.g., using sampling [17]) and/or by reducing the data generated per node per query through aggregation. By having greater control over the transport and by being able to add extra intelligence into it, the *ToS* system can allow easy implementation of the above strategies. Other directions along which we would like to take this work include experimentation on very large scale platforms such as BG/L (already ported and functional), and adding new custom transports to TAU such as MRNET.

References

1. Shende, S., Malony, A.D.: The TAU parallel performance system. The International Journal of High Performance Computing Applications 20(2), 287–331 (2006)
2. Sottile, M., Minnich, R.: Supermon: A high-speed cluster monitoring system. In: CLUSTER'02: International Conference on Cluster Computing (2002)
3. Bailey, D.H., et al.: The nas parallel benchmarks. The International Journal of Supercomputer Applications 5(3), 63–73 (1991)

4. Nataraj, A., Malony, A., Shende, S., Morris, A.: Kernel-Level Measurement for Integrated Parallel Performance Views: the KTAU Project. In: CLUSTER'06: International Conference on Cluster Computing, IEEE Computer Society Press, Los Alamitos (2006)
5. de St. Germain, J.D., Parker, S.G., McCorquodale, J., Johnson, C.R.: Uintah: A massively parallel problem solving environment. In: HPDC'00: International Symposium on High Performance Distributed Computing, pp. 33–42 (2000)
6. Gu, W., et al.: Falcon: On-line monitoring and steering of large-scale parallel programs. In: 5th Symposium of the Frontiers of Massively Parallel Computing, McLean, VA, pp. 422–429 (1995)
7. Ribler, R., Simitci, H., Reed, D.: The Autopilot performance-directed adaptive control system. Future Generation Computer Systems 18(1), 175–187 (2001)
8. Tapus, C., Chung, I.H., Hollingworth, J.: Active harmony: Towards automated performance tuning. In: SC'02: ACM/IEEE conference on Supercomputing (2002)
9. Eisenhauer, G., Schwan, K.: An object-based infrastructure for program monitoring and steering. In: 2nd SIGMETRICS Symposium on Parallel and Distributed Tools (SPDT'98), pp. 10–20 (1998)
10. Miller, B., Callaghan, M., Cargille, J., Hollingsworth, J., Irvin, R., Karavanic, K., Kunchithapadam, K., Newhall, T.: The paradyn parallel performance measurement tool. Computer 28(11), 37–46 (1995)
11. Roth, P., Arnold, D., Miller, B.: Mrnet: A software-based multicast/reduction network for scalable tools. In: SC'03: ACM/IEEE conference on Supercomputing (2003)
12. Roth, P., Miller, B.: On-line automated performance diagnosis on thousands of processes. In: 11th ACM SIGPLAN Symposium on Principles and Practice of Parallel Programming, pp. 69–80. ACM Press, New York (2006)
13. Huck, K.A., Malony, A.D., Shende, S., Morris, A.: TAUg: Runtime Global Performance Data Access Using MPI. In: Mohr, B., Träff, J.L., Worringen, J., Dongarra, J. (eds.) PVM/MPI 2006. LNCS, vol. 4192, pp. 313–321. Springer, Heidelberg (2006)
14. Ludwig, T., Wismüller, R., Sunderam, V., Bode, A.: Omis – on-line monitoring interface specification (version 2.0). LRR-TUM Research Report Series 9 (1998)
15. Wismuller, R., Trinitis, J., Ludwig, T.: Ocm – a monitoring system for interoperable tools. In: 2nd SIGMETRICS Symposium on Parallel and Distributed Tools (SPDT'98), pp. 1–9 (1998)
16. Gerndt, M., Fürlinger, K., Kereku, E.: Periscope: Advanced techniques for performance analysis. In: In: Parallel Computing: Current & Future Issues of High-End Computing, In the International Conference ParCo 2005, 13-16 September 2005, pp. 15–26. Department of Computer Architecture, University of Malaga, Spain (2005)
17. Mendes, C., Reed, D.: Monitoring large systems via statistical sampling. International Journal of High Performance Computing Applications 18(2), 267–277 (2004)

Practical Differential Profiling[*]

Martin Schulz and Bronis R. de Supinski

Center for Applied Scientific Computing
Lawrence Livermore National Laboratory
PO Box 808, L-560, Livermore, CA 94551, USA
{schulzm,bronis}@llnl.gov

Abstract. Comparing performance profiles from two runs is an essential performance analysis step that users routinely perform. In this work we present *eGprof*, a tool that facilitates these comparisons through differential profiling inside *gprof*. We chose this approach, rather than designing a new tool, since *gprof* is one of the few performance analysis tools accepted and used by a large community of users.

eGprof allows users to "subtract" two performance profiles directly. It also includes callgraph visualization to highlight the differences in graphical form. Along with the design of this tool, we present several case studies that show how *eGprof* can be used to find and to study the differences of two application executions quickly and hence can aid the user in this most common step in performance analysis. We do this without requiring major changes on the side of the user, the most important factor in guaranteeing the adoption of our tool by code teams.

1 Motivation

Users can choose from a large variety of performance analysis tools to optimize their parallel applications, like TAU [1], Vampir [7], Paradyn [6], or Open|SpeedShop [9] to name just a few. However, it is our experience from working with code teams both at Lawrence Livermore National Laboratory (LLNL) and beyond that those tools are best suited to performance analysis experts. They provide rich functionality that naturally implies a somewhat steep learning curve. Since members of the code teams focus on adding functionality to their applications that increase the science they can achieve, they only have limited time to spend on performance optimization (often only 1 or 2 weeks of effort per year across the entire team). Thus, they encounter the learning curve with each use and choose to use relatively simple tools for that reason.

In this work we therefore pursue a different approach to improve performance tools for these occasional performance tool users. Rather than developing a new

---

[*] This work was performed under the auspices of the U.S. Department of Energy by University of California Lawrence Livermore National Laboratory under contract No. W-7405-Eng-48 (UCRL-CONF-227812).

A.-M. Kermarrec, L. Bougé, and T. Priol (Eds.): Euro-Par 2007, LNCS 4641, pp. 97–106, 2007.
© Springer-Verlag Berlin Heidelberg 2007

tool with more advanced functionality, we analyzed the usage patterns of those tools that are already accepted by the code teams and determined their main deficiencies with respect to their common usage pattern based on user feedback. Based on this analysis, we are augmenting these tools to suit those usage patterns better while retaining their ease-of-use. Our new toolset can be deployed transparently to the user without requiring any workflow or instrumentation changes on their side since it merely provides additional functionality. It will therefore easily find acceptance by a large group of users.

In this work, we apply this strategy to profiling of parallel applications with *gprof*, a command line driven profiler that is installed on almost all systems. While it is limited in its scope, it has found a large acceptance among our code teams because of its simplicity, wide installation base, and existing support by almost any compiler. However, *gprof* (like most other profiling tools) does not provide any direct support for the most common analysis step used in profiling: the comparison of executions, e.g., before and after coding changes intended to improve performance. Instead the user is left with having to compare large text logs of profiles manually, which is both tedious and error-prone.

We therefore extend the *gprof* toolset to include differential profiling allowing the user to directly compare two execution profiles as well as callgraphs from two different application executions. In addition, we provide a graphical representation of both individual and differential callgraphs to visualize the often complex information encoded in *gprof*'s callgraph results.

Combined, this provides an easy way to study the impact of code optimizations and parameter changes, as well as code properties both within one rank and across ranks. We will demonstrate this using four case studies covering various scenarios for single and multi-node performance analysis. In all cases, our extensions concisely present the key differences between individual executions in a few lines without the need for long manual searches.

2   Related Work

Only few tools support differential or comparative performance analysis. One of the exceptions is Open|SpeedShop [9], a recently developed performance toolset for Linux clusters, which includes the ability to align and contrast results from multiple runs. Further, Karavanic has investigated difference operators for performance event maps in Paradyn [6] as part of her Ph.D. thesis [4].

Both PerfDMF [3] and PerfTrack [5] provide a base infrastructure capable of supporting differential performance analysis. They both deploy relational databases to store the results of performance analysis across multiple runs. This data can later be queried and then compared using external tools.

Most other tools, however, only have the ability to work with data gathered during a single run and leave the user with the task to manually contrast the individual results. Due to the complexity and size of performance data, in particular from large scale parallel applications, this often tedious task risks missing key aspects.

3   eGprof: Differential Profiling with gprof

Differential profiling is a useful tool in many scenarios. However, to make it practical and accepted by code teams, it is important to provide a tool that is familiar to users as well as easy to use. To achieve these goals we decided to integrate the concept of differential profiling into *gprof*, which is part of the *GNU binutils* package. Thus, it is available on almost any UNIX based system, familiar to users, and already in use by most code teams.

Our design follows two main guidelines: any extension must a) be optional and cannot alter the default behavior of the tool and b) follow the main philosophy of the original *gprof*. These guidelines ensure the tool can be transparently deployed and is easy to use, which ensures acceptance by existing *gprof* users.

eGprof, our extended *gprof*, provides two major new features that aid differential profiling: the ability to subtract two profiles from each other and thereby to show their differences; and the visualization of both single and differential callgraphs.

3.1   Profile Subtraction

To use the unmodified *gprof* users must instrument their code using the *-pg* switch. The resulting binary then produces a *gmon.out* file containing the execution profile. *gprof* reads this profile along with the symbol information contained in the binary itself and produces both an execution profile and a dynamic callgraph in a textual representation.

To enable differential profiling, we allow the user to "subtract" a second profile from a baseline. Both profiles are collected in the same way and specified as command line options[1]. In this case, we first load both profiles individually, create the performance histogram and the callgraph information as in the original *gprof*, and associate the performance data (both the "positive" data from the baseline profile and the "negative" data from the second one) with the corresponding symbols.

We then align the two histograms, subtract the respective timing information for each symbol, and then store the newly created histogram. To generate the common callgraph, we first start with the baseline or "positive" graph. For each edge in the "negative" graph that is already present in the baseline graph we subtract the number associated with this edge, i.e., the number of times this edge was traversed at runtime, from the value associated with the existing edge. If the edge does not already exist, we add it, but record the negative of the associated number for this edge. In both cases, histogram and callgraph, we must take special care of zero values: in contrast to the original *gprof* in which zero means no performance information and hence can be removed, a zero in *eGprof* means that the performance in the two profiles is equal, which is relevant information that must be retained.

This basic approach works as long as both input profiles originate from executions of the same binary and hence are associated with the same symbol

---

[1] Note if only a baseline profile is specified, *eGprof* behaves exactly like the original *gprof* ensuring that this addition is fully transparent.

% time	cumulative seconds	self seconds	calls	self s/call	total s/call	name
50.37	5.51	5.51	31	0.18	0.18	f_d
29.16	8.70	3.19	1	3.19	8.52	f_c
10.51	9.85	1.15				main
9.96	10.94	1.09	1	1.09	1.27	f_b

% time	cumulative seconds	self seconds	calls	self s/call	total s/call	name
35.28	3.08	3.08				main
23.37	5.12	2.04	22	0.09	0.09	f_d
21.88	7.03	1.91	1	1.91	3.76	f_c
19.47	8.73	1.70	1	1.70	1.89	f_b

% time	cumulative seconds	self seconds	calls	self s/call	total s/call	name
47.60	3.47	3.47	9	0.39	0.39	f_d
-26.47	5.40	-1.93				main
17.56	6.68	1.28	0	-1.28	-4.76	f_c
-8.37	7.29	-0.61	0	0.61	0.62	f_b

% impact	+self seconds	-self seconds	diff seconds	+self calls	-self calls	diff calls	sym +-	name
47.60	5.51	2.04	3.47	31	22	9	XX	f_d
-26.47	1.15	3.08	-1.93	-	-	0	XX	main
17.56	3.19	1.91	1.28	1	1	0	XX	f_c
-8.37	1.09	1.70	-0.61	1	1	0	XX	f_b

Fig. 1. Histograms for two executions (top two); differential histograms in old (third) and new format (bottom)

information. In many usage scenarios, e.g., for the evaluation of code optimizations, however, each profile is generated from a separate executable. In such a case, we require the user to specify the second binary along with its profile. We then load the symbol tables from each binary, but keep them in separate data structures since each binary potentially has different code regions associated with the same symbol. Instead we sort and align the two symbol tables against each other, and then link matching symbols.

In both cases we print the resulting differential profile in the existing *gprof* format, as shown in Figure 1. The top two histograms show two profiles of a demo program with different parameters and the third histogram shows the generated differential analysis. It shows, for example, that 47.6% of all performance changes were caused by a runtime decrease in routine *f_d* of 3.47 seconds, while *f_b* and *main* ran slower in the second execution by 1.93 and 0.61 seconds respectively.

By maintaining the output format, users can continue to use any evaluation scripts that parse the *gprof* output or external visualization tools that postprocess the output (e.g., *xprofiler*). Since this format is, however, not easily readable we also provide the option to print the resulting analysis in a new, more detailed format that further eases the manual analysis of the results and also includes the information from the separate profiles for easy comparison. The bottom histogram in Figure 1 shows the results of the above profile in the new format. For the remainder of the paper we will use this new format for illustration purposes.

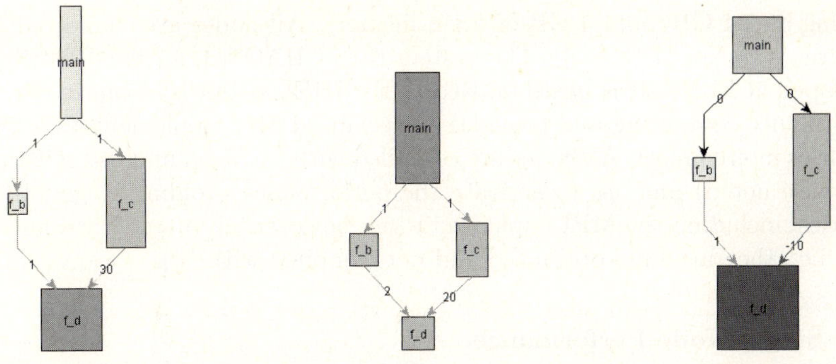

Fig. 2. Callgraphs of two executions (left, middle); differential callgraph (right)

3.2   Visualizing Callgraphs

gprof also provides dynamic callgraph information for all instrumented code pieces. However, this format is often difficult to read and adding differential information makes it even more complex. We therefore added support for callgraph visualization. On request, *eGprof* exports the callgraph as a GML (Graphic Markup Language) file. Figure 2 shows an example using the same data as above. The nodes represent the individual functions in the executable, and the edges show all calls between functions executed by the program together with the call frequency. Optionally, the size of the node can show both the scaled inclusive (height) and exclusive (width) time the code spent in this routine.

The left and the middle graph show the original execution profiles of the sample program and the right graph shows the differential callgraph. The latter is color coded to illustrate the direction of changes between the two original callgraphs. Again we see the large changes in *f_d* as well as in the exclusive time of *main*. Additionally, the graph shows an unchanged call frequency from *main* to *f_b* and *f_c*, while the second execution called *f_d* one more time from *f_b*, but ten times less from *f_c*.

By enabling *eGprof* to write GML files rather then visualizing the data itself, we ensure that the existing gprof maintains its command line philosophy and remains simple and easy to use. The created GML files can be visualized with any GML capable viewer. We use the freely available tool *yed* [2] for all graphs in this paper.

4   Case Studies

Differential profiles can be used in many scenarios. In the following we present four case studies on parallel applications showing the use of eGprof for both single and multi-node performance analysis. All experiments were conducted on *mcr*, a large scale cluster installed at LLNL. Each node consists of two Intel Xeon CPUs

---

[2] http://www.yworks.com/en/products_yed_about.htm

running at 2.4 GHz and 4 GB of main memory. All nodes are connected using Quadrics's QsNet II (Elan3). The system runs CHAOS 3, a Linux distribution developed at LLNL. It is based on Red Hat's RHEL 4, but is optimized for high performance computing and provides a specialized MPI implementation for the Quadrics interconnect. All codes are compiled with gcc 3.4.4 and use -*O2* (unless otherwise noted) and -*pg* to activate the performance profiling in gcc. System libraries, including the MPI implementation, however, are unmodified and used as is, i.e., they are fully optimized and not compiled with -*pg*.

4.1   Single Node Performance

Performance Optimization: Performance optimization is usually an iterative process of selecting an appropriate optimization method, applying it to the code and then analyzing the performance of the newly generated code in comparison to the original one. The latter step is essential for understanding the impact of the chosen optimization and for selecting the one for the next step. This process is greatly aided by our differential profiling approach.

We show an example of this use on SMG2000, a Semicoarsining Multigrid Solver based on the hypre library [2]. We compile the base version of the application using the -*O2* flag and then optimize by allowing inlining using -*O2 -finline*. Figure 3 shows the resulting output for the top ten routines. While the performance only marginally improves, we clearly see that the 69536 invocations of *hypre_ExchangeLocalData* were inlined by the compiler eliminating all calls to this routine in the second version. The callgraph (omitted due to space constraints) shows that this routine was originally called by *hypre_InitializeCommunication* and the histogram correspondingly shows an increase in time spent in this routine due to the inlining. The top 2 routines, however, benefit from inlining, most likely due to compiler optimizations enabled by the inlined routines.

Understanding Parameter Impact: Many applications have a large set of tunable algorithm parameters that enable the adjustment of codes to new platforms or problems. However, complex interactions among the parameters as well

```
   %     +self  -self   diff   +self  -self   diff   sym
impact   sec.   sec.    sec.   calls  calls   calls  +-    name
 21.65   55.25  54.83   0.42    6247   6247       0   XX    hypre_SMGResidual
  7.73   33.68  33.53   0.15    4881   4881       0   XX    hypre_CyclicReduction
  5.15    0.10   ---    0.10   69536      -   69536   XX    hypre_ExchangeLocalData
  5.15    2.89   2.79   0.10    1176   1176       0   XX    hypre_SemiInterp
  5.15    0.82   0.72   0.10     980    980       0   XX    hypre_StructAxpy
  4.12    0.65   0.57   0.08    1014   1014       0   XX    hypre_CycRedSetupCoarseOp
  4.12    0.48   0.40   0.08     168    168       0   XX    hypre_StructVectorClearGhostValues
 -4.12    0.66   0.74  -0.08     336    336       0   XX    hypre_StructVectorSetConstantValues
 -3.61    0.05   0.12  -0.07   69536  69536       0   XX    hypre_InitializeCommunication
  2.58    0.42   0.37   0.05      84     84       0   XX    hypre_SMGSetupInterpOp
```

Fig. 3. Comparing performance of SMG2000 with and without inlining

% impact	+self sec.	-self sec.	diff sec.	+self calls	-self calls	diff calls	sym +-	name
60.72	536.56	---	536.56	–	–	0	XX	ATL_dupKBmm10_10_2_b1
-22.51	---	198.88	-198.88	–	–	0	XX	ATL_dJIK80x80x80TN80x80x0_a1_b1
-10.80	---	95.46	-95.46	–	–	0	XX	ATL_dJIK0x0x20TN20x20x0_a1_bX
1.60	14.12	---	14.12	–	–	0	XX	ATL_dJIK0x0x10TN10x10x0_a1_bX
0.60	20.15	14.86	5.29	500	50	450	XX	HPL_dlaswp00N
-0.43	---	3.79	-3.79	–	–	0	XX	ATL_dJIK0x0x20TN1x1x20_a1_bX
-0.40	---	3.57	-3.57	–	–	0	XX	ATL_dJIK0x0x50TN50x50x0_a1_bX
-0.34	4.51	7.48	-2.97	–	–	0	XX	ATL_dtrsmKLLNU
-0.33	---	2.89	-2.89	–	–	0	XX	ATL_dupMBmm0_1_0_b1_loopa
0.23	4.22	2.23	1.99	–	–	0	XX	ATL_dcol2blk_a1
-0.22	---	1.95	-1.95	–	–	0	XX	ATL_dJIK0x0x26TN26x26x0_a1_bX
-0.17	0.47	1.97	-1.50	–	–	0	XX	ATL_drow2blkT_a1
0.16	1.41	---	1.41	–	–	0	XX	ATL_dJIK0x0x10TN1x1x10_a1_bX
0.13	8.98	7.82	1.16	2	2	0	XX	HPL_pdmatgen
0.12	14.22	13.15	1.07	55042118	50541398	4500720	XX	HPL_lmul

Fig. 4. Comparing two executions of HPL with different blocksizes

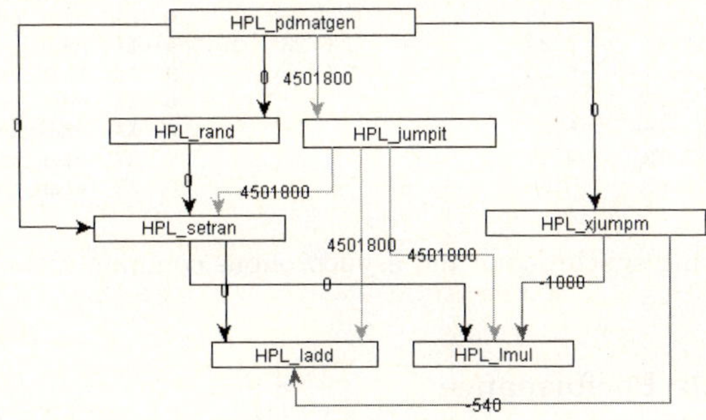

Fig. 5. Differential callgraph for HPL with varying block sizes

as with the target architecture often makes understanding their impact difficult. Differential profiling can help analyze the impact of parameter variations.

The High Performance LINPACK (HPL) [8] benchmark from the University of Tennessee uses a two-dimensional, block-cyclic data distribution and LU factorization with row partial pivoting featuring multiple look-ahead depths. Details of the algorithm can be fine-tuned with a large range of parameters. However, only rough guidelines exist on how to choose the best settings for a given target architecture, forcing the user to rely on hand-tuning for each platform.

In our experiment we use a constant problem size of N=5000 and vary one of the most significant parameters, the blocking size NB between 10 and 100, which causes more than a factor of two difference in performance. Figure 4 presents the differential profile for NB=10 and NB=100. It shows that HPL uses a drastically different set of routines in the underlying BLAS library for its computation depending on the value of NB[3]. This difference is further illustrated in the excerpt of the callgraph shown in Figure 5.

---

[3] The missing call number information is caused by the ATLAS library (all routines starting with *ATL_*) since this library was preinstalled and not compiled with *-pg*.

% impact	+self seconds	-self seconds	diff seconds	+self calls	-self calls	diff calls	sym +-	name
-98.69	975.14	981.94	-6.80	12	12	0	XX	sweep_
-0.58	2.20	2.24	-0.04	12	12	0	XX	flux_err_
-0.15	0.00	0.01	-0.01	1	1	0	XX	initgeom_
0.15	0.96	0.95	0.01	1	1	0	XX	initialize_
0.15	0.01	0.00	0.01	1	1	0	XX	inner_auto_
-0.15	0.00	0.01	-0.01	1440	1440	0	XX	snd_real_
0.15	18.06	18.05	0.01	12	12	0	XX	source_
0.00	---	---	0.00	2	-	2	XX	timers
0.00	0.00	0.00	0.00	1	1	0	XX	MAIN__

Fig. 6. Comparing rank 0 and 1 of Sweep3D

% impact	+self seconds	-self seconds	diff seconds	+self calls	-self calls	diff calls	sym +-	name
-19.09	614.20	640.62	-26.42	263328	263328	0	XX	hypre_SMGResidual
7.70	15.90	5.24	10.66	-	-	0	XX	elan_tportBufFree_locked
5.27	12.11	4.82	7.29	-	-	0	XX	elan_tportGCBufPool
5.19	55.36	48.18	7.18	-	-	0	XX	elan_progressFragLists
4.94	331.77	324.93	6.84	14144	205728	-191584	XX	hypre_CyclicReduction
-4.54	44.23	50.51	-6.28	-	-	0	XX	MPIR_Unpack2
-4.39	---	6.08	-6.08	-	-	0	XX	PMPI_Irecv
3.87	5.35	---	5.35	-	-	0	XX	PMPI_Recv
3.37	70.38	65.72	4.66	-	-	0	XX	elan_pollWord
2.65	28.65	24.99	3.66	-	-	0	XX	elan_tportRxStart

Fig. 7. Comparing synchronous and asynchronous communication in SMG2000

4.2   Cross-Node Performance

Understanding Load Balancing: The efficiency of parallel applications depends to a large degree on correct load balancing, i.e., ensuring that all tasks have roughly the same amount of work and hence do not incur long wait times at synchronization points. Comparing performance profiles of different ranks is one way to check for correct load balancing.

To illustrate this point we use Sweep3D from the ASCI Blue Benchmark suite. This code solves a 1-group time-independent discrete ordinates 3D cartesian geometry neutron transport problem. We run this code with 32 processes and create a separate profile for each rank. Using *eGprof* we then compare two representative tasks[4], in this case 0 and 1.

Figure 6 shows that the execution profiles of the individual ranks of Sweep3D are very similar indicating a well balanced code. All routines are called the same number of times (except for the timing routine, which is only called on the master node, rank 0) and the time spent in each routine is nearly identical.

Synchronous vs. Asynchronous Communication: Synchronous communication in parallel programs can lead to long blocking delays. In some cases this delay can be hidden using non-blocking communication calls: these routines allow send and receive operations to start without immediately waiting for their

---

[4] All rank combinations exhibit similar behavior.

completion. Instead the call returns and the application can overlap computation with the message transfer.

We show the effects of non-blocking communication by contrasting the performance of a synchronous and an asynchronous version of SMG2000. For this experiment we statically link the MPI library to the application (to give *eGprof* access to its symbols) and execute both versions using 32 processes. We then add the data from all ranks into a single profile for each run using *gprof*'s sum option and contrast the two global profiles.

The resulting differential profile in Figure 7 shows that the first version uses blocking receives (*MPI_Recv*), while the second version uses the non-blocking counterpart (*MPI_Irecv*). The performance difference between those two calls is, however, negligible since these routines merely start the receive. The actual operation is conducted inside Quadric's Elan library, for which the first version shows longer execution times due to the blocking operations. The main computation routine (*hypre_SMGResidual*), however, executes slower in the non-blocking version. This is caused by the concurrent message transfer in the Elan library. Overall, however, the non-blocking version performs better showing that the code benefits from this style of communication.

4.3   Discussion

The case studies discussed above show that differential profiling can help users to quickly identify the key difference between two application executions, typically by just looking at the top ten or fifteen routines in the profile. Without this feature, users must manually examine the entire profile for both runs and match up the symbols, which may be presented in a different order, to understand execution time differences completely. For SMG2000, a relatively simple benchmark, the *gprof* output has around 370 lines for the histogram and over 2700 for the callgraph. Realistic applications with millions of lines of code, as are typical in environments like national laboratories, generate significantly more output so manual comparison is intractable. Together with the callgraph visualization, our *eGprof* implementation automates this activity to support quick comparison of any two performance profiles.

5   Conclusion

Code teams demand easy solutions in familiar environments for their performance analyses. They usually do not have the time to learn complex and new tools, which instead are normally used by a few selected users specializing in performance optimization. We therefore take the approach to analyze those simple tools that are accepted and in use by our code teams and add missing functionality that directly aids their typical workflow.

Following this approach, we designed and implemented *eGprof*, an extension of the popular and widely available *gprof* profiler. Our version adds support for differential profiling to enable a quick and efficient comparison of the performance observed during two executions, the most common step in performance analysis. Further we support callgraph visualization to give users a quick graphical overview of the performance of their codes and the differences between two executions. In our cases studies we showed how these new features can significantly aid both single and multi-node performance analysis. In all cases, the output of *eGprof* was able to point to the key difference between application executions within the top 15 routines of the resulting profile.

Our *eGprof* implementation gives code teams access to powerful performance analysis techniques in a familiar environment. It does not require any new setup or performance measuring techniques on their side and therefore guarantees the lowest possible learning curve and easy acceptance.

References

1. Bell, R., Malony, A., Shende, S.: ParaProf: A Portable, Extensible, and Scalable Tool for Parallel Performance Profile Analysis. In: Kosch, H., Böszörményi, L., Hellwagner, H. (eds.) Euro-Par 2003. LNCS, vol. 2790, pp. 17–26. Springer, Heidelberg (2003)
2. Falgout, R., Yang, U.: hypre: a Library of High Performance Preconditioners. In: Sloot, P.M.A., Tan, C.J.K., Dongarra, J.J., Hoekstra, A.G. (eds.) Computational Science - ICCS 2002. LNCS, vol. 2331, pp. 632–641. Springer, Heidelberg (2002)
3. Huck, K., Malony, A., Bell, R., Morris, A.: Design and Implementation of a Parallel Performance Data Management Framework. In: Proceedings of the 2005 International Conference on Parallel Processing (August 2005)
4. Karavanic, K.: Experiment Management Support for Parallel Performance Tuning. PhD thesis, Department of Computer Science, University of Wisconsin (1999)
5. Karavanic, K., May, J., Mohror, K., Miller, B., Huck, K., Knapp, R., Pugh, B.: Integrating Database Technology with Comparison-Based Parallel Performance Diagnosis: The PerfTrack Performance Experiment Management Tool. In: Proceedings of IEEE/ACM Supercomputing '05 (November 2001)
6. Miller, B., Callaghan, M., Cargille, J., Hollingsworth, J., Irvin, R., Karavanic, K., Kunchithapadam, K., Newhall, T.: The Paradyn Parallel Performance Measurement Tool. IEEE Computer 28(11), 37–46 (1995)
7. Nagel, W.E., Arnold, A., Weber, M., Hoppe, H.C., Solchenbach, K.: VAMPIR: Visualization and analysis of MPI resources. Supercomputer 12(1), 69–80 (1996)
8. Petitet, A., Whaley, R.C., Dongarra, J., Cleary, A.: Hpl - a portable implementation of the high-performance linpack be nchmark for distributed-memory computers. Available at `http://www.netlib.org/benchmark/hpl/`.
9. The Open|SpeedShop Team. Open|SpeedShop for Linux (November 2006), `http://www.openspeedshop.org/`

Decision Trees and
MPI Collective Algorithm Selection Problem

Jelena Pješivac-Grbović, George Bosilca, Graham E. Fagg, Thara Angskun,
and Jack J. Dongarra

Innovative Computing Laboratory,
The University of Tennessee Computer Science Department
1122 Volunteer Blvd., Knoxville, TN 37996-3450, USA
{pjesa,bosilca,fagg,angskun,dongarra}@cs.utk.edu

Abstract. Selecting the close-to-optimal collective algorithm based on
the parameters of the collective call at run time is an important step
for achieving good performance of MPI applications. In this paper, we
explore the applicability of C4.5 decision trees to the MPI collective
algorithm selection problem. We construct C4.5 decision trees from the
measured algorithm performance data and analyze both the decision tree
properties and the expected run time performance penalty.

In cases we considered, results show that the C4.5 decision trees can be
used to generate a reasonably small and very accurate decision function.
For example, the broadcast decision tree with only 21 leaves was able
to achieve a mean performance penalty of 2.08%. Similarly, combining
experimental data for reduce and broadcast and generating a decision
function from the combined decision trees resulted in less than 2.5% relative performance penalty. The results indicate that C4.5 decision trees
are applicable to this problem and should be more widely used in this
domain.

1   Introduction

The performance of MPI collective operations is crucial for good performance of
MPI applications that use them [1]. For this reason, significant efforts have gone
into the design and implementation of efficient collective algorithms for both homogeneous and heterogeneous cluster environments [2,3,4,5,6,7]. Performance of
these algorithms varies with the total number of nodes involved in communication, system and network characteristics, size of data being transferred, current
load and, if applicable, the operation that is being performed, as well as the segment size that is used for operation pipelining. Thus, selecting the best possible
algorithm and segment size combination (*method*) for every instance of collective
operation is important.

To ensure good performance of MPI applications, collective operations can be
tuned for the particular system. The tuning process often involves detailed profiling of the system, possibly combined with communication modeling, analyzing
the collected data, and generating a *decision function*. During run-time, the decision function selects the close-to-optimal method for a particular collective

A.-M. Kermarrec, L. Bougé, and T. Priol (Eds.): Euro-Par 2007, LNCS 4641, pp. 107–117, 2007.
© Springer-Verlag Berlin Heidelberg 2007

instance. As the amount of the system performance information can be significant, the decision function building mechanism must be efficient both in terms of storage and the time-to-decision performance. We are interested in different approaches to storing and accessing the large amount of performance data.

This paper studies the applicability of C4.5 decision trees [8] to the MPI collective algorithm/method selection problem. We assume that the system of interest has been benchmarked and that detailed performance information exists for each of the available collective communication methods.[1] With this information, we focus our efforts on investigating whether the C4.5 algorithm is a feasible way to generate static decision functions.

The paper proceeds as follows: Section 2 discusses existing approaches to the decision making/algorithm selection problem; Section 3 provides background information on the C4.5 algorithm; Section 4 discusses the mapping of performance measurement data to C4.5 input, Section 5 presents experimental results; and Section 6 concludes the paper with discussion of the results and future work.

2   Related Work

The MPI collective algorithm selection problem has been addressed in many MPI implementations. In FT-MPI [10], the decision function is generated manually using a visual inspection method augmented with Matlab scripts used for analysis of the experimentally collected performance data. This approach results in a precise, albeit complex, decision function. In the MPICH-2 MPI implementation, the algorithm selection is based on bandwidth and latency requirements of an algorithm, and the switching points are predetermined by the implementers [6]. In the tuned collective module of Open MPI [11], the algorithm selection can be done in either of the following three ways: via a compiled decision function; via user-specified command line flags; or using a rule-based run-length encoding scheme that can be tuned for a particular system.

Our previous work [12] used quadtree encoding to store the information about the optimal collective algorithm performance on a particular system. This structure was used either to generate a decision function or as an in-memory decision system for selecting a close-to-optimal method at run-time.

Alternatively, data mining techniques can be applied to the algorithm selection problem with replacing the original problem by an equivalent classification problem. The new problem is to classify collective parameters *(collective operation, communicator size, message size)* into a correct category, a method in our case, to be used at run-time.

Vuduc et al. construct statistical learning models to build different decision functions for the matrix-matrix multiplication algorithm selection [13]. In their work, they consider three methods for decision function construction: parametric

---

[1] Detailed benchmarking of all possible methods takes a significant amount of time. If this is not an option, performance profiles can be generated using a limited set of performance measurements coupled with performance modeling [9].

modeling; parametric geometry modeling; and non-parametric geometry modeling. The non-parametric geometry modeling uses statistical learning methods to construct implicit models of the boundaries/switching points between the algorithms based on the actual experimental data. To achieve this, Vuduc et al. use the *support vector machines* method[14].

Conceptually, the work presented in this paper is close to the non-parametric geometry modeling work done by Vuduc et al. However, our problem domain is different: MPI collective operations are used instead of matrix-matrix multiplication, and we use the C4.5 algorithm instead of *support vector machines* methods. To the best of our knowledge, we are the only group that has approached the MPI collective tuning process in this way.

3   C4.5 Algorithm

C4.5 is a supervised learning classification algorithm used to construct decision trees from the data [8]. C4.5 can be applied to the data that fulfills the following requirements:

- *Attribute-value description*: information about a single entry in the data must be described in terms of attributes. The attribute values can be discrete or continuous and, in some cases, the attribute value may be missing or can be ignored.
- *Predefined classes*: the training data has to be divided into predefined classes or categories. This is a standard requirement for supervised learning algorithms.
- *Discrete classes*: the classes must be clearly separated and a single training case either belongs to a class or it does not. C4.5 cannot be used to predict continuous class values such as the cost of a transaction.
- *Sufficient data*: the C4.5 algorithm utilizes an inductive generalization process by searching for patterns in data. For this approach to work, the patterns must be distinguishable from random occurrences. What constitutes the "sufficient" amount of data depends on a particular data set and its attribute and class values, but in general, statistical methods used in C4.5 to generate tests require reasonably large amount of data.
- *"Logical" classification models*: generated classification models must be represented as either decision trees or a set of production rules [8].

The C4.5 algorithm constructs the initial decision tree using a variation of the Hunt's method for decision tree construction (Figure 1). The main difference between C4.5 and other similar decision tree building algorithms is in the test selection and evaluation process (last case in Figure 1). The C4.5 utilizes information *gain ratio* criterion, which maximizes normalized information gain by partitioning T in accordance with a particular test [8].

Once the initial decision tree is constructed, a pruning procedure is initiated to decrease the overall tree size and decrease the estimated error rate of the tree[8].

Given a set of training cases, T, and set of classes $C = \{C_1, C_2, ..., C_k\}$,
the tree is constructed recursively by testing for the following cases:

1) T contains one or more cases which all belong to the same class C_j:
   A leaf node is created for T and is denoted to belong to C_j class;
2) T contains no cases:
   A leaf node is created for T and is assigned the most most frequent class at the parent node;
3) T contains cases that belong to more than one class:
   Find a test that splits T set to a single-class collections of cases.
   This test is based on a single attribute value, and is selected such that it results in one or more mutually exclusive outcomes $\{O_1, O_2, ...O_n\}$.
   The set T is then split into subsets $\{T_1, T_2, ...T_n\}$ such that the set T_i contains all cases in T with outcome O_i.
   The algorithm is then called recursively on all subsets of T.

Fig. 1. Hunt's method for decision tree construction [8]

Additional parameters that affect the resulting decision tree are:

- *weight*, which specifies the minimum number of cases of at least two outcomes of a test. This prevents near-trivial splits that would result in almost flat and really wide trees.
- *confidence level*, which is used for prediction of tree error rates and affects the pruning process. The lower the confidence level, the greater the amount of pruning that takes place.
- *attribute grouping*, which can be used to create attribute value groups for discrete attributes and possibly infer patterns occurring in sets of cases with different values of an attribute, but do not occur for other values of that attribute.
- *windowing*, which enables construction of multiple trees based on a portion of the test data and then selects the best performing tree [8].

4 MPI Collectives Performance Data and C4.5

We use the collective algorithm performance information on a particular system to extract the information about the *optimal methods*. The optimal method on the particular system is the method that achieves the lowest duration for a particular set of input parameters.

The collected performance data can be described using the collective name, communicator and message size attributes. The collective name attribute has discrete values such as broadcast, reduce, etc. Communicator and message size attributes have continuous values. Additionally, constructive induction can be used to create composite attributes that can capture additional system information. For example, a *total data per node* attribute can be used to distinguish between a single-process-per-node and two-processes-per-node run. Moreover, such attributes can potentially indirectly capture information about the system bottlenecks. In this paper, however, we focus on performance data that is fully described by the collective name, communicator and message size attributes.

The predefined set of classes in our case contains methods that were optimal for some of the data points. The class names consist of the algorithm name and segment size used, for example, Linear_0KB or SplitBinary_16KB. The classes are well defined, and by construction, the data with the same input parameters can belong to a single class only.

As far as the "sufficient" data requirement is concerned, the performance measurement data contains a considerable number of data points in the communicator - message size range. We do not cover every single possible communicator or message size, but our training data set usually contains around 1000 data points, so we feel that for this type of problem, collected data is sufficient to give reasonable results.

The goal of this work is construction of decision functions, so we provide the functionality to generate the decision function source code in C from the constructed decision trees: the internal nodes are replaced by a corresponding *if* statement, and leaf nodes return the decision method index/name. We did not utilize the `c4.5rules` program for this purpose.

5   Experimental Results and Analysis

In this work, we used release 8 of the C4.5 implementation by J.R. Quinlan [15] to construct decision trees based on existing performance data for broadcast and reduce collectives collected on the Grig cluster at the University of Tennessee, Knoxville.

The Grig cluster has 64 dual Intel(R) Xeon(TM) processor nodes at 3.2 GHz and Fast Ethernet and MX interconnects. The experimental data from the Grig cluster in this paper was gathered using the Fast Ethernet interconnect.

The performance data in this paper was collected using the MPICH-2 [16] version 1.0.3 and OCC library [17]. The OCC library implements a number of collective algorithms on top of MPI point-to-point operations and can be used with any MPI implementation. The OCC benchmark measures collective operation performance by repeating the operation a number of times. To avoid pipelining effects, a balanced barrier (such as Bruck) is inserted between every collective call, and the time to execute the barrier is subtracted from the total running time. More advanced benchmarks, such as SKaMPI [18], can be used as well. The only requirement is to convert the benchmark results to C4.5 input file format.

In our experiments, we tested decision trees constructed using different weight and confidence level constraints. We did not use windowing because our data was relatively sparse in comparison to the complete communicator - message size domain size, so we did not expect that there would be a benefit by not utilizing all available data points. Also, since communicator and message sizes were described as continuous attributes, we were not able to use the grouping functionality of C4.5.

We constructed decision trees both per-collective (e.g., just for broadcast or alltoall) and for the set of collectives that have similar or the same set of available implementations (e.g., both have Linear, Binary, and Pipeline algorithms) and for which we expected to have similar decision functions (e.g., broadcast and reduce).

5.1 Analysis of Broadcast Decision Trees

Figure 2 shows three different decision maps[2] for a broadcast collective on the Grig cluster. We considered five different broadcast algorithms (Linear, Binomial, Binary, Split Binary, and Pipeline)[3] and four different segment sizes (no segmentation, 1KB, 8KB, and 16KB). The measurements covered all communicator sizes between two and 28 processes and message sizes in the 1B to 384KB range with total of 1248 data points. The original performance data set contains $1248 \times 4 \times 5$ data points.

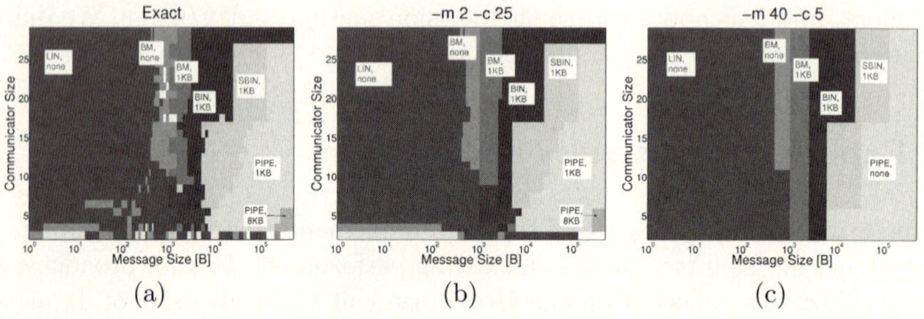

Fig. 2. Broadcast decision maps from the Grig cluster: (a) Measured (b) '-m 2 -c 25' (c) '-m 40 -c 5'. X-axis corresponds to message sizes, Y-axis represents the communicator sizes. Different colors correspond to different method indices. In this Figure, "LIN" stands for Linear, BM for Binomial, "BIN" for Binary, "SBIN" for Split Binary, and "PIPE" for Pipeline algorithm. Also, "none", "1KB", and "8KB" refer to the corresponding segment size.

Figure 2 (a) shows an exact decision map generated from experimental data. The subsequent maps were generated by C4.5 decision trees constructed by specifying different values for weight ("-m") and confidence level ("-c") parameters (See Section 3). The statistics about these and additional trees can be found in Table 1.

The exact decision map in Figure 2 (a) exhibits trends, but there is a considerable amount of information for intermediate size messages (between 1KB and 10KB) and small communicator sizes. The decision maps generated from different C4.5 trees capture general trends very well. The amount of captured detail depends on weight, which determines how the initial tree will be built, and confidence level, which affects the tree pruning process. "Heavier" trees require that branches contain more cases, thus limiting the number of fine-grained splits.

---

[2] Decision map is a 2D representation of the decision tree output for a particular communicator and message size ranges.

[3] For more details on these algorithms, refer to [9].

Table 1. Broadcast decision tree statistics corresponding to the data presented in Figure 2. Size refers to the number of leaf nodes in the tree. Errors are in terms of misclassified training cases. The data set had 1248 training cases. The median performance penalty was 0% in all cases.

Command line	Before pruning		After pruning			Performance penalty		
	Size	Errors	Size	Errors	Predicted Error	Min	Max	Mean
-m 2 -c 25	133	7.9%	127	7.9%	14.6%	0%	75.41%	0.66%
-m 4 -c 25	115	8.8%	95	9.4%	15.0%	0%	316.97%	1.16%
-m 6 -c 15	99	10.4%	65	11.5%	17.6%	0%	316.97%	3.24%
-m 8 -c 5	73	12.0%	47	12.8%	21.0%	0%	316.97%	1.66%
-m 40 -c 5	21	17.8%	21	17.8%	21.9%	0%	316.97%	2.08%

A lower confidence level allows for more aggressive pruning, which also results in coarser decisions.

Looking at the decision tree statistics in Table 1, we can see that the default C4.5 tree ('-m 2 -c 25') has 127 leaves and a predicted misclassification error of 14.6%. Using a slightly "heavier" tree '-m 4 -c 25' gives us a 25.20% decrease in tree size (95 leaves) and maintains almost the same predicted misclassification error. As we increase tree weight and decrease the confidence level, we produce the tree with only 21 leaves (83.46% reduction in size) with a 50% increase in predicted misclassifications (21.9%).

In this work, the goal is to construct reasonably small decision trees that will provide good run-time performance of an MPI collective of interest. Given this goal, the number of misclassified training examples is not the main figure of merit we need to consider. To determine the "quality" of the resulting tree in terms of collective operation performance, we consider the performance penalty of the tree. The performance penalty is the relative difference between the performance obtained using methods predicted by the decision tree and the experimentally optimal ones.

The last three columns in Table 1 provide performance penalty statistics for the broadcast decision trees we are considering. The minimum, mean, and median performance penalty values are rather low - less than 4%, even as low as 0.66%, indicating that even the simplest tree we considered should provide good run-time performance. Moreover, the simplest tree, "-m 40 -c 5", had a lower performance penalty than the "-m 6 -c 15," which indicates that the percent of misclassified training cases does not translate directly into a performance penalty of the tree.

In all cases, the mean and median performance penalty values are excellent, but the maximum performance penalty of 316.97% requires explanation. At communicator size 25 and message size 480, the experimentally optimal method is Binary algorithm without segmentation (1.12 ms), but most decision trees select Binomial algorithm without segmentation (4.69 ms). However, the Binomial algorithm performance in the neighborhood of this data point is around and less than 1 ms, which implies that the 4.69 ms result is probably affected by external factors. Additionally, in the "-m 40 -c 5" tree, only six data points had a performance penalty above 50%.

5.2 Combined Decision Trees

It is reasonable to expect that similar MPI collective operations have similar decision functions on the same system. To test this hypothesis, we decided to analyze the decision trees generated from the experimental data collected for broadcast and reduce collectives on the Grig system. Our implementations of these collectives are symmetric; each of them has Linear, Binomial, Binary, and Pipeline based implementations. Broadcast supports the Split Binary algorithm for which we do not have an equivalent in reduce implementation, but we expect that C4.5 should be able to handle these cases correctly.

The training data for this experiment contains three attributes (collective name, communicator size, and message size) and the same set of predetermined classes as in the broadcast-only case.

Figure 3 shows the decision maps generated from the combined broadcast and reduce decision tree. The leftmost maps in both rows are the exact decisions for each of the collectives based on experimental data. The remaining maps are generated by querying the combined decision tree. Figures 3 (b) and (e) were generated using a "-m 2 -c 25" decision tree, while (c) and (f) were generated by a "-m 20 -c 5" decision tree. Table 2 provides the detailed information about

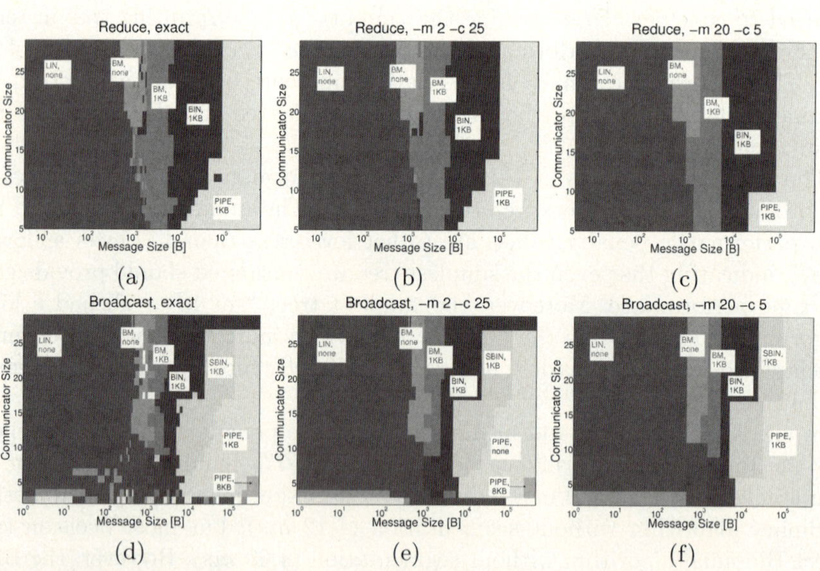

Fig. 3. Combined broadcast and reduce decision maps from the Grig cluster: (a) reduce, Exact (b) reduce, '-m 2 -c 25' (c) reduce, '-m 20 -c 5' (d) broadcast, Exact (e) broadcast, '-m 2 -c25' (f) broadcast, '-m 20 -c 5'. Color has the same meaning as in Figure 2.

Table 2. Statistics for combined broadcast and reduce decision trees corresponding to the data presented in Figure 3. Size refers to the number of leaf nodes in the tree. Errors are in terms of misclassified training cases. The data set had 2286 training cases.

Command line	Before Pruning		After Pruning			Mean performance penalty	
	Size	Errors	Size	Errors	Predicted error	Broadcast	Reduce
-m 2 -c 25	239	137	221	142	12.6%	0.66%	0.41%
-m 6 -c 25	149	205	115	220	14.0%	1.62%	0.71%
-m 8 -c 25	127	225	103	235	14.4%	1.64%	0.72%
-m 20 -c 5	63	310	55	316	20.6%	2.40%	0.93%
-m 40 -c 25	33	392	33	392	19.6%	2.37%	1.53%

the combined decision trees of interest including the mean performance penalty of the trees.

The structure of combined broadcast and reduce decision trees reveals that the test for the type collective occurs for the first time on the third level of the tree. This implies that the combined decision tree is able to capture the common structure of the optimal implementation for these collectives, as one would expect based on decision maps in Figure 3.

5.3   C4.5 Decision Trees vs. Quadtree Encoding

Quadtree encoding is an alternative method for storing performance information and generating decision functions based on the performance data. We explored this approach in [12].

The quadtree results for broadcast collective on the Grig cluster showed that the 6-level quadtree can represent experimental data fully. The 5-level quadtree for this data set, incurred around 5.41% mean performance penalty, while the 3-level quadtree introduced 8.83% mean performance penalty. In comparison, the C4.5 decision trees we considered incurred less than 3.5% mean performance penalty.

The main benefit of the quadtree encoding is the fact that the size of the generated quadtree can be easily manipulated. This allows us to limit the maximum number of expressions that need to be evaluated to reach the decision. The depth of the C4.5 decision tree is hard to estimate, making it impossible to set an a priori limit on the maximum number of expressions to be evaluated in the final decision function.

The main benefit of C4.5 decision trees is the ability to handle multi-dimensional data automatically. In this paper, we added collective name as a third dimension in Section 5.2. The composite attributes or ordinal attributes that describe system information can be automatically handled by C4.5. The quadtree encoding is restricted to two-dimensional data (communicator and message sizes), and cannot be easily extended to include additional attributes. Moreover, one-dimensional decisions (such as "for this communicator size and all message sizes use method A, but do not use this method for neighboring communicator sizes") cannot be captured with size-restricted quadtrees, while C4.5 does not have this problem.

6 Discussion and Future Work

In this paper, we studied the applicability of C4.5 decision trees to the MPI collective algorithm/method selection problem. We assumed that the system of interest has been benchmarked and that detailed performance information exists for each of the available collective communication methods. Using this information, we focused on investigating whether C4.5 decision trees are a feasible way to generate static decision functions.

Using a publicly available C4.5 implementation, we constructed decision trees based on existing performance data for broadcast and reduce collectives. We evaluated decision trees constructed using different weight and confidence level parameters.

Our results show that C4.5 decision trees can be used to generate a reasonably small and very accurate decision function: the mean performance penalty on existing performance data was within the measurement error for all trees we considered. For example, the broadcast decision tree with only 21 leaves was able to achieve a mean performance penalty of 2.08%. Moreover, using this tree, only six points in the communicator - message size ranges we tested would incur more than 50% performance penalty. Similar results were obtained for reduce and alltoall.

Additionally, we combined the experimental data for reduce and broadcast to generate the combined decision trees. These trees were also able to produce decision functions with less than a 2.5% relative performance penalty for both collectives. This indicates that it is possible to use information about one MPI collective operation to generate a reasonably good decision function for another collective, under the assumption that the two are similar.

In the direct comparison to the decision functions generated by the quadtrees from [12], C4.5 trees produced decision functions with lower mean performance penalties. However, the size and structure of a C4.5 decision tree is less predictable than the one of the corresponding quadtree. More detailed comparison of both methods is planned for future work.

Our findings demonstrate that the C4.5 algorithm and decision trees are applicable to this problem and should be more widely used in this domain. In the future, we plan to use C4.5 decision trees to reevaluate decision functions in FT-MPI and the tuned collective module of Open MPI. We also plan to integrate C4.5 decision trees with our MPI collective testing and performance measurement framework, OCC.

References

1. Rabenseifner, R.: Automatic MPI counter profiling of all users: First results on a CRAY T3E 900-512. In: Proceedings of the Message Passing Interface Developer's and User's Conference, pp. 77–85 (1999)

2. Worringen, J.: Pipelining and overlapping for MPI collective operations. In: 28th Annyal IEEE Conference on Local Computer Network, Bonn/Königswinter, Germany, pp. 548–557. IEEE Computer Society Press, Los Alamitos (2003)
3. Rabenseifner, R., Träff, J.L.: More efficient reduction algorithms for non-power-of-two number of processors in message-passing parallel systems. In: Kranzlmüller, D., Kacsuk, P., Dongarra, J.J. (eds.) Recent Advances in Parallel Virtual Machine and Message Passing Interface. LNCS, vol. 3241, Springer, Heidelberg (2004)
4. Chan, E.W., Heimlich, M.F., Purkayastha, A., van de Geijn, R.M.: On optimizing of collective communication. In: Proceedings of IEEE International Conference on Cluster Computing, 145–155 (2004)
5. Bernaschi, M., Iannello, G., Lauria, M.: Efficient implementation of reduce-scatter in MPI. Journal of Systems Architure 49(3), 89–108 (2003)
6. Thakur, R., Rabenseifner, R., Gropp, W.: Optimization of Collective Communication Operations in MPICH. International Journal of High Performance Computing Applications 19(1), 49–66 (2005)
7. Kielmann, T., Hofman, R.F.H., Bal, H.E., Plaat, A., Bhoedjang, R.A.F.: MagPIe: MPI's collective communication operations for clustered wide area systems. In: Proceedings of the ACM SIGPLAN symposium on Principles and Practice of Parallel Programming, pp. 131–140. ACM Press, New York (1999)
8. Quinlan, J.R.: C4.5: Programs for Machine Learning. Morgan Kaufmann Publishers, San Mateo, California (1993)
9. Pješivac-Grbović, J., Angskun, T., Bosilca, G., Fagg, G.E., Gabriel, E., Dongarra, J.J.: Performance analysis of MPI collective operations. In: Proceedings of IPDPS'05 - PMEO-PDS Workshop, p. 272.1. IEEE Computer Society Press, Los Alamitos (2005)
10. Fagg, G.E, Gabriel, E., Bosilca, G., Angskun, T., Chen, Z., Pješivac-Grbović, J., London, K., Dongarra, J.: Extending the MPI specification for process fault tolerance on high performance computing systems. In: Proceedings of the International Supercomputer Conference (ISC) 2004, Primeur (2004)
11. Fagg, G.E., Bosilca, G., Pješivac-Grbović, J., Angskun, T., Dongarra, J.: Tuned: A flexible high performance collective communication component developed for Open MPI. In: Proccedings of DAPSYS'06, Innsbruck, Austria, pp. 65–72. Springer, Heidelberg (2006)
12. Pješivac-Grbović, J., Fagg, G.E., Angskun, T., Bosilca, G., Dongarra, J.J.: MPI collective algorithm selection and quadtree encoding. In: Mohr, B., Träff, J.L., Worringen, J., Dongarra, J. (eds.) Recent Advances in Parallel Virtual Machine and Message Passing Interface. LNCS, vol. 4192, pp. 40–48. Springer, Heidelberg (2006)
13. Vuduc, R., Demmel, J.W., Bilmes, J.A.: Statistical Models for Empirical Search-Based Performance Tuning. International Journal of High Performance Computing Applications 18(1), 65–94 (2004)
14. Vapnik, V.N.: Statistical Learning Theory. Wiley, New York, NY (1998)
15. Quinlan, J.R.: C4.5 source code (2006), http://www.rulequest.com/Personal
16. MPICH-2: Implementation of MPI 2 standard (2005),
    http://www-unix.mcs.anl.gov/mpi/mpich2
17. OCC: Optimized Collective Communication Library (2005),
    http://www.cs.utk.edu/~pjesa/projects/occ
18. SKaMPI: Special Karlsruher MPI Benchmark (2005),
    http://liinwww.ira.uka.de/~skampi

Profiling of Task-Based Applications on Shared Memory Machines: Scalability and Bottlenecks

Ralf Hoffmann and Thomas Rauber

Department for Mathematics, Physics and Computer Science
University of Bayreuth, Germany
`{ralf.hoffmann,rauber}@uni-bayreuth.de`

Abstract. A sophisticated approach for the parallel execution of irregular applications on parallel shared memory machines is the decomposition into fine-grained tasks. These tasks can be executed using a task pool which handles the scheduling of the tasks independently of the application. In this paper we present a transparent way to profile irregular applications using task pools without modifying the source code of the application. We show that it is possible to identify critical tasks which prevent scalability and to locate bottlenecks inside the application. We show that the profiling information can be used to determine a coarse estimation of the execution time for a given number of processors.

1 Introduction

Due to the dynamic computation structure of irregular applications, task pools have been shown to be a well suited execution environment for these types of applications [1]. Applications decomposed into a large number of fine-grained tasks managed by task pools can be executed efficiently on a wide variety of parallel shared memory machines. This includes large systems like IBM p690 or SGI Altix2 with more than 32 processors but also smaller dual or multi-core SMP systems. The task pool implementation takes care of executing available task and storing newly created tasks to realize a dynamic task structure.

The granularity of the different tasks is a major factor for the resulting performance, since it partially determines the overhead introduced by the task management. It is however difficult to predict the performance and scalability of a specific application. It has been shown in our previous work [1] that this overhead can be reduced by using hardware operations for the task pool management, but the performance also depends on other parameters which cannot be improved easily. A limited number of available tasks, and therefore a larger waiting time for new runnable tasks, is an example of a limiting factor beyond the scope of the task pool implementation.

A detailed analysis of the internal task structure of an application is required to determine bottlenecks, to find scalability problems, and to suggest code improvements. For programs with a static task structures this can be done by analyzing the directed acyclic graphs (DAGs). But for irregular applications the task graph is usually not known before execution, and even then it may only be

A.-M. Kermarrec, L. Bougé, and T. Priol (Eds.): Euro-Par 2007, LNCS 4641, pp. 118–128, 2007.

known partially. Analyzing the source code and predicting the performance can be difficult for complex and irregular multi-threaded applications.

Applications designed to utilize task pools are completely independent from the actual task pool implementation. Therefore, a profiling task pool can be used to analyze performance characteristics of the application in the background. Possible information gathered in the profiling process include the task creation scheme (i.e., the task graph), statistics about the execution time of each individual task, task stealing operations for load balancing, and so on. In this paper we concentrate on gathering information about the task runtime and the time each processor waits for new tasks to identify bottlenecks in the application. The contribution of the paper is to propose a method to analyze the task structure of arbitrary applications utilizing task pools and to show that it is possible to predict the performance for a large number of processors by using profiling results from runs with a small number of participating processors. As case study the method is applied to the hierarchical radiosity from the SPLASH-2 suite [2].

The rest of the paper is organized as follows. Section 2 introduces the profiling method and Section 3 describes the data analysis. Section 4 presents a detailed case study. Section 5 discusses related work and Section 6 concludes the paper.

2   A Profiling Task Pool

For a task based execution the application defines tasks by providing the operations executed by the tasks as functions. Figure 1 shows the generic structure of a task-based application. After creating initial tasks, depending on the actual input of the application, each processor executes the main loop and asks for tasks to execute until all tasks are finished. Each task is a sequential function, and it can create an arbitrary number of new tasks. The number of executable tasks varies over time, but usually there is a much larger number of tasks than processors. The internal implementation of the task pool is hidden but it is accessible by all threads via an application programming interface.

The granularity of the tasks depends on the actual computations done by the corresponding functions and it can have a large influence on the resulting performance. For example, many small tasks lead to frequent access to the task pool with a potentially larger overhead due to mutual exclusion while a small number of larger tasks can limit the available degree of parallelism.

The actual execution time (in the following referred to as *task size*) of each task instance is hard to predict since it depends on hardware parameters like CPU speed or memory bandwidth as well as input-dependent parameters like loop bounds. Another important factor is the time which threads have to wait for new runnable tasks (referred to as *task waiting time*). It is not easy to predict whether there are enough tasks available at any given time.

If an application does not scale well there are several issues to consider to improve the performance of the application. A limiting factor may be the task structure of the application, but also the task pool implementation may limit the

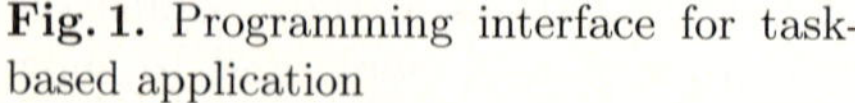

```
struct Task { Function, Argument };
// 1. Initialization phase (processor 1)
for (each work unit U of the input data)
    TaskPool.create_initial_task(U.Function,
        U.Argument);
// 2. Working phase
processor 1...p:
    loop:
        Task T ← TaskPool.get();
        if (T = ∅) exit;
        T.execute(); // may create new tasks
        T.free();
```

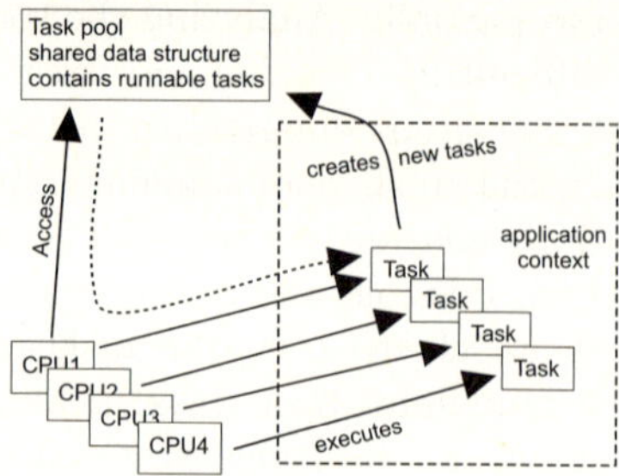

Fig. 1. Programming interface for task-based application

Fig. 2. Parallel execution scheme of a task-based application

scalability. But even if an application scales well on a given machine, it remains unclear whether it will achieve good speedups when using a larger number of processors or faster machines. To obtain the profiling information required to address these issues we select a well performing yet simple task pool implementation from previous work [1] and a modified version is used to gather statistical information about the task execution. Because the task pool appears as a black box to the application this does not require code changes in the application.

Figure 2 illustrates the parallel execution of a task-based application. Each thread accesses the task pool to obtain executable tasks. The code of the task is executed in the application context which can create new tasks for later execution. Besides the application context there is a task pool context which covers the task management, i.e. searching or waiting for new tasks. In this paper we concentrate on profiling the time spent in both contexts and show how this information can be used to identify problems within the application. The time spent in the application context is the execution time of the task code which we call the *task size*. This time represents the work done by the actual application. The larger the task size the more work is done in a single task, so this value helps to identify the task granularity of the application. The task size does not only provide information about the task granularity, it can also be used to obtain information about possible bottlenecks due to parallel execution. For example, if the granularity of a task is independent of the number of threads, but the actual execution times increase with the number of threads then this indicates non-obvious scalability problems. Reasons for such a behavior can be a higher number of cache misses, increased memory access time due to remote access, or higher lock contention. For each single task the profiling mechanism stores which function has been executed along with the corresponding execution time.

The time spent in the task pool context is the time needed for the task management including waiting for new executable tasks. We refer to the sum of these times as *waiting time*. This is the time which is not spent executing the actual application, so it indicates an overhead. An increase in the waiting time also indicates a scalability problem. Reasons can be: the threads access the task pool too often causing mutual exclusion or there are not enough executable task available, so some threads need to wait. Some of the reasons for scalability problems can be addressed at the task pool level, e.g., by modifying the task pool implementation (as shown, for example, in [1]); other problems need to be addressed at the application level. In any case, detailed information about the waiting time can be used to find bottlenecks. For the waiting time the profiling mechanism measures the time spent in the task pool after finishing a task and before executing a new task. This waiting time is associated with the new task indicating that this task was not available early enough for execution.

3   Profiling Methodology

The execution of a task based application using the profiling task pool generates a large data set which needs to be analyzed for detailed information. In the first step we determine global statistical values which include the number of tasks executed, the total task size (i.e., total time spent in the application), the average task size, the total waiting time (i.e., total time spent outside the application), and the average waiting time.

This information allows first overall conclusions about the task pool usage for a specific application. If the number of tasks is small compared to the number of processors the load balancing effect of the task pool is limited. The waiting time can be considered as overhead as this is the time spent in the task pool and not in the application. If the waiting times are long compared to the task size, then this indicates that too few tasks are available for execution at some times.

For a detailed analysis we create task histograms (see Figure 3 for an example) which count the number of occurrences for every task size. Together with the waiting time these plots allow detailed statements about the performance impacts of the interaction between the application and the task pool. The important observations from the histogram can be summarized as follows:

- Large tasks mean a low overhead but possibly indicate limited parallelism or load imbalance.
- Many medium sized tasks suggest a good balance between high overhead and load imbalance.
- Many small tasks (i.e. a high occurrence of tasks on the left side of the histogram) indicate a large overhead in the task pool.
- An similar shape of the task size histogram for different number of processors indicate a suitable task structure, as the execution time of a task does not depend on the number of processors. Otherwise, memory or lock contention or cache invalidations are possible reason for unsatisfactory performance.
- The waiting time should always be as low as possible, i.e., most of the occurrences should be on the left-hand side of the histogram.
- The majority of the waiting times should be below the task size curve. Otherwise, the waiting time is more significant than the actual computation indicating a serious problem in the parallel application.
- Large waiting times, even if they occur rarely, indicate limited parallelism at some time inside the application.

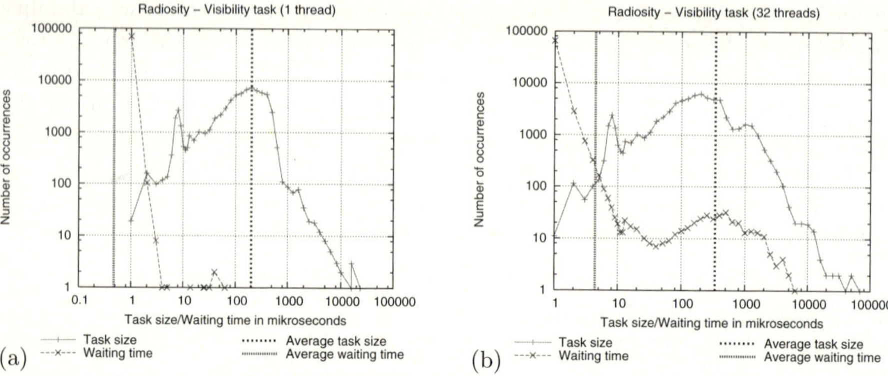

Fig. 3. Task histogram for the "visibility" task of the radiosity application using 1 (a) and 32 (b) threads on the Power4 system

In the following, we discuss the analysis of the profiling information for a selected task type to show how this information can be evaluated for an actual task. The information are gathered from an IBM p690 server which is a symmetric multiprocessor (SMP) with 32 Power4 processors running at 1.7 GHz. A more detailed analysis of the complete application is given in Section 4. Figure 3 shows the results of the data analysis for the hierarchical radiosity application for a specific task type using 1 and 32 threads. For single threaded execution the waiting time for most of the tasks is 1 (0 is counted as 1 due to the logarithmic scale), i.e., there is almost no time spent in the task pool. On average, a task takes $201\,\mu s$ to finish and there are only several hundreds out of about 70,000 tasks with a task size $\leq 5\,\mu s$. Taking all zero waiting times into account, the average waiting time is $\approx 0.5\,\mu s$ so the task size is roughly 400 times larger than the waiting time in the task pool on average. The maximum waiting time of $63\,\mu s$ is also much smaller than the average task size.

For 32 threads, Figure 3b indicates that there are no significant dependencies on the number of threads for this task type as the shape of the task size curve is similar to the single thread curve. All tasks take nearly the same time to complete when executing the application with 32 threads so the execution is not limited by memory bandwidth or cache size. On the other hand, the waiting time increases from 1 to 32 threads. The average waiting time is $4.4\,\mu s$ which is around 10 times more than for the single thread run, but the average task size increases only from $201\,\mu s$ to $340\,\mu s$. Even more important, the largest waiting time is $6319\,\mu s$ which is around 100 times more than for the single thread run. However, the majority of the waiting time is $\leq 1\,\mu s$ and the average task size is still almost 80 times larger than the waiting time on average. On this system, the large increase in the waiting time is not reflected by a performance decrease as the maximum waiting time of $\approx 7\,ms$ is much smaller than the largest task which took almost $70\,ms$ to complete. It can be expected that on a larger or a faster system the problem may limit the scalability especially because such large

Algorithm 1. Framework of the hierarchical radiosity application

```
Phase 1:                              Task REFINEMENT(P):
  for all input patches P do            compute form factor and refine recursively;
    insert P into BSP tree;           Task RAY(P):
    create task REFINEMENT(P);          for all interactions I do
Phase 2:                                  if error(I) too large then
  repeat                                    refine element(I) and create interactions;
    for all patches P in BSP tree do      else if unfinished(I) then
      create task RAY(P);                   create task VISIBILITY(I);
    execute tasks;                        else
  until error small enough;                 gather energy;
Phase 3:                                    if P is leaf then propagate to parent;
  for all patches P in BSP tree do          else for each child C do
    create task AVERAGE(P,average);           create task RAY(C);
  execute tasks;                      Task VISIBILITY(P):
  for all patches P in BSP tree do      compute visibility for given interactions;
    create task AVERAGE(P,normalize);    continue task RAY(P);
  execute tasks;                      Task AVERAGE( P, mode ):
                                        if P is leaf then average or normalize values;
                                        else for each child C do
                                          create task AVERAGE(C,mode);
```

waiting times occur several times. For this task type we can draw the following conclusions using the profiling information:

1. large task sizes in contrast to waiting times even for 32 processors indicate good scalability;
2. the overhead of the task pool is negligible;
3. there is a small increase in task size and a bigger increase in waiting time when more processors are used, so perfect scalability will not be reached especially for a larger number of processors;
4. the small number of very small tasks also indicates a suitable task structure;

4   A Case Study for Performance Prediction

In this section we describe how the task profiling can be used to predict the performance of an application. The case study is done by considering the *hierarchical radiosity application* [3] which is a global illumination algorithm that renders geometrical scenes by computing the equilibrium distribution of light. A hierarchical subdivision of the object surfaces is performed dynamically at runtime, and interactions between the surface elements representing the transport of light are evaluated. Parallelism is exploited across interactions and subdivision elements. The application is subdivided into four task types, see Algorithm 1 for an overview of the application. At the beginning the surfaces (or patches) of the initial scene are divided into sub-patches. This is done by the "refinement" task which can be executed in parallel for different patches. The computation is done by the "ray" task which calculates the energy exchange with other patches. This task can issue new "visibility" tasks and "ray" tasks to evaluate sub-patches. The "average" task post-processes the computed values.

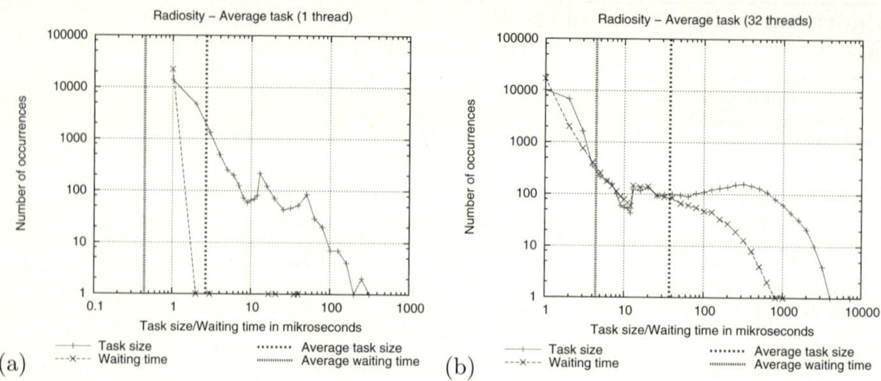

Fig. 4. Task size histogram for the "average" task of the radiosity application using 1 (a) and 32 (b) threads on the Power4 system

4.1 Application Evaluation

The implementation uses four different task types for different stages of calculating the resulting images of a given scene.

Visibility task. We already have investigated the "visibility" task in the previous section, see Figure 3. We have seen that this particular task performs well also for a large number of processors (32).

Average task. Figure 4 shows the task histogram for the "average" task. This task performs a post-processing step for averaging and normalizing the calculated radiosity values. The majority of the task sizes is very small, less than $10\,\mu s$. For one thread (Figure 4a), the average task size is $2.67\,\mu s$ and there are only a few tasks larger than $100\,\mu s$. The average waiting time is $0.44\,\mu s$.

For 32 threads, the shape of the task size graph changed much more than for the "visibility" task shown in Figure 3. There is now a significant number of task sizes larger than $100\,\mu s$. The average task size is with $37.35\,\mu s$ almost 14 times larger than in the single threaded run. Because the number of participating threads does not influence the granularity of this task type, this significant increase indicates a performance problem. The average waiting time is 10 times larger ($4.44\,\mu s$) in contrast to a single thread execution. This is 12% of the average task size which is a significantly larger fraction than for the "visibility" task considered in the previous section.

The conclusion is that this particular task type does not scale very well. The increase in waiting time is not extremely large but the fraction of the waiting time on the task size is large enough to influence the performance. More important is the significant increase of the task size which needs attention to improve the performance of the application.

Ray task. The "ray" task actually calculates the energy of the patches. The shape of the task histogram is similar to the histogram for the "average" task.

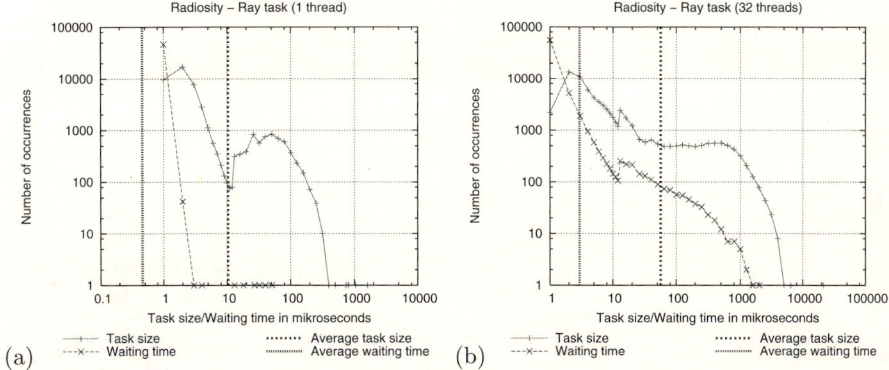

Fig. 5. Task histogram for the "ray" task of the radiosity application using 1 (a) and 32 (b) threads on the Power4 system

When using a single thread (Figure 5a) the majority of the tasks are small (less than $10\,\mu s$), but there are also several larger tasks leading to an average task size of $\approx 10.3\,\mu s$. The average waiting time is very small ($\approx 0.46\,\mu s$, $\approx 4\%$ of the task size). For 32 threads we see the expected increase in the average waiting time which is with $\approx 2.94\,\mu s$ only around 6.5 times larger. The average task size is ≈ 5.47 times larger ($56.43\,\mu s$) which is less than the increase in the task size for the "average" task, but it indicates a similar problem. The conclusion is that this task type works slightly better than the "average" task, but still has scalability problems due to the majority of small tasks and an increase in the task size when using more processors.

Refinement task. The "refinement" task is used to divide the patches of the scene into smaller sub-patches. We observe a different behavior (Figure 6) than for the other task types. For a single thread, no task is smaller than $11\,\mu s$ and the majority is $11 - 13\,\mu s$ but there are several larger tasks ($50 - 500\,\mu s$). The average task size is $12.24\,\mu s$. The waiting time is small, mostly less than $1\,\mu s$. The peak at around $10\,\mu s$ and the few larger waiting times up to $4500\,\mu s$ represent overhead in the task pool implementation. This task type is created and executed first, so the internal data structures to store the large number of tasks need to be created. The average waiting time is very small ($0.61\,\mu s$).

For 32 threads we observe major scalability problems. The average task size is more than 14 times larger ($172.77\,\mu s$), and the average waiting time is almost 27 times larger ($16.45\,\mu s$). The absence of very small tasks should be a good sign for good scalability, but the significant increase of the task size and waiting time indicates serious scalability problems for this task.

4.2   Performance Prediction

For a coarse estimation of the execution time of the application for a specific number of processors, we use the profiling information from measurements with

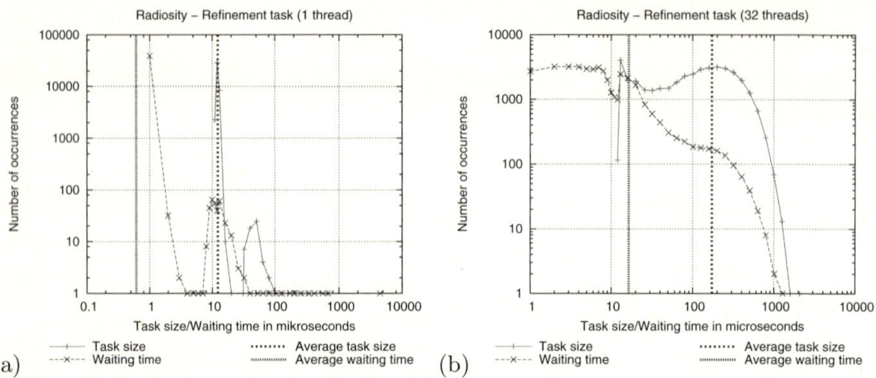

Fig. 6. Task histogram for the "refinement" task of the radiosity application using 1 (a) and 32 (b) threads on the Power4 system

Table 1. Estimated combined execution and waiting time for 32 threads

Radiosity	Computation time		Waiting time	
Task type	Estimated time in μs	Actual time measured in μs	Estimated time in μs	Actual time measured in μs
RAY	2,922,294.15	5,794,241.25	125,116.07	296,063.50
REFINEMENT	20,249,703.33	12,141,060.70	299,718.71	703,022.05
AVERAGE	1,751,265.41	1,556,440.05	158,188.50	123,334.20
VISIBILITY	22,136,753.20	38,105,390.00	51,363.556	361,316.50
Sum	47,060,016.06	57,597,132.00	634,386.84	1,483,736.25

a smaller number of processors to extrapolate the task sizes and waiting times for a larger number of processors. The information for each pair of consecutive number of processors is used to linearly extrapolate the task size and waiting time to the target number of processors. Several extrapolation values are combined to the final value by weighting each value by its distance to the target number of processors. The prediction is done by using only information gathered from the profiling process without considering details about the application. As tasks are arbitrary complex functions, the actual performance depends on many parameters unknown to the profiling task pool (like lock contention inside the task, memory bandwidth requirements, and so on). The estimated performance cannot be expected to be exact, but it can give an indication how well the application will scale.

Table 1 shows the estimated results and the actual results on the Power4 system for the computation time and the waiting time both for 32 processors. The estimated values are calculated by taking measured results for 1, 2, 4, 8 and 16 processors into account. Most values are underestimated or slightly overestimated, and the only specific value which is significantly overestimated is the execution time for the refinement tasks. It is around 66 percent above the actual value which is still acceptable considering the low assumptions made and the few information available.

Using these values the predicted speedup for 32 processors is 9.75 whereas the actual speedup is 7.87. The speedup for 16 processors is ≈ 9.5 so the estimation suggests that using twice as many processors does not give a significant benefit for the execution time. The predicted increase of the task sizes, waiting times and number of tasks helps pointing out which tasks need more attention than others.

5   Related Work

Application profiling is a well known technique to analyze the performance of an application. Modern compilers support adding profiling code to the application to be able to find time consuming parts of an application. Tools like VAMPIR [4] allow the evaluation of such information.

The method presented in [5] does not depend on compile time instrumentation but uses the binary code to profile the application and predict the performance. The method profiles memory accesses to predict the estimated execution time for larger inputs for sequential execution. [6] tries to predict the performance of a selected application for parallel execution even on future architectures but requires detailed analysis of the actual application.

[7] proposes a framework to automatically tune an application. Similar to ATLAS [8], but more generic, the framework is able to select an efficient implementation of certain library functions used by the application. A small amount of source code changes are required and the optimizations are application specific while we are trying to optimize generic task-parallel applications.

As a similar approach to analyze an application and identify performance problems, [9] proposes a method to use a simulator to obtain memory access information (cache misses etc.) and suggest improvements. In our work we are trying to avoid the overhead of a simulator and source code modifications and we also consider the impact of contention which are not modeled by cache statistics. [10] proposes a method to profile parallel applications using TAU. Similar to our work, TAU profiles different contexts (phases) but this requires instrumentation of the source code.

6   Conclusion

The profiling methods proposed in this paper allow to study the behavior of irregular applications and identify scalability problems without code changes or even recompilations of the actual application. Splitting the execution time into the application context and the task pool context makes it possible to evaluate different task types separately and even to consider single tasks. The application context models the actual computations but also covers possible contention inside single tasks. The task pool context covers the available parallelism and overhead of the task based execution.

The proposed method allows the investigation of specific tasks for a given number of processors to identify possible scalability problems inside the task and to indicate missing parallelism inside the application or a too large overhead from the task pool implementation. The isolated examination of single tasks allows us to point out specific tasks which indicates problems and also to propose changes to improve the scalability.

Similar profiling information are otherwise only available by changing the application or recompile it to use profile information from compilers or utilize hardware counters which is not always available or wanted.

Acknowledgments. We thank the NIC Jülich for providing access to their computing systems.

References

1. Hoffmann, R., Korch, M., Rauber, T.: Performance Evaluation of Task Pools Based on Hardware Synchronization. In: Proceedings of the 2004 Supercomputing Conference (SC'04), Pittsburgh, PA (2004)
2. Woo, S.C., Ohara, M., Torrie, E., Singh, J.P., Gupta, A.: The SPLASH-2 programs: Characterization and methodological considerations. In: Proceedings of the 22nd International Symposium on Computer Architecture, Santa Margherita Ligure, Italy, pp. 24–36 (1995)
3. Hanrahan, P., Salzman, D., Aupperle, L.: A Rapid Hierarchical Radiosity Algorithm. In: Proceedings of SIGGRAPH (1991)
4. Brunst, H., Kranzlmüller, D., Nagel, W.E.: Tools for Scalable Parallel Program Analysis - Vampir VNG and DeWiz. In: Juhasz, Z., Kacsuk, P., Kranzlmüller, D. (eds.) DAPSYS. Kluwer International Series in Engineering and Computer Science, vol. 777, pp. 93–102. Springer, Heidelberg (2004)
5. Marin, G., Mellor-Crummey, J.: Cross-Architecture Performance Predictions for Scientific Applications Using Parameterized Models. In: Proceedings of Joint International Conference on Measurement and Modeling of Computer Systems - Sigmetrics 2004, New York, NY, pp. 2–13 (June 2004)
6. Kerbyson, D.J., Alme, H.J., Hoisie, A., Petrini, F., Wasserman, H.J., Gittings, M.: Predictive performance and scalability modeling of a large-scale application. In: Proceedings of the 2001 Supercomputing Conference (SC'01), IEEE/ACM SIGARCH, p. 37 (2001)
7. Tapus, C., Chung, I.H., Hollingsworth, J.K.: Active Harmony: Towards Automated Performance Tuning. In: Supercomputing '02: Proceedings of the 2002 ACM/IEEE conference on Supercomputing, Los Alamitos, CA, USA, pp. 1–11. IEEE Computer Society Press, Los Alamitos (2002)
8. Whaley, R.C., Dongarra, J.J.: Automatically Tuned Linear Algebra Software. Technical report, University of Tennessee (1999)
9. Faroughi, N.: Multi-Cache Profiling of Parallel Processing Programs Using Simics. In: Arabnia, H.R. (ed.) Proceedings of the PDPTA, pp. 499–505. CSREA Press (2006)
10. Malony, A., Shende, S.S., Morris, A.: Phase-Based Parallel Performance Profiling. In: Joubert, G.R., Nagel, W.E., Peters, F.J., Plata, O.G., Tirado, P., Zapata, E.L. (eds.) Proceedings of the PARCO. John von Neumann Institute for Computing Series, vol. 33, pp. 203–210. Central Institute for Applied Mathematics, Jülich, Germany (2005)

Search Strategies for Automatic Performance Analysis Tools*

Michael Gerndt and Edmond Kereku

Technische Universität München, Fakultät für Informatik I10,
Boltzmannstr.3, 85748 Garching, Germany
`gerndt@in.tum.de`

Abstract. Periscope is a distributed automatic online performance analysis system for large scale parallel systems. It consists of a set of analysis agents distributed on the parallel machine. This article presents the architecture of the node agent and its central part, the search strategy driving the online search for performance properties. The focus is on strategies used to analyze memory access-related performance properties in OpenMP programs.

1   Introduction

Performance analysis tools help users in writing efficient codes for current high performance machines. Since the architectures of today's supercomputers with thousands of processors expose multiple hierarchical levels to the programmer, program optimization cannot be performed without experimentation.

To tune applications, the user has to carefully balance the number of MPI processes vs the number of threads in a hybrid programming style, he has to distribute the data appropriately among the memories of the processors, has to optimize remote data accesses via message aggregation, prefetching, and asynchronous communication, and, finally, has to tune the performance of a single processor.

Performance analysis tools can provide the user with measurements of the the program's performance and thus can help him in finding the right transformations for performance improvement. Since measuring performance data and storing those data for further analysis in most tools is not a very scalable approach, most tools are limited to experiments on a small number of processors. To investigate the performance of large experiments, performance analysis has to be done online in a distributed fashion, eliminating the need to transport huge amounts of performance data through the parallel machine's network and to store those data in files for further analysis.

Periscope [4] is such a distributed online performance analysis tool. It consists of a set of autonomous agents that search for performance bottlenecks in a

---

* This work is being funded by the German Science Foundation under contract GE 1635/1-3.

A.-M. Kermarrec, L. Bougé, and T. Priol (Eds.): Euro-Par 2007, LNCS 4641, pp. 129–138, 2007.

subset of the application's processes and threads. The agents request measurements of the monitoring system, retrieve the data, and use the data to identify performance bottlenecks. The types of bottlenecks searched are formally defined in the APART Specification Language (ASL) [2,1].

The focus of this paper is on the agent's architecture and the search strategies guiding the online search for performance properties. We present the search strategies not for analyzing MPI programs on large-scale machines, but for analyzing the memory access behavior of OpenMP programs on a single shared memory node.

The next section presents work related to the automatic performance analysis approach in Periscope. Section 3 presents Periscope's architecture. The detailed description of the agent architecture and the role of the search strategy is discussed in Section 4. Search strategies implemented for memory access properties are presented in Section 5. Results from experiments are given in Section 6 and a summary and outlook in Section 7.

2   Related Work

Several projects in the performance tools community are concerned with the automation of the performance analysis process. Paradyn's [7] Performance Consultant automatically searches for performance bottlenecks in a running application by using a dynamic instrumentation approach. Based on hypotheses about potential performance problems, measurement probes are inserted into the running program. Recently MRNet [8] has been developed for the efficient collection of distributed performance data. However, the search process for performance data is still centralized.

The Expert [10] tool developed at Forschungszentrum Jülich performs an automated post-mortem search for patterns of inefficient program execution in event traces. Potential problems with this approach are large data sets and long analysis times for long-running applications that hinder the application of this approach on larger parallel machines.

Aksum [3], developed at the University of Vienna, is based on a source code instrumentation to capture profile-based performance data which is stored in a relational database. The data is then analyzed by a tool implemented in Java that performs an automatic search for performance problems based on JavaPSL, a Java version of ASL.

Periscope goes beyond those tools by performing an automatic online search in a distributed fashion via a hierarchy of analysis agents.

3   Architecture

Periscope consists of a user interface, a hierarchy of analysis agents and two separate monitoring systems (Figure 1).

The user interface allows the user to start up the analysis process and to inspect the results. The agent hierarchy performs the actual analysis. The node

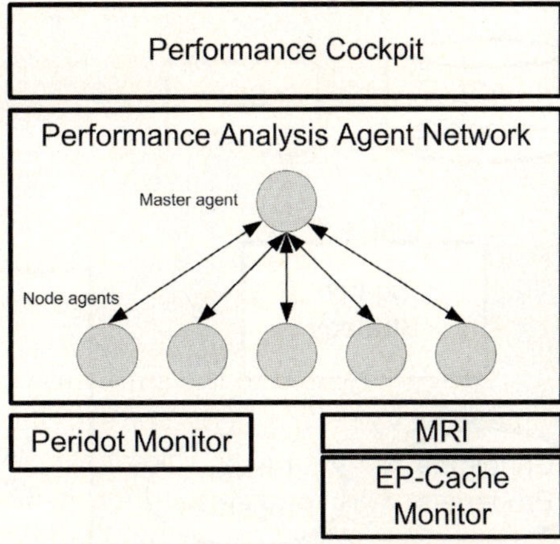

Fig. 1. Periscope currently consists of a frontend, a hierarchy of analysis agents, and two separate monitoring systems

agents autonomously search for performance problems which have been specified with ASL. Typically, a node agent is started on each SMP node of the target machine. This node agent is responsible for the processes and threads on that node. Detected performance problems are reported to the master agent that communicates with the performance cockpit.

The node agents access a performance monitoring system for obtaining the performance data required for the analysis. Periscope currently supports two different monitors. The work described in this article is mainly based on the *EP-Cache monitor* [6] developed in the EP-Cache project focusing on memory hierarchy information in OpenMP programs.

Detected performance bottlenecks are reported back via the agent hierarchy to the frontend.

4   Node Agent Architecture

The node agents search autonomously for performance bottlenecks in the processes/threads running on a single SMP node. Figure 2 presents the architecture and the sequence of operations executed within a node agent.

The figure consists of three main parts, the agent, the monitor, and the data structures coupling the agent and the monitor. These data structures reside in a shared memory segment and are used to configure the monitoring as well as to return performance data.

The agent's main components are the agent control, the search strategy, and the experiment control. As presented in Section 3, the node agent is part of an agent hierarchy. The **master agent** starts the bottleneck search via the **Agent Control and Command (ACC) message** `ACC check` marked as (1) in the diagram. Before the message is sent, the application was started and suspended in

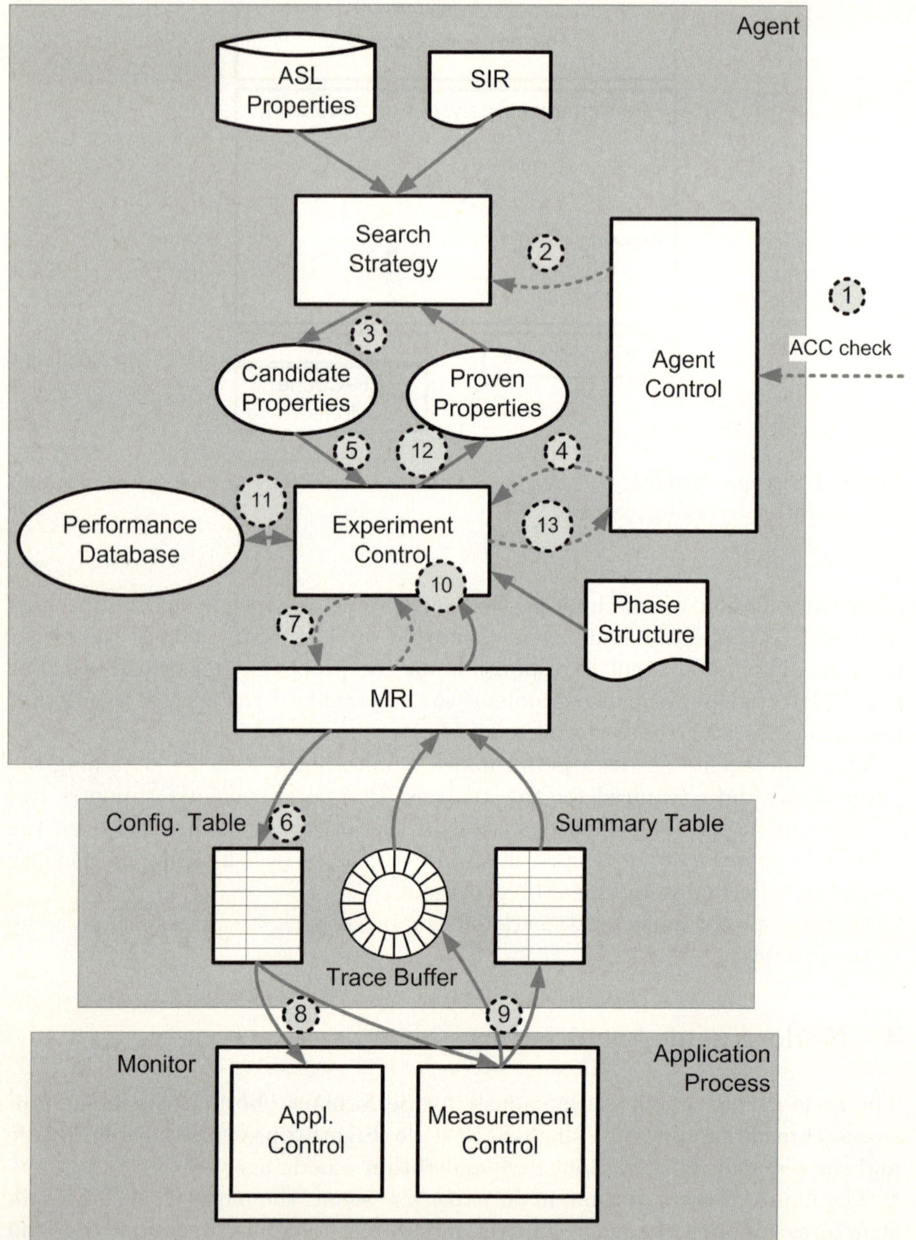

Fig. 2. The agent's search is triggered by a control message `ACC check` from its parent in the agent hierarchy. Performance data are obtained from the monitor linked to the application via the Monitor Request Interface (MRI).

the initialization of the monitoring library linked to the application. In addition the node agent was instructed to attach to the application via the shared memory segment.

The agent performs a **multistep bottleneck search**. Each search step starts (2) with determining the **set of candidate performance properties** that will be checked in the next experiment. This candidate set is determined by a **search strategy** based on the set of properties found in the previous step (3). At the beginning, the set of evaluated properties is empty. The applied search strategy is determined when the agent is started. Most of the strategies take also the program's structure into account.

The **source code instrumenter** used for insertion of the monitor library calls [5] generates information about the program's regions (main routine, subroutines, loops, etc.) and the data structures used in the program in the **Standard Intermediate Program Representation (SIR)** developed in the European American APART working group on automatic performance analysis tools. The SIR is an XML-based language defined for C++, Java, and Fortran [9].

After the candidate set was determined, the agent control starts a new experiment (4). The **experiment control** accesses all the properties in the candidate set and checks whether the required performance data for proving the property are available. If not, it configures the monitor via new **MRI measurement requests**. The requests, such as `measure the number of cache misses in the parallel loop in line 52 in file foo.f`, are inserted into a **configuration table** (6).

Once all the properties were checked for missing performance data, the experiment is started (7). The MRI provides an **application control interface** that allows the experiment control to release a suspended application and to specify when the application is suspended again to retrieve the performance data.

This approach is based on the concept of **program phases**. Usually programs consist of several phases, e.g., an initialization phase, a computation phase, and a termination phase. Frequently the computation phase is repetitive, i.e., it is executed multiple times. Such repetitive phases can be used to perform the multistep search of the node agent. Program phases need to be specified by the user. We provide two ways for specification. The user can mark program parts as a **user region** via directives that are replaced by calls to the monitoring library by the source instrumenter. As an extension, phases can also be marked by manual insertion of **phase boundary function calls** that specify additional properties of the phase, e.g., whether it is a repetitive phase or a execute-once phase. Currently, our prototype supports the specification of a single user region which is assumed to be repetitive. If no such region is specified, the whole program is taken as a phase and is restarted to perform additional experiments.

During program execution, the **monitoring library** checks whether the end of the current phase is reached (8) and configures **hardware and software sensors** for measuring the requested performance data. These data are inserted into a **trace buffer** (9) if trace information is requested or into a **summary table** if aggregated information is to be computed.

When the application is suspended (10) the experiment control is informed by the MRI and it retrieves the measured performance data via the MRI into the internal **performance database** (11). Trace data are handled differently. If the trace buffer is filled during execution, a callback function of the agent is triggered to extract the data into the performance database. Currently our node agent does not use this feature. All the performance properties are based on summary data.

The experiment control evaluates the candidate performance properties and inserts the proven properties into the **proven properties set** (12). At the end of this search step, the control is returned to the agent control (13).

5   Search Strategies

Periscope currently supports a number of search strategies. The strategies provide a simple interface which consists of a routine creating an initial candidate set and a refinement routine that determines from the set of proven properties

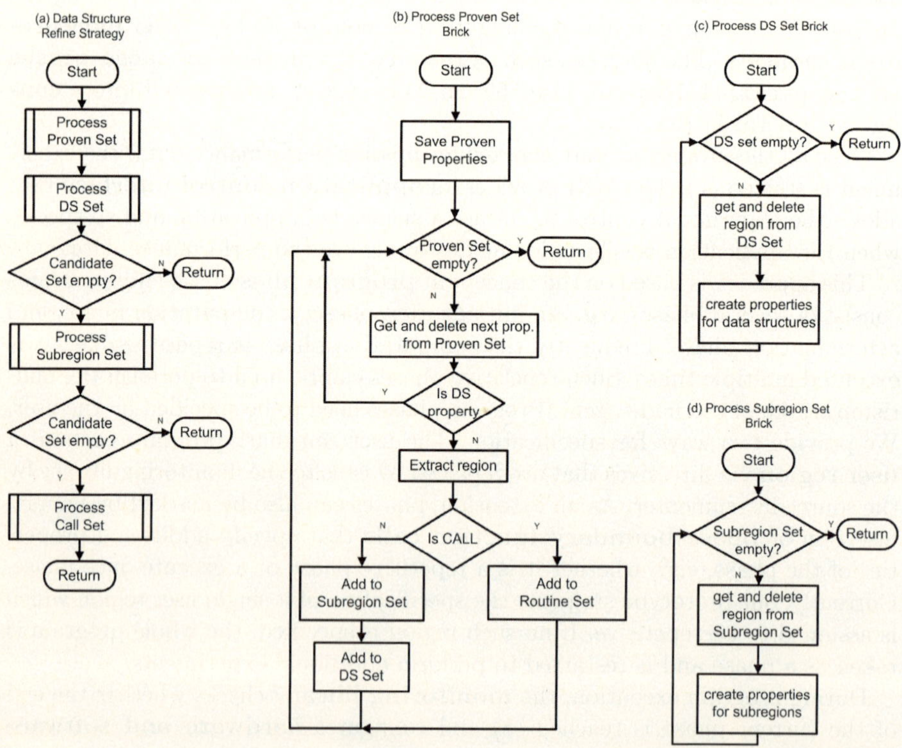

Fig. 3. Implementation of the strategy that refines the search with respect to data structures, subregions, and called subroutines

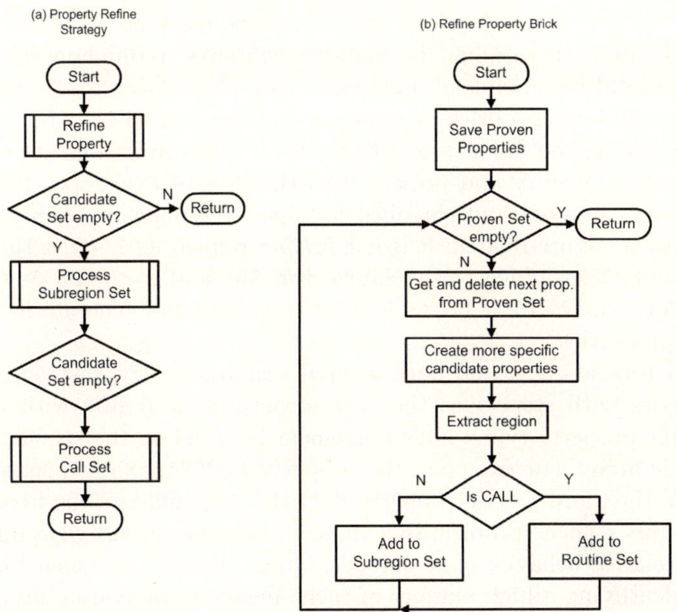

Fig. 4. Strategy that refines proven properties with respect to more specific properties in the property hierarchy

a new candidate set for the next search step. Since the node agent only accesses those routines, search strategies can be implemented as classes that are dynamically loaded and thus, without having to recompile the agent if a new strategy is available.

Search strategies are based on a number of strategy bricks supporting reuse of code. Figure 3 introduces the *Data Structure Refine Strategy* which was designed to perform a search for memory access inefficiencies. It takes a memory access-related property and searches for occurences of this property in the program's execution. The refinement is based on a novel feature of the EP-Cache monitor. It not only provides measurements of cache misses etc. via hardware counters for program regions, but allows to restrict measurements to address regions. Thus, the agent can check properties that are related to individual data structures, e.g., `high number of cache misses for array A in loop 20`.

The refinement routine in Figure 3.a first processes the set of proven properties from the previous search step. It then refines proven properties with respect to the data structures accessed in the program region (Fig. 3.c). After analyzing the current set of program regions with memory access inefficiencies with respect to the data structures, these regions are further analyzed with respect to their subregions (Fig. 3.d) in the next search step. If no refinement of the current set of proven properties with respect to data structures and subregions is possible, properties found for individual subroutine calls are further investigated. The strategy brick `Process Call Set` is not shown in the figure. It is very similar to the two bricks discussed before, but also keeps track of the already checked subroutines. Since there might be multiple call sites, redundant searches of a subroutine would be possible otherwise.

The missing strategy brick is `Process Proven Set` (Fig. 3.b). This brick starts with saving the found properties since all the properties are ranked according to their severity and presented to the user of Periscope. Then all the properties are analyzed and classified for further refinement. If a property is already data structure-related, it is not further refined. Otherwise the region is extracted and either added to the `Subregion Set` and to the `DS Set` or to the `Routine Set`. The strategy bricks process these sets and generate more precise candidate properties.

Figure 4 presents a second implemented search strategy. Instead of refining the properties with respect to the data structures, it refines with respect to more specific property types. This refinement is based on the specification of a property hierarchy. For example, the property `LC2DMissRateInSomeThread` is refined into the more precise property `UnbalLC2DMissRate`. The first property only highlights a cache problem in a thread while the second gives information about the relative behavior of all threads. Other obvious refinements are from a property identifying a high number of cache misses to individual properties for read and write misses and for local vs remote misses on ccNUMA architectures.

6   Experiments

We tested the search strategies with several OpenMP examples. Here, we present the results for the SWIM benchmark from the SPEC benchmark suite. The first experiment analyzes a sequential run of SWIM with the `Data Structure Refine Strategy`. Periscope was used to search for severe LC3 miss rate. The results of the search are presented in form of search paths which show the refinements on region level and on data structures.

```
        The results of the automatic search for SWIM


                    Region                    LC3MissesInThread
Application Phase( USER_REGION, swim.f, 84 )
    calc2( CALL_REGION, swim.f, 92 )                0.022
    calc2( SUB_REGION, swim.f, 315 )
      ( PARALLEL_REGION, swim.f, 332              0.028
          ( DO_REGION, swim.f, 336 )              0.028


Application Phase( USER_REGION, swim.f, 84 )
    calc2( CALL_REGION, swim.f, 92 )                0.022
    calc2( SUB_REGION, swim.f, 315 )
      ( DO_REGION, swim.f, 354 )                   0.302
          unew( DATA_STRUCTURE, swim.f, 3 )        0.279
          vnew( DATA_STRUCTURE, swim.f, 3 )        0.281
          pnew( DATA_STRUCTURE, swim.f, 3 )        0.281
```

```
Application Phase( USER_REGION, swim.f, 84 )
   calc2( CALL_REGION, swim.f, 92 )                    0.022
   calc2( SUB_REGION, swim.f, 315 )
      ( LOOP_REGION, swim.f, 360 )                     0.053


Application Phase( USER_REGION, swim.f, 84 )
   ( DO_REGION, swim.f, 116 )                          0.046
         unew( DATA_STRUCTURE, swim.f, 3 )             0.046
         vnew( DATA_STRUCTURE, swim.f, 3 )             0.046
         pnew( DATA_STRUCTURE, swim.f, 3 )             0.046


Application Phase( USER_REGION, swim.f, 84 )
   calc3z( CALL_REGION, swim.f, 145 )                  0.043
```

The severity shown is simply the miss rate. If the severity is redefined to take into account also the amount of time spent in the code region, the last found property for routine CALC3Z is the most critical.

We also tested SWIM on the SGI Altix Bx2 at Leibniz Computing Centre. We run it with 16 and 32 threads on this ccNUMA architecture and applied the Refine Property Strategy for cache problems on the level two data cache.

```
The results for SWIM running with 16 threads


       Region        LC2DMissRateInSomeThread    UnbalC2DMissRate
(DO_REGION,swim.f,437)       0.36                      3.24
(DO_REGION,swim.f,294)       0.29                     14.91
(DO_REGION,swim.f,354)       0.29                      5.67
(DO_REGION,swim.f,116)       0.08                       ---


The results for SWIM running with 32 threads


(DO_REGION,swim.f,437)       0.30                     11.97
(DO_REGION,swim.f,354)       0.24                     16.46
(DO_REGION,swim.f,116)       0.08                      0.74
```

SWIM has cache problems on almost the same regions in both configurations. What we observe is that the problem of unbalanced cache misses is aggravated when running with 32 threads.

7   Summary

Periscope is an automatic performance analysis tool for high-end systems. It applies a distributed online search for performance bottlenecks. The search is executed in an incremental fashion by either exploiting the repetitive behavior of program phases or by restarting the application several times.

The search strategies defining the refinement of found properties into new candidate properties are loaded dynamically so that new strategies can be integrated without recompilation of the tool.

This article presented the architecture of the agents and the integration of the search strategy with the agent's components and the monitoring system. Search strategies are assembled from building blocks called search bricks. The presented strategies have been developed for searching memory access inefficiencies in OpenMP codes.

Future work will focus on developing search strategies that take into account instrumentation overhead, the limited number of resources in the monitor, the progress of the search in other agents etc. The work presented here is a starting point for the development of more intelligent automatic performance analysis tools.

References

1. Fahringer, T., Gerndt, M., Riley, G., Träff, J.: aff. Knowledge specification for automatic performance analysis. APART Technical Report (2001), http://www.fz-juelich.de/apart
2. Fahringer, T., Gerndt, M., Riley, G., Träff, J.L.: Specification of performance problems in MPI-programs with ASL. In: International Conference on Parallel Processing (ICPP'00), pp. 51–58 (2000)
3. Fahringer, T., Seragiotto, C.: Aksum: A performance analysis tool for parallel and distributed applications. In: Getov, V., Gerndt, M., Hoisie, A., Malony, A., Miller, B. (eds.) Performance Analysis and Grid Computing, pp. 189–210. Kluwer Academic Publisher, Dordrecht (2003)
4. Gerndt, M., Fürlinger, K., Kereku, E.: Advanced techniques for performance analysis. In: Joubert, G.R., Nagel, W.E., Peters, F.J., Plata, O., Tirado, P., Zapata, E. (eds.) Parallel Computing: Current&Future Issues of High-End Computing (Proceedings of the International Conference ParCo 2005), NIC Series, vol. 33, pp. 15–26a (2006)
5. Gerndt, M., Kereku, E.: Selective instrumentation and monitoring. In: International Workshop on Compilers for Parallel Computers (CPC 04) (2004)
6. Kereku, E., Gerndt, M.: The EP-Cache automatic monitoring system. In: International Conference on Parallel and Distributed Systems (PDCS 2005) (2005)
7. Miller, B.P., Callaghan, M.D., Cargille, J.M., Hollingsworth, J.K., Irvin, R.B., Karavanic, K.L., Kunchithapadam, K., Newhall, T.: The Paradyn parallel performance measurement tool. IEEE Computer 28(11), 37–46 (1995)
8. Roth, P.C., Arnold, D.C., Miller, B.P.: MRNet: A software-based multicast/reduction network for scalable tools. In: SC2003, Phoenix (November 2003)
9. Seragiotto, C., Truong, H., Fahringer, T., Mohr, B., Gerndt, M., Li, T.: Standardized Intermediate Representation for Fortran, Java, C and C++ programs. APART Working Group Technical Report, Institute for Software Science, University of Vienna (Octorber 2004)
10. Wolf, F., Mohr, B.: Automatic performance analysis of hybrid MPI/OpenMP applications. In: 11th Euromicro Conference on Parallel, Distributed and Network-Based Processing, pp. 13 –22 (2003)

Experiences Understanding Performance in a Commercial Scale-Out Environment

Robert W. Wisniewski[1], Reza Azimi[2], Mathieu Desnoyers[3],
Maged M. Michael[1], Jose Moreira[1], Doron Shiloach[1], and Livio Soares[2]

[1] IBM T.J. Watson Research Center
[2] University of Toronto
[3] École Polytechnique de Montréal

Abstract. Clusters of loosely connected machines are becoming an important model for commercial computing. The cost/performance ratio makes these scale-out solutions an attractive platform for a class of computational needs. The work we describe in this paper focuses on understanding performance when using a scale-out environment to run commercial workloads. We describe the novel scale-out environment we configured and the workload we ran on it. We explain the unique performance challenges faced in such an environment and the tools we applied and improved for this environment to address the challenges. We present data from the tools that proved useful in optimizing performance on our system. We discuss the lessons we learned applying and modifying existing tools to a commercial scale-out environment, and offer insights into making future performance tools effective in this environment.

1   Introduction

For the past decade mainstream commercial computing has been moving from uniprocessor computing systems to multiprocessor ones. During the first phase of the commercial multiprocessor revolution, shared-memory multiprocessors (SMPs) became pervasive. SMPs of increasing size, with processors of increasing clock rate, offered ever more computing power to handle the needs of even large corporations. SMPs currently represent the mainstream of commercial computing. Companies like IBM, HP, and Sun invest heavily in building bigger and faster SMPs. These large SMPs are also called *scale-up* systems.

Computational needs, however, have continued to rise, and more recently, there has been increased interest in clusters of loosely coupled systems for commercial computing. These clusters are also called *scale-out* systems. Computer manufacturers have made it easier to deploy scale-out solutions with rack-optimized and blade servers. Scale-out has been the only viable alternative for large scale scientific computing for several years, as observed in the evolution of the TOP500 systems [1], and they are now becoming more popular for commercial computing. For the past year, we have been working on the problem of demonstrating a scale-out system with superior performance in commercial

A.-M. Kermarrec, L. Bougé, and T. Priol (Eds.): Euro-Par 2007, LNCS 4641, pp. 139–149, 2007.

applications. In order to optimize such systems it is important to understand their performance.

A commercial scale-out (CSO) environment presents several challenges to understanding the performance of applications. We describe these in Section 2. In Section 3 we describe the tools we developed, ported, and used to understand the performance of our environment. Using these tools we obtained a better performance of our system and used that to guide a directed optimization. We present this data in Section 4. During the work of understanding performance and tailoring the tools for our commercial scale-out environment we learned valuable lessons and gained insights into what future performance tools should offer to be effective in this environment. We present these in Section 5. We present related work in Section 6 and conclude in Section 7.

2   Understanding Performance in Commercial Scale Out

Understanding performance in a commercial scale-out environment has two challenges similar to large parallel scientific environments. The first is that the large number of processing elements generate a large number of trace streams with a tremendous amount of data. It is thus important to develop automatic techniques for determining which parts of which traces should receive manual attention. For longer running applications it is important to have techniques to limit the amount of data.

The second challenge is that to correlate events from different machines a synchronized time is needed. Some parallel tool developers perform a linear shift between the first and last event for each trace. This is only sufficient for traces of smaller time windows collected on higher-end systems with more accurate clocks. In the CSO space, with commodity hardware the clocks often do not drift linearly and these techniques are not sufficient.

The two additional challenges that commercial workloads introduce are the complexity of the software stack and a large number of simultaneous threads of execution on a single processing element. The stack complexity manifests itself by requiring multiple loosely related coordinating processes. When multiple coordinating processes are executing it becomes difficult to understand the causality relationship between observed system behavior and the process or group of processes causing the behavior. As a simple example, if both a database and web server are running on a machine, and the observed behavior is that we are running short of page cache space, given our knowledge of these applications, we can probably conclude that it is the database causing the problem. However, if the characteristics of the running applications are not clearly understood in isolation, or if there is any interaction between them, it becomes much more difficult to understand particular system behavior. Related to this issue is understanding the interrelations between processes on different machines. For example, why did process X on machine 1 start process Y on machine 2.

Typical commercial workloads generate hundreds or thousands of concurrently executing threads in each machine. The threads do not necessarily all belong to a single process or logical unit of computation. It is therefore not meaningful to combine the performance of all the threads, nor is it practical to analyze them on a cases-by-case basis. In the scope of this work we made modifications to the tools described in Section 3 to handle these issues.

3   Performance Tools

A major difficulty in a commercial scale-out environment is the number of different software pieces interacting together. Profiling tools like `gprof`, `oprofile`, and `nmon` show only a breakdown of the CPU time, but do not identify which resource a program is waiting for when it does not have the CPU. More information can be extracted by using an instrumentation approach.

The Linux Trace Toolkit Next Generation (LTTng) [2], is a trace-based tool that extracts information in a unified manner from all execution layers in the software stack (from hypervisor through user space) with minimal impact on the application performance and system behavior. The Linux Trace Toolkit Viewer (LTTV) uses the data output from LTTng, merges the data collected from each software stack layer, and organizes them in data structures that permit the identification of the producer of these events (which node, process, thread) and classification of the execution context in which the event occurs (process context or in which system call, trap, interrupt, softirq).

We have added features to LTTng to support commercial scale-out environments. We added PowerPC-specific instrumentation and tracing support for the Java language. This support was implemented using a JNI interface that calls a C handler which in turn calls the LTTng user space tracer to record the information. Once the information is available, it is important to be able to identify the information source. This has been made possible by adding *thread branding* events that are triggered when a new Java thread starts. The thread brand event records information about the name of the main thread function and specific thread information. The analysis and visualization tool, LTTV, has been extended to include thread brands into its representation of the system state, and allows filtering on them. To ameliorate start-up time for the many large files, we added a new precomputation module.

To complement the functionality of LTT, we designed a performance monitoring facility that provides an easy-to-use interface to the hardware Performance Monitoring Unit (PMU) of the PowerPC processors. This facility uses statistical sampling to continuously identify microprocessor bottlenecks. It has been implemented as a kernel module that performs hardware performance counter (HPC) multiplexing, does PC and data sampling, and calculates a stall breakdown model.

The key to achieve acceptable overhead during run-time monitoring is to minimize the frequency of user-kernel protection boundary crossings. In our implementation, the sampling module is fully implemented inside the Linux 2.6 kernel. As a result, except for infrequent control operations (such as initialization or reset), there is no interaction between the user code and the performance monitoring module.

Table 1. Types of miss events with their potential effect in the micro-architecture function

Miss Event	Effect	Description
I-cache miss	Empty reorder buffer	In-flight instructions after mispredicted branch are flushed
Branch misprediction	Empty reorder buffer	In-flight instructions after branch are flushed
Data cache miss	Retirement stops	Delay in the Load-Store-Unit (LSU) due to data cache miss.
Address translation misses	Retirement stops	Address translation (e.g., TLB) miss
LSU basic latency	Retirement stops	Delay in one of the LSUs to finish execution of an instruction
Rejections	Retirement stops	One of the units rejects an instruction – it must be reissued
FPU latency	Retirement stops	Delay in one of the FPUs to finish the computation
Integer latency	Retirement stops	Delay in one of the integer units to finish the computation
Other sources	Retirement stops	Usually results in flushing the pipeline

We alleviate the problem of having limited number of physical counters by dynamically multiplexing [3] the set of hardware events counted by the HPCs using fine-grained time slices. The sampling module assigns each group of events a fraction of g cycles out of a *multiplexing round* R that is the time period in which all HPC groups have a chance to be scheduled. At the end of each HPC group's time slice, the sampling engine automatically assigns another HPC group to be counted by the hardware PMU. The value that is read from an HPC after g cycles is scaled linearly, as if that event had been counted during the entire R-cycle period. As a result, the user program is presented with N *logical* HPCs on top of n *physical* HPCs, where N can be an order of magnitude larger than n. Our earlier experimental evaluation [3] demonstrated that the statistical distance between the sampled and real rates of hardware events is small.

We use the hardware performance counters to calculate a Cycle-Per-Instruction (CPI) breakdown that attributes CPU cycles to the different hardware components that caused them. We restrict our approach to a breakdown of *stall cycles*. A stall cycle is a processor cycle in which no instruction completes (retires). When no stall occurs, the CPU throughput, in terms of IPC, is fairly close to the pipeline width and is fairly application-independent.

The key idea behind the stall breakdown model is that most bottlenecks can be detected by speculatively attributing a *source* to each stall. There are two major categories of such stalls:

– *Empty Reorder Buffer:* This implies that the front-end has not been able to feed the back-end in time. Assuming the micro-architecture is designed and tuned properly, such situations happen mostly when there is an I-Cache miss, or when a branch misprediction occurs.

– *Completion Stops:* In this case, the reorder buffer is not empty, but the oldest instruction bundle in the reorder buffer cannot retire. This happens mainly because one or more of its instructions in the bundle have not yet finished (i.e., they are waiting for an functional unit to provide the results).

We call the hardware events that can cause a stall *long-latency events*. The long-latency events we consider in this study are listed in Table 1 along with the type of stalls they cause and the potential effect they may have. By taking all sources of stalls into account, the following formula can be used to speculatively characterize the potential CPU bottlenecks at each phase in the program execution:

$$\text{CPI}_{\text{real}} = \sum_{i=0}^{n} \text{Stall}_i + \text{CPI}_C$$

where, Stall_i is the number of stalls caused by long-latency event i in the monitoring period, and CPI_C is number of *completion cycles* in each of which at least one instruction is completed. In fact, CPI_C can be used as an estimate for the CPI that can be achieved by an *ideal* hardware in which all the long latency events are removed and performance is solely determined by the program dependences and the width of the pipeline.

4   Experimental Results

Our Commercial Scale-Out (CSO) environment was built using PowerPC blades to run the Nutch [4] web search application workload. The basic building block of the cluster is a BladeCenter-H (BC-H) chassis. Each BC-H chassis has 14 blade slots and is coupled with one DS4100 storage controller with a 2 Gbs Fiber Channel. Of the 8 chassis in our cluster, 4 are filled with JS21 blades. These are quad-processor (dual-socket, dual-core) PowerPC 970 blades, running at 2.5 GHz, with 8 GiB of memory each. The chassis are interconnected through two nearest-neighbor networks: a 4 Gbs Fiber Channel network and a 1 Gbs Ethernet network. Each DS4100 consists of dual redundant RAID controllers and 14 SATA drives of 400GB each.

Nutch is an open-source distributed web search application built on top of the Lucene search library [5]. The query engine consists of one or more front-ends and one or more back-ends, as shown in Figure 1. The front-ends provide a web interface for queries. Each back-end is responsible for a data partition, a subset of the complete set of documents. When a front-end receives a query, it broadcasts the query to all back-ends. Each back-end responds to the front-end with the top n (e.g., 10) documents in its data partition that match the query. The front-end collects the responses from all the back-ends to produce a single list of the top n documents with the best overall matches. For each document in the top n list, the front-end asks the corresponding back-end to return representative snippets of the document text. The front-end then responds to the requester with the query results. The Nutch architecture is similar to the Google search engine [6].

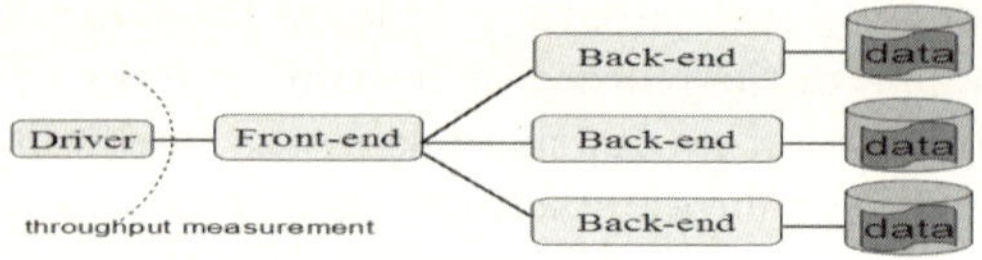

Fig. 1. Nutch distributed query environment

We took two approaches in understanding the performance of our application. The first was to use HPCs to understand the CPI-breakdown of our application. The second approach was to use LTT to examine and categorize time spent in different OS services.

Our measurements with HPCs show that the CPI for query fluctuates between 1.5 and 2.0 during a run, with an average of 1.7. This is far from the PowerPC 970 peak CPI of 0.2, but it is within what would be expected from previous experience with SPECcpu2000 benchmarks where we observed a CPI of 1.53. To better understand what aspects of the hardware were limiting the CPI we used the tool we described earlier to classify stalls. Figure 2(a) shows the stall breakdown with each bar representing a 1 second analysis period. In each second there are a total of 10 billion processor cycles executed by 4 processors running at 2.5GHz. Figure 2(b) shows the average over time of the data in Figure 2(a), separated for cycles in user mode, kernel mode, and total.

Instructions complete on only 20% of the cycles. Because instructions in the PowerPC 970 complete in bundles, multiple instructions can complete per cycle. The number of instructions executed per second (10 billion cycles/second + 1.7 cycles per instruction = 5.9 billion instructions/second), shows that the average bundle size is approximately 3 instructions (out of a maximum of 5). The non-stall CPI for query, computed by dividing the number of non-stall (completed) cycles by the number of instructions is 0.34. Again, this is similar to the non-stall CPI for SPECcpu2000 where we observed a CPI of 0.35.

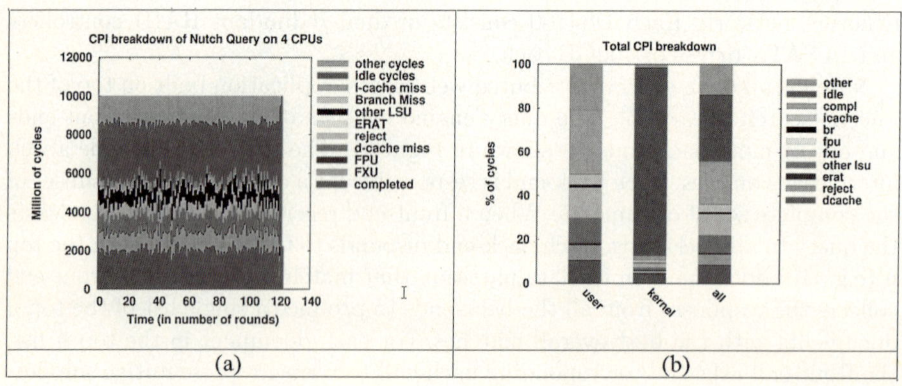

Fig. 2. Hardware performance counter data

The data shows that for a significant number of cycles (25%), the processor is idle. In principle, we should be able to reduce that idle time by increasing the load on the node. This is accomplished by increasing the number of tasks on the node. There is of course a balance, because an excessive number of threads causes a slowdown. In more extensive experiments, we found that we could keep the idle time down to between 10-15%, and are investigating how to drive this even lower.

Finally, we observe that the number of cycles wasted on the I-cache or on branch mispredictions is relatively small, stalls due to the fixed-point units account for 10% of the cycles, and stalls because of the memory hierarchy (D-cache, reject, ERAT, and other LSU) represent approximately 20% of the cycles. The fraction of cycles wasted in the memory hierarchy is similar to the fraction of cycles doing useful work. Thus, a perfect memory system would at most double the performance of the existing processors. Alternatively, for an equivalent amount of chip area, a commensurate benefit could be obtained by doubling the number of processors and maintaining the memory hierarchy per processor.

Using LTT to sample this workload at various points throughout its execution we determined that its behavior characteristics are stable over time. The following figures (Figures 3-4) are taken over a ten second snapshot during the portion of query running on a JS21 back-end blade as described earlier. Each of the data points was sampled several times and the below figures represent the median point. Figure 3(a) shows an overall breakdown of time in the system. As shown, the application (user mode) takes 70% of the CPU time, the system 20%, and 10% remains idle. One of our goals was to find bottlenecks in the system stack and to optimize for those. Thus, we examined the system closer with LTT. Figure 3(b) shows the breakdown into the major categories of system time. Figure 3(a) indicates that system call time dominates system activity for this application.

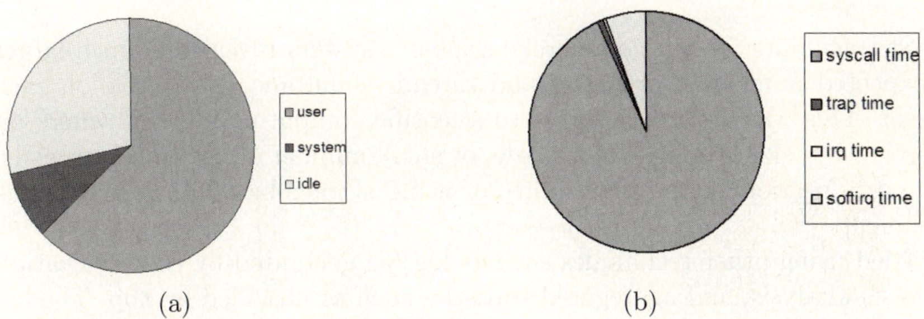

Fig. 3. Time breakdown for overall system (a) and for kernel (b)

As system call time was critical to this application, we wanted to ascertain which system call was having the greatest impact on system performance. We thus used LTT to breakdown the system call time into its various components. As seen in Figure 4(a) `read()` is one of the largest components of system time. Based on the impact of `read()` time, we optimized the `read()` system call in our libOS environment. The optimization targeted for `read()` was to perform a specialized caching. We were able to achieve this because the libOS environment can be tailored to a specific application. This capability demonstrates the strength of a libOS environment [7]. Figure 4(b) shows a comparison of the modified `read()` libOS versus original Linux system. Using performance monitoring as provided by LTT, we were able to identify system calls having the significant impact on performance and target them for optimization. After optimizing `read()` we were able to improve the overall performance of our Nutch/Lucene search application.

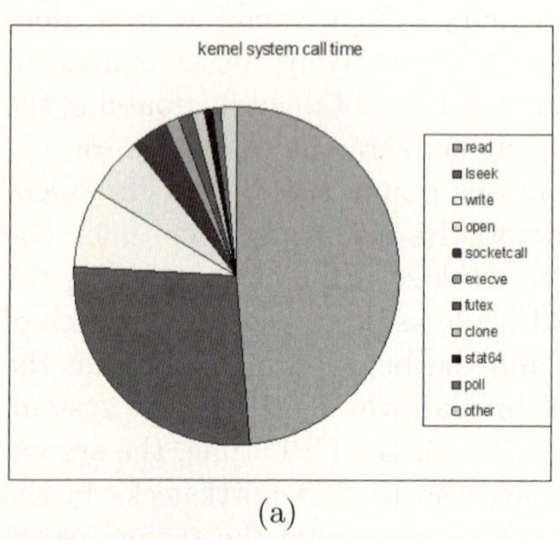

(a)

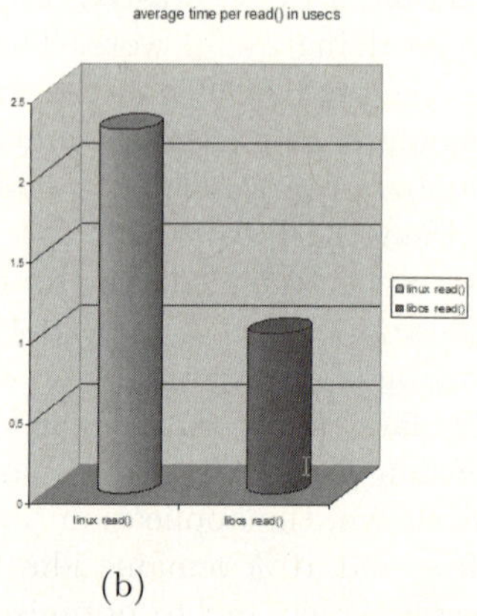

(b)

Fig. 4. Time breakdown for system calls (a) and comparison between original Linux and modified libOS (b)

5   Lessons Learned and Future Tools Requirements

The first trace data for our commercial scale-out system revealed a much larger than expected number of processes and threads simultaneously active on each processor. This was a stark contrast to scientific computing systems where on each processing element there is a single, or small number of, executing threads. We started questioning the plausibility of using a trace-based analysis for this environment.

We tried using other techniques such as log (as generated by the application of interest) analysis, and aggregated statistics such as provided by `top`. The log based approach applies just to the application. It does not help significantly with system understanding, and there is a tremendous amount of data generated by the logs. In tracking down one particular problem we could not use the logs because they perturbed the system too much and generated so much data we could not store it. There was no tool, other than `grep` and `split` (the files were too big to load into any editor) designed to process the logs.

In light of these issues, we decided to make modifications to our trace-based analysis to make it more useful for commercial scale-out. The first capability added was process branding to identify how and why a process is created. We could now identify processes that were started to create other worker JVMs upon receiving requests from the master node. This clarified one of the puzzles as to why some Java processes did nothing but wake up occasionally and go back to sleep without doing anything. We now had a good indication that it was not because they were blocked waiting for some resource, rather they just didn't have anything to do and were waiting for work. This highlights another difference in programming models from scientific to commercial computing. It is important that tools be able to recognize why a process is not computing, even when the reason occurred outside the time window over which the trace was collected. Another feature we added was the ability to only view a single process tree. This allowed us to examine threads of execution that were related to a particular higher-level computation. These few features made a meaningful difference in the understanding we could get from a trace-based analysis of a commercial workload, but still much more could be done. While there are some commercial workloads that behave similar to scientific workloads, and vice-versa, the experiences described above represent general tendencies for scientific and commercial computing.

Based on our experience, we have developed a set of recommendations for future performance tools:

- **Process branding:** An automatic way to determine how and why processes were started. For example, for a daemon that forks processes in response to requests from a master or other cooperating processes, we need the ability to track that relationship.
- **Time synchronization:** Add events into the trace generation and put analysis capability into the post-processing tools allowing timestamps collected on different machines to be adjusted on a fine granularity table-based manner. This would allow all events in a CSO system to have consistent timestamps and be displayed on a single timeline.
- **Automatic idle thread determination:** The inability to overlap I/O, poor use of synchronization primitives, etc., causes performance loss because the processor becomes idle. We need the ability to identify why the processor is not fully utilized, and in particular, to understand why a thread is unable to run at a given time.
- **Cross-machine logical causality:** Allow tools to display on a single timeline processes from one machine that have been created as a result of a process running on a different machine. This requires time synchronization and thread branding.
- **Tree-based causality:** Provide the ability to select a given process and see all processes that have been created because of actions taken by this process. Should work across machines.
- **Selective aggregation of different performance data:** Currently tools such as LTT can provide statistical information about how much time a

processor spent executing particular system calls. This capability is provided on a process by process basis. What is needed is the ability to provide it on a tree-based causality basis.

6   Related Work

Barroso *et. al.* [6] describe Google's search architecture. This is similar to Nutch distributed query in partitioning data and designating web servers (front-ends) and index and document servers (back-ends). However, the former maintains separate index and document servers, while in the Nutch architecture, back-ends are responsible for both searching the index and providing the relevant snippets from the document data partitions.

The original LTT project [8] has been followed by LTTng [9]. The latter project reused work done in IBM's K42 [10,11] on atomic tracing algorithms. The current work on with LTTng targets the Xen hypervisor and supports analysis of the interactions between the hypervisor and operating system.

Software multiplexing of HPCs is implemented for PAPI [12] at user-level using OS signals [13,14]. Due to the large overhead of switching HPCs at user level (signal delivery plus kernel/user context switches), the granularity of multiplexing must be large which in turn results in high sampling error. DCPI [15] uses statistical sampling of the HPCs to identify system-wide hot spots and pipeline stalls at the instruction level. A major simplifying assumption in DCPI is that there is a fixed distance between the instructions that causes HPCs to overflow and the overflow exception. However, this assumption has been shown not to be realistic in modern processors with deep and wide pipelines. A simpler alternative for stall breakdown is to use fixed penalties for long-latency operations such as cache misses [16]. This approach though, significantly overestimates the actual penalties as it does not take potential overlapping of concurrent operations into account. Our approach is more accurate because it exploits existing hardware support to measure the stalls that actually occur.

7   Conclusion

In this paper we described the challenges a commercial scale-out environment poses for performance understanding, including a complex execution stack with multiple loosely related processes executing concurrently and the requirement for a large number of threads of execution per single processing element. We used LTTng and hardware performance counters to investigate performance in such an environment. This process was enlightening in terms of what the needs are for performance tools in the commercial scale-out space. We shared the lessons we learned as part of going through this process and then provided a set of recommendations for future performance tools for commercial scale-out environments.

References

1. Top 500 supercomputer sites, http://www.top500.org/
2. Lttng home web page, http://ltt.polymtl.ca
3. Azimi, R., Stumm, M., Wisniewski, R.W.: Online performance analysis by statistical sampling of microprocessor performance counters. In: ICS International Conference on Supercomputing, Cambridge, Massachusetts (June 2005)
4. Cafarella, M., Cutting, D.: Building Nutch: Open source search. j-QUEUE 2(2), 54–61 (2004)
5. The lucene home page, http://lucene.apache.org/
6. Barroso, L.A., Dean, J., Hölzle, U.: Web search for a planet: The Google Cluster Architecture. IEEE Micro 23(2), 22–28 (2003)
7. Ammons, G., Appavoo, J., Butrico, M., Silva, D.D., Grove, D., Kawachiya, K., Krieger, O., Rosenburg, B., Hensbergen, E.V., Wisniewski, R.W.: Libra: A library operating system for a jvm in a virtualized execution environment. In: VEE. Virtual Execution Environments, June 13-15,2007, San Diego, CA, 2007
8. Yaghmour, K., Dagenais, M.R.: The linux trace toolkit. Linux Journal (May 2000)
9. Desnoyers, M., Dagenais, M.R.: The lttng tracer: A low impact performance and behavior monitor for gnu/linux. In: OLS, 2006. Ottawa Linux Symposium (July 2006)
10. The K42 operating system, http://www.research.ibm.com/k42/
11. Wisniewski, R.W., Rosenburg, B.: Efficient, unified, and scalable performance monitoring for multiprocessor operating systems. In: Supercomputing, November 17-21, 2003, Phoenix Arizona (2003)
12. Dongarra, J., London, K., Moore, S., Mucci, P., Terpstra, D., You, H., Zhou, M.: Experiences and lessons learned with a portable interface to hardware performance counters. In: Proceedings of Workshop Parallel and Distributed Systems: Testing and Debugging (PATDAT), joint with the 19th Intl. Parallel and Distributed Processing Symposium (IPDPS), Niece, France (April 2003)
13. Mathur, W., Cook, J.: Improved estimation for software multiplexing of performance counters. In: Proceedings of the 13th Intl. Symposium on Modeling, Analysis, and Simulation of Computer and Telecommunication Systems (MASCOTS), Atlanta, GA (September 2005)
14. May, J.M.: MPX: Software for multiplexing hardware performance counters in multithreaded systems. In: Proceedings of the Intl. Parallel and Distributed Processing Symposium (IPDPS), San Francisco, CA (April 2001)
15. Anderson, J., Berc, L., Dean, J., Ghemawat, S., Henzinger, M., Leung, S., Sites, D., Vandervoorde, M., Waldspurger, C., Weihl, W.: Continuous profiling: Where have all the cycles gone? In: Proceedings of the 16th ACM Symposium of Operating Systems Principles (SOSP), Saint-Malo, France (October 1997)
16. Wassermann, H.J., Lubeck, O.M., Luo, Y., Bassetti, F.: Performance evaluation of the SGI Origin2000: a memory-centric characterization of lanl asci applications. In: Proceedings of ACM/IEEE Conference on Supercomputing (SC), San Jose, CA (November 1997)

Detecting Application Load Imbalance on High End Massively Parallel Systems

Luiz DeRose, Bill Homer, and Dean Johnson

Cray Inc.
Mendota Heights, MN, USA
`{ldr,homer,dtj}@cray.com`

Abstract. Scientific applications should be well balanced in order to achieve high scalability on current and future high end massively parallel systems. However, the identification of sources of load imbalance in such applications is not a trivial exercise, and the current state of the art in performance analysis tools do not provide an efficient mechanism to help users to identify the main areas of load imbalance in an application. In this paper we discuss a new set of metrics that we defined to identify and measure application load imbalance. We then describe the extensions that were made to the Cray performance measurement and analysis infrastructure to detect application load imbalance and present to the user in an insightful way.

1   Introduction

The current trend in high performance computing is to have systems with very large number of processors. In the latest list of Top 500 Supercomputing Sites [1], the smallest system in the top 10 has more than 9,000 processing elements, and the top 3 systems have more than 25,000 processors each. Moreover, with the recent shift in the microprocessor industry, which stopped riding the frequency curve and started increasing the number of processor cores in a chip, we will see a faster growth in the number of processing elements in these high end massively parallel systems. With the arrival of the "many-core" processors, we are going to see several massively parallel systems with tens and hundreds of thousands of processing elements in the near future. However, in order to perform at scale on these massively parallel systems, applications will have to be very well balanced. Thus, users will need performance analysis tools that can identify and display sources of load imbalance in an intuitive way.

A variety of performance measurement, analysis, and visualization tools have been created to help programmers tune and optimize their applications. These tools range from simple source code profilers [2], to sophisticated tracers and binary analysis tools. However, HPC performance tools currently tend to focus on processor performance, which is normally measured with hardware performance counters (e.g., Perfctr [3], PAPI [4], SvPablo [5], HPM Toolkit [6], HPCView [7]); analysis of communication (e.g., Vampir [8], VampirGuideView [9], Paraver [10], Jumpshot [11]); analysis of the memory subsystem (e.g., Sigma [12]), performance prediction (e.g., dimemas [13], Metasim [14]), or a combination of the above (e.g., TAU [15], KOJAK [16], Paradyn [17]). However, in general these performance tools do not focus on detection of load imbalance.

A.-M. Kermarrec, L. Bougé, and T. Priol (Eds.): Euro-Par 2007, LNCS 4641, pp. 150–159, 2007.

In order to address this problem, we extended the Cray performance measurement and analysis infrastructure [18], which consists of the CrayPat Performance Collector and the Cray Apprentice[2] Performance Analyzer, to automatically identify sources of performance imbalance, and present to the user in an insightful way. The main innovations presented in this paper include the definition of new metrics for evaluation of load imbalance in an application, and new insightful approaches for presenting load balance information in both textual and graphical forms.

The remainder of this paper is organized as follows: In Section 2 we briefly describe the Cray performance measurement and analysis infrastructure. In Section 3 we discuss our load balance metrics and demonstrate their use in textual form, using as an example the ASCI Sweep3d benchmark [19]. In Section 4 we discuss approaches for visualization of load imbalance, also using as example the Sweep3d benchmark. Finally, we present our conclusions in Section 5.

2   The Cray Performance Measurement and Analysis Infrastructure

The Cray performance measurement and analysis infrastructure consists of the CrayPat Performance Collector and the Cray Apprentice[2] Performance Analyzer. CrayPat provides an infrastructure for automatic program instrumentation at the binary level with function granularity. Users can select the functions to be instrumented by name or by groups, such as MPI, I/O, memory. CrayPat also provides an API for fine grain instrumentation. When instrumenting at a function level, users do not need to modify the source code, the makefile, or even recompile the program. CrayPat uses binary rewrite techniques at the object level to create an instrumented application, which is generated with a single static re-link, managed by CrayPat. When using the CrayPat API to instrument code regions source code modification and recompilation are needed, but other than the instrumentation differences, CrayPat and Cray Apprentice[2] treat code regions as user functions. Thus, for simplicity, in this paper we will refer to any instrumented section of the code (code regions, user functions, MPI functions, etc) as functions.

The second main component of the CrayPat Performance Collector is its runtime performance data collection library, which can be activated by sampling or by interval timers. Performance data can be generated in the form of a profile or a trace file, and its selection is based on an environment variable.

A third main CrayPat component is the report generator *(pat_report)*, which is a utility that reads the performance file that was created by the runtime library and generates text reports, presented in the form of tables.

Finally, the Cray Apprentice[2] Performance Analyzer is a multi-platform, multi-function performance data visualization tool that takes as input the performance file generated by CrayPat and provides the familiar notebook-style tabbed user interface, displaying a variety of different data panels, depending on the type of performance experiment that was conducted with CrayPat and the data that was collected. Cray Apprentice[2] provides call-graph based profile information and timeline based trace visualization, supporting the traditional parallel processing and communication mechanisms, such as MPI, OpenMP, and SHMEM, as well as performance visualization for I/O.

3   Load Imbalance Metrics

The first step in order to be able to report load imbalance in an insightful way is to define load imbalance metrics that are intuitive. Thus, we defined a couple of metrics: *"imbalance percentage"* and *"imbalance time"*. Assuming that n is the number of processing elements, we define the imbalance percentage of a parallel application[1] as:

$$\text{imbalance percentage} = \frac{\text{maximum time} - \text{average time}}{\text{maximum time}} \times \frac{n}{n-1} \qquad (1)$$

The goal of imbalance percentage is to provide an idea of the "badness" of the imbalance. Thus, to make it intuitive, we defined it to be in the range of 0 to 100, where a perfectly balanced code segment would have 0 imbalance percentage and a serial portion of a code segment on a parallel application (for example a serial I/O) would have imbalance percentage of 100. The imbalance percentage corresponds to the percentage of time that the rest of the team, excluding the slowest processing element, is not engaged in useful work on the given function. If they are idle, this is the "percentage of resources available for parallelism" that is wasted. In the worst case (as in the example above), one processing element does all the work, so that the others are 100% wasted.

Notice, however, that a section of code that has a high imbalance percentage should not necessarily be the main target of performance optimization. Following the example above, a serial I/O will always have imbalance percentage of 100, independently of the amount of time that it takes. If the fraction of time spent on that particular I/O operation is small, its impact on the overall performance of the application will not be significant. Thus, we also need a metric that is related to execution time, in order to identify regions of the program that should be considered for optimization. We opted to define a metric that would provide an estimation to the user of how much time in the overall program would be saved, if the corresponding section of code had a perfect balance. Thus we defined the imbalance time as:

$$\text{imbalance time} = \text{maximum time} - \text{average time} \qquad (2)$$

The imbalance time for a code section represents an upper bound on the *potential saving* that could result from perfectly balancing that particular code section. It is only an upper bound because it assumes that other processing elements are simply waiting without doing any useful work while the slowest member finishes.

Figure 1 shows a pat_report table displaying the profile output[2] with load imbalance information from a 48 processors execution of the Sweep3d benchmark, running 20 iterations with a $150 \times 150 \times 150$ grid, on a Cray XT3. As described in Section 2, CrayPat provides functionality for automatic performance instrumentation at the function level, and provides an API for hand instrumentation at a smaller granularity. The imbalance metrics described above are automatically calculated at the level that the code was instrumented. Pat_report splits the profile by the different instrumentation groups (user

---

[1] As defined, the imbalance percentage is only valid for parallel programs. Serial programs would have an imbalance percentage of 0.

[2] This is the summary version of the output, which only shows the lines where the percentage of time is at least 0.05% of the total time.

```
Table 1:   Profile by Function Group and Function

 Time % |        Time |Imb. Time |  Imb.  |   Calls |Group
        |             |          |Time %  |         | Function
        |             |          |        |         |  PE='HIDE'


100.0% | 3.661935 |        -- |     -- | 675604 |Total
|----------------------------------------------------------------
|  72.2% | 2.644437 |        -- |     -- | 245380 |USER
||---------------------------------------------------------------
||  97.1% | 2.568997 | 0.126365 |   4.8% |    576 |sweep_
||   1.8% | 0.047655 | 0.001252 |   2.6% |    576 |source_
||   0.3% | 0.008992 | 0.000212 |   2.4% |    576 |flux_err_
||   0.3% | 0.007960 | 0.000940 |  10.8% | 118080 |snd_real_
||   0.1% | 0.003182 | 0.001944 |  38.7% |     48 |MAIN_
||   0.1% | 0.002816 | 0.000591 |  17.7% | 118080 |rcv_real_
||   0.1% | 0.001637 | 0.039748 |  98.1% |     48 |inner_
||   0.1% | 0.001386 | 0.000055 |   3.9% |     48 |initialize_
||===============================================================
|  27.8% | 1.016999 |        -- |     -- | 238224 |MPI
||---------------------------------------------------------------
||  79.9% | 0.812923 | 0.232522 |  22.7% | 118080 |mpi_recv_
||  12.8% | 0.129678 | 0.125148 |  50.2% |   1536 |mpi_allreduce_
||   5.8% | 0.059340 | 0.010950 |  15.9% | 118080 |mpi_send_
||   0.8% | 0.007631 | 0.000221 |   2.9% |    192 |mpi_bcast_
||   0.7% | 0.007423 | 0.000347 |   4.6% |    144 |mpi_barrier_
|===============================================================
```

Fig. 1. CrayPat profile with load balance by function group and function

functions, MPI, I/O, Memory, etc), which provides an idea of the balance of the various phases of the application (computation, communication, I/O, and memory allocation). In the example shown in Figure 1, only MPI and user functions were instrumented, and we observe that 72.2% of the total time was spent in users functions, while 27.8% of the time was spent in MPI functions. We also observe that the function that has the highest potential savings ("Imb. Time") is the "MPI_recv", which is consistent with the Sweep3d application, since it communicates using a wave front approach, which creates a communication imbalance, since the higher ranks tend to wait longer on receives. Although not shown in the example, CrayPat can also display the call path, as well as source code and line number information for each function.

Notice that in the profile shown in Figure 1, by default, "Time" is the average time per processing element. Thus, the imbalance time will be greater than the average time if the maximum time is more than twice the average. So it would not be unusual to see imbalance times that are larger than the average time for some functions, as is the case of function "inner" in Figure 1, but typically these would not be near the top of the profile.

```
    Table 2: Load Balance across PE's by Function
      This table shows only lines with Time% > 5.0.

   Time % |   Cum.  |    Time  |  Calls  |Group
          | Time %  |          |         | Function
          |         |          |         |  PE[mmm]

  100.0% | 100.0% | 3.661935 | 675604  |Total
|-------------------------------------------------
|  72.2% |  72.2% | 2.644437 | 245380  |USER
||------------------------------------------------
||  97.1% |  97.1% | 2.568997 |     576 |sweep_
|||-----------------------------------------------
3||   2.2% |   2.2% | 2.695363 |      12 |pe.0
3||   2.1% |  52.7% | 2.590278 |      12 |pe.31
3||   2.0% | 100.0% | 2.487526 |      12 |pe.37
||================================================
|  27.8% | 100.0% | 1.016999 | 238224  |MPI
||------------------------------------------------
||  79.9% |  79.9% | 0.812923 | 118080  |mpi_recv_
|||-----------------------------------------------
3||   2.7% |   2.7% | 1.045446 |    1440 |pe.47
3||   2.0% |  56.9% | 0.786257 |    2880 |pe.16
3||   1.4% | 100.0% | 0.563250 |    1440 |pe.0
|||================================================
||  12.8% |  92.7% | 0.129678 |    1536 |mpi_allreduce_
|||-----------------------------------------------
3||   4.1% |   4.1% | 0.254826 |      32 |pe.0
3||   2.1% |  71.1% | 0.128529 |      32 |pe.21
3||   0.0% | 100.0% | 0.002649 |      32 |pe.47
|||================================================
||   5.8% |  98.5% | 0.059340 | 118080  |mpi_send_
|||-----------------------------------------------
3||   2.5% |   2.5% | 0.070290 |    2880 |pe.27
3||   2.0% |  58.6% | 0.056842 |    2160 |pe.6
3||   1.2% | 100.0% | 0.033176 |    1440 |pe.0
|================================================
```

Fig. 2. CrayPat profile with Load Balance across PE's by Function

The point of the imbalance metric is to reveal cases in which the average time underestimates the "contribution" of a function to the elapsed time of the program. The functions with the best opportunities for reducing runtime by improving load balance will be among those for which the maximum time spent by a single processing element exceeds the average time over all processing elements by an amount that is a significant fraction of the program runtime. Our "imbalance time" is precisely the excess of maximum over average. Notice that in the statement above, we have to say "are among" instead of just "are" because of cases like the following:

```
void F1()
{
    if (rank%2) G1()
    else        G2();
}
```

Here `G1()` and `G2()` will be very unbalanced individually, but if `F1()` has balanced inclusive time, then this section of the code will probably not have a load balance problem. Currently, CrayPat only shows the imbalance metrics for exclusive times. We are in the process of extending it to also provide imbalance metrics for inclusive times.

An alternative approach to display load balance information, which is also provided by CrayPat, is shown in Figure 2. This pat_report table displays the maximum, median, and minimum values for each function and corresponding PEs[3]. If desired, one can display the complete PE distribution for each function, but for runs with a large processor count, such table would not be practical. Another option is available, where the report presents a distribution with the top three and the bottom three PEs, in addition to the median.

4   Visualization of Load Imbalance

As described in Section 2 users can visualize the performance data generated by CrayPat with the Cray Apprentice[2] Performance Analyzer. One of the multiple views provided by Cray Apprentice[2] is the call graph profile shown in Figure 3. The call graph profile has a similar approach to the call graph visualization described in [20], where the size of the boxes are relative to the execution time of the function. In our case, the height of a rectangle represents the exclusive time of the function, while the width represents its children time (i.e., the inclusive time minus the exclusive time, which is the same as the sum of the time of all its children). Thus, when searching for the code segments that take most of the execution time, users should look for large tall boxes in the call graph.

As described above, when looking at the overall call graph, users have an idea of how each of the code segments contributes to total execution time of the application. We extended this call graph representation by adding a second level of abstraction, such that users could quickly observe load balance information when looking at a single rectangle. With this second level of abstraction, the height of each individual rectangle in the call graph represents the maximum execution time across all processing elements. The size of the darker blue bar in the left side of the rectangle is proportional to the average time, while the size of the light blue bar in the right side is proportional to the minimum time across all processing elements. The rest of the rectangle is filled with yellow. Thus, when looking at a single rectangle, users have an idea of the load balance of the code segment considering all processing elements. In order to identify functions that have a high percentage of load imbalance users should look for rectangles with a large yellow area. In particular, a large amount on yellow in the left side of a tall rectangle will indicate a high potential saving, since the amount of yellow in the left side of a rectangle represents the difference between the maximum time and the average

---

[3] In the interest of brevity, this table only shows lines where the percentage of time is at least 5.0% of the total time. This threshold can be selected by the user when running pat_report. The default is to show only functions that execute at least 0.05% of the time.

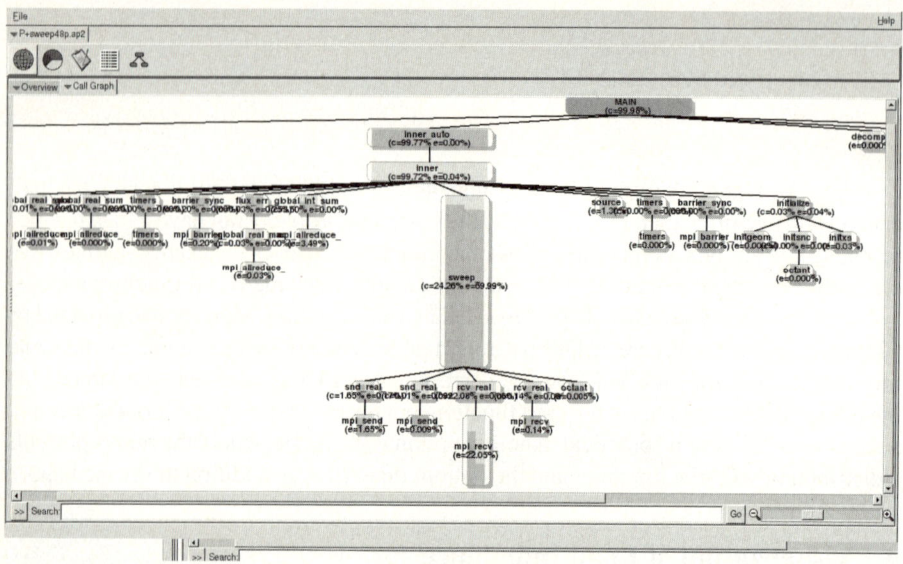

Fig. 3. Cray Apprentice[2] call graph view

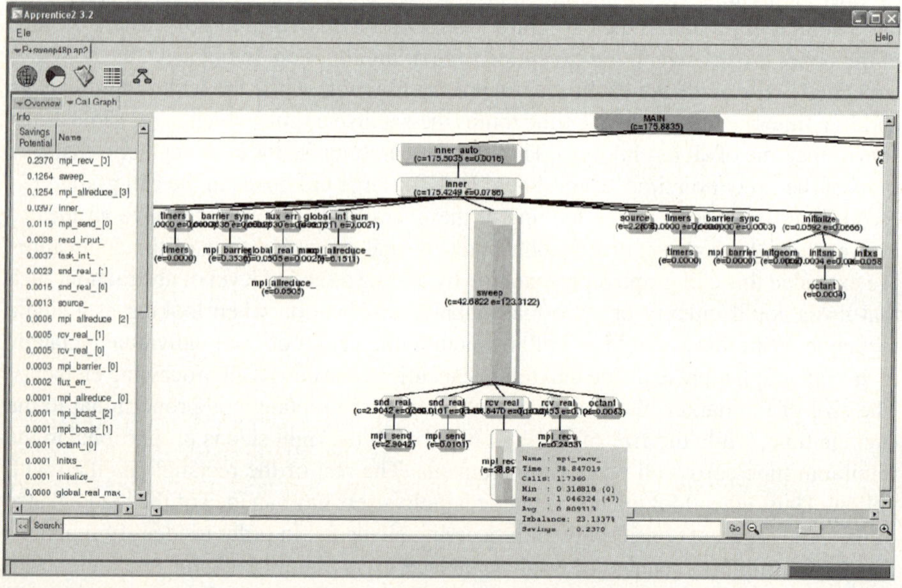

Fig. 4. Load balance view

time for the function, as defined in Equation 2. A rectangle almost completely filled with yellow (e.g., the function "`inner`" in Figure 3) normally indicates a function that is executing on a single PE.

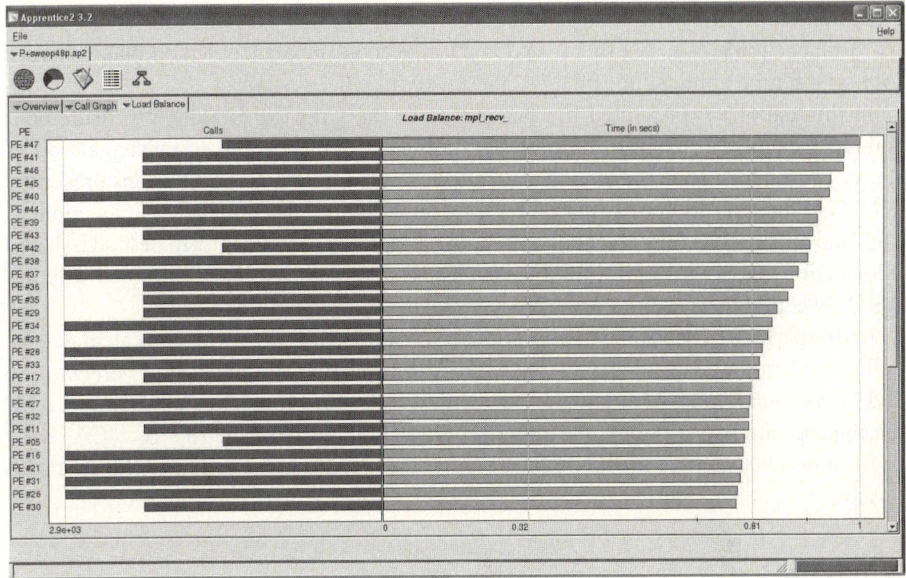

Fig. 5. Load balance view

The call graph view also provides a list of functions, which can be sorted by exclusive time, imbalance percentage, or imbalance time, as shown in Figure 4. Functions that have multiple call sites will appear multiple times in the list (with numbers added to the names for disambiguation). The sorted list helps users to quickly locate the main sources of load imbalance, which is helpful especially with large call graphs. When clicking on a function name the call graph pane will focus on the corresponding rectangle and highlight it. Also, when the user puts the cursor on top of a rectangle in the call graph, a popup window will appear, as shown for the function "`mpi_recv`" in Figure 4. The popup window displays the measured information for the function, including the corresponding PE that had the minimum and maximum times, as well as the average time, imbalance percentage, and imbalance time. Whenever collected during runtime, hardware counters data is also displayed in the popup window.

Cray Apprentice[2] also provides a view with the load balance distribution for any function in the call graph, as shown in Figure 5. The PEs in the load balance view can be sorted by time or number of calls. In addition, this load balance view display lines indicating the minimum, average, and maximum times for the function, as well as marks indicating the range of plus and minus one standard deviation from the mean, which can be used for a better understanding of the load balance distribution.

5   Conclusion

Applications will need to be well balance in order to achieve high scalability on current and future high end massively parallel systems. However, there is no standard for the measurement of load imbalance in an application and the current state of the art in performance tools does not focus on detection of load imbalance, which makes harder for users to tune applications on these massively parallel systems.

In this paper we presented the extensions to the Cray performance measurement and analysis infrastructure to support measurement, identification, and visualization of sources of performance imbalance in an application. The main contributions presented here were the definition of new metrics for evaluation of load imbalance in an application, and approaches for presentation in both textual and graphical form of load balance information that are insightful for the user. These approaches were exemplified with the ASCI Sweep3d benchmark.

In this paper we focus on the discussion of imbalance for time, which is calculated by default on CrayPat. However, CrayPat extends the concepts of *"imbalance percentage"* and *"imbalance (time)"* for any metric of interest. Using Figure 1 as an example, with the appropriate user options, pat_report could calculate these metrics for "CALLS", or for any other metric, including hardware counter metrics, such as FLOPS or cache misses.

References

1. Top500 Supercomputer Sites: The 28th TOP500 List (2006),
   http://www.top500.org/
2. Graham, S., Kessler, P., McKusick, M.: gprof: A Call Graph Execution Profiler. In: Proceedings of the SIGPLAN '82 Symposium on Compiler Construction, Boston, MA, pp. 120–126. Association for Computing Machinery (June 1982)
3. Pettersson, M.: Linux X86 Performance-Monitoring Counters Driver. Computing Science Department, Uppsala University - Sweden (2002),
   http://user.it.uu.se/~mikpe/linux/perfctr/
4. Browne, S., Dongarra, J., Garner, N., Ho, G., Mucci, P.: A Portable Programming Interface for Performance Evaluation on Modern Processors. The International Journal of High Performance Computing Applications 14(3), 189–204 (2000)
5. DeRose, L., Reed, D.: Svpablo: A Multi-Language Architecture-Independent Performance Analysis System. In: Proceedings of the International Conference on Parallel Processing, pp. 311–318 (August 1999)
6. DeRose, L.: The Hardware Performance Monitor Toolkit. In: Sakellariou, R., Keane, J.A., Gurd, J.R., Freeman, L. (eds.) Euro-Par 2001. LNCS, vol. 2150, pp. 122–131. Springer, Heidelberg (2001)
7. Mellor-Crummey, J., Fowler, R., Marin, G., Tallent, N.: HPCView: A tool for top-down analysis of node performance. The Journal of Supercomputing 23, 81–101 (2002)
8. Nagel, W., Arnold, A., Weber, M., Hoppe, H.C., Solchenbach, K.: Vampir: Visualization and Analysis of MPI Resources. Supercomputer 12, 69–80 (1996)
9. Kim, S., Kuhn, B., Voss, M., Hoppe, H.C., Nagel, W.: VGV: Supporting Performance Analysis of Object-Oriented Mixed MPI/OpenMP Parallel Applications. In: Proceedings of the International Parallel and Distributed Processing Symposium (April 2002)
10. European Center for Parallelism of Barcelona (CEPBA): Paraver - Parallel Program Visualization and Analysis Tool - Reference Manual (November 2000),
    http://www.cepba.upc.es/paraver
11. Wu, C., Bolmarcich, A., Snir, M., Wootton, D., Parpia, F., Chan, A., Lusk, E., Gropp, W.: From trace generation to visualization: A performance framework for distributed parallel systems. In: Proceedings of Supercomputing 2000 (November 2000)

12. DeRose, L., Ekanadham, K., Hollingsworth, J.K., Sbaraglia, S.: SIGMA: A Simulator Infrastructure to Guide Memory Analysis. In: Proceedings of SC2002, Baltimore, Maryland (November 2002)
13. Labarta, J., Girona, S., Cortes, T.: Analyzing scheduling policies using Dimemas. Parallel Computing 23(1–2), 23–34 (1997)
14. Snavely, A., Carrington, L., Wolter, N., Labarta, J., Badia, R., Purkayastha, A.: A framework for performance modeling and prediction. In: Proceedings of SC2002, Baltimore, Maryland (November 2002)
15. Bell, R., Malony, A.D., Shende, S.: A Portable, Extensible, and Scalable Tool for Parallel Performance Profile Analysis. In: Kosch, H., Böszörményi, L., Hellwagner, H. (eds.) Euro-Par 2003. LNCS, vol. 2790, pp. 17–26. Springer, Heidelberg (2003)
16. Wolf, F., Mohr, B.: Automatic performance analysis of hybrid mpi/openmp applications. Journal of Systems Architecture, Special Issue 'Evolutions in parallel distributed and network-based processing' 49(10–11), 421–439 (2003)
17. Miller, B.P., Callaghan, M.D., Cargille, J.M., Hollingsworth, J.K., Irvin, R.B., Karavanic, K.L., Kunchithapadam, K., Newhall, T.: The Paradyn Parallel Performance Measurement Tools. IEEE Computer 28(11), 37–46 (1995)
18. DeRose, L., Homer, B., Johnson, D., Kaufmann, S.: The New Generation of Cray Tools. In: Proceedings of Cray Users Group Meeting – CUG 2005 (May 2005)
19. Lawrence Livermode National Laboratory: the ASCI sweep3d Benchmark Code (1995), `http://www.llnl.gov/asci_benchmarks/asci/limited/sweep3d/`
20. DeRose, L., Pantano, M., Aydt, R., Shaffer, E., Schaeffer, B., Whitmore, S., Reed, D.A.: An Approach to Immersive Performance Visualization of Parallel and Wide-Area Distributed Applications. In: Proceedings of 8th International Symposium on High Performance Distributed Computing - HPDC'99 (August 1999)

A First Step Towards Automatically Building Network Representations

Lionel Eyraud-Dubois[1], Arnaud Legrand[2], Martin Quinson[3],
and Frédéric Vivien[4,1]

[1] LIP, Université de Lyon - CNRS - INRIA, Lyon, France
[2] LIG - MESCAL, UJF - INPG - CNRS - INRIA, Grenoble, France
[3] Nancy University - LORIA, Nancy, France
[4] INRIA
Lionel.Eyraud-Dubois@ens-lyon.fr, Arnaud.Legrand@imag.fr,
Martin.Quinson@loria.fr, Frederic.Vivien@ens-lyon.fr

Abstract. To fully harness Grids, users or middlewares must have some knowledge on the topology of the platform interconnection network. As such knowledge is usually not available, one must uses tools which automatically build a topological network model through some measurements. In this article, we define a methodology to assess the quality of these network model building tools, and we apply this methodology to representatives of the main classes of model builders and to two new algorithms. We show that none of the main existing techniques build models that enable to accurately predict the running time of simple application kernels for actual platforms. However some of the new algorithms we propose give excellent results in a wide range of situations.

Keywords: Network model, topology reconstruction, Grids.

1   Introduction

Grids are parallel and distributed systems that result from the sharing and aggregation of resources distributed between several geographically distant organizations [1]. Unlike classical parallel machines, Grids present heterogeneous and sometimes even non-dedicated capacities. Gathering accurate and relevant information about them is then a challenging issue, but it is also a necessity. Indeed, the efficient use of Grid resources can only be achieved through the use of accurate network information. Qualitative information such as the network topology is crucial to achieve tasks such as running network-aware applications [2], efficiently placing servers [3], or predicting and optimizing collective communications performance [4].

However, the description of the structure and characteristics of the network interconnecting the different Grid resources is usually not available to users. This is mainly due to security (fear of Deny Of Service attacks) and privacy reasons (ISP do not want you to know where their bottlenecks are). Hence a need for tools which automatically construct models of platform networks. Many tools

A.-M. Kermarrec, L. Bougé, and T. Priol (Eds.): Euro-Par 2007, LNCS 4641, pp. 160–169, 2007.

and projects provide some network information. Some rely on simple ideas while others use very sophisticated measurement techniques. Some of these techniques, though, are sometimes ineffective in Grid environments due to security issues. Anyway, to the best of our knowledge, these different techniques have never been compared rigorously in the context of Grid computing platforms. Our aim is to define a methodology to assess the quality of network model building tools, to apply it to representatives of the main classes of model builders, to identify weaknesses of existing approaches, and to propose new model building algorithms.

The main contributions of this paper is the design of two new reconstruction algorithms, and some evaluations that highlight the weaknesses of classical algorithms and demonstrate the superiority of one of our new algorithms.

The rest of this article is organized as follows. In Section 2, we review the main observation techniques and we identify some that are effective in Grid environments. In Section 3 we review existing reconstruction algorithms and we identify a few representative ones. Based on the analysis of potential weaknesses of these algorithms, we propose two new algorithms. We recall in Section 4 a few quality metrics we proposed in [5] to assess the quality of reconstruction algorithms. In Section 5, we evaluate through simulation the quality of the studied reconstruction algorithms with respect to all the proposed metrics. This evaluation is performed on models of real platforms and on synthetic models.

2   Related Work

Network discovery tools have received a lot of attention in recent years. However, most of them are not suited to Grids. Indeed, much of the previous work (e.g., Remos [6]) rely on low-level network protocols like SNMP or BGP, whose usage is generally restricted for security reasons (it is indeed possible to conduct Deny Of Service attacks by flooding the routers with requests). As a matter of fact, in a Grid, even traceroute or ping-based tools (e.g., TopoMon [7], Lumeta [8], IDmaps [9], Global Network Positioning [10]) are getting less and less effective. For example, ICMP is more and more often disabled, once again to avoid Deny Of Service attacks based on flooding: the Skitter project [11] reports that over 5 years the number of hosts replying to ICMP requests decreased by 2 to 3% per month. Even if recent works have proposed similar or even better functionalities without relying on ICMP, some of them (e.g., pathchar [12]) require specific privileges, which make them unusable in our context. It is mandatory to rely on tools that only use *application-level measurements*, i.e., measurements that can be done by any application running on a computing Grid without any specific privilege. This comprises the common end-to-end measurements, like bandwidth and latency, but also interference measurements (i.e., whether a communication between two machines A et B has non negligible impact on the communications between two machines C et D). Many projects rely on this type of measurements. An example is the NWS (Network Weather Service) [13] software, which constitutes a *de facto* standard in the Grid community as it is used by major *Grid middlewares* like Globus [14], or *Problem Solving Environments* (PSEs) like

DIET [15], to gather information about the current state of a platform and to predict its evolutions. NWS is able to report the end-to-end bandwidth, latency, and connection time, which are typical application-level measurements. However, the NWS project focuses on quantitative information and does not provide any kind of topological information. It is however natural to address this issue by aggregating all NWS information in a single clique graph and use this labeled graph as a network model. In another example, interference measurements have been used in ENV [16] and enabled to detect, to some extent, whether some machines are connected by a switch or a hub. A last example is ECO [17], a collective communication library, that uses plain bandwidth and latency measurements to propose optimized collective communications (e.g., broadcast, reduce, etc.). These approaches have proved to be very effective in practice, but they are generally very specific to a single problem and we are looking for a general approach.

3   Studied Reconstruction Algorithms

We are thus looking for a tool based on application-level measurements that would enable any network-aware application to benefit from reasonably accurate information on the network topology. In most previous works, the underlying network topology is either a clique [13,17] or a tree [18,16]. Our reference reconstruction algorithms are thus clique, minimal spanning tree on latencies, and maximal spanning tree on bandwidths. As our experiments show (Section 5), these methods produce very simple graphs, and often fail to provide a realistic view of platforms. We thus designed two new reconstruction algorithms, as a first step towards a better reconstruction. The first algorithm aims at improving an already built topology and is meant to be used to improve an existing spanning tree. The second one reconstructs a platform model from scratch, by growing a set of connected nodes. Both algorithms keep track of the routing while building their model, to be able to correct a route connecting two nodes whose latency was previously inaccurately predicted. We focus on latency rather than on bandwidth as bandwidths are less discriminant.

Algorithm IMPROVING. Algorithm IMPROVING is based on the observation that if the latency between two nodes is badly over-predicted by the current route connecting them, an extra edge should be inserted to connect them through an alternate and more accurate route. Among all pairs of "badly connected" nodes, we pick the two nodes with the smallest possible measured latency, and we add a direct edge between them. Each time IMPROVING adds an edge, for each pair of nodes whose latency is over-predicted, we check whether that pair cannot be better connected through the just introduced edge, and we update the routing if needed. This edge addition procedure is repeated until all predictions are considered sufficiently accurate. The accuracy of predictions is necessarily arbitrary. In our implementation, it corresponds to a deviation of less than 10% from actual measurements.

Algorithm AGGREGATE. Algorithm AGGREGATE uses a more local view of the platform. It expands a set of already connected nodes, starting with the two closest nodes in terms of latency. At each step, AGGREGATE connects a new *selected* node to the already connected ones. The selected node is the one closest to the connected set in terms of latency. AGGREGATE iteratively adds edges so that each route from the selected node to a connected node is sufficiently accurate. Added edges are greedily chosen starting from the edge yielding a sufficiently accurate prediction for the largest number of routes from the selected node to a connected node. We slightly modified this scheme to avoid adding edges that will later become redundant. A new edge is added only if its latency is not significantly larger (meaning less than 50% larger) than that of the first edge added to connect the selected node. Because of this change, we may move to a new selected node while not all the routes of the previous one are considered accurate enough. We thus keep a list of *inaccurate* routes. For each edge addition we check whether the new edge defines a route rendering accurate an inaccurate route. When all nodes are connected, we add edges to correct all remaining inaccurate routes, starting with the route of lowest latency.

4   Assessing the Quality of Reconstructions

We want to thoroughly assess the quality of reconstruction algorithms. To fairly compare various topology mapping algorithms, we have developed ALNeM (Application Level Network Mapper). ALNeM is developed with GRAS [19] that provides a complete API to implement distributed application on top of heterogeneous platforms. Thanks to two different implementations of GRAS, ALNeM can work seamlessly on real platforms as well as, with SimGrid [20], on simulated platforms. ALNeM is made of three main parts: 1) a measurement repository; 2) a distributed collection of sensors performing bandwidth, latency, and interference measurements; 3) a topology builder which uses the repository.

To assess the quality of model builders, we use two different and complementary approaches. For both approaches, we consider a series of original platforms; and for each platform we compare the original platform and the models built from it. The two approaches can be seen as different points of view on models: a communication-level one and an application-level one.

4.1   End-to-End Metric

A platform model is "good" if it allows to accurately predict the running time of applications. The prediction accuracy depends on the model capacity to render different aspects and characteristics. Often, researchers only focus on bandwidth prediction. However, latencies and interferences can also greatly impact an application performance. Therefore, we consider the three following characteristics:

Bandwidth. We need to know the available bandwidths as soon as the applications send messages of different sizes.

Latencies. Latencies are often overlooked in Grid computing, but Casanova presented an example [21] where one third of the time needed to transfer a 1 GByte of data would be due to latencies. Therefore, we must model them.

Interferences. Many distributed applications use collective communications (e.g., broadcasts or all-to-all) or independent communications between disjoint pairs of processors. The only knowledge of latencies and bandwidths does not allow to predict the time needed to realize two communications between two disjoint pairs of processors: this depends on whether the two communications use a same physical link[1]. Also, knowing the network topology and being able to predict communication interferences enables to derive more efficient algorithms [2].

Methodology. Our evaluation methodology is based on simulations. Given one original platform, we measure the end-to-end latencies and bandwidths between any two processors. We also measure the end-to-end bandwidths obtained when any two pairs of processors simultaneously communicate. We then perform the same measurement on the reconstructed models. To compare the results, we build an accuracy index for each reconstruction algorithm, each graph, and each studied network characteristic. For latencies and bandwidths, following [22], we define *accuracy* as the maximum of the two ratios x_R/x_M and x_M/x_R, where x_R is the reconstructed value and x_M is the original measured one. We compute the accuracy of each pair of nodes, and then the geometric mean of all accuracies.

4.2   Application-Level Measurements

To simultaneously analyze a combination of the characteristics studied with end-to-end measurements, we compare, through simulations, the performance of several classical distributed routines when run on the original graph and on each of the reconstructed ones. This evaluates the predictive power of the reconstruction algorithms for applications with more complex but realistic communication patterns. We study the following simple distributed algorithms (listed from the simplest communication pattern to the most complicated one):

- **Token ring:** a token circulates three times along a randomly built ring (the ring structure is not correlated to that of the interconnection network).
- **Broadcast:** a randomly picked node sends a message to all the other nodes.
- **All-to-all:** all the nodes simultaneously perform a broadcast.
- **Parallel matrix multiplication (pmm):** a matrix multiplication is realized using ScaLAPACK outer product algorithm [23].

This evaluation must be done through simulations. Indeed, the measurements on the reconstructed models can obviously not be done experimentally. Furthermore, the comparison of experimental (original platform) and simulated (reconstructed models) measurements would introduce a serious bias in the evaluation framework, due to the differences between the actual world and the simulator.

5   Experimental Results

We present two types of experiments: the first one is based on a modeling of a real network architecture, while for the second one we generated synthetic platforms

---

[1] It also depends whether the shared communication link is a backbone [21], etc.

using GridG [24]. As stated in Section 3, we evaluate several reconstruction algorithms. In addition to our three reference reconstruction methods (CLIQUE, minimal spanning tree on latencies (TREELAT), and maximal spanning tree on bandwidths(TREEBW)), we analyze the performance of the AGGREGATE algorithm, and of the IMPROVING procedure applied to both spanning trees: IMPTREELAT and IMPTREEBW.

5.1   Renater

Renater[2] is the French public network infrastructure that connects major universities. Thanks to a collection of accounts in several universities, we were able to measure latencies and bandwidths between the corresponding hosts. For security reasons, these measurements were performed using the most basic tools, namely `ping` for latency and `scp` of bandwidths. Thanks to the topology information available on the Renater website we created a model of this network, that we annotated with the bandwidths and latencies we measured. We then executed our reconstruction algorithms on the obtained model.

Figure 1 shows the evaluation of the reconstructed topologies. For end-to-end metrics, we plotted the average accuracy for both latency and bandwidth, and we also detailed the average accuracy for over- and under-predicted values. Unsurprisingly, CLIQUE has excellent end-to-end performances whereas TREE-LAT and TREEBW have poor ones. AGGREGATE over-estimates bandwidth for a few pairs of hosts, but both IMPTREELAT and IMPTREEBW have excellent end-to-end performances.

Regarding applicative performance, CLIQUE is unsurprisingly good for TOKEN and BROADCAST where there is always at most one communication at a time and very bad for ALL2ALL and PMM. IMPTREELAT and IMPTREEBW are once again equivalent and now clearly have much better results than any other heuristics. They are actually within 10% of the optimal solution for all applicative performances. Last, the interference evaluation (Figure 1c) enables us to distinguish IMPTREEBW and IMPTREELAT. IMPTREEBW accurately predicts more than 95% of interferences whereas IMPTREELAT overestimates 50% of interferences!

This experiment shows that our reconstruction algorithms are able to yield platforms with good predictive power. It also suggests that our IMPTREEBW algorithm can provide very good reconstructions. The good performance of IMPTREEBW may be explained by the fact that this is the only algorithm which builds a non trivial graph (i.e., not a clique) while using both the information on latencies and bandwidths. However, these encouraging results obtained on a realistic platform must be confirmed by a more comprehensive set of experiments, using a large number of different platforms, which we do in the next section.

5.2   GridG

For a thorough validation of our algorithms, we used the GridG platform generator [24] to study realistic Internet-like platforms, which may be different from

---

[2] `http://www.renater.fr`

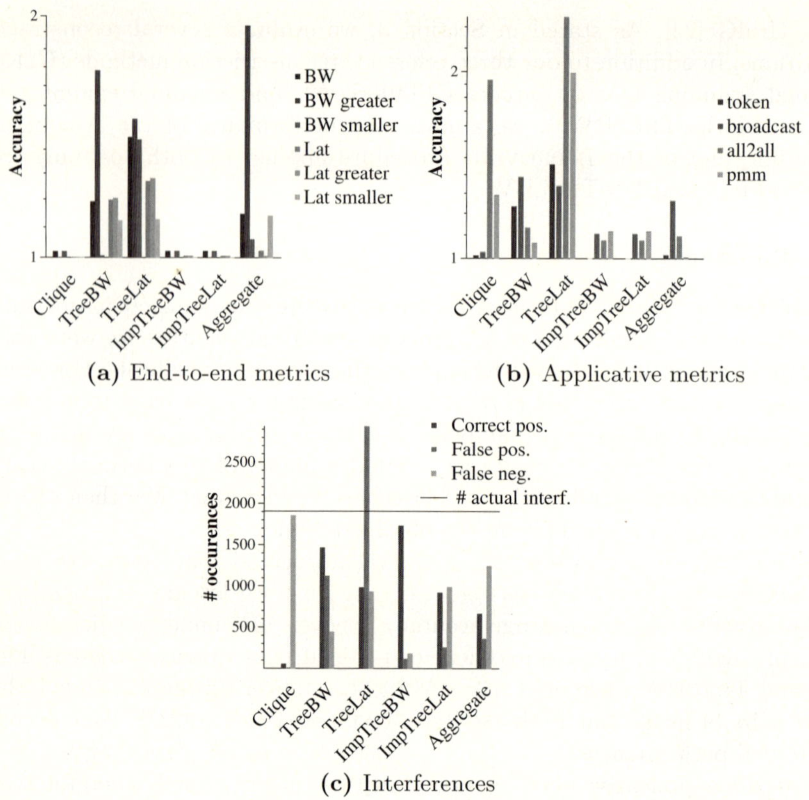

(a) End-to-end metrics

(b) Applicative metrics

(c) Interferences

Fig. 1. Simulated tests on the Renater platform

the very few platforms we can access and thus test directly. In this experiment, we generated two different kinds of platforms: in the first group, all of the hosts are known to the measurement procedure, which means that it is possible to deploy a process on all internal routers of the platform. In the second group, only the external hosts are known to the algorithms. For each group, we generated 40 different platforms, each of them containing about 60 hosts.

The results are shown on Figures 2 and 3. For end-to-end metrics, we plotted the average accuracy for both latency and bandwidth, and we also detailed the average accuracy for over- and under-predicted values. We have also indicated the minimum and maximum values obtained over all 40 platforms.

Figure 2 confirms the results of the previous section: the improved trees have very good predictive power, especially IMPTREEBW, with an average error of 3% on the most difficult application, namely ALL2ALL. The results of CLIQUE would be very good too. But as it fails to take interferences into account, it fails to accurately predict the running time of ALL2ALL. (Note that the fact that CLIQUE over-estimates the bandwidth for a few pairs of hosts is due to routing asymmetry in the original platform.) We can also see that the basic spanning trees have better results than in the previous experiment. This is due

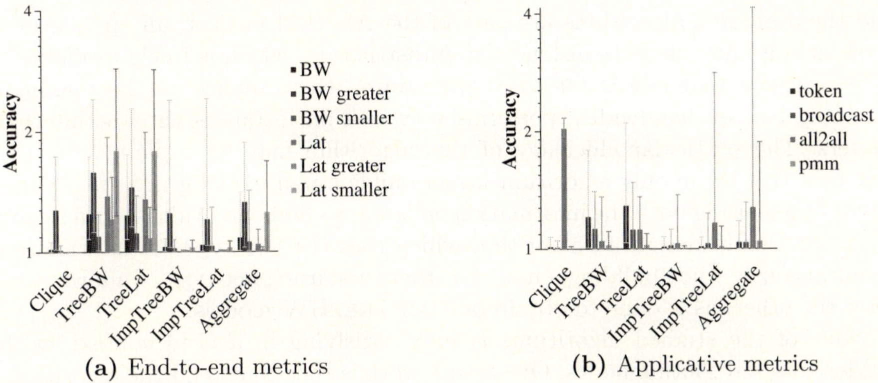

(a) End-to-end metrics
(b) Applicative metrics

Fig. 2. Simulated tests on the GridG platforms, with processes on every host

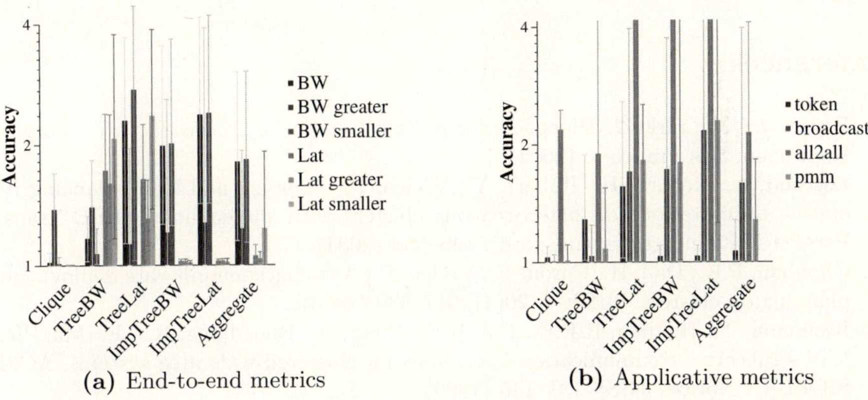

(a) End-to-end metrics
(b) Applicative metrics

Fig. 3. Simulated tests on the GridG platforms, with hidden routers

to the fact that GridG platforms contain parts that are very tree-like, which these algorithms are able to reconstruct easily.

However, Figure 3 shows that platforms with hidden routers are much more difficult to reconstruct. The performance of the clique platform remains the same as before, but all other algorithms suffer from a severe degradation. It is not clear yet whether this degradation comes from a wrong view of the topology of the platform, or from the wrong bandwidth predictions which we can see on Figure 3a.

6   Conclusion

In this work, we proposed two new reconstruction algorithms and we compared them with classical reconstruction algorithms (namely spanning trees and cliques) through a thorough evaluation framework. This evaluation framework and the evaluated algorithms are part of the ALNeM project, an application-level measurement and reconstruction infrastructure, which is freely available[3].

We showed that our IMPROVING procedure, when applied to the maximal spanning tree on bandwidth, performs very well on instances without internal routers. The particular efficiency of this algorithm may be explained by the fact that this is the only algorithm which builds a non trivial graph (i.e., not a clique) while using both the information on latencies and bandwidths. As a future work, we should design an algorithm which uses the two types of information simultaneously when building a model, rather than using one type of information after the other, as is done to obtain our IMPTREEBW models.

None of the studied algorithms is fully satisfying in a Grid context, with hidden internal routing nodes. Our future work is thus to extend the algorithms to enable them to cope with such a situation. So far, no algorithm is using any information on interferences. This should also be addressed as this information should enable us to design even more efficient network model building tools.

References

1. Foster, I.: The Grid 2: Blueprint for a New Computing Infrastructure. Morgan Kaufmann, San Francisco (2004)
2. Legrand, A., Renard, H., Robert, Y., Vivien, F.: Mapping and load-balancing iterative computations on heterogeneous clusters with shared links. IEEE Trans. Parallel Distributed Systems 15(6), 546–558 (2004)
3. Chouhan, P.K., Dail, H., Caron, E., Vivien, F.: Automatic middleware deployment planning on clusters. IJHPCA 20(4), 517–530 (2006)
4. Kielmann, T., Hofman, R.F.H., Bal, H.E., Plaat, A., Bhoedjang, R.A.F.: MagPIe: MPI's collective communication operations for clustered wide area systems. ACM SIGPLAN Notices 34(8), 131–140 (1999)
5. Eyraud, L., Quinson, M.: Assessing the quality of automatically built network representations. In: 1st Workshop on Programming Models for Grid Computing (in Proceedings of CCGrid07 (2007)
6. Dinda, P., Gross, T., Karrer, R., Lowekamp, B., Miller, N., Steenkiste, P., Sutherland, D.: The architecture of the remos system. In: HPDC-10 (2001)
7. den Burger, M., Kielmann, T., Bal, H.E.: TOPOMON: A monitoring tool for grid network topology. In: Sloot, P.M.A., Tan, C.J.K., Dongarra, J.J., Hoekstra, A.G. (eds.) Computational Science - ICCS 2002. LNCS, vol. 2330, pp. 558–567. Springer, Heidelberg (2002)
8. Burch, H., Cheswick, B., Wool, A.: Internet mapping project, `http://www.lumeta.com/mapping.html`
9. Francis, P., Jamin, S., Jin, C., Jin, Y., Raz, D., Shavitt, Y., Zhang, L.: Idmaps: A global internet host distance estimation service. IEEE/ACM Transactions on Networking (October 2001)
10. Ng, T., Zhang, H.: Predicting internet network distance with coordinates-based approaches. In: INFOCOM. 1, 170–179 (2002)
11. The cooperative association for internet data analysis, `http://www.caida.org/`

---

[3] `http://gforge.inria.fr/plugins/scmcvs/cvsweb.php/contrib/ALNeM/?cvsroot=simgrid`

12. Downey, A.B.: Using pathchar to estimate internet link characteristics. In: Measurement and Modeling of Computer Systems, pp. 222–223 (1999)
13. Wolski, R., Spring, N.T., Hayes, J.: The Network Weather Service: A distributed resource performance forecasting service for metacomputing. Future Generation Computing Systems, Metacomputing Issue 15(5–6), 757–768 (1999)
14. Foster, I., Kesselman, C.: Globus: A Metacomputing Infrastructure Toolkit. International Journal of Supercomputing Applications 11(2), 115–128 (1997)
15. Caron, E., Desprez, F.: DIET: A scalable toolbox to build network enabled servers on the grid. IJHPCA 20(3), 335–352 (2006)
16. Shao, G., Berman, F., Wolski, R.: Using effective network views to promote distributed application performance. In: PDPTA (June 1999)
17. Lowekamp, B., Beguelin, A.: ECO: Efficient collective operations for communication on heterogeneous networks. IPDPS'96 (1999)
18. Byers, J.W., Bestavros, A., Harfoush, K.: Inference and labeling of metric-induced network topologies. IEEE TPDS 16(11), 1053–1065 (2005)
19. Quinson, M.: GRAS: A research & development framework for grid and P2P infrastructures. In: PDCS2006 (2006)
20. Legrand, A., Quinson, M., Fujiwara, K., Casanova, H.: The SimGrid project - simulation and deployment of distributed applications. In: HPDC-15, IEEE Computer Society Press, Los Alamitos (2006)
21. Casanova, H.: Modeling large-scale platforms for the analysis and the simulation of scheduling strategies. In: IPDPS (April 2004)
22. Tsafrir, D., Etsion, Y., Feitelson, D.G.: Backfilling using system-generated predictions rather than user runtime estimates. IEEE TPDS 18(6), 789–803 (2007)
23. Blackford, L.S., Choi, J., Cleary, A., D'Azevedo, E., Demmel, J., Dhillon, I., Dongarra, J., Hammarling, S., Henry, G., Petitet, A., Stanley, K., Walker, D., Whaley, R.C.: ScaLAPACK Users' Guide. SIAM (1997)
24. Lu, D., Dinda, P.: Synthesizing realistic computational grids. In: Proceedings of ACM/IEEE Supercomputing 2003 (SC 2003) (November 2003)

Topic 3
Scheduling and Load-Balancing

Henri Casanova, Olivier Beaumont, Uwe Schwiegelshohn, and Marek Tudruj

Topic Chairs

While scheduling and load-balancing problems have been studied for several decades, the dramatic multi-scale shifts in distributed systems and their usage in the last few years have raised new and exciting challenges. These challenges span the entire spectrum from theory to practice, as demonstrated by the selection of papers in the scheduling and load-balancing topic this year at Europar. Out of the twenty-three submissions to the topic we accepted six papers. The topic organizers would like to thank all reviewers whose work made it possible for each paper to receive at least three reviews.

In "Cooperation in Multi-Organization Scheduling", the authors demonstrate at a theoretical level that when multiple organizations share compute resources with selfish goals it is always preferable for these organizations to collaborate with respect to job scheduling.

The paper "A Framework for Scheduling with Online Availability" obtains new complexity results for classical scheduling problems with the assumption that availability of compute resources is stochastic throughout application execution.

In "A Parallelisable Multi-Level Banded Diffusion Scheme for Computing Balanced Partitions with Smooth Boundaries", the author proposes a novel and practical technique for domain partitioning that yields partitions of higher quality than previously proposed approaches, and in particular that yields smooth partition boundaries.

The authors of "Toward Optimizing Latency under Throughput Constraints for Application Workflows on Clusters" proposes pipelining and task replication strategies for minimizing the latency of a scientific workflow on a homogeneous cluster, while enforcing that the workflow's throughput be above a predetermined threshold.

The paper "Scheduling File Transfers for Data-Intensive Jobs on Heterogeneous Clusters" presents both an algorithm based on Integer Programming and a heuristic to schedule collective file transfers between multiple storage devices distributed within a multi-cluster platform.

The author of "Load Balancing on an Interactive Multiplayer Game Server" presents an experimental study of multi-threaded multiplayer game servers, and highlights several practical solutions to improve the scalability of such servers including the use of an efficient load-balancing approach.

A.-M. Kermarrec, L. Bougé, and T. Priol (Eds.): Euro-Par 2007, LNCS 4641, p. 171, 2007.
© Springer-Verlag Berlin Heidelberg 2007

Toward Optimizing Latency Under Throughput Constraints for Application Workflows on Clusters[*]

Nagavijayalakshmi Vydyanathan[1], Umit V. Catalyurek[2], Tahsin M. Kurc[2],
Ponnuswamy Sadayappan[1], and Joel H. Saltz[2]

[1] Dept. of Computer Science and Engineering
{vydyanat,saday}@cse.ohio-state.edu
[2] Dept. of Biomedical Informatics
{umit,kurc,saltz}@bmi.osu.edu
The Ohio State University

Abstract. In many application domains, it is desirable to meet some user-defined performance requirement while minimizing resource usage and optimizing additional performance parameters. For example, application workflows with real-time constraints may have strict throughput requirements and desire a low latency or response-time. The structure of these workflows can be represented as directed acyclic graphs of coarse-grained application tasks with data dependences. In this paper, we develop a novel mapping and scheduling algorithm that minimizes the latency of workflows that act on a stream of input data, while satisfying throughput requirements. The algorithm employs pipelined parallelism and intelligent clustering and replication of tasks to meet throughput requirements. Latency is minimized by exploiting task parallelism and reducing communication overheads. Evaluation using synthetic benchmarks and application task graphs shows that our algorithm 1) consistently meets throughput requirements even when other existing schemes fail, 2) produces lower-latency schedules, and 3) results in lesser resource usage.

1 Introduction

Complex application workflows can often be modeled as directed acyclic graphs (DAGs) of coarse-grained application components with data dependences. The quality of execution of these workflows is often gauged by two metrics: latency and throughput. Latency is the time to process an individual data item through the workflow, while throughput is a measure of the aggregate rate of processing of data. It is often desirable or necessary to meet a user-defined requirement in one metric, while achieving higher performance value in the other metric and minimizing resource usage. Workflows with real-time constraints, for example, can have strict throughput requirements, while interactive query processing may have strict latency constraints. To be able to meet requirements and minimize resource usage is also important in settings such as Supercomputer centers, where resources (e.g., a compute cluster) have an associated cost and are contended for by multiple clients.

[*] This research was supported in part by the National Science Foundation under Grants #CCF-0342615 and #CNS-0403342.

A.-M. Kermarrec, L. Bougé, and T. Priol (Eds.): Euro-Par 2007, LNCS 4641, pp. 173–183, 2007.

Workflows in domains such as image processing, computer vision, signal processing, parallel query processing, and scientific computing often act on a stream of input data [1,2]. Each task in the workflow repeatedly receives input data items from its predecessor tasks, computes on them, and writes the output to its successors. Multiple data items can be processed in a parallel or pipelined manner and independent tasks can be executed concurrently. In this paper, we present a novel approach for the scheduling of such workflows on clusters of homogeneous processors. Our algorithm employs pipelined, task and data parallelism in an integrated manner to meet strict throughput constraints and minimize latency. *Pipelined parallelism* is the concurrent execution of dependent tasks in the workflow on different instances of the input data stream, *data parallelism* is the concurrent processing of multiple data items by replicas of a task, and *task parallelism* is the concurrent execution of independent tasks on the same instance of the data stream.

We compare our approach against two existing schemes: Filter Copy Pipeline (FCP) [3] and EXPERT (EXploiting Pipeline Execution undeR Time constraints) [2]. Evaluations are done using synthetic benchmarks and application task graphs in the domains of Image Analysis, Video Processing and Computer Vision [1,2,4]. We show that our algorithm is able to 1) consistently meet throughput requirements even when the other schemes fail, 2) generate schedules with lower latency, and 3) reduce resource usage.

2   Related Work

Several researchers have addressed the problem of minimizing the parallel completion time (latency) of applications modeled as DAGs. As this problem is NP-complete [5], heuristics have been proposed and a survey of these can be found in [6]. Researchers have also proposed the use of pipelined scheduling for maximizing the throughput of applications. Hary and Ozguner [7] discussed heuristics for maximizing the throughput of application DAGs, while Yang [8] presented an approach for resource optimization under throughput constraints. Benoit and Robert [9] have addressed the problem of maximizing the throughput of pipeline skeletons of linear chains of tasks on heterogeneous systems. These techniques, however, do not consider replication of tasks.

Though many papers focus on optimizing latency or throughput in isolation, very few address both. Subhlok and Vondran [10] have proposed a dynamic programming solution for optimizing latency under throughput constraints for applications composed of a chain of data-parallel tasks. Benoit and Robert [11] study the theoretical complexity of latency and throughput optimization of pipeline and fork graphs with replication and data-parallelism under the assumptions of linear clustering and round-robin processing of input data items. In [3], Spencer et al. presented the Filter Copy Pipeline (FCP) scheduling algorithm for optimizing latency and throughput of data analysis application DAGs on heterogeneous resources. FCP computes the number of copies of each task that is necessary to meet the aggregate production rate of its predecessors and maps the copies to processors that yield their least completion time. Another closely related work is [2], where Guirado et al. have proposed a task mapping algorithm called

EXPERT (EXploiting Pipeline Execution undeR Time constraints) that minimizes latency of streaming applications, while satisfying a given throughput constraint. EXPERT identifies maximal clusters of tasks that can form synchronous stages that meet the throughput constraint and maps tasks in each cluster to the same processor so as to reduce communication overheads and minimize latency.

3   Task Graph and Execution Model

A workflow can be modeled as a connected, weighted DAG $G = (V, E)$, where V, the set of vertices, represents non-homogeneous sequential tasks and E, the set of edges, represents data dependences. The task graph G acts on a stream of data, where each task repeatedly receives input data items from its predecessors, computes on them, and writes the output to its successors. The weight of a vertex (task) $t_i \in V$, is its execution time to process a single data item, $et(t_i)$. The weight of an edge $e_{i,j} \in E$, $wt(e_{i,j})$, is the communication cost measured as the time taken to transfer a single data item of size $d_{i,j}$ between t_i and t_j. The length of a path in G is the sum of the weights of the tasks and edges along that path. The *critical path* of G, denoted by $CP(G)$, is the longest path in G. The *bottom level* of a task t in G, $bottomL(t)$, is defined as the length of the longest path from t to the exit task, including the weight of t.

In this paper, we target homogeneous compute clusters for execution of the task graph G. Our algorithm assumes that the execution behavior of the tasks in G is not strongly dependent on the properties of the input data items and that profiling G on several representative data sets gives a reasonable measure of the task execution times. The system model assumes overlap of computation and communication.

The latency of a schedule of task graph G on P processors is the time taken to process a single data item through G. G', the DAG that represents the dependences in the schedule, can be constructed from G by adding zero-weight *pseudo-edges* between concurrent tasks in G that are mapped to the same processor. These pseudo-edges denote induced dependences. The latency is defined to be the critical path length of G'.

Let a task-cluster denote the group of all tasks that are mapped to the same processor. The time taken by a task-cluster C_i to process a single data item is given by the sum of the execution times of its constituent tasks, i.e $et(C_i) = \sum_{\forall t \in C_i} et(t)$. If the workflow is assumed to act on a stream of independent data items (i.e processing of each data item is independent of the processing of other data items), replicas of a task/task-cluster can be executed concurrently. If $nr(C_i)$ denotes the number of replicas of task-cluster C_i, the aggregate processing rate of C_i, $pr(C_i)$ is given by $\frac{nr(C_i)}{et(C_i)}$ data items per unit time. Each replica of a task-cluster is assumed to be executed on a separate processor. For example, assume that tasks t_1 and t_2 are mapped to task-cluster C and $bottomL(t_1) > bottomL(t_2)$ in G'. Let $nr(C)$ be 2, the replicas be mapped to processors P_1 and P_2, $et(t_1) = 10$, and $et(t_2) = 20$. Then, on each of these processors, t_1 processes a data item followed by t_2. The processing rate of C is $\frac{2}{(10+20)}$.

The data transfer rate of an edge $e_{i,j}$, $dr(e_{i,j})$, is $\frac{1}{\frac{d_{i,j}}{bw_{i,j}}}$ data items per unit time, where $bw_{i,j} = \min(nr(t_i), nr(t_j)) \times bandwidth$. Here, bandwidth corresponds to the minimum of disk or memory bandwidth of the system depending on the location of data and the network bandwidth. $nr(t_i)$ denotes the number of replicas of task t_i. As we assume that computation and communication can overlap, the overall processing rate or throughput of the workflow is determined by the slowest task-cluster or edge, and is given by $\min(\min_{\forall C_i} pr(C_i), \min_{\forall e_{i,j}} dr(e_{i,j}))$.

4 Workflow Mapping and Scheduling Heuristic

Given a workflow-DAG G, P homogeneous processors and a throughput constraint T, our workflow mapping and scheduling heuristic (WMSH) generates a mapping and schedule of G on P that minimizes the latency while satisfying T. The algorithm consists of three main heuristics, which are executed in sequence: the *Satisfy Throughput Heuristic* (STH) to meet the user-defined throughput requirements, the *Processor Reduction Heuristic* (PRH) to ensure that the resulting schedule does not require more processors than available, and the *Latency Minimization Heuristic* (LMH) to minimize the workflow latency. In this section, we describe each of these heuristics. Details on the proofs for theorems can be found in the technical report [12].

Theorem 1. *Given a workflow-DAG $G = (V, E)$ that acts on a stream of independent data items, the maximum achievable throughput T_{max}, on P homogeneous processors is given by $\frac{P}{\sum_{t \in V}(et(t))}$, where $et(t)$ is the time taken by t to process a single data item.*

T_{max} can be achieved by mapping all tasks in G to a single task-cluster and making P replicas, each mapped to a unique processor. However, this mapping suffers from a large latency as it fails to exploit parallelism between concurrent tasks in G. For the sake of presentation, the rest of this section assumes that G acts on a stream of independent data items and hence all tasks can be replicated. However, the heuristics described here can be applied when processing of a data item is dependent on the processing of certain other data items (i.e replication of tasks is not allowed), by enforcing the weight of every task-cluster to be $\leq \frac{1}{T}$, for a given throughput constraint $T \leq T_{max}$. T_{max} in this case, is the reciprocal of the weight of the largest task in G.

Given a throughput constraint $T \leq T_{max}$, STH verifies whether a non-pipelined low latency schedule, generated by priority-based list-scheduling [6], meets the throughput requirement. The tasks in G are prioritized in the decreasing order of their bottom-levels and scheduled in priority order to processors that yield their least completion time. If the throughput of this schedule (which is the reciprocal of the latency) is $\geq T$, STH returns this schedule. Otherwise, the following steps are executed to obtain a low-latency pipelined schedule that satisfies T. To generate a pipelined schedule, each task $t_i \in V$ is mapped to a separate task-cluster C_i. Let M denote the set of all the task-clusters. The number of replicas of C_i, $nr(C_i)$, required to satisfy T is computed as $nr(C_i) = T \times et(C_i)$. When there is no throughput constraint, $nr(C_i) = 1$. For all edges $e_{i,j} \in E$, whose data transfer rate is $< T$, STH avoids the communication overhead by merging the task-clusters containing the incident tasks. When two task-clusters are merged, the DAG G' representing the dependences in the schedule is constructed from G by adding zero weight *pseudo-edges* between concurrent tasks in G that are mapped to the same task-cluster. The pseudo-edges originate from the task with the

---

Algorithm 1. PRH: Processor Reduction Heuristic

1: **function** PRH(G', M)   ▷ $G' \leftarrow$ schedule DAG returned by STH, $M \leftarrow$ set of task-clusters returned by STH
2:     $P' = \sum_{C_i \in M}(\lceil nr(C_i)\rceil)$

3:     **repeat**
4:         $\mathcal{C}' \leftarrow \{(C_i, C_j) \mid C_i \in M \wedge C_j \in M \wedge \lceil nr(C_i) + nr(C_j)\rceil < (\lceil nr(C_i)\rceil + \lceil nr(C_j)\rceil)\}$
5:         **while** $\mathcal{C}'$ **not** empty $\wedge (P' > P)$ **do**
6:             Pick the task-cluster pair (C_i, C_j) from $\mathcal{C}'$ that yields the largest decrease in latency when merged. Preference is given to task-clusters that are connected, not concurrent and which produce the largest resource wastage when merged.
7:             For all task-pairs $(t_a, t_b) \in C_i \times C_j \mid t_a$ concurrent to t_b in G, add a *pseudo-edge* in G' originating from the task with the larger bottom-level.
8:             For all edges $e_{a,b} \in G \mid (t_a, t_b) \in C_i \times C_j, wt(e_{a,b}) \leftarrow 0$ in G'
9:             Merge C_i and C_j and update M
10:            $P' \leftarrow P' - 1$
11:            Update $\mathcal{C}'$
12:        **if** $P' > P$ **then**
13:            Pick the task-cluster pair (C_i, C_j) that yields the maximum value of $\lceil (nr(C_i) + nr(C_j))\rceil - (nr(C_i) + nr(C_j))$ and the largest decrease in latency when C_i and C_j are merged.
14:            For all task-pairs $(t_a, t_b) \in C_i \times C_j \mid t_a$ concurrent to t_b in G, add a *pseudo-edge* in G' originating from the task with the larger bottom-level.
15:            For all edges $e_{a,b} \in G \mid (t_a, t_b) \in C_i \times C_j, wt(e_{a,b}) \leftarrow 0$ in G'
16:            Merge C_i and C_j and update M
17:    **until** $P' \leq P$
18:    **return** $< G', M >$

---

larger bottom-level. Edges between tasks mapped to the same task-cluster have zero weight in G'. An example to illustrate this is given in the technical report [12].

Following STH, PRH is executed. The total number of processors required to execute $nr(C_i)$ copies of each task-cluster C_i, where each copy is mapped to a unique processor, is $P' = \sum_{C_i \in M} \lceil nr(C_i)\rceil$. If $P' > P$, PRH merges certain task-clusters and obtains a schedule that uses $\leq P$ processors. Once a feasible schedule is obtained, LMH is called to optimize the latency. PRH and LMH output a set of task-clusters and the pipelined schedule is obtained by mapping each replica of a task-cluster to a unique processor. Tasks within a task-cluster are run in the decreasing order of their bottom-levels and iterate over the instances of the data stream. We now present PRH and LMH in greater detail.

4.1 Processor Reduction Heuristic (PRH)

PRH recursively merges pairs of task-clusters based on some metric until we get a mapping that uses $\leq P$ processors.

Theorem 2. *If task-clusters C_i and C_j are merged and P_i and P_j are the number of processors required to run the replicas of C_i and C_j respectively, i.e $P_i = \lceil nr(C_i)\rceil$ and $P_j = \lceil nr(C_j)\rceil$, the number of processors required to run the replicas of the new task-cluster formed that meets the throughput constraint is either $P_i + P_j$ or $P_i + P_j - 1$.*

The pseudo code of PRH is illustrated in Algorithm 1. Step 4 of the algorithm considers all pairs of task-clusters that when merged would reduce the number of processors used by 1. Among these, PRH picks the task-cluster pair that yields the largest decrease in latency when merged. To break ties, preference is given to task-clusters that are connected, not concurrent, and which produce the largest resource wastage, in that order (step 6). Task-clusters C_i and C_j are "connected" if there exists some task t_a in C_i and some task t_b in C_j such that $e_{a,b}$ is an edge in G. Task-clusters C_i and C_j are "not concurrent" if for all pairs of tasks (t_a, t_b), $t_a \in C_i$ and $t_b \in C_j$, t_a is not concurrent to t_b in G. Resource wastage of a task-cluster C is defined as $\lceil nr(C) \rceil - nr(C)$. Giving preference to task-cluster pairs that yield a larger resource wastage reduces the possibility of fragmentation. Steps 5-11 are repeated as long as there are task-cluster pairs that reduce the processor count and $P' > P$. After all possible task clusterings, if the resource usage is still greater than P at step 12, defragmentation is done in steps 13-16 where the task-clusters that produce the largest resource wastage are merged. To break ties, the one that causes the largest decrease in latency is chosen. The outer-loop (steps 3-17) are repeated until the resource usage is lesser than or equal to P. At the end of the processor reduction phase, a mapping M and schedule G' is obtained that meets the throughput constraint and uses $\leq P$ processors.

4.2   Latency Minimization Heuristic (LMH)

LMH is called to refine the mapping obtained by PRH to further optimize the latency by reducing communication overheads. The task-clusters in M are considered by LMH as indivisible macro-tasks. A macro-task therefore, may contain one or more tasks. The incoming and outgoing edges of a macro-task is the union of the incoming and outgoing edges, respectively, of the tasks that it contains, without considering edges between tasks belonging to the macro-task. Hence, the term task in Theorem 3 is the same as macro-task in the case where multiple tasks are mapped to same task-cluster by PRH.

Theorem 3. *Let G' and M denote a schedule and mapping of G that meets the throughput constraint and uses $\leq P$ processors. Let $e_{i,j}$ be an edge in G' from task/macro-task t_i to t_j such that the in-degree$(t_i) = $ in-degree$(t_j) = 1$ and the out-degree$(t_i) = $ out-degree$(t_j) = 1$ (i.e. t_i and t_j are connected along a linear chain in that order). Let t_k be the parent of t_i and t_l be the child of t_j. If $wt(e_{i,j}) > wt(e_{k,i}) + wt(e_{j,l})$, it is optimal to merge t_i and t_j to a single task-cluster, assuming that all tasks can be replicated. If replication is not allowed, t_i and t_j can be merged to a single task-cluster only if $et(t_i) + et(t_j) \leq \frac{1}{T}$ and $e_{i,j}$ satisfies the above condition.*

This theorem can be extended to the case where the tasks/macro-tasks are not connected in a linear chain. Details of this can be found in the technical report [12].

Algorithm 2 describes LMH. LMH identifies the set E^* of edges where it is optimal to merge the incident tasks (theorem 3 and its extensions) (step 2) and merges the task-clusters of the incident tasks (steps 6-8). After merging, E^* is updated (step 9). Steps 4-9 are repeated until E^* is empty. In steps 10-14, among the edges along $CP(G')$ that do not cause an increase in latency when zeroed-in, LMH zeroes-in the edge with the largest maximum possible decrease in latency. To break ties, the edge $e_{i,j}$ with the minimum value of the sum of number of critical edges to t_i and number of critical edges to t_j is chosen. The outer-loop of steps 3-15 is repeated until all edges in $CP(G')$ cause an increase in latency when zeroed-in. Details regarding the order of complexity of WMSH can be found in [12].

Algorithm 2. LMH: Latency Minimization Heuristic

1: **function** LMH(G', M)   ▷ $G' \leftarrow$ schedule DAG returned by PRH, $M \leftarrow$ mapping returned by PRH.
2:     $E^* \leftarrow$ set of all edges in G' where it is optimal to merge the incident tasks (theorem 3)
3:     **repeat**
4:         **while** E^* **not** empty **do**
5:             $e_{i,j}$ is an edge in E^*
6:             For all task-pairs $(t_a, t_b) \in$ clusterOf$(t_i)\times$clusterOf$(t_j) \mid t_a$ concurrent to t_b in G, add a *pseudo-edge* in G' originating from the task with the larger bottom-level. ▷ clusterOf(t_i) is the task-cluster that contains task t_i.
7:             For all edges $e_{a,b} \in G \mid (t_a, t_b) \in$ clusterOf$(t_i)\times$clusterOf(t_j), $wt(e_{a,b}) \leftarrow 0$ in G'
8:             Merge clusterOf(t_i) and clusterOf(t_j), update M
9:             Update E^*
10:            Pick edge $e_{i,j}$ in $CP(G')$ that does not increase the latency when clusterOf(t_i) and clusterOf(t_j) are merged and has maximum value of $\min(wt(e_{i,j}), CPL(G') - LBL(G))$ and minimum value of $(|\text{critical-edges}(t_i)| + |\text{critical-edges}(t_j)|)$ ▷ $CPL(G') \leftarrow$ Critical Path Length of G', $LBL(G) \leftarrow$ Lower Bound on Latency of G
11:            For all task-pairs $(t_a, t_b) \in$ clusterOf$(t_i)\times$ clusterOf$(t_j) \mid t_a$ concurrent to t_b in G, add a *pseudo-edge* in G' originating from the task with the larger bottom-level.
12:            For all edges $e_{a,b} \in G \mid (t_a, t_b) \in$ clusterOf$(t_i)\times$clusterOf(t_j), $wt(e_{a,b}) \leftarrow 0$ in G'
13:            Merge clusterOf(t_i) and clusterOf(t_j) and update M
14:            Update E^*
15:        **until** For all edges $e_{i,j}$ in $CP(G')$, latency increases when clusterOf(t_i) and clusterOf(t_j) are merged
16:        **return** $< G', M >$

---

5   Performance Analysis

This section evaluates the performance of WMSH against previously proposed schemes: Filter Copy Pipeline (FCP) [3] and EXPERT (EXploiting Pipeline Execution undeR Time constraints) [2], and FCP-e and EXPERT-e, their modified versions. When FCP fails to utilize all processors and does not meet the throughput requirement T, FCP-e recursively calls FCP on the remaining processors until T is satisfied or all processors are used. EXPERT-e replicates the task-clusters by dividing the remaining processors among them in the ratio of their weights. The performance of these algorithms is evaluated using both synthetic task graphs and those derived from applications, using simulations.

5.1   Synthetic Task Graphs

Two sets of synthetic benchmarks were used in the evaluations: 1) Benchmark-I: randomly generated task graphs with communication delays [13], and 2) Benchmark-II: synthetic graphs generated using the DAG generation tool in [14]. More details on the benchmarks can be found in the technical report [12]. Figure 1 plots the performance on benchmark-I on 32 and 64 processors. The x-axis is the throughput constraint, which is decreased from the maximum achievable throughput (T_{max}) in steps of 0.25. The symbol ≈ 0 denotes the case when there is no throughput constraint (or negligibly small). The y-axis is the average latency ratio. Latency ratio is the ratio of the latency of the schedule generated by an algorithm to that of WMSH. The results show that WMSH consistently generates schedules that meet the throughput constraint, while FCP and EXPERT fail at large throughput requirements ((T_{max} and $0.75*T_{max}$). Though FCP replicates tasks, it computes the number of replicas independent of the number of processors and fails to refine the number of replicas when it maps multiple tasks to the

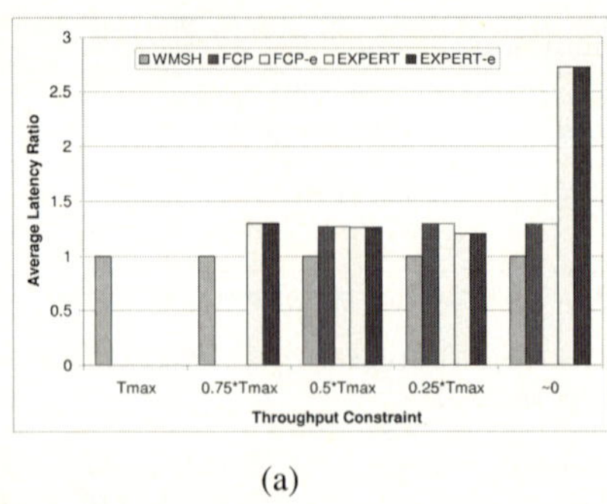

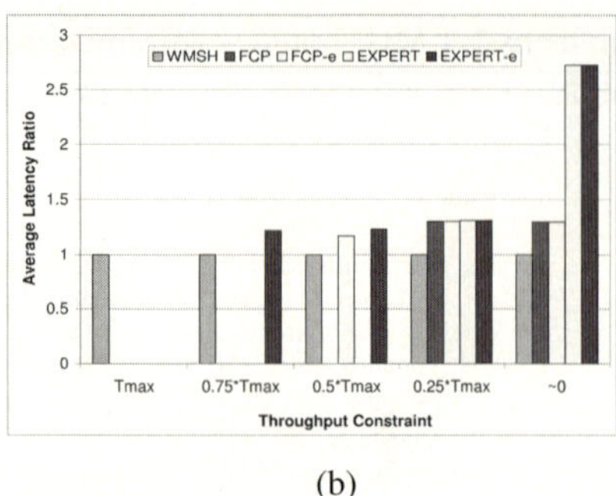

(a)   (b)

Fig. 1. Performance on Benchmark-I on (a) 32 processors, (b) 64 processors. (The missing bars indicate that the corresponding algorithm could not meet the throughput requirement).

Table 1. Performance on Benchmark-I on 64 processors (a) Average Throughput Ratio, (b) Average Utilization Ratio. (The missing values in (b) indicate that the corresponding algorithm could not meet the throughput requirement).

(a)

T	WMSH	FCP	FCP-e	EXPERT	EXPERT-e
T_{max}	1.00	0.31	0.35	0.40	0.68
$0.75*T_{max}$	1.00	0.41	0.55	0.53	1.00
$0.50*T_{max}$	1.00	0.59	1.00	0.80	1.00

(b)

T	WMSH	FCP	FCP-e	EXPERT	EXPERT-e
T_{max}	1.00	-	-	-	-
$0.75*T_{max}$	0.91	-	-	-	0.94
$0.50*T_{max}$	0.73	-	1.00	-	0.94

same processor. EXPERT does not replicate tasks. The modified versions are designed to overcome some of these limitations and hence, meet the constraint in some of the cases where FCP or EXPERT fail.

With respect to latencies, we find that WMSH generates lower latency schedules. On 32 processors, FCP generates 27%-29% longer latencies than WMSH, while EXPERT generates 20%-30% longer latencies when throughput constraint is relaxed upto $0.25*T_{max}$. As EXPERT creates maximal task-clusters with weights $\leq \frac{1}{T}$, for negligible throughput constraint, it groups all tasks to a single task-cluster resulting in large latencies. For FCP-e, we used the smallest of the latencies of all the workflow instances it creates and hence it is similar to that of FCP. Latency in EXPERT-e is similar to EXPERT, since EXPERT-e only replicates tasks; this improves the throughput but does not alter the latency. As P is increased, T_{max} increases, and hence, there are more instances where FCP and EXPERT do not satisfy T.

Table 1(a) shows the average throughput ratio for the schemes for Benchmark-I on 64 processors. The throughput ratio is the ratio of the throughput achieved by an algorithm to the throughput constraint. If the achieved throughput is greater than the constraint, the ratio is taken to be 1. Beyond $0.5*T_{max}$, all schemes meet the constraint. When FCP and EXPERT fail, they generate schedules with throughput atleast 40% and 20% less than the constraint, respectively. Table 1(b) shows the average utilization ratio for the schemes. The utilization ratio is given by the ratio of the number of processors used by an algorithm to the total number of available processors. Among schemes that satisfy T, WMSH produces lower-latency schedules while using fewer processors. For example, when T is $0.5*T_{max}$, utilization of WMSH is 27% lower than that of FCP-e and 19%

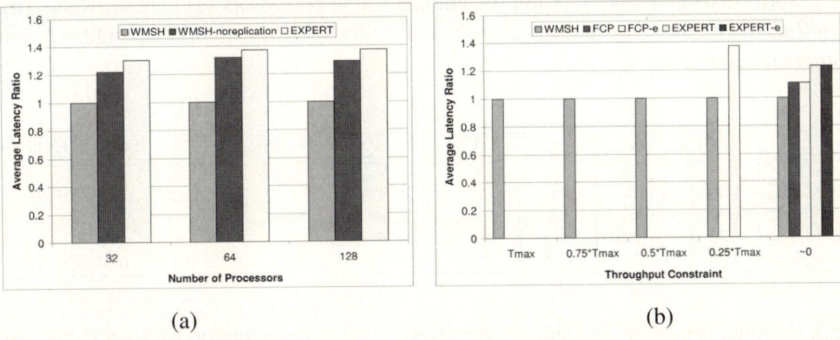

Fig. 2. (a) Relative Performance of WMSH, WMSH with replication disabled and EXPERT when throughput constraint is $\frac{1}{\max_{t \in V}(et(t))}$, (b) Performance on Benchmark-II on 32 processors and CCR=10. (The missing bars in (b) indicate that the corresponding algorithm could not meet the throughput requirement).

lower than EXPERT-e, and it produces latencies 15% and 19% shorter than FCP-e and EXPERT-e respectively.

As EXPERT does not replicate tasks, we compared its performance with that of WMSH with replication disabled (Fig. 2(a)). Even with no replication, WMSH produces lower latencies than EXPERT. WMSH with replication shows the least latency as tasks connected by edges with heavy communication cost can be mapped to the same task-cluster and replicated to meet the throughput constraint. Thus replication not only helps in improving throughput but also minimizing the latency.

To study the impact of communication costs, we evaluated the schemes using Benchmark-II by varying the communication to computation ratio (CCR) as 0.1, 1 and 10. Figure 2(b) shows the performance when CCR=10. Due to space constraints, we have not included results for CCR=0.1,1. These can be found in the technical report [12]. For larger CCR values, we find more instances where FCP, EXPERT and their modified versions do not meet the throughput constraint, while WMSH always does. WMSH intelligently zeroes-in heavy edges by mapping the incident tasks to the same task-cluster and replicating this cluster to meet the throughput constraint. Though FCP minimizes communication costs in some capacity by mapping replicas to processors that yield their least completion time, it still incurs the cost when the processor to which the parent task is mapped is heavily loaded (as mapping the task to this processor would cause a larger completion time). EXPERT does not replicate and hence cannot cluster heavy tasks that also have a huge communication cost. The modified versions of the schemes do not completely avoid the communication overheads as they only replicate tasks. As for Benchmark-I, WMSH generates the lowest latency schedules that use lesser resources.

5.2 Application Task Graphs

Evaluations were done using task graphs from computer vision, multimedia and imaging domains. Due to space limitations, we present results for only two applications; detailed evaluation can be found in the technical report [12]. Table 2 shows the

Table 2. Performance of Darpa Vision Benchmark on 32 processors (a) Latency Ratio, (b) Utilization Ratio. (The missing values indicate that the corresponding algorithm could not meet the throughput requirement).

(a)

T	WMSH	FCP	FCP-e	EXPERT	EXPERT-e
T_{max}	1.00	-	-	-	-
$0.75*T_{max}$	1.00	-	-	-	-
$0.50*T_{max}$	1.00	-	-	-	-
$0.25*T_{max}$	1.00	-	1.04	-	-
≈ 0	1.00	1.04	1.04	1.15	1.15

(b)

T	WMSH	FCP	FCP-e	EXPERT	EXPERT-e
T_{max}	1.00	-	-	-	-
$0.75*T_{max}$	0.75	-	-	-	-
$0.50*T_{max}$	0.53	-	-	-	-
$0.25*T_{max}$	0.31	-	1.00	-	-
≈ 0	0.25	0.47	0.47	0.03	1.00

Table 3. Performance of MPEG video compression on 32 processors (a) Latency Ratio, (b) Utilization Ratio. (The missing values indicate that the corresponding algorithm could not meet the throughput requirement).

(a)

Segments	WMSH	FCP	EXPERT
2	1.00	1.00	1.21
4	1.00	1.00	1.36
8	1.00	1.00	1.41
16	1.00	1.00	1.24

(b)

Segments	WMSH	FCP	EXPERT
2	0.13	0.13	0.09
4	0.25	0.41	0.22
8	0.50	0.78	0.47
16	1.00	1.00	1.00

performance for the Darpa Vision Benchmark (DVB) [4], which performs model-based object recognition of a hypothetical object. We find that FCP, EXPERT and their modified versions do not meet the throughput requirement T, in many instances. In cases where they satisfy T, WMSH produces schedules with shorter latencies and lower resource utilization than FCP. When T is negligible, the schedule generated by WMSH uses 22% fewer processors than that of FCP and has 4% lower latency. WMSH also produces latencies 15% lower than that of EXPERT.

Table 3 shows results for an MPEG video compression application [2]. Due to frame encoding dependences, MPEG frames have to be processed in-order and hence task replication is not allowed. However, the input frames can be divided into N segments, which can be processed in parallel. We assumed T_{max} to be the reciprocal of the weight of the largest task and varied N from 2 to 16. We find that FCP and WMSH generate schedules with similar latencies, but WMSH has upto 28% lower utilization. Though EXPERT shows lower utilization, it generates schedules with 21%-41% longer latencies than WMSH or FCP. The scheduling times in these experiments were less than a second suggesting that scheduling is not a time critical operation for these applications.

6   Conclusion

This paper presents a mapping and scheduling heuristic that minimizes the latency of workflows that operate on a stream of data, while satisfying strict throughput requirements. Our algorithm meets the throughput constraints through pipelined parallelism and replication of tasks. Latency is minimized by exploiting task parallelism and reducing communication overheads. Evaluation using synthetic and application task graphs indicate that our heuristic is always guaranteed to meet the throughput requirement and hence can be deployed for scheduling workflows with real-time constraints. Further, it produces lower latency schedules and utilizes lesser resources.

Acknowledgments. We would like to thank Dr. Yves Robert and Dr. Anne Benoit for their valuable discussions and constructive reviews on the paper.

References

1. Kumar, V.S., Rutt, B., Kurc, T., Catalyurek, U., Saltz, J., Chow, S., Lamont, S., Martone, M.: Large image correction and warping in a cluster environment. In: Supercomputing Conf. p. 79 (2006)
2. Guirado, F., Ripoll, A., Roig, C., Luque, E.: Optimizing latency under throughput requirements for streaming applications on cluster execution. In: Cluster Computing Conf. (2005)
3. Spencer, M., Ferreira, R., Beynon, M., Kurc, T., Catalyurek, U., Sussman, A., Saltz, J.: Executing multiple pipelined data analysis operations in the grid. In: Supercomputing Conf. pp. 1–18 (2002)
4. Shukla, S.B., Agrawal, D.P.: Scheduling pipelined communication in distributed memory multiprocessors for real-time applications. SIGARCH Comput. Archit. News 19(3) (1991)
5. Garey, M.R., Johnson, D.S.: Computers and Intractability; A Guide to the Theory of NP-Completeness. W. H. Freeman & Co, New York, USA (1990)
6. Kwok, Y.K., Ahmad, I.: Static scheduling algorithms for allocating directed task graphs to multiprocessors. ACM Comput. Surv. 31(4), 406–471 (1999)
7. Hary, S.L., Ozguner, F.: Precedence-constrained task allocation onto point-to-point networks for pipelined execution. IEEE Trans. Par. Distrib. Syst. 10(8), 838–851 (1999)
8. Yang, M.T., Kasturi, R., Sivasubramaniam, A.: A pipeline-based approach for scheduling video processing algorithms on now. IEEE Trans. Par. Distrib. Syst. 14(2), 119–130 (2003)
9. Benoit, A., Robert, Y.: Mapping pipeline skeletons onto heterogeneous platforms. Technical Report LIP RR-2006-40 (2006)
10. Subhlok, J., Vondran, G.: Optimal latency-throughput tradeoffs for data parallel pipelines. In: 8th ACM Symp. on Parallel Algorithms and Arch, pp. 62–71. ACM Press, New York (1996)
11. Benoit, A., Robert, Y.: Complexity results for throughput and latency optimization of replicated and data-parallel workflows. Technical Report LIP RR-2007-12 (2007)
12. Vydyanathan, N., Catalyurek, U., Kurc, T., Sadayappan, P., Saltz, J.: An approach for optimizing latency under throughput constraints for application workflows on clusters. Technical Report OSU-CISRC-1/07-TR03, The Ohio State University (2007)
13. Davidovic, T., Crainic, T.G.: Benchmark-problem instances for static scheduling of task graphs with communication delays on homogeneous multiprocessor systems. Computers & OR 33(8), 2155–2177 (2006)
14. Vallerio, K.: Task graphs for free, http://ziyang.ece.northwestern.edu/tgff/maindoc.pdf

Load Balancing on an Interactive Multiplayer Game Server

Daniel Cordeiro[1,*], Alfredo Goldman[1], and Dilma da Silva[2]

[1] Department of Computer Science, University of São Paulo
`danielc,gold@ime.usp.br`
[2] Advanced Operating System Group, IBM T. J. Watson Research Center
`dilma@watson.ibm.com`

Abstract. In this work, we investigate the impact of issues related to performance, parallelization, and scalability of interactive, multiplayer games. Particularly, we study and extend the game QuakeWorld, made publicly available by *id Software* under GPL license. We have created a new parallelization model for Quake's distributed simulation and implemented this model in QuakeWorld server. We implemented the model adapting the QuakeWorld server in order to allow a better management of the generated workload. We present in this paper our experimental results on SMP computers.

1   Introduction

Game developers recently started a deep discussion about the uses of multiple processors in computer games. The new generation of video game consoles provides a multicore processor environment, but the current game developing technology does not use the full potential of the new video game consoles yet.

Abdelkhalek et al. [1,2,3] started an investigation of the behavior of interactive, multiplayer games in multi-processed computers. They characterized the computing needs and proposed a parallelization model for QuakeWorld, an important computer game optimized for multiplayer games.

The results presented by Abdelkhalek showed that his multithreaded server implementation suffered from performance problems because of their misuse of game semantics in order to create a lock strategy and because of the use of a static division of work between the available processors.

We present a dynamic load balancing and scheduling mechanism that utilizes the semantics of the simulation executed by the server. The performance analysis shows that we are able to improve the parallelism rate from 40% obtained by Abdelkhalek to 55% of the total execution time.

This paper is organized as follows. Section 2 gives an overview of the QuakeWorld server. Section 3 describes previous efforts in parallelizing the Quake server. Section 4 presents the proposed parallelization model for Quake and Section 5 presents the performance analysis. Section 6 presents our conclusions.

* The author would like to thank The State of São Paulo Research Foundation (FAPESP, grant no. 03/10064-4) for the financial support.

A.-M. Kermarrec, L. Bougé, and T. Priol (Eds.): Euro-Par 2007, LNCS 4641, pp. 184–194, 2007.

2   Quake Server

The Quake game is an interactive, multiplayer action game developed and distributed by *id Software* [4]. Its release in 1996 was a important mark in the game industry because it was the first time that a game was developed using three-dimensional models in the simulation and graphics engines. Later that year, QuakeWorld was released with enhancements for Internet games[1].

The Quake multiplayer game follows the client-server architecture. One game session has a single, centralized server that is responsible for the execution of all physics simulations, for evolving the game model (that gives the semantics of the game to the simulation), and for the propagation of state modifications to all connected clients. Up to thirty-two clients can join an open game session. It is the client responsibility to collect and send the input events from the player to the server and to do the graphics rendering of the current state of the simulation.

2.1   Quake Server Architecture

The Quake server is a single process, event-driven program. The server-side simulation consists of execution of frames. The frame task is composed by three distinct stages: updating the world physics, receiving and processing events sent by clients and creating and sending replies to all active players.

During the simulation of the world physics, each solid game entity (players, bullets, blocks, etc.) is simulated by the engine. The engine simulates the effects of gravity, velocity, acceleration, etc. The request and response processing is the more computational intensive phase. It reads all messages from the server socket, simulates the events sent in the message (movement commands, jump command, text messages, etc.), applies the command to the game state and replies to the clients with the changes in the game state. Only active clients, i.e. clients that sent a message in this frame receives a reply.

3   Multithreaded Version

Abdelkhalek et al. started an effort to characterize the behavior and performance of the original version of Quake [1]. They increased the limit of players from the original 32 to 90 simultaneous players. Their experiments showed that the incoming bandwidth is practically constant and is low (a few KBytes/s) and the outgoing bandwidth depends on the number of simultaneous clients, but does not exceed a few KBytes/s. The actual performance limitation was the processing power bottleneck. With more than 90 simultaneous users, server performance started to degrade.

In his following works [2,3], Abdelkhalek presented a multithreaded version of the Quake server. In order to avoid semantic and correctness errors that could

---

[1] We will use the names "Quake" and "QuakeWorld" interchangeably in this text to refer this new improved version.

arise from reordering the frame computation stages or overlapping them, two invariants were imposed: (i) each server phase is distinct and should not overlap with other phases; and, (ii) each phase should execute in the original order: world processing, request processing, and finally reply processing.

Also, at this first parallelization attempt, there is no load balancing mechanism. Each client is assigned to a server thread at connection time. The server uses one thread per available CPU and clients are assigned to the threads in a round-robin fashion. All computations related to one client are executed by the same thread and each client sends their messages directly to the assigned thread.

3.1   Tasks Decomposition

The frame tasks were decomposed as follows. The authors noted that the world processing stage takes less than 5% of the total execution time of the original version regardless of the quantity of players [2] and excluded this execution from the parallelization effort. So, there is a synchronization barrier before and after the world processing stage and only the coordinator thread executes this phase.

After that, each thread starts to read and execute all received messages at the thread's socket. Each thread does not proceed to the next synchronization barrier until all messages are consumed. The last phase is the reply processing, where each thread determines which entities are of interest to each client and sends out information only for those, i.e. notifies a client only about changes in entities that are visible to him.

3.2   Synchronization

Threads must synchronize the access to shared data structures between the synchronization barriers. During the world processing stage – which is sequential – and the reply processing stage there is no need to do any synchronization. In the last stage, the thread reads from shared data structures, but only writes in thread local buffers and sockets. However, during the request processing, the execution of the event from one client can alter the state of another entity, possibly being simulated by another thread.

The synchronization mechanism utilized by Abdelkhalek uses the semantics of the data structure known as Areanode tree. The Areanode tree is a binary-tree that allows the server to quickly find the list of entities that a single entity can interact using the location of this particular entity.

A node in an Areanode tree is defined by a tuple (3D region, plane). This plane is perpendicular to one of the three axes and divides the 3D region in two subregions. Each subregion generated by this division is used to define each one of the two tuples that will be children of this node. The 3D region of the root is defined as the entire virtual game map. The planes are calculated when server starts and are based on the Binary Space Partitioning (BSP) [5] representation of the game map. The maximum depth of this tree is hard-coded to be four.

Using the Areanode tree, the synchronization implements a region-based locking scheme. For each message consumed by a thread, the server finds the client that sent the message, locates the leaf in the Areanode tree that contains this client (based in its position) and acquires a lock in each node from leaf to root until the region determined by the current node contains all the entities that can be modified by the execution of this message. This method guarantees that two different threads will not change the state of the same entity nor the global data structures of the server in a undesirable manner.

3.3   Analysis

The performance analysis presented in [2] showed that in up to 70% of the execution time of the server, at least one thread is waiting at a synchronization point: 30% in lock contention and 40% in synchronization barriers.

Using the semantics of the game objects, Abdelkhalek did some performance optimizations in the mechanism used to acquire locks and improved the time spent with lock contention from 30% to 20%. We believe that there are two major problems with this approach:

- **lack of a dynamic load balancing method.** The static allocation of clients in threads at connection time may create a large load unbalance and may increase the time spent by the threads at synchronization barriers;
- **misuse of the semantics of game objects.** Locking by node in the Areanode creates a coarse grained mechanism that penalizes performance. Also, using the proposed lock mechanism makes the lock acquire operation time-consuming, since the Areanode must be traversed from leaf to root.

4   Parallelization Methodology

We propose a new parallelization model for the Quake server based in the Bulk Synchronous Parallel model [6] implemented using an event-driven architecture [7] with only one synchronization barrier.

4.1   Motivation

A preliminary analysis indicated that the original Quake server has low memory footprint and indicated also that only 10% of the time is spent with network-related system calls. This means that all the remaining time is spent on the execution of the server frames. According to the classification proposed by Pai et al. in [7], the original Quake server is implemented as a single-process event-driven (SPED) server. The server uses non-blocking system calls to perform asynchronous I/O operations. The server is able to overlap CPU and networking operations, achieving an efficient utilization of computing resources.

The nature of Quake simulation and the fact that it is written as an event-driven program makes Quake a good candidate for parallelism. The good results achieved by Zeldovitch et al. with his libasync-smp [8] showed that it is possible to have good performance improvements using coarse-grained parallelism in event-driven programs that have natural opportunities for parallel speedup.

Abdelkhalek's work with Quake, described in Section 3, introduced a multi-threaded version for Quake server that uses a static distribution of work between the available processors. Lack of load balancing and a locking strategy that misuse game objects semantics leaded to suboptimal speedups.

Using the ideas proposed by Zeldovitch and inspired by Abdelkhalek's work we created a multi-process event-driven system that is composed by multiple SPED process, as shown in next section.

4.2   Methodology

The nature of the simulations done by the Quake server determines which events can be executed concurrently. In order to keep the correct game semantics, all events originated in the same client must be executed sequentially and in the same order as in the original server. However, if the simulation of one entity does not interfere on the simulation of another entity, then the execution of both entities' events can be multiplexed by the server without any side effect to the simulation.

Our model decomposes the frame execution in four phases: world processing phase; task scheduling; parallel request and reply processing; synchronization barrier. Phases 1 and 2 runs in a coordinator process and does not run in parallel. The third phase is executed in various processors and the last phase is a synchronization barrier that waits the completion of each process frame. Each server frame is a super-step in the BSP model where phases 1 and 2 compound the input phase, local computation occurs in phase 3 and global communication and a synchronization barrier (output-phase) takes place in phase 4.

World Processing. The world processing phase is the first phase of this model and runs sequentially on the coordinator process. There is an inversion of control in this point of the server execution. The server runs the interpreter for the script programming language called QuakeC used by *id Software* to implement the game semantics. Each entity in the game has its own QuakeC functions associated to them and these functions implements the game semantics for actions and physical simulation for this entity.

Task Scheduling. Our task scheduling uses the entity position update mechanism employed by Quake server.

During the request processing, for every client α that sent an event during the current frame, a list L_α with all entities whose distance from α is not bigger than 256 pixels (α's action range) is computed by the Quake server. This list is used by the server to predict what entities (L_α) can be affected by the event sent by the client (α) being processed.

The game semantics of this list guarantees that if we use the same processor to simulate the events from client α and from all entities in L_α, then we do not need to do any kind of communication between the processors during the request and reply processing. We just need to schedule the simulation of all entities in the same list to the same processor. Our scheduling and load balancing algorithm works as follows.

First, we compute the set of connected components of the undirected graph $G = (V, E)$, where V is the list of all active entities in the current frame and $E = \{(\alpha, \beta) \mid \alpha \in V \wedge \beta \in L_\alpha\}$, i.e. an edge in (α, β) in E exists if the execution of events sent by α can potentially affect the entity β. It is important to note that the original quake code must compute the L_α for all active α clients. We use the already created lists to compute the connected components set using a union-find with path compression algorithm. The overhead to compute this connected components is not significative.

Then, we use the Longest Processing Time First (LPT) [9] heuristic to distribute the jobs among the available processors. Each job is defined as the task of simulating all the events related to the entities in one connected component and the weight of the job is the number of vertices in this connected component. In the context of the Quake simulation, the load produced by one active client is proportional to the number of entities that are near that client. Using the total number of entities in the same connected component as the weight of the job produced good load balancing results, as showed in Section 5.

The use of LPT allowed a lightweight implementation and guarantee good results with many simultaneous clients [9]. Note, however, that this definition of job weight can lead to load unbalance in cases where many entities are concentrated in one spot of the map (a special region of interest in the game map, for instance). We analyzed the influence of the scenario in load balancing in Section 5.2. After the scheduling, each processor starts the execution of request and reply processing.

Request and Reply Processing. The request and reply processing is the most computing intensive phase during the execution of a server frame. It comprehends reading the messages from the network, executing all events sent during the previous frame and computing and sending the replies to the clients. In the sequential version, this phase represents 80% of the server execution time.

Using the ideas proposed in libasync-smp [8], we create one worker process for each available CPU via the `fork()` system call. Each worker process runs as an independent SPED program that executes only the entities in the connected components assigned by the scheduler algorithm to this CPU.

The performance overhead caused by the creation (with `fork()`) of the worker processes is reduced because of the low memory footprint of each independent SPED process and mainly because of the copy-on-write (COW) [10] semantics of `fork()` implemented by the Linux Kernel. The impact of the use of `fork()` will be analyzed in Section 5.

Other implementation detail is related to the operating system scheduler. Using the Linux Trace Toolkit [11], a kernel tracer that generates traces of the execution of programs in an instrumented Linux kernel, we found that the default Linux scheduler does not work well with the workload generated by our parallel Quake server. The worker threads were sometimes scheduled to run in the same processor used by the coordinator process.

Using the new `sched_setaffinity()` and `sched_getaffinity()`, introduced in the Linux kernel 2.6, we could implement a more strict behavior in the kernel scheduler, that forced every new process to start in an idle processor. The performance improvements are shown in Section 5.

Inter-process synchronization. After sending the replies to the clients, the server starts the global communication phase. Each worker thread writes the updated entities state in a pre-allocated shared memory region and the coordinator reads and updates the new state.

Global data structures that depend on entity states, like the Areanode tree, are updated automatically during the next frame, at the world processing phase, so there is no need to update these structures at this point.

Instead of using the usual `wait()` system call to wait for the termination of the worker process, we use an alternative Linux mechanism that allows the coordinator process to proceed while the operating system destroys the auxiliary processes. The world processing phase – which is sequential – runs in parallel with the destruction of the worker process. In order to do that, we used the `clone()` system call without the option `SIGCHLD`. This means that the new child process is independent from the parent process and does not need to wait for a call to `wait()` to be destroyed. In the next section we present the performance analysis of our new parallel server.

5   Experimental Results

In order to evaluate the workload using several clients running the game simultaneously, we implemented[2] an automated Quake player in the Quake client code. Every ten client frames, the player starts to walk. After five client frames, the player chooses randomly one action event to send to the server. The action events available to the automated player are *jump*, a *gun shot* and *change direction* (that changes the direction of the movement being executed). After five more client frames, the client stops, and the procedure is repeated. We believe that this simple scheme are well suited for our analysis purposes. The event send rate is very close to the event send rate of an actual human player.

All results presented in this section are measured during the execution of game sessions with 160 simultaneous clients and all experiments were repeated at least thirty times.

5.1   Environment

The experimental environment that has been used to run the clients is made up of five AMD Athlon™ XP 2800+ (2.25 GHz), 1.0 GB of RAM and connected through a 100 Mbps Ethernet network. We utilized the Linux kernel version 2.6.16.2.

---

[2] Client and server source code can be downloaded from: `http://grenoble.ime.usp.br/~danielc/quake.tar.bz2` under GPL license.

The server was tested in two different systems. One has two Intel Xeon 3.0 GHz processors and 2.0 GB of RAM running Linux kernel version 2.6.17.7 with the patches required by the Linux Trace Toolkit version 0.5.76. The other has 8 UltraSPARC 900 MHz processors and 32.0 GB of RAM running Solaris 9.

5.2   Load Balancing

In order to analyze the results achieved by our load balance algorithm, we must understand the influence of the scenario in the server performance.

Scenario Influence. The scenario influence in our implementation showed up as an important issue since earlier tests. There are three issues related to the scenario that impacts our parallelization scheme:

- **quantity of map entities:** Quake maps have special entities that implements some functionalities related to map elements like elevators;
- **size of reply messages:** if the map defines more small rooms, the size of the reply message tend to be small, but the reply processing tends to be more processor intensive. If the map defines big rooms, the size of messages tends to increase;
- **concentration areas:** some areas of the map can have more importance to the game semantics and tend to concentrate a large number of players.

Small rooms and concentration areas can induce the creation of big connected components in our load balancing algorithm. We used two different scenarios in our analysis. A regular map created to games with 32 simultaneous users distributed by *id Software* called *death32c*[3] and one created by us, that are free of concentration areas.

Job size. In a 160 clients session, using the map *death32c* our load balance algorithm creates in average nine jobs with weight equals to eight in a single frame. With our own created map, without concentration areas, there are twenty jobs with weight equals to four in average. The normalized distribution of this jobs among the processors are shown in Fig. 1.

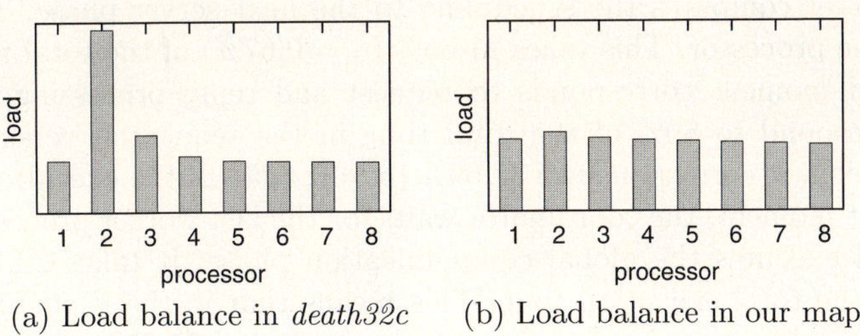

(a) Load balance in death32c   (b) Load balance in our map

Fig. 1. Load balance among processors in the Solaris environment

---

[3] Available at `ftp://ftp.idsoftware.com/idstuff/quakeworld/maps/`.

Performance Results. Section 4.2 describes some implementation details that allowed us to control the operating system scheduling and how we parallelized the destruction of worker process.

Table 1. Performance of implementations evaluated in Linux environment

Evaluated implementation	frames/s	messages/frame	messages/s
using `fork()` and `wait()`	95	33	3.135
using `clone()` and `wait()`	198	18	3.564
using `clone()` without `wait()`	350	11	3.850

By analyzing the traces, we discovered that when the server uses `fork()` and `wait()` primitives just one of the available processors is used almost all the time. The results presented in Table 1 show that using `clone()` and controlling the operating system processor allocation, we increased the frame rate from 95 to 198 frames/s. Parallelizing the destruction of the worker process improved the frame rate from 198 to 350 frames/s.

Preliminary results using a computer with four processors showed that we achieved performance improvements using up to three processors. With more than three the server frame rate starts to drop.

Workload Analysis. The analysis of the execution trace generated by the Linux Trace Toolkit brought some interesting results.

The creation of new worker processes in each server frame represents 2.12% of the execution time in *death32c* and 1.49% in our map. The copy of the memory pages through the Linux copy-on-write mechanism corresponds to 4.69% of the execution time in *death32c* and 5.05% in our map.

We can classify the execution in four different moments according to the server parallelism. During the first moment, the frame starts and the world processing phase runs in parallel with the destruction of the worker process from the last frame. This corresponds to 17,93% of the execution time (with standard deviation $\sigma = 0.51\%$).

During a second moment, the server finishes the world processing phase and then the server computes the scheduling to the next server phase. The server uses only one processor. This takes 21.66% ($\sigma = 0.67\%$) of the total time.

The third moment corresponds to request and reply processing. This two phases correspond to 80% of the total time in the sequential version. In our parallel version, it corresponds to 37.69% ($\sigma = 0.42\%$) of the execution time.

In the last moment, the coordinator waits for the last worker process to finish its jobs and executes the global communication phase. It takes 22.70% ($\sigma = 22.70\%$) of the total execution time. This means that in the *death32c* map we utilize all available processors' processing power during 55.65% of the time.

The results obtained with our map are similar. The first moment takes 19.13% ($\sigma = 0.83\%$) of total time, the second takes 22.77% ($\sigma = 0.97\%$) and the third 34.43% ($\sigma = 0.63\%$) of the total time. The last moment uses 23.65% ($\sigma = 0.97\%$). We use all available processor' processing power in 53.65% of the total time.

6 Conclusion

In this work we investigate the parallelization and a dynamic load balance strategy to interactive, multiplayer game servers. We used QuakeWorld to describe the implementation issues of the proposed methodology in a real world system. Quake is an important representative of this class of applications.

Previous works from other researchers presented a multithreaded architecture with static distribution of work among available processors. This work distribution and the shared memory synchronization scheme did not take advantage from the semantics of the objects being simulated by the game. Their implementation obtained full parallelism only in 40% of the time, mainly because lock contention and load unbalance.

We present a new dynamic load balancing and scheduling mechanism that utilizes the semantics of the simulation executed by the server. The division of the tasks allowed the creation of a Bulk Synchronous Parallel model for QuakeWorld, in which the execution of the computational phase is lockless.

Our experiments analyze the impact of certain external parameters of the simulation such as the virtual map utilized. With kernel instrumentation we measured a total parallelization time improvement from around 40% from previous work to 55% in our implementation.

References

1. Abdelkhalek, A., Bilas, A., Moshovos, A.: Behavior and performance of interactive multi-player game servers. In: Proceedings of the International IEEE Symposium on the Performance Analysis of Systems and Software (ISPASS-2001), Arizona, USA, November 2001, IEEE Computer Society Press, Los Alamitos (2001)
2. Abdelkhalek, A., Bilas, A., Moshovos, A.: Behavior and performance of interactive multi-player game servers. Cluster Computing 6(4), 355–366 (2003)
3. Abdelkhalek, A., Bilas, A.: Parallelization and performance of interactive multi-player game servers. In: Proceedings of 18th International Parallel and Distributed Processing Symposium, apr 2004, p. 72a. IEEE Computer Society Press, Los Alamitos (2004)
4. id Software homepage (2006), http://www.idsoftware.com
5. Shimer, C.: Binary space partition trees (1997) Available at http://www.cs.wpi.edu/~matt/courses/cs563/talks/bsp/bsp.html
6. Valiant, L.: A bridging model for parallel computation. Communications of the ACM 33(8), 103–111 (1990)
7. Pai, V.S., Druschel, P., Zwaenepoel, W.: Flash: An efficient and portable Web server. In: Proceedings of the USENIX 1999 Annual Technical Conference, California, EUA, June 1999, pp. 199–212 (1999),

8. Zeldovich, N., Yip, A., Dabek, F., Morris, R., Mazieres, D., Kaashoek, F.: Multiprocessor support for event-driven programs. In: Proceedings of the 2003 USENIX Annual Technical Conference, June 2003, pp. 239–252 (2003)
9. E.C. Jr., Sethi, R.: A generalized bound on LPT sequencing. In: SIGMETRICS '76: Proceedings of the 1976 ACM SIGMETRICS conference on Computer performance modeling measurement and evaluation, pp. 306–310. ACM Press, New York (1976)
10. Bovet, D.P., Cesati, M.: Understanding the Linux Kernel. 3rd edn. O'Reilly Media (November 2005)
11. Yaghmour, K., Dagenais, M.R.: Measuring and characterizing system behavior using kernel-level event logging. In: Proceedings of the 2000 USENIX Annual Technical Conference, Berkeley, CA, June 2000, pp. 13–26. USENIX Association (2000)

A Parallelisable Multi-level Banded Diffusion Scheme for Computing Balanced Partitions with Smooth Boundaries

François Pellegrini

ENSEIRB, LaBRI and INRIA Futurs
Université Bordeaux I
351, cours de la Libération, 33405 TALENCE, France
`pelegrin@labri.fr`

Abstract. Graph partitioning algorithms have yet to be improved, because graph-based local optimization algorithms do not compute smooth and globally-optimal frontiers, while global optimization algorithms are too expensive to be of practical use on large graphs. This paper presents a way to integrate a global optimization, diffusion algorithm in a banded multi-level framework, which dramatically reduces problem size while yielding balanced partitions with smooth boundaries. Since all of these algorithms do parallelize well, high-quality parallel graph partitioners built using these algorithms will have the same quality as state-of-the-art sequential partitioners.

1 Introduction

Graph partitioning is an ubiquitous technique which has applications in many fields of computer science and engineering, such as workload balancing in parallel computing, database storage, VLSI design or bio-informatics. It is mostly used to help solving domain-dependent optimization problems modeled in terms of weighted or unweighted graphs, where finding good solutions amounts to computing, eventually recursively in a divide-and-conquer framework, small vertex or edge cuts that balance evenly the weights of the graph parts.

Many algorithms have been proposed to compute efficient partitions of any graphs, such as graph or evolutionary algorithms, spectral methods, or linear optimization methods. Basically, all of these methods belong to two distinct classes: global methods, which consider all of the graph data, and local optimization heuristics, which try to improve locally a preexisting partition. Global methods often yield better results, but their costs dramatically increases along with problem size, which makes them practically impossible to use for graphs comprising several tens million vertices, which are the graphs now being considered in many scientific engineering problems.

The multi-level approach [1,2] has been a quite successful attempt to combine both approaches. It consists in repeatedly computing a set of increasingly coarser albeit topologically similar versions of the graph to partition, by finding matchings which collapse vertices and edges, until the coarsest graph obtained is no

A.-M. Kermarrec, L. Bougé, and T. Priol (Eds.): Euro-Par 2007, LNCS 4641, pp. 195–204, 2007.

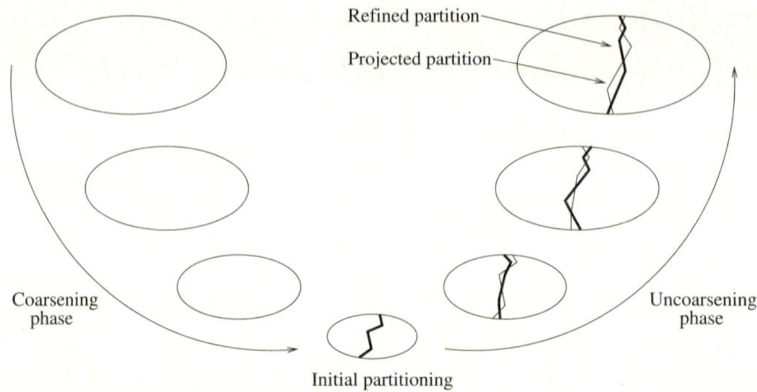

Fig. 1. Multi-level framework for computing a bipartition of a graph

larger than a few hundreds of vertices, then computing a separator on this coarsest graph, and projecting back this separator, from coarser to finer graphs, up to the original graph. Most often, a local optimization algorithm, such as Kernighan-Lin [3] or Fiduccia-Mattheyses [4] (FM), is used in the uncoarsening phase to refine the partition that is projected back at every level, such that the granularity of the solution is the one of the original graph and not the one of the coarsest graph, as illustrated in Figure 1. This approach improves quality over plain graph algorithms, and speed over plain global optimization algorithms, by taking the best of both worlds. Global optimization algorithms can be used on small graphs to give the general direction of the partition to set, and inexpensive local optimization algorithms can be used at low cost on finer graphs with tens of million vertices.

However, the quality of partitions produced by this approach is not as good as the one that would be yielded by plain global optimization algorithms. Coarsening artifacts, as well as the meshing topology of the original graphs, trap local optimization algorithms in local optima of their cost functions, such that frontiers are often made of non-optimal sets of segments, as illustrated in Figure 5.a.

This paper describes an efficient way to integrate diffusion schemes into a multi-level framework, so as to compute partitions with small and smooth frontiers in a time equivalent in magnitude to the one of state-of-the-art local optimization algorithms. It is organized as follows. After presenting related works in Section 2, we introduce in Section 3 our multi-level banded diffusion scheme, and show some partitioning and mapping results, obtained with SCOTCH 5.0, in Section 4. Then comes the conclusion.

2   Related Works

Many authors had already noticed that partitions yielded by local optimization algorithms were not optimal. One of the most vocal communities was the one of the users of iterative linear system solving methods [5], which experienced that such partitions were not fitted for their purpose, as subdomains with longer frontiers or irregular shapes resulted in a larger number of iterations to achieve convergence. To measure the quality of each of the parts, several authors defined a metric called *aspect ratio*, which can be thought in 2D as a measure of the perimeter of a part with respect to the square root of its area. The more compact a part is, the smaller its aspect ratio value is, as ideal parts are of circular shape in the Euclidean space.

In [6], Diekmann *et al.* evidenced such a behavior, and proposed both a measure of the aspect ratio of the parts, as well as a set of heuristics to create and refine the partitions, with the objective of decreasing their aspect ratio. Among these algorithms is a "bubble-growing" algorithm. This algorithm is based on the observation that sets of soap bubbles self-organize so as to minimize the surface of their interfaces, which is indeed what is expected from a partitioning algorithm. Consequently, the authors' idea was to grow, from as many seed vertices as the desired number of parts, a collection of expanding bubbles, by performing breadth-first traversals rooted at these seed vertices. Once every graph vertex has been assigned to some part, each part computes its center based on the graph distance metric. These center vertices are taken as new seeds and the expansion process is started again, until it converges, that is, until centers of subdomains no longer move. An important drawback of this method is that it does not guarantee that all parts will hold the same number of vertices, which requires to call other heuristics in turn to perform load balancing. Also, all of the graph vertices must be visited many times, which makes this algorithm quite expensive, all the more it is combined with costly algorithms such as simulated annealing, and the computation of the aspect ratio requires some knowledge on the geometry of the graphs, which is not always available.

In [7], Meyerhenke and Schamberger further explore the bubble model, and devise a way to grow the bubbles by solving, possibly in parallel, systems of linear equations, instead of iteratively computing bubble centers. This method yields partitions of high quality too, but is very slow, even in parallel [8], and the load balancing problem is also not addressed, which requires to resort to a greedy load balancing algorithm afterwards.

In [9], Wan *et al.* explore a diffusive model, called the *influence model*, where vertices impact their neighbors by diffusing them information on their current state. This model also does not handle load balancing properly.

3   Multi-level Banded Diffusion Scheme

In spite of their better quality, all of the above diffusion schemes have two drawbacks: first, they do not naturally balance loads between parts and second, they are expensive as they involve all of the graph vertices. The method that we propose in this paper addresses both of these problems.

3.1   The Jug of the Danaides

The diffusion scheme that we propose can apply to an arbitrary number of parts, but for the sake of clarity we will describe it in the context of graph

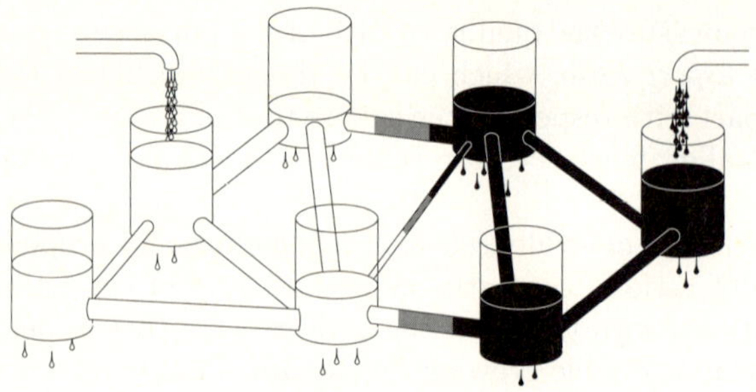

Fig. 2. Sketch of our diffusion model

bipartitioning, that is, with two parts only. We model the graph to bipartition
in the following way, depicted in Figure 2. Nodes are represented as barrels of
infinite capacity, which leak such that one unit of liquid at most drips per unit
of time. When graph vertices are weighted, always with integer weights, the
maximum quantity of liquid to be lost per unit of time is equal to the weight of
the vertex. Graph edges are modeled by pipes of section equal to their weight.
In both parts, a source vertex is chosen, to which a source pipe is connected,
which flows in $\frac{|V|}{2}$ units of liquid per unit of time. Two sorts of liquids are in fact
injected in the system: scotch in the first pipe, and anti-scotch in the second pipe,
such that when some quantity of scotch mixes with the same quantity of anti-
scotch, both vanish. To ease the writing of the algorithm in the bipartitioning
case, scotch is represented by positive quantities and anti-scotch is represented
by negative ones, so that mutual destruction naturally takes place when adding
any two quantities of opposite signs.

The diffusion algorithm performs as outlined in Figure 3. For each time step,
and for each vertex, the amount of liquid (whether scotch or anti-scotch) which
remains after some has leaked is spread across the connecting pipes towards
the neighboring barrels, according to their relative sections. This process could
be iterated until convergence, but in fact it is only performed for a number of
steps sufficient to achieve sign stability. Indeed, we are not interested in complete
convergence, but in the stability of the signs of all content quantities borne by
graph vertices, which indicate whether scotch or anti-scotch dominates in the
barrels, that is, if some vertex belongs to part 0 or 1.

Since $|V|$ units of both liquids are injected on the whole per unit of time, and
since all of the barrels can leak the same overall amount in the same time, the
system is bound to converge, all the more that liquid can disappear by collision
of scotch and anti-scotch. As in the bubble schemes, what is expected is that a
smooth front will be created between the two parts. The purpose of the algorithm
is more to have a global smoothing of the frontier than a strict minimization of
the cut. In fact, unlike all of the algorithms presented in the previous section, our
method privileges load balancing over cut minimization. For this latter criterion,
we rely on an additional feature of our scheme, as explained below.

```
while (number of passes to do) {
    reset contents of new array to 0;
    old[s_0] ← old[s_0] − |V|/2;                /* Refill source barrels */
    old[s_1] ← old[s_1] + |V|/2;
    for (all vertices v in graph) {
        c ← old[v];                             /* Get contents of barrel              */
        if (|c| > weight[v]) {                  /* If not all contents have leaked     */
            c ← c − weight[v] * sign(c);        /* Compute what will remain            */
            σ ← Σ_{e=(v,v')} weight[e];         /* Sum weights of all adjacent edges   */
            for (all edges e = (v, v')) {       /* For all edges adjacent to v         */
                f ← c * weight[e]/σ;            /* Fraction to be spread to v'         */
                new[v'] ← new[v'] + f;          /* Accumulate spreaded contributions   */
            }
        }
    }
    swap old and new arrays;
}
```

Fig. 3. Sketch of the jug-of-the-Danaides diffusion algorithm. Scotch, represented as positive quantities, flows from the source of part 1, while anti-scotch, represented as negative quantities, flows from the source of part 0. For each step, the current and new contents of every vertex are stored in arrays old and new, respectively.

3.2   Band Graphs in a Multi-level Scheme

Our diffusion algorithm, as such, presents two weaknesses: nothing is said about the selection of the seed vertices, and performing such iterations over all of the graphs vertices is very expensive compared to local optimization algorithms which only consider vertices in the immediate vicinity of the frontiers.

To address these two problems concurrently, we use a method we have developed in [10], illustrated in Figure 4. It consists in using a multi-level scheme in which refinement algorithms are not applied to the full graphs but to band graphs that contain vertices that are at most at some small distance, typically 3, from the projected separator. In these band graphs, two additional "anchor" vertices represent all of the removed vertices of each part, and are connected to the last band layers of vertices of each of the parts. The vertex weight of the anchor vertices is equal to the sum of the vertex weights of all of the vertices they replace, to preserve the balance of the two band parts.

The underlying reasoning of this pre-constrained banding scheme is that since every refinement is classically performed by means of a local algorithm, which perturbs only in a limited way the position of the projected separator, local refinement algorithms need only to be passed a subgraph that contains the vertices that are very close to the projected separator. We have experimented that, when performing Fiduccia-Mattheyses refinement on band graphs that contain only vertices that are at distance at most 3 from the projected separators, the quality of the finest separator not only remains constant, but even significantly

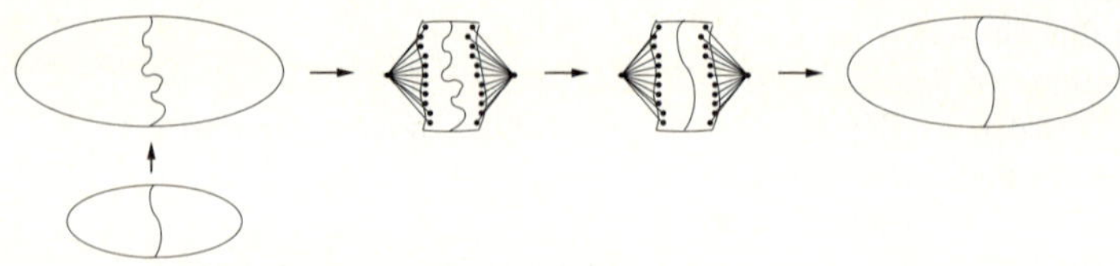

Fig. 4. Multi-level banded refinement scheme. A band graph of small width is created around the projected finer separator, with anchor vertices representing all of the removed vertices in each part. After some optimization algorithm (whether local or global) is applied, the refined band separator is projected back to the full graph, and the uncoarsening process goes on.

improves in most cases. Our interpretation is that this pre-constrained banding prevents local optimization algorithms from exploring and being trapped in local optima that would be too far from the global optimum sketched at the coarsest level of the multi-level process.

Such a banded scheme is ideal for using our diffusion scheme, as anchor vertices represent a natural choice to be taken as seed vertices. Indeed, the most important problem for bubble-growing algorithms is the determination of the seed vertices from which bubbles are grown, which requires expensive processes involving all of the graph vertices [6,7]. Since anchor vertices are connected to all of the vertices of the last layers, the diffused liquids flow as a front as if they originated from the farthest vertices from the frontier, which is indeed what would happen if they flowed from the center of a bubble having the frontier as its perimeter.

3.3   Parallelization

Our diffusion algorithm has the additional interest of being highly scalable. If we assume that full graphs, as well as band graphs, are distributed across processors such that every processor holds a fraction of the graph vertices along with their adjacency lists, like what is done for instance in PT-SCOTCH [11], the parallel version of SCOTCH, the parallel version of the algorithm is straightforward. Every processor performs its local update and computes the contributions it has to spread to distant neighbors, after which these contributions are sent to their destination processors in order to be aggregated. In order to cover communication by computations, vertices that have distant neighbors can be processed first, then communications are started, and vertices with purely local adjacency lists can be processed in the mean time, before received contributions are aggregated.

4   Experimental Results

The diffusion algorithm discussed above has been implemented, as a sequential graph bipartitioning method, in version 5.0 the SCOTCH [12] graph partitioning and static mapping software. Its k-way implementation is not yet available, because it requires more coding, including a k-way band extraction algorithm

Table 1. Description of the test graphs that we use, which all relate to 3D problems, except *thread*. $|V|$ and $|E|$ are the vertex and edge cardinalities, in thousands.

| Graph | Size ($\times 10^3$) | | Average | Graph | Size ($\times 10^3$) | | Average |
	$|V|$	$|E|$	degree		$|V|$	$|E|$	degree
altr4	26	163	12.50	conesphere1m	1055	8023	15.21
audikw1	944	38354	81.28	ocean	143	410	5.71
auto	449	3315	14.77	oilpan	74	1762	47.77
bmw32	227	5531	48.65	pwt	37	145	7.93
body	45	164	7.26	thread	30	2220	149.32
bracket	63	367	11.71				

which does not exist to date. All of the necessary floating-point arithmetic has been implemented in single precision.

The tests were run on a Lenovo ThinkPad T60 laptop, with an Intel dual-core T2400 processor running at 1.8 MHz and 1 Gb of memory. As we ran sequential tests only, the dual-core feature of the processor is not relevant. The test graphs we have used in our experiments are listed in Table 1. These graphs were partitioned into 2 to 128 parts, and the three quality metrics that we consider are the number of cut edges, called **Cut**, a load imbalance ratio equal to the size of the largest part divided by the average size, called **MaCut**, and the maximum diameter of the parts, referred to as **MDi**, which is an indirect metric of the shape of the partition, and is usable even in the case of graphs of unknown or nonexistent geometry. This latter metric is insufficient, as it does not really capture the smoothness of the interfaces, since irregularly shaped parts can still have small diameters; the best proof would have been to run an iterative solver and measure convergence rates basing on the numbers of iterations. This work is in progress.

Three diffusion heuristics were compared against the classical strategy implemented in SCOTCH 4.0, referred to as RMF in the following, which performs recursive bipartitioning with bipartitions computed in a multi-level way, using FM refinement.

The first method, RMBD, uses the same recursive bipartitioning and multilevel strategy, but banded diffusion is performed during the multi-level refinement steps. The results achieved with this method validate our approach: the obtained partitions have very smooth boundaries (see Figure 5.b), and are adequately balanced if the number of diffusion iterations is sufficiently high, as shown in Table 2.

When performing 100 diffusion steps, the average **MaCut** value for RMBD is 1.046, only 1.80 % higher than the one of RMF. However, the maximum diameter **Mdi** is not significantly reduced, and is even increased on average by 4.69% with respect to RMF. This method is also 5.33 times slower than RMF and increases the cut by about 20%, which makes it of little practical use.

We have therefore experimented a second method, RMBDF, where the classical FM algorithm is applied to the band graph after the diffusion algorithm. The idea of this strategy is to benefit from the global optimization capabilities brought by the diffusion algorithm, while locally optimizing the frontier

Table 2. Evolution of the cut size (Δ**Cut**), of the load imbalance ratio (Δ**MaCut**) and of the maximum diameter of the parts (Δ**MDi**) produced by various partitioning heuristics with respect to the RMF strategy, averaged over all test graphs and numbers of parts. Figures below partitioning strategy names indicate the number of diffusion steps performed.

Method	RMBD				RMBDF		RMBaDF
	500	200	100	40	500	40	40
Δ**Cut** (%)	+19.51	+20.01	+18.15	+21.49	+2.26	+3.10	-3.17
Δ**MaCut** (%)	+0.58	+1.12	+1.80	+9.76	-0.95	-0.29	-0.21
Δ**MDi** (%)	+3.86	+1.92	+4.69	+5.43	+2.26	+3.10	-3.24
ΔTime ($\times$)	21.31	9.33	5.33	2.93	21.47	2.99	3.07

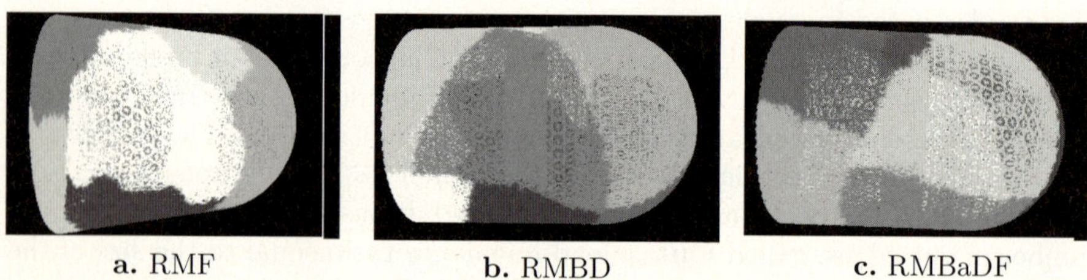

a. RMF   b. RMBD   c. RMBaDF

Fig. 5. Partition of graph **altr4** into 8 parts using three different strategies. The segmented frontiers produced by FM-like algorithms are clearly evidenced in Figure **a**. RMBD produces the smoothest boundaries, as shown in Figure **b**. RMBaDF takes the best of both worlds, in Figure **c**.

afterward. Even when performing 40 diffusion steps only, the smoothness of the boundaries is preserved and parts are more balanced, while the cut is only increased by 3.10% with respect to RMF. This strategy is also only three times slower than RMF, which is extremely fast for a diffusion-based algorithm.

In order to favor the minimization of diameters, we have modified our diffusion method so as to double at each step the amount of liquid borne by every vertex, in an "avalanche"-like process. This method is referred to as "aD". It is no longer bound to converge, and indeed causes overflows for large numbers of diffusion steps, but gives good results for small numbers of iterations. As a matter of facts, we can see in Table 2 that the RMBaDF method is the most efficient one on average, and yields better results than the classical RMF method while still providing smooth boundaries, as evidenced in Figure 5.c.

For the sake of comparison, we compare in Table 3 some of our results against the ones obtained with K-MEΠS. K-MEΠS uses direct k-way partitioning instead of recursive bipartitioning, which usually makes it more efficient when the number of parts increases, and also much faster (from 10 to 20 times). As analyzed in [13], the performance of recursive bipartitioning methods tends to decrease when the number of parts increases, which should limit the efficiency of RMBDF methods for large numbers of parts. A full k-way diffusion algorithm is therefore required.

Table 3. Comparison of the results, in terms of cut size (**Cut**) and maximum diameter of the parts (**MDi**), between three heuristics: multi-level with FM refinement (RMF, as implemented in SCOTCH 4.0), multi-level with banded diffusion and FM refinements (RMBaDF), and K-METiS

Test case		Number of parts						
		2	4	8	16	32	64	128
altr4								
RMF	**Cut**	1688	**3197**	**4978**	**7788**	**11905**	17656	24478
	MDi	50	52	**40**	33	**25**	21	**14**
RMBaD(40)F	**Cut**	**1621**	3203	5017	7776	11980	17669	24831
	MDi	**48**	46	41	**30**	**25**	**18**	14
KMeTiS	**Cut**	1670	3233	4981	8115	12147	**17355**	**24058**
	MDi	**48**	**45**	41	34	26	22	**14**
bmw32								
RMF	**Cut**	17271	**54424**	84222	**120828**	**181844**	**267427**	**394418**
	MDi	93	116	130	106	74	120	68
RMBaD(40)F	**Cut**	16032	54446	**83422**	124945	183454	275594	411154
	MDi	91	130	**96**	**84**	**68**	63	**56**
KMeTiS	**Cut**	**15529**	55506	92658	125686	193169	286111	420965
	MDi	87	**108**	99	87	70	**61**	68

5 Conclusion and Future Work

In this paper, we have presented a diffusion algorithm which, used in a multi-level banded framework, results in smoother partition frontiers and more compact parts. Used in our banded context, this algorithm is fast enough to be used on very large graphs, as it is only about three times slower than classical local optimization schemes. The 2-way sequential version has been integrated in version 5.0 of SCOTCH.

This algorithm is also easily parallelizable and highly scalable, which makes it a very good candidate for the realization of a fast and efficient parallel graph partitioner, taking advantage of the parallel multi-level and band graph extraction routines already developed in PT-SCOTCH in the context of sparse matrix reordering.

Even more than classical FM-like algorithms, this algorithm is constrained by the greedy nature of the recursive bipartitioning scheme, which prevents the global improvement of frontiers computed at previous stages. A full k-way version of the algorithm is therefore under development, which extends the 2-way model by considering k different liquids having the same mutual annihilation properties, such that when p different liquids are mixed in the same barrel, only the most abundant one remains. This behavior is equivalent to the one of our algorithm in the 2-way case. Using a native k-way scheme should also significantly reduce running times compared to recursive bipartitioning. A parallel version is also being developed.

References

1. Hendrickson, B., Leland, R.: A multilevel algorithm for partitioning graphs. In: Proceedings of Supercomputing (1995)
2. Karypis, G., Kumar, V.: A fast and high quality multilevel scheme for partitioning irregular graphs. SIAM J. on Scientific Computing 20(1), 359–392 (1998)
3. Kernighan, B.W., Lin, S.: An efficient heuristic procedure for partitionning graphs. BELL System Technical Journal 49, 291–307 (1970)
4. Fiduccia, C.M., Mattheyses, R.M.: A linear-time heuristic for improving network partitions. In: Proc. 19th Design Automat. Conf. pp. 175–181. IEEE Computer Society Press, Los Alamitos (1982)
5. Vanderstraeten, R., Keunings, R., Farhat, C.: Beyond conventional mesh partitioning algorithms. In: SIAM Conf. on Par. Proc. pp. 611–614 (1995)
6. Diekmann, R., Preis, R., Schlimbach, F., Walshaw, C.: Aspect ratio for mesh partitioning. In: Pritchard, D., Reeve, J.S. (eds.) Euro-Par 1998. LNCS, vol. 1470, pp. 347–351. Springer, Heidelberg (1998)
7. Meyerhenke, H., Schamberger, S.: Balancing parallel adaptive FEM computations by solving systems of linear equations. In: Cunha, J.C., Medeiros, P.D. (eds.) Euro-Par 2005. LNCS, vol. 3648, pp. 209–219. Springer, Heidelberg (2005)
8. Meyerhenke, H., Schamberger, S.: A parallel shape optimizing load balancer. In: Nagel, W.E., Walter, W.V., Lehner, W. (eds.) Euro-Par 2006. LNCS, vol. 4128, pp. 232–242. Springer, Heidelberg (2006)
9. Wan, Y., Roy, S., Saberi, A., Lesieutre, B.: A stochastic automaton-based algorithm for flexible and distributed network partitioning. In: Proc. Swarm Intelligence Symposium, pp. 273–280. IEEE, Los Alamitos (2005)
10. Chevalier, C., Pellegrini, F.: Improvement of the efficiency of genetic algorithms for scalable parallel graph partitioning in a multi-level framework. In: Nagel, W.E., Walter, W.V., Lehner, W. (eds.) Euro-Par 2006. LNCS, vol. 4128, pp. 243–252. Springer, Heidelberg (2006)
    http://www.labri.fr/~pelegrin/papers/scotch_efficientga.pdf
11. Chevalier, C., Pellegrini, F.: PT- SCOTCH: A tool for efficient parallel graph ordering. Parallel Computing (submitted), http://www.labri.fr/~pelegrin/papers/scotch_parallelordering_parcomp.pdf
12. SCOTCH: Static mapping, graph partitioning, and sparse matrix block ordering package, http://www.labri.fr/~pelegrin/scotch/
13. Simon, H.D., Teng, S.H.: How good is recursive bipartition. SIAM J. Scientific Computing 18(5), 1436–1445 (1997)

A Framework for Scheduling
with Online Availability

Florian Diedrich[*,**] and Ulrich M. Schwarz[**]

Institut für Informatik,
Christian-Albrechts-Universität zu Kiel,
Olshausenstr. 40, 24098 Kiel, Germany
{fdi,ums}@informatik.uni-kiel.de

Abstract. With the increasing popularity of large-scale distributed computing networks, a new aspect has to be considered for scheduling problems: machines may not be available permanently, but may be withdrawn and reappear later. We give several results for completion time based objectives: 1. we show that scheduling independent jobs on identical machines with online failures to minimize the sum of completion times is $(8/7 - \epsilon)$-inapproximable, 2. we give a nontrivial sufficient condition on machine failure under which the SRPT (shortest remaining processing time) heuristic yields optimal results for this setting, and 3. we present meta-algorithms that convert approximation algorithms for offline scheduling problems with completion time based objective on identical machines to approximation algorithms for the corresponding preemptive online problem on identical machines with discrete or continuous time. Interestingly, the expected approximation rate becomes worse by a factor that only depends on the probability of unavailability.

1   Introduction

Since the advent of massive parallel scheduling, machine unavailability has become a major issue. Machines can be damaged and thus are not operational until some maintainance is undertaken, or idle time is donated by machines. In either case, it is a realistic assumption that the time of availability of the individual machines can neither be controlled nor foreseen by the scheduler.

Related Problems and Previous Results. In classical scheduling, dynamic machine unavailability has at best played a minor role; however, unreliable machines have been considered as far back as 1975 [1] in the offline setting. Semi-online adversarial variants of the makespan problem were studied by Sanlaville [2] and Albers & Schmidt [3]. In the semi-online setting, the next point in time when machine

---

[*] Research supported in part by a grant "DAAD Doktorandenstipendium" of the German Academic Exchange Service; part of this work was done while visiting the LIG, Grenoble.

[**] Research supported by EU research project AEOLUS, Algorithmic Principles for Building Efficient Overlay Computers, EU contract number 015964.

A.-M. Kermarrec, L. Bougé, and T. Priol (Eds.): Euro-Par 2007, LNCS 4641, pp. 205–213, 2007.

availability may change is known. The discrete time step setting considered here is a special case (i.e. we assume that at time $t + 1$, availability changes) that is closely linked to the unit execution time model. Sanlaville & Liu [4] have shown that *longest remaining processing time* (LRPT) is an optimal strategy for minimizing the makespan even if there are certain forms of precedence constraints on the jobs.

Albers & Schmidt [3] also give results on the true online setting which are obtained by imposing a "guessed" discretization of time.

The general notion of solving an online problem by re-using offline solutions was used by Hall et al. [5] and earlier by Shmoys et al. [6], where $\sum w_j C_j$ and $\min \max C_j$ objectives, respectively, with online job arrivals were approximated using corresponding or related offline algorithms.

Applications. We focus mainly on a setting where there is a large number of individual machines which are mainly used for other computational purposes and donate their idle periods to perform fractions of large computational tasks. The owner of these tasks have no control of the specific times of availability; there even might be no guarantee of availability at all. Our model provides a formal framework to deal with various objective functions in such a situation.

New Results. We first study a setting of discrete time steps where the jobs are known in advance and availability is revealed on-line and give three main results: we prove $(8/7 - \epsilon)$-inapproximability for any online algorithm that tries to minimize the average completion time; we show that the *shortest remaining processing time first* heuristic SRPT solves this problem optimally if the availability pattern is increasing zig-zag; and finally, we present a meta-algorithm for a stochastic failure model that uses offline approximations and incurs an additional approximation factor that depends in an intuitive way on the failure probability. Our approach holds for any offline preemptive algorithm that approximates the makespan or the (possibly weighted) sum of completion times, even in the presence of release dates and in-forest precedence constraints, and slightly weaker results are obtained for general precedence constraints. We also show how our results can be adapted to a semi-online model of time, i.e. breakdowns and preemptions can occur at any time, but the time of the next breakdown is known in advance.

In our probabilistic model of machine unavailability, for every time step t each machine is unavailable with probability $f \in [0, 1)$; this can also be seen as a failure probability of f where there is a probability of $1 - f$ that the machine will be available again at the next time. Notice that in the setting studied in [7], no such probability assumptions are made; here machines just are statically available or unavailable.

2   Definitions

We will consider the problem of scheduling n jobs $J_1, \ldots, J_n$, where a job J_i has processing time p_i which is known *a priori*. We denote the completion time of J_i

in schedule σ with $C_i(\sigma)$ and drop the schedule where it is clear from context. We denote the number of machines available for a given time t as $m(t)$ and the total number of machines $m = \max_{t \in \mathbb{N}} m(t)$. The SRPT algorithm simply schedules the jobs of shortest remaining processing time at any point, preempting running jobs if necessary.

3   Lower Bounds

In the following, we assume that at any time, at least one machine is available; otherwise, no algorithm can have bounded competitive ratio unless it always minimizes the makespan as well as the objective function.

Theorem 1. *For* $Pm, fail|pmtn|\sum C_j$, *no online algorithm has competitive ratio* $\alpha < 8/7$.

Proof. Consider instance I with $m = 3$ and 3 jobs, where $p_1 = 1$ and $p_2 = p_3 = 2$. Let ALG be any online algorithm for our problem. The time horizon given online will start with $m(1) = 1$. We distinguish two cases depending on the behaviour of the algorithm. In *Case 1*, ALG starts J_1 at time 1, resulting in $C_1(\sigma_1) = 1$. Then the time horizon is continued by $m(2) = 3$ and finally $m(3) = m(4) = 1$. It is easy to see that the best sum of completion times the algorithm can attain is 8. However, starting J_2 at time 1 yields a schedule with a value of 7.

In *Case 2*, ALG schedules J_2 at time 1; the same argument works if J_3 is scheduled here. The time horizon is continued by $m(2) = m(3) = 2$. Enumerating the cases shows that the algorithm can only get $\sum C_j \geq 8$, however, by starting J_1 at time 1, an optimal schedule gets $\sum C_j = 7$. $\square$

Theorem 2. *For* $Pm, fail|pmtn|\sum C_j$, *the competitive ratio of SRPT is* $\Omega(m)$.

Proof. For $m \in \mathbb{N}$, m even, consider m machines and m small jobs with $p_1 = \cdots = p_m = 1$ and $m/2$ large jobs with $p_{m+1} = \cdots = p_{m+m/2} = 2$. We set $m(1) = m(2) = m$ and $m(t) = 1$ for every $t > 2$.

SRPT generates a schedule σ_2 by starting the m small jobs at time 1 resulting in $C_j(\sigma_2) = 1$ for each $j \in \{1, \ldots, m\}$. At time 2 all of the $m/2$ large jobs are started; however, they cannot be finished at time 2 but as time proceeds, each of them gets executed in a successive time step. This means that $C_{m+j}(\sigma_2) = 2 + j$ holds for each $j \in \{1, \ldots, m/2\}$. In total, we obtain $\sum C_j = m + \sum_{j=1}^{m/2}(2 + j) = \Omega(m^2)$.

A better schedule will start all long jobs at time 1 and finishes all jobs by time 2, for $\sum C_j \leq 3m$. $\square$

4   SRPT for Special Availability Patterns

Throughout this section, we assume the following availability pattern which has been previously studied for min-max objectives [7]:

Definition 1. *Let* $m : \mathbb{N} \to \mathbb{N}$ *the machine availability function; m forms an* increasing zig-zag pattern *iff the following condition holds:*

$$\forall t \in \mathbb{N} : m(t) \geq \max_{t' \leq t} m(t') - 1 .$$

Intuitively, we may imagine that machines may join at any time and that only one of the machines is unreliable.

Lemma 1. $p_i < p_j$ *implies* $C_i \leq C_j$ *for each* $i, j \in \{1, \ldots, n\}$ *in some suitable schedule σ that minimizes* $\sum C_j$.

Proof. Fix a schedule σ that minimizes $\sum C_j$. Let $i, j \in \{1, \ldots, n\}$ with $p_i < p_j$ but $C_i > C_j$. Let I_i, I_j be the sets of times in which J_i, J_j are executed in σ, respectively. We have $0 < |I_i \setminus I_j| < |I_j \setminus I_i|$ since $p_i < p_j$, $C_i > C_j$. Let $g : I_i \setminus I_j \to I_j \setminus I_i$ be an injective mapping; construct a schedule σ' from σ in the following way: for all $t \in I_i \setminus I_j$ exchange the execution of J_i at time t with the execution of J_j at time $g(t)$. Then we have $C_k(\sigma) = C_k(\sigma')$ for every $k \neq i, j$, furthermore $C_i(\sigma) = C_j(\sigma')$ and $C_j(\sigma) \geq C_i(\sigma')$. Iterating the construction yields the claim. $\square$

Theorem 3. *SRPT is an optimal algorithm if machine availabilities form an increasing zig-zag pattern.*

Proof. Assume a counterexample $J_1, \ldots, J_n$, m such that $\sum_{j=1}^{n} p_j$ is minimal. Fix an optimal schedule σ_{OPT} and an SRPT schedule σ_{ALG} such that the set D of jobs that run at time 1 in only one of $\sigma_{\mathrm{OPT}}, \sigma_{\mathrm{ALG}}$ is of minimal size.

If $|D| = 0$, then σ_{OPT} and σ_{ALG} coincide at time 1 up to permutation of machines. Denote with I the set of jobs running at time 1. By definition of SRPT, $I \neq \emptyset$. We define a new instance by setting

$$\forall j = 1, \ldots, n : p'_j := \begin{cases} p_j - 1, & J_j \in I, \\ p_j, & J_j \notin I \end{cases}$$

$$\forall t \in \mathbb{N} : m'(t) := m(t+1) .$$

It is obvious that $\sum p'_j < \sum p_j$, and so it would be a smaller counterexample. Hence, $D \neq \emptyset$.

We will now argue that there must be some job run by σ_{OPT} that is not run by σ_{ALG} at time 1 and then show that we can exchange these jobs in σ_{OPT} without increasing the objective function value, leading to a counterexample of smaller $|D|$.

Assume that all jobs run by σ_{OPT} also run in σ_{ALG}. Since $|D| > 0$, there is some job in σ_{ALG} that is not in σ_{OPT}, hence σ_{OPT} contains an idle machine. Hence, all n available jobs must run in σ_{OPT} at time 1 by optimality, a contradiction to $|D| > 0$.

Thus there is a job J_j run by σ_{OPT} which is not run by σ_{ALG}. Since not all n jobs can run in σ_{ALG} at time 1 and SRPT is greedy, there must be a different job J_i which is run in σ_{ALG}, but not in σ_{OPT}. By definition of SRPT, we know $p_i < p_j$, and $C_i(\sigma_{\text{OPT}}) \leq C_j(\sigma_{\text{OPT}})$ by Lemma 1.

We will now show that it is always possible to modify σ_{OPT} to execute J_i at time 1 instead of J_j. Preferring J_i will decrease its completion time by at least 1, so we need to show that the total sum of completion times of the other jobs is increased by at most 1.

Case 1: if J_j does not run at time C_i in σ_{OPT}, we have $C_j > C_i$ and we can execute J_i at time 1 and J_j at time C_i. This does not increase the completion time C_j, and any other job's completion time remains unchanged.

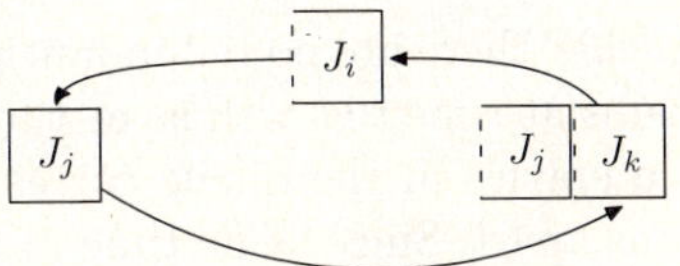

Fig. 1. Case 2 in the proof of Theorem 3

Case 2: The following construction is sketched in Fig. 1. J_j runs at time C_i. We will execute J_i at time 1 and J_j at time $C_j + 1$ for a total change of $\sum C_j$ of at most 0. This can trivially be done if there is an idle machine in σ_{OPT} at time $C_j + 1$. Otherwise, there are $m(C_j + 1)$ jobs running at that time. We still have an idle machine at time C_i, freed up by moving J_i to time 1, and want to displace one of the $m(C_j + 1)$ jobs into this space. We note that we may not choose jobs that are already running at time C_i. There are at most $m(C_i) - 2$ such jobs, since we know J_i and J_j are running at that time. By the increasing zig-zag condition and $C_i \leq C_j < C_j + 1$, we know that

$$m(C_j + 1) > m(C_i) - 1 > m(C_i) - 2,$$

so at least one job, say J_k, is not excluded. Since no part of J_k is delayed, C_k does not increase. $\qquad\square$

5   Algorithm MIMIC

The basic idea of algorithm MIMIC is to use an offline approximation for reliable machines and re-use this given schedule as far as possible. More precisely, let us assume that we already have an α-approximate schedule σ for the offline case for an objective in $\{\sum w_j C_j, \sum C_j, C_{\max}\}$. We will first convert the schedule into a queue Q in the following way; we note that this is for expository reasons and not needed in the implementation.

> For any time t and any machine $i \in \{1, \ldots, m\}$, the job running at
> time t on machine i in schedule σ is at position $(t-1)m + i$ in the   (1)
> queue.

Setup: calculate Q.

Online Execution: if at time t, $m(t)$ machines are available, preempt all currently running jobs, remove the first $m(t)$ entries from Q and start the indicated jobs. Terminate when Q becomes empty.

Fig. 2. Algorithm MIMIC for independent jobs

Note that this means "idle" positions may occur in the queue; this is not exploited. We can now use the queue in our online scheduling algorithm in Fig. 2.

Remark 1. In the generated schedule, no job runs in parallel to itself.

Proof. We assume w.l.o.g. that there are no redundant preemptions in the offline schedule σ, i.e. if a job j runs at time t as well as at time $t+1$, it remains on the same machine. Hence, two entries in the queue corresponding to the same job must be at least m positions apart. Since at no time in the online schedule, more than m machines are available, no two entries of the same job can be eligible simultaneously. $\qquad\square$

We now take a different view upon machine failure: instead of imagining failed machines, we consider that "failure blocks" are inserted into the queue. Since there is a one-to-one correspondence of machine/time positions and queue positions given by (1), this is equivalent to machine failures. We recall an elementary probabilistic fact:

Remark 2 (Expected run length). In our setting (machines failing for a time step independently with probability f), the expected number of failure blocks in front of each non-failure block is $f/(1-f)$.

We can now bound how long the expected completion of a single job is delayed in the online schedule σ':

Lemma 2. *For any job j, we have $E[C_j(\sigma')] \leq \frac{1}{1-f}C_j(\sigma) + 1$.*

Proof. We note that since there are always m machines in the offline setting, there cannot be any blocks corresponding to j in the queue after position $mC_j(\sigma)$ before failure blocks are inserted. This means that after random insertion of the failure blocks, the expected position of the last block of job j is at most $\left(1 + f/(1-f)\right)mC_j(\sigma)$. In light of (1), this yields

$$E[C_j(\sigma')] = \lceil \frac{1}{m}(mC_j(\sigma)\frac{1}{1-f})\rceil \leq \frac{1}{1-f}C_j(\sigma) + 1\,,$$

which proves the claim. $\qquad\square$

Theorem 4. *MIMIC has asymptotic approximation ratio $1/(1-f)$ for independent unweighted jobs and $(1+\epsilon)/(1-f)$ for sum of weighted completion times with release dates.*

This is achieved by exploiting known offline results for different settings (cf. Tab. 1). We should note in particular that since machine failure at most delays a job, our model is applicable to settings with non-zero release dates. We list the result of Kawaguchi & Kyan mainly because it is obtained by a very simple *largest ratio first* heuristic, as opposed to the more sophisticated methods of Afrati et al. [8], which gives it a very low computational complexity.

Table 1. Selection of known offline results that can be used by MIMIC

Setting	Source		ax. ratio
$P\|pmtn\|\sum C_j$	McNaughton	[9]	1
$P\|\|\sum w_j C_j$	Kawaguchi and Kyan	[10]	$(1+\sqrt{2})/2$
$P\|r_j, pmtn\|\sum w_j C_j$	Afrati et al.	[8]	PTAS
$P\|r_j, prec, pmtn\|\sum w_j C_j$	Hall et al.	[5]	3

5.1 Handling Precedence Constraints

We note that our results stated so far cannot be simply used if there are general precedence constraints, as the following example shows:

Example 1. Consider four jobs $J_1, \ldots, J_4$ of unit execution time such that $J_1, J_2 \prec J_3, J_4$. The queue $J_1 J_2 J_3 J_4$ corresponds to an optimal offline schedule. If a failure occurs during the first time step, we have $C_2 = 2$ and MIMIC schedules one of J_3, J_4 in parallel to J_2.

The main problem is that jobs J_2 and J_3 have a distance of $1 < m = 2$ in the queue, so they may be scheduled for the same time step online even though there is a precedence constraint on them. Conversely, if the distance is at least m, they are never scheduled for the same time step.

Since our setting allows free migration of a job from one machine to another, we can sometimes avoid this situation: if the precedence constraints form an *in-forest*, i.e. every job has at most one direct successor, we can rearrange the jobs in the following way: if at a time t, a job J_0 is first started, and $J_1, \ldots, J_k, k \geq 1$ are those of J_0's direct predecessors that run at time $t - 1$, w.l.o.g. on machines $1, \ldots, k$, we assign J_0 to machine k. This ensures that the distance in the queue from J_0 to J_k and hence also to $J_1, \ldots, J_{k-1}$ is at least m.

If we have general precedence constraints, we cannot guarantee that all jobs are sufficiently segregated from their predecessors, as seen above. We can, however, use a parameterized modification to extend results to the case of general precedence constraints and only incur an additional factor of $1 + \epsilon$: fix $k \in \mathbb{N}$ such that $k^{-1} \leq \epsilon$. We will use time steps of granularity k^{-1} in the online schedule.

For the offline instance, we first scale up all execution times by k, essentially shifting them to our new time units, and then increase all execution times by 1. Since all scaled execution times were at least k to begin with, this gives at most an additional factor of $1 + k^{-1} \leq 1 + \epsilon$ for the objective function value. In the queue, we leave the additional time slot empty. This forces the queue distance of jobs which are precedence-constrained to be at least m, hence, the generated schedule is valid.

6   Continuous Time and the Semi-online Model

In this section, we will adapt the idea of algorithm MIMIC—reusing an offline approximation—to the more general semi-online setting by methods similar to Prasanna & Musicus' continuous analysis [11]. In the semi-online setting, changes of machine availability and preemptions may occur at any time whatsoever, however, we know in advance the next point in time when a change of machine availability will take place. We can use this knowledge to better convert an offline schedule into an online schedule, using the algorithm MIMIC′ in Fig. 3: during each interval of constant machine availability, we calculate the area $m(t)\delta$ we can schedule. This area will be used up in time $m(t)\delta/m$ in the offline schedule. We take the job fractions as executed in the offline schedule and schedule them online with McNaughton's wrap-around rule [9]. Precedence constraints can be handled by suitable insertion of artificial interruptions.

Setup: Calculate offline schedule σ_{offline}; $t_{\text{offline}} := 0$
Online Execution: if $m(t)$ machines are available during the interval $[t, t + \delta)$:

- Set $\delta_{\text{offline}} := \min\{m(t)\delta/m, \min\{C_j(\sigma_{\text{offline}}) - t_{\text{offline}}|t_{\text{offline}} \leq C_j(\sigma_{\text{offline}})\}\}$.
- Set $\delta_{\text{online}} := m\delta_{\text{offline}}/m(t)$.
- Schedule all job fractions that run in the interval $[t_{\text{offline}}, t_{\text{offline}} + \delta_{\text{offline}})$ in σ_{offline} in the online interval $[t, t + \delta_{\text{online}})$ using McNaughton's wrap-around rule.
- Set $t_{\text{offline}} := t_{\text{offline}} + \delta_{\text{offline}}$ and proceed to time $t + \delta_{\text{online}}$.

Fig. 3. Algorithm MIMIC′

Since at time $C_j(\sigma_{\text{offline}})$, a total area of $mC_j(\sigma_{\text{offline}})$ is completed, we have the following bound on the online completion times $C_j(\sigma_{\text{online}})$:

$$\int_0^{C_j(\sigma_{\text{online}})} m(t)dt \leq mC_j(\sigma_{\text{offline}}). \tag{2}$$

If we set $\forall t : E[m(t)] = (1 - f)m$ to approximate our independent failure setting above, equation (2) simplifies to $C_j(\sigma)(1 - f)m \leq mC_j(\sigma_{\text{offline}})$, which again yields a $1/(1 - f)$-approximation as in Lemma 2, thus we obtain the following result.

Theorem 5. *Algorithm MIMIC′ non-asymptotically matches the approximation rates of MIMIC for the continuous semi-online model.*

7   Conclusion

In this paper we have presented a simple yet general framework permitting the transfer of results from offline scheduling settings to natural preemptive online extensions. We have studied the min-sum objectives $\sum C_j$ and $\sum w_j C_j$; we have also considered the behaviour of $C_{\max}$, which permits the transfer of bicriteria results. We remark that algorithm MIMIC permits a fast and straightforward implementation which indicates its value as a heuristic for practice where real-time data processing is important; this holds in particular when the underlying offline scheduler has low runtime complexity.

Acknowledgements. The authors thank Jihuan Ding for many fruitful discussions on lower bounds, and the anonymous referees for their valuable comments which helped improve the quality of the exposition.

References

1. Ullman, J.D.: NP-complete scheduling problems. J. Comput. Syst. Sci. 10(3), 384–393 (1975)
2. Sanlaville, E.: Nearly on line scheduling of preemptive independent tasks. Discrete Applied Mathematics 57(2-3), 229–241 (1995)
3. Albers, S., Schmidt, G.: Scheduling with unexpected machine breakdowns. Discrete Applied Mathematics 110(2-3), 85–99 (2001)
4. Liu, Z., Sanlaville, E.: Preemptive scheduling with variable profile, precedence constraints and due dates. Discrete Applied Mathematics 58(3), 253–280 (1995)
5. Hall, L.A., Shmoys, D.B., Wein, J.: Scheduling to minimize average completion time: Off-line and on-line algorithms. In: Proceedings of the Seventh Annual ACM-SIAM Symposium on Discrete Algorithms, pp. 142–151. ACM Press, New York (1996)
6. Shmoys, D.B., Wein, J., Williamson, D.P.: Scheduling parallel machines on-line. SIAM J. Comput. 24(6), 1313–1331 (1995)
7. Sanlaville, E., Schmidt, G.: Machine scheduling with availability constraints. Acta Informatica 35(9), 795–811 (1998)
8. Afrati, F.N., Bampis, E., Chekuri, C., Karger, D.R., Kenyon, C., Khanna, S., Milis, I., Queyranne, M., Skutella, M., Stein, C., Sviridenko, M.: Approximation schemes for minimizing average weighted completion time with release dates. In: Proceedings of FOCS '99, pp. 32–44 (1999)
9. McNaughton, R.: Scheduling with deadlines and loss functions. Mgt. Science 6, 1–12 (1959)
10. Kawaguchi, T., Kyan, S.: Worst case bound of an LRF schedule for the mean weighted flow-time problem. SIAM Journal on Computation 15(4), 1119–1129 (1986)
11. Prasanna, G.N.S., Musicus, B.R.: The optimal control approach to generalized multiprocesor scheduling. Algorithmica 15, 17–49 (1996)

Scheduling File Transfers for Data-Intensive Jobs on Heterogeneous Clusters[*]

Gaurav Khanna[1], Umit Catalyurek[2], Tahsin Kurc[2],
P. Sadayappan[1], and Joel Saltz[2]

[1] Dept. of Computer Science and Engineering
{khannag,saday}@cse.ohio-state.edu
[2] Dept. of Biomedical Informatics
The Ohio State University
{umit,kurc}@bmi.osu.edu, Joel.Saltz@osumc.edu

Abstract. This paper addresses the problem of efficient collective scheduling of file transfers requested by a batch of tasks. Our work targets a heterogeneous collection of storage and compute clusters. The goal is to minimize the overall time to transfer files to their respective destination nodes. Two scheduling schemes are proposed and experimentally evaluated against an existing approach, the Insertion Scheduling. The first is a 0-1 Integer Programming based approach which is based on the idea of time-expanded networks. This scheme achieves the minimum total file transfer time, but has significant scheduling overhead. To address this issue, we propose a maximum weight graph matching based heuristic approach. This scheme is able to perform as well as insertion scheduling and has much lower scheduling overhead. We conclude that the heuristic scheme is a better fit for larger workloads and systems.

1   Introduction

Data centers consisting of collections of storage and compute clusters provide a viable environment for hosting large scientific datasets and providing analysis services. Scientific datasets are typically stored as a set of files, distributed across multiple storage nodes. Data analysis is carried out by downloading subsets of datasets from storage systems to compute systems. Analysis tasks are then executed on local data. A data center should be able to support efficient execution of batches of analysis tasks, in which a task requests a set of files and the sets of files requested by different tasks may overlap (i.e., tasks may share files). Efficient execution of such a batch of tasks involves addressing two key problems. The first problem is the mapping of tasks to compute nodes such that the volume of overall data transfer is minimized. The second one is the transfer of files from storage nodes to compute nodes. The staging of files should be carefully scheduled and executed to minimize the contention, while accounting for the topology and the heterogeneity of bandwidths in the system.

---

[*] This research was supported in part by the National Science Foundation under Grants #CCF-0342615 and #CNS-0403342.

A.-M. Kermarrec, L. Bougé, and T. Priol (Eds.): Euro-Par 2007, LNCS 4641, pp. 214–223, 2007.

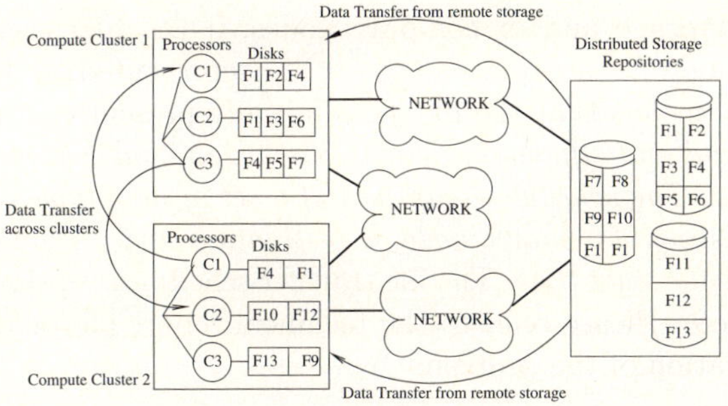

Fig. 1. Scheduling problem

In our earlier work, we looked at the problem of scheduling and mapping a batch of data-intensive tasks [1]. This paper addresses the file transfer scheduling problem, given a mapping of tasks to nodes. In other words, it focuses on the second phase of the overall problem. We propose two approaches. The first one formulates the problem using 0-1 Integer Programming (IP) by employing the concept of time-expanded networks [2]. The second approach employs max-weighted graph matching to yield a schedule which tries to minimize contention and maximize the parallelism in the system. We carry out an experimental evaluation of these algorithms, comparing them against the insertion scheduling heuristic [1]. Our results show that the IP formulation results in better schedules, but introduces high scheduling overhead. The second approach performs as well as the insertion scheduling and also takes much less time to compute a schedule thereby making it a good choice for larger workloads and systems.

2   Problem Definition and Related Work

Problem Definition: We target batches consisting of independent sequential tasks. Each task requests a subset of files from a dataset and can be executed on any of the compute nodes. The files required by a task should be staged to the node where the task is allocated. We assume a single port model wherein multiple requests to the same node are serialized. A heterogeneous multi-cluster environment consisting of compute and storage nodes is represented by graph $G = (V, E)$, referred to here as a *platform graph*. Here, V is the set of nodes and E is the set of edges. We assume that the graph G is connected. We employ a store and forward model of file transfer which implies that if a file f_ℓ needs to be transferred from a node v_i to a non-adjacent node v_j, the file is routed along one of the multiple possible paths between v_i and v_j. A copy of the file is left in each intermediate node thereby increasing the number of replicas of each file leading to potentially higher parallelism for other requests.

The input to the scheduler is a set of two tuples $R = \{< f_\ell, v_i >\}$ representing that the file f_ℓ needs to be transferred to the node v_i. The initial mapping of files to nodes (storage and/or compute nodes, if the file has been replicated on a compute node for a previous request) is represented by the set $D = \{<f_\ell, v_j >\}$, which means that the file f_ℓ is initially present on the node v_j. *Our objective is to find and efficiently execute a schedule that will minimize the total file transfer time. The schedule comprises of a set of four tuples $< v_i, v_j, f_\ell, t >$, each tuple consisting of a source node, a destination node, a file to be transferred and the file transfer start time.* The file transfer scheduling optimization problem is $NP - complete$. Please refer to the technical report [3] for the proof. Fig. 1 shows an illustration of the problem.

Related Work: Giersch et al. [4] address scheduling of a collection of tasks sharing files onto heterogeneous clusters. Their work proposed extensions to the MinMin heuristic [5] to lower the scheduling cost. In our past work, we looked at the problem of scheduling a batch of data-intensive tasks on homogeneous clusters [1]. Our prime focus was to address the first phase of the overall problem that is to accomplish task mapping. GridFTP [6] is a protocol which enables high performance data movement by employing techniques like multiple TCP streams per transfer and striped transfers. In contrast, our work is complementary to GridFTP and can be applied in conjunction with it.

3   Scheduling Schemes

3.1   Insertion Scheduling Based Approach

Giersch et al. [4] employ an insertion scheduling scheme to schedule file transfers. In our past work [1], we developed a Gantt chart based heuristic based on a similar idea which is applied in the conjunction with the task mapping schemes. The basic idea was to memorize the duration and the start time of file transfers for each link and use this information to generate schedules for pending requests.

The transfer completion time (TCT) to transfer a file f_ℓ from a node v_i to a node v_j, $TCT_{\ell ij}$, is estimated as the sum of the earliest time a transfer can start and the actual transfer time. At each step, the algorithm chooses a file, destination node pair $< f_\ell, v_k >$ and schedules the transfer of file f_ℓ to node v_k. To accomplish this, it finds the expected transfer completion time TCT of each file in the input request set on its respective destination node and among them chooses the $< f_\ell, v_k >$ pair with the minimum expected transfer completion time. This process is then repeated until all the file transfers have been scheduled. For a platform graph $G = (V, E)$ and an input request set R, the complexity of insertion scheduling is $O(|R|^2 \times |V| \times (|E| + |V|log(|V|)))$ [3].

3.2   0-1 Integer Programming-Based Approach

In the following discussion we use subscripts i and j for nodes, e for edges, ℓ for files and t for time. We represent time in discrete units and the smallest unit of time represents the least time taken to transfer a file from a source node to a destination node among all files and node pairs. In our formulation, we make use of the concept of time-expanded networks [2]. A time-expanded network captures the temporal aspects of network flow such that flows over time in the original network can be treated as flows in the time-expanded network. Let T^* denote the upper bound on the total completion time of all the file transfers. For each file f_ℓ to be transferred, we construct a time expanded network $G'_l = (V'_l, E'_l)$ as follows. For each node v_i in the system and each time $t = 0, ..., T^*$, we add a vertex v_{it} to the graph G'_l. For an edge $e = \{v_i, v_j\}$ connecting any two nodes v_i and v_j, $Time_{lij}$ represents the transfer time of file f_ℓ on the link $e = \{v_i, v_j\}$. We add a directed edge $(v_{it}, v_{jt'})$ to the time expanded network G'_l if $t' \leq T^*$, where $t' = t + Time_{lij}$. The objective function of the 0-1 IP scheme is to the minimize the overall file transfer time $FileTransferTime = \sum_{(\forall t)} Busy_t$ under a set of constraints. It solves for the following set of variables: 1) $Busy_t$, which is a binary variable. $Busy_t = 1$, if there is a file transfer which is finished at time t or a later point in time. 2) $X_{\ell it}$, which is a binary variable. $X_{\ell it} = 1$, if file f_ℓ is available on node v_i at time t, and 0 otherwise. 3) $Y_{\ell e}$, which is a binary variable. $Y_{\ell e} = 1$, if the edge e in the time expanded network G'_l is used to transfer the file f_ℓ, and 0 otherwise. The constraints for 0-1 IP are:

At t=0, certain files are present on certain nodes.

$$(\forall \ell)(\forall i, < f_\ell, v_i >\in D)X_{\ell i 0} = 1 \tag{1}$$

A file f_ℓ is present on a node v_i at time t either if it is already present on the node at time $t - 1$ or due to the file transfer of the file f_ℓ to the node v_i from one of the nodes v_j such that the file transfer is finished at time t. $I_{\ell it}$ is the set of directed edges incident on the node v_{it} in the time-expanded network G'_l.

$$(\forall \ell)(\forall i)(\forall t)X_{\ell it} = (\sum_{(\forall e, e \in I_{\ell it})} Y_{\ell e}) + X_{\ell it-1} \tag{2}$$

At time $t = T^*$, each file must be present at its respective destination nodes.

$$(\forall \ell)(\forall i, < f_\ell, v_i >\in R)X_{\ell i T^*} = 1 \tag{3}$$

A file f_ℓ can be transferred from the node v_i at time t only if its present on the node v_i at time t. In addition, at most one outgoing arc is allowed from a node v_i at time t. $O_{\ell it}$ is the set of directed edges outgoing from the node v_{it} in the time-expanded network G'_l.

$$(\forall \ell)(\forall t)(\sum_{(\forall e, e \in O_{\ell it})} Y_{\ell e}) \leq X_{\ell it} \tag{4}$$

A file f_ℓ once staged to a node v_i remains available on the node.

$$(\forall \ell)(\forall i)(\forall t)X_{\ell it} \leq X_{\ell it+1} \tag{5}$$

Each node v_i can be involved in at most one send or receive at a time t. Let $C_{\ell it}$ be the set of all incoming and outgoing arcs of the time-expanded network G'_l that would make the node v_i busy during the time $[t, t + 1)$. Note that this

---

Algorithm 1. Maximum Weighted Matching based Scheduling Heuristic

---

Require: Platform $G = (V, E)$ and an input request set consisting of $< f_\ell, v_i >$ pairs
1: **while** there exists a pending request **do**
2:     **for** each pending request $< f_\ell, v_i >$ **do**
3:         Run the Modified Dijkstra's algorithm on Graph G for the request $< f_\ell, v_i >$. Let $Path_{\ell i}$ denote the file transfer path which yields the earliest completion time for the request.
4:     Create a file transfer graph $G' = (V', E')$ as follows.
5:     **for** each pending request $< f_\ell, v_i >$ **do**
6:         Let nodes v_{i1} and v_{i2} comprise the first hop of the file transfer path $Path_{\ell i}$.
7:         $V' = V' \cup \{v_{i1}, v_{i2}\}$.
8:         Add an edge with weight $\frac{1}{TCT}$ between v_{i1} and v_{i2} in G'. Here, TCT denotes the minimum completion time of the request
9:     Run the Max-weighted matching algorithm on the Graph G' to get a Matching
10:    Schedule the chosen set of edges belonging to the Matching

---

includes all arcs that start at time $t' \leq t$, end at a time $t' \geq (t+1)$, and having v_i as its source or target node.

$$(\forall t)(\forall i)(\sum_{(\forall \ell)(\forall e, e \in C_{\ell it})} Y_{\ell e}) \leq Busy_t \tag{6}$$

The objective function is such that the network may be busy for, say, 5 time steps with $Busy_1 = ... = Busy_5 = 1$, be idle for the next 10 time steps, $Busy_6 = ... = Busy_{15} = 0$, and finishing the transfer in the next 2 time steps, $Busy_{16} = Busy_{17} = 1$. This would lead to objective value 7, which is seemingly wrong since the network is busy even at time $t = 17$. To address this problem, we introduce the following constraint.

$$(\forall t)Busy_t \geq Busy_{t+1} \tag{7}$$

3.3 Max-Weighted Matching Based Scheduling Scheme (MMSS)

The MMSS is an iterative algorithm and employs max-weighted matching as illustrated in Algorithm 1. For a graph $G = (V, E)$, we define the set $M \in E$ as a matching of Graph G, if no two edges in M have a common vertex. The weight of the matching is the sum of the weights of the edges which form the matching. A maximum weighted matching is defined as the matching of maximum weight.

In each iteration, the algorithm creates a file transfer graph $G' = (V, E')$ whose vertices $v' \in V$ correspond to the nodes in the system and whose edges e' correspond to file transfers. Each input request can possibly consist of multiple hops, i.e., a set of intermediate nodes can be used to transfer the file to its final destination. An input request $< f_\ell, v_i >$ is considered as pending, if the file f_ℓ is not yet present on the node v_i. For each such pending request, the algorithm computes the path $Path_{\ell i}$ of file transfer which yields the minimum transfer time for the file f_ℓ onto the node v_i. This step requires running a variant of

Dijkstra's shortest path algorithm on G to find which one of the multiple possible sources to stage the file from. The file transfer corresponding to the first hop of the path $Path_{\ell i}$ is then added as an edge to G' between the corresponding pair of vertices in G'. Note that for a multi-hop request, the first hop changes with time as the file gets closer to its destination node. The weight of an edge in the file transfer graph corresponding to an input request is $\frac{1}{TCT}$ where TCT is the expected minimum completion time of the request. The idea behind this weight assignment is to give higher priority to file transfers which can finish early. Finally, the algorithm employs max-weighted matching on the file transfer graph to obtain a set of non-contending ready file transfers and schedules them. In this work, we modify the Dijkstra's algorithm to take into account the wait times of the source and the destination nodes as well as the link bandwidths. Once a file transfer is scheduled between a source and a destination node, the wait time on both the nodes is incremented by the expected file transfer time, which is simply the size of the file divided by the bandwidth. Therefore, the transfer completion time for an unscheduled transfer between a source-destination pair is the sum of the earliest idle time (which is simply the maximum of the wait times on the two nodes) and the expected file transfer time. This procedure works iteratively until all the file transfers have been scheduled.

We employ Gabow's $O(|V|^3)$ implementation of the Edmond's algorithm for computing maximal matching on graphs [7]. The worst case complexity of the matching based heuristic is $O(|R| \times (|V|^4))$. For furthur details, please refer to the technical report [3]. The number of input requests $|R|$ is typically orders of magnitude higher than the number of vertices $|V|$. Therefore, in practice, the matching based heuristic is expected to perform much faster than the insertion scheduling approach presented in Section 3.1.

4   Experimental Results

For experimental evaluation, we used both randomly generated workloads as well as workloads derived from two application classes: satellite data processing (**SAT**) and biomedical image analysis (**IA**) [1,8]. For **IA**, we implemented a program to emulate studies that involve analysis on images obtained from MRI and CT scans. A 1 Terabyte dataset was created which emulates a study involving 2000 patients and images acquired over several days from MRI and CT scans. The sizes of images were 10 MB and 100 MB for MRI and CT scans, respectively. Images were distributed among all the storage nodes in a round robin fashion. To generate datasets for **SAT**, we employed an emulator developed in [8]. For **SAT**, the 250GB dataset was distributed across the storage nodes using a Hilbert-curve based declustering method. Each file in the dataset was 50 MB. To generate the input file request set for the two application domains, we apply our task-mapping technique [1] to map a batch of tasks onto a set of compute nodes. Since each task is associated with a set of files, the task mapping provides information about the destination nodes for each file.

In addition to the three schemes; the integer programming (IP), the graph matching based approach (*Matching*), and the insertion scheduling approach (*Insertion*), we implemented a base scheme, referred to here as *Indep_local*. This is a relatively simpler scheduling scheme where each destination node knows the set of files it needs and makes requests for each of them one by one. The destination nodes acting as clients do not interact with each other before making their respective requests. In the experiments, IP uses the feaspump solver [9], available through the NEOS Optimization Server [10] to compute the schedule. The upper bound T^* defined in Section 3.2 was set to be value obtained by *Matching*. Since the feaspump solver gives feasible solutions which may not be optimal, we apply binary search in conjunction with the solver to get the optimal value of the objective function.

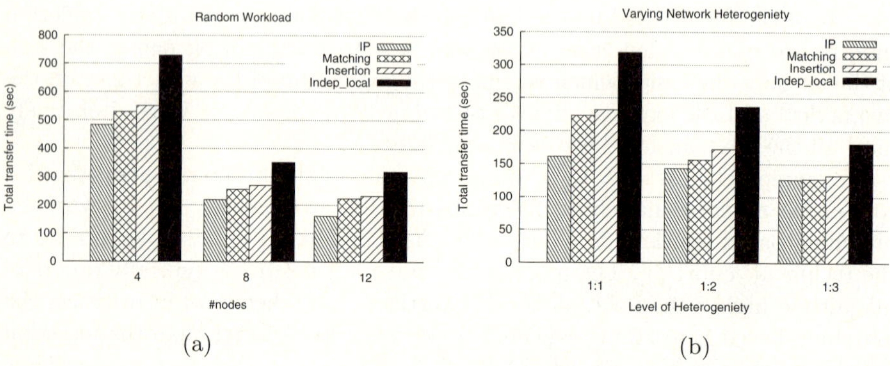

(a)   (b)

Fig. 2. (a) Performance of all schemes for a randomly generated workload, (b) Performance of all schemes with varying network heterogeneity

Fig 2(a) compares the different schemes in terms of the overall file transfer time (in seconds). These experiments were conducted on randomly generated workloads. The initial distribution of files on the nodes was also chosen randomly. The input request set consisted of 50 file transfers each involving 1GB files. The results show that IP results in the best schedule. This is because IP is able to integrate the global information of the input request set and the platform topology into the objective function. *Matching* performs quite similar to *Insertion*; it yields a schedule that minimizes end-point contention, since the graph matching ensures that each at step, a set of non-conflicting transfers are chosen. *Indep_local* performs the worst as expected.

Fig 2(b) shows the performance of the schemes when network heterogeneity is varied. This experiment was conducted using 12 nodes by employing the workload used in Figure 2(a). Since the workload was random, each of the 12 nodes could possibly act as sources for some files and destinations for others. We abstracted the platform graph (see Section 2) as a fully-connected network and emulated heterogeneity by randomly choosing half of the links to have double

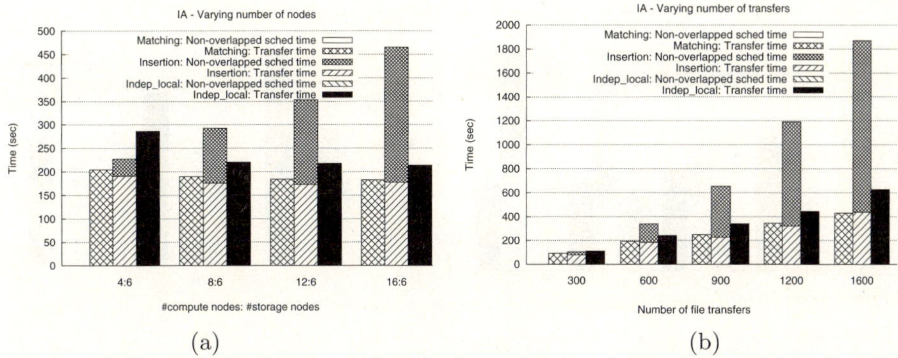

Fig. 3. Performance of different schemes for IA workload with (a) varying number of nodes, (b) varying number of file transfers

and triple the communication bandwidth as compared to the remaining links. These are denoted by $(1:2)$ and $(1:3)$ in the results; $(1:1)$ corresponds to a homogeneous network case. On the cluster machine used for the experiments, the network heterogeneity is achieved by transferring proportionally smaller amounts of data on the faster links followed by locally padding the rest of bytes to the file. The experimental results show that the performance gap between IP and the other approaches decreases with increasing heterogeneity. At low heterogeneity, IP performs better because it explores a much larger search space thereby achieving a better global solution. However, as the extent of heterogeneity increases, the search space of efficient solutions becomes more and more restricted to faster links and all the schemes take that into account.

Figure 3 shows the scalability results with varying number of compute nodes and varying number of input requests. Since IP takes too long to execute even for moderately-sized workloads, we show results for the other three schemes only. To analyze the scalability of *Matching* with respect to the number of compute nodes, we ran experiments with an **IA** workload consisting of around 250 tasks over 4, 8, 12, 16 compute nodes and 6 storage nodes. Note that the Figure 3 shows the performance in terms of two metrics, namely the total file transfer time and the non-overlapped scheduling time. The non-overlapped scheduling time is the difference between the end-to-end execution time and the total file transfer time. The end-to-end execution time is defined as the elapsed time between the instant when the scheduler accepts a batch of requests to the instant when all the requests have been completed. In other words, the non-overlapped scheduling time is the perceived scheduling overhead.

For *Insertion*, the end-to-end execution time is simply the sum of the scheduling time and the total file transfer time. *Insertion* generates the entire schedule once at the beginning followed by the transfer of files. For *Matching*, on the other hand, the schedule is generated iteratively while the file transfers are taking place. Therefore, the non-overlapped scheduling time is negligible and the end-to-end execution time closely matches the overall file transfer time.

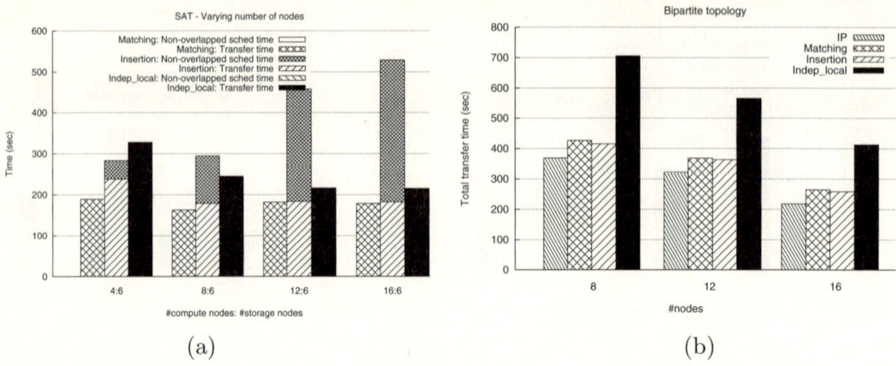

(a)   (b)

Fig. 4. (a) Performance of different schemes for SAT workload with varying number of nodes, (b) Performance of all schemes by employing a bipartite platform graph

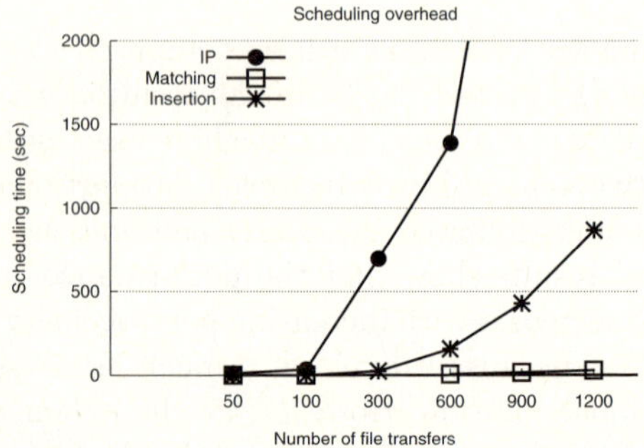

Fig. 5. Scheduling overhead for all schemes

Figure 3(a) shows that *Matching* performs significantly better than *Insertion* in terms of the end-to-end execution time. This is because, non-overlapped scheduling time in *Matching* is very small. In terms of the total file transfer time, the performance of *Matching* is quite close to *Insertion*. Figure 3(b) shows the results with increasing number of requests for an **IA** workload. We observe that *Matching* is able to perform much better than *Insertion*. This is because *Insertion* has a quadratic dependence of its complexity on the number of requests as opposed to *Matching* which has a linear dependence.

Fig 4(a) shows the performance results for a **SAT** workload in terms of the total file transfer time and the non-overlapped scheduling time. We observe that *Matching* outperforms *Insertion* by upto 20% in terms of the total file transfer time. In terms of the end-to-end execution time, *Matching* does significantly better. Fig 4(b) shows the results in terms of total file transfer time on a bipartite platform graph for all the schemes. The bipartite topology was emulated by having two distinct subsets of nodes with links only across the two sets. We employed a randomly generated workload with multiple destination node mappings for each file. The results show expected trends except that the performance of *Indep_Local* is much worse than the other approaches. This is because each file needs to be sent to multiple different destinations, thereby leading to increased contention due to multiple simultaneous requests for the same file. Fig 5 shows the scheduling times for various schemes. The scheduling time shown is the actual time spent in generating the schedule. *IP* has a high scheduling overhead for larger configurations, due to its exponential complexity. The scheduling time of *Insertion* is higher than that of *Matching*, as expected.

5   Conclusion

We proposed two strategies for collectively scheduling a set of file transfer requests made by a batch of data-intensive tasks on heterogeneous systems - one approach employs 0-1 Integer Programming and the other employs max-weighted matching. The results show that the IP formulation results in the best overall file transfer time. However, it suffers from high scheduling time. The matching based approach results in slightly higher file transfer times, but is much faster than the IP based approach. Moreover, the matching based approach is able to match the performance of Insertion scheduling with a much lower scheduling overhead. Our conclusion is that the IP based approach is attractive for small workloads, while the matching based approach is preferable for large scale workloads.

References

1. Khanna, G., Vydyanathan, N., Kurc, T., Catalyurek, U., Wyckoff, P., Saltz, J., Sadayappan, P.: A hypergraph partitioning based approach for scheduling of tasks with batch-shared I/O. In: Proc. of CCGrid'05, vol. 2, pp. 792–799 (2005)
2. Ford, L.R., Fulkerson, D.R.: Constructing maximal dynamic flows from static flows. Operations Research 6, 419–433 (1958)
3. Khanna, G., Catalyurek, U., Kurc, T., Sadayappan, P., Saltz, J.: Scheduling file transfers for data-intensive jobs on heterogeneous clusters. Technical Report OSU-CISRC-1/07-TR05, CSE Dept, The Ohio State University (2007)
4. Giersch, A., Robert, Y., Vivien, F.: Scheduling tasks sharing files from distributed repositories. In: Danelutto, M., Vanneschi, M., Laforenza, D. (eds.) Euro-Par 2004. LNCS, vol. 3149, pp. 246–253. Springer, Heidelberg (2004)
5. Ibarra, O.H., Kim, C.E.: Heuristic algorithms for scheduling independent tasks on nonidentical processors. J. ACM 24, 280–289 (1977)
6. Allcock, W., Bresnahan, J., Kettimuthu, R., Link, M.: The globus striped gridftp framework and server. In: Proc. of SuperComputing'05 (2005)
7. Gabow, H.N.: An efficient implementation of edmonds' algorithm for maximum matching on graphs. J. ACM 23, 221–234 (1976)
8. Uysal, M., Kurc, T.M., Sussman, A., Saltz, J.: A performance prediction framework for data intensive applications on large scale parallel machines. In: O'Hallaron, D.R. (ed.) LCR 1998. LNCS, vol. 1511, pp. 243–258. Springer, Heidelberg (1998)
9. Fischetti, M., Glover, F., Lodi, A.: The feasibility pump. Math. Program. 104, 91–104 (2005)
10. Czyzyk, J., Mesnier, M.P., Moré, J.J.: The neos server. IEEE Comput. Sci. Eng. 5, 68–75 (1998)

Cooperation in Multi-organization Scheduling*

Fanny Pascual[1,2], Krzysztof Rzadca[2,3], and Denis Trystram[2]

[1] INRIA Rhône-Alpes, France
[2] LIG Grenoble University, France
[3] Polish-Japanese Institute of Information Technology, Warsaw, Poland

Abstract. The distributed nature of the grid results in the problem of scheduling parallel jobs produced by several independent organizations that have partial control over the system. We consider systems composed of n identical clusters of m processors. We show that it is always possible to produce a collaborative solution that respects participant's selfish goals, at the same time improving the global performance of the system. We propose algorithms with a guaranteed worst-case performance ratio on the global makespan: a 3-approximation algorithm if the last completed job requires at most $m/2$ processors, and a 4-approximation algorithm in the general case.

1   Introduction

The grid computing paradigm [1] introduces new and difficult problems in scheduling and resource management. A grid can be viewed as an agreement to share resources between a number of independent organizations (such as laboratories, or universities), with little, or no, central, administrative control [2], forcing them to interact. An organization is an administrative entity grouping users and computational resources. Organizations are free to join or to leave the system, if the gain experienced is lower than the cost of participation. Therefore, in order to sustain the grid, the resource management system must achieve an acceptable performance not only at the level of the community of users (as in classic, monocriterion scheduling), but also on the between-organizations level. Some globally-optimal approaches may be unacceptable because they implicitly favor jobs produced by one organization, therefore reducing the performance experienced by the others.

In this paper, we study the problem of scheduling parallel jobs [3] produced by several *organizations*. Each organization owns and controls a cluster, that together form a computational grid. The global goal is to minimize the makespan [3], the time moment when all the jobs are finished. However, each organization is only concerned with the makespan of its own jobs. An organization can always quit the grid and compute all its jobs on its local cluster. Therefore, a solution which extends the makespan of an organization in comparison with such a local solution is not *feasible*, even if it leads to a better global makespan. Such an organization would

* This research was partly supported by the FP6 Network of Excellence CoreGRID funded by the European Commission (Contract IST-2002-004265). Krzysztof Rzadca is partly supported by the French Government Grant number 20045874.

A.-M. Kermarrec, L. Bougé, and T. Priol (Eds.): Euro-Par 2007, LNCS 4641, pp. 224–233, 2007.

prefer to quit the grid, to compute all its jobs locally and not to accept any other jobs on its cluster. The considered scheduling problem is therefore an extension of the the classic, parallel job scheduling [3] by a series of constraints stating that that no organization's makespan can be increased.

The main contribution of the paper is the demonstration that several independent organizations have always interest to collaborate in a load-balancing grid system. We propose an algorithm producing solutions that guarantee that no organization's makespan is increased, at the same time having guaranteed approximation ratio (worst-case performance) regarding the globally-optimal solution. Assuming that each cluster has m of processors, the proposed algorithm is a 3-approximation if the last finished job is *low* (requires at most half of the available processors), and a 4-approximation in the general case.

This paper is organized as follows. Section 2 introduces some notations, formally defines the model and the problem and presents some motivating examples. Section 3 considers a problem of scheduling local and foreign jobs on a single multiprocessor cluster with guaranteed performance for local jobs. Section 4 presents the algorithms for n multiprocessor clusters and proves the approximation ratios. Related work is discussed in Section 5. Section 6 discusses the results obtained and concludes the paper.

2   Preliminaries

2.1   Notation and the Model of the Grid

By $\mathcal{O} = \{O_1, \ldots, O_n\}$ we denote the set of independent organizations forming the grid. Each organization O_k owns a cluster M_k. Each cluster M_k has m identical processors. By $\mathcal{M}$ we denote the set of all clusters.

The set of all the jobs *produced* by O_k is denoted by $\mathcal{I}_k$, with elements $\{J_{k,i}\}$. By $\mathcal{J}_k$ we denote the set of jobs *executed* on O_k's cluster M_k. If $J_{k,i} \in \mathcal{J}_k$, the job is executed *locally*, otherwise it is *migrated*. Job $J_{k,i}$ must be executed in parallel on $q_{k,i}$ processors of exactly one cluster during $p_{k,i}$ time units. It is not possible to divide a job between two, or more, clusters. We denote by $p_{\max} = \max p_{k,i}$ the maximum length of job. $J_{k,i}$ is *low* if it needs no more than a half of cluster's processors ($q_{k,i} \leq \frac{m}{2}$), otherwise it is *high*.

By $C_{k,i}$ we denote the completion (finish) time of job $J_{k,i}$. For an organization O_k, we may compute the maximum completion time (makespan) as $C_{\max}(O_k) = \max_{k,i}\{C_{k,i} : J_{k,i} \in \mathcal{I}_k\}$. The global makespan $C_{\max}$ is the maximum makespan of organizations, $C_{\max} = \max_k C_{\max}(O_k)$.

For cluster M_k, a *schedule* is a mapping of jobs $\mathcal{J}_k$ to processors and start times in such a way that, at each time, no processor is assigned to more than one job. We can define the *makespan $C_{\max}(M_k)$ of cluster M_k* as the maximum completion time of jobs $\mathcal{J}_k$ assigned to that cluster, $C_{\max}(M_k) = \max_i\{C_{j,i} : J_{j,i} \in \mathcal{J}_k\}$. At any time t, *utilization $U_k(t)$* of M_k is the ratio of the number of assigned processors to the total number of processors m. A *scheduler* is an application which produces schedules, given the sets of jobs produced by each organization.

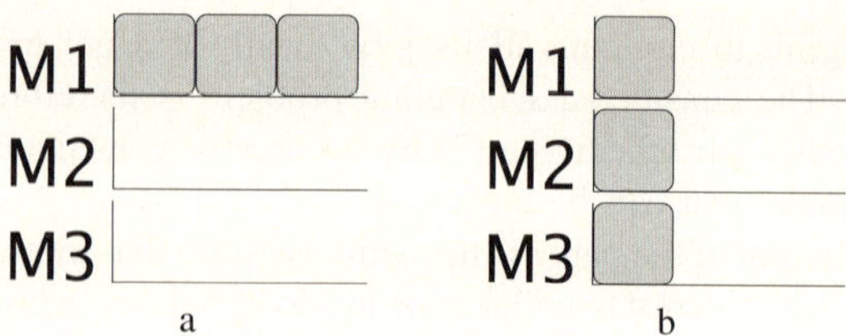

Fig. 1. Executing all the jobs locally (a) may lead to n approximation ratio regarding the globally-optimal solution (b). All the jobs were produced by organization O_1, the owner of M_1.

2.2   Problem Statement

We consider off-line, clairvoyant scheduling with no preemption on time-sharing processors. Those assumptions are fairly realistic in most of the existing scheduling systems, which use batches [4] and which require the user to define the run-time of the posted jobs. Each organization O_k wants to minimize the date $C_{\max}(O_k)$ at which all the locally produced jobs $\mathcal{I}_k$ are finished. Organization O_k does not care about the performance of other organizations, nor about the actual makespan $C_{\max}(M_k)$ on local cluster M_k, if the last job to be executed is not owned by O_k. However, $C_{\max}(O_k)$ takes into account jobs owned by O_k and executed on non-local clusters, if there are any.

The *Multi-Organization Scheduling Problem* (MOSP) is the minimization of the makespan of all the jobs (the moment when the last job finishes) with an additional constraint that no makespan is increased compared to a preliminary schedule in which all the clusters compute only locally produced jobs. More formally, let us denote $C_{\max}^{loc}(O_k)$ as a makespan of O_k when $\mathcal{J}_k$, the set of jobs executed by M_k is equal to the set of locally produced jobs, i.e. $\mathcal{J}_k = \mathcal{I}_k$. MOSP can be defined as:

$$\min C_{\max} \text{ such that } \forall_k \ C_{\max}(O_k) \leq C_{\max}^{loc}(O_k). \tag{1}$$

By restricting the number of organizations to $n = 1$, the size of the cluster to $m = 2$ and the jobs to sequential ones ($q_{k,i} = 1$), we obtain the classic, NP-hard problem of scheduling sequential jobs on two processors $2|p_j|C_{max}$ [5]. Therefore, MOSP is also NP-hard.

2.3   Motivation

A number of instances motivate organizations to cooperate and accept non-local jobs, even taking into account the fact that the resulting configuration is not necessary globally optimal. A non-cooperative solution (without the grid) is that all the organizations compute their jobs on their local clusters. However, such a solution can be as far as n times worse than the optimal one (see Figure 1). Note also that careful scheduling offers more than simple load balancing of the previous example. By matching certain types of jobs, bilaterally profitable solutions are also possible (see Figure 2). Nevertheless, a certain price must be paid in order to produce solutions in which all the organizations have incentive to participate. Figure 3 presents an instance in which the globally-optimal solution extends the makespan of one of the organizations. Consequently, all

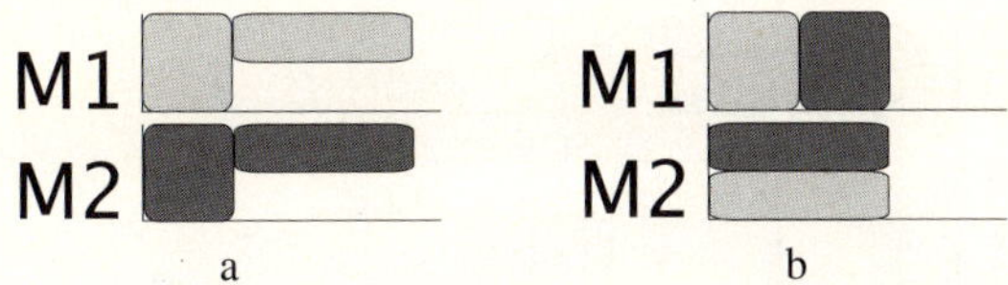

Fig. 2. By matching certain types of jobs, cooperative solution (b) delivers better makespans for both organizations than a solution scheduling all the jobs locally (a). The light gray jobs were produced by organization O_1, the dark gray ones by O_2.

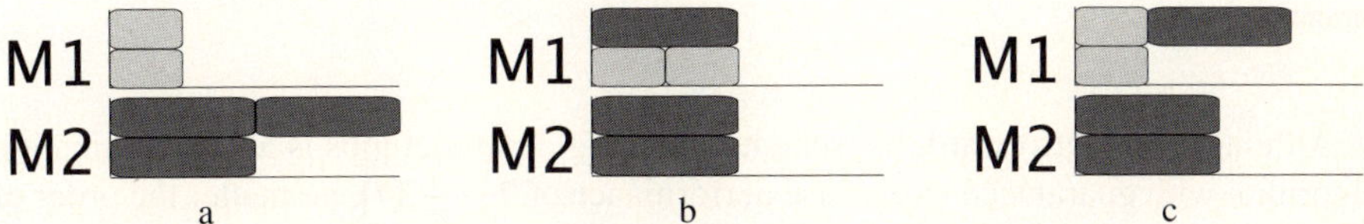

Fig. 3. Globally-optimal solution (b) is inadmissible, as it extends the makespan of organization O_1 (the producer of light gray jobs) in comparison with the local solution (a). The best solution not extending O_1's makespan (c) is $\frac{3}{2}$ from the global optimum.

the algorithms that meet the constraint have at least $\frac{3}{2}$ approximation ratio regarding the globally-optimal solution.

3 Scheduling on One Cluster

Let us first focus on the simple case of scheduling rigid parallel jobs on one cluster consisting of m identical processors. Note that as in this section there is no reason to distinguish between a cluster and an organization, we will simply use $C_{\max}$ to denote the makespan and omit the index of the organization in other notations (e.g. $q_{k,i}$ becomes q_i). We will use here the classic list scheduling algorithm, which has an approximation ratio equal to $2 - \frac{1}{m}$. We show that if the jobs are ordered according to decreasing number of required processors, the resulting schedule achieves fairly homogeneous utilization. Preliminary results established in this section will be later used to solve the general problem of multi-organization scheduling in Section 4.

3.1 List Scheduling

List scheduling [6] is a class of heuristics which work in two phases. In the first phase, jobs are ordered into a list. In the second phase, the schedule is constructed by assigning jobs to processors in a greedy manner. Let us assume that at time t, m' processors are free in the schedule under construction. The scheduler chooses from the list the first job J_i requiring no more than m' processors, schedules it to be started at t, and removes it from the list. If there is no such job, the scheduler advances to the earliest time t' when one of the scheduled jobs finishes.

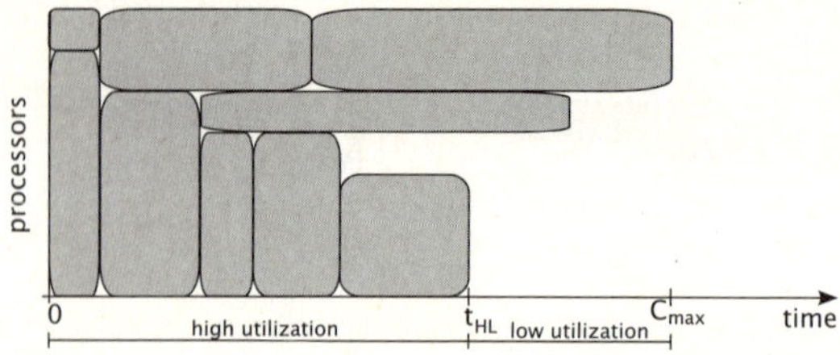

Fig. 4. When jobs are presorted according to number of required processors, the schedule can be divided into two regions with utilization $U(t) > \frac{1}{2}$ (up to t_{HL}) and $U(t) \leq \frac{1}{2}$ (after that moment)

Although straightforward, list scheduling of rigid parallel jobs is an approximation algorithm with guaranteed worst case performance of $2 - \frac{1}{m}$ [7], no matter the order of jobs in the first phase. A polynomial time algorithm with better approximation ratio is not known.

3.2   Highest First (HF) Job Order

The $2 - \frac{1}{m}$ approximation ratio of list scheduling does not depend on the particular order of jobs in the list. Therefore, we may choose a criterion which gives some interesting properties of the resulting schedule without loosing the approximation ratio.

Let us consider jobs ordered according to the Highest First (HF) rule, i.e. by non-increasing q_i. The following proposition holds:

Proposition 1. *All HF schedules have the same structure consisting of two consecutive regions of high ($t \in [0, t_{HL}) : U(t) > \frac{1}{2}$) and low ($t \in [t_{HL}, C_{max} : U(t) \leq \frac{1}{2}$) utilization, where $0 \leq t_{HL} \leq C_{max}$ (Figure 4).*

Proof: First, note that no high job is scheduled after a period of low utilization. Indeed, as soon as a high job is completed, the following highest job is scheduled (according to the HF rule). Thus there is no low utilization period before that all the high jobs have been completed. The proof is now by contradiction. Let us assume that at time t the utilization is low ($U(t) \leq \frac{1}{2}$), and that at time $t' > t$ the utilization is high ($U(t') > \frac{1}{2}$). Let us consider a job J_i scheduled at time t'. It is not possible that J_i is a high job because no high job can be scheduled after a period of low utilization, as noted before. If J_i is low ($q_i \leq \frac{m}{2}$) then it could have been scheduled at time t, and scheduling it after t contradicts the greedy principle of the list scheduling algorithm.   □

4   Multi-organization Scheduling

In this section we present an algorithm that address the Multi-Organization Scheduling Problem. This algorithm has a guaranteed approximation ratio regarding to the global makespan, at the same time not worsening the local solutions that are produced by the organizations while computing independently.

The algorithm that we propose computes a lower bound of the global makespan and then moves all the jobs which start after twice this date to the end of the schedules of less-loaded clusters. Details follow.

Let us denote by $W = \sum p_{k,i} q_{k,i}$ the total work in the system, or the total *surface* of the jobs. As all the jobs must fit into available processors, the global makespan $C_{\max}$ is not less than the lower bound $LB = \frac{W}{mn}$.

Let us assume that all the organizations list-scheduled their jobs on their local machines according to HF order. The *Multi-Organization Load Balancing Algorithm (MOLBA)* is the following one. It starts to compute LB. All the clusters with local makespans between $2LB$ and $2LB + p_{\max}$ are ignored (we do not move their local jobs). For the rest of the clusters, all the jobs that start after time $2LB$ are moved from their local clusters to a migration queue. Finally, the jobs from the migration queue are list-scheduled onto all available clusters. The jobs are scheduled sequentially in a greedy manner. No migrated job can delay a local job: a job $J_{k,i}$ is scheduled before the original makespan of the host cluster M_j ($t < C_{\max}^{loc}(O_j)$) only if at least $q_{k,i}$ processors are free on M_j from time t to time $t + p_{k,i}$. Such a strategy is similar to the well-known conservative backfilling in FCFS (First Come First Serve).

We prove in Sections 4.1 and 4.2 that this algorithm is a $3-$approximation of the global makespan $C_{\max}$ when the last completed task is a low task, and that is it a 4-approximate algorithm in the general case. We also show that this algorithm does not increase the local makespans of the organizations (therefore holding the constraint in Eq. 1). We start with a lemma that characterizes the structure of all the clusters' schedules.

In the schedule returned by MOLBA, on each cluster, we denote by t_L^{start} the first moment when the utilization is lower than or equal to $\frac{1}{2}$. Similarly, t_L^{end} is the last moment when the utilization is larger than 0 and lower than or equal to $\frac{1}{2}$. We first prove the following lemma:

Proposition 2. *In the schedule returned by MOLBA, on each cluster, the length of the time interval between t_L^{start} and t_L^{end} (denoted by P_L) is shorter than or equal to $p_{\max}$.*

Proof: Each cluster schedules its local jobs with HF. Then, it may add jobs from other organizations, also in HF order. Proposition 1 shows that, in a schedule returned by HF, the only zone of low utilization is at the end of the schedule. Thus, on each cluster, there are at most two zones of low utilization: possibly one at the end of the schedule of the local jobs, and also possibly one at the end of the schedule.

Let $J_{k,i}$ be the *low* job that finishes last on cluster M_j. After $J_{k,i}$ finishes, utilization is either high, or zero. Thus, by P_L definition, $J_{k,i}$ cannot finish before t_L^{end}. $J_{k,i}$ does not start after t_L^{start}, as utilization at t_L^{start} is low, so there are enough free processors to execute a *low* job. Thus, the length of P_L is smaller than or equal to the length of $J_{k,i}$, which is not longer than p_{max}. $\qquad\square$

Proposition 3. *After MOLBA finishes, there is at least one cluster whose $t_L^{start} \leq 2\,LB$.*

Proof: The proof is by contradiction. Suppose that there exists $\epsilon > 0$ such that all the clusters have high utilization until time $2\,LB + \epsilon$. Then, the total surface of jobs computed by all the clusters is greater than $2\,LB \cdot mn \cdot 0.5 = W$, i.e. greater than the total work available, which leads to a contradiction. $\qquad\square$

4.1   Low Jobs

We show in this section that, in the schedule returned by MOLBA, if the last completed job is a *low* job, then MOLBA is a $3-$approximate algorithm.

Proposition 4. *The makespan of the schedule returned by MOLBA is a 3-approximation of the optimal makespan, if the last completed job is* low. *Moreover, all the organizations have incentive to cooperate.*

Proof: The proof uses two well known lower bounds of the optimal makespan $C^*_{\max}$. Firstly, the longest job (of length $p_{\max}$) must be completed, so $C^*_{\max} \geq p_{\max}$. Secondly, all the jobs must fit onto available processors, so $C^*_{\max} \geq LB = \frac{W}{mn}$. The last job $J_{k,i}$ finishes at $C_{\max}$. Recall that this job is a low job. Proposition 3 guarantees that there is at least one cluster with low utilization before or at time $2\,LB$. Thus, job $J_{k,i}$ does not start after $2\,LB$, since we use a list scheduling algorithm. Hence, $C_{\max} \leq 2LB+p_{k,i} \leq 2LB + p_{\max} \leq 3C^*_{\max}$.

As no migrated job can delay a local job, makespans of organizations that were receiving tasks are not modified. The organizations that were sending tasks have their makespan reduced because of the global approximation ratio. The schedule of the rest of organizations is not modified. Thus, the constraint in Equation (1) is satisfied and all the organizations have incentive to cooperate. $\qquad\square$

4.2   General Case

Let us now consider the case where the last completed job can have any height. We now show that MOLBA achieves an approximation ratio of 4 on the global makespan. We suppose here that we "cut" the schedule where each organization schedule its local jobs at time $3\,LB$ (and not $2\,LB$ as in the previous case). We do not move the local tasks of organizations with local makespans smaller than $4\,LB$. For the rest of the clusters, all the jobs which start after time $3\,LB$ before the load balancing procedure are moved into a queue and then scheduled using the HF list algorithm.

Proposition 5. *MOLBA is a 4-approximate algorithm and all the organizations have incentive to cooperate.*

Proof: Let us prove this Proposition by contradiction. Let C^*_{max} be the makespan of an optimal schedule, and let us suppose that a job starts after time $3\,C^*_{max}$ in the schedule returned by MOLBA. This means that this job could not have been started before : for all $i \in \{1,\ldots,n\}$, $C_{max}(M_i) \geq 3\,C^*_{max}$. Proposition 2 shows that, for each cluster, the zone where at most half of the processors are busy is smaller than or equal to C^*_{max}. Thus, on each cluster, the zone where at least half of the processors are busy is larger than or equal to $2\,C^*_{max}$. As we have seen it previously, $C^*_{max} \geq \frac{W}{n\,m}$, where $W = \sum p_{k,i} q_{k,i}$. Thus, the total work which is done before $3\,C^*_{max}$ in the zones of high utilization is larger than or equal to $\frac{n\,m}{2}(2\,\frac{W}{n\,m}) = W$. This is not possible since the total work which has to be done is equal to W. Thus, no job starts after time $3\,C^*_{max}$, and no job is completed after time $4\,C^*_{max}$.

The proof that all the organizations have incentive to cooperate is analogous to the proof of Proposition 4. $\qquad\square$

There are some other special cases in which the presented approximation ratio can be improved. When there are $n = 2$ organizations, the original version of the algorithm ("cutting" the schedules at $2\,LB$) is $3-$approximate. We omit the proof because of the lack of space. For n clusters, and when all the jobs are *low*, the algorithm is also $3-$approximate, since this special case is included in the proof presented in Section 4.1. Finally, when all the tasks are high, no two tasks can be scheduled in parallel on one cluster. Thus, the problem corresponds to scheduling sequential tasks on n processors. Any list scheduling algorithm is, in this case, $2 - \frac{1}{n}$ approximate. It is straightforward to guarantee that all the organizations have incentive to cooperate. Each task is scheduled on its local processor, unless there is a free processor that already scheduled its local tasks.

5   Related Work

In this paper we have studied the interest of collaboration between independent parties. We have claimed that if a proposed, collaborative solution does not deteriorates any participant's selfish goal, it will be adopted by all the participants.

Using a reasonable set of assumptions, we have demonstrated that it is always possible to produce such collaborative solutions. Moreover, we have developed an algorithm which has a worst-case guarantee on the social goal (the makespan of the system), at the same time respecting selfish goals of participants.

In this section we will briefly summarize how the concept of collaboration and the distributed nature of systems has been understood by and used in other works.

Non-cooperative game theory studies situations in which a set of selfish agents optimize their own objective functions, which also depend on strategies undertaken by other agents. The central notion is the Nash equilibrium [8], a situation in which no agent can improve its own objective function by unilaterally changing his/her strategy. It can be useful to define a *social* (global) objective function, which expresses the performance of the system as a whole. The ratio between the values of this function in the worst Nash equilibrium and in an optimal solution is called the Price of Anarchy (PoA)[9]. This can be interpreted as the cost of no cooperation and can be high. In the context of scheduling, [10] measures PoA when selfish sequential jobs choose one of the available processors. A related measure, Price of Stability (PoS) [11,12] compares the socially-best Nash equilibrium with the socially-optimal result. Usually, in order to find such an equilibrium, a centralized protocol gathers information from, and then suggests a strategy to, each participant. Since the proposed solution is a Nash equilibrium, the participants do not have incentive to unilaterally refuse to follow it. [13] computes PoS in the same model as [10], but relaxes the selfishness of jobs by a factor of α and studies the trade-off between α and the approximation ratio of the global makespan. The collaborative solution proposed by our algorithm approximates the socially-best Nash equilibrium, because it optimizes the global goal with a guarantee that no participant has the incentive to deviate from the proposed solution.

Cooperative game theory studies similar situations, but assumes that players can communicate and form coalitions. The members of a coalition split the sum of their payoffs after the end of the game. Note that this requires that the payoffs are transferable, which is not the case in our problem.

Papers proposing *distributed* resource management or distributed load balancing usually solve the problem of optimizing a common goal with a decentralized algorithm. [14] shows a fully decentralized algorithm that always converges to a steady state. [15] presents a similar algorithm with the divisible load job model. Those approaches contrast with our algorithm. Although the algorithm is centralized, it respects the decentralized goals of participants. We are, however, aware that a load balancing algorithm in large scale systems must be decentralized. In [16] a fully distributed algorithm balances selfish *identical* jobs on a network of identical processors. The aim of each job is to be on the least loaded machine. The work focus on the time needed to converge towards a Nash equilibrium. Alternative approaches propose to balance the load by an implicit barter trade of CPU power [17], or explicit computational economy [18].

6   Conclusion and Perspectives

In this work we have considered the problem of cooperation between selfish participants of a computational grid. More specifically, we studied a model of the grid in which selfish organizations minimize the maximum completion time of locally-produced jobs. Under some basic assumptions (off-line, clairvoyant system, idle time of machines is free) we have demonstrated that it is always possible to respect the selfish goals at the same time improving the performance of the whole system. The cooperative solutions have a constant worst case performance, a significant gain compared to selfish solutions that can be arbitrary far from the optimum. We deliberately focused on the analysis of the worst-case performance in order to avoid the plethora of problems of the experimental methodology in grid systems.

Our aim was not to find an algorithm solving the general problem of grid resource management, which complexity is overwhelming for any kind of mathematical modeling. However, we claim that the positive results given by this paper proves that cooperation achieved at the algorithmic (as opposed to e.g. economic) level is possible. Note that it should be fairly straightforward to relax some of our assumptions, e.g. to use on-line scheduling in batches instead of off-line. An interesting direction would also be to consider this mutiorganization scheduling problem with heterogeneous clusters.

In our future work, we would like to study the effect of the increased effort of individuals on the global goal. More specifically, we would like to relax the hard constraint of "not being worse than the local solution" to an approximation of "not being worse than α times local solution".

References

1. Foster, I., Kesselman, C.: The Grid 2. Blueprint for a New Computing Infrastructure. Elsevier, Amsterdam (2004)
2. Foster, I.: What is the grid, `http://www-fp.mcs.anl.gov/~foster/Articles/WhatIsTheGrid.pdf`
3. Blazewicz, J.: Scheduling in Computer and Manufacturing Systems. Springer, Heidelberg (1996)
4. Shmoys, D., Wein, J., Williamson, D.: Scheduling parallel machines on-line. SIAM Journal on Computing 24 (1995)
5. Garey, M.R., Johnson, D.: Computers and Intractability: A Guide to the Theory of NP-Completeness. WH Freeman & Co, New York, USA (1979)
6. Graham, R.: Bounds on multiprocessor timing anomalies. SIAM J. Appl. Math 17(2) (1969)
7. Eyraud-Dubois, L., Mounie, G., Trystram, D.: Analysis of scheduling algorithms with reservations. In: Proceedings of IPDPS, IEEE Computer Society Press, Los Alamitos (2007)
8. Osborne, M.J., Rubinstein, A.: A Course in Game Theory. MIT Press, Cambridge (1994)
9. Koutsoupias, E., Papadimitriou, C.: Worst-case equilibria. In: Meinel, C., Tison, S. (eds.) STACS 99. LNCS, vol. 1563, pp. 404–413. Springer, Heidelberg (1999)
10. Christodoulou, G., Koutsoupias, E., Nanavati, A.: Coordination mechanisms for selfish scheduling. In: Baeten, J.C.M., Lenstra, J.K., Parrow, J., Woeginger, G.J. (eds.) ICALP 2003. LNCS, vol. 2719, pp. 345–357. Springer, Heidelberg (2003)
11. Schultz, A.S., Stier Moses, N.: On the performance of user equilibria in traffic networks. In: Proceedings of SODA, 86–87 (2003)
12. Anshelevich, E., Dasgupta, A., Kleinberg, J.M., Tardos, É., Wexler, T., Roughgarden, T.: The price of stability for network design with fair cost allocation. In: Proceedings of FOCS, pp. 295–304 (2004)
13. Angel, E., Bampis, E., Pascual, F.: The price of approximate stability for a scheduling game problem. In: Nagel, W.E., Walter, W.V., Lehner, W. (eds.) Euro-Par 2006. LNCS, vol. 4128, Springer, Heidelberg (2006)
14. Liu, J., Jin, X., Wang, Y.: Agent-based load balancing on homogeneous minigrids: Macroscopic modeling and characterization. IEEE TPDS 16(7), 586–598 (2005)
15. Rotaru, T., Nageli, H.H.: Dynamic load balancing by diffusion in heterogeneous systems. J. Parallel Distrib. Comput. 64(4), 481–497 (2004)
16. Berenbrink, P., Friedetzky, T., Goldberg, L., Goldberg, P., Hu, Z., Martin, R.: Distributed Selfish Load Balancing. In: Proceedings of SODA, pp. 354–363. ACM Press, New York (2006)
17. Andrade, N., Cirne, W., Brasileiro, F., Roisenberg, P.: Ourgrid: An approach to easily assemble grids with equitable resource sharing. In: Feitelson, D.G., Rudolph, L., Schwiegelshohn, U. (eds.) JSSPP 2003. LNCS, vol. 2862, pp. 61–86. Springer, Heidelberg (2003)
18. Buyya, R., Abramson, D., Venugopal, S.: The grid economy. Special Issue on Grid Computing 93, 698–714 (2005)

Topic 4
High-Performance Architectures and Compilers

Michael O'Boyle, François Bodin, Jose Gonzalez, and Lucian Vintan

Topic Chairs

Parallelism is now a central concern for architecture designers and compiler writers. Instruction-level parallelism and increasingly multi-cores are present in all contemporary processors. Furthermore, we are witnessing a convergence of interests with architects and compiler writers addressing large scale parallel machines, general-purpose platforms and specialised hardware designs such as graphic co-processors or low-power embedded systems. Modern systems require system software and hardware to be designed in tandem, hence this topic is concerned with architecture design and compilation. Twenty-four papers were submitted to the track of which five were accepted split over two sessions.

In "Starvation-Free Transactional Memory System", the authors focus on starvation effects that show up in transactional memory. Simple protocols are prone to failure if a first-come first-served policy is used. Low complexity hardware solutions are proposed that do not affect overall performance. In "Towards Real-Time Compression of Hyperspectral Images Using Virtex-II FPGAs" the authors develop an FPGA-based data compression technique. This approach depends on the concept of spectral un-mixing of pixels and sub-pixel targets in hyperspectral analysis. The authors use a two-stage approach which is implemented on an existing FPGA allowing on-board near real-time data compression.

The paper "Optimizing Chip Multiprocessor Work Distribution using Dynamic Compilation" examines how sequential applications benefit from chip multiprocessors. It develops an automatically parallelizing approach based on dynamic compilation which adaptively tunes work distribution. This is experimentally evaluated on the Jamaica chip multi-processor. In " Program Behavior Characterization through Advanced Kernel Recognition" the authors examine how to develop automatic techniques that summarize the behavior of real applications and hide implementation details. This is achieved by describing applications in terms of computational kernels using the XARK compiler. This is thoroughly evaluated on a large set of benchmarks.

In "Compositional Approach applied to Loop Specialization" the authors examine specialised code generation techniques that consider the impact of data sizes on code quality This can lead to code expansion and decision tree overhead. The authors develop a new folding method to specialize code at the assembly level while reducing overhead. They demonstrate the need for specialization on small loops and provide a detailed experimental evaluation.

A.-M. Kermarrec, L. Bougé, and T. Priol (Eds.): Euro-Par 2007, LNCS 4641, p. 235, 2007.
© Springer-Verlag Berlin Heidelberg 2007

Program Behavior Characterization Through Advanced Kernel Recognition[*]

Manuel Arenaz, Juan Touriño, and Ramón Doallo

Computer Architecture Group
Department of Electronics and Systems
University of A Coruña, A Coruña, Spain
{arenaz,juan,doallo}@udc.es
http://gac.des.udc.es

Abstract. Understanding program behavior is at the foundation of program optimization. Techniques for automatic recognition of program constructs (from now on, computational kernels) characterize the behavior of program statements, providing compilers with valuable information to to guide code optimization. Our goal is to develop automatic techniques that summarize the behavior of full-scale real applications by building a high-level representation that hides the complexity of implementation details. The first step towards this goal is the description of applications in terms of computational kernels such as induction variables, reductions, and array recurrences. To this end we use XARK, a compiler framework that recognizes a comprehensive collection of frequently used kernels. This paper presents detailed experiments that describe several benchmarks from different application domains in terms of the kernels recognized by XARK. More specifically, the SparsKit-II library for the manipulation of sparse matrices, the Perfect benchmarks, the SPEC CPU2000 collection and the PLTMG package for solving elliptic partial differential equations are characterized in detail.

1 Introduction

Automatic code optimization hinges on advanced symbolic analysis to gather information about the behavior of programs. Compiler techniques for automatic kernel recognition carry out symbolic analysis in order to discover program constructs that are frequently used by software developers. Such techniques were shown to be a powerful mechanism to improve the performance of optimizing and parallelizing compilers. Well-known examples are the substitution of induction variables with closed-form expressions, the detection of reduction operations to raise the effectiveness of dependence analysis, the characterization of the access patterns of array references to predict program locality, or the automatic replacement of sequential algorithms with platform-optimized parallel versions.

---

[*] This research was supported by the Ministry of Education and Science of Spain and FEDER funds of the European Union (Project TIN2004-07797-C02), and by the Galician Government (Projects PGIDT05PXIC10504PN and PGIDIT06PXIB105228PR).

A.-M. Kermarrec, L. Bougé, and T. Priol (Eds.): Euro-Par 2007, LNCS 4641, pp. 237–247, 2007.

XARK [2] is an extensible compiler framework for automatic recognition of computational kernels. Unlike previous approaches that focus on specific and isolated kernels [6,8,11], XARK provides a general solution that detects a comprehensive collection of kernels that appear in real codes with regular and irregular computations. The recognition algorithm analyzes data dependences and control flow altogether, and handles scalar and array variables in a unified manner. The kernels are organized in families that share common syntactical properties. Some well-known examples are induction variables, scalar reductions, irregular reductions and array recurrences.

The rest of the paper is organized as follows. Section 2 gives a general overview of the XARK compiler and describes the families of computational kernels. Section 3 shows detailed experimental results for the benchmarks SparsKit-II, Perfect, SPEC CPU2000 and PLTMG. Finally, Section 4 concludes the paper and outlines future work.

2   The XARK Compiler

2.1   Overview

XARK [2] is a compiler framework that provides a general solution to the problem of automatic kernel recognition. Three key characteristics distinguish XARK from previous approaches: (1) completeness, as it recognizes a comprehensive collection of computational kernels that involve integer-valued and floating-point-valued scalar and array variables, as well as if-endif constructs that introduce complex control flows; (2) robustness against different versions of a computational kernel; and (3) extensibility, as its design enables the addition of new recognition capabilities with little programming effort.

XARK internals consist of a two-phase demand-driven classification algorithm that analyzes the data dependences and the control flow graph of a program through its Gated Single Assignment (GSA) representation [12], which is an extension of the well-known Static Single Assignment (SSA) form where reaching definition information of scalar and array variables is represented syntactically. For illustrative purposes consider the example code presented in Figure 1. For the sake of clarity, the details about the GSA form and the loop index variable h have been omitted. The code consists of a loop do_h that computes a kernel called *consecutively written array* (see Section 2.2 later in this paper). At run-time, consecutive entries of the array a are written in consecutive memory locations determined by the value of the linear induction variable i. The complexity of this loop comes from the fact that i is incremented in one unit in those iterations where the condition $c(h)$ is fulfilled. In general, the condition is not loop-invariant, so the value of i in each iteration cannot be calculated as a function of the loop index variable h.

The framework is built on top of an intermediate representation where the source code statements are represented as abstract syntax trees, and the data dependences between statements are captured as use-def chains between the trees. In the first phase, XARK identifies the strongly connected components

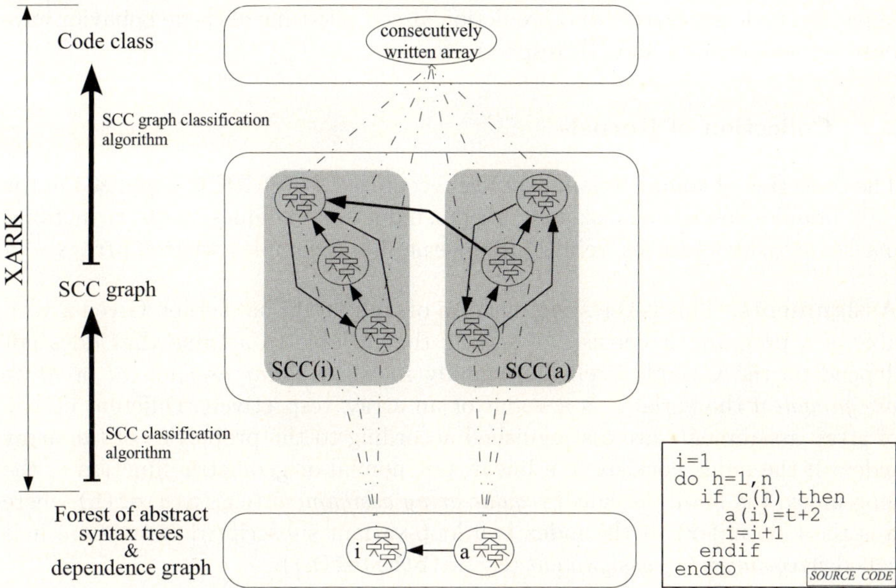

Fig. 1. Overview of the XARK compiler

(SCCs) of the dependence graph and carries out an intra-SCC analysis that determines the type of kernel computed during the execution of the statements of each SCC. As a result of this intra-SCC analysis, the code is decomposed into a set of mutually dependent kernels that capture the run-time behavior of each source code variable. In the example, the code is decomposed into two SCCs (namely, SCC(i) and SCC(a)) that enable the recognition of the induction variable and the array assignment (see Section 2.2) computed as a result of executing the statements i=i+1 and a(i)=t+2, respectively. The dependence relationships between a and i are represented as use-def chains between the SCCs.

In the second phase, XARK focuses on the use-def chains between statements of different SCCs in order to recognize more complex kernels that result from combining simpler kernels in the same code. In the example of Figure 1, the isolated detection of the induction variable i and the array assignment a does not provide enough information to recognize the consecutively written array. Thus, during the inter-SCC analysis, XARK checks that array a is indexed with the induction variable i; it analyzes the control flow graph to prove that every time a(i)=t+2 is executed, i=i+1 is also executed; and checks that i is incremented in one unit in every loop iteration where c(h) is fulfilled. Under such conditions, the consecutively written array a is recognized successfully. The results of this stage provide the compiler with high level information about the behavior of the program, hiding the complexity of implementation details. This information is very useful for the construction of other passes of a parallelizing or optimizing compiler. Examples of successful application of XARK in the scopes

of parallel code generation and prediction at compile-time of cache behavior have
been presented in [3] and [1], respectively.

2.2   Collection of Kernels

The collection of computational kernels recognized by XARK is organized in the
eight families described next: assignments, induction variables, maps, reductions,
masks, array recurrences, reinitialized kernels, and complex written arrays.

Assignments. This is the simplest form of computational kernel. Given a variable of a program, it consists of setting the variable to a value that does not
depend on the variable itself. The family is called *scalar assignment* or *array
assignment* if the variable is a scalar or an array, respectively. Different classes
of array assignments are distinguished according to the properties of the array
index. If the index consists of a linear, polynomial or geometric function of the
loop index, the family is called *regular array assignment* (e.g., `a(h)=f(h)` where
`h` is the loop index). If the index is a loop-variant subscripted expression, it is
called *irregular array assignment* (e.g., `a(b(h))=f(h)`).

Induction Variables. This family represents the type of scalar, integer-valued
variables that are updated in the iterations of a loop, either in every iteration
or in those iterations that fulfill a given condition. Different classes of induction
variables (IVs) are distinguished: *linear*, if an integer-valued loop-invariant value
is added to the value of the IV in each iteration (e.g., `i=i+1` is an IV of step
one); *polynomial*, if it is the value of another IV that is added (e.g., `i=i+j` where
`j=j+1`); and *geometric*, if the IV is multiplied by a loop-invariant (e.g., `k=2*k+1`).

Maps. A distinguishing characteristic of IVs is that there is a closed form function that allows the computation of the next value of the variable starting from its
initial value or from its current value. A *map* represents a sequence of values that
do not have such a closed form. In each loop iteration, the variable is assigned the
value of an array reference whose subscript expression contains an occurrence
of the variable (e.g., `i=next(i)`). When the variable is an array, different types
of regular and irregular access patterns are considered (e.g., `a(h)=next(a(h))`,
with loop index `h`, is an array map with a regular access pattern).

Reductions. A *scalar reduction* is a kernel with one scalar variable that is
defined in terms of itself and at least one loop-variant subscripted expression
(e.g., `r=r+a(h)`). An *array reduction* is defined in a similar manner, the reduction
variable being an array (e.g., `r(h)=r(h)+a(h)`). The characteristics of the index
expression lead to distinguish between regular and irregular array reductions.
Well-known examples of this family of kernels are adding the elements of a vector,
and finding the minimum/maximum element in each row of a matrix. A variant
of a minimum/maximum reduction consists of gathering additional information
about the reduction variable, for instance, the position of the minimum (or the
maximum) value within each row.

```
flag=true
do h=1,n
   if flag then
      ...
      flag=false
   endif
   ...
enddo
```

```
do h=1,n
   i=0
   do hh=1,m
      i=i+1
   enddo
enddo
```

Fig. 2. Example of loop that computes a mask kernel

Fig. 3. Example of loop that computes a reinitialized induction variable kernel

Masks. Masks are kernels that modify the value of a variable if its content fulfills a boolean condition. This family is called either *scalar find&set* or *array find&set*. A typical example is a loop that contains a set of statements that are executed only in the first loop iteration (see Figure 2). When the condition is *true* (`flag` in the example), such statements are executed and the condition is set to *false* to avoid the execution in the subsequent loop iterations. Array masks with regular and irregular access patterns are also considered.

Array Recurrences. Array recurrences are kernels that compute the value of the elements of the array using the values of other elements of the array (e.g., `a(h)=a(h-1)+1`). Unlike array reductions, array recurrences use different index expressions to access to the elements of the array. Regular and irregular access patterns are also considered.

Reinitialized Kernels. Real codes may contain more elaborate program constructs built from the kernels described above. From a graphical point of view, they can be interpreted as a point in a multidimensional space where the syntactical kernels are the values represented in the axes. Thus, a reinitialized kernel is as follows: first, an assignment that sets a scalar/array variable to a given value at the beginning of every iteration of a loop; and second, an induction variable, a map, a reduction, a mask or an array recurrence that updates the value of the scalar/array variable during the execution of an inner loop. The example shown in Figure 3 contains a reinitialized IV `i`.

Complex Written Arrays. Another interesting family is called *complex written array* [7]. It consists of a scalar kernel (e.g., induction variable, reinitialized IV, scalar reduction) that defines the array entries to be modified during the execution of the code, and an array assignment whose left-hand side subscript is a linear function of the scalar variable. When the scalar kernel is an IV of step one, the kernel is called *consecutively written array* (see the example code of Figure 1). When it is a reinitialized IV of step one, it is called *segmented consecutively written array*. Other variants of complex written arrays recognized by XARK involve an array reduction or an array recurrence instead of an array assignment. Then, they are called *(segmented) consecutively reduced array* and *(segmented) consecutively recurrenced array*, respectively.

3 Experimental Results

3.1 Benchmarks

Four benchmark suites have been used in the experiments: the Fortran routines included in SPEC CPU2000 [10], the Perfect benchmarks [5], the SparsKit-II library [9] and the PLTMG (Piecewise Linear Triangle Multi-Grid) code [4]. SPEC2000 and Perfect are well-known benchmarks that have been extensively used in the literature. SparsKit-II and PLTMG have been selected because their source codes contain plenty of irregular computations that cover the typical kernels found in full-scale applications. Table 1 shows the size of the benchmarks in terms of number of routines and number of code lines, and presents the percentage of loops recognized successfully by the XARK compiler.

Table 1. Summary of characteristics of the benchmark suite

	SPEC2000	Perfect	SparsKit-II	PLTMG	Totals
#Routines	273	608	103	258	1242
#Code lines	53173	60136	8286	27530	149125
#Loops analyzed	769	1245	293	651	2958
#Loops recognized	609	955	224	502	2290
%Loops recognized	79%	77%	76%	77%	77%

SparsKit-II [9] contains routines for the manipulation of sparse matrices. It is organized in four modules: MATVEC, devoted to basic matrix-vector operations (e.g., matrix-vector products and triangular system solvers); BLASSM, which covers basic linear algebra operations (e.g., matrix-matrix products and sums); UNARY, to carry out unary operations with sparse matrices (e.g., extract a submatrix); and FORMATS, for the conversion of sparse matrices between different types of sparse storages.

SPEC CPU2000 [10] consists of six Fortran codes: a program in the area of quantum chromodynamics (WUPWISE), two weather prediction programs (SWIM and APSI), a very simple multi-grid solver for computing a three dimensional potential field (MGRID), coupled nonlinear partial differential equations solver in the scope of computational fluid dynamics and computational physics (APPLU), and a program in the area of high energy nuclear physics accelerator design (SIXTRACK).

The Perfect Benchmarks [5] are a collection of thirteen scientific and engineering Fortran programs that are representative of applications executed on high-performance computers and that have been used extensively in research on parallelizing and restructuring compilers. It covers different application areas: fluid dynamics (ADM, ARC2D, FLO52, OCEAN and SPEC77), chemical and physical modeling (BDNA, MDG, QCD and TRFD), engineering design (DYFESM and SPICE), and signal processing (MG3D and TRACK).

PLTMG (Piecewise Linear Triangle Multi-Grid) [4] is a Fortran-77 code that consists of an adaptive multi-grid solver for two dimensional problems in general domains.

Table 2. Kernel families recognized by XARK in Sparskit-II and PLTMG

Kernel family	MATVEC				BLASSM				UNARY				FORMATS				PLTMG			
	N	S	C	L	N	S	C	L	N	S	C	L	N	S	C	L	N	S	C	L
Assignments																				
Scalar assignment	16	1-1	0-0	2-4	37	1-1	1-1	1-3	61	1-1	0-4	1-3	82	1-2	1-4	1-4	503	1-4	1-9	1-4
Regular assignment	12	1-1	0-0	1-2	23	1-2	1-2	1-3	39	1-1	0-1	1-3	55	1-2	1-2	1-2	456	1-9	1-4	1-3
Irregular assignment					13	1-1	1-1	1-2	11	1-4	1-2	1-3	47	1-2	1-1	1-3	67	1-3	1-2	1-2
Induction variables																				
Linear	1	1-1	0-0	4-4	6	1-3	1-2	1-3	12	1-1	1-2	1-2	17	1-5	1-2	1-3	100	1-6	1-5	1-4
Polynomial																	5	1-1	0-0	1-2
Geometric									1	1-1	0-0	3-3	1	1-1	0-0	4-4	38	1-1	0-2	1-2
Maps																				
Linked list																	44	1-2	1-9	1-4
Regular map																	2	1-1	0-0	2-2
Reductions																				
Scalar reduction	5	1-1	0-0	1-1					12	1-1	0-1	1-3	4	1-1	0-0	1-1	105	1-2	1-3	1-2
Regular reduction	2	1-1	0-0	2-2	2	1-1	0-0	1-1	4	1-1	0-0	1-1	2	1-2	0-1	1-2	86	1-3	0-1	1-3
Irregular reduction	9	1-1	0-0	1-4	1	1-1	1-1	1-1	7	1-3	1-1	1-2	13	1-1	1-1	1-2	24	1-3	0-1	1-2
Scalar min/max									2	1-2	1-2	1-1	4	1-2	1-2	1-2	6	1-2	1-3	1-2
Masks																				
Scalar find-and-set					2	1-1	1-1	1-1									13	1-2	1-4	1-3
Regular find-and-set									4	1-2	1-1	1-1					6	1-2	1-2	1-2
Irregular find-and-set									2	1-1	1-1	1-2					5	2-4	1-3	1-3
Recurrences																				
Regular recurrences									7	1-1	0-0	1-1	18	1-1	0-0	1-2	51	1-3	0-1	1-2
Irregular recurrences									2	1-1	1-1	2-2	2	1-1	1-2	1-2	15	1-1	0-0	1-3
Reinitialized kernels																				
Induction variables									2	2-2	1-1	2-3	2	2-2	1-1	2-2				
Maps																	9	2-3	2-2	2-4
Reductions	1	2-2	0-0	2-2					5	2-2	1-1	2-2	2	2-2	0-0	2-3	4	2-2	0-0	2-2
Complex written arrays																				
C. written ar. (CwriA)					13	2-6	1-2	1-2	17	2-2	1-1	1-3	31	2-2	1-2	1-4	31	2-6	1-3	1-2
C. recurrenced ar. (CrecA)													1	2-2	0-0	2-2	5	2-2	1-1	2-2
Segmented CwriA									2	3-3	0-0	2-2								
TOTALS	46	1-2	0-0	1-4	97	1-6	0-2	1-3	190	1-4	0-4	1-3	281	1-5	0-4	1-4	1575	1-9	0-9	1-4
Kernels with irregular computations	21	46%			62	64%			96	51%			157	56%			642	41%		

3.2 Recognition Results

The first step towards the characterization of the behavior of programs is the recognition of the computational kernels that appear in the code. The last row of Table 1 measures the effectiveness of XARK in terms of the percentage of loops whose body has been decomposed into a set of kernels recognized by the compiler. The percentage of *recognized loops* is 77% on average, ranging from 76% in SparsKit-II up to 79% in SPEC2000.

The experiments revealed that the loops of the benchmarks can be described in terms of the eight kernel families recognized by XARK, which were introduced in Section 2.2. Tables 2-4 summarize the number of kernels N found in each module of SparsKit-II, PLTMG, Perfect and SPEC2000. The last rows show the totals for each module, including the percentage of computational kernels that contain irregular computations. Measurements of the complexity of the kernels are also presented: S is the range of statements that compose the kernels; C is the range of conditions checked in if-endif statements; and L is the range of nested loops that contain the statements of the kernels. The ranges are displayed in the format m-M, the numbers m and M being the minimum and the maximum, respectively.

Sparskit-II and PLTMG are codes that contain a high percentage of irregular computations, ranging from 41% in PLTMG up to 64% in the BLASSM module of Sparskit-II (see last row of Table 2). Irregularity is due to the presence of kernels with irregular access patterns (e.g., irregular assignment, irregular reduction, irregular find-and-set and irregular recurrence), array references in the conditions of if-endif constructs, or read-only subscripted array references

244   M. Arenaz, J. Touriño, and R. Doallo

Table 3. Kernel families recognized by XARK in Perfect

Kernel family	ADM N	S	C	L	SPICE N	S	C	L	QCD N	S	C	L	MDG N	S	C	L	TRACK N	S	C	L	BDNA N	S	C	L
Assignments																								
Scalar assignment	278	1-3	1-6	1-3	42	1-1	0-0	1-2	14	1-2	1-2	1-2	20	1-1	0-0	1-3	2	1-1	0-0	1-1	360	1-3	0-2	1-4
Regular assignment	169	1-10	1-4	1-3	23	1-1	0-0	1-2	22	1-1	1-1	1-3	23	1-1	0-0	1-2	45	1-2	1-1	1-2	113	1-4	0-1	1-2
Irregular assignment																					1	1-1	0-0	2-2
Induction variables																								
Linear	62	1-2	2-2	1-3	20	1-1	0-0	1-2	2	1-1	0-0	4-4	23	1-1	1-1	1-3	6	1-1	1-1	1-2	43	1-1	0-0	1-3
Polynomial	4	1-1	0-0	2-3																				
Geometric	12	1-1	1-1	1-2	1	1-1	0-0	1-1					2	1-1	0-0	1-2	9	1-1	0-0	1-3	7	1-1	0-0	1-2
Reductions																								
Scalar reduction	17	1-1	0-0	1-3	6	1-1	0-0	1-2	2	1-1	0-0	1-1	22	1-2	0-0	1-2	4	1-1	0-0	1-2	33	1-1	0-0	1-2
Regular reduction	13	1-2	0-0	1-2	5	1-1	0-0	1-1	5	1-1	0-0	1-2	7	1-3	0-0	1-2	8	1-2	1-1	1-3	49	1-6	0-0	1-4
Irregular reduction									1	1-1	1-1	2-2									14	1-12	0-0	1-3
Masks																								
Regular find-and-set	1	2-2	1-1	1-1																				
Recurrences																								
Regular recurrences	7	1-1	0-0	1-3	5	1-1	0-0	1-1					1	1-1	0-0	1-1	6	1-1	0-0	1-1	20	1-3	0-0	1-2
Irregular recurrences					10	1-2	0-0	1-1																
Reinitialized kernels																								
Induction variables	7	2-2	0-0	2-3	1	2-2	0-0	2-2	3	2-2	0-0	4-4	10	2-2	0-0	2-4					1	2-2	0-0	3-3
Reductions					1	2-2	0-0	2-2					4	2-2	0-0	2-2					2	2-2	0-0	2-4
Complex written arrays																								
C. written ar. (CwriA)	4	2-2	0-0	1-2					45	2-28	1-1	1-4	10	2-2	0-0	1-4					23	2-2	0-0	1-1
C. reduced ar. (CredA)	8	2-2	0-0	2-3									4	2-2	0-0	2-2					45	2-2	0-0	1-2
Segmented CwriA													1	3-3	0-0	2-2								
Segmented CredA													1	3-3	0-0	2-2								
TOTALS	582	1-10	0-6	1-3	114	1-2	0-0	1-2	94	1-28	0-2	1-4	128	1-3	0-1	1-4	80	1-2	0-1	1-3	711	1-12	0-2	1-4
Kernels with irregular computations	29	5%			22	19%			6	6%			22	17%			8	10%			118	17%		

Kernel family	OCEAN N	S	C	L	DYFESM N	S	C	L	MG3D N	S	C	L	ARC2D N	S	C	L	FLO52 N	S	C	L	TRFD N	S	C	L	SPEC77 N	S	C	L
Assignments																												
Scalar assignment	14	1-2	3-3	1-2	32	1-1	1-4	1-4	747	1-2	1-1	1-3	158	1-3	0-3	1-3	40	1-1	0-0	1-2	1	1-1	0-0	2-2	64	1-3	1-2	1-3
Regular assignment	29	1-2	0-0	1-2	55	1-3	0-1	1-4	65	1-5	0-0	1-3	65	1-4	0-0	1-3	56	1-4	0-0	1-3	5	1-2	0-0	1-2	120	1-2	0-1	1-2
Irregular assignment					6	1-1	1-1	2-2																				
Induction variables																												
Linear	27	1-1	0-0	1-2	10	1-1	0-0	1-2	194	1-2	0-0	1-3	24	1-2	1-1	1-3	11	1-1	1-1	1-2	10	1-2	1-1	1-4	62	1-2	1-1	1-3
Polynomial	8	1-1	0-0	1-2					52	1-1	0-0	2-3													2	1-1	1-1	2-2
Geometric	9	1-1	0-0	1-2					6	1-1	0-0	1-1					2	1-1	0-0	1-1					9	1-1	0-0	1-2
Reductions																												
Scalar reduction	4	1-1	0-0	1-1	18	1-2	0-1	1-4	2	1-1	0-0	1-1	1	1-1	0-0	2-2	4	1-1	0-0	1-1	4	1-1	0-0	2-4	33	1-1	1-1	1-3
Regular reduction	30	1-1	0-0	1-2	18	1-4	1-1	1-3	13	1-2	0-0	1-2	30	1-4	0-0	1-3	18	1-1	0-0	1-3	4	1-2	0-0	1-1	59	1-2	1-1	1-3
Irregular reduction					4	1-3	1-1	1-2																				
Scalar min/max																									3	1-1	1-1	1-2
Masks																												
Scalar find-and-set																									1	1-1	1-1	2-2
Recurrences																												
Regular recurrences	6	1-1	0-0	1-2	5	1-1	1-1	1-2	2				69	1-2	0-0	1-2	11	1-2	0-0	1-2					7	1-1	1-1	1-2
Reinitialized kernels																												
Induction variables	4	2-2	0-0	2-2	2	2-2	0-0	2-4	36	2-2	0-0	2-3					10	2-2	0-0	2-3	1	2-2	0-0	2-2	16	2-2	0-0	2-3
Reductions					11	2-2	0-0	2-4																	14	2-2	0-0	2-2
Complex written arrays																												
C. written ar. (CwriA)	2	2-2	0-0	1-1					2	2-2	0-0	1-1									2	2-2	0-0	1-1	2	2-2	0-0	1-1
C. recurrenced ar. (CrecA)																	2	2-2	0-0	1-1								
Segmented CwriA																									15	4-10	0-0	2-3
TOTALS	133	1-2	0-3	1-2	161	1-4	0-4	1-4	1119	1-5	0-1	1-3	347	1-4	0-3	1-3	154	1-4	0-1	1-3	27	1-2	0-1	1-4	407	1-10	0-2	1-3
Kernels with irregular computations	7	5%			30	19%			23	2%			29	8%			8	5%			0	0%			24	6%		

used in the computations of other kernels. Sparskit-II consists of small routines
with complex computations. Apart from the high percentage of irregular computations, this complexity is reflected in the maximum number of statements
(6 and 5 in BLASSM and FORMATS, respectively), conditions (4 in UNARY
and FORMATS) and nested loops (4 in MATVEC, FORMATS and PLTMG).
The experiments also reveal that PLTMG is a complex application. It consists
of 1575 kernels that involve up to 9 statements, 9 conditions and 4 nested loops.
Note that PLTMG is the unique benchmark that contains kernels of the eight
families presented in Section 2.2. It contains 46 maps, which is a family that
does not appear in any other benchmark. This fact shows that the recognition

Table 4. Kernel families recognized by XARK in SPEC CPU2000

Kernel family	WUPWISE				SWIM				MGRID				APPLU				SIXTRACK				APSI			
	N	S	C	L	N	S	C	L	N	S	C	L	N	S	C	L	N	S	C	L	N	S	C	L
Assignments																								
Scalar assignment	6	1-2	1-2	1-3					1	1-1	0-0	3-3	24	1-1	0-0	1-4	32	1-2	1-2	1-3	276	1-3	1-6	1-3
Regular assignment	2	1-1	0-0	1-2	23	1-1	0-0	2-2	4	1-1	0-0	3-3	26	1-25	0-1	1-4	432	1-29	1-4	1-4	167	1-10	1-4	1-3
Irregular assignment									1	2-2	0-0	1-1					3	1-1	1-1	1-1				
Induction variables																								
Linear	4	1-1	0-0	1-1													22	1-2	1-1	1-4	61	1-2	2-2	1-3
Polynomial																					4	1-1	0-0	2-3
Geometric									3	1-1	0-0	3-3	6	1-1	0-0	2-4	30	1-1	0-0	1-2	12	1-1	1-1	1-2
Reductions																								
Scalar reduction	2	1-1	0-0	1-1	3	1-1	0-0	2-2	1	1-1	0-0	3-3	3	1-1	0-0	2-2	8	1-1	0-0	1-1	17	1-1	0-0	1-3
Regular reduction	12	1-3	1-2	1-3	4	1-1	0-0	2-2	5	1-4	0-0	1-3	8	1-2	0-0	1-4	36	1-2	1-2	1-4	10	1-1	0-0	1-2
Irregular reduction																	4	1-1	1-1	1-1				
Scalar min/max									1	1-1	1-1	3-3												
Regular min/max													2	1-1	1-1	1-4								
Masks																								
Scalar find-and-set	1	2-2	4-4	1-1													1	1-1	1-1	2-2				
Regular find-and-set																	1	2-2	2-2	1-1	1	2-2	1-1	1-1
Recurrences																								
Regular recurrences					30	1-1	0-0	1-1	5	1-2	0-0	1-2	9	1-1	0-0	1-3	86	1-3	1-2	1-3	8	1-2	0-0	1-3
Reinitialized kernels																								
Induction variables																					7	2-2	0-0	2-3
Reductions	6	2-2	0-0	3-3													2	2-2	0-0	2-2				
Complex written arrays																								
C. written ar. (CwriA)	1	2-2	0-0	1-1													1	4-4	0-0	1-1	4	2-2	0-0	1-2
C. reduced ar. (CredA)	2	2-2	0-0	1-1																	6	2-2	0-0	2-3
TOTALS	36	1-2	0-4	1-3	60	1-1	0-0	1-2	21	1-4	0-1	1-3	78	1-25	0-1	1-4	658	1-29	0-4	1-4	573	1-10	0-6	1-3
Kernels with irregular computations	9	25%			7	12%			2	10%			8	10%			171	26%			27	5%		

of all the kernel families is essential in order to fully characterize the behavior of real applications.

Perfect and SPEC2000 are codes characterized by the regularity of their computations. On average, only 11% of the kernels contain some type of irregularity, ranging from 0% in TRFD up to 26% in SIXTRACK. The complexity of the analysis of the Perfect benchmarks is mainly due to the existence of plenty of linear, polynomial and geometric induction variables. SPEC2000 includes large applications such as APPLU, SIXTRACK and APSI. They contain complex implementations of the well-known kernels assignments, induction variables and reductions. Note that the maximum number of statements is 29 in SIXTRACK, the maximum number of conditions is 6 in APSI, and the maximum number of nested loops is 4 in APPLU and SIXTRACK. These results demonstrate that the representation of programs in terms of computational kernels hides the complexity of the implementation, and eases the understanding of program behavior by the compiler.

Overall, the three families assignments, induction variables and reductions cover, on average, 83% of the computations of the recognized loops. In MATVEC, ADM, TRACK, OCEAN, MG3D and APSI, they cover more than 90% of the computations, being almost 100% in MATVEC and MG3D. The family of complex written arrays covers 6% of the kernels on average. This percentage raises up to 13%, 14% and even 48% in MDG, BLASSM and QCD, respectively. Note that, implicitly, these kernels involve the computation of IVs and reinitialized IVs. Finally, it should be noted that the experiments enabled to identify new computational kernels not studied in the literature so far, in particular, array maps, irregular array recurrences, consecutively reduced (and recurrenced) arrays and segmented consecutively reduced (and recurrenced) arrays.

4   Conclusion

This article has demonstrated that a significant amount of the regular and irregular computations carried out in full-scale real applications can be characterized using the families of computational kernels recognized by the XARK compiler. The representation of programs in terms of kernels hides the complexity of implementation details, providing optimizing compilers with a promising tool to reason about programs and, thus, to guide program optimizations.

In addition, the experiments have shown that full-scale real applications require the recognition of all the kernel families detected by XARK. Finally, note that new kernel families that had not been studied in the literature have been found in the benchmarks.

As future work we intend to give a step forward by describing program behavior using a higher-level of abstraction that consists of dependence relationships between the computational kernels. It is also intended to use such high-level representation for code generation using a stream programming model.

References

1. Andrade, D., Arenaz, M., Fraguela, B.B., Touriño, J., Doallo, R.: Automated and Accurate Cache Behavior Analysis for Codes with Irregular Access Patterns. Concurrency and Computation: Practice and Experience (in press)
2. Arenaz, M., Touriño, J., Doallo, R.: A GSA-Based Compiler Infrastructure to Extract Parallelism from Complex Loops. In: 17th International Conference on Supercomputing, San Francisco, CA, 193–204 (2003)
3. Arenaz, M.: Compiler Support for Parallel Code Generation through Kernel Recognition. In: 18th International Parallel and Distributed Processing Symposium, Santa Fe, NM (2004)
4. Bank, R.E.: PLTMG Package. Available at `http://cam.ucsd.edu/~reb/software.html` [Last accessed May 2007]
5. Berry, M., Chen, D., Koss, P., Kuck, D., Pointer, L., Lo, S., Pang, Y., Roloff, R., Sameh, A., Clementi, E., Chin, S., Schneider, D., Fox, G., Messina, P., Walker, D., Hsiung, C., Schwarzmeier, J., Lue, K., Orzag, S., Seidl, F., Johnson, O., Swanson, G., Goodrum, R., Martin, J.: The Perfect Club Benchmarks: Effective Performance Evaluation of Supercomputers. International Journal of Supercomputer Applications 3(3), 5–40 (1989)
6. Gerlek, M.P., Stoltz, E., Wolfe, M.: Beyond Induction Variables: Detecting and Classifying Sequences using a Demand-Driven SSA. ACM Transactions on Programming Languages and Systems 17(1), 85–122 (1995)
7. Lin, Y., Padua, D.A.: On the Automatic Parallelization of Sparse and Irregular Fortran Programs. In: O'Hallaron, D.R. (ed.) LCR 1998. LNCS, vol. 1511, pp. 41–56. Springer, Heidelberg (1998)
8. Pottenger, W.M., Eigenmann, R.: Idiom Recognition in the Polaris Parallelizing Compiler. In: 9th International Conference on Supercomputing, Barcelona, Spain, pp. 444–448 (1995)
9. Saad, Y.: SPARSKIT: A Basic Tool Kit for Sparse Matrix Computations (Version 2). Available at `http://www.cs.umn.edu/~saad/software/SPARSKIT/sparskit.html` [Last accessed May 2007]

10. SPEC. SPEC CPU2000. Standard Performance Evaluation Corporation. Available at `http://www.spec.org/cpu2000/` [Last accessed May 2007]
11. Suganuma, T., Komatsu, H., Nakatani, T.: Detection and Global Optimization of Reduction Operations for Distributed Parallel Machines. In: 10th International Conference on Supercomputing, Philadelphia, PA, pp. 18–25 (1996)
12. Tu, P., Padua, D.A.: Gated SSA-Based Demand-Driven Symbolic Analysis for Parallelizing Compilers. In: 9th International Conference on Supercomputing, Barcelona, Spain, pp. 414–423 (1995)

Towards Real-Time Compression of Hyperspectral Images Using Virtex-II FPGAs

Antonio Plaza

Department of Computer Science, University of Extremadura
Avda. de la Universidad s/n, E-10071 Caceres, Spain
aplaza@unex.es

Abstract. Hyperspectral imagery is a new type of high-dimensional image data which is now used in many Earth-based and planetary exploration applications. Many efforts have been devoted to designing and developing compression algorithms for hyperspectral imagery. Unfortunately, most available approaches have largely overlooked the impact of mixed pixels and subpixel targets, which can be accurately modeled and uncovered by resorting to the wealth of spectral information provided by hyperspectral image data. In this paper, we develop an FPGA-based data compression technique which relies on the concept of spectral unmixing, one of the most popular approaches to deal with mixed pixels and subpixel targets in hyperspectral analysis. The proposed method uses a two-stage approach in which the purest pixels in the image (endmembers) are first extracted and then used to express mixed pixels as linear combinations of end-members. The result is an intelligent, application-based compression technique which has been implemented and tested on a Xilinx Virtex-II FPGA.

1   Introduction

Due to significantly improved spectral resolution provided by latest-generation hyper-spectral imaging sensors, hyperspectral imagery expands the capability of multispectral imagery in many ways, such as subpixel target detection, object discrimination, mixed pixel classification and material quantification [1]. Each pixel in a hyperspectral image is composed of hundreds of reflectance values which define a 'spectral signature' for each pixel. By realizing the importance of hyperspectral data compression, many efforts have been devoted to designing and developing compression algorithms for hyperspectral imagery [2]. Two types of data compression can be performed, lossless and lossy, in accordance with redundancy removal. More specifically, lossless data compression is generally considered as data compaction which eliminates unnecessary redundancy without loss of information. By contrast, lossy data compression removes unwanted redundancy or insignificant in-formation which results in entropy reduction. Which type of compression should be used depends heavily upon the application under study. For example, in medical imaging, lossless compression is preferred. However, in this case only small compression ratios can be achieved (typically, 3:1

A.-M. Kermarrec, L. Bougé, and T. Priol (Eds.): Euro-Par 2007, LNCS 4641, pp. 248–257, 2007.

or below). On the other hand, video processing such as high definition television (HDTV) can greatly benefit from lossy compression. For remotely sensed imagery, both types of compression have been investigated in the past [2].

Our main focus in this work is to design compression techniques able to reduce significantly the large volume of information contained in hyperspectral data while, at the same time, being able to retain information that is crucial to deal with mixed pixels and subpixel targets. These two types of pixels above are essential in many hyperspectral analysis applications, including military target detection and tracking, environmental modeling and assessment at sub-pixel scales, etc. A subpixel target is a mixed pixel with size smaller than the available pixel size (spatial resolution) [3]. So, it is embedded in a single pixel and its existence can only be verified by using the wealth of spectral information provided by hyperspectral sensors. A mixed pixel is a mixture of two or more different substances present in the same pixel [4]. In this case, spectral information can greatly help to effectively characterize the substances within the mixed pixel via spectral unmixing techniques [5]. When hyperspectral image compression is performed, it is critical and crucial to take into account these two issues, which have been generally overlooked in the development of lossy compression techniques in the literature [6].

The possibility of real-time, onboard data compression is a highly desirable feature to overcome the problem of transmitting a sheer volume of high-dimensional data to Earth control stations via downlink connections. An exciting new development in the field of specialized commodity computing is the emergence of hardware devices such as field programmable gate arrays (FPGAs), which can bridge the gap towards onboard and real-time analysis of remote sensing data [7,8]. FPGAs are now fully reconfigurable, which allows one to adaptively select a data processing algorithm (out of a pool of available ones) to be applied onboard the sensor from a control station on Earth. The ever-growing computational demands of remote sensing applications can fully benefit from compact, reconfigurable hardware components and take advantage of the small size and relatively low cost of these units as compared to clusters or networks of computers [9].

In this work, we explore a solution based on mapping the proposed compression algorithm on FPGA hardware. The remainder of the paper is organized as follows. Section 2 develops a new application-oriented lossy compression algorithm which utilizes a two-stage approach: first, a pixel purity index (PPI) algorithm is used to extract the purest pixels (endmembers) in the image, and then a linear spectral unmix-ing (LSU) procedure is used to express mixed pixels as linear combinations of endmembers, weighted by their respective abundance fractions. Section 3 maps the proposed compression algorithm in hardware using systolic array design. Section 4 provides experimental evidence about the algorithm performance using a real image data set collected by the NASA Jet Propulsion Laboratory's Airborne Visible Infra-Red Imaging Spectrometer (AVIRIS). Parallel performance data are given using a Xilinx Virtex-II FPGA. Finally, Section 5 concludes with some remarks and hints at plausible future research.

2   Hyperspectral Data Compression

2.1   Compression Algorithm

The idea of the proposed data compression algorithm is to represent a hyperspectral image cube by a set of fractional abundance images [10]. More precisely, for each N-dimensional pixel vector $\mathbf{f}_i$, its associated abundance vector $\mathbf{a}_i$ of E dimensions is used as a fingerprint of $\mathbf{f}$ with regards to E endmembers obtained by the pixel purity index (PPI) algorithm. The implementation of the proposed data compression algorithm can be summarized by the following steps:

1. Use the PPI algorithm to generate a set of E endmembers $\{\mathbf{e}_e\}_{e=1}^{E}$.
2. For each pixel vector $\mathbf{f}_i$ in the input scene, use the LSU algorithm to estimate the corresponding endmember abundance fractions $\mathbf{a}_i = \{a_{i1}, a_{i2}, \cdots, a_{iE}\}$ and approximate $\mathbf{f}_i = \mathbf{e}_1 \cdot a_{i1} + \mathbf{e}_2 \cdot a_{i2} + \cdots + \mathbf{e}_E \cdot a_{iE}$. Note that this is a reconstruction of $\mathbf{f}_i$.
3. Construct E fractional abundance images, one for each endmember.
4. Apply lossless predictive coding to reduce spatial redundancy within each of the E fractional abundance images, using Huffman coding to encode predictive errors.

2.2   Pixel Purity Index (PPI)

The PPI is a well-known approach to deal with the problem of mixed pixels in hyperspectral imaging. In this work, we use an improved version of the PPI algorithm [4] as the first step of our compression algorithm. Due to the algorithm's propriety and limited published results, we provide an outline of the algorithm which is based on limited published results and our own interpretation [3].

The PPI generates a large number of random, N-dimensional unit vectors called 'skewers' through the dataset. Every data point is projected onto each skewer, and the data points that correspond to extrema in the direction of a skewer are identified and placed on a list. As more skewers are generated, the list grows, and the number of times a given pixel is placed on this list is also tallied. The pixels with the highest tallies are considered the final endmembers.

The inputs to PPI are a hyperspectral data cube $\mathbf{F}$ with N dimensions; a maximum number of endmembers to be extracted, E; the number of random skewers to be generated during the process, K; a cut-off threshold value, t_v, used to select as final endmembers only those pixels that have been selected as extreme pixels at least t_v times throughout the process; and a threshold angle, t_a, used to discard redundant endmembers. The output of the algorithm is a set of E final endmembers $\{\mathbf{e}_e\}_{e=1}^{E}$. The algorithm is summarized as follows:

1. Produce a set of K randomly generated unit vectors $\{\mathbf{skewer}_j\}_{j=1}^{K}$.
2. For each $\mathbf{skewer}_j$, all sample pixel vectors $\mathbf{f}_i$ in the original data set $\mathbf{F}$ are projected onto $\mathbf{skewer}_j$ via dot products of $|\mathbf{f}_i \cdot \mathbf{skewer}_j|$ to find sample vectors at its extreme (maximum and minimum) projections, thus forming an extrema set for $\mathbf{skewer}_j$ which is denoted by $S_{extrema}(\mathbf{skewer}_j)$. Despite

the fact that a different $\mathbf{skewer}_j$ would generate a different extrema set $S_{extrema}(\mathbf{skewer}_j)$, it is very likely that some sample vectors may appear in more than one extrema set. In order to deal with this situation, we define an indicator function of a set S, denoted by $I_S(\mathbf{x})$, to denote membership of an element $\mathbf{x}$ to that particular set as follows:

$$I_S(\mathbf{f}_i) = \begin{cases} 1 \text{ if } \mathbf{x} \in S \\ 0 \text{ if } \mathbf{x} \notin S \end{cases} \tag{1}$$

3. Calculate the PPI score associated to the pixel vector $\mathbf{f}_i$ using the following equation:

$$N_{PPI}(\mathbf{f}_i) = \sum_{j=1}^{K} I_{S_{extrema}(\mathbf{skewer}_j)}(\mathbf{f}_i) \tag{2}$$

4. Find the pixels with value of $N_{PPI}(\mathbf{f}_i)$ above t_v, and form a unique set of endmembers $\{\mathbf{e}_e\}_{e=1}^{E}$.

2.3 Linear Spectral Unmixing (LSU)

For each sample pixel vector $\mathbf{f}_i$ in $\mathbf{F}$, a set of abundance fractions specified by $\mathbf{a}_i = \{a_{i1}, a_{i2}, \cdots, a_{iE}\}$ is obtained using the set of endmembers $\{\mathbf{e}_e\}_{e=1}^{E}$, so that $\mathbf{f}_i$ can be expressed as a linear combination of endmembers as follows:

$$\mathbf{f}_i = \mathbf{e}_1 \cdot a_{i1} + \mathbf{e}_2 \cdot a_{i2} + \cdots + \mathbf{e}_E \cdot a_{iE} \tag{3}$$

In order to achieve the decomposition above, we multiply each pixel $\mathbf{f}_i$ by $(\mathbf{M}^T\mathbf{M})^{-1}\mathbf{M}^T$, where $\mathbf{M} = \{\mathbf{e}_e\}_{e=1}^{E}$ and the superscript 'T' denotes the matrix transpose operation. In the expression above, abundance sum-to-one and non-negativity constraints are imposed, i.e., $\sum_{e=1}^{E} a_{ie} = 1$ and $a_{ie} \geq 0$ for all $i = \{1 \cdots T\}$, where T is the total number of pixels in the image $\mathbf{F}$, and for all $e = \{1 \cdots E\}$, where E is the total number of endmembers extracted by PPI.

3 FPGA-Based Hardware Implementation

In this subsection, we describe a hardware-based parallel strategy for implementation of the hyperspectral data processing chain which is aimed at enhancing replicability and reusability of slices in FPGA devices through the utilization of systolic array de-sign [11]. One of the main advantages of systolic array-based implementations is that they are able to provide a systematic procedure for system design that allows for the derivation of a well defined processing element-based structure and an interconnection pattern which can then be easily ported to real hardware configurations [12].

The rationale behind our systolic array-based parallelization can be summarized as follows. The PPI algorithm consists of computing a very large number

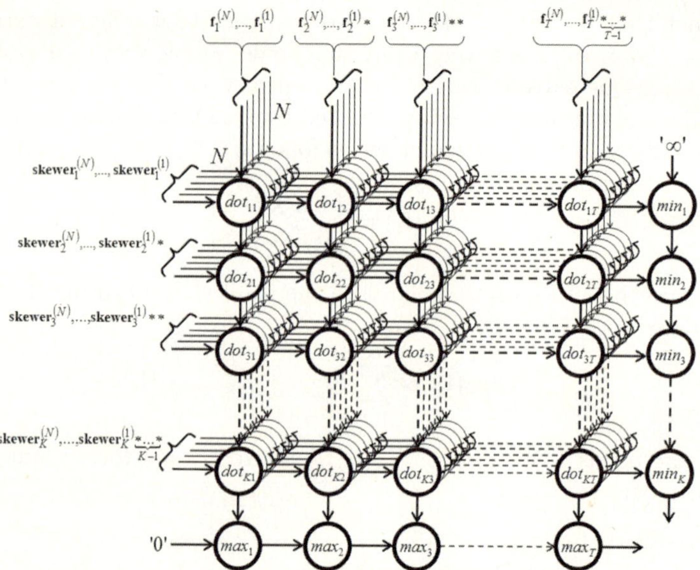

Fig. 1. Systolic array design for the proposed FPGA implementation of the PPI

of dot-products, and all these dot-products can be performed simultaneously. As a result, a possible way of parallelization is to have a hardware system able to compute K dot-products in the same time against the same pixel $\mathbf{f}_i$, where K is the number of skewers and $i = \{1 \cdots T\}$, with T being the total number of pixels in the input scene. Now, if we suppose that we cannot simultaneously compute K dot-products but only a fraction K/P, where P is the number of available processing units, then the PPI algorithm can be split into P passes, each performing dot-products, where T is the total number of input pixels to be fed to the systolic. From an architectural point of view, each processor receives T pixels, computes T dot-products, and keeps in memory the two pixels having produced the *min* and the *max* dot-products. In this scheme, each processor holds a different skewer which must be input before each new pass.

Fig. 1 illustrates the above principle, in which local results remain static at each processing element, while pixel vectors are input to the systolic array from top to bottom and skewer vectors are fed to the systolic array from left to right. In Fig. 1, asterisks represent delays while $\mathbf{skewer}_j^{(n)}$ denotes the value of the n-th band of the j-th skewer, with $j \in \{1, \cdots, K\}$ and $n \in \{1, \cdots, N\}$, being N the number of bands of the input hyperspectral scene. Similarly, $\mathbf{f}_i^{(n)}$ denotes the reflectance value of the n-th band of the i-th pixel, with $i \in \{1, \cdots, T\}$, being T is the total number of pixels in the input image. The processing nodes labeled as *dot* in Fig. 1 perform the individual products for the skewer projections. On the other hand, the nodes labeled as *max* and *min* respectively compute the maxima and minima projections after the dot product calculations have been completed. In fact, the *max* and *min* nodes can be respectively seen as part of a 1-D systolic array which avoids broadcasting the pixel while simplifying the collection of the results.

The main advantage of the systolic array described in Fig. 1 is its scalability. Depending of the resources available on the reconfigurable board, the number of processors can be adjusted without modifying the control of the array. In order to reduce the number of passes, we decide to allocate the maximum number of processors in the available FPGA components. In other words, although in Fig. 1 we represent an ideal systolic array in which T pixels can be processed, this is not the usual situation, and the number of pixels usually has to be divided by P, the number of available processors. In this scenario, after T/P systolic cycles, all the nodes are working. When all the pixels have been flushed through the systolic array, T/P additional systolic cycles are thus required to collect the results for the considered set of P pixels and a new set of P different pixels would be flushed until processing all T pixels in the original image. Finally, to obtain the vector of endmember abundances $\{a_{i1}, a_{i2}, \cdots, a_{iE}\}$ for each pixel $\mathbf{f}_i$, the multiplication of each $\mathbf{f}_i$ by $(\mathbf{M}^T\mathbf{M})^{-1}\mathbf{M}^T$, where $\mathbf{M} = \{\mathbf{e}_e\}_{e=1}^{E}$, is done as described in [13], i.e. using a simple parallel block algorithm.

The algorithm described above was synthesized using Handel-C, a hardware design and prototyping language that allows using a pseudo-C programming style. The source code in Handel-C corresponding to step 2 of our FPGA implementation of the PPI algorithm is shown in Algorithm 1. The implementation was compiled and transformed into an EDIF specification automatically by using the DK3.1 software package. We also used other tools such as Xilinx ISE 6.1i[1] to carry out automatic place and route (PAR), and to adapt the final steps of the hardware implementation to the Virtex-II FPGA used in experiments.

Algorithm 1. Handel-C implementation of the PPI for FPGAs

```
void main(void) {
    unsigned int 16 max[E]; //E is the number of endmembers
    unsigned int 16 end[E];
    unsigned int 16 i;
    unsigned int 10000 k; //k denotes the number of skewers
    unsigned int 224 N; //N denotes the number of bands
    par (i = 0; i < E; i++) max[i] = 0;
    par (k = 0; k < E; k++) {
        par (k = 0; k < E; k++) {
            par (j = 0; j < N; j++) {
                Proc_Element[i][k](pixels[i][j],skewers[k][j],0@i,0@k);}}}
    for (i = 0; i < E; i++) {
        max[i]=Proc_Element[i][k](0@max[i], 0, 0@i, 0@k); }
    phase_1_finished=1
    while (!phase_2) { //Waiting to enter phase 2 }
    for (i = 0; i < E; i++) end[i]=0;
    for (i = 0; i < E; i++) {
        par (k = 0; k < E; k++) {
            par (j = 0; j < N; j++) {
                end[i]=end[i]&&Proc_Element[i][k](pixels[i][j],skewers[k][j],0,0);}}}
    phase_2_finished=1
    global_finished=0
    for (i = 0; i < E; i++)   global_finished=global_finished&&end[i];
```

---

[1] http://www.xilinx.com

4   Experimental Results

This section provides an assessment of the effectiveness of the hardware-based compression algorithm described in sections 2 and 3. The algorithm was implemented on a Virtex-II XC2V6000-6 FPGA, which contains 33,792 slices, 144 Select RAM Blocks and 144 multipliers (of 18-bit x 18-bit). The algorithm was applied to a real hyperspectral scene collected by an AVIRIS flight over the Cuprite mining district in Nevada, which consists of 614×512 pixels and 224 bands. The site has several exposed minerals of interest. Fig. 2(left) shows a spectral band of the image, and Fig. 2(right) plots the spectra of five minerals measured in the field by U.S. Geological Survey (USGS).

In order to explore the quality of the compressed images produced by the proposed compression method, Table 1 reports the spectral angle similarity scores [1,3] among the USGS reference signatures in Fig. 2 and the PPI-extracted endmembers from the resulting images after data compression (the lowest the scores, the highest the similarity), using compression ratios of 20:1, 40:1 and 80:1 (given by different tested values of input parameter E). Spectral similarity scores below 0.1 are widely considered as a requirement in many applications [4].

As expected, the highest-quality endmembers were extracted from the original data set. As the compression ratio was increased, the quality of extracted

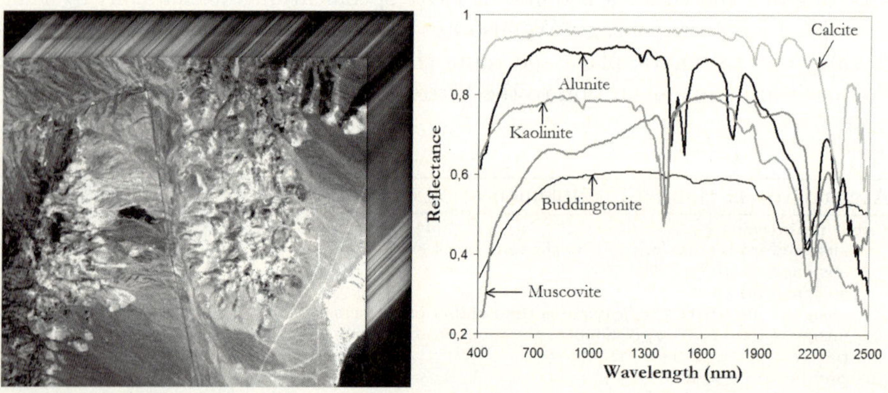

Fig. 2. AVIRIS hyperspectral image (left) and USGS mineral signatures (right)

Table 1. Spectral similarity scores among USGS spectra and endmembers extracted from the original image, and from several reconstructed versions of the image after applying PPI/LSU, JPEG2000 and SPIHT algorithms with different compression ratios

Mineral signature	Original image	PPI/LSU: 20:1	40:1	80:1	JPEG2000: 20:1	40:1	80:1	SPIHT: 20:1	40:1	80:1
Alunite	0.063	0.069	0.078	0.085	0.112	0.123	0.133	0.106	0.119	0.129
Buddingtonite	0.042	0.053	0.061	0.068	0.105	0.131	0.142	0.102	0.125	0.127
Calcite	0.055	0.057	0.063	0.074	0.102	0.128	0.139	0.097	0.122	0.134
Kaolinite	0.054	0.059	0.062	0.071	0.114	0.140	0.151	0.110	0.134	0.146
Muscovite	0.067	0.074	0.082	0.089	0.123	0.145	0.167	0.116	0.139	0.152

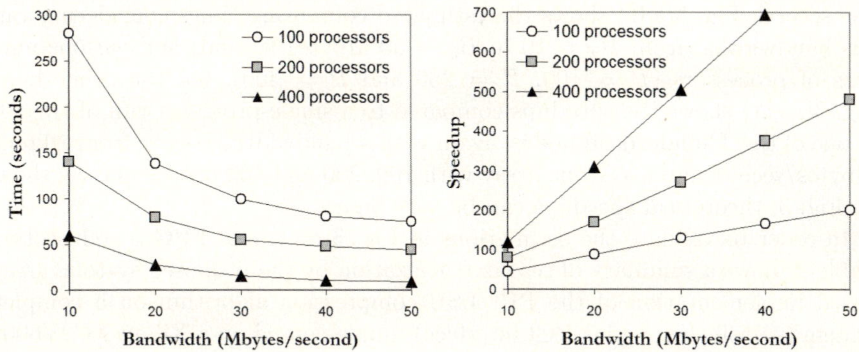

Fig. 3. Time in seconds for computing the full hyperspectral data compression algorithm using a reconfigurable board connected to a PC through the I/O bus (left) and speedup compared to a single-processor version running on a single AMD PC (right)

Table 2. Resource utilization for the FPGA implementation (operation frequency given in MHz)

Number of processors	Total gates	Total slices	% of total	Frequency
100	97,443	1,185	3%	29,257
200	212,412	3,587	10%	21,782
400	526,944	12,418	36%	18,032

endmembers was decreased. For illustrative purposes, we have also included the results provided by two standard methods in our comparison, i.e., the wavelet-based JPEG2000 multi-component [14] and the set partitioning in hierarchical trees (SPIHT) [15]. The JPEG2000 implementation used for our experiments was the one available in kakadu software[2]. Both techniques are 3-D compression algorithms that treat the hyperspectral data as a 3-D volume, where the spectral information is the third dimension. Results in Table 1 show that such 3-D techniques, which enjoy great success in classical image processing, may not necessarily find equal success in hyperspectral image compression. Specifically, techniques able to preserve the spectral information are required to characterize mixed pixels and subpixel targets. As demonstrated by Table 1, for the same compression ratio, a 3-D lossy compression may result in significant loss of spectral information which can be preserved much better, in turn, by an application-oriented algorithm such as the proposed PPI/LSU. It should be noted that the serial versions of the algorithms in Table 1 required several minutes of computation to compress the AVIRIS Cuprite data set in a PC with AMD Athlon 2.6 GHz processor and 512 MB of RAM (specifically, the PPI/LSU algorithm required almost one hour).

The average performance of the systolic array is mainly determined by the dot-product capacity, that is the number of additions/subtractions executed in

---

[2] http://www.kakadusoftware.com

one second. Fig. 3(left) shows the estimated computing times considering various bandwidths (from $B_w = 10$ to $B_w = 50$ Mbytes/second) and various numbers of processors ($P = 100$, $P = 200$ and $P = 400$). On the other hand, Fig. 3(right) shows the speedups compared to a single-processor run of the PPI in one of the Thunderhead nodes, again with a bandwidth ranging from 10 to 50 Mbytes/second and a systolic array with 100, 200 and 400 processors. As shown by Fig. 3, theoretical speedups can be very high.

In order to validate the estimations in Fig. 3 on a real FPGA architecture, Table 2 shows a summary of resource utilization by the proposed systolic array-based implementation of the PPI/LSU compression algorithm on a complete system (systolic array plus PCI interface), implemented on a Xilinx XC2V6000-6 board, using different numbers of processors. We measured an average PCI bandwidth of 15 Mbytes between the PC and the board, leading to a speedup of 120 when running the PPI/LSU with a maximum number of $P = 400$ processors. The optimum trade-off between the achieveable parallelism versus clockrate was found for the maximum number of processors used since the balance between the speedup found and the operation frequency (around 18 MHz) was satisfactory while at the same leaving enough room in the FPGA for implementation of additional algorithms (only 36% of the FPGA resources were used).

It should be noted that, in our experimentation, the performance was seriously limited by the transfer rate between the PC and the board: the array is able to absorb a pixel flow of above 40 Mbytes/second, while the PCI interface can only provide a flow of 15 Mbytes. This experiment, however, demonstrated that the considered board, even with a non-optimized PCI connection (with no DMA), can still yield very good speedup for the PPI/LSU, with a final execution time for all the compression procedure of only 7.94 seconds for $P = 400$ processors. This response is not strictly in real-time since the cross-track scan line in AVIRIS, a pushbroom instrument [1], is quite fast (8.3 msec). This introduces the need to process a full image cube (614×512 pixels with 224 bands) in no more than 5 seconds to fully achieve real-time performance. However, we anticipate that the proposed FPGA design can still be significantly optimized to fulfill real-time requirements (even without increasing the number of processors) by improving the communication bandwidth between the PC and the FPGA board, which seems feasible given the limited flow of the considered PCI interface.

5   Conclusion

The wealth of spectral information provided by hyperspectral sensors is essential in many applications, and needs to be retained by compression algorithms. Standard 3-D lossy compression techniques may cause significant loss of crucial information that is provided by mixed pixels and subpixel targets, which are essential in hyperspectral imaing applications. In order to satisfy (near) real-time requirements, we have developed an FPGA-based algorithm for onboard data compression. A major goal is to overcome the bottleneck introduced by the bandwidth of the downlink connection from the observatory platform. Experimental results demonstrate that our hardware version makes appropriate use of computing resources in the considered FPGA, and further provides a response in (near) real-time which is believed to be acceptable in most applications. It should be noted that efficient onboard compression has been a long-awaited goal by the remote sensing community. In this regard, the reconfigurability of FPGA systems opens many innovative perspectives from an application point of view. Although the experimental results presented in this paper are encouraging, further work is still needed to arrive to optimal parallel design and implementations for the proposed and other hyperspectral compression algorithms.

References

1. Chang, C.-I.: Hyperspectral imaging: Detection & classification. Kluwer Academic Publishers, Dordrecht (2003)
2. Motta, G., Rizzo, F., Storer, J.A.: Hyperspectral data compression. Springer, New York (2005)
3. Plaza, A., Chang, C.-I.: Impact of initialization on design of endmember extraction algorithms. IEEE Trans. Geoscience and Remote Sensing 44, 3397–3407 (2006)
4. Plaza, A., Martinez, P., Perez, R., Plaza, J.: A quantitative and comparative analysis of endmember extraction algorithms from hyperspectral data. IEEE Trans. Geoscience and Remote Sensing 42, 650–663 (2004)
5. Chang, C.-I., Plaza, A.: A Fast Iterative Implementation of the Pixel Purity Index Algorithm. IEEE Geoscience and Remote Sensing Letters 3, 63–67 (2006)
6. Du, J., Chang, C.-I.: Linear Mixture Analysis-Based Compression for Hyperspectral Image Analysis. IEEE Trans. Geoscience and Remote Sensing 42, 875–891 (2004)
7. El-Araby, E., El-Ghazawi, T., Le Moigne, J.: Wavelet spectral dimension reduction of hyperspectral imagery on a reconfigurable computer. In: Proc. of the 4th IEEE International Conference on Field-Programmable Technology, vol. 1, pp. 861–867 (2004)
8. Fry, T., Hauck, S.: Hyperspectral image compression on reconfigurable platforms. In: Proc. of the 10th IEEE Symposium on Field-Programmable Custom Computing Machines, vol. 1, pp. 305–312 (2002)
9. Plaza, A., Valencia, D., Plaza, J., Martinez, P.: Commodity cluster-based parallel processing of hyperspectral imagery. Journal of Parallel and Distributed Computing 66, 345–358 (2006)
10. Ramakhrishna, B., Plaza, A., Chang, C.-I., Ren, H.: Spectral/spatial hyperspectral image compression. In: Motta, G., Rizzo, F., Storer, J.A. (eds.) Hyperspectral data compression, pp. 309–346 (2005)
11. Valero-Garcia, M., Navarro, J., Llaberia, J., Valero, M., Lang, T.: A method for implementation of one-dimensional systolic algorithms with data contraflow using pipelined functional units. Journal of VLSI Signal Processing 4, 7–25 (1992)
12. Zhang, D., Pal, S.K.: Neural Nets & Systolic Array Design. World Scientific (2002)
13. Dou, Y., Vassiliadis, S., Kuzmanov, G., Gaydadjiev, G.: 64-bit floating-point FPGA matrix multiplication. In: Proc. of the 13th ACM/SIGDA International Symposium on FPGAs, vol. 1, pp. 123–129 (2005)
14. Taubman, D.S., Marcellin, M.W.: JPEG2000: Image Compression Fundamentals, Standard and Practice. Kluwer, Boston (2002)
15. Said, A., Pearlman, W.A.: A New, Fast, and Efficient Image Codec Based on Set Partitioning in Hierarchical Trees. IEEE Transactions on Circuits and Systems 6, 243–350 (1996)

Optimizing Chip Multiprocessor Work Distribution Using Dynamic Compilation

Jisheng Zhao, Matthew Horsnell, Ian Rogers, Andrew Dinn, Chris Kirkham, and Ian Watson

University of Manchester, UK
`{jishengz,horsnell,irogers,adinn,chris,watson}@cs.man.ac.uk`

Abstract. How can sequential applications benefit from the ubiquitous next generation of chip multiprocessors (CMP)? Part of the answer may be a dynamic execution environment that automatically parallelizes programs and adaptively tunes the work distribution. Experiments using the Jamaica CMP show how a runtime environment is capable of parallelizing standard benchmarks and achieving performance improvements over traditional work distributions.

Keywords: Automatic parallelization, feedback-directed optimization, dynamic execution.

1 Introduction

In most traditional optimizing compilers, each optimization has a corresponding performance prediction. These predictions are often based on abstract metrics, with the assumption that there is a direct correlation between the metric and the runtime performance. However, a given program may have markedly different characteristics when run with different input data or on different architectures, significantly impacting the performance of an optimization.

A runtime compilation environment has the potential to take into consideration the most promising optimizations and pick a good choice based on runtime profiling data to maximize performance, and avoid reapplying optimizations shown to incur performance degradation.A compiler-enabled virtual machine framework is presented capable of collecting runtime performance information and automatically reconfiguring the executing code. Using this *Online Tuning Framework* (OTF), a loop-based program can be parallelized and tuned at runtime, with acceptable overheads, increasing the performance when compared to traditional parallelization schemes.

This paper is organized as follows. Section 2 introduces the online tuning framework and its interaction with the Jikes Research Virtual Machine. Section 3 describes the experimental methodology. Section 4 presents and discusses the results from experimental evaluation of the OTF, Section 5 describes how this work compares to related research, and finally Section 6 concludes this paper.

A.-M. Kermarrec, L. Bougé, and T. Priol (Eds.): Euro-Par 2007, LNCS 4641, pp. 258–267, 2007.

2   Online Tuning Framework

The *Online Tuning Framework* (OTF) consists of three distinct elements: the loop parallelizing compiler (see Section 2.1), the adaptive optimization component (see Section 2.2), and the runtime profiler (see Section 2.3).

The OTF is embedded within the adaptive optimization system (AOS) of the Jikes Research Virtual Machine (RVM) [1,2]. The Jikes RVM captures runtime information by instrumenting the running code at the method-level. Once the instrumentation indicates that a given method is *hot* (i.e. number of times the method is executed is above a threshold), the AOS decides whether to compile it using an optimizing compiler[3]. The OTF hijacks this decision, so that any hot method is also considered for parallelizing optimizations. The following sections describe in detail the internal elements of the OTF and how they interact with the AOS.

2.1   Loop Parallelizing Compiler

The Loop Parallelizing Compiler (LPC) searches for fine-grain parallel code within amenable loop structures. The LPC works within two phases of the Jikes RVM optimizing compiler's workflow:

Loop Analysis and Annotation. Loop analysis and annotation occurs in the high-level optimization phase. In this phase the LPC detects loop structures, analyses the data dependencies within them, creates parallel loops where these dependencies can be maintained, and annotates the loops with high-level pseudo code. By performing the analysis of loops at this high-level compiler phase the LPC benefits from Java's strong typing and single static assignment (SSA) form [4].

In order to determine whether array accesses within the loops are amenable to parallelization the Banerjee Test [5] is performed. This allows *do-all* and *do-across* loops to be created when presented with loop carried dependencies on arrays with affine indices. The code for the parallel loop body is placed at the end of the method containing the parallel loop. All parallel loop bodies have a prologue to set up their state and load loop-constant values, and an epilogue to join them back with the parent thread.

Parallel Thread Creation. Parallel thread creation occurs in the machine-level optimization phase. In this phase the previously inserted pseudo-code is replaced by machine specific code, enabling the code to fork new threads on idle processor contexts[1], as well as applying different adaptively optimizing distribution policies.

---

[1] A processor context is defined as a hardware supported context within a chip multiprocessor architecture.

2.2 Adaptive Optimization Component

The Adaptive Optimization Component (AOC) inserts one or more optimizations deemed to be appropriate for optimizing a given loop, identified by the LPC, into the code. The AOC is invoked by the LPC to place the optimizations around the identified parallel loops. Presently the AOC supports three adaptive optimizations for parallelizable code (see Optimizations 1–3 below). These three optimizations vary either the number of loop iterations inside a block[2] or tile, the number of threads created, or the manner in which the blocks or tiles are distributed. By varying these factors the OTF is able to find strategies that best balance the costs associated with threading, the cache performance, and the system load.

Optimization 1 – Adaptive Block Division (ABD). For a given loop the total number of iterations is divided into blocks. Each block is then distributed through the creation of a parallel thread. The parallel threads can be run on any available processor context. In all optimizations if a thread cannot be invoked on a remote processor context the generator thread must consume its work before continuing to distribute subsequent threads. This optimization uses two simple hill-climbing like algorithms [6] to adaptively divide the total number of iterations in a loop into blocks. The first algorithm, searches for an optimal divisor in the range $1 \leq$ optimal $\leq$ number of processor contexts. An extension to this algorithm is used to increase the search to the range $1 \leq$ optimal $\leq m$, where m is a multiple of the number of processor contexts.

Optimization 2 – Adaptive Tile Division (ATD). As loops can be tiled to take advantage of data resuse [7], selecting a suitable tile size is a common technique for improving performance. This optimization is applied when a perfect nested loop is identified by the LPC. The 2-dimensional loop traversal of the iteration space is divided into tiles which are then distributed by the creation of parallel threads. Each tile has a corresponding divisor pair. Given a divisor pair (D_i, D_j), D_i is the divisor corresponding to the outer loop iterator and D_j corresponds to the inner loop iterator. Adaptive searching, again using simple hill-climbing, starts from the divisor pair $(M, 1)$, where M is the total number of processor contexts, this is equivalent to naïve ABD. D_i is incrementally decreased and D_j is increased. Two algorithms are used for ATD, one adaptively optimizes regular 2-dimensional loops and another optimizes triangular 2-dimensional loops. The division of both dimensions of the iteration space is configurable, but currently, the total number of tiles created is restricted to the total number of processor contexts. For regular 2D loops: $D_i \times D_j = M$, and for triangular 2D loops: $D_i \times \frac{D_j+1}{2} = M$.

Optimization 3 – Adaptive Version Selection (AVS). This optimization performs runtime selection between block/tile based loop distribution, as described in optimizations 1 and 2, and distribution using a cyclic recursive distribution (CRD), shown by Figure 1(b). CRD divides the tiles/blocks, using

---

[2] The term *block* is used to mean a contiguous sequence of loop iterations.

divisor pairs generated using ABD/ATD, such that half of them remain with the generator thread and half are distributed by the creation of a parallel generator thread to another processor context. This happens recursively until all the blocks/tiles are distributed. A variable cyclic stride pattern is also applied to shuffle the division of the blocks.

2.3   Runtime Profiler

To evaluate the performance of the selected adaptive optimizations, the OTF needs to be able to calculate each optimizations runtime profile. This is achieved by inserting two additional code stubs at the start and end of the parallelized loop being profiled. The first stub extracts from the architecture the cycle count[3] prior to the loops execution and the second stub extracts the cycle count after the loop has executed. The second stub is also responsible for reporting the total execution time, and the number of loop iterations, back to the AOS, running in a parallel thread, as shown in Figure 1(a).

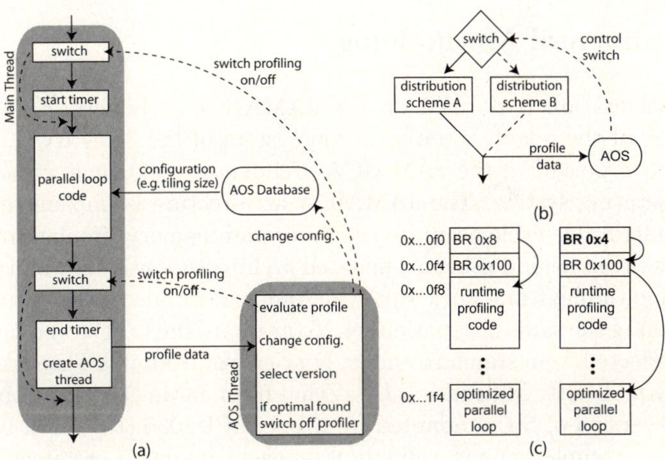

Fig. 1. Runtime profiling and adaptive version switching

The OTF is then able to calculate the execution time per iteration of each invocation of the loop and can make a decision about the comparative performance with other invocations of the loop under different optimizations and divisors. Once an optimal divisor is found for a given optimization the AOS stops profiling it and either switches to a different distribution scheme using AVS, as shown in Figure 1(b), or having assessed all optimizations the AOS switches off the runtime profiler and runs any subsequent executions of the loop using the

---

[3] Although this mechanism is machine specific; instructions exist in the main architectures: RDTSC (x86), mftb (PPC), TICK register (SPARC).

best optimization found. The code stub that previously invoked the runtime profiling is modified, so that future execution of the code no longer needs to execute any code inside the profiling phase, Figure 1(c). It should also be noted that the AOS instrumentation code for loop back-edges is removed from the parallel code sections which prevents their interruption at runtime.

The precision of the execution time metric is a major factor in getting good results from the presented optimizations, and there are two issues that affect the precision. The first is that not all loops are of static length or duration, it is possible that both the number of iterations and the loops content will vary per invocation. The second issue is that the execution timings are affected by system noise, for example cold caches and other unrelated thread activity. To overcome these issues, the execution time for a given optimization on a parallelized loop is calculated, as an arithmetic mean of the cycles per iteration for four invocations of that loop. Loops that exhibit large profile deviations, defined as having a *coefficient of variation*[4] (CV) greater than a configurable threshold, for this work set at 0.1, are deemed unstable, the profiling code is switched off and the current best optimization is used.

3   Experimental Methodology

The experiments are performed on the JAMAICA architecture [8] using the OTF as part of the adaptive optimization system of the Jikes RVM. The Jikes RVM has been ported to the JAMAICA architecture and runs without an underlying operating system. The JAMAICA architecture is implemented within a highly configurable cycle-accurate processor and memory simulation platform. The main decision behind using a simulated architecture was the ability to evaluate the online tuning framework on a wide range of simulated hardware configurations all using the same instruction set. To evaluate the OTF seven benchmarks have been selected from standard suites, FourierTest from jBYTEmark [9], Euler from JavaGrande [10], MatrixMul, LU, Zchol from JaMa [11], Java Linpack [12] and a Java version of Swim adapted from SpecCPU2000 [13]. Each benchmark is executed to completion and validation on each simulated architecture configuration. The configurations assess the performance of the OTF in the presence of varying cache sizes, the number of cores and the number of contexts per core.

4   Results and Discussion

Figure 2 shows the OTF searching for an optimal divisor in the inner loop for the matrix multiply benchmark using ABD. By the third invocation of the parallelized loop the initial overhead of the runtime profiling code is amortized by the optimized performance, and by iteration 6 a local optimal divisor for this loop has been found. It should be noted that by the very nature of the

---

[4] Coefficient of variation (CV) is the ratio of the standard deviation to the arithmetic mean.

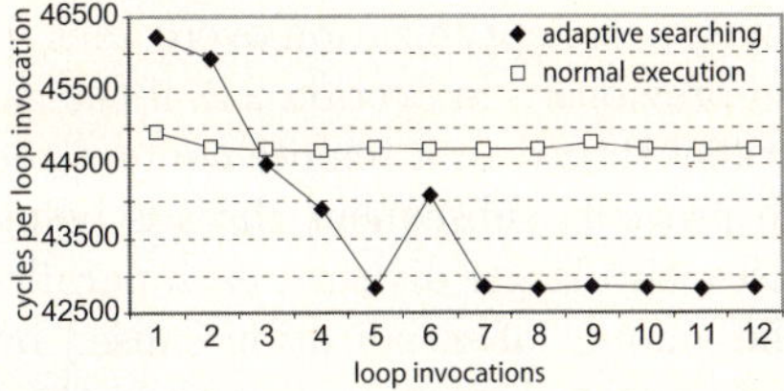

Fig. 2. Searching profile using ABD for the matrix multiply benchmark

hill-climbing algorithms used, the adaptive searching finishes after finding local-optimal solutions.

Figure 3 presents the results of optimizing the benchmarks using the ABD optimization. The results show the speedup achieved using the adaptively found local-optimal divisor, listed in the table, compared to dividing the loop iterations equally to a fixed number of threads, in this case equal to the total number of processor contexts.

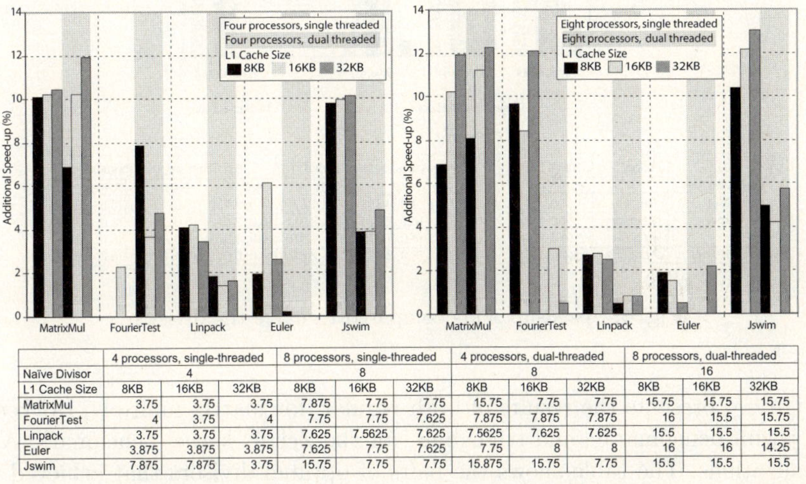

	4 processors, single-threaded			8 processors, single-threaded			4 processors, dual-threaded			8 processors, dual-threaded		
Naïve Divisor	4			8			8			16		
L1 Cache Size	8KB	16KB	32KB	8KB	16KB	32KB	8KB	16KB	32KB	8KB	16KB	32KB
MatrixMul	3.75	3.75	3.75	7.875	7.75	7.75	15.75	7.75	7.75	15.75	15.75	15.75
FourierTest	4	3.75	4	7.75	7.75	7.625	7.875	7.875	7.875	16	15.5	15.75
Linpack	3.75	3.75	3.75	7.625	7.5625	7.625	7.5625	7.625	7.625	15.5	15.5	15.5
Euler	3.875	3.875	3.875	7.625	7.75	7.625	7.75	8	8	16	16	14.25
Jswim	7.875	7.875	3.75	15.75	7.75	7.75	15.875	15.75	7.75	15.5	15.5	15.5

Fig. 3. Speedup of ABD compared to naïve division

For the majority of cases shown in figure 3, the optimal divisor is a value less than the naïve divisor. This is due to the nature of the distribution scheme. The processor context responsible for distributing the parallel threads, the generator, is always the last available for processing any of the loop iterations. In the case of a smaller divisor, 3.75 as opposed to 4, a loop with 100 iterations will be distributed such that the first three distributed threads contain a block of 26 iterations, and the fourth contains only 22. This scheme is therfore able to trade-off the overhead on the generator thread. In the cases where the optimal divisor is larger than the naïve divisor, the optimization overcomes the context contention overhead. As mentioned previously, in Section 2.2, if the generator thread is not able to distribute a parallel thread to a remote context the work must be done by the generator which prevents subsequent threads being distributed. As the size of the block decreases with larger divisors, each parallel thread contains less work, which reduces the amount of serialization caused by context contention. For this reason in some cases a divisor greater than the number of processor contexts performs better.

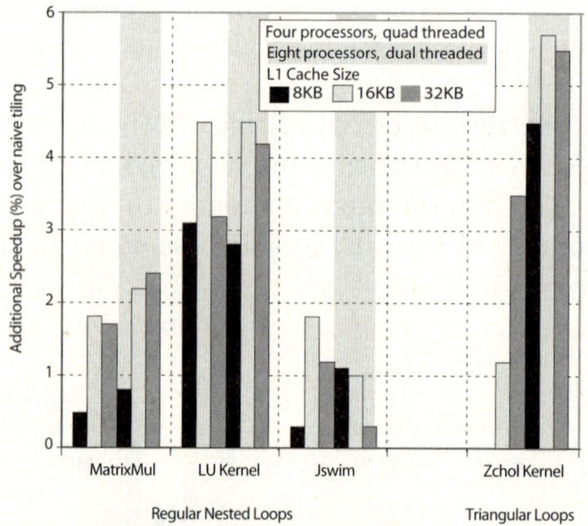

	Loop	Naïve Divisor	4 processors, quad threaded			8 processors, dual threaded		
L1 Cache Size			8KB	16KB	32KB	8KB	16KB	32KB
MatrixMul	Regular	(4.0, 4.0)	(4.0, 3.875)	(4.0, 3.75)	(8.0, 1.875)	(4.0, 3.75)	(8.0, 1.875)	(8.0, 1.875)
LU Kernel	Regular	(4.0, 4.0)	(4.0, 6.5)	(8.0, 2.0)	(8.0, 2.0)	(8.0, 2.0)	(8.0, 2.0)	(8.0, 2.0)
Jswim	Regular	(4.0, 4.0)	(4.0, 3.75)	(8.0, 1.875)	(8.0, 1.875)	(8.0, 1.875)	(8.0, 1.875)	(8.0, 1.875)
Zchol Kernel	Tringular	(4.0, 7.0)	(4.0, 6.0)	(8.0, 3.0)	(8.0, 2.875)	(4.0, 6.0)	(8.0, 3.0)	(8.0, 2.875)

Fig. 4. Speed-up ATD compared to naïve tile division

Figure 4 presents the results of both regular, for matrix multiply, LU kernel and jswim benchmarks, and triangular, for the Zchol[5] kernel benchmark, ATD optimizations. The table shows the pairs of optimal divisors that achieved the speedups shown, compared to the performance using naïve tile divisor pairs. It should be noted that naïve tile divisor pairs performed better than ABD on the benchmarks shown. Both the regular and the triangular tile division algorithms, optimize the performance of nested loops to achieve more efficient cache use. The table illustrates the variation amongst the optimal divisor pairs given variations in the architecture and cache size. The OTF is able to increase the performance for almost all configurations presented. The ATD optimizations search for an optimal tile size which can take advantage of data resuse and therefore improve cache behaviour. Each architecture used to assess the ATD optimizations contained 16 processor contexts, and the results for the optimal

---

[5] Zchol implements the Cholesky decomposition of a positive definite matrix.

divisors are normalized against the results gathered using the divisor pair (4.0, 4.0) for regular 2D loops, and (4.0, 7.0) for triangular 2D loops. Both of these divisor pairs generate equal sized square tiles. In the Java programming language, operations on multi-dimensional arrays can be optimized. For example, a loop iteration that loads a value from a 2D array $A[i][j]$, where j is the inner loop iterator, needs two memory load instructions: load $A[i]$ to ref_i and load $ref_i[j]$ into the target register operand. The load operation for $A[i]$ can be moved outside of the loop, reducing the total load operations. For this reason the divisor pair $(8.0, 2.0)$ achieves performance benefits over the simple $(4.0, 4.0)$ divisor pair.

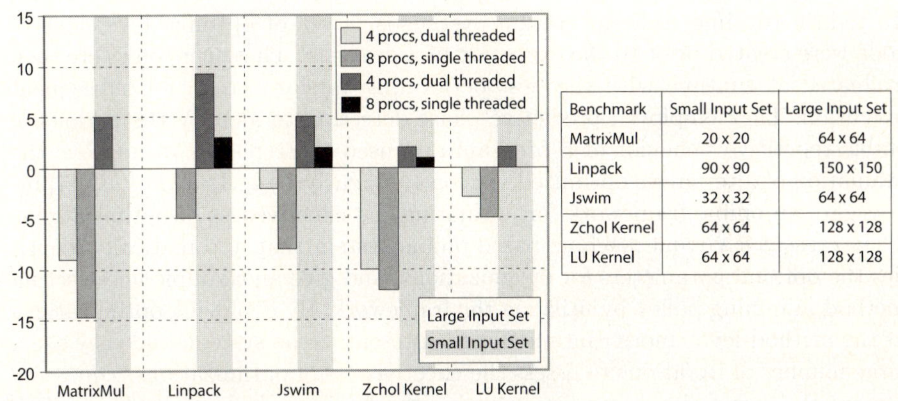

Benchmark	Small Input Set	Large Input Set
MatrixMul	20 x 20	64 x 64
Linpack	90 x 90	150 x 150
Jswim	32 x 32	64 x 64
Zchol Kernel	64 x 64	128 x 128
LU Kernel	64 x 64	128 x 128

Fig. 5. Speed-up from CRD on top of initial ABD/ATD gains

By using the AVS optimization, the OTF can select between a traditional and a cyclic recursive distribution (CRD) policy. To show the affect of AVS the performance of traditional distribution, using optimal divisors located using the ABD/ATD optimizations, and CRD, using the same divisors, was compared. Figure 5, shows the additional performance speedups gained by using CRD with the ABD/ATD optimizations. For each benchmark both a large and small data set was evaluated. Clearly for larger data sets the CRD scheme degrades the performance of the best ABD/ATD using traditional distribution, however, for smaller datasets additional performance increases are achieved for most of the benchmarks. CRD gains performance for smaller data sets as it uses a tree-like distribution policy to create parallel threads. This reduces the overhead of thread creation in the initial generator thread. It also employs a cyclic block distribution, which has been shown to improve the cache performance on multi-threading processors [14], especially for loops working on contiguous memory segments. Furthermore, when CRD is used on triangular loops, this cyclic distribution leads to less variation in the total amount of work received by each processor context. The drawback of CRD, however, is that it increases the number of cache misses when used in a multi-processor environment. This is because a contiguous memory segment will be mapped to different processors' caches.

5   Related Work

Voss and Eigenmann [15] established an adaptive optimization framework named ADAPT which performs dynamic optimization on hot spots through empirical search. The ADAPT uses dynamic recompilation to evaluate different optimizations and a domain-specific language to drive the search on the optimization space for a specific optimization (e.g. for loop unrolling, each level of unrolling will be compiled, run and timed, and the fastest version will be kept and used for the hot spot). The compiler used for recompilation was run on a parallel processor which reduced the recompilation overhead at runtime. *Fursin et al.* [16] explored online empirical searches for scientific benchmarks. To reduce runtime code generation overheads, a set of optimized versions of code were created prior to the execution of a program. These versions were then evaluated at runtime with the best performing version chosen for subsequent execution. They employed predictive phase detection to identify the periods of stable repetitive behavior of a program and used these phases to improve the evaluation of alternative optimized versions. Similarly *Lau, Arnold et al.* [17] investigate an online framework for evaluating the effectiveness of optimizations. They present a virtual machine based online system that automatically identifies the optimal parameters for optimizations, and give an example for selecting method inlinining policy by utilizing the framework. By deploying optimizations at the method-level, more runtime noise is present in the system, and they use a large number of iterations to assess the effectiveness of optimizations. *Diniz and Rinard* [18] use a simple version selection mechanism which reacts to runtime inputs and loop parameters. Their dynamic optimization system also generates code to provide dynamic feedback, allowing automated selection of the best synchronization policy for parallel execution.

In contrast with the above work, the method presented in this paper combines a loop-level parallel compiler and an adaptive optimization framework within a Java virtual machine that works on a CMP architecture. By employing the Jamaica CMP's capability of distributing fine-grain parallelization, the adaptive optimization system can perform online adaptive tuning to improve the performance of parallelized code for smaller data sets with acceptably low overheads.

6   Conclusion and Futher Work

This paper has presented an Online Tuning Framework, capable of locating and parallelizing loops. The framework is able through runtime profiling, to search for an optimal division of a parallelizable loop into blocks and an optimal distribution of the loop blocks across the parallel resources of a chip multiprocessor. This system realises additional performance speedups, up to 12% on the benchmarks assessed, over a traditional parallelization system, including the initial overheads involved with the profiling system.

The OTF is currently directed to optimize work distribution within parallel code sections which exhibit stability during profiling. As part of further studies the framework will be extended to *time-out*, whereupon previously unstable code sections, and those previously optimized but performing poorly, will be re-evaluated. Additionally the framework is being extended to automatically optimize work distribution across larger and more scalable CMP architectures.

References

1. IBM: Jikes^{TM} Research Virtual Machine (2005), `http://jikesrvm.sourceforge.net/`
2. Arnold, M., Fink, S.: Adaptive optimization in the Jalapeño JVM. In: Proceedings of the 15th ACM SIGPLAN conference on Object-oriented programming, systems, languages, and applications, pp. 47–65 (2000)
3. Burke, M., et al.: The Jalapeeño dynamic optimizing compiler for Java. In: Proceedings ACM 1999 Java Grande Conference, San Francisco, CA, United States, pp. 129–141. ACM Press, New York (1999)
4. Knobe, K., Sarkar, V.: Array SSA form and its use in parallelization. In: Symposium on Principles of Programming Languages, pp. 107–120 (1998)
5. Banerjee, U.: Loop Transformations for Restructuring Compilers: The Foundations. Springer, Heidelberg (1993)
6. Mitchell, M.: An Introduction to Genetic Algorithms. MIT Press, Cambridge (1996)
7. Wolfe, M., Shanklin, C., Ortega, L.: High Performance Compilers for Parallel Computing. Addison-Wesley Longman Publishing Co., Inc, Boston, MA, USA (1995)
8. Wright, G.: A single-chip multiprocessor architecture with hardware thread support. PhD thesis, The University of Manchester (2001)
9. Grehan, R., Rowell, D.: jBYTEMark Benchmark (1998)
10. Bull, J., et al.: A benchmark suite for high performance Java. Concurrency Practice and Experience 12(6), 375–388 (2000)
11. Hicklin, J., Moler, C.: **Ja**va **Ma**trix Package Benchmarks (July 2005)
12. Dongarra, J., Wade, R.: Linpack Benchmark - Java Version
13. Henning, J.: SPEC CPU2000: measuring CPU performance in the New Millennium. Computer 33(7), 28–35 (2000)
14. Lo, J., Eggers, S.: Tuning compiler optimizations for simultaneous multithreading. In: International Symposium on Microarchitecture, pp. 114–124 (1997)
15. Voss, M., Eigenmann, R.: High-level adaptive program optimization with ADAPT. In: PPOPP, pp. 93–102 (2001)
16. Fursin, G., Cohen, A., O'Boyle, M., Temam, O.: A practical method for quickly evaluating program optimizations. In: Conte, T., Navarro, N., Hwu, W.-m.W., Valero, M., Ungerer, T. (eds.) HiPEAC 2005. LNCS, vol. 3793, pp. 29–46. Springer, Heidelberg (2005)
17. Lau, J., Arnold, M.: Online performance auditing: using hot optimizations without getting burned. In: PLDI, pp. 239–251 (2006)
18. Diniz, P., Rinard, M.: Dynamic feedback: an effective technique for adaptive computing. In: Proceedings of the ACM SIGPLAN 1997 conference on Programming language design and implementation, pp. 71–84 (1997)

Compositional Approach Applied to Loop Specialization

Lamia Djoudi, Jean-Thomas Acquaviva, and Denis Barthou

Université de Versailles, France
{lamia.djoudi,denis.barthou,jean-thomas.acquaviva}@uvsq.fr

Abstract. An optimizing compiler has a hard time to generate a code which will perform at top speed for an arbitrary data set size. In general, the low level optimization process must take into account parameters such as loop trip count for generating efficient code. The code can be specialized depending upon data set size ranges, at the expense of code expansion and decision tree overhead.

We propose for loop structures a new method to specialize code at the assembly level, cutting drastically the overhead cost with a new folding approach. Our technique can generate and combine sequentially at the assembly level several versions, tuned for small, medium and large iteration number. We first show on the SPEC benchmarks the need for specialization on small loops. Then we demonstrate the benefit of our method on kernels with detailed results.

1   Introduction

An optimizing compiler has a hard time to generate a code which will perform at top speed for an arbitrary data set size. In general, Schwiegelshohn *et al.*[1] have shown there is no one best scheduling function for a loop, for all possible data sets. Even for regular programs, the best latency is only reached asymptotically [2,3], for large iteration counts. Splitting loop index to obtain better schedules[4], or tiling iteration domains are well known techniques that improve latency. These transformations are driven according to source code features such as dependencies or memory reuse. On the other hand, low level optimizations must take into account parameters such as loop trip count for generating efficient code: for example, short loop trip count would favor full unrolling while very large loop trip counts will favor deep software pipelining. To some extent, the code generated has to be specialized depending upon data set size ranges and then has to use extensive versioning to apply these different specialized versions. The classical drawback of such an optimization scheme is code expansion and decision tree overhead. It usually puts a hard limit on the total number of different specialized versions generated.

We propose, for loop structures, a new method to specialize code at the assembly level and to drastically cut the overhead cost with a new folding approach. Taking the assembly code, we are able for instance to generate three versions

A.-M. Kermarrec, L. Bougé, and T. Priol (Eds.): Euro-Par 2007, LNCS 4641, pp. 268–279, 2007.

tuned for small, medium and large iteration number. We combine all these versions into a code that switches smoothly from one to the other while the iteration count increases. Hence, the resulting code achieves the same level of performance as each version on its specific iteration interval.

We first show on the SPEC benchmarks the need for specialization on small loops. Then we demonstrate the benefit of our method on kernels optimized with software pipeline, with experimental results.

1.1   Motivating Example

Loop optimization is a critical part in the compiler optimization chain. A routinely stated rule is that 90% time of the execution time is spent in 10% part of the code. Another rule, implicitly used by the community, is that the number of iterations for loops in scientific code is *large*. Consequently, loops are often unrolled, pipelined deeply and data streams aggressively prefetched.

However, optimizations for asymptotic behavior involve a part of risk. For instance in software pipeline, depth is always increased if it can reduce the Initiation Interval. This yields to codes which deliver poor performance when the number of iterations is limited. Figure 1 clearly illustrates the trade-off that the compiler has to handle on a simple vector loop named Tridia. ICC 8.1, first unrolls this loops two times and generates a software pipeline of depth 2. While ICC 9.0 unrolls this loops 8 times, then applies software pipeline. The corresponding tail code is also software pipelined.

The corresponding performance evaluation is:

- ICC 9.0: $65 \times \frac{N}{8} + 130$ (unrolled 8 times) and $10 \times (N \bmod 8) + 20$ for tail code.
- ICC 8.1: $24 \times \frac{N}{2} + 48$ (unrolled 2 times) and $14 \times (N \bmod 2)$ for tail code.

As illustrated by figure 1, ICC 9.0 choice is justified for asymptotic performance but is doubtful when the number of iterations is small.

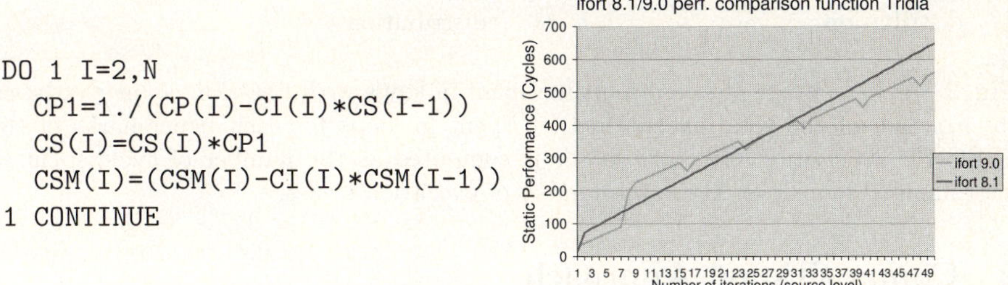


```
DO 1 I=2,N
   CP1=1./(CP(I)-CI(I)*CS(I-1))
   CS(I)=CS(I)*CP1
   CSM(I)=(CSM(I)-CI(I)*CSM(I-1))
 1 CONTINUE
```

Fig. 1. Tridia code and its performance with ICC version 8.1 and 9.0. Moving from version 8.1 to 9.0, ICC has changed part of its code generation policy. None of the two different versions is optimal over the whole possible range of iterations.

To evaluate the importance of the short loops we have performed a set of measurements on the SPEC FP 2000 benchmarks. Using MAQAO tool [5], all loops are instrumented. Instrumentation is done at the assembly level to prevent distortions in the compiler optimization chain. This instrumentation simply measures the number of iterations executed per loop and the number of CPU cycles spent within the loop. At the end, histograms are built according to the number of iterations of these loops weighted by their execution time. We use ICC v9.0 with flags reported for SPEC results, including profiling guided optimization, on a 1.6 GHz / 9 MB L3 Itanium 2 system. The only instrumented loops are counted loops, software pipelined loops, but loops driven by conditional branches are currently not caught by our tool.

Additionally the numbers provided should be considered knowing that ICC performs aggressive unrolling (most of the time by a factor of 8), consequently reducing the number of loops iterations.

Figure 2(a) details measurements made on the number of iterations for loops of CFP2000 codes using the `ref` data set. The answer is surprising: 25% of loop time is spent in loops with less than 8 iterations. A more detailed analysis shows that 6 over 14 benchmarks form the CFP2000 spend half of their loop time in loops with less than 16 iterations. Therefore, loop tuning based on infinite number of iteration is missing real performance opportunities, and compilers should not over-simplify loop behavior. In order to back the idea that short loops are an important problem, Figure 2(b) details the fraction of the execution time spent in loops benchmarks in SPEC.

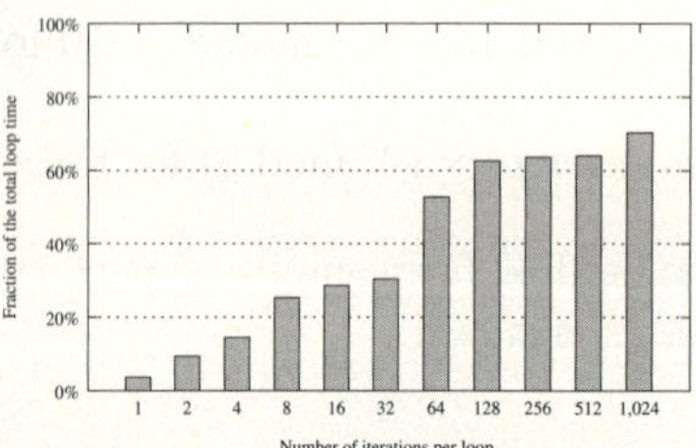

(a) SPEC FP overall cumulative loop distribution

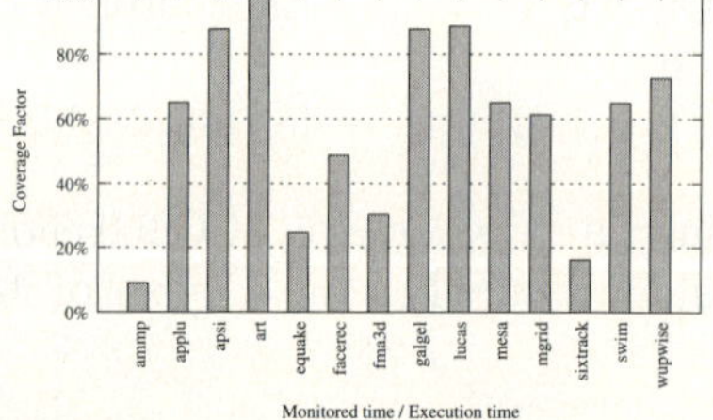

(b) SPEC FP overall cumulative loop distribution

Fig. 2. (a) Percentage of execution time spent in loops with iteration count $\leq$ x-label. (b) Fraction of the total execution time spent in loops for each benchmarks of the SPEC FP 2000 suite. Coverage factor is computed as the number of cycle spent in instrumented loops over the total number of execution cycles.

2   Compositional Approach

The idea of the approach, given different codes (and schedules) for the same loop, to combine or compose them into one code that achieves the same performance as the best code, for any iteration count. The purpose of this method is to assemble together in a costless and smart way different versions of the same loops. It is understated that the quality of the resulting composed version will depends on the quality of the different individual versions available at the origin.

Iterative compilation is an approach that relies on the generation of many different versions of the same code to find out the best among them. Loops generated by iterative compilation are therefore good candidates.

2.1 Iterative Compilation Framework

Iterative compilation is decomposed usually into two steps: (i) Generate multiple optimized versions of the same code. The goal is to generate a small number of versions that have good performance while covering a wide range of iterations. (ii) Compare the performance of these different versions (either by a model or by a dynamic evaluation) and build a program combining them. Most of the research effort has been on the first step, and for the second one, it usually boils down to generate a decision tree. This means that for one execution, only one of the specialized code is executed. In this paper, we focus on this second step and generalize the previous case by enabling one execution to execute several specialized codes in sequence. Indeed, optimization may be beneficial only on a given range of iterations. For each of the following optimizations, we describe its limitations and the conditions for which it applies

1. *peeling*; Peeling enables a rescheduling of the first iterations of a loop. It generates more opportunities for a better resource usage, with a free schedule, at the expense of code expansion.
2. *unrolling*; Unrolling a loop body offers the opportunity for better ILP. The higher the unrolling factor, the higher the impact on performance of the tail code for small loops.
3. *data prefetching*; Data prefetching cuts by a large amount the read/write latency of memory accesses. Tuning the prefetch distance is highly dependent on the total number of iterations. For small loops, the prefetch distance is too large for the prefetching to be effective. In this case, removing the prefetches may free resources for a better ILP, therefore increasing performance.
4. *software pipeline*; The Initiation Interval (II) is usually the value minimized by software pipeline algorithms, and represents the amount of time between two successive start of iterations. This comes at the expense of the latency required to execute a complete iteration, which is important for small loops.

This shows that there are many opportunities in which it would be interesting to combine different optimized codes according to the iteration count.

2.2 Performance Model

Consider two different optimized versions of the same loop, called L_1 and L_2. This can be generalized to any number of versions. We assume that these loops are inner loops (they do not include other loops). The cycle count of L_1 is given by the formula: $c_1(i) = \alpha_1.i + \beta_1$, where α_1 is an rational number, β_1 an integers and the cycle count is rounded down. Similarly for L_2, $c_2(i) = \alpha_2.i + \beta_2$. For instance, for Figure 1, the cycle count for the loop L_1 generated by ICC 8.1 is defined with $\alpha_1 = 12, \beta_1 = 48$ and for the loop L_2 generated by ICC 9.0, $\alpha_2 = 8.125$, and $\beta_2 = 130$, without the tail code. Tail code is considered for our purposes as another version of the code, within the range of 7 or 1 iteration (for the loops unrolled 8 and 2 times respectively).

We consider the case where the two loops are such that: $\alpha_2 < \alpha_1$ and $\beta_1 < \beta_2$, meaning that L_1 is faster than L_2 when $i < \frac{\beta_2 - \beta_1}{\alpha_1 - \alpha_2}$ and L_2 outperforms L_1 for larger number of iterations. We would like to build a *best* code such that:

$$\forall n, c_{best}(n) = \min_k(c_k(n)).$$

This *best* code is built by an optimization function min: $\min(L_1, L_2) = best$. This function min defines a minimum on codes with respect to the performance, for all iteration values. Due to the difficulty of building the minimum of two codes without introducing any overhead, we propose to tackle a more pragmatic problem. We want to build a code $\min(L_1, L_2)$ with a level of performance very close to the performance of the best of the two codes. The following constraints are applied to the loop to build:

1. Asymptotic performance (in cycle/iteration, when iteration count grows) is the same than the best asymptotic performance of L_1 and L_2.
2. Each loop is possibly called many times, each time with a possible different loop trip count. Given a loop count distribution, the average gain in cycle/iteration compared to the best asymptotic performance of L_1 and L_2 is positive.
3. When performance of both loops L_1 and L_2 meet, the loop built moves to the best code for asymptotic performance.

Note that the second constraint does not compel the new loop to outperform L_2 and L_1 for each iteration count, but in general, for all the execution of the loop, some cycles have been gained. The reason is that some overhead may appear when switching from one version to the other. The best code would have the following cycle count:

$$c_{12}(i) = \begin{cases} i \le B : c_1(i) \\ i > B : c_2(i) - c_2(B) + c_1(B) + \gamma \end{cases}$$

where B is the integer $\frac{\beta_2 - \beta_1}{\alpha_1 - \alpha_2}$ and γ represents the overhead necessary when going from one version to the other. This overhead represents register initializations, branch mispredicts,.... The difference in cycles/iteration between the asymptotic best loop and the new loop $\min(L_1, L_2)$ is, for an iteration count i:

$$dpi(i) = \begin{cases} i \le B : (c_2(i) - c_1(i))/i \\ i > B : (c_2(B) - c_1(B) - \gamma)/i \end{cases}$$

The difference in cycle/iteration is asymptotically 0, meaning that this new version is as fast as L_2. When the loop iteration count is uniformly distributed among iterations $[1..N]$, the average difference in cycle/iteration is obtained by:

$$adpi(N) = \sum_{i=1}^{N} \frac{dpi(i)}{N}.$$

This definition can easily be adapted to other distributions. In particular, distribution of values caught during profiled execution can be used. When $adpi(N)$ is positive, it means that for a uniform distribution of loop trip counts in $[1, N]$, the new loop $\min(L_1, L_2)$ is in average faster than the best asymptotic loop L_2. This value is positive when $N < B$ since each difference/iteration is positive for all iterations $i < B$. For higher values of N, $adpi(N) > 0$ if:

$$\gamma < \frac{B(\alpha_2 - \alpha_1)(1 + H(N) - H(B)) + (\beta_2 - \beta_1)H(N)}{H(N) - H(B)}, \tag{1}$$

where $H(N)$ is the harmonic number $H(N) = \sum_{k=1}^{N} \frac{1}{k}$. As H is a strictly increasing function, when N asymptotically grows, γ must be such that:

$$\gamma < c_2(B) - c_1(B).$$

This constraint implies that the new loop $\min(L_1, L_2)$ takes less cycles than the best asymptotic version, for any value of the iteration count. From this constraint we can deduce the basic steps to build the code $\min(L_1, L_2)$:

1. Compare $c(L_1)$ and $c(L_2)$, in order to compute B
2. Assuming L_1 outperforms L_2 for the first B iterations, evaluate the code in-between necessary for the transition and the overhead β generated.
3. If inequality 1 is satisfied, then build the minimum of the two codes. Otherwise the overhead is too significant w.r.t. the total execution time.

For the example in Figure 1, $B = 21$, inequality (1) entails that for $N > 293$, γ can no longer be strictly positive. For $N = 200$, γ can be up to 15 cycles.

2.3   Scopes and Limits

As with all other versioning schemes, our approach improves performance at the expense of code size. If the number of iteration remains constant or at least in a single range the extra code size and some instruction overhead will penalize the execution time. However if we considere the SPEC benchmark as representative of the average code complexity it be can safely stated that iteration range is varying a lot and that specialization on iteration number will mostly inscreases performance.

3   Assembly to Assembly Transformation

We first present an assembly code dependence analysis then describe two particular transformations, loop peeling and transformation of prefetching, as well as their composition. These two steps are for an Itanium architecture, but we believe this can be generalized to other platforms. Such post-compiler optimization is already a hot topic of research [6,7,8].

3.1   Code Flattening and Dependence Graph

In order to preserve code semantics, the validity of the transformations applied is checked by computing a *data dependence graph* (DDG) on the assembly codes. Dependencies considered can be either intra-iteration or inter-iteration and dependence analysis is required by the peeling and jam transformations described in the following section.

In the case of pipelined loops on Itanium, dependence analysis is more complex and loop flattening is a preliminary transformation. IA64 Hardware support for software pipelining includes a rotating register bank and predicate registers [9]. Loop flattening is a transformation that removes the effect of software pipeline: it renames carefully registers according to their rotating status or not, and predicates are used to retrieve the iteration number when the instruction becomes valid.

A data dependence graph is then built between the different instructions in the loop. Register dependences are built by reaching definition analysis. For memory accesses, the alias analysis performed relies on two techniques: We apply a conservative approach, based on the schedule generated by the compiler. The base rule is that all memory accesses are interdependent (read or write with write). If the compiler schedules two instructions within less cycles than the minimum latency of the first instruction (the one being the possible dependency anchor) then we assume that the compiler did this schedule on purpose and therefore that the two instructions are independent. For instance, if a load `ld f32,[r31]` is scheduled 3 cycles before a store `st [r33],f40` and the minimum latency of a load is 6 cycles, then both statements are independent.

We also resort to a partial symbolic computation of addresses, using induction variable analysis on address registers. The value of address registers can often be computed with respect to some initial parametric values coming from registers defined outside of the loop (parameters of a function for instance). In this case, our de-ambiguation policy depends on the original compilation flags (either with or without no-alias flag). More independent statements can be found that way.

3.2   Peeling and Prefetching Transformations

We show the composition approach using loop peeling and prefetching.

Peeling is the process of 'taking off' a number of iterations from a loop body, and consequently explicitly express them at the beginning or end of the loop. This is often done in order to match two different bounds of two subsequent loops. Generally, the positive effect of this technique is better understood if explained in conjunction with loop fusion. In our approach, the peeling has also a positive effect if explained in conjunction with software pipelining. Compared to warming up stages of software pipelined loop, an interleaving scheme does not increase latency but increases the number of iterations simultaneously in-flight. This does not yield to excessive register pressure. In fact, the global register pressure depends on the number of iterations simultaneously alive. Our peeling techniques is careful enough to keep this number below the software pipelined loop asymptotic behavior. The initial schedule of peeled iterations is the schedule obtained after a possible flattening. Then iterations are jammed (or interleaved) with a list scheduling algorithm with priority to the first iterations. The statements of the first iteration peeled are scheduled first and have higher priority over the statements of the second iteration. Indeed, this schedule improves over the initial schedule, w.r.t. the difference in cycle per iteration, as presented in Section 2.2. If the initial (flatten) loop has a cycle count function of the form $\alpha.i+\beta$, a jammed version of the peeled iterations takes $c_{peeled}(i) = \alpha'.i+\beta$ cycles with $\alpha' \leq \alpha$ where α' is a rational number. The list scheduling algorithm ensures that the longest dependence chain in one iteration is not increased, the latency α' is lower or equal to α. Finally, a mechanism is needed to ensure program correctness if the number of total iterations is smaller than the number of peeled iterations. One calculated branch is used for a late entry into the peeled code, and predicate registers guard interleaved instructions. The branch uses a branch register to store the address to jump in. Setting a value to a branch register is a 6 cycle long operation. If the execution time of interleaved iterations exceeds 6 cycles we use this kind of register to minimize the overall latency. Moreover, instructions guarded by predicates prevent from executing interleaved instructions that do not belong to the desired peeled iterations. $\log_2(N)$ comparisons, and $\log_2(N)/6$ cycles are necessary to set the predicate registers of N peeled iterations. The overhead is limited to a couple of cycles.

For prefetching, the prefetch distance is estimated from the symbolic computation performed before and from the increment of the address registers, we assess the number n of first loop iterations that do not take advantage of the prefetch. The loop is then split into a sequence of two similar loops. The first loop has no prefetches (they are replaced by nops) and has n iterations. The second loop is the initial loop, performing the remaining iterations.

4   Related Work

Specialization is a well known technique to obtain high performance programs. Compiled time specialization often boils down to the generation of codes that are in mutually exclusive execution paths. Splitting iteration space in order to apply different optimizations for each fragment has been proposed by Griebl et al.[4]: their goal is to partition the iteration space according to the dependence pattern for each statement. This increases control but increases the number of affine schedules that can be computed for each code. Tiling is another transformation that changes the iteration domain for better scheduling. However, very few works resort to loop versioning in order to explicitly reduce the overall latency of the loop. This is due to the intractability of general performance models (finding best latency affine schedules is still a difficult problem). That is one reason why asymptotic loop counts are generally considered for optimization. In this paper, we do not consider memory models (or cache model) and assume that the compiler or a tool evaluates accurately the performance of inner loops. This is easier on assembly than on source code.

Software pipelining[10] is a key optimization for VLIW and EPIC architectures. In particular, modulo scheduling, as used by the ICC compiler, exhibits instruction parallelism, even in the presence of data dependencies, that greatly improves performance. Modulo scheduling targets large iteration counts and tries to find an initiation interval (II) as small as possible, defining the throughput of the loop. However, when the iteration count is small, this may increase the loop latency (in particular when the II is essentially constrained by resources). Loop peeling is a well known technique for improving the behavior of the code for small iteration count. As it comes at the expense of code size, compiler heuristics usually prefer to not use it. With our approach, it is possible to decide, according to the awaited iteration count distribution, whether peeling is worth or not. Moreover, our technique would take advantage of profile information since the distribution is then more accurate.

Prefetch works by bringing in data from memory well before it is needed by a memory operation. This hides the cache miss latencies for data accesses and thus improves performance. Typically, when a cache line is brought in from memory, it contains several elements of an array that is accessed in the loop. This means that a new cache line is needed for this array access only once in several iterations. This depends on several factors such as the stride of the array access pattern, the cache line size, etc. In fact, if redundant prefetch instructions are issued every iteration, it may degrade performance. Prefetch instructions require extra issue slots and thereby increase the cycle per iteration ratio. Redundant prefetches can overload the memory subsystem and thereby adversely affect the performance, and prefetch too much in advance can also result in cache pollution and performance degradation [11]. Prefetches are interesting only when the iteration trip count is large enough to make data access at the prefetch distance. This implies that for medium iteration numbers, prefetch instructions can be removed.

5   Experiments

We consider three benchmarks: a DAXPY loop ($Y[i] = \alpha \times X[i] + Y[i]$) and two benchmarks from the CFP2000: GALGEL and MGRID. The DAXPY illustrates the combination of both unrolling and prefetch specialization.

DAXPY: Prefetch instructions must be generated for both X and Y arrays. It appears, that using prefetch degrades the initiation interval of the software pipelined loop due to extra pressure on memory slot.

Based on our performance model, there are three versions of the initial code:

- First zone: peeling, each block of 8 iterations costs $30 + N \bmod 9$. Peeling degree is set to 8 since is corresponds to the minimal latency in the DDG (4 cycles) which is just enough to schedule 8 floating point instructions.
- Second zone: disable prefetch, the formula is in $1 \times N + \alpha$[1]
- Third zone: enable prefetch, formula: $2 \times N + \alpha$

From the prefetch version, we know that the prefetch distance is set to 800B. Therefore, considering that every iteration is consuming 8 Bytes, it means that the loop needs to iterate at least 100 times before accessing the first prefetched data. So for this 100 first elements, it can used a loop without prefetch instructions. Performance results are detailed in Figure 3.

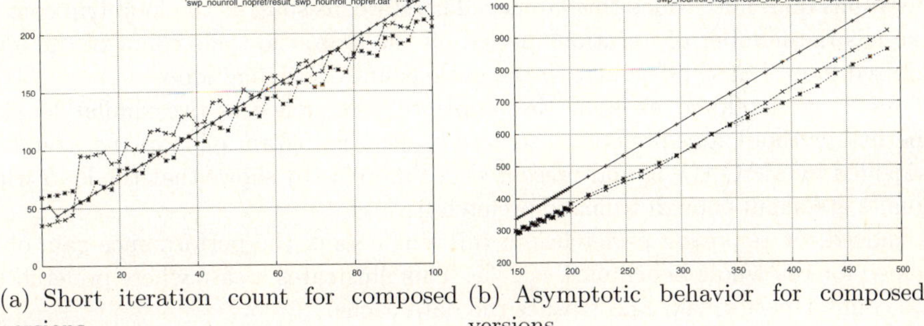

(a) Short iteration count for composed versions

(b) Asymptotic behavior for composed versions

Fig. 3. (a) is a close-up of the relative performance between composed versioning, prefetch and no prefetch. For the first 8 iterations, composed versioning (referred as *DAXPY composed peeling/no prefetch* on the graph) outperforms all the other versions. However for the 9th iteration, composed versioning suffers from the overhead of filling up pipeline, while they are already filled-up for both other versions. This is consistent with our policy: overheads should be postponed as far as possible. Therefore even if these overheads still account for the same number of cycles their *relative cost* is smaller. Notice, that clearly composed versioning follows the same behavior than no prefetch version. In this example we chose to stick with no pretech up to 128 iterations. (b) details the behavior for large number of iterations. Composed versioning sticks with the no prefetch slope outperforming the prefetch version up to a hundred iterations. Beyond, it sticks to the original version, for the best asymptotic performance.

GALGEL, loop b1_20: The loop *b1_20* is a pipelined loop, and is one of the many versions generated by the compiler for one source loop. The loop has 8 iterations in the **train** input data set, 11 in the **ref** input data set. For this loop, we performed a peeling of one iteration. Table 4-(a) sums up the results of the peeling transformation.

---

[1] The Itanium architecture can not sustain one branch per cycle without inserting a stall cycles, the real formula is $1.7 \times N + \alpha$.

Iteration count	Cycles		Performance Model		Gain
	Orig.	Peeling	Orig.	Peeling	
8	37760118	33984118	16xN+32	16xN+28	2.5 %
11	373744918	344995318	16xN+32	16xN+28	1.92 %

Fig. 4. Peeling one iteration out of loop b1_20, for each iteration count are given: the cycle count of the original loop and of the peeled loop (excluding peeled iteration), the cycle count according to a static performance model, and the performance gain of the peeling in % w.r.t. the original version

MGRID, loop b7_81: The loop $b7\_81$ in the assembly code is memory access intensive, since it performs in two cycles two load-pairs (equivalent to four loads) and four stores. The loop uses one prefetch instruction and is pipelined. Peeling the loop does not bring significant performance gain, according to the performance model. Indeed each peeled iteration takes 2 cycles and interleaving peeled iterations does not reduce this latency. Therefore, as soon as the loop trip count exceeds the number of iterations peeled off the loop, the cycle count of the optimized loop should be similar to the cycle count of original loop.

As for prefetching, we split the loop into a sequence of two similar loops, the first without any prefetch instruction. The histogram of loop trip counts, provided by MAQAO [5] and presented in Figure 4-(b) shows that the loop trip counts are small enough to make prefetch useless.

Indeed, by removing prefetches in this single loop, the performance gain obtained for the whole benchmark is 25%. This illustrates a case where prefetches are counter-performant and trashes the data cache.

6   Conclusion

The stem of our work is the diagnosis that in scientific computing a consequent fraction of execution time is spent in loops with a small number of iterations. However, even modern compilers seem to bet everything on asymptotic performance. Clearly there are performance opportunities for non-asymptotic behaviors and optimization must be adapted to the size of data, and for loop, to the iteration range.

Therefore, we come out with a novel method to version codes. This compositional versioning limits the overhead, reduces costly decision tree height and exploits and executes as much as possible of the generated code. This new technique is based on loop versioning, according to the iteration count distribution. This is a generalization of simple asymptotic evaluations. Given a loop count distribution, either coming from static analysis of the code, provided by the user through pragmas, or observed by profiling, we propose a smart loop versioning scheme. In particular, we split index sets so that each iteration range can be optimized more aggressively. The proposed optimizations are, for short range: peeling and for medium range: turning prefetching off, in addition to any versions proposed by the compiler. The first results on SPEC benchmarks show up to 25% speed up for one benchmark.

From an implementation point of view, our work is still in progress and while we are currently able to handle limited pieces of code and vector loops, we are now building the infrastructure to address the whole SPEC benchmark. One of the main issue to address is the switching overhead. In order to reduce it, we are investigating a way to peel off not only complete iteration but also software pipeline prologue and epilogue. Where the goal is to reschedule and interleave all these instructions allowing to switch directly from one version to a fully loaded pipeline.

References

1. Schwiegelshohn, U., Gasperoni, F., Ebcioglu, K.: On Optimal Parallelization of Arbitrary Loops. Journal of Parallel and Distributed Computing 11, 130–134 (1991)
2. Darte, A., Robert, Y.: Affine-by-statement scheduling of uniform and affine loop nests over parametric domains. Journal of Parallel and Distributed Computing 29(1), 43–59 (1995)
3. Rau, B.R.: Iterative modulo scheduling: an algorithm for software pipelining loops. In: Int. Symp. on Microarchitecture, San Jose, California, United States, pp. 63–74. ACM Press, New York (1994)
4. Griebl, M., Feautrier, P., Lengauer, C.: Index set splitting. Int. Journal of Parallel Programming 28(6), 607–631 (2000)
5. Djoudi, L., Barthou, D., Carribault, P., Lemuet, C., Acquaviva, J.T., Jalby, W.: Exploring application performance: a new tool for a static / dynamic approach. In: LACSI Los Alamos Computer Science Institute Symposium (2005)
6. Cooper, K., Dasgupta, A., Kennedy, K.: Vizer: A system to vectorize intel x86 binaries. In: LACSI Los Alamos Computer Science Institute Symposium (December 2002)
7. Merten, M., Thiems, M.: An overview of the IMPACT x86 binary reoptimization framework. Technical report (July 1998)
8. Larus, J., Schnarr, E.: EEL: Machine-independent executable editing. In: Int. Conf. on Programming Language Design and Implementation, pp. 291–300 (1995)
9. McNairy, C., Soltis, D.: Itanium 2 processor microarchitecture. IEEE Micro 23(2), 44–55 (2003)
10. Allan, V.H., Jones, R.B., Lee, R.M., Allan, S.J.: Software pipelining. ACM Computing Surveys 27(3), 367–432 (1995)
11. Doshi, G., Krishnaiyer, R., Muthukumar, K.: Optimizing software data prefetches with rotating registers. In: Int. Conf. on Parallel Architectures and Compilation Techniques, Barcelona, Catalunya, Spain, IEEE Computer Society Press, Los Alamitos (2001)

Starvation-Free Transactional Memory-System Protocols*

Mridha Mohammad Waliullah and Per Stenstrom

Department of Computer Science and Engineering
Chalmers University of Technology
SE-412 96, Göteborg, Sweden
{waliulla,pers}@ce.chalmers.se

Abstract. Transactional memory systems trade ease of programming with run-time performance losses in handling transactions. This paper focuses on starvation effects that show up in systems where unordered transactions are committed on a demand-driven basis. Such simple commit arbitration policies are prone to starvation. The design issues for commit arbitration policies are analyzed and novel policies that reduce the amount of wasted computation due to roll-back and, most importantly, that avoid starvation are proposed. We analyze in detail how to incorporate them in a TCC-like transactional memory protocol. The proposed schemes have no impact on the common-case performance and add quite modest complexity to the baseline protocol.

Keywords: Multiprocessors, transactional memory, starvation.

1 Introduction

As multi-core architectures are becoming commonplace, the need to make parallel programming easier is becoming acute. Transactional memory (TM) [1,2,3,4,7] promises to reduce the programming effort by relieving the programmer from resolving complex, fine-grain, inter-thread dependences by classical synchronization primitives such as locks and event synchronizations. Instead, coarse program segments form transactions that will either execute atomically or not at all. If transactions run by different threads have no dependencies, they can run concurrently. On the other hand, if a data dependency or a conflict appears, one of the transactions is squashed and must re-execute. Therefore, transactional memory trades programming simplicity for wasted execution at run-time.

Conflicts can be detected eagerly or lazily [4]. Under lazy conflict detections, such as TCC [2], the modifications done by a transaction are isolated until the point when the transaction is to commit. When the transaction commits, other transactions that have speculatively read data modified by the committing transaction will be squashed. Squashing does not only waste useful work. We show in this paper that it can cause starvation.

---

* This research is sponsored by the SARC project funded by the EU under FET. The authors are members of HiPEAC - a Network of Excellence funded by the EU under FP6.

A.-M. Kermarrec, L. Bougé, and T. Priol (Eds.): Euro-Par 2007, LNCS 4641, pp. 280–291, 2007.

This paper makes several contributions. Firstly, it analyzes in detail how to implement feasible commit arbitration schemes for TCC-like TM protocols. Secondly, and most importantly, it contributes with two novel starvation-free commit arbitration policies. Our overall approach to detect and remedy a potential starvation problem is to track how many times a certain transaction has been squashed. At the time a thread is ready to commit, it will not be allowed to do so if there is an ongoing transaction that has been squashed more times than the committing one. Then, the committing thread is stalled until that transaction has committed. Apart from avoiding devastating starvation situations, we show experimentally, using eight applications from SPLASH-2, that our starvation-free policies have virtually no impact on common-case performance and that they can be implemented with minor modifications to a TCC-like TM protocol.

As for the rest of the paper, we first introduce the architectural framework and frame the problem in more detail in Section 2. Section 3 is devoted to the novel arbitration schemes and especially how they are incorporated in the architectural framework. We then move on to the experimental results in Sections 4-5 by first describing the methodology. Section 6 puts our work in context of the TM literature and we conclude in Section 7.

2 System Framework

In this section, we define the framework of our study including the software assumptions in Section 2.1 and the architectural framework in Section 2.2. Finally, we frame the problem addressed by this research in detail in Section 2.3.

2.1 Software Assumptions

We assume that parallel programs uses transactions only and that a transaction is annotated by the programmer using start transaction (tx_begin) and end transaction (tx_end) constructs. In case of parallel applications using critical sections and barriers transactions are formed so that the following simple rules are followed: 1) critical sections are guaranteed to be encapsulated within a transaction and 2) a transaction is terminated and a new transaction starts at a barrier. However, a transaction can be terminated and a new one can start between two barriers as long as it happens outside critical sections [6,8,9,10].

In the assumed TM system, threads can execute beyond a barrier as long as they do not conflict with a thread that has not reached the barrier. This is supported using the notion of ordered as well as unordered transactions. A *phase number* is associated with each transaction which is incremented when a barrier is passed. All transactions that are started after the dynamic invocation of a certain barrier get the same phase number. If two transactions have the same phase number they are unordered; otherwise they are ordered and must commit in the ascending order of their phase numbers. Let's next consider the system model that supports this software model.

2.2 Architectural Framework

We consider a multi-core system that consists of n nodes where each node consists of a processor core (or core for simplicity) with its private L1 (optionally L2) cache connected to a shared L2 (optionally L3) cache via a bus or other broadcast medium according to Figure 1. Each core can be optionally (simultaneously) multithreaded with k hardware threads, where k is typically a small number (four or less).

Each L1 cache is extended with meta data to keep track of which blocks have been speculatively read and written using a read (R) and a write (W) bit, respectively, by setting the corresponding bits. When a transaction is finished, it will try to commit by requesting the bus. If a block that has been speculatively written is replaced, it is placed in a victim cache (VC) attached to each L1 cache. VC overflow is treated as described in [2].

In the baseline system, multiple commit requests are arbitrated through a central arbitration unit (denoted as CAU in Figure 1). To adhere to the semantics of ordered transactions, it attempts to select a committing transaction among the ones with the lowest phase number. Among these unordered transactions, it selects a candidate using FIFO.

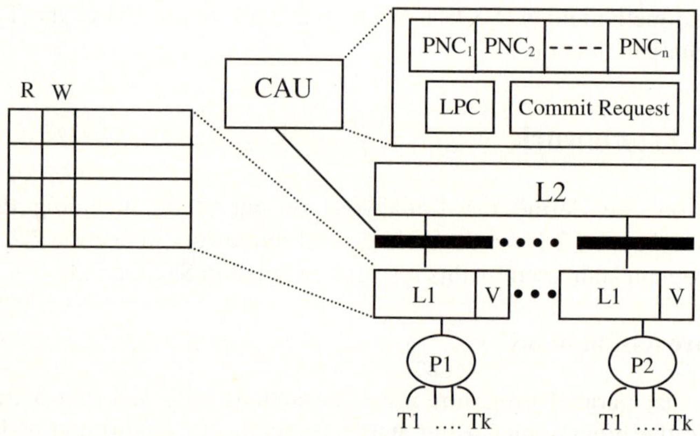

Fig. 1. Baseline architecture framework

To implement the baseline arbitration policy, the CAU uses three components: phase number counters (PNC), a lowest phase counter (LPC), and a FIFO with all commit-requests. The PNCs keep track of the current phase number of each thread using $N = n \times k$ phase-number counters, given n nodes and k hardware threads per node.

There are two types of commit requests: ordered and unordered. When a thread passes a barrier, an ordered commit request is sent. When an ordered commit request is granted by the CAU, the corresponding PNC is incremented. The LPC keeps track of the lowest phase number of any thread, i.e., $\min(PNC_i)$, i=1,...,N. Finally, the FIFO simply keeps all pending commit requests on a first-come, first serve basis. Given these components, the CAU uses the LPC to filter out the requests from the FIFO that can commit, i.e., the transactions having the lowest phase numbers. It then picks the first one of these in the FIFO queue. A node with a granted commit request broadcasts its write set (the set of blocks having a W-bit set) to all L1 caches. All nodes with a block belonging to the write set and with its R-bit set will be notified to squash their ongoing transactions. Squashing a transaction involves the following steps: 1) invalidate all blocks having either the R or W-bit set; 2) gang-clear all R and W bits; and, 3) restart the transaction by reinstalling the architectural state.

2.3 A Starvation Scenario

A major limitation of any transactional memory system is the performance lost due to squashes. More seriously, the simple arbitration policy assumed in the original TCC proposal is actually prone to starvation as the example in Figure 2 clearly demonstrates.

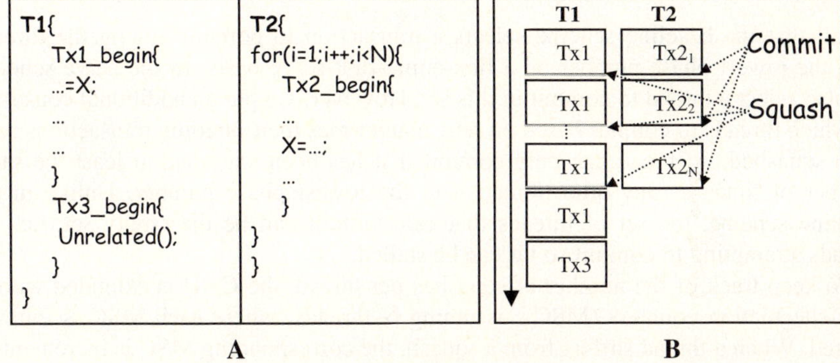

Fig. 2. A starvation scenario

Let's consider two threads (T1 and T2) that execute the code in Figure 2A. T1 executes a transaction (Tx1) that reads variable X followed by another transaction that does not conflict with any (Tx3). On the other hand, T2 executes a transaction (Tx2) that modifies X N times. Further, the execution time of Tx1 is assumed to be longer than that of Tx2.

Now consider the execution scenario in Figure 2B in which the execution of Tx1, Tx2, and Tx3 is tracked along the vertical time axis and where the commit and squash points are marked. As Tx1 and Tx2 obviously conflict, Tx2 will successfully commit whereas Tx1 will be squashed. As T2 will invoke Tx2 again, while T1 attempts to re-execute Tx1, the same scenario may repeat N times. As a result, the execution of all transactions will be serialized.

In another scenario, assuming the same code, Tx1 could have successfully committed before the first invocation of Tx2. Then, the execution of Tx3, and possibly subsequent transactions, could overlap the execution of the N-1 invocations of Tx2. Clearly, although the commit arbitration mechanism assumed in TCC meets fairness objectives at the level of commit requests between ordered and unordered transactions, it may result in starvation at the software level.

We have observed that a nearly as bad scenario showed up for Raytrace in SPLASH-2 (to be discussed in Section 5). The key reason for the devastating scenario of Figure 2B is that the CAU is completely unaware of the history by which transactions get squashed. In the next section, we propose novel arbitration policies that address this shortcoming.

3   Starvation-Free Arbitration Policies

To avoid starvation, the overall approach is to give priority to ongoing transactions that have already suffered from being squashed. We consider two schemes that differ in the way transactions are given priority to be selected and in implementation complexity. They are referred to as the naïve and the elaborate schemes. Both schemes extend the already existing central arbitration unit in the baseline system.

3.1   The Naïve Scheme

Recall that the baseline scheme selects a transaction to commit among the threads with the lowest phase number on a first-come-first-serve basis. In the naïve scheme, we also select a thread to commit in this set. However, we put an additional constraint on which threads to commit based on how many times their ongoing transactions have been squashed. A thread can only commit if it has been squashed at least the same number of times as any other thread with the lowest phase number. Unlike in the baseline scheme, the set of threads that can commit can be the empty set, i.e., all threads attempting to commit so far can be stalled.

To keep track of the number of squashes per thread, the CAU is extended with N miss-speculation counters (MSC), assuming N threads, where each MSC is initially cleared. When a thread suffers from a squash, the corresponding MSC is incremented. Conversely, when a thread commits its transaction, the corresponding MSC is cleared. In Figure 3, we show the extra mechanisms needed by the naïve scheme.

The CAU obviously knows when a thread can commit its ongoing transaction but lacks knowledge about when a transaction is squashed. Hence, it needs to know when to increment the MSCs. This is accomplished by requiring that each time a node selects to squash a transaction, it notifies the CAU so that the corresponding MSC can be incremented. Hence, the commit protocol involving broadcasting the write set and matching it against the read set in each node, must be extended with a final phase to notify the CAU about the hardware threads on each node that were squashed.

One solution is to let all nodes serially report to the CAU about squashed transactions. This would obviously cause a lot of overhead so we propose the following solution. All hardware threads have a SQUASH signal that is raised when a transaction has been squashed. Further, the bus is extended with N dedicated lines; one for each SQUASH-signal. These lines are used as increment-enable signals to the set of MSCs maintained by the CAU as shown in Figure 3.

In summary, in terms of structures, the naïve scheme is extended with as many miss-speculation counters as the number of hardware threads. Additionally, it needs a

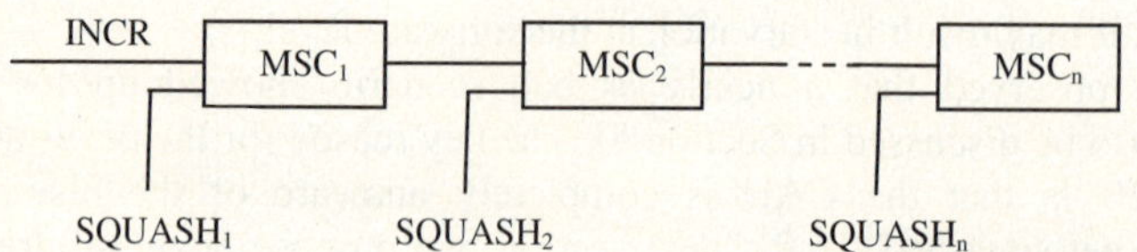

Fig. 3. Mechanisms used by the naive scheme not part of the baseline

special-purpose bus with as many lines as the number of hardware threads. Given the fairly limited number of nodes and hardware threads per core in a multi-core chip for a foreseeable future, this solution appears to be reasonable.

3.2 The Elaborate Scheme

In the naïve scheme, a committing transaction is stalled when any transaction with the lowest phase number and a higher MSC exists in the system disregarding the fact that these transactions could be data independent of each other. Intuitively, it would be unwise to stall a transaction that has no conflict with a transaction that has a higher MSC because, then, it is not the cause of a potential starvation scenario.

In the elaborate scheme, a committing transaction that has the lowest phase number and has no conflict with any transaction with the same phase number and a higher MSC is allowed to commit. Conversely, the CAU will stall a transaction either if the committing transaction has not the lowest phase number (as in the baseline) or it has the lowest phase number and it conflicts with another transaction that has the same phase number and a higher MSC.

Unfortunately, the CAU has no knowledge about what transactions would be squashed if it grants a commit request. Therefore, we need to extend the naïve scheme by breaking up a commit operation into two phases: 1) check-for-data-races 2) make-commit-decision where the first phase is not part of the naïve and the baseline scheme.

To check for data conflicts, the committing node first broadcasts the write set of its transaction. Each node checks locally whether its ongoing transaction conflicts with the committing transaction. If this is the case, it activates the SQUASH signal introduced in the naïve scheme (see Figure 3). The CAU will check the MSCs of the transactions with their SQUASH signals activated and if any of them is higher than the MSC of the committing node and they have the same (lowest) phase number, the committing node is not allowed to commit. On the other hand, if the committing transaction will be allowed to commit, the CAU will notify the nodes that have their SQUASH signals activated to squash their ongoing transactions.

In summary, in comparison with the naïve scheme, the elaborate scheme must extend the commit protocol with a phase that checks whether the transaction that requests to commit can do so without squashing a thread that has the same phase number but a higher MSC value. Note that even if a committing transaction will stall, a bus transaction that checks for conflicts is still needed. Thus, the elaborate scheme causes more traffic than the naïve scheme.

4 Experimental Methodology

In order to analyze whether the new arbitration schemes result in fewer useless cycles in terms of delays (waiting time for commit decision) or miss-speculations, we have built a simulation model of the baseline system and augmented it with the naïve and the elaborate scheme.

As we are only concerned with the number of cycles lost due to stalling a committing thread and a thread that miss-speculates, we have opted for a simple model of the memory system. Hence, we do not charge any cycles for cache misses at any level of the memory hierarchy. We model a 16-core system with a single hardware thread per core. Each node has a 128-Kbyte, 4-way L1 cache with a block size of 16 bytes. This would correspond to multi-cores with a combined L1/L2 inclusive cache hierarchy of the same capacity. Further, we maintain an infinite buffering space for speculatively modified blocks by assuming an infinite victim cache.

The system model is driven by traces generated by the eight SPLASH-2 applications [11] in Table 1. We use a trace-driven methodology to drive the system model. Each of the eight benchmarks is first run on Simics [5] with Sun Sparc as the target machine. Benchmarks are run with the original synchronization primitives and we collect statistics only in the parallel phase of each benchmark.

Table 1. Benchmarks and their data sets

Applications	Inputs	Applications	Inputs
Barnes	2048 particles	Ocean	34x34 grid
FFT	2^{16}-data points, cache line size-16, number of cache line-8K	Radix	1024 radix, 256K keys
		Raytrace	teapot.env
FMM	2048 particles	Volrend	Head
LU-Non contig.	512x512 matrix		

Transactions are marked by using the Simics magic instructions in-lined in the implementation of locks and barriers in the ANL-macros. There is a magic instruction at the beginning and at the end of these macros. All memory references between these magic instructions are filtered out of the trace.

A new transaction starts at a barrier and will terminate at the next barrier or after 1000 instructions, which of them happens first, unless that point occurs inside a critical section. If so, the transaction is continued until the corresponding unlock construct is executed. For nested locks, we flatten them out with the outermost lock, i.e., no transaction can start or terminate within the boundary of this outermost critical section.

It is well-known that a trace-driven approach that we use will not reflect the correct interleaving of events. However, as our goal is not to assess absolute performance but rather to compare the number of useless cycles using the different arbitration policies, we feel that the methodology adopted is adequate for this purpose.

5   Experimental Results

In Section 5.1, we experimentally confirm that the starvation scenario in Section 2.3 can happen. Then, we analyze the tradeoffs between the baseline, the naïve scheme, and the elaborate scheme in detail in Section 5.2

5.1   The Case for Starvation of the Baseline Protocol

Figure 4 shows the execution time of each of the 16 threads normalized to the slowest running thread when running the application Raytrace on three systems. For each thread, the three bars correspond to the baseline system, the naïve scheme, and the elaborate scheme, from left to right. Further, each bar is decomposed into three

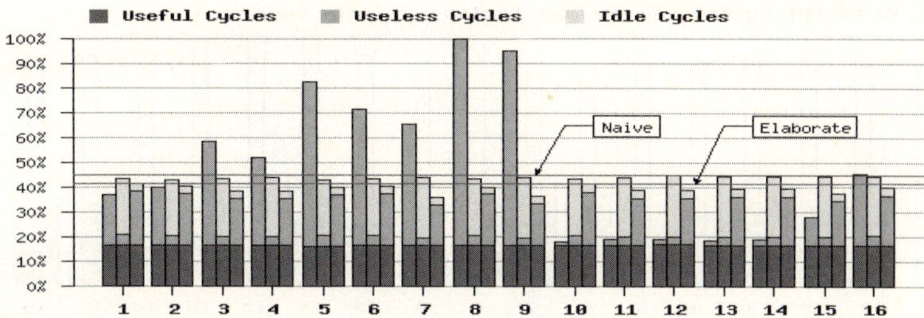

Fig. 4. Execution time of individual threads in Raytrace normalized to the slowest thread. The first, second, and third bars in each cluster correspond to the baseline, the naive and the elaborate arbitration scheme respectively.

sections that break down the execution time into useful cycles, useless cycles, and idle cycles (waiting for commit decision), from bottom to top.

As can be seen in Figure 4, there is a huge difference in execution time between different threads under the baseline arbitration scheme. For example, the baseline system execution time for thread 8 is five times longer than that of thread 10! However, the number of useful cycles across the threads is about the same so the effect is not attributed to load imbalance. The difference stems from the number of useless cycles. Trace inspection has revealed that Raytrace suffers from a similar starvation situation as described in Section 2.3 which degrades the performance.

Considering the execution times across threads for the naïve and the elaborate schemes (the two rightmost bars in each cluster), we can see that the difference between the slowest and the fastest threads is small. This suggests that these schemes successfully eliminate starvation. In fact, the execution time, which is dictated by the slowest thread, is cut down by as much as between 55% and 59% using the naïve and the elaborate schemes.

Continuing with the difference between the naïve and the elaborate schemes, it is clear from Figure 4 that the naïve scheme suffers a lot from stalling. However, it is interesting to note that the elaborate scheme also suffers from useless cycles which are attributed to miss-speculations. A close inspection of Figure 4 reveals that the naïve scheme tends to reduce the number of cycles lost for miss-speculations at the expense of cycles lost for stalling threads. However, the elaborate scheme performs slightly better than the naïve scheme – the execution time is around 4% shorter.

5.2   Tradeoffs Between the Naïve and the Elaborate Scheme

In Figure 5, we present the same results as in Figure 4 but for the ocean application. A striking observation is that the naïve scheme suffers from a huge idle time. This is caused by transactions that have to wait for unrelated transactions with a higher miss-speculation count to commit. This is unfortunate, as the waiting transactions could have been committed in the first place.

On the other hand, the elaborate scheme manages to keep the number of stall cycles low but at the expense of extending a commit transaction with a "check-for-data-races" phase. It is interesting to understand whether a simple modification of the naïve scheme could reduce the number of idle cycles.

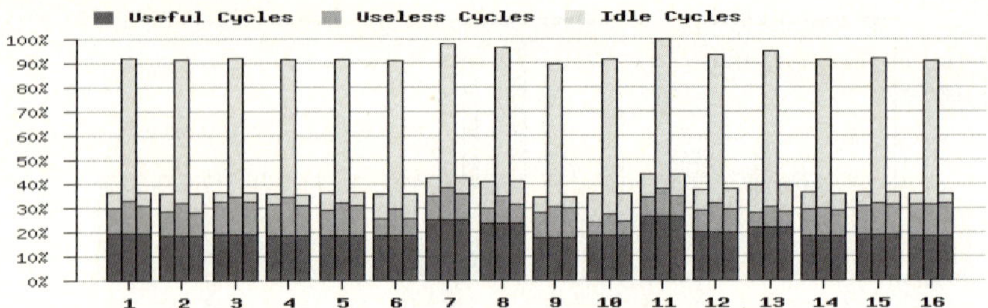

Fig. 5. Execution time of individual threads in Ocean normalized to the slowest thread. The first, second, and third bars in each cluster correspond to the baseline, the naive, and the elaborate arbitration scheme respectively.

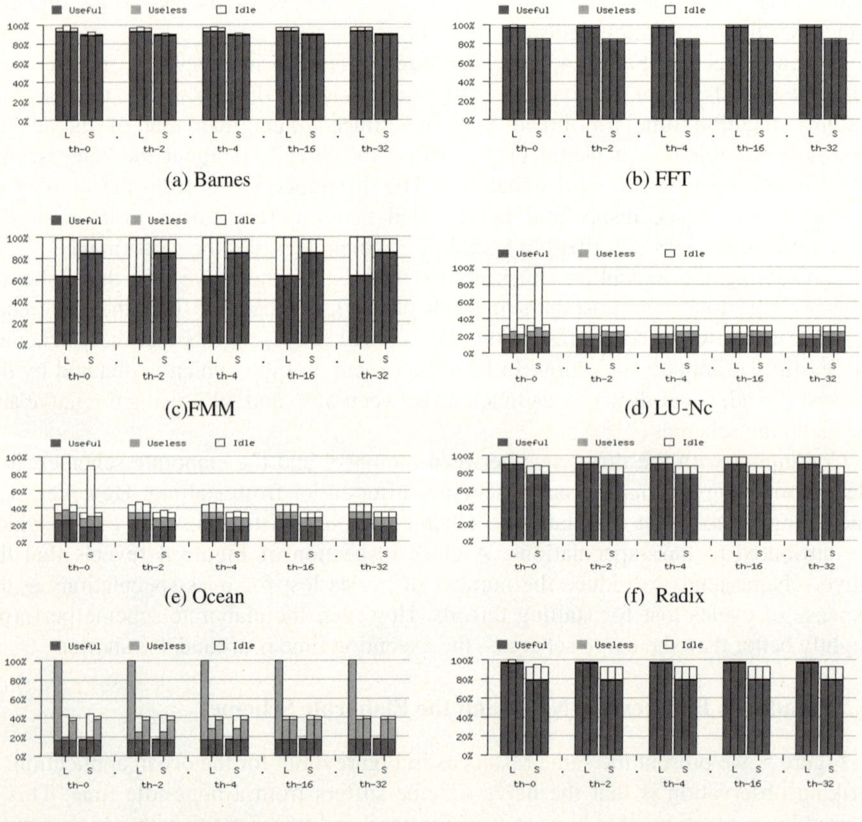

(a) Barnes

(b) FFT

(c)FMM

(d) LU-Nc

(e) Ocean

(f) Radix

(g) Raytrace

(h) Volrend

Fig. 6. Relative performance of the commit arbitration schemes for the eight applications using threshold values 0, 2, 4, 16, and 32 plotted through diagrams (a) to (h). The 1st and 2nd clusters under each threshold represent the execution time for the slowest and fastest running threads respectively. Each cluster uses three bars that represent the relative execution time for the same thread using all three arbitration schemes.

This encouraged choosing threads that have to stall under the naïve scheme more selectively. Recalling the example starvation scenario in Section 2.3, the chief observation is that a symptom of a starvation situation is that a request has been denied several times in a row. In our improved naïve scheme, we allow a transaction (with the lowest phase number) to commit if its MSC count is lower than another one (with the lowest phase number) minus an offset, which we refer to as a threshold.

Obviously, a key issue is to select an appropriate threshold. Therefore, we experimented with several threshold values. Figure 6 shows the performance of the eight applications under the naïve scheme using threshold values of 0, 2, 4, 16, and 32 for the naïve scheme which correspond to the different bar clusters (each consisting of six bars) in each diagram. In the figure, we have depicted the execution time for the longest (L) and the shortest (S) running thread for each application. This enables us to pinpoint a starvation scenario if it exists.

As we go from lower to higher threshold values, the idle time for the naïve scheme is reduced significantly. It is important to note that it becomes as low as the baseline arbitration for a threshold value of 16 for the applications which do not encounter any starvation.

When an application uses barrier synchronizations, the performance of an application that is run under a transactional memory paradigm will suffer from idle cycles lost due to ordered commits. This is because a transaction with a higher phase number must wait for a transaction with a lower phase number to commit. As a result, also the baseline scheme suffers from idle cycles in all applications that use barriers to synchronize as can be seen from Figure 6. One exception is Raytrace which does not use barriers.

A striking observation is that as we increase the threshold, the performance difference between the naïve scheme and the elaborate scheme diminishes. Therefore, our recommendation is that the naïve scheme with a fairly high threshold (16 for example) can safely eliminate starvations.

6 Related Work

We have proposed arbitration schemes for TM systems to avoid starvation. Hill et al. have classified TM systems in two groups depending on lazy/eager conflict detection [4]. While [1,4,7] are referred to as eager, [2] does lazy conflict resolution. This paper has focused on the starvation problem for TM systems using lazy conflict resolution.

Rajwar et al. [9] proposed to timestamp a transaction when it is instantiated for prioritizing transactions and in a conflicting scenario lower priority transactions rollback or wait. The policy ensures forward progress and starvation freedom in a system where conflicting scenario is visible before making a commit decision. The author also assumed a system configuration where conflict is visible at the time of the conflicting memory access. On the other hand, in a lazy conflict detecting system, conflict is visible only after a transaction commit and memory modification is revealed to others. However, we can use the timestamp based prioritization scheme to serialize all commits disregarding the fact that many transactions don't have any conflict among them (independent transactions). Nevertheless, a concern is that it would unnecessarily force independent transactions to wait.

TM systems that detect conflict lazily, know about the conflict at the commit point of any of the involving transactions. The straightforward first-come first-serve arbitration policy used in TCC may lead to a starvation as described in Section 2.3. Our proposed arbitration scheme targets a TM system that detects conflicts lazily and especially uses a central arbitration unit to select a thread to commit.

Hammond et al. [2] propose the use of pseudo-barriers to make forward progress for all processors. The idea is to arbitrarily insert a barrier implicitly so that a thread that is subject to starvation will get a chance to catch up. Unfortunately, the paper does not elaborate on how to make use of the idea. In fact, we have shown in this paper that it is important to monitor when a starvation scenario is about to happen. We do this by using miss-speculation counters. By arbitrarily inserting barriers, there is simply no such monitoring mechanism to provide feedback.

7   Concluding Remarks

In this paper, we have proposed novel commit arbitration schemes for TM systems that avoids starvation. We show how they can be implemented in a framework based on TCC. As a general approach to avoid starvation, we propose that the commit arbitration policy should be provided with feedback on squash counts of other threads. To this end, we propose the naïve scheme and the elaborate scheme. Through a detailed implementation and performance evaluation study we found the following. Both proposed schemes can avoid starvation but do it at the expense of lost cycles due to delaying the point at which a thread can commit.

We analyzed the design space of the simplest policy – the naïve policy – and found that it can eliminate most of the stall cycles by introducing a threshold value between the committing thread's and the other ongoing thread's miss-speculation count. By doing this, we have found that naïve manages to remove most of the stall cycles at the expense of some quite modest structure and protocol extensions. Overall, this paper shows that it is possible to avoid starvation of transactional memory protocols at a modest implementation cost.

References

1. Ananian, C.S., Asanovi'c, K., Kuszmaul, B.C., Leiserson, C.E., Lie, S.: Unbounded Transactional Memory. In: Proceedings of the 11th International Symposium on High-Performance Computer Architecture, San Franscisco, CA, February 2005, pp. 316–327 (2005)
2. Hammond, L., Wong, V., Chen, M., Hertzberg, B., Carlstrom, B., Davis, J., Prabhu, M., Wijaya, H., Kozyrakis, C., Olukotun, K.: Transactional Memory Coherence and Consistency. In: Proc. of the 31st Annual International Symposium on Computer Architecture, München, Germany, June 19-23, pp. 102–113 (2004)
3. Herlihy, M., Moss, J.E.B.: Transactional Memory: architectural support for lock-free data structures. In: Proceedings of the 20th International Symposium on Computer Architecture, pp. 289–300 (1993)
4. Moore, K.E., Bobba, J., Moravan, M.J., Hill, M.D., Wood, D.A.: LogTM: Log-based Transactional Memory. In: Proceedings of the 12th Annual International Symposium on High Performance Computer Architecture (HPCA-12), Austin, TX (February 11-15, 2006)

5. Magnusson, P.S., Christianson, M., Eskilson, J., et al.: Simics: A full system simulation platform. IEEE Computer 35(2), 50–58 (2002)
6. Martinez, J., Torrellas, J.: Speculative synchronization: Applying thread-level speculation to parallel applications. In: Proceedings of the 10th International Conference on Architectural Support for Programming Languages and Operating Systems (October 2002)
7. Rajwar, R., Herlihy, M., Lai, L.: Virtualizing transactional memory. In: Proceedings of the 32nd International Symposium on Computer Architecture, June 2005, pp. 494–505 (2005)
8. Rajwar, R., Goodman, J.: Speculative Lock Elision: enabling highly concurrent multithreaded execution. In: MICRO 34: Proceedings of the 34th ACM/IEEE International Symposium on Microarchitecture, pp. 294–305. IEEE Computer Society Press, Los Alamitos (2001)
9. Rajwar, R., Goodman, J.: Transactional Lock-free Execution of Lock-Based Codes. In: Proceedings of 10th International Conference on Architectural Support for Programming Languages and Operating Systems (ASPLOS) (October 2002)
10. Rundberg, P., Stenstrom, P.: Reordered speculative execution of critical sections. In: Proceedings of the 2002 International Conference on Parallel Processing (February 2002)
11. Woo, S.C., Ohara, M., Torrie, E., Singh, J.P., Gupta, A.: The SPLASH-2 Programs: Characterization and Methodological Considerations. In: Proceedings of the 22nd International Symposium on Computer Architecture, Santa Margherita Ligure, Italy, June 1995, pp. 24–36 (1995)

Topic 5
Parallel and Distributed Databases

Marta Patiño-Martinez, Genoveva Vargas-Solar, Elena Baralis,
and Bettina Kemme

Topic Chairs

Advances in data exploitation (access, query, retrieval, analysis, mining) are inherent to current and future information systems. Today, accessing great volumes of information is reality; tomorrow data intensive management systems will enable huge user communities to transparently access multiple pre-existing autonomous, distributed and heterogeneous resources (data, documents, services). Existing data management solutions do not provide efficient techniques for exploiting and mining tera-datasets available in clusters, peer to peer and grid architectures. Parallel and distributed databases are a key element for achieving scalable, efficient systems that will both cost-effectively manage and extract knowledge from huge amounts of highly distributed and heterogeneous digital data repositories.

The track received 16 submissions from which 6 papers were accepted for publication.

The first paper is devoted to P2P data management. "Topology-Aware Approach for Distributed Data Reconciliation in P2P Networks" presents an algorithm for data reconciliation that exploits knowledge about the topology of the underlying P2P network to reduce the communication cost. Two other papers are devoted to parallel and distributed algorithms for data mining. "Efficient Distributed Data Condensation for Nearest Neighbor Classification" presents a distributed algorithm for computing a consistent subset of the training datasets for the nearest neighbor classification whilst "Parallel Nearest Neighbour algorithms for Text Categorization" explores the parallelization of the nearest neighbour problem.

"Handling Request Variability for QoS-max Measures" focuses on how optimizing the multi-programming level of the database to maximize the throughput.

"Search Engine Accepting On-Line Updates" discusses how to address updates in parallel search web engines. Finally, "A Multi-layer Collaborative Cache for Question Answering" addresses how to distribute a cache for answering queries efficiently.

A.-M. Kermarrec, L. Bougé, and T. Priol (Eds.): Euro-Par 2007, LNCS 4641, p. 293, 2007.

A Multi-layer Collaborative Cache for Question Answering[*]

David Dominguez-Sal[1], Josep Lluis Larriba-Pey[1], and Mihai Surdeanu[2]

[1] DAMA-UPC, Universitat Politècnica de Catalunya,
Jordi Girona 1,3 08034 Barcelona, Spain
{ddomings,larri}@ac.upc.edu
[2] TALP, Universitat Politècnica de Catalunya,
Jordi Girona 1,3 08034 Barcelona, Spain
surdeanu@lsi.upc.edu

Abstract. This paper is the first analysis of caching architectures for Question Answering (QA). We introduce the novel concept of multi-layer collaborative caches, where: (a) each resource intensive QA component is allocated a distinct segment of the cache, and (b) the overall cache is transparently spread across all nodes of the distributed system. We empirically analyze the proposed architecture using a real-world QA system installed on a cluster of 16 nodes. Our analysis indicates that multi-layer collaborative caches induce an almost two fold reduction in QA execution time compared to a QA system with local cache.

1 Introduction

Question Answering (QA) systems are tools that extract short answers from a textual collection in response to natural language questions. For example, the answer to the question "How many countries are part of the European Union?" is "25". This functionality is radically different from that of typical information retrieval systems, e.g. Google or Yahoo, which return a set of URLs pointing to documents that the user has to browse to locate the answer.

A Question Answering service, like some other of today's applications, demands huge amounts of CPU and I/O resources that a single computer can not satisfy. The usual approach to implement systems with such requirements consists of networks of computers that work in parallel. On the other hand, the use of caching techniques reduces significantly the computing requirements thanks to the reuse of previously accessed data. Our target is to add the infrastructure to get the benefits of parallelism and caching to a QA system.

[*] David Dominguez is supported by a grant from the *Generalitat de Catalunya* (2005FI00437). Mihai Surdeanu is a research fellow within the Ramón y Cajal program of the Ministerio de Educación y Ciencia. The authors want to thank *Generalitat de Catalunya* for its support through grant number GRE-00352 and Ministerio de Educación y Ciencia of Spain for its support through grant TIN2006-15536-C02-02, and the European Union for its support through the Semedia project (FP6-045032).

A.-M. Kermarrec, L. Bougé, and T. Priol (Eds.): Euro-Par 2007, LNCS 4641, pp. 295–306, 2007.

In this paper we propose a collaborative caching architecture for distributed QA. We propose a new multi-layer cache where we store the data according to its semantics. We identify two types of data used by the two resource-intensive processes of a QA application, and according to it, we implement a two layer cache in our system. We show that these data sets can be cached cooperatively, with an increase in the application's performance over the caching of one or the other data sets alone. Furthermore, we do not limit the architecture to local data accesses, we provide and evaluate techniques that make it possible to find and retrieve data from remote computers in order to save disk accesses and computing cycles.

We tested all our results in a real Question Answering system. We show that the use of cooperative multi-layer distributed caches improves over the use of local multi-layer caches alone. Furthermore, the results obtained in a distributed environment reach from almost two, up to five fold improvements of the hit rates; the execution time is reduced by 50% to 60% on a 16 processor paralel computer, without parallelizing the QA task.

This paper is organized as follows. In Section 2 we describe our QA system. Section 3 explains the cache architecture for the QA system. Finally, Section 4 contains the empirical analysis of the proposed cache architecture. We end with a review of related work and conclusions.

2   Question Answering

The QA system implemented uses a traditional architecture, consisting of three components linked sequentially: *Question Processing* (QP), *Passage Retrieval* (PR), and *Answer Extraction* (AE). We describe these components next.

Question Processing: The QP component detects the type of the expected answer for each input question, i.e. a numeric answer for the question above. We model this problem as a supervised Machine Learning (ML) classification problem: (a) from each question we extract a rich set of features that include lexical, syntactic, and semantic information, and (b) using a corpus of over 5,000 questions annotated with their expected answer type, we train a multi-class Maximum Entropy classifier to map each question into a known answer type taxonomy [1].

Passage Retrieval: Our PR algorithm uses a two-step query relaxation approach: (a) in the first step all non-stop question words are sorted in descending order of their priority using only syntactic information, e.g. a noun is more important than a verb, and (b) in the second step, the set of keywords used for retrieval and their proximity is dynamically adjusted until the number of retrieved documents is sufficient, e.g. if the number of documents is zero or too small we increase the allowable keyword distance or drop the least-significant keyword from the query. The actual information retrieval (IR) step of the algorithm is implemented using a Boolean IR system[1] that fetches only documents that contain *all* keywords in the current query. PR concludes with the elimination of the documents whose relevance is below a certain threshold. Thus the system avoids applying the next CPU-intensive module for documents that are unlikely to contain a good answer.

Answer Extraction: The AE component extracts candidate answers using an in-house *Named Entity Recognizer* [2], and ranks them based on the properties of the context where they appear. We consider as candidate answers all entities of the same type as the answer type detected by the QP component. Candidate answers are ranked using a formula inspired from [3].

This QA system obtained state-of-the-art performance in one recent international evaluation. More details on the system architecture and the evaluation framework are available in [4].

QA systems are excellent solutions for very specific information needs, but the additional functionality provided, i.e. query relaxation and answer extraction, does not come cheap: using the set of questions from previous QA evaluations[2] we measured an average response time of 75 seconds/question. The execution time is distributed as follows: 1.2% QP, 28.8% PR, and 70.0% AE. This brief analysis indicates that QA has two components with high resource requirements: PR, which needs significant I/O to extract documents from the underlying document collection, and AE, which is CPU-intensive in the extraction and ranking of candidate answers.

3   A Multi-layer Distributed Cache for Question Answering

As previously mentioned, QA has two resource intensive components that can benefit from caching: PR and AE which are I/O and CPU intensive respectively. This observation motivates the architecture proposed in this paper: we introduce a two-layer distributed caching architecture for QA, with one layer dedicated to each resource-intensive component.

Figure 1 depicts the architecture of the distributed QA system extended with multi-layer caching on each node. We consider that each node has a finite amount of memory for the cache. Figure 1 shows that, in each node, the cache is divided into two layers: the first dedicated to caching PR results and the second used to cache AE information. The PR layer caches only the raw text documents retrieved from disk. The AE layer stores passages, which are the retrieved documents as well as their syntactic and semantic analysis. An entry is stored either in the PR layer or in the AE layer because the cache architecture is inclusive: a document cached in the PR layer will hit in the PR phase but miss in the AE phase, while a passage cached in the AE layer will hit in both.

---

[1] http://lucene.apache.org
[2] http://trec.nist.gov/data/qa.html

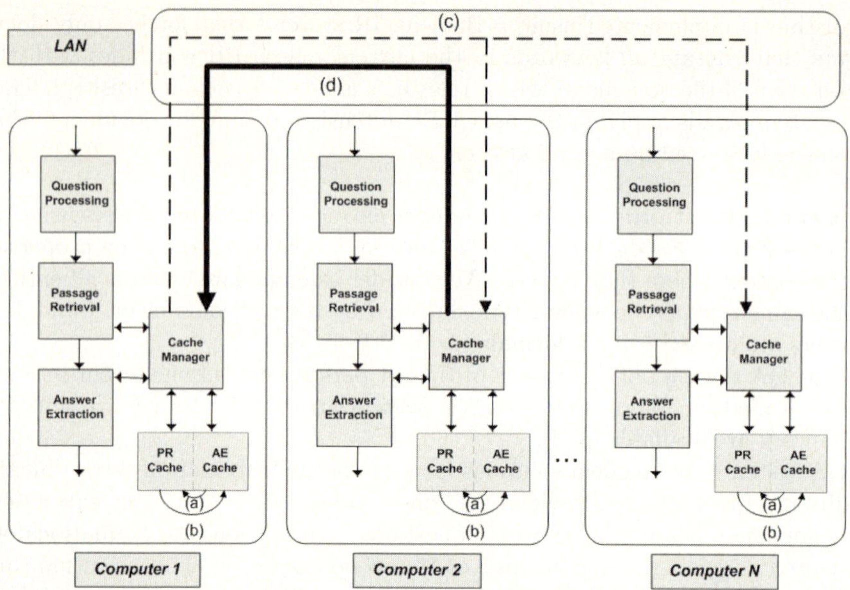

Fig. 1. Diagram of the QA system. QA is composed of three sequential blocks QP, PR and AE. PR and AE can request and put information into the cache using the cache manager. The cache manager implements several operations: (a) Demote: An entry from the AE cache is moved to the PR cache and the natural language analysis is reset; (b) Promote: An entry from the PR cache is moved to the AE cache; (c) Petition: broadcast request message to other caches; (d) Response: the node that can resolve the petition replies with the requested information.

Note that: (a) due to the filtering at the end of PR, only a subset of the documents accessed in PR reach AE and (b) an AE cache unit is significantly larger than the corresponding PR cache unit (approximately five times larger in our setup). These observations justify the need for the multi-layer cache.

The caching unit is a collection document or a passage. This approach increases the probability of a hit because: (a) even though the same natural language question can be reformulated in many distinct variants, they all require the same documents/passages, and (b) the documents/passages needed by questions that are different but on the same topic are likely to overlap significantly.

3.1   The Cache Manager

Both layers are managed by the same cache manager, which implements the local intra and inter-layer operations and the collaboration protocol between nodes. The operations of the cache manager are logically grouped in two protocols: the *local protocol*, which defines the operations within a node, and the *distributed protocol*, which defines the collaboration between nodes.

Local Operations

With respect to the cache management protocol, we employ two policies:

- One for the internal management of a layer, which defines the *intra-layer* operations. We use a least recently used (LRU) algorithm. This means that entries in a layer are ordered in decreasing order of their usage time stamps.
- A second policy, which defines the *inter-layer* operations. We use a promotion/demotion algorithm. When a document is read from disk for the first time it is promoted in the PR layer; then, the oldest document is demoted from PR (i.e. it is completely removed from the cache). When we miss a passage in the AE layer but the document exists in the PR layer, we promote the entry to the AE layer. Then, the oldest entry in the AE layer is demoted to the PR layer. When we demote a unit from the AE to the PR layer the syntactico-semantic information is deleted.

Distributed Protocol

Question Answering is an expensive task. In other words, a single computer can not process the amount of user queries that a web search engine receives. The solution, as with information retrieval engines, is distributed computing [5]. We present an extension of our cache to a collaborative cache for a cluster based architecture, where each node runs a copy of the QA system. We compare several alternatives for implementing the distributed system:

Local: A multi-layer cache is deployed locally in each cluster node. We call this implementation *Local*. There is no communication between nodes.

*Broadcast Petition (*BP*):* We build a collaborative cache using a protocol that asks for remote work units using a network broadcast (operation (c) in Figure 1). If another node has already processed that work unit, it answers with the information stored in its cache (operation (d) in Figure 1). The response includes information of subsequent layers if the remote node knows it. The QA process continues when all requests for a query have been fullfilled or the request times out, in which case the request is processed locally. Results received are locally cached using the local caching protocol.

For example, if a node asks for a document (that is missing in the local cache) and another node has the document and the passage, the remote node will answer with the information from both the AE and PR layer because this information will be probably needed later. Sending all available information once reduces the number of network connections and the response time.

*Broadcast Petition Recently (*BPR*):* This model is similar to *BP*, but it locally caches work units received from other nodes only if they were repeatedly accessed in the same node (at least twice) in the past x seconds. For example: the first time that a node retrieves remotely the document A, the node will not store it locally. The next time the node must retrieve A, if it occurs before x seconds, a duplicate copy will be cached locally because the entry is considered to be frequently accessed. Otherwise, the local cache will not store it locally. The protocol is efficiently implemented with a queue with the ids and the timestamps of petitions. When a node receives the information remotely, it checks if it has previously requested the same data recently with a lookup to a queue of the most recent petitions.[3]

This method minimizes the chance that documents that are not frequently accessed have multiple copies in the network. This is an important feature for QA where storing a cache unit is expensive, e.g. documents may be large and the natural language information takes significant space.

Our collaborative solutions avoid failure problems because all nodes are equal: if one machine crashes, the system continues to run. Furthermore, we can increase the processing power of our QA cluster simply by warm plugging more computers. To balance the load, we dispatch questions in a round robin fashion among the nodes in the system. However, note that in a real-world implementation the dispatcher does not have to be part of the distributed QA system itself. It can be implemented within a Domain Name Service (DNS) server, which maps a unique name to all the IP addresses in the QA cluster.

4   Experiments

In this section we test the performance of the multi-layer cache and the different collaborative protocols using the previously described QA system.

We used as textual repository the TREC document collection which has approximately 4GB of text in approximately 1 million documents. The questions sent to the system were selected from the questions that were part of former TREC-QA evaluations (700 different questions).

We tested the multi-layer cache in our QA system for Zipf question distributions. [6,7,8] show that web servers and IR systems follow zipfian distributions with estimations of parameter α between 0.60 and 1.45. Considering that QA systems are similar to IR systems from the usage point of view, it is reasonably safe to assume that question distributions for real-world QA systems will also follow a similar Zipf distribution.

4.1   Experiments with Multi-layer Cache

Setup: One single PC dual Xeon 2.4Ghz with 1.5GB of RAM. The question set contains 150 queries randomly selected following a $\text{Zipf}_{\alpha=1.0}$ distribution.

Experiments: In this experiment we show that multi-layer caches perform better than single level caches. We assigned to the QA system a fixed pool of memory for storing the PR and the AE layers. We executed several runs with different partitions of the available memory between the two layers. We set the size of an

---

[3] This can be implemented efficiently in several ways. For example, we can add a hash if the number of recent requests is known to be large.

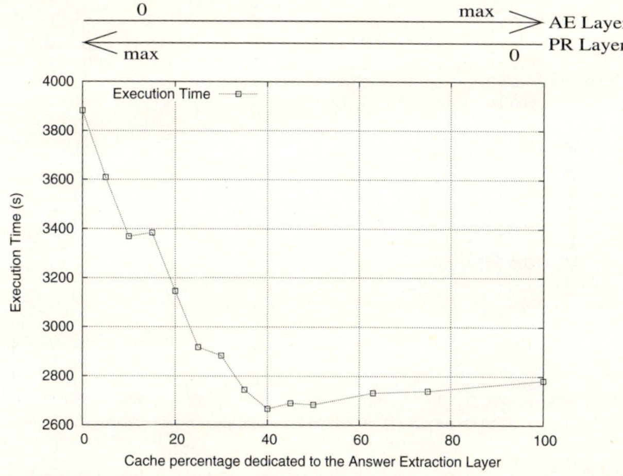

Fig. 2. Execution time of the QA system in a single computer with different partitions

entry in the AE layer as 5 times the size of the entry in the PR layer to accommodate the additional information generated by the natural language analysis. A QA system with a single cache layer corresponds to partitioning all the available memory for either PR or AE.

In Figure 2 we plot the execution times for different partitions of the cache. The vertical axis is the total execution time. The horizontal axis represents the partitioning of the available memory between cache layers. The extremes of the graph correspond to single level caches: the left end is a PR only cache, and the right end is an AE only cache. Our experiments show that the best execution time corresponds to partitioning about 40% of the cache to the AE layer and the rest to the PR layer. All other partitions yield higher execution times. Figure 2 also proves that single layer caches, i.e. when the cache is complely dedicated to either PR or AE (left and right-most points in Figure 2), are less efficient than multi layer caches.

4.2 Distributed Protocol Experiments

Setup: A cluster with 16 nodes, each equipped with two Xeon 2-Ghz processors and 2GB RAM. All nodes were interconnected with a 1GB Ethernet. We dispatched the questions to the QA system using a separate client process that runs in the same LAN. The client process sends questions following the chosen Zipf distribution with a pause of 10 ms between questions. The question set contains 700 questions distributed in a round robin fashion among the nodes.

We performed several experiments with different cache sizes. In each node we set the maximum cache size as $\frac{1}{16}$ of the size required to cache all the work units in a single node[4]. Hence, when all memory is used, the overall cache can store all

---

[4] This value corresponds to the right end in Figures 3 and 4.

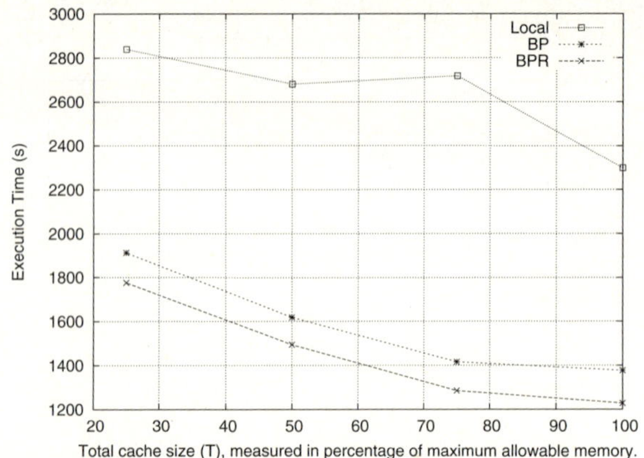

Fig. 3. Execution time of the distributed QA system with 16 nodes. Question distribution is Zipf $_{\alpha=1.0}$.

the data generated in the experiment if there were no duplicates. We partitioned the cache memory between the PR and AE layers using the best configuration from the experiment in Figure 3 (60% for PR and 40% for AE).

We experimented empirically with several values of recency for the *BPR* algorithm, between 30 and 180 seconds, but we obtained similar results in this range. This is caused by the exponential nature of the Zipf distribution, which yields large gaps in access frequencies between the very frequent questions and the remaining ones. For example, the most frequent questions in our experiments repeated up to eight times per node, sometimes even consecutively. On the other hand, there is a big set of questions that will appear more than once in the question set but are rarely accessed twice by a node. Due to this large gap in access frequencies, *BPR* achieves its goal of avoiding caching the latter class of questions without fine tuning the recency parameter. In the experiments reported here the value is set to 120 seconds.

Experiments: In Figure 3 we plot the execution time for different memory size configurations and cache algorithms. The vertical axis represents execution time, and the horizontal is the size of the memory pool allocated to the system. This experiment is done using a Zipf$_{\alpha=1.0}$ distribution. Figure 3 confirms that both collaborative approaches have better performance than the *Local* method. The execution time of the *local* protocol (top line in Figure 3) is up to two times slower than the execution time of the collaborative methods.

Figure 4 explains in detail the better performance of the collaborative caching algorithms (*BP* and *BPR*). The figure shows the cache hit rates for the PR and the AE layers for the three models proposed. We show both the local cache hits (lines with clear marks) and the total hits, i.e. local + remote (lines with solid marks). At a first glance, one can see that collaborative caches have better performance, i.e. their hit ratio is approximately 25% larger than the hit ratio of local protocols. If

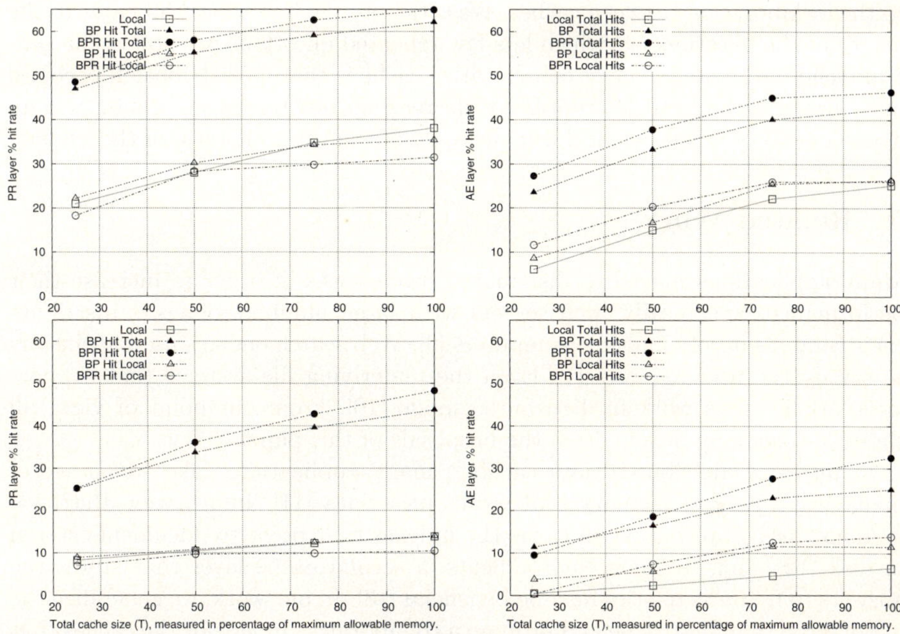

Fig. 4. Hit rates for a distributed QA system with 16 nodes. The left column shows the PR layer and the right column is the AE layer. First row plots results for a Zipf$_{\alpha=1.0}$, second row for a Zipf$_{\alpha=0.6}$.

we compare *BP* and *BPR* in the first row of Figure 4, we see that *BPR* has a higher ratio of total hits. In the PR layer, *BPR* has a 2-7% lower hit ratio than *BP* because when the system asks for a remote document it must be requested more than once in order to cache it locally. On the other hand, *BPR* yields a more efficient global cache by avoiding useless copies in the network, so the system has a higher number of total hits. In the AE layer, collaborative algorithms perform slightly better than the local protocol for local accesses. The explanation is the inclusive nature of the proposed caching architecture: if a node requests a document from the PR cache and a remote node node has the only corresponding passage in the AE cache, it replies with the complete passage. Hence the original node will have a remote hit in the PR phase and a local hit in AE. Figure 3 confirms that the better hit rate makes the *BPR* protocol up to 12% faster than *BP*.

As expected, when we decrease the cache size, the number of local hits becomes considerably smaller. Nevertheless, even if we can not allocate sufficient memory for caching results, the use of a collaborative cache mantains a large number of total hits in the system.

We tested the algorithms with another distribution, setting $\alpha = 0.6$, which is a lower bound for typical usages of web search engines and other IR systems. The results are shown in the last row of Figure 4. The trend is the same as for $\alpha = 1.0$ as we saw above. The most notable difference is that the local hit rates are lower but the number of remote hits does not decrease at the same rate due to the collaboration between caches. We attribute the lower local hit rates to the length of the question set, which lets few repeated questions per node. The general trend of the double layer collaborative cache is that as distribution becomes more uniform the local hit rate keeps decreasing, but the cache hits on remote documents or passages smooth the decrease of the total hit ratio of the system.

5   Related Work

Many applications moved to distributed frameworks in order to increase their performance. Specifically, web servers with large numbers of visits distribute their contents across multiple computers [9]. Web search engines use big clusters to satisfy the user requests [5]. Even the underlying file systems that support large applications work in distributed mode [10]. From our point of view, all these systems may benefit from the proposals of this paper.

In terms of distributed data caching, many applications take advantage of managing a pool of memory that stores past data [11]. Information Retrieval systems, for instance, save in cache the indexes that point to documents stored in disk [12], and, in some environments, a second cache layer that stores the answers to frequent queries may be beneficial [13]. In our work, we generalize the use of more than one cache and propose a collaborative model for the caches that helps improve their performance. Furthermore, we show that a correct allocation of the available memory is crucial if we want to get the maximum performance of our application.

Our approach should not be confused with multi-level caches [14,15]. A multi-level cache is oriented towards a client-server architecture where the cached data can be stored in different computers in the route between the client and the server. Our multi-layer cache proposes a management protocol for an application with several pipelined stages. Our extensions for a distributed environment also differ from multi-level because the computers do not have to be organized in a hierarchical topology, which is inherent of multi-level caches.

Question Answering has been a prolific research field mainly devoted to improving the accuracy of the results obtained [16,17,18]. The work on improving the quantitative performance of the QA problem is scarce. [19] proposed a distributed QA architecture but their focus is on load-balancing algorithms rather than caching strategies. Our work can be seen as a necessary complement to their load distribution techniques. Some QA systems implement caches for previous answers [20] or take advantage of the caches of IR systems to reduce the Passage Ranking execution time [21]. In this paper, we give a distributed solution for caching the different consuming phases in QA, and the granularity of our solution allows for the efficient reuse of the results from different previous questions.

6   Conclusion

To our knowledge, this paper is the first analysis of caching architectures for Question Answering. QA, usually implemented with a pipeline architecture, has resource requirements that are distinct and much larger than its historical predecessor, Information Retrieval: Passage Retrieval, which extracts the content of relevant documents, has high I/O usage, and Answer Extraction, which identifies the exact answer in the relevant documents, is CPU intensive. Driven by these observations we proposed and analyzed a caching architecture that: (a) is multi-layer, i.e. each resource intensive pipeline stage is allocated a segment of the cache, and (b) is collaborative, i.e. the overall system cache is transparently spread across all nodes of the distributed system. We proposed two different protocols for the management of the distributed cache.

We empirically analyzed the proposed cache architecture using a real-world QA system and question distributions in both single-processor and distributed environments (16 nodes). Our first experiments show that the multi-layer approach is better than a single-layer architecture where all cache memory is allocated to either of the resource-intensive components. The QA system with the best multi-layer cache configuration has a response time up to a 30% smaller than the system with a single-layer. Secondly, we showed that, apart from allowing for warm plugs and unplugs, the distributed approach improves the hit ratio of the multi-layer collaborative caching, and also reduces the amount of communication by keeping information that is frequently reused locally. Overall, the distributed QA system enhanced with our best cache architecture has almost twofold gains in execution times from the system with local caching.

References

1. Xin, L., Roth, D.: Learning question classifiers: the role of semantic information. Natural Language Engineering 12(03), 229–249 (2005)
2. Surdeanu, M., Turmo, J., Comelles, E.: Named Entity Recognition from Spontaneous Open-Domain Speech. In: Proceedings of Interspeech (2005)
3. Pasca, M.: Open-Domain Question Answering from Large Text Collections. CSLI Publications Stanford, Calif (2003)
4. Surdeanu, M., Dominguez-Sal, D., Comas, P.: Performance Analysis of a Factoid Question Answering System for Spontaneous Speech Transcriptions. Interspeech (2006)
5. Barroso, L., Dean, J., Hölzle, U.: Web search for a planet: The google cluster architecture. IEEE Micro 23(2), 22–28 (2003)
6. Breslau, L., Cao, P., Fan, L., Phillips, G., Shenker, S.: Web caching and zipf-like distributions: Evidence and implications. In: INFOCOM'99. Eighteenth Annual Joint Conference of the IEEE Computer and Communications Societies. Proceedings. pp. 126–134, IEEE (1999)
7. Markatos, E.P.: On caching search engine query results. Computer Communications 24(2), 137–143 (2001)
8. Scime, A., Baeza-Yates, R.: Web Mining: Applications and Techniques. In: Idea Group (2005)
9. Cardellini, V., Casalicchio, E., Colajanni, M., Yu, P.S.: The state of the art in locally distributed Web-server systems. ACM Computing Surveys 34(2), 263–311 (2002)

10. Ghemawat, S., Gobioff, H., Leung, S.T.: The Google file system. In: Proceedings of the nineteenth ACM symposium on Operating systems principles, pp. 29–43. ACM Press, New York (2003)
11. Raunak, M.: A survey of cooperative caching. Technical report (1999)
12. Alonso, R., Barbara, D., Garcia-Molina, H.: Data caching issues in an information retrieval system. ACM Transactions on Database Systems (TODS) 15(3), 359–384 (1990)
13. Saraiva, P., de Moura, E., Ziviani, N., Meira, W., Fonseca, R., Riberio-Neto, B.: Rank-preserving two-level caching for scalable search engines. ACM SIGIR, 51–58 (2001)
14. Zhou, Y., Philbin, J., Li, K.: The multi-queue replacement algorithm for second level buffer caches. In: Proceedings of the General Track: 2002 USENIX Annual Technical Conference, Berkeley, CA, USA, pp. 91–104. USENIX Association (2001)
15. Wong, T., Wilkes, J.: My cache or yours? making storage more exclusive. In: Proceedings of the General Track: 2002 USENIX Annual Technical Conference, Berkeley, CA, USA, pp. 161–175. USENIX Association (2002)
16. Chu-Carroll, J., Czuba, K., Duboue, P., Prager, J.: Ibm's piquant ii in trec2005. In: Proceedings of the Fourthteen Text REtrieval Conference (TREC) (2005)
17. Robert, E.N.: Javelin i and ii systems at trec 2005. In: Proceedings of the Fourthteen Text REtrieval Conference (TREC) (2005)
18. Moldovan, D., Pasca, M., Harabagiu, S., Surdeanu, M.: Performance issues and error analysis in an open-domain question answering system. ACM Transactions in Information Systems 21(2), 1–22 (2003)
19. Surdeanu, M., Moldovan, D., Harabagiu, S.: Performance analysis of a distributed question/answering system. Transactions on Parallel and Distributed Systems (2002)
20. Lam, S., Ozu, M.: Querying Web data-the WebQA approach. In: Web Information Systems Engineering. WISE 2002, pp. 139–148 (2002)
21. Zheng, Z.: AnswerBus Question Answering System. In: Human Language Technology Conference (HLT 2002), pp. 24–27 (2002)

Handling Request Variability for QoS-Max Measures

Pedro Furtado

University of Coimbra
`pnf@dei.uc.pt`

Abstract. We denote as QoS-max the control of a request processing system to try to maximize QoS qualities and we focus on external, non-intrusive approaches with statistics on readily measurable quantities. In order to do this, the controller characterizes requests in terms of response times (or resource use) and uses that characterization to try to achieve QoS-max. However, measures vary both between different requests and for different runs of the same request. In this paper we show how we incorporated these for robust statistical QoS-max control. We use a simulator and requests with varied arrival and duration distributions to show the effectiveness of the variability handling approach.

Keywords: Performance, Transactional Systems, Autonomic.

1 Introduction

As we move from statically controlled systems to systems offering autonomic functionality and from standalone systems to virtualized environments where services can run in any platform, it becomes necessary to devise mechanisms that can learn from runtime conditions and provide QoS features. In traditional systems, control was enforced by static configuration parameters such as a limit on the number of sessions or transactions that could be accepted simultaneously. With the recent trend towards embedding autonomic functionality into systems, adaptability enables the system to adjust more tightly to working conditions, which may also change with time. Expressive constraints can be used for QoS control, like for instance, requests should be answered in less than 5 seconds, the miss rate on that deadline should not be above 15% and the throughput should be as large as possible. We emphasize a special requirement for the QoS control system: that it should be readily deployable in any request processing system without any change. Autonomic QoS Control requires a model to predict what would be the response to possible parameter adaptations. Prediction has been proposed using tools that include analytic models, resource capacities and request demands for those resources [2, 14] or response time statistics [8]. Whatever the approach that is used, variability is a practical fact, which is not due to the types of measures taken, but rather to characteristics of requests, systems and running conditions, so that total precision is not possible in most practical environments and approaches should account for variability. Some causes of variability in response time or resource demands by requests include system properties such as caches (e.g. part of indexes or data may be in the cache of a DBMS or not), locks, system overheads that are external to the request processing system,

A.-M. Kermarrec, L. Bougé, and T. Priol (Eds.): Euro-Par 2007, LNCS 4641, pp. 307–317, 2007.

request properties (e.g. a transaction that submits an order may include a single or twenty order lines), or multiple request run patterns (combinations of multiple requests running simultaneously, each at a different execution point). The characteristics of variability are such that no practical model with reasonable overhead may account for each (e.g. it would not be practical to account for which items are cached or the resource usage patterns at the precise running point of each running request).

Execution statistics from TPCC [19] are used to show the variability within and among requests. After presenting a basic control design, we show how statistics can be collected and processed to have effective control with highly variable workload characteristics. A simulation is used to test the robustness of the approach against varied workload conditions, with different arrival and duration distributions.

The paper is structured as follows: in section 2 we analyze the response time variability within a TPCC workload. After presenting the QoS control approach in section 3, we discuss how the collected statistics are used in a robust, high-variation resilient QoS controller in section 4. In section 5 we present experimental results which show, using multiple distributions, that the robust approach is able to deal with variability satisfactorily. In section 6 we review related work section 7 concludes the paper.

2 Workload Variability

One of the most important control knobs for performance and QoS in a transactional system is the number of requests running simultaneously, which we denote as Execution Cardinality (EC), because the response times of requests are very influenced by that variable, as shown in Figure 1a for the TPC-C workload [19]. From the point-of-view of response time of a request arriving and executing alone, the best EC would be 1. However, because requests arriving have to wait in a queue until they can execute, a much higher value of EC than 1 typically allows the system to explore its parallelism and resource utilization features better (e.g. while one thread waits for I/O the other one computes or does I/O on another disk). The best EC value actually varies with conditions (requests being submitted, database size, system utilization, among others). With QoS objectives (execution of all requests should take less than 2 seconds) and statistic information we can have a dynamic control of EC, but then the statistic information becomes very relevant.

Our objective in this section is to analyze experimentally the variability of the TPC-C [19] workload in a test case. We used a system running on a Pentium 4 with 1GB of memory and PostgreSQL version 8.1. The experiment consisted in logging response times for several runs of the transaction mix with incrementing EC values, from 1 to 50. Figure 1a shows the average response time versus EC and Figure 1b shows the standard deviation of response times for the two slowest transactions. From Figure 1a we can see that average response times increase steadily and in some cases (e.g. NEW_ORDER) almost linearly as EC increases. Looking at Figure 1b for two of the transactions and in Figure 2 for all transactions (with EC=9), we can see that the standard deviation is quite large.

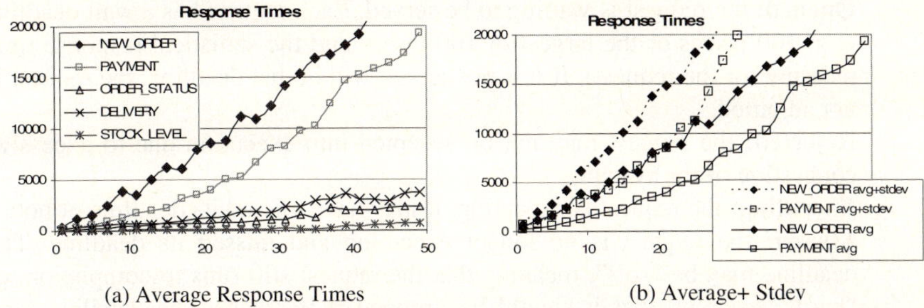

(a) Average Response Times   (b) Average+ Stdev

Fig. 1. Response Times versus EC

In Figure 2 the standard deviation was actually larger than the average for all transactions except NEW_ORDER.

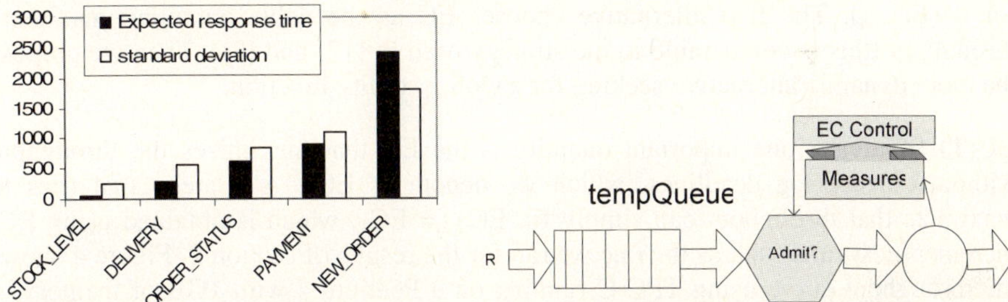

Fig. 2. Response Time Statistics EC=9   **Fig. 3.** Controller Architecture

The possibility of very disparate response times between different requests is obvious and expected, as some requests have to process a lot of information while others implement simple and fast tasks. This depends on the application scenario. But the observed disparate response times for different runs of the same request are also to be expected, because there are many physical details that determine the response time. For instance, the backend DBMS may have the required data already in its cache or not (for performance reasons, DBMSs have large caches and both table and index data maybe or not be in the cache), an index may be used or not (selectivity issues), there may be more or less contention for resources or locks over data. The transaction itself may process more or less data (e.g. a new order may process one or 100 order lines). Given the objectives of having an external, non-intrusive and readily deployable control that does not go into system or application details, what would be a robust prediction procedure? Due to the high variability of the measured item, the model must account for that variability using conservative statistical parameters.

3 Basic QoS$_{max}$ Controller

Figure 3 shows the controller architecture, which includes a temporary queue, an admission procedure, a monitor collecting measures and a control element. In this system, requests can be at any of the following states:

Queued: the request is waiting to be served. Each request has a wait deadline (e.g. 100 msecs or the largest of 100 msecs and the statistical response time measure for the request). If it is not served within that deadline, the request is not admitted.

Rejected: the request may not be accepted into execution due to excessive congestion of the system;

Executing: the request is executing. It may have missed its deadline or not;

Missed: the request is no longer executing and missed its deadline. The deadline may be "soft", meaning that the request still runs to completion, or "hard", meaning that it should be dropped if it reaches the deadline (and possibly rollbacked). We consider mostly soft deadlines, but both alternatives can be used;

Ended Within Deadline (EWD): requests that executed to completion and did not miss their deadline.

We now describe two alternatives to control EC by finding a maximum utility value for it (EC_{max}). The first alternative chooses EC as the value obtaining maximum throughput (this is comparable to the strategy used in [17] and [8]). Then we propose the more dynamic alternative seeking for a QoS_{max} utility function.

EC(T) Strategy: one important quantity is the EC that maximizes the throughput without considering deadlines, which we denote as EC_T. A strategy that tries to maximize that throughput can simply fix $EC_{max}= EC_T$, which is obtained using EC-increment test runs such as the one we ran for the results of section 2. Figure 4 shows the throughput curve using TPC-C running on a Pentium 4 with 1GB of memory, 4 disks and PostgreSQL version 8.1. In this case $EC_T=20$.

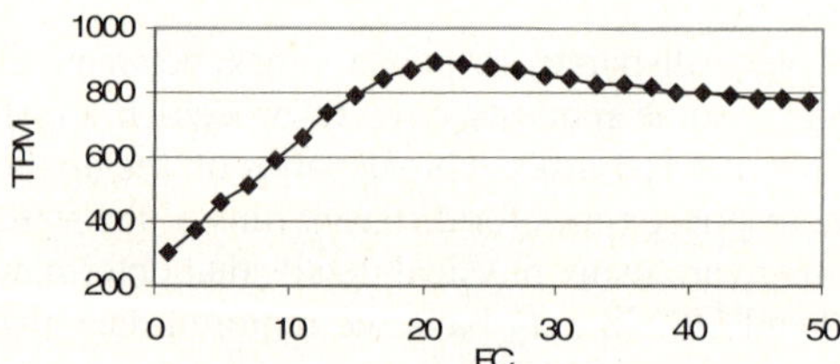

Fig. 4. Throughput Curve to determine EC_T

The EC(T) strategy is adequate when we do not consider that requests should not take more than a certain time to run (deadlines).

E(ES) Strategy: this approach searches the EC value that allows more requests to end within their time constraint (deadline). It tests the utility function for each candidate EC within a pre-defined interval [EC_{low}, EC_{high}] and chooses the one that maximizes the utility function. We used default values $EC_{low} = 4$ (as it should not be too low to avoid rejecting too many requests) and $EC_{high}= 50$. The initial EC_{max} value ($EC_{default}$) is set to EC_T.

The approach assumes statistics are collected in the form $RT(r_i,EC_i)$: request r_i has response time statistic value $RT(r_i,EC_i)$ when the execution cardinality is EC_i.

Consider the current (within a recent interval) workload $\{r_i, f_i\}$, where r_i is a request and f_i is its frequency in the workload. Consider also that each request r_i should execute within time t_i. For an EC_{max} candidate EC_j we compute the utility function U_j value as $EC_j \times \sum ft_i$, where ft_i is 0 if the statistic response time for request r_i is above t_i when EC_j is considered or the request frequency f_i otherwise.

Algorithm:
```
For each candidate ECj
  Uj=0;
  For each Request ri in workload W,
        If ( RT(ri,ECi)<ti ) Uj+=  ECj × fi
  Choose max { Uj}
```

The test runs that obtain $RT(r_i, EC_i)$ are to happen offline (e.g. during the night at weekends) (some EC_j values are tested, the remaining interpolated). If a request does not have offline test values yet, it counts as if its deadline is just above its response time with $EC=EC_T$ (so that EC_T would be a good choice for it).

Queue handling: the time waiting in the queue adds to the total response time of requests, even though the request is not being served while in the queue. The primary purpose of the queue is to hold requests temporarily during request arrival bursts, but the maximum amount of time waiting should be constrained. The queue may implement any policy, with FIFO often being the policy of choice, but there needs to be an additional control on the size of the queue to reject requests based on the waiting time. In our proposals this is implemented by means of an Admittance Decision Time limit (ADT) that is contracted for all requests, for an individual application or for a specific request. For instance, ADT<=100ms or ADT<=max{100ms,10%RT}, where RT is the expected runtime of the request. Given this parameter, the request waits in the queue for at most ADT and if it is not accepted until then, it is immediately rejected.

Control rate: the frequency of control (number of times the algorithm runs) can be set as desired, but in our experimental work we observed that the overhead is relatively modest if the workload is not too large, so the control can run for instance every 100 msecs (this is the value we used in the experiments) or every second. The size of the workload window depends on this choice. If the frequency of control is large, the system will be able to tune itself to current workload conditions.

4 Accounting for Variability

Intra-request variability refers to the response time variability among different runs of a request. It may lead to inadequate setting of EC_{max} and consequently to much higher miss rates than desired. We propose a simple conservative estimation to deal with this problem. Inter-request variability also poses a relevant issue concerning response time and throughput when EC is constrained: fast executing requests may be blocked by slower ones executing. In this case we propose a simple scheme to have a balanced execution between slow and fast requests.

Intra-request Variability: consider a miss rate (default=0%). Deadlines are specified for requests and the QoS_{max} Control functionality should try to achieve those while maximizing the throughput. This means that the EC_{max} value should not be such that, due to request variability, many more requests miss their deadlines than desired. Given that very precise prediction of actual running times is impractical for reasons we discussed before, the solution is for the control approach to be sufficiently conservative to avoid excessive miss rates. In control algorithm, the response time was taken from values $RT(r_i,EC_i)$, which is a statistic response time value. While it could be the expected response time (average of observed values), the more conservative estimate requires other measures to be used instead. If $E(RT(r_i,EC_i))$ is the expected response time and the standard deviation is $S(RT(r_i,EC_i))$, a simple more conservative estimate is to use $E(RT(r_i,EC_i)) + S(RT(r_i,EC_i))$. In this case the offline test runs must collect these two quantites, as depicted in Figure 1b; A much more conservative statistic would be obtained by using the maximum observed response time $max(RT(r_i,EC_i))$. The most flexible solution is for a control configuration parameter to specify a desired percentile in the observed response times distribution. That solution requires much more information, a histogram of response times for each pair $<r_i,EC_i>$.

Inter-request Heterogeneity: given strict deadlines, the QoS control may constrain EC to a small value (e.g. $EC_{max}<10$) to achieve QoS_{max}. This in turn may lead to a blocking behavior, as slow running requests may "block" fast running ones. For instance, if EC=4, fast requests enter and leave execution fast, while slow requests enter execution and are left executing for a much longer time. As a consequence, at certain moments there will be too few or no execution slots available to maintain the high throughput of fast requests. We show this effect in the experimental section.

We wanted to provide a counterbalance mechanism to avoid blocking. For that we formulate the objective that the runtime-to-arrival_rate ratio be similar for all types of request. Requests are divided into request groups RG_i, defined by expected response time intervals (this can be done automatically) and seeing that RG_is have similar runtime-to-arrival rates. Given arrival rate λ_i and runtime β_i, the ratio is β_i/λ_i. When a request arrives and is classified into a group i, we compare the runtime-to-arrival rate of that group β_i/λ_i to the one of the remaining groups β_r/λ_r,

```
If(EC<ECmax)
        if(θh × βr/λr < βi/λi)
                reject request
        else
                admit request
```

Where $\theta_h >1$ is a threshold. For instance, if $\beta_r/\lambda_r=30\%$, $\beta_i/\lambda_i= 65\%$ and $\theta_h =1.5$, the request group has ran more than 1.5 times the remaining groups, so the request is rejected. As a practical implementation for this work we defined two groups based on the response times being above or below $\mu-\sigma$ (overall average and standard deviation of response times) and $\theta_h =1.5$. As part of our future work we intend to design automated strategies to determine these parameters for best performance.

5 Experimental Results

Our experimental objective was to analyze the effects of variance and the robustness of the approach in responding to variable conditions (represented here by distribution variance), and trying to provide the QoS_{max} objective of maximizing the number of requests that End-Within-the-Deadline under variable workloads and congestion.

In order to test this, we have built an event-driven simulator that uses the throughput curve of Figure 4 (obtained from tests with TPCC) to model the effects of multiple simultaneous executions. We have set request deadlines to 5 seconds and varied either duration (request duration when running alone, between 100 ms and 1.5 seconds) or arrival rates (between 10 ms and 500 ms), fixing the other parameters (the duration at 1 second, the inter-arrival time at 100ms). In every case, these were always mean values for each of the distributions depicted in Figure 5 (Uniform, Exponential, Gaussian and Poisson, with varied rates). We included low variance and high variance distributions on purpose, so that we could test the effects of variability.

DISTR.	avg	stdev	DISTR.	avg	stdev
Poisson	1013	92	Gaussian	1000	33
Exponential	1038	976	Uniform	1024	578

Fig. 5. Characteristics of distributions used in the experiments

We have engaged in extensive experimentation with the arrival and duration distributions from Figure 5, then focused the analysis to reach conclusions on the variability issue. The results were consistent across distributions, with the variability being the major factor. Two major experiments are shown: The first one tests the response of the system to different variances, by successively increasing that parameter in a Gaussian request duration distribution with mean 1000 msecs. The second experiment tests the system response while varying the mean duration of requests in an exponential request duration distribution. In both cases a uniform arrival distribution was used. Each simulation run submitted 10000 transactions according to the arrival rate distribution and statistics.

Our results concern the following control alternatives:

ACTUAL: this is the control that would result if the controller could guess the exact duration of the request (which it cannot);

EC(T): the Maximum Throughput strategy, that is, EC_{max} is fixed at the maximum throughput value obtained from the test runs (Figure 4);

EXP: QoS_{max} control using the average of previous request run durations;

EXP+STD: QoS_{max} control using the average+stdev quantity from previous request run durations;

5.1 Variance, With and Without Inter-request Handling

This experiment tests the behavior of the system with different variances and the capability of the approach to handle those different variances. A Gaussian request duration distribution is submitted with expected duration 1000 msecs and successively larger variance (from 100^2 to 800^2 msecs). Figure 6 shows the EWD throughput (a) and the Miss Rate (b) against variability, and for comparison purposes it also shows the results when inter-request blocking avoidance is off (c and d). It is clear from Figure 6a that EXP+STD achieves much better T_{EWD} than EXP as soon as variability appears and also that T_{EWD} of EXP+STD is only slightly less than T_{EWD} if the actual request duration values were known by the controller (ACTUAL). Both EXP+STD and EXP are much better than E(T). This can also be seen by looking at the miss rates of Figure 6b.

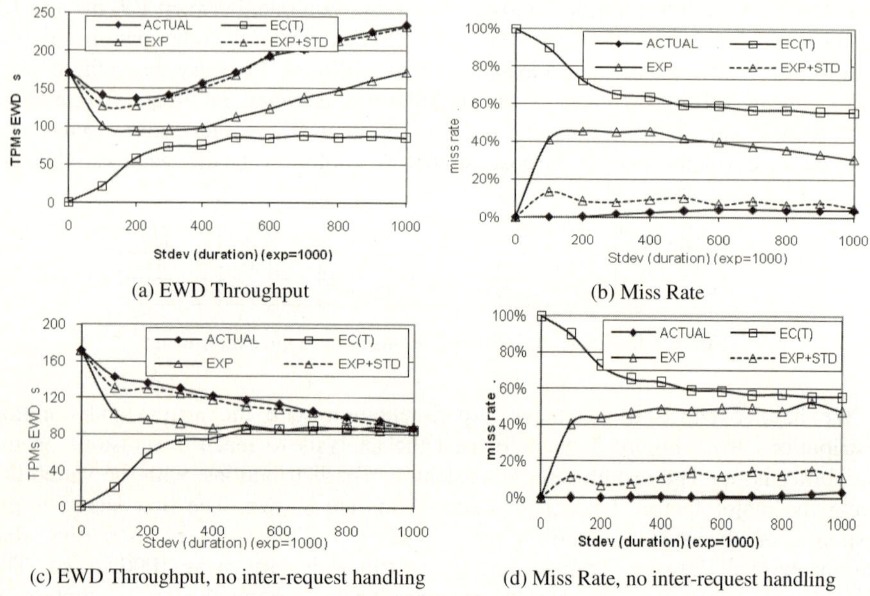

(a) EWD Throughput

(b) Miss Rate

(c) EWD Throughput, no inter-request handling

(d) Miss Rate, no inter-request handling

Fig. 6. T_{EWD} and Miss Rate vs Variability

The results in Figure 6c and 6d prove that, as we expected, inter-request variability handling to avoid fast request blocking is very important, because without that strategy the T_{EWD} throughput of ACTUAL, EXP and EXP+STD converge to that of EC(T) when variability is high (Figure 6c), even though miss rates are similar to those of Figure 6b.

5.2   Exponential with Varied Durations

In this experiment we tested varying expected durations over an Exponential distribution for durations and uniform arrival distribution. Figure 7 shows T_{EWD} (7a) and the Miss Rates (7b). When the durations are small (e.g. below 500 msecs) EC(T), EXP and EXP+STD have the similar behavior. There is no congestion, expected duration is still relatively small and therefore these strategies fix the same EC_{max} at the maximum throughput value of EC(T). However, as the expected duration increases, the miss rate also starts increasing, as a fraction of requests has much larger durations and miss the deadlines. As a result, the throughput of ACTUAL is larger than the one of EXP or EXP+STD for this part of the Figure spectrum. From 500 msecs on, EXP+STD senses the larger expected durations and decreases EC_{max}, the throughput catches up. A similar adjustment also happens for EXP but it reacts much later. The miss rate of EC(T) increases significantly as we increase the expected duration, so its T_{EWD} is much worse than the other alternatives.

This result shows that QoS_{max} adapts to varied durations, but while EXP exhibits a large miss rate in some cases (up to 30%) and not-so-good throughput, EXP+STD miss rate was always less than 10% and its throughput was almost always as good as the one QoS_{max} would get if it could know actual durations in advance (ACTUAL).

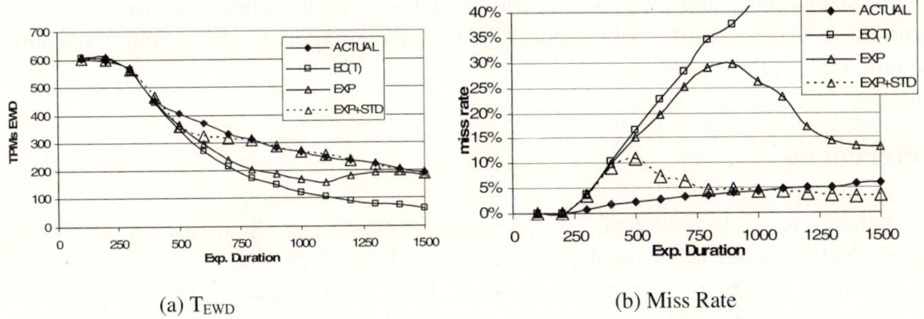

(a) T_{EWD} (b) Miss Rate

Fig. 7. T_{EWD} and Miss Rate vs Duration

6 Related Work

There are several related works on Quality-of-service in Web-servers or computer-systems in general. The works in [9, 10, 11] focus on QoS for request processing in real-time databases, especially aimed at applications with strict time requirements such as military ones [1, 10]. They propose strategies to meet deadlines for individual requests, but the authors consider an all-in-memory context, as this renders the system completely predictable and therefore controllable with precision. In [13, 18] the authors propose feedback control to adapt the admission procedure to guarantee deadlines in real time systems, using a control theory approach. In [2, 3, 14] control is based in a specification of QoS targets. There, prediction is based on an analytic model using Markov Chains and Queuing Networks, taking resource demands and capacities as inputs. Quality-of-Service was also studied for Web Servers. In [7] the authors propose session-based Admission Control (SBAC), noting that longer sessions may result in purchases and therefore should not be discriminated in overloaded conditions. They propose self-tunable admission control based on hybrid or predictive strategies. [6] uses a rather complex analytical model to perform admission control. There are also approaches proposing some kind of service differentiation: [4] proposes architecture for Web servers with differentiated services; [5] defines two priority classes for requests, with corresponding queues and admission control over those queues. [16] proposes an approach for Web Servers to adapt automatically to changing workload characteristics and [8] proposes a strategy that improves the service to requests using statistical characterization of those requests and services. Comparing to our own work, all the works referenced above except [8] are either intrusive, require extensive modifications to systems, or use analytical models that may not provide guarantees against real system. The work in [8] does not consider deadlines and does not adapt constantly as ours does, instead it fixes the maximum throughput capacity when it runs the test runs.

7 Conclusion

In this paper we proposed variability-solving solutions for robust QoS control. We have discussed the fact that variability is inherent to systems and no practical acceptable overhead control can account for all the variability factors. We presented the control strategy and then the variability-handling approaches, with the objective of turning the strategy into a robust one. Our extensive experimental results have shown the advantage of our proposals.

References

1. Abbott, R., Garcia-Molina, H.: Scheduling Real-Time Requests: A Performance Evaluation. ACM Trans. Database System 17, 513–560 (1992)
2. Bennani: Autonomic Computing Through Analytic Performance Models, Ph.D in CS dissertation, George Mason University (May 2006)
3. Bennani, Menasce: Assessing the Robustness of Self-Managing Computer Systems under Highly Variable Workloads. In: Proc. International Conf. Autonomic Computing (ICAC-04), May 17-18, 2004, New York, NY (2004)
4. Bhatti, N., Friedrich, R.: Web server support for tiered services. IEEE Network 13(5), 64–71 (1999)
5. Bhoj, Rmanathan, Singhal: Web2K: Bringing QoS to Web servers. Tech. Rep. HPL-2000-61, HP Labs (May 2000)
6. Chen, X., Mohapatra, P., Chen, H.: An admission control scheme for predictable server response time forWeb accesses. In: Proceedings of the 10th World Wide Web Conference, May 2001, Hong Kong (2001)
7. Cherkasova, Phaal: Session-based admission control: A mechanism for peak load management of commercial Web sites. IEEE Req. on Computers, 51(6) (June 2002)
8. Elnikety, S., Nahum, E., Tracey, J., Zwaenepoel, W.: A Method for Transparent Admission Control and Request Scheduling in E-Commerce Web Sites. In: WWW2004: The Thirteenth International World Wide Web Conference, May 2004, New York City, NY, USA (2004)
9. Kang, K.D., Son, S.H., Stankovic, J.A.: Managing Deadline Miss Ratio and Sensor Data Freshness in Real-Time Databases. IEEE Trans. on Knowledge and Data Engineering 16(10), 1200–1216 (2004)
10. Kang, K.: QoS-Aware Real-Time Data Management. PhD thesis, U. Virginia (May 2003)
11. Kang, K.D., Son, S.H., Stankovic, J.A., Abdelzaher, T.F.: A QoSSensitive Approach for Timeliness and Freshness Guarantees in Real-Time Databases. In: Proc. 14th Euromicro Conf. Real-Time Systems (June 2002)
12. Kanodia, V., Knightly, E.W.: Ensuring latency targets in multiclass Web servers. IEEE Requests on Parallel and Distributed Systems 13(10) (October 2002)
13. Lu, C., Stankovic, J., Tao, G., Son, S.: Feedback Control Real-Time Scheduling. Framework, Modeling and Algorithms, special issue of Real-Time Systems Journal on Control-Theoretic Approaches to Real-Time Computing 23(1/2), 85–126 (2002)
14. Menasce, Bennani.: On the Use of Performance Models to Design Self-Managing Computer Systems. In: Proc. 2003 Comp. Measur. Group Conf., December 7-12, 2003, Dallas, TX (2003)
15. Mogul, J.C., Ramakrishnan, K.K.: Eliminating receive livelock in an interrupt-driven kernel. ACM Requests on Computer Systems 15(3), 217–252 (1997)

16. Pradhan, P., Tewari, R., Sahu, S., Chandra, A., Shenoy, P.: An observation-based approach towards self managing Web servers. In: International Workshop on Quality of Service, May 2002, Miami Beach, FL (2002)
17. Schroeder, Harchol-Balter, M., Iyengar, Nahum, Wierman: How to determine a good multi-programming level for external scheduling. In: 22nd International Conference on Data Engineering (ICDE 2006) (2006)
18. Stankovic, J., Lu, C., Son, S., Tao, G.: The Case for Feedback Control Real-Time Scheduling. In: EuroMicro Conference on Real-Time Systems (June 1999)
19. TPCC: http://www.tpc.org

A Topology-Aware Approach for Distributed Data Reconciliation in P2P Networks

Manal El Dick[1], Vidal Martins[1,2], and Esther Pacitti[1]

[1] ATLAS Group, INRIA and LINA, University of Nantes, France
[2] PPGIA/PUCPR - Pontifical Catholic University of Paraná, Brazil
`manal.el-dick@univ-nantes.fr`, `firstname.lastname@univ-nantes.fr`

Abstract. A growing number of collaborative applications are being built on top of Peer-to-Peer (P2P) networks which provide scalability and support dynamic behavior. However, the distributed algorithms used by these applications typically introduce multiple communications and interactions between nodes. This is because P2P networks are constructed independently of the underlying topology, which may cause high latencies and communication overheads. In this paper, we propose a topology-aware approach that exploits physical topology information to perform P2P distributed data reconciliation, a major function for collaborative applications. Our solution (P2P-Reconciler-TA) relies on dynamically selecting nodes to execute specific steps of the algorithm, while carefully placing relevant data. We show that P2P-Reconciler-TA introduces a gain of 50% compared to P2P-Reconciler and still scales up.

1   Introduction

Collaborative applications are getting common as a result of rapid progress in distributed technologies (grid, P2P, and mobile computing). There are currently many projects aimed at constructing these applications on top of P2P networks because of their properties: decentralization, self-organization, scalability and fault-tolerance. As an example of such applications, consider a community of scientists working on the same project while geographically dispersed. Scientists collaborate by sharing and processing data without relying on a central server.

A P2P network is an overlay network built over a physical network. Each node is logically connected to a set of nodes, referred to as its neighbors. Normally, the neighborhood of a node is set without much knowledge of the underlying topology, causing a mismatch between the P2P overlay and the physical network. Thus, communications between nodes incur high latencies and overload the network. Distributed P2P algorithms, used by collaborative applications and built on top of P2P networks, introduce large data transfers and frequent interactions between the nodes involved. The performance of such algorithms may degrade drastically because of two orthogonal problems: inefficient overlay and nodes selection without taking into account topology information.

Approaches have been focusing on the P2P overlay network. The goal is to construct an overlay that reflects the underlying topology. Some proposals such

A.-M. Kermarrec, L. Bougé, and T. Priol (Eds.): Euro-Par 2007, LNCS 4641, pp. 318–327, 2007.

as [2] group nodes into clusters based on network distance or IP addresses. Others [3,7,8,9] try to improve nodes neighborhood in terms of proximity. Regarding the selection issue, studies such as [7] are typically limited to finding one nearby node wrt. an origin node when searching for data replicated at multiple nodes. In this paper, we propose a solution at the application level (over a P2P network). We focus on P2P semantic reconciliation, used within optimistic replication.

Optimistic replication is largely used as a solution to provide data availability for dynamic collaborative applications. It allows the asynchronous updating of replicas such that applications can progress even though some nodes are disconnected or have failed. This enables asynchronous collaboration among users. However, concurrent updates may cause replica divergences and conflicts, which should be reconciled. The P2P-Reconciler approach [4] performs distributed semantic reconciliation over a P2P network structured as a distributed hash table (DHT) [6,8]. Thus, the mismatch between the P2P overlay and the physical network may introduce poor performance.

In this paper, we propose P2P-Reconciler-TA, an approach used to improve P2P-Reconciler response times. P2P-Reconciler-TA dynamically takes into account the physical network topology combined with the DHT properties when executing reconciliation. Our approach is designed for distributed reconciliation. However the metrics, cost functions and the general approach can be useful in different contexts.

The main contributions of this paper are: (1) metrics and cost functions that rely on characteristics of both P2P and underlying networks; (2) a distributed algorithm for dynamically selecting the best nodes to execute specific reconciliation steps while considering dynamic data placement; and (3) experimental results that show that P2P-Reconciler-TA yields excellent scalability, with very good performance and limited overhead.

The rest of this paper is organized as follows. Section 2 discusses the most relevant related work. Section 3 provides preliminaries that constitute the basis of our work. Section 4 presents P2P-Reconciler-TA, detailing our metrics, cost functions and topology-aware approach. Section 5 gives a performance evaluation based on our implementation. Section 6 concludes.

2   Related Work

Many efforts have been made to exploit topological information in order to improve the performance of P2P environments. In [3], the authors describe a measurement-based technique to dynamically connect physically close nodes and disconnect physically distant nodes. A design improvement of the P2P overlay CAN [6] aims at constructing it in a way congruent to the underlying topology. Pastry [8] and Tapestry [9] are both P2P overlay routing infrastructures that take into account network locality to establish nodes neighborhoods. In comparison with these works, we focus mainly on the algorithms, not on the overlay structure. Given a set of nodes, we exploit topological information to select the "best" nodes to participate in the different steps of an algorithm, in a way that achieves optimal performance. As such, those approaches can be used to complement our work and improve our results.

Location-aware clustering is proposed in [2] where physically close nodes are grouped into clusters. However, the control of the topology relies on a centralized server, which introduces high overheads due to the dynamic changes which are frequent in P2P environments. In contrast, our solution is based on network information gathered and refreshed in a scalable and inexpensive manner.

3   Preliminaries

In this section, we introduce some preliminaries that constitute the basis of our work. We briefly describe the P2P-Reconciler and CAN necessary to understand our topology-aware approach.

3.1   P2P-Reconciler

P2P-Reconciler [4] performs distributed semantic reconciliation over a P2P network structured as a distributed hash table (DHT) [6,8]. A DHT provides a hash table abstraction over multiple computer nodes. Data placement in the DHT is determined by a hash function which maps data identifiers into nodes.

P2P-Reconciler takes advantage of the action-constraint framework of Ice-Cube [1] to perform semantic reconciliation. According to this framework, the application semantics can be described by means of constraints between actions. A *constraint* is an application invariant, e.g. *a parcel* constraint establishes the "all-or-nothing" semantics, i.e. either all *parcel*'s actions execute successfully in any order, or none does. For instance, consider a user that improves the content of a shared document by producing two *related* actions a_1 and a_2 (e.g. a_1 changes a document paragraph and a_2 changes the corresponding translation); in order to assure the "all-or-nothing" semantics, the application should create a parcel constraint between a_1 and a_2. These actions can conflict with other actions. Therefore, the aim of reconciliation is to take a set of actions with the associated constraints and produce a schedule, i.e. a list of ordered actions that do not violate constraints.

With P2P-Reconciler, reconciliation is executed in 6 distributed steps in order to maximize parallel processing. Each step is performed simultaneously and independently by a subset of nodes referred to as *reconciler nodes*. Data produced or consumed during reconciliation are held by different *reconciliation objects*. A reconciliation object is stored in one particular node called *provider node*, based on its object identifier. We restrict the reconciliation work to a subset of nodes (the reconciler nodes) in order to maximize performance.

P2P-Reconciler's steps proceed as follows. First, nodes execute local actions to update replicated data while respecting user-defined constraints. Then, these actions and constraints are inserted into the appropriate reconciliation objects. When the reconciliation is launched, P2P-Reconciler selects the best reconcilers according to communication costs. Once reconcilers are chosen, they retrieve actions and constraints from their corresponding provider nodes and produce a global schedule by resolving conflicting updates based on the application semantics. This schedule is locally executed at every node; thereby assuring eventual consistency (i.e. all replicas eventually achieve the same final state when users stop submitting updates).

3.2 CAN

Basic CAN [6] is a virtual Cartesian coordinate space to store and retrieve data as *(key, value)* pairs. At any point in time, the entire coordinate space is dynamically partitioned among all nodes in the system, so that each node owns a distinct zone that represents a segment of the entire space. To store (or retrieve) a pair (k_1, v_1), key k_1 is deterministically mapped onto a point P in the coordinate space using a uniform hash function. Then (k_1, v_1) is stored at the node that owns the zone to which P belongs. Intuitively, routing in CAN works by following the straight line path through the Cartesian space from source to destination coordinates.

Optimized CAN aims at constructing its logical space in a way that reflects the topology of the underlying network. It assumes the existence of well-known landmarks spread across the network. A node measures its round-trip time to the set of landmarks and orders them by increasing latency (i.e. network distance). The coordinate space is divided into bins such that each possible landmarks ordering is represented by a bin. Physically close nodes are likely to have the same ordering and hence will belong to the same bin.

4 P2P-Reconciler-TA

P2P-Reconciler-TA is a distributed protocol for reconciling conflicting updates in *topology-aware* P2P networks. A P2P network is classified as topology-aware if its topology is established by taking into account the physical distance among nodes (e.g. in terms of latency times). P2P-Reconciler-TA aims at exploiting the physical proximity of nodes to improve the reconciliation performance. Briefly, P2P-Reconciler-TA works as follows. Based on the network topology, it selects the best provider and reconciler nodes. These nodes then reconcile conflicting updates and produce a schedule, which is an ordered list of non-conflicting updates. In this work, we focus on node allocation by proposing a dynamic distributed algorithm for efficiently selecting provider and reconciler nodes. We first introduce some definitions. Then, we present the allocation algorithm in details.

4.1 Metrics

P2P-Reconciler-TA uses the following reconciliation objects: action log (L_R), action summary (AS), clusters set (CS), and schedule (S). The action log contains update actions to be reconciled; the action summary holds constraints among actions; the clusters set stores clusters of conflicting actions; and the schedule holds an ordered list of actions that do not violate constraints. For availability reasons, we produce k replicas of each reconciliation object and store these replicas into different providers. We note these terms as follows:

- **RO:** set of reconciliation objects $\{LR, AS, CS, S\}$
- **ro:** a reconciliation object belonging to RO (e.g. CS, LR, etc.)
- **ro$_i$:** the replica i of the reconciliation object ro (e.g. CS_1 is the replica 1 of CS), where $1 \leq i \leq k$; the coordinates (x_i, y_i) are associated with ro_i and determines the ro_i placement over the CAN coordinate space; ro_i is stored at the provider node p_{roi} whose zone includes (x_i, y_i)
- **P$_{ro}$:** set of k providers p_{roi} that store replicas of the reconciliation object ro
- **best(P$_{ro}$):** the most efficient provider node holding a replica of ro

We apply various criteria to select the best provider nodes. One of these criteria establishes that a provider node should not be isolated in the network, i.e. it should be close to a certain number of neighbors that can become reconcilers, and therefore are called *potential reconcilers*. The physical proximity in terms of latency is not enough; a potential reconciler should also be able to access provider's data at an acceptable cost. Thus, such a potential reconciler is considered a *good neighbor* of the associated provider node. We now present metrics and terms applied in provider node selection:

- **accessCost(n, p):** cost for a node n accessing data stored at the provider node p in terms of latency and transfer times. The transfer time relies on the message size, which is usually variable. For simplicity, we consider a message of fixed size (e.g. 4 Kb). Equation (1) shows that the $accessCost(n, p)$ is computed as the latency between n and p (noted $latency(n, p)$) plus the time to transfer the message msg from p to n (noted $tc(p, n, msg)$

$$accessCost(n, p) = latency(n, p) + tc(p, n, msg) \tag{1}$$

- **maxAccessCost:** maximal acceptable cost for any node accessing data stored in provider nodes; if $accessCost(n, p) > maxAccessCost$, n is considered far away from p, and thus is not a good neighbor of p
- **potRec(p):** number of potential reconcilers that are good neighbors of p
- **minPotRec:** minimal number of potential reconcilers required around a provider node p in order to accept p as a candidate provider; if $potRec(p) < minPotRec$, p is considered isolated in the network
- **candidate provider:** any provider node p with $potRec(p) \geq minPotRec$ is considered a candidate in the provider selection
- **cost provider:** node that stores costs used in node selection.
- **QoN(p):** quality of network around the provider node p. It is defined as the average access cost associated with good neighbors of p, and it is computed by (2). In this equation, n_i represents a good neighbor of p

$$QoN(p) = \frac{1}{potRec(p)} \sum_{i=1}^{potRec(p)} accessCost(n_i, p) \tag{2}$$

Another criterion for selecting a provider node is its proximity of other providers. During a reconciliation step, a reconciler node often needs to access various reconciliation objects. By approximating provider nodes, we reduce the associated access costs. We also need the following terms applied in reconciler selection:

- **candidate reconcilers:** set of nodes that are candidate to become reconcilers. This set includes all good neighbors of selected providers
- **step:** a reconciliation stage

4.2   Detailed Algorithm

P2P-Reconciler-TA selects provider nodes and candidate reconcilers as follows. Every provider node regularly evaluates its network quality and, according to the number of potential reconcilers around it, the provider announces or cancels its candidature to the cost provider node. The cost provider, in turn, manages candidatures by monitoring which providers have the best network quality. Whenever the best providers change, the cost provider performs a new selection and notifies its decision to provider nodes. Following this notification, provider nodes inform their good neighbors whether they are candidate reconcilers or not. With the selection of new providers, current estimated reconciliation costs are discarded and new estimations are produced by the new candidate reconcilers. Thus, selected provider nodes and candidate reconcilers are dynamically changing according to the evolution of the network topology. We now detail each step of node allocation.

Computing Provider Node's QoN. A provider node computes its network quality by using equation (2) and the input data supplied by its good neighbors. Good neighbors introduce themselves to the provider nodes as follows. Consider that node n has just joined the network. For each reconciliation object $ro \in RO$, n looks for the closest node that can provide ro, noted p_{ro}, and if $accessCost(n, p_r o)$ is acceptable, n introduces itself to p_{ro} as a good neighbor by informing $accessCost(n, p_{ro})$. Node n finds the closest p_{ro} as follows. First, n uses k hash functions to obtain the k coordinates (x_i, y_i) corresponding to each replica ro_i. Then, n computes the Cartesian distance between n's coordinates and each (x_i, y_i). Finally, the closest p_{ro} is the one whose zone includes the closest (x_i, y_i) coordinates. The closest p_{ro} is called the n's *reference provider* wrt. ro. Figure 1 illustrates how node n finds its reference provider wrt. the action summary reconciliation object (AS).

Provider nodes and the associated potential reconcilers cope with the dynamic behavior of the P2P network as follows. A provider node dynamically refreshes its QoN based on its good neighbors' joins, leaves, and failures. Joins and leaves are notified by the good neighbors whereas failures are detected by the provider node based on the expiration of a ttl (time-to-live) field. On the other hand, a good neighbor dynamically changes a reference provider p_{ro} whenever p_{ro} gives up the responsibility for ro. If p_{ro} disconnects or transfers ro to another provider, p_{ro} notifies these events to its good neighbors. However, if p_{ro} fails its good neighbors detect such failure and change the corresponding reference provider.

324   M.E. Dick, V. Martins, and E. Pacitti

Managing Provider Candidature. The network quality associated with a provider node dynamically changes as its potential reconcilers join, leave, or fail. Thus, a provider node often refreshes its candidature as follows. When the neighborhood situation of a provider p switches from isolated (i.e. p has a few of potential reconcilers around it) to surrounded (i.e. $potRec(p) \geq minPotRec$), p announces its candidature to the cost provider. In contrast, when p switches from surrounded to isolated, p cancels its candidature. Finally, if p's QoN varies while it remains surrounded by potential reconcilers, p updates its QoN. Figure 2 illustrates AS candidate providers for $minPotRec = 4$.

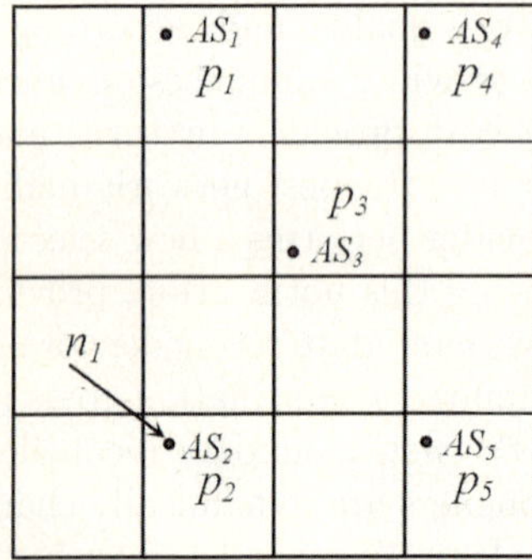

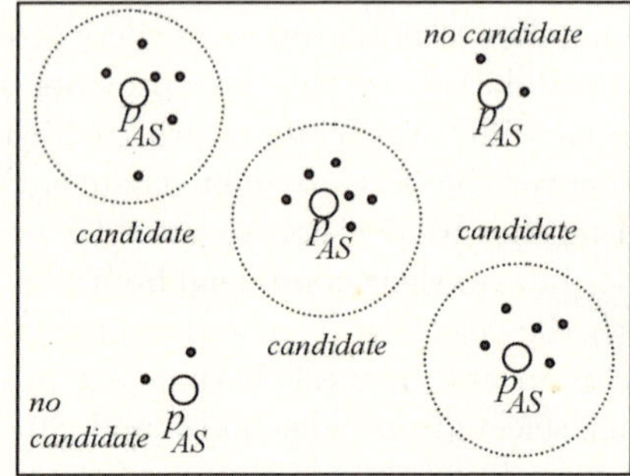

Fig. 1. Finding the AS reference provider

Fig. 2. Managing provider candidature

Selecting Provider Nodes. For each reconciliation object, P2P-Reconciler-TA must select the best provider node. This selection should take into account the proximity among providers since different providers are accessed in the same reconciliation step. We reduce the search space of best providers by applying the heuristic illustrated in Fig. 3. First, we select $best(P_{AS})$ and $best(P_{CS})$ (Fig. 3a). These nodes must be as close as possible from each other because AS and CS are the most accessed reconciliation objects and both are often retrieved in the same step. Next, we select $best(P_{LR})$ and $best(P_S)$ based on the pair $(best(P_{AS}), best(P_{CS}))$ previously selected (Fig. 3b); $best(P_{LR})$ must be as close as possible to $best(P_{AS})$ since a reconciler accesses both $best(P_{LR})$ and $best(P_{AS})$ in the same step whereas $best(P_S)$ must be as close as possible to $best(P_{CS})$ for the same reason. Figure 3c shows the selected providers of our illustrative scenario (i.e. p_{AS1}, p_{CS3}, p_{S1}, and p_{LR5}).

All candidate providers have at least $minPotRec$ potential reconcilers around them. However, the network quality (QoN) may vary a lot from one provider to another. Therefore, instead of consider all candidates we begin the selection by filtering, for each reconciliation object, the k best providers in terms of QoN. Afterwards, we evaluate only the distances among these filtered candidates. For instance, in Fig. 3 only 2 candidates per reconciliation object were filtered (i.e. $\{(p_{AS1}, p_{AS3}), (p_{CS3}, p_{CS4}), (p_{S1}, p_{S2}), (p_{LR1}, p_{LR5})\}$). For

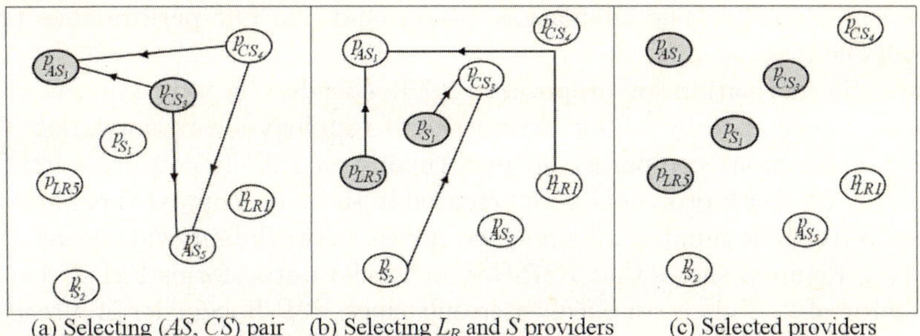

(a) Selecting (AS, CS) pair   (b) Selecting L_R and S providers   (c) Selected providers

Fig. 3. Selecting provider nodes

selecting the pair $(best(P_{AS}), best(P_{CS}))$, the cost provider sends the set of filtered CS providers (i.e. $FCS = \{p_{CS3}, p_{CS4}\}$) to each filtered AS provider (i.e. $FAS = \{p_{AS1}, p_{AS3}\}$). Afterwards, each $p_{ASi} \in FAS$ computes the latency between p_{ASi} and each $p_{CSj} \in FCS$, noted $latency(p_{ASi}, p_{CSj})$, and returns these latencies to the cost provider in the following tuple format: $< p_{ASi}, p_{CSj}, latency(p_{ASi}, p_{CSj}) >$. The cost provider merges such tuples arranging them in ascending order of latency. Finally, the cost provider retrieves the first tuple (i.e. the one with the smallest latency) and designates the associated pair of providers (i.e. (p_{ASi}, p_{CSj})) as selected providers. The same approach is used to select $best(P_{LR})$, which should be close to $best(p_{AS})$, as well as to select $best(P_S)$, which should be close to $best(P_{CS})$.

The candidate providers filtered to participate of the provider selection vary with time. To face this dynamic behavior of candidatures, the cost provider automatically launches a new provider selection whenever the set of filtered candidates change.

Notifying Provider Selection. Changing the selected provider leads to changes in the set of candidate reconcilers and invalidates all estimated reconciliation costs. As a result, the cost provider discards estimated costs and notifies the result of provider selection to provider nodes. The provider nodes, in turn, proceed as follows. If the provider p switches from selected to unselected, p notifies its good neighbors that from now on they are no longer candidate reconcilers. In contrast, if the provider p switches from unselected to selected, p notifies its good neighbors that from now on they are candidate reconcilers.

5   Performance Evaluation

To validate P2P-Reconciler-TA and study its performance, we implemented it on top of a simulated overlay P2P network based on topology-aware CAN. We performed many tests but, for space reasons, we present only part of the experimental results. The simulation background and our performance model are available in [5].

The main motivation for proposing P2P-Reconciler-TA is to improve reconciliation performance by taking advantage of topology-aware networks. Thus, our first experiment compares the performance of P2P-Reconciler with P2P-Reconciler-TA. Both protocols were executed in the same context (i.e. number of actions to reconcile number of connected nodes, network bandwidths and latencies, etc.). Figure 4 shows that P2P-Reconciler-TA outperforms P2P-Reconciler by a factor of 2. This is an excellent result since P2P-Reconciler is already an efficient protocol and CAN is not the most efficient topology-aware P2P network (e.g. Pastry and Tapestry are more efficient than CAN).

The second experiment aims at observing the scalability of P2P-reconciler-TA by studying the impact of the number of connected nodes on the reconciliation time (the larger the number of nodes is, the larger the average number of hops needed to lookup an identifier in the P2P network). We varied the number of connected nodes from 64 to 4000 whereas the number of reconciled actions was 106. Although most collaborative applications have limited numbers of participants, we considered large numbers of nodes to prove that P2P-reconciler-TA adapts perfectly to several other contexts. Figure 5 represents the reconciliation time with a straight line, which means an excellent scalability wrt. the number of connected nodes. This happens because provider and reconciler nodes are as close as possible independently of the network size.

Recall that for each reconciliation object, P2P-Reconciler-TA must select the best provider node. Even with a limited number of replicas, the search space is quite large as the combination of provider nodes must be taken into account. We aim at drastically reducing the search space of best providers while preserving the best alternatives in the reduced space. This allows us to efficiently select provider nodes. So, our third experiment studies the selection of provider nodes by varying the number of candidate providers per reconciliation object. The candidates are chosen according to their network quality. Figure 6 shows that P2P-Reconciler-TA achieves the best performance with small numbers of candidates (e.g. 3 or 4). This is an excellent result since the smaller the number of candidates is, the smaller the search space.

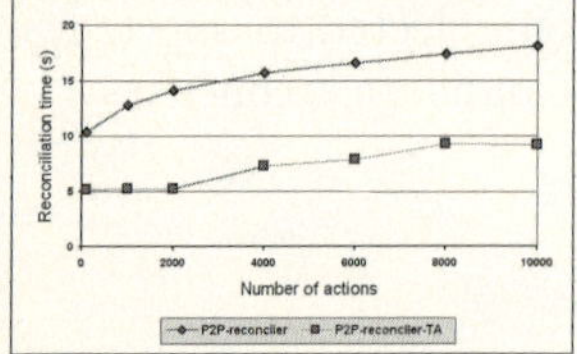
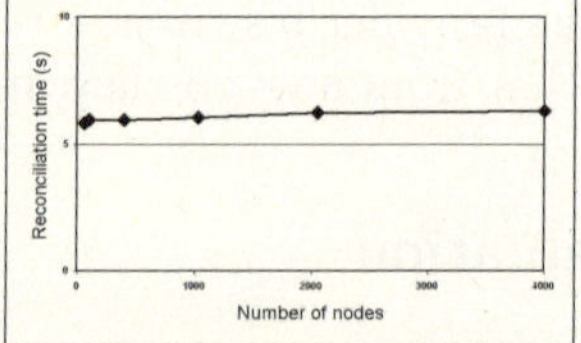
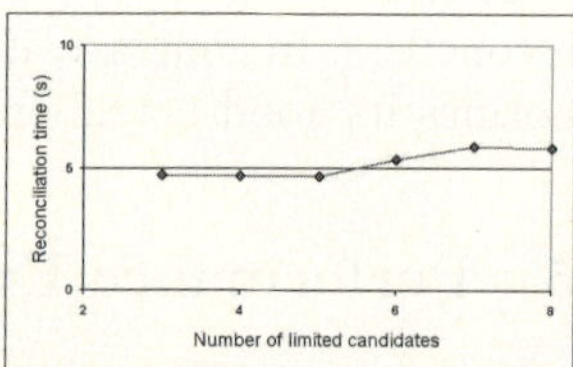

Fig. 4. Varying number of actions   **Fig. 5.** Varying number of nodes   **Fig. 6.** Varying number of candidate providers

6   Conclusion

In this paper, we proposed a topology-aware approach to improve response times in P2P distributed semantic reconciliation. The P2P-Reconciler-TA algorithm dynamically takes into account the physical network topology combined with the DHT properties when executing reconciliation. We proposed metrics and cost functions to be used for dynamically selecting the best nodes to execute reconciliation, while considering dynamic data placement. We validated P2P-Reconciler-TA through implementation and simulation. The experimental results show that our approach achieves a performance improvement by a factor of 2 in comparison with P2P-Reconciler. In addition, P2P-Reconciler-TA has proved to be scalable with limited overhead and thereby suitable for P2P environments. Our approach is conceived for distributed reconciliation; however our metrics, costs functions as well as our selection approach can be useful in several contexts.

References

1. Kermarrec, A-M., Rowstron, A., Shapiro, M., Druschel, P.: The IceCube approach to the reconciliation of diverging replicas. In: Proc. of ACM PODC, ACM Press, New York (2001)
2. Krishnamurthy, B., Wang, J., Xie, Y.: Early measurements of a cluster-based architecture for P2P systems. In: Proc of ACM SIGCOMM (2001)
3. Liu, Y., Xiao, L., Liu, X., Ni, L.M., Zhang, X.: Location awareness in unstructured P2P pystems. IEEE Trans. Parallel Distrib. Syst. 16(2) (2005)
4. Martins, V., Pacitti, E.: Dynamic and distributed reconciliation in P2P-DHT networks. In: Proc. of Euro-Par (2006)
5. P2P-Reconciler-TA. `http://www.sciences.univ-nantes.fr/lina/gdd/members/vmartins/`
6. Ratnasamy, S., Francis, P., Handley, M., Karp, R., Shenker, S.: A scalable content-addressable network. In: Proc. of ACM SIGCOMM (2001)
7. Ratnasamy, S., Handley, M., Karp, R.M., Shenker, S.: Topologically-aware overlay construction and server selection. In: Proc. of IEEE INFOCOM (2002)
8. Rowstron, A., Druschel, P.: Pastry: scalable, distributed object location and routing for large-scale P2P systems. In: Proc. of IFIP/ACM Middleware, ACM Press, New York (2001)
9. Zhao, B.Y., Huang, L., Stribling, J., Rhea, S.C., Joseph, A.D., Kubiatowicz, J.D.: Tapestry: a resilient global-scale overlay for service deployment. IEEE Journal on Selected Areas in Communications (JSAC) 22(1) (2004)

Parallel Nearest Neighbour Algorithms for Text Categorization

Reynaldo Gil-García[1], José Manuel Badía-Contelles[2], and Aurora Pons-Porrata[1]

[1] Center of Pattern Recognition and Data Mining, Universidad de Oriente, Cuba
{gil,aurora}@csd.uo.edu.cu
[2] Dpt. Computer Science and Engineering, Universitat Jaume I, Castellón, Spain
badia@icc.uji.es

Abstract. In this paper we describe the parallelization of two nearest neighbour classification algorithms. Nearest neighbour methods are well-known machine learning techniques. They have been successfully applied to Text Categorization task. Based on standard parallel techniques we propose two versions of each algorithm on message passing architectures. We also include experimental results on a cluster of personal computers using a large text collection. Our algorithms attempt to balance the load among the processors, they are portable, and obtain very good speedups and scalability.

1   Introduction

Text Categorization (also known as text classification) is the task of assigning documents to one or more predefined categories. This task, which falls at the crossroads of Information Retrieval and Machine Learning, has witnessed a booming interest in the last years from researchers and developers alike. Text Categorization is an important component in many information management tasks such as spam detection [1], real time sorting of email or files into folders, document filtering [2], document dissemination, document routing [3], topic identification, classification of Web pages and automatic building of Yahoo!-style catalogs. Different learners have been applied in the Text Categorization literature, including probabilistic methods, decision tree and decision rule learners, example-based methods, support vector machines and classifier committees.

Most sequential text categorization algorithms have large running times. On the other hand, the volume of data available for analysis is growing rapidly. The huge size of text collections and their high dimension make classification a highly computationally-demanding application, to the point that parallel computing is an essential tool for its solution. Parallelism offers a natural and promising approach to cope with the problem of efficient categorization in large text collections.

Nearest neighbour (NN) classifiers are well-known methods for text categorization problems. This approach classifies an unknown sample into the categories of its nearest neighbours, according to some similarity measure. A particular case of NN classifiers is the k-nearest neighbour rule (k-NN), which assigns each unknown sample to the category most frequently represented among the k nearest training

A.-M. Kermarrec, L. Bougé, and T. Priol (Eds.): Euro-Par 2007, LNCS 4641, pp. 328–337, 2007.

samples. The NN classifiers include the following features: 1) conceptual simplicity, 2) easy implementation, 3) they can be designed even if there are few training samples, 4) known error rate bounds, 5) they can be implemented when categories are overlapped with each other, 6) good accuracy, 7) they have no design phase and simply store the training set, and 8) they can be performed in linear time with respect to the cardinality of the training set. The last feature can be a serious problem in very large text collections, because the computation of similarities is very time-consuming.

Several parallel versions of the NN learning method can already be found in the literature. Li [4] proposed several parallel nearest neighbour classification algorithms on two different types of SIMD computers. In [5] a parallel nearest neighbour classifier which uses attribute weights in the computation of distance is presented. This algorithm was implemented on the Connection Machine CM-2. It partitions the training set into p mutually exclusive data subsets that are distributed across the p processors. Then, each processor computes the similarity between the training samples in its local data subset and the new sample to be classified. Jin et al. [6] apply the same data distribution on a cluster parallel computer. Each node processes its training samples to calculate locally the k-nearest neighbours to the new sample and then, a global reduction is carried out to compute the overall k-nearest neighbours. The experimental results showed good speedup up to 8 nodes.

Finally, Jin et al. [7] also proposed a parallel version of k-NN algorithm on a shared memory computer. Each training sample is processed by one processor. After processing the sample, each processor updates the list of k current nearest neighbours. They used a full-replication scheme to avoid the race conditions. The obtained speedup was moderate up to four processors.

In this paper, we introduce scalable and efficient parallel versions of two nearest neighbour classification methods for distributed memory computers. Our versions use the standard parallel techniques: master-slave and pipelining. The algorithms were implemented using the MPI library on a cluster parallel computer. The performance of the proposed algorithms is evaluated on the Reuters Corpus Volume I [8], which is the new standard benchmark for text categorization tasks.

The remainder of this paper is organized as follows. In Section 2 we briefly review the nearest neighbour classification methods addressed in this paper. Section 3 describes our parallel algorithms. In Section 4 we show the experimental results and compare the proposed parallel versions. Finally, Section 5 contains a summary of the work.

2 Sequential Algorithms

As mentioned above, the nearest neighbour methods classify an unknown document d into the categories of its nearest neighbours in the training set. The training set is a collection of m documents labelled with their categories.

The nearest neighbour classifiers usually involve three phases: (i) the nearest neighbour finding from the training documents, (ii) a voting phase, in which each category assigns a vote to d, and (iii) a decision rule, in which a decision is made from these votes to classify the new document. During the last years, a large number of NN algorithms have aroused from various scientific communities. Many of them focus on increasing classification rates, either changing the method to find the nearest neighbours or varying the voting scheme.

In this paper we focus on two NN methods, which assume different neighbourhood definitions. The first one, the traditional k-NN, starts at d and grows a spherical region until it encloses k training documents, where k is a user defined parameter.

The second method [9], considers a kind of neighbourhood which inspects a sufficiently small and near area to d. In this method, the number of neighbours is not fixed, but rather the neighbourhood radius is automatically adjusted from the nearest neighbour of d. This neighbourhood takes into account instead all neighbours enclosed within a spherical region defined from the nearest neighbour. The method uses two parameters α and β, which provide a convenient way of obtaining such a neighbourhood. From now on, we will refer to this method as *braNN*. In Figure 1 both neighbourhoods are graphically depicted. The shady regions represent the neighbourhoods in both methods.

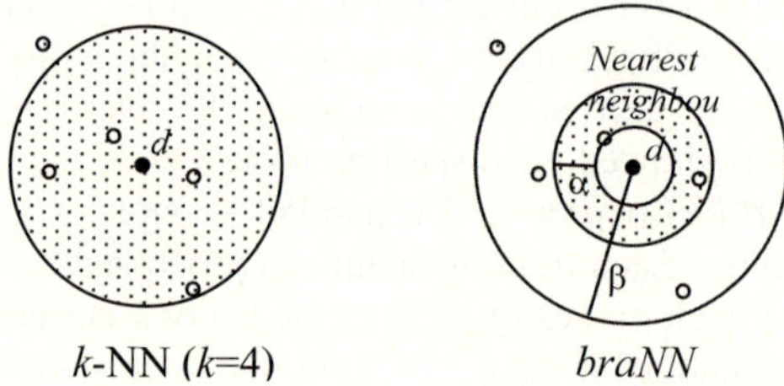

k-NN (k=4)   *braNN*

Fig. 1. Neighbourhoods of k-NN and *braNN* methods

Different methods have been used to calculate the votes of each category (second phase). One of these methods considers the similarity of the nearest neighbours and their category association to calculate the votes [10], i.e.,

$$V(c_i,d) = \frac{\sum_{d_j \in N(c_i)} \cos(d,d_j)}{\sum_{d_j \in N} \cos(d,d_j)}, \tag{1}$$

where N is the set of all nearest neighbours of the document d, $N(c_i)$ is the set of the nearest neighbours labelled with the category c_i, and $\cos(d,d_j)$ is the cosine between the two document vectors, which is the similarity measure commonly used in text categorization tasks.

From these category votes, several rules can be applied for deciding whether d should be classified under c_i (third phase). In this paper, we used the thresholding decision rule. According with this rule, the document is assigned to categories with the score greater than a certain threshold value γ. Notice that this decision rule allows a multi-label categorization.

Both nearest neighbour methods, k-NN and *braNN*, employ the same voting scheme and decision rule. To sum up, the steps of NN classifiers we use are shown in Algorithm 1. The step 3 depends on the neighbourhood definition. Algorithms 2 and 3 show the construction of the neighbourhood in k-NN and *braNN* methods, respectively.

Algorithm 1. NN classifiers

1. Let *Tr* be the training set.
2. Let *d* be the document to classify.
3. Build the set *N* of neighbours of *d* according with a certain neighbourhood definition.
4. Compute the votes $V(c_i, d)$ for each category c_i using formula (1).
5. For each category c_i:
     (a) If $V(c_i, d) \geq \gamma$:
          i. Assign *d* to c_i.

Algorithm 2. Construction of the k-NN neighbourhood

1. Let *N* be the list of neighbours of *d*, decreasingly ordered by its similarity with *d*.
2. $N = \varnothing$.
3. For each training document $d_j \in Tr$:
     (a) Insert d_j in *N*.
     (b) If $|N| > k$:
          i. Remove the last element of *N*.

Algorithm 3. Construction of the *braNN* neighbourhood

1. Build the set $N_\beta = \{d_j \in Tr \,/\, \cos(d, d_j) \geq \beta\}$
2. Let *max* be the similarity between *d* and its nearest neighbour in N_β.
3. Let *N* be the list of neighbours of *d*, $N = \varnothing$.
4. For each training document $d_j \in N_\beta$:
     (a) If $\cos(d, d_j) \geq max - \alpha$:
          i. Add d_j to *N*.

The space complexity of both algorithms is $O(m)$, because the whole training set is stored. On the other hand, its time complexity is $O(nm)$, where n is the number of documents to classify, because for each unknown sample the similarity with each training document is calculated.

3 Parallel Algorithms

Our parallel algorithms assume that we have p processors each with a local memory. These processors are connected using a communication network. We do not assume a specific interconnection topology for the communication network, but the access time to the local memory of each processor must be cheaper than time to communicate the same data with other processor.

We present two parallel versions of the algorithms k-NN and *braNN*. The first parallel algorithm uses a master-slave scheme whereas the second one uses the pipeline technique.

3.1 Master-Slave Algorithms

In the distributed memory setting, the nearest neighbour classifiers can be parallelized by dividing the data to be classified among the processors and replicating the training set. Thus, each processor can classify the data items it owns. A uniform data distribution can produce a load imbalance, since the computation of similarities of the sample to the training documents is the most time-consuming task, and the documents can have different sizes. An easy way to achieve load balancing is to use a master-slave scheme.

Our parallel algorithm can be described as follows. First, the master process broadcasts the training set. Then, the master sends a sample to each slave process. The slaves determine independently the nearest neighbours of their samples and classify them. Each time a slave finishes its work with a sample it asks for a new one to the master.

It is important to notice that no significant differences between k-NN and *braNN* exist in this parallel version. In the first case, each processor determines the k nearest neighbours (see Algorithm 2), whereas in the second one the processors determine the neighbourhood defined by α and β parameters (see Algorithm 3).

The time complexity of this scheme is $O(nm/(p-1))$, because the n unknown documents are classified independently by the $p-1$ slave processors and the communications and waiting time can be overlapped with the computations. The space complexity of this parallel algorithm is $O(m)$, because each processor stores the whole training set. Thus, this version does not have space scalability.

3.2 Pipeline Algorithms

Instead of dividing the data to be classified, in this parallel version, the training set is partitioned into p mutually exclusive data subsets that are distributed among the p processors. Once this data is distributed, the problem of finding the nearest neighbours of each document to classify can be broken up into several stages.

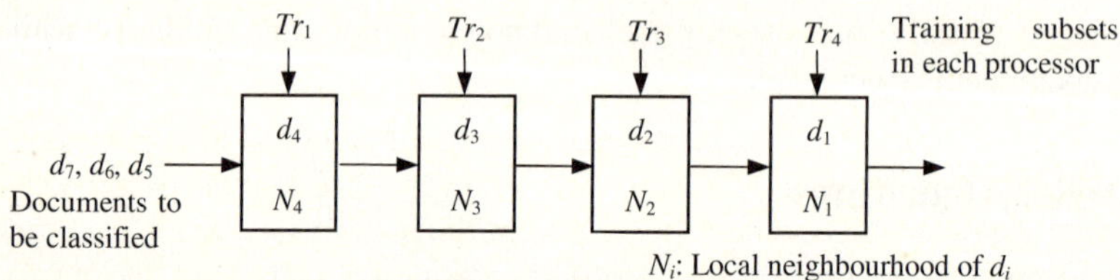

Fig. 2. Pipeline scheme

One pipeline solution could have a separate stage in each processor for updating the neighbourhood of each unknown sample, as shown in Figure 2. Each stage of the pipeline computes the similarities between the training documents in its local data subset and the document to be classified, and updates the list of nearest neighbours provided by the previous stage. Each unknown document and its partial neighbourhood are sent from one processor to the next. Thus, the last processor obtains the global neighbourhood and it also classifies the sample.

The steps of this parallel algorithm are shown in Algorithm 4.

Algorithm 4. Pipelining

---

1. Scatter the training documents among the processors.
2. While True:
   (a) If processor 0:
      i. Load the unknown document *d*.
      ii. Build the local neighbourhood of *d*.
      iii. Send *d* and its neighbours to processor 1.
   (b) Else:
      i. Receive *d* and its neighbours from the previous processor.
      ii. Update the local neighbourhood of *d*.
      iii. If it is the last processor:
         A. Classify the document *d*.
      iv. Else:
         A. Send *d* and its neighbours to the next processor.

The neighbourhood updating (step 2(b)ii) depends on the nearest neighbour classifier. In parallel k-NN algorithm, each processor updates the neighbour list associated to each unknown sample it receives. This list is ordered by similarity to the sample and must only contain k documents. When a processor receives the sample and the local list of neighbours from the previous one, it inserts its training documents in the list according with its similarity to the sample.

On the other hand, in parallel *braNN* algorithm, each processor updates the nearest neighbour to the unknown sample and the list of those training documents whose similarity is greater than both β and *max*-α (*max* is the similarity between the nearest neighbour and the new sample).

The time complexity is $O(nm/p)$, because each processor computes the similarities between each sample and its m/p training documents. In this parallel algorithm the communications can be overlapped with the computations. The space complexity of this parallel version in both algorithms is $O(m/p)$, because the training set is scattered among the p processors. Thus, this parallel algorithm has space scalability.

Notice that both, this parallel version and the algorithm proposed by Jin [6], distribute the training set among the processors and that local neighbourhoods are calculated in each processor. However, both algorithms differ on the procedure to compute the global neighbourhood. In Jin's algorithm a global reduction is carried out whereas we use a pipeline.

4 Performance Evaluation

The target platform for our experimental analysis is a cluster of SMP processors connected through an Infinitband network. The cluster consists of 8 nodes, each containing 4 Intel Itanium-2 processors and 4 Gbyte of RAM. The proposed parallel algorithms have been implemented on a Linux operating system, and we have used a specific implementation of the MPI message-passing library that offers small latencies and high bandwidths on the Infinitband network. This library also optimizes the communications between processors on the same node by exploiting the shared memory as communication mechanism. We have executed the parallel algorithms varying the number of processors from 4 to 32.

We used the Reuters Corpus Volume I (RCV1-v2) [8] as our testing medium. RCV1-v2 collection consists of over 800000 newswire stories that have been manually classified in 103 categories. This collection is partitioned (according to the LYRL2004 split we have adopted) into a training set of 23149 documents and a test set of 781265 documents. The documents are represented using the traditional vector-space model. Terms are statistically weighted using Cornell *ltc* term weighting [11].

Due to the fact that we focus on a parallelization scheme, no accuracy results for the data are given in this paper. We neither tune the parameters of the algorithms, but we fix the values that report good results in the literature.

In this paper, we compare our two parallel versions of k-NN and *braNN* algorithms. We also implemented the parallel version proposed by Jin [6] and compare it with them. The performance comparison using different number of processors for k-NN and *braNN* algorithms is shown in Figures 3 and 4, respectively.

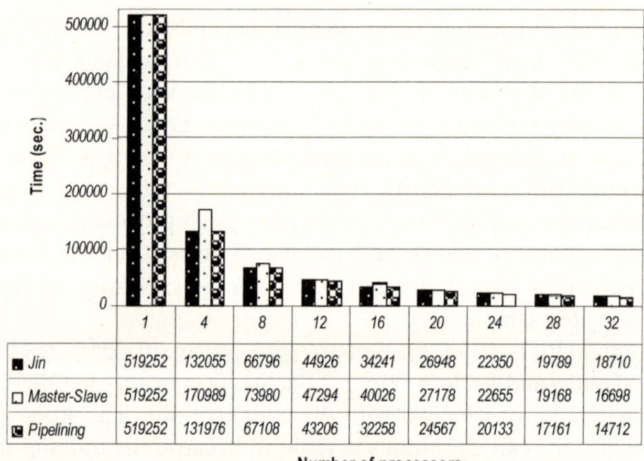

	1	4	8	12	16	20	24	28	32
Jin	519252	132055	66796	44926	34241	26948	22350	19789	18710
Master-Slave	519252	170989	73980	47294	40026	27178	22655	19168	16698
Pipelining	519252	131976	67108	43206	32258	24567	20133	17161	14712

Fig. 3. Running times of parallel k-NN versions

Several observations can be made by analyzing the results in these figures. First, both sequential algorithms spent a lot of time classifying the documents. Second, all parallel versions clearly reduce the sequential time. Notice that the sequential k-NN algorithm takes about 6 days to classify this collection, while the pipeline parallel version reduces this time to 4.08 hours on 32 processors. In case of *braNN* algorithm, the time decreases from 5.26 days down to 3.58 hours.

A third observation that also emerges clearly from the figures is that all parallel versions of *braNN* algorithm are less time-consuming than the corresponding parallel versions of k-NN algorithm independently of the number of processors used. This fact is in agreement with the results presented earlier in [9]. This can be explained because *braNN* algorithm does not need to create a sorted list of k nearest neighbours.

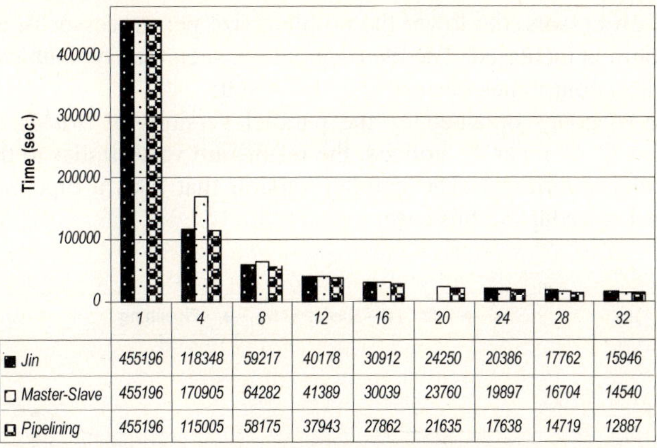

	1	4	8	12	16	20	24	28	32
■ Jin	455196	118348	59217	40178	30912	24250	20386	17762	15946
□ Master-Slave	455196	170905	64282	41389	30039	23760	19897	16704	14540
▨ Pipelining	455196	115005	58175	37943	27862	21635	17638	14719	12887

Fig. 4. Running times of parallel *braNN* versions

Figure 5 shows the speedups obtained by the parallel k-NN algorithms. As we can see, Jin's algorithm slightly outperforms our parallel version that uses the master-slave model for a small number of processors. However, our algorithm obtains larger speedups while increasing the number of processors. This can be explained because our algorithm uses only p-1 processors to classify the documents and, therefore, its speedup is always near p-1. On the contrary, as Jin's algorithm uses a global reduction operation, its efficiency goes down as the number of processors increases. Thus, we can conclude that our parallel version is more scalable than Jin's algorithm.

On the other hand, Figure 5 shows clearly that pipeline parallel version achieves superlinear speedup. We think that it is due to the data distribution used. The higher

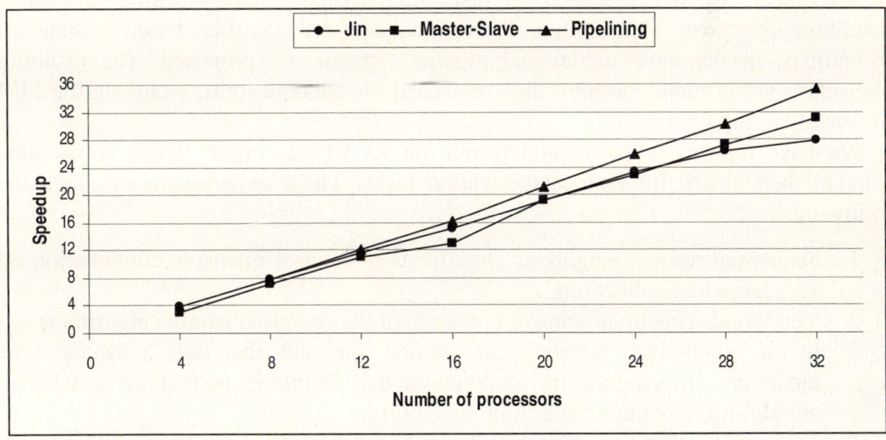

Fig. 5. Speedup of parallel k-NN versions

the number of processors, the lower the problem size per processor is, and therefore, the cache hit ratio is increased. The overlapping between the communications and the computations also contributes positively to this result.

Finally, the speedups obtained by the parallel versions of *braNN* algorithm are shown in Figure 6. As it can be noticed, the results are very similar to those obtained by *k*-NN parallel algorithms. The parallel version that uses a pipeline model also achieves the best speedups in this case.

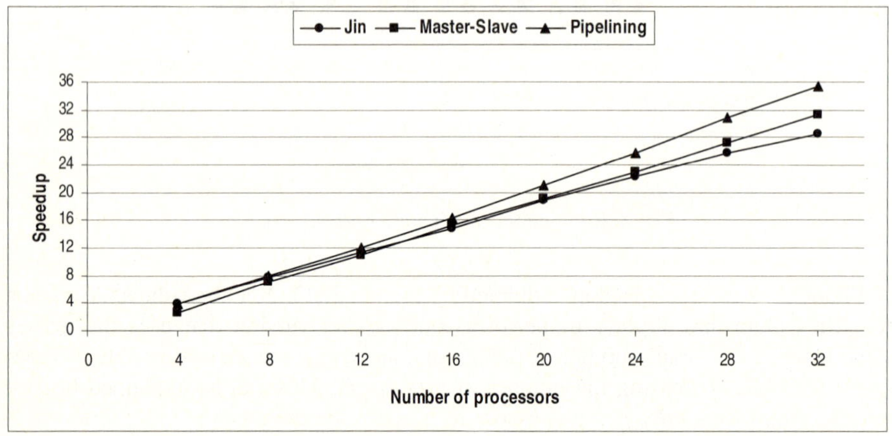

Fig. 6. Speedup of parallel *braNN* versions

5   Conclusion

In this paper, we have focused on distributed memory parallelization of nearest neighbour classifiers. Two new parallel versions of these classifiers based on standard techniques: master-slave model and pipeline scheme are proposed. The resulting algorithms are portable, because they are based on standard tools, including the MPI message-passing library.

We have reported experimental results on RCV1-v2 corpus, which is the new standard benchmark for text categorization tasks. These experiments establish the following:

1. Sequential nearest neighbour classification requires intensive computation on very large text collections.
2. Very good speedup is achieved for each of the proposed parallel algorithms.
3. In the comparison between our parallel approach that uses a master-slave model and Jin's algorithm, we conclude that despite its poor space scalability, our algorithm obtains better time scalability.
4. The parallel algorithm that uses a pipeline model offers clear advantages over the parallel versions we experimented with in that it is both spatially and temporally scalable. This method achieves the same space scalability as Jin's

algorithm and the best time scalability. The pipeline version also obtains superlinear speedup.

The proposed parallel algorithms can be used in many highly computationally demanding-applications such as information organization, filtering, routing, topic tracking and text categorization tasks, allowing to process huge text collections.

Acknowledgments. This work was partially supported by the Research Promotion Program 2006 of Universitat Jaume I, Spain, the CICYT project TIN2005-09037-C02-02 and FEDER.

References

[1] Drucker, H., Wu, D., Vapnik, V.N.: Support Vector Machines for Spam Categorization. IEEE Transactions on Neural Networks 10(5), 1048–1054 (1999)

[2] Eichmann, D., Srinivasan, P.: Adaptive Filtering of Newswire Stories using Two-Level Clustering. Information Retrieval 5, 209–237 (2002)

[3] Iyer, R.D., Lewis, D.D., Schapire, R.E., Singer, Y., Singhal, A.: Boosting for Document Routing. In: Proceedings of the Ninth International Conference on Information and Knowledge Management (2000)

[4] Li, X.: Nearest neighbor classification on two types of SIMD machines. Parallel Computing 17, 381–407 (1991)

[5] Creecy, R.H., Masand, B.M., Smith, S.J., Waltz, D.L.: Trading MIPS and memory for knowledge engineering. Comm. of the ACM 35(8), 48–63 (1992)

[6] Jin, R., Agrawal, G.: A Middleware for Developing Parallel Data Mining Implementations. In: Proceedings of the First SIAM Conference on Data Mining (2001)

[7] Jin, R., Yang, G., Agrawal, G.: Shared memory parallelization of data mining algorithms: Techniques, programming interface and performance. IEEE Transactions on Knowledge and Data Engineering 17, 71–89 (2005)

[8] Lewis, D., Yang, Y., Rose, T., Li, F.: Rcv1: A new benchmark collection for text categorization research. Machine Learning Reseach 5, 361–397 (2004)

[9] Gil-García, R., Pons-Porrata, A.: A new nearest neighbor rule for text categorization. In: Martínez-Trinidad, J.F., Carrasco Ochoa, J.A., Kittler, J. (eds.) CIARP 2006. LNCS, vol. 4225, pp. 814–823. Springer, Heidelberg (2006)

[10] Yang, Y.: Expert network: Effective and efficient learning from human decisions in text categorization and retrieval. In: SIGIR'94, 17th ACM International Conference on Research and Development in Information Retrieval, Ireland, pp. 13–22. ACM Press, New York (1994)

[11] Buckley, C., Salton, G., Allan, J.: The effect of adding relevance information in a relevance feedback environment. In: Proceedings of the Seventeenth Annual International ACM-SIGIR Conference on Research and Development in Information Retrieval, pp. 292–300. ACM Press, New York (1994)

Efficient Distributed Data Condensation for Nearest Neighbor Classification

Fabrizio Angiulli[1] and Gianluigi Folino[2]

[1] DEIS, Università della Calabria
Via P. Bucci 41C, 87036, Rende (CS), Italy
`f.angiulli@deis.unical.it`
[2] Institute of High Performance Computing and Networking (ICAR-CNR)
Via P. Bucci 41C, 87036 Rende (CS), Italy
`folino@icar.cnr.it`

Abstract. In this work, PFCNN, a distributed method for computing a consistent subset of very large data sets for the nearest neighbor decision rule is presented. In order to cope with the communication overhead typical of distributed environments and to reduce memory requirements, different variants of the basic PFCNN method are introduced. Experimental results, performed on a class of synthetic datasets revealed that these methods can be profitably applied to enormous collections of data. Indeed, they scale-up well and are efficient in memory consumption and achieve noticeable data reduction and good classification accuracy. To the best of our knowledge, this is the first distributed algorithm for computing a training set consistent subset for the nearest neighbor rule.

1   Introduction

Even though collecting data capabilities of organizations is dramatically increasing, often they cannot take advantage of these collections of potential useful information, since ad-hoc data mining algorithms may be unavailable, while traditional machine learning and data analysis tools are practicable only on small data sets.

A very useful task is to build a model over it so as to obtain a classifier for prediction purposes. The *nearest neighbor rule* [2,8,4] is one of the most extensively used nonparametric classification algorithms, simple to implement yet powerful, due to its theoretical properties guaranteeing that for all distributions its probability of error is bounded above by twice the Bayes probability of error. The naive implementation of this rule has no learning phase, in that it uses all the training set objects in order to classify new incoming data. But, a number of training set condensation algorithms have been proposed that extract a *consistent subset* of the overall training set, namely CNN, MCNN, NNSRM, FCNN, and others [6,7,3,1], i.e. a subset that correctly classifies all the discarded training set objects through the nearest neighbor rule. These algorithms have been shown to achieve in some cases condensation ratios corresponding to a small percentage of the overall training set.

However, the performances of these algorithms may degrade considerably, both in terms of memory and time, when they have to cope with huge data sets, consisting of a very large number of objects. Indeed, this amount of data can be too large to fit

A.-M. Kermarrec, L. Bougé, and T. Priol (Eds.): Euro-Par 2007, LNCS 4641, pp. 338–347, 2007.

into the main memory. Furthermore, execution time may become burdensome or even prohibitive.

Parallel and distributed computation can be exploited in order to manage efficiently these enormous collections of data. Furthermore, recently, the new emerging paradigm of grid computing [5] has chiefly provided the access to large resources of computing power and storage capacity. Typically, a user can harness the unused and idle resources that organizations share in order to solve very complex problems. Moreover, data reduction through the partitioning of the data set into smaller subsets seems to be a good approach. But, to the best of our knowledge, *no parallel or distributed consistent subset learning algorithm for the nearest neighbor rule has been proposed in literature.*

In this paper, a distributed training set consistent subset learning algorithm for the nearest neighbor rule exhibiting high efficiency both in terms of time and of memory usage is presented. The algorithm, called PFCNN, for Parallel Fast Condensed Nearest Neighbor Rule, is a distributed version of the sequential algorithm FCNN [1], which has been shown to outperform all the other training set consistent subset methods. Distribution of data and their consequent handling raise many problems that can be faced in different ways if we privilege the usage of memory rather than the scalability or the execution time. Thus, different clever variants of the basic distributed method are proposed that keep in count these aspects.

The rest of the paper is organized as follows. Section 2 recalls the sequential FCNN rule. Section 3 describes the architecture of the PFCNN method and its variants. Subsequent Section 4 describes the PFCNNs algorithms. Section 5 reports experimental results on class of synthetic very large data sets. Finally, Section 6 reports conclusions and depicts future work.

2   The FCNN Rule

In this section, the sequential FCNN rule [1] is recalled. First of all, some preliminary definitions are provided. We define T as a labelled training set from a metric space with distance metrics d. Let x be an element of T. Then we denote $nn(x, T)$ as the nearest

---

Algorithm FCNN(T : training set)

1. Initialize the set S to the empty set
2. Initialize the set ΔS to the set $Centroids(T)$
3. While the set ΔS is not empty:
   (a) Augment the set S with the set ΔS
   (b) Initialize the set ΔS to the empty set
   (c) For each object y in the set S, insert into ΔS the representative object of the Voronoi enemies of y in T w.r.t. S
4. Return the set S

Fig. 1. The (sequential) FCNN rule

neighbor of x in T according to the distance d. $l(x)$ will be the label associated to x. Given a labelled data set T and an element y of M, the *nearest neighbor rule* $NN(y, T)$ assigns to y the label of the nearest neighbor of y in T, i.e. $NN(y, T) = l(nn(y, T))$ [2]. A subset S of T is said to be a *training set consistent subset of T* if, for each $x \in T$, $l(x) = NN(x, S)$ [6]. Let S be a subset of T, and let y be an element of S. By $Vor(y, S, T)$ it is denoted the set $\{x \in T \mid \forall y' \in S, d(y, x) \leq d(y', x)\}$, that is the set of the elements of T that are closer to y than to any other element y' of S, called the *Voronoi cell* of y in T w.r.t. S. Furthermore, by $Voren(y, S, T)$ it is denoted the set $\{x \in Vor(y, S, T) \mid l(x) \neq l(y)\}$, whose elements are called *Voronoi enemies* of y in T w.r.t. S. $Centroids(T)$ is the set containing the centroids of each class label in T. Given a set of points S having the same class label, the *centroid* of S is the point of S which is closest to the geometrical center of S. The Fast Condensed Nearest Neighbor Rule [1], FCNN for short, relies on the following property: a set S is a training set consistent subset of T for the nearest neighbor rule if for each element y of S, $Voren(y, S, T)$ is empty.

The FCNN algorithm is shown in Figure 1. The algorithm initializes the consistent subset S with a seed element from each class label of the training set T. In particular, the seeds employed are the centroids of the classes in T. The algorithm is incremental. During each iteration the set S is augmented until the stop condition, given by the property above, is reached. For each element of S, a *representative* element of $Voren(y, S, T)$ w.r.t. y is selected and inserted into S. The behavior of two different definitions of representative were investigated. The first definition, FCNN1 is the name of the implementation of the FCNN rule using this definition, selects as representative the nearest neighbor of y in $Voren(y, S, T)$, that is the element $nn(y, Voren(y, S, T))$ of T. The second definition, FCNN2, selects as representative the class centroid in $Voren(y, S, T)$ closest to y, that is the element $nn(y, Centroids(Voren(y, S, T)))$ of T.

As far as the comparison between the two methods in the sequential scenario [1], the FCNN2 rule appears to be little sensitive to the complexity of the decision boundary, since it rapidly covers regions of the space far from the centroids of the classes and always performs about a few batches of ten iterations. The FCNN1 is slightly slower than the FCNN2 since it may require more iterations, up to a few hundreds. Conversely, the FCNN1 is likely to select points very close to the decision boundary, and hence may return a subset smaller than that of the FCNN2. As for the time complexity of the method, let N denote the size of the training set T and let n denote the size of the consistent subset S computed, then the FCNN rule requires at most Nn distance computations to compare the elements of T with the elements of S.

3   The PFCNN Architecture

Despite the FCNN algorithm is fast, its time requirements grow with the size of the dataset. When huge collections of data have to be handled, it is of interest to scale-up the method. It will be shown that time and memory requirements of large data sets can be coped with a distributed implementation of the FCNN algorithm, called PFCNN, whose architecture is introduced next.

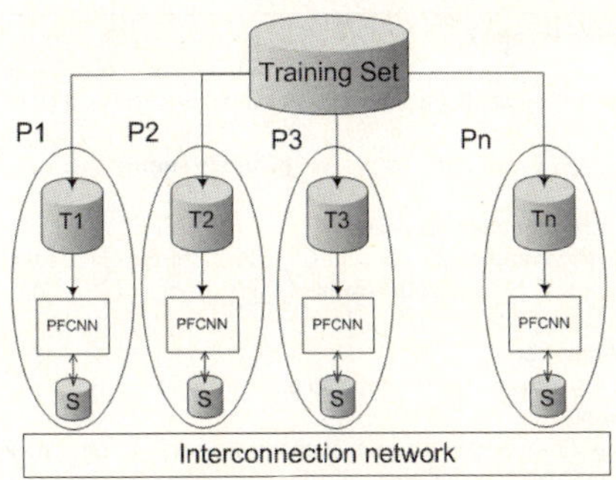

Fig. 2. PFCNN architecture

The general architecture of the PFCNNs algorithms is illustrated in Figure 2. The architecture is composed of n nodes $P_1, \ldots, P_n$. The original training set T is partitioned in n disjoint partitions $T_1, \ldots, T_n$, each assigned to a distinct node. PFCNN can also be used when the data set is already distributed among nodes and cannot be moved (i.e. for privacy reasons). Each node i computes, in parallel, the overall condensed set S using only its partition T_i of the training set. Note that the condensed data set S is replicated on each node. However, it amounts to a very small percentage of the training set (usually, it is some order of magnitude smaller).

The following section describes the two basic PFCNNs strategies, that is the PFCNN1 and PFCNN2 rules, and then introduces different variants, namely the PFCNN-t, PFCNN-p, and PFCNN-b, that further improve time and memory consumption of the two basic rules.

4   PFCNN Rules

PFCNN1 rule. Figure 3 shows the Parallel FCNN1 algorithm. We recall that the PFCNN1 rule is the variant of the PFCNN rule using the nearest neighbor as representative of the Voronoi enemies of a consistent subset element.

Let p the number of nodes available. Each node is identified by i such that $1 \leq i \leq p$. The pseudo-code reported in Figure 3 is executed on the generic node i. Variables there employed are local to the node i, except for those handled by parallel functions, which instead come from different nodes. When it will be needed to distinguish the node i from which a variable v comes from, then the notation v^i will be used.

Next we describe the data structures employed and how data is located on the different nodes. The overall training set T, containing N objects, is randomly partitioned into p equally sized disjoint blocks $T_1, \ldots, T_p$ and then each node i receives in input the block T_i. Differently from the training set T, each node maintains a local copy of the entire consistent subset S. Furthermore, each node maintains two arrays: *nearest* and *rep*. The array *nearest*, having size $\frac{N}{p}$, contains for each point x in T_i its closest point *nearest*$[x]$ in the set S. The array *rep* contains, for each point y in S, its representative *rep*$[y]$ of the misclassified points lying in the Voronoi cell of y in T_i w.r.t. S.

Algorithm PFCNN1(T_i : a training set block)

1. For each class $j = 1, \ldots, m$: compute the sum $s[j]$ of all the elements of T_i of the class j, together with their number $N[j]$
2. For each class $j = 1, \ldots, m$: $s[j] = \textbf{parallel–sum}(s^1[j], \ldots, s^P[j])$, $N[j] = \textbf{parallel–sum}(N^1[j], \ldots, N^P[j])$
3. For each class $j = 1, \ldots, m$: compute the center $c[j] = s[j]/N[j]$
4. For each class $j = 1, \ldots, m$: compute the element $C[j]$ in T_i of the class j which is closest to $c[j]$
5. For each class $j = 1, \ldots, m$: $C[j] = \textbf{parallel–min}(\langle C^1[j], \mathrm{d}(c[j], C^1[j])\rangle, \ldots, \langle C^P[j], \mathrm{d}(c[j], C^P[j])\rangle)$
6. Initialize the set ΔS to the set $\{C[1], \ldots, C[m]\}$
7. Initialize the set S to the empty set
8. For each element x in T_i: set $nearest[x]$ to undefined
9. While the set ΔS is not empty:
    (a) For each element x in $T_i - S$, and for each element y in ΔS: if the distance between x and y is less than the distance from x and $nearest[x]$ then set $nearest[x]$ to y
    (b) For each element y in S: set $rep[y]$ to undefined
    (c) For each element x in $T_i - S$: if the class of x is different from the class of $nearest[x]$ and the distance from x to $nearest[x]$ is less than the distance from $nearest[x]$ to $rep[nearest[x]]$ then set $rep[nearest[x]]$ to x
    (d) Augment the set S with the set ΔS
    (e) For each y in S, $rep[y] = \textbf{parallel–min}(\langle rep^1[y], \mathrm{d}(y, rep^1[y])\rangle, \ldots, \langle rep^P[y], \mathrm{d}(y, rep^P[y])\rangle)$
    (f) Initialize the set ΔS to the empty set
    (g) For each element y is S: if $rep[y]$ is defined then insert $rep[y]$ into ΔS
10. Return the set S

Fig. 3. The PFCNN1 rule

Now we are in the position of commenting on the code. First of all, steps 1-3 compute the geometrical center of each training set class, while steps 4-5 compute the centroids $C[1], \ldots, C[m]$ of each class. Two communication functions are employed in these steps, that is **parallel–sum** and **parallel–min**. The **parallel–sum**$(v^1, \ldots, v^P)$ is a parallel function which gathers the p (arrays of) integer or real numbers $v^1, \ldots, v^P$ from the p nodes and then returns the sum $v^1 + \ldots + v^P$ of these values. The **parallel–min**$(\langle u^1, v^1 \rangle, \ldots, \langle u^P, v^P \rangle)$ is a parallel function gathering the p values $u^1, \ldots, u^P$, together with the p integer or real numbers $v^1, \ldots, v^P$, and then returning the value u^i associated to the smallest number v^i among $v^1, \ldots, v^P$.

Once the centroids $C[1], \ldots, C[m]$ of the training set classes are computed, the set ΔS is initialized to $\{C[1], \ldots, C[m]\}$, the consistent subset S is initialized to the empty set, the closest element $nearest[x]$ in S of each element x in T_i is set to undefined (steps 6-8), and then the iterative part of the algorithm starts.

During each iteration, the array $nearest$ and rep must be updated since they represent, respectively, the partitioning of the points of T_i into Voronoi cells and the points in the new set ΔS. Let ΔS be the set of points to be added to the set S during the current iteration (at the first iteration this set coincides with the class centroids). To update the array $nearest$, the training set points in $(T_i - S)$ are compared with the points in the set ΔS (step 9.(a)). Clearly, it is not needed to compare the points in $(T_i - S)$ with the points in S, since this comparison was already done in the previous iterations and nearest neighbors so far computed are currently stored in $nearest$.

After having computed the closest point $nearest[x]$ in ΔS, of the points x in $(T_i - S)$, the array rep is updated efficiently (step 9.(c)) as follows: if the class of x is different from the class of $nearest[x]$, then x is misclassified. In this case, if the distance from

---

Algorithm PFCNN2(T_i : a training set block)

1-8. The same as the PFCNN1 rule
  9. While the set ΔS is not empty:
   (a) For each element x in $T_i - S$, and for each element y in ΔS: if the distance between x and y is less than
        the distance from x and $nearest[x]$ then set $nearest[x]$ to y
   (b) Augment the set S with the set ΔS
   (c) For each element y in S, and for each class $j = 1, \ldots, m$: compute the sum $s[y, j]$ of all the elements x
        in T_i of the class j such that $nearest[x] = y$, together with their number $N[y, j]$
   (d) For each element y in S, and for each class $j = 1, \ldots, m$: $s[y, j] = $ **parallel–sum**$(s^1[y, j], \ldots,$
        $s^p[y, j])$, $N[y, j] = $ **parallel–sum**$(N^1[y, j], \ldots, N^p[y, j])$
   (e) For each element y in S, and for each class $j = 1, \ldots, m$: compute the center $c[y, j] = s[y, j]/N[y, j]$
   (f) For each element y in S, and for each class $j = 1, \ldots, m$: compute the element $C[y, j]$ in T_i of the class
        j such that $nearest[C[y, j]] = y$ which is closest to $c[y, j]$
   (g) For each element y in S, and for each class $j = 1, \ldots, m$: $C[y, j] = $ **parallel–**
        min$(\langle C^1[y, j], \mathrm{d}(C^1[y, j], c^1[y, j])\rangle, \ldots, \langle C^p[y, j], \mathrm{d}(C^p[y, j], c^p[y, j])\rangle)$
   (h) Initialize the set ΔS to the empty set
   (i) For each element y in S: set $rep[y]$ to undefined
   (j) For each element y is S: set $rep[y]$ to the point among $C[y, 1], \ldots, C[y, m]$ which is closest to y
   (k) For each element y is S: if $rep[y]$ is defined then insert $rep[y]$ into ΔS
 10. Return the set S

Fig. 4. The PFCNN2 rule

$nearest[x]$ to x is less than the distance from $nearest[x]$ to its current representative $rep[nearest[x]]$, then $rep[nearest[x]]$ is set to x.

At the end of each iteration, for each y in S, the elements $rep^i[y]$ of each node i are exploited to find the representative of the Voronoi enemies of y in the overall training set T (step 9.(e)). Indeed, for each y in S, its nearest enemy in T w.r.t. S is the closest point among its nearest enemies $rep^1[y]$, ..., $rep^p[p]$ w.r.t., respectively, $T_1, \ldots, T_p$. This closest point can be retrieved efficiently by using the parallel function **parallel–min** as shown in Figure 3.

Once the true representatives of the Voronoi enemies of each point in the current consistent subset S are computed, and stored into the array rep, the set ΔS is built with the points stored into the entries of the array rep. Notice that not all the entries of the array rep will be defined, since it might there exist points in S whose Voronoi cell contains only points of the same class.

PFCNN2 rule. Figure 4 shows the Parallel FCNN2 algorithm. We recall that this rule differs from the PFCNN1 for the definition of representative of the Voronoi enemies. In particular, the representative is defined as the closest class centroid. As for the data structures there employed, the training set block T_i, the consistent subset S, and the arrays $nearest$ and rep have the same semantics described in the previous section.

Steps 1-8 are the same as the PFCNN1 rule, while subsequent step 9 represents the main iteration of the algorithm. During each iteration, first of all, each element x in $(T_i - S)$ is compared with the elements y of ΔS, and the entry $nearest[x]$ of the array $nearest$ is updated to contain the element of S which is closest to x (step 9.(a)). Once the elements in ΔS have been compared with all the elements in $T_i - S$, the array rep can be updated. To this aim, steps 9(c)-(e) compute the centers $c[y, j]$ of the points of the Voronoi cell of y in T w.r.t. S having class label j, while subsequent steps 9(f)-(g) compute the centroids $C[y, j]$ of the points of the Voronoi cell of y in T w.r.t. S having class label j. Finally, steps 10(h)-(k) set the entries $rep[y]$ of rep to the centroid among $C[y, 1], \ldots, C[y, m]$ which is closest to y, and then build the new set ΔS.

In the following, variants of the two above described basic rules, namely the PFCNN-t, PFCNN-p, and PFCNN-b rules, are introduced.

PFCNN-t. If the distance employed satisfies the *triangular inequality*, then the number of distances computed by the PFCNN rules can be reduced. Indeed, since at the beginning of each iteration the distance from each object x of T_i to its current closest element $nearest[x]$ in S is known, this information can be exploited to compare each object x of T with a subset of ΔS instead of the entire set ΔS, thus saving distance computations. This subset will be composed only by the elements of ΔS candidate to be closer than $nearest[x]$ to x.

To this aim, for each y in S, the distances from y to the elements of ΔS are computed, and then these elements are sorted in order of increasing distance from y. Then, the elements of the Voronoi cell of y in T_i w.r.t. S, that is the elements of T_i such that $nearest[x] = y$, are compared with the elements in ΔS having distance from y less than twice the distance from x and y. Indeed, by the triangular inequality, they are all and the only elements of ΔS candidate to be closer to x than y. Notice that this strategy does not need to store together all the distances in the set $D = \{d(y, z) \mid y \in S, z \in \Delta S\}$. Indeed, while visiting the Voronoi cell of $y \in S$, only the distances among y and the elements of the set ΔS are needed.

We call PFCNN-t rule the method obtained by augmenting the PFCNN rule with the strategy above depicted. The PFCNN1-t and PFCNN2-t rules may reduce the number of distances computed w.r.t. the PFCNN1 and PFCNN2 rules, respectively, and thus accelerating their execution time. However, since the sets S and ΔS are identical in each node, it is the case that the same computation, i.e. the calculation of all the pairwise distances in the set D, will be carried out in each node. Though this strategy has the advantage of not requiring additional communications, this replicated computation may deteriorate the speed-up of the algorithm.

PFCNN-p and PFCNN-b. The PFCNN-t rules can be scaled-up by parallelizing even the computation of the distances in the set D and their sorting. To this aim, each node i can compute a disjoint subset of the distances in D, sort them, and then it can gather in a *single* communication the distances computed by any other node. We call PFCNN-p rule the PFCNN-t rule augmented with the strategy depicted above. Differently from the PFCNN-t rule, the PFCNN-p rule stores together all the distances in the set D, and, hence, depending on the characteristics of the dataset, it could require a huge amount of memory. As an example, if $|S| = 10^5$ and $|\Delta S| = 10^4$, then D is composed by one thousand million floating point numbers.

Memory consumption of the PFCNN-p rule can be alleviated, even if at the expense of *multiple* communications. To this purpose, S can be partitioned into b_n blocks, say them $B_1, \ldots, B_{b_n}$, having size b_s each. Then, the strategy of the PFCNN-p rule can be applied iteratively to each block B_h, $h = 1, \ldots, b_n$, and at the end of each iteration, i.e. after having used them, the distances $\{dist(y, z) \mid y \in B_h, z \in \Delta S\}$ can be discarded. We call PFCNN-b rule the PFCNN-p rule modified as described above. The effect of varying the size of the buffer on the two strategies will be discussed in the experimental results section.

5 Experiments

All the experiments were performed on a Linux cluster with 16 Itanium2 1.4GHz nodes each having 2 GBytes of main memory and connected by a Myrinet high performance network.

In order to compare the behavior of the different strategies, we considered a family of synthetically generated data sets, called *Checkerboard*. Each data set of the family is composed by two dimensional points into the unit square. A 4×4 checkerboard, ideally drawn onto the unit square, partitions the points in two classes associated to white and black cells of the board. We take into account data sets composed by one million points (each point is encoded with three words, two representing point coordinates and the last one representing the class label, for a total of 11MB), ten millions (114MB), twenty millions (229MB), and fifty millions of points (573MB).

The cost of the methods depends on the size of the set ΔS during the various iterations. Thus, it is interesting to preliminarily examine the course of ΔS. As for the PFCNN1, the number of iterations increased from about one hundred, for one million of points (1M), to about one thousand, for fifty millions of points (50M), while the size of ΔS remained always below 200. As for the PFCNN2, on the contrary, the number of iterations remained almost the same regardless the data set size, while the maximum size of ΔS increased sensibly: about $1,000$ for 1M, $4,000$ for 10M, $5,000$ for 20M, and $8,000$ for 50M.

Now we comment on the speedup curves of Figure 5. It is worth to notice that the PFCNN1 and PFCNN2 scale almost linearly. This confirms that the parallelization is very efficient. As for the triangular inequality based strategies, for all the dataset sizes, the PFCNN2-p outperforms the PFCNN2-t. This is due to the parallelization of the comparison among the elements of S and ΔS. The same is not true for the PFCNN1-t and PFCNN1-p that require almost the same amount of time (except for the smallest dataset, where the PFCNN1-p is sensibly better than the PFCNN1-t). This behavior is due to the fact that the size of ΔS is small over all the iterations and distance computation savings do not reward the additional communication overhead to be paid by the PFCNN1-p. Except for the PFCNN2, the different PFCNN2 strategies scale worst than the corresponding PFCNN1 strategies. Since on average the set ΔS computed by the PFCNN2 is much greater than the same set computed by the PFCNN1, then we

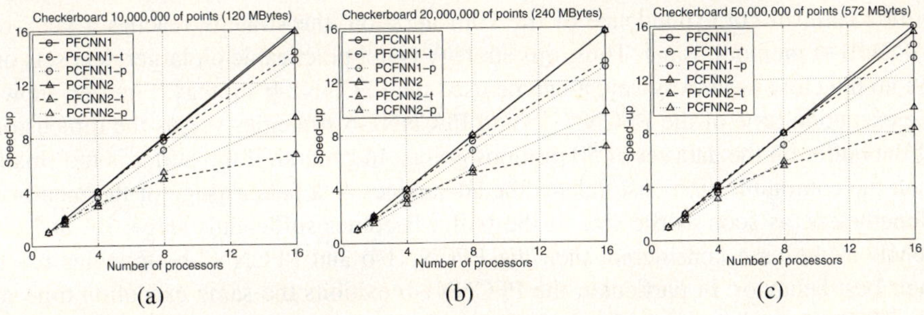

Fig. 5. Checkerboard dataset: speedup

Table 1. Checkerboard data set: execution time and memory usage

(a) Execution time for 50 millions of points.

	Seq	2	4	8	16
PFCNN1	134457.5	67229.2	33658.0	16839.6	8434.7
PFCNN1-t	39792.6	19896.7	9657.6	4978.6	2737.2
PFCNN1-p	–	23156.3	11154.8	5623.9	2939.6
PFCNN2	172798.6	86395.7	43201.5	21724.2	11161.2
PFCNN2-t	8253.3	4128.6	2204.3	1489.0	991.5
PFCNN2-p	–	5385.8	2639.9	1411.3	838.5

(b) Maximum (average in bracket) memory usage per node (MBytes) with $p = 16$.

	1M	10M	20M	50M
FCNN-Seq	19 (19)	191 (191)	382 (382)	954 (954)
PFCNN1-t	1 (1)	12 (12)	24 (24)	60 (60)
PFCNN1-p	3 (2)	15 (13)	35 (28)	75 (66)
PFCNN2	1 (1)	13 (13)	25 (25)	61 (61)
PFCNN2-t	1 (1)	13 (13)	25 (25)	61 (61)
PFCNN2-p	35 (8)	362 (73)	682 (141)	1788 (353)

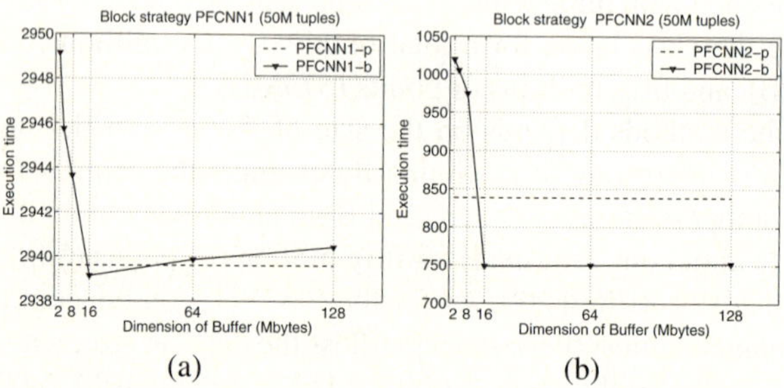

(a)   (b)

Fig. 6. Checkboard dataset: execution time vs buffer dimension

expect that the triangular inequality guarantees great savings on the PFCNN2 rule, and hence that the PFCNN2-t and PFCNN2-p strategies are faster than the PFCNN1-t and PFCNN1-p strategies, respectively. This behavior is confirmed by the execution time reported in Table 1(a). The same behavior cannot be observed on the PFCNN1 and PFCNN2. It can be concluded that, without the time savings guaranteed by the triangular inequality, the PFCNN2 is slower than the PFCNN1. As for the size of the condensed set, the PFCNN1 algorithm (the size of the subset S computed amounts to 0.08% for the 50M data set) shows a higher compression ratio than the PFCNN2 (the size of the subset S computed amounts to 0.10% for the 50M data set).

Table 1(b), shows the memory usage per node, assuming that 16 nodes are used. Interestingly, memory becomes critical only for the PFCNN2-p and when the data set consists of 50 millions of points. In fact, as memory depends on the factor $|S| \cdot |\Delta S|$, in this case, in correspondence of the 19th iteration, the strategy reaches a peak of 1788MB of memory usage. Thus, this strategy is not practicable on larger data sets on the architecture used. Anyway, it can be used the PFCNN2-b strategy. Figure 6 shows the execution time of the PFCNN1-b and PFCNN2-b strategies versus the dimension of the buffer on the data set of 50 millions points. In general, if the buffer is too small, then the communication cost deletes the advantages of a better usage of the memory. Nonetheless, as soon as the size of the buffer becomes sufficiently large, i.e. at least 16MB in the case considered, then the PFCNN1-b and PFCNN2-b strategies reach their best behavior. In particular, the PFCNN1-b exhibits the same execution time of the PFCNN1-p, since the buffer is sufficient to store all the distances between the elements of S and the elements of ΔS, being the latter set very small. Surprisingly, the

PFCNN2-b performs better than the PFCNN2-p. This can be explained since the efficient memory usage rewards the little overhead due to additional communications. As a result, the PFCNN2-b strategy terminates in about 750 seconds, which is the fastest time scored on this dataset, with a buffer of 16MB and a total memory usage of 67MB.

6 Conclusions and Future Work

A distributed algorithm for computing a consistent subset of a very large data set for the nearest neighbor decision rule has been presented and it is shown that it scales perfectly. To the best of our knowledge, this is the first distributed algorithm for computing a training set consistent subset for the nearest neighbor rule.

We briefly point out the strengths and weakness of the different strategies. The two basic strategies, PFCNN1 and PFCNN2, scale almost linearly and are suitable to distributed environment as computational grids. Triangular inequality based strategies (PFCNN-t, PFCNN-p, and PFCNN-b) further reduce execution time. The PFCNN-t is advantageous in grid environments, since it limits communication overhead, the PFCNN-p is more adapt to parallel architectures, while the PFCNN-b uses more efficiently the memory. Experiments performed on a parallel architecture, showed that the algorithm scales well both in terms of memory consumption and execution time. The algorithms were able to manage very large collections of data in a small amount of time; e.g. about 12 minutes to process a data set of about 0.6GB composed by fifty millions objects.

As a future work we are planning to validate the algorithm on real data sets and on a grid environments, to apply it to distributed sources of data, and to enhance the strategies with the management of disk resident data (actually, the basic PFCNN rules are very suitable for this kind of data).

References

1. Angiulli, F.: Fast condensend nearest neighbor rule. In: Proc. of the 22nd International Conference on Machine Learning, Bonn, Germany, pp. 25–32 (2005)
2. Cover, T.M., Hart, P.E.: Nearest neighbor pattern classification. IEEE Trans. on Inform. Th. 13(1), 21–27 (1967)
3. Devi, F.S., Murty, M.N.: An incremental prototype set building technique. Pat. Recognition 35(2), 505–513 (2002)
4. Devroye, L.: On the inequality of cover and hart in nearest neighbor discrimination. IEEE Trans. on Pat. Anal. and Mach. Intel. 3, 75–78 (1981)
5. Foster, I., Kesselman, C.: The Grid2: Blueprint for a New Computing Infrastructure. Morgan Kaufmann, San Francisco (2003)
6. Hart, P.E.: The condensed nearest neighbor rule. IEEE Trans. on Inform. Th. 14(3), 515–516 (1968)
7. Karaçali, B., Krim, H.: Fast minimization of structural risk by nearest neighbor rule. IEEE Trans. on Neural Networks 14(1), 127–134 (2002)
8. Stone, C.: Consistent nonparametric regression. Annals of Statistics 8, 1348–1360 (1977)

A Search Engine Accepting On-Line Updates

Mauricio Marin[1], Carolina Bonacic[2], Veronica Gil Costa[3], and Carlos Gomez[1]

[1] Yahoo! Research, Santiago, University of Chile
[2] ARTECS, Complutense University of Madrid, Spain
[3] DCC, University of San Luis, Argentina

Abstract. We describe and evaluate the performance of a parallel search engine that is able to cope efficiently with concurrent read/write operations. Read operations come in the usual form of queries submitted to the search engine and write ones come in the form of new documents added to the text collection in an on-line manner, namely the insertions are embedded into the main stream of user queries in an unpredictable arrival order but with query results respecting causality. The search engine is built upon distributed inverted files for which we propose generic strategies for load balance and concurrency control.

1   Introduction

The inverted file is a well-known index data structure for supporting fast searches on very large text collections. Web search engines use this strategy to index huge collections of text. A number of papers have been published reporting experiments and proposals for efficient parallel query processing upon inverted files which are distributed on a set of P processor-memory pairs [1,2,3,4,5,6]. It is clear that efficiency on clusters of computers is only achieved by using strategies devised to reduce communication among processors and maintain a reasonable balance of the amount of computation and communication performed by the processors to solve the search queries.

An inverted file is composed of a vocabulary table and a set of posting lists. The vocabulary table contains the set of relevant terms found in the collection. Each of these terms is associated with a posting list which contains the document identifiers where the term appears in the collection along with additional data used for ranking purposes. To solve a query, it is necessary to get the set of documents associated with the query terms and then perform a ranking of these documents so as to select the top K documents as the query answer.

The two dominant approaches to distributing the inverted file onto P processors are (**a**) the document partitioned strategy in which the documents are evenly distributed onto the processors and an inverted file is constructed in each processor using the respective subset of documents, and (**b**) the term partitioned strategy in which a single inverted file is constructed from the whole text collection to then distribute evenly the terms with their respective posting lists onto the processors. In addition, some strategies which are hybrids between these two schemes have been proposed to improve load balance. The way the inverted file is

A.-M. Kermarrec, L. Bougé, and T. Priol (Eds.): Euro-Par 2007, LNCS 4641, pp. 348–357, 2007.

partitioned onto the processors dictates the way in which the parallel processing of queries is performed.

Query operations over parallel search engines are usually read-only requests upon the distributed inverted file. This means that one is not concerned with multiple users attempting to write information on the same text collection. All of them are serviced with no regards for consistency problems since no concurrent updates are performed over the data structure. Insertion of new documents is effected off-line in Web search engines. However, it is becoming relevant to consider mixes of read and write operations.

For example, for a large news service we want users to get very fresh texts as the answers to their queries. Certainly we cannot stop the server every time we add and index a few news into the text collection. It is more convenient to let write and read operations take place concurrently. This becomes critical when one think of a world-wide trade system in which users put business documents and others submit queries using words like in Web search engines. Notice that the pageRank based ranking and lack of support for on-line updates promoted by Web search engines do not apply in this context. Also in this case it is feasible to keep all posting lists in main memory.

In this paper we present strategies and performance results for a search engine devised for this purpose. The search engine can be built upon the document or term partitioned approaches, and variations. We focus on the two most important factors affecting its performance, namely the load balancing strategies used to achieve good parallelism in query processing, and the concurrency control strategy used to avoid R/W conflicts in the underlying distributed inverted file. The contribution of this paper is that in both cases we propose efficient strategies and evaluate their comparative performance against alternative strategies. Another contribution of this paper is the design of the search engine itself, a design that (**i**) decouples ranking from posting list treatment in order to achieve a high rate of queries per second and (**ii**) is able to accommodate an efficient concurrency control algorithm by ways of the bulk-synchronous organization of the ranking and posting lists fetching processes. To the best of our knowledge no search engine which is able to cope efficiently with concurrent read/write operations has been proposed in the literature.

The remainder of this paper is organized as follows. Section 2 describes our search engine. Section 3 discusses load balancing strategies and section 4 concurrency control algorithms. Section 5 presents conclusions.

2   Search Engine

The parallel processing of queries is basically composed of a phase in which it is necessary to fetch parts of all of the posting lists associated with each term present in the query, and perform a ranking of documents in order to produce the results. After this, additional processing is required to produce the answer to the user. This paper is concerned with the fetching+ranking part and we are interested in situations in which a high traffic of queries is arriving at the search

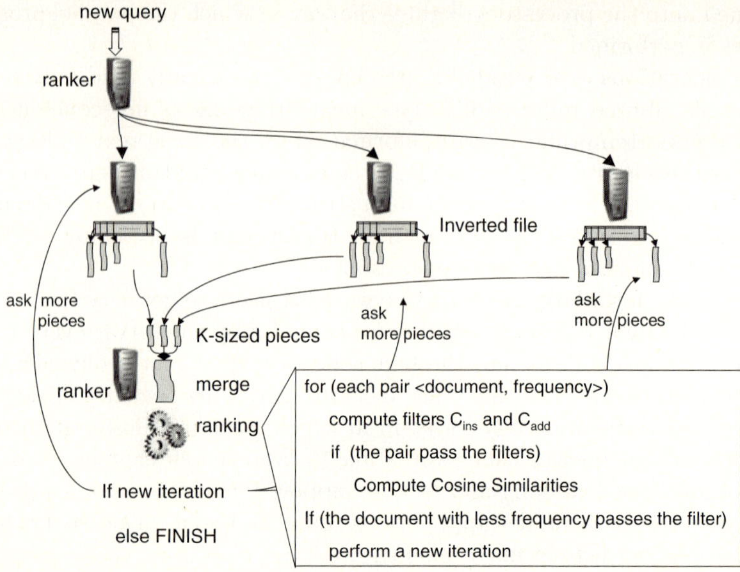

Fig. 1. Search engine: list fetching, ranking and iterations

engine and it is relevant to optimize the query throughput. At the parallel server side, queries arrive from a receptionist machine that we call the *broker*.

The broker machine is in charge of routing the queries to the cluster's processors and receiving the respective answers. It decides to which processor routing a given query by using a load balancing heuristic. The particular heuristic depends on the approach used to partition the inverted file. Overall the broker tends to evenly distribute the queries on all processors.

The processor in which a given query arrives is called the *ranker* for that query since it is in this processor where the associated document ranking is performed. Every query is processed using two major steps: the first one consists on fetching a K-sized piece of every posting list involved in the query and sending them to the ranker processor.

In the second step, the ranker performs the actual ranking of documents and, if necessary, it asks for additional K-sized pieces of the posting lists in order to produce the K best ranked documents that are passed to the broker as the query results. We call this *iterations*. Thus the ranking process can take one or more iterations to finish. In every iteration a new piece of K pairs (doc_id, frequency) from posting lists are sent to the ranker for every term involved in the query. See figure 1.

Under this scheme, at a given interval of time, the ranking of two or more queries can take place in parallel at different processors along with the fetching of K-sized pieces of posting lists associated with other queries. We assume a situation in which the query arrival rate in the broker is large enough to let the broker distribute QP queries onto the P processors.

We use the vectorial method for performing the ranking of documents along with the filtering technique proposed in [7]. Consequently, the posting lists are kept sorted by frequency in descending order. Once the ranker for a query receives all the required pieces of posting lists, they are merged into a single list and passed throughout the filters. If it happens that the document with the least frequency in one of the arrived pieces of posting lists passes the filter, then it is necessary to perform a new iteration for this term and all others in the same situation. We also provide support for performing the intersection of posting lists for boolean AND queries. In this case the ranking is performed over the documents that contain all the terms present in the query.

The search engine is implemented on top of the BSP model of parallel computing [8] as follows. In BSP the computation is organized as a sequence of *supersteps*. During a superstep, the processors may perform computations on local data and/or send messages to other processors. The messages are available for processing at their destinations by the next superstep, and each superstep is ended with the barrier synchronization of the processors. The underlying communication library ensures that all messages are available at their destinations before starting the next superstep. Thus at the beginning of each superstep the processors get into their input message queues both new queries placed there by the broker and messages with pieces of posting lists related to the processing of queries which arrived at previous supersteps. The processing of a given query can take two or more supersteps to be completed. All messages are sent at the end of every superstep and thereby they are sent to their destinations packed into single messages to reduce communication overheads. In addition, in the input message queues are requests to index new documents and merge them into the distributed inverted file the resulting pairs (id_doc,frequency).

2.1   Experiments

We use two text databases. The first one is a 2GB (and 12GB) sample of the Chilean Web taken from the `www.todocl.cl` search engine. The text is in Spanish. Using this collection we generated a 1.5GB index structure with 1,408,447 terms. Queries were selected at random from a set of 127,000 queries taken from the todocl log. The second collection is a set of crawled documents along with a query log from an experimental search engine at Yahoo! Labs. We took 300,000 random queries from the query log making sure that all terms are in the document collection. In practice we did not observe any significant differences in the results from both text collections and thereby in this paper we report results from the first one.

The experiments were performed on a cluster with dual processors (2.8 GHz) that use NFS mounted directories. This system has 2 racks of 6 shelves each with 10 blades to achieve 120 processors.

In every run we process 10,000 queries in *each* processor. That is the total number of queries processed in each experiment reported below is 10,000 P. For our collection the values of the filters C_{ins} and C_{add} were both set to 0.1 and we set K to 1020. On average, the processing of every query finished with

$0.6K$ results after 1.5 iterations. Before measuring running times and to avoid any interference with the file system, we load into main memory all the files associated with queries and the inverted file.

3   Load Balance

Key to the efficiency of search engines is the data structure used to support queries over the text collection. Inverted files are used for this purpose, in particular most search engines distribute the inverted file onto the processors using the document partitioned approach. The main advantages of this scheme are its suitability for boolean AND queries and simplicity of maintenance since inserting a new document requires updating the piece of inverted file stored in one processor only. Typically a certain number of documents are accumulated and inserted into a copy of the inverted file to then replace it for the new version.

However, the term partitioned approach has the advantage of being able to achieve a better performance and has deserved attention in the literature. Its main disadvantage lays in the problem of calculating the intersection of posting lists for AND queries. This because as terms are distributed onto the processors, the pairs (doc_id, frequency) tend to be in different processors with high probability, demanding as a result more communication to calculate intersections. How costly can be this increase in communication is matter of term distribution as we show in the following.

In figure 2 [left] we show a term distribution which makes more likely term co-residence for intersections using 32 processors. We obtained the normalized frequencies for each term distribution. From the query log, we obtain all possible pairs and compute a "hit" for each term doing as follows. If terms t_i and t_j are terms of a query q, we add 1 to the frequency of each term if t_i and t_j are in the same processor. The figure 2 [left] shows that the distribution of terms on processors labeled by v3 increases the probability of co-residence. In this case from the query log we detect the P^2 most frequent terms and distribute them circularly onto the processors.

For the rest of the terms we determine the processor in which they are stored by using the following strategy. From a query log we count the frequency in which pairs of terms (t_1, t_2) appear in queries (all permutations are considered for queries with three or more terms). We form a list sorted by decreasing frequency order. Then the pairs (t_1, t_2) are removed one by one from the front of the list and are placed on the processors using the following rules. If both t_1 and t_2 have not been assigned to a processor, place t_1 and t_2 in the processor with the least number of pairs. If one of the terms is already assigned to a processor, its companion goes to the same processor. This defines the initial allocation of terms to processors. We then use *tabu* search to improve on that. The method moves terms among processors by trying to reduce the BSP cost of executing the query log and using an optimization criteria to prune the search space. Distribution v1 applies this method without first distributing circularly the most frequent terms and distribution v2 distributes circularly the only first P most frequent terms.

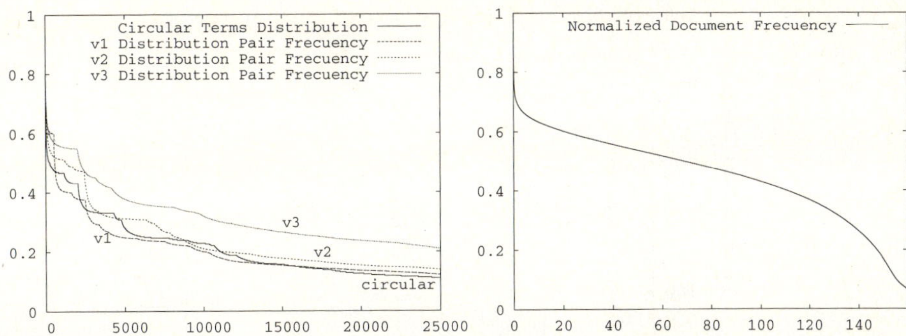

Fig. 2. Probability of processor co-residence for the most frequent query terms and documents containing terms in the query log. [left] the x-axis shows the normalized frequency of co-residence for the 25K most frequent pairs of terms in queries (0 is the most frequent one). [right] the x-axis shows documents sorted from the one containing most of the query terms to the one with least ones (values in thousands of docs).

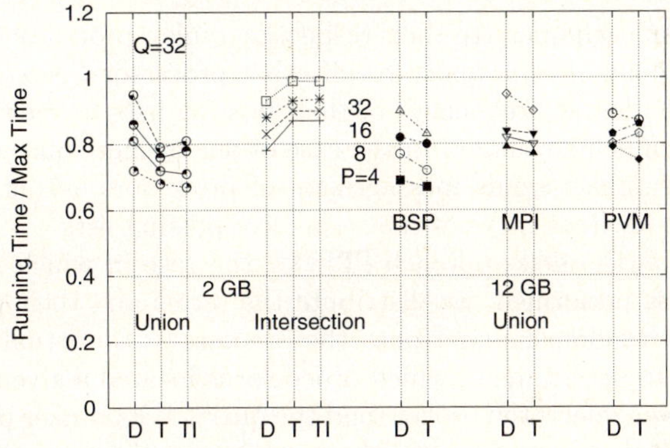

Fig. 3. Normalized running times for 4, 8, 16 and 32 processors under queries requiring union and intersection of posting lists. For the term partitioned index the intersection operation is performed under circular distribution of terms ('T') and the v3 distribution of terms (TI). The document partitioned approach is denoted by D. The first part shows results for BSP over a 2GB text collection whereas the second part shows results for a 12GB text collection and inverted files implemented using the BSP, MPI and PVM communication libraries. Top curves are for $P= 32$ and bottom ones are for $P= 4$.

Figure 2 [right] is more useful in the context of the next section as it shows for each document the amount of relevant terms that are present in the query log. This indicates that there is a high probability that during the insertion of a new document in the collection some of its terms are likely to be identical to those coming from the user queries.

For the document partitioned approach (**D**) the load balancing heuristic is simple. The ranker is selected in a circular manner among the P processors. The broker performs a broadcast of every query to all the processors. But it does so in two steps. First and exactly as in the term partitioned approach, the broker

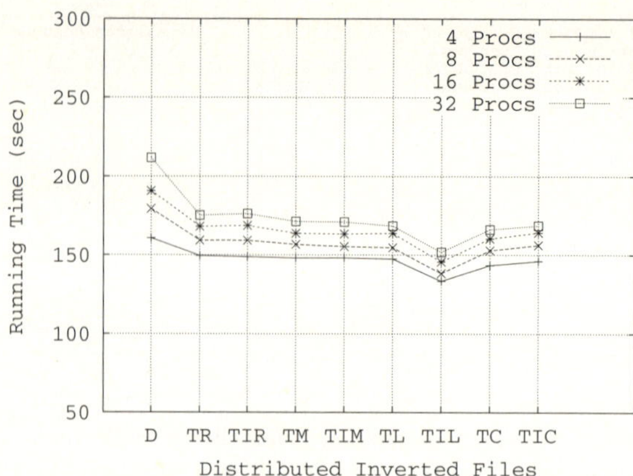

Fig. 4. Running times for 4, 8, 16 and 32 processors for different alternatives for load balancing computation and communication

sends one copy of each query to their respective ranker processors. Secondly, the ranker sends a copy of each query to all other processors. Next, all processors send K/P pairs (doc_id, frequency) of their posting lists to their rankers which perform the documents ranking. In the case of one or more query terms passing both filters, the ranker sends messages to all processors asking for additional K/P pairs (doc_id, frequency) of the respective posting lists.

In the term partitioned approach (**T**), for the case in which no intersection of posting lists is performed, we distribute the terms and their posting lists in an uniformly at random manner onto the processors (we actually use the rule id_term mod P to determine in which processor is located a given term). In this scheme, for a given query, the broker send the query to its ranker processor which upon reception sends messages to others asking for the first K pairs (doc_id, frequency) of every term present in the query. The same is repeated if one or more terms pass the filters.

In figure 3 we show results for the document and term partitioned approaches for distributed inverted files. We show running times for union and intersection of posting lists during ranking. For the term partitioned index if the terms are co-resident their posting lists are intersected locally. When the terms are not co-resident the complete smallest posting list is sent to the other processor. If the documents in the intersection pass the filters new pieces of posting lists are requested. In the document partitioned approach the intersections are always effected locally. We also present a comparison with inverted files implemented by using the asynchronous message passing approach to parallel computing upon MPI and PVM. Both indexes achieve competitive performance.

In figure 4 we show different strategies we devised to improve load balance in the term (T) partitioned approach and compare its running against the document (D) partitioned strategy. No intersections are performed. The letter R stands for ranker distributed circularly among the processors. Letter I indicates distribution v3 to improve locality for intersection operations. Letter M stands for upper limits to the number of ranking operations performed in every processor in each superstep. Letter L indicates limits only to the number of list fetches allowed in each processor and superstep. For both cases we used limit= $2.5\,Q$. Letter C indicates a case in which rankers keep in cache memory the pieces of posting lists arrived from other processors. The results show that the term partitioned strategy achieves good efficiency when we decouple ranking from list fetching. For ranking a good strategy is to distribute rankers circularly among the processors. The figure also shows that imposing upper limits to the number of list fetching performed in each superstep and processor can further improve running time. Performance also improves from the fact that in the strategies "I" we distribute circularly the most frequent terms onto the processors.

4 Concurrency Control

An important component of the proposed search engine is its support for concurrent write and read operations as new documents are expected to be arriving from the broker machine along with queries in an unpredictable manner. In the following we describe a very efficient strategy to support concurrent R/W operations over the document and term partitioned bulk-synchronous inverted files.

A first point to note is that the semantics of supersteps tell us that all messages are in their target processors at the start of each superstep. That is, no messages are in transit at that instant and all the processors are barrier synchronized. If the broker assigns a unique integer timestamp to every R/W operation that it sends to the processors, then it suffices to process all messages in timestamp order in each processor to avoid R/W conflicts. These timestamps are also used as the ids for the respective queries. To this end, every processor maintains its input message queue organized as a priority queue with keys given by id_query integer values. The broker selects the processors to send new documents in a circular manner. Upon reception, a document is parsed to extract all its relevant terms.

In the document partitioned inverted file the posting lists of all the parsed terms are updated locally in the same processor. However, this is not effected in the current superstep but in the next one in order to wait for the arrival of broadcasted terms which could have timestamps smaller than the one associated to the current document insertion operation. Instead the processor sends to itself a message in order to wait one superstep to proceed with the updating of the posting lists.

In the term partitioned inverted file the arrival of a new document to a processor is much more demanding in communication. This because once the document is parsed a pair (id_doc,frequency) for each term has to be sent to every processor holding the posting list of the respective terms. However, insertion of new document is expected to be comparatively less frequent than queries in real-life settings.

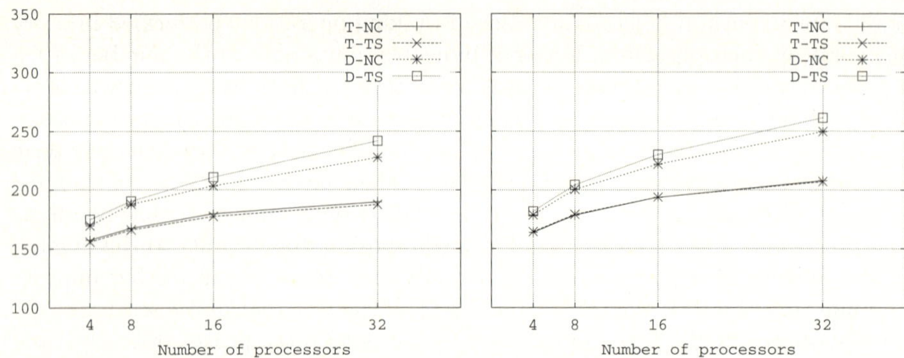

Fig. 5. Running times (sec) for processing $10,000$ queries in each processor by inserting $Q = 32$ new queries (read operations) in each superstep and processor respectively. With probability p a write operation is generated and new the document is formed with 100 words randomly generated from the *same* query log to increase R/W conflicts. D and T stand for document and term partitioned inverted files, and NC and TS stand for no-concurrency and timestamped-concurrency control respectively. Figure [left] shows results for a probability of document arrival per user query of $p=0.25$ whereas figure [right] shows results for probability $p=0.5$ (y-axis is the same to [left]).

A complication arises from the fact that the processing of a given query can take several iterations (supersteps). A given pair (id_doc,frequency) cannot be inserted in its posting list by a write operation with timestamp larger than the query being solved throughout several iterations. We solve this by keeping aside this pair for those queries and logically including the pair in the posting list for queries with timestamps larger than the pair one. In practice the pair is physically inserted in the posting list in frequency descending order but it is not considered by queries with smaller timestamps.

In figure 5 we show performance results for the concurrency control strategy. The figure shows the running time for both the document and term partitioned inverted files under a situation of doing nothing to prevent read/write conflicts and the case in which these conflicts are avoided using the bulk-synchronous timestamp based strategy. The results shows that the running time of the strategy is very similar to the running time of doing nothing, that is, a case in which R/W conflicts never take place and thereby it represents the "fastest" concurrency control algorithm.

5   Conclusion

We have presented the design of a search engine which provides an efficient support for the on-line arrival of new documents. The design can accommodate both the document and term partitioned approaches to distributing inverted files onto the processors. We have proposed efficient implementations of both approaches and found that even for intersection operations the term partitioned approach performs comparatively well whereas for cases in which this operation is not a requirement this strategy achieves a very high rate of queries per second.

It is not clear that for the type of application our search engine is intended for the intersection operations are a firm requirement for all queries. Moreover, an half-the-way case can be to perform the ranking of the K-sized pieces of posting lists without performing any intersection and once a sufficient amount of well ranked documents is reached (say $2K$), the ranker assigns the highest scores to the document that contain all the query terms. This is an issue we plan to explore in the near future.

Also, the adoption of the BSP model of computing allowed us to solve in a trivial manner the problem of controlling concurrency for read and write operations. The experimental results show that this strategy is very efficient.

Acknowledgment. Partially funded by Cyted Project 505PI0058.

References

1. Badue, C., Baeza-Yates, R., Ribeiro, B., Ziviani, N.: Distributed query processing using partitioned inverted files. Eighth Symposium on String Processing and Information Retrieval (SPIRE'01), pp. 10–20 (November 2001)
2. Buttcher, S., Clarke, C.: Indexing time vs. query time trade-offs in dynamic information retrieval systems. In: International Conference on Information and Knowledge Management, pp. 317–318 (2005)
3. MacFarlane, A., McCann, J., Robertson, S.: Parallel search using partitioned inverted files. In: 7th International Symposium on String Processing and Information Retrieval, pp. 209–220. IEEE Computer Society Press, Los Alamitos (2000)
4. Moffat, W., Webber, J., Zobel, Baeza-Yates, R.: A pipelined architecture for distributed text query evaluation. Information Retrieval (October 5, 2006)
5. Orlando, S., Perego, R., Silvestri, F.: Design of a parallel and distributed web search engine. In: Proc. 2001 Parallel Computing Conf., pp.197–204 (2001)
6. Zobel, J., Moffat, A.: Inverted files for text search engines. ACM Computing Surveys 38(2) (2006)
7. Persin, M., Zobel, J., Sacks-Davis, R.: Filtered document retrieval with frequency-sorted indexes. Journal of the American Society for Information Science 47(10), 749–764 (1996)
8. Valiant, L.G.: A bridging model for parallel computation. Comm. ACM 33, 103–111 (1990)

Topic 6
Grid and Cluster Computing

Rosa M. Badia, Christian Pérez, Artur Andrzejak, and Alvaro Arenas

Topic Chairs

With some years of research and development Grid computing is starting to be a mature subject with relevant methodologies, software and tools available both for the academia and industry. Grid computing represents the culmination of truly general distributed computing across various resources in a ubiquitous, open-ended infrastructure to support a wide range of different application areas. However, there is still many areas in which further research is required to achieve a user-friendly, efficient, secure and reliable grid.

For this edition of the EuroPar conference, the Grid topic called for papers that present research and results in the areas of: Grid middleware, Resource/Service/Information discovery, Resource management and scheduling, Grid programming models, tools, and algorithms, Dependability, adaptability, and scalability, Security for grids, Monitoring and event notification for grids, Workflow management, Grid accounting and economics, Innovative Applications for Grid and Cluster Computers, Desktop Grids, and Automated management of resources and applications.

The topic received a large amount of papers (48) from which 11 where selected for the program of the conference following a high quality review process where all received at least 3 reviews. From these lines, we would like to thank all the reviewers that helped in this review process.

The selected papers feature topics from the areas of workflows, desktop grids, fault-tolerance, grid scheduling, symbolic computation, characterization of groups of jobs, simple and efficient execution of distributed applications, decission making in cross-grid environments, resource ranking, and P2P Data-dissemination techniques.

The chairs of the topic invite you to read and enjoy this collection of papers.

A.-M. Kermarrec, L. Bougé, and T. Priol (Eds.): Euro-Par 2007, LNCS 4641, p. 359, 2007.
© Springer-Verlag Berlin Heidelberg 2007

Characterizing Result Errors
in Internet Desktop Grids

Derrick Kondo[1], Filipe Araujo[2], Paul Malecot[1], Patricio Domingues[3],
Luis Moura Silva[2], Gilles Fedak[1], and Franck Cappello[1]

[1] INRIA Futurs, France
[2] University of Coimbra, Portugal
[3] Polytechnic Institute of Leiria, Portugal

Abstract. Desktop grids use the free resources in Intranet and Internet environments for large-scale computation and storage. While desktop grids offer a high return on investment, one critical issue is the validation of results returned by participating hosts. Several mechanisms for result validation have been previously proposed. However, the characterization of errors is poorly understood. To study error rates, we implemented and deployed a desktop grid application across several thousand hosts distributed over the Internet. We then analyzed the results to give quantitative and empirical characterization of errors stemming from input or output (I/O) failures. We find that in practice, error rates are widespread across hosts but occur relatively infrequently. Moreover, we find that error rates tend to not be stationary over time nor correlated between hosts. In light of these characterization results, we evaluated state-of-the-art error detection mechanisms and describe the trade-offs for using each mechanism.

1   Introduction

Desktop grids use the free resources in Intranet and Internet environments for large-scale computation and storage. For over 10 years, desktop grids have been one of the largest distributed systems in the world providing TeraFlops of computing power for applications from a wide range of scientific domains, including climate prediction, computational biology, and physics [1]. Despite the huge computational and storage power offered by desktop grids and their high return on investment, there are several challenges in using this volatile and shared platform effectively.

One critical issue is the validation of results computed by volatile and possibly malicious hosts. In large and complex distributed systems, errors in results are inevitable, and errors can stem from different sources. Some sources can be computational. For example, an error could result from a CPU miscalculation due to overclocking and overheating [2]. Other sources can be related to failures during application input or output (I/O). For example, if a machine crashes when the application is writing to an output file or checkpoint, only a partial number in-memory data blocks could have been flushed to disk (not necessarily

A.-M. Kermarrec, L. Bougé, and T. Priol (Eds.): Euro-Par 2007, LNCS 4641, pp. 361–371, 2007.
© Springer-Verlag Berlin Heidelberg 2007

in order) [3], which would lead to an erroneous result[1]. Thus, effective error detection mechanisms are essential, and several methods have been proposed previously [7,8].

However, little is known about the nature of errors in real systems. Yet, the trade-offs and efficacy among different error detection mechanisms are dependent on how errors occur in real systems. Thus, we focus on characterizing errors, specifically I/O errors, in a real system by addressing the following critical questions. What is the frequency and distribution of host I/O error rates? How stationary are host I/O error rates? How correlated are I/O error rates between hosts? In light of the error characterization, what is the efficacy of state-of-the-art error detection mechanisms?

To help answer those questions, we deployed an Internet desktop grid application across several thousand desktop hosts. We then validated the results returned by hosts, and we analyzed the invalid results to characterize quantitatively the I/O error rates in a real desktop grid project.

2   Background

At a high-level, a typical desktop grid system consists of a server from which **workunits** of an **application** are distributed to a **worker** daemon running on each participating host. The workunits are then executed when the CPU is available, and upon completion, the **result** is returned back to the server. We define a result **error** to be any result returned by a worker that is not the correct value or within the correct range of values. We call any host that has or will commit at least one error an **erroneous host** (whether intentionally or unintentionally).

Workunits of an application are often organized in groups of workunits or **batches**. To achieve overall low rates for a batch of tasks, the individual error rate per host must be made small. Consider the following scenario described in [7] where a computation consists of 10 batches, each with 100 workunits. Assuming that any work unit error would cause the entire batch to fail, then to achieve an overall error rate of 0.01, the probability of a result being erroneous must be no greater than 1×10^{-5}. Many applications (for example, those from from computational biology [2] and physics [1]) require (low) bounds on error rates as the correctness of the computed results is essential for making accurate scientific conclusions.

3   Related Work

To the best of our knowledge, there has been no previous study that gives quantitative estimates of error rates from empirical data. Nevertheless, several mechanisms for reducing errors in desktop grids have been proposed. We discuss three

---

[1] While a number of mechanisms exist to ensure atomic writes or to detect file corruption, in practice, few if any are provided by desktop grid systems [4,5,6] or utilized by desktop grid applications themselves.

of the most common state-of-the-art methods [7,8,2] namely spot-checking, majority voting, and credibility-based techniques, and emphasize the issues related to each method.

The **majority voting** method detects erroneous results by sending identical workunits to multiple workers. After the results are retrieved, the result that appears most often is assumed to be correct. In [7], the author determines the amount of redundancy for majority voting needed to achieve a bound on the frequency of voting errors given the probability that a worker returns a erroneous result. Let the error rate φ be the probability that a worker is erroneous and returns an erroneous result unit, and let ε be the percentage of final results (after voting) that are incorrect.

Let m be the number of identical results out of $2m - 1$ required before a vote is considered complete and a result is decided upon. Then the probability of an incorrect result being accepted after a majority vote is given by:

$$\varepsilon_{majv}(\varphi, m) = \sum_{j=m}^{2m-1} \binom{2m-1}{j} \varphi^j (1 - \varphi)^{2m-1-j} \tag{1}$$

The redundancy of majority voting is $\frac{m}{1-\phi}$.

A more efficient method for error detection is **spot-checking**, whereby a workunit with a known correct result is distributed at random to workers. The workers' results are then compared to the previously computed and verified result. Any discrepancies cause the corresponding worker to be **blacklisted**, i.e., any past or future results returned from the erroneous host are discarded (perhaps unknowingly to the host).

Erroneous workunit computation was modelled as a Bernoulli process [7] to determine the error rate of spot-checking given the portion of work contributed by the host, and the rate at which incorrect results are returned. The model uses a work pool that is divided into equally sized batches.

Allowing the model to exclude coordinated attacks, let q be the frequency of spot-checking, let n be the amount of work contributed by the erroneous worker, let f be the fraction of hosts that commit at least 1 error, and let s be the error rate per erroneous host. $(1 - qs)^n$ is the probability that an erroneous host is not discovered after processing n workunits. The rate which spot-checking with blacklisting will fail to catch bad results is given by:

$$\varepsilon_{scbl}(q, n, f, s) = \frac{sf(1 - qs)^n}{(1 - f) + f(1 - qs)^n} \tag{2}$$

The amount of redundancy of spot-checking is given by $\frac{1}{1-q}$.

To address the potential weaknesses of majority voting and spot-checking, **credibility-based systems** were proposed [7], which use the conditional probabilities of errors given the history of host result correctness. The idea is based on the assumption that hosts that have computed many results with relatively few errors have a higher probability of errorless computation than hosts with a history of returning erroneous results. Workunits are assigned to hosts such that more attention is given to the workunits distributed to higher risk hosts.

To determine the credibility of each host, any error detection method such as majority voting, spot-checking, or various combinations of the two can be used. The credibilities are then used to compute the conditional probability of a result's correctness.

Note that the methods described in this section were designed to detect errors from any source, including a computational source (for example, CPU miscalculations) or an I/O source (for example, a crash during a checkpoint). In the sections that follow, we determine the methods' efficacy in detecting errors caused by I/O failures.

4   Method

We studied the error rates of a real Internet desktop grid project called XtremLab [9]. XtremLab uses the BOINC infrastructure [4] to collect measurement data of desktop resources across the Internet. The XtremLab application currently gathers CPU availability information by continuously computing floating point and integer operations, and every 10 seconds, the application will write the number of operations completed to file. Every 10 minutes, the output file is uploaded to the XtremLab server.

In this study, we analyze the outputs of the XtremLab application to characterize the rate at which errors can occur in Internet-wide distributed computations. In particular, we collected traces between April 20, 2006 to July 20, 2006 from about 4400 hosts. From these hosts, we obtained over 1.3×10^8 measurements of CPU availability from 2.2×10^6 output files. We focused our analysis on about 600 hosts with more then 1 week worth of CPU time in order to ensure the statistic significance of our conclusions .

Errors in the application output are determined as follows. Output files uploaded by the workers are processed by a validator. The validator conducts both syntactical and semantics checks of the output files returned by each worker. The syntactical checks verify the format of the output file (for example, that the time stamps recorded were floating numbers, the correct number of measurements were made, and each line contains the correct number of data). The semantic checks verify the correctness of the data to ensure that the values reported fall in the range of feasible CPU availability values. Any output files that failed these checks were marked as erroneous. After careful inspection of output files that failed synatatic or semantic checks, we found that most of these output files were truncated prematurely or had scrambled output (perhaps due to an out-of-order flush of in-memory blocks [3], for example). As such, we assume that any output file that fails a syntactic or semantic check would have corresponded to an I/O error of an application (for example, an erroneous checkpoint) that would lead to an erroneous result[2]. To date, there has been little specific

---

[2] We do not believe the errors are due to network failures during transfers of output files. This is because BOINC has a protocol to recover from failures (of either the worker of server) during file transmission that ensures the integrity of files transfers [4].

data about error rates in Internet desktop environments, and we believe that
the above detection method gives a first-order approximation of the I/O error
rates for a real Internet desktop grid project.

5   Error Characterization

5.1   Frequency of Errors

The effectiveness of different methods by which errors are detected is heavily
dependent on the frequency of errors among hosts. We measured the fraction of
workunits with errors per host, and show the cumulative distribution function
(CDF) of these fractions in Figure 1.

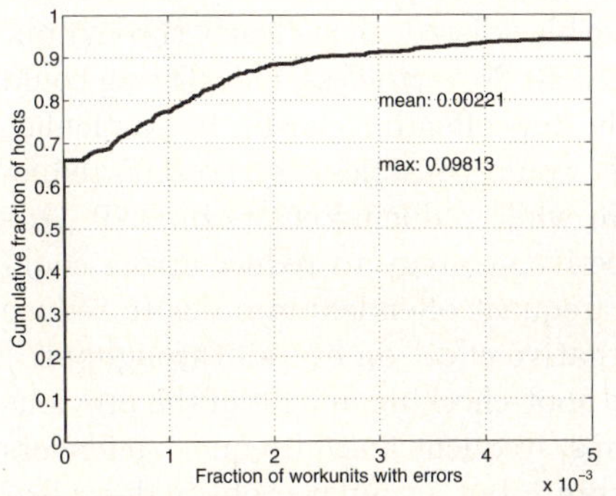
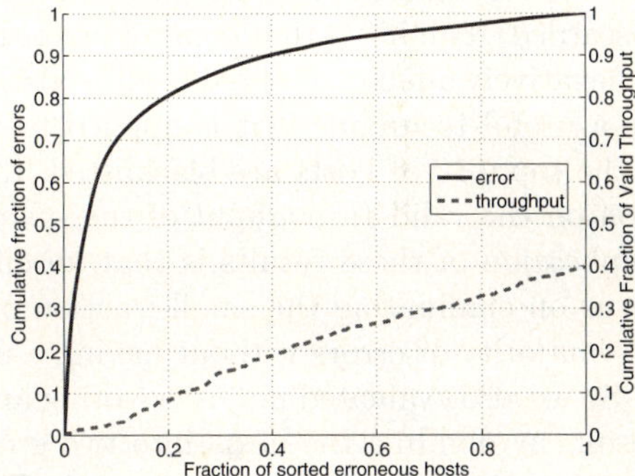

Fig. 1. Error Rates of Hosts in Entire
Platform

Fig. 2. Cumulative Error Rates and Effect
on Throughput

We find that a remarkably high percentage of hosts (about 35%) returned
at least one corrupt result in the 3 month time frame. Given that the overall
error rate is low and a significant fraction of hosts result in at least one error,
blacklisting all erroneous hosts may not be an efficient way of preventing errors.

An error rate of 0.002 may seem so low that error correction, detection and
prevention are moot, but consider the scenario in Section 2 again where the
desired overall error rate is 0.01. In that case, the probability of a result being
erroneous must be no greater than 1×10^{-5}. If $\varphi = 0.002$ as shown in Figure 1,
we can simply conduct a majority vote where $m = 2$ to achieve an error rate of
about 1×10^{-5} and with a redundancy of about 2.0.

While spot-checking can achieve a similar error rate of about 1×10^{-5}, spot-
checking requires a large number of workunits to be processed before achieving
it. For example, to achieve a similar error rate of 1×10^{-5} via spot-checking
where $q = 0.10$, $f = 0.35$ (from Figure 1), $s = 0.003$ (as shown in Table 1),
Equation 2 requires that the number of workunits (n) processed by each worker
be *greater* than 5300. While redundancy is lower at 1.11 compared to majority
voting, if each workunit requires 1 day of CPU time (which is a conservative

estimate as shown in [1]), it would require *at least* 14.5 years of CPU time *per worker* before the desired rate could be achieved. Even if we increase q to 0.25 (and redundancy is 1.33), spot-checking requires $n = 3500$ (or at least 9.5 years of CPU time per worker assuming a workunit is 1 day of CPU time in length).

Figure 2 shows the skew of the frequency of errors among those erroneous hosts. In particular, we sort the hosts by the total number of errors they committed, and the blue, solid plot in Figure 2, shows the cumulative fraction of errors. For example, the point (0.10, 0.70) shows that the top 0.10 of erroneous hosts commit 0.70 of the errors. Moreover, the remaining 0.90 of the hosts cause only 0.30 of the errors. We refer to the former and latter groups as **frequent** and **infrequent offenders**, respectively.

Figure 2 also shows the effect on throughput if the top fraction of hosts are blacklisted, assuming than an error is detected immediately and that after the error is detected, all workunits that had been completed previously by the host are discarded. If all hosts that commit errors are blacklisted, then clearly throughput is negatively affected and reduced by about 0.40. Nevertheless, blacklisting could be a useful technique if it is applied to the top offending hosts. In particular, if the top 0.10 of hosts are blacklisted, this would cause less than a 0.05 reduction on the valid throughput of the system while reducing errors by 0.70. One implication of these results is that an effective strategy to reduce errors could focus on eliminating the small fraction of frequent offenders in order to reduce the majority of errors without having a negative effect on overall throughput.

So we also evaluated majority voting and spot-checking in light of the previous result, by dividing the hosts into two groups, frequent and infrequent offenders based on the knee of the curve shown in Figure 2, but a similar problem described earlier occurs. The error rate for majority voting ε_{majv} is given by Equation 1, where $\varphi = f_{all} \times s_{frequent} \times f_{frequent} + f_{all} \times s_{infrequent} \times f_{infrequent}$. Note that f_{all} is simply the fraction of workers that could result in at least one error (0.35). $s_{frequent}$ (0.0335) and $s_{infrequent}$ (0.001) (see Table 1) are the error rates for frequent and infrequent offenders respectively. $f_{frequent}$ (0.10) and $f_{infrequent}$ (0.90) are the fraction of erroneous workers in the frequent and infrequent groups respectively.

We plot ε_{majv} as a function of m in Figure 3(a). We find that the error rate ε_{majv} decreases exponentially with m, beginning at about 1×10^{-5} for $m = 2$.

We also compute the error rate for spot-checking with blacklisting when dividing the hosts in terms of frequent and infrequent offenders. The error rate ε_{scbk} is given by the sum of the error rates for each grouping, $\varepsilon_{scbk,frequent}$ and $\varepsilon_{scbk,infrequent}$. Note that $\varepsilon_{scbk,infrequent}$ is given by substituting $f_{all} \times f_{frequent}$ for f and $s_{frequent}$ for s in Equation 2. $\varepsilon_{scbk,infrequent}$ can be calculated similarly.

Then we plot in Figure 3(b) $\varepsilon_{scbk,frequent}$, $\varepsilon_{scbk,infrequent}$, and ε_{scbk} as a function of n (the number of workunits that must be computed by each worker) where $q = 0.10$. The plot for the frequent offenders decreases exponentially; this is because the error rate for the hosts is relatively high, and so after a series of workunit computations, the erroneous hosts are rapidly detected. The plot for the infrequent offenders decreases very little even as n increases significantly.

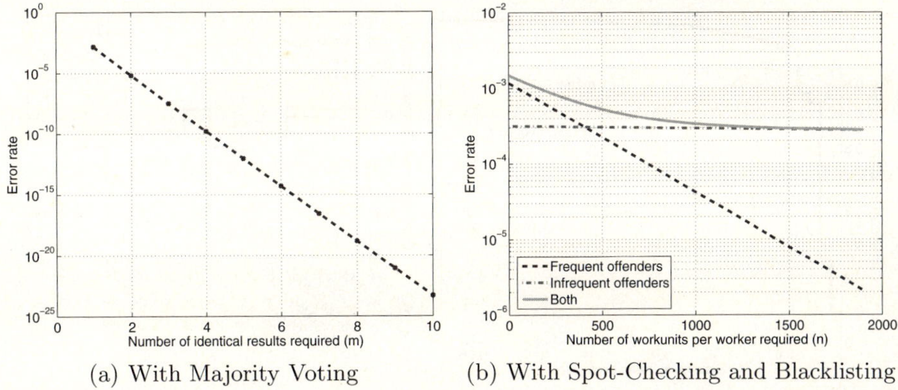

(a) With Majority Voting   (b) With Spot-Checking and Blacklisting

Fig. 3. Error Rate Bounds

This is because the error rates for the infrequent offenders are relatively low, and thus, increasing n does not improve detection nor reduce errors significantly. The effect of the net error rate ε is that it initially decreases rapidly for n in the range $[0, 1000]$. Thereafter, the error rate decreases little.

Thus, spot-checking acts as a low-pass filter in the sense that hosts with high error rates can be easily detected (and can then be blacklisted); however, hosts with low error rates remain in the system. If all frequent offenders are detected by spot-checking and blacklisted, then by Figure 2, this will reduce error rates by 0.70 (or equivalently, an error rate of 63×10^{-5}) and cause only a 0.05 reduction in throughput due to blacklisting. However, to reduce the error rate down to 1×10^{-5}, spot-checking must detect errors from both frequent *and* infrequent offenders. As shown by Figure 3(b), spot-checking will not efficiently detect errors from infrequent offenders because it requires a huge number workunits to be processed by each worker. From Figure 3(b), we conclude that spot-checking can reduce error rates down to about 2×10^{-4} quickly and efficiently. To achieve lower error rates, one should consider using majority voting. In the next section, we show that spot-checking may have other difficulties in real-world systems.

5.2 Stationarity of Error Rates

Intuitively, a process is stationary if its statistical properties do not change with time. In this section, we investigate how stationary the mean of the host error rate s is over time, and describe the implications for error detection mechanisms given our findings.

We measured the stationarity of error rates by determining the change in mean error rates over 96 hour periods for each host. That is, for every 96 hours of wall-clock time during which the worker had been active, we determined the mean error rate on each host, and measured the change in error rates from one period to the next. After close inspection of the results, we found that hosts often have long periods with no errors, and that when errors occurred, they occurred

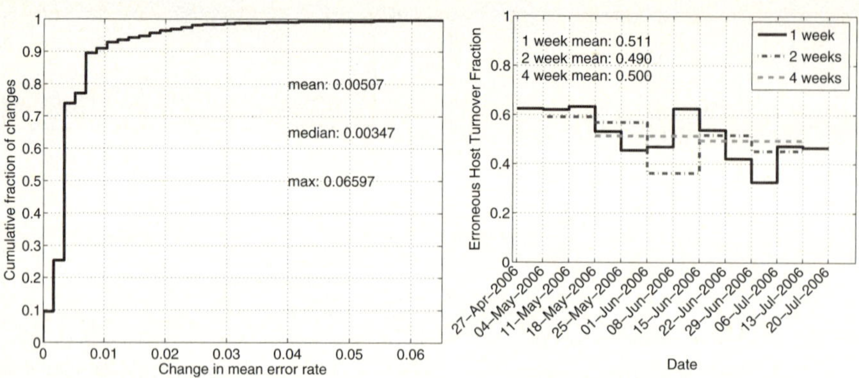

Fig. 4. Error Rate Stationarity

Fig. 5. Turnover Rate of Erroneous Hosts

sporadically. Figure 4 shows the cumulative distribution function of error rate changes over all hosts. This results in Figure 4 show that workunit errors are not very stationary, and in fact, the error rate fluctuates significantly over time.

We also computed statistics for *host* error rates over 96 hour periods. This characterizes s as defined in Section 3. Table 1 shows the mean, standard deviation, and coefficient of variation (which is the standard deviation divided by the mean) for all hosts , the top 10% of erroneous hosts, and the bottom 90% of erroneous hosts. We find that even for relatively long 96 hour periods, the host error rate is quite variable. In particular, the coefficients of variation for all hosts, the top 10%, and the bottom 90% are 3.48, 0.89, and 2.01 respectively.

Table 1. Statistics for Host Error Rates over 96 hour Periods

Host Group	Statistic		
	μ	σ	σ/μ
All erroneous	0.0034	0.018	3.48
Top 10% erroneous	0.0335	0.030	0.89
Bottom 90% erroneous	0.001	0.002	2.01

To investigate the seasonality of errorless periods, we determined whether the set of hosts that err from time period to time period are usually the same hosts or different. In particular, we determined the erroneous host turnover rate as follows. For a specific time period, we determine which set of hosts erred, and then compared this set with the set of the hosts that erred in the following time period. The erroneous host turnover fraction is then the fraction of hosts in the first set that do not appear in the second set. We computed the erroneous host turnover fraction for time periods of 1 week, 2 weeks, and 4 weeks (see Figure 5). For example, the first segment at about 0.62 corresponding to the 1 week period between April 27 and May 4 means that only 0.62 of the hosts that erred between April 20 and April 27 also erred between April 27 and May 4 . On average, the turnover rate is about 0.50 for all periods, meaning that from time period to time period, 0.50 of the erred hosts will be newly erred hosts. That is, the 0.50 of erred hosts had *not* erred in the previous period.

One explanation for the lack of stationarity is that desktop grids exhibit much host churn, as users (and their hosts) often participate in a project for a while and then leave. In [10], the authors computed host lifetime by considering the time interval between entry and its last communication to the project. The host was considered "dead" if it had not communicated to the project for at least 1 month. They found that a host lifetime in Internet desktop grids was on average 91 days.

One implication is that mechanisms that depend on the consistency of error rates, such as spot-checking and credibility-based methods, may not be as effective as majority voting. Spot-checking depends partly on the consistency of error rates over time. Given the high variability in error rates and the intermittent periods without any errors, a host could pass a series of spot-checks, and thereafter or in between spot-checks, the host could produce a high rate of error. Conversely, an infrequent offender could have a burst of errors, be identified as an erroneous host via spot-checking, and then blacklisted. If this occurs with many infrequent offenders, this could potentially have a negative impact on throughput as shown in Figure 2. The same is true for credibility-based systems as a host with variable error rates could build a high credibility, and then suddenly, cause high error rates. Thus, the estimated bounds resulting from spot-checking or credibility-based methods may not be accurate in real-world systems. By contrast, majority voting is not as susceptible to fluctuations in error rates, as the error rate (and confidence bounds on the error rate) decrease exponentially with the number of votes. Nevertheless, the effectiveness of majority voting could be hampered by correlated errors, which we investigate in the next section.

5.3   Correlation of Error Rates

Using the trace of valid and erroneous workunit completion times, we computed the empirical probability that any two hosts had an error at the same time. That is, for each 10 minute period between April 20 to July 20, 2006, and for each pair of hosts, we counted the number of periods in which both hosts computed an erroneous workunit, and the total number of periods in which both hosts computed a workunit (erroneous or correct). Using those counts, we then determined the empirical probability that any two hosts would give an error simultaneously, i.e., within the same 10 minute period. In this way, we determined the empirical joint probability of two hosts having an error simultaneously.

We then determined the "theoretical" probability of two hosts having an error simultaneously by taking the product of their individual host error rates. The individual host error rates are given by dividing the number of erroneous workunits per host by the total number of workunits computed per host (as described in Section 5.1). After determining the "theoretical" probabilities for each host pair, we then determined the difference between the theoretical and empirical probabilities for each host pair. If the error rates for each pair of hosts are not positively correlated, then the theoretical probability should be greater than or equal to the empirical, and the difference should be nonnegative.

Figure 6 shows the cumulative distribution for the differences between theoretical and empirical pairwise error rates. We find most (0.986) of the theoretical pairwise error rates were greater than the empirical. This suggests that the error rates between hosts are not positively correlated. Moreover, only 0.01443 of the pairings had differences less than 0. After carefully inspecting the number of workunits computed by these host pairs, we believe these data points are in fact outliers due a few common errors made by both hosts over a relatively low number of workunits.

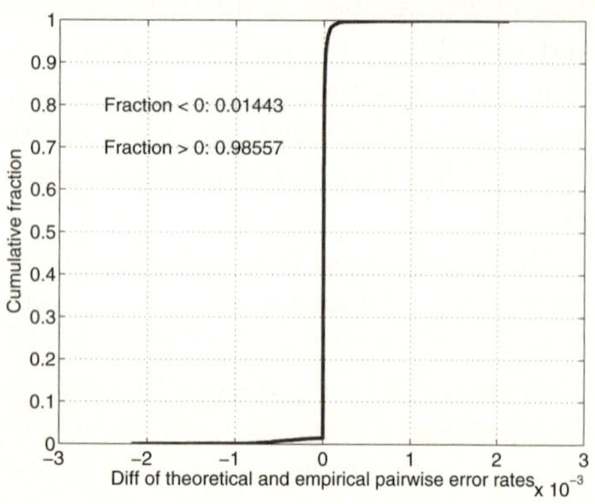

Fig. 6. Pairwise Host Error Rates

6   Summary

We characterized quantitatively the I/O error rates in a real Internet desktop grid system with respect to the distribution of errors among hosts, the stationarity of error rates over time, and correlation among hosts. In summary, the characterization findings were as follows. First, a significant fraction of hosts (about 35%) will commit at least a single error over time. Second, the mean error rate over all hosts (0.0022) is low. Third, a large fraction (e.g. about 70%) of errors result from a small fraction (e.g. 10%) of hosts. Fourth, error rates over time vary greatly and do not seem stationary. Error rates can vary as much as 3.48 over time. The turnover rate for erroneous hosts can be as high as 50%. Fifth, error rates between two hosts often seem uncorrelated. While correlated errors could occur during a coordinated attack or after worm propagation, we do not believe it is the most common source of errors in practice.

In light of these characterization findings, we showed the effectiveness of several error prevention and detection mechanisms namely blacklisting, majority voting, spot-checking, and credibility-based methods. We concluded the following. First, if one can afford redundancy or one needs an error rate to be less then 2×10^{-4}, then majority voting should be strongly considered. Second, if one can afford an error rate greater than 2×10^{-4} and can make batches relatively long (ideally with at least 1000 work units and at least 1 week of CPU time per worker), then spot-checking with blacklisting should be strongly considered. Third, fluctuations in error rates over time may limit the effectiveness of credibility-based systems.

For future work, we plan to address a limitation of this study. Our method measures the I/O error rates of only a single, compute-intensive application.

While we believe this application is representative of most Internet desktop grid applications in terms of its high ratio of computation compared to communication, applications with different I/O or computation patterns could potentially differ in error rates. Thus, to study I/O and computational errors of another application, we will deploy a real scientific application that has easily verifiable results . Nonetheless, we believe this is the first study of a real project to give quantitative estimates of I/O error rates.

References

1. Catalog of boinc projects. `http://boinc-wiki.ath.cx/index.php?title=Catalog_of_BOINC_Powered_Proje%cts`
2. Taufer, M., Anderson, D., Cicotti, P., Brooks III, C.L.: Homogeneous redundancy: a technique to ensure integrity of molecular simulation results using public computing. In: Proceedings of the International Heterogeneity in Computing Workshop (2005)
3. Oltean, A.: How to do atomic writes in a file (December 2005), `http://blogs.msdn.com/adioltean/archive/2005/12/28/507866.aspx`
4. Anderson, D.: Boinc: A system for public-resource computing and storage. In: Proceedings of the 5th IEEE/ACM International Workshop on Grid Computing, Pittsburgh, USA (2004)
5. Fedak, G., Germain, C., N'eri, V., Cappello, F.: XtremWeb: A Generic Global Computing System. In: Proceedings of the IEEE International Symposium on Cluster Computing and the Grid (CCGRID'01) (May 2001)
6. Chien, A., Calder, B., Elbert, S., Bhatia, K.: Entropia: Architecture and Performance of an Enterprise Desktop Grid System. Journal of Parallel and Distributed Computing 63, 597–610 (2003)
7. Sarmenta, L.: Sabotage-tolerance mechanisms for volunteer computing systems. In: Proceedings of IEEE International Symposium on Cluster Computing and the Grid, IEEE Computer Society Press, Los Alamitos (2001)
8. Zhao, S., Lo, V.: Result Verification and Trust-based Scheduling in Open Peer-to-Peer Cycle Sharing Systems. In: Proceedings of IEEE Fifth International Conference on Peer-to-Peer Systems, IEEE Computer Society Press, Los Alamitos (2001)
9. Malecot, P., Kondo, D., Fedak, G.: Xtremlab: A system for characterizing internet desktop grids (abstract). In: Proceedings of the 6th IEEE Symposium on High-Performance Distributed Computing, IEEE Computer Society Press, Los Alamitos (2006)
10. Anderson, D., Fedak, G.: The Computational and Storage Potential of Volunteer Computing. In: Proceedings of the IEEE International Symposium on Cluster Computing and the Grid (CCGRID'06), IEEE Computer Society Press, Los Alamitos (2006)

Evaluation of a Utility Computing Model Based on the Federation of Grid Infrastructures[*]

Tino Vázquez, Eduardo Huedo, Rubén S. Montero, and Ignacio M. Llorente

Departamento de Arquitectura de Computadores y Automática
Facultad de Informática
Universidad Complutense de Madrid, Spain
{tinova,ehuedo}@fdi.ucm.es, {rubensm,llorente}@dacya.ucm.es

Abstract. Utility computing is a service provisioning model which will provide adaptive, flexible and simple access to computing resources, enabling a pay-per-use model for computing similar to traditional utilities such as water, gas or electricity. On the other hand, grid technology provides standard functionality for flexible integration of diverse distributed resources. This paper describes and evaluates an innovative solution for utility computing, based on grid federation, which can be easily deployed on any infrastructure based on the Globus Toolkit. This solution exhibits many advantages in terms of security, scalability and site autonomy, and achieves good performance, as shown by results, mainly with compute-intensive applications.

1   Introduction

Utility is a computing term related to a new paradigm for an information technology (IT) provision which exhibits several potential benefits for an organization [1]: reducing fixed costs, treating IT as a variable cost, providing access to unlimited computational capacity and improving flexibility, thereby making resource provision more agile and adaptive. Such valuable benefits may bring the current fixed-pricing policy of IT provision to an end, where computing is carried out within individual corporations or outsourced to external service providers [2].

The deployment of a utility computing solution involves a full separation between the provider and the consumer. The consumer requires a uniform, secure and reliable functionality to access the utility computing service and the provider requires a scalable, flexible and adaptive infrastructure to provide the service. Moreover, the solution should be based on standards and allow a gradual deployment in order to obtain a favorable response from application developers and IT staff.

[*] This research was supported by Consejería de Educación of Comunidad de Madrid, Fondo Europeo de Desarrollo Regional (FEDER) and Fondo Social Europeo (FSE), through BioGridNet Research Program S-0505/TIC/000101, by Ministerio de Educación y Ciencia, through research grant TIN2006-02806, and by European Union, through BEinGRID project EU contract IST-2005-034702.

A.-M. Kermarrec, L. Bougé, and T. Priol (Eds.): Euro-Par 2007, LNCS 4641, pp. 372–381, 2007.

In a previous work [3], we have proposed a solution for federating grids which can be deployed on a grid infrastructure based on the Globus Toolkit (GT). Such a solution demonstrates that grid technology overcomes utility computing challenges by means of its standard functionality for flexible integration of diverse distributed resources.

A similar approach for the federation of grid infrastructures has been previously applied to meet LCG and GridX1 infrastructures [4], hosting a GridX1 user interface in a LCG computing element. However, this solution imposes software, middleware and network requirements on worker nodes. The Globus project is also interested in this kind of *recursive* architectures, and is working on Bouncer [5], which is a Globus job forwarder initially conceived for federating TeraGrid and Open Science Grid infrastructures. There are other approaches to achieve middleware interoperability, for example between UNICORE and Globus [6] and between gLite and UNICORE [7].

On the other hand, MOAB Grid Suite from Cluster Resources[1] is a grid workload management solution that integrates scheduling, management, monitoring and reporting of workloads across independent clusters. Moreover, Condor's Flocking and GlideIn mechanisms [8] provide similar functionality, allowing job transfers across Condor pools' boundaries or the deployment of remote Condor daemons, respectively. Nevertheless, these solutions are not based on standards and require the same workload manager to be installed in all resources. In this sense, the Open Grid Forum's Grid Scheduling Architecture research group[2] is working on a standard architecture for the interaction between different metaschedulers.

Finally, other projects, like Gridbus [9] or GRIA [10], are developing components for accounting, negotiation and billing, in order to provide end-to-end quality of service driven by computational economy principles.

In this work, we use a technology for the federation of grid infrastructures to build a utility model for computing, which provides full metascheduling functionality, and it is flexible, scalable and based on standards. The rest of this paper is as follows. Section 2 presents a solution for building utility grid infrastructures based on the Globus Toolkit and the GridWay Metascheduler by means of GridGateWays. The architecture of a GridGateWay is described and evaluated in Section 3, while Section 4 shows some experimental results. Finally, Section 5 presents some conclusions and our plans for future work.

2   Utility Grid Infrastructures

A grid infrastructure offers a common layer to integrate non-interoperable computational platforms by defining a consistent set of abstraction and interfaces for access to, and management of, shared resources [11]. Most current grid infrastructures are based on the Globus Toolkit [12], that implements a collection of high level services at the grid infrastructure layer. These services include,

---

[1] http://www.clusterresources.com
[2] http://forge.gridforum.org/projects/gsa-rg

among others, resource monitoring and discovery (Monitoring and Discovery Service, MDS), resource allocation and management (Grid Resource Allocation and Management, GRAM), file transfer (Reliable File Transfer, RFT, and GridFTP) and a security infrastructure (Globus Security Infrastructure, GSI). The Globus layer provides a uniform interface to many different Distributed Resource Manager (DRM) systems, allowing the development of grid workload managers that optimize the use of the underlying computing platforms.

The technological feasibility of the utility model for computing services is established by using a novel grid infrastructure based on Globus Toolkit components and the GridWay Metascheduler[3] [13]. Globus Toolkit services provide a uniform, secure and reliable interface to heterogeneous computing platforms managed by different DRM systems. The main innovation of our model is the use of Globus Toolkit services to recursively interface to the services available in a federated Globus based grid.

A set of Globus Toolkit services hosting a GridWay Metascheduler, what we call a *GridGateWay*, provides the standard functionality required to implement a gateway to a federated grid. Such a combination allows the required virtualization technology to be created in order to provide a powerful abstraction of the underlying grid resource management services and meets the requirements imposed by a utility computing solution [3]. The GridGateWay acts as the utility computing service, providing a uniform standard interface based on Globus interfaces, protocols and services for the secure and reliable submission and control of jobs, including file staging, on grid resources.

Application developers, portal builders, and ISVs may interface with the utility computing service by using the Distributed Resource Management Application API (DRMAA)[4] [14]. DRMAA is an Open Grid Forum (OGF)[5] standard that constitutes a homogeneous interface to different DRM systems to handle job submission, monitoring and control, and retrieval of finished job status.

The grid hierarchy in our utility computing model is clear. An enterprise grid, managed by the IT Department, includes a GridGateWay to an outsourced grid, managed by the utility computing service provider. The outsourced grid provides pay-per-use computational power when local resources are overloaded. This hierarchical grid organization may be extended recursively to federate a higher number of partner or outsourced grid infrastructures with consumer/provider relationships. Figure 1 shows one of the many grid infrastructure hierarchies that can be deployed with GridGateWay components. This hierarchical solution, which resembles the architecture of the Internet (characterized by the end-to-end argument [15] and the IP hourglass model [16]), involves some performance overheads, mainly higher latencies, which will be quantified in Section 4.

The access to resources, including user authentication, across grid boundaries is under control of the GridGateWay service and is transparent to end users. In fact, different policies for job transfer and load balancing can be defined in

---

[3] http://www.gridway.org
[4] http://www.drmaa.org
[5] http://www.ogf.org

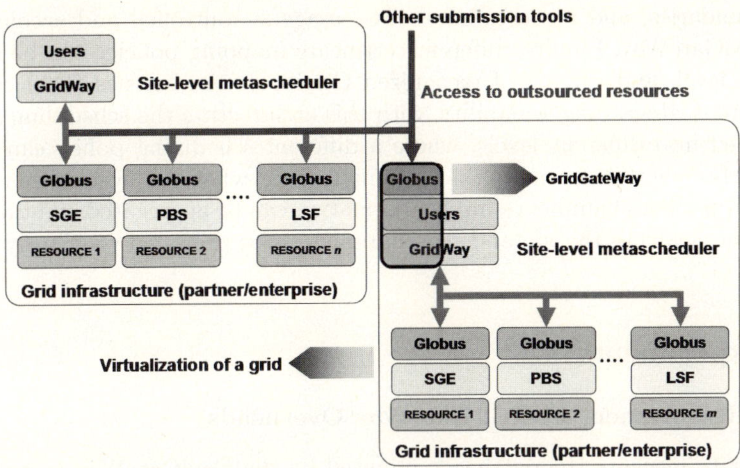

Fig. 1. Utility computing solution based on the Globus Toolkit and the GridWay Metascheduler

the GridGateWay. The user and resource accounting and management could be performed at different aggregation levels in each infrastructure.

On the other hand, the cultural and business model changes required for adopting the utility model should be gradual, starting with access to a local workload manager, followed by an in-house enterprise grid and finally moving onto outsourced services. These adoption steps are transparent to the end user in the proposed solution, since he always interfaces with a given GridGateWay using DRMAA standard.

3 Architecture of the GridGateWay

To interface GridWay through GRAM, a new scheduler adapter has been developed along with a scheduler event generator. Also, a scheduler information provider has been developed in order to feed MDS with scheduling information. Therefore, the main functionality of the GridGateWay is provided by the GridWay Metascheduler, although accessed through the standard interfaces provided by the Globus Toolkit. Moreover, the GridGateWay takes advantage of recent improvements in the GridWay Metascheduler, like multiple user support, accounting, client and resource fault-tolerance, and new drivers to access different grid services. It is expected that new functionality related to the GridGateWay will be added to GridWay, like new scheduling policies for enterprise, partner and utility grid infrastructures, security and certificate management policies, billing, etc.

The presented architecture has a number of advantages in terms of security and scalability. Regarding security, only the GridGateWay should be accessible from the Internet, and only Globus firewall requirements are imposed. Moreover, it is possible to restrict the dissemination of system configuration outside site boundaries, and resource access and usage is controlled and accounted in the GridGateWay. Finally, different certificate mapping policies can be defined at each level, and the set of recognized Certificate Authorities (CA) can also be different. Regarding scalability, with this architecture the scheduling process is divided into different levels, where a different scheduling policy can be applied. Also, there is no need to disseminate everywhere all system monitoring information. For example, resource information can be aggregated and then published by means of the r_∞ and $n_{1/2}$ parameters, as proposed previously by the authors [17].

4   Experiments

4.1   Measurement of GridGateWay Overheads

In order to measure the overheads imposed by the GridGateWay architecture, we set up a simple infrastructure where a client runs an instance of the GridWay Metascheduler interfacing with a GridGateWay (running another instance of the GridWay Metascheduler accessed through Globus Toolkit services) which, in turns, interfaces with a Globus resource managed by the Fork job manager. The application used was a simple `/bin/echo` (using the executable in the remote resource), so the required computational time was negligible and the file transfer costs were minimum (basically the standard input/output/error streams).

The average Globus overhead in both the GridGateWay and the resource was 6 seconds. The average scheduling overhead in the client machine was 15 seconds, since the scheduling interval was set to 30 seconds, while in the GridGateWay it was only 5, since the scheduling interval was set to 10 seconds in order to improve the response time. Therefore, the difference in execution time between a direct execution and an execution through a GridGateWay can be as low as 11 seconds. This difference could be higher when more file transfers are involved.

4.2   Enterprise Grid Resource Provision from a Partner Grid Through a GridGateWay

Now, in order to evaluate the behavior of the proposed solution, we set up a more realistic infrastructure where a client run an instance of the GridWay Metascheduler interfacing local resources in an enterprise (UCM, in this case), based on GT4 Web Services interfaces, and a GridGateWay that gives access to resources from a partner grid (*fusion* VO of EGEE, in this case), based on GT pre-Web Services interfaces. The simultaneous use of different Middleware Access Drivers (MAD) to access multiple partner grid infrastructures has been demonstrated before [18,19]. In fact, that could be an alternative for the coexistence of different grid infrastructures, although based on distinct middleware (GT2, GT4, LCG, gLite...).

In this configuration, `draco` is the client machine, providing access to the enterprise grid, and `cepheus` is the GridGateWay, providing access to the partner grid (see Figure 2). Notice that, in this case, the GridGateWay is hosted in the

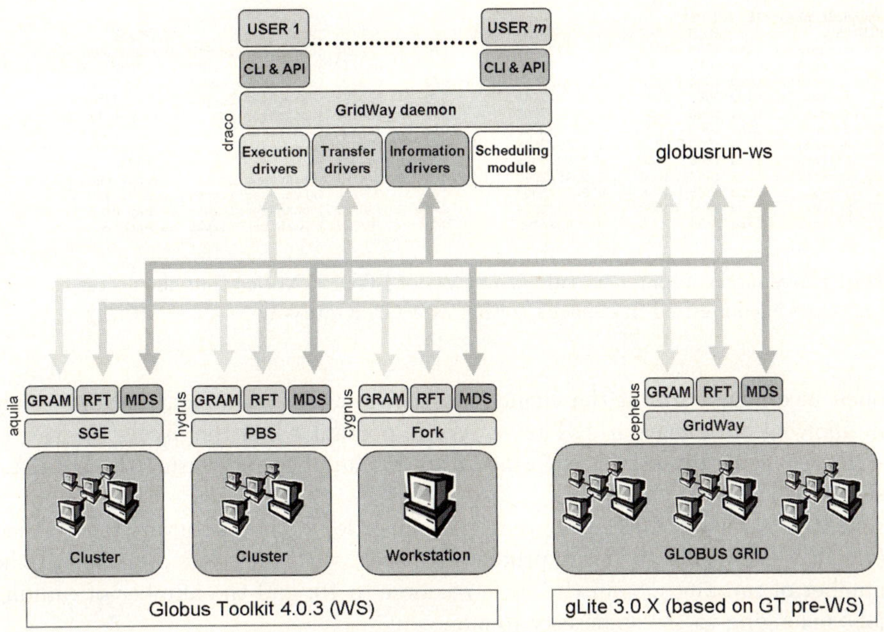

Fig. 2. Experimental environment

```
ehuedo@draco:~$ gwhost
HID OS              ARCH  MHZ %CPU  MEM(F/T)    DISK(F/T)    N(U/F/T) LRMS  HOSTNAME
0   Linux2.6.16-2-6 x86   3216   0  831/2027 114644/118812    0/0/1 Fork  cygnus.dacya.ucm.es
1   Linux2.6.16-2-a x86_6 2211 100  671/1003  76882/77844     0/2/2 SGE   aquila.dacya.ucm.es
2   Linux2.6.16-2-6 x86   3215   0  153/2027 114541/118812    0/0/1 Fork  draco.dacya.ucm.es
3   Linux2.6.16.13- x86   3200 200   10/512  148855/159263    0/0/2 SGE   ursa.dacya.ucm.es
4   Linux2.6.16-2-a x86_6 2211 100  674/1003  76877/77844     0/2/2 PBS   hydrus.dacya.ucm.es
5   NULLNULL        NULL    0   0    0/0            0/0 6/665/1355 GW    cepheus.dacya.ucm.es
```

Fig. 3. Resources in enterprise grid (UCM). Notice that cepheus, acting as a GridGate-Way, appears as another resource with GW (GridWay) as LRMS (Local Resource Management System) and provides no information about its configuration, but only about free and total nodes.

enterprise. However, in a typical business situation it would be hosted in the partner infrastructure. The characteristics of resources from UCM are shown in Figure 3, as provided by GridWay command gwhosts, while those from EGEE (*fusion* VO) are shown in Figure 4.

Notice also that, with the GridGateWay architecture, it is possible to access resources with non WS (Web Services) interfaces, like those present in the LCG computing element of gLite 3, using WS interfaces, like those present in the new services provided by GT4, which are based on the Web Services Resource Framework (WSRF). Therefore, the interface the client sees is more homogeneous. Moreover, the access pattern of EGEE resources requires the client to start a GridFTP or GASS server. Nevertheless, GridWay doesn't need such server to access WS-based resources. With the proposed architecture, there is a GridFTP server already started in the GridGateWay, so there is no need to

```
ehuedo@cepheus:~$ gwhost
HID OS             ARCH   MHZ %CPU   MEM(F/T)  DISK(F/T)    N(U/F/T) LRMS                HOSTNAME
0   Scientific Linu i686  1001   0    513/513        0/0  3/103/224 jobmanager-lcgpbs  lcg02.ciemat.es
1   ScientificSL3.0 i686   551   0    513/513        0/0     0/3/14 jobmanager-lcgpbs  ce2.egee.cesga.es
2   Scientific Linu i686  1000   0  1536/1536        0/0     0/2/26 jobmanager-lcgpbs  lcgce01.jinr.ru
3   Scientific Linu i686  2800   0  2048/2048        0/0     0/0/98 jobmanager-lcgpbs  lcg06.sinp.msu.ru
4   Scientific Linu i686  1266   0  2048/2048        0/0    0/26/58 jobmanager-pbs     ce1.egee.fr.cgg.com
5   Scientific Linu i686  2800   0  2048/2048        0/0  0/135/206 jobmanager-lcgpbs  node07.datagrid.cea.fr
6   ScientificSL3.0 i686  3000   0  2048/2048        0/0  0/223/352 jobmanager-lcgpbs  fal-pygrid-18.lancs.ac
7   Scientific Linu i686  2400   0  1024/1024        0/0  0/139/262 jobmanager-lcgpbs  heplnx201.pp.rl.ac.uk
8   Scientific Linu i686  3000   0  2048/2048        0/0     0/0/60 jobmanager-pbs     cluster.pnpi.nw.ru
9   Scientific Linu i686  1098   0  3000/3000        0/0     0/5/16 jobmanager-lcgpbs  grid002.jet.efda.org
10  Scientific Linu i686  2800   0  1024/1024        0/0    0/30/32 jobmanager-lcgpbs  ce.hep.ntua.gr
11  Scientific Linu i686  2000   0    492/492        0/0      0/2/7 jobmanager-lcgpbs  ce.epcc.ed.ac.uk
```

Fig. 4. Resources in partner grid (*fusion* VO of EGEE). Notice that the access to these resources is configured in cepheus, acting as a GridGateWay.

open incoming ports in the client, and the firewall requirements results solely in allowing outgoing ports. This is very important when the access to grid resources is generalized and performed from ISV applications, using the DRMAA standard.

In the case of EGEE resources, and in order to not saturate the testbed (which is supposed to be at production level) with our tests, we limited the number of running jobs in the same resource to 10, and the number of running jobs belonging to the same user to 30.

First of all, we want to compare the direct access to EGEE resources and the access through a GridGateWay. In order to do that, in a first experiment we submitted the jobs directly to the GridWay instance running on cepheus. And, in a second experiment, we submitted the jobs to the GridWay instance running on draco with cepheus, acting as a GridGateWay, as the unique resource.

In this case, the application used was the distributed calculation of the π number as $\int_0^1 \frac{4}{1+x^2} dx$. Each task computes the integral in a separate section of the function and all results are finally added to obtain the famous 3.1415926435... Therefore, the required computational time is now higher (about 10 seconds on a 3.2 GHz Pentium 4 for each task) and file transfer costs include the executable and the standard input/output/error streams (about 10 KB per task).

Figure 5 shows the throughput achieved in EGEE (*fusion* VO) resources when accessed directly and when they are accessed through the GridGateWay. The number of tasks submitted was 100. As expected, there are differences in latency (response time) and throughput when directly accessing the resources and when using the GridGateWay. A throughput of 284.8 jobs/hour was achieved with direct access, versus 253.9 jobs/hour with the access through the GridGateWay. Therefore, the use of a GridGateWay supposes a performance loss of only 10.85%. Notice that this performance loss has been obtained with an application requiring only 10 seconds to execute. Since the overheads are independent on the computational time required by the application, they will suppose a smaller fraction of the total time when more demanding applications are used.

Figure 6 shows the throughput achieved in UCM when provisioning partner resources from EGEE (*fusion* VO) through a GridGateWay. The number of tasks submitted was 100 again. In this case, there are also differences in latency between in-house and partner resources. Besides network connection and the use

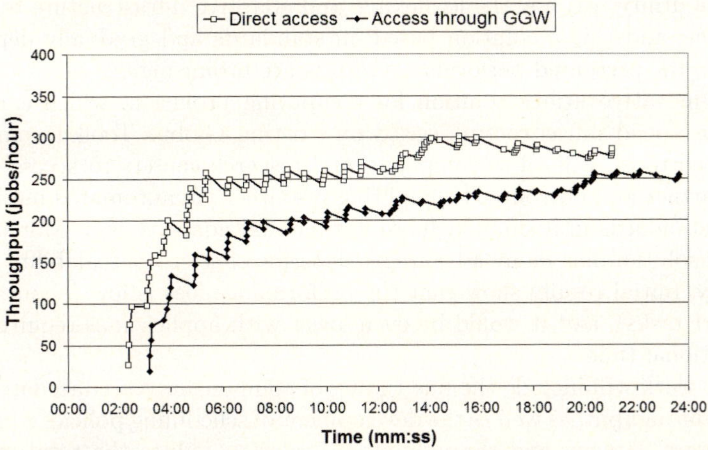

Fig. 5. Throughput achieved in EGEE (*fusion* VO) when accessed directly and when accessed through a GridGateWay

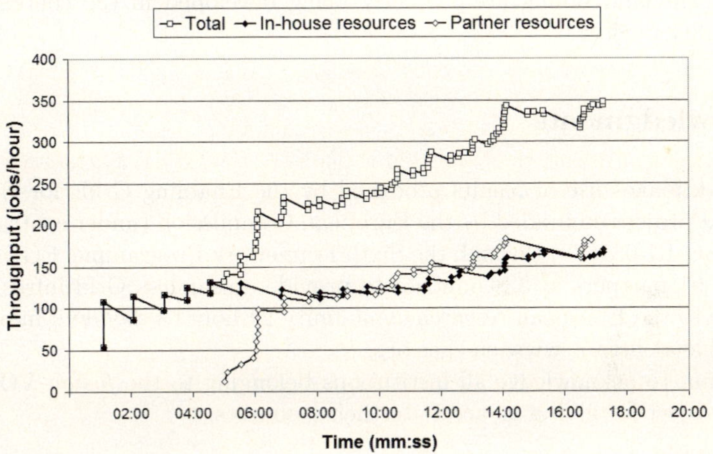

Fig. 6. Throughput achieved in UCM when provisioning resources from EGEE (*fusion* VO) through a GridGateWay

of a GridGateWay, it is also due to the production status of partner resources, as they are under heavy usage. The aggregated throughput achieved was 347.5 jobs/hour. In-house and partner resources contributed almost equally (170.3 and 181.4 jobs/hour, respectively) to the aggregated throughput.

5   Conclusions and Future Work

We have presented and evaluated a solution that meets the requirements for utility computing, namely: (i) a uniform, secure and reliable functionality to access the utility; (ii) a scalable, flexible and adaptive infrastructure to provide the service; and (iii) a solution based on standards and gradually deployable. Moreover, the presented performance results are promising.

This innovative utility solution for computing provision, which can be deployed on a grid infrastructure based on existing Globus Toolkit components and related tools, will allow companies and research centers to access their in-house, partner and outsourced computing resources via automated methods using grid standards in a simpler, more flexible and adaptive way. Moreover, the proposed solution has many advantages in terms of security, scalability and site autonomy. Initial results show that the performance loss is low (about 10% for very short tasks), and it would be even lower with applications requiring more computational time.

Future work will include the fine tuning of components to reduce latency and increase throughput, as well as the development of scheduling policies considering these factors, latency and throughput, in order to reduce the total execution time of a whole workload, and also taking into account resource ownership to reduce the associated cost (in terms of real money, resource usage or partner satisfaction). New components for negotiation, service level agreement, credential management, and billing are currently being developed in the context of the Grid4Utility project[6].

Acknowledgments

This work makes use of results produced by the Enabling Grids for E-sciencE project, a project co-funded by the European Commission (under contract number INFSO-RI-031688) through the Sixth Framework Programme. EGEE brings together 91 partners in 32 countries to provide a seamless Grid infrastructure available to the European research community 24 hours a day. Full information is available at http://www.eu-egee.org.

We want to acknowledge all institutions belonging to the *fusion* VO[7] of the EGEE project for giving us access to their resources.

References

1. Murch, R.: Introduction to Utility Computing: How It Can Improve TCO. Prentice-Hall, Englewood Cliffs (2005)
2. Carr, N.G.: The End of Corporate Computing. MIT Sloan Management Review 46(3), 67–73 (2005)
3. Llorente, I.M., Montero, R.S., Huedo, E., Leal, K.: A Grid Infrastructure for Utility Computing. In: Proc. 15th IEEE Intl. Workshops on Enabling Technologies: Infrastructures for Collaborative Enterprises (WETICE 2006), pp. 163–168. IEEE Computer Society Press, Los Alamitos (2006)

---

[6] http://www.grid4utility.org
[7] http://grid.bifi.unizar.es/egee/fusion-vo

4. Agarwal, A., Ahmed, M., Berman, A., et al.: GridX1: A Canadian Computational Grid. Future Generation Computer Systems 23(5), 680–687 (2007)
5. Baumbauer, C., Goasguen, S., Martin, S.: Bouncer: A Globus Job Forwarder. In: Proc. 1st TeraGrid Conf (TeraGrid 2006) (2006)
6. Breuer, D., Wieder, P., van den Berghe, S., von Laszewski, G., MacLaren, J., Nicole, D., Hoppe, H.C.: A UNICORE-Globus Interoperability Layer. Computing and Informatics 21, 399–411 (2002)
7. Mallmann, D., Riedel, M., Twedell, B., Streit, A.: Interoperability Plan for UNI-CORE. Technical Report MSA3.3, EGEE-II (2006)
8. Frey, J., Tannenbaum, T., Livny, M., Foster, I., Tuecke, S.: Condor-G: A Computation Management Agent for Multi-Institutional Grids. Cluster Computing 5(3), 237–246 (2002)
9. Buyya, R., Venugopal, S.: The Gridbus Toolkit for Service Oriented Grid and Utility Computing: An Overview and Status Report. In: Proc. 1st IEEE Intl. Workshop on Grid Economics and Business Models (GECON 2004), pp. 19–36. IEEE Computer Society Press, Los Alamitos (2004)
10. Surridge, M., Taylor, S., Roure, D.D., Zaluska, E.: Experiences with GRIA Industrial Applications on a Web Services Grid. In: Proc. 1st Intl. Conf. e-Science and Grid Computing (e-Science05), pp. 98–105. IEEE Computer Society Press, Los Alamitos (2005)
11. Foster, I., Tuecke, S.: Describing the Elephant: The Different Faces of IT as Service. Enterprise Distributed Computing 3(6), 26–34 (2005)
12. Foster, I.: Globus Toolkit Version 4: Software for Service-Oriented Systems. In: Jin, H., Reed, D., Jiang, W. (eds.) NPC 2005. LNCS, vol. 3779, pp. 2–13. Springer, Heidelberg (2005)
13. Huedo, E., Montero, R.S., Llorente, I.M.: A Framework for Adaptive Execution on Grids. Software - Practice and Experience 34(7), 631–651 (2004)
14. Herrera, J., Huedo, E., Montero, R.S., Llorente, I.M.: GridWay DRMAA 1.0 Implementation – Experience Report. Document GFD.E-104, DRMAA Working Group – Open Grid Forum (2007)
15. Carpenter, E.B.: Architectural Principles of the Internet. RFC 1958 (June 1996)
16. Deering, S.: Watching the Waist of the Protocol Hourglass. In: 6th IEEE International Conference on Network Protocols (1998)
17. Montero, R.S., Huedo, E., Llorente, I.M.: Benchmarking of High Throughput Computing Applications on Grids. Parallel Computing 32(4), 267–279 (2006)
18. Vázquez-Poletti, J.L., Huedo, E., Montero, R.S., Llorente, I.M.: Coordinated Harnessing of the IRISGrid and EGEE Testbeds with GridWay. J. Parallel and Distributed Computing 66(5), 763–771 (2006)
19. Huedo, E., Montero, R.S., Llorente, I.M.: A Modular Meta-Scheduling Architecture for Interfacing with Pre-WS and WS Grid Resource Management Services. Future Generation Computer Systems 23(2), 252–261 (2007)

The Characteristics and Performance of Groups of Jobs in Grids

Alexandru Iosup, Mathieu Jan, Ozan Sonmez, and Dick Epema

Department of Electrical Engineering, Mathematics and Computer Science
Delft University of Technology, Delft, The Netherlands
Members of the CoreGRID European Virtual Institute on Grid Scheduling
{A.Iosup,M.Jan,O.O.Sonmez,D.H.J.Epema}@tudelft.nl

Abstract. Even though with few exceptions, grid workloads are dominated by single-node jobs, not all of these jobs are necessarily independent or unrelated. For instance, sets of jobs may be grouped because they are submitted by users in batches, e.g., to perform parameter sweeps. However, there is no reported data to confirm the presence and structure of these groupings, despite the large potential impact of such information. To address this lack of information, in this work we present a first investigation into the characteristics of groups of jobs present in grid workloads. First, we define three types of job groupings: batch, continued, and bursty submissions. Then, we analyze the characteristics of these groupings for three long-term traces from currently deployed grid environments. Notably, our results show that the various groupings are responsible for up to 96% of the total CPU time consumption. Finally, we present insights into the performance of real grids in dealing with grouped jobs.

1 Introduction

It has become evident that many currently deployed grids, such as Grid'5000 [1], the NorduGrid [2] and the GLOW[1], are running almost exclusively single-node jobs [3]. However, this does not preclude some jobs in such workloads to be in some way related. For instance, users may submit batches of jobs in order to perform parameter sweeps, or they may submit workflows that consist of sequential jobs with precedence constraints. As a consequence, single-node grid jobs may be logically grouped in some way, e.g., in batches of identical jobs [4]. However, the presence and the structure of such groups of jobs have yet to be analyzed, despite their potentially tremendous impact on the efficiency and the functionality of grid resource management solutions. To address this gap, in this work we first characterize various job groupings that occur in production and research grid environments. Then, we present an assessment of their impact on the performance of real grid systems.

---

[1] Grid Laboratory of Wisconsin, [Online] cs.wisc.edu/condor/glow/, February 2007.

A.-M. Kermarrec, L. Bougé, and T. Priol (Eds.): Euro-Par 2007, LNCS 4641, pp. 382–393, 2007.

Over the last decade, much work has focused on enabling parallel applications in grids [5,6]. In parallel, scientists have exploited grids for running completely independent jobs. Combining these two extremes, a third direction focuses on running sets of single-node jobs with convenient groupings, e.g., parameter sweeps [4]. With the performance of grid systems being highly dependent on the structure of their workloads [7,8,9,10], it is imperative to establish which types of job groupings exist in real grid workloads, to what extent they exist, and what is the impact of their structure on the performance of real grid systems. Therefore, in this work we answer the following set of questions:

- **What are the dependencies among the jobs submitted by a single user?** Do users submit single, independent jobs, or batches of jobs such as parameter sweeps? How can we characterize groups of related jobs?
- **What is the physical structure of such job groupings?** How many jobs appear in batches? And what are their characteristics, e.g., their duration and resource requirements?
- **What is the impact of the job groupings on the performance of grids?** What is the total processing time of grouped vs. non-grouped jobs? Do job groupings "average out" the waiting time, and achieve a lower slowdown, compared to their non-grouped counterparts?

2   Grid Environments

In order to answer our questions, we have obtained long-term traces from three large grid systems. Each trace records the jobs that have been submitted to the grid, and includes for each job at least the identity of the user submitting it, and the job's time information (the time of the job submission, job start, and job end). In Table 1, we summarize the content of the traces considered in this work: Grid'5000, NorduGrid, and GLOW. Our workload analysis covers almost two million jobs submitted over a period of three years by more than 800 users. Note that the data sources have been selected to represent different types of grids: research for Grid'5000 and production for NorduGrid and GLOW, with GLOW a much more dynamic system than NorduGrid.

The first testbed used in our work is Grid'5000 [1], an experimental grid platform consisting of 9 sites geographically distributed in France. Each site

Table 1. Summary of the content of the studied traces. The $\star$ sign marks restrictions due to data scarcity (see text). UJM and LRM/GRM are acronyms for User Job Manager and Local/Grid Resource Manager jobs log, respectively. GRP and USR represent the number of unique groups/VOs and of users in the workload, respectively.

Trace ID	System	Source	Period	Number of observed					Consumed
				Sites	CPUs	Jobs	GRP	USR	CPUTime
T-1	Grid'5000	LRM	05/'04-11/'06	15	~2500	951K	10	473	651y
T-2	NorduGrid	GRM	05/'04-02/'06	~75	~2000	781K	106	387	2443y
T-3	GLOW	UJM	09/'06-01/'07	1$\star$	~1400	216K	1$\star$	18	55y

comprises one or several clusters, for a total of 15 clusters inside Grid'5000. We have analyzed traces recorded by all batch schedulers handling Grid'5000 clusters (OAR [11]), from the beginning of the Grid'5000 project up to November 2006. Note that most clusters of Grid'5000 were made available during the first half of 2005.

The second testbed considered is NorduGrid [2], a large scale production grid. In NorduGrid, non-dedicated resources are connected using the Advanced Resource Connector (ARC) as Grid middleware [12]. Over 75 different clusters have been added over time to the infrastructure. We have obtained the ARC logs of NorduGrid for a period spanning from 2003 to 2006. In these logs, the information concerning the grid jobs is logged locally, then transferred to a central database voluntarily. The logging service can be considered fully operational only since mid-2004.

The third source of data is the Grid Laboratory of Wisconsin (GLOW), a campus-wide distributed computing environment that serves the computing needs of the University of Wisconsin-Madison's scientists. This Condor-based pool consists of over 1400 machines shared temporarily by their rightful owners [4]. We have obtained a trace comprising all the jobs submitted by one Virtual Organization (VO) in the Condor-based GLOW pool, in Madison, Wisconsin. The trace spans four months, from September 2006 to January 2007.

3   Groups of Jobs

In this section, we introduce definitions for the three types of groups of jobs that we distinguish in this paper.

Let W be the workload of a grid, which we consider as a set $\{J_i | i = 1, ..., |W|\}$ of jobs ordered according to increasing submission time, so $ST(J_i) < ST(J_j)$ if $i < j$, where $ST(\cdot)$ returns the submission time of a job. A *group of jobs* G is a subset of W, which we assume again to be ordered according to submission time. The *seed* J_G of a group G of jobs is the job in G with the earliest submission time. We define the *arrival time of a group* G of jobs as the submission time of its seed, $ST(J_G)$, and we define the *inter-arrival time between two groups* G, G' as the difference between their arrival times $IAT(G, G') = ST(J_{G'}) - ST(J_G)$. We also define the *duration of a group* G as the difference between $ST(J_G)$ and $LFT(G)$, where $LFT(\cdot)$ returns the finish time of the last job in G.

If the function $user(J)$ returns the identify of the user who has submitted job J, then we denote by $W_u = \{J_i | user(J_i) = u\}$ the *group of jobs submitted by the same user* u. A *batch submission with time parameter* Δ *of user* u is a maximal contiguous subsequence G of W_u such that for any two successive jobs J, J' in G, $ST(J') \leq ST(J) + \Delta$. Similarly, we define a *continued submission with time parameter* Δ *of user* u as a maximal contiguous subsequence G of W_u such for any two successive jobs J, J' in G, $ST(J) \leq FT(J') + \Delta$, where the function $FT(\cdot)$ returns the finish time of a job. Similarly, a *bursty submission with time parameter* Δ is a batch submission of all users jointly. We further define a *Parameter Sweep Application* (PSA) as *batch submission with time parameter* Δ *of user* u with the additional constraint that all jobs execute the same application.

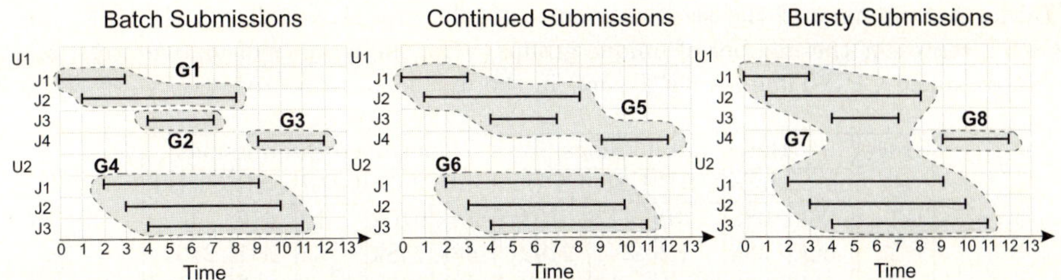

Fig. 1. The structure of a group of jobs: (left) batch submission, (center) continued submission, and (right) bursty submission

We denote by $\mathbf{G}_u^B(\Delta)$ the *ordered set of batch submissions with time parameter Δ of user u* containing all the batch submissions with time parameter Δ of user u, ordered according to their arrival times. Note that for any two successive G, G' in $\mathbf{G}_u^B(\Delta)$, $ST(J_{G'}) > LFT(G) + \Delta$. Similarly, we denote by $\mathbf{G}_u^C(\Delta)$ the *ordered set of continued submissions with time parameter Δ of user u*, and by $\mathbf{G}^S(\Delta)$ (S comes from spike) the *ordered set of bursty submissions with time parameter Δ*. From hereon, we call collectively the elements of $\mathbf{G}_u^B(\Delta)$ of any user u in the workload as batch submissions, or simply as *batches*.

We define as *non-batch, non-continued* and *non-bursty submissions* (collectively called *non-grouped submissions*) as the individual jobs from W that do not belong to any group of jobs in $\mathbf{G}_u^B(\Delta)$, $\mathbf{G}_u^C(\Delta)$, or $\mathbf{G}^S(\Delta)$.

We now define several performance metrics associated with the concept of job grouping. We define the *runtime of a group* (RT) as the amount of time during which at least one group job is running. We define the *duration of a group G* as $LFT(G) - ST(J_G)$. Then, we define the *idle time of a group* (IT) as the difference between the duration and the runtime of the group. We further define the *slowdown of a group* as the ratio between its duration and its runtime. Last, we define the *average group run time* (ART), the *average group idle time* (AIT), and the *average group slowdown* (ASD) as the average group runtime, group idle time, and group slowdown across all groups.

Figure 1 illustrates these definitions using the jobs submitted by two users, $U1$ and $U2$, over some period of time. According to our definition of batches, user $U1$ submits three batches ($G1$, $G2$ and $G3$), whereas user $U2$ submits only one batch ($G4$). Following the definition of continued submissions, user $U1$ submits only one group of jobs, $G5$. Finally, the definition of bursty submission divides the jobs into two groups $G7$ and $G8$, where $G7$ contains jobs from both $U1$ and $U2$.

4   The Characteristics of Jobs Groupings

In this section, we analyze the structure of batch, continued and bursty submissions. We first report an analysis of grouped submissions at the workload level. Then, we present the characteristics of grouped submission. Finally, we target the characteristics of jobs in grouped submissions. These results are clearly dependent on the value of Δ. Considering the overhead of existing grid middleware,

Table 2. Summary of the sizes of groups of jobs, for $\Delta = 120s$. GRP, USR, and CPUT represent the number of unique groups/VOs, the number of unique users, and the consumed CPU Time (in CPUyears) in the workload, respectively.

Trace	Batch Submissions					Continued Submissions				
ID	Sub.	Jobs	GRP	USR	CPUT	Sub.	Jobs	GRP	USR	CPUT
T-1	26k	808k	10	417	193y	14k	910k	10	417	462y
T-2	50k	738k	82	341	2192y	48k	738k	82	341	2192y
T-3	13k	205k	1	17	53y	13k	205k	1	17	53y

Trace	Bursty Submissions				
ID	Sub.	Jobs	GRP	USR	CPUT
T-1	6k	930k	9	163	581y
T-2	34k	759k	97	338	2325y
T-3	13k	204k	1	17	53y

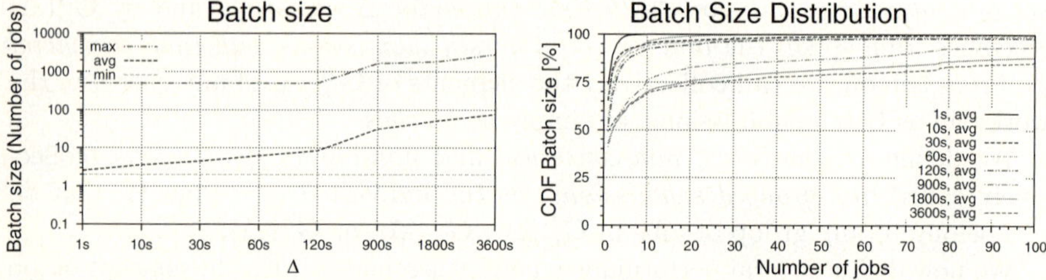

Fig. 2. The impact of parameter Δ on batch size, for trace T-3

which is commonly in the order of tens of seconds [13], we show throughout this section values obtained for $\Delta = 120s$, unless otherwise specified.

For brevity, we show graphical results only for batch and bursty submissions, as these are expected to form the majority of load arriving in the grid in a relatively short interval of time, and therefore potentially have the maximum impact on the grid's performance. We comment on the results obtained for continued submissions. We have also investigated the presence of PSAs; for T-1, 75% of batches are in fact PSAs. This result indicates that jobs are mostly submitted as bags-of-tasks, and not as workflows, with therefore waiting times mainly not being the consequences of dependencies and synchronizations between jobs. Furthermore, this result also shows that jobs, grouped according to our definitions (see Section 3), are principally related between them. Note that in trace T-1, the main differences between PSAs and batches are an increased group size by 9 in average, and a duration time divided by 5.7, but with an almost double coefficient of variation (1.7).

4.1 Workload-Level Analysis

We first investigate the grouped submissions at the workload level.

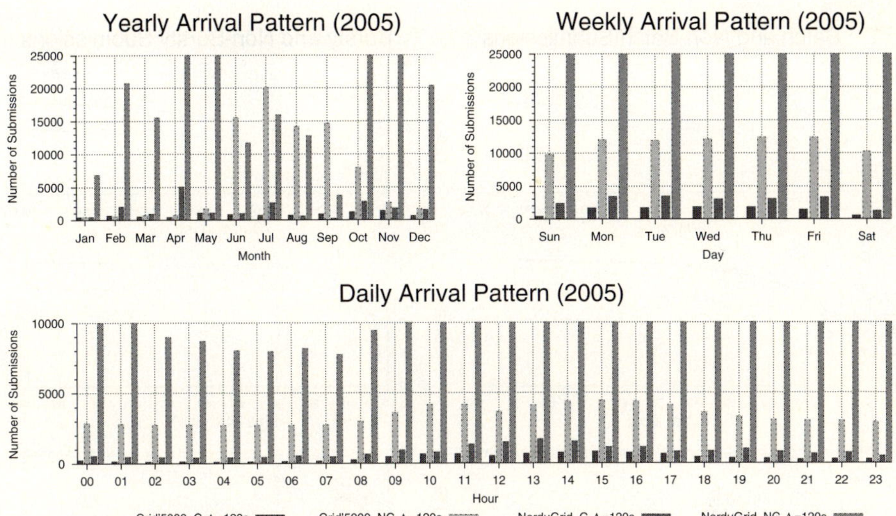

Fig. 3. The yearly, weekly, and daily arrival patterns of batch (caption **G**) and non-batch (caption **NG**) submissions, for all traces; $\Delta = 120s$

Figure 2 shows the impact of Δ on the batch size, for trace T-3. The curve displaying the maximum batch size has a breaking point around $\Delta = 120s$. We have obtained similar results for the other traces. Furthermore, the CDF of batch size for $\Delta = 10s$, $30s$ and $60s$ are almost identical. Therefore, we select $\Delta = 120s$ as the basis for the reported results in this section.

Table 2 shows the summary of the sizes of grouped jobs, for batch, continued, and bursty submissions. Relative to the complete traces, batch submissions are responsible for 85%–95% of the jobs, and for 30% 96% of the total consumed CPU time. Similar values for the number of jobs, and much higher values (from 70% upwards) for the total consumed CPU time can be observed for continued and bursty submissions. The columns *Sub.* and *Jobs* show the number of submissions and of jobs for each type of groups in the trace, respectively. We have further investigated the averages, standard deviation and the extremes for the size of the batches. For T-1, T-2 and T-3, the average size of the batches is 31 ± 110, 15 ± 33 and 15 ± 38, respectively. The maximum size of a batch submission is, however, rather large: 1993, 2000 and 608 for T-1, T-2 and T-3, respectively. The average size for continued submissions are 64, 16 and 16 (for T-1, T-2 and T-3, respectively), and 162, 22 and 16 for bursty submissions (again for T-1, T-2 and T-3, respectively).

Figure 3 shows the yearly, weekly, and daily arrival patterns of batch grouped and non-batch submissions, for both T-1 and T-2 traces, and for 2005. Note that T-3 trace is not shown due to its short length (see Section 2). Note that a single batch submission includes several jobs. The yearly arrival pattern shows large variations, with 4-9 times more jobs submitted during peak months (e.g., October for both traces and April for T-2) vs. low-activity months (e.g., January for both traces). For trace T-1, the presence of peak months late in the yearly

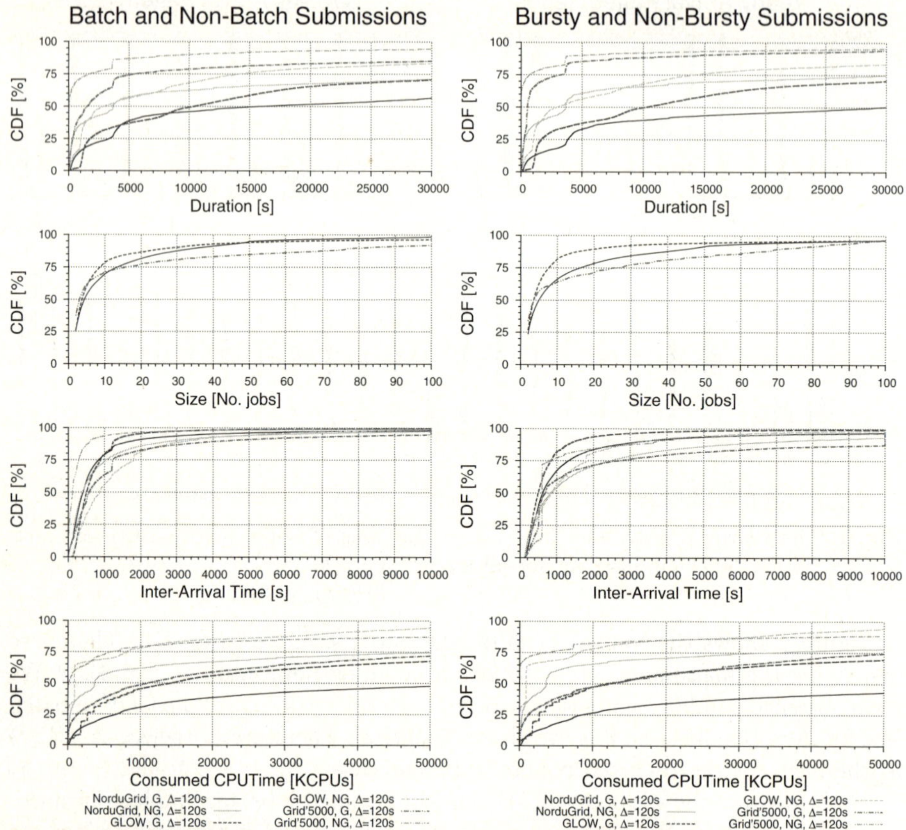

Fig. 4. The characteristics of the grouped (caption **G**) and of the non-grouped (caption **NG**) submissions, for all traces. Row 1: duration per submission. Row 2: the size of grouped submissions. Row 3: the Inter-Arrival Time distributions. Row 4: the consumed CPUTime.

pattern is explained by full availability of Grid'5000 after the first half of 2005 (see Section 2). The weekly pattern is less variable, with only 20-50% more jobs during high- vs. low-intensity days. The high- and low-intensity days correspond to weekday and weekend, respectively. There is a clear variation with daytime, with twice as many jobs submitted during peak vs. low hours. We have obtained similar results for the continued and bursty submissions.

4.2 Group-Level Analysis

We now analyze the characteristics of grouped submissions.

Figure 4 plots the cumulative distribution function (CDF) of the characteristics of the grouped and non-grouped batch and bursty submissions. The duration of batch submissions, shown in row 1, is statistically higher than that of their non-batch counterparts. For instance, for T-2 the average duration of a batch submission is 1.5 days vs. 1 day for non-batch submissions. The average duration for batch submissions for traces T-1, T-2 and T-3 is 3.85, 1.5 and 0.34 days, respectively.

To complement the results presented in Table 2, we show in row 2 the sizes of batch and of bursty submissions; this metric is not shown in row 2 as it is useless for non-grouped submissions (the value is always 1 (job)!). Surprisingly, these sizes are relatively low: 75% of the batch submissions are size 15-20 (T-1 and T-2), or <10 (T-3); the results for bursty submissions are only slightly larger (except for T-2). This is an indication of a heavy-tail distribution of values: despite the majority of submissions being of low size, there are a few jobs that are oversized (outside the graphic, to the right).

As shown in row 3, the inter-arrival time between consecutive groups (see Section 3) is high: 50% of the values are between 400 and 700s, and 25% between 200 and 500s, for all traces. This is expected, given that the interarrival time between two groups represents the size of the group and, for the same user, a period of time with no submissions of at least $\Delta = 120s$.

Finally, the CPU time consumed by grouped jobs is much higher than that consumed by non-grouped jobs. Note that the CPU time provided by one node during one day of continuous work is 86.4 KCPUs (or roughly 100 KCPUs). We detail in Section 4.4 the average CPU time consumption per grouped and non-grouped jobs, for each trace.

4.3 Job-Level Analysis

Our third analysis part targets the characteristics of jobs in grouped submissions.

Figure 5 shows the characteristics of individual jobs, per submission type. Unexpectedly, there is no clear distinction between the consumed CPU time of batch and bursty submissions jobs, and that of their non-grouped counterparts, shown in row 1. For T-1, grouped jobs consume statistically less resources than their non-grouped counterparts, however note that grouped jobs consume negligible time (more than 90% are below 10 KCPUs). For T-2, grouped jobs consume more resources, and for T-3 they consume the same amount. The CDFs of CPU time for both grouped and non-grouped submissions converge on the high spectrum (over 100 KCPUs, or roughly 1.25 days). The average runtime of batch, continued and bursty submissions jobs (CDFs shown in row 3) are 0.66±6.65, 0.66±6.56 and 0.73±6.97 for T-1 respectively; 1.04±3.16, 1.04±3.16 and 1.04±3.18 days for T-2 respectively, and 2.27±5.59 hours in all cases for T-3. The average wait time (CDFs shown in row 2) of group jobs are higher than the runtime: between 30% to 45% for T-1 and 71% for T-3; with also increased standard deviations: from 6 for T-1 up to 16 for T-3, for durations expressed in hours. Note that trace T-2 does not provide any information about the waiting time of jobs.

We further answer the question: *do parallel jobs inside batches exist, or every grouping is conveniently parallel?* We define a *node* as a schedulable resource unit, e.g., processor for NorduGrid and Grid'5000, or virtual node in Condor. There exist parallel jobs in grouped submissions only for T-1 and T-2; T-3 is

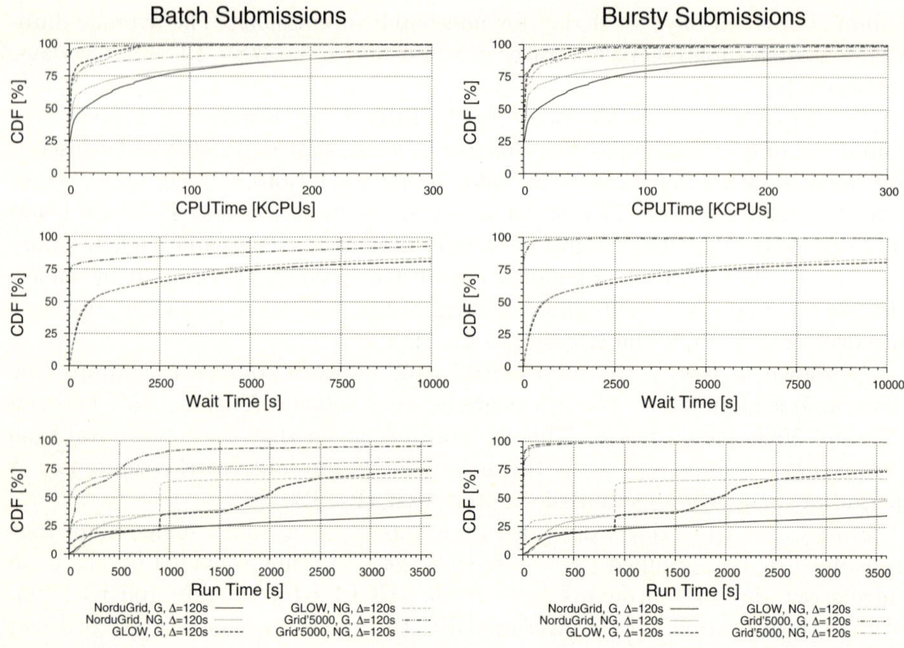

Fig. 5. The characteristics of individual jobs inside grouped (caption **G**) and non-grouped (caption **NG**) submissions, for all traces. Row 1: consumed CPU Time. Row 2: wait time. Row 3: run time.

composed entirely from single-node jobs. The average parallelism of all grouped submissions are 1±1, 2±7 and 1 for T-1, T-2 and T-3 respectively. Looking at the CDF function of the individual job sizes, 37% of the batch submissions in T-1 are of size 2, and 9% are of a size higher than 2, with a maximum of 325 nodes. The maximum size for T-2 is 64.

4.4   The Performance Impact of Grouped Submissions

We now turn our attention to investigate performance impact of grouped submissions, compared to that of non-grouped submissions. Expectedly, the average CPU time consumed by batch submissions for trace T-1 is much higher than their non-batch counterparts. The coefficient of variation is relatively high: 9.18. Other traces follow a same pattern, but are not shown for brevity.

Table 3. Summary of performance characteristics of the Grid'5000 for batches and non-batches

Trace ID	Batch Submissions				Non-Batch Submissions			
	CPU Time [KCPUs]	ART [s]	AIT [s]	ASD [s]	CPU Time [KCPUs]	ART [s]	AIT [s]	ASD [s]
T1	236	14181	568483	2156	101	4127	4233	280

By submitting enough jobs in the same batch, the waiting time of the individual jobs can be "averaged out", that is, at least one job in the batch is active at any given moment throughout the duration of the batch [14]. This leads to an ideal AIT of 0. Surprisingly, in the studied trace the batches display an high AIT value: over 4000% (!) of the ART. This situation also adversely impacts the ASD value: it is above 2000, and much higher than the ASD of non-batches. We would like to draw the attention of the research community to the problem of minimizing the AIT for batch submissions, by taking into account, to some extend, these groups of submissions in the design of scheduling policies for instance.

5   Related Work

In the past decades, a great deal of attention has been given to understanding and detailing the characteristics of workloads of large-scale multi-processor systems, and on the evaluation of their impact on existing and potential architectural solutions [7,8,9,10,15]. In particular, the seminal work of Feitelson et al. [7,8,10] lead to the comprehension of the workload modeling importance, and to a community-accepted model for rigid jobs in parallel environments [8,10]. However, there is relatively little work on characterizing grouped submissions in large-scale environments [9,14,15,16]. To the best of the authors' knowledge, this work is the first attempt to fully characterize grouped submissions in grids.

The results most closely related to our work are reported in [9], which characterize the size of batch submissions for $\Delta = 7200s$. However, our grid traces would include just a few batches at this interval, as most jobs would be grouped together. In [15], authors characterize applications that can be found in several scientific job batches. Still, they do not estimate the characteristics investigated in this work. Finally, the performance of running PSAs of genomics (i.e., BLAST) and of high-energy physics applications in grid environments have been explored in [14] and [16], respectively.

6   Conclusion and Ongoing Work

In this paper, we have presented a first investigation into the characteristics of groups of jobs present in grid workloads as well as insights into their impact on the performance of grids. First, we have formally defined three types of job groups: batch (and PSAs), continued and bursty submissions. Then, we have analyzed traces coming from one large research grid, namely Grid'5000, and two large production grids, NorduGrid and GLOW. Our analysis 1) shows the presence of arrival time patterns for groups of jobs and 2) reveals the characteristics of the groups and their composing jobs. Notably, we found that batch submissions account for up to 96% of the CPU time consumed in the analyzed traces. Finally, we have contrasted the performance obtained by grouped submissions with that of their non-grouped counterparts. Our results show that batch submissions have high idle times. This points to the need for new, more practical, results in an important area of research: the minimization of the idle time for grouped jobs. We plan to investigate more thoroughly the impact of grouped submissions on grid performance, both with simulations and with experiments in our DAS grid environment in the Netherlands.

Acknowledgements

This research work is carried out in the context of the Virtual Laboratory for e-Science project (`www.vl-e.nl`), which is supported by a BSIK grant from the Dutch Ministry of Education, Culture and Science (OC&W), and which is part of the ICT innovation program of the Dutch Ministry of Economic Affairs (EZ).

We gratefully acknowledge and thank Dr. Franck Cappello and the Grid'5000 team, Dr. Balász Kónya and the NorduGrid team, and Dr. Miron Livny, Dan Bradley, and the Condor team, for providing us with the grid traces used in this work.

References

1. Bolze, R., Cappello, F., Caron, E., Daydè, M., Desprez, F., Jeannot, E., Jègou, Y., Lanteri, S., Leduc, J., Melab, N., Mornet, G., Namyst, R., Primet, P., Quetier, B., Richard, O., Talbi, E.G., Irena, T.: Grid'5000: a large scale and highly re-configurable experimental grid testbed. International Journal of High Performance Computing Applications 20(4), 481–494 (2006)
2. Eerola, P., Kónya, B., Smirnova, O., Ekelöf, T., Ellert, M., Hansen, J.R., Nielsen, J.L., Wäänänen, A., Konstantinov, A., Herrala, J., Tuisku, M., Myklebust, T., Ould-Saada, F., Vinter, B.: The nordugrid production grid infrastructure, status and plans. In: The 4th International Workshop on Grid Computing (Grid2003), Phoenix, AZ, USA (2003) 158–165
3. Iosup, A., Dumitrescu, C., Epema, D., Li, H., Wolters, L.: How are real grids used? the analysis of four grid traces and its implications. In: IEEE/ACM, pp. 262–269. IEEE Computer Society Press, Los Alamitos (2006)
4. Thain, D., Tannenbaum, T., Livny, M.: Distributed computing in practice: the Condor experience. Concurrency and Computation: Practice and Experience 17(2-4), 323–356 (2005)
5. Karonis, N.T., Toonen, B.R., Foster, I.T.: MPICH-G2: A grid-enabled implementation of the message passing interface. J. Parallel Distrib. Comput. 63(5), 551–563 (2003)
6. van Reeuwijk, K., van Niewpoort, R.V., Bal, H.E.: Developing Java Grid applications with Ibis. In: Cunha, J.C., Medeiros, P.D. (eds.) Euro-Par 2005. LNCS, vol. 3648, pp. 411–420. Springer, Heidelberg (2005)
7. Feitelson, D.G., Rudolph, L., Schwiegelshohn, U., Sevcik, K.C., Wong, P.: Theory and Practice in Parallel Job Scheduling. In: Feitelson, D.G., Rudolph, L. (eds.) Job Scheduling Strategies for Parallel Processing. LNCS, vol. 1291, pp. 1–34. Springer, Heidelberg (1997)
8. Feitelson, D.G.: Packing schemes for gang scheduling. In: Feitelson, D.G., Rudolph, L. (eds.) JSSPP. LNCS, vol. 1162, pp. 89–110. Springer, Heidelberg (1996)
9. Squillante, M., Yao, D., Zhang, L.: Analysis of job arrival patterns and parallel scheduling performance. Perform. Eval. 36-37(1-4), 137–163 (1999)

10. Lublin, U., Feitelson, D.G.: The workload on parallel supercomputers: modeling the characteristics of rigid jobs. Journal of Parallel and Distributed Computing 63(11), 1105–1122 (2003)
11. Costa, G.D., Georgiou, Y., Huard, G., Martin, C., Mouniè, G., Neyron, P., Richard, O.: A batch scheduler with high level components. In: Proceedings of the 5th IEEE/ACM International Symposium on Cluster Computing and the Grid (CC-Grid '05), Cardiff, UK, pp. 776–783. IEEE Computer Society Press, Los Alamitos (2005)
12. Ellert, M., et al.: Advanced Resource Connector middleware for lightweight computational Grids. FGCS 23(2), 219–240 (2007)
13. Dumitrescu, C., Raicu, I., Ripeanu, M., Foster, I.: DiPerF: An automated distributed performance testing framework. In: Proceedings of the 5th IEEE/ACM International Workshop on Grid Computing (GRID'04), Pittsburgh, PA, USA, pp. 289–296. IEEE Computer Society Press, Los Alamitos (2004)
14. Dumitrescu, C., Raicu, I., Foster, I.: Experiences in Running Workloads over Grid3. In: Zhuge, H., Fox, G.C. (eds.) GCC 2005. LNCS, vol. 3795, pp. 274–286. Springer, Heidelberg (2005)
15. Thain, D., Bent, J., Arpaci-Dusseau, A.C., Arpaci-Dusseau, R.H., Livny, M.: Pipeline and batch sharing in grid workloads. In: HPDC, pp. 152–161. IEEE Computer Society, Los Alamitos (2003)
16. Bonacorsi, D., et al.: Running CMS software on GRID Testbeds. Computing in High Energy and Nuclear Physics, 24-28, March 2003, La Jolla, CA, USA (2003)

Vigne: Executing Easily and Efficiently a Wide Range of Distributed Applications in Grids

Emmanuel Jeanvoine[1,3], Christine Morin[2,3], and Daniel Leprince[1]

[1] EDF R&D
[2] INRIA
[3] IRISA – PARIS project-team

Abstract. Using grid resources to execute scientific applications requiring a large amount of computing power is attractive but not easy from the user's point of view. Vigne is a Grid system designed to provide users with a simplified view of a grid. This paper presents a set of system services that allow to run a wide range of distributed applications in a simple and efficient manner. A running prototype has been implemented as a proof of concept and experiments on the Grid'5000 testbed show the efficiency of our approach.

1 Introduction

Numerical simulation has an important place in several scientific fields where experimentation is impossible. Grids have become attractive for the execution of numerical simulations since they provide a large computational power. However, using a grid where resources are spread all around the world and have a dynamic behavior is still a challenge.

Several systems are designed to execute applications on a grid and in particular the distributed ones. However they suffer from several issues. Sometimes, they are dedicated to some kinds of applications. They often require some modifications in the code of the application, some re-linking with external libraries or even a specific implementation. Finally, few systems are able to deal with particular application constraints like the need for an efficient network between the application tasks.

We propose Vigne, a system to simplify the use of a grid from the users' and the administrators' point of view. We have designed this system in order to deal with the dynamic behavior of resources in a large scale environment. It is self-healing and self-organizing [1]. Thanks to an architecture built on peer-to-peer overlay networks [2], Vigne can handle several simultaneous failures and does not suffer from any single point of failure. Vigne is composed of a set of services that provide facilities for resource discovery and allocation, memory sharing, file transfers, resource interfaces and application management. With these services, Vigne supports a wide range of applications, even the legacy ones, and does not require their modification. Vigne takes into account the specific constraints of the distributed applications and is able to perform synchronized and spatial

A.-M. Kermarrec, L. Bougé, and T. Priol (Eds.): Euro-Par 2007, LNCS 4641, pp. 394–403, 2007.
© Springer-Verlag Berlin Heidelberg 2007

resource allocations by providing a simple application description formalism. Finally, Vigne is designed to harness all the grid models like desktop grids, peer-to-peer computing or federated clusters.

Experiments conducted on the Grid'5000 testbed [3] with a large number of nodes validate our approach and show its efficiency.

The rest of the paper is as follows. In Section 2 we present the background with some distributed application models and related works; we also introduce our approach. In Section 3, we present the Vigne features related to the execution of distributed applications on a grid. In Section 4 we present experiments of Vigne on the Grid'5000 testbed. Finally, we conclude and present future works in Section 5.

2   Background

Several kinds of applications can take advantage of the computational power of a grid. In particular, the distributed applications can use several nodes of the grid to reduce the computation time. The applications composed of a single task are not taken into account in this paper since executing them is not so challenging.

We present in this section several models of distributed applications and some related works on the execution of distributed applications in grids.

2.1   Distributed Application Models

Several paradigms can be used to develop distributed applications according to the nature of the parallelism that can be introduced in those applications. First of all, the simplest model of distributed application is the "bag of tasks". In this model, a lot of independent tasks are executed and the goal is to minimize the makespan. Neither synchronization, nor close resource allocation is required between tasks.

Master-worker applications are also composed of a large set of independent tasks. With this model, the master assigns the tasks to the workers and collects the results. The resources required for the execution must be allocated at the same time, which represents a synchronization dependency between the tasks.

Some applications, inherited from the cluster computing world, are composed of communicating tasks. Such applications are usually based on message passing environments like MPI and are used for instance to solve linear algebra computations. Other applications, designed to solve problems related to multi-physic domains, use code coupling methods. These two kinds of applications may communicate intensively and require an efficient network connection. With this model, the resources required for the execution must be allocated at the same time and must be close, from the network point of view, in order to provide high performance network communications. This represents synchronization and spatial dependencies between the tasks. Other applications have a more complex architecture and are composed of sub-applications. The sub-applications can have several architectures like: sequential, parallel, code-coupling or master-worker.

Table 1. Dependencies between tasks according to the application models

	Synchronization	Spatial	Precedence
Bag of tasks			
Master-Worker	X		
Parallel/Code-coupling	X	X	
Workflow	X	X	X

The particularity of these complex applications is that data may be exchanged between the sub-applications. These applications are often represented with a workflow. With this complex model, synchronization and spatial dependencies may be required according to the sub-application models. Furthermore, a precedence dependency is added because some sub-applications may require the output of another one to be executed.

So, it is possible to identify three kinds of dependencies between the tasks of a distributed applications that are: spatial, synchronization and precedence dependencies. This is summarized in the table 1.

2.2   Running Distributed Applications in Grids

Several grid middlewares support the execution of distributed applications since they rely on a batch scheduler. These middlewares are not presented here and we only focus on those that natively support the execution of distributed applications and we explain how they address the issues of dependencies presented before.

Globus [4] is a toolkit that provides a set of low-level features to federate resources on a grid. It allows to run distributed applications and provides a partial solution for the synchronization dependency without requiring the modification of the applications. The problem is that Globus does not provide facilities to synchronize the allocation of several resources for a parallel or a code-coupling application. Zorilla [5] is a system based on an overlay network and provides locality-aware co-allocation with a mechanism called *flooding scheduling*. However, due to the nature of the scheduling, a lot of resources may not be used and the resulting allocation might be inefficient.

Some middlewares offer attractive features to handle synchronization and/or spatial dependencies between the tasks but they impose the modification of the applications. [6] presents an architecture for co-allocation in Globus. Primitives must be added in the applications to perform an initial barrier to synchronize all the resources used before an execution. No spatial dependency is taken into account in this system. Legion [7] provides similar features but also implies the modification of the applications in order to give them access to the Legion object-oriented programming model. Vishwa [8] provides a framework to execute distributed applications with synchronization dependencies by using the concept of Distributed Pipes that requires the modification of the applications. KOALA [9] is a meta-scheduler built on top of Globus. It provides features for spatial dependencies by placing tasks close to specified data like a file server for instance. XtremWeb [10] is a platform for global peer-to-peer computing. It targets master-worker and workflow applications and provides facilities to create distributed applications with JavaRMI or XML-RPC.

Several systems handle precedence dependencies between the application tasks. These systems are mainly workflow engines like Triana [11].

3 Vigne, a Grid System Designed to Simplify the Execution of Distributed Applications

Our contribution is the design and the implementation of the Vigne system that allows to execute distributed applications with the possibility to define synchronization, spatial and precedence dependencies between the tasks and without modifying legacy applications.

In this section, we present the Vigne features to simplify the execution of distributed applications. In particular we present the Vigne features to handle the three dependencies between the tasks mentioned in Section 2.1 and the support for legacy applications.

3.1 Full Application Example

Figure 1 presents an example of application that is used in the following. The application is composed of 7 tasks. Task 1 produces a result that is used by two task groups ({2,3}and {4,5}). The result of the group {2,3}is used by 7. The result of the group {4,5}is used by 6. Finally the result of 6 is used by 7. We can notice that the two groups represent two parallel applications. Figure 2 shows the application description in an XML-like syntax that can be used to submit the example application to the Vigne system. In order to save space, the things related to the type of resources required for the execution are not included in this description.

3.2 Handling Synchronization Dependencies

Vigne allows to define several tasks in an application like one can see the 7 tasks in the example. If no constraint is given to the system, the tasks are executed independently. This means that the system performs a resource discovery and a resource allocation for each task and then it executes those tasks on their respective allocated resources. In the application example, there is no synchronization constraints between the tasks 1, 6 and 7.

If a synchronization constraint is given between a subset of tasks, Vigne performs a resource discovery for each task and a synchronized resource allocation for the tasks impacted by the constraint. When a resource is discovered for an application, Vigne ensures that this resource will not be used for another application at the same time by temporary locking it. At the end of the allocation, the unused resources are unlocked. As a consequence, the resources synchronously allocated with Vigne are available and free.

In our example, the groups {2,3} and {4,5} have a synchronization constraint. With Vigne, we describe that with the property `basic_synchro`. Then the value

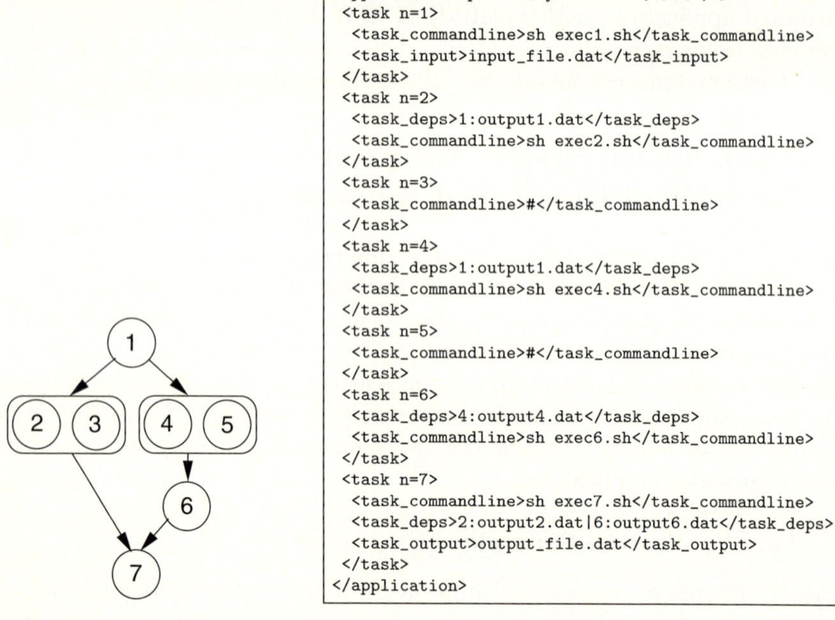

```
<application spatial_synchro="{2,3}|{4,5}">
 <task n=1>
  <task_commandline>sh exec1.sh</task_commandline>
  <task_input>input_file.dat</task_input>
 </task>
 <task n=2>
  <task_deps>1:output1.dat</task_deps>
  <task_commandline>sh exec2.sh</task_commandline>
 </task>
 <task n=3>
  <task_commandline>#</task_commandline>
 </task>
 <task n=4>
  <task_deps>1:output1.dat</task_deps>
  <task_commandline>sh exec4.sh</task_commandline>
 </task>
 <task n=5>
  <task_commandline>#</task_commandline>
 </task>
 <task n=6>
  <task_deps>4:output4.dat</task_deps>
  <task_commandline>sh exec6.sh</task_commandline>
 </task>
 <task n=7>
  <task_commandline>sh exec7.sh</task_commandline>
  <task_deps>2:output2.dat|6:output6.dat</task_deps>
  <task_output>output_file.dat</task_output>
 </task>
</application>
```

Fig. 1. Architecture of the ap- **Fig. 2.** Vigne description for the example application
plication example

of the property defines which tasks are impacted by the synchronization constraint. In the Vigne formalism, tasks that have a dependency are defined in a
set and several sets can be defined by separating them with a pipe.

3.3   Handling Spatial Dependencies

To handle spatial dependencies between tasks, Vigne uses the same mechanisms
that the ones used to handle synchronization dependencies but in addition, the
resource allocator minimizes the distance, from the network point of view, between the resources synchronously allocated to the tasks.

Several methods can be used to measure the distance between resources in a
network. We present here two methods that can be used in the Vigne system.
First, it is possible to create the network matrix of the latencies between the
discovered resources of all the tasks concerned by a spatial constraint. Even if
that kind of method provides good results because it is really accurate, constructing the network matrix induces a significant network traffic, in particular
if the grid nodes have a dynamic behavior. Indeed, in this case the network matrix cannot be created once and for good. That is why we choose to use a more
simple method that uses the DNS name of the resources. We assume that the
resources that share a bigger DNS suffix have a high probability to be close. For
instance, the node **node22.rennes.grid5000.fr** should be closer to the node
node51.rennes.grid5000.fr than to the node **node13.lyon.grid5000.fr**.

Then, we use a backtracking algorithm that performs successive tries to find a good combination of the resources.

To specify spatial dependencies in Vigne, one must use the property spatial_synchro of the tag application and the same pattern that the one defined in Section 3.2. In Vigne, a spatial constraint induces a synchronization constraint. In our example, the groups {2,3}and {4,5}have a spatial constraint because they represent two parallel applications with intensive network communications.

3.4 Handling Precedence Dependencies

In order to run workflow applications or specific application models, Vigne provides users with features to formulate precedence relations between tasks.

When an application is submitted to Vigne, each task of the application is analyzed. If some tasks do not have precedence dependency with other tasks, Vigne performs a resource discovery for them. This is the case in our example for the task 1. For the other tasks, Vigne waits for the dependencies between these tasks to be satisfied. As a consequence, when an task completes its execution, Vigne checks if other tasks depends on this task and notifies the given tasks. When a task receives a notification, Vigne checks if all the dependencies are satisfied and if it is the case, it launches the resource discovery for this task. In our example, the tasks 2, 3, 4, 5 are executed only at the end of the task 1. The task 6 is executed at the end of the tasks 4 and 5. Finally, the task 7 is executed at the end of the tasks 2, 3 and 7. As far as resources are allocated only when a subset of the tasks is ready to be executed, no resource is wasted. However, this approach has a major drawback. If an application is composed of some tasks that require resources that are not in the grid, the execution application would be interrupted and some tasks would have been previously executed uselessly. A solution would be to reserve the resources at the submission time, but in this case, a lot of resources may be wasted. If we consider a large grid, the probability that tasks require resources that are not in the grid is quite low. So we prefer the first solution, which is best to avoid resource wasting.

When precedence dependencies between tasks are formulated, Vigne also offers the facility to transfer directly some files between the resources of the dependent tasks. This operation is possibly performed at the end of the execution of a task.

In this paper, we only focus on precedence dependencies induced by file exchanges between tasks, i.e. we deal with the case where one task produces some files that are used as an input for other tasks. Another application model where the tasks use shared memory to exchange data has been studied in [1].

To specify precedence dependencies in Vigne, one must use the tag task_deps of the concerned tasks. The value of the tag is as follows (in a regular expression syntax):

```
task_number:[file][,file]*[|task_number:file[,file]*]*
```

In our example, one can see that the task 7 needs the output files output2.dat of the task 2 and output6.dat of the task 6 to be executed.

We can notice that the description of our application example does not specify precedence dependencies between the tasks 1 and 3 because the tasks 2 and 3 are synchronized with a spatial, and implicitly with a synchronization, dependency. As a consequence, the task 3 will not be executed before the end of the task 1. The same occurs between the tasks 1 and 5.

3.5   Legacy Application Support

Some applications need to know the location of all their tasks when they are running. For instance, one must provide a *machine-file* to run an MPI application. As far as the location of the resources used by the application is not known at the submission time, this kind of information cannot be given by users.

Vigne provides several environment variables in the execution environment of all the resources used by the application.

These variables are $VIGNE_MACHINEFILE and $VIGNE_TASK_n (where n is the number of the task). $VIGNE_MACHINEFILE contains the path to a file where the location of all the resources used by the application are written. $VIGNE_TASK_n contains the DNS name of the resource allocated to the task n. As a consequence, the application does not need to be modified to be executed.

For instance, the script exec2.sh used by the task 2 in our example (see Fig. 2) may contain something like:

```
echo $VIGNE_TASK_2 > machines && echo $VIGNE_TASK_3 >> machines
lamboot -d machines && mpirun -np 2 mpi_program && lamhalt
```

The first line of the script is used to create the machine-file with the resources used by the tasks 2 and 3. The second line is used to launch mpi_program on two resources using the LAM MPI environment.

4   Experiments

4.1   Vigne Prototype

A working prototype of Vigne has been implemented with all the features presented in Section 3. It runs on Linux based systems and it does not rely on any middleware. Vigne is written in C and the prototype is materialized by a daemon that must be run on each grid node and a client used by grid users to perform queries like: application submission, information about execution, execution cancellation, manual file transfers between tasks, etc.

To submit an application, a user must provide an XML description like it is shown in Figure 2. To simplify the submission of applications, a GUI could be developed to generate easily the XML description.

4.2   Evaluation of the Co-allocation Mechanisms

In these experiments, we evaluate the co-allocation mechanisms of Vigne. To do that, we have used a parallel MPI application computing the number π. Implicitly, an MPI application has synchronization dependencies between its tasks

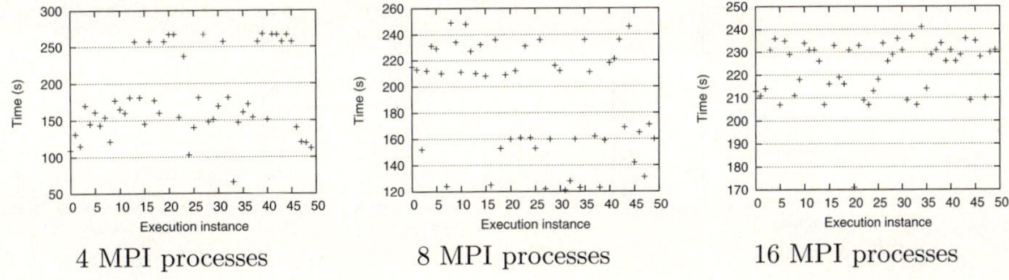

Fig. 3. Execution time of the application without spatial optimizations

since each task must be run at the same time. Furthermore, this application performs intensive network communications, so it has spatial dependencies between its tasks.

The precedence dependency is not evaluated here since it can only be evaluated qualitatively and the real difficulties rely in the management of the two other dependencies.

We perform two types of experiments. In the first one, we do not use the spatial synchronization, so we use the `basic_synchro` property in the application description. In the second one, we use the `spatial_synchro` property in order to minimize the distance between the resource allocated to the tasks. For each experiment, we launch 50 instances of the application (the instances are launched every two seconds). Each experiment is launched three times, with a different value for the number of nodes used by the MPI application. The values are 4, 8 and 16 nodes.

4.3   Setup

To evaluate the Vigne prototype, we have used the Grid'5000 [3] testbed with a large number of nodes. Grid'5000 is a French experimental testbed for research in Grid computing composed of several hundreds of nodes spread over 9 sites in France. We have used 376 nodes spread over 6 sites: Lille (31), Lyon (45), Nancy (42), Orsay (93), Rennes (115), Sophia (50). Vigne has been deployed over all these nodes.

4.4   Results

On Figure 3, one can see the execution time of each instance of the MPI application for the experiment where the spatial optimizations have not been used. On Figure 4, one can see the execution time of each instance of the application for the experiment where the spatial optimizations have been used.

If we observe Figures 3 and 4, we can see that the spatial optimization seriously reduces the execution time of the tasks. We can also observe that an average speedup of 40% is obtained with the spatial optimizations when we use 8 nodes instead of 4 nodes. Without spatial optimization, there is no real speedup on average. With 16 nodes, the spatial optimizations do not provide as good results

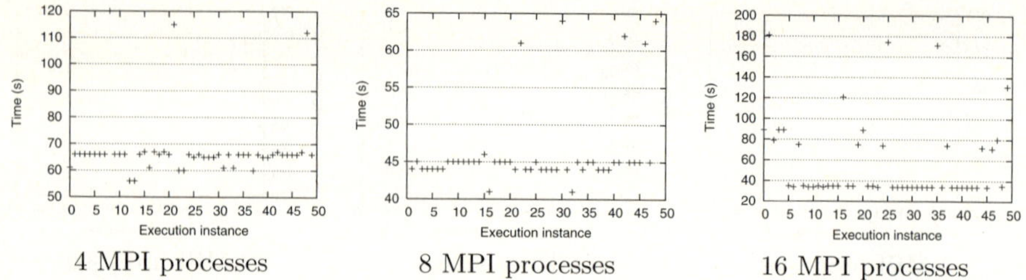

4 MPI processes   8 MPI processes   16 MPI processes

Fig. 4. Execution time of the application with spatial optimizations

Table 2. Occurrences of the number of sites used in a computation according to the number of nodes per application

	Without spatial optimization			With spatial optimization		
	4 nodes	8 nodes	16 nodes	4 nodes	8 nodes	16 nodes
1 site	1	0	0	47	45	33
2 sites	14	0	0	3	5	12
3 sites	22	9	0	0	0	2
4 sites	13	24	1	0	0	3
5 sites	0	14	19	0	0	0
6 sites	0	3	30	0	0	0

as with 4 or 8 nodes because the resources are much more overloaded and in this case it is hard to find 16 resources on the same site. However, the results with the spatial optimization are better than those without optimization. To confirm the results shown on Figures 3 and 4, the Table 2 shows the dispersion of the executions according to the number of sites used each time. Here, we can observe that with the spatial optimization, the great majority of the executions occurs with a resource allocation on a single site per application. Without optimization, the resources were mainly allocated on 2 sites and more. The results show that the co-allocation features of Vigne are working and are really efficient. The application used in the experiments performs intensive network communication, but this represents the worst case for a co-allocation mechanism. Indeed, if the resources used by the tasks are randomly allocated in the grid, large latencies between the nodes induce to much overhead for the overall performance, as shown in Figure 3.

5   Conclusion

This paper presents Vigne, a Grid system to execute simply and efficiently distributed applications in large scale grids. In particular, we propose a set of system features implemented in the system Vigne that allow to execute distributed applications without modification of the code and in an efficient way thanks to spatial optimizations in the resource allocation service. The Vigne system and the proposed mechanisms have been evaluated on the Grid'5000 testbed with a large number of nodes. The results show the efficiency of our approach.

Future works will be dedicated to the ability to execute distributed applications on a grid where resources may be located behind firewalls or private networks. Vigne will also be evaluated with industrial applications, in particular in the framework of the SALOME numerical platform.

References

1. Rilling, L.: Vigne: Towards a Self-Healing Grid Operating System. In: Nagel, W.E., Walter, W.V., Lehner, W. (eds.) Euro-Par 2006. LNCS, vol. 4128, pp. 437–447. Springer, Heidelberg (2006)
2. Jeanvoine, E., Rilling, L., Morin, C., Leprince, D.: Using overlay networks to build operating system services for large scale grids. In: Proceedings of the 5th International Symposium on Parallel and Distributed Computing (ISPDC 2006), Timisoara, Romania (July 2006)
3. Grid'5000 website. Web Page: http://www.grid5000.fr
4. Foster, I., Kesselman, C.: Globus: A metacomputing infrastructure toolkit. International Journal of Supercomputer Applications 11(2), 115–128 (1997)
5. Drost, N., van NieuwPoort, R.V., Bal, H.E.: Simple locality-aware co-allocation in peer-to-peer supercomputing. In: Proceedings of the Sixth International Workshop on Global and Peer-2-Peer Computing (GP2P), Singapore, vol. 2, p. 14 (May 2006)
6. Czajkowski, K., Foster, I., Kesselman, C.: Resource co-allocation in computational grids. In: Proceedings of High Performance Distributed Computing (HPDC-8 '99), pp. 219–228. IEEE Computer Society Press, Los Alamitos (1999)
7. Chapin, S.J., Katramatos, D., Karpovich, J.F., Grimshaw, A.S.: The legion resource management system. In: Proceedings of the Job Scheduling Strategies for Parallel Processing, London, UK, pp. 162–178 (1999)
8. Venkateswara Reddy, M., Vijay Srinivas, A., Gopinath, T., Janakiram, D.: Vishwa: A reconfigurable peer-to-peer middleware for grid computations. In: Proceedings of the 35th International Conference on Parallel Processing, Ohio, USA, pp. 381 390. IEEE Computer Society Press, Los Alamitos (2006)
9. Mohamed, H., Epema, D.: The design and implementation of the koala co-allocating grid scheduler. In: Sloot, P.M.A., Hoekstra, A.G., Priol, T., Reinefeld, A., Bubak, M. (eds.) EGC 2005. LNCS, vol. 3470, pp. 640–650. Springer, Heidelberg (2005)
10. Cappello, F., Djilali, S., Fedak, G., Herault, T., Magniette, F., Nri, V., Lodygensky, O.: Computing on large-scale distributed systems: Xtremweb architecture, programming models, security, tests and convergence with grid. Future Generation Computer Systems 21, 417–437 (2005)
11. Taylor, I., Shields, M., Wang, I., Harrison, A.: Visual grid workflow in triana. Journal of Grid Computing 3(3-4), 153–169 (2005)

Are P2P Data-Dissemination Techniques Viable in Today's Data-Intensive Scientific Collaborations?

Samer Al-Kiswany[1], Matei Ripeanu[1], Adriana Iamnitchi[2],
and Sudharshan Vazhkudai[3]

[1] University of British Columbia
`{samera,matei}@ece.ubc.ca`
[2] University of South Florida
`anda@cse.usf.edu`
[3] Oak Ridge National Laboratory
`vazhkudaiss@ornl.gov`

Abstract. The interest among a geographically distributed user base to mine massive collections of scientific data propels the need for efficient data dissemination solutions. An optimal data distribution scheme will find the delicate and often application-specific balance among conflicting success metrics such as minimizing transfer times, minimizing the impact on the network, and uniformly distributing load among participants. We use simulations to explore the performance of classes of data-distribution techniques, some of which successfully deployed in large peer-to-peer communities, in the context of today's data-centric scientific collaborations. Based on these simulations we derive several recommendations for data distribution in real-world science collaborations.

1 Introduction

Modern science is data-intensive. Large-scale simulations, new scientific instruments, and large-scale observatories generate massive volumes of data that need to be analyzed by large, geographically dispersed user communities. These trends are emerging in fields as diverse as bioinformatics and high-energy physics. Examples include the Large Hadron Collider (LHC) experiment at CERN and the DØ experiment at Fermi Lab. Aiding in the formation of these collaborative data federations are ever increasing network capabilities including high-speed optical interconnects (e.g., Lambda-Grid) and highly optimized bulk transfer tools and protocols (e.g., GridFTP).

Data dissemination in such federations involves dynamic distribution of subsets of data, available at one site, to one or many collaborating locations for real-time analysis and visualization. For instance, the PetaBytes of data from the LHC experiment have to be distributed world-wide, across national and regional centers.

Two conflicting arguments compete to shape the one-to-many delivery of large scientific data over well provisioned networks. On one side, there is the intuition that the well provisioned networks are sufficient to guarantee good data-delivery performance; sophisticated algorithms that adapt to unstable or limited-resource environments are superfluous and add unjustified overheads in these environments. The counterargument is that advanced data dissemination systems are still required as the size of data and the relatively large collaborations create contention and bottlenecks on

A.-M. Kermarrec, L. Bougé, and T. Priol (Eds.): Euro-Par 2007, LNCS 4641, pp. 404–414, 2007.

shared resources, which hinder efficient usage. Additionally, even if contention for shared resources is not a serious concern, the question if networks are over provisioned and thus generate unnecessary costs remains.

These two arguments motivate this study: we explore, experimentally, the space of solutions for one-to-many, large-scale data delivery in today's environments via simulations. We consider solutions typically associated with peer-to-peer (P2P) applications (such as BitTorrent) and evaluate them under our target scenario of large data federations. To this end, we used both generated as well as real production Grid testbed topologies in our evaluations.

Our contribution is twofold. First, we quantitatively evaluate and compare a set of representative data-delivery techniques applied to a grid environment. The quantitative evaluation is then used to derive well-supported recommendations for choosing data-dissemination solutions and for provisioning the Grid networking infrastructure. Further, our study contributes to a better understanding of the performance tradeoffs in the data-dissemination space. To the best of our knowledge, this is the first, head-to-head comparison of alternative solutions using multiple performance metrics. Second, we propose a simulation framework that can be used to explore optimal solutions for specific deployments or can be extended for new dissemination solutions.

To derive our recommendations, we identify a relevant set of candidate solutions from different domains (Section 3), build a simulator (presented in Section 4) and evaluate the candidate solutions on key metrics such as time-to-delivery, generated overhead, and load balance (Section 5). We summarize our findings in Section 6.

2 Data in Scientific Collaborations

Today, Grids are providing an infrastructure that enables users to dynamically distribute and share massive datasets. However, most data distribution strategies currently in place involve explicit data movement through batch jobs that are seldom sympathetic to changing network conditions, congestion and latency, and rarely exploit the collaborative nature of modern-day science [1].

On the other hand, P2P file sharing and collaborative caching efficiently exploit patterns in users' data sharing behavior. However, such techniques are not directly adaptable to Grid settings because of different usage scenarios. In particular, three key differences make it difficult to predict the behavior of P2P techniques in scientific data federations: scale of data, data usage characteristics, and resource characteristics.

The *scale of data* poses unique challenges: scientific data consists of massive collections comprising of hundreds to thousands of files. For instance, of the more than one million files accessed in DØ between January 2003 and May 2005, more than 5% are larger than 1GB and the mean file size is larger than 300MB [2].

Usage of data in scientific communities varies in *intensity* compared to other communities. For example, 561 scientists from DØ processed more than 5PB of data in 29 months, which translates to accessing more than 1.13 million distinct data files [2]. However, *popularity distributions* for scientific data are more uniform than in P2P systems or in the Web. Further, in scientific environments, files are often used in groups and not individually.

Finally, *resource availability* in grids poses smaller challenges than in P2P systems. Computers stay connected for longer, with significantly lower churn rate and higher availability due to hardware and software configurations. At the same time, data federations are overlays built atop well-provisioned networks (e.g., TeraGrid) as opposed to the commercial Internet. Additionally, *resource sharing is often enforced by out-of-band means*, such as agreements between institutions or between institutions and funding agencies. For this reason, mechanisms that enforce participation and fair sharing, such as the tit-for-tat scheme of BitTorrent, are often unnecessary.

To summarize, most Grid data distribution strategies currently in place fail to exploit the characteristics and data usage patterns emerging in today's scientific collaborations. The properties (huge data volumes, well provisioned networks, stable resources, and cooperative environments) of these data collaborations, however, invite the *question whether P2P data-distribution strategies result in tangible gains on well-endowed network infrastructures on which today's Grids are deployed*. A careful study is necessary to derive recommendations for building and provisioning future testbeds and choosing efficient dissemination approaches for science collaborations.

3 Data Distribution: Solutions and Metrics

We have identified a number of techniques as potential candidates for our comparison. We provide a classification of data distribution techniques (Section 3.1), detail the techniques we explore in this paper (Section 3.2), and present the criteria over which they are typically evaluated (Section 3.3).

3.1 Classification of Approaches

We identify three broad categories of techniques to optimize data distribution: data staging, data partitioning, and exploiting orthogonal bandwidth. In this section we describe these techniques and discuss them in the context of our target environment.

Data Staging. With data staging, participating nodes are used as intermediate storage points in the data distribution solution. Such an approach is made feasible due to the emergence of network overlays. For instance, it is becoming increasingly common practice for application-specific groups to build collaborative networks, replete with their application-level routing infrastructure. This is based on the premise that sophisticated applications are more aware of their resource needs, deadlines, and constraints and can thus perform better resource allocation and scheduling. In this vein, P2P file-sharing systems can be viewed as data-sharing overlays with sophisticated application-level routing performed atop traditional Internet.

In data grids, data staging is encouraged by the increasing significance of application-level tuning of large transfers. For instance, collaborating sites often use path information to make informed decisions to access data from preferred locations, based on a delivery constraint schedule [3]. A logical extension is thus to utilize the participating sites as intermediary data staging points for efficient data dissemination. Additionally, a data distribution infrastructure can include a set of intermediary, strategically placed resources to stage data (e.g., IBP [4]).

Data Partitioning. To add flexibility, various P2P data distribution solutions split files into blocks that are transferred independently (e.g., BitTorrent[5]). Much like the aforementioned application-level routing, this approach allows applications a greater degree of control over data distribution. Further, it enables application-level error correction. For example, when downloading a file from multiple replicas, partitioning can be coupled with erasure coding to achieve fault tolerance. Partitioning techniques have significant value in a data grid setting. Bulk data movement in the Grid is usually long-haul transfers that have to survive a wide range of failures (e.g., network outage, security proxy expiration). Thus, there is a genuine need to provide application-level resilience for data transfers.

Orthogonal Bandwidth Exploitation. Once a basic file partitioning mechanism is in place, it can then be used to accelerate data distribution by exploiting "orthogonal bandwidth", i.e., the bandwidth that cannot be used by a traditional, source-routed data-distribution tree. This is the underlying premise in a number of commercially deployed (e.g., BitTorrent) or academically designed (e.g., Bullet [6]) data-distribution systems that owe much of their success to such optimizations.

Intuitively, it seems these techniques will have commensurate gains when applied to data grids. However, several of these optimizations are designed to work in a non-cooperative environment, where peers contend for scarce resources (e.g., bandwidth). One question to address is how this intuition translates when the bandwidth is plentiful and users are cooperative, as is the case with current scientific data collaborations.

3.2 Candidate Solutions for Evaluation

For our experimental study, we selected representative solutions from each of the categories presented above. We also include other traditional, well understood techniques for comparison. A brief description of these solutions follows (Our technical report [7] has a complete discussion).

Application-level multicast (ALM) solutions organize participating nodes into a source-rooted distribution tree overlay. What differentiates various ALM solutions is the algorithm used to build and maintain the distribution tree. For our experiments, we chose ALMI [8], a solution offering near optimal trees built using global views.

BitTorrent [5] is a popular data distribution system that exploits the upload bandwidth of participating peers for efficient data dissemination. Participating nodes build transitory pair-wise relationships and exchange missing file blocks. BitTorrent assumes a non-cooperative environment and employs a tit-for-tat incentive mechanism to discourage free riders. Each node selects its peers to minimize its own time to acquire content disregarding the overall efficiency of the data distribution operation.

Bullet [6] offers a way to exploit orthogonal bandwidth by distributing disjoint subsets of data using an initial distribution tree (we use the ALM-built tree in this study). Informed delivery techniques [9] are then used to reconcile these data sets stored at destination nodes, which exchange the necessary blocks.

Logistical multicast [4] employs strategically placed nodes in an overlay to expedite data distribution. We evaluate an idealized variation of this approach that associates storage with each router in our topologies.

Spider [10] offers a set of heuristics that enables fast content distribution by building multiple source-rooted trees (assuming global views).

3.3 Success Metrics

Multiple categories of success metrics can be defined for most data management problems. We note that the relative importance of these metrics is dependent on the application context. Thus, no data distribution solution is optimal for all cases and a careful evaluation of various techniques is required to choose a solution appropriate for a specific application context and deployment scenario. The most representative performance objectives include:

- *Minimizing transfer times*: The application focus may be to minimize the average, median, N^{th} percentile, or the highest transfer time to destination.
- *Minimizing the overall impact on the network*: This involves minimizing the load on bottleneck links, the volume of generated traffic, or the overall network 'effort'.
- *Load balancing*: Enlisting all participating sites in the data dissemination effort makes spreading the load among them crucial.
- *Fairness*: Being fair in the presence of concurrent transfers can be an important concern depending on the lower-layer network and the protocols used.

4 Simulating Data Dissemination

To evaluate the techniques above, we have built a block-level simulator. This section presents the set of decisions that guided our simulator design. For a detailed description of the simulator we refer the reader to Al Kiswany et al. [7].

As with most simulators, the main tradeoff is between the amount of resources allocated for simulation and simulation fidelity. At one end of the possible design spectrum are packet-level simulators and emulators. These require significant hardware resources, but model application performance faithfully by running unmodified application code and simulating/emulating network transfers at the IP-packet level. At the other end of the spectrum are high-level simulators that abstract the application transfer patterns and employ only coarse network modeling. Our simulator sits in between these two extremes. From an application perspective, the granularity is file block transfer, a natural choice since many of the data dissemination schemes we investigate use file blocks as their data management unit. From a network perspective, while we do not simulate at the packet level, we do, however, simulate link level contention between application flows.

Additionally, our simulator design is guided by the following decisions:

- *Ignore control overheads.* For our target scenario, protocol control overheads are often orders of magnitude lower than the effort to transfer the actual data payload. Moreover, since the control messages overlap or are piggybacked on data transfers, the latency introduced by the control channel is often negligible.
- *Use of global views.* Our simulator uses a global view of the system in order to hide algorithmic details that are not relevant to our investigation.
- *Ignore competing traffic.* To increase simulator scalability we do not directly model competing traffic. Competing traffic can be modeled, however, by varying the available bandwidth of the links in the simulated network topology.

We experiment with the four solutions for data dissemination solutions described above: application-level multicast (ALM), BitTorrent, Bullet, and logistical multicast. To study their performance, we compare them with two base cases: IP multicast (and its improvement using Spider heuristics) and the naïve, yet popular, solution that sends separate copies of the data from the source to each destination.

For the more complex protocols, Bullet and BitTorrent, the simulator models each block transfer. This is necessary due to the non-deterministic nature of these data dissemination solutions. The simulations use a default block size of 512KB, as in deployed BitTorrent systems. We experimented with different block sizes and noticed that the block size does not significantly affect the results as long as files can be split in a large number of blocks. All our simulations explore the performance of distributing a 1GB file over different topologies.

5 Experimental Results

We use the physical network topologies of two real-world grid testbeds EGEE [11] and GridPP [12]. The EEGE topology is presented in Figure 1. Our technical report [7] presents the detailed GridPP topology. Additionally, to increase the confidence in our results, we use BRITE to generate two sets of larger (hundreds of nodes) Waxman topologies [13]. These sets have the same number of nodes and constant overall bandwidth, and differ only in the density of core links. Our goal is to compare results on these two sets of

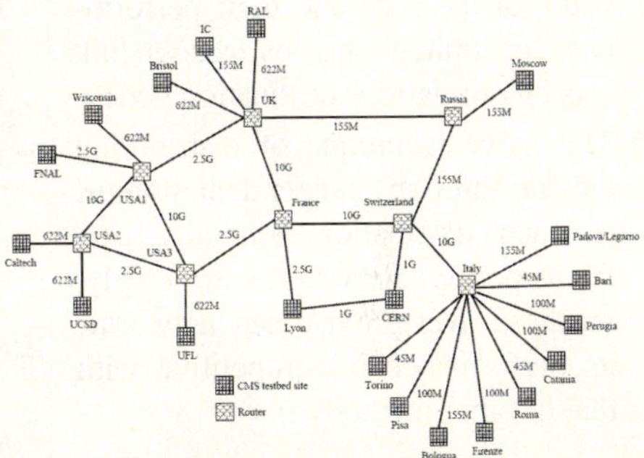

Fig. 1. EGEE topology

topologies to obtain a more direct measure of the degree to which various protocols are able to exploit network path diversity. Due to space limitation, we present the key observations and direct the interested reader to our detailed results, for all of these topologies, in our technical report [7].

5.1 Performance: File Transfer Time

Depending on the application context, the performance focus can be on minimizing the average, median, N^{th} percentile, or the largest transfer time to destination. To cover all these criteria, for each data dissemination technique we present the evolution in time of the number of destinations that have completed the file transfer.

Figure 2 presents this evolution for the original EGEE topology. Note that here Spider builds only one dissemination tree and is thus equivalent to IP-multicast. In spite of the slightly different results for various topologies, the following observations are common:

- IP-multicast and Logistical Multicast are the best solutions to deliver a file to the slowest node as they optimally exploit the bandwidth on bottleneck links.
- Intermediate progress with IP-multicast is poor. The reason is that multicasting schemes do not include buffering at intermediate points in the network and limit their data distribution rate to the rate of the bottleneck link.
- Logistical Multicast is among the first to complete the file dissemination and also offers one of the best intermediate progress performance. This is partially a result of the bandwidth distribution in these two topologies: the bottlenecks are the site access links and not the links at the core of the network. As a result, Logistical Multicast is

410   S. Al-Kiswany et al.

able to push the file fast through the core routers that border the final access link and thus offer near optimal distribution times.

- Application-level multicast (ALM), Bullet and BitTorrent are worse but comparable to Logistical Multicast, both in terms of finishing time as well as intermediate progress. They are able to exploit the plentiful bandwidth at the core and their performance is limited only by access link capacity of various destination nodes.

- The naïve technique of distributing the file through independent streams to each destination generally performs poorly. However, surprisingly, on these over-provisioned networks, its performance is competitive with that of other methods.

The surprisingly good performance of parallel independent transfers in these topologies clearly indicates that the network core is over-provisioned. We are interested in exploring the performance of data dissemination techniques at different core-to-access link capacity ratios for the following two reasons. First, if the core is over-provisioned, we would like to understand how much bandwidth (and eventually money) can be saved by reducing the core capacity without significantly altering the dissemination performance. Second, we aim to understand whether independent transfers perform similarly well when compared to more sophisticated techniques under different network conditions.

With these two goals in mind we ran the same simulations on a set of hypothetical topologies. These topologies are similar to the original EGEE and GridPP topologies except that the bandwidth of the core links (the links between the routers) is $^1/_2$, $^1/_4$, $^1/_8$, or $^1/_{16}$ of the original core link bandwidth.

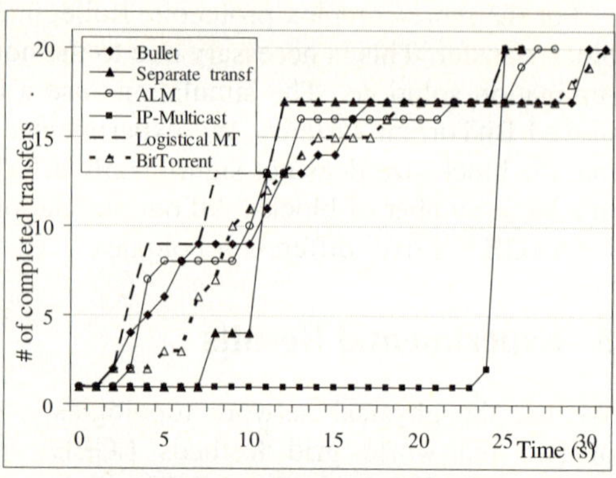

Fig. 2. Number of destinations that have completed the file transfer (original EGEE topology)

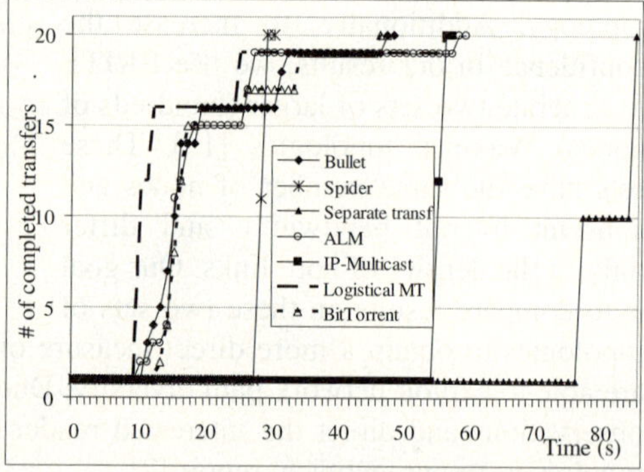

Fig. 3. Number of destinations that have completed the file transfer (EGEE topology with core bandwidth reduced to $^1/_8$ of the original)

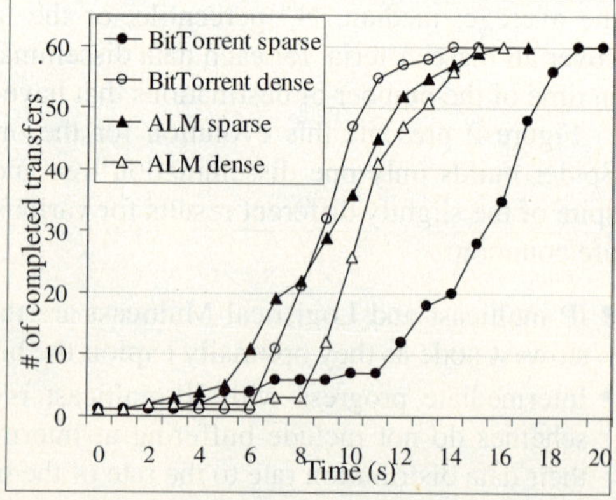

Fig. 4. Number of destinations that have completed the file transfer with two generated topologies. The dense topology has four times more links in the core with 4 times less average bandwidth per link.

Figure 3 presents the case of core link bandwidth equal to $^1/_8$ of the original bandwidth in the EGEE. The performance of the parallel independent transfers degrades much faster than the performance of any other technique. Additionally, for our topologies the performance of the more sophisticated dissemination schemes does not degrade significantly when reducing the core capacity to ½ or ¼ of the original one. This is testament to their ability to exploit orthogonal bandwidth. Furthermore, it is an indication that similar performance can be obtained at lower network core budgets by employing sophisticated data distribution techniques.

To further investigate the ability to exploit alternate network paths, we generate a set of topologies in which the aggregate core bandwidth is maintained constant, but the number of core links is varied. Figure 4 compares the intermediate progress of the BitTorrent and ALM protocols on two topologies: the 'dense' topology has four times more links in the core (and four times lower link bandwidth). The results underline BitTorrent's ability to exploit all available transport capacity (Bullet shows similar behavior) unlike ALM whose relative performance degrades for denser networks.

5.2 Overheads: Network Effort

The traditional method to compare overheads for tree-based multicast solutions is to compare maximum link stress (or link stress distributions), where, link stress is defined as the number of identical logical flows that traverse the link. However, the same metric cannot be applied to Bullet or BitTorrent as these protocols dynamically adjust their data distribution patterns and, therefore, link stress varies continuously. For this reason, we propose a new metric to estimate network effort. We estimate the volume of duplicate traffic that traverses each physical link and aggregate it over all links in the testbed.

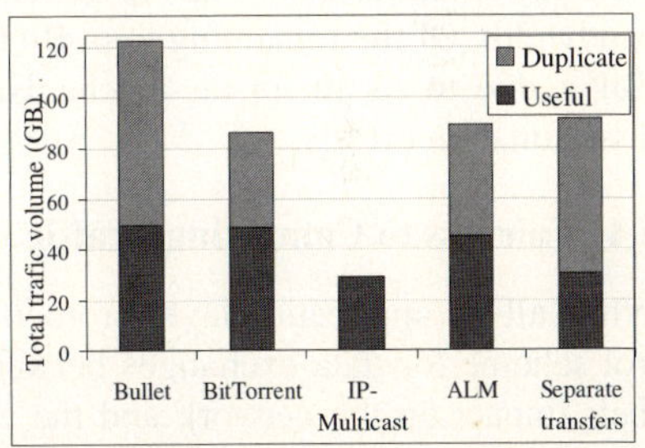

Fig. 5. Overhead of each protocol on EGEE topology

Figure 5 presents the generated useful and overhead traffic for each protocol for the original EGEE topology. We consider *useful*, the data traffic that remains after excluding all link-level duplicates. The following observations can be made from Figure 5, and can be generalized, as there is little variance across all topologies. First, as expected, IP-layer solutions do not generate any duplicates and thus are optimal in terms of total traffic. Second, Bullet, BitTorrent and ALM require significantly higher network effort. Bullet emerges as the largest bandwidth consumer. This is because it uses approximate representations of the set of blocks available at each node. False negatives on these data representations generate additional traffic overhead. BitTorrent generates slightly smaller overheads as nodes employ exact representations to denote the set of blocks available locally. Finally, ALM trees also introduce considerable overhead as the tree construction algorithm is optimized for high-bandwidth dissemination and ignores node location in the physical topology.

5.3 Load Balance

Another metric to evaluate the performance of data dissemination schemes is load balancing. To this end, we estimate the volume of data processed (both received and sent) at each end-node. Obviously, IP-layer techniques that duplicate packets at routers or storage points inside the network, will offer ideal load balancing. Sending

data through independent connections directly from the source will offer the worst load balance as the source load is proportional to the number of destinations.

Figure 6 presents the load balancing performance of the remaining techniques: ALM, BitTorrent, and Bullet. These results are obtained for the GridPP topology, but again, the relative order of these techniques in terms of load balance does not change across all our experiments.

Apart from independent transfers, ALM has the worst load balance among the three solutions as it tends to increase the load on the nodes with ample access-link bandwidth. Of the remaining two, BitTorrent offers slightly better load balancing than Bullet due to its tit-for-tat mechanism that implicitly aims to evenly spread data-dissemination efforts.

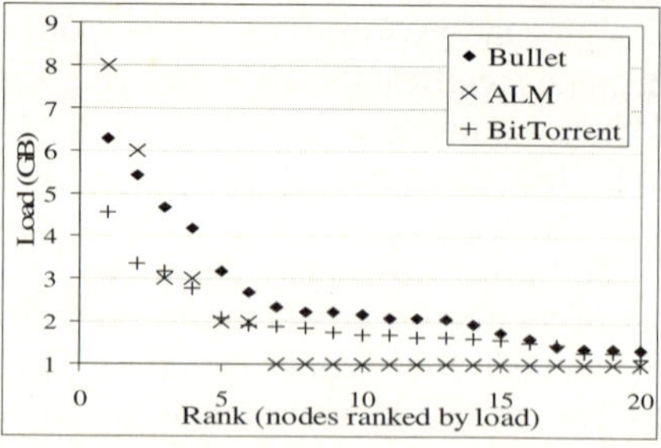

Fig. 6. Load balancing for ALM, BitTorrent and Bullet. Nodes are ranked in decreasing order of their load.

5.4 Fairness to Competing Traffic

While all the application layer protocols we analyze use TCP-friendly congestion control scheme for data exchanges between each individual pair of nodes, they differ in their impact on the network and the competing traffic. In spite of this, there is little related work on analyzing the relative fairness of data distribution schemes. We use link stress as a metric to estimate impact on competing traffic: the higher the number of flows a distribution scheme maps on a physical link, the higher the impact on competing traffic. This impact is non-negligible. The average link stress generated by Bullet can be as high as 12, while the maximum link stress can be as high as 23. This implies that, if a unicast transfer shares its bottleneck link with a link on which Bullet generates such stress, its allocated bandwidth is drastically reduced.

6 Summary

This study focuses on the problem of one to many data dissemination in the context of today's science grids. Data dissemination in these environments is characterized by relatively small collaborations (tens to hundreds of participating sites), large data files, and well-provisioned networks. This study provides an experimentally-supported answer to the question: "Given the characteristics of deployed grids, what benefits can P2P solutions for one-to-many data dissemination offer?"

Our simulation-based comparison of seven solutions drawn from traditional data delivery systems and P2P networks shows the following:

- Some of *today*'s Grid testbeds are over-provisioned. In this case, the deployment is scalable with the size of the user community, and P2P solutions that adapt to dynamic and under-provisioned networks do not bring significant benefits. While they improve load balancing, they add significant overheads and, more importantly, do not offer significant improvements in terms of distribution time.

- Application-level schemes such as BitTorrent, Bullet and application-level multicast perform best in terms of file-delivery time. However, they introduce high-traffic overheads, even higher than independent parallel transfers. On the other hand, BitTorrent and Bullet are designed to deal with dynamic environment conditions, which might be desirable in some scenarios.
- The naive solution of separate data transfers from source to each destination yields reasonable performance on well-provisioned networks but its performance drops dramatically when the available bandwidth decreases. In such cases, adaptive P2P-like techniques that are able to exploit multiple paths existing in the physical topology can offer good performance on a network that is less well provisioned.

In short, the P2P solutions that offer load balancing, adaptive data dissemination, and participation incentives, lead to unjustified costs in today's scientific data collaborations deployed on over-provisioned network cores. However, as user communities grow and these deployments scale (as already seen in Open Science Grid or TeraGrid) P2P data delivery mechanisms will outperform other techniques.

In any case, network provisioning has to progress hand-in-hand with improvements and the adoption of intelligent, adaptive data dissemination techniques. In conjunction with efficient data distribution techniques, appropriate network provisioning will not only save costs while building/provisioning collaborations, but also derive optimal performance from deployed networks.

References

1. Iamnitchi, A., Ripeanu, M., Foster, I.: Small-World File-Sharing Communities. In: Infocom2004, Hong Kong (2004)
2. Iamnitchi, A., Doraimani, S., Garzoglio, G.: Filecules in High-Energy Physics: Characteristics and Impact on Resource Management. In: HPDC 2006, France (2006)
3. Vazhkudai, S., Tuecke, S., Foster, I.: Replica Selection in the Globus Data Grid. In: IEEE International Conference on Cluster Computing and the Grid (CCGRID 2001) (2001)
4. Beck, M., Moore, T., Plank, J.S., Swany, M.: Logistical Networking: Sharing More Than the Wires. In: Active Middleware Services Workshop, Norwell, MA (2000)
5. Cohen, B.: BitTorrent web site (2005), http://www.bittorrent.com
6. Kostic, D., Rodriguez, A., Albrecht, J., Vahdat, A.: Bullet: High Bandwidth Data Dissemination Using an Overlay Mesh. In: SOSP'03, Lake George, NY (2003)
7. Al-Kiswany, S., Ripeanu, M., Iamnitchi, A., Vazhkudai, S.: Are P2P Data-Dissemination Techniques Viable in Today's Data Intensive Scientific Collaborations? University of British Columbia (2007)
8. Pendarakis, D., Shi, S., Verma, D., Waldvogel, M.A.: An Application Level Multicast Infrastructure. In: USITS'01 (2001)
9. Byers, J., Considine, J., Mitzenmacher, M., Rost, S.: Informed Content Delivery Across Adaptive Overlay Networks. In: SIGCOMM2002, Pittsburg, PA (2002)
10. Ganguly, S., Saxena, A., Bhatnagar, S., Banerjee, S., et al.: Fast Replication in Content Distribution Overlays. In: IEEE INFOCOM, Miami, FL (2005)

11. Enabling Grids for E-sciencE Project (2006)
12. Britton, D., Cass, A.J., Clarke, P.E.L., Coles, J.C., et al.: GridPP: Meeting the Particle Physics Computing Challenge. In: UK e-Science All Hands Conference (2005)
13. Medina, A., Lakhina, A., Matta, I., Byers, J.: BRITE: An Approach to Universal Topology Generation. In: International Workshop on Modeling, Analysis and Simulation of Computer and Telecommunications Systems- MASCOTS '01, Cincinnati, Ohio (2001)

Increasing Parallelism for Workflows in the Grid[*]

Jonathan Martí[1], Jesús Malo[1], and Toni Cortes[1,2]

[1] Barcelona Supercomputing Center
{jonathan.marti,jesus.malo,toni.cortes}@bsc.es
[2] Universitat Politècnica de Catalunya
http://www.ac.upc.edu

Abstract. Workflow applications executed in Grid environments are not able to take advantage of all the potential parallelism they might have. This limitation in the usage of parallelism comes from the fact that when there is a producer/consumer situation communicating using files, the consumer does not start its execution till the producer has finished creating the file to be consumed, and the file has been copied to the consumer (if needed).

In this paper, we propose a publish/subscribe mechanism that allows consumers to read the file at the same time it is being produced. In addition, this mechanism is implemented in a transparent way to the application, so does not require any special feature from the local filesystems.

Finally, we show that our mechanisms can speedup applications significantly. In our best test we divided by two the execution time of some applications, but other applications may have even higher benefits.

Keywords: Grid Workflow Parallelism Storage.

1   Introduction

Parallel applications are one of the main targets for the Grid, since they are built by several tasks that perform work mostly independent of the other ones. Sometimes data has to flow from one task to another and some mechanisms, such as message passing or shared files, can be used to achieve this. The most clear example of Grid applications that communicate using files can be found in workflows, where a workflow engine divides the application into tasks and decides what dependencies these tasks have among them. In this environment, most of the dependencies are due to shared files, and the general case is that a given task cannot be executed till its input file has been written. This will only happen when predecessor task has finished.

The way current Grid and workflow systems handle these dependencies is very simple. Let's assume a workflow with only two tasks where *task2* depends on the output of *task1*. First *task1* is executed and once it finishes, the system decides the node where *task2* will be executed and copies the file (if needed) to

---

[*] This work has been partially supported by the Spanish Ministry of Science and Technology under the TIN2004-07739-C02-01 grant.

A.-M. Kermarrec, L. Bougé, and T. Priol (Eds.): Euro-Par 2007, LNCS 4641, pp. 415–424, 2007.

this node. This copy is normally done using GridFTP or a similar mechanism. Finally, *task2* is executed and can use the file produced by *task1*.

We can see that this mechanism does not exploit all potential parallelism we could achieve. If *task2* could start processing the file before it is completed, then we could start both tasks in parallel and allow the second one to process the file as it is being generated. Unfortunately, this kind of mechanisms are not available in Grid environments, and this is the problem we address.

In this paper we propose a publish/subscribe mechanism that allows a file to be generated by one task and consumed (while it is being generated) by as many tasks as needed in the Grid. This solution will be integrated in GRID superscalar [1], a well-known workflow engine.

2   Design and Implementation

We have implemented a publish/subscribe middleware that allows subscribers to read remote files, while they are being generated by producers. Figure 1 shows a simple scenario where our approach is running. As you can see, the daemon on the publisher side is reading the file contents while it is being generated by *task1*. In parallel, the publisher daemon is sending these data to the subscribed daemons which write the contents to a named pipe and optionally to a temporary file, so that *task2* could read the data concurrently.

It is also important to notice that all this mechanism is completely transparent to the application and that no modification is done on the application code due to the proposed data forwarding.

Using named pipes enables an application running on the subscriber side to read the contents while they are being written. The problem in using regular files would be that appending data to a file produces the consequent *EOF*, and a typical application based on regular reads would fail trying to read the contents from that file. On the other hand, if the daemon places the named pipe where the application hopes to find the file causes the application to block till data it needs is read.

Using temporary files is optional, but on one hand it allows eventual transparency, since if a file copy is left in the subscriber side, the middleware is finally acting as a traditional file-transfer protocol; in the other hand it allows dealing with replica management mechanisms.

Certainly, our mechanism could introduce CPU overhead on nodes while monitoring the published files. This is the reason behind our decision to monitor files using a dynamic polling mechanism that enables the adaption of the event-based communication protocol to file generation rate (this is detailed later in the Implementation issues). However, our approach is designed for large applications that produce large files (e.g. hundreds of MB or GB), so the overhead introduced by monitoring files is negligible compared with benefits obtained.

Furthermore, taking into account the middleware design, it would be easy to integrate it as a Globus web-service. Doing it, would enable to start the daemons/service as soon as Globus does.

Operations provided by the daemon

- Publish file: asks the daemon to monitor a specific local file. The daemon is then prepared to accept incoming subscriptions for that file.
- Subscribe to remote file: asks the daemon to subscribe to a file located at a remote host. The daemon generates a new named pipe to attach the incoming remote contents (i.e. incoming update events - event-based protocol is explained later). Furthermore, these incoming events may be written to a temporary file to eventually keep a replica.
- Process subscription: asks the daemon to process an incoming subscription, coming from a remote host, to a local file.
- Process event: asks the daemon to process an incoming event related to a remote file which is subscribed to.
- Worker is over: informs the daemon that worker has finished generating some published file. This is broadcast to each subscriber to that file which proceed closing the named pipe and placing the temporary file instead.

Event-based communication protocol

Files are transferred using a communication protocol based on two types of events: *UPDATE* and *END*.

While the published file is being generated, each subscriber to this file receives *UPDATE* events coming from the daemon responsible of that file. These events contain the data corresponding to pieces of the file that each subscriber has not received yet. So that, these data is processed by subscribed daemons appending them to the associated named pipe and temporary file.

Eventually, when the remote file is closed, so the generation is over, it is produced the *END* event from remote daemon to the subscribers. When subscribers receive it, they proceed closing the named pipe and placing the temporary file instead of it, so the file is saved on subscriber too as if transferred in a classical way.

Implementation of the daemon

Figure 2 shows the interaction between the main daemon components. The box in the middle shows these components: *EventProducer*, *EventDispatcher*, and *EventProcessor*. *EventDispatcher* and *EventProcessor* manage an event queue to dispatch and process events respectively. The daemon interacts on the one hand with remote daemons, shown with the upper boxes, and on the other hand with local files and named pipes.

The components introduced before are implemented as the following threads:

- EventProducer: thread that monitors a local file declared to be published, and generates events for each subscriber to that file. These events are enqueued at the EventDispatcher event queue.
- EventDispatcher: thread that sends the events enqueued by *FIFO* policy.
- EventProcessor: thread that process the events received from a daemon to which it is subscribed (*UPDATE* or *END*).

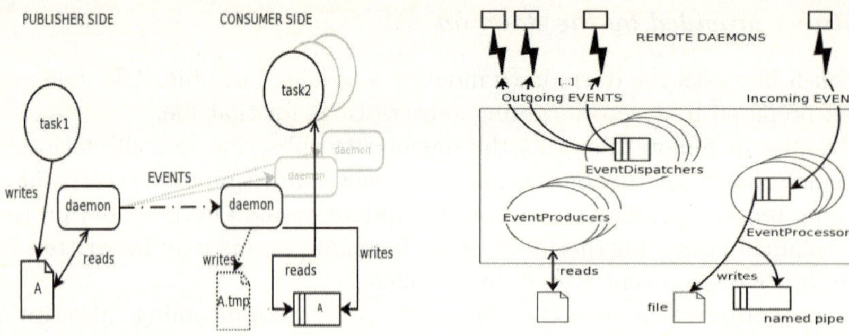

Fig. 1. Essential scenario   **Fig. 2.** Daemon components

Adaptive file monitoring

As we have introduced in section 2, we implemented an adaptive monitoring algorithm to reduce the CPU overhead produced by the *EventProducer*. This thread is in charge of monitoring one published file. To know whether the file has changed, so new events must be produced, *EventProducer* polls the file size periodically. Polling the file size too much often, obviously increase the CPU overhead, and even more if there are many *EventProducers* monitoring several files. So to deal with this situation, we decided to correct the polling time dynamically with a minimum of 10 ms, i.e. the minimum elapsed time between each polling request must be 10 ms. The period time increases when *EventProducer* polls the file size and it has not changed, otherwise the period time decreases.

3   Integration in GRID Superscalar

To validate our approach, we integrated it with GRID superscalar [1]. GRID superscalar is a programming framework that enables the easy development of applications to be run in a computational Grid.

GRID superscalar uses a master/worker schema to map the application workflow. That is, the master is responsible to control the application course by means of delegating its corresponding tasks to several workers. Specifically, the master evaluates task dependencies in terms of the input/output data arguments they use. Then, based on this evaluation, it calculates how to parallelize the application in the Grid, i.e. how to map the workflow among the nodes.

When the master detects a dependency between two tasks *task1(out fileA)* and *task2(in fileA)*, it solves it by executing first *task1*, then transferring *fileA* to worker responsible of *task2* and eventually executing *task2*.

Here it is where our approach comes into play. The idea is that using our middleware enables the master to delegate the execution of *task2* while *fileA* is being generated by *task1*. Figure 3 shows an abstract comparison between both behaviours in terms of wasted time. With the original GRID superscalar, tasks with data dependencies are executed consecutively, whereas using our middleware they could be executed concurrently.

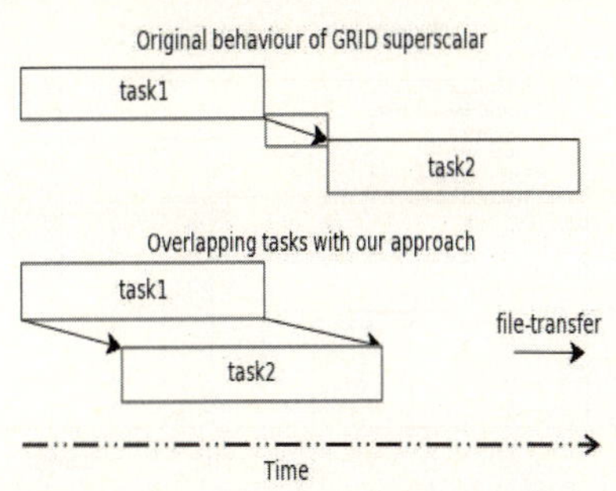

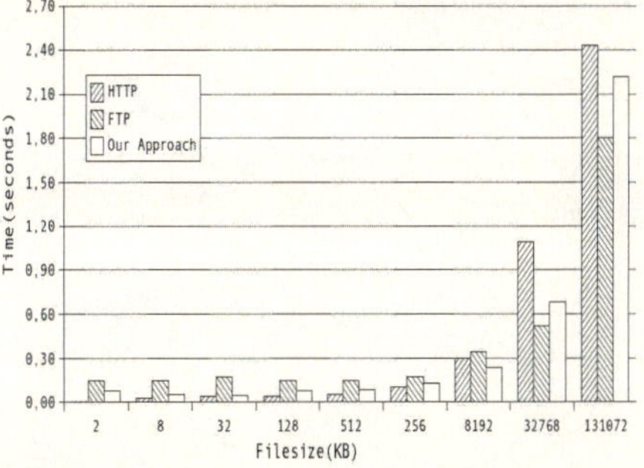

Fig. 3. Task overlapping   **Fig. 4.** FTP vs HTTP vs Our protocol

4 Performance Evaluation

Case 1: The protocol

First of all, we compared our communication protocol based on events versus HTTP and FTP. We did it in terms of file-transfer bandwidth, since we wanted to test if the performance of our approach is acceptable according to protocols specifically designed for point-to-point file-transfer.

Our middleware is actually designed for support N to M file content distribution, but as it must be implemented at application level since we are working in Grid environments, you can see our middleware as a derivative from several point-to-point file-transfers. Therefore, doing this test we compare the performance between the point-to-point basis of our protocol versus the existing ones.

The environment where we performed the test consists of two Pentium M 1.73Gh, with 1GB of memory, and Gigabit Ethernet cards.

To obtain the HTTP results, we developed a shell-script based on the *wget* command to transfer files from 1KB to 128MB from one laptop to the other. To obtain the FTP results we developed a shell-script based on *ftpupload* to transfer the same files. And finally to obtain the results about our approach we subscribed the daemon from one of the laptops to the other and transferred files already generated using our event-based communication protocol. Furthermore, in order to be on equals terms as far as possible, we disabled the temporary file creation for our approach, which obviously carries out an overhead while it supposes two writes per actual write.

To deal with interferences in our LAN, we executed the tests several times, but actually we did not observed any significant difference among executions.

Figure 4 shows the results obtained, from what we conclude that on one hand FTP is between 30-40% faster than the others for large files, what is expected due to it is designed for point-to-point communications and optimized for transferring large files. However, our approach is close to HTTP, which is a good result considering that *wget* is designed for point-to-point communications, although it is slower than FTP because it is not optimized for transferring large files as FTP does.

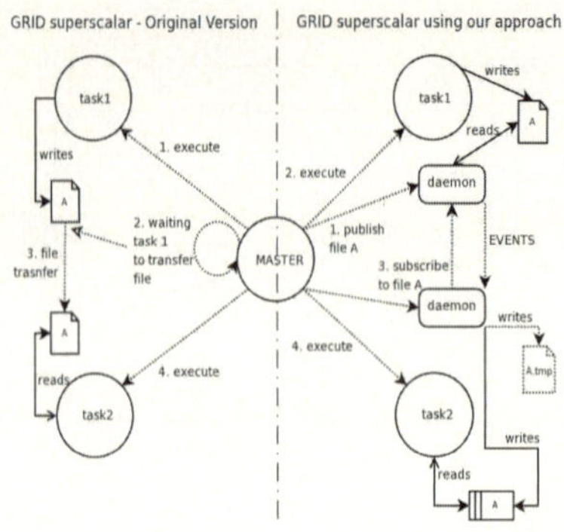

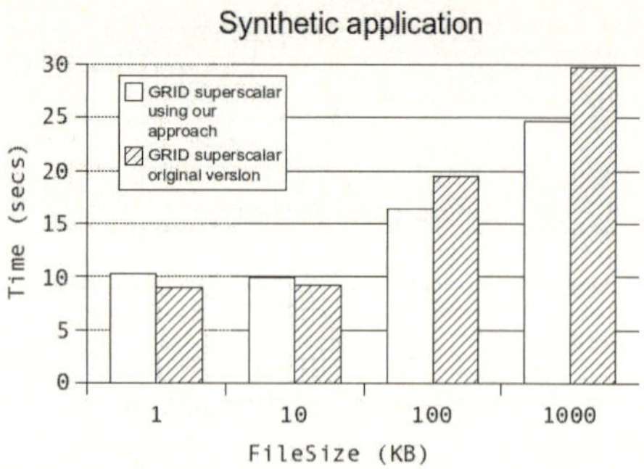

Fig. 5. GRID superscalar integration   **Fig. 6.** Synthetic case results

As a global conclusion,our approach is acceptable for file-transfer communications in terms of bandwidth because taking account of our approach allows N to M communications among daemons, as a publish/subscribe approach, it carries out an inevitable overhead, but even so we obtained results comparable to well-known protocols that do not deal with goals like our approach does.

Case 2: Synthetic load

We checked the benefits of our approach integrating it with GRID superscalar. Figure 5 shows the comparison between GRID superscalar using our approach, and the original version. Depending on the version used, the behaviour for running the basic scenario of two data dependent tasks is different. In terms of time, we already introduced the difference in figure 3 commented before.

We have checked what happened when GRID superscalar manages an application based on two functions *task1(out fileA)* as the producer and *task2(in fileA)* as the consumer. So *task1* is executed as a worker in one of the nodes, and *task2* is executed as a worker on the other node. These nodes are: Khafre (PIII 700MHz,2.5GB memory) Khepri (Power4 1000MHz, 2GB memory).

We have executed one hundred tests for this experiment to deal with possible interferences with another users either in *Khafre* or *Khepri*. For each test, the *fileA* used in the application varies from 1KB to 10MB. We decided to stop at 10MB because in fact, the benefits appear since file is larger than 100KB and they could be obviously extrapolated to larger applications which generate larger files. Figure 6 shows the results obtained.

Case 3.1: Matrix multiply

The application called *MatMul* is based on matrix multiplications, specifically it does (AxB)x(CxD) where A,B,C,D are input matrices. These matrices are passed as files. So first task, *task1*, is responsible for computing AxB, the second task, *task2*, computes CxD, and finally the third task, *task3*, computes (AxB)x(CxD) with the results obtained from *task1* and *task2*. So there is an obvious data dependency between *task3* and its predecessors *task1* and *task2*.

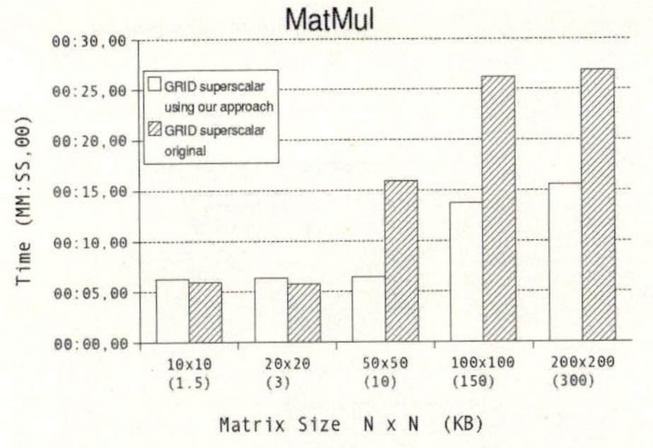

Fig. 7. MatMul results   Fig. 8. MatMul behaviour through time

We have executed the application ten times for both cases, with and without using our approach. Furthermore, we performed the tests with different matrix sizes, from 10x10 (1.5KB) to 200x200 (300KB). Results are shown in figure 7.

The significance of this test is the behaviour obtained with both cases commented before, since the master scheduling, and consequent assignments of task to workers, are different depending on whether GRID superscalar is using our approach or not. Figure 8 shows the utilization of each worker for both cases.

Specifically, the original version does not take advantage of a third node, because file F is already generated and accessible from node 2, whereas using our middleware does it, overlapping execution of tasks. Then, benefits appear with input matrices from 20x20 on. The improvement increases from 100x100 up to 200x200 matrices, reaching a 42% of time reduction.

Case 3.2: Double cryptography

In this test we performed the evaluation of our approach using an application based on cryptology we called GRIDCypher. This is a basic application where a file must be encrypted in two steps from sender A to the receiver B, so that it is possible to ensure two things: first, the file will be only read by the receiver B since it has been encrypted with its public key and could only be decrypted with B private key; second, the receiver could trust the file was generated by the sender A, since it could be only decrypted with A's public key.

We have executed the application ten times for both cases, with and without using our approach. Furthermore, we performed the tests with different file sizes, from 100 bytes to 100KB. Results are shown in figure 9.

Figure 10 shows the utilization of each worker while this execution is performed. Using our approach, workers run concurrently while files are transferred step by step. On the other hand, the original GRID superscalar does not take advantage of more than one worker, since as soon as each step is finalized the same worker is able to do the next one.

The original version saves all the file transfers, but for each step the current task must wait the entire file from previous one, whereas using our approach, GRID superscalar could deal with parallelism between more than one worker.

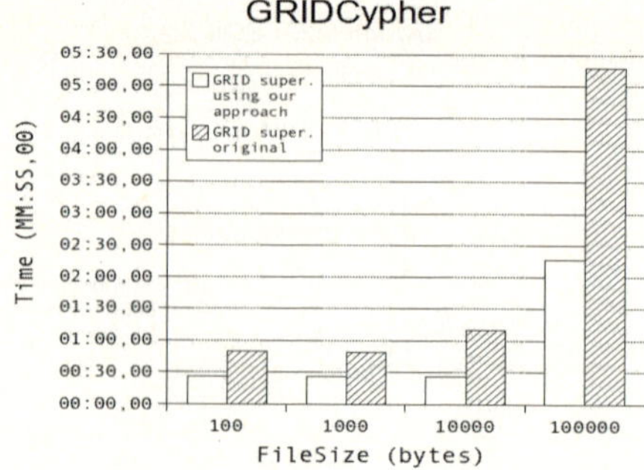

Fig. 9. GRIDCypher results

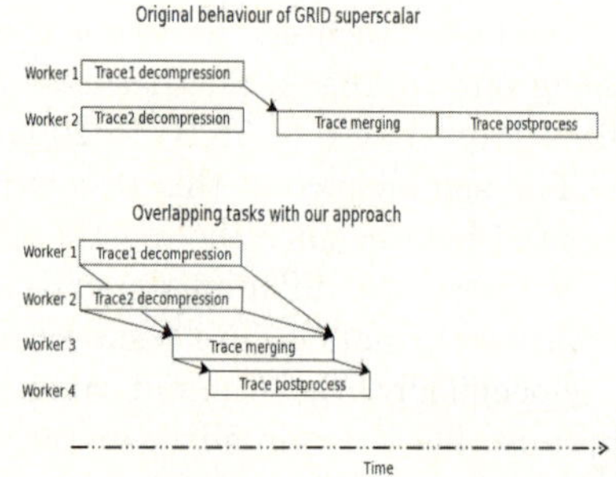

Fig. 10. GRIDCypher through time

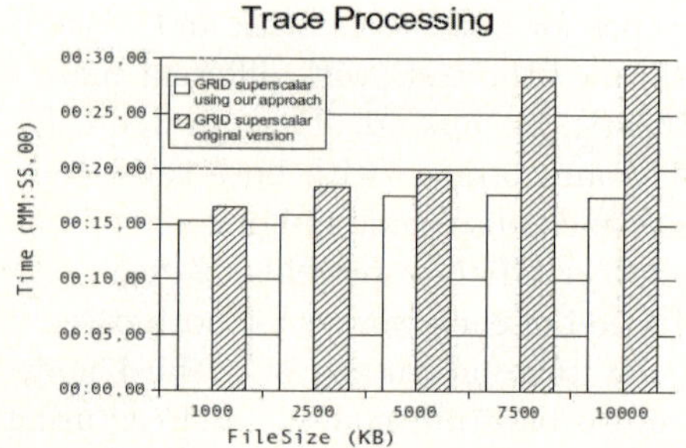

Fig. 11. Trace case results

Fig. 12. Traces case through time

This behaviour is reflected in the results which shown that using our middleware benefits increase since the first case, reaching a 57% of time reduction.

Case 3.3: Multitrace processing

In this test we performed the evaluation of our approach using an application based on processing PARAVER [2] traces. The scenario consists of two tasks *task1* and *task2* which decompress two PARAVER traces in parallel. Afterwards, the decompressed traces are merged with the Unix *merge* command executed by a third task *task3*. Finally, the merged trace is filtered via *grep* command.

There are interesting data-dependencies between the first two decompression tasks and the responsible for merging them. Eventually, there is another data-dependency between this merging task and the final one about post-processing.

We have executed the application ten times for both cases, with the original GRID superscalar and using our approach. Furthermore, we ran the tests with different file sizes, from 1MB to 10MB. Results are shown in figure 11.

Benefits are shown in all the cases, but significantly ones come from 7,5MB on. Figure 12 shows the utilization of each worker while the application is running. The original GRID superscalar version does not take advantage of more than two workers due to data dependencies among defined tasks, whereas using our approach, the benefits coming from overlapping tasks are important enough to reach a time reduction from 7% to 41%.

5   Related Work

Mainly, the are two types of publish/subscribe systems: topic-based and content-based. Topic-based systems are much like newsgroups. The idea is that subscribers express their interest by joining a group (related to a topic), so that all messages/events related to that topic are broadcast to all users subscribed to that group. Content-based systems are designed to allow subscriptions to specific contents, i.e. users express their interests specifying predicates over the values of a number of well-defined attributes. Therefore the matching of published events to subscriptions is done based on the content, i.e. the values of the attributes.

Our approach could be all part of the topic-based publish/subscribe systems taking into account that published files could be seen as topics. Currently it exists several topic-based approaches, like Scribe [3] or Bayeux [4], but mostly of them are not designed for file distribution.

There are other topic-based publish/subscribe systems commonly used in the world wide web, like RSS [5] which is based on broadcasting XML messages. But they are implemented using client-side polling to ask for the contents.

Current research lines on file-sharing distributed systems are mostly focused on using P2P [6] techniques. Peers transfer files using protocols that allow file chunks to be retrieved from remote peers in parallel. In a sense, this could address the problem that our middleware does, since peers could download each file from just one remote peer. But P2P systems are not designed to enable peers to read the contents sequentially and while they are being generated.

To address the entire issue this paper is focused on, we know of a system called Distributed File System (DFS) [7]. DFS is designed for enable file distribution in publish/subscribe fashion as our approach does. DFS allows the files to be read while they are being remotely generated thanks to a special file communication protocol based on events as our middleware does.

But the main problem in DFS is that when a file has to be broadcast to several subscribers, everyone has to receive the data before going on. This restriction introduces two basic problems. First, if a subscriber host is down, DFS enters into a loop trying to send the event to this host and making the other subscribers to wait for it. Second, regarding the integration of the middleware with a system like GRID superscalar, if the middleware used was DFS, an application based on task schema as shown in figure 13 would suffer an unnecessary delay in *task3* causing an unnecessary virtual dependency between *task2* and *task4*. On the other hand, using our middleware, the *EventDispatcher* thread responsible of sending events to the daemons subscribed to a specific local file, do not wait for any subscribers' response to continue sending other possible enqueued events. So that using our approach, the application commented could be executed faster.

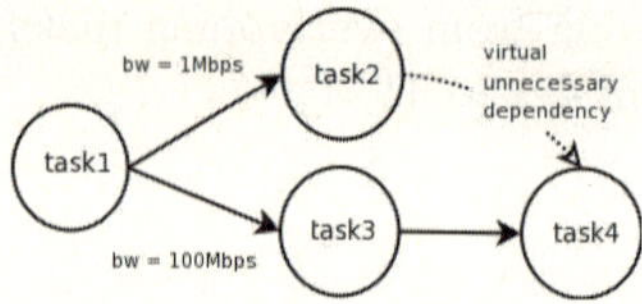

Fig. 13. Virtual unnecessary dependency using DFS instead of our approach

Furthermore, DFS is a framework with its own I/O API, so that it is necessary to change the application code to use DFS *open, close, read, write* I/O calls, whereas with our approach the applications keep their original code.

6   Conclusion

In this paper we have presented a publish/subscribe mechanism that can be used to improve parallelism in workflow applications by allowing the processing of a file at the same time as it is being generated in a different node in the Grid.

We have also integrated this mechanism into GRID superscalar to show that workflow engines can use this mechanism very easily and improve parallelism without requesting any special feature from the file system used.

Finally, we have measured the performance benefit of such a mechanism and have conclude that even for small files (less than 100KB), a significant performance (up to a speedup of 2) can be achieved.

References

1. Badia, R., Labarta, J., Sirvent, R., Prez, J., Cela, J., Grima, R.: Programming Grid Applications with GRID superscalar. Journal of Grid Computing 1(2), 151–170 (2003)
2. Obtain detailed information from raw performance traces, http://www.cepba.upc.es/paraver/
3. Castro, M., Druschel, P., Kermarrec, A., Rowstron, A.: SCRIBE: A large-scale and decentralized application-level multicast infrastructure. IEEE Journal on Selected Areas in communications (JSAC) 20(8), 1489–1499 (2002)
4. Zhuang, S.Q., Zhao, B.Y., Joseph, A.D., Katz, R.H., Kubiatowicz, J.D.: Bayeux: An architecture for scalable and fault-tolerant wide-area data dissemination. In: Proceedings of NOSSDAV (June 2001)
5. Rss advisory board announcements and really simple syndication news, http://www.rssboard.org/
6. Lua, K., Crowcroft, J., Pias, M., Sharma, R., Lim, S.: A survey and comparison of peer-to-peer overlay network schemes. Communications Surveys & Tutorials, IEEE, 72–93 (2005)
7. Chen, K., Huang, Z., Li, B., Huang, E., Rajic, H., Kuhn, R., Chen, W.: Distributed File Streamer: A Framework for Distributed Application Data Coupling. In: 7th IEEE/ACM Grid, pp. 168–175. ACM Press, New York (2006)

Persistent Fault-Tolerance for
Divide-and-Conquer Applications on the Grid

Gosia Wrzesinska, Ana-Maria Oprescu, Thilo Kielmann, and Henri Bal

Vrije Universiteit Amsterdam
{gosia,amo,kielmann,bal}@cs.vu.nl

Abstract. Grid applications need to be fault tolerant, malleable, and migratable. In previous work, we have presented *orphan saving*, an efficient mechanism addressing these issues for divide-and-conquer applications. In this paper, we present a mechanism for writing partial results to checkpoint files, adding the capability to also tolerate the *total loss* of all processors, and to allow suspending and later resuming an application.

Both mechanisms have only negligible overheads in the absence of faults, even with extremely short checkpointing intervals like one minute. In the case of faults, the new checkpointing mechanism outperforms orphan saving by 10 % to 15 %. Also, suspending/resuming an application has only little overhead, making our approach very attractive for writing grid applications.

1 Introduction

In grid environments, the availability of computing resources changes constantly. Processor crashes are more likely to occur than in traditional parallel environments. Also, processors may be taken away from an application because they are claimed by another, higher-priority application, or because a processor reservation has ended. At the same time, new processors may become available.

A grid application must be able to adapt to such changes in order to survive in a grid environment and to achieve good performance. In particular, three issues have to be dealt with: *Fault tolerance,* which is the ability of an application to operate in the presence of hardware and software failures. *Malleability,* which is the ability of an application to handle processors joining and leaving an on-going computation. And *migratability,* which is the ability of an application to relocate to a different set of computational resources during the run.

These three issues are closely related to each other. For example, if an application can handle crashing processors (fault tolerance) and continue working on the diminished number of processors, it can also handle leaving processors (partial malleability). Even more, if the processors are leaving *gracefully* (i.e., after a prior notice) handling it may be more efficient than handling crashing processors. Further, if an application is malleable, it is also migratable: it can be migrated from one set of resources to another by first adding the new processors to the computation and then removing the old ones.

A.-M. Kermarrec, L. Bougé, and T. Priol (Eds.): Euro-Par 2007, LNCS 4641, pp. 425–436, 2007.

In previous work, we have presented an efficient mechanism supporting fault-tolerance, malleability, and migration of divide-and-conquer applications [1], in the following called *orphan saving*. It is based on re-executing jobs done by processors that have either crashed or left voluntarily, while preserving as many partial results as possible in the application's main memory. With this mechanism, applications are guaranteed to complete successfully, as long as at least one processor is alive, at any moment during the execution.

In this paper, we complement orphan saving by a mechanism for writing partial results to checkpoint files, adding the capability to also tolerate the *total loss* of all processors, and to allow suspending an application and to resume it later, whenever CPU's become available again.

Our performance evaluation shows that both fault-tolerance mechanisms have negligible overhead in the absence of faults. Suspending and resuming the tested applications has only $1\% - 11\%$ overhead, compared to uninterrupted runs. In the case of faults, the new checkpointing mechanism even outperforms orphan saving.

In Section 2, we briefly introduce the divide-and-conquer model, along with the *orphan saving* fault-tolerance mechanism. Section 3 presents our new, checkpointing fault-tolerance mechanism. In Section 4, we evaluate both mechanisms using two application programs. Related work is discussed in Section 5. In Section 6, we draw our conclusions.

2 Divide-and-Conquer and the Orphan Saving Mechanism

Divide-and-conquer applications operate by recursively dividing a problem into subproblems. The recursive subdivision goes on until the subproblems become trivial to solve. After solving subproblems, their results are recursively combined until the final solution is assembled. This leads to an *execution tree* of nested tasks. The excellent suitability of the divide-and-conquer paradigm for writing grid applications has been shown many times before [2,3,4,1].

We have implemented our fault-tolerance mechanisms within Satin, a Java framework for creating grid-enabled divide-and-conquer applications. With Satin, the programmer annotates the sequential code with divide-and-conquer primitives (marker interfaces and a synchronization method). Satin's byte-code rewriter generates the necessary communication and load-balancing code. In the following, we are using Satin to discuss and evaluate our fault-tolerance mechanisms.

In Satin, invocations of annotated divide-and-conquer methods lead to the creation of entries in the processor's local *work queue*. Work is distributed across the processors by work stealing: when a processor runs out of work, it picks another processor at random and steals a job from its work queue. After computing the job, the result is returned to the originating processor. Satin uses a very efficient, grid-aware load balancing algorithm which hides wide-area latencies by overlapping local and remote stealing [3]. Though the combination of local and remote

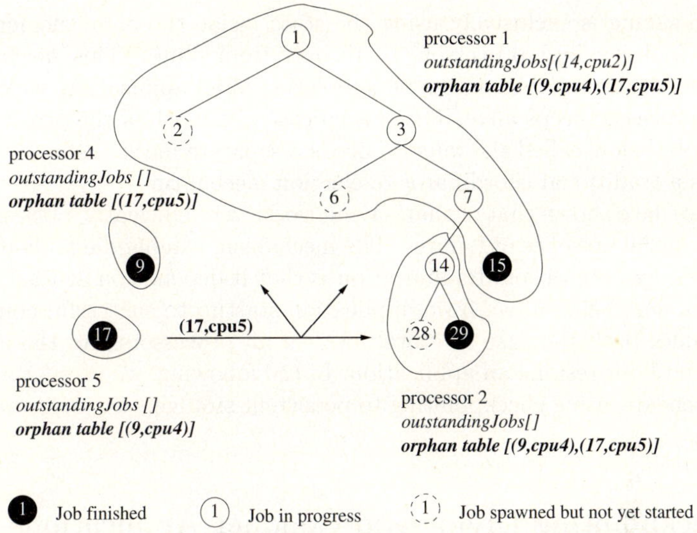

Fig. 1. *Orphan saving* re-using orphan results

stealing could allow for a job to be stolen *from* two or more processors in turn, the problem of livelock does not arise, simply because the job will eventually be stolen by a processor with an empty *work queue*, where the job will get executed immediately.

Adding a new machine to a divide-and-conquer computation is straightforward: the new machine simply starts stealing jobs from other machines. Removing processors (both voluntarily and in case of faults) can be handled by recomputing the work stolen by leaving processors. Implementing such a recomputation scheme efficiently, however, is not trivial, due to the problem of *orphan jobs*: those jobs that have been stolen *from* a leaving processor. The novelty of our *orphan saving* mechanism [1] is its ability to efficiently re-use the results of orphan jobs in case of a fault. This is achieved by carefully inserting the results of orphan jobs into the execution tree once the lost jobs have been resubmitted.

An example is sketched in Figure 1. Here, processor 3 has crashed. In response, processor 1 has re-submitted jobs 2 and 6 (once stolen *by* processor 3) to its work queue, while processors 4 and 5 have notified all processors that they have results for jobs 9 and 17 (once stolen *from* processor 3) available. As it is not known beforehand which processor will need an orphan result, all processors have to be notified where it could be found. As soon as an orphan job is up for recomputation, the processor in charge will check its orphan table first, thus re-using the orphan results.

Divide-and-conquer is extremely robust because utilizing orphan results is merely a performance optimization, albeit an important one. In case of several errors, leading to the loss of some or all orphan results, the divide-and-conquer application will still compute the correct result, only its performance will be impacted, depending on the amount of losses.

Orphan saving is exclusively using the state inside the main memory of the surviving (not leaving) processors to recover from faults. This mechanism is sufficient to guarantee the successful completion of an application, as long as at least one processor keeps alive. The special case of a crash of the processor with the root job, below called the *master*, needs a separate mechanism; in this case, Satin uses a traditional coordinator re-election mechanism.

In [1], we have shown that orphan saving works very efficiently, both in the absence and in the presence of failures. The mechanism can handle fault-tolerance, malleability, and migration. Its limitation is that it depends on at least one processor at a time being alive, from application startup to successful completion. This excludes both the case of "total loss" of all processors and the ability to suspend and later resume an application. In the following, we present a complementary scheme, using checkpointing to persistent storage, that overcomes these limitations.

3   Checkpointing Divide-and-Conquer Applications

To overcome the limitation of storing all orphan results in volatile main memory, we have developed a mechanism that stores intermediate results, just like orphan results, but on stable storage. Using the orphan result data structure allows us to build a very light-weight checkpointing scheme, comparable to application-level checkpointing, except that the Satin runtime system is taking care of storing the relevant data. This combines the advantages of system-level checkpointing (user transparency) and of application-level checkpointing (efficiency due to low data volume.)

All processors of a divide-and-conquer application periodically save their partial results in a *checkpoint file*. Along with a job's result, both *jobID* and originating *processorID* are stored. Each processor writes its own checkpoint asynchronously from the others, avoiding synchronization overheads. This is possible due to the robustness of divide-and-conquer and orphan saving, as explained in Section 2. The checkpointing interval is user defined.

Processors do not access the checkpoint file directly. Instead, they send their data to a centralized *coordinator* processor that is in charge of reading and writing the actual file. The coordinator's role is twofold:

1. **Fault tolerance:** If a processor crashes, the coordinator searches the checkpoint file for results computed by that processor. Those results are fetched into the coordinator's memory. Next, the orphan saving mechanism is used to re-use those results; they are treated just like orphan jobs. For each of those results, the coordinator forwards the *jobID* along with *its own processorID* to the other processors, allowing them to integrate these results into the job tree, as necessary.
2. **Suspend/resume:** When a computation is started, the coordinator checks whether the user-specified checkpoint file already exists. If so, the coordinator assumes that the computation has been restarted. All results from the

checkpoint file are read into the memory of the coordinator and for each of them, the *jobID* and the coordinator's *processorID* are sent to the other processors.

3.1   The Checkpoint File

The checkpoint file is accessed by the coordinator, but it need not necessarily be located on the coordinator's local filesystem. In fact, the user may specify an arbitrary location for the checkpoint file. Access to the checkpoint file is implemented using the Java implementation of the Grid Application Toolkit (GAT) [5]. The GAT provides transparent access to various grid middleware systems. For accessing remote files, a programmer only needs to specify a URI referring to the location. The GAT then takes care of selecting and using the appropriate protocol, like, for example, FTP, SSH, HTTP, or GridFTP.

The results stored in a checkpoint file are partially redundant. This is caused by the fact that many jobs are stored in the checkpoint file, along with their direct parent or another ancestor job. In such cases, only the ancestor is useful. Depending on the checkpoint interval, there may be more or less redundancy. We use *checkpoint compression* to reduce the number of such redundant jobs in a checkpoint file.

3.2   The Coordinator

Initially, the master is elected as checkpointing coordinator. To achieve good I/O performance, however, the coordinator is re-elected from among the processors taking part in the computation, based on actual I/O performance with the checkpoint file. For this purpose, each processor measures the time it takes to write a small file to the location of the checkpoint file. The master collects these results and selects the processor with the shortest file write time as the new coordinator.

If the coordinator crashes, a new coordinator has to be elected. The new election is initiated by the master, which sends a *coordinator reelection* message to all processors. Then, the above coordinator election procedure is performed. The processors postpone checkpointing until the election is completed.

If the coordinator has crashed while another process was sending checkpoint data to it, this data will be lost. Because loss of checkpoint data may only affect performance of the application but never its correctness, we do not take any action to avoid such situations.

The coordinator may also crash while writing to the checkpoint file and the checkpoint file may be corrupted. Therefore, each time a coordinator is initialized, it inspects the checkpoint file for possible errors. If errors are found, a new checkpoint file containing all non-damaged results is created and used.

We are using two different mechanisms to detect coordinator crashes: one implemented in the communication layer (Ibis), another one implemented in the Ibis Registry. If one of both mechanisms would lead to false positives, still this would only affect the *performance* of our system, but not its *correctness*.

To minimize the overhead of checkpointing, we use *concurrent checkpointing* [6]. The results are written to the checkpoint file by a separate thread in the coordinator process. This thread runs concurrently with the Satin computation.

There is no synchronization between the workers and the coordinator in terms of sending/receiving the checkpoint data.

4   Evaluation

We will now evaluate the performance of our fault-tolerance mechanisms, both in the absence and in the presence of faults. We compare the orphan-saving algorithm to the new checkpointing mechanism, the latter with various checkpointing intervals.

The experiments were carried out on the Distributed ASCI Supercomputer (DAS-2), consisting of five clusters that are located at universities in the Netherlands. We have used a total of 32 nodes on two DAS-2 clusters (16 nodes each). Each node contains two 1-GHz Pentium-III's and at least 1 GB RAM. All nodes run RedHat Linux. Within a cluster, nodes are connected by 2 Gb/s Myrinet. For intra-cluster communication we have used 100 Mb/s Ethernet and SurfNet, the Dutch academic Internet backbone. The bandwidth between the clusters is about 700 Mb/s. The round-trip latencies are around 2 ms.

In order to gain broad insights about our checkpointing mechanism, we have selected one computation-intensive application (TSP) and one communication-intensive code (Raytracer). The code for the Traveling Salesman Problem (*TSP*) searches for a shortest path connecting a set of cities. TSP is a well-known, NP-complete problem that has many applications in science and engineering. TSP was parallelized by evaluating different paths in parallel. TSP is a computation-intensive application and sends only little data. *Raytracer* renders a bitmap image from an abstract scene description. Raytracer has been parallelized by recursively subdividing the bitmap into smaller parts, and rendering the parts in parallel. Raytracer is a relatively communication-intensive application.

4.1   Performance Overhead in the Absence of Crashes

First, we assess the overhead on application performance, caused by either orphan saving or checkpointing, when no processors are leaving or crashing. For both applications, we compare the plain Satin system (without any fault tolerance) to Satin with orphan saving, and to Satin with checkpointing, using checkpointing intervals of one, two, and five minutes. Note that these checkpointing intervals are rather short, compared to traditionally used values, like 30 minutes or one hour.

Figure 2 shows runtimes of the two applications. The overhead of orphan saving is negligible. Likewise, checkpointing has small overhead and the overhead does not seem to depend on the checkpointing interval. This is because our checkpointing is completely asynchronous, both on the nodes sending the data to the coordinator, and on the coordinator itself. Table 1 lists the maximal sizes

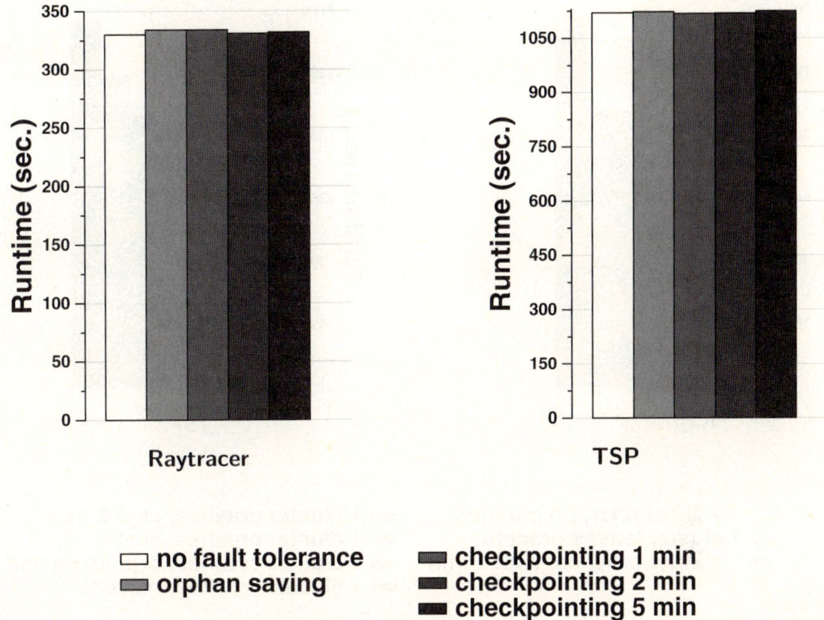

Fig. 2. Application overhead during crash-free execution

Table 1. Checkpoint file sizes

interval	1 min	2 min	5 min
Raytracer	28.0 MB	13.8 MB	16.6 MB
TSP	217 KB	128 KB	55 KB

of the checkpoint files for different checkpoint intervals. The checkpoint files produced by TSP are small, since TSP does not process much data. Raytracer is more data intensive, and therefore produces bigger checkpoint files.

4.2   Performance in the Presence of Crashes

Next, we evaluate the performance of both orphan saving and checkpointing in the presence of crashes. Again, we have run the two applications on 32 nodes in 2 clusters. We removed one of the clusters in the middle of the computation, that is, after half of the time it would take on 2 clusters without processors leaving. The case when half of the processors leave is the most demanding, as the biggest number of orphan jobs is created in this case. On average, the number of orphans does not depend on the moment when processors leave, except for the initial and final phase in the computation.

For our analysis, we compare to two extreme cases. One is *naive fault toler-ance*, where orphan results are always discarded. The other extreme is running

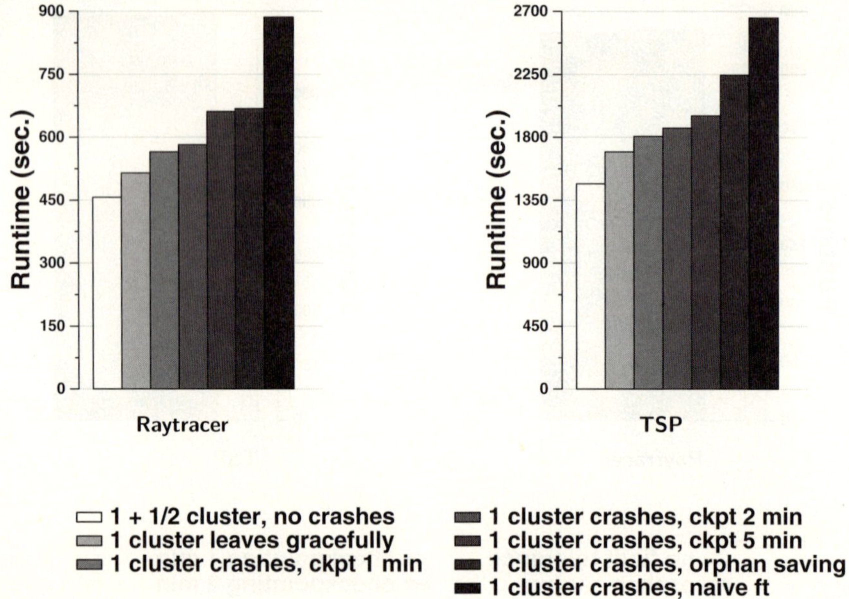

Fig. 3. Application performance in the presence of crashes

on a smaller number of processors right from the beginning, without processors leaving, denoted as the *1+1/2 cluster* case. We also compare to the situation in which nodes are leaving *gracefully* (after a prior notification) where all orphan results are used via the orphan saving mechanism.

We used checkpointing intervals of 1, 2, and 5 minutes. To allow a fair comparison between the checkpointing intervals, we enforced that the crash always occurred exactly in the middle of a checkpointing interval. We achieved this by adjusting the time the *first* checkpoint during the computation was taken.

The graphs in Figure 3 show the runtimes of both applications. The runtimes shown are averages taken over 4 runs. In 50% of the runs, the crashing (or leaving gracefully) cluster contained the master.

Figure 3 shows that orphan saving outperforms the *naive* approach by 15% to 25%. Checkpointing improves the performance of the system by further 10% to 15%. The performance improvement is the largest with small checkpointing intervals. This is a consequence of the (quasi-) constant checkpointing overhead, as shown in Figure 2. If nodes are leaving gracefully, the orphan saving algorithm provides up to 40% performance improvement over the *naive* algorithm.

Figure 3 also presents runtimes for the situation in which, instead of nodes leaving or crashing, a smaller number of nodes is available to the application right from the beginning. This smaller number of nodes is chosen to be equivalent to the accumulated number of nodes in case of the investigated crashes and graceful leaves, denoted as the *1+1/2 cluster* case. This case is best being compared to the situation of graceful leaves, indicating a small but noticeable cost (12% for Raytracer and 15% for TSP) of the orphan saving algorithm which is to be paid

Table 2. Runtimes and checkpoint file size for suspending/resuming applications

application	uninterrupted run	suspend/resume	file size
Raytracer	322 s	360 s	12 MB
TSP	1106 s	1116 s	19 KB

for moving the orphan results and some amount of redundant computation of orphan jobs that get recomputed before their results arrive at the other nodes.

4.3   Performance of Suspending/Resuming an Application

To evaluate the performance of the suspend/resume mechanism, again we ran both applications on 32 nodes in 2 clusters. In the middle of the computation, we stopped the applications, which, in turn, checkpointed their results and exited. Next, we have restarted the applications on the same processor set, using the checkpoint files just created.

The runtimes of both uninterrupted and suspended/resumed execution, along with the checkpoint file sizes, are listed in Table 2. These times are application runtimes only, they do *not* include the overhead of scheduling and (re-)starting the applications. The overhead of suspending and resuming an application is 11% for the data intensive Raytracer application and only 1% for the compute-bound TSP. This overhead is caused by the need to write and read the checkpoint file, The size of the checkpoint file is determined by the checkpoint data structures (including data serialization) and the size of checkpoint intervals. Practically no work is lost while suspending and resuming an application.

5   Related Work

Checkpointing is used in grid computing by systems such as Condor [7] and Cactus [8]. Dynamite [9] uses checkpointing to support load balancing through the migration of tasks for PVM and MPI applications.

Unfortunately, checkpointing causes execution time overhead, even if there are no crashes, mainly caused by writing the state of the processes to stable storage. This overhead might be reduced by using concurrent checkpointing [10]. Another problem of most checkpointing schemes is the complexity of the crash recovery procedure, especially in dynamic and heterogeneous grid environments where rescheduling the application and retrieving and transferring the checkpoint data between nodes is non-trivial. The final problem of checkpointing is that in most existing implementations, the application needs to be restarted at the same number of processors as before the crash, so it does not support malleability. An exception is SRS [11], a library for developing malleable data-parallel applications.

The checkpointing overhead can be reduced by application-level checkpointing, as, e.g., done by Cactus [8]. Here, the application itself determines which data to checkpoint, allowing to reduce the data to a minimum. Satin pushes this idea even further by not only writing small data sets, but also by writing the data asynchronously, without interrupting the ongoing computation. This leads to a very efficient checkpointing scheme, albeit being restricted to divide-and-conquer applications.

Several fault tolerance mechanisms for divide-and-conquer applications have been proposed. In the DIB system [12], processors redo work of other processors even if no crash has been detected. Redoing occurs while a processor waits for its steal request being granted. Instead of staying idle, the processor starts redoing work that was stolen from it earlier but the result of which has not yet been received. This approach is robust since crashes can be handled even without being detected. However, this strategy can lead to a large amount of redundant computation.

Another approach was proposed in [13]. Here, the problem of orphan jobs is partially addressed by storing not only the identifier of the parent processor (the processor from which the job was stolen), but also the identifier of its *grandparent* processor. When the parent processor crashes, the orphaned job is directed to the grandparent instead. Obviously, if both ancestor processors crash, the orphaned job cannot be reused anymore. While this mechanism can be extended further, the price to pay is higher overhead for the additional control data.

Atlas [2] is yet another divide-and-conquer system, based on CilkNOW [14], an extension of Cilk [15], a C-based divide-and-conquer system, to networks of workstations. Atlas was designed with heterogeneity and fault tolerance in mind but aims only at moderate performance. Its fault tolerance mechanism is also based on redoing work. The problem of orphan jobs is not addressed in Atlas.

6   Conclusion

Grid applications need to be fault tolerant, malleable, and migratable. In previous work, we have presented *orphan saving*, an efficient mechanism addressing these issues for divide-and-conquer applications. With orphan saving, applications are guaranteed to complete successfully, as long as at least one processor is alive, at any moment during the execution.

In this paper, we have presented a mechanism for writing partial results to checkpoint files, adding the capability to also tolerate the *total loss* of all processors, and to allow suspending an application and to resume it later, whenever CPU's become available again.

Our performance evaluation has shown that both fault-tolerant mechanisms have only negligible overheads in the absence of faults. This allows us to use very short checkpointing intervals, such as one minute. In the case of faults, the new checkpointing mechanism outperforms orphan saving by 10 % to 15 %. Due to the short checkpointing intervals, our mechanism is recovering from crashes very efficiently. In our tests, suspending and later resuming an application has only between 1 % and 11 % overhead, compared to uninterrupted runs. We also use a special technique, *checkpoint file compression*, to control the size of the checkpoint file.

Divide-and-conquer lends itself very well for fault-tolerant, malleable, or migratable execution, because any job of an application's execution tree can always be recomputed in case its result was lost. Both our mechanisms, orphan saving and the new checkpointing scheme, can execute very efficiently due to effective re-use of partial (orphan) results in the case of crashes, malleability, or migration.

The new checkpointing mechanism also adds the capability of suspending the execution of an application, and to resume it later from a checkpoint file. Due to the robustness of divide-and-conquer, the computation can even be resumed from (the valid parts of) a damaged checkpoint file which might be the result of a crash of the processor writing the file. With this added value, divide-and-conquer becomes an even more attractive paradigm for implementing grid applications.

Acknowledgements

This work has been partially supported by the *CoreGRID* Network of Excellence, funded by the European Commission's FP6 programme (contract IST-2002-004265). We would like to thank our alumnus student Kris Borg who implemented the first version of the checkpointing mechanism for his M.Sc. thesis project.

References

1. Wrzesinska, G., van Nieuwport, R.V., Maassen, J., Bal, H.E.: Fault-tolerance, Malleability and Migration for Divide-and-Conquer Applications on the Grid. In: 19th IEEE International Parallel and Distributed Processing Symposium (IPDPS'05), IEEE Computer Society Press, Los Alamitos (2005)
2. Baldeschwicler, J., Blumofe, R., Brewer, E.: ATLAS: An Infrastructure for Global Computing. In: Seventh ACM SIGOPS European Workshop on System Support for Worldwide Applications, Connemara, Ireland, pp. 165–172 (September 1996)
3. van Nieuwpoort, R.V., Kielmann, T., Bal, H.: Efficient Load Balancing for Wide-Area Divide-and-Conquer Applications. In: ACM SIGPLAN Symposium on Principles and Practice of Parallel Programming, Snowbird, Utah, USA, pp. 34–43 (June 2001)
4. van Nieuwpoort, R.V., Maassen, J., Wrzesinska, G., Kielmann, T., Bal, H.E.: Satin: Simple and Efficient Java-based Grid Programming. Scalable Computing: Practice and Experience 6(3), 19–32 (2005)
5. Allen, G., Davis, K., Goodale, T., Hutanu, A., Kaiser, H., Kielmann, T., Merzky, A., van Nieuwpoort, R., Reinefeld, A., Schintke, F., Schütt, T., Seidel, E., Ullmer, B.: The Grid Application Toolkit: Towards Generic and Easy Application Programming Interfaces for the Grid. Proceedings of the IEEE 93(3), 534–550 (2005)
6. Li, K., Naughton, J.F., Plank, J.S.: Real-time, concurrent checkpointing for parallel programs. In: 2nd ACM SIGPLAN Symposium on Principles and Practice of Parallel Programming (PPoPP'90), pp. 79–88. ACM Press, New York (1990)

7. Litzkow, M., Livny, M., Mutka, M.: Condor - A Hunter of Idle Workstations. In: Proceedings of the 8th International Conference of Distributed Computing Systems, San Jose, California, pp. 104–111 (June 1988)
8. Allen, G., Benger, W., Goodale, T., Hege, H.C., Lanfermann, G., Merzky, A., Radke, T., Seidel, E., Shalf, J.: The Cactus Code: A Problem Solving Environment for the Grid. In: Proceedings of the Ninth IEEE International Symposium on High Performance Distributed Computing (HPDC9), Pittsburgh, Pennsylvania, USA, pp. 253–260 (August 2000)
9. Iskra, K.A., Hendrikse, Z.W., van Albada, G.D., Overeinder, B.J., Sloot, P.M.A., Gehring, J.: Experiments with migration of message-passing tasks. In: Buyya, R., Baker, M. (eds.) GRID 2000. LNCS, vol. 1971, pp. 203–213. Springer, Heidelberg (2000)
10. Plank, J.: Efficient Checkpointing on MIMD architectures. PhD thesis, Princeton University (1993)
11. Vadhiyar, S.S., Dongarra, J.J.: SRS – a framework for developing malleable and migratable parallel applications for distributed systems. Parallel Processing Letters 13(2), 291–312 (2003)
12. Finkel, R., Manber, U.: DIB – A Distributed Implementation of Backtracking. ACM Transactions of Programming Languages and Systems 9(2), 235–256 (1987)
13. Lin, F.C.H., Keller, R.M.: Distributed Recovery in Applicative Systems. In: Proceedings of the 1986 International Conference on Parallel Processing, University Park, PA, USA, pp. 405–412 (August 1986)
14. Blumofe, R., Lisiecki, P.: Adaptive and Reliable Parallel Computing on Networks of Workstations. In: USENIX 1997 Annual Technical Conference on UNIX and Advanced Computing Systems, Anaheim, California, pp. 133–147 (January 1997)
15. Blumofe, R.D., Joerg, C.F., Kuszmaul, B.C., Leiserson, C.E., Randall, K.H., Zhou, Y.: Cilk: An Efficient Multithreaded Runtime System. Journal of Parallel and Distributed Computing 37(1), 55–69 (1996)

Adaptable Distance-Based Decision-Making Support in Dynamic Cross-Grid Environment

Julien Gossa[1], Jean-Marc Pierson[2], and Lionel Brunie[1]

[1] LIRIS INSA-Lyon, UMR5205 F-69621, France
`firstname.lastname@liris.cnrs.fr`
[2] IRIT University Paul Sabatier UMR5505 F-31062, France
`lastname@liris.cnrs.fr`

Abstract. The grid environment presents numerous opportunities for business applications as well as for scientific ones. Nevertheless the current trends seem to lead to several independent specialized grids in opposition to the early visions of one generic world wide grid. In such a cross-grid context, the environment might be harder to manipulate whereas more decisions must be handled from user-side. Our proposal is a distance-based decision-making support designed to be usable, adaptable and accurate. Our main contribution is to ensure the profitability of classical monitoring solutions by improving their usability. Our approach is illustrated and validated with experiments in a real grid environment.[1]

1 Introduction

The term Grid Islands has been proposed by GridBus project [1] to describe the situation of several grid solutions cohabiting without any collaboration. In their paper, De Assunao and al. analyze the lack of interoperability between grid islands and the necessity of transparent and secure interlinks to build a large World Wide Grid. Our vision of grid evolution is slightly different: Each specialized domain has so specific needs and constraints that it seems very difficult to satisfy them in a generic environment. Consequently, the perspective of several specialized and adapted grids is more realistic. In such a context the users interact with several grids middlewares. This interaction is allowed by Service Oriented Architectures, as WSRF-based grids are now usable with the standard software equipment. Many proposals already address the security requirement. But few works are interesting in the requirement of a usable decision-making support. It presents numerous challenges and requirements that must be mandatory addressed, as illustrated in the next section.

1.1 Use Case

The figure 1 shows a cross-grid environment composed of nine sites distributed throughout France. One VO is composed of two grids: *Grid*1 and *Grid*2. One

---

[1] This work is supported by the French ministry of research within the ACI "masse de données" (Grid for Geno-Medicine Project).

A.-M. Kermarrec, L. Bougé, and T. Priol (Eds.): Euro-Par 2007, LNCS 4641, pp. 437–446, 2007.

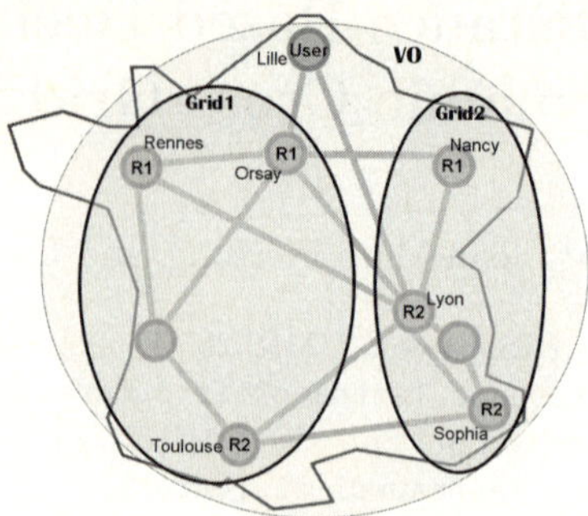

Fig. 1. Example of user decision-making in cross-grid context

member of this grid is a biologist located at Lille. He has to compute the signature of a medical image. This classical task implies generally two resources: $R1$ an image normalization service and $R2$ a signature computation service. These are replicated on both grids. He knows the size of initial image, the size of the intermediate one and the size of signature. Moreover, he knows that signature computation is CPU expensive.

The task can be submitted either to *Grid1* or to *Grid2*. Assuming that grid middlewares are able to automatically compose resources, $R1$ and $R2$ are linked automatically. Another alternative is to handle the composition from user side and to use $R1$ and $R2$ from different grids. This last possibility seems less efficient as the user must retrieve the output from $R1$ and resend it to $R2$.

To make this decision, the user has access to the monitoring information about the infrastructure: CPU speed and load of each resource, latency and bandwidth between the resources... But using these data is difficult as the task implies both computation and communication with different magnitude: While the normalization is mainly communication expensive, the signature computation is mainly CPU expensive. Moreover the possible combinations of resources are numerous and consequently confusing. Unfortunately in a cross-grid environment the users can not rely on the classical grid decision-making units, such as resource brokers or schedulers, as they are not designed to collaborate. Some decisions must be handled from the user side.

In such a case, the user needs to rely on a usable decision-making support.

1.2   Grid Decision-Making Support

The decisions in grid environment concern all grid resources: data, databases, services, hosts... And all users: end-users selecting resources to submit tasks, administrators planning deployment, developers sizing software...

In SOA the number of services is meant to dramatically increase. In this context, the optimization of the whole architecture relies on the optimization of each service. Consequently the decision-making support will be needed at each level of the architecture and must be adaptable to several contexts.

Several aspects must be taken into account: (1) *The infrastructure topology and condition*: communication, computation and storage capabilities... (2) *The characteristics of the task*: communication, computation and storage needs... And (3) *the objectives to be achieved*: End-user response delay, load balancing, optimization of financial costs...

Moreover, decision-making supports have some mandatory requirements:

Usability, Accuracy, and Profitability: They must be available under the best conditions and the provided information must be as close as possible of user concerns, otherwise it will not be used. Furthermore, the ratio profit/cost must be satisfactory.

Adaptability, Flexibility and Extensibility: They must be suitable whatever the user needs are or the infrastructures evolution is.

Scalability: They must be able to handle world-scale decisions as easily as simple ones, particularly from the user point of view.

This article is organized as follows. First in Section 2 we present the existing decision-making supports provided by grid environments and their limitations. Second we describe our proposal in Section 3 and analyse some experimentations in Section 4. Finally, we conclude with perspectives in Section 5.

2   Related Work

Decision-making in existing grid environment is mainly supported by monitoring systems. The identification of relevant metrics for grid environment and measurement methods has been made by the Network Measurements Working Group of the Global Grid Forum in [2]. Recent developments in grid infrastructure have lead to effective tools providing these measurements, such as the Monitoring and Discovery Service of Globus [3], R-GMA [4], and SCALEA-G [5]. Most of them are based on the Grid Monitoring Architecture (GMA). Another approach is adopted by the Network Weather Service [6]. It is able to capture the condition of both network and hosts. It can provide the raw measurements of the classical metrics as well as forecasts.

An alternative to monitoring system providing several raw metric measurements, is the concept of *distance*. Several Distance Vector protocols (RIP, IGRP, EIGRP, OSPF,...) have been implemented for routing in packet-switched networks. Here, the concept of *distance* is mostly the number of hops between two end-points. Since then, this concept has often been reused for very different purposes such as data management, network topology discovering, resource brokering, nodes clustering, etc. We do think that the popularity of this concept comes from its similarity with our real world. So it constitutes a precious help in the understanding of the network and in the elaboration of decision-making processes. The Distance Map Services (DMS) such as IDMaps in [7] and Global Network Positioning (GPN) [8] aim at providing an estimation of the distance between all the hosts of a network while minimizing the number of measurements. Most of them are limited to the latency which is the easiest and most inexpensive metric to measure.

Both monitoring systems and DMS aim at efficiency and scalability. They are too low-level to be really usable as a full support for decision-making: Their users have to deal with raw metrics such as latencies or CPU loads which are far from their actual concerns. In grid environment, the tasks and goals are more complex. Furthermore, while DMS provide too limited information, monitoring systems are highly resource expensive. Producing and providing the monitoring information consume the monitored resource as well as communication resources. It might be a waste if the monitoring data are not fully exploited.

Consequently, they are inadequate to fully support decision-making in grid environment. There is a need for an advanced decision-making support, improving the usability while achieving a good adaptability and accuracy.

3   NDS: The Network Distance Service

Our proposal is a distance-based decision support. It provides distances adaptable to any given *task*. We call *task* an interaction between hosts of a network. Generally, it corresponds to the invocation of one service. But it may be more basic tasks such as data retrieval or storage. Such distances are meant to be more usable than raw metrics and more relevant than the distance provided by Distance Map Services. Basically, they are based on the composition of different raw metrics provided by external monitoring systems. This computation is embedded in a Web Service developed with Globus Toolkit 4: the Network Distance Service.

3.1   Metric Model

According to the GGF NM-WG in [2], *a metric is a quantity related to the performance and reliability of the Internet.* We call *measures* the actual values related to metrics.

We note M the set of metrics. It includes the *bandwidth* (BW), the *latency* (L), the *CPU speed* ($CPUs$), and the *CPU availability* ($CPUa$). The measure of the network metric m from the host i to the host j is noted $m_{i,j}$. The measure of the host metric m for the host i is noted m_i.

NDS must be able to use the measures provided by any monitoring tools. Thus monitoring tools are accessed through command lines execution. Then, no wrapper has to be developed and new metrics and tools are integrated through a simple JNDI configuration file. The only requirement is that the monitoring tools must by queryable from the execution host of NDS.

3.2   Compound Metric Model

In order to come closer to user concerns and to improve its usability, NDS embeds two compound metrics related to the two basic kinds of grid task: data transfers or computations.

DTC for *Data Transfer Cost*. It assesses the cost to transfer a piece of data of size *data_size* from one host i to another j. According to [9], "the Raw Bandwidth model using NWS forecasts can be used effectively to rank alternative candidate schedules". Actually, we observe that data size does not influence only distance, but also relevance of the different metrics: Latency is the key factor about small size data, whereas Bandwidth is the key factor about large ones. Thus *DTC* includes the 3 RTT needed to open and close TCP/IP connections.

$$DTC_{i,j}(data\_size) = 3 \times (L_{i,j} + L_{j,i}) + \frac{data\_size}{BW_{i,j}}$$

CTC for *Computation Task Cost* assesses of the cost to execute *nb_cycles* CPU cycles on the host i. *nb_cycles* can be obtained either by calibration or compilation technics. Its automatic extraction will be investigated in future work.

$$CTC_i(nb\_cycles) = \frac{nb\_cycles}{CPUs_i \times CPUa_i}$$

One can note that theses compound metrics are rather basic. Some advanced characteristics are not taken into account, such as host architecture, OS, bus frequency, buffer size, scheduler configuration, protocol, MTU, TCP/IP configuration... Nevertheless, tools like NWS measure the metrics from the application layer. Consequently they already take all the influencing characteristics into account without having to identify them.

Moreover compound metrics are declared as easily as raw metrics: It is easy to refine them or to add new ones assessing of any goals. Even non-functional aspects like financial costs can be integrated as soon as the raw data are available.

3.3   NDS Queries

The NDS queries include three sections:

Set of hosts: The set of names of hosts involved in the decision. It can include several named subsets and is noted $\mathcal{H}$.

Task Properties Set: The set of task parameters involved in the distance computation which are not monitoring metrics. It is a set of named values noted TPS.

Distance Function: The real-valued function that gives the final value of the distance between two nodes. We note it df.

It is a nonlinear combination of monitoring metrics and values in TPS. This function is the core of distance computation. It must be relevant to the aspects the user wants to assess. Generally its result must tend to zero when the performances tend to perfection, as distances generally represent costs.

The next section shows some examples of NDS queries in concrete use cases.

4   Experiments

The experiments are made on Grid5000 [10]. This experimental platform allows to reserve nodes to conduct distributed experiments. Its topology is shown in the Figure 1. Its network is high-performance (GB/s) and shared among all users. The nodes are not shared during reservation and heterogeneous architectures are represented (AMD and Intel, 32 bits and 64 bits, simple and double core). The raw metrics used by NDS in distance computations are provided by NWS.

Our approach is validated by comparing the execution time for all possible decisions. These times are retrieved with a Fake Service: The Fake Client sends *in* random bytes to the Fake Service; Then this one makes *comp* divisions of double typed variables; Finally, *out* random bytes are returned to either the client or another Fake Service. This allows to emulate a large range of resource. All presented times are means of 10 experiments.

4.1   Selection

The problem presented in section 1.1 is to decide which instances of $R1$ and $R2$ optimize the execution time. This problem can be solved by computing and sorting the distance representing the task performance. This distance must include $s1 = 10MB$ the size of the initial image, $s2 = 100KB$ the size of the intermediate image, $s3 = 1KB$ the size of the signature, $c1 = 100\ 000$ the number of CPU cycles needed by the normalization, and $c2 = 10\ 000\ 000$ the number of CPU cycles needed by the signature computation.

The corresponding NDS query is:

$$
\begin{cases}
\mathcal{V} & = \{client = \{\text{node-36.lille}\}, \\
& \qquad R1 = \{\text{gdx0039.orsay, grillon-20.nancy, helios51.sophia}\}, \\
& \qquad R2 = \{\text{node-2.lyon, paravent74.rennes, node-25.toulouse}\}\} \\
TPS & = \{s1\text{=}10\ 000\ 000,\ s2\text{=}100\ 000,\ s3\text{=}1\ 000,\ c1\text{=}100\ 000,\ c2\text{=}10\ 000\ 000\} \\
df & = DTC_{client,R1}(s1) + CTC_{R1}(c2) \\
& \quad + (sameVO_{R1,R2})?DTC_{R1,R2}(s2) : DTC_{R2,client}(s2) + DTC_{client,R2}(s2) \\
& \quad + CTC_{R2}(c2) + DTC_{R2,client}(s3)
\end{cases}
$$

Where *client*, $R1$, and $R2$ into *df* represent the client host and the instances of resources declared in $\mathcal{V}$. In this example *df* represents the costs to achieve the whole task using $R1$ and $R2$. Please note the use of the special test *sameVO* returning *true* if the given parameters are members of a same VO. It is integrated in NDS as a raw metric and implemented into a fake external tools that can be replaced by a real one if available. It shows the expressivity of NDS which uses the library Java math Expression Parser (JEP [11]) to parse *df*. It supports classical mathematical operations and functions as well as Boolean tests and character strings.

The figure 2 shows both the experimental execution times (in seconds) and the distances given by NDS (in arbitrary units) according to the selection of $R1$ and $R2$. One can note that the 9 alternate possibilities are almost perfectly ranked by NDS and that the best solution was hard to guess as `gdx0039.orsay`

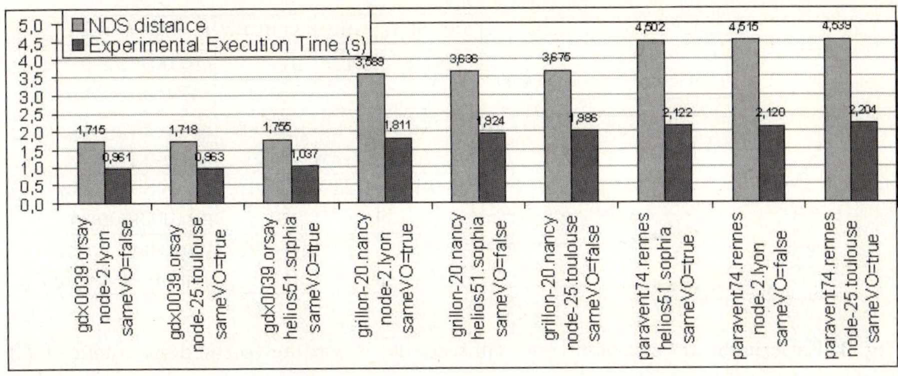

Fig. 2. Experimental execution times and NDS distances according to $R1$ and $R2$

and `node-2.lyon` are not in the same grid. Furthermore the important decision is actually the selection of $R1$. Indeed the results are sensibly equal whatever the selection of $R2$ is. This shows that first data transfer is the key factor of the efficiency of the whole execution, which was hard to guess too.

An important observation is that NDS distances are not exactly directly proportional to the real execution times. Consequently they can not be expressed in physical units like seconds and they are not previsions and must be used for ranking exclusively. Nevertheless, NDS distances can be used to support decisions in a wide range of applications, for instance in deployment tasks.

4.2   Deployment

Another problem is to decide how many instances of a resources must be deployed and where. In graph theory, this problem is called the *k-medians problem*:

Given a set V of points in a metric space endowed with a metric distance function df, and given a desired number k of resulting clusters, partition S into non-overlapping clusters $C_1, \ldots, C_k$ and determine their "centres"

$$\mu = \{\mu_1, \ldots, \mu_k\} \subset V \text{ so that } criterion = \sum_{i \in \mathcal{V}} min_{j=1}^{k} df(i, \mu_j) \text{ is minimized.}$$

In our scenario, k is the number of instances of the resource, while $\mu_1, \ldots, \mu_k$ are their optimal locations. We show in [12] how the distance produced by NDS can be validated to be "metric distance". NDS embeds an algorithm to compute and compare the *criterion* of each possible solution. It was used to make decisions about the placement of $R1$ and $R2$.

We assume that the resources clients have been identified on 19 nodes of 7 sites of Grid5000: `Lyon` (3 nodes), `Rennes` (2 nodes), `Orsay` (2 nodes), `Toulouse` (2 nodes), `Sophia-Antipolis` (4 nodes), `Nancy` (2 nodes) and `Lille` (1 nodes). Moreover we assume the client requests distribution uniform and the resource deployable on any of the 19 nodes.

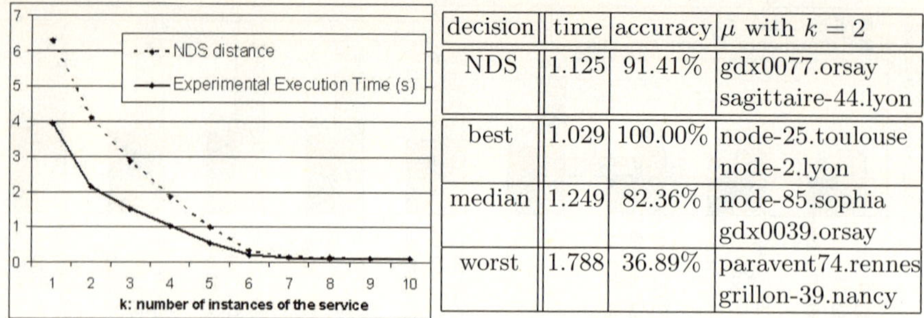

decision	time	accuracy	μ with $k = 2$
NDS	1.125	91.41%	gdx0077.orsay sagittaire-44.lyon
best	1.029	100.00%	node-25.toulouse node-2.lyon
median	1.249	82.36%	node-85.sophia gdx0039.orsay
worst	1.788	36.89%	paravent74.rennes grillon-39.nancy

Fig. 3. Experimental execution times (in seconds) according to the deployment of $R1$

The corresponding NDS query is:

$$\begin{cases} \mathcal{V} & = \{client = \{\text{sagittaire-18.lyon}, \text{ ... (all of the 19 hosts) }\}, \\ & \quad location = \{\text{sagittaire-18.lyon}, \text{ ... (all of the 19 hosts)}\}\} \\ TPS = \{in, out, comp\} \\ df & = DTC_{client,location}(in) + CTC_{location}(comp) + DTC_{location,client}(out) \end{cases}$$

The provided distances assesses a resource invocation from all nodes to all nodes with *in* bytes of input, *comp* CPU cycles of computation, and *out* bytes of output. It allows executing the k-medians algorithm and making the deployment decisions to optimize global resource access time, as stated below.

Deployment of $R1$: $TPS = \{in = 10\,000\,000, out = 100\,000, comp = 100\,000\}$.

The Figure 3 shows k-medians algorithms results in two forms: First and according to k the best case of (1) the mean experimental execution times from all of the 19 nodes to their closest resource and (2) the distance-based k-server criteria. Obviously as $R1$ is communication expensive, deploying numerous instances allows to come closer to the clients and consequently to improve the performance. But NDS highlights two particular values $k = 2$ and $k = 6$ where the tangent line changes: The speed-up is very impressive from $k = 1$ to $k = 2$, good from $k = 2$ to $k = 6$, and null afterwards. This is perfectly assessed by NDS and allows the user to decide how many instances he will deploy according to the gain and cost to add a new resource.

Second, the table shows the experimental result with $k = 2$ of four placements: recommended by NDS thanks to the k-medians algorithm, and experimental best, median and worst. The mean execution time from all the 19 nodes shows that the NDS recommendation achieves a good accuracy, especially regarding the median and worst decisions which might possibly be made by intuitive means.

Deployment of $R2$: $TPS = \{in = 100\,000, out = 1\,000, comp = 10\,000\,000\}$.

In Figure 4 one can see that the replication of $R2$ does not lead to any speed-up, which is perfectly assessed by NDS. Indeed as $R2$ is mainly computation expensive, it does not need to be brought close to each client. Then one can decide

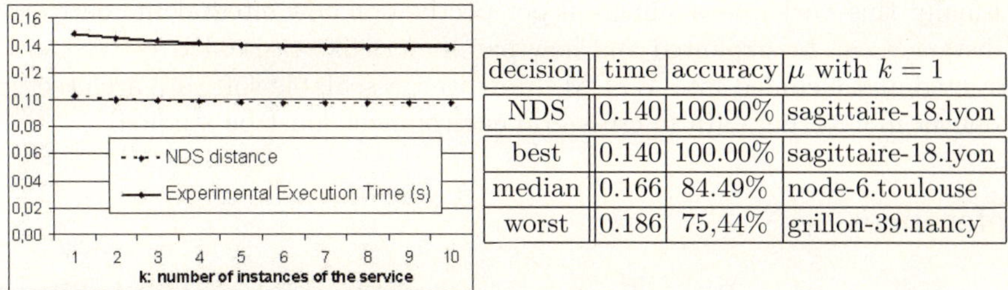

decision	time	accuracy	μ with $k = 1$
NDS	0.140	100.00%	sagittaire-18.lyon
best	0.140	100.00%	sagittaire-18.lyon
median	0.166	84.49%	node-6.toulouse
worst	0.186	75,44%	grillon-39.nancy

Fig. 4. Execution times (in seconds) according to the deployment of $R2$

to deploy only one instance. Moreover, the NDS placement recommendation is perfect while a average decision might lead to a loss of more than 15%.

Global deployment results and discussion: We have tested the NDS recommendations with $\{in, out, comp\} \in \{\{1\ 000, 100\ 000, 10\ 000\ 000\}^3\}, \forall k \in [1, 19]$. They achieve a global mean accuracy of 94.26%.

The main limitation of NDS in deployment problems, is the necessity to fix the TPS values. This can be avoided by computing the integral of df over the values of TPS according to their distribution. Moreover, the change in speed-up can be highlighted by the derivative of the k-medians criterion with respect to k. This is a part of our future work.

The purpose of this experiment is to show the adaptability and usability of our service: A large variety of complex and concrete user problems can be accurately solved with actually simple queries. NDS does not directly address the scalability. Actually NDS can handle as many host as necessary, but the underlying monitoring system has to be exhaustive and thus might present scalability issues. Nevertheless, NDS improve the usability of this system and thus ensure its profitability.

5 Conclusion and Future Work

We have presented a novel distance-based decision-making support called the *Network Distance Service*. It is designed for dynamic SOA-based grids and is particularly useful in cross-grid environments where some decisions must be handled from user-side.

The NDS provides an uniform and high level access to monitoring information, making the expression of the need at same time easy and accurate. It allows the computation of distances adapted to any task in a wide range of application. Its accuracy and usability have been shown with concrete experiments for classical decision-making problems in a real platform. Our main contribution is to ensure the profitability of monitoring systems by improving the usability of the produced data. NDS can be used either by users through a JAVA GUI or by other services through WSDL API. Moreover NDS can easily solve complex problems thanks to embedded algorithms.

Finally, this work opens numerous perspectives on how integral and derivative of distances can be exploited and how parameters like task CPU cycles can be extracted and included into WSDM. Moreover, a scalable software architecture to deliver monitoring data in cross-grid environment must be studied.

References

1. de Assuncao, M.D., Buyya, R.: A case for the world wide grid. Technical Report GRIDS-TR-2006-1, GRIDS Laboratory, Mel bourne University, Australia (February 2006)
2. Lowekamp, B., Tierney, B., Cottrell, L., Hughes-Jones, R., Kielmann, T., Swany, M.: A hierarchy of network performance characteristics for grid applications and services. In: Global Grid Forum (June 2004)
3. Globus_Alliance: Monitoring and discovery service, http://www.globus.org/mds/
4. Cooke, A., Gray, A., et al.: The relational grid monitoring architecture: Mediating information about the grid (2004)
5. Truong, H.L., Fahringer, T.: Scalea-g: a unified monitoring and performance analysis system for the grid, 12(4), 225–237 (2004)
6. Wolski, R., Spring, N.T., Hayes, J.: The Network Weather Service: a distributed resource performance forecasting service for metacomputing. Future Generation Computer Systems 15(5–6), 757–768 (1999)
7. Francis, P., Jamin, S., Jin, C., Jin, Y., Raz, D., Shavitt, Y., Zhang, L.: IDMaps: A global Internet host distance estimation service. IEEE/ACM Transactions on Networking 9(5), 525–540 (2001)
8. Ng, T.S.E., Zhang, H.: Predicting internet network distance with coordinates-based approaches. In: Proceedings IEEE INFOCOM 2002, vol. 1, pp. 170–179. IEEE, Los Alamitos (2002)
9. Faerman, M., Su, A., Wolski, R., Berman, F.: Adaptive performance prediction for distributed data-intensive applications. In: ACM/IEEE SC99 Conference on High Performance Networking and Computing, Portland, OR, USA, IEEE Computer Society Press, Los Alamitos (1999)
10. Grid 5000, https://www.grid5000.fr/
11. JEP - Java Math Expression Parser. Singular Systems, www.singularsys.com/jep/
12. Gossa, J.: Evaluation of network distances properties by nds, the network distance service. In: 3th International Workshop on Networks for Grid Applications (GRIDNETS'06), IEEE/Create-Net (October 2006)

Negotiation Strategies Considering Opportunity Functions for Grid Scheduling

Jiadao Li[1], Kwang Mong Sim[2], and Ramin Yahyapour[1]

[1] Members of the Core Grid Institute on Resource Management and Scheduling
Institute for Robotics Research - Information Technologies
University of Dortmund, 44221 Dortmund, Germany
`jiadao.li@uni-dortmund.edu,`
`ramin.yahyapour@uni-dortmund.edu`
[2] Department of Computer Science, Hong Kong Baptist University
Kowloon Tong, KLN, Hong Kong
`prof_sim_2002@yahoo.com`

Abstract. In Grid systems, nontrivial qualities of service have to be provided to users by the resource providers. However, resource management in a decentralized infrastructure is a complex task as it has to cope with different policies and objectives of the different parties: providers and consumers/users. Agreement-based resource management is considered to solve many of these problems as the conflicts between the users and resource providers can be reconciled in a negotiation process. Such negotiation processes must be automated with no or minimal human interaction, considering the potential scale of Grid systems and the amount of necessary transactions. Therefore, strategic negotiation models and strategic negotiation strategies play important roles. In this paper, negotiation strategies considering time and opportunity functions for Grid scheduling are proposed and examined. The simulation results demonstrate that the negotiation strategies are suitable and effective for Grid environments.

1   Introduction

Grid computing [1,2] is a Service Oriented Architecture (SOA [3,2]) in which a resource user typically requires a certain service quality to be provided by the resource owners. Here, agreement based resource management [4] is typically considered as a suitable approach for this scenario. Negotiation is the process towards creating suitable agreements between different parties in a Grid. The negotiation process in a Grid computing environment should be done automatically and transparently with the growing scale in Grids [5]. In order to automate the negotiation process, suitable negotiation models are required that take the different policies and objectives of the resource providers and resource users into account and produce suitable service level agreements in reasonable time with minimized or even no user and provider interference.

In our previous work [6,7], a strategic negotiation model which supports the automatic negotiation in Grid computing was proposed and evaluated. In that model, the user, or more precisely some meta/grid-scheduling agent or job broker on his behalf, will contact different resource providers, negotiate with several of them and make a

A.-M. Kermarrec, L. Bougé, and T. Priol (Eds.): Euro-Par 2007, LNCS 4641, pp. 447–456, 2007.

decision to commit to a particular agreement with one resource provider. This is considered as the *one to many* negotiation type [8]. A *Concurrent bilateral negotiation model* [9] is suitable for this problem. This paper is an extension of the previous work. In this paper, the strategic negotiation model proposed in [6] is adopted, while we now add negotiation strategies considering the opportunity functions. An opportunity function determines the bargaining position of a negotiation agent based on available outside options [10]. A Grid resource management system needs to continuously adapt to changes in the availability of computing resources (i.e., outside options). These strategies have been implemented and evaluated.

2   Related Work

There are many approaches proposed for the Grid resource management problem, for example, economic methods. An overview of such methods can be found in [11] by Buyya et al., or in in [12] by Ernemann et al. , or by Wolski et al. in [13], or by Lai et al. in [14,15]. In these papers, economic based resource management in Grid computing are investigated and several economic models are evaluated. To this end, a lot of effort has been made on Grid resource management with support of service level agreements (SLAs). In [4], the concepts of agreement-based resource management in the Grid computing environment are introduced and a general agreement model is presented; in [16], a Grid resource broker supporting SLAs called GRUBER is presented and evaluated in a real grid. In [17], the very few existing research initiatives on applying bargaining as a mechanism for managing Grid resources are reviewed and compared.

In this paper, we now adopt the negotiation strategies considering the time and opportunity functions [10] in the earlier proposed negotiation model for Grid scheduling and present an evaluation with discrete event-based simulation.

3   Negotiation Model

As a bilateral negotiation model is the building block of concurrent negotiation model, we will briefly introduce this first in this section.

3.1   Bilateral Negotiation Model

There are three parts in the bilateral negotiation model [18]: 1) the negotiation protocol, 2) the used utility functions or preference relationships for the negotiating parties, and 3) the negotiation strategy that is applied during the negotiation process. In this paper, we adopted Rubinstein's sequential alternating offer protocol in Grids, see [19]. In the negotiation process, when one negotiation side times out or an agreement is created, the negotiation process will end. Disagreement is considered as the worst outcome, therefore, the negotiation party will always try to avoid opting out of the negotiation. In this negotiation model, the negotiation parties do not know the opponents' private reservation information and their preferences/utility functions.

Typically, the objectives of a user for a computational job are to obtain a shorter response time and/or to get cheaper resources, while the resource providers expect to gain higher profit and/or higher utilization. However, our model is not restricted to particular objectives and can be flexibly defined for different scenarios. In real Grid systems, there can be many different negotiation objectives, which are interdependent and should be simultaneously dealt with which yields to a multi-criteria optimization problem [20]. In this presented research work, without limiting generality we restrict our analysis on considering the expected waiting time of the job and the expected cost per cpu time. However, the model can be applied and extended to other criteria as well. For the resource providers, we also assume two corresponding negotiation issues: the expected waiting time until a job can be started $T_s^t(Job)$, and the expected price $P_s^t(Job)$. The expected waiting time for the newly incoming job can be obtained from the current resource status and the future schedule plan considering other created agreements which have to be fulfilled. The expected price will be obtained via the negotiation process. In paper [6], more details about the utility functions are given.

3.2   Negotiation Strategies Considering the Time and Opportunity Functions

In the negotiation process, it is assumed that both of the negotiation agents behave according to *the good-faith bargaining principles* which means that it is usually not easily reversed [21]. Here, on the basis of the initial offer values, successive offers by sellers are monotonically decreasing while successive offers by the buyers are monotonically increasing. In order to create the agreement, both of the negotiation parties want to narrow the difference between the offers and counter offers with respect to different negotiation issues. In the strategic negotiation model, the negotiation agents can take different kinds of negotiation strategies developed in the agent community [22] to create the negotiation offer at different negotiation times. In [10], the negotiation strategies are proposed and analyzed for market-driven agents to make prudent compromises taking into account factors such as time preference, opportunity functions, competition factors.

In a multilateral negotiation, having outside options may give a negotiator more bargaining "power". However, negotiations may still break down if the proposals between two negotiators are too far apart.

Suppose agent B (the job agent) engages S_j (the resource provider) in round t. At any negotiation round t, $B's$ last proposal (bid) is represented by a utility vector $(V_t^{B \to S_j}, W_t^{B \to S_j})$ and $S_j's$ proposal (offer) is a utility vector $(V_t^{S_j \to B}, W_t^{S_j \to B})$.

The opportunity function

$$O(n_t^B, v_t^{B \to S_j}, \langle W_t^{S_j \to B} \rangle) = 1 - \prod_{j=1}^{n_t^B} \frac{V_t^{B \to S_j} - W_t^{S_j \to B}}{(V_t^{B \to S_j} - c^B)} \tag{1}$$

determines the amount of concession based on 1) trading alternatives (number of trading parties n_t^B) and differences in utilities $V_t^{B \to S_j}$ generated by the proposal of the job agent and the counter-proposal of its trading party $W_t^{S_j \to B}$. c^B is the worst possible utility for agent B. Space limitation precludes detailed derivations from being included here, but they can be found in [10].

As explained before, the negotiation party will also modify the negotiation offer with the negotiation time going on. There are many ways of defining the function $\alpha_j^a(t)$ to model the effects of the remaining negotiation time. We also use the following function to calculate the $\alpha_j^a(t)$, see [6]:

$$\alpha_j^a(t) = k_j^a + (1 - k_j^a)(\frac{t}{t_{max}^a})^{1/\beta}, \tag{2}$$

where t_{max}^a is the deadline of the negotiation party a for the completion of the negotiation, t denotes the current time instant in the available negotiation time set, the parameter β is the degree of convexity that determines the type of the negotiation party in the time dependent strategy. Different β values yield different negotiation strategies. For the initial bargaining value k_j^a is used, for which the following relation holds $0 \le k_j^a \le 1$.

As pointed out in [10], there are several means of combining the time and the opportunity function effects to create the offers for the negotiation parties, for instance, $0.5 * (T(t) + O(t))$, or $T(t) * O(t)$. Here we use the former one.

It is assumed that P_c^t is the offered price at time t by the user, P_s^t is the offered price at time t by the resource provider; $T_c^t(job)$ is the proposed waiting time at time t by the user, $T_s^t(job)$ is the acceptable waiting time for the specific job at time t according to the current resource status considering the future reserved resource as well.

We assume that V_j is the utility function of the negotiation party which associates with the negotiation issue j and the $x_{a \to b}^t[t]$ is the offer provided by one party (denoted by a) to another negotiation party (denoted by b).

If V_j is decreasing:

$$x_{a \to b}^t[t] = a_j^t + 0.5((min(max_j^a, b_j^t) - a_j^a) * (O(t) + \alpha_j^a(t)), \tag{3}$$

if V_j is increasing:

$$x_{a \to b}^t[t] = a_j^t + (1 - 0.5 * (\alpha_j^a(t) + O(t))(min(max_j^a, b_j^t) - a_j^t), \tag{4}$$

Equations 3 and 4 represents the job user's strategy and the resource provider's strategy respectively.

As there are two negotiation issues involved in this negotiation process, we assume that if the offer in which one of the negotiation issues from the opponent is satisfied, then it will accept this value and not further change it but only change the value of the remaining other issue in the following negotiation process. This is just a first heuristic to analyze the behavior of the negotiation strategies. In real life, negotiation issues will not be independent and thus acceptable deal is not easily reached. This will require more complex negotiation strategies that need to be considered in future work. For now, we accept that the negotiation issues are modified according to the previously made assumption on the monotonous increase/decreas by the parties.

3.3   Concurrent Bilateral Negotiation Model

As mentioned above, in the Grid environment a number of resources will typically be available which are capable of fulfilling the job constraints after the resource discovery phase. The user or a corresponding scheduling component will contact different resource providers and initiate the negotiation process for the actual resource allocation. The negotiation relationship is of the "one to many" type, which can be treated as a concurrent bilateral model. In the concurrent negotiation threads in which a single user is involved, the reservation value of the negotiation issues and preferences are the same. However, the user may adopt different strategies with respect to different negotiation opponents. Furthermore, they might change the negotiation strategies during the negotiation process based on available information from different negotiation threads.

Because these negotiation threads are executed concurrently, it is very difficult to predict whether the user might achieve a better offer from another negotiation thread if there is already a suitable offer that could be committed to an agreement. In our model, we assumed that once an agreement is available, it will be created and committed. Of course, in a real life scenario the job agent might actually exploit the available time to find several offers and decide at the end on the best offer. In paper [23], we analyze the results of tradeoff between the "best" and the "first available" agreement. In this paper, for simplicity, we restricted our examination to accepting the first available agreement. If one negotiation thread is successfully negotiated, all of the other negotiation threads will be terminated. The agreement can then be used by provisioning and execution service to actually start a job on the local resource management system.

4 Evaluation

Discrete event simulation has been used to evaluate the proposed negotiation model. Currently, there is no real data from Grid computing environments that include suitable information for negotiation models. However, high performance computing is still the typical application scenario for Grid technology. For this scenario, workload traces are available which were recorded on actual machine installations [24]. Therefore, our first evaluations are based on such traces. However, negotiation information are not included in this data as none of the real systems supported negotiation models yet. To this end, the missing information can only be modeled based on first assumptions. In the following the simulation configuration is described and the simulation results are analyzed.

4.1 Simulation Configuration

At the beginning of the negotiation, the negotiation parties will always make the offers which are most favorable to themselves. So we assume initial values of 0 for k_j^a of all the negotiation parties. For performance analysis we assume a negotiation interval of 1s between each negotiation round. In the following we describe the models of the users and the resource providers. In order to evaluate the learning-based negotiation algorithms, we will compare simulation results in different simulation cases.

User Model. In our simulation we consider parallel batch jobs in an online scenario. Typically, users will behave quite differently in the negotiation process. For our simulation, we assume two different kinds of user objectives: time-optimization and cost-optimization. The actual behavior of real users will be investigated in future research work. Below are the parameters of the user modeling which have been applied for the simulation.

- Negotiation span is uniformly distributed in $[0, 30]$s.
- Maximum price of the different job user is uniformly distributed in $[4.0, 9.0]$.
- Acceptable waiting time for the job users are uniformly distributed in $[0, 36000]$s.
- For the tough negotiator, β value is uniformly distributed in $[0.02, 0.2]$.
- For the conceder negotiator, β value is uniformly distributed in $[20, 40]$.
- Weights of time and price for the time-optimization are 0.8 and 0.2, while the weights of the time and price for the cost-optimization are 0.2 and 0.8.

Resource Provider. For the local resource management system an FCFS scheduling strategy with backfilling [25] is adopted which is common for parallel computers. There is no preemption allowed in our scenario. To this end, in this evaluation we do not consider the co-allocation of resources from different providers. The resources are all homogeneous and only differ in the number of available CPU nodes at each site. The simulated hardware configurations of the resource providers are consistent with actual configurations of the systems from which the real traces are originated. In this paper, we present results for traces from the Cornell Theory Center [24] which had 512 CPU nodes. In our simulation we assumed a Grid scenario with 6 different machines (parallel computer or cluster with a given set of CPU nodes) and therefore 6 resource providers. However, to stay consistent with the available workload from the CTC traces, the total number of nodes for all simulated machines is again 512 nodes. The number of nodes on each machine and the negotiation parameters for each resource provider are given below.

- The numbers of CPU nodes for the machines are $\{384,64,16,16,16,16\}$.
- Their different maximum prices per CPU time are $\{8.2, \ 8.0, \ 7.5, \ 7.6, \ 7.4, \ 7.5\}$.
- Their different minimum prices per CPU time are $\{2.4, \ 2.3, \ 2.0, \ 1.95, \ 1.90, \ 1.80\}$.
- Negotiation deadlines of different resource providers are all 30s, which means that usually the resource provider will not opt out of the negotiation once the negotiation thread is created.
- For the conceding negotiator, β value is $\{32, \ 35, \ 34, \ 38, \ 40, \ 40\}$.
- For the tough negotiator, β value is $\{0.03, 0.05, 0.04, 0.10, 0.05, 0.06\}$

4.2 Evaluation Criteria

In the following we provide some first evaluation remarks which give some qualitative information about the performance of the model. The actual quality will have to be verified with better workload models and real implementations.

- Comparison between the negotiation result and the reference point [21], which is the middle of the zone of possible agreement of user and resource provider: $[C_j^{max}, S_j^{min}]$. The reference point is computed by the following function:

$$U_j^{ref} = \frac{C_j^{max} + S_j^{min}}{2} \tag{5}$$

– The rate of successfully created agreement for all jobs.
– The negotiation overhead to create the agreement measured by the time taken to create the agreement. In our case, we use the final negotiation rounds which represents the required number of messages exchanged. The actual network overhead will depend on the actual network speed for this message exchange.
– In computational Grids, the users will concern about the response time and waiting time of a job, while for the resource providers the utilization and the profit will probably be the main objectives. We also compare these criteria to get some feedback about the feasibility of the negotiation model.

Simulation Results. We used the first 10000 jobs from the CTC workload traces [24] to conduct our simulation. We compare the on average required number of negotiation rounds for the successful creation of an agreement, and the rate of successfully created agreements in comparison to the total number of job requests. Other criteria are the average weighted response time (AWRT), the average weighted waiting time (AWWT), the average price difference between the agreement price (AP) and the reference price (RP). For the weight in AWRT and AWWT, the job resource consumption is used, see [26]. This weight prevents any favor of jobs with high or low resource consumption over each other.

In order to evaluate the simulation results, we compared the simulation cases with the associated simulation cases that we did in previous publications [6]. The following simulation cases are considered. Case 1: Both of them use the conceding strategy [6]; Case 2: Both of them use the conceding strategy and opportunity functions; Case 3: Both of them use the linear strategy [6]; Case 4: Both of them use linear strategy and opportunity functions; Case 5: Both of them use the tough strategy [6]; Case 6: Both of them use tough strategy and opportunity functions;

The success rate of negotiations in these 6 cases are as the following tables. The successful rate of negotiations in these 10 cases are 94.65%, 94.03%, 62.87%, 53.49%, 1.95%, 43.40% respectively, so we can see that using the time and opportunity together can yield higher creation rate in Case 6 than Case 5; in other simulation cases the creation rate are comparable.

Figure 1 shows a selection of results. R1 to R6 stands for the resources from 1 to resource 6 respectively. From these simulation results, we can see that the negotiation agents using the time and opportunist functions to narrow the differences between the offers and counter offers can achieve higher utilities than using negotiation strategies in our former work in [6]. In the simulation cases which use the time and opportunity functions, the AWWT is less than the cases in [6]. Except in the Case 5, there are no agreements created in R1, so the AWWT is 0, in other resources (except R2 and R4), the jobs can be started immediately due to the quite lower utilization rate. But the users usually pay more for the needed resources than in the simulation cases we did before as shown in the result figure. AWRT is comparable and in the same range as for Grid models which do not use negotiation models but conventional queuing systems. That means, the presented model can be considered feasible for real Grid infrastructure as it does not lead to any drawbacks in the performance results. To the contrary the negotiated waiting time of the jobs will be guaranteed by the resource providers which is the anticipated quality of service level and can be seen as a major asset of

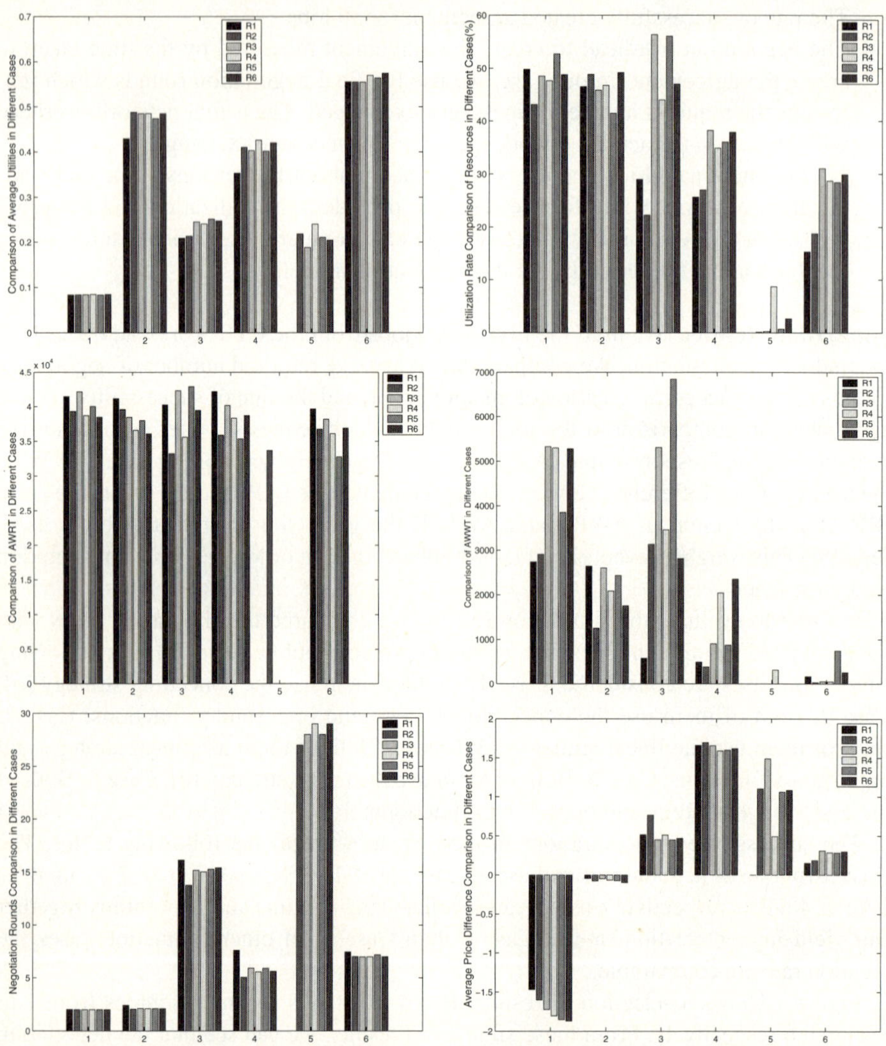

Fig. 1. Comparison between different negotiation cases including the results for the individual six resources

such an approach. Also we can see that the simulation cases using the opportunity functions can work with less negotiation rounds than using the pure negotiation tactic. An agent using opportunity function is more likely to reach a quicker agreement because it has higher chance of exploring more negotiation options. From these simulations, we can see that negotiation strategies considering time and opportunity functions are quite flexible and effective, and can actually be used in the dynamically changing Grid infrastructure.

5   Conclusions and Future Work

In this paper, we proposed and evaluated a strategic negotiation model in the Grid scenario. This model has been evaluated using discrete event based simulation. The results show that it can be applied in the practical use in automatic job scheduling. The presented results can be seen as first steps in analyzing the features and requirements for automatic negotiation strategies. However, the actual evaluation of the obtained service quality is difficult to obtain as there is no valid user job and preference workload model for Grids available which takes economic functionality into account. The presented results indicate that the negotiation overhead in terms of exchanged messages is manageable for practical application. In the future work, we will try other combinations of opportunity and time functions which can be used as the alternatives for the negotiation strategies for Grid scheduling.

Acknowledgement

This research work is partially carried out under the FP6 Network of Excellence CoreGRID funded by the European Commission (Contract IST-2002-004265). K. M. Sim acknowledges financial support from the Hong Kong Research Grant Council (project code: RGC/HKBU210906). The authors would like to thank the anonymous referees for the comments and suggestions.

References

1. Foster, I., Kesselman, C., Tuecke, S.: The anatomy of the Grid: Enabling scalable virtual organizations. In: Sakellariou, R., Keane, J.A., Gurd, J.R., Freeman, L. (eds.) Euro-Par 2001. LNCS, vol. 2150, Springer, Heidelberg (2001)
2. Foster, I., Kesselman, C.: The Grid: Blueprint for a New Computing Infrastructure. Morgan Kaufmann, San Francisco (2003)
3. The w3c web services architecture working wroup public draft (February 2004), http://www.w3.org/TR/ws-arch/
4. Czajkowski, K., Foster, I., Kesselman, C.: Agreement-based resource management. Proceedings of the IEEE 93(3), 631–643 (2005)
5. Foster, I., Jennings, N.R., Kesselman, C.: Brain meets brawn: Why grid and agents need each other. In: Kudenko, D., Kazakov, D., Alonso, E. (eds.) AAMAS II, 2005. LNCS (LNAI), vol. 3394, pp. 8–15. Springer, Heidelberg (2005)
6. Li, J., Yahyapour, R.: Negotiation strategies for grid scheduling. In: GPC 2006. LNCS, vol. 3947, pp. 42–52. Springer, Heidelberg (2006)
7. Li, J., Yahyapour, R.: Learning-based negotiation strategies for grid scheduling. In: Proceedings of the International Symposium on Cluster Computing and the Grid (CCGRID2006), pp. 567–583. IEEE Computer Society Press, Los Alamitos (2006)
8. Lomuscio, A.R., Wooldridge, M., Jennings, N.R.: A classification scheme for negotiation in electronic commerce. International Journal of Group Decision and Negotiation 12(1), 31–56 (2003)
9. Nguyen, T.D., Jennings, N.R.: A heuristic model of concurrent bi-lateral negotiations in incomplete information settings. In: Proc. 18th Int. Joint Conf. on AI, Acapulco, Mexico, pp. 1467–1469. IEEE Computer Society Press, Los Alamitos (2003)

10. Sim, K.M.: Equilibria, prudent compromises, and the "waiting" game. IEEE Transactions on Systems, Man, and Cybernetics, Part B 35(4), 712–724 (2005)
11. Buyya, R.: Economic-based Distributed Resource Management and Scheduling for Grid Computing. PhD thesis, Monash University, Melbourne, Australia (2002)
12. Ernemann, C., Yahyapour, R.: Applying Economic Scheduling Methods to Grid Environments. In: Grid Resource Management - State of the Art and Future Trends, pp. 491–506. Kluwer Academic Publishers, Dordrecht (2003)
13. Wolski, R., Plank, J.S., Brevik, J., Bryan, T.: Analyzing market-based resource allocation strategies for the computational grid. International Journal of High Performance Computing Applications 15(3) 258–281 (2001)
14. Lai, K., Rasmusson, L., Adar, E., Sorkin, S., Zhang, L., Huberman, B.A.: Tycoon: an Implemention of a Distributed Market-Based Resource Allocation System. Technical Report arXiv:cs.DC/0412038, HP Labs, Palo Alto, CA, USA (December 2004)
15. Kevin Lai, B.A.H., Fine, L.: Tycoon: A Distributed Market-based Resource Allocation System. Technical Report arXiv:cs.DC/0404013, HP Labs, Palo Alto, CA, USA (April 2004)
16. Dumitrescu, C., Foster, I.T.: Gruber: A grid resource usage sla broker. In: Cunha, J.C., Medeiros, P.D. (eds.) Euro-Par 2005. LNCS, vol. 3648, pp. 465–474. Springer, Heidelberg (2005)
17. Sim, K.M.: A survey of bargaining models for grid resource allocation. ACM SIGecom Exch. 5(5), 22–32 (2006)
18. Kraus, S.: Strategic Negotiation in Multi-Agent Environments. MIT Press, Cambridge (2001)
19. Rubinstein, A.: Perfect equilibrium in a bargaining model. Econometrica 50, 97–110 (1982)
20. Kurowski, K., Nabrzyski, J., Oleksiak, A., Weglarz, J.: Multicriteria Aspects of Grid Resource Management. In: Grid Resource Management - State of the Art and Future Trends, pp. 271–295. Kluwer Academic Publishers, Dordrecht (2003)
21. Raiffa, H.: The Art and Science of Negotiation. Harvard Universtiy Press, Cambridge (1982)
22. Jennings, N.R., Faratin, P., Lomuscio, A.R., Parsons, S., Sierra, C., Wooldridge, M.: Automated negotiation: Prospects, methods and challenges. Int. J. of Group Decision and Negotiation 2(10), 199–215 (2001)
23. Li, J., Yahyapour, R.: A strategic negotiation model for grid scheduling. International Transactions on Systems Science and Applications 1(4), 411–421 (2006)
24. Standard workload format (July 2005), http://www.cs.huji.ac.il/labs-parallel/workload/index.html
25. Lifka, D.: The ANL/IBM SP scheduling system. In: Feitelson, D.G., Rudolph, L. (eds.) IPPS 1995. LNCS, vol. 949, pp. 295–303. Springer, Heidelberg (1995)
26. Schwiegelshohn, U., Yahyapour, R.: Analysis of first-come-first- serve parallel job scheduling. In: Proceedings of the 9th SIAM Symposium on Discrete Algorithms, pp. 629–638 (January 1998)

SymGrid: A Framework for Symbolic Computation on the Grid

Kevin Hammond[1], Abdallah Al Zain[2], Gene Cooperman[3],
Dana Petcu[4], and Phil Trinder[2]

[1] School of Computer Science, University of St Andrews, St Andrews, UK
`kh@cs.st-and.ac.uk`
[2] Dept. of Mathematics and Comp. Sci., Heriot-Watt University, Edinburgh, UK
`{ceeatia,trinder}@macs.hw.ac.uk`
[3] College of Computer Science, Northeastern University, Boston, USA
`gene@ccs.neu.edu`
[4] Institute e-Austria, Timişoara, Romania
`petcu@info.uvt.ro`

Abstract. This paper introduces the design of **SymGrid**, a new Grid framework that will, for the first time, allow multiple invocations of symbolic computing applications to interact via the Grid. **SymGrid** is designed to support the specific needs of symbolic computation, including computational steering (greater interactivity), complex data structures, and domain-specific computational patterns (for irregular parallelism). A key issue is heterogeneity: **SymGrid** is designed to orchestrate components from different symbolic systems into a single coherent (possibly parallel) Grid application, building on the **OpenMath** standard for data exchange between mathematically-oriented applications. The work is being developed as part of a major EU infrastructure project.

1 Introduction

Symbolic Computation is often distinguished from Numerical Analysis by the observation that symbolic computation deals with exact computations, while numerical analysis deals with approximate quantities, including issues of floating point error. Examples of problems that motivated the original development of symbolic computation include symbolic differentiation, indefinite integration, polynomial factorization, simplification of algebraic expressions, and power series expansions. Because of the focus on exactness, symbolic computation has followed a very different evolutionary path from that taken by numerical analysis. In contrast to notations for numerical computations, which have emphasised floating point arithmetic, monolithic arrays, and programmer-controlled memory allocation, symbolic computing has emphasized functional notations, greater interactivity, very high level programming abstractions, complex data structures, automatic memory management etc. With this different direction, it is not surprising that symbolic computation has different requirements from the Grid than traditional numerical computations. For example, the emphasis on interactivity

A.-M. Kermarrec, L. Bougé, and T. Priol (Eds.): Euro-Par 2007, LNCS 4641, pp. 457–466, 2007.

for symbolic computation leads to a greater stress on *steering* of computations, and the emphasis on functional programming has led to the identification of sophisticated *higher-order* computational patterns. Similar patterns have recently found commercial use in, for example, Google's **MapReduce** system [1,2].

In order to support the specific needs of symbolic computation on the Grid, as part of the EU Framework VI SCIEnce project (Symbolic Computation Infrastructure in Europe, RII3-CT-2005-026133), we have designed a new Grid framework, **SymGrid**, which builds on and extends standard Globus middleware capabilities, providing support for Grid Services and for orchestration of symbolic components into high-performance Grid-enabled applications forming a computational Grid. The project will initially integrate four major computational algebra systems into **SymGrid**: Maple [3], GAP [4], Kant [5] and MuPad [6]. In this way, heterogenous symbolic components can be composed into a single large-scale (possibly parallel) application. Our work builds on earlier, long-standing work on adaptive parallel systems [7], parallel symbolic computations [8,9] and distributed/Grid-enabled symbolic computations [10].

The key novelties of the **SymGrid** approach are:

1. we integrate several represenatative symbolic computation systems (GAP [4], Maple [3], Kant [5] and MuPad [6]) into a single, generic and non-exclusive middleware framework;
2. we provide a sophisticated interactive *computational steering* interface integrating seamlessly into the interactive front-ends provided by each of our target symbolic computation systems, and providing simple, transparent and high-level access to Grid services;
3. by defining common data and task interfaces for all systems, we allow complex computations to be constructed by orchestrating heterogeneous distributed components into a single symbolic application;
4. by exploiting well-established adaptive middleware that we have developed to manage complex irregular parallel computations on clusters and shared-memory parallel machines, and that has recently been ported to the Grid, we allow a number of advanced autonomic features that are important to symbolic computations including automatic control of task granularity, dynamic task creation, implicit asynchronous communication, automatic sharing-preserving data-marshalling and unmarshalling, ultra-lightweight work stealing and task migration, virtual shared memory, and distributed garbage collection;
5. we identify new domain-specific patterns of symbolic computation that may be exploited to yield platform-specific implementations for a wide range of classical symbolic computations, and which may be combined dynamically to yield complex and irregular parallel computation structures; and
6. we target a new user community which may have massive computational demands yet where exposure to parallelism/Grids is not common; this may serve as a template for extending Grid technologies to similar user bases.

This paper introduces the design of the **SymGrid** framework (Section 2), describes middleware components that support both Grid services (Section 2.1) and the composition of heterogeneous symbolic components into large-scale Grid-enabled applications (Section 2.2), identifies new and important computational patterns that are specific to symbolic computation (Section 3), and places our work in the context of previous work on parallel and Grid-enabled symbolic computations (Section 4).

2   The SymGrid Design

SymGrid comprises two main middleware components (Figure 1): **SymGrid-Services** provides a generic interface to Grid services, which may be engines of computational algebra (CA) packages; **SymGrid-Par** links multiple instances of these engines into a coherent parallel application that may be distributed across a geographically-distributed computational Grid. **SymGrid-Par** may itself be registered as a Grid service. The engines are linked by well-defined interfaces that use the standard **OpenMath** protocol for describing mathematical objects [11], developed as part of EU Framework V project MONET (IST-2001-34145). We have constructed prototype implementations of both components and are in the process of robustifying these for general use.

2.1   SymGrid-Services

Modern Grid and Web technologies allow simple access to remote applications and other services. Services are exposed through standard ports, which provide mechanisms to allow the discovery of new services and to support interaction with those services. **SymGrid-Services** provides a set of WSRF-compliant interfaces from symbolic computations to both Grid and Web services, and allows straightforward encapsulation of symbolic computations as Grid service components, including automatic client generation. Mathematical objects are communicated between users and services using **OpenMath** protocols.

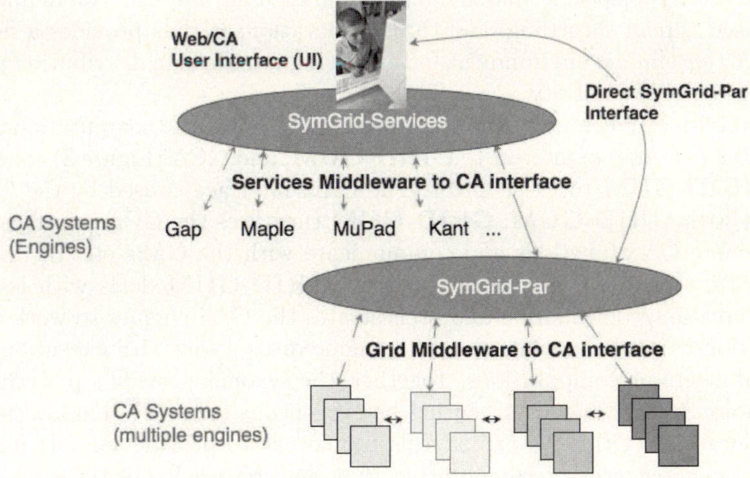

Fig. 1. The SymGrid Design

While Grid services have different goals from pure Web services (sharing computing power and resources such as disk storage databases and software applications, compared with simply sharing information), a Grid service is essentially a Web service with a few new additions: stateful services, service instantiation, named service instances, two-level naming scheme, a base set of service capabilities, and lifetime management. These new capabilities improve user interaction with remote symbolic computing services: prior computations can be stored in the service state, using instantiation, personalized services can be created, the naming schemes allow services to be easily modified, standard service data elements allow standard search facilities to be implemented, and, due to the transient character of the services, resource management can easily be performed. All of these features are supported through **SymGrid-Services**.

2.2   SymGrid-Par

SymGrid-Par orchestrates symbolic components into a (possibly parallel) Grid-enabled application. Each component executes within an instance of a Grid-enabled engine, which can be geographically distributed to form a wide-area computational Grid, built as a loosely-coupled collection of Grid-enabled clusters. Components communicate using the same **OpenMath** data-exchange protocol that is also used by **SymGrid-Services**. In this way, a high degree of integration is achieved between services and high-performance computations.

SymGrid-Par is built around **GRID-GUM** [7], a system designed to support parallel computations on the Grid, and which has been adapted to interface with symbolic engines. **GRID-GUM** builds on the basic MPICH-G2 transport protocols and task management capabilities to provide a range of very high-level facilities aimed at supporting large-scale, complex parallel systems. It includes support for ultra-lightweight thread creation, distributed virtual shared-memory management, multi-level scheduling support, automatic thread placement, automatic datatype-specific marshalling/unmarshalling, implicit communication, load-based thread throttling, and thread migration. It thus provides a flexible, adaptive, *autonomic* environment for managing parallelism/distribution at various degrees of granularity.

SymGrid-Par comprises two interfaces: CAG links the computational algebra systems (CASs) of interest to **GRID-GUM**; and GCA (Figure 2) conversely links **GRID-GUM** to these CASs. The CAG interface is used by the CAS to interact with **GRID-GUM**. **GRID-GUM** then uses the GCA interface to invoke remote CAS functions and communicate with the CASs etc. In this way, we achieve a clear separation of concerns: **GRID-GUM** deals with issues of thread creation/coordination and orchestrates the CAS engines to work on the application as a whole; while the CAS engine deals solely with execution of individual algebraic computations. Together, the systems provide a powerful, but easy-to-use, framework for executing heterogeneous symbolic computations on a computational Grid. We exploit this framework to provide support for commonly found patterns of computation that may be used directly from within symbolic programs.

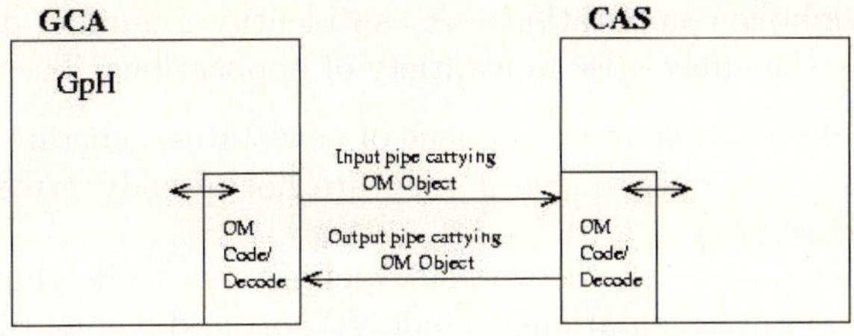

Fig. 2. GCA Design

3   Patterns of Symbolic Computation

We have identified a number of common patterns used in symbolic computation applications. Each of these patterns is potentially amenable to specific kinds of parallel execution. **SymGrid** will support these patterns as dynamic *algorithmic skeletons* [12], which may be called directly from within the computational steering interface. In general (and unlike most previous skeletons approaches), these patterns may be nested or composed dynamically as required to form the Grid computation. They may also mix computations taken from different computational algebra systems (in which case the computations are steered to an appropriate engine using **SymGrid-Par**).

The standard patterns that we have identified are listed below. These patterns are a subset of those identified by Gorlatch and Cole as appropriate algorithmic skeletons for the Grid [13,14], and are also similar to those previously identified for the **ParGap** parallel implementation of GAP [15].

```
parMap::       (a->b) ->                          [a] ->    [b]
parZipWith::   (a->b->c) -> [a] ->                [b] ->    [c]
parReduce::    (a->b->b) ->  b ->                 [a] ->    b
parMapReduce:: (a->b->b) -> (c->[(d,a)]) -> c ->           [(d,b)]
```

Here each argument to the pattern is separated by an arrow (->), and may operate over lists of values ([..]), or pairs of values ((..,..)). All of the patterns are *polymorphic*: a, b etc. stand for (possibly different) concrete types. The first argument in each case is a function of either one or two arguments that is to be applied in parallel. For example, `parMap` applies its function argument to each element of its second argument (a list) in parallel, and `parReduce` will reduce its third argument (a list) by applying the function between pairs of elements, ending with the value supplied as its second argument. The parallel semantics and implementations of these standard patterns are well established [15,13,14] and we will therefore not describe these in detail here.

3.1   Domain-Specific Patterns for Symbolic Computation

Parallel symbolic computation often requires irregular parallelism, which in turn leads to non-traditional patterns of parallel behaviour. Based on our experience with parallel symbolic computations, we can identify a number of new *domain-specific* patterns that may arise in a variety of applications. These include:

1. *orbit calculation:* given an open queue of new states, generate neighbors and place them in the open queue if they are not already present (related to breadth-first search).
2. *duplicate elimination:* given two lists, merge them while eliminating duplicates. Typical implementations include: (i) sort and merge; (ii) hash one list and check if the other list contains a duplicate
3. *completion algorithm:* given a set of objects, new objects can be *generated* from any pair of objects. Each new object is *reduced* against existing objects according to some rules, and if an object is thereby reduced to the trivial object, it is discarded. The newly reduced object is then used to reduce other existing objects.
4. *chain reduction:* given a chain of objects, (for example, the rows of a matrix during a Gaussian elimination), any later object in the chain must be reduced against each of the earlier objects in the chain. The algorithm terminates when no further reductions are possible.
5. *partition backtracking:* given a set of objects such as permutations acting on a set of points the algorithm searches for some *basis objects*. Any known basis objects allow additional pruning during the algorithm. As an optimization to backtracking search, the set of points may be partitioned into a disjoint union of two subsets, and the algorithm then finds all basis objects that respect this partition. The latter objects allow the remaining search to be more aggressively pruned.

Each of these patterns gives rise to highly-parallel computations with a high degree of inter-task interactions, and complex patterns of parallelism that may involve a mixture of task- and data-parallel computations.

4   Related Work

Work on parallel symbolic computation dates back to at least the early 1990s. Roch and Villard [16] provide a good general survey as of 1997. Significant research has been undertaken for specific parallel computational algebra algorithms, notably term re-writing and Gröbner basis completion (e.g. [17,18]). A number of one-off parallel programs have also been developed for specific algebraic computations, mainly in representation theory [19]. While several symbolic computation systems include some form of operator to introduce parallelism (e.g. parallel GCL, which supports Maxima [8], parallel Maple [20], or parallel GAP [9]), there does not exist a large corpus of production parallel algorithms. This is at least partly due to the complexities involved in programming such algorithms using explicit parallelism, and the lack of generalised support for communication, distribution etc. By abstracting over such issues, and, especially, through the identification of the domain-specific patterns noted above, we anticipate that **SymGrid** will considerably simplify the construction of such computations.

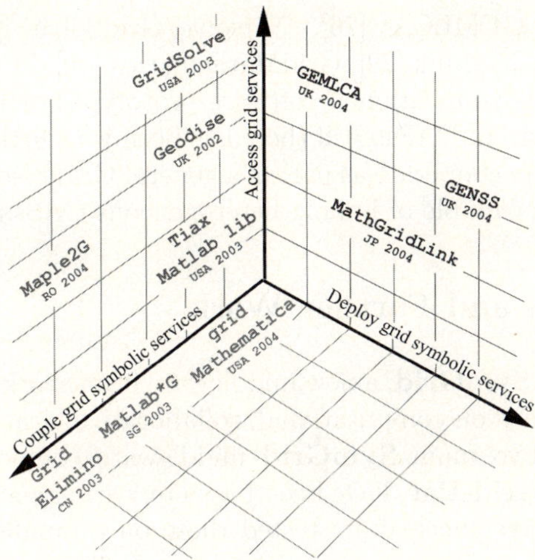

Fig. 3. Existing Grid-enabled Symbolic Computing Systems

4.1 Symbolic Computation for Computational Grids

While, as outlined above, parallel symbolic systems exist that are suitable for either shared-memory or distributed memory parallel systems (e.g. [21,22,15,23,24]), there is relatively little work on Grid-based symbolic computation. Figure 3 shows the relationship between the main existing systems (including **Maple2G**, developed by one of the authors [10]). The axes represent the degree in which a tool or a package satisfies the following requirements: deployment of symbolic Grid services (x-axis); composition of different grid symbolic services (y-axis); and accessing external grid services (z-axis). While the vision of Grid computing is that of simple and low-cost access to computing resources without artificial barriers of physical location or ownership, none of these Grid-enabled systems conforms entirely to the basic requirements of this vision. **SymGrid** is therefore highly novel in aiming both to allow the construction of heterogeneous symbolic computations and in allowing seamless access to the Grid from within symbolic systems. In order to achieve such a vision, we must be able to:

1. *deploy* symbolic Grid services;
2. allow *access* to available Grid services from the symbolic computation; and,
3. *combine* different symbolic Grid services into a coherent whole.

This is the challenge that we have attempted to meet in the design of the **SymGrid** system.

4.2 Symbolic Computations as Grid Services

Even less work has so far been carried out to interface CASs to the Grid. A number of projects have considered the provision of CASs as Grid services, e.g. GENSS [25], GEMLCA [26], Netsolve/GridSolve [27], Geodise [28], Maple2G [10], MathGridLink [29], GridMathematica [30]. More details are given in [31]. Using the above mentioned software prototypes or tools, Grid services can be called from within CASs. Still there has been very little work on adapting CASs so that they can cooperate as part of a general Grid resource. None of these systems is, however, capable of linking heterogeneous CASs as in **SymGrid**.

5   Conclusions and Further Work

We have introduced **SymGrid**, a new middleware framework supporting generic symbolic computations on computational/collaborative Grids. Prototype implementations of the two main **SymGrid** middleware components, **SymGrid-Services** and **SymGrid-Par**, have been based on earlier successful but system-specific work. We have successfully tested these on a number of computations running on the Grid, including a parallel orbit calculation, which has delivered promising speedup results on a test Grid linking the UK, Germany and Romania. We are in the process of enhancing these implementations to provide improved robustness, full heterogeneity and good support for the patterns of computation that we have identified above. In collaboration with our research partners, we are also identifying good and realistic symbolic computation exemplars, including ones with novel heterogeneous aspects.

While we have focused in this paper on the four computational algebra systems that form the immediate target of the SCIEnce project (Maple, Gap, Kant and MuPad), the majority of our work is highly generic, and we intend in due course to explore the inclusion of other symbolic systems. In the longer term, we also intend to work on providing highly autonomic scheduling and work management resource managers that will take account of information about future execution behaviours of the parallel computation. Unlike many traditional (often numeric) Grid computations, which have a regular (and therefore relatively easily predicted behaviour), the dynamic and highly irregular nature of the workload means that we must explore new and previously untried approaches based, for example, on statistical prediction or static analysis. We will also explore a new lightweight mechanism for supporting automatic task recovery to improve fault tolerance for long running computations [32], which has not yet been applied in a Grid setting. This will provide us with increased robustness in (potentially) unreliable settings, such as widely-distributed computational Grids built on standard network communication protocols.

Acknowledgements

This research is partially supported by European Union Framework 6 grant RII3-CT-2005-026133 SCIEnce: Symbolic Computing Infrastructure in Europe, by a one-year SOED support fellowship from the Royal Society of Edinburgh, and by National Science Foundation Grant ACIR-0342555.

References

1. Dean, J.: Experiences with MapReduce, an abstraction for large-scale computation. In: Proc. PACT 2006 – Intl. Conf. on Parallel Architectures and Compilation Techniques, p. 1 (2006)
2. Dean, J., Ghemawat, S.: MapReduce: Simplified data processing on large clusters. In: Proc. OSDI '04 – Sixth Symp. on Operating System Design and Implementation, pp. 137–150 (2004)
3. Char, B.W., et al.: Maple V Language Reference Manual. Maple Publishing, Waterloo, Canada (1991)
4. Group, T.G.: Gap – groups, algorithms, and programming, version 4.2 St Andrews, (2000) http://www.gap-system.org/gap
5. Daberkow, M., Fieker, C., Klüners, J., Pohst, M., Roegner, K., Schörnig, M., Wildanger, K.: Kant v4. J. Symb. Comput. 24(3/4), 267–283 (1997)
6. Morisse, K., Kemper, A.: The Computer Algebra System MuPAD. Euromath Bulletin 1(2), 95–102 (1994)
7. Al Zain, A., Trinder, P., Loidl, H.W., Michaelson, G.: Managing Heterogeneity in a Grid Parallel Haskell. J. Scalable Comp.: Practice and Experience 6(4) (2006)
8. Cooperman, G.: STAR/MPI: Binding a parallel library to interactive symbolic algebra systems. In: Proc. Intl. Symp. on Symbolic and Algebraic Computation (ISSAC '95). Lecture Notes in Control and Information Sciences, vol. 249, pp. 126–132. ACM Press, New York (1995)
9. Cooperman, G.: GAP/MPI: Facilitating parallelism. In: Proc. DIMACS Workshop on Groups and Computation II. DIMACS Series in Discrete Maths. and Theoretical Comp. Sci., vol. 28, pp. 69–84. AMS, Providence, RI (1997)
10. Petcu, D., Paprycki, M., Dubu, D.: Design and Implementation of a Grid Extension of Maple. Scientific Programming 13(2), 137–149 (2005)
11. The OpenMath Format (2007), http://www.openmath.org/
12. Cole, M.I.: Algorithmic Skeletons. In: Hammond, K., Michaelson, G. (eds.) Research Directions in Parallel Functional Programming; ch.13, pp. 289–304. Springer, Heidelberg (1999)
13. Alt, M., Dünnweber, J., Müller, J., Gorlatch, S.: HOCs: Higher-Order Components for grids. In: Getov, T., Kielmann, T. (eds.) Components Models and Systems for Grid Applications. CoreGRID, pp. 157–166 (2004)
14. Cole, M.: Algorithmic Skeletons: Structured Management of Parallel Computation. The MIT Press, Cambridge, MA (1989)
15. Cooperman, G.: Parallel GAP: Mature interactive parallel. Groups and computation, III (Columbus, OH, 1999) de Gruyter, Berlin (2001)
16. Roch, L., Villard, G.: Parallel computer algebra. In: ISSAC'97, IMAG Grenoble, France (1997) (Preprint)
17. Amrhein, B., Gloor, O., Küchlin, W.: A case study of multithreaded Gröbner basis completion. In: In Proc. of ISSAC'96, pp. 95–102. ACM Press, New York (1996)
18. Bündgen, R., Göbel, M., Küchlin, W.: Multi-threaded AC term re-writing. In: Proc. PASCO'94, vol. 5, pp. 84–93. World Scientific, Singapore (1994)
19. Michler, G.O.: High performance computations in group representation theory. Institut fur Experimentelle Mathematik, Univerisitat GH Essen (1998) (Preprint)
20. Bernardin, L.: Maple on a massively parallel, distributed memory machine. In: Proc. PASCO '97, pp. 217–222. ACM Press, New York (1997)
21. Char, B.: A user's guide to Sugarbush — Parallel Maple through Linda. Technical report, Drexel University, Dept. of Mathematics and Comp. Sci. (1994)

22. Chan, K., Díaz, A., Kaltofen, E.: A Distributed Approach to Problem Solving in Maple. In: Proc. 5th Maple Summer Workshop and Symp., pp. 13–21 (1994)
23. Küchlin, W.: PARSAC-2: A parallel SAC-2 based on threads. In: Sakata, S. (ed.) AAECC-8 1990. LNCS, vol. 508, pp. 341–353. Springer, Heidelberg (1991)
24. The GpH-Maple Interface (2007), `http://www.risc.uni-linz.ac.at/software/ghc-maple/`
25. GENSS (2007), `http://genss.cs.bath.ac.uk/index.htm`
26. Delaitre, T., Goyeneche, A., Kacsuk, P., Kiss, T., Terstyanszky, G., Winter, S.: GEMLCA: Grid Execution Management for Legacy Code Architecture Design. In: Proc. 30th EUROMICRO Conference, pp. 305–315 (2004)
27. Agrawal, S., Dongarra, J., Seymour, K., Vadhiyar, S.: NetSolve: past, present, and future; a look at a Grid enabled server. In: Making the Global Infrastructure a Reality, pp. 613–622. Wiley, Chichester (2003)
28. Geodise (2007), `http://www.geodise.org/`
29. Tepeneu, D., Ida, T.: MathGridLink – Connecting Mathematica to the Grid. In: Proc. IMS '04, Banff, Alberta (2004)
30. GridMathematica2, `http://www.wolfram.com/products/gridmathematica/`
31. Petcu, D., Tepeneu, D., Paprzycki, M., Ida, I.: In: Engineering the Grid: status and perspective, pp. 91–107, American Scientific Publishers (2006)
32. Cooperman, G., Ansel, J., Ma, X.: Transparent adaptive library-based checkpointing for master-worker style parallelism. In: Proc. IEEE Intl. Symp. on Cluster Computing and the Grid (CCGrid06), pp. 283–291. IEEE Computer Society Press, Los Alamitos (2006)

Grid Resource Ranking Using Low-Level Performance Measurements*

George Tsouloupas and Marios D. Dikaiakos

Dept. of Computer Science,
University of Cyprus
1678, Nicosia, Cyprus
{georget,mdd}@cs.ucy.ac.cy

Abstract. This paper outlines a feasible approach to ranking Grid resources based on an easily obtainable application-specific performance model utilizing low-level performance metrics. First, Grid resources are characterized using low-level performance metrics; Then the performance of a given application is associated to the low-level performance measurements via a *Ranking Function*; Finally, the Ranking Function is used to rank all available resources on the Grid with respect to the specific application at hand. We show that this approach yields accurate results.

1   Introduction

Matching between resource requests and resource offerings is one of the key considerations in Grid computing infrastructures. Currently, the implementation of matching is based on the *matchmaking* approach introduced by the Condor project [6], adapted to multi-domain environments and Globus, and extended to cover aspects such as data access and work-flow computations, interactive Grid computing, and multi-platform interoperability. Matchmaking produces a ranked list of resources that are compatible to the submitted resource requests. Ranking decisions are based on published information regarding the number of CPU's of each resource, their nominal speed, the nominal size of main memory, the number of free CPU's, available bandwidth, etc. This information is retrieved from Grid information services such as the Monitoring and Discovery Service of Globus.

This approach works well in cases where the main consideration of end-users is to allocate sufficient numbers of idle CPU's in order to achieve a high job-submission throughput with opportunistic scheduling. In several scenarios, however, reliance to matchmaking is not sufficient; for instance, when end-users wish to "shop around" for Grid computing resources before deciding where to deploy a high-performance computing application, or when Virtual Organization (VO) operators want to audit the real availability and configuration status of their providers' computing resources [4]. In such cases, the information published by

---

* This work was supported in part by the European Commission through projects EGEE (contract INFSO-RI-031688) and g-Eclipse (contract 034327).

A.-M. Kermarrec, L. Bougé, and T. Priol (Eds.): Euro-Par 2007, LNCS 4641, pp. 467–476, 2007.

resource providers and Grid monitoring systems does not provide sufficient detail and accuracy. Grid users need instead the capability to interactively administer benchmarks and tests, retrieve and analyze performance metrics, and select resources of choice according to their application requirements. To provide Grid users with such a *test-driving* functionality, we designed and implemented *Grid-Bench*, a framework for evaluating the performance of Grid resources interactively. GridBench facilitates the definition of parameterized execution of various probes on the Grid, while at the same time allowing for archival, retrieval, and analysis of results [9,10]. GridBench comes with a suite of open-source micro-benchmarks and application kernels, which were chosen to test key aspects of computer performance, either in isolation or collectively (CPU, memory hierarchy, network, etc.) [11].

In this paper, we present *SiteRank*, a component that we developed on top of GridBench to support the user-driven ranking of computational Grid resources. SiteRank enables Grid users to easily construct and adapt ranking functions that: (i) Take as arguments performance metrics derived with the low-level benchmarks of GridBench [11]; the selection of these metrics can be done manually or semi-automatically by the end-user, through the user interface of GridBench. (ii) Combine the selected metrics into a linear model that takes into account the particular requirements of the application that the user wishes to execute on the Grid (e.g., memory vs. floating-point performance bound). Using a ranking function, Grid users can derive rankings of Grid resources that are tailored to their specific application requirements.

In this paper, we describe the methodology followed by SiteRank to develop ranking functions. Furthermore, we demonstrate the use of SiteRank in the ranking of the computational resources of EGEE, which is the largest production-quality Grid in operation today [1]. To this end, we examine two alternative applications running on EGEE: povray, a ray-tracing application, and SimpleScalar, a simulator used for hardware-software co-verification and micro-architectural modelling. Our results show that SiteRank functions can provide an accurate ranking of EGEE resources, in accordance to the different requirements that each application has. Furthermore, that the careful selection of the low-level metrics used in the linear model is very important for the construction of accurate ranking functions.

The rest of this paper is organized as follows: Section 2 introduces SiteRank and its ranking methodology. Section 3 describes the use of SiteRank in the ranking of EGEE resources for the two applications of choice: povray and SimpleScalar. We conclude in Section 4.

2   SiteRank

Computational resources on the Grid exhibit considerable variance in terms of different performance characteristics. This leads to non-uniform application performance that significantly varies between applications.

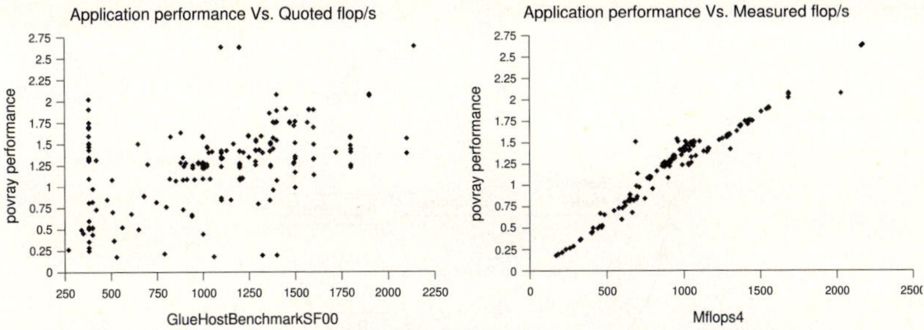

Fig. 1. The relationship of application performance to **quoted** and **measured** metrics

One approach for ranking resources in terms of performance is the one taken by the current (EGEE) infrastructure, which is to publish GlueHostBench-markSF00 (SPEC-Float 2000 floating point performance metric) and GlueHost-BenchmarkSI00 (SPEC-Int 2000 integer performance metric) values for each site. Unfortunately, values quoted by site administrators cannot be relied upon; This is evident in Figure 1 which compares the effectiveness of a **quoted** metric (Figure 1 left) in contrast to a measured metric(Figure 1 right). The charts speak for themselves; Clearly, the **quoted** metric does a very poor job in justifying application performance[1]. It is important to note that this would be *inadequate* even if the quoted values were correct, since application performance depends on much more than just two metrics (see Section 3).

2.1   The Ranking Methodology

The GridBench tool provides a *SitcRank module* that allows the user to interactively and semi-automatically build a *ranking model*. A *ranking model* consists of *filtering*, *aggregation* and *ranking functions* (Figure 2).

Filtering refers to a user selection regarding which results will be included or excluded in the ranking process. *Attribute filtering* allows the user to limit the selected set of measurements to the ones that match certain criteria in the benchmark description. E.g. limiting the selection to a specific VO , type of CPU, or the date and time results were obtained.

Aggregation allows the user to specify grouping of the measurements. The user can specify whether each measurement will count equally, irrespective of which worker-node it was executed on. In this case, the reported metric may possibly be less representative of the resource as a whole because some worker-nodes may be over-represented. On the other hand, this will tend to be more representative of what the user actually experiences once the resource's policy is applied. The *Aggregation* step produces a set of statistics for each metric: *mean*, *standard-deviation*, *min*, *max*, *average-deviation* and *count*.

---

[1] Similar results are obtained with GlueHostBenchmark**SI**00 just as with GlueHost-Benchmark**SF**00.

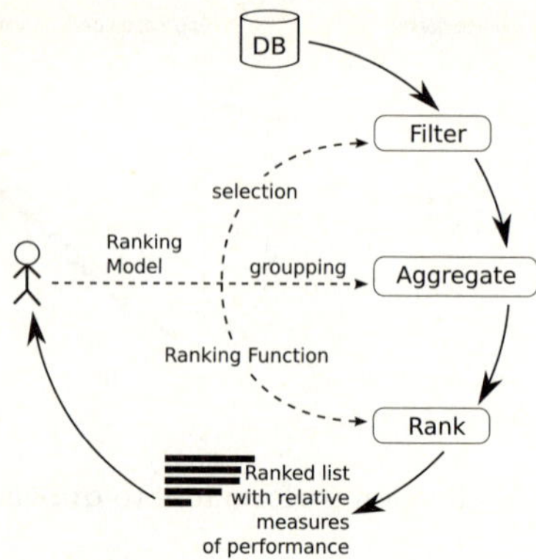

Fig. 2. The ranking process

During the aggregation step, the raw metrics are normalized according to a base value. The base values are configurable and in our experiments we used values from a typical 3.0GHz Xeon worker-node. For example, we used the value of 1050.0 to normalize the Mflops4 metric. The aggregation step is also important for the conversion of vector-type metrics, such as the ones produced by CacheBench into scalars (see later description on the *c512k* metric) so that they can be used in ranking functions.

Ranking Function Construction: The end goal of this methodology is a ranked list of computational resources that reflects the performance that users will experience running a specific application. It involves establishing a relationship between application performance and a set of low-level measurements. The process is illustrated in Figure 3, and it is outlined by the following steps (see Section 3 for an example):

1. **Sampling:** Obtain low-level performance metrics m for a small sample of resources – typically 10-15% of the full-set of resources. For the same sample of resources also obtain application performance measurements, i.e. application completion times. The application performance of this sample is denoted α where each $\alpha = 1/(completion\ time)$.

2. **Ranking Function Generation:** Determine a *Ranking Function R* based on the low-level metric data m and application performance α, so that $\alpha = R(m)$. This involves the selection of the low-level metrics that closely correlate to this application's performance, followed by a linear fit of the data, i.e. multivariate regression.

3. **Estimation:** For the set of the remaining resources, obtain only low-level performance metrics M, and apply the ranking function in order to obtain an estimate of the application performance A_{est} such that $A_{est} = R(M)$. Sorting A_{est} produces the *Rank Estimation*.

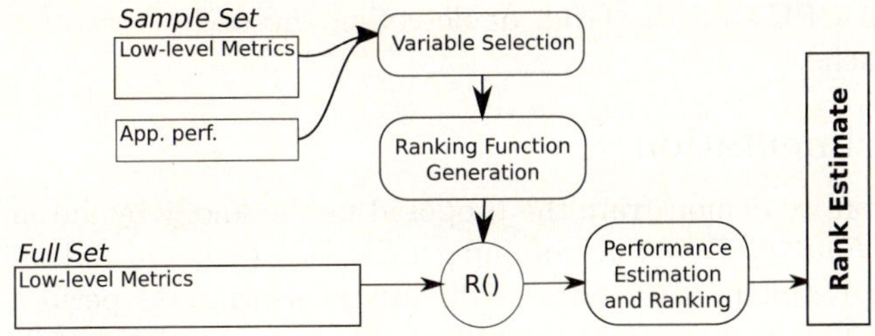

Fig. 3. *Rank Estimate* generation process outline

Table 1. Metrics and Benchmarks

Factor	Metric	Delivered By
CPU	Floating-Point operations per second	Flops
CPU	Integer operations per second	Dhrystone
Main memory	sustainable memory bandwidth in MB/s	Stream
Main memory	Available physical memory in MB	Memsize
Cache	memory bandwidth by varying array sizes in MB/s	CacheBench
Disk	Disk bandwidth for read/write/rewrite	bonnie++
Interconnect	latency, bandwidth and bisection bandwidth	MPPTest

2.2 Metrics

Selecting the *right* metrics to characterise the resources is of utmost importance in order to adequately characterize the major computational characteristics that affect application performance. In fact, we consider a *good set of metrics* one that can adequately explain the performance of several distinct applications. In the process of picking the right metrics and the right benchmarks to deliver these metrics, we limited ourselves to freely available tools that we could widely deploy and run. We also aimed at keeping the number of metrics low and we favored well-known metrics. A more detailed discussion can be found in [11].

Table 1 shows a list of low-level metrics and the associated benchmarks. The Flops benchmark yields 4 metrics, *Mflops1*, *Mflops2*, *Mflops3* and *Mflops4*, each consisting of different mixes of floating-point additions, subtractions multiplications and divisions. Dhrystone yields the *dhry* integer performance metric. The STREAM memory benchmark yields the *copy*, *add*, *multiply* and *triad* metrics which measure memory bandwidth using different operations. For cache metrics, measuring memory (cache) bandwidth B by allocating and accessing progressively larger array sizes s, the CacheBench benchmark produces a series of values B_s where $s = 2^8, 2^9, 2^{10} \ldots 2^n$. By summing up the product of the bandwidths and respective sizes we derive a metric that takes into account both the cache size *and* the cache speed : $\sum_{s=8}^{n} s \times B_s$. For example, summing up to 512kb, i.e. $\sum_{s=8}^{19} s \times B_s$ yields the *c512k* metric. This is done for sizes up to 512kb, 1Mb, 2Mb, 4Mb, 8Mb yielding the metrics *c512k*, *c1M*, *c2M*, *c4M* and *c8M* respectively. This approach alleviates the problem of looking up the cache size for the multitude of CPU's on the Grid, or detecting the cache sizes of a potentially multilevel cache.

3 Experimentation

In this section we demonstrate the proposed methodology by automatically determining a *Ranking Function*, obtaining a *Ranking Estimate* and validating that the Ranking Estimate is accurate by directly measuring the performance of the application. This is done for two applications, on a set of about 230 sites that belong to the EGEE infrastructure. We user two serial applications:

povray: The Povray v3.6 ray-tracing application using the a 40x40 scene.
sisc: The SimpleScalar, computer architecture simulation.[2]

For this experiment, we aimed at having between 2 and 3 measurements from each computational resource. One noteworthy fact is that we could only obtain results for about 160 out of the 230 sites. This was partly due to errors and site unavailability, but also due to exhausted quotas at some resources. We used the GridBench framework to obtain our measurements. The process of integrating the two applications into GridBench including the compilation took less than one hour and only needs to be performed once. The process of actually running all the experiments took less than 10 minutes, although we did have to wait for a few hours until the results from all the queued jobs were in. We then exported these results into an open-source statistics software package[3] ("R").

The data-set obtained by running the benchmarks on all the available computational resources will be referred to from now on as the *full-set*. Out of the full-set, we obtained a random sample, henceforth referred to as the *sample-set*, with results from 24 resources (15% of the full-set). A *correlation matrix* indicates which metrics are most correlated to application performance; this is shown in Figure 4. The problem of collinearity must be taken into consideration when narrowing down the selection of metrics. As shown in Figure 4 some metric groups are highly collinear, in such cases we eliminate the collinear metrics by selecting one metric out of the group, i.e. the one with the highest correlation to the application. In this example we kept *Mflops4* and discarded *Mflops2*, *Mflops3* and *dhry*. Selecting the *Mflops4* and *c512k* metrics for building the Ranking Function, leads to the next step, i.e. calculating the a and b coefficients in order to best satisfy:

$$\alpha_{povray} = a \times Mflops4 + b \times c512k$$

Outlier removal is achieved by performing a linear regression, and data-points that fall more than two standard deviations away from the rest are filtered out. In our specific example, 2 out of the 18 points were dropped. Linear regression is

---

[2] Limited execution privileges for the Virtual Organization through which we performed our experiments, dictated that we use parameters resulting in short application completion times. This applied both to **povray** and to **sisc**.

[3] Use of the R software was limited to establishing the relationship between the low-level metrics and application performance, and the validation of the results. All charts included in the paper we created using GridBench.

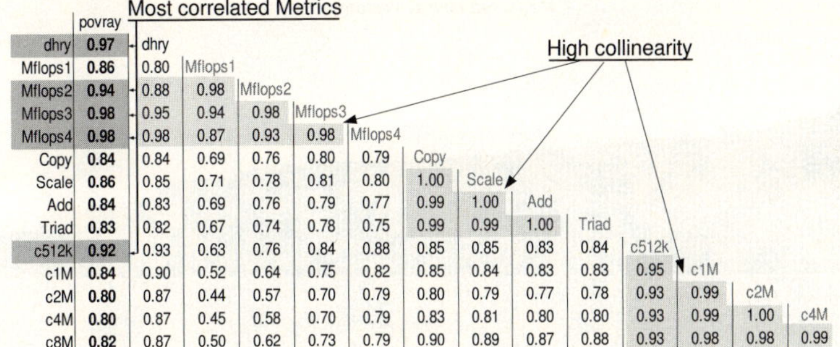

Fig. 4. Correlation Matrix for the *povray* application

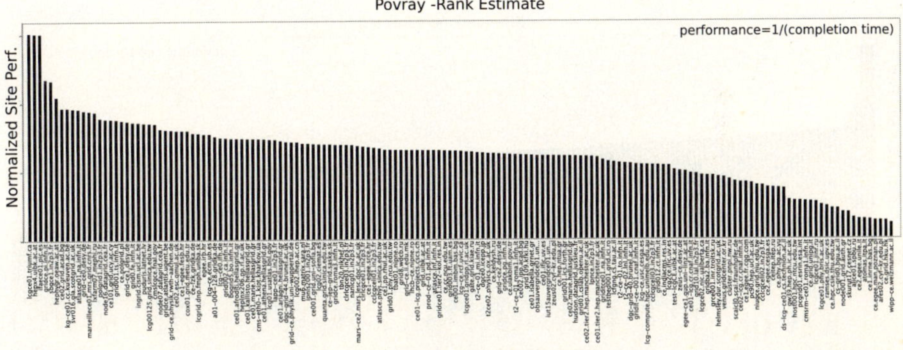

Fig. 5. Rank Estimate for the **povray** application

performed once again using the filtered sample-set, which yields the coefficients $a = 0.94$ (for *Mflops4*) and $b = 0.46$ (for *c512k*). Finally, we apply this model on the *full-set* in order ·estimate the performance of the application:

$$A_{povray} = 0.94 M_{Mflops4} + 0.46 M_{c512k}$$

Ordering the list of resources by A_{povray} gives the *Rank Estimate*. The Rank Estimate is shown in Figure 5. In order to test that the Ranking Estimate is accurate the performance of the application was directly measured for the whole infrastructure. This is only necessary in order to validate the model and not part of the methodology. The measured performance is shown in Figure 6. The agreement between the Rank Estimate and the measured ranking can be statistically tested by calculating the *rank correlation*. There are several ways of doing this, such as Kendall's τ, which ranges from -1 to 1 and is also known as the "bubble-sort distance". Kendall's τ yielded $\tau = 0.90$. Spearman's ρ, which again ranges from -1 to 1, yielded $\rho = 0.977$. Finally, Pearson's correlation coefficient yielded 0.98. All three of the statistics show that the two rankings are quite similar. The τ statistic appears considerably lower that the other two, due to the fact that our data-set contains a lot of resources that are of almost identical performance.

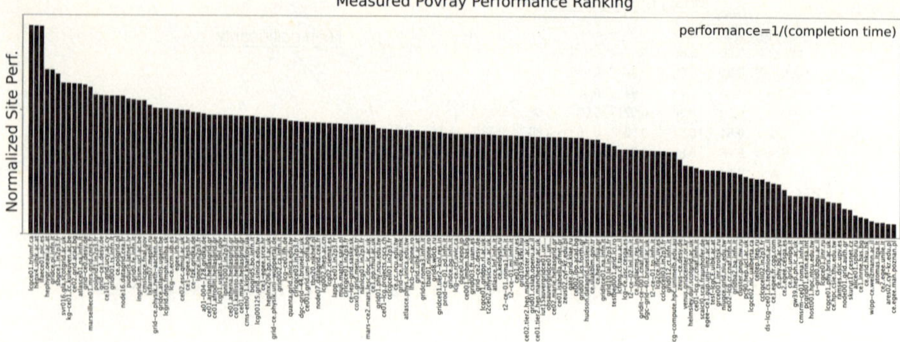

Fig. 6. Measured **povray** performance on 159 resources of the EGEE infrastructure

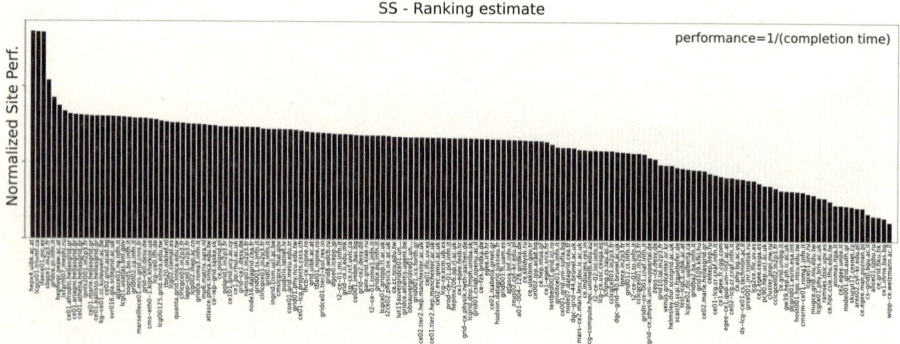

Fig. 7. Rank Estimate for the **sisc** application on the EGEE infrastructure

Extremely small fluctuations in measurement are enough to change the ordering. Yet,the performance of the resources is nearly identical, so the reordering is not very significant. For this reason the authors are inclined to take $\rho = 0.977$ as the more representative measure.

For the second application **sisc** we used the same methodology and the same sample-set that was used in the previous case. The metrics dictated by the correlation matrix are *dhry* and *c512k*. Performing the regression, outlier removal and then estimating the metric coefficients yields:

$$\boldsymbol{A}_{sisc} = 0.27\boldsymbol{M}_{dhry} + 0.18\boldsymbol{M}_{c512k}$$

The Ranking estimate is given in Figure 7. The correlation of estimated and actual is again quire high with a value of $\rho = 0.959$. Thus, for both applications the ranking of resources based on low-level measurements provides results that are very close to the ranking produced by running the application itself.

4   Conclusion

The work presented here, i.e. Ranking based on derived models of low-level metrics, describes an alternative way of choosing and ranking resources. We propose a semi-automated user-driven approach to ranking Grid resources that employs user-specified metrics and *ranking functions*.

The process of running benchmarks collecting and analysing results and generating ranked lists, would simply not be feasible if it had to be done manually, especially if it had to be done by the end user. Eventually, resource performance information will be coupled with resource pricing information. Users will then be able to "shop around" and pick the right resources (e.g. black-listing or white-listing) in order to influence the matchmaking process is a way that benefits them. The SiteRank module of the GridBench tool allows the user to interactively construct and modify ranking functions based on the collected measurements. The *Ranking Estimate* has proven to be quite accurate with a very high correlation to measured application performance for at least two applications, *povray* and *SimpleScalar*.

We have illustrated that current approaches to expressing the performance of resources, such as publishing the *quoted*, not measured, GlueHostBenchmarkSF00 and GlueHostBenchmarkSI00 metrics into the information system are not satisfactory, since they do not correlate well with at least the two applications that we have investigated.

Other tools in the general area of Grid testing and benchmarking include the Grid Assessment Probes [3], DiPerF [5] and the Inca test harness and reporting framework [7]. These are testing/benchmarking frameworks that provide functionality ranging from testing of Grid services to the monitoring of service agreements. In contrast, we focus on user-driven performance exploration and ranking. Benchmarking as a data-source for resource-brokering is explored in [2]. This work suggests the application of *weights* to different resource attributes and the use of *application benchmarks* to obtain a ranking that can eventually be used for resource brokering; we have also suggested this in our previous work [8].

Choosing the right metrics to collect is of vital importance, as an incomplete set of metrics will yield poor characterization. For example, our initial experiments did not include metrics that characterize the memory cache. While we had been collecting measurements about the cache, the data was in a form that was rather difficult to integrate into a regular function. Also, we had falsely assumed that the cache effects would be largely accounted for in other metrics. The initial results were not at all encouraging; but including the cache metrics, i.e. *c512k*, completely changed the situation. Indicative was the improvement of the ρ rank correlation statistic from approximately $\rho = 0.8$ to $\rho = 0.96$ for the *SimpleScalar* application. This also confirms the importance of a well-sized, fast cache to computational applications.

Further plans include the investigation of more applications, especially applications that are not CPU/memory bound, in order to evaluate the extent to which the metrics that we collect provide sufficient characterization.

References

1. Enabling Grids for E-SciencE project, `http://www.eu-egee.org/`
2. Afgan, E., Velusamy, V., Bangalore, P.V.: Grid resource broker using application benchmarking. In: Sloot, P.M.A., Hoekstra, A.G., Priol, T., Reinefeld, A., Bubak, M. (eds.) EGC 2005. LNCS, vol. 3470, pp. 691–701. Springer, Heidelberg (2005)

3. Chun, G., Dail, H., Casanova, H., Snavely, A.: Benchmark probes for grid assessment. In: 18th International Parallel and Distributed Processing Symposium (IPDPS 2004), CD-ROM / Abstracts Proceedings, 26-30 April 2004, Santa Fe, New Mexico, USA. IEEE Computer Society (2004)
4. Coles, J.: Grid Deployment and Operations: EGEE, LCG and GridPP. In: Proceedings of the UK e-Science All Hands Meeting 2005, (accessed October 2005) (2005), http://www.allhands.org.uk/proceedings/2005
5. Dumitrescu, C., Raicu, I., Ripeanu, M., Foster, I.: Diperf: an automated distributed performance testing framework. In: Proceedings of the 5th International Workshop on Grid Computing (GRID2004), IEEE Computer Society Press, Los Alamitos (2004)
6. Raman, R., Livny, M., Solomon, M.: Matchmaking: An extensible framework for distributed resource management. Cluster Computing 2(2), 129–138 (1999)
7. Smallen, S., Olschanowsky, C., Ericson, K., Beckman, P., Schopf, J.: The inca test harness and reporting framework. In: SC '04: Proceedings of the 2004. ACM/IEEE conference on Supercomputing, Washington, DC, USA, 2004, p. 55. IEEE Computer Society Press, Los Alamitos (2004)
8. Tiramo-Ramos, A., Tsouloupas, G., Dikaiakos, M.D., Sloot, P.: Grid Resource Selection by Application Benchmarking: a Computational Haemodynamics Case Study. In: Sunderam, V.S., van Albada, G.D., Sloot, P.M.A., Dongarra, J.J. (eds.) ICCS 2005. LNCS, vol. 3514, pp. 534–543. Springer, Heidelberg (2005)
9. Tsouloupas, G., Dikaiakos, M.D.: GridBench: A Tool for Benchmarking Grids. In: Proceedings of the 4th International Workshop on Grid Computing (Grid2003), pp. 60–67. IEEE Computer Society, Los Alamitos (2003)
10. Tsouloupas, G., Dikaiakos, M.D.: GridBench: A Workbench for Grid Benchmarking. In: Sloot, P.M.A., Hoekstra, A.G., Priol, T., Reinefeld, A., Bubak, M. (eds.) EGC 2005. LNCS, vol. 3470, pp. 211–225. Springer, Heidelberg (2005)
11. Tsouloupas, G., Dikaiakos, M.D.: Characterization of Computational Grid Resources Using Low-level Benchmarks. In: Second IEEE International Conference on e-Science and Grid Computing (e-Science'06), pp. 70–77. IEEE Computer Society, Los Alamitos (2006)

Topic 7
Peer-to-Peer Computing

Alberto Montresor, Fabrice Le Fessant, Dick Epema, and Spyros Voulgaris

Topic Chairs

Distributed systems have experienced a shift of scale in the past few years. This evolution has generated an interest in peer-to-peer systems and resulted in much interesting work. Peer-to-peer systems are characterized by their potential to scale due to their fully decentralized nature. They are self-organizing, adapting automatically to peer arrivals and departures, and are highly resilient to failures. They rely on a symmetric communication model where peers act both as servers and clients. As the peer-to-peer concepts and technologies become more mature, many distributed services and applications relying on this model are envisaged in the context of large-scale distributed and parallel systems. This topic examines peer-to-peer technologies, applications, and systems, and also identifies key research issues and challenges. Twenty-six papers were submitted to the track and we accepted six. We organized two sessions, the first devoted to the problem of query management in structured and unstructured overlay networks, the second containing a broader selection of topics.

In "Path Query Routing in Unstructured Peer-to-Peer Networks", a solution is proposed to the problem of storing an XML database over an unstructured peer-to-peer network. The system combines multi-level bloom filters for path queries with exponentially decaying bloom-filters for neighbourhood knowledge, to decide to which neighbours queries should be forwarded in the network. In "Processing Top-k Queries in Distributed Hash Tables", a new algorithm is presented, to extract only the first k replies matching a given query in a Distributed Hash Table. Multiple parameters can be taken into account in the query, provided each parameter domain can be split uniformly among some of the peers. In "Multi-dimensional Range Queries Over Structured Overlays" the authors suggest SOMA, a CAN-like overlay for executing range queries on multiple attributes. SOMA is based on a virtual d-dimensional Cartesian coordinate space, where each dimension corresponds to one attribute. SOMA processes range queries in $\log(N)$ routing steps, N being the number of nodes.

In "Asynchronous Distributed Power Iteration with Gossip-based Normalization", the authors have designed a fully distributed and robust algorithm based on gossip for finding the dominant eigenvector of large and sparse matrices. In "Capitalizing on Free Riders in P2P Networks", instead of fighting free riders by shutting them out, a mechanism is presented to benefit from them by letting them handle the forwarding of search queries. This mechanism shifts the load of search queries to the free riders, turning them into an asset instead of a menace.

A.-M. Kermarrec, L. Bougé, and T. Priol (Eds.): Euro-Par 2007, LNCS 4641, pp. 477–478, 2007.
© Springer-Verlag Berlin Heidelberg 2007

The paper "Content-Based Publish/Subscribe Using Distributed R-Trees", introduces a class of distributed R-trees and their applicability to building content-based publish/subscribe overlays. The study focuses on the properties of the resulting topology and the accuracy of event dissemination.

Path Query Routing in Unstructured Peer-to-Peer Networks

Nicolas Bonnel, Gildas Ménier, and Pierre-Francois Marteau

Valoria, Université de Bretagne-Sud
Campus de Tohannic, Bat. Yves Coppens
56000 Vannes
{nicolas.bonnel,gildas.menier,pierre-francois.marteau}@univ-ubs.fr

Abstract. In this article, we introduce a way to distribute an index database of XML documents on an unstructured peer-to-peer network with a flat topology (i.e. with no super-peer). We then show how to perform content path query routing in such networks. Nodes in the network maintain a set of Multi Level Bloom Filters that summarises structural properties of XML documents. They propagate part of this information to their neighbor nodes, allowing efficient path query routing in the peer-to-peer network, as shown by the evaluation tests presented.

1   Introduction

There is a growing need for large databases of semi-structured documents and this raises management and querying problems that are specific to this kind of data. XML [1] is a standard for semi-structured document encoding. Building an index database allows to speed up the querying process [2] and in general this database is larger than the database itself. Distributing the index database allows to store and manage a larger amount of information.

Peer-to-peer (P2P) systems can be used to manage XML data [3]. They can have different architectures. Napster [4] for instance is a centralized P2P network that was very popular in the early 00'. The use of a central repository to answer queries makes this system poorly scalable and vulnerable to failure. Others P2P systems don't rely on a central server : they are decentralized, thus they are very scalable, and fault tolerant. They are divided in two categories.

First, structured P2P networks link network topology and location of data. Most of them implement a Distributed Hashtable (DHT) [5,6,7,8,9] and provide one basic operation : given a key, they map the key to a node. This is performed by using a distributed hash function. They use content routing to forward the key to the corresponding node. They are very well suited to retrieve rare information (i.e. with a low number of replicas). Their main limitation is that it is very costly to perform range and approximative queries, because hashing destroys the order on keys. Data clustering on those systems is mostly delicate, as data placement is given by the hash value of keys.

A.-M. Kermarrec, L. Bougé, and T. Priol (Eds.): Euro-Par 2007, LNCS 4641, pp. 479–488, 2007.

Second, unstructured P2P networks have no constraint between location of data and network topology. Gnutella [10] is an example of such working system. Query forwarding can be achieved either by flooding [11] - consuming a lot of bandwidth - or by random walk : a path is selected randomly according to a uniform distribution. They are suited to retrieve highly replicated data, but have limitations for rare information retrieval. Data clustering on such systems can be achieved because there is no constraint on data placement. Nowadays, most of the P2P systems used have a decentralized unstructured topology.

Our current work focuses on indexing XML data and distributing the index database. As we need to be able to perform approximative path queries, we choose an unstructured P2P architecture. The primary purpose of the nodes on the network is not to manage this index database. They all may have to set aside the management of the database if a user would need the resources on this node for his own purpose. For this reason our P2P network don't have super peers. The use of an unstructured P2P architecture allows us to have a network topology as close as possible to the physical topology and can lead to use less network resources. As we need to be able to perform path queries from all nodes of the network, we need a mecanism to route path queries. In this article, we describe a scheme based on the use of exponentially decaying multi level Bloom filters to address this problem.

Section 2 recalls previous work, section 3 presents our system, section 4 shows our experiments and section 5 talks about possible future work.

2   Related Work

2.1   Unstructured P2P Routing Mechanisms

Many studies have already dealt with this problem. For instance, the problem of routing path queries in unstructured peer to peer network has been tackled by [12,13,14]. However they all consider hierarchical topologies with super-peers, which we want to avoid for the reasons stated previously.

A hybrid search mechanism has been proposed [15] to achieve routing for both common and rare items. This system uses flooding to locate highly replicated information and a DHT to locate rare information.

Random walk efficiency can be improved while launching k similar queries. k-random walk has been proposed by [11] that improves query response rate with little network trafic.

Distributed-index mechanisms can improve the efficiency of content routing. Different Routing Indices (RI) have been proposed by [16] : compound, hop-count and exponential. Various kinds of Bloom filters [17] have been used to encode these distributed indexes.

Attenuated Bloom filters have been proposed by [18]. Exponentialy Decaying Bloom Filter (EDBF) has been proposed by [19] and Multi-level Bloom Filter (MLBF) has been introduced by [12]. Both are presented later in this article.

2.2   Bloom Filters

A Bloom filter [17] is a data structure that answers approximatively to set membership queries. It is made of an array of m bits and k hash functions $h_1, ..., h_k$, as shown in figure 1.

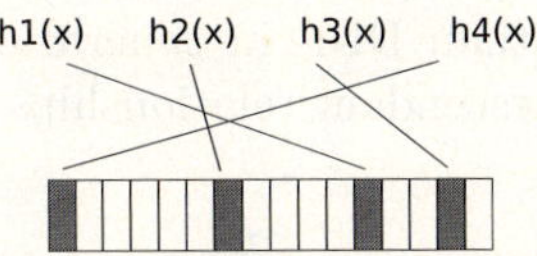

Fig. 1. Insertion of an element x in a Bloom filter of length m=16 and k=4 hash functions

When no element has been inserted into the filter, all bits are set to 0. When an element x is inserted into the filter, all bits given by the k hash functions $h_i(x)$ are set to 1 with a classical bitwise-or operation. A test membership of an element to a set is answered by checking that all the bits given by the k hash functions are set to 1.

False positives are possible but false negatives are not. The probability of having a positive answer for an element not belonging to the filter, with n being the number of elements inserted in the filter, is $(1 - (1 - \frac{1}{m})^{nk})^k$.

More elaborated Bloom filters have been proposed [18,20,21,22]. For instance Counting Bloom Filter (CBF) [23] allows the removing of elements from the filter. In those filters, the array of bits is replaced by an array of integer, each of them acting as a counter. Inserting an element is performed by increasing integers given by the k hash functions. Removing an element is performed by decreasing those integers. A set membership query is answered by checking that all integers are strictly positives.

Exponentialy Decaying Bloom Filter (EDBF). This filter has been proposed by [19]. Introducing an element is performed as in a classical Bloom filter. However, querying for an element x gives the number $\theta(x)$ of bits equal to 1. Thoses filters are then used to encode probabilistic routing tables, in which $\frac{\theta(x)}{k}$ is the probability to find element x among a given link in the network (Each node maintains a filter for each neighbor he has). This probability decays exponentially with the number of *hops* (or node transitions) from the node where the element x is stored.

Nodes update their filters periodically. The filter for one neighbor is updated with the information attenuated from all other neighbors and the information without attenuation from the local EDBF of the node. The attenuation is performed by resetting each bit to 0 with a probability $1/d$, where d is the decay of the filter. The aggregation of information corresponds to a bitwise-or operation.

Multi Level Bloom Filters (MLBF). It has been introduced by [12]. The main idea they propose is to use a set of Bloom filters to describe structural properties of set of XML documents. There are 2 sets of Bloom filters : Breadth Bloom Filters (BBF) and Depth Bloom Filters (DBF), as shown in figure 2. A BBF is composed of i Bloom filters $\{BBF_1, ..., BBF_i\}$. Inserting an XML document is performed by inserting all nodes of level l in BBF_l. A DBF is composed of j Bloom filters $\{DBF_1, ..., DBF_j\}$. Inserting an XML document is performed by inserting all subpath of length l in DBF_l. According to the authors, BBF works better than DBF in general case, but the last can drive path queries with ancestor-descendant relationships.

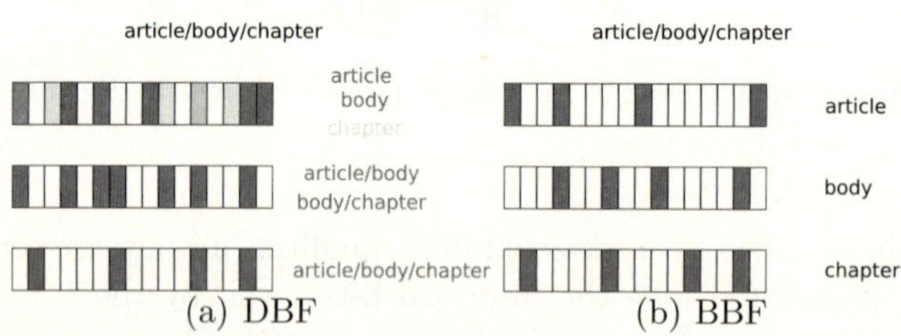

Fig. 2. Multi level Bloom filter with a BBF and DBF of size 3. Each subfilter has a length m=16 and k=4 hash functions. This figure illustrates the insertion of the path "article/body/chapter" in both filters.

To perform routing, nodes are then clustered in a hierarchical organization. A set of nodes are designated as *root nodes*, according to their storage and processing capabilities, and are connected to a main channel that allows them to communicate between themselves. Each root node has a merged filter that contains all the elements in its children peers and a local filter that contains elements hosted on this node. When a root node receives a query, it first checks its local filter. It then propagates the query to its children if there is a hit in its merged filter.

A hierarchical organization is adapted to a network of nodes with different processing and storage capabilities. However, our system is designed to work on a network in which nodes will have varying processing and storage capabilities. For this reason, we can't have a hierarchical topology and need a mechanism to drive path queries in a random graph like topology. EDBF have proved to be efficient in such topologies but can't drive path queries. Next section presents our system that perform path query routing in a random graph like network topology using an exponentially decaying version of MLBF.

3   System Design

3.1   Path Indexing Scheme

Each node of the network performs the indexing of a part of the set of XML documents. For each XML document, paths leading to content are indexed in a hashtable, using the QDBM library [24].

Each node carries a BBF of size l and a Reverse Breadth Bloom Filter (RBBF) composed also of l Bloom filters $\{RBBF_1, ..., RBBF_l\}$ of size l. For each path P of length k, $P = /e_1/.../e_k/$ and $\forall i \in [1, ..., k]$ we insert e_i in BBF_i and $RBBF_{k-i}$. For each node, BBF and RBBF are made of CBF, so that elements can easily be removed.

We don't use DBF for two reasons. First, according to [12], they are less efficient than BBF. Second, they cannot be used efficiently to drive path queries containing unknown elements. For instance, let us consider a query such as $/A/?/C/$. This query can be driven using information in the first and third level of the BBF, whereas only information in the first level of the DBF (paths of length 1) can be used.

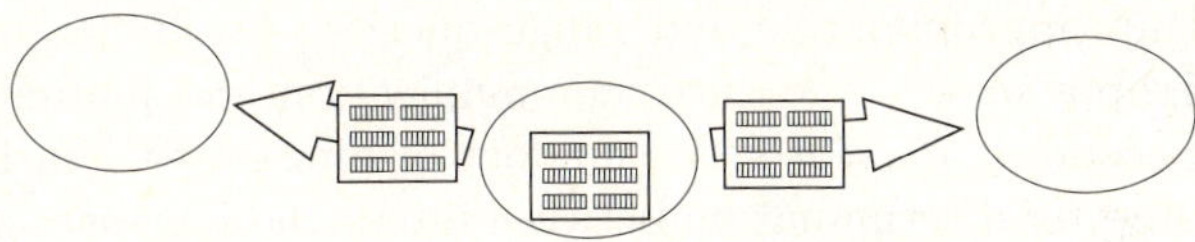

Fig. 3. Example of filter settings on a node with 2 neighbors. BBF and RBBF have a size of 3.

We encode probabilistic routing table using an exponentially decaying version of BBF and RBBF. Thus, each node has two filters for each neighbor, as shown in figure 3. Updating the filters is performed as described in [19].

3.2   Querying Mechanism

The query language we consider is a subset of the XML Path Language (XPath) [25]. Let E be the set of all XML element names. Let "?" be a don't care element and $E^? = E \cup \{\text{"?"}\}$. Let S be the set of relationship symbols, $S = \{\text{"/"}, \text{"//"}\}$. "/" describes a parent-child relationship, and "//" describes an ancestor-descendant relationship. "/" is a particular case of "//". A path query P of length k is defined as a word on $E^? \cup S$, with $P = s_0\, e_1\, s_1\, ...\, e_k\, s_k$, where $e_i \in E^?$ and $s_i \in S$. $/e_1$ stands for the root element. We assume each local node has a mechanism to answer queries exactly. Either the begining or the end of the searched paths must be known, thus s_0 and s_k cannot be at the same time both "//".

Let P be a path query containing no "//" relationship, $P = /e_1/.../e_k/$. Let $Q(F, x)$ be the query response of the presence of the element x in the filter F. Querying MLBF is performed as follows :

$$Q(BBF, P) = Q(BBF_1, e_1).Q(BBF_2, e_2) \ldots Q(BBF_k, e_k) \tag{1}$$

$$Q(RBBF, P) = Q(RBBF_1, e_k).Q(RBBF_2, e_{k-1}) \ldots Q(BBF_k, e_1) \tag{2}$$

Querying counting filters return *true* or *false* (exact query), and the "." in equations 1 and 2 stands for the logical AND. Querying exponentially decaying filters return a result in $[0, 1]$ (approximative query), and the "." in equations 1 and 2 stands for the product.

MLBF querying is performed by querying BBF and RBBF. If the path query contains "//", the subpath before the first "//" is answered by BBF and the subpath after the last "//" is answered by RBBF. If the path query contains only parent-child relationships, the whole query is answered by both filters. The global result is then given by the product (exponentially decaying filters) or the logical AND (counting filters) of the results given by BBF and RBBF.

3.3   Clustering

Clustering similar data increases routing efficiency for two reasons. First, it allows to speed up approximative and range queries. As similar information is located on neighbor nodes, it reduces communication and minimizes resources requirement. Second, it decreases the number of bits set to 1 in Bloom filters. This leads to a better discrimination between filters and increases gradient routing performance.

There are two ways to perform clustering : node and data clustering. The former aggregates nodes with similar content, while the latter aggregates data that are closed on the same node. As we would like to have as few constraints as possible on nodes, and especially on the network topology, we perform data clustering. Our algorithm clusters similar XML documents on the same nodes or on neighbor nodes.

Let $P = /e_1/.../e_k/$ a path of length k. We define for the node N:

$$\psi(N, P) = |\{i \in [1, \ldots, k]/BBF_i(e_i) = true\}|$$
$$+ |\{i \in [1, \ldots, k]/RBBF_{k-i}(e_i) = true\}|$$
$$\omega(N, P) = \psi(N, P) + \sum_{x \in neighborhood\ N} \psi(x, P)$$

$S(N)$: a linear function of the space available on node N

D_N : the set of XML paths stored in N

Our clustering algorithm works as follows : each node N launches periodically an agent. This agent contains a copy of an XML path P taken at random from D_N and the indexing information related to this path, a Time To Live (TTL), a reference to the node N, and the score of the node N for this path P. This score is given by the product $\omega(N, P).S(N)$.

The agent moves to a random neighbor until his TTL reaches 0 or until it finds a node with a better score for the path P. He keeps track of the sequence of visited nodes. If a better node is found, the path P and related indexing information is moved from the agent to that node, and then a deletion message for this path and the associated information is sent to the original node N using routing information stored by the agent. If the node already stores the path P, indexing information is merged. If the TTL reaches 0 the agent is discarded, and the path P remains on the node N.

3.4   Query Forwarding

Query forwarding is performed using the Scalable Query Routing algorithm (SQR) proposed by [19]. If the query is forwarded for the first time from a node, it is sent to the neighbor with the highest score when querying the exponentially decaying MLBF of the link to this neighbor. If the query has already been seen, it is forwarded to a random neighbor.

As we only use Breadth Bloom Filters, there is no relation between the elements that compose a path. For instance, if the paths /book/chapter/ and /article/abstract/ are inserted into the MLBF, it will answer true for the paths /book/abstract/ and /article/chapter/, even if they are not in the filter. However, because of our clustering algorithm, if such paths exist, they will likely be in the same node, or in the neighborhood of the node.

4   Experiments

Experiments have been made with a simulation written in java. The settings of these experiments can be found in figure 4. In those experiments, information was not replicated. A subset of the wikipedia collection from INEX 2006 [26,27] containing 260000 XML documents has been used. We used a stoplist containing commons elements, such as <article>, and small length elements, like <b> used for bold text. We removed all those elements in the paths we indexed. For these experiments, we made comparisons with random walk rather than with flooding, because it would be too costly to retrieve information located far away from the query source.

Parameter	Value
Number of nodes	200
Node degree	3 - 8
XML documents per node	1300
Length of filters	2^{13}
Size of BBF and RBBF	3
Number of hash functions	32
Decay	8

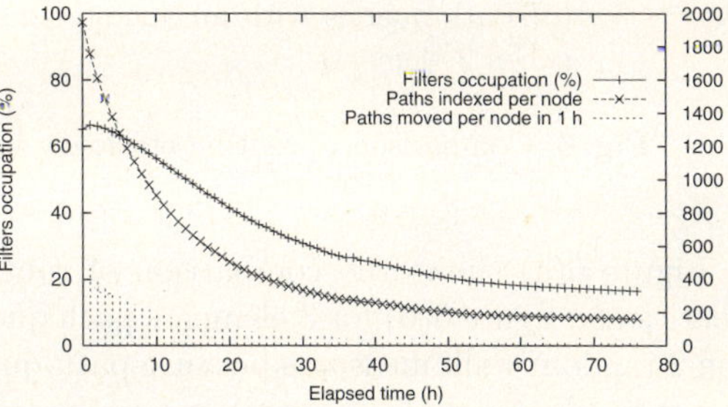

Fig. 4. Experiment settings   **Fig. 5.** Impact of clustering

Figure 5 show the evolution of the number of paths moved, the number of indexed paths and the filters occupation. Merging similar paths on the same nodes allows to lower the number of bits set to 1 in filters by a factor of 3. Our algorithm converge to a minimum, as the number of paths moved gradually get closer to 0. The comparison of routing efficiency for path queries between SQR and random walk has been performed by averaging 1000 different queries launched 20 time each. We used an hop count measure instead of a TTL one because the computer used could have varying CPU or memory resources (for these experiments).

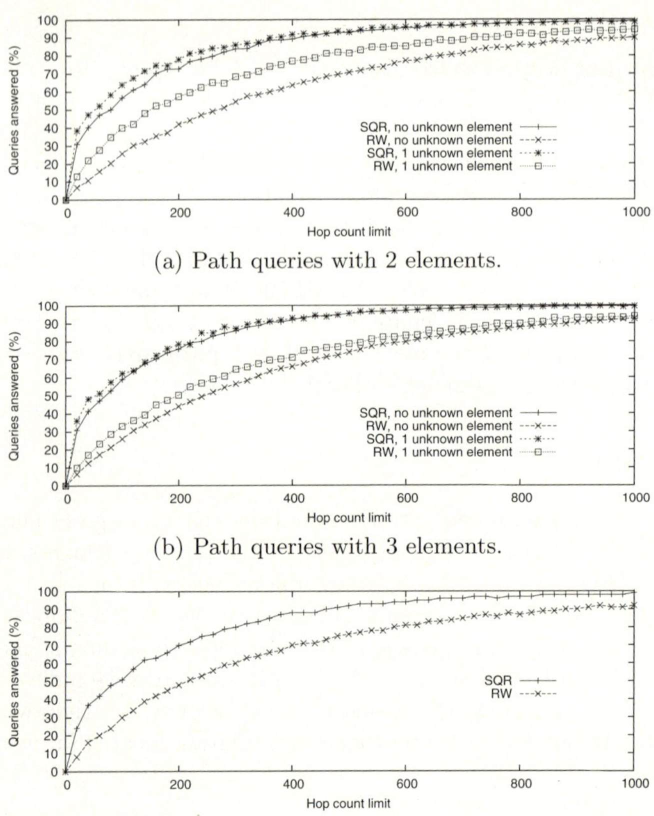

(a) Path queries with 2 elements.

(b) Path queries with 3 elements.

(c) Path queries with an ancestor-descendant relationship between 2 elements.

Fig. 6. Comparison of routing efficiency between SQR and random walk

Figure 6(a) shows the comparison of routing efficiency for 2 elements path queries and figure 6(b) for 3 elements path queries. There is a difference between the 2 random walk measures because path queries with 1 unknown element (for instance */document/?/*) cover more paths than path queries with no unknown element (for instance */document/book/*). However, as the hop count limit increase, the 2 SQR measure get closer, as having no unknown element provide more information for routing.

Figure 6(c) shows results with 2 elements path queries, linked by an ancestor-descendant relationship (for instance */document//paragraph/*). Although performances are a bit lower than those with 2 elements path queries containing only parent-child relationships, there still a good increase of routing efficiency. For instance, 70% of queries can be answered with a *hop* count limit two times lower than with random walk.

5   Conclusion and Future Work

We have introduced a way to perform stochastic approximative path query routing in an unstructured P2P network. We have shown that with our clustering algorithm, SQR outperforms random walk to forward simple approximative path queries. Filters used for routing can be maintained at low cost, as we can control the frequency of the updates, and because Bloom filters use few memory.

Further research efforts are required to be able to drive more elaborated path queries. Furthermore in our experiments information was not replicated. We will study the impact of information replication to see how SQR performs with path query through multiple data replica. For the time being, our simulator works on a single computer. We plan to distribute our application on a cluster to be able to perform testing on much larger XML databases.

References

1. Bray, T., Paoli, J., Sperberg-McQueen, C.M.: Extensible Markup Language (XML) 1.0 - W3C recommendation 10-february-1998. Technical Report REC-xml-19980210 (1998)
2. McHugh, J., Widom, J., Abiteboul, S., Luo, Q., Rajamaran, A.: Indexing semistructured data (1998)
3. Koloniari, G., Pitoura, E.: Peer-to-peer management of xml data: issues and research challenges. SIGMOD Rec. 34(2), 6–17 (2005)
4. Napster: Napster homepage (2001), http://www.napster.com/
5. Morris, R., Karger, D., Kaashoek, F., Balakrishnan, H.: Chord: A Scalable Peer-to-Peer Lookup Service for Internet Applications. In: ACM SIGCOMM 2001, San Diego, CA, ACM Press, New York (2001)
6. Ratnasamy, S., Francis, P., Handley, M., Karp, R., Shenker, S.: A scalable content addressable network. Technical Report TR-00-010, Berkeley, CA (2000)
7. Rowstron, A., Druschel, P.: Pastry: Scalable, decentralized object location, and routing for large-scale pocr-to-peer systems. In: Guerraoui, R. (ed.) Middleware 2001. LNCS, vol. 2218, Springer, Heidelberg (2001)
8. Zhao, B.Y., Kubiatowicz, J.D., Joseph, A.D.: Tapestry: An infrastructure for fault-tolerant wide-area location and routing. Technical Report UCB/CSD-01-1141, UC Berkeley (April 2001)
9. Clarke, I., Sandberg, O., Wiley, B., Hong, T.W.: Freenet: A distributed anonymous information storage and retrieval system. In: Federrath, H. (ed.) Designing Privacy Enhancing Technologies. LNCS, vol. 2009, Springer, Heidelberg (2001)
10. Clip2: The gnutella protocol specification v0.4 (2002)
11. Lv, C., Cao, P., Cohen, E., Li, K., Shenker, S.: Search and replication in unstructured peer-to-peer networks (2001)
12. Koloniari, G., Pitoura, E.: Content-based routing of path queries in peer-to-peer systems. In: Bertino, E., Christodoulakis, S., Plexousakis, D., Christophides, V., Koubarakis, M., Böhm, K., Ferrari, E. (eds.) EDBT 2004. LNCS, vol. 2992, Springer, Heidelberg (2004)
13. Dragan, F., Gardarin, G., Yeh, L.: Mediapeer: A safe, scalable p2p architecture for xml query processing. In: DEXA Workshops, pp. 368–373 (2005)

14. Wang, Q., Jha, A.K., Ozsu, M.T.: An xml routing synopsis for unstructured p2p networks. waimw 0, 23 (2006)
15. Loo, B.T., Huebsch, R., Stoica, I., Hellerstein, J.M.: The case for a hybrid p2p search infrastructure (2004)
16. Crespo, A., Garcia-Molina, H.: Routing indices for peer-to-peer systems (2002)
17. Bloom, B.H.: Space/time trade-offs in hash coding with allowable errors. Communications of the ACM 13(7), 422–426 (1970)
18. Rhea, S.C., Kubiatowicz, J.: Probabilistic location and routing. In: Proceedings of INFOCOM 2002 (2002)
19. Kumar, A., Xu, J., Zegura, E.W.: Efficient and scalable query routing for unstructured peer-to-peer networks. In: Proc. of IEEE Infocom, IEEE Computer Society Press, Los Alamitos (2005)
20. Cohen, S., Matias, Y.: Spectral bloom filters. In: SIGMOD '03: Proceedings of the 2003 ACM SIGMOD international conference on Management of data, New York, NY, USA, pp. 241–252. ACM Press, New York (2003)
21. Chazelle, B., Kilian, J., Rubinfeld, R., Tal, A.: The Bloomier filter: An efficient data structure for static support lookup tables
22. Kumar, A., Xu, J., Wang, J., Spatscheck, O., Li, L.: Space-code bloom filter for efficient per-flow traffic measurement. In: Infocom (2004)
23. Li, F., Pei, C., Jussara, A., Andrei, B.: Summary cache: A scalable wide-area web cache sharing protocol (Technical Report 1361). In: Proceedings of SIGCOMM'98, Computer Sciences Department, Univ. of Wisconsin-Madison (February 1998)
24. Hirabayashi, M.: Quick database manager (2006), `http://qdbm.sourceforge.net/`
25. Robie, J., Fernández, M.F., Boag, S., Chamberlin, D., Berglund, A., Kay, M., Siméon, J.: XML path language (XPath) 2.0. W3C proposed reccommendation, W3C (November 2006), `http://www.w3.org/TR/2006/PR-xpath20-20061121/`
26. INEX: Initiative for the evaluation of xml retrieval (2006), `http://inex.is.informatik.uni-duisburg.de/2006/`
27. Denoyer, L., Gallinari, P.: The wikipedia xml corpus. SIGIR Forum 40(1), 64–69 (2006)

Processing Top-k Queries in Distributed Hash Tables*

Reza Akbarinia[1,2], Esther Pacitti[1], and Patrick Valduriez[1]

[1] INRIA and LINA, University of Nantes, France
[2] Shahid Bahonar University of Kerman, Iran
FirstName.LastName@univ-nantes.fr, Patrick.Valduriez@inria.fr

Abstract. Distributed Hash Tables (DHTs) provide a scalable solution for data sharing in large scale distributed systems, *e.g.* P2P systems. However, they only provide good support for exact-match queries, and it is hard to support complex queries such as top-k queries. In this paper, we propose a family of algorithms which deal with efficient processing of top-k queries in DHTs. We evaluated the performance of our solution through implementation over a 64-node cluster and simulation. Our performance evaluation shows very good performance, in terms of communication cost and response time.

1 Introduction

Distributed Hash Tables (DHTs), *e.g.* CAN [20] and Chord [23], provide an efficient solution for data location and lookup in large-scale P2P systems. While there are significant implementation differences between DHTs, they all map a given key onto a peer p using a hash function and can lookup p efficiently, usually in $O(log\ n)$ routing hops where n is the number of peers [13]. DHTs typically provide two basic operations [13]: *put(key, data)* stores a pair *(key, data)* in the DHT using some hash function; *get(key)* retrieves the data associated with *key* in the DHT. These operations enable supporting exact-match queries only. Recently, much work has been devoted to supporting more complex queries in DHTs such as range queries [11] and join queries [15]. However, efficient evaluation of more complex queries in DHTs is still an open problem [5].

An important kind of complex queries is top-k queries. Given a dataset D and a scoring function f, a top-k query retrieves the k tuples in D with the highest scores according to f. Top-k queries have attracted much interest in many different areas such as network and system monitoring [2][6], information retrieval [3][16], sensor networks [22][24], multimedia databases [7][12][19], spatial data analysis [14], data streams [18], etc. The main reason for such interest is that they avoid overwhelming the user with large numbers of uninteresting answers which are resource-consuming. Most of the efficient approaches for top-k query processing in centralized and distributed systems, *e.g.*[4][6][8][17], are based on the Threshold Algorithm (TA) [10][12][19]. TA is applicable for queries where the scoring function is monotonic, *i.e.*, any increase in the value of the input does not decrease the value of the output.

---

* Work partially funded by ARA "Massive Data" of the French ministry of research and the European Strep Grid4All project.

A.-M. Kermarrec, L. Bougé, and T. Priol (Eds.): Euro-Par 2007, LNCS 4641, pp. 489–502, 2007.

In a large-scale P2P system, top-k queries can be very useful [3]. For example assume a community of car dealers who want to take advantage of a DHT to share some data about the used cars which they are willing to sell. Assume they agree on a common Car description in relational format. The Cars relation includes attributes such as car-id, price, mileage, mark, model, picture, etc. Suppose a user wants to submit the following query to obtain the 10 top answers ranked by a scoring function over price and mileage:

> SELECT car-id, price, mileage FROM Cars
> WHERE price < 3000 AND mileage < 60000
> **ORDER BY** scoring-function(price, mileage) **STOP AFTER** 10

The user specifies the scoring function according to the criteria of interest. For instance, in the query above, the scoring function could be (- (20*price + mileage)).

The problem of top-k queries has been addressed in unstructured P2P networks, *e.g.* in [1], and also super-peer networks, *e.g.* in [3]. However, the specific nature of DHTs, *i.e.* data storage and retrieval based on hash functions, makes it quite challenging to support top-k queries [5]. A simple solution for supporting top-k queries in DHTs is to retrieve all tuples of the relations involved in the query, compute the score of each retrieved tuple, and finally return the k tuples whose scores are the highest. However, this solution cannot scale up to a large number of stored tuples. Another solution is to store all tuples of a relation in the DHT by using the same key (*e.g.* relation's name), thus all tuples are stored at the same peer. Then, top-k queries can be processed at the central peer using well-known centralized algorithms. However, the central peer becomes a bottleneck and a single point of failure.

What we need is an efficient solution which can scale up to large numbers of peers and avoids any centralized data storage. In this paper, we propose such a solution for top-k query processing in DHTs. Our main contributions are the followings:

- We propose a data storage mechanism that not only provides good support for exact-match queries, but also enables efficient execution of top-k queries using our algorithms. It stores relational data in the DHT in a fully decentralized way, and avoids skewed distribution of data among peers.
- We propose a family of three algorithms which deal with efficient processing of top-k queries in DHTs. The first algorithm efficiently supports top-k queries with monotonic scoring functions. The second one supports top-k queries with a much larger class of scoring functions. We propose two optimizing strategies that reduce significantly the communication cost of the latter algorithm. We analytically prove that the algorithm finds correctly the k highest scored tuples. We propose a third algorithm for the cases where only a small set of relations' attributes are used in scoring functions. At the expense of incurring a small amount of redundancy on the DHT, the third algorithm yields much performance gains in terms of response time and communication cost.
- We evaluated the performance of our algorithms through implementation over a 64-node cluster and simulation using SimJava up to 10,000 peers. The results show the effectiveness of our solution for processing top-k queries in DHTs.

The rest of this paper is organized as follows. In Section 2, we present our mechanism for storing the shared data in a DHT. In Section 3, we present our algorithms for processing top-k queries in DHTs. Section 4 describes a performance evaluation of our algorithms through implementation over a 64-node cluster and simulation using SimJava. Section 5 concludes.

2 Data Storage Mechanism

In this section, we propose a mechanism for storing relational data in the DHT. This mechanism not only provides good support for exact-match queries, it also enables efficient execution of our top-k query processing algorithms. In our data storage mechanism, peers store their relational data in the DHT with two complementary methods: tuple storage and attribute-value storage. In this paper, we assume that the data which are stored in the DHT are highly available by using yet proposed approaches, *e.g.* using multiple hash functions as in [20].

2.1 Tuple Storage

With the tuple storage method, each tuple of a relation is entirely stored in the DHT using its tuple identifier (*e.g.* its primary key) as the storage key. This enables looking up a tuple by its identifier. Let R be a relation name and A be the set of its attributes. Let T be the set of tuples of R and $id(t)$ be a function that denotes the identifier of a tuple $t \in T$. Let h be a hash function that hashes its inputs into a DHT key, *i.e.* a number which can be mapped by the DHT onto a peer. For storing relation R, each tuple $t \in T$ is entirely stored in the DHT where the storage key is $h(R, id(t))$, *i.e.* the hash of the relation name and the tuple identifier. Hereafter, the key by which we store a tuple in the DHT is called *tuple storage key*.

Tuple storage allows us to answer exact-match queries on the tuple identifier. For example, consider relation Car(<u>car-id</u>, price, mileage, ...) in which *car-id* is the primary key. If we store the tuples of this relation in the DHT using the tuple storage method, we are able to answer to exact match queries on *car-id* attribute, *e.g.* "Is there any car whose *car-id* is equal to 20?". But, it does not help answering exact-match queries on other attributes, *e.g.* "Is there any car whose price is 2000?". Attribute-value storage helps answering such queries.

A straightforward extension to tuple storage is to partition (fragment) the relation horizontally and store all tuples of each partition with the same key, thereby at the same peer. The key for storing the tuples of each partition can be constructed as for attribute-value storage which we describe in the next section. For very large numbers of tuples, this extension can be much more efficient than storing each tuple in the DHT with a different key.

2.2 Attribute-Value Storage

Attribute-value storage stores individually the attributes that may appear in a query's equality predicate or in a query's scoring function in the DHT. Thus, like database secondary indices, it allows checking for the existence of tuples using attribute values.

Our attribute-value storage method has two important properties. 1) after retrieving an attribute value from the DHT, peers can retrieve easily the corresponding tuple of the attribute value; 2) attribute values that are relatively "close" are stored at the same peer. To satisfy the first property, the key used for storing the entire tuple, *i.e.* tuple storage key, is stored along with the attribute value. The second property is satisfied by using the concept of domain partitioning as follows. Consider an attribute a and let D_a be its domain of values. Assume there is a total order $<$ on D_a, *e.g.* D_a is numeric, string, date, etc. D_a is partitioned into n nonempty sub-domains d_1, d_2, ..., d_n such that their union is equal to D_a, the intersection of any two different sub-domains is empty, and for each $v_1 \in d_i$ and $v_2 \in d_j$, if $i<j$ then we have $v_1<v_2$. For example, the attribute "mileage", whose domain is integer values in [0..300K], can be partitioned into 30 sub-domains [0..10K), [10K..20K), ..., [290K..300K]. Given a value v, the sub-domain to which v belongs is denoted by *sd(a, v)*. The number of sub-domains of an attribute and the lower bound of each sub-domain are known to all peers of the DHT. Therefore, given an attribute a and a value v, any peer can locally compute *sd(a, v)*.

The key which is used for storing an attribute value in the DHT is constructed as follows. Let R be a relation, a be an attribute of R, and v be the value of a in a tuple t, then the key for storing v in the DHT is *h(R, a, sd(a, v))*, *i.e.* the hash of the relation name, attribute name and the sub-domain to which v belongs. Therefore, the attribute values that belong to the same sub-domain are stored with the same key. Thus, they are maintained at the same peer.

The values of the attributes, for which we do an attribute-value storage, are stored two times in the DHT, *i.e.* once upon tuple storage and once upon attribute-value storage. However, this controlled redundancy allows us to answer exact-match queries on these attributes.

2.3 Uniform Distribution of Attribute-Values

The method which we use for partitioning attribute domains should avoid skewed distribution of attribute values within sub-domains, which may yield load unbalance among peers. For instance, simply dividing the domain into n equal-width sub-domains, as we did for attribute "mileage" above, may yield attribute storage skew. Using histogram-based information that describes the distribution of the values of an attribute, we can do a better partitioning that uniformly distributes the values within the sub-domains. Formally, let $p_a(v)$ be the probability density function that describes the probability that attribute a takes a value equal to v. Let $lb(d)$ be a function that denotes the lower bound of a sub-domain d, to obtain a uniform partitioning we choose the sub-domains d_1, d_2, ..., d_n such that:

$$\int_{lb(d_i)}^{lb(d_{i+1})} p_a(v)dv = \frac{1}{n} \quad , \text{ for } 1 \leq i \leq n\text{-}1 \tag{1}$$

By these *n-1* equations, the lower bounds are chosen in such a way that the sub-domains have equal cumulative numbers of values. We know that the lower bound of d_1 is equal to the lower bound of D_a, so $lb(d_1)$ is determined. Thus, we have *n-1* equations with *n-1* variables, *i.e.* $lb(d_2),..., lb(d_n)$. By solving the equations, we can determine the value of $lb(d_2),..., lb(d_n)$.

3   Top-k Query Processing Algorithms

In this section, we first propose DHTop1, an algorithm for efficient executions of top-k queries whose scoring function is monotonic. Then, based on DHTop1, we propose the DHTop2 algorithm which supports a much larger group of scoring functions. Finally, we propose the DHTop3 algorithm which exploits a small amount of redundancy in the DHT to yield high performance gains in the execution of top-k queries.

3.1   DHTop1 Algorithm

Let Q be a given top-k query, f be its scoring function, and p_{int} be the peer at which Q is issued. Let *scoring attributes* be the attributes that are used in f. We assume that the values of the scoring attributes are stored in the DHT by attribute-value storage. DHTop1 starts at p_{int} and proceeds in two phases as follows:

1. For each scoring attribute α do

    Create a list L_α and add all sub-domains of α to it;

    Remove from L_α the sub-domains which do not satisfy Q's condition;

    Sort L_α in descending order of its sub-domains;

2. end-condition := false;

    For each scoring attribute α do in parallel

    $i := 1$;

    $n :=$ number of sub-domains in L_α;

    While (end-condition = false) and ($i \leq n$) do

    > Send Q to the peer p that maintains the α values whose sub-domain is $L_\alpha[i]$. p returns to p_{int} its values of α which satisfy Q's condition, one by one in descending order, along with their corresponding tuple storage key;

    > $v :=$ the first α value returned by p;

    > While ($v \neq$ null) and (end-condition = false) do
    > - Retrieve the corresponding tuple of v and compute its score. If it is one of the k highest scores, then record the tuple in a list Y;
    > - If there are k tuples in Y whose scores are higher than the threshold then set end-condition to true and return to the user these k tuples;
    > - If end-condition is false then set v to the next α value returned by p;
    > - If v is null (*i.e.* all values returned by p have been received) then set $i := i + 1$;

The threshold we use is inspired from the TA algorithm [10] and is computed as follows. Let $\alpha_1, \alpha_2, ..., \alpha_m$ be the scoring attributes. Let $v_1, v_2, ..., v_m$ be the last values received respectively for attributes $\alpha_1, \alpha_2, ..., \alpha_m$. The threshold is defined as $\delta = f(v_1, v_2, ..., v_m)$. After receiving each attribute value, the threshold value is recomputed.

Let N be the number of stored tuples and m be the number of scoring attributes. In a way similar to [9], we can prove that, in the worst case, the number of tuples which should be retrieved by DHTop1 is $O(N^{(m-1)/m} * k^{1/m})$, which is sub-linear in the number of tuples. Due to space limitations we omit the proof. In average case, the execution time of DHTop1 is much better than an algorithm that retrieves all tuples (see Section 4).

We have proved that the DHTop1 works correctly for top-k queries with monotonic scoring functions. The proof is similar to the proof which we give for the correctness of the DHTop2 algorithm in the following section.

3.2 DHTop2

DHTop1 efficiently executes top-k queries whose scoring function is monotonic, *i.e.* any increase in the value of the input does not decrease the value of the output. For example the function $f(x_1, x_2) = x_1 + x_2$ is monotonic, but $f(x_1, x_2) = x_1 - x_2$ is not monotonic because an increase in the value of x_2 decreases the output of the function. Many of the popular aggregation functions, *e.g.* Min, Max, Average, are monotonic. However, there are many useful functions that are not monotonic including most of linear functions. In this section, we propose the algorithm DHTop2 which works with a super-set of monotonic functions. We call these functions *IODEV scoring functions*.

3.2.1 IODEV Scoring Functions

To define IOD-EV scoring functions, we need the two following definitions.

Definition 1 (Increasing wrt variable x): *A scoring function is increasing wrt variable x if any increase in the value of x does not decrease the output of the scoring function.*

Definition 2 (Decreasing wrt variable x): *A scoring function is decreasing wrt variable x if any increase in the value of x does not increase the output of the scoring function.*

For example the function $f(x_1, x_2) = x_1 - x_2$ is increasing wrt x_1 and decreasing wrt x_2. Now we can define IOD-EV scoring functions.

Definition 3 (Increasing Or Decreasing wrt Each Variable (IOD-EV)): *A scoring function f is IOD-EV if for each variable x, f is increasing wrt x or decreasing wrt x.*

For example, the functions $f(x_1, x_2) = x_1 - x_2$ and $f(x_1, x_2) = (x_1)^3 - (x_2)^3$ are IOD-EV. The set of monotonic scoring functions is a subset of IOD-EV functions because a monotonic scoring function is increasing wrt every variable, thus it is *IOD-EV*. The linear functions are also IOD-EV. This can be demonstrated as follows. A linear function can be written as $f(x_1, x_2, ..., x_m) = a_0 + a_1x_1 + a_2x_2 + ...+ a_mx_m$ where $a_0, a_1, ..., a_m$ are constant values. For each variable x_i if its coefficient a_i is positive then f is increasing wrt x_i, otherwise it is decreasing wrt x_i. Thus, all linear functions are IOD-EV.

3.2.2 Algorithm

Let Q be a given top-k query, and f be its scoring function. Assume f is IOD-EV, then DHTop2 algorithm is obtained from DHTop1 by performing the following modifications:

- In Phase 1, the sorting of L_α is done as follows. If the scoring function f is increasing wrt α then L_α is sorted in descending order of the lower bound of its sub-domains. Otherwise L_α is sorted in ascending order. Let $lb(d)$ denote the lower bound of a sub-domain d, and $L_\alpha[i]$ denote the ith sub-domain of L_α. At the end of this step, L_α is as follows. If f is increasing wrt α then $lb(L_\alpha[i]) \geq lb(L_\alpha[i+1])$ for $i \geq 1$. And if f is decreasing wrt α then $lb(L_\alpha[i]) \leq lb(L_\alpha[i+1])$ for $i \geq 1$.
- In Phase 2, the peer p, which maintains the α values whose sub-domain is $L_\alpha[i]$, returns the attribute values to p_{int} in the following order. If f is increasing wrt α then p returns the α values in descending order. Otherwise, it returns them in ascending order.

Example. Consider a relation $R(a, b, c)$ such that the domain of the attributes a and b is the real values in [0..10]. Assume that the domain of attribute a is partitioned into 4 sub-domains d_1=[0..3), d_2=[3..5), d_3=[5..8) and d_4=[8..10]. Assume the domain of attribute b is partitioned into 5 sub-domains d'_1=[0..3), d'_2=[3..5), d'_3=[5..6), d'_4=[6..8) and d'_5=[8..10). Consider the following query Q which is issued at p_{int}:

```
SELECT  *  FROM  R
WHERE  a < 6  AND b < 5
ORDER BY (a-b)    STOP AFTER  5
```

In the above query, the scoring attributes are a and b. The scoring function, *i.e.* $f=a-b$, is increasing wrt a and decreasing wrt b. Thus, in Phase 1, L_a is sorted in descending order and L_b in ascending order. At the end of Phase 1, we have L_a=<d_3, d_2, d_1> and L_b=<d'_1, d'_2>. Notice that some of the sub-domains of a (and b) are removed from L_a (and L_b) due to Q's condition, *e.g.* d_4 is removed from L_a because of a<6 in Q's condition.

3.2.3 Proof of Correctness

Let us now prove the correctness of the DHTop2 algorithm for top-k queries with IOD-EV scoring functions. For this, we prove the following lemma.

Lemma 1: *Let f be an IOD-EV scoring function, v_i and v'_i be two values of a scoring attribute a_i such that v_i is retrieved by DHTop2, and v'_i is retrieved after v_i or it is not retrieved by DHTop2, then we have $f(x_1, x_2,..., x_{i-1}, v'_i, x_{i+1},..., x_m) \leq f(x_1, x_2,..., x_{i-1}, v_i, x_{i+1},..., x_m)$ for any value $x_j ,1 \leq j \leq m$ and $j \neq i$. In other words, if we only change the value of a_i by replacing v_i with v'_i, the output of the scoring function decreases or does not change.*

Proof: With respect to the possible values v_i and v'_i, there are two cases to consider. In the first case, v_i and v'_i belong to two different sub-domains, *e.g.* d_1 and d_2 respectively. Thus, d_1 is before d_2 in L_{ai}. If f is increasing wrt a_i, then considering the first phase of the algorithm, we have $lb(d_1) \geq lb(d_2)$. Thus, we have $v_i \geq v'_i$ and since f is increasing wrt a_i, we have $f(x_1, x_2,..., x_{i-1}, v'_i, x_{i+1},..., x_m) \leq f(x_1, x_2,..., x_{i-1}, v_i, x_{i+1},..., x_m)$, *i.e.* increasing the value of a_i does not decrease the output of the function. Now, if f is decreasing wrt a_i, then considering the first phase of the algorithm, we have $lb(d_1) \leq lb(d_2)$. Thus $v_i \leq v'_i$ and since f is decreasing wrt a_i, we have $f(x_1, x_2,..., x_{i-1}, v'_i$

$x_{i+1},..., x_m) \leq f(x_1, x_2,..., x_{i-1}, v_i, x_{i+1},..., x_m)$. The second case is when v_i *and* v'_i belong to the same sub-domain. If f is increasing wrt a_i, then considering the second phase of the algorithm, we have $v_i \geq v'_i$ and thus $f(x_1, x_2,..., x_{i-1}, v'_i, x_{i+1},..., x_m) \leq f(x_1, x_2,..., x_{i-1}, v_i, x_{i+1},..., x_m)$. If f is decreasing wrt a_i then we have $v_i \leq v'_i$ and thus $f(x_1, x_2,..., x_{i-1}, v'_i, x_{i+1},..., x_m) \leq f(x_1, x_2,..., x_{i-1}, v_i, x_{i+1},..., x_m)$. $\square$

The following theorem provides the correctness of our algorithm.

Theorem 1: *If f is an IOD-EV scoring function, then DHTop2 finds the k top tuples correctly.*

Proof: The proof is by contradiction. Let Y be the set of k top tuples obtained by DHTop2, and t' be the tuple in Y whose score is the lowest. We assume there is a tuple $t'' \notin Y$ such that its score is greater than t', and we show that this assumption yields to a contradiction. Let $a_1, a_2, ..., a_m$ be the scoring attributes. Let $v_1, v_2, ..., v_m$ be the last values, *i.e.* before ending the algorithm, retrieved respectively for attributes $a_1, a_2, ..., a_m$. Let $v'_1, v'_2, ..., v'_m$ be the values of the attributes $a_1, a_2, ..., a_m$ in t', respectively. Let $v''_1, v''_2, ..., v''_m$ be the values of attributes $a_1, a_2, ..., a_m$ in t'', respectively. Since t'' is not in Y, it was not retrieved during the execution of our algorithm. Thus, none of its values, *i.e.* $v''_1, v''_2, ..., v''_m$, was retrieved by p_{int}, because if the value of any attribute of a tuple was retrieved, the entire tuple would have been retrieved by the algorithm. By applying Lemma 1 on attribute a_1 we have $f(v_1, v_2, ..., v_m) \geq f(v''_1, v_2, ..., v_m)$. By applying Lemma 1 on attribute a_2, we have $f(v''_1, v_2, v_3,..., v_m) \geq f(v''_1, v''_2, v_3,..., v_m)$. By continuing the application of Lemma 1 on attributes $a_3,..., a_m$, we have $f(v_1, v_2, ..., v_m) \geq f(v''_1, v_2, ..., v_m) \geq f(v''_1, v''_2, ..., v_m) \geq... \geq f(v''_1, v''_2, ..., v''_{m-1}, v_m) \geq f(v''_1, v''_2, ..., v''_{m-1}, v''_m)$. Therefore, we have $f(v_1, v_2, ..., v_m) \geq f(v''_1, v''_2, ..., v''_m)$. According to the end condition of the algorithm, we have $f(v'_1, v'_2, ..., v'_m) \geq f(v_1, v_2, ..., v_m)$, and by comparing this inequality with the former one, we have $f(v'_1, v'_2, ..., v'_m) \geq f(v''_1, v''_2, ..., v''_m)$. In other words, the score of tuple t' is greater than that of t'', which yields to a contradiction. $\square$

3.2.4 Optimizations

In order to further reduce the communication cost of the DHTop2 algorithm, we propose two optimizing strategies: batch retrieval of attribute values and retrieving each tuple at most once.

Batch retrieval of attribute values (BRAV). In Phase 2 of the basic version of DHTop2, the values of the scoring attributes are returned to p_{int} one by one, *i.e.* each value in a message. Since each message has its own overhead, *e.g.* latency, returning only one value per message is very costly. To reduce such overhead, we modify Phase 2 of the algorithm such that the peer, which maintains the values of a sub-domain, sends the attribute values to p_{int} in a batch fashion, *e.g.* k values per message.

Retrieving each tuple at most once (RTO). In Phase 2 of the basic version of DHTop2, after retrieving each value of a scoring attribute, the corresponding tuple of that value is retrieved. Since there may be several scoring attributes, a tuple may be retrieved several times. However, after the first retrieval of the tuple and comparing its score with the k highest scores, there is no need to retrieve it again because either the tuple is in the set Y or its score cannot be one of the k highest scores. Thus, to optimize our algorithm, we change Phase 2 such that p_{int} maintains in a list the identifiers of all tuples which have yet been retrieved. Before retrieving a tuple, p_{int} checks the list, and if the identifier of the tuple is in the list, it does not retrieve the tuple.

3.3 DHTop3

By analyzing many scoring functions in useful queries, we observed that only a small set of a relation's attributes are likely to be used for scoring. For instance, the attributes typically used for scoring used cars are price and mileage. In addition to their small number, these attributes typically have small size, *e.g.* numerical. Based on this observation, we modify our storage mechanism such that the value of the attributes typically used for scoring are stored upon each attribute-value storage, *e.g.* upon storing the price value, we store the value of both price and mileage in the DHT.

Formally, let R be a relation, and A_{sf} be the set of attributes of R which are typically used for scoring. Let t be a tuple of R, and V be the set of values of attributes of A_{sf} in tuple t. Let $v \in V$ be the value of $a \in A_{sf}$ in tuple t. Upon storing v by the modified attribute-value storage, we store all values involved in V in the DHT. Like with the basic attribute-value storage, we also store the storage key of t along with V.

With the modified attribute-value storage, the values of the attributes involved in A_{sf} are stored $/A_{sf}/$ times in the DHT. At the expense of this redundancy, which is usually low because $/A_{sf}/$ is small, we can now compute a tuple score at peers that maintain the attribute values without having to retrieve the tuple.

Relying on the modified attribute-value storage, we develop a top-k query processing algorithm, called DHTop3, based on DHTop2 by performing the following modifications in the second phase. The peer p, which maintains the attribute values of sub-domain $L_\alpha[i]$, computes the scores of the corresponding tuple of each attribute value, and returns the scores along with the attribute values to p_{int}. Thus, p_{int} no longer needs to retrieve the corresponding tuple of the received values; it only keeps the storage keys of the k highest scored tuples in the set Y and continues until the end condition holds. Finally, p_{int} retrieves from the DHT the k tuples whose storage keys are maintained in Y.

4 Performance Evaluation

We evaluated the performance of the algorithms, which we proposed in the pervious section, through implementation and simulation. The implementation over a 64-node cluster was useful to validate our algorithms and calibrate the simulator. The simulator allows us to study scale up to high numbers of peers (up to 10,000 peers).

In this section, we first describe our experimental setup. Then, we investigate the scalability of our algorithms by increasing the number of peers and also by increasing the number of tuples which, we store for each relation, in the DHT. Finally, we evaluate the performance of our algorithm by varying other parameters such as the number of requested tuples, *i.e.* k, the distribution of attribute values, and the number of sub-domains of each attribute.

4.1 Experimental Setup

Our implementation and simulation are based on Chord [23] which is an efficient DHT. We tested our algorithms over a cluster of 64 nodes connected by a 1-Gbps network. Each node has two Intel Xeon 2.4 GHz processors, and runs the Linux operating system. We make each node act as a peer in the DHT.

To study the scalability of our algorithm far beyond 64 peers, we also implemented a simulator using SimJava. After calibration of the simulator, we obtained simulation results similar to the implementation results up to 64 peers. Thus, since the simulator allows us to study larger systems and due to space limitations, we only report simulation results for most of our tests.

Our default settings for different experimental parameters are shown in Table 1. Most of these settings are the same as in [7]. In our tests, we use a synthetically generated relation with six attributes a_i, $1 \le i \le 6$ and the domain of the attributes is numeric. The default number of tuples of the relation is 100,000 and they are randomly generated in two different ways: (1) Uniform data set, and (2) Gaussian data set. With (1), the values of attributes are independent of each other, and the distribution of the values of each attribute is uniform. This is our default setting. With (2), the values of different attributes are independent of each other, and the values for each attribute are generated via overlapping multidimensional Gaussian belles.

In our tests, the top-k query Q is delivered to a randomly selected peer. The selectivity of Q over the generated data is 10% and the scoring function specified in Q is the linear function $f(a_1, a_2, a_3, a_4, a_5, a_6) = a_1 + a_2 + a_3 + a_4 + a_5 + a_6$. Typically, users are interested in a small number of top answers, thus we set $k=10$. In our storage mechanism, the domain of each attribute is uniformly partitioned into n sub-domains and the default value for n is 100. The network parameters of the simulator are shown in Table 2. We use parameter values which are typical of P2P systems [21]. The simulator allows us to perform tests up to 10,000 peers, after which the simulation data no longer fit in RAM. This is quite sufficient for our tests.

Table 1. Default setting of experimental parameters

Parameter	Default values
Number of tuples	100,000
K	10
Number of attributes	6
Data set	Uniform
Data selectivity	10 %
Number of attribute's sub-domains	100

Table 2. Network parameters of the simulator

Parameter	Default values
Bandwidth	Normally distributed random, Mean = 56 Kbps, Variance = 32
Latency	Normally distributed random, Mean = 150 ms, Variance = 100
Number of peers	10,000 peers

To evaluate the performance, we measure the following metrics. 1) Response time: the time elapsed between the delivery of Q to p_{int} and the end of the algorithm. 2) Communication cost: the total number of bytes which are transferred over the network for executing a given top-k query.

We evaluated the performance of our three algorithms DHTop1, DHTop2, and DHTop3. We also compared our algorithms with the simple algorithm which we introduced in the introduction of this paper. The simple algorithm, which we denote as Simple_DHT, retrieves all tuples of the relations involved in the query, computes the score of each retrieved tuple, and finally returns the k tuples whose scores are the highest.

4.2 Scale Up

In this section, we investigate the scalability of our algorithm. For this, we study the effect of the number of peers and also the number of stored tuples on performance.

Effect of the Number of Peers

We used both our implementation and our simulator to study the response time while varying the number of peers. Using our implementation over the cluster, we ran experiments to study how response time increases with the addition of peers. Figure 1

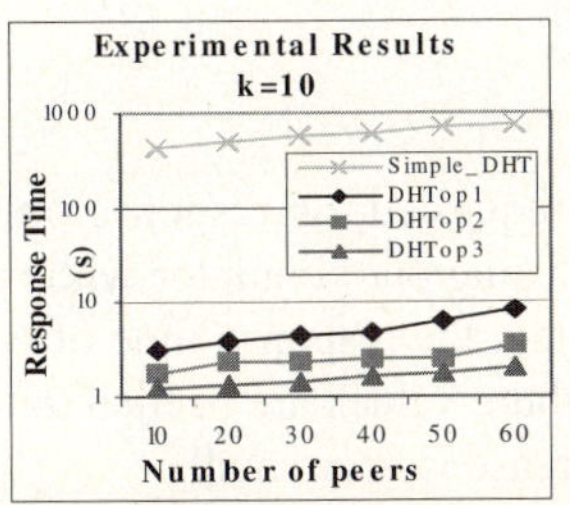

Fig. 1. Response time vs. number of peers

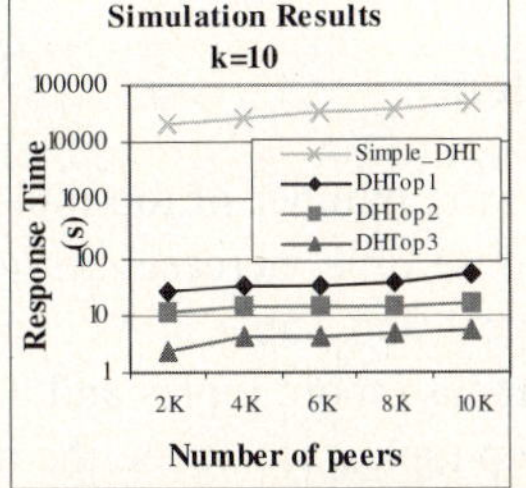

Fig. 2. Response time vs. number of peers

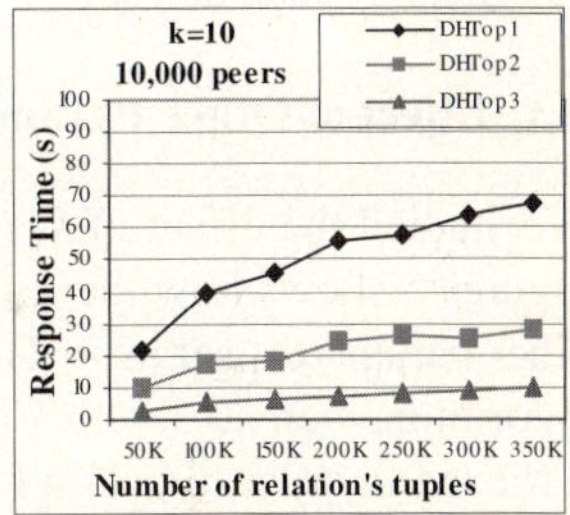

Fig. 3. Response time vs. number of relation's tuples

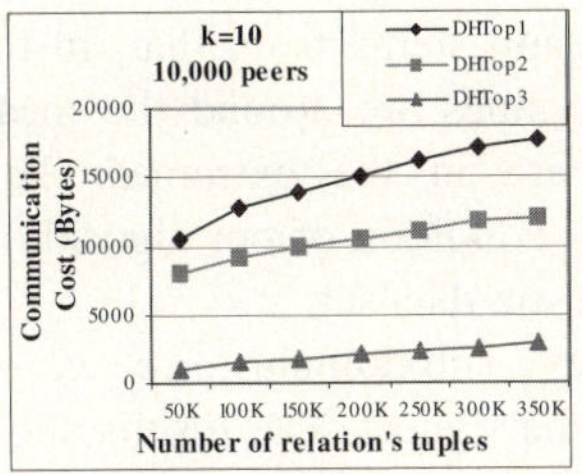

Fig. 4. Communication cost vs. number of relation's tuples

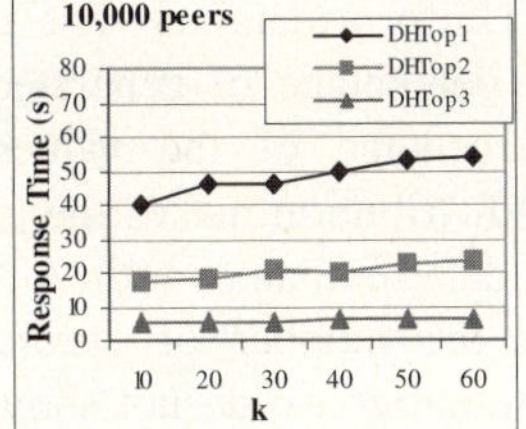

Fig. 5. Response time vs. k

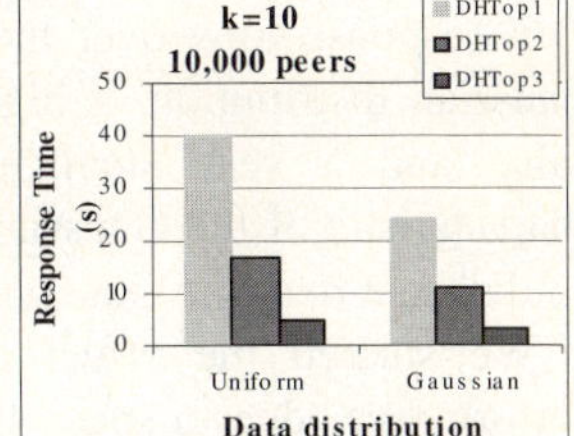

Fig. 6. Response time over uniform & Gaussian data sets

shows the response time of our three algorithms and the Simple_DHT algorithm with the addition of peers up to 64. In all four algorithms, the response time grows logarithmically with the number of peers. However, the response time of our algorithms is much better than Simple_DHT.

Using simulation, Figure 2 shows the response times of the four algorithms with the number of peers increasing up to 10000 and the other parameters set as in Table 1 and Table 2. Overall, the experimental results correspond qualitatively with the simulation results. However, we observed that the response time gained from our experiments over the cluster is slightly better than that of simulation, simply because of faster communication in the cluster.

Effect of the Number of Tuples

We used our simulator to study the response time while varying the number of tuples which we store for each relation in the DHT. Figure 3 shows how the response time of our three algorithms increases with the number of tuples, using our simulator with the other parameters set as in Table 1 and Table 2. The number of tuples has a very small impact on DHTop3, because it does not need to retrieve stored tuples for computing the scores. The impact of the number of tuples on the response time of DHTop2 is less than DHTop1 due to the optimizations which we proposed for DHTop2. We also tested the communication cost of our algorithms. Using the simulator, Figure 4 depicts the number of bytes with increasing numbers of tuples, with the other parameters set as in Table 1 and Table 2.

4.3   Effect of Other Parameters

We studied the effect of k, *i.e.* the number of top tuples requested, on response time. Figures 5 shows how the response time increases with k, using our simulator with the other parameters set as in Table 1 and Table 2. As expected, the response time of our algorithms increases with k because more tuples and attribute values are needed to be retrieved in order to obtain k top tuples. However, the increase is very small.

We investigated the response time of our algorithms over two Uniform and Gaussian data sets that we have generated synthetically, as described in Section 4.1. Using our simulator, Figure 6 shows the response time of our algorithms over Uniform and Gaussian data sets, with the other parameters set as in Table 1 and 2. The response time of our algorithms over the Gaussian data set is much better than their response time over the Uniform data set. The reason stems from that, in the Gaussian distribution, a high percentage of generated values are around the mean value and a very small percentage of the values are in the extremes. This characteristic of the Gaussian distribution makes the end condition of our algorithms hold sooner over the Gaussian data set than over the Uniform data set.

We studied the effect of the number of attributes' sub-domains, *i.e.* n, on performance (due to space limitations we do not show the figure). The results show that for the case of issuing a lot of simultaneous queries, increasing n reduces the average response time because it increases the number of peers that are responsible for maintaining the values of an attribute, so the load of each peer decreases.

5 Conclusion

In this paper, we addressed the problem of efficient top-k query processing in DHTs. We first proposed a mechanism for data storage in DHTs which provides good support for exact-match queries and enables efficient execution of our top-k query processing algorithm. Then, we proposed a family of algorithms which deal with efficient processing of top-k queries in DHTs. We evaluated the performance of our algorithms through implementation over a 64-node cluster and simulation using SimJava. The results showed the effectiveness of our solution for processing top-k queries in DHTs.

References

[1] Akbarinia, R., Pacitti, E., Valduriez, P.: Reducing Network Traffic in Unstructured P2P Systems Using Top-k Queries. Distributed and Parallel Databases 19(2) (2006)

[2] Babcock, B., Olston, C.: Distributed Top-K Monitoring. In: SIGMOD Conf. (2003)

[3] Balke, W.-T., Nejdl, W., Siberski, W., Thaden, U.: Progressive Distributed Top k Retrieval in Peer-to-Peer Networks. In: ICDE Conf. (2005)

[4] Bast, H., Majumdar, D., Schenkel, R., Theobald, M., Weikum, G.: IO-Top-k: Index-access Optimized Top-k Query Processing. In: VLDB Conf. (2006)

[5] Blanco, R., Ahmed, N., Hadaller, D., Sung, L.G.A., Li, H., Soliman, M.A.: A Survey, of Data Management in Peer-to-Peer Systems. Technical Report CS-2006-18, University of Waterloo (2006)

[6] Cao, P., Wang, Z.: Efficient Top-K Query Calculation in Distributed Networks. In: PODC Conf. (2004)

[7] Chaudhuri, S., Gravano, L., Marian, A.: Optimizing Top-K Selection Queries over Multimedia Repositories. IEEE Trans. on Knowledge and Data Engineering 16(8) (2004)

[8] Das, G., Gunopulos, D., Koudas, N., Tsirogiannis, D.: Answering Top-k Queries Using Views. In: VLDB Conf. (2006)

[9] Fagin, R.: Combining fuzzy information from multiple systems. J. Comput. System Sci. 58(1) (1999)

[10] Fagin, R., Lotem, J., Naor, M.: Optimal aggregation algorithms for middleware. J. of Computer and System Sciences 66(4) (2003)

[11] Gao, J., Steenkiste, P.: An Adaptive Protocol for Efficient Support of Range Queries in DHT-Based Systems. In: IEEE Int. Conf. on Network Protocols (ICNP) (2004)

[12] Güntzer, U., Kießling, W., Balke, W.T.: Optimizing Multi-Feature Queries for Image Databases. In: VLDB Conf. (2000)

[13] Harren, M., Hellerstein, J.M., Huebsch, R., Loo, B.T., Shenker, S., Stoica, I.: Complex Queries in DHT-based Peer-to-Peer Networks. In: Int. Workshop on Peer-to-Peer Systems (IPTPS) (2002)

[14] Hjaltason, G.R., Samet, H.: Index-driven similarity search in metric spaces. ACM Transactions on Database Systems (TODS), 28(4) (2003)

[15] Huebsch, R., Hellerstein, J., Lanham, N., Loo, B.T., Shenker, S., Stoica, I.: Querying the Internet with PIER. In: VLDB Conf. (2003)

[16] Kimelfeld, B., Sagiv, Y.: Finding and approximating top-k answers in keyword proximity search. In: PODS Conf. (2006)

[17] Michel, S., Triantafillou, P., Weikum, G.: KLEE: A Framework for Distributed Top-k Query Algorithms. In: VLDB Conf. (2005)

[18] Mouratidis, K., Bakiras, S., Papadias, D.: Continuous monitoring of top-k queries over sliding windows. In: SIGMOD Conf. (2006)

[19] Nepal, S., Ramakrishna, M.V.: Query Processing Issues in Image (Multimedia) Databases. In: ICDE Conf. (1999)

[20] Ratnasamy, S., Francis, P., Handley, M., Karp, R.M., Shenker, S.: A scalable content-addressable network. In: SIGCOMM Conf. (2001)

[21] Saroiu, S., Gummadi, P.K., Gribble, S.D.: A Measurement Study of Peer-to-Peer File Sharing Systems. Proc. of Multimedia Computing and Networking (MMCN) (2002)

[22] Silberstein, A., Braynard, R., Ellis, C.S., Munagala, K., Yang, J.: A Sampling-Based Approach to Optimizing Top-k Queries in Sensor Networks. In: ICDE Conf. (2006)

[23] Stoica, I., Morris, R., Karger, D.R., Kaashoek, M.F., Balakrishnan, H.: Chord: A scalable peer-to-peer lookup service for internet applications. In: Proc. of SIGCOMM (2001)

[24] Wu, M., Xu, J., Tang, X., Lee, W.-C.: Monitoring Top-k Query in Wireless Sensor Networks. In: ICDE Conf. (2006)

A Structured Overlay for Multi-dimensional Range Queries[*]

Thorsten Schütt, Florian Schintke, and Alexander Reinefeld

Zuse Institute Berlin

Abstract. We introduce SONAR, a structured overlay to store and retrieve objects addressed by multi-dimensional names (*keys*). The overlay has the shape of a multi-dimensional torus, where each node is responsible for a contiguous part of the data space. A uniform distribution of keys on the data space is not necessary, because denser areas get assigned more nodes. To nevertheless support logarithmic routing, SONAR maintains, per dimension, fingers to other nodes, that span an exponentially increasing number of *nodes*. Most other overlays maintain such fingers in the *key-space* instead and therefore require a uniform data distribution. SONAR, in contrast, avoids hashing and is therefore able to perform range queries of arbitrary shape in a logarithmic number of routing steps—independent of the number of system- and query-dimensions.

SONAR needs just one hop for updating an entry in its routing table: A longer finger is calculated by querying the node referred to by the next shorter finger for its shorter finger. This doubles the number of spanned nodes and leads to exponentially spaced fingers.

1   Introduction

The efficient handling of multi-dimensional range queries in Internet-scale distributed systems is still an open issue. Several approaches exist, but their lookup schemes are either expensive (space-filling curves) [2] or use probabilistic approaches like consistent hashing [10] to build the overlay.

We propose a system for storing and retrieving objects with d-dimensional keys in a peer-to-peer network. SONAR (Structured Overlay Network with Arbitrary Range-queries) directly maps the multi-dimensional data space to a d-dimensional torus. It supports range queries of arbitrary shape, which are useful, for example, in geo-information systems where objects in a given distance of a position are sought. SONAR can also be employed in Internet games with millions of online-players who concurrently interact in a virtual space and need quick access to the local surroundings of their avatars. In a broader context, SONAR can be employed as a hierarchical publish/subscribe system, where published events are categorized by several independent attributes. The category of published events addresses a data point in the d-dimensional space and consumers subscribing to subareas will receive all events published in their subarea.

---

[*] Part of this work was carried out under the SELFMAN and XtreemOS projects funded by the European Commission.

A.-M. Kermarrec, L. Bougé, and T. Priol (Eds.): Euro-Par 2007, LNCS 4641, pp. 503–513, 2007.

The paper is organized as follows: First, we discuss related work. Then, in Section 3, we introduce SONAR. In Section 4, we present empirical results and in Section 5 we conclude the paper with a brief summary.

2   Related Work

Several systems [1] have been proposed that support complex queries with multi-dimensional keys and ranges. They can be split into two groups.

a) Space Filling Curves. These systems [2,9,16] use locality preserving space-filling curves to map multi-dimensional to one-dimensional keys. They provide less efficient range queries than the space partitioning schemes described below, because a single range query may cover several parts of the curve, which have to be queried separately (Fig. 5a). Chawathe et al. [7] present performance results of a real-world application using Z-curves on top of OpenDHT. The query performance (≈ 2 sec. for ≤ 30 nodes) is rather low due to the layered approach.

b) Space Partitioning. The schemes using space partitioning split the key-space among the nodes. SONAR belongs to this group of systems. The proposed systems mainly differ by their routing strategies.

CAN [14] was one of the very first DHTs. It hashes the key-space onto a multi-dimensional torus. While the topology resembles that of SONAR and MURK (see below), CAN uses just the neighbors for routing and it does not support range queries.

SWAM [4] employs a Voronoi-based space partitioning scheme and uses a small-world graph overlay with routing tables of size $O(1)$. The overlay is not built by some regular partitioning scheme (e.g. kd-tree [5]) but uses a sample technique to place the fingers.

Multi-attribute range queries were also addressed by Mercury [6] which needs a large number of replicas per item to achieve logarithmic routing performance. SWORD [12] uses super-peers and query-caching to allow multi-attribute range queries on top of the Bamboo-DHT [15].

Ganesan *et al.* [9] proposed two systems for multi-dimensional range queries in peer-to-peer systems: SCRAP and MURK. SCRAP uses the traditional approach of mapping multi-dimensional to one-dimensional data with space-filling curves which destroys the data locality. Consequently, each single multi-dimensional range-query is mapped to several one-dimensional queries. MURK is more similar to our approach, as it divides the data space into hypercuboids with each partition assigned to one node. In contrast to SONAR, MURK uses a heuristic approach based on skip graphs [3] to set routing fingers.

3   System Design

We first present the overlay topology of SONAR and then discuss its routing and lookup strategy. Thereafter we present mechanisms that make SONAR robust under churn.

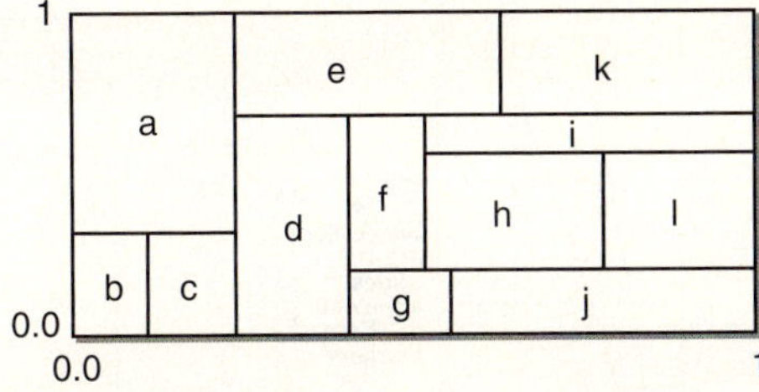

Fig. 1. Example two-dimensional overlay with attribute domains $[0, 1]$

SONAR is used to store and retrieve *objects*. It works on a d-dimensional torus, the *key-space*. Objects have a name, the *key*, which is a vector of d components, the *attributes* of the key. Each *dimension* of the torus is responsible for one *attribute domain*. Figure 1 illustrates a two-dimensional key-space ($[0, 1]^2$). Arbitrarily located computers, the *nodes*, are each responsible for a dedicated area (hypercuboid) in the key-space of the overlay (rectangles in Fig. 1). The *node-space* has the same extent as the key-space, but is completely filled with nodes. Two nodes in the node-space are *adjacent* (or *neighbors*) when their key-space is adjacent. The direct mapping between key-space and node-space guarantees adjacent keys to be stored on the same or adjacent nodes, which enables efficient range queries across node boundaries by local query propagation.

Nodes are dynamically assigned to the key-space such that each node serves roughly the same number of objects. Load-balancing is done by changing the responsibility of nodes instead of moving around objects in the key-space. That becomes necessary when the number of objects or nodes in the system changes (Sect. 3.4).

3.1   Overlay Topology

As illustrated in Figure 1, the two-dimensional key-space is covered by rectangles, each of them containing about the same number of objects. Because the keys are generally not uniformly distributed, the rectangles have different sizes and thus may have more than one neighbor per direction. The neighbors are stored in *neighbor lists*, one per dimension.

The overlay described so far resembles that of CAN [14] except for the hashing in CAN, which prevents efficient range queries. Consequently, SONAR would also need $O(\sqrt[d]{N})$ network hops if it would just use the neighbors for routing. In the following, we introduce routing tables to achieve logarithmic routing performance.

3.2   Routing

For routing, SONAR uses separate *routing tables*, one per dimension. Each routing table contains *fingers* spanning an exponentially increasing number of nodes (Fig. 2). With a total of $\log N$ routing fingers, the average number of hops is reduced to $O(\log N)$ [17].

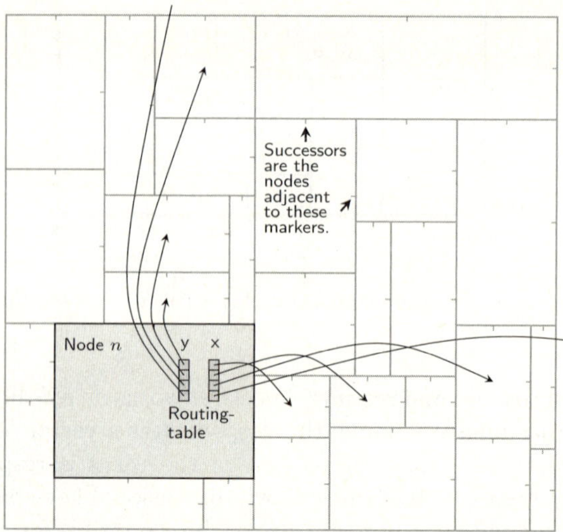

Fig. 2. Routing table for the two-dimensional case

To calculate its ith finger in the routing table, a node looks at its $(i-1)^{th}$ finger and asks the remote node listed there for the $(i-1)^{th}$ finger. At the lowest level, the fingers point to the successor.

$$finger_i = \begin{cases} successor & : i = 0 \\ finger_{i-1}.getFinger(i-1) & : i \neq 0 \end{cases}$$

This update process works in a running system, but also during startup. Initially, all fingers are set to *unknown* except for the finger to the successor. Filling the second entry will always succeed, because the successor knows its successor. Filling further entries may fail (result *unknown*), because the remote node may not have determined the corresponding entry yet. But with subsequent periodic updates, eventually all nodes will get their entries filled. The resulting structure is similar to skip lists [13], but the behavior is more deterministic.

Successor Used for Routing. A node may have more than one neighbor per direction. We define the node adjacent to the middle of the respective side to be the *successor*. Successors are marked by small ticks in Figure 2.

Due to the different box sizes and the calculation of longer fingers from shorter ones, fingers are not necessarily straight in one direction. Slight deviations in y-direction might occur when following the fingers of the x-direction (and vice versa), as shown in Figure 2. Our empirical results indicate, however, that this does not affect the logarithmic routing performance (Sect. 4).

Routing Table Size. Each node holds approximately $\log N$ fingers in its routing tables. However, not knowing the total number of nodes N, how many fingers should a node put into each of its d routing table so that the total is $\log N$?

```
// calculates the entries of a routing table
void updateRoutingTable(int dim) {
  int  i = 1;
  bool done = false;

  rt[dim][0] = this.Successor[dim];

  while (!done) {
    Node candidate = rt[dim][i - 1].getFinger(dim, i - 1);
    if (IsBetween(dim, rt[dim][i - 1].Key, candidate.Key, this.Key)){
      rt[dim][i] = candidate;
      i++;
    } else
      done = true;
  }
}

// checks whether the resp coordinate of pos lies between start and end
bool IsBetween(Dim dim, Key start, Key pos, Key end);
```

Fig. 3. Finger calculation for dimension dim

Mercury [6] predicts the system size N by estimating the key density. SONAR uses a simpler, deterministic solution with less overhead.

For each dimension dim, SONAR's finger update algorithm (Fig. 3) inserts an additional finger $finger_i$ as long as its position is between that of the last routing table entry $finger_{i-1}$ and that of the node itself. Otherwise the new finger circles around the ring and is not inserted.

Our results in Section 4 confirm that each node holds indeed $\log N$ fingers. The construction process guarantees—in contrast to Chord [18]—that no two fingers point to the same node. Since the fingers in the routing tables span an exponentially increasing number of nodes, the routing table of each dimension has a total of $\lceil \log D \rceil$ entries on the average, where D is the number of nodes in this direction on the torus.

Cost of a Finger Update. Our periodically running finger update algorithm needs just one network hop to determine an entry in the routing table. Chord in contrast needs $O(\log n)$ for the same operation, because it performs a DHT lookup to calculate a finger.

3.3 Lookup and Range Queries

As in other DHTs, SONAR uses greedy routing. In each node the finger that maximally reduces the Euclidean distance to the target in the key-space is followed, independently of the dimension (see Fig. 4).

SONAR supports range queries with multiple attributes. In its most basic form a range query is defined by d intervals for the d attribute domains. The range query finds all keys whose attributes match the respective intervals and returns the corresponding objects. Because of their shape, such range queries are called d-dimensional *rectangular* range queries.

In practice, users sometimes need to define circles, polygons, or polyhedra in their queries. Figure 5 illustrates a two-dimensional circular range query defined

```
// find the responsible node for a given key
Node find(Point target) {
  Node nextHop = findNextHop(target);
  if (nextHop == this)
    return this;
  else
    return nextHop.Find(target);
}

double getDistance(Node a, Point b);

Node findNextHop(Point target) {
  Node candidate = this;
  double distance = getDistance(this, target);

  if (distance == 0.0)
    // found target
    return this;

  for (int d = 0; d < dimensions; d++) {
    for (int i = 0; i < rt[d].Size; i++) {
      double dist = getDistance(rt[d][i], target);
      if (dist < distance) {
        // new candidate
        candidate = rt[d][i];
        distance = dist;
      }
    }
  }
  // will never happen:
  Assert(candidate != this);
  return candidate;
}
```

Fig. 4. Lookup for a target

by a center and a radius. Here, we assume a person located in the governmental district of Berlin searching for a hotel in 'walking distance' (circle around the person). The query is first routed to the node responsible for the center of the circle and then forwarded to all neighbors that partially cover the circle (Fig. 5b). The query is checked against the local data and the results are returned to the requesting node. Figure 6 shows the pseudocode of this algorithm. Note that redundant messages are eliminated. op is an additional check for objects in the queried area—in this case for type hotel.

SONAR performs a range query with a single lookup. When the target node does not hold the complete key range, the query is locally forwarded. Systems with space-filling curves, in contrast, usually require more than one lookup for a single range query because they map connected areas to multiple independent line segments, see Figure 5a.

3.4   Topology Maintenance to Handle Churn

Node Join. When a node joins the system, the key-space of a participating node has to be split and the key responsibilities subdivided. To achieve this, two things must be done (we first describe random splitting and then include load balancing):

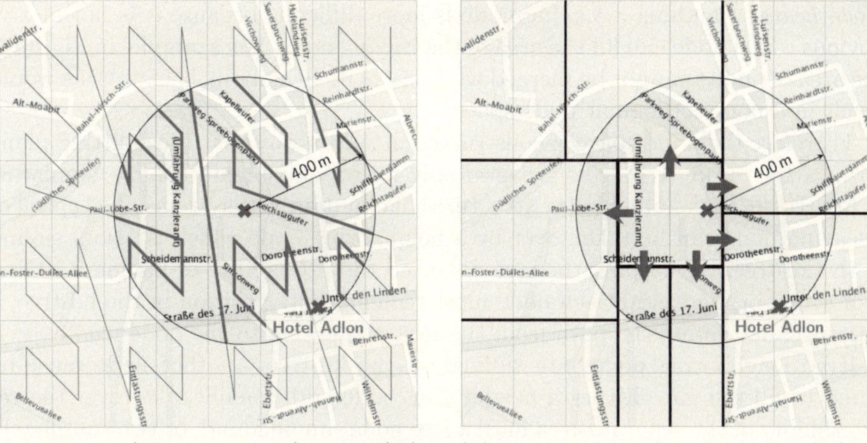

(a) Z-curve (8 line segments): $8 \cdot log_2(N)$ (b) Neighborhood broadcast: $log_2(N)+6$

Fig. 5. Circular range query

```
// perform a range query
void queryRange(Range r, Operation op) {
  Node center = Find(r.Center);
  center.doRangeQuery(r, op, newId());
}

void doRangeQuery(Range r, Operation op, Id id) {
  // avoid redundant executions
  if (pastQueries.Contains(id))
    return;
  pastQueries.add(id);

  foreach (Node neighbor in this.Neighbors)
    if (r ∩ neighbor.Range != ∅)
      neighbor.doRangeQuery(r \ this.Range, op, id);

  // execute operation locally
  op(this, r);
}
```

Fig. 6. Range query algorithm

1. Select a random target node: A random position in the key-space is routed
   to and a random walk is started from there. The final target node of this is
   the candidate to be split. The random walk ensures that nodes responsible
   for larger areas of the key-space are not preferred over smaller ones.
2. Split the key-space and transfer one part to the new node: Splits are parallel
   to one of the coordinate system axes. The selection of the axis to be split
   should not strictly favor one dimension over the others because the number
   of nodes to be contacted for a range query could become disproportionately
   high, when a large interval for the favored dimension is specified in the query.
   Also, node leaves could become more expensive.

Node Leave. Handling a leaving node is more difficult, because it is not always obvious which node can fill the area of the leaving node. For example, in Figure 1, the area of node f cannot be merged with any of its neighbors, because this would result in a non-rectangular node-space.

Therefore the node-space is constructed in such a way that the splitting plane forms a *kd-tree* [5]. KD-trees are used only for topology maintenance, similar as in MURK [9], but not as index structures like in database systems. The space of a leaving node can be taken over by a neighboring node which is also a sibling in the kd-tree. By keeping the tree balanced the probability of having a sibling as a neighbor increases. Each node must remember its position in the kd-tree, a bit-string describing the path from the root of the tree to the node itself.

If no neighboring nodes are siblings in the kd-tree, another node must be found to fill the gap. Either a neighboring node additionally takes over the responsibility of the separate area until a free node can be found, or two completely independent nodes that are siblings in the kd-tree have to be found to merge them and thus free a node that takes over the free area. The former concept, called *virtual nodes*, is also used for load-balancing in other systems.

Load Balancing. Load-balancing can be implemented by either adjusting the boundaries of the responsibilities locally or freeing nodes in underloaded areas and moving them to overloaded areas. The former has similar issues as a node leave—the boundaries are interlocked with limited room for adjustments. The latter was shown to be converging [11] with predictable performance.

The balancing can be based on different metrics for load, like object or query load or a combination of both. To avoid thrashing effects a threshold for performing a load-balancing round must be introduced.

4   Empirical Results

For testing the performance of SONAR we used a traveling salesman data set with 1,904,711 cities[1]. The cities' geographical locations follow a Zipf distribution [19] which is also common in other scenarios.

We assigned the responsibility of nodes by recursively splitting the key-space at the longer side, so that each part gets half of the cities until enough rectangles are created. Figure 7 shows a sample splitting for 256 nodes.

The coordinates were mapped onto a doughnut-shaped torus rather than a globe, because in a globe all vertical rings meet at the poles. This would not only cause a routing bottleneck at the poles but would also result in different ring directions for the western and eastern hemisphere (southwards vs. northwards).

Figure 8 shows the results for various all-to-all searches in networks of different sizes. The routing performance, depicted by the '+' ticks, almost perfectly matches the expected $0.5 \log_2 N$ hops. Only in the larger networks the expected value slightly deviates.

---

[1] http://www.tsp.gatech.edu/world

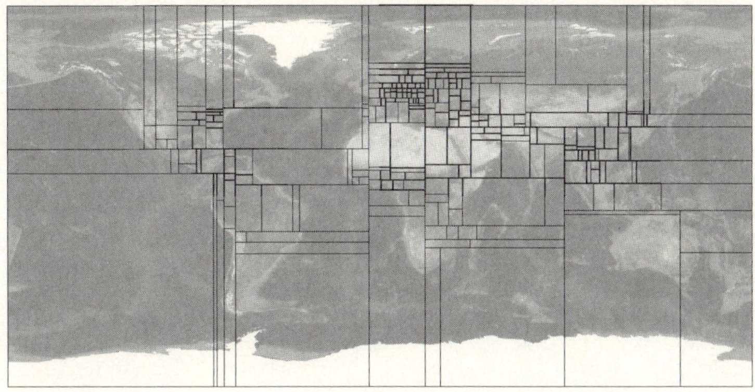

Fig. 7. 1,904,711 cities split evenly into 256 rectangular nodes

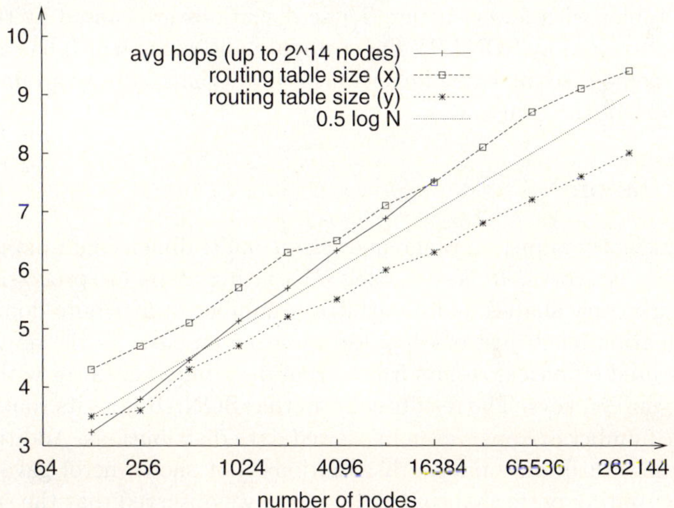

Fig. 8. SONAR results for increasing system sizes (2-dimensional)

We also checked whether the number of fingers in the routing tables, which are calculated without global information (Fig. 4), meets our expectations. The '□' ticks give the routing table sizes in horizontal direction, and the '∗' ticks represent the sizes in vertical direction. As expected, both graphs have the same slope of $0.5 \log_2 N$: One lies consistently above, the other below. This is attributed to the different domain sizes of the coordinate system (360 versus 180 degrees) and to the uneven number of splitting planes.

Figure 9 gives further insight into the characteristics of SONAR's routing tables. It shows—again for various network sizes—the deviation of the table sizes from their expected size $\log_2 N$ (denoted by '0'). As can be seen, the same pattern applies for all network sizes: About 50% of the tables contain one extra

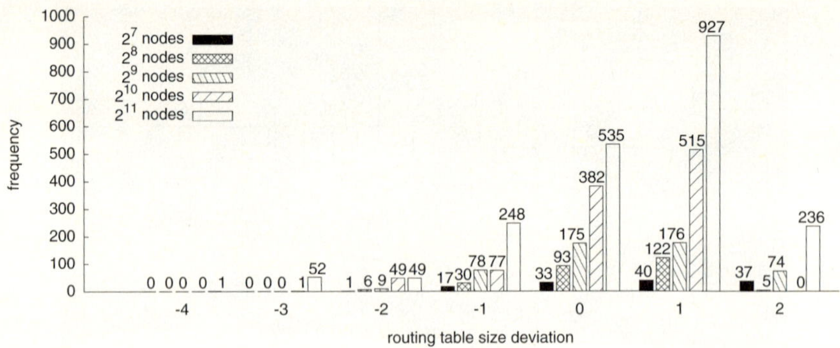

Fig. 9. Routing table size deviation from the expected value

entry, about 25% meet the expected size of $\log_2 N$, while there is a decreasing number of tables with fewer entries. These deviations are caused by the uneven key distribution and by SONAR's finger update algorithm which has a tendency to insert in some cases an extra finger that is more than halfway around the ring (but still 'left' of the own node).

5   Conclusion

SONAR efficiently supports range-queries on multi-dimensional data in structured overlay networks. It needs $O(\log N)$ routing steps for processing range-queries of arbitrary shapes and an arbitrary number of attribute domains. The finger calculation needs just one hop for updating an entry in the routing table.

We presented empirical results from a Zipf distributed data set with approximately two million keys. The results confirm that SONAR does its routing with a logarithmic number of hops—even in skewed data distributions. Additional tests with other practical and uniform distributions (not shown here) gave the same logarithmic routing performance. Furthermore, we observed that the sizes of the distributed routing tables are always $O(\log N)$ although they are autonomously maintained by the nodes with local information only.

Acknowledgements

Thanks to the anonymous reviewers for their valuable comments. The topographic images were taken from the 'Blue Marble next generation' project of NASA's Earth Observatory. Thanks to Slaven Rezić for the street map of Berlin.

References

1. Aberer, K., Onana Alima, L., Ghodsi, A., Girdzijauskas, S., Haridi, S., Hauswirth, M.: The essence of P2P: A reference architecture for overlay networks. In: P2P 2005 (2005)

2. Andrzejak, A., Xu, Z.: Scalable, efficient range queries for Grid information services. In: P2P 2002 (2002)
3. Aspnes, J., Shah, G.: Skip graphs. In: SODA (January 2003)
4. Banaei-Kashani, F., Shahabi, C.: SWAM: A family of access methods for similarity-search in peer-to-peer data networks. In: CIKM (November 2004)
5. Bentley, J.: Multidimensional binary search trees used for associative searching. Communications of the ACM 18(9) (1975)
6. Bharambe, A., Agrawal, M., Seshan, S.: Mercury: Supporting scalable multi-attribute range queries. In: ACM SIGCOMM 2004 (August 2004)
7. Chawathe, Y., Ramabhadran, S., Ratnasamy, S., LaMarca, A., Shenker, S., Hellerstein, J.: A Case Study in building layered DHT applications. In: SIGCOMM'05 (August 2005)
8. Gaede, V., Günther, O.: Multidimensional access methods. ACM Computing Surveys, 30(2) (1998)
9. Ganesan, P., Yang, B., Garcia-Molina, H.: One torus to rule them all: Multi-dimensional queries in P2P systems. In: WebDB2004 (2004)
10. Karger, D., Lehman, E., Leighton, T., Panigrah, R., Levine, M., Lewin, D.: Consistent hashing and random trees: Distributed caching protocols for relieving hot spots on the World Wide Web. In: ACM Sympos. Theory of Comp. (May 1997)
11. Karger, D., Ruhl, M.: Simple efficient load balancing algorithms for peer-to-peer systems. In: Voelker, G.M., Shenker, S. (eds.) IPTPS 2004. LNCS, vol. 3279, Springer, Heidelberg (2005)
12. Oppenheimer, D., Albrecht, J., Patterson, D., Vahdat, A.: Design and implementation tradeoffs for wide-area resource discovery. In: 14th IEEE Symposium on High Performance Distributed Computing (HPDC-14) (July 2005)
13. Pugh, W.: Skip lists: A probabilistic alternative to balanced trees. Communications of the ACM (June 1990)
14. Ratnasamy, S., Francis, P., Handley, M., Karp, R., Shenker, S.: A scalable content-addressable network. In: ACM SIGCOMM 2001 (August 2001)
15. Rhea, S., Geels, D., Roscoe, T., Kubiatowicz, J.: Handling churn in a DHT. In: Proceedings of the USENIX Annual Technical Conference (June 2004)
16. Schmidt, C., Parashar, M.: Enabling flexible queries with guarantees in P2P systems. IEEE Internet Computing, 19–26 (May/June 2004)
17. Schütt, T., Schintke, F., Reinefeld, A.: Structured overlay without consistent hashing: Empirical results. In: GP2PC'06 (May 2006)
18. Stoica, I., Morris, R., Kaashoek, M.F., Karger, D., Balakrishnan, H.: Chord: A scalable peer-to-peer lookup service for Internet application. In: ACM SIGCOMM 2001 (August 2001)
19. Zipf, G.: Relative frequency as a determinant of phonetic change. Harvard Studies in Classical Philiology (1929)

Asynchronous Distributed Power Iteration with Gossip-Based Normalization*

Márk Jelasity[1], Geoffrey Canright[2], and Kenth Engø-Monsen[2]

[1] University of Szeged, HAS Research Group on AI, Hungary
[2] Telenor R&D, Fornebu, Norway

Abstract. The dominant eigenvector of matrices defined by weighted links in overlay networks plays an important role in many peer-to-peer applications. Examples include trust management, importance ranking to support search, and virtual coordinate systems to facilitate managing network proximity. Robust and efficient asynchronous distributed algorithms are known only for the case when the dominant eigenvalue is exactly one. We present a fully distributed algorithm for a more general case: non-negative square matrices that have an arbitrary dominant eigenvalue. The basic idea is that we apply a gossip-based aggregation protocol coupled with an asynchronous iteration algorithm, where the gossip component controls the iteration component. The norm of the resulting vector is an unknown finite constant by default; however, it can optionally be set to any desired constant using a third gossip control component. Through extensive simulation results on artificially generated overlay networks and real web traces we demonstrate the correctness, the performance and the fault tolerance of the protocol.

1 Introduction

The calculation of the dominant eigenvector of a matrix has long been a fundamental tool in almost all areas of science. In recent years, eigenvector calculation has found new and important applications in fully distributed environments such as peer-to-peer (P2P) overlay networks.

For example, the PageRank algorithm [1] calculates the *importance ranking* for hyperlinked web pages. The calculated ranks are given by the dominant eigenvector of a matrix that can be derived from the adjacency matrix of the graph defined by the hyperlinks. Fully distributed algorithms have already been proposed to implement PageRank [2,3,4]. As another example, *trust assignment* is a key problem in P2P networks. In [5] a method was proposed, that assigns a global trust value to each peer, through calculating the dominant eigenvector of the matrix containing local (pairwise) trust values. Finally, the eigenvectors that belong to the largest few absolute eigenvalues also play a role in esthetic low

---

* This work was completed while Márk Jelasity was with the University of Bologna. This work was partially supported by the Future and Emerging Technologies unit of the European Commission through Project DELIS (IST-2002-001907).

A.-M. Kermarrec, L. Bougé, and T. Priol (Eds.): Euro-Par 2007, LNCS 4641, pp. 514–525, 2007.

dimensional *graph layout* [6]. This application is relevant in virtual coordinate assignment that allows to map the actual delays among all pairs of nodes onto the distance in the n-dimensional Euclidian space [7].

Motivated by these applications, and firmly believing that new ones will keep emerging, we identify fully distributed eigenvector calculation as an important P2P service that should be studied in its own right.

The environments we target impose special requirements. We assume, that there is a large number of nodes, the connections are volatile and unreliable and the eigenvector needs to be continuously updated and maintained in a decentralized way. Communication is implemented through message passing, where messages can be dropped or delayed. However, nodes have access to a local clock that measures the passage of real time with a reasonable accuracy. We do not assume that the local clocks at different nodes are synchronized.

In this model, we propose a protocol that involves three components. The first is an instantiation of the asynchronous iteration model described in [8]. This algorithm requires that the dominant eigenvalue is exactly one. We extend this protocol with a gossip-based control component that allows the iteration algorithm to converge even if the dominant eigenvalue is less than or greater than one. A third gossip component can be applied to explicitly control the exact value of the vector norm (which is an unspecified finite value without this third component).

These extensions make the asynchronous iteration robust to dynamic change and errors. Traditional methods are very sensitive to the dominant eigenvalue being exactly one: the slightest deviation results in misbehavior on the long run. Besides, the protocol is able to implement algorithms that assume a dominant eigenvalue different from one. A recent promising example is a ranking method using unnormalized web-graphs [9].

We demonstrate the correctness, the performance and the fault tolerance of the protocol through extensive simulation results on artificially generated overlay networks and real web traces.

2   Chaotic Asynchronous Power Iteration

Given a square matrix A, vector x is an *eigenvector* of A with *eigenvalue* λ, if $Ax = \lambda x$. Vector x is a *dominant* eigenvector if there are no other eigenvectors with an eigenvalue larger than $|\lambda|$ in absolute value. In this case λ is a *dominant eigenvalue* and $|\lambda|$ is the *spectral radius* of A.

Motivated by the various application areas mentioned in the Introduction, in this paper we concentrate of the abstract problem of calculating the dominant eigenvector of a weighted neighborhood matrix of some large network, in a fully distributed way. By "fully distributed" we mean the worst case, when the elements of the vector are held by individual network nodes, one vector element per one node. The matrix A is defined by physical or overlay *links* between the network nodes, more precisely, the *weights* assigned to these links: let matrix

```
1: loop {Active Thread}
2:    wait(Δ)                                1: loop {Passive Thread}
3:    for each j ∈ out-neighborsᵢ do         2:    x ← receive(*)
4:        send A_{ji}w_i to j                3:    k ← sender(x)
5:    b_i ← Σ_{k∈in-neighbors_i} b_{ki}      4:    b_{ki} ← x
6:    w_i ← b_i
```

Fig. 1. Asynchronous iteration executed at node i

element A_{ij} be the weight of the link from node j to node i. If there is no link from j to i then $A_{ij} = 0$.

In [8], Lubachevsky and Mitra present a chaotic asynchronous family of message passing algorithms to calculate the dominant eigenvector of a non-negative irreducible matrix, that has a spectral radius of one. Figure 1 shows an instantiation of this framework, that we will apply in this paper.

In the algorithm, the values w_i represent the elements of the vector that converges to the dominant eigenvector. The values b_{ki} are buffered incoming weighted values from incoming neighbors in the graph. These values are not necessarily up-to-date, but, as shown in [8], the only assumption about message failure is that there is a finite upper bound on the age of these values. The age of value b_{ki} is defined by the time that elapsed since k sent the last update successfully received by i. This bound can be very large, so delays and message drop are tolerated extremely well. In addition, the values b_{ki} have to be initialized to be positive.

In dynamic scenarios, when nodes or network links are added or removed, the algorithm is still functional. Temporary node failures, churn, and link failures are all regarded as message failures, and are therefore covered by the assumption of the finite upper bound on update delay. Permanent changes can be dealt with as well: after the change the vector will start converging to the new eigenvector, provided simple measures are taken to make sure nodes remove dead links and take new ones into consideration. Due to lack of space, we omit further details on the dynamic scenarios.

3   Adding Normalization

Let λ_1 be a dominant eigenvalue of A. We can assume that $\lambda_1 \geq 0$ since A was assumed to be non-negative. The asynchronous method described above is known to work correctly if $\lambda_1 = 1$, but if $\lambda_1 > 1$ or $\lambda_1 < 1$, then the vector elements will grow indefinitely or tend to zero, respectively. This motivates us to propose a control component that continuously approximates the *average growth rate* of the vector elements, and normalizes each updated component with this value. Note that after we achieve convergence, the growth rate of every single vector element becomes λ_1. This suggests that approximating the global average using local, limited information is a viable plan.

We adopt the gossip protocol described in [10] to approximate the average growth rate. More precisely, we will use this algorithm to approximate the

```
1: loop {Active Thread}
2:    wait($\Delta_r$)                          1: loop {Passive Thread}
3:    $j \leftarrow$ GETRANDOMPEER()            2:    $r_j \leftarrow$ receive(*)
4:    send $r_i$ to $j$                         3:    send $r_i$ to sender($r_j$)
5:    $r_j \leftarrow$ receive($j$)             4:    $r_i \leftarrow (r_i + r_j)/2$
6:    $r_i \leftarrow (r_i + r_j)/2$
```

Fig. 2. Gossip based averaging protocol executed by node i. The local state of i is denoted as r_i.

geometric mean of the local growth rates $b_i^{(m+1)}/w_i^{(m)}$ over all nodes i, where $b_i^{(m+1)}$ is the value calculated in line 5 in Figure 1 and $w_i^{(m)}$ is the value of w_i before executing line 6. The geometric mean is a more natural choice since we average multiplicative factors.

The averaging protocol (shown in Figure 2) is run by all nodes in parallel with the distributed power iteration. The local state r_i of node i is the current approximation of the average at node i. As a result of the protocol, at all nodes these approximations quickly converge to the average of the initial values of the local approximations. The protocol relies on a *peer sampling service*, accessed by GETRANDOMPEER, that returns a random node from the system. We use NEWSCAST to implement this service, a detailed description can be found in [11]. The details of the NEWSCAST protocol are not required for understanding the present work, the only important aspect is the incurred communication cost. Fortunately, the averaging protocol and newscast can be implemented to use the same connections (sending messages to the same nodes at the same time). This means the NEWSCAST adds no extra cost.

To calculate the geometric mean, each node i, when updating w_i, overwrites the local approximation of the growth rate by the *logarithm* of the locally observed growth rate of the vector element held by the node. That is, node i sets $r_i = \log(b_i^{(m+1)}/w_i^{(m)})$. The approximation of the growth rate is therefore $e^{r_i(t)}$ at node i at time t. This value is used to normalize b_i, that is, we replace line 6 by $w_i = b_i/e^{r_i(t)}$ in the active thread of the iteration algorithm.

A cycle length $\Delta_r < \Delta$ is chosen so that a sufficiently accurate average is calculated, in spite of the continuous updates of r_i external to the averaging protocol. According to preliminary experiments, setting $\Delta_r = \Delta/5$ is already a safe choice on all the problems we examined. This is because, based on the results from [10], the approximation error decreases exponentially fast, besides, the growth rate is similar at all vector elements, as mentioned before.

4   Controlling the Vector Norm

The iteration component combined with gossip-based normalization is designed to achieve convergence, however, the norm of the converged vector is not known in advance. In some applications this might not be sufficient, since interpreting a single vector element becomes impossible, only relative values carry information.

Besides, in scenarios when the matrix A constantly and frequently changes, the vector norm can grow without bounds or can tend to zero without explicitly controlling the vector norm. Finally, knowing a suitable vector norm makes it possible to implement some algorithms that require global knowledge. We will describe the random surfer operator of the PageRank algorithm as an example.

To address these issues, we apply a second gossip component for calculating two measures: the maximum and the average of the absolute value of the vector elements. The maximum is also known as the norm $\|.\|_\infty$, while the average is equal to $\|.\|_1/N$, where N is the dimension of the vector.

To get the maximum, in the protocol in Figure 2 we simply have to replace lines 6 and 4 in the active and passive threads, respectively, with calculating the maximum instead: $r_i \leftarrow \max(r_i, r_j)$.

It must be noted that in the case of norm calculation, $\Delta_r = \Delta/30$ appears to be necessary according to preliminary experiments, since, unlike growth rates, the vector elements themselves are not guaranteed to be similar.

Let us now assume that $n_i(t)$ is the approximation of either the maximum or the average of the vector at node i. To push this value towards one, we propose the following heuristic control factor to modify the normalization factor to introduce a bias towards the vector of which the average or maximum, respectively, is one. Intuitively, if $n_i(t)$ is too large, we decrease the local value a little more, and if it is too low, we increase a little more. More formally, we calculate a factor c as

$$c = e^{r_i(t)} \cdot \left(\frac{0.2}{1 + 1/n_i(t)} + 0.9 \right) \tag{1}$$

and subsequently replace line 6 with $w_i = b_i/c$. The factor c in (1) is a sigmoid function over the logarithm of $n_i(t)$, transformed to have range $[0.9, 1.1]$. This means that the growth rate approximation is never altered by more than 10% no matter how far the average is from one.

As a relatively more complex example of the possibilities this framework offers, we present the implementation of the random surfer operator used in the PageRank algorithm [1]. This operator will in turn allow us to implement PageRank as well.

The PageRank algorithm is concerned with the normalized adjacency matrix of a directed graph (e.g., the WWW link graph). Apart from this directed graph, the PageRank algorithm uses a "random surfer" operator R as well, defined as $R_{ij} = 1/N$, for all $i,j = 1,\ldots,N$, where N is the number of nodes. This corresponds to the definition of R as being a uniform random walk on the fully connected graph (hence the name "random surfer"). The net effect of R is to add a constant weight to each node at each propagation step. In other words, R times any vector gives a vector which is uniform, and whose value may be known if the average of the vector is known [1]. Hence, we can effectively replace matrix A with the PageRank operator $(1 - \epsilon)A + \epsilon R$ where the second term involving R may be known as long as the average of $\mathbf{w}$ is known. Note that ϵ is a parameter of the PageRank algorithm, and defines the weight of the random surfer operator.

As described above, we can in fact obtain an approximation of the vector average. Then we can implement the PageRank R operator—a global operator—using purely local operations: node i now has the update rule

$$w_i = (1 - \epsilon)\frac{b_i}{e^{r_i(t)}} + \epsilon n_i(t) \tag{2}$$

where $n_i(t)$ is the locally known converged approximation of the average at time t.

Finally, note that controlling the average of the vector and applying the random surfer operator can be done simultaneously as well, using the update rule

$$w_i = (1 - \epsilon)\frac{b_i}{c} + \epsilon n_i(t). \tag{3}$$

where c is the same as in (1).

5   Experimental Results

We performed extensive event-based simulation experiments using the PEERSIM simulator developed at the University of Bologna [12]. The goal of the experiments was to demonstrate that our method, is both efficient and robust to failure.

5.1   Notes on the Implementation

In the case of one of the components—the gossip-based protocol that continuously calculates the *average* of the current vector approximation, described in Section 4—we applied two modifications to increase robustness.

First, instead of the protocol in Figure 2, we applied a variant presented in [13]. This variant is very similar; the main difference is that it is slightly modified so that it can apply the "push only" communication model, while the original version is based on the "push-pull" model. In the push model, the nodes only send messages, but need not answer them. In the push-pull model all messages must be answered immediately. We apply the push variant because it is more robust to message delays: while in the push-pull version the state of the nodes are inconsistent for a short time (between the sending and reception of the answer), this problem does not exist in the push version.

The second modification of this component is, that we apply the *restarting* technique described in [10]. According to this technique the computation of the average is performed in consecutive *epochs*. During one epoch the local values of the nodes (r_i) are not allowed to be updated to allow for convergence. The start of the new epoch is fully automatic and distributed. All messages are tagged with an epoch identifier, that is increased when a given node has completed a specified number of cycles. The node also updates its epoch counter if it detects an incoming message that is tagged by a larger epoch identifier than its own. This way the start of the new epoch is propagated very quickly in an epidemic fashion. The length of the epoch is defined as 30 cycles in our experiments. Recall that one cycle is Δ_r time units, and $30\Delta_r = \Delta$.

5.2 Artificially Generated Matrices

For evaluating the protocol we applied a set of artificially generated matrices with controlled properties. To model real applications mentioned in the Introduction, all matrices are sparse and are derived from the adjacency matrix of a link graph. First let us define the graphs that were used to define the matrices.

All graphs have 5000 nodes. The baseline case is a directed random graph, according to the k-out model. In this model, k random out-links are added to each node. We generated an instance of this model with $k = 8$.

The second graph is a scale free graph generated by the Barabási-Albert model [14]. Most importantly, the degree distribution of this graph follows a power law, that is extremely unbalanced with many low degree nodes and a few high degree nodes, and that is known to describe many interesting emergent networks such as the WWW or social relationships [14]. The parameter of the model was set to two. In this case the Barabási-Albert model defines an undirected graph by starting with two disconnected nodes, and subsequently adding nodes one by one, linking each new node to two existing nodes. These two nodes are selected with a probability proportional to their degree (preferential attachment). The average degree in the graph is thus four.

The third graph was generated starting with an undirected ring, and adding two random out-links from all nodes (note that this procedure follows a modified version of the Watts-Strogatz model [15]). The motivation behind using this graph is that, as we will see, its adjacency matrix has a small eigenvalue gap, which results in a slow convergence of the power iteration. This graph was chosen to test whether our method is sensitive to a small eigenvalue gap.

The matrices were derived from the adjacency matrices of these graphs. We note for completeness that the specific instances of the directed graphs we used (the random k-out and the small gap graphs) were all strongly connected. This is not a sufficient condition for convergence but makes it very likely in the case of random graphs. Since our convention in this paper has been that the element A_{ij} describes the weight for network link (j, i)—so that matrix vector multiplication can be defined by sending messages along the outgoing (and not incoming) links—the adjacency matrices were first transposed. The first set of matrices consists of the transposed adjacency matrices. The second set contains the column normalized versions of the matrices in the first set. The normalized versions are such that the weights of the outgoing links sum up to one for all nodes, therefore these matrices describe random walks on the graphs.

Table 1 shows the first two largest magnitude eigenvalues for all the problem instances. Note that the eigenvalue gap (the difference between the first and second largest eigenvalues) determines the convergence speed of the power iteration [16], and thus it is a good indicator of the speed of our method as well. For a small gap, convergence is slow. With a zero gap, the power iteration does not converge at all. Since all matrix elements are real and non-negative, the largest eigenvalue is real and non-negative as well.

Table 1. The first and second largest magnitude eigenvalues. Note that the largest magnitude eigenvalue is guaranteed to be real and positive.

	random k-out		scale free		small gap	
	normalized	unnormalized	normalized	unnormalized	normalized	unnormalized
λ_1	1.0000	8.0000	1.0000	1.3981	1.0000	4.1938
$|\lambda_2|$	0.3573	2.8345	0.8373	1.1737	0.9754	3.9976

5.3   Results

Each experiment was carried out as follows. First, each node i was initialized to have $w_i = 1$ and $b_i = 1$. In the simulations we started each node at a random time within the first Δ time units counted from the first snapshot time t_0.

Two versions of the method were run for each problem. In the first, we do not apply the algorithm described in Section 4. In this case we expect the vector norm to converge to a previously unknown value, given that we do not change the underlying matrices in these experiments. In the second version we do apply vector normalization. In particular, we apply the *maximum* of the vector for this purpose, and therefore we expect the maximum to converge to one.

The evaluation metrics were as follows. We first computed the correct dominant eigenvector ($\mathbf{x}$) using a centralized algorithm. Following general practice in matrix computations, we measured the angle of the actual approximation and the correct vector to characterize convergence. That is, we computed the cosine of the angle $\cos\alpha^{(i)} = \|\mathbf{x}^T\mathbf{w}^{(i)}\|_2 / \|\mathbf{x}\|_2 / \|\mathbf{w}^{(i)}\|_2$, and used the angle $\alpha^{(i)}$ as a metric, which tends to zero during the iteration, as i increases. As a second metric, we measured the maximum of the vector elements to verify normalization.

The failure scenarios involved varying message drop probabilities, and varying message delays. Message drop was modeled by dropping all messages with a given probability, and message delay and delay jitter was modeled by drawing a delay value from a specified interval uniformly, for all messages. Obviously, these settings were applied for all messages sent by any of the components of the protocol equally.

Figure 3 shows the results of the experiments. First of all, even the more moderate failure scenario can be considered pessimistic, not to mention the more severe scenario. This is because in the application scenarios we envision, the interval Δ can be rather long, in the range of ten to thirty seconds, so a delay of 10% of Δ is already large. Most importantly, from the point of view of the averaging and maximum finding protocols, that have a much shorter cycle length of $\Delta_r = \Delta/30$, these delay values are extreme.

From the experiments we can conclude that when the vector norm is not controlled explicitly, then convergence is fast, comparable to that of the centralized power iteration. Our preliminary experiments (not shown) suggest that message delay has virtually no effect on the convergence results, when $P_{drop} = 0$. Higher drop rates slow down convergence but do not change its characteristics significantly.

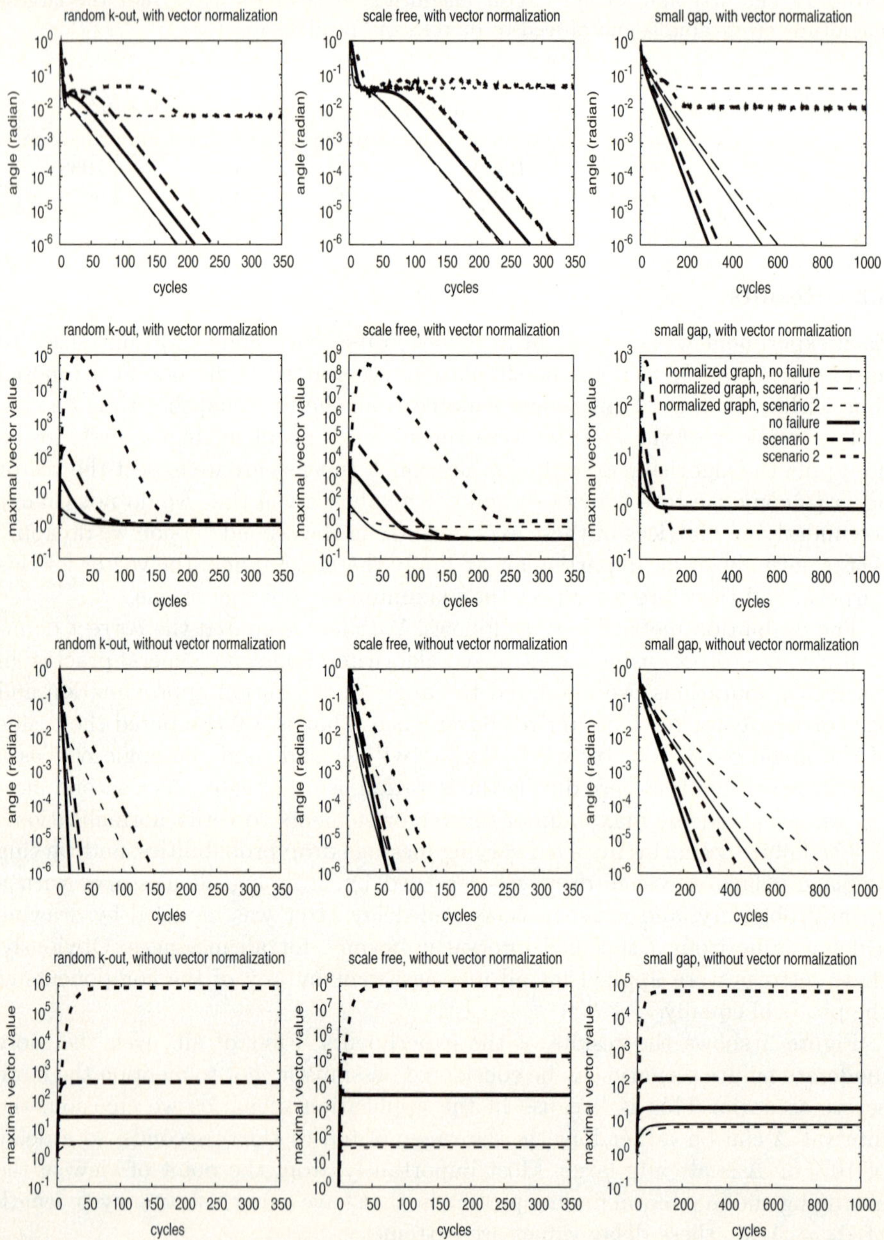

Fig. 3. Simulation results. Scenario 1 involves $P_{drop} = 0.1$, and a random message delay drawn from $[0, \Delta/10]$ uniformly. In scenario 2, $P_{drop} = 0.3$ and the message delay is drawn from $[\Delta/10, \Delta/2]$.

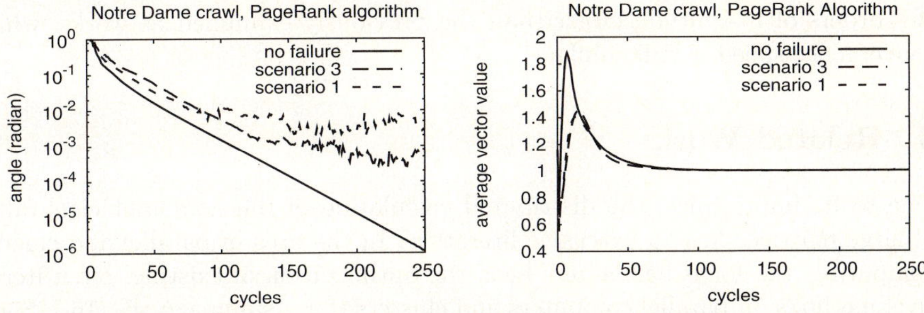

Fig. 4. Simulation results with the PageRank algorithm. Scenario 1 involves $P_{drop} = 0.1$, and a random message delay drawn from $[0, \Delta/10]$ uniformly. In scenario 3, $P_{drop} = 0$ and the message delay is the same as in scenario 1.

When we do apply vector normalization, convergence slows down somewhat due to the interference of vector normalization and asynchronous iteration. In the extreme failure scenario we don't achieve full convergence. The reason is that the extremely high delay and message drop rate prevents the propagation of the current maximum of the vector to all nodes during the interval Δ, and so different nodes might normalize with a different value. As a side effect, the maximum does not converge to one, and there is a constant noise factor in the approximation of the eigenvector. However, in the less severe, but still pessimistic scenario we do achieve convergence.

5.4   PageRank on WWW Crawl Data

As a more realistic case study, we tested our method on the same dataset used in [17], available from the authors. It was generated by a crawler, starting from one page within the domain of the University of Notre Dame. This sample has 325729 nodes. On this dataset we executed the PageRank algorithm, as described in Section 4. The weight of the random surfer operator was $\epsilon = 0.2$. This way, for the complete linear operator of the PageRank algorithm, we have $\lambda_1 = 0.84648$, $\lambda_2 = 0.8$.

The results of the method are shown in Figure 4 in various failure scenarios. We can observe that the protocol is now more sensitive to failure than in the case of the previous experiments, although the achieved accuracy is still satisfactory (note the logarithmic scale of the plots). The reason is that in this case the vector average is used for controlling the norm of the vector, that is, it is guaranteed that the average of the vector stays one. The average is used to implement the random surfer operator as well. However, the calculation of the average is more sensitive to failure than the calculation of the maximum. This way, the approximation of the actual average of the vector has a small noise factor, that is inherited by the approximation of the ranks.

We can also note that the protocol scales well: the network examined here is two orders of magnitude larger than the previously examined networks, while convergence speed is still similar.

6   Related Work

Due to its importance, the distributed calculation of the dominant eigenvalue of large matrices has an extensive literature. In the area of parallel and cluster computing, the focus has largely been the optimization of existing, often iterative, methods on parallel computers and clusters (for a summary, see [16]). Such optimizations include partitioning; for example, different parts of the vector can be freely assigned to different processors in order to minimize message exchange and to maximize speedup. Besides, due to the reliable computing platform, synchronization can be efficiently implemented. This model is radically different from ours: in our case the assignment is fixed and given *a priory*, and the main goal is to achieve robustness to high rates of message delivery failures.

Asynchronous protocols have also been proposed for implementing iterative methods, and important convergence results are available as well (see [18] for a summary). These protocols are extremely fault tolerant and also efficient, but so far no algorithms are known that can deal with the case when the dominant eigenvalue is different from one. This introduces a certain sensitivity to dynamic environments even if $\lambda_1 \approx 1$, besides, many interesting applications where $\lambda_1 \neq 1$ cannot be tackled, for example [9].

Finally, in the context of P2P systems the main focus is on distributed PageRank implementations, where in all cases $\lambda_1 = 1$ is assumed, for example, [2,4,3]. The EigenTrust protocol in [5] also applies a similar implementation, but the authors assume all values are updated in each round, presumably unaware of the advantages of the long existing asynchronous version of the protocol, and thereby offering a rather fragile algorithm.

7   Conclusion

In this paper we have addressed the problem of designing a fully distributed and robust algorithm for finding the dominant eigenvector of large and sparse matrices, that are represented as weights of links between nodes of a network. Our contribution can be summarized as follows. First of all, our algorithm does not require the dominant eigenvalue to be one. This is an important feature even if the problem involves a dominant eigenvalue of one (like PageRank does). In PageRank, sophisticated techniques for "fixing" the graph are required to make sure the dominant eigenvalue is one, which are not needed in our case, as we demonstrated. Besides, the protocol opens the door for applications where the dominant eigenvalue is known to be different from one [9].

Second, the norm of the approximation of the dominant eigenvector can be controlled as well. In other words, in addition to guaranteeing that the norm of the vector converges to a finite value, we can define this value explicitly using an additional gossip-component. This also means that the algorithm can be run indefinitely in a continuously changing environment.

Finally, we demonstrated the robustness of the algorithm through event-based simulation experiments, both on artificially generated graphs and on web-crawl data.

References

1. Page, L., Brin, S., Motwani, R., Winograd, T.: The pagerank citation ranking: Bringing order to the web. Technical report, Stanford Digital Library Technologies Project (1998)
2. Sankaralingam, K., Sethumadhavan, S., Browne, J.C.: Distributed pagerank for p2p systems. In: Proc. HPDC-12, pp. 58–69 (2003)
3. Shi, S., Yu, J., Yang, G., Wang, D.: Distributed page ranking in structured p2p networks. In: Proc. ICPP03, pp. 179–186 (October 2003)
4. Parreira, J.X., Donato, D., Michel, S., Weikum, G.: Efficient and decentralized PageRank approximation in a peer-to-peer web search network. In: Proc. VLDB, pp. 415–426 (2006)
5. Kamvar, S.D., Schlosser, M.T., Garcia-Molina, H.: The eigentrust algorithm for reputation management in p2p networks. In: Proc. WWW, ACM Press, New York (2003)
6. Koren, Y.: On spectral graph drawing. In: Warnow, T.J., Zhu, B. (eds.) COCOON 2003. LNCS, vol. 2697, pp. 496–508. Springer, Heidelberg (2003)
7. Dabek, F., Cox, R., Kaashoek, F., Morris, R.: Vivaldi: A decentralized network coordinate system. In: Proc. SIGCOMM 2004, ACM Press, New York (2004)
8. Lubachevsky, B., Mitra, D.: A chaotic asynchronous algorithm for computing the fixed point of a nonnegative matrix of unit radius. J. of the ACM 33(1), 130–150 (1986)
9. Burgess, M., Canright, G., Engø-Monsen, K.: Importance-ranking functions derived from the eigenvectors of directed graphs. Technical Report DELIS-TR-0325, DELIS Project (2006)
10. Jelasity, M., Montresor, A., Babaoglu, O.: Gossip-based aggregation in large dynamic networks. ACM Transactions on Computer Systems 23(3), 219–252 (2005)
11. Jelasity, M., Guerraoui, R., Kermarrec, A.M., van Steen, M.: The peer sampling service: Experimental evaluation of unstructured gossip-based implementations. In: Jacobsen, H.A. (ed.) Middleware 2004. LNCS, vol. 3231, pp. 79–98. Springer, Heidelberg (2004)
12. PeerSim: http://peersim.sourceforge.net/
13. Kempe, D., Dobra, A., Gehrke, J.: Gossip-based computation of aggregate information. In: Proc. FOCS'03, pp. 482–491. IEEE Computer Society Press, Los Alamitos (2003)
14. Albert, R., Barabási, A.L.: Statistical mechanics of complex networks. Reviews of Modern Physics 74(1), 47–97 (2002)
15. Watts, D.J., Strogatz, S.H.: Collective dynamics of 'small-world' networks. Nature 393, 440–442 (1998)
16. Bai, Z., Demmel, J., Dongarra, J., Ruhe, A., van der Vorst, H. (eds.): Templates for the Solution of Algebraic Eigenvalue Problems: a Practical Guide, SIAM, Philadelphia (2000)
17. Albert, R., Jeong, H., Barabási, A.L.: Diameter of the world wide web. Nature 401, 130–131 (1999)
18. Frommer, A., Szyld, D.B.: On asynchronous iterations. Journal of Computational and Applied Mathematics 123(1-2), 201–216 (2000)

Capitalizing on Free Riders in P2P Networks

Yuh-Jzer Joung[1], Terry Hui-Ye Chiu[1,2], and Shy Min Chen[1,3]

[1] Dept. of Information Management,
National Taiwan University, Taipei, Taiwan
[2] Dept. of Industrial Engineering and Management,
MingChi Institute of Technology, Taishan, Taipei, Taiwan
[3] Dept. of Business and Management, MingChi Institute of Technology,
Taishan, Taipei, Taiwan

Abstract. Free riding is a common phenomenon in P2P networks. Although several mechanisms have been proposed to handle free riding—mostly to exclude free riders, few of them have actually been adopted in practical systems. This may be attributed to the fact that they are nontrivial, and that completely eliminating free riders could jeopardize the sheer power created by the huge volume of participants in a P2P network. Rather than excluding free riders, in this paper we incorporate and utilize them to provide global index service to the files shared by other peers. The simulation results indicate that our model can significantly boost the search efficiency of a plain Gnutella, and the model is quite resistant to system churn rate and the ratio of free riding.

1 Introduction

Free riding is a common phenomenon in peer-to-peer (P2P) systems. The study by Adar et al. [1] in 2000 found that 66 percent of peers in Gnutella were free riders. Moreover, the top 1 percent of peers were responsible for 47 percent of requests, while the top 25 percent of peers provided for 98 percent. Follow up study by Hughes et al. [2] in 2005 found that free riders had increased to 85 percent, whereas the top 1 percent of peers were responsible for 50 percent of requests and the top 25 percent of peers still provided for 98 percent of requests.

Most people believe that free riding may degrade system performance and add vulnerability, possibly corruption, to the system. Therefore, several mechanisms have been proposed to handle free riding. These include *payment-based* [3,4], where each resource has a price so that a peer needs to "pay" for downloading the resource; *reciprocal-based* [5,6], where a peer's ability to consume resources is based on its contribution to the system;[1] *reputation-based* [7,8], where a peer's ability to consume resources is based on it's reputation, which is evaluated by other peers based on past interactions with the peer; and *punishment* [9], where

---

[1] In addition, BitTorrent's "tit-for-tat" peer selection strategy, which has been proven quite successfully in limiting free riding, may also be classified into this category. The BitTorrent platform, however, is very different from the Gnutella-like unstructured P2P networks we will be studying in the paper.

A.-M. Kermarrec, L. Bougé, and T. Priol (Eds.): Euro-Par 2007, LNCS 4641, pp. 526–536, 2007.

a free rider's request is eventually ignored to force it to disconnect from the network.

However, few of these proposals have been implemented in a practical system. For example, Gnutella, which has attracted more than two million users by Dec. 2005, is still reluctant to implement any anti-free-riding measures. In fact, the above findings on Gnutella also show that even though the level of free riding has increased, the system can still sustain it. Moreover, even though free riders do exist, there are always some peers willing to share due to an altruistic or the so called *"glow worm"* effect. This has been supported by an empirical study on human behavior conducted by Gu et. al [10]. Another reason may be attributed to that existing measures for combatting free riders often have high overhead on communication, storage, and/or computation, and some even require an external centralized authority.

By observing human behaviors, we believe that free riders may always exist regardless of any action taken against them. On the other hand, a P2P network is often appraised by the sheer power created by the huge volume of its participants (for example, such a network is particularly useful for disseminating information, for, say, marketing and advertising). Completely eliminating free riders could jeopardize its power. Therefore, why not to make free riders part of the network and try to utilize them, rather than expelling them from the network?

Based on the studies in [1,2,11,12], we can see that free riders have the following characteristics: (a) short connection time, (b) often requesting popular files, and (c) obviously, they do not like to share—if they have a choice. This motivates us to think of making them to contribute what they cannot control (e.g., the platform and the protocol) and, usually, do not care much.

In this paper we propose a Gnutella-like unstructured P2P network that utilizes free riders (those that share none) to serve as indexing nodes to normal peers (those that share some files). Since free riders often request popular files, they can easily point to the peers that have the target files one is looking for. Free riders' short connection time can be overcome by the large number of their pals. Our simulation results indicate that the network can significantly boost the search efficiency of a plain Gnutella. Moreover, routing and search costs are effectively shifted to free riders and, with a proper setting, the search success rate of normal peers can be much higher than that of free riders—all of these provide some incentive for peers to contribute files to the network. More importantly, our network is quite resistant to system churn rate and the increasing ratio of free riding. Below we present our system.

2   The System

2.1   Neighbor Tables

Like ordinary P2P networks, each peer u in our model maintains a *neighbor table* of contact information about peers to which u may forward a query. The contact information consists of node IP and port. We say that v is a *neighbor* of u if u maintains v's contact information in its neighbor table.

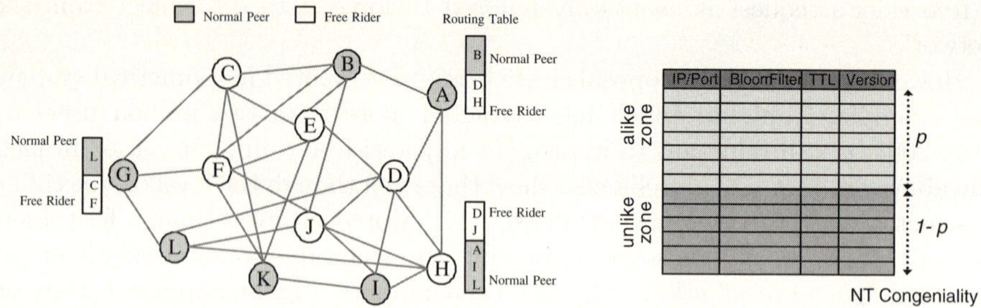

Fig. 1. Overview of network architecture (left) and the neighbor table structure (right)

In addition, each entry in the neighbor table also contains a Bloom filter [13,14]
summarizing the files shared by the corresponding peer (see Fig. 1). With the
Bloom filters in the neighbor table, a peer receiving a query can check if any peer
it maintains in the neighbor table may have the queried file. If so, the query can
be sent directly to the peer for a final check. Otherwise, the query will be resolved
using our search mechanism that is to be described shortly in Section 2.3. The
fields TTL and Version in a neighbor table entry are for maintenance of the
entry. This will also be clear shortly.

A more important role of the Bloom filters is for a peer to select neighbors
in its neighbor table, which allows some form of clusters to be established in
the network, thereby speeding up search process. For our purpose, a peer u's
neighbor table is partitioned into two zones: *alike* and *unlike*. Intuitively, the
alike zone is for peers that are "akin" to u in terms of whether or not they wish
to share, while the unlike zone is for peers that are "different" from u. Thus, a
normal peer's neighbors in the alike zone are all normal peers, and its neighbors
in the unlike zone are all free riders. On the other hand, a free rider's neighbors
in the alike zone are all free riders, while its neighbors in the unlike zone are all
normal peers.

When there are more candidates than a zone can fit, a replacement scheme
is needed to help select better neighbors in the zone. There are several possible
schemes. Here we select peers based on how "similar" their Bloom filters are
(i.e., based on the keywords they have in their files). Specifically, let B_1 and
B_2 be two Bloom filters, and let $B_1 \wedge B_2$ denote the Bloom filter B such that
$B[i] = B_1[i] \wedge B_2[i]$. Then

$$sim(B_1, B_2) = \left| \{i \mid B[i] = 1, \text{where } B = B_1 \wedge B_2\} \right|$$

That is, $sim(B_1, B_2)$ is the number of bit-1 the two Bloom filters have in common.

The replacement scheme for the alike zone is as follows: u prefers v to w in
filling u's alike zone if their Bloom filters B_u, B_v satisfy the following condition:
$sim(B_u, B_v) > sim(B_u, B_w)$. If $sim(B_u, B_v) = sim(B_u, B_w)$, then the one that
is *least recently used (LRU)* is replaced. In addition, we require that a peer v
can fill in u's alike zone only if $One(B_u) > 0 \Rightarrow One(B_v) > 0$, and $One(B_u) =$

$0 \Rightarrow One(B_v) = 0$, where $One(B)$ denotes the number of bit-1 in B. Because a normal peer's Bloom filter is a nonzero vector, the condition implies that only normal peers can fill in a normal peer's alike zone, and only free riders can fill in a free rider's alike zone.

On the other hand, a peer that cannot fill in u's alike zone is a candidate for u's unlike zone, and if there is more than one candidate, the replacement scheme is simply based on LRU. Note that by the criterion given for the alike zone, only free riders can fill in a normal peer's unlike zone, and only normal peers can fill in a free rider's unlike zone.

We call the percentage p of a neighbor table that is allocated to the alike zone the *NT congeniality*. Thus, $1 - p$ is the percentage of the unlike zone in a neighbor table.

2.2   Network Construction and Maintenance

Our network, as shown in Fig. 1, is a typical Gnutella like unstructured P2P network. A peer u that wishes to join the network first prepares a Bloom filter to summarize the files it wishes to share. Then it sends PING-PONG messages via some hook-up node to the network. Like the Gnutella protocol, the purpose of the messages is to learn a list L of nodes in the network. ¿From the list, u can build its neighbor table according to the rules described in Section 2.1, thereby connecting itself to the network. Note that during the operation of the network, u may also learn more peers in the network, e.g., when other peers join the network and send PING-PONG messages to u, or when u involves in a search process in which query messages from other nodes arrive at u. The new peer information allows u to modify its neighbor table entries so as to improve search efficiency. Again, neighbor table entries are replaced according to the rules described in Section 2.1.

Because a peer may leave the network without notifying other peers that maintain it as a neighbor, neighbor table entries may become stale. Moreover, when peers change the files they wish to share, their Bloom filters must also be updated. So in the network each peer must periodically refresh its neighbor table entries. The TTL field in a neighbor table entry (see Fig. 1) allows the owner of the entry to determine when the entry needs to be verified with the peer specified in the entry. The version field allows both peers to verify if the Bloom filter in the entry is up to date.

2.3   Search Scheme

The basic idea of our system is to let free riders collectively provide index service to normal peers. To do so, we recall that by the neighbor selection mechanism, a normal peer will have normal peers in its alike zone and free riders in its unlike zone; while a free rider will have free riders in its alike zone and normal peers in its unlike zone. If the size of unlike zone is relatively large compared to the size of alike zone, then a normal peer connects to a lot of free riders, while each of the free riders connects to a lot of normal peers. By storing normal peers'

Bloom filters at free riders, the free riders can collectively provide index service to normal peers.

To make use of the index service, the search scheme is simply as follows: Let u be a peer that receives a query. If u has the target file, then it returns the file to the querier and the search process ends. Otherwise, u checks the Bloom filters in its neighbor table to see if any of its neighboring peers may have the file. If so, it forwards the query message to the neighbor, say v, to confirm this. If v does have the file, it sends the result to u, which in turn forwards the result to the querier and then ends the search process. Note that v will not forward the query message for u if it does not have the file. In general, in the absence of false positives, normal peers do not involve in the search process unless they have the target files.

If none of u's neighboring peers has the queried file, u forwards the query message to all free riders in its neighbor table, and waits for the result. Each of u's free rider neighbors, upon receiving the query message from u, processes the query in the same way as u does, and so on, until a certain depth of search path has been reached. If the target file is located at some peer w, then w returns the file along the search path back to u that initiates the query. Each peer in the returning path learns of w (and its Bloom filter), and may add u to its neighbor table if w can replace an existing entry.

3   Experimental Result

3.1   Evaluation Metrics

Observe that if objects are randomly distributed to a network of N nodes, then the probability for locating a unique object o from a randomly selected set of P nodes is $\frac{P}{N}$. By comparing to this random search, we can evaluate whether a system or a search mechanism can offer a better success rate than a blind search. Therefore, we define a metric called *search condensity* as follows: Let $\mathcal{A}$ be a search mechanism over a network of N nodes. Suppose that the search success rate is s when the search space size (number of nodes visited) is P. Then the search condensity SC of $\mathcal{A}$ is defined by

$$SC = \frac{s}{\frac{P}{N}} = s \times \frac{N}{P}$$

That is, SC is a measure of how effective a mechanism can "boost" the search success rate of a network as compared to a random blind search. Note that SC may vary with P and N, and so it is sometimes more meaningful to write SC as a function of P/N, i.e., $SC(P/N)$. Also note that $SC(1) = 1$.

The second parameter is the workload for a peer to process a query. We measure the workload of a peer by the number of query messages it has received during some observation interval.

The third parameter is to measure the duplication ratio of query messages. Observe that because search in an unstructured network is more or less a blind

Table 1. Statistics of the dataset obtained from the Boeing Proxy Log

owner	owns	max	min	avg
Peer	File	103,880	1	87.28
Peer	Keyword	981	1	35.87
File	Keyword	137	1	4.47

process, duplicated query messages may arrive at the same node. The *query duplication ratio* measures how many query messages are duplicated during a search process. It is defined as follows: Let P be the search space size of a query, and let M be the total number of query messages sent for the query. Then the query duplication ratio is defined by $\frac{M}{P}$.

3.2 Dataset

Since real P2P dataset is hard to obtain, we use Web proxy logs collected from Boeing[2] to simulate file distribution in a P2P environment. The logs basically record which clients have requested which URLs. We use the file path in a log to represent a queried file owned by the host IP in the URL. The dataset we used consists of 3,761,054 URLs belonging to 31,743 Web hosts, which then allows us to model 3,761,054 files distributed over 31,743 peers.

Because query in our system is by keywords, we need to extract keywords from each queried file to represent the file. To do so, we extracted words from the file path using the following characters '/', '?', '-', '_', and '&' as delimiters. After the process, each word may contain some non-alphanumeric characters such as '%' and could have a length more than 20. For simplicity, we further removed non-alphanumeric characters, and truncated words that have length more than 20. The remaining words are then treated as keywords associated with the file. To avoid having a long Bloom filter, a host that has more than 1,000 keywords in its files was also removed from our dataset. In total, 61 out of 31,743 hosts were removed. Table 1 gives some statistics about the dataset. Moreover, on average the peer support of a file (number of peers having the file) is nearly 1.

3.3 Simulation Settings

Our simulator is written in Java. It uses a single-process event-driven architecture to simulate concurrent activities of peers, such as joining, querying, and message delivery. All time measured in the simulation will be relatively to the time unit of the simulator (which is calibrated to 0.1 second.) The default simulation settings are shown in Table 2.

To simulate the dynamics of P2P networks, each peer will be given a session time when it connects to the network. We use the method presented in [15] to model peer session time so that peers' behaviors are close to the empirical

---

[2] See `http://www.web-caching.com/traces-logs.html`

Table 2. Default simulation settings

Parameter	Default Value
Network Size	50,000
Free Rider Ratio	0.85
Routing Table Size	20
Bloom Filter Size	8,000
Median Session Time	36,000
Average Message Delay	2.84
Search TTL	3
Query Frequency	0.5 per time unit

results studied in [16]. Moreover, free riders typically have shorter session time than normal peers [11]. Our network dynamics model will also reflect this.

In the simulation, when we need a network of size N, the network is constructed and maintained as follows. Initially, we let N nodes concurrently join the network. After the join process is completed, we start to time each peer's session, and remove peers from the network when their sessions end. To keep the network size roughly stable, when a peer u leaves the network, a random peer that is currently offline is "awakened" to rejoin the network. Similarly, u becomes offline until it is awakened by another peer. Note that when a peer first joins the network, it will randomly contact 3-5 hook-up nodes to build its neighbor table. When a peer leaves and rejoins the network, it refreshes its neighbor table by first checking if its previous neighbors are still alive, and also randomly contacts 3-5 hook-up nodes to find new peers that may fill in its neighbor table. Entries in the neighbor table are replaced in the same way as described in Section 2.1.

Free riders are chosen randomly from the network according to their population. For normal peers, we let each node randomly map to a unique Web host in our Boeing dataset so that the nodes simulate normal peers that own some files. All querying activities are performed after the join process is completed. To simulate the querying process, we randomly select an alive peer in the network as the originator of the query. The query target is again randomly selected from some existing file, and is represented by the keyword set of the selected file.

3.4 Performance Evaluation

We measure the performance of the system from various perspectives. Note that it takes time for each node to learn of other peers to fill in its routing table so as to meet the peer selection requirement in its alike and unlike zones. According to our preliminary test, for most parameters we shall be studying, the system becomes stable approximately after one query has been issued per peer. So, unless stated, otherwise all measurements in the experiments are taken after 50,000 queries have been issued to the system.

We first measure search condensity with respect to NT congeniality p. We vary $p = 0.0$ to 1.0 on a scale of 0.1. The results are shown in Fig. 2(a). We see that except for $p = 0.0$, small p generally implies high search condensity. Observe that search condensity is a measure of how "effective" query messages are issued to search the target. In general, high search condensity may be obtained when the search space size is small. However, search success rate may also be compromised due to this small search space. In Fig. 2(b) and (c) we have also drawn the success rate and the actual search space (the number of peers that have been queried per search) measured in the experiment for each p. We see that the overall success rate reaches its high 0.58 when $p = 0.5$, but drops to 0.31 when $p = 0.1$, and to 0.44 when $p = 0.9$ on the other side. The actual search space size, however, grows in polynomial with p. To see the trend of a search space size, recall that in our search scheme, query messages will only be forwarded to free riders to expand the search scope. So when a normal peer initiates a query, if the neighboring normal peers do not have the target file, then $(1 - p)n$ queries will be forwarded to the neighboring free riders, where n is the neighbor table size. In contrast, if a query is initiated at a free rider, then $p \times n$ queries may be forwarded to its neighboring free riders. For TTL=3, a normal peer's query may expand to $(1 - p)p^2 n^3$ peers, while a free rider's query may expand to $p^3 n^3$ peers. Taking the percentage of free riders and normal peers into account, on average a query should have a search space of size

$$(0.15(1 - p) + 0.85p)p^2 n^3. \tag{1}$$

So the size grows in polynomial with p. By combining (b) and (c), it is not surprising to see that search condensity drops almost in polynomial with p.

Note that $p = 0$ and $p = 1$ represent two extreme cases in our model. When $p = 0$, a normal peer's neighbors are all free riders, while a free rider's neighbors are all normal peers. So a query initiated by a normal peer will be forwarded to its neighboring free riders. If they happen to have indexed the target file, the search succeeds; otherwise, the search fails as no more free riders can be used to forward the query. On the other hand, if a query is initiated by a free rider, then the query is resolved if the free rider's neighboring normal peers have the target file; otherwise the search fails as the free rider has no neighboring free riders to forward the query. For $p = 1$, the situation is rather opposite. A normal peer's neighbors are all normal peers, while a free rider's neighbors are all free riders. So a query initiated by a normal peer can be resolved if the peer or its neighbors happen to have the target file; and a query initiated by a free rider will have no chance to be resolved, as all peers within the search range are free riders and they do not index any normal peers.

In Fig. 2(b), we have also identified the search success rate of normal peers and free riders. We see that when $p \leq 0.5$, normal peers have higher search success rate than free riders, and the difference is significant when p is small. Moreover, when $p = 0.2 - 0.5$, normal peers' search success rate is even higher than the maximum overall success rate. This suggests that when setting NT congeniality to these values, our mechanism can also create some incentive to normal peers, as they will have a higher search success rate. All together, we see that $p = 0.2$ appears to be an attractive setting, as it can yield quite high search condensity (206.6) without compromising too much in search success rate for normal peers.

The search space size in Eq. (1) assumes that no two peers within the search range have a common neighbor. Otherwise, the actual search space will be

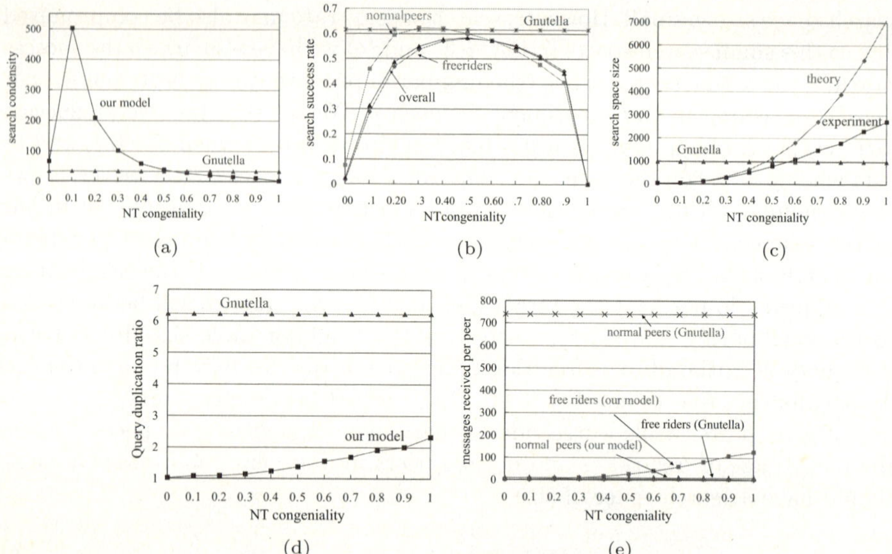

Fig. 2. NT congeniality vs. (a) search condensity; (b) success rate; (c) search space size; (d) query duplication ratio; and (e) workload

smaller, and query message duplication ratio will increase. From Fig. 2(c), it can be expected that the query duplication ratio will increase as p increases. The measured query duplication ratio is shown in Fig. 2(d).

The workloads of normal peers and free riders are shown in Fig. 2(e). The workload is measured by the number of query messages received per normal peer/free rider for every consecutive 1,000 queries issued to the system. We see that the workload of free riders grows in polynomial with p, while the workload of normal peers is barely observable. This is because in our search scheme, a normal peer x can receive a query forwarded from another peer y only when x's Bloom filter maintained by y indicates that x may have the target file. So the workload of normal peers is solely due to a successful search or false positive of the Bloom filter. On the other hand, free riders are responsible for forwarding and processing queries. As shown in Fig. 2(c), the search space grows in polynomial with p. Therefore, the workload of free riders also grows in polynomial with p.

For comparison, the search performance of the plain Gnutella is also shown in the figures. Since in the plain Gnutella a peer selects its neighbors based only on LRU, the NP congeniality parameter has no effect on the system performance. So all measured values remain constant at the y-axis. In Fig. 2(a), Gnutella's search condensity is about 32.5, which is considerably low as compared to our scheme at small p. Note that because a successful search will also cause all peers on the query-result returning path to update their neighbor tables, search in the plain Gnutella is not exactly a random search. So its search condensity is higher than 1. In Fig. 2(b), Gnutella's search success rate is about 0.61, which is slightly higher than ours at small p. (There is virtually no difference between normal peers and free riders in the success rate.) However, this high success rate is at the cost of large search space size. For example, in Fig. 2(c), Gnutella's search space is 958, as opposed to 118 needed in our scheme at $p = 0.2$. Moreover, Gnutella's query duplication ratio in Fig. 2(d) is considerably high. In Fig. 2(e), the workload of normal peers in Gnutella is also very high, while there is virtually no workload to free riders. This is because as time goes by, both normal peers and free riders will maintain only normal peers as their neighbors in their routing tables, thus yielding all query loads to normal peers. Because there are only 15% of normal peers, each normal peer may be connected to by a number of peers, and hence the high query duplication ratio.

4   Concluding Remarks

We have proposed a Gnutella-like unstructured P2P network that utilizes free riders to serve as indexing nodes to normal peers. To do so, when building the network, we partitioned a peer's neighbors into two groups: one that is "congenial" to the peer, and the other that has an opposite characteristic. As a result, a normal peer's congenial neighbors are normal peers, while its uncongenial neighbors are free riders. On the other hand, a free rider's congenial neighbors are free riders, while its uncongenial neighbors are normal peers. In addition, we used Bloom filters to summarize the files of the peers, and let each peer keep a copy of its neighboring peer's Bloom filter. Let p be the portion of the congenial neighbors within a peer's neighbor table. Then when p is small, free riders collectively can provide index service to the files shared by normal peers. Likewise, the search scheme is simply to forward queries to free riders, as they have more information about which normal peers may have the target files. This effectively shifts the routing and search costs to free riders. Hence, a peer either contributes some files to other peers, or provides indexing service to them. Note that the indexing service is part of the network protocol, and so it is out of a peer's control—unless it wishes to hack the protocol. Security threats are a general problem in P2P networks, and they are beyond the scope of the paper.

Our simulation results indicate that with a network of size 50,000, 85% of free riders, a neighbor table of size 20, and a search TTL of 3, setting $p = 0.2$ can significantly boost the search condensity of a plain Gnutella—206.6 vs. 32.5! Moreover, the search success rate of normal peers is also higher than that of free riders (0.59 vs. 0.47), thereby offering some incentive to normal peers. More importantly, our network is quite resistant to system churn rate and the increasing ratio of free riders (although, due to space limitation, this part of simulation result is not shown here). In fact, the higher the free riding ratio, the more the number of peers to provide indexing service, and therefore the higher the search success rate and search condensity.

In our model, we distinguish normal peers and free riders simply based on whether or not they have contributed any file to the network. Although this is a typical distinguishing criterion, a more practical (but sophisticated) approach is to judge a peer based on its real contribution to the others: for example, how many files it offers, how many of them have actually been requested, and how frequent of the requests. In the future we wish to modify our system so that a peer's search efficiency and workload can truly reflect how much it actually contributes.

References

1. Adar, E., Huberman, B.A.: Free riding on Gnutella. First Monday 5(10) (October, 2000), Available from `http://firstmonday.org/issues/issue5_10/adar/index.html`
2. Hughes, D., Coulson, G., Walkerdine, J.: Free riding on Gnutella revisited: the bell tolls? IEEE Distributed Systems Online 6(6) (2005)
3. Golle, P., Leyton-Brown, K., Mironov, I.: Incentives for sharing in peer-to-peer networks. In: ACM EC, pp. 264–267. ACM Press, New York (2001)
4. Figueiredo, D., Shapiro, J., Towsley, D.: Incentives to promote availability in peer-to-peer anonymity systems. In: ICNp, 110–121 (2005)
5. Vishnumurthy, V., Chandrakumar, S., Sirer, E.G.: KARMA: A secure economic framework for peer-to-peer resource sharing. In: Proceedings of the 1st Workshop on Economics of Peer-to-Peer Systems (2003)
6. Kamvar, S.D., Schlosser, M.T., Garcia-Molina, H.: Incentives for combatting freeriding on P2P networks. In: Kosch, H., Böszörményi, L., Hellwagner, H. (eds.) Euro-Par 2003. LNCS, vol. 2790, Springer, Heidelberg (2003)
7. Dutta, D., Goel, A., Govindan, R., Zhang, H.: The design of a distributed rating scheme for peer-to-peer systems. In: Proceedings of the 1st Workshop on Economics of Peer-to-Peer Systems (2003)
8. Hales, D.: From selfish nodes to cooperative networks - emergent link-based incentives in peer-to-peer networks. In: P2P, 151–158 (2004)
9. Karakaya, M., Korpeoglu, I., Ulusoy, Ö.: A distributed and measurement-based framework against free riding in peer-to-peer networks. P2P, 276–277 (2004)
10. Gu, B., Jarvenpaa, S.: Are contributions to P2P technical forums private or public goods? – An empirical investigation. In: Proceedings of the 1st Workshop on Economics of P2P Systems (2003)
11. Yang, M., Zhang, Z., Li, X., Dai, Y.: An empirical study of free-riding behavior in the maze P2P file-sharing system. In: Castro, M., van Renesse, R. (eds.) IPTPS 2005. LNCS, vol. 3640, pp. 182–192. Springer, Heidelberg (2005)
12. Beverly, I.V., R.E.: Reorganization in network regions for optimality and fairness. Master's thesis, MIT EECS (2004)
13. Bloom, B.H.: Space/time trade-offs in hash coding with allowable errors. CACM 13(7), 422–426 (1970)
14. Cheng, A.H., Joung, Y.J.: Probabilistic file indexing and searching in unstructured peer-to-peer networks. Computer Networks 50(1), 106–127 (2006)
15. Joung, Y.J., Wang, J.C.: Chord^2: A two-layer chord for reducing maintenance overhead via heterogeneity. Computer Networks 51(3), 712–731 (2007)
16. Saroiu, S., Gummadi, P.K., Gribble, S.D.: A measurement study of peer-to-peer file sharing systems. In: MMCN (2002)

Content-Based Publish/Subscribe
Using Distributed R-Trees

Silvia Bianchi[1], Pascal Felber[1], and Maria Gradinariu[2]

[1] University of Neuchâtel, Switzerland
[2] LIP6, INRIA-Université Paris 6, France

Abstract. Publish/subscribe systems provide a useful paradigm for selective data dissemination and most of the complexity related to addressing and routing is encapsulated within the network infrastructure. The challenge of such systems is to organize the peers so as to best match the interests of the consumers, minimizing false positives and avoiding false negatives.

In this paper, we propose and evaluate the use of R-trees for organizing the peers of a content-based routing network. We adapt three well-known variants of R-trees to the content dissemination problem striving to minimize the occurrence of false positives while avoiding false negatives. The effectiveness and accuracy of each structure is analyzed by extensive simulations.

1   Introduction

Publish/subscribe is an appealing communication primitive for large scale dynamic networks due to the loosely coupled interaction between the publishers and subscribers. In this paradigm, publishers produce *events* and subscribers express their interests through *subscriptions*; any *event* matching the subscription is delivered to the corresponding subscriber. The matching procedure is performed by brokers, which are also responsible for the event delivery. In this way, publishers and subscribers are completely desynchronized in time and space. Many applications such as stock quotes, network management systems, RSS feed monitoring, already benefit from this paradigm.

Publish/subscribe systems designed so far follow two main directions: *topic-based* and *content-based* systems [1]. The *topic-based* systems are similar to group communication where events published on a specific topic are forwarded to all clients subscribed in this topic. The *content-based* systems provide a finer granularity, where subscribers specify their interests based on event contents.

Traditional solutions for content routing are usually based on a fixed infrastructure of reliable brokers. While subscriptions are dynamic, the event routing structure remains mostly static. This approach limits the scalability and routing accuracy with the increase and dynamism of subscription populations. Moreover, this solution introduces single point of failures and bottlenecks.

Another approach to content routing is to design it free of brokers infrastructure, and organize subscribers and publishers in a peer-to-peer overlay through which messages flow to interested parties. By using an adequate structure and gathering subscribers with similar interests to form *semantic communities*, events can be quickly disseminated within a community without incurring significant filtering cost [2]. Obviously,

A.-M. Kermarrec, L. Bougé, and T. Priol (Eds.): Euro-Par 2007, LNCS 4641, pp. 537–548, 2007.
© Springer-Verlag Berlin Heidelberg 2007

for such techniques to be efficient, one needs to properly structure the overlay to: avoid *false negatives* (a subscriber failing to receive an event it is interested in); minimize the occurrence of *false positives* (a subscriber receiving an event that it is not interested in); *self-adapt* to the dynamic nature of the systems, with peers joining, leaving, and failing; and maintaining the *overlay balanced* in order to provide a publication service time logarithmic in the size of the network similar to the DHT-based implementations. Our challenge was to propose an efficient overlay that positively responds to the above mentioned requirements.

In this paper, we present a novel approach, called distributed R-trees, to address the limitations of content routing in publish/subscribe systems. Distributed R-trees are a class of content-based publish/subscribe overlays where subscribers and publishers are organized in peer-to-peer balanced structures based only on their interests. Our overlays are derived from R-trees [3] and R*-trees [4] that are well-known indexing structures that are specially designed to support spatial database queries. We have implemented the distributed, scalable, and fault tolerant version of these particular data structures and analyzed their impact on the false positives/false negatives via extensive simulations.

Our overlays achieve the efficiency through: 1) organizing subscribers in a distributed and completely decentralized virtual balanced tree, based only on their interests; 2) providing a zero risk of false negatives and maintaining a low level of false positives; 3) masking faults via self-stabilization techniques. The self-stabilization of our structure and its correctness analysis are proposed in a companion paper [5].

The rest of the paper is organized as follows: Section 2 reviews some related work in this domain. Section 3 introduces the considered publish/subscribe model and revisits the R-tree characteristics and its variants. Section 4 presents our distributed R-tree overlays. Section 5 evaluates the effectiveness and accuracy of distributed R-trees for publish/subscribe and Section 6 concludes the paper.

2   Related Work

Content-based publish/subscribe over peer-to-peer systems has been widely addressed in recent years (e.g., [6,7,2,8,9]). Surprisingly, most of these systems aim at providing scalability and fault-tolerance but very few of them address the central problem in publish/subscribe systems: the presence of false positives and false negatives.

One of the techniques used in publish/subscribe is the *rendezvous*. This technique steams in identifying particular nodes where subscriptions meet publications [10,11,12]. The main drawback in such solutions is the high load reported on the *rendezvous* nodes since they centralize all the filtering performed in the system. In contrast, our overlay is totally distributed (i.e., decentralized) and every peer in the system participates in the matching and event dissemination.

Another popular technique in the design of publish/subscribe systems is the use of DHT-based overlays (e.g., Pastry or CAN). The advantage of using overlays is the logarithmic guaranties on the hit time for the publication. HOMED [13] presents a peer-to-peer hypercube overlay for distributed publish/subscribe systems. Also, the peers are organized based on their interests. Similarly, Meghdoot [14] uses the CAN infrastructure. In this system, the subscriptions composed by multiple predicates are partitioned and distributed onto CAN nodes. These approaches have two main drawbacks: the lack of scalability for publish/subscribe systems that require complex subscriptions and the large number of false positives/negatives. The first problem was addressed in [15] by using multi-dimensional spaces. Terpstra *et al.* [15] partition the event space among the peers in the system, but they broadcast the events and the subscriptions to all the peers in the system. Consequently, the number of false positives is in the order of the number of subscriptions in the system.

One of the techniques that focused on minimizing the false positives/false negatives is the organisation of subscribers based on their similarity [2,8,9]. In the first two systems, subscriptions form unbalanced trees and the publication complexity is strongly dependent on the subscription distribution. Contrary to this approach, our structure is balanced and offers guarantees comparable to the DHT-based implementations. Sub2Sub [9] is constructed on top of an epidemic semantic-based group membership. Nodes that share the same subscription are linked together in a ring. The impact of this architecture on the level of false positives is studied only inside the similarity groups. Our work does not only provide logarithmic guarantees with respect to the publication hit time but also extends the study of false positives to the whole trajectory of events.

Another approach is the subscription merging [16,17], which also groups subscriptions based on their similarity and creates a new subscription containing the set. This new subscription is similar to the MBR[1] and it is used in the matching procedure. The merging algorithm used to identify the groups implies in the number of false positives generated in the system. However, the merging problem was proved to be NP-hard [16]. An optimization of subscription merging is presented by Ouksel *et al.* [18] where they propose a Monte Carlo type algorithm. Contrary to our approach, the algorithm introduces false negatives due to the probabilistic nature of the algorithm.

3   Background and System Model

3.1   Content-Based Publish/Subscribe Systems

Data Model. As most other publish/subscribe systems, we assume that an event is a set of attribute-value pairs. Each attribute has a name and a numeric or string value. A subscription is a conjunction of predicates over the attribute values, i.e., $S = f_1 \wedge \ldots \wedge f_j$, where f_i is defined as a tuple $f_i = (n_i \quad op_i \quad v_i)$ with n_i the name of the attribute, op_i an operator ($<, >, \leq, \geq, =, \neq$, etc.), and v_i a constant value. For example, a subscription expressed on the attributes a and b may be of the form $(v_i < a < v_j) \wedge (v_k < b < v_l)$.

A schema composed by n attributes may be represented in a cartesian space with n dimensions. The subscriptions correspond to poly-rectangles and events as points. Figure 1 shows a set of subscriptions defined on 2-dimensional space with two attributes. Note that, if one attribute is undefined, then the corresponding rectangle is unbounded in the associated dimension.

---

[1] MBR is usually represented by the coordinates of the upper left corner and the bottom right corner of the corresponding rectangle.

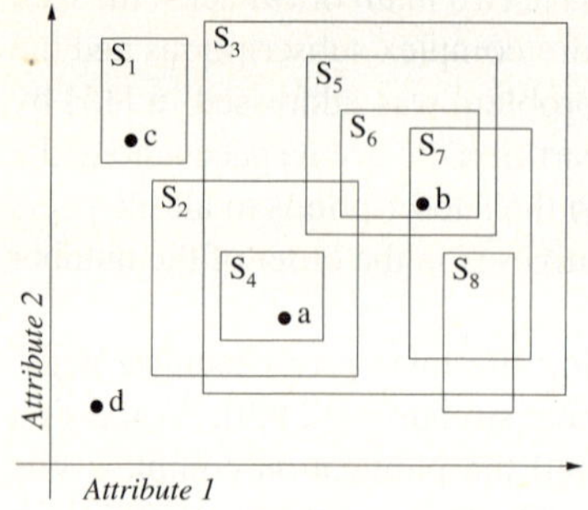

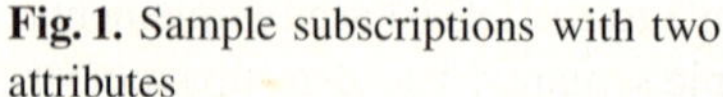

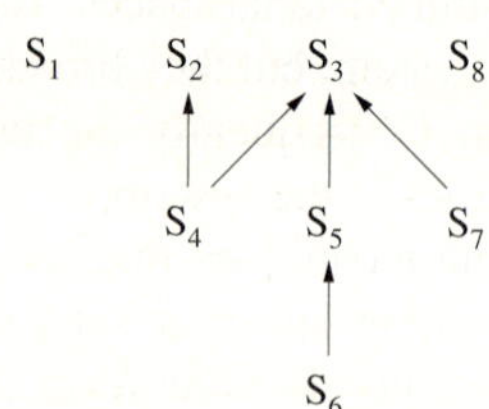

Fig. 1. Sample subscriptions with two attributes

Fig. 2. Containment graph for the subscriptions of Figure 1

Publish/subscribe systems can take advantage of the property of *subscription containment* in order to improve the filtering procedure. A subscription S_1 contains another subscription S_2, or $S_1 \supseteq S_2$ if S_1 has a larger scope than S_2. This means that any event E_1 that matches S_2 also matches S_1. Conversely, we say that S_2 is contained by S_1, or $S_2 \subseteq S_1$. Note that the containment relationship is transitive and defines a partial order. Geometrically, the containment corresponds to the enclosure relationships between the poly-space rectangles (see Figure 1).

Content-based Routing Protocols. An efficient publish/subscribe overlay should minimize the occurrence of false positives (a peer receiving a message that it is not interested in) and avoid false negatives (a peer failing to receive a message that it is interested in).

A straightforward approach for avoiding false positives and false negatives is to organize the subscribers in a tree structure according to containment relationships [19], such that the subscription of a peer contains the subscriptions of its descendants. Indeed, if an event matches the containee, it *has* to match the container (this guarantees no false negatives); conversely, if it does not match the container, it cannot match the containee (this guarantees no false positives). Figure 2 illustrates the containment graph from the mapping of the example in Figure 1.

A direct mapping of the containment graph to a tree structure [2] is often inadequate. First, it requires a virtual root with as many children as subscriptions that are not contained in any other subscription. Second, depending on the subscription workload, the resulting tree might be heavily unbalanced with a high variance in the degree of internal nodes. Another approach consists in building one containment tree per dimension and adding a subscription to each tree for which it specifies an attribute filter [8]. This solution tends to produce flat trees with high fan-out and generates a significant number of false positives.

Our objective in this paper is to improve these approaches by using bounded-degree height-balanced trees, while preserving the containment relationships that ensure accurate content dissemination. To that end, we propose distributed extensions of the R-tree index structures.

3.2   R-Tree Index Structures

R-trees were first introduced by Guttman [3]. An R-tree is a height-balanced tree where each node in the tree is represented by the smallest poly-space rectangle enclosing all the rectangles in its subtree, called *minimum bounding rectangle* (MBR).

An R-tree is characterized by the following properties: *(a)* Every non-leaf node has between m and M children, except for the root that has at least two children; *(b)* The height of an R-tree containing N objects is $\lceil log_m(N) \rceil - 1$; *(c)* The worst space utilization for each node except the root is m/M.

In a classical R-tree structure, the actual objects are only stored in the leaves of the tree and internal nodes only maintain MBRs. An R-tree constructed from the sample subscriptions of Figure 1 is shown in Figure 3 (for $m = 1$ and $M = 3$) and its spatial representation in Figure 4. Note that all subscriptions are stored in the leaves and the role of internal nodes $B_1, \ldots, B_5$ is to keep track of the bounding rectangles that contain their descendants. In distributed settings, obviously, internal nodes must be managed by specific peers in the system.

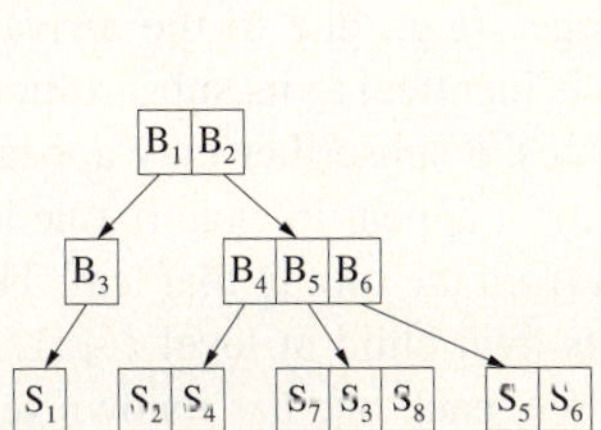

Fig. 3. R-tree for the subscriptions of Figure 1

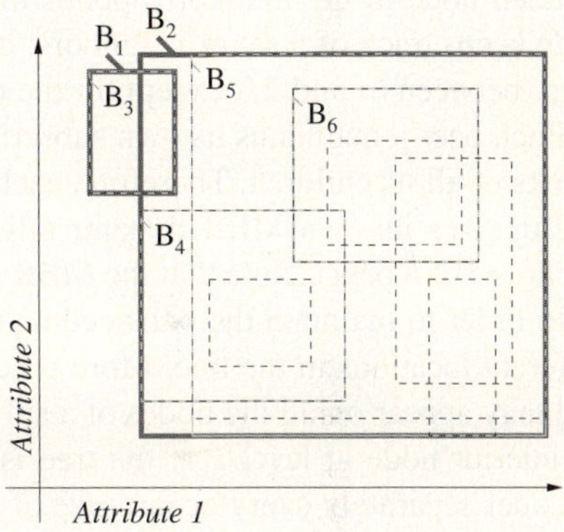

Fig. 4. Spatial representation of the R-tree of Figure 3

Upon a join of a new node, if the children set becomes bigger than M entries, the children set must be split. There are several well-known methods for splitting an overflowing node during the join. In the following, we present three classical methods, which are supported by our distributed R-tree structures: *(i)* The *linear* method [3] chooses two children from the overflowing node such that the union of their MBRs waste the most area and places each one in a separate node. The remaining children are assigned to the nodes whose MBR is increased the least by the addition. This method takes linear time. *(ii)* The *quadratic* method [3] chooses two children from the overflowing node that would waste the most area if they were in the same node, and place each one in a separate node. The remaining MBRs are examined and the one whose addition maximizes the difference in coverage between the MBRs associated with each node is added to the node whose coverage is minimized by the addition. This method takes quadratic time. *(iii)* The R^*-*tree* splitting method [4] attempts to reduce not only the coverage, but also the overlap. Instead of just splitting the node when it overflows, it also tries to allocate some entries to a better suited node through reinsertion.

In the next section, we extend R-trees and show how the resulting *distributed R-tree* variants can be used to produce efficient content dissemination networks.

4   Distributed R-Trees for Content-Routing

Our distributed R-tree is a virtual height-balanced tree for content-routing in publish/ subscribe systems, where the peers are organized according to their interests (subscriptions). Unlike traditional R-trees, each node in the tree maps to one peer in the network.

Every peer has two roles: publisher/subscriber and router. A peer that registers a subscription will participate in the overlay (the nature of the subscription will influence the position of the peer in the tree) and may or not publish events. Also, every non-leaf node acts as a forwarder during event dissemination. Node positions may change due to the dynamism of the system: a leaf node may become an internal node and inversely.

Each node in the tree corresponds to a subscription. The peer associated with the node keeps track of a set of neighbors: its parent and, for non-leaf nodes, a set of children (between m and M, except for the root).

Each peer p maintains its own subscription S_p, as well as an MBR that encloses the MBRs of all its children. Therefore, each node keeps track of the MBRs of its children and updates its own MBR dynamically when it changes (e.g., due to the arrival or departure of a peer). Note that the MBR of a leaf node is identical to its subscription.

In order to maintain the balanced nature of the R-trees, a subscriber may appear at different locations in the tree. More precisely, a node must appear in exactly one leaf, and may appear on all the nodes of a suffix of the path from the root to that leaf. Thus, an interior node at level l of the tree is recursively its own child at level $l + 1$. We consider separately every occurrence of a peer in the tree: each one has its own set of neighbors and related data.

The choice of which peers are promoted as internal nodes is performed according to existing containment relationships so as to minimize the likeliness for false positives. We typically choose the node whose current MBR is largest. Hence, if a peer whose MBR covers all the other MBRs in the children set, then it trivially becomes the new parent: the containment relationship is preserved and there is no occurrence of false positives. If MBRs in the children set intersect or are disjoint, the peer with the largest MBR is chosen in order to minimize the size of the area corresponding to false positives.

Figure 5 illustrates a possible configuration of the distributed R-tree structure for the subscriptions of Figure 1; and Figure 6 shows the associated communication graph.

Selective Event Dissemination. Event filtering and dissemination in distributed R-trees are fully distributed among the peers in the system. This process is very simple and only relies on local information.

Upon receiving an event, a peer forwards it to each child whose MBR contains the event, unless the message was received by that child. If the message originated from a descendant (i.e., it is propagated upward the tree), then the peer forwards it to its parent as well. If the event matches the local subscription, it is delivered to the user.

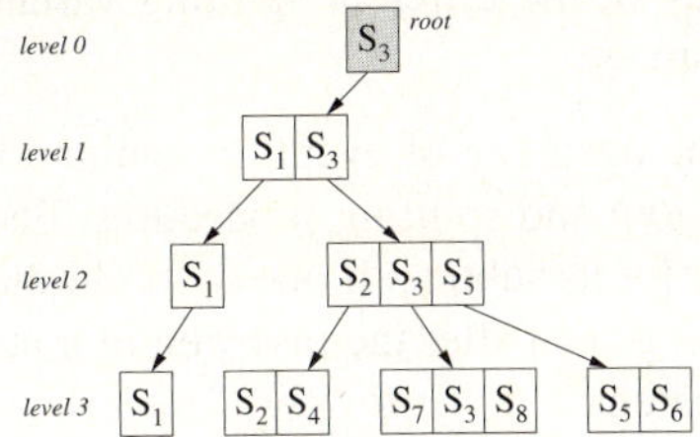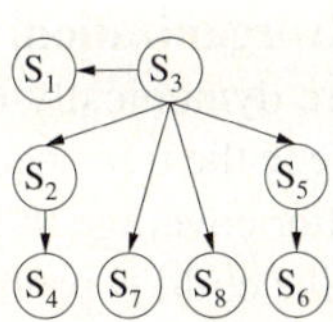

Fig. 5. Distributed R-tree for the subscriptions of Figure 1

Fig. 6. Communication graph for the distributed R-tree of Figure 5

An event E_i matches a subscription iff the subscription's attributes are satisfied by the event, i.e., the event falls in the attribute range for each of the dimensions. If a node receives an event that does not match its subscription, the event is considered a false positive. As leaf nodes have an MBR equal to their subscriptions, only internal nodes can experience false positives. Moreover, if the subscriptions of all the descendants of a subscriber p are contained within p's subscription, then p's MBR is identical to its subscription and it does not experience false positives.

By construction, our R-tree structures cannot produce false negatives during dissemination, i.e., all the subscribers that have subscribed for an event eventually receive it (unless there is a failure).

Registering a Subscription. A peer joining the network by registering a subscription can contact any existing peer. The subscription request is redirected upward the tree until it reaches the root. Then, it is pushed downwards to the last non-leaf node whose interests are closest to those of the new subscriber (as determined by comparing MBRs). Having neighbors with similar interests helps minimize the occurrence of false positives. Our distributed R-tree structures support two variants for selecting the best branches when traversing down the tree to register a new subscription: 1) R: we choose the subtree that needs the least enlargement of its MBR to insert the new subscription; upon tie, we select the subtree with the smallest MBR [3]; 2) R^*: we proceed as above until we reach the last non-leaf nodes; then, we insert the new subscription in the node that needs the least overlap enlargement; upon tie, we select the node whose MBR needs the least area enlargement [4].

The concepts of coverage and overlap are important to minimize the occurrence of false positives. Each of these variants attempts to minimize the coverage or/and the overlap of the MBRs at the same level. Minimizing the coverage of a node's MBR permits to minimize the dead space between its MBR and the MBR of its children. Minimizing the overlap of MBRs at the same level avoids an event being disseminated to all the subtrees.

As previously discussed, each node has between m and M entries by level (except for the root node). If the new parent has less than M children, it inserts the new subscription in its children set. Otherwise, the parent creates two children sets with at least m subscriptions, effectively creating a new subtree. Splits propagate upward the tree: a split at a node may trigger a split at its parent when inserting the newly created subtree.

We have implemented and compared the efficiency of the different splitting variants (quadratic, linear, R^*) applied to our distributed R-trees.

Dynamic Reorganization. In order to improve the accuracy of event dissemination, the nodes are dynamically reorganized during the join and splitting procedures. Each internal node in the tree checks if it is the best cover for its subtree. If one of its children provides better coverage (e.g., because its MBR has grown after the insertion of a new node), then the child is promoted and replaces its parent.

Canceling a Subscription. We only discuss controlled departures here, i.e., peers explicitly unregistering their subscriptions from the system. Fault tolerance is supported by the means of a self-stabilizing tree maintenance protocol discussed in [5].

A peer leaving the system notifies its parent(s). When receiving such a notification, the parent removes its departing child from its children set. After the peer leaves the system, the whole branch that used to contain the departing peer, must be repaired.

If the children set drops below m after the peer leaves the system, the children set is reinserted in the tree. Thus, for each child, the whole subtree rooted by the child is reinserted in the tree at the same level. Conversely, if the children set remains between m and M, there are two scenarios: if the parent is the departing peer, then a new parent is promoted to occupy the vacant position; otherwise, the MBR of the parent is updated since it may become smaller.

5   Evaluation

This section describes the results of our evaluation and comparision of the different distributed R-tree structures presented earlier.

Table 1. Parameters used for the experiments

Parameter	Values
Splitting Method	[quadratic], linear, R*
Number of subscriptions	$1,000$, $2,500$, $5,000$, $[10,000]$, $25,000$, $50,000$
Number of events	$[2,500]$
Subscription distribution	uniform, uniform-25:75, uniform-10:90, Zipf [Zipf-25:75], Zipf-10:90
Event distribution	[uniform]
Number of dimensions	2, $[4]$, 6, 8, 10, 12
Degree of the tree (m, M)	$(2,5)$, $[(5,10)]$, $(10,20)$, $(15,30)$, $(20,40)$

Experimental Setup. Subscriptions are defined as a set of d attribute-range pairs, each of which corresponds to a dimension. The range specifies the set of values that the consumer is interested in. Without loss of generality, we used range values between 0 and $1,000$. Note that a range may represent a single value. Events are points in the d-dimensional space.

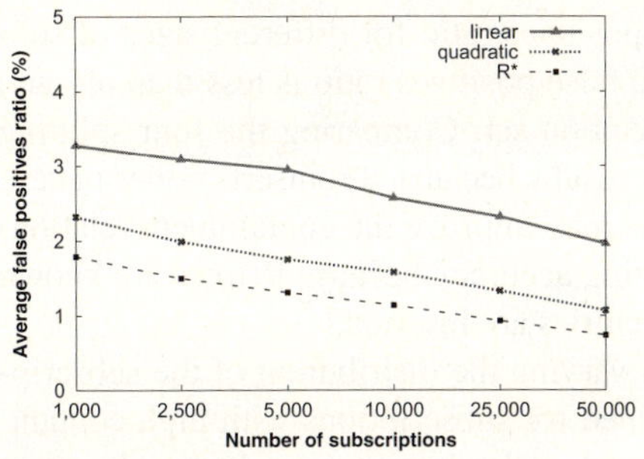

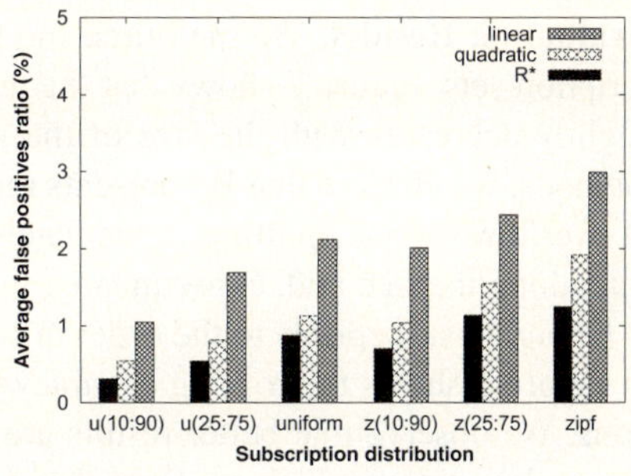

Fig. 7. False positives ratio for different subscription set sizes

Fig. 8. False positives ratio for different subscription distributions

We analyzed the performance of the system under uniform and skewed subscription workloads; and with a uniform event distribution. Skew is simulated using a power-law distribution (Zipf with $\alpha = 1$) and is applied to the origin of subscriptions only: their size is always chosen according to a uniform distribution.

To model and observe the influence of containment relationships, we have generated some subscription sets with a given ratio of container/containee subscriptions. Given a ratio of X:Y, we have first generated $X\%$ of the subscription population according to the current distribution. For each subscription in the remaining $Y\%$, we have taken the following steps: select a random subscription S from the current set; generate a uniform random subscription S' such that $S \supseteq S'$; and insert S' in the set. This method guarantees that at least $Y\%$ of the subscriptions are containees. We considered uniform and Zipf distributions, as well as two X:Y ratios: 25%:75% and 10%:90%.

In the experiments, we evaluated the efficiency of our approach in terms of *false positives ratio*, i.e., the percentage of the nodes in the system that receive events that does not match their interests. For simplicity, we assume that events are injected at the root. Note that this assumption is equivalent to having each event with at least one interested consumer being produced by a publisher with a matching subscription, i.e., producers never experience false positives locally.

Obviously, an event that does not match a single subscription is expected to show a lower false positive ratio than an event with many interested subscribers, because the latter is likely to be propagated deeper in the tree. Therefore, we shall also observe the effect of event popularity in our study. As leaves have an MBR equal to their subscriptions and a node forwards an event to each of its children whose MBR contains the event, only interior nodes can experience false positives. We do not consider false negatives since our distributed R-tree structures do not produce any.

The number of events was fixed in all simulations to $2,500$ for computing false positive ratios. We have used the parameters shown in Table 1 (default values are in brackets). For each simulation, we have varied the values of the parameter to be observed and fixed the remaining ones to their default value.

Evaluation Results. We measured the false positives ratio for different sizes of subscription sets. Figure 7 shows that the average false positives ratio is less than 5% and slightly decreases with the size of the subscription set. Comparing the four splitting methods, we observe that R* presents the best results because it reinserts nodes in case of overflow instead splitting immediately. This may improve the containment relationship along the tree and, consequently, the routing accuracy because R-trees are known to be highly susceptible to the order in which entries are inserted.

Figure 8 shows the routing accuracy when varying the distribution of the subscriptions. We observe that better results are obtained for subscriptions with high containment relationship, which confirms that our trees do indeed preserve and take advantage of containment relationships. Accuracy is also slightly better with a uniform subscription distribution.

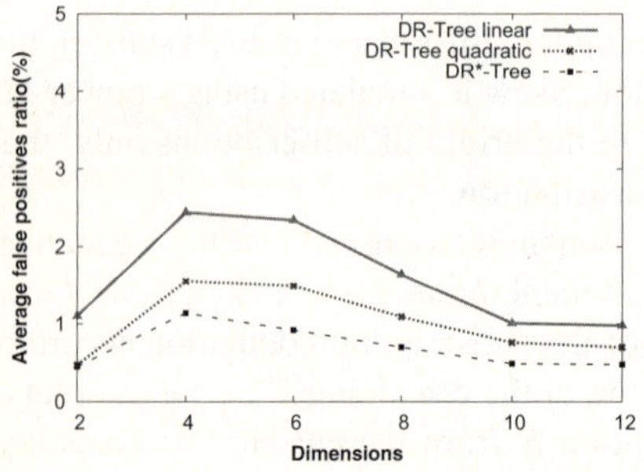

Fig. 9. False positives ratio for different dimensions

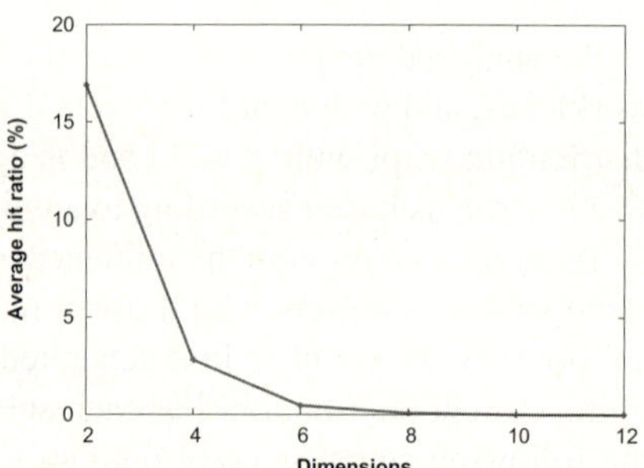

Fig. 10. Hit ratio for different dimensions

Figure 9 illustrates the average false positives ratio for different dimensions. Surprisingly, accuracy improves with the number of dimensions. This is due to the fact that less nodes are interested in the event, as can be seen in Figure 10. We have, therefore, plotted the false positives ratio as a function of the number of hits for each experiment with three dimensions. The results are shown in Figure 11. We observe now that the average false positives ratio actually increases with the dimension but remains reasonably small, never reaching 10% even for linear. The same general trends are exhibited by all three splitting method.

As discussed before, our approach differs from traditional R-trees in that some subscriptions may appear at different levels in the logical tree. Thus, the degree of a node varies depending on its position and number of occurrences in the tree, in addition to the values of m and M. Figure 12 presents simulation results for different degrees between $M = 5$ and $M = 40$, where $m = M/2$; and Table 2 shows the maximum, average, and variance of the degree of internal nodes. We observe a clear trade-off between accuracy and nodes degree: increasing the degree improves accuracy. In the studied scenario, a value of $M = 20$ appears to be a good compromise.

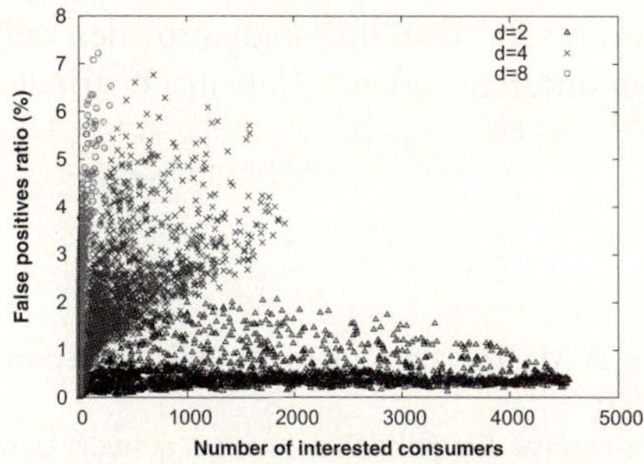

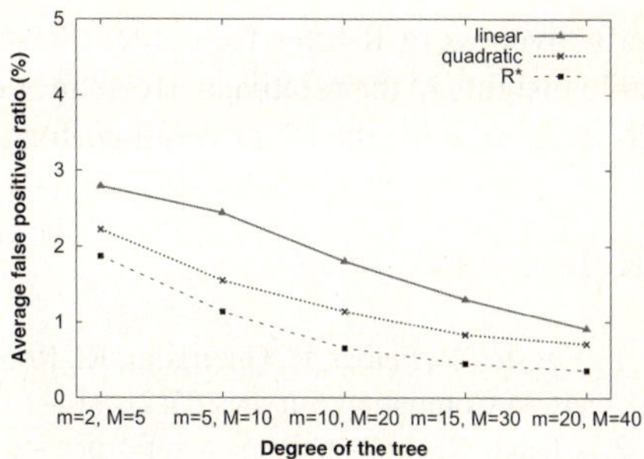

Fig. 11. False positives ratio vs. hit ratio for different dimensions (one point corresponds to one experiment)

Fig. 12. False positives ratio for different degrees of the tree

For comparison purposes, a tree built as a direct mapping of the containment graph using the same $10,000$ subscriptions (Zipf-25:75) would have a virtual root node with approximately $2,000$ children and would obviously not be height-balanced.

Table 2. Degree statistics

Degree	quadratic			linear			R*		
	Max	*Avg*	*Var*	*Max*	*Avg*	*Var*	*Max*	*Avg*	*Var*
m=2, M=5	20	4.59	5.09	19	4.61	5.08	25	3.91	8.47
m=5, M=10	29	7.99	10.98	28	7.93	10.66	30	7.49	16.27
m=10, M=20	51	14.92	25.40	37	14.82	23.15	42	15.05	32.87
m=15, M=30	51	21.91	35.70	50	21.69	33.52	57	22.28	56.29
m=20, M=40	68	28.02	55.38	60	28.39	54.02	77	30.00	81.10

6   Conclusion

In this paper we studied a class of distributed R-trees and their applicability to build content-based publish/subscribe overlays. Our study focused on the properties of the resulting topology and the accuracy of event dissemination (occurrence of false positives and false negatives). Distributed R-trees, proposed in this paper, are a decentralized implementation of the R-tree structure and its variants, which are widely used in database systems. These overlays are fully adapted to embed a publish/subscribe system with complex subscriptions (multi-dimensional) and cope with the dynamism of the system. The overlays are designed such that they eradicate false negatives and drastically drop the number of false positives. Moreover, organizing the peers based on their interests minimizes both the amount of matchings in the system and the latency during event dissemination (the worst case being logarithmic in the size of the network). We have implemented and analyzed via simulations the evolution of false positives applying the different variants of insertion and splitting methods. Independently of the absolute effectiveness of R-trees for content-based publish/subscribe, this study provides valuable insights in the relative performance of their different variants. Note that distributed R*-trees provide the best overall performance.

References

1. Eugster, P., Felber, P., Guerraoui, R., Kermarrec, A.M.: The many faces of publish/subscribe. ACM Computing Surveys 35(2), 114–131 (2003)
2. Chand, R., Felber, P.: Semantic peer-to-peer overlays for publish/subscribe networks. In: Cunha, J.C., Medeiros, P.D. (eds.) Euro-Par 2005. LNCS, vol. 3648, Springer, Heidelberg (2005)
3. Guttman, A.: R-trees: A dynamic index structure for spatial searching. In: Proceedings of ACM SIGMOD, pp. 47–57. ACM Press, New York (1984)
4. Beckmann, N., Kriegel, H., Schneider, R., Seeger, B.: The R*-tree: An efficient and robust access method for points and rectangles. In: Proceedings of ACM SIGMOD, ACM Press, New York (1990)
5. Bianchi, S., Datta, A.K., Felber, P., Gradinariu, M.: Stabilizing dynamic R-tree based spatial filters. In: Proceedings of 27th ICDCS (2007)
6. Perng, G., Wang, C., Reiter, M.: Providing content-based services in a peer-to-peer environment. In: Proceedings of 3rd DEBS, pp. 74–79 (2004)
7. Triantafillou, P., Aekaterinidis, I.: Content-based publish/subscribe over structured P2P networks. In: Proceedings of 3rd DEBS, pp. 104–109 (2004)
8. Anceaume, E., Gradinariu, M., Datta, A.K., Simon, G., Virgillito, A.: A semantic overlay for self-* peer to peer publish/subscribe. In: Proceedings of 26th ICDCS, p. 22 (2006)
9. Voulgaris, S., Riviere, E., Kermarrec, A.M., van Steen, M.: Sub-2-sub: Self-organizing content-based publish subscribe for dynamic large scale collaborative networks. In: Proceedings of 5th IPTPS (2006)
10. Aekaterinidis, I., Triantafillou, P.: Pastrystrings: A comprehensive content-based publish/subscribe DHT network. In: Proceedings of 26th ICDCS, p. 23 (2006)
11. Baldoni, R., Marchetti, C., Virgillito, A., Vitenberg, R.: Content-based publish-subscribe over structured overlay networks. In: Proceedings of 25th ICDCS, pp. 437–446 (2005)
12. Renesse, R., Bozdog, A.: Willow: DHT, aggregation, and publish/subscribe in one protocol. In: Proceedings of 3rd IPTPS (2004)
13. Choi, Y., Park, K., Park, D.: HOMED: a peer-to-peer overlay architecture for large-scale content-based publish/subscribe system. In: Proceedings of 3rd DEBS, pp. 20–25 (2004)
14. Gupta, A., Sahin, O.D., Agrawal, D., Abbadi, A.E.: Meghdoot: Content-based publish/subscribe over P2P networks. In: Proceedings of 5th Middleware (2004)
15. Terpstra, W., Behnel, S., Fiege, L., Zeidler, A., Buchmann, A.P.: A peer-to-peer approach to content-based publish/subscribe. In: Proceedings of 2nd DEBS, pp. 1–8 (2003)
16. Crespo, A., Buyukkokten, O., Garcia-Molina, H.: Query merging: Improving query subscription processing in a multicast environment. IEEE TKDE 15(1), 174–191 (2003)
17. Li, G., Hou, S., Jacobsen, H.: A unified approach to routing, covering and merging in publish/subscribe systems based on modified binary decision diagrams. In: Proceedings of 25th ICDCS, pp. 447–457 (2005)
18. Ouksel, A., Jurca, O., Podnar, I., Aberer, K.: Efficient probabilistic subsumption checking for content-based publish/subscribe systems. In: Proceedings of 7th Middleware (2006)
19. Carzaniga, A., Rosenblum, D.S., Wolf, A.L.: Design and evaluation of a wide-area event notification service. ACM Transactions on Computer Systems 19(3), 332–383 (2001)

Topic 8
Distributed Systems and Algorithms

Luís Rodrigues, Achour Mostefaoui, Christof Fetzer, and Philippas Tsigas

Topic Chairs

Parallel computing is increasingly exposed to the development and challenges of distributed systems, such as the lack of load balancing, asynchrony, long latencies, network partitions, failures, disconnected operations, heterogeneity and protocol standardization. Furthermore, distributed systems are becoming larger, more diverse and more dynamic. This Euro-Par topic provides a forum for research and practice, of interest to both academia and industry, about distributed systems, distributed computing, distributed algorithms, and parallel processing on distributed systems. Submission was encouraged across the whole area, with emphasis on the following: design and practice of distributed algorithms and data structures, analysis of the behaviour of distributed systems and algorithms, distributed operating systems, parallel processing on distributed systems, resource and service discovery, resource sharing and in distributed systems, distributed fault tolerance, security in distributed systems, scalability, concurrency and performance in distributed systems, middleware for parallel computations, web services, interoperability and standards, self-organised and self-adjusting distributed systems.

Thirty-two papers were submitted in this topic, eight of which have been accepted for publication. The accepted papers cover a wide range of aspects in the distributed system and algorithms topic. Three papers are related with the problem of data sharing in distributed and parallel systems, two papers address issues related with mobile agents in graphs, and, finally, three papers address fundamental distributed algorithm problems, such as failure detection, consensus, and spanning tree construction.

A.-M. Kermarrec, L. Bougé, and T. Priol (Eds.): Euro-Par 2007, LNCS 4641, p. 549, 2007.
© Springer-Verlag Berlin Heidelberg 2007

Accelerate Data Sharing in a Wide-Area Networked File Storage System[*]

Kun Zhang, Hongliang Yu, Jing Zhao, and Weimin Zheng

Department of Computer Science and Technology
Tsinghua University, Beijing 100084, China

Abstract. Up to now, more and more people use Internet storage services as a new way of sharing. File sharing by a distributed storage system is quite different from a specific sharing application like BitTorrent. And as large file sharing becomes popular, the data transmission rate takes the place of the response delay to be the major factor influencing user experience. This paper introduces strategies used to accelerate file sharing with low bandwidth consumptions in a deployed wide-area networked storage system - *Granary*. We use a popularity and locality sensitive replication strategy to put files closer to users that request it frequently. The *Hybrid* server selection scheme and the *Remote Boosting* replication mechanism are also presented. Experimental results show these methods offer better sharing speed and cost less network bandwidth than conventional caching schemes.

1   Introduction

The improvement of computer hardware and network infrastructure is offering more bandwidth and storage spaces, making it possible to distribute files as large as gigabytes via the Internet. While people storing more data through network, the need for sharing data like music, home video or picture albums is growing too. Email was used as a sharing tool before, which was inconvenient for its notable delay and limitation on file sizes.

Up to now, more and more people use Internet storage services as a new way of sharing. For example, commercial on-line storage services like Box.net and Omnidrive is offering more than 1GB storage space for each user. Sharing based on a storage system is quite different from specific file-sharing applications like BitTorrent and eMule. By uploading the file to storage servers, users does not need to keep on-line to share the data. The quality of service is also guaranteed by the infrastructure of the storage system, not end users. This lead to different choices of data replication and caching schemes.

As audio and video sharing becomes usual and popular, the data transmission rate takes the place of the response delay to be the major factor influencing user experience. From another aspect, as the file size grows, bandwidth consumption

[*] This work is supported by the National Natural Science Foundation of China under Grant No. 60603071 and 60443040.

A.-M. Kermarrec, L. Bougé, and T. Priol (Eds.): Euro-Par 2007, LNCS 4641, pp. 551–562, 2007.

is becoming more critical than before. OceanStore [1] propose a introspection mechanism that can be used for cluster recognition or replica management, which could be a potential solution to this issue, but unfortunately there is not any further study or implementation for this mechanism. As OceanStore's prototype, Pond [2] uses a simple caching mechanism to accelerate archive data retrieval. Similarly, when a PAST node routes a file for a insertion or retrieval operation, it caches the file on its local storage [3]. However, caching is a passive scheme and is confined by the routing path of the original file. In Pangaea [4], nodes that have low network latency to each other are clustered into groups, and an aggressive replication mechanism is used to put replicas in the client's local nodes. However, it is not explained how nodes are clustered or how a client find a local node, which is very important. Moreover, the systems above do not tackle the performance problem of sharing large files sizing megabytes or even gigabytes, especially the transfer rate of downloading them. Therefore, it is still challenging to accelerate sharing speed with low bandwidth consumption in such systems.

In this paper, we introduce a distributed storage system which provides reliable storage service to over 500 users at present. Users backup their files and share some of them with others. We implement a replication mechanism to solve the problems brought by file sharing. It evaluates the popularity of files, clusters clients that are close to each other, makes more copies of popular files on the nodes that are closest to the most demanding users, thus decreases unnecessary network traffic, reduces network congestion on the critical links and finally improves download speed. It has two characteristics:

- *It's popularity sensitive.* It records the requests for every file, analyzes popularity, and propagates more replicas for more popular files. Popularity sensitivity can help relieve the impact of requests for popular files.
- *It's locality sensitive.* It detects and analyzes the location of storage nodes and clients, places file replicas closer to the users that request it frequently, thus decrease unnecessary inter-domain traffics and reduce network congestion on the critical links.

2   Background

Our distributed storage system is named as *Granary.*It is composed of dedicated nodes at a global scale and provides paid storage services to end users. It utilizes *PeerWindow* [5] as the node collection algorithm and *Tourist* [6] to route messages to proper nodes, on top of which a highly available DHT [7,3] is built. The DHT stores DHT objects. A DHT object consists of a key and a value, and comes with a 128-bit hash value, which is the MD5 hash of its key. Every node is assigned a unique 128-bit key called nodeId, which can be the hash value of a per-machine signature such as MAC address. A DHT object is replicated and stored on the closest nodes in terms of nodeId, which are probably disperse in terms of physical location. This provides high availability to DHT objects.

Unlike PAST, *Granary* doesn't directly store files in DHT. Instead, it stores and replicates files in the upper layer and puts only the locations of replicas (which is called the *replica list*) and the meta data of files in the DHT. The high availability of DHT guarantees that the meta data can be accessible despite of network disconnections or node failures. A *Granary* server acts not only as a DHT node, which routes messages for other nodes and stores DHT objects that are mapped to it, but also as an upper-level storage server which clients access directly for file uploading and downloading. This choice of design allows us to develop a more flexible replication algorithm that are not constrained by the binding of nodeIds and objectIds in the DHT. Since trivial operations like message transmission and meta data read/write are all handled by the DHT layer, we are able to focus on the replication strategy itself.

Most contributive systems [7,3,1,2] divide files into smaller blocks and spread blocks among nodes. The client connects to the closest node, which routes queries and data for it. This is a reasonable choice because they are made of contributive and weak nodes which tend to go off-line at any time. Dividing files into blocks and replicating these blocks ensures the availability of file even if some nodes become unavailable. However, *Granary* stores full copies of files on dedicated nodes. These nodes are more powerful and stable, but much less in quantity, compared with contributive systems. So *Granary* doesn't need to divide files to ensure availability. Additionally, storing full copies eliminates the need to communicate with many nodes simultaneously for file retrieval, thus reduces the traffic for block searching overhead.

Files in *Granary* are immutable, but a user can upload a new version of his file with the same file name. Different versions of a file are distinguished by version number. The servers do not delete the older versions immediately after a new version has been uploaded, because some other clients may be downloading the old one at the moment. The coexistence of older and newer versions will only long for a while, because only the newest version is visible to newly arriving clients, so that the older versions can be deleted after remaining download sessions have finished. When the user deletes a file, its meta data is removed from DHT, and the accommodating servers will delete their local replicas after remaining download sessions have finished.

In following sections, we will first describe some fundamental designs which will be later used. Then, strategies to accelerate sharing in *Granary* are presented. Furthermore, experimental results of these strategies will be shown at last.

3   Fundamental Design

3.1   Caching vs. Replication

Both caching and replication are widely used to improve the performance of distributed systems [8,7,3]. Although they are very similar to each other, they do have differences. Firstly, caching passively retains data that may be used for the next time, while replication may actively create copies of data; Secondly, the data cached is volatile while replica is persistent; Finally, caching usually introduces no extra cost besides space occupation, while replication may bring extra cost, e.g. network traffic. We choose replication to accelerate the file retrieval in *Granary* for the following reasons:

- Replication is active, so we can do predictions and make copies proactively;
- Even if we use a caching method, replication is still needed to provide availability, and it would be a waste of space to keep both replicas and cache;
- Caching is used in systems where the data is transmitted via more than one nodes before reaching the client. Although caching does not bring extra network traffic, data routing and forwarding do. In *Granary*, clients download data directly from servers. If properly designed, it will not cause more network traffic with replication than a routing-and-forwarding system with caching.

3.2   Network Distance

Sharing in *Granary* is boosted by the optimal placement of data replicas. To achieve it, we must know network distances in the system first. Usually, network distance between two hosts can be measured by round-trip time (RTT), trace-route hops, DNS name and IP address.

Trace-route reveals the underlying topology of the routers. However, it will brings too much extra traffic if we trace-route every requesting client. M. Andrews [9] grouped hosts into different autonomous systems (*ASes*) that have similar round-trip time (RTT) between them. He organized all known hosts by an *IP tree*, in which each node represents an IP prefix. With each server logs every file transmission to and from any other hosts, the similarities of RTT between leaf nodes are calculated. If the difference is below a threshold, these leaf nodes are merged and folded up to their parent. The result is a tree with its leaves representing different autonomous systems.

This method only considered IP address which may encountered a problem when an autonomous system includes several IP prefixes. For example, our campus network includes two network prefixes: (59.66.0.0/16) and (166.111.0.0/16), but they will never be clustered in such an IP tree. However, they are both assigned to the same DNS name suffix (*.ip.tsinghua.edu.cn*). To solve it, we introduce another data structure called the *domain tree*, with a similar structure as the DNS system. The *domain tree* records the DNS names of every clients. Except for using DNS name suffixes instead of IP prefixes, the *domain tree* is the same as the *IP tree*. Therefore the same clustering process can be applied to it.

In *Granary*, The IP distance $d_{ip}^{1,2}$ between hosts H_1 and H_2 is defined as the height of the smallest subtree that contains both the cluster leaf of H_1 and that of H_2. The height is zero based, and if the cluster leaf of a host doesn't exist, the default distance will be set to 24, as if a leaf is created for the host. If H_1 and H_2 both have DNS names, the domain distance $d_{domain}^{1,2}$ is defined in the same way in the domain tree. Otherwise, $d_{domain}^{1,2}$ will be set to the height of the domain tree. We divide d_{ip} and d_{domain} by 24 and the height of the domain tree respectively, and get the normalized distances $\hat{d}_{ip}$ and $\hat{d}_{domain}$ which are in the range $[0, 1]$. We define the network distance $d^{1,2}$ between H_1 and H_2 as:

$$d^{1,2} = \hat{d}_{ip}^{1,2} \times \hat{d}_{domain}^{1,2} \tag{1}$$

If H_1 and H_2 are in the same group, either from the IP tree or the domain tree, $d^{1,2}$ will be zero. Because transmissions between servers are also logged, we are able to evaluate the distance between two servers or between a server and a client.

4   Accelerate Sharing

4.1   Replica Generation

To accelerate sharing in a distributed storage system, it is very important to decide the number of data replications. In *Granary*, we use a simple thought that more replicas should be allocated for data that is more popular. Then the problem here is how we decide the replication number according to the popularity.

Researches on web caching reveal that the query frequency follows a Zipf-like distribution [10]. Researches on video-on-demand systems also show that replication distribution should follow the Zipf-like distribution [11]. Zipf-like distribution claims that the query frequency q_i of the ith popular object is proportional to $1/i^\alpha$, where α is a distribution-dependent constant. Based on Zipf-like distribution, we estimate the required number of replicas q_i for the ith popular file by

$$q_i = \frac{\beta M}{i^\alpha} \tag{2}$$

where M is the total query frequency of all files on all servers, and β is a constant configured according to the expected disk usage of the system. We assume that the servers are evenly loaded, so that M can be estimated by the total query frequency on a server (m) and the number of servers (N).

$$M = N \times m \tag{3}$$

Granary stores the logs of file access for a duration (e.g., 6 hours), and keeps a list of files sorted by request frequency. When a file is being requested, the server updates the list. Let r_i be the number of replicas of this file, if $r_i < q_i$, a new replica will be generated automatically.

However, the composition of the queries on a node in a distributed system is different from that in a single-server system. Since a client may have multiple servers to choose, the topological distribution of clients will influence the query distribution on a certain server.

For a certain client, we define a server as a *local server* if the network distance between them is below a predefined threshold (e.g. 0.5). To the opposite, we get a *remote server*. Then, requests sent to *local/remote server* can be named as *local/remote requests*. Although the server selection scheme described in Section 4.3 tends to choose a *local server*, a server may receive *remote requests* in two cases:

- *Local servers* are overloaded and the client have to choose a *remote server*;
- *Local servers* don't have a copy of the requested file.

Therefore, *remote requests* usually imply a misplacement of replicas and that we need to create a closer replica for these clients, while *local requests* better comply to the popularity distribution in the server's local domain.

For example, suppose Tom, a user at *CampusA*, has uploaded a video to *Granary*, and shares it to his friends. Then he notifies his friends about it by Email. It is possible that the friends in *CampusA* get Tom's message first and become the first client to download it. If the storage system finds out that this file is popular near *CampusA*, it will place all replicas near *CampusA*. Some days later, his friends in *CampusB*, which is far from *CampusA*, get Tom's Email and try to download the data. Because this file doesn't get more popular, based on the replication distribution strategies above, no more replicas will be generated. So there will never be replicas near *CampusB*, even though it's quite slow for the friends in *CampusB* to download from servers near *CampusA*.

In *Granary*, the problem is fixed by a mechanism called *Remote Boosting* (RB). We compare these two kinds of requests. If a file has more *remote requests* than *local requests*, even if its replica number is enough according to its popularity, the node will create a new replica for it. The replica placement scheme described in Section 4.2 will ensure that the new replica to be created near *CampusB*. After the new replica has been successfully created, the initial replica in *CampusA* will be deleted to accord with the file's popularity. This deletion does not actually remove the replica from the storage server. Instead, it is just removed from the file's replica list and marked as deleted (trashed) on the storage server. A trashed replica will be really deleted when the server needs to spare disk space for new replicas, but if the file is decided to replicate on this server again, the trashed replica (if survived) could be recycled so that the cost of replica propagation would be saved.

4.2   Replica Propagation

The optimal placement of a new replica is to place it near users demanding it. We quantize the benefit of placing a replica of file f on a certain server a by the average transmission distance $\bar{D}(a)$.

$$\bar{D}(a) = \frac{\sum_j d^{j,a} \times q_j}{\sum_{\cdot j} q_i} \tag{4}$$

where $d^{j,a}$ is the network distance from client j to server a, and q_j indicates the request frequency from client j for file f. We will select the server with the minimal $\bar{D}(a)$ as the *destination server*, on which we make a new replica.

The data of new replicas is distributed in a pulling manner. Let's suppose an initiating server A has decided to propagate file f to *destination server B*. Firstly, server A adds the ID of B in the *replica list* of file f. In *Granary*, the replica list is stored in its DHT layer which ensures consistency and availability, and can be globally accessed. Then server A sends a message to B to notify about it. After that, server B downloads the file content from some server, which is not necessarily A, but is chosen by the server selection scheme described in the next section.

This method ensures the replicas to be propagated as soon as possible. Additionally, since the replica list is already updated, clients can start to download from server B even before the replica propagation is finished. In fact, they often complete the download almost simultaneously with the propagating process. This is because that the network speed from a client to a server is often slower than that between two servers in *Granary*. Apparently, the effect is the same as setting up a proxy server on B. It is just another benefit from our strategies.

If there is no enough space on *destination server* to store the new replica, some file on the node will be deleted to spare the storage space. We use the *least recently used* (LRU) scheme to decide which file to delete. However, it means a reduce of replicas for some file. To ensure data availability, a minimal number replicas (e.g. 2) is persisted for every file. If the node is unable to allocate enough disk space for the file content, it will remove itself from the replica list in the meta data thus cancels this replication. However, if such kind of things happens frequently, what we really need to do is put more servers in this area.

4.3   Selecting a Proper Server

In *Granary*, a client logs in to a server he knows first, which is called the *portal*. The *portal* can be selected from the server list stored on the client, which will be updated every time he logs in. With this design, client can get the latest *portal* list, and access the system even when portal list update is unavailable.

After accessing the selected *portal*, a list of available servers will be sent back to the client. It includes nodes with enough disk space when uploading. As for downloading, it is composed of those accommodating the file's replicas, which can be learned from the meta data stored in the DHT layer. After that, the client broadcasts a request message to all available servers. Thus each server will send back a reply message, including an estimated *weighted mean data rate* $(\bar{R})$ and the *network distance* between the server and the client.

The $\bar{R}$ is calculated from the logs recorded in the requesting client's cluster leaf in the IP tree or domain tree, as indicated below. In such formula, j is the index of the log; $size_j$ indicates the size of the file transmitted; and age_j shows how many hours it has been since the transmission finished; while $time_j$ describes the time the transmission takes.

$$\bar{R} = \frac{\sum_j size_j \times 0.5^{age_j}}{\sum_j time_j \times 0.5^{age_j}} \tag{5}$$

The $\bar{R}$ reflects the most recent transfer speed to and from the requesting client or other clients near it. We use it to prevent the convergence of client's selection. If every client simply chooses the closest server, it may happen that most clients converge to a small fraction of servers and thus overload them, while other servers are still idle [12]. The client picks half of the available servers with higher $\bar{R}$, from which it then selects the one with the shortest distance. Because an overloaded server concedes a low $\bar{R}$, it will be rejected in the first step. This server selection scheme is named as the *Hybrid* selection.

5   Evaluation

Granary has been deployed in 5 universities across China. It has attracted about 500 registered users to use it. Each user stores about 800 megabytes data in average in the system. The scale of the current system is still too small for a strategy evaluation, so we use two kinds of experiments to demonstrate the effectiveness of our replication schemes. The first one is a network simulation using logs from two heavy loaded sharing systems. And the second is an emulation carried out on PlanetLab.

5.1   Simulation Results

In this section, two simulations were done for evaluating the effectiveness of the strategies used under heavy workload. The first one was carried using a transfer log from a FTP server mainly used for multimedia file sharing in our campus, and the other was finished with request logs from a large scale wide-area network video on demand(VoD) system [13]. Apparently, FTP is a traditional way of file sharing. As for VoD, it can be looked as a kind of real time sharing method. Both systems are used for multimedia sharing, which is also targeted by *Granary*. The details of these two workloads are shown in Table 1. We developed a simulation program that simulates a *Granary* system running in a bandwidth-limited wide-area environment. We ran simulations on a network topology generated by GT-ITM[1]. The network consisted of 9 transit domains, each of which owned 12 transit nodes in average. A transit node was connected with 1 stub domain which had 10 stub nodes commonly. Thus in total there were 108 transit nodes and 1080 stub nodes. The bandwidth of inter-domain links was 1Gbps and that of intra-domain links was 100Mbps. Each client got a bandwidth of 10Mbps. Finally, forty stub nodes were chosen randomly to be the storage servers, while the others would be assigned as client hosts.

We have evaluated several combinations of techniques in the simulations. Before introducing the experimental results, we firstly explain the terminology that would be used below.

- *Caching* – the caching scheme used by Pond [2] and PAST [3]. The client requests the nearest server for a file, which retrieves the file from the accommodating servers, sends it to the client and keeps it in its local cache.

---

[1] GT-ITM project, `http://www-static.cc.gatech.edu/projects/gtitm/`

Table 1. Characteristics of the simulation workloads

	FTP	VoD
Duration	136 days	30 days
Number of files	8,540	7,525
Average file size	56MB	125MB
Number of requests	97,239	3,021,401
Requests per hour	30	4,196
Size of downloaded data	12,573GB	264,562GB

Table 2. The *average transmission rate of downloads* (Avg. rate) and the *average traffic on network links* (Avg. traffic) in the simulations using the FTP and VoD workloads

	Avg. rate (byte/s)		Avg. traffic (MB)	
	FTP	VoD	FTP	VoD
Caching	773,588	634,381	13,220	191,400
Granary (C)	773,338	828,779	13,280	163,700
Granary (C + RB)	773,655	830,062	13,260	163,000
Granary (H)	773,728	838,942	13,250	162,200
Granary (H + RB)	774,172	840,365	13,220	161,400
Full mirror	900,269	892,462	5,997	120,700

- *Granary* – the replication scheme described in this paper and used in the *Granary* system. It may come with one or two of the following strategies:
  - *Closest* (C) – the client always choose the server with the shortest distance, using the network distance metric defined in Section 3.2.
  - *Hybrid* (H) – the client choose half of the available servers with higher *weighted mean data rate* ($\bar{R}$), from which it chooses the one with the shortest distance. It is explained in Section 4.3.
  - *Remote Boosting* (RB) – the server propagates extra replicas for files that are not very popular but have more remote requests than local requests. See Section 4.2 for details.
- *Full mirror* – all files have already been mirrored on every server. It gets the optimum performance at the cost of maximum disk usage.

Table 2 shows that with the VoD workload the average download rate under *Granary* (H + RB) is 32.5% higher than under *Caching*, while the improvement is not so obvious with the FTP workload. After further analysis, we found that in FTP workload the requests for a single file is evenly scattered over a long period of time, so that every file is requested at a rather low frequency. Cache and replication are usually ineffective when files are not frequently requested. This can explain why the result of FTP workload is not as good as that of the VoD's.

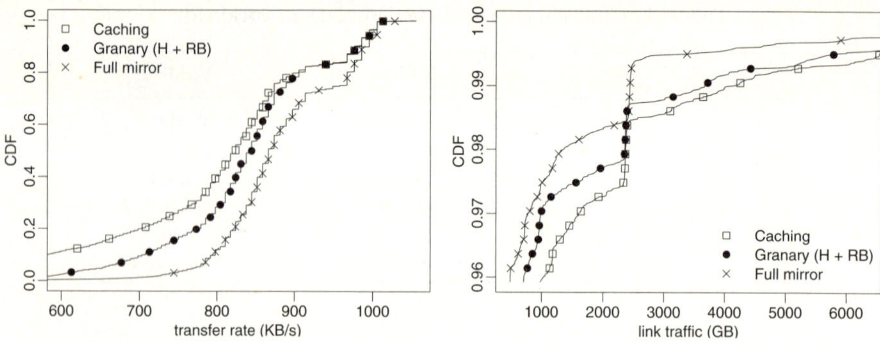

Fig. 1. The CDF of transmission rate of downloads (VoD workload)

Fig. 2. The CDF of link traffic on the most critical links (VoD workload)

From Figure 1, we can find that *Granary* (H + RB) has reduced the fraction of downloads with transfer rate below 700KB/s from 20% to 10%, which is an obvious improvement of user experience.

Then let's have a look at the bandwidth consumption. We recorded the total throughput on every network link that connects two hosts. Table 2 demonstrates that *Granary* replication schemes consume less network bandwidth than *Caching*. It is clear from Figure 2 that network traffic on the most critical links, which are the ones with the biggest throughput, has been notably reduced by *Granary*.

5.2 Experiment Results on Planetlab

PlanetLab[2] is an Internet-scale testbed for distributed systems. To test the performance of *Granary* in a wide-area network, we deployed *Granary* on 40 nodes of PlanetLab, which are located dispersively in different countries all over the world including China, Singapore, US, Poland, UK, Spain, Denmark and Germany, and carried out two experiments: the *REPLAY* experiment and the *SHARE* experiment.

The *REPLAY* experiment is based on a request log from the *Granary* system that has been running as a public service in our campus for a few months. The log includes 746 download requests for 448 different files, from 60 different IP addresses. We doubled the density of the requests and assigned them to client simulators deployed on 120 PlanetLab nodes. The client simulators replayed the requests, and recorded the transmission rate of every download.

The *SHARE* experiment is to simulate a typical scenario of sharing, that a user uploads a file and tells his friends all over the world to download it. We set up client simulators on 193 randomly selected PlanetLab nodes, and made them download a 5MB-sized file one by one.

---

[2] PlanetLab, `http://www.planet-lab.org/`

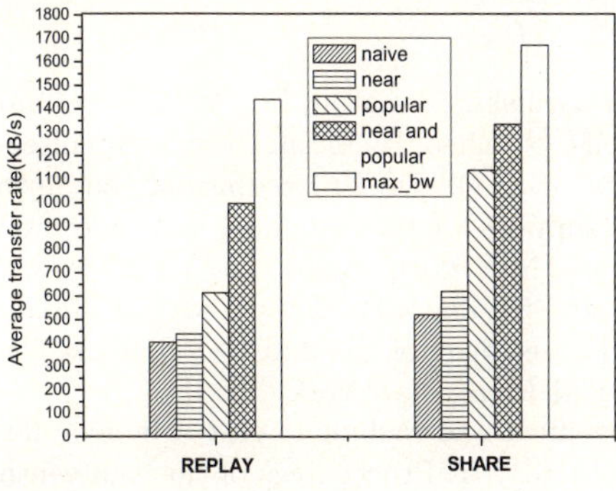

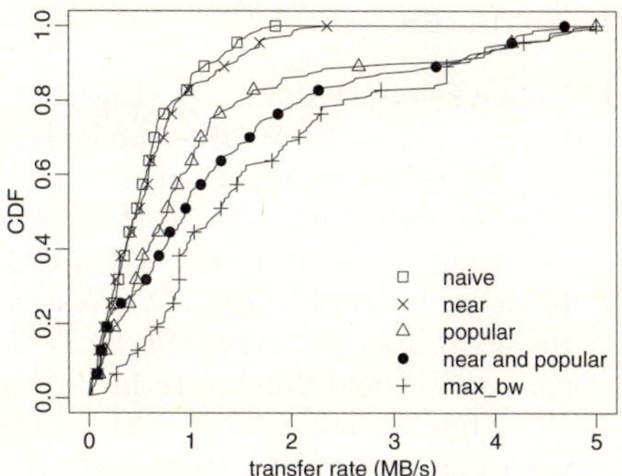

Fig. 3. Comparison of average transmission rate of downloads from experiments on PlanetLab

Fig. 4. CDF of transmission rate of downloads from the SHARE experiment on PlanetLab

Four replication configurations are tested. The *naive replication* places 2 replicas on random servers and keeps them unmoved; the *near replication* places replicas close to the requesters; the *popular replication* decides the number of replicas by the file's popularity; the *near and popular replication* is the combination of the two, which is the default setting of *Granary*. In addition, we measure the maximal bandwidth of every client, and refer it as *max_bw* in the results.

The average transmission rate of downloads can be found in Figure 3. We can see that after applying *near and popular replication*, downloads have been accelerated by nearly 60% in the *REPLAY* experiment, and more than 100% in the *SHARE* experiment. The *SHARE* often performed better than the *REPLAY*, because a file is shared by more users in the *SHARE* than in the *REPLAY*.

Finally, we can have a further look to the *SHARE* experiment in Figure 4. The performance gap between different replication method is clearly shown. The CDF of *near and popular* is quite close to that of the *max_bw*, indicating it as a very optimized method.

6 Conclusion

More and more people store and share their files, including large files such as audio and video, on the Internet. There has been a lot of work improving specific file sharing systems such as BitTorrent, but improving the performance of file sharing on a distributed storage system still remains a challenge. In this paper, we introduced how we tried to accelerate the sharing of files on *Granary*, a distributed file storage system deployed in several campuses across China. *Granary* replicates popular files near to the users that request it for reducing its bandwidth consumption and improving the transmission rate of file downloads. Combined with the *Hybrid* server selection scheme and the *Remote Boosting* mechanism, our replication scheme performs notably better than conventional caching schemes in file sharing according to the experimental results.

References

1. Kubiatowicz, J., Bindel, D., Chen, Y., Czerwinski, S., Eaton, P., Geels, D., Gummadi, R., Rhea, S., Weatherspoon, H., Wells, C., Zhao, B.: Oceanstore: an architecture for global-scale persistent storage. In: ASPLOS-IX: Proceedings of the ninth international conference on Architectural support for programming languages and operating systems, pp. 190–201. ACM Press, New York (2000)
2. Rhea, S., Eaton, P., Geels, D., Weatherspoon, H., Zhao, B., Kubiatowicz, J.: Pond: the oceanstore prototype. In: FAST '03: Proceedings of the 2nd USENIX Conference on File and Storage Technologies, ACM Press, New York (2003)
3. Rowstron, A., Druschel, P.: Storage management and caching in past, a large-scale, persistent peer-to-peer storage utility. In: SOSP '01: Proceedings of the eighteenth ACM symposium on Operating systems principles, pp. 188–201. ACM Press, New York (2001)
4. Saito, Y., Karamanolis, C., Karlsson, M., Mahalingam, M.: Taming aggressive replication in the pangaea wide-area file system. In: OSDI '02: Proceedings of the 5th symposium on Operating systems design and implementation, Boston, Massachusetts, pp. 15–30. ACM Press, New York (2002)
5. Hu, J., Li, M., Yu, H., Dong, H., Zheng, W.: Peerwindow: An efficient, heterogeneous, and autonomic node collection protocol. In: Proceedings of the 2005 International Conference on Parellel Processing (ICPP-05), IEEE Computer Society Press, Los Alamitos (2005)
6. Zheng, W., Hu, J., Li, M.: Granary: Architecture of object oriented internet storage service. In: CEC-EAST '04: Proceedings of the E-Commerce Technology for Dynamic E-Business, IEEE International Conference on (CEC-East'04), pp. 294–297. IEEE Computer Society Press, Los Alamitos (2004)
7. Dabek, F., Kaashoek, M.F., Karger, D., Morris, R., Stoica, I.: Wide-area cooperative storage with cfs. In: SOSP '01: Proceedings of the eighteenth ACM symposium on Operating systems principles, pp. 202–215. ACM Press, New York (2001)
8. Clarke, I., Sandberg, O., Wiley, B., Hong, T.W.: Freenet: A distributed anonymous information storage and retrieval system. In: Proceedings of Designing Privacy Enhancing Technologies: Workshop on Design Issues in Anonymity and Unobservability, Berlin/Heidelberg, pp. 46–66. Springer, Heidelberg (2001)
9. Andrews, M., Shepherd, B., Srinivasan, A., Winkler, P., Zane, F.: Clustering and server selection using passive monitoring. In: INFOCOM '02: Proceedings of the 21th Annual IEEE Conference on Computer Communications, pp. 1717–1725. IEEE Computer Society Press, Los Alamitos (2002)
10. Breslau, L., Cao, P., Fan, L., Phillips, G., Shenker, S.: Web caching and zipf-like distributions: Evidence and implications. In: INFOCOM '99: Proceedings of the 18th Annual IEEE Conference on Computer Communications, pp. 126–134. IEEE Computer Society Press, Los Alamitos (1999)
11. Chervenak, A.L., Patterson, D.A., Katz, R.H.: Choosing the best storage system for video service. In: Multimedia '95: Proceedings of the third ACM international conference on Multimedia, San Francisco, California, United States, pp. 109–119. ACM Press, New York (1995)
12. Mogul, J.C.: Emergent (mis)behavior vs. complex software systems. In: EuroSys '06, Proceedings of the 1st EuroSys Conference, Leuven, Belgium, pp. 293–304. ACM Press, New York (2006)
13. Yu, H., Zheng, D., Zhao, B.Y., Zheng, W.: Understanding user behavior in large scale video-on-demand systems. In: EuroSys '06, Proceedings of the 1st EuroSys Conference, Leuven, Belgium, pp. 333–344. ACM Press, New York (2006)

Esodyp+: Prefetching in the Jackal Software DSM

Michael Klemm[1], Jean Christophe Beyler[2], Ronny T. Lampert[1],
Michael Philippsen[1], and Philippe Clauss[2]

[1] University of Erlangen-Nuremberg, Computer Science Department 2
Martensstr. 3 91058 Erlangen Germany
{klemm,philippsen}@cs.fau.de, ronny.lampert@gmail.com
[2] Université de Strasbourg, LSIIT/ICPS
Pôle API, Bd Sébastian Brant
67400 Illkirch-Graffenstaden France
{beyler,clauss}@icps.u-strasbg.fr

Abstract. Prefetching transfers a data item in advance from its storage location to its usage location so that communication is hidden and does not delay computation. We present a novel prefetching technique for object-based Distributed Shared Memory (DSM) systems and discuss its implementation. In contrast to page-based DSMs, an object-based DSM distributes data on the level of objects, rendering current prefetchers for page-based DSMs unsuitable due to more complex data streams. To predict future data accesses, our prefetcher uses a new predictor (Esodyp+) based on a modified Markov model that automatically adapts to program behavior. We compare our prefetching strategy with both a stride prefetcher and the prefetcher of the Delphi DSM system. For several benchmarks our prefetching strategy reduces the number of network messages by about 60 %. On 8 nodes, runtime is reduced by 15 % on average. Hence, network-bound programs benefit from our solution. In contrast to the other predictors, Esodyp+ achieves a prediction accuracy above 80 % with only 8 % of unused prefetches for the benchmarks.

1 Introduction

Today's high-performance computing landscape mainly consists of Symmetric Multi-Processors (SMPs) and clusters [1]. SMPs provide a hardware-managed global address space; clusters are assembled from nodes with private memory. Thus, programmers must explicitly communicate (e. g. with MPI's send/receive) to access remote data. An Software-based Distributed Shared Memory system (S-DSM), e. g. TreadMarks [2], Delphi [3], or Jackal [4], provides a shared-memory illusion on top of the distributed memory. The DSM adds a high-latency level (the remote memory) to the memory hierarchy of the nodes (registers, caches, main memory); performance drops due to the growing data access latencies.

Prefetching provides a solution to this problem by prematurely requesting data before it is needed. Most current hardware (e. g. Intel Itanium, IBM

A.-M. Kermarrec, L. Bougé, and T. Priol (Eds.): Euro-Par 2007, LNCS 4641, pp. 563–573, 2007.
© Springer-Verlag Berlin Heidelberg 2007

POWER) offer prefetch instructions, that let the CPU asynchronously fetch data from main memory. Instead of adding such instructions at compile time, Esodyp [5] inserts prefetches into the program *while it is executed*; the inserted prefetches are then executed by the CPU as usual. Esodyp continuously monitors all accesses to local memory in a Markov-like model, allowing it to predict future accesses. Esodyp+ extends Esodyp for use in an S-DSM on a cluster. It is implemented in Jackal, an object-based DSM, whose compiler prefixes each object access with an *access check* to test whether or not the object is cached on the local node. If not, the runtime system (RTS) requests the object from its home node. Esodyp+ extends the checks to monitor data access patterns in order to predict future accesses and prefetches the objects by bulk transfers.

Obviously, a dynamic prefetcher is superflous for embarrassingly/pleasingly parallel applications or if data access patterns can be analyzed and optimized statically. In contrast, network-bound applications with complex data access patterns benefit from our prefetching solution.

Section 2 covers related work. Section 3 introduces the Jackal DSM protocol. Section 4 presents the modified Markov model and the necessary extensions of the Esodyp+ approach for an S-DSM. Section 5 discusses issues of the Esodyp+ integration into the Jackal DSM. Section 6 shows the performance gain that Esodyp+ achieves and compares it to two known predictors.

2   Related Work

The related work can roughly be divided into a hardware axis and a size axis. With respect to hardware, prefetching occurs either (1) for single CPUs/SMPs or (2) in clusters, especially in S-DSMs. With respect to size, either (A) a single data item is prefetched or (B) a bulk transfer is used.

Almost all current CPUs provide prefetch instructions for single data items (1A). There are also projects, e. g. ADORE [6] or Chilimbi/Hirzel's work [7], that dynamically insert prefetches into programs. Modern hardware platforms with caches show a simple form of bulk prefetching (category 1B) since a whole cache line is "prefetched" upon memory accesses. On clusters, prefetching single data items (2A) is prohibitively expensive. As our proposal is in category 2B, we present this related work in more detail.

JIAJIA [8] uses a history of past accesses to predict. Delphi [3] predicts by means of a third order differential finite context method. Adaptive++ [9] relies on lists of past accesses to predict. In [10], an OpenMP DSM uses the inspector/executor pattern to determine future accesses. All these projects asynchronously request predicted pages one by one. For an object-based DSM this is not viable, as prefetching small objects cannot hide high network latencies.

Stride predictors [11] often separate data streams by using the instruction address as a hash key. While this works for stable access patterns/regular strides, it does not for the more complex access patterns of object-based DSM systems.

Whereas most prefetchers in S-DSMs request page by page, our approach bulk-transfers sets of objects. MPI programmers often use this technique to manually transfer large amounts of data by one MPI send/receive pair. Similarly, Java RMI [12] ships a deep copy of arguments to remote method invocations even if some of the shipped objects are not needed at the remote side. Jackal statically groups associated objects into larger ones [13] and, aggregating access checks, it requests a set of objects and ships it in bulk fashion [14]. Esodyp+ is not limited to *for* loops, but can be applied to arbitrary sequences of code, and it dynamically decides which objects to combine for transfer.

Instead of object combining, TreadMarks dynamically builds page groups, i. e. larger prefetching units [2]. A predictor continuously monitors page faults and decides which pages to group. The predictor is also capable of ungrouping pages if too much false sharing is caused. Our approach is similar in that it requests a set of objects during a prefetching request. However, the to-be-prefetched objects are not grouped and un-grouped, but the set is dynamically formed at each prediction step as necessary.

3   The Jackal DSM Protocol

Jackal [4] implements an object-based DSM for Java on clusters. Instead of cache lines or pages, data transfer and memory consistency are implemented on the granularity of Java objects to reduce false sharing. Jackal respects the Java Memory Model (JMM) [15] and provides a single system image to programmers.

```
1   int foo(SomeObject o) {          1   int foo(SomeObject o) {
2                                     2       if (!readable(o))
3                                     3           fetch(o, readable);
4       return o.field;              4       return o.field;
5   }                                5   }
```

Fig. 1. A function before (left) and after the insertion of an access check (right)

Jackal compiles to a native executable, e. g. for IA-32 or Itanium, and inserts access checks into the Java code to prepare it for the usage in the DSM environment. Fig. 1 shows a bit of original Java code (left) and the added check as Java-like pseudo-code (right). If the accessed object is not locally available, a message is sent to the *home-node* that stores the object's master copy. This request is answered with message that contains the requested data. Thus, the delay caused is roughly two times the network latency plus the cost of object serialization and deserialization. A prefetcher suffers from about the same latency for its request, but reduces the number of blocking waits by requesting the next accessed objects at the same time.

To uniquely identify each object that is allocated in the DSM, a Global Object Reference (GOR) is created. A GOR is a tuple *(alloc-node, alloc-address)*, where *alloc-node* is the logical rank of the node that created the object and *alloc-address* is the object's address on that node. The GOR is fixed during an object's life time; only if the object is reclaimed by the garbage collector, the GOR is released and may be recycled. When a non-local object is received, it is stored in a local caching heap and it is assigned a local address in the local address space for efficient object access. The DSM system maps these local addresses to their corresponding GORs and hence to the home-nodes (for status or data updates).

Whereas Jackal mostly transfers individual objects, arrays are handled differently for performance reasons. Since transferring an array as a whole is not an option, the RTS partitions it into *regions* of 256 bytes. Thus, a failing access check requests only one region from the array's home-node.

4   Esodyp+

Esodyp+ extends Esodyp, the Entirely Software DYnamic data Prefetcher, a dynamic predictor that models address sequences with a variation of the classic Markov model [5]. This section briefly presents the model, how it reacts to the addresses passed by the DSM framework, and how it can help to prefetch objects. A complete explanation can be found in [5].

Classical Markov predictors [16] use two major parameters: *depth* and *prefetching distance*. The *depth* defines the number of past items used to calculate the predictions. The *prefetching distance* defines how much items will be predicted; a value of 1 means that one next element is predicted. Esodyp+ is more flexible by creating and applying a graph instead of these parameters. With this graph, the model can define a maximum depth and can handle all smaller depths simultaneously. When predicting the next N accesses, Esodyp+ uses counters to prioritize the predictions. This is a major difference to other table-driven models [16].

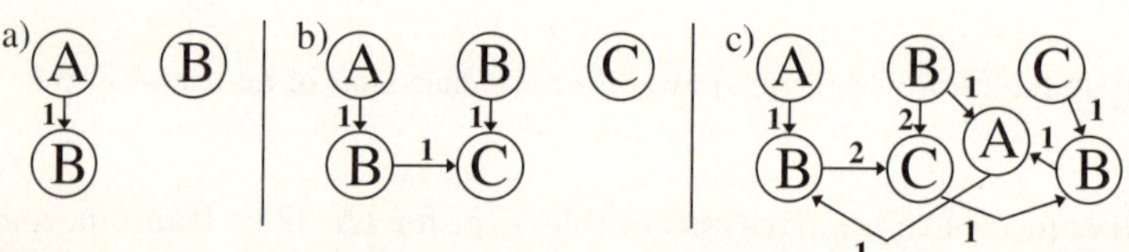

Fig. 2. The Esodyp+ graph (with a depth of 2) after two accesses (a), three accesses (b), and after the full sequence has been passed to the model (c)

To illustrate graph creation, let us assume Esodyp+ sees the following sequence of addresses: A, B, C, B, A, B, C. Fig. 2(a) shows the depth-2 graph after the first two accesses. For depth 2, the model takes into account the last two accesses. The arrow signifies that, after an access to A, an access to B occurred. For nodes without a successor, the predictor cannot predict anything. The single node B symbolizes that, if all we know is there was an access to B, nothing can be predicted. The edge label 1 indicates that Esodyp+ has seen the sequence (A, B) once. Graph construction evolves to Fig. 2(b) when the next address is seen. Two more nodes are added to the graph; C is attached to both Bs. This symbolizes that C occurs both after accessing B and after a sequence (A, B). After the whole sequence has been processed the full graph has 7 nodes, see Fig. 2(c). There, the edge label 2 indicates that an edge has been followed twice. By keeping in memory the current position in the graph, if the next access maps to a known pattern, Esodyp+ predicts in constant time. This is a major benefit compared to other predictors that perform calculations in order to predict.

Moreover, Esodyp+ does not just handle simple addresses but it can store any type of information in its prediction model. Hence, we could directly map GORs to the model and would get an exact prediction. However, for each monitored GOR, at least one node in the graph would be created. Since this would make the model unmanageable for large working sets, it has to be kept compact without losing prediction accuracy. Our solution to this problem is similar to a stride prefetcher, as we compute the differences between two subsequent GORs and store only the difference in the model.

Originally, Esodyp was implemented using a *construction phase* to create the model that later is used in the *prediction phase*. Esodyp+ merges both phases and emits predictions even while it is constructing. This helps to reduce the overhead of the model by starting to predict earlier. This means that prediction strides must be recalculated if changes are made in the graph. Every time a change to the most probable child of a node is more recent than the last change of the current node, the prediction is recalculated. This makes Esodyp+ more dynamic and lightweight than other predictors. Like Esodyp, Esodyp+ triggers a flushing mechanism as soon as there are too many mispredictions [5].

5   Integration of a Predictor into Jackal

There are several ways a predictor for an object-based DSM must be special. Since predictors cannot guarantee accuracy, the DSM runtime must still check the validity of predicted GORs. Except for an object's creator, nodes do not know GORs without previously having accessed the object. Hence, only the home-node of an object can check a GOR for correctness. Thus, the predicted GORs are sent to the home-nodes and an object is only sent back for valid GORs; otherwise the request is safely ignored.

Predictors often only predict the next probable address. For each address a prefetch instruction is emitted into the instruction stream [7,6]. Because of high network latencies in the DSM, prefetching of single objects does not give enough overlap of communication and computation. Hence, Esodyp+ emits predictions to the next N objects. If N objects are bulk-transferred, the program can continue without delay until the $(N+1)$th object is needed. In addition, the message count is reduced from $2N$ to 2 by serializing together N objects into a single message. While predicting the next object can be fairly accurate, the farther away a predicted access is, the smaller is the accuracy. This causes predictions of unused objects. In addition, a growing N increases the chance of false sharing, which in turn increases DSM protocol activity. Our measurements have shown that, on average, $N = 10$ is a good trade-off between false sharing and message reduction for the Jackal DSM.

Since prefetching alters the sequence of memory accesses (accesses that would regularly happen later in the execution are now performed earlier when a node prefetches data), it interferes with memory consistency. As a data item might be updated while a prefetch is outstanding, depending on the memory model, a prefetch either can continue or has to be canceled. Jackal implements the JMM [15] even in the presence of a prefetcher. Simplified, each thread owns a private memory to cache its working set of data items that must be flushed when a synchronization point is reached. Hence, an update is usually not visible to other threads until all of them have reached a synchronization point as well and have flushed their caches. Prefetching blends well with such a weak memory consistency model. An active prefetch is not affected by concurrent updates that happen in the private cache of the updater. If the update hits a synchronization, it still does not affect the prefetch, since the JMM requires a synchronization for a cache refresh as well. As the requestor waits for the prefetch request to complete, it cannot reach a synchronization point in the meantime.

6   Performance

To evaluate the performance of Esodyp+ in Jackal, we compare it to both a stride predictor and the Delphi predictor on a Gigabit Ethernet cluster of Xeon 3.20 GHz nodes with 2 GB of memory and Linux (kernel 2.6.14.2). Since all prefetchers use the same RTS interfaces and the same maximum of N objects for each prefetching request, the measured differences are only caused by the overhead incurred and by the prediction quality. In all tests, each thread uses its own predictor model (since a global predictor does not make sense). Hence, the memory consumptions below are per thread and not globally.

We present four benchmarks and discuss the effects of the prefetchers. Since a predictor is useless in an embarrassingly parallel program, the benchmarks represent classes of applications that communicate with different access patterns. We feel that the selection is representative for applications that make use of general purpose DSMs. The results are the averages over 5 runs of each benchmark.

6.1   Predictors

To validate our predictor, we compare it to two other predictors.

Our multi-stream **stride predictor** implementation [11] with a table of 128 bytes calculates a new stride as the difference between the current GOR and the last GOR. Using a *confidence counter*, predictions are only made if the same stride occurs a certain number of times in sequence. To give the stride prefetcher a better chance to keep recurring strides, we use a *dirty counter* that enables the predictor to ignore a different stride as long as the recurring stride reappears quickly enough in the sequence of strides. The next data accesses are predicted by adding the active stride to the current address. Stride predictors can only

Table 1. Runtimes and message counts for the benchmarks (best in bold)

	Nodes	Runtime (in seconds)				Messages (in thousands)			
		w/o	Stride	Delphi	Esodyp+	w/o	Stride	Delphi	Esodyp+
SOR	2	24.0	**23.7**	23.7	23.9	27.6	7.1	**5.4**	6.0
	4	13.2	12.3	**12.3**	12.4	83.1	22.7	17.9	**17.6**
	6	9.4	**8.6**	8.6	8.7	139.2	36.2	**29.9**	30.0
	8	8.0	**7.2**	7.3	7.3	196.0	53.9	**40.6**	42.3
Water	2	113.8	113.7	114.5	**102.3**	1593.0	1593.0	1593.0	**962.4**
	4	78.0	78.2	78.5	**62.1**	2651.6	2651.6	2651.5	**1593.9**
	6	64.4	64.4	64.5	**52.6**	3260.4	3260.5	3260.6	**1967.9**
	8	58.7	58.7	59.1	**49.6**	3756.5	3756.7	3758.4	**2248.9**
Blur2D	2	10.6	5.9	9.3	**3.6**	223.9	223.9	134.6	**33.4**
	4	6.9	6.6	5.9	**4.0**	385.4	385.4	190.0	**99.8**
	6	7.9	10.0	6.1	**5.0**	483.4	483.4	230.6	**166.2**
	8	8.8	13.5	7.1	**6.9**	581.5	581.5	277.2	**232.3**
Ray	2	**69.2**	71.7	71.8	69.9	9.2	5.9	8.7	**5.9**
	4	35.4	36.1	36.5	**35.2**	27.3	17.4	25.8	**17.1**
	6	24.1	24.2	24.6	**23.8**	45.4	29.0	42.1	**28.4**
	8	18.4	18.5	18.9	**18.3**	57.3	42.7	59.9	**41.1**

Table 2. Accuracy of different prefetchers and their unused but prefetched objects

	Accuracy (in %)			Unused objects (in %)		
	Stride	Delphi	Esodyp+	Stride	Delphi	Esodyp+
SOR	81,09 %	**97,13 %**	95,39 %	2,48 %	**0,75 %**	6,61 %
Water	8,91 %	31,81 %	**67,81 %**	79,42 %	**7,88 %**	12,83 %
Blur2D	0,00 %	49,94 %	**76,81 %**	**0,00%**	1,96 %	10,36 %
Ray	79,99 %	0,00 %	**83,12 %**	6,31 %	100,00 %	**1,93 %**
Average	42,50 %	44,72 %	**80,78 %**	22,05 %	27,65 %	**7,93 %**

predict very regular accesses. As a GOR not only consists of a memory address but also contains a node rank, the Stride often mispredicts in non-array programs or for complex data distributions.

Our **Delphi predictor** implementation [3] continuously updates its information and uses a constant memory cache. As it cannot detect the frequency of sequences, rare sequences cause it to forget earlier sequences and to emit mispredictions. Delphi uses a hash function to map sequences to its table. In the best case, each sequence receives a unique index. However, the number of conflicts depends on the memory access pattern. Our implementation employs a 4096-entry hash table that uses the last three accesses as the hash key. Each entry contains a pointer to a structure containing a GOR and access check information. Hence, the total size of the table is 96 KB per thread. In a certain sense, Delphi is closest to a Markov model, as it uses a fixed depth of 3. But it cannot handle depths of 1 or 2 simultaneously. For n sequences $(A, B, *)$, Delphi needs n entries, each of which stores A, B, and the last element. Esodyp+, however, handles all the depths 1, 2, and 3, and stores the subsequence (A, B) only once. A prioritized linked list then covers the n possibilities.

6.2   Benchmarks

SOR iteratively solves discrete Laplace equations on a 2D grid ($4,100 \times 4,100$; 50 iterations). It computes new values of a grid point as the average value of four neighboring points. Each thread of SOR receives a contiguous set of rows of the grid. SOR only communicates at the boundaries of the partitions, at which the threads read the values of the grid points of other threads.

Although restructuring compilers for array-based languages may handle such highly regular programs better, it is instructive how a prefetcher affects SOR. With a model of 2.2 KB, Esodyp+ reduces SOR's runtime by about 9 % and the message count by about 78 %. As can be seen in Table 1, all prefetchers roughly achieve the same gain; Stride is slightly faster due to its lower internal overhead. Table 2 shows that Delphi is most accurate closely followed by Esodyp+. Stride cannot outperform the others as the access sequence is not simple enough.

Water [17] simulates moving water molecules by means of an (N-square) N-body simulation. Work is divided by assigning molecules to different threads. After a thread has finished computing new directions and accelerations for its molecules in the current time step, it publishes these results for the other threads by means of a special class that implements both synchronization and simultaneous exchange of updates with other threads.

Water has a large working set (1,792 molecules) that is communicated after each time step, making it highly network-bound. The Stride cannot correctly predict, as the objects are scattered over the DSM. Delphi suffers from a number of conflicts in its database and from the high number of objects being accessed. In contrast, Esodyp+ builds an almost perfect model of 14.8 KB and reduces runtime by up to 20 % (for 4 nodes). The message count is decreased by about 40 %. No static analysis can achieve similar results for such irregular applications. Esodyp+ achieves 68 % accuracy (Table 2), which is twice as good as Delphi and about 8 times better than Stride. Stride predicts many unused objects since, once a decision is made, the program behavior has already changed.

Blur2D implements a 2D convolution filter that softens a picture of 400×400 gray-scale points (20 iterations). It uses a 2D array of double precision values that describe the pixels' gray values. The value of a pixel is computed as the average of itself and its eight neighbors. For such a stencil computation (like SOR) accesses are difficult to predict because the parallelization does not fit to the data distribution of the DSM. While Jackal favors a *row-wise* distribution scheme, Blur2D uses a *column-wise* work distribution. Hence, false sharing and irregular access patterns make Blur2D highly network-bound.

In contrast to Stride and Delphi, Esodyp+ (5.4 KB model size) saves roughly 20 % runtime on 8 nodes and the number of messages drops by 60 %. Esodyp+ predicts more accurately (almost 77 %) due to the additional information in its model. In contrast to Delphi and Stride, when constructing the request message,

Esodyp+ first tries the most probable next access. If this leads to an object that has already been requested, it uses another edge of its graph to predict a less probable but still possible access. Hence, Esodyp+ prefetches not only the most likely sequence but also less frequently ocurring data access sequences. For increasing node counts, it cannot compensate the false sharing. Stride's simplistic model is the reason for its poor behavior. Due to a high number of consecutive accesses to the same DSM region, Stride does not emit any good predictions. As of the confidence counter, Stride does not prefetch when a pattern change occurs. Disabling this counter causes a 50 % slow-down due to mispredictions. Delphi's hash table is not able to provide enough slots to store the complex data access sequences of Blur2D.

Ray renders a 3D scene constructed from 2, 000 randomly placed spheres. The image is stored as a 2D array (500×500) of RGB values. For parallelization, the image is partitioned into independent sub-images that are assigned to the threads. Raytracing is inherently parallel: it has a read-only working set (the spheres) and a thread-local working set that is written (the sub-images).

On average, the prefetchers do not gain any performance, although Stride and Esodyp+ (1.5 KB model) save roughly 35 % of the messages. Delphi is unable to make correct predictions due to conflicts in the hash table. This leads to an accuracy of 0 % and an unused percentage of 100 %. Stride is able to keep a score of 79 % compared to the 83 % of Esodyp+. However, Stride makes more unused predictions (6,31 % vs. 1.93%). Transferring read-only data at initialization, the threads work on local data that is only transferred at the end of the computation. Obviously, for embarrassingly parallel applications prefetching does not improve performance. Variations in the cluster load cause the fluctuations in Table 1.

To **summarize**, let us compare the overheads of the predictors. When learning, Stride has the lowest overhead as it only considers strides relative to the last access. Delphi causes a higher overhead by computing the hash index for each access. Esodyp+ updates a few nodes and counters in the model, which also results in a higher overhead. When predicting, Delphi still calculates the hash index of the current access sequence, whereas Esodyp+ predicts by just following pointers to the most probable prefetch candidates. Hence, for stabilized models, Esodyp+ reaches the low overhead of a stride predictor. This is a major advandage of the Esodyp+ model, as it is able to modelize complex memory access behavior with a low prediction overhead once the model has been created.

7   Conclusions and Future Work

In this paper, we have shown that it is worthwhile to integrate a predictor into an object-based DSM, since generally predictors can compensate their overheads. Esodyp+, a novel Markov-based prefetcher performs better than two existing prefetchers. It is more precise and more efficient in predicting and emitting prefetches and reduces the message count by about 60 %. It reduces runtime by 15 % on an 8-node computation. On average, it achieves an accuracy of 80 % compared to 45 % of the other predictors. Hence, our prefetcher is well-suited for network-bound applications with complex data access patterns.

Because of its Markov-like models, Esodyp+ automatically adapts to various access patterns and is only limited when a pattern is completely irregular. Therefore, instead of passing GORs, we work on adding structural data about object associations (i. e. connections between individual objects) to Esodyp+. We will also try to automatically choose an optimal prefetching distance at runtime, such that the distance best fits the program, reduces the potential of false-sharing, and speeds up the program. Another avenue of work is program phase detection. By using Jackal's internal profilers, the predictor's impact on the program can be assessed. If prefetching is not efficient, it may be temporarily switched off to get rid of its runtime overhead.

References

1. TOP500 List (2006), http://www.top500.org/
2. Keleher, P., Dwarkadas, S., Cox, A.L., Zwaenepoel, W.: TreadMarks: Distributed Shared Memory on Standard Workstations and Operating Systems. In: Proc. Winter 1994 USENIX Conf., San Francisco, CA, pp. 115–131 (1994)
3. Speight, E., Burtscher, M.: Delphi: Prediction-Based Page Prefetching to Improve the Performance of Shared Virtual Memory Systems. In: Proc. Intl. Conf. on PDPTA, Las Vegas, NV, pp. 49–55 (2002)
4. Veldema, R., Hofman, R.F.H., Bhoedjang, R.A.F., Bal, H.E.: Runtime Optimizations for a Java DSM Implementation. In: Proc. ACM-ISCOPE Conf. on Java Grande, Palo Alto, CA, pp. 153–162 (2001)
5. Beyler, J.C., Clauss, P.: ESODYP: An Entirely Software and Dynamic Data Prefetcher based on a Markov Model. In: Proc. 12th Workshop on Compilers for Parallel Computers, A Coruna, Spain, pp. 118–132 (2006)
6. Lu, J., Chen, H., Fu, R., Hsu, W., Othmer, B., Yew, P., Chen, D.: The performance of runtime data cache prefetching in a dynamic optimization system. In: Proc. 36th IEEE/ACM Intl. Symp. on Microarchitecture, San Diego, CA, pp. 180–190 (2003)
7. Chilimbi, T.M., Hirzel, M.: Dynamic Hot Data Stream Prefetching for General-Purpose Programs. In: Proc. ACM Conf. on PLDI, Berlin, Germany, pp. 199–209 (2002)
8. Liu, H., Hu, W.: A Comparison of Two Strategies of Dynamic Data Prefetching in Software DSM. In: Proc. 15th Intl. Parallel & Distributed Processing Symp., San Francisco, CA, pp. 62–67 (2001)
9. Bianchini, R., Pinto, R., Amorim, C.: Data Prefetching for Software DSMs. In: Proc. Intl. Conf. on SC, Melbourne, Australia, pp. 385–392 (1998)
10. Jeun, W.C., Kee, Y.S., Ha, S.: Improving Performance of OpenMP for SMP Clusters through Overlapping Page Migrations. In: Proc. Intl. Workshop on OpenMP, Reims, France, CD-ROM (2006)
11. Fu, J.W.C., Patel, J.H., Janssens, B.L.: Stride Directed Prefetching in Scalar Processors. SIGMICRO Newsletter 23(1-2), 102–110 (1992)
12. SUN Microsystems: RMI Specification (1998) http://java.sun.com/products/jdk/1.2/docs/guide/rmi/spec/rmi-title.doc.html
13. Veldema, R., Philippsen, M.: Using Object Combining for Object Prefetching in DSM Systems. In: Proc. 11th Workshop on Compilers for Parallel Computers, Seeon, Germany (2004)

14. Veldema, R., Hofman, R.F.H., Bhoedjang, R.A.F., Jacobs, C.J.H., Bal, H.E.: Source-Level Global Optimizations for Fine-Grain Distributed Shared Memory Systems. In: ACM Symp. on PPoPP, Snowbird, UT, pp. 83–92 (2001)
15. Manson, J., Pugh, W., Adve, S.V.: The Java Memory Model. In: Proc. 32nd ACM Symp. on PoPL, Long Beach, CA, pp. 378–391 (2005)
16. Joseph, D., Grunwald, D.: Prefetching Using Markov Predictors. IEEE Transactions on Computers 48(2), 121–133 (1999)
17. Woo, S.C., Ohara, M., Torrie, E., Singh, J.P., Gupta, A.: The SPLASH-2 Programs: Characterization and Methodological Considerations. In: Proc. 22nd Intl. Symp. on Computer Architecture, St. Margherita Ligure, Italy, pp. 24–36 (1995)

Modeling and Validating the Performance of Atomic Broadcast Algorithms in High Latency Networks

Richard Ekwall and André Schiper

École Polytechnique Fédérale de Lausanne (EPFL), 1015 Lausanne, Switzerland
{nilsrichard.ekwall,andre.schiper}@epfl.ch

Abstract. The performance of consensus and atomic broadcast algorithms using failure detectors is often affected by a trade-off between the number of communication steps and the number of messages needed to reach a decision.

In this paper, we model the performance of three consensus and atomic broadcast algorithms using failure detectors in the oft-neglected setting of wide area networks and validate this model by experimentally evaluating the algorithms in several different setups.

1 Introduction

1.1 Context

Chandra and Toueg introduced the concept of failure detectors in [1]. Since then, several atomic broadcast [2] and consensus [1,3] algorithms based on failure detectors have been published.

The performance of these algorithms is affected by a trade-off between the number of communication steps and the number of messages needed to reach a decision. Some algorithms reach decisions in few communication steps but require more messages to do so. Others save messages at the expense of additional communication steps (to diffuse the decision to all processes in the system for example). This trade-off is heavily influenced by the message transmission and processing times. When deploying an atomic broadcast algorithm, the user must take these factors into account to choose the algorithm that is best adapted for the given network environment.

The performance of these algorithms has been evaluated in several environments, both real and simulated [4]. However, these evaluations are limited to a symmetrical setup: all processes are on the same local area network and have identical peer-to-peer round-trip times. Furthermore, they only consider low round-trip times between processes (and thus high message processing costs), which is favorable to algorithms which limit the number of sent messages, at the expense of additional communication steps.

A.-M. Kermarrec, L. Bougé, and T. Priol (Eds.): Euro-Par 2007, LNCS 4641, pp. 574–586, 2007.
© Springer-Verlag Berlin Heidelberg 2007

1.2   Contributions

In this paper, we model and evaluate the performance of three atomic broadcast algorithms using failure detectors with three different communication patterns (the first based on reduction to a centralized consensus algorithm [1], the second based on reduction to a decentralized consensus algorithm [3] and the third one, a ring based algorithm [2]) in wide area networks. We specifically focus on the case of a system with three processes, — i.e., supporting one failure — where either (i) all three processes are on different locations and (ii) the three processes are on two locations only (and thus one of the locations hosts two processes). The system with three processes is interesting as it (1) has no single point of failure, (2) represents the case in which the group communication algorithms reach their best performance and (3) can be well modeled analytically. The algorithms are experimentally evaluated with a large variation in link latency (e.g., round-trip times ranging from 4 to 300 ms).

We propose a simple model of the wide area network to analytically predict the performance of the three algorithms. The experimental evaluation confirms that the model correctly predicts the performance for average system loads and for all round-trip times that we considered.

The experimental evaluation of the algorithms leads to the following conclusions. First, the number of communication steps of the algorithms is the predominant factor in wide area networks, whether the round-trip time is high (300 ms) or, more surprisingly (since message processing times are no longer negligible), if it is low (4 ms). The performance ranking of the three algorithms is the same in all the wide area networks considered, despite the two orders of magnitude difference between the smallest and largest round trip times. Second, the performance of each of the algorithms heavily depends on setup issues that are orthogonal to the algorithm (typically the choice of the process that starts each iteration of the algorithm, which can be always the same process, or which can shift from one process to another at each iteration). These setup issues also determine the maximum achievable throughput. Finally, measurements referenced in the paper show, as expected, that the performance ranking of the three algorithms is fundamentally different in a wide area network than in a local area network.

2   Motivation and Related Work

In [5], the authors show that consensus cannot be solved in an asynchronous system with a single crash failure. Several extensions to the asynchronous model, such as failure detectors [1], have circumvented this impossibility and agreement algorithms [1,3,2] have been developed in this extended model. The performance of these atomic broadcast algorithms is evaluated in different ways. Usually, the formal presentation of the agreement algorithms is accompanied by analytical bounds on the number of messages and communication steps that are needed to solve the problem [1,3,6]. This coarse-grained evaluation of the performance of the algorithms is however not sufficiently representative of the situation in a real environment.

To get a more accurate estimation of the performance of the atomic broadcast algorithms, they have often been evaluated in local area networks [2], simulated in a symmetrical environment where all links between processes have identical round-trip times [4] or evaluated in hybrid models that introduce artificial delays to simulate wide area networks [6]. Although these performance evaluations do provide a representative estimate of the performance of atomic broadcast on a local area network, they cannot be used to extrapolate the performance of the algorithms on a wide area network, where the ratio between communication and processing costs is completely different. Furthermore, evaluating the performance of atomic broadcast on wide area networks is not only of theoretical interest. As [7] shows, it is feasible to use atomic broadcast as a service to provide consistent data replication on wide area networks. In this paper, we model the performance of these algorithms *and* validate this analysis by experimentally evaluating the algorithms in wide area networks.

We now discuss the central trade-off that explains the impact of network latency on the performance of atomic broadcast algorithms.

2.1 The Trade-Off Between Number of Messages and Communication Steps

The processes executing the atomic broadcast algorithms that we consider in this paper communicate with each other to agree on a common message delivery sequence. To do so, they need to exchange a minimum number of messages in a number of communication steps. There is here a trade-off on the number of communication steps and the number of sent messages. Usually, a higher number of messages enables the algorithm to reach a decision in fewer communication steps and vice-versa.

Each communication step has a cost. Indeed, each additional communication step induces a delay on the solution to the problem. This cost is typically low in a local area network, whereas it increases with the latency in a wide area network.

Sending messages also has a cost. Whenever a message is sent, it has to be handled by the system. This handling includes costs related to algorithmic computations on its content, serialization (i.e. transforming the message to and from an array of bytes that is sent on the network) and bandwidth used for the transmission.

These costs characterize the trade-off between the number of messages sent and communication steps needed by the algorithm. If a communication step costs nothing, then the algorithm that sends the least number of messages performs the best. If, on the other hand, a communication step is very expensive, the algorithm that sends most messages (and thus saves on the number of communication steps) has the best performance. In this paper, several network latencies are studied to evaluate their impact on this trade-off.

2.2   Related Work

In [8], the authors study the influence of network loss on the performance of two atomic broadcast algorithms in a wide area network. To do this, the authors combine experimental results obtained on a real network with an emulation of the atomic broadcast algorithms. The scope of the work in [8] is different from ours: they evaluate the impact that message loss has on the performance of atomic broadcast algorithms whereas we model and evaluate the impact of *network latency* on the relative performance of different algorithms.

Bakr and Keidar evaluate the duration of a communication round on the Internet in [9]. Their work focuses on the running time of four distributed algorithms with different message exchange patterns, and in particular, the effect of message loss on these algorithms. Their experiments are run on a large number of hosts (10) and the algorithms that they examine do not allow messages to be lost (i.e. an algorithm waits until it has received *all* messages it is expecting). The scope of [9] is similar to ours in that they analyze the relative performance of algorithms with different communication patterns on a wide area network. However, their algorithms are *not* representative of failure detector based atomic broadcast algorithms. Indeed, in the three algorithms we consider, processes never need to wait for messages from *all* the other processes. Thus, if messages from one process are delayed because of a high-latency link, it does not necessarily affect the performance of the atomic broadcast algorithm (whereas it would in [9]).

In [6], an atomic broadcast algorithm that is specifically targeted towards high latency networks is presented. The authors also evaluate the performance of the algorithm in a local area network with added artificial delays (to simulate the high latency links). The artificial delay is however not sufficient to adequately represent the network links of a wide area network. Indeed, such links are also characterized by a lower bandwidth than local area network links. In our performance measurements, we show that in some cases, the low bandwidth of the wide area links strongly limits the performance of the algorithms that are considered.

Several other papers (e.g. [4,2,10,11]) have studied the performance of atomic broadcast algorithms, but these papers either study the performance of the algorithms in a local area network or through simulation.

3   System Model and Algorithms

We consider an asynchronous system of n processes $p_0, \ldots, p_{n-1}$. The processes communicate by message passing over reliable channels and at most f processes may fail by crashing (i.e. we do not consider Byzantine faults). The system is augmented with unreliable failure detectors [1,2].

In the following paragraphs, we informally present reliable broadcast, consensus and atomic broadcast [1]. Reliable broadcast and consensus are building blocks for solving atomic broadcast in two of the atomic broadcast algorithms that we consider.

3.1   Reliable Broadcast, Consensus and Atomic Broadcast

In the *reliable broadcast* problem, defined by the primitives *rbroadcast* and *rdeliver*, all processes need to agree on a common set of delivered messages. In this paper, we consider the reliable broadcast algorithm presented in [1], which requires $O(n^2)$ messages and a single communication step to *rbroadcast* and *rdeliver* a message m.

Informally, in the *consensus* problem, defined by the two primitives *propose* and *decide*, a group of processes have to agree on a common decision. In this paper, we consider two consensus algorithms that use the $\Diamond \mathcal{S}$ failure detector [1] (presented in Section 3.2).

In the atomic broadcast problem, defined by the two primitives *abroadcast* and *adeliver*, a set of processes have to agree on a common total order delivery of a set of messages. It is a generalization of the reliable broadcast problem with an additional ordering constraint. In this paper, we consider two atomic broadcast algorithms which are described in Section 3.3.

3.2   Two Consensus Algorithms

The first consensus algorithm, proposed by Chandra and Toueg [1] and noted CT, is a centralized algorithm that requires 3 communication steps, $O(n)$ messages and 1 reliable broadcast for all processes to reach a decision in *good* runs (i.e. runs without any crashes or wrong suspicions). The behavior of the CT algorithm in good runs is detailed in [12].

The second consensus algorithm, proposed by Mostéfaoui and Raynal [3] and noted MR, is a decentralized algorithm that requires 2 communication steps and $O(n^2)$ messages[1] for all processes to reach a decision in good runs. The behavior of the MR consensus algorithm in good runs and in a system with $n = 3$ processes is detailed in [12].

On the choice of a coordinator: Both the CT and MR consensus algorithms use a *coordinator* that proposes the value that is to be decided upon. This coordinator can be any process in the system, as long as it can be deterministically chosen by all processes (based only on information that is locally held by each process). In the analytical and experimental evaluations of these algorithms, we examine how the choice of the first coordinator influences the performance of the algorithms. We also study the case where the first coordinator changes between instance number k of consensus and the next instance $k + 1$.

3.3   Two Atomic Broadcast Algorithms

Chandra-Toueg atomic broadcast [1]: The Chandra and Toueg atomic broadcast algorithm requires at least one reliable broadcast and a consensus execution for all processes to *abroadcast* and *adeliver* messages.

---

[1] The MR consensus algorithm does not use reliable broadcast as a building block. Instead, reliable diffusion of the decision is ensured by an ad-hoc protocol using n^2 messages.

Whenever a message m is *abroadcast*, it is reliably broadcast to all processes (first communication step in good runs). The processes then execute consensus on the messages that haven't been *adelivered* yet (using the CT or MR algorithm in our case). If m is in the decision of consensus, then m is *adelivered*. A waiting period happens if a consensus execution is already in progress and therefore prevents m from being proposed at once for a new consensus.

Token using an unreliable failure detector [2]: The token-based atomic broadcast algorithm (noted *TokenFD*) solves atomic broadcast by using an unreliable failure detector noted $\mathcal{R}$ and by passing a token among the processes in the system. It requires three communication steps in a system with $n = 3$ processes and $O(n)$ messages for all processes to *abroadcast* and *adeliver* messages.

Whenever a message m is *abroadcast*, it is sent to all processes (first communication step). Message m is then added to the token that circulates among the processes. After the second communication step, m is *adelivered* by the token-holder which sends an update to all other processes about this delivery (third communication step). Again, a waiting period only happens if the token is already being sent on the network (without m) and therefore prevents m from being ordered immediately.

4   Performance Metrics and Workloads

The following paragraphs describe the benchmarks (i.e., the performance metrics and the workloads) that were used to experimentally evaluate the performance of the three atomic broadcast algorithms (reduction to CT consensus; reduction to MR consensus; TokenFD algorithm). The benchmarks in [4,2] are similar to the ones we use here.

Performance metric – latency vs. throughput: The performance metric that was used to evaluate the algorithms is the latency of atomic broadcast. For a single atomic broadcast, the latency L is defined as follows. Let t_a be the time at which the *abroadcast*(m) event occurred and let t_i be the time at which *adeliver*(m) occurred on process $p_i \in \{p_0, \ldots, p_{n-1}\}$. The latency L is then defined as $L \stackrel{def}{=} (\frac{1}{n} \sum_{i=0}^{n-1} t_i) - t_a$. In our experimental performance evaluation, the mean for L is computed over many messages and for several executions. 95% confidence intervals are shown for all the results.

Workloads: The workload specifies how the *abroadcast* events are generated. We chose a simple symmetric workload where all processes send atomic broadcast messages (without any payload) at the same constant rate and the *abroadcast* events follow a Poisson distribution. The global rate of atomic broadcasts is called the *throughput* T.

Furthermore, we only consider the system in a stationary state (when the rate of *abroadcast* messages is equal to the rate of *adelivered* messages) and we only evaluate the performance of the algorithms in good runs (i.e., without any process failures or wrong suspicions).

We specifically focus on the case of a system with three processes, supporting one failure. This system size might seem small. However, atomic broadcast provides strong consistency guarantees (that can be used to implement active replication for example [13]) and is limited to relatively small degrees of replication. If a large degree of replication is needed, then alternatives that provide weaker consistency should be considered [14].

5   Modeling the Performance of the Algorithms

This section discusses the analytical performance evaluation of the two atomic broadcast (and consensus) algorithms in a wide area network. We present the two wide area network models that are considered. Due to lack of space, the derivation of the latencies of the algorithms in this model is not further presented, but can be found in [12]. The predictions of the model (i.e. the numerical applications of the model to the four experimental setups) are shown alongside the experimental evaluation of the algorithms in Section 6.

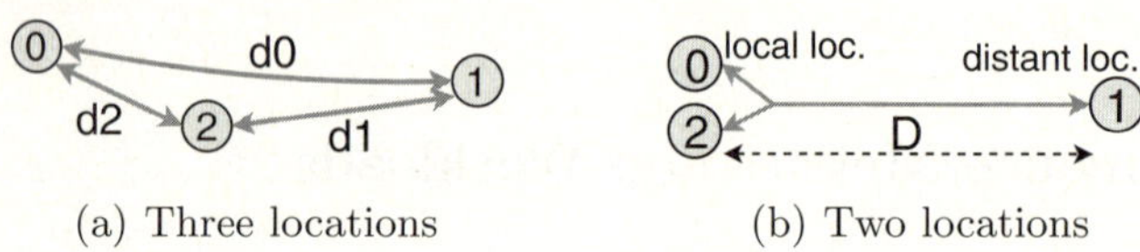

(a) Three locations   (b) Two locations

Fig. 1. Theoretical model of a wide area network

Figure 1(a) presents the model of a wide area network system with three processes on three different locations. The network latency between location i and location $i + 1$ is noted d_i. Without loss of generality, we assume that $d_0 \geq d_1 \geq d_2$. The model is simplified, in the sense that the processing costs of the messages are considered negligible.

Furthermore, the model does not take other factors into account, such as the bandwidth of the links or message loss.

Figure 1(b) presents the model of a system with three processes, one of which is on a distant location. The network latency between the distant location and the local location is noted D. The *two-location model* is a special case of the previous model, with $d_0 = d_1 = D$ and $d_2 = 0$.

The average latency of the three atomic broadcast algorithms in this model can be found in [12].

6   Experimental Performance Evaluation

In the following section, the experimental performance of the atomic broadcast algorithms presented in Section 3 are compared. First, we briefly present the evaluation environments that were considered and then the results that were

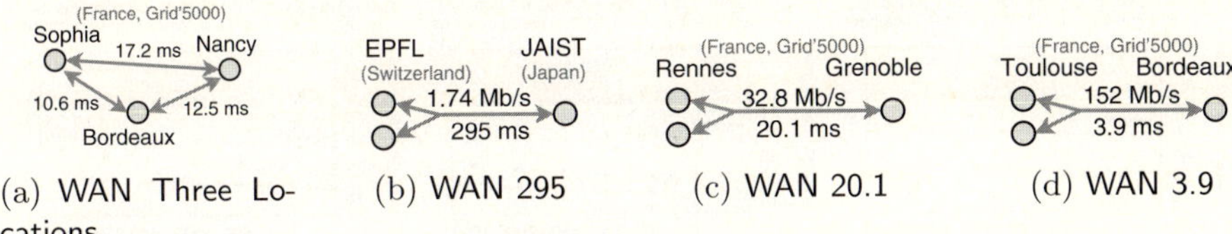

Fig. 2. Wide area network environments in decreasing order of round trip times

obtained are presented, analyzed and compared to the analytical evaluation of Section 5. The algorithms presented in this paper are all implemented in Java, using the Neko framework [15].

6.1   Evaluation Environments

Four wide area network environments were used to evaluate the performance of the three atomic broadcast and consensus algorithms (see Fig. 2) All machines run a Linux distribution (2.6.8 to 2.6.12 kernels) and a Sun Java 1.5.0 virtual machine. The following paragraphs describe the different wide area network environments in which the atomic broadcast algorithms are evaluated.

Three-location wide area network: The first evaluation environment (noted WAN Three Locations, Figure 2(a)) is a system with three locations on Grid'5000 [16], a French grid of interconnected clusters designed for the experimental evaluation of distributed and grid computing applications. The round-trip times of the links between the three processes are respectively $2d_0 = 17.2$ ms, $2d_1 = 12.5$ ms and $2d_2 = 10.6$ ms. The observed bandwidth of the three links are respectively 30.1 Mbits/s, 41.4 Mbits/s and 48.7 Mbits/s.

Two-location wide area networks: Three environments were used to evaluate the performance of atomic broadcast on wide area networks with two different locations:

  – WAN 295 (Figure 2(b)): The first two-location environment consists of one location in Switzerland and one in Japan. The round-trip time between the locations is $2D = 295$ ms and the bandwidth of the connecting link is 1.74 Mb/s.
  – WAN 20.1 and WAN 3.9: The two following environments are systems with both locations on Grid'5000. The WAN 20.1 system (Figure 2(c)) features a round-trip time between locations of $2D = 20.1$ ms and a link bandwidth of 32.8 Mb/s. The WAN 3.9 system (Figure 2(d)) features a round-trip time between locations of $2D = 3.9$ ms and a link bandwidth of 152 Mb/s. The performance characteristics of the three algorithms are similar in WAN 20.1 and WAN 3.9.

6.2   Validation of the Model with the Experimental Results

We now discuss the validation of the model presented in Section 5 by the experimental evaluation of the three atomic broadcast algorithms. As mentioned

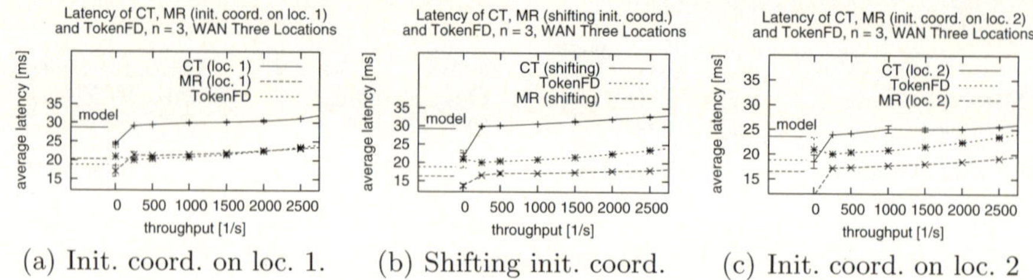

(a) Init. coord. on loc. 1.   (b) Shifting init. coord.   (c) Init. coord. on loc. 2

Fig. 3. Average latency of the three algorithms as a function of the throughput in the WAN Three Locations setting

in Section 4, the performance graphs present the average latency as a function of the throughput in the system. Furthermore, for the CT and MR consensus algorithms, the results are given for an initial coordinator that is fixed in one location or shifting with each new consensus execution. The TokenFD algorithm has no concept of coordinator and its results are the same for all three settings (they are repeated to give a point of comparison with respect to CT and MR). The modeled performance of the algorithms is shown on the far-left of each graph (noted "model").

Table 1. Latency (in milliseconds) of the three algorithms for a throughput of 1000 msgs/s in the WAN Three Locations setting

(a) Init. Coord. on location 1

Algorithm	Measured	Modeled	Diff.
CT	29.92	28.66	+4.4%
MR	21.55	20.32	+6.1%
TokenFD	20.81	18.82	+10.6%

(b) Shifting init. coord.

Algorithm	Measured	Modeled	Diff.
CT	30.57	29.24	+4.5%
MR	17.09	16.28	+5.0%
TokenFD	20.81	18.82	+10.6%

In all experimental setups, the measurements confirm the estimations of the model (Figures 3 and 4), especially in the case of moderate throughputs. When the throughput increases, the load on the processors and on the network (which is not modeled) affects the latency of the algorithms (illustrated in particular in Figure 4(c)), which decreases the accuracy of the model.

Furthermore, when the throughput is very low, the measured latencies of CT and MR are lower than what the model predicts. Indeed, our analysis assumes a load in which messages are *abroadcast* often enough that there is always a consensus execution in progress. In the low throughput executions however, there is a pause between the consensus executions. An unordered message that is received during this pause is immediately proposed in a new consensus execution and thus, the *waiting* phase presented in Section 5 does not apply to that message.

Finally, the point that was not predicted by the analytical model is the result for high throughputs when the initial coordinator of CT and MR is on a local location, illustrated by Figure 4(c). Indeed, in this setting, the system never

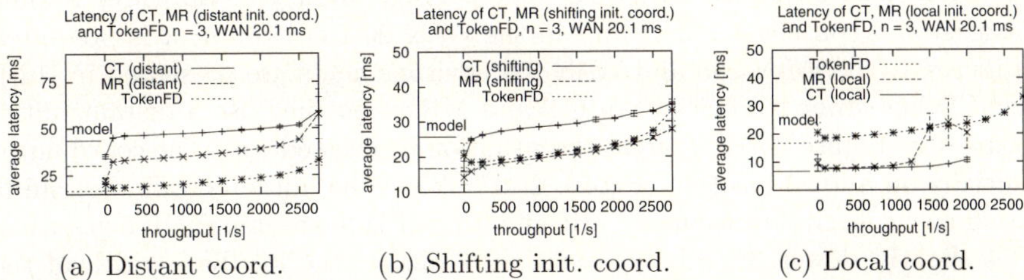

(a) Distant coord.   (b) Shifting init. coord.   (c) Local coord.

Fig. 4. Average latency of the three algorithms as a function of the throughput in the WAN 20.1 setting

reaches a stationary state given a sufficiently high throughput. The processes on the local location reach consensus decisions very fast without needing any input from the distant location. The updates that are then sent to the distant location saturate the link between both locations (its bandwidth is only 32.8 Mbits/s in WAN 20.1). The process on the distant location thus takes decisions slower than the two local processes and prevents the average latency of atomic broadcast from stabilizing. This problem does not affect the settings with a distant or shifting initial coordinator, since the distant location periodically acts as a consensus coordinator, providing a natural flow control. Setup issues, such as the choice of the initial coordinator, thus affect the maximum achievable throughput of the algorithms.

Table 1 presents the latency of the three algorithms in the WAN Three Locations setting, with a throughput of 1000 messages per second. The model predicts the experimental measurements with an error of 5 to 10%.

6.3   Comparing the Performance of the Three Algorithms

WAN Three Locations: The average latency of the three algorithms in the WAN Three Locations environment is presented in Figure 3. TokenFD and MR outperform CT for all locations of the initial coordinator and for all throughputs, due to the additional communication step that is needed by CT. TokenFD and MR perform similarly when the initial MR coordinator is on site 1 (which is the worst-case scenario for MR), whereas MR achieves slightly better latencies than TokenFD for both other initial coordinator locations.

Surprisingly enough, the result of using a shifting initial coordinator in the CT and MR algorithms are opposite: in the case of MR, the latency is lower using a shifting initial coordinator than a fixed initial coordinator on any location, whereas in CT it is higher. The explanation is the following: MR and CT both start a new consensus execution after two communication steps if the coordinator is on a fixed location. If the coordinator shifts, a new execution can start as soon as the next non-coordinator process decides. This is done after *one* communication step in MR (if $n = 3$), but after *three* steps in CT, as explained in Section 5.

WAN 295, WAN 20.1 **and** WAN 3.9: The average latency of the three atomic broadcast and consensus algorithms in the WAN 20.1 environment is presented in Figures 4 (the WAN 295 and WAN 3.9 environements are presented in [12]). TokenFD has lower latencies than CT and MR when they use a distant initial coordinator (Figure 4(a)), whereas the situation is reversed when the coordinator is initially on a local location (Figure 4(c)). When the initial coordinator shifts at each new consensus execution, MR and TokenFD have similar latencies while CT is slightly slower. Finally, as mentioned earlier, the low bandwidth of the link between both locations prevents MR and CT from reaching stable average latencies when the initial coordinator is on the local locations and the throughput is high.

Communication steps versus number of messages: As expected, the performance results presented above show that *communication steps have the largest impact on performance in wide area networks, whereas the number of sent messages is a key to the performance in a local area network [12].* The validity of this statement however varies with the round-trip time of the network that is considered. As the network latency decreases, the impact of the additional messages that need to be sent and processed increases. In the case of networks with 3.9 ms or even 20.1 ms round-trip times, this impact is clearly observable.

However, for a given set of parameters, the algorithm with the best performance is generally the same (whether a wide area network with a 3.9 ms round-trip time is considered or one with a 295 ms round-trip time) and it is correctly predicted by the model.

Finally, we also saw that choosing a CT and MR coordinator on the local location (without implementing an additional flow control mechanism) is not necessarily the best solution performance-wise, since the system cannot reach a stationary state as the total throughput increases. Shifting the initial coordinator between locations at each new consensus execution or choosing the TokenFD algorithm results in a natural flow control which enables the system to remain in a stationary state even for high throughputs (at the expense of a higher average *adelivery* latency).

7   Conclusion

This study confirms that the relative performance between the algorithms is fundamentally different between a local area network and a wide area network (even in wide area networks with small round-trip times): in the former case, the number of sent messages (i.e. the number of messages that need to be processed) largely determines the performance of the algorithms, whereas the communication steps have the most impact in the latter case.

Within wide area networks on the other hand, the performance ranking of the three algorithms remains the same, despite the (two order of magnitude) difference in the round-trip time between the smallest and largest wide area networks. Furthermore, this ranking is correctly predicted by our model. The study also showed that algorithms or parameters which provide a natural flow control (such as the TokenFD atomic broadcast algorithm or the Chandra-Toueg and Mostéfaoui-Raynal consensus algorithms with an initial coordinator that shifts between locations at each new consensus) are effective in reaching higher throughputs in wide area networks.

References

1. Chandra, T., Toueg, S.: Unreliable failure detectors for reliable distributed systems. Journal of ACM 43(2), 225–267 (1996)
2. Ekwall, R., Schiper, A., Urbán, P.: Token-based atomic broadcast using unreliable failure detectors. In: Proc. of the 23rd Symposium on Reliable Distributed Systems (SRDS 2004), Florianópolis, Brazil (October 2004)
3. Mostefaoui, A., Raynal, M.: Solving Consensus using Chandra-Toueg's Unreliable Failure Detectors: A Synthetic Approach. In: Jayanti, P. (ed.) DISC 1999. LNCS, vol. 1693, Springer, Heidelberg (1999)
4. Urbán, P., Shnayderman, I., Schiper, A.: Comparison of failure detectors and group membership: Performance study of two atomic broadcast algorithms. In: Proc. of the Int'l Conf. on Dependable Systems and Networks (DSN), pp. 645–654 (June 2003)
5. Fischer, M., Lynch, N., Paterson, M.: Impossibility of Distributed Consensus with One Faulty Process. Journal of ACM 32, 374–382 (1985)
6. Vicente, P., Rodrigues, L.: An Indulgent Total Order Algorithm with Optimistic Delivery. In: 21st IEEE Symp. on Reliable Distributed Systems (SRDS-21), Osaka, Japan, pp. 92–101 (October 2002)
7. Lin, Y., Kemme, B., Patiño-Martínez, M., Jiménez-Peris, R.: Consistent data replication: Is it feasible in WANs?. In: Proc. 11th International Euro-Par Conference, Lisbon, Portugal, pp. 633–643 (September 2005)
8. Anker, T., Dolev, D., Greenman, G., Shnayderman, I.: Evaluating total order algorithms in WAN. In: Proc. International Workshop on Large-Scale Group Communication, Florence, Italy (October 2003)
9. Bakr, O., Keidar, I.: Evaluating the running time of a communication round over the internet. In: Proc. of the 21st ACM Ann. Symp. on Principles of Distributed Computing, pp. 243–252 (2002)
10. Sousa, A., Pereira, J., Moura, F., Oliveira, R.: Optimistic Total Order in Wide Area Networks. In: 21st IEEE Symp. on Reliable Distributed Systems (SRDS-21), Osaka, Japan, pp. 190–199 (October 2002)
11. Guerraoui, R., Levy, R.R., Pochon, B., Quéma, V.: High Throughput Total Order Broadcast for Cluster Environments. In: IEEE International Conference on Dependable Systems and Networks (DSN 2006) (June 2006)
12. Ekwall, R., Schiper, A.: Comparing Atomic Broadcast Algorithms in High Latency Networks. Technical Report LSR-REPORT-2006-003, École Polytechnique Fédérale de Lausanne, Switzerland (July 2006)
13. Schneider, F.: Replication Management using the State-Machine Approach. In: Mullender, S. (ed.) Distributed Systems, 2nd edn. ACM Press Books, pp. 169–198. Addison-Wesley, London, UK (1993)
14. Alvisi, L., Marzullo, K.: Waft: Support for fault-tolerance in wide-area object oriented systems. In: Proc. of the 2nd Information Survivability Workshop – ISW '98, October 1998, pp. 5–10. IEEE Computer Society Press, Los Alamitos (1998)

15. Urbán, P., Défago, X., Schiper, A.: Neko: A single environment to simulate and prototype distributed algorithms. Journal of Information Science and Engineering 18(6), 981–997 (2002)
16. Cappello, F., Caron, E., Dayde, M., Desprez, F., Jeannot, E., Jegou, Y., Lanteri, S., Leduc, J., Melab, N., Mornet, G., Namyst, R., Primet, P., Richard, O.: Grid'5000: a large scale, reconfigurable, controlable and monitorable Grid platform. In: Grid'2005 Workshop, Seattle, USA, IEEE/ACM (November 13-14 2005)

A Joint Data and Computation Scheduling Algorithm for the Grid

Fangpeng Dong and Selim G. Akl

School of Computing, Queen's University
Kingston, ON, Canada K7L 3N6
{dong,akl}@cs.queensu.ca

Abstract. In this paper a new scheduling algorithm for the Grid is introduced. The new algorithm (JDCS) combines data transfer and computation scheduling by a back-trace technique to reduce remote data preloading delay, and is aware of resource performance fluctuation in the Grid. Experiments show the effectiveness and adaptability of this new approach in the Grid environment.

Keywords: Workflow, Scheduling, Algorithm, Grid Computing, Data.

1 Introduction

The development of Grid infrastructures now allows workflow to be submitted and run on remote Grid resources. At the same time, fledging data Grid projects enable remote data access, replication, and transfer. These advances make it possible to run data-intensive workflows in the Grid. In this paper, we consider the problem of assigning metatasks in a workflow, which is assumed to be represented by an acyclic directed graph (DAG), to Grid resources. The objective is to minimize the total schedule length of the whole workflow (also known as the makespan). As the general form of this optimization problem is NP-Complete [6], heuristic approaches are usually adopted. In the Grid environment, the problem becomes even more challenging: first, the performance of a Grid resource is usually dynamically changing, which makes it harder to get an accurate estimate in advance of a task's execution time; second, input data of a task may cross a long distance from a data storage site to the computational resource. If computational resources and data storage sites are dynamically selected by a Grid scheduler, communication cost for input data transfer may vary according to different data storage and computational resource combinations. This situation is different from the traditional cases where data and computation are located on the same site, or data sites and computational sites are fixed prior to scheduling so that communication cost is a known constant.

Based on the above observations, we propose a workflow scheduling algorithm called JDCS (Joint Data and Computation Scheduling). It considers the possibility of overlapping the time of input data preloading and that of computation, thus reducing the waiting time of a task to be scheduled. This is achieved by using a back-trace technique. To overcome the difficulties brought about by performance fluctuation, JDCS takes advantage of mechanisms such as Grid performance prediction and

A.-M. Kermarrec, L. Bougé, and T. Priol (Eds.): Euro-Par 2007, LNCS 4641, pp. 587–597, 2007.

resource reservation [1], which can capture the resource performance information and provide some kind of guaranteed performance to users. These approaches make it possible for Grid schedulers to get relatively accurate resource information prior to making a schedule, and according to the obtained information, JDCS updates metatask priorities dynamically when it makes a schedule.

The remainder of this paper is organized as what follows. Section 2 introduces models used by JDCS. Section 3 describes JDCS in detail, while Section 4 presents experimental results and a performance analysis. In Section 5, related research is reviewed. Conclusions and suggestions for future work are provided in Section 6.

2 Definitions and Models

A *Grid scheduler* running JDCS receives submitted workflows from Grid users, retrieves resource information from Grid Information Services, creates schedules and commits scheduled metatasks to Grid resources. The Grid has a set of computational resources $\{p_1,..., p_n\}$ and a set of data storage sites $\{d_1,..., d_m\}$. The performance of a computational resource is not only heterogeneous, but also dynamically fluctuating. In a resource management system supporting advance reservation, available resource performance at a specific time can be known by calculating the workload generated by jobs that have reserved resources at that time, as Fig. 1 (a) indicates. Thus, by referring resource predictors or resource management components supporting reservation, the performance fluctuation can be caught. Theoretically, if the time axis can be divided into fine granular periods, the performance within a period can be approximated as a constant. So, to describe the fluctuation, a sequence of time slots $s_1,..., s_k$ is introduced. The computational power of p_i in s_j is denoted as a constant $c_{i,j}$, which can be known from Grid Information Services. To decide the number of time slots which will be used to complete the workflow, an optimistic estimation strategy is used: The scheduler estimates the serial executing time of the entire workflow on each resource, and chooses the smallest one. This strategy is based on the expectation that a parallel processing, even in the worst case, is not worse than the best sequential one.

The network connection between p_i and p_j is denoted as $clink_{i,j}$ and one between p_i and d_j is denoted as $dlink_{i,j}$. The bandwidth of $clink_{i,j}$ or $dlink_{i,j}$ is denoted as $cw_{i,j}$ or $dw_{i,j}$, respectively. For $cw_{i,j}$, if $i = j$, $cw_{i,j} = +\infty$, which implies the communication delay within p_i is 0. Fig. 1 (b) gives an example of the system model. The input data for a metatask is located on a data storage d_j. Computational resources themselves can also provide limited storage capacity that allows input data to be precached prior to the start of the computation. The capacity of the input data cache on p_i is denoted as c_i. Once a metatask is assigned to a computational resource, its input data preloading can start, and once a metatask starts, the input data will be deleted and occupied space will be released. Due to the limitation of cache capacity, it may not be possible for all metatasks to pre-upload their data at the same time to the computational site. Available data cache space on computational resource p_j, at time T is denoted as $ACS(p_j, T)$, which initially equals c_i. To avoid potential conflicts resulting from simultaneous data preloading for different metatasks on the same resource, when a data transfer starts, required space on the destination will be reserved.

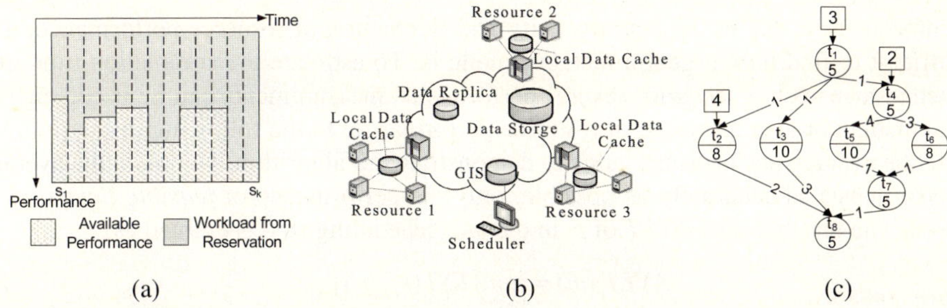

(a)   (b)   (c)

Fig. 1. (a) Performance fluctuation resulted from advance reservation; (b) A Grid Model. (c) A DAG depicting a metatask workflow with raw input data

We assume that a workflow can be represented by a directed acyclic graph (DAG) G (Fig. 1 (c)). A circular node t_i in G represents a metatask, where $1 \leq i \leq v$, and v is number of metatasks in the workflow. The computational power consumed to finish t_i is denoted as q_i. An edge $e(i,j)$ from t_i to t_j means that t_j need an intermediate result from t_i, and $t_j \in succ(t_i)$, where $succ(t_i)$ is the set of all immediate successors of t_i. Similarly, we have $t_j \in pred(t_i)$, where $pred(t_i)$ is the set of immediate predecessors of t_i. The weight of $e(i,j)$ gives the size of intermediate results transferred from t_i to metatask t_j. A metatask may also take raw data as input, which originally resides on data storage sites. The raw input data size of metatask t_i is denoted as r_i. We have $D_{i,j} = 1$, if the raw input data of metatask t_i is on storage d_j, and $D_{i,j} = 0$, otherwise. In the example, a square node with an arrow to a metatask node represents the raw data input, and the number indicates its size.

3   JDCS Algorithm

To achieve a feasible schedule with a small makespan, the JDCS algorithm has the following features: (1) It considers the possibility of raw input data preloading that can be overlapping with computation to reduce the waiting time of a metatask. (2) It is aware of data cache capacity and schedules data preloading accordingly in order to avoid conflicts. (3) It updates the rank of a metatask in each scheduling step so that critical metatasks will be recognized dynamically. (4) To avoid myopic decisions, it looks ahead along the current longest unscheduled path to select a resource for the current metatask. (5) In order to use idle time slots on a resource, it can insert an unscheduled metatask before a scheduled metatask on that resource if the insertion does not violate precedence orders and cache space limitation.

3.1   Metatask Ranking

To schedule a workflow efficiently, it is important to identify the *critical tasks* in each scheduling step. A delay of critical tasks may result in the extension of the schedule length. Usually, the priority of a metatask node can be obtained by finding the maximum "distance" from this node to the starting or exiting node. Here, distance means the sum of computational and communication costs along a certain path.

Unfortunately, due to the heterogeneity and fluctuation of resource performance, it is difficult to find how urgent a metatask node is. To estimate the completion time of a metatask in such a scenario, several performance measurements can be used, such as the median or average value of resource performance. In the following discussion, we use the average performance value to demonstrate our algorithm. To obtain the average performance as accurately as possible, only the performance of *feasible time slots* is used. The feasible time $AVLT$ of p_i in the mth scheduling step is defined as:

$$AVLT_i(m) = \min_{t_j \in RQ}\{EST(t_j, p_i)\} \tag{1}$$

Here $EST(t_j, p_i)$ is earliest start time of t_j on p_i (See (3)). So, $AVLT_i(m)$ is the earliest time when a metatask in the ready queue (RQ) can start on p_i at the mth scheduling step. Time slots after this time will be considered feasible and the corresponding performance within these slots will be used to update priorities of task nodes. For simplicity, we remove m from all expressions, without losing generality. The average performance of p_i along time avg_c_i and the average performance of all available computational resources avg_c are given by the following two equations:

$$avg\_c_i = \frac{1}{k - AVLT_i} \sum_{AVLT_i \le j \le k} c_{i,j} \quad avg\_c = \frac{1}{n} \sum_{1 \le i \le n} avg\_c_i$$

Similarly, the average unit communication cost between a computational resource p_i and others avg_cd_i, the overall average unit communication cost between any two resources avg_cd and the average unit communication cost between a data storage site d_j and any computational resources avg_dd_j are given by the following equations:

$$avg\_cd_i = \frac{1}{n} \sum_{1 \le j \le n} 1/cw_{i,j} \quad avg\_cd = \frac{1}{n} \sum_{1 \le i \le n} avg\_cd_i \quad avg\_dd_j = \frac{1}{n} \sum_{1 \le i \le n} 1/dw_{i,j}$$

We assumed that the time required to complete a metatask on different resources is uniformly related to the performance of resources. So, if the performance of a processing node p_j is a constant K, it will finish metatask t_i within time q_i/K. We have the same assumption for the communication cost. These assumptions lead to following equation which relates computational cost, resource performance and time for the performance fluctuating scenario:

$$q_i = (s_s^{end} - s_{start}) \times c_{j,s}/u + \sum_{e=s+1}^{c-1} c_{j,e} + (s_{complete} - s_c^{start}) \times c_{j,c}/u \tag{2}$$

where s_i^{start} and s_i^{end} are the start and end of a time slot, u is the length of a time slot, s_{start} and $s_{complete}$ are start time and completion time of t_i, and s and c are indices of time slots in which t_i starts and completes respectively. Now we can express the priority of a metatask $rank_u$ using a recursive definition:

$$rank_u(t_i) = \frac{q_i}{avg\_c} + \max_{t_j \in succ(t_i)} (TransTime(i, j) + rank_u(t_j))$$

$$TransTime(i, j) = \max(r_j \times \min_{1 \le x \le m \wedge D_{j,x}=1} (avg\_dd_x), e(i, j) \times avg\_cd)$$

$TransTime(i, j)$ gives the estimated time needed to transfer intermediate results or raw input data to metatask t_j, which is an immediate successor of t_i. Since t_i is not scheduled yet, the time for intermediate results and that for raw input are both estimated using average values. The meaning of $rank_u$ and *TransTime* implies that a conservative policy is used in the ranking method. It assumes that raw input data preloading of t_i will not start from the data storage until t_i's predecessors finish. This conservative policy may assign a higher rank to metatasks with large raw input data, thus giving them higher priority and more chances to get cache space reservation.

As scheduling proceeds, available performance of computational resources will change. Instead of using a static rank value computed at the beginning of the scheduling, JDCS applies a dynamic iterate ranking strategy, that is, once a metatask node is scheduled, the ranks for all of the metatask nodes yet to schedule will be updated using the resource performance in new feasible time slots.

A metatask node cannot start until it receives all of the intermediate results and raw input data. The value of $EST(t_i, p_j)$ is related to two factors: the latest intermediate result from t_i's predecessors, and the ready time of pre-loaded raw input data. According to the restrictions, the raw input data pre-loading of metatask t_i is not relevant to the finish time of t_i's predecessors directly. Ideally, t_i should start as soon as possible when its last predecessor finishes. We denote the time when t_i's last intermediate result can be ready on p_j as $LIRT(t_i, p_j)$, and we can obtain:

$$LIRT(t_i, p_j) = \max_{t_x \in pred(t_i)} (CT(t_x) + e(x, j)/cw_{PA(t_x),j})$$

$CT(t_x)$ is the real complete time of t_x. Raw input data preloading of t_i should finish before $LIRT(t_i, p_j)$, otherwise, t_i will suffer from additional delay. So, intuitively, raw data preloading should start as early as possible. Unfortunately, due to the limited data cache capacity, if raw input data of t_i is preloaded too early, it may occupy space that should be given to metatasks scheduled earlier than t_i. To find out the proper time to start preloading, JDCS uses a trace-back technique. Starting from time $LIRT(t_i, p_j)$, it traces back to check and reserve the earliest available cache space on p_j for the input data of t_i. If there is no space available backward, it will change the search direction to forward. Procedure *Find_Preloading_ StartTime*$(t_i, p_j, time)$ in Fig. 2 describes how this works. It returns the earliest preloading start time of t_i on p_j $EPST(t_i, p_j)$. Thus, the earliest raw input data ready time $ERRT(t_i, p_j)$ and $EST(t_i, p_j)$ can be formulated as:

$$ERRT(t_i, p_j) = EPST(t_i, p_j) + \min_{1 \le x \le m \wedge D_{j,x}=1} (r_i / dw_{j,x})$$

$$(3)$$

$$EST(t_i, p_j) = \max(LIRT(t_i, p_j), ERRT(t_i, p_j))$$

For simplicity, the *while* loop iterates using integer time, but actually, the value of ACS only changes when a new data transfer is started or space is released. In the worst case, the *while* loop will run $O(v*L)$ times, where L is a constant indicating the maximum degree of a metatask graph. According to the restrictions and cache management policy, when a metatask node is scheduled, all of its predecessors have been already scheduled, so the scheduler can know precisely what the cache space status was at any time in the past. Therefore, tracing back and inserting input data is not going to conflict with any other scheduled metatasks, or result in deadlock.

With $EST(t_i, p_j)$ and performance of p_j in different times slots, it is straightforward to get the earliest complete time of t_i on p_j $ECT(t_i, p_j)$, according to (2) where $s_{start} = EST(t_i, p_j)$ and $s_{complete} = ECT(t_i, p_j)$. At each step, the scheduler chooses the unscheduled metatask which has the highest dynamic rank.

3.2 Resource Assignment

In the second phase, the scheduler selects a resource for a metatask. Intuitively, a "good" assignment for a metatask should be the resource that can complete it the earliest. But the problem of this intuitive way is that it may lead to a local optimum. For example, we can consider the following case with a sequence of metatasks that forms a path P from metatask t_a to an exit node. It can happen that after t_a is assigned to a resource p_x that can finish t_a the earliest, the rest of metatask nodes on P also have to be assigned to p_x because the communication cost between any of them might delay their earliest complete time otherwise. But, if they are scheduled to another resource, say p_y, it is possible that the execution of the remaining metatasks on P can make up the communication delay because p_y has a faster computational speed. So instead of using the simple earliest complete time strategy, JDCS adopts a look-ahead approach, which is described by *Rule* 1 to avoid a biased schedule.

Rule 1: If t_i is the current metatask to be scheduled and t_j is the direct child of t_i on the longest path P_{t_i} measured by task rank from t_i to an exit node, then t_i should be scheduled to resource p_x which satisfies:

$$\min_{1\le x,y\le n} \{PEST(t_j) + EPT(P_{t_i}, p_y, PEST(t_j))\}$$

where $PEST(t_j) = ECT(t_i, p_x) + \max(e(i,j)/cw_{x,y}$ and $EPT(P, p_i, T) = (\sum_{t_i \in P} q_j) / \sum_{T \le j \le k} c_{i,j} / (k-T).$

We cannot know the real earliest start time of t_j, since at least one predecessor of t_j has not been scheduled yet. Function *EPT* computes the estimated total execution time of metatasks on path P_{t_i} using the average performance of each resource after $PEST(t_j)$. Therefore, *Rule* 1 states that instead of finding the processing node that can finish t_i the earliest, we are trying to find a pair of resources, p_x and p_y, so that execution time for the metatasks on the longest path from t_i to an exit node will be minimized.

To utilize idle time slots on resources, JDCS allows task insertion, which is formalized by *Rule* 2. It states that a task can be assigned to a resource only if there are time slots large enough to accommodate it without delaying metatasks already scheduled or violating precedence orders among the metatasks.

Rule 2: A metatask t_i can be assigned to resource p_j which is already assigned with a sequence of metatasks $\{ t_{j_1}, t_{j_2}, \ldots, t_{j_n} \}$ at time s, if there is some m such that for every metatask t_{j_x} in $\{ t_{j_1}, t_{j_2}, \ldots, t_{j_m} \}$, $ECT(t_{j_x}, p_j) \le EST(t_i, p_j)$, and for every metatask t_{j_y} in $\{ t_{j_m}, t_{j_{m+1}}, \ldots, t_{j_n} \}$, $ECT(t_i, p_j) \le EST(t_{j_y}, p_j)$.

The pseudo code of the JDCS algorithm is shown in Fig.2. The total complexity of JDCS is in the order of $O(n^2 v^2 + nkv)$.

4 Experiments

Simulation experiments are conducted to evaluate the performance of JDCS in the Grid. The performance metric we used for the comparison is the Scheduled Length

Ratio (SLR), defined as the ratio of the real makespan to the lower bound of any possible scheduling, which is the minimum length of the critical path. In the experiments, the basic topology of Grid resources is generated using GridG1.0 [3]. This tool allows us to get realistic computational resources and networking settings for simulation, such as processing capability, data cache size, bandwidth and delay for LAN and WAN. Grid data storage sites are assumed to follow a uniform geographic distribution in the Grid. Performance of a computational resource is assumed to follow a Poisson distribution having the average value given by GridG as the average. TGFF [4] is used to generate basic DAGs. TGFF allows customized settings such as average number of metatask nodes in the graph, average out-degree and in-degree for a node, and range of computation and communication costs. However, TGFF cannot generate raw input data for a metatask. As a result, a graph generated by TGFF is reprocessed and raw data input is inserted randomly to metatasks in the graph. The size of raw input data is also uniformly distributed in the same range given to TGFF for cost generation.

```
Find_Preloading_StartTime(t_i, p_j,      5.    Call Select_Resource(t);
time)                                     6.    Update rank_u;
1.  EPST = time;                          7.  }
2.  if (ACS(p_j, PST) > = r_i){
3.    While (ACS(p, EPST) > = r_i)        Select_Resource (metatask t)
4.       EPST--;                          1. Find the longest path P from t;
5.  }else                                 2. For available resources p_i{
6.      while (ACS(p, EPST) < r_i)        3.   Get EST(t,p_j) and ECT(t,p_i);
7.         EPST++;                        4.   For available resources p_j
8.  Return EPST                           5.     Call EAT(P, p_j, PEST(t));
                                          6. }
Algorithm JDCS(DAG G)                     7. Assign t to p_i that satisfies
1. Compute initial task rank_u;              Rule 1;
2. Initialize the ready queue RQ          8. Set Data transfer time for t
   with the entry metatask;                  and write the cache log of p_i;
3. While(there are unscheduled            9. For all available resources p_i,
   metatasks){                               update AVLT_i of p_i.
4.    Select the highest rank
      metatask t in RQ;
```

Fig. 2. Pseudo code of JDCS

To test JDCS in different resource settings and workflow patterns, six groups of experiments are conducted to test the influence of the following parameters to schedule results: 1) the number of input data replicas (DR) in the Grid; 2) the ratio of the average data input size to the average data cache size (ICR) on computational resources, 3) the ratio of the average degree of a node to the total number of metatasks in a graph (DTR); 4) the percentage of metatasks having raw input data (RIP) in a workflow, 5) the average ratio of computation cost to communication cost a metatask graph (CCR) (a high CCR value means a metatask graph is computation-intensive); and 6) the resource background workload factor (BWF) which decides the percentage that the performance of a computational resource can increase or drop in different time slots. For every group of experiments, we used five sets of workflows whose average metatask number varied from 20 to 100.

To test how the number of replicas of input data in the Grid will influence scheduling results, three different settings are compared, namely, (1) no replica, which means each metatask only has one copy of raw input data on data storage sites, (2) one replica, and (3) two replicas. As expected (Fig. 3 (a)), data replication benefits the schedule because the scheduler has more choices to reduce the preloading cost. But the small margin between the curves of one replica and two replicas implies that the gain from increasing the numbers of replicas is limited if they are evenly distributed in the network. We also test the outcome of the original HEFT [3] algorithm, which only considers intermediate results but not raw input data preloading when it makes a schedule, and a revised HEFT algorithm using the trace-back method introduced in this paper. The results show that the original approach will bring significant lag to the performance, which supports the basic motivation of our work in this paper.

Another element impacting scheduling results is how large the data cache of computational resources is. To make results comparable, we use a normalized ratio of input data size to cache size (Fig. 3(b)). It can be observed that as the ratio increases, SLR is increased dramatically and slopes of curves turn from sub-linear to super-linear. The explanation is that as the ratio is higher, it becomes more difficult for a metatask to get a free data cache space to preload its input data. In particular, when the ratio is higher than 0.5, in most cases, simultaneous preloading is impossible because a data cache can only hold input data for one metatask.

Intermediate results in a workflow are also competing for data cache space on computational resources. So the ratio of intermediate data size to the data cache size also influences a schedule. Given a uniform distribution of intermediate result size when a metatask graph is generated, the total size of intermediate results is proportional to the number of edges in the graph, or the degrees of metatask nodes. To describe the edge density in a graph, the ratio of the average degree of each node to the total number of nodes is used in our experiments (Fig. 3 (c)). Results show that the performance of JDCS is stable in different ratios. Intuitively, SLR will be higher as the size of intermediate results increases. But increasing the degree of metatasks implies the task graph is more connected and the length of the path to an exit node might become shorter. As JDCS uses a look-forward strategy in the resource selecting phase, a shorter path means a more accurate estimation of the execution time, which will benefit the task assignment. Another reason is given in the following analysis.

The third element relating to data cache competition is the number of metatasks having raw input data. Different percentages of the metatasks having raw input data in a workflow are tested (Fig. 3 (d)). It can be observed that, as the percentage increases, gaps between different curves increase, which implies that JDCS is more sensitive to changes in the number of raw input data than that of intermediate results. The reason behind this is that the timing of intermediate results transfer is more restricted by both precedence orders among metatasks and cache size limitation, while the raw input data preloading solely depends on cache availability.

In our experiments, we assume CCR only influences the ratio of computation to intermediate data, but not the raw input data. Fig. 3 (e) shows that, as CCR increases, the SLR drops significantly. The reason behind is that the more communication in the workflow, the higher SLR is going to be. But as CCR reaches a point in our model, the main contribution of the communication cost will come from raw input data preloading, which will then limit the drop of SLR.

The last group of experiments is to discover how JDCS adapts to performance fluctuation of computational resources. In the experiments, we allow the background workload of a computational resource to increase up to a certain percentage of its full performance in different time slots (Fig. 3 (f)). It can be observed that, as the fluctuation in performance grows, *SLR* is going to be extended. However in every workflow set, the increase remains stable and moderate. This implies that JDCS can work well even with resource performance changes in a wide range of up to 80%.

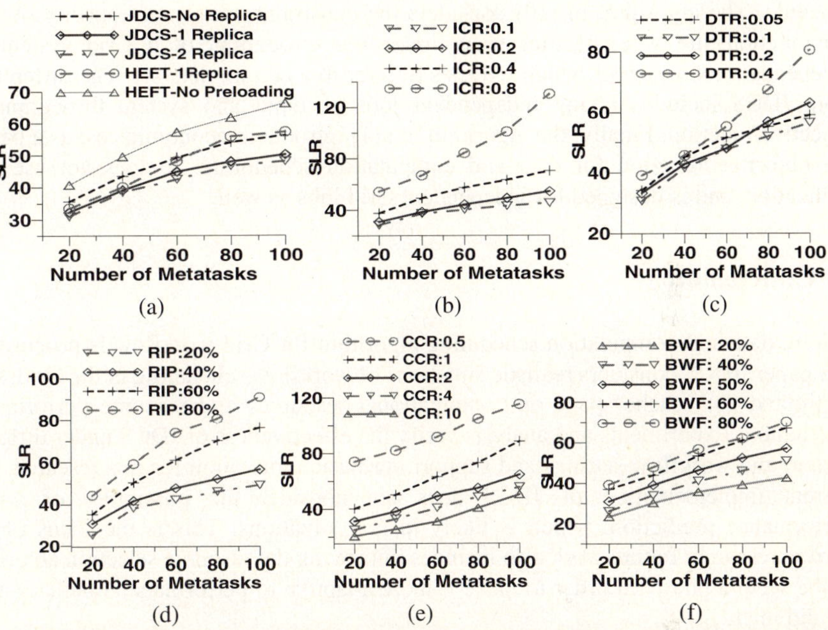

Fig. 3. Performance on different settings of: (a) DR (ICR=0.2, DTR =0.2, RIP=40%, CCR = 2, BWF=50%). (b) ICR (DR=1, DTR =0.2, RIP=40%, CCR = 2, BWF=50%). (c) DTR (DR=1, ICR=0.2, RIP=40%, CCR = 2, BWF=50%); (d) RIP (DR=1, DTR =0.2, ICR=0.2, CCR = 2, BWF=50%), (e) CCR (DR=1, DTR =0.2, ICR=0.2, RIP=40%, BWF=50%) and (f) BWF (DR=1, DTR =0.2, ICR=0.2, RIP=40%, CCR =2).

5 Related Works

In [7], Park classifies different scenarios about locations of data and computation in the Grid into five different categories. **JDCS belongs to the category of Remote Data and Different Remote Execution, because both computation and data sites are selected dynamically and not necessarily close to each other. We believe this is a more general case compared with the other four patterns in the Grid.** There are two different points of view on the relationship between data scheduling and computation scheduling, namely, decoupling the two kinds of scheduling or combining them. In a decoupled method given in [8], data replication strategies are independent from task scheduling strategies, and a complete scheduling algorithm is an arbitrary combination of the two. The primary goal of that approach is to maximize the system throughput. By contrast, the goal of JDCS is to optimize the finish time of each Grid workflow, and the algorithm does not generate new data replication in scheduling procedures. Scheduling of computation is combined with data replica selection in [9] to reduce the execution time of a collection of applications which can be unified to compose a DAG. The goal of the approach is the same as JDCS, but main differences include: 1) JDCS focuses on finding the right time to start a data input preloading other than selecting a replica; and 2) JDCS does not partition a DAG graph to make a schedule. The algorithm in [10] considers the constraint of storage capacity of each computation site when it tries to optimize the throughput of a Grid system for independent jobs each of which requests to refer to a certain data set. The differences from JDCS includes taking independent jobs as input and system throughput as objective function. Finally, the algorithm in [11] introduces economic cost as a part of the objective function for data and computation scheduling. It does not use data replication, and is designed for independent Grid jobs as well.

6 Conclusion

A joint data and computation scheduling algorithm for Grid workflow is proposed in this paper. JDCS considers realistic situations of workflow scheduling in the Grid, such as limited data cache space on computational resources and resource performance fluctuation. Experiments and analysis verify the effectiveness of JDCS under different system and workflow settings and support our basic motivation for this research. The current implementation of JDCS does not consider the possibility of wrong performance prediction, which is likely in real situations. This is the focus of our current research. Future work also includes improving the resource selection algorithm in the second phase, in order to make it more adaptive to performance fluctuations as we did in [12].

References

[1] Yang, L., Schopf, J.M., Foster, I.: Conservative Scheduling: Using Predicted Variance to Improve Scheduling Decisions in Dynamic Environments. Super-Computing (2003)

[2] Aggarwal, K., Kent, R.D.: An Adaptive Generalized Scheduler for Grid Applications. In: The 19th HPCS (2005)

[3] Topcuoglu, H., Hariri, S., Wu, M.Y.: Performance-Effective and Low-Complexity Task Scheduling for Heterogeneous Computing. IEEE Trans. on Parallel and Distributed Systems 13(3), 260–274 (2002)

[4] Lu, D., Dinda, P.: Synthesizing Realistic Computational Grids. Super-Computing (2003)

[5] Dick, R.P., Rhodes, D.L., Wolf, W.: TGFF Task Graphs for Free. In: The 6th International Workshop on Hardware/Software Co-design (1998)

[6] El-Rewini, H., Lewis, T., Ali, H.: Task Scheduling in Parallel and Distributed Systems. PTR Prentice Hall, Englewood Cliffs (1994)

[7] Park, S., Kim, J.: Chameleon: a Resource Scheduler in a Data Grid Environment. In: The 3rd CCGrid (2003)

[8] Ranganathan, K., Foster, I.: Decoupling Computation and Data Scheduling in Distributed Data-Intensive Applications. In: The 11th HPDC (2002)

[9] Alhusaini, A.H., Prasanna, V.K., Raghavendra, C.S.: A Unified Resource Scheduling Framework for Heterogeneous Computing Environments. In: The 8th HCW Workshop (1999)

[10] Desprez, F., Vernois, A.: Simultaneous Scheduling of Replication and Computation for Bioinformatic Applications on the Grid. In: Proc. of CLADE2005 (2005)

[11] Venugopal, S., Buyya, R.: A Deadline and Budget Constrained Scheduling Algorithm for eScience Applications on Data Grids. In: Hobbs, M., Goscinski, A.M., Zhou, W. (eds.) ICA3PP 2005. LNCS, vol. 3719, Springer, Heidelberg (2005)

[12] Dong, F., Akl, S.: PFAS: A Resource Performance Fluctuation Aware Workflow Scheduling Algorithm for Grid Computing. In: The 16th HCW Workshop (2007)

Distributed Computation of All Node Replacements of a Minimum Spanning Tree[*]

Paola Flocchini[1], Toni Mesa Enriquez[2], Linda Pagli[3],
Giuseppe Prencipe[3], and Nicola Santoro[4]

[1] University of Ottawa, Canada
`flocchin@site.uottawa.ca`
[2] Universidad de La Habana, Cuba
`tonymesa@matcom.uh.cu`
[3] Università di Pisa, Italy
`{pagli,prencipe}@di.unipi.it`
[4] Carleton University, Canada
`santoro@scs.carleton.ca`

Abstract. In many network applications the computation takes place on the *minimum-cost* spanning tree (MST) of the network; unfortunately, a single link or node failure disconnects the tree.

In this paper we consider for the first time the problem of computing all the replacement minimum-cost spanning trees *distributively*, and we efficiently solve the problem. We design a solution protocol and we prove that the total amount of data items communicated during the computation is $O(n^2)$. This communication can be achieved either transmitting $O(n)$ long messages, if the system so allows, or $O(n^2)$ standard messages. Even in systems that do not allow long messages, the proposed protocol constitutes a significant improvement over the individual computation of the replacement trees.

Keywords: Minimum Spanning Tree, Replacement Tree, Node Failure, Distributed Algorithms.

1 Introduction

1.1 The Framework

In most network applications, the computation takes place not on the entire network but solely on a spanning subnet. There are several reasons for this fact; first and foremost, it is done to reduce the amount of communication and thus the associated costs; it is done also for security reasons, e.g. to minimize the exposure of messages to external eavesdroppers. The subnet used is typically a special spanning tree of the network G; in particular, the *minimum-cost* spanning tree (MST) is used for basic network tasks such as broadcasting, multicasting, leader election and synchronization. The major drawback of using a MST is the

[*] Research partially supported by NSERC Canada.

A.-M. Kermarrec, L. Bougé, and T. Priol (Eds.): Euro-Par 2007, LNCS 4641, pp. 598–607, 2007.

high vulnerability of its tree structure to link and/or node failures: a single failure disconnects the spanning tree, interrupting the message transmission. Hence it is crucial to update the MST after changes in network topology. In this paper we update MST after single node deletions. In a graph $G = (V, E)$, with n nodes there are n possible instances of a single node deletions. Let T be the MST of G. Informally, the *All Node Replacements* (ANR) problem is to update T in each of the instances of single node deletion. Observe that this problem is much more difficult than the related *All Link Replacement* (ALR) problem where the goal is to update T in each of the instances of single *edge* deletion. In fact, the deletion of a single node u is equivalent to the simultaneous deletion of all its $deg(u)$ incident edges.

The re-computation of the new MST in each instance is rather expensive. This is particularly true if the re-computation is done distributively in the network after a failure; in addition, if the failures in the system are mostly temporary, the usefulness of these re-computations is limited and the rational for affording their cost becomes questionable. For these reasons, to solve the All Node Replacements problem in reality means to *pre-compute* the n replacement minimum spanning trees, one for each possible node failure in the tree [4,10,13]; the computed information is then used only if a node fails, and only as long as the failure persists. The computational challenge is to be able to combine work among the n different pre-computations, so that the the total cost is much less than that incurred by computing each replacement tree individually. This problem has been extensively investigated, and efficient solutions have been developed for both the sequential and parallel settings (e.g., see [3,4,10,13,15]).

In this paper we consider the *distributed* version of this problem. That is we investigate the All Node Replacements problem when the computational entities are nodes of G themselves, and each can only communicate by exchanging messages with its neighbours. The network itself must pre-compute the n replacement minimum spanning trees; the information so obtained is then stored (distributively) together with the original MST tree T, and used whenever a node failure is detected; the original minimum spanning tree T is reactivated once the network has recovered from the transient fault.

The repeated application of a distributed MST construction protocol (e.g., [7,11]) will cost at least $O(nm + n^2 \log n)$ messages, where m denotes the number of edges. Surprisingly, no more efficient distributed solutions exist for this problem, prior to this work. As stated in [4] (where efficient serial and parallel solutions were presented): *Designing an efficient distributed algorithm for ANR remains an open problem.*

1.2   Main Result

In this paper we consider the problem of computing all the replacement minimum-cost spanning trees *distributively*, and we efficiently solve the problem.

We design a distributed algorithm for computing all the replacement MSTs of the minimum cost spanning tree T of the network G, one for each possible node failure, and we show how to store the computed information in order to restore the tree's connectivity when the temporary fault occurs.

We prove that the total amount of data items communicated during the computation (the data complexity) is $O(n^2)$. This communication can be achieved transmitting only $O(n)$ long messages between neighbours, if the system so allows; otherwise $O(n^2)$ standard messages suffice. In other words, with this complexity, our protocol constructs a MST that maintains its minimum-cost properties even after a single (but arbitrary) link or node failure.

Even in systems that do not allow long messages, the proposed protocol constitutes a significant improvement over the individual computation of the replacement trees. Indeed, for dense graphs, our protocol constructs all the n replacement MSTs of the minimum spanning tree T with the same number of messages required just to compute T.

The communication structure of the algorithm is surprisingly simple, as it consist of a single broadcast phase followed by a convergecast phase. The difficulty is to determine what information is locally needed, which items of data have to be transmitted in these two phases, and how the communicated information must be locally employed. This schema is reminescent of the one used for computing all the swap-edges of a shortest-path tree [8,9], but the similarity is limited to the structure. In fact, since the failure of a single node u is equivalent to the simultaneous deletion of all its $deg(u)$ incident edges, the nature of the problem changes dramatically, and those approaches can not be used here.

They can however be employed, as we show, to solve the simpler All Link Replacement (ALR) problem where the goal is to update T in each of the instances of single *edge* deletion.

1.3   Related Work

The All Node Replacements (ANR) problem was first studied in a *serial* environment by Chin and Houck [3]. A more efficient solution has been developed by Das and Loui [4], and later improved by Nardelli, Proietti and Widmayer [13]. When G is *planar*, improved bounds have been obtained by Gaibisso, Proietti and Tan [10]. The simpler All Edge Replacements (AER) problem is implicitly solved by Dixon, Rauch and Tarjan [5]; an improved solution was later developed by Nardelli, Proietti and Widmayer [13].

In the *parallel* setting, Tsin presented an algorithm to update a MST after a single node deletion [15]; thus, concurrent use of this algorithm solves ANR in parallel. A subsequent parallel solution to ANR is obtained by combining the parallel algorithms presented by Johnson and Metaxas [12]. A more efficient parallel technique has been designed by Das and Loui [4]. The simpler All Edge Replacements (AER) problem is efficiently solved by using the parallel verification algorithm of Dixon and Tarjan [6].

In the *distributed* setting, the *construction* of the MST of a network has received considerable attention. The well known protocol by Gallager, Humblet and Spira uses $O(m + n \log n)$ messages, where m denotes the number of edges [11]. This protocol is not only elegant but also optimal, since $\Omega(m + n \log n)$

messages are needed regardless of their size [14]. In fact, all subsequent work (e.g., [7]) has been dedicated to reducing the time needed in synchronous executions.

To solve AER and ANR, one may use repeated applications of a distributed MST construction protocol; this brute-force approach will cost at least $O(nm + n^2 \log n)$ messages. The more complex problem of updating a MST with multiple node and edge deletions was considered by Cheng, Cimet and Kumar [2]; however, when used in the ANR and in the AER problems, their solution would not yield any improvement over the brute-force approach (it would actually be worse). Indeed, prior to this work, no efficient distributed solutions exist for either problems.

2   Terminology and Problems

2.1   Definitions

Let $G = (V, E)$ be an undirected graph, with $n = |V|$ vertices and $m = |E|$ edges. A *label* of length $l \leq \log n$ is associated to each vertex of G. A non negative real *weight* $w(e)$ is associated to each edge e. A *subgraph* $G' = (V', E')$ of G is such that $V' \subseteq V$ and $E' \subseteq E$. If $V' \equiv V$ and G' is connected, then G' is a *spanning* subgraph. A graph G is *2-edge connected* or *2-node connected* if it remains connected after the removal of any one of its edges (or any one of its nodes). Let $T = (V, E(T))$ be a spanning tree of graph G rooted in r, arbitrary node of T. A spanning tree $T = (V, E(T))$ is called *minimum spanning tree MST* of G if the sum of tree edge weights is minimum over all spanning trees.

A subtree rooted at some node x is denoted by T_x. The *parent* of a node x is indicated as *parent*(x) and its *children* as *children*(x). Consider an edge $e = (x, y) \in E(T)$ with y closer to r, the root of T; if such an edge is removed, the tree is disconnected in two subtrees: T_x and $T \setminus T_x$. A *swap* edge for $e = (x, y)$ is any edge $e' = (u, v) \in E \setminus \{e\}$ that connects the two subtrees. It can be easily seen that the MST of $G - e$, called the replacement tree T_{G-e} can be computed by selecting the swap edge of minimum weight connecting T_x and $T \setminus T_x$.

We consider a *distributed computing system* with communication topology G. Each computational entity x is located at a node of G, has local processing and storage capabilities, has a unique label $\lambda_x(e)$ from a totally ordered set associated to each of its incident edges e, knows the weight of its incident edges, and can communicate with its neighboring entities by transmission of bounded sequences of bits called messages. The nodes do not know the topology G, but only their incident edges with their labels. The communication time includes processing, queueing, and transmission delays, and it is finite but otherwise unpredictable. In other words, the system is *asynchronous*. All the entities execute the same set of rules, called *distributed algorithm* (e.g., see [14]).

In the following, when no ambiguity arises, we will use the terms entity, node and vertex as equivalent; analogously, we will use the terms link, arc and edge interchangeably.

2.2   The All Edges Replacement Problem and Its Solution

Let G be 2-edge connected. The *All Edges Replacement* problem, denoted as $AER(G,T)$ with input G and T is that of finding T_{G-e} for every edge $e \in E(T)$.

The $AER(G,T)$ problem can be solved distributively by applying one of the algorithmic shells of [9], where the input tree is now an MST of G, instead of a shortest-path tree, and where the *best swap edge e'* for e is the one leading to the minimal total weight; hence, this function can be computed locally by each node by simply summing the weight of e' and subtracting the weight of e from the total MST's weight. The overall message complexity is then the same as in [9] amounting to $O(n_r^*)$, where n_r^* is the number of edges of the transitive closure of $T \setminus \{r\}$ and $0 \le n_r^* \le (n-1)(n-2)/2$, which is of $O(n^2)$.

2.3   The All Nodes Replacement Problem

Let $G = (V, E)$ be 2-node connected. Consider a node $x \in V$; if such node is removed from T together with its incident edges, the tree is disconnected into the subtrees $T_{x_1}, \ldots, T_{x_k}$, where $x_1, ..., x_k$ are the children of x; let $T' = T \setminus \{T_{x_1}, ..., T_{x_k}, \{x\}\}$. Let x_0 be the parent of x, and E' be the set of non tree edges; we will call $\mathcal{U}_x = \{e = (u,v) \in E' | u \in T_{x_i}, 1 \le i \le k, v \in T'\}$ the set of *upwards edges* of x and $\mathcal{H}_x = \{e = (u,v) \in E' | u \in T_{x_i}, v \in T_{x_j}, 1 \le i,j \le k, i \ne j\}$ the set of *horizontal edges* of x. For node x, the set of the *best upward edges* $\mathcal{U}'_x \subseteq \mathcal{U}_x$ is the set containing the edges of minimum weight (if any) connecting $T_{x_i}, 1 \le i \le k$ and T', and the set of the *best horizontal edges* is the set $\mathcal{H}'_x \subseteq \mathcal{H}_x$ containing the edges of *minimum weight* connecting T_{x_i} and $T_{x_j}, 1 \le i \ne j \le k$ (if any). In the following, we will use also the notation $\mathcal{U}, \mathcal{U}', \mathcal{H}$, and $\mathcal{H}'$, when the reference to the removed node is clear from the context.

From [13] we know that the MST of $G - x$ can be computed through the computation of the MST of the *contracted graph* $G_x = (V_x, E_x)$, where $V_x = x_0, x_1, ... x_k$ and $E_x = \mathcal{H}' \cup \mathcal{U}'$, obtained contracting to a single vertex each subtree $T_{x_i}, 1 \le i \le k$, and T'. The edges of the obtained MST, say T_{G-x}, are the *replacement set* of edges for x.

The computation of all the replacement sets for each node failure will be called the *All Nodes Replacement* or simply $ANR(G,T)$ problem in the following. We are interested in the distributed solution of the $ANR(G,T)$ problem.

3   Solving the *ANR* Problem

Consider the problem of computing the replacement edges for the failure of node x of T; the computation is performed simultaneously for all possible node failures. We first present a distributed algorithm described at high level, while the details of each module will be discussed later. At high level the algorithm consists of a *broadcast* phase started by the children of the root, followed by a *convergecast* phase started by the leaves. The idea is that each node x is able to compute its replacement set, when all its children have already computed their replacement sets in the convergecast phase. Node x determines also a set of edges, useful to compute the replacement sets for all its ancestors (except for the root), that is for $a_i, 2 \leq i \leq s$, where a_2 is the parent of x in T and a_s a child of r.

Once node x has computed its replacement set, composed of edges having at least one endpoint in its subtrees, it sends them back to its children, each one to the root of the proper subtree. In the case node x fails, each child knows which edges have to be activated in its subtree.

ALL NODES REPLACEMENT (*ANR(G,T)*)

[Broadcast.]

1. Each child x of the root starts the broadcast by sending to its children a list containing its name.
2. Each node y, receiving a list of names from its parent, appends its name to the received list and sends it to its children.

[Convergecast.]

1. Each leaf z selects, among its non tree incident edges, the best upwards edge and the best horizontal edges for each ancestor a in the received list. Then sends the lists of those edges to its parent (if different from r).
2. An internal node y waits until it receives the information computed from each of its children: this information contains the set of the upwards edges $\mathcal{U}'$ and the set of horizontal edges $\mathcal{H}'$ for y.
   (a) y computes the MST of the graph $G_y = (V_y, E_y)$ where $V_y = \{parent(y), children(y)\}$ and $E_y = \{\mathcal{U}' \cup \mathcal{H}'\}$ and sends the edges of T_{G_y}, that is the *replacement set RS_y* for y to its proper subtrees.
   (b) y then selects, among its incident non tree edges and the information received from its children, the best upwards edge and the best horizontal edges for each of its ancestor.
   (c) y finally sends the lists of these edges to its parent (if different from r).

To show how this high level algorithmic structure works we must specify in more details the convergecast phase and, in particular, the operations executed by each node. First of all, let us define the structure of the information received by a node x from each of its children: it is composed by s lists, one for x and one for each of the other $s - 1$ ancestors a_j, $2 \leq j \leq s$ (except for the root). For each $x_i, 1 \leq i \leq k$ let $L_j^i, 1 \leq j \leq s$ be the list from x_i for x and for the other ancestors a_j. Each L_j^i is composed of two fields, called UP and HOR. For $L_1^i, 1 \leq i \leq k$, the field UP, denoted as $UP(L_1^i)$ will contain the best upwards edge from T_{x_i} for x. The set composed by $UP(L_1^1)$, ... , $UP(L_1^k)$ are used to compute $\mathcal{U}'$; $UP(L_j^i), 1 \leq i \leq k, 2 \leq j \leq s$, will contain the best upwards edge encountered until now for a_j, that is the best upwards edge for a_j outgoing from T_{x_i}. Note that every edge is always stored together with its weight.

The field HOR of each list L_1^i, $1 \leq i \leq k$, denoted as $HOR(L_1^i)$, is a pointer to a possibly empty list of at most $k - 1$ best horizontal edges connecting T_{x_i} and T_{x_h}, $1 \leq h \leq k$, $h \neq i$. The edges in the lists $HOR(L_1^1), \ldots, HOR(L_1^k)$ form the set $\mathcal{H}'$. Let $d(a_j)$ be the degree of a_j in T; the size of the lists $HOR(L_j^i)$, $1 \leq i \leq k$ and $1 \leq j \leq s$, is at most equal to $d(a_j) - 1$. For $j > 1$, such lists contain the best horizontal edges found until now for a_j, that is the best horizontal edges outgoing from T_{x_i} for a_j.

Some of the information sent to a node from its children is shown in Figure 1(a). Note that, since the horizontal edges are computed independently by each subtree, each edge will appear twice in the lists.

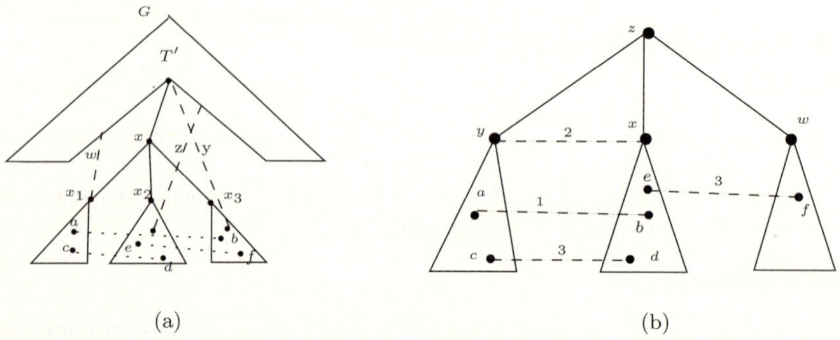

(a)   (b)

Fig. 1. (a) The upwards and horizontal edges sent to x by its children x_1, x_2, x_3 used for the computation of the replacement set for x. $UP(L_1^1)$ contains the upwards edge w, $UP(L_1^2)$ contains z and $UP(L_1^3)$ contains y. $\mathcal{U}' = \{w, y, z\}$ $HOR(L_1^1)$ contains the edges (a, b) and (c, d); $HOR(L_1^2)$ contains (d, c) and (e, f) and $HOR(L_1^3)$ contains (b, a) and (f, e). $\mathcal{H}' = \{(a, b)(c, d)(d, c)(e, f)(b, a)(f, e)\}$. (b) Selection of the horizontal edges for z in algorithm $MyAUH$ executed by x.

Once a node has received the sets $\mathcal{U}'$ and $\mathcal{H}'$ from its children, it has to compute the MST of the contracted graph G_x. This can be done locally with an optimal sequential algorithm, with no exchange of additional messages. The only problem is that in the sets $\mathcal{U}'$ and $\mathcal{H}'$, the edges are indicated by their endpoints, while the nodes of the contracted graph G_x are the children and the parent of x. For this purpose, the endpoints of these edges must be relabeled.

Let us describe in detail the operations executed by node x. First of all x computes the new MST for the contracted graph $G_x = (V_x, E_x)$ by considering the lists transmitted to it from its children (Algorithm $MyMST$).

MyMST

(* Algorithm for node x*)

1. Construct the contracted graph $G_x = (V_x, E_x)$ of $G - x$ where $V_x = \{parent(x), children(x)\}$. E_x is obtained by the union of the sets $\{UP(L_1^i), HOR(L_1^i)\}, 1 \le i \le k$, relabeled as follows: any edge $e = (a, b) \in UP(L_1^i)$ becomes $(x_i, parent(x))$. For any edge $e = (a, b) \in HOR(L_1^i)$ search the list $HOR(L_1^j), 1 \le i, j \le k, i \ne j$, containing the edge $e = (b, a)$ and rename (a, b) as (x_i, x_j).
2. Compute the MST of G_x locally with an optimal algorithm.
3. Reassign to the set of edges ET_{G_x} of T_{G_x} their original names besides the new names. ET_{G_x} is the replacement set for x.
4. Send any edge $e = (a, b) \in ET_{G_x}$, relabelled as (x_i, x_j) to child x_i.

Note that the relabeling operation is needed because even if node x knows the label i of the child from which it receives the information, an edge (a, b) coming from x_i does not explicitly specify to which subtree of x the node b belongs.

We now describe the algorithm of x which computes the *best upwards edge* for each ancestor $a_j, 2 \leq j \leq s$, among its incident upwards edges and the edges in $UP(L_j^i), 1 \leq i \leq k$. In addition x computes the *best horizontal edges* among its incident edges that are horizontal with respect to a_j and the edges in $HOR(L_j^i), 1 \leq i \leq k, 2 \leq j \leq s$ (Algorithm *MyAUH*).

Node x will produce the new $s - 1$ lists $L_j^x, 2 \leq j \leq s$ to send to its parent. Note that while the best upward and horizontal edges that x computes for its parent are the final ones, the edges computed for all the other ancestors can be worse than the final ones; they will be ultimately computed for each node when *their children* execute Algorithm *MyAUH*.

Algorithm *MyAUH* makes use of the boolean function *anc(x,y)* which is *true* if and only if node x is an ancestor of y, and of the function *nca(x,y)* which returns the nearest common ancestor of x and y in a given tree, that is the common ancestor of x and y, whose distance from x and y is smaller than the distance of any other ancestor. Let $In(x)$ be the set of non tree edges incident to x. With respect to a node x, the horizontal edges connecting the same pair of subtrees of x will be called *analogous* in the following.

MyAUH

(* Algorithm for node x *)

1. Among the edges in $In(x)$: select those for which $nca(x, y) = z$, $z \neq x$ and $z \neq y$; let *min* be the one of minimum weight; For each ancestor node $a_j, 2 \leq j \leq s$: compute the *best upwards edge* as the one of minimum weight among $UP(L_j^i), 1 \leq i \leq k$, *min*, and the edges belonging to $In(x)$ such that $anc(a_j, x) = true$; store the *best upwards edge* in $UP(L_j^x)$.

2. Among the edges $e = (x, y) \in In(x)$: select those for which $nca(x, y) = a_j, 2 \leq j \leq s$. For each j if there is a set of analogous edges, then choose the one of minimum weight. For each ancestor node $a_j, 2 \leq j \leq s$, $d(a_j) = d$, consider the selected incident edges $e = (x, y)$ such that $nca(x, y) = a_j$ and the edges $e = (h, h') \in HOR(L_j^i), 1 \leq i \leq k$; if there is a set of analogous edges then choose the one of minimum weight. All the selected edges are then stored in $HOR(L_j^x)$.

4   Correctness and Complexity

4.1   Basic Properties

We first introduce some properties needed to show how a node x can locally efficiently perform the operations in Algorithm *MyAUH*.

In order for a node to decide if the other endpoint of an incident edge is its ancestor it is sufficient to check the information collected in the broadcast phase.

Property 1. *Given $e = (x, y) \in In(x)$, $anc(y, x)$ can be checked at node x and no communication is needed.*

Property 1 derives from the fact that, after the broadcast phase, x knows all of its ancestors, and if y does not belong to the list of ancestors the function is false.

The *nearest common ancestor* of pairs of nodes $x, y \in T$, $nca(x, y)$ must be also computed. In a recent work [1], it has been shown that this information can be locally computed in constant time, through a proper labeling of the tree that requires labels of $O(log n)$ bits, denoted as $l(x)$, that can be precomputed by a depth first traversal of the tree. Therefore, our basic algorithm $ANR(G, T)$ has to be slightly modified to transmit, for each node x, $l(x)$ instead of x. Once such labeling is computed for T, each node can be distinguished by its label. Then, from [1] and since $l(y)$ is accessible at x, we have:

Property 2. *Let $e = (x, y) \in In(x)$. $nca(x, y)$ can be computed at x and no communication is needed.*

In the selection of the horizontal edges we need to check whether two edges, having the same nearest common ancestor z, connect the same pair of subtrees of z, that is they are *analogous*: only the one with minimal weight, must be selected. In this way, node x selects at most one edge from T_x to any other subtree rooted in its siblings and this is important to bound the size of the information sent by every node.

The situation is depicted in Figure 1(b), where all horizontal edges (x, y), (b, a), (d, c), and (e, f) have the same nearest common ancestor z, but (x, y), (b, a), and (d, c) are analogous since they connect the same pair of subtrees T_x and T_y; only the one of minimum weight (b, a) is chosen; edge (e, f) is the unique connecting T_x and T_w, then is directly chosen. Besides the other information, x will then send to z the list $HOR(L_z^x)$ containing $(b, a)(e, f)$.

The problem is now how to detect the *analogy* between two horizontal edges. We have the following:

Lemma 1. *Let (a, b) and (c, d) be two edges such that $a \in T_y$, $c \in T_y$, and $nca(a, b) = nca(c, d) = z$. These edges are analogous if $nca(b, d) = x$, $x \neq z$. The condition can be checked at y for each z and no communication is needed.*

The proof of Lemma 1 can be followed observing Figure 2, where, for the edges (x, y) and (d, c), $nca(y, c)$ is different from z, hence they are analogous. Viceversa, for (x, y) and (e, f), $nca(y, f)$ is equal to z, hence the condition does not hold.

4.2   Analysis

We now prove the correctness of our basic algorithm ALL NODES REPLACEMENT $ANR(G, T)$. We have:

Theorem 1. *In algorithm ANR(G,T) each node $z \neq r$:*

(i) *correctly computes the best upwards edge and the best horizontal edges for its parent.*

(ii) *determines for each ancestor a, different from the parent and the root, the best upward edges and the best horizontal edges for a in T_z.*

We now establish the data complexity required by the algorithm. We recall that the preprocessing phase consists of a depth first search of the tree requiring $O(n)$ messages. We have:

Theorem 2. *The data complexity of algorithm* ANR(G,T)*is* $O(n^2)$.

The algorithm ALL NODES REPLACEMENT terminates leaving, in the children of each node, the edges to activate in case of failure. Let x be the node which fails, $x_1, ... x_k, 1 \leq i \leq k$ its children, and let RS_x be the replacement set of edges for x_i. Every x_i will contain the subset $RS_{x_i} \subseteq RS_x$ of edges having an endpoint in T_{x_i}; it starts a broadcast phase sending RS_{x_i} down in its subtree; in this phase the nodes that discover to be incident to one edge $e \in RS_{x_i}$ activate the edge. This activation phase requires a data complexity of order $O(d_{x-1} \times n)$, since at most d_{x-1} edges have to reach $O(n)$ nodes.

References

1. Alstrup, S., Gavoille, C., Kaplan, H., Rauhe, T.: Nearest common ancestor: A survey and a new distributed algorithm for a distributed environment. Theory of Computing System 37, 441–456 (2004)
2. Cheng, C., Cimet, I.A., Kumar, S.P.R.: A protocol to maintain a minimum spanning tree in a dynamic topology. Comput. Commun. Rev. 18(4), 330–338 (1988)
3. Chin, F., Houck, D.: Algorithms for updating minimal spanning trees. J. Comput. System Sci. 16(3), 333–344 (1978)
4. Das, B., Loui, M.C.: Reconstructing a minimum spanning tree after deletion of any node. Algorithmica 31, 530–547 (2001)
5. Dixon, B., Rauch, M., Tarjan, R.E.: Verification and sensitivity analysis of minimum spanning trees in linear time. SIAM J. Computing 21(6), 1184–1192 (1992)
6. Dixon, B., Tarjan, R.E.: Optimal parallel verification of minimum spanning trees in logarithmic time. Algorithmica 17(1), 11–17 (1997)
7. Faloutsos, M., Molle, M.: A linear-time optimal-message distributed algorithm for minimum spanning trees. Distributed Computing 17(2), 151–170 (2004)
8. Flocchini, P., Mesa Enriques, A., Pagli, L., Prencipe, G., Santoro, N.: Point of failure shortest-path rerouting. IEICE Trans. Inf. Syst. E89-D (2), 700–708 (2006)
9. Flocchini, P., Pagli, L., Prencipe, G., Santoro, N., Widmayer, P., Zuva, T.: Computing all the best swap edges distributively. In: Higashino, T. (ed.) OPODIS 2004. LNCS, vol. 3544, pp. 154–168. Springer, Heidelberg (2004)
10. Gaibisso, C., Proietti, G., Tan, R.B.: Optimal MST maintenance for transient deletion of every node in planar graphs. In: Warnow, T.J., Zhu, B. (eds.) COCOON 2003. LNCS, vol. 2697, pp. 404–414. Springer, Heidelberg (2003)
11. Gallager, R.G., Humblet, P.A., Spira, P.M.: A distributed algorithm for minimum spanning tree. ACM Trans. Prog. Lang. and Systems 5(1), 66–77 (1983)
12. Johnson, D.B., Metaxas, P.: A parallel algorithm for computing minimum spanning trees. J. Algorithms 19, 383–401 (1995)
13. Nardelli, E., Proietti, G., Widmayer, P.: Nearly linear time minimum spanning tree maintenance for transient node failures. Algoritmica 40, 119–132 (2004)
14. Santoro, N.: Design and Analysis of Distributed Algorithms. Wiley, Chichester (2007)
15. Tsin, Y.H.: On handling vertex deletion in updating minimum spanning trees. Information Processing Letters 27(4), 167–168 (1988)

Locating a Black Hole in an Un-oriented Ring Using Tokens: The Case of Scattered Agents

Stefan Dobrev[1], Nicola Santoro[2], and Wei Shi[2]

[1] University of Ottawa, Canada
sdobrev@site.uottawa.ca
[2] Carleton University, Canada
{santoro,swei4}@scs.carleton.ca

Abstract. Black hole search in a ring network has been studied in a *token model*. It is known that locating the black hole in an anonymous ring using tokens is feasible, if the team of agents is initially *co-located*. When dealing with the *scattered* agents, the problem was so far solved only when the orientation of the ring is known.

In this paper, we prove that a black hole can be located in a ring using tokens with scattered agents, even if the ring is un-oriented. More precisely, first we prove that the black hole search problem can be solved using only three scattered agents. We then show that, with k ($k \geq 4$) scattered agents, the black hole can be located fewer moves. Moreover, when k ($k \geq 4$) is a constant number, the move cost can be made optimal. These results hold even if both agents and nodes are *anonymous*.

Keywords: Black Hole, Mobile Agent, Token, Ring, Anonymous, Asynchronous, Scattered, Un-oriented.

1   Introduction

1.1   The Problem and Related Work

The computational issues related to the presence of a harmful agent have been explored in the context of intruder capture and network decontamination. In the case of a harmful host, the focus has been on the *black hole*, a node that disposes of any incoming agent without leaving any observable trace of this destruction [1,2,3,4,5,6].

A *black hole* (BH) models a network site in which a resident process (e.g., an unknowingly installed virus) deletes visiting agents or incoming data; furthermore, any undetectable crash failure of a site in an asynchronous network transforms that site into a BH. In presence of a BH, the first important goal is to determine its location. To this end, a team of mobile system agents is deployed; their task is completed if, within finite time, at least one agent survives and knows the links leading to the BH. The research concern is to determine under what conditions and at what cost mobile agents can successfully accomplish this task, called the *black hole search* (BHS) problem.

A.-M. Kermarrec, L. Bougé, and T. Priol (Eds.): Euro-Par 2007, LNCS 4641, pp. 608–617, 2007.

The computability and complexity of Bhs depend on a variety of factors, first and foremost on whether the system is *synchronous* [1,2,6] or *asynchronous* [3,4]. Indeed the nature of the problem changes drastically and dramatically: the former is a great simplification of the later. For example, if there is a doubt on whether or not there is a Bh in the system, in absence of synchrony, this doubt can *not* be removed. In fact, in an asynchronous system, it is *undecidable* to determine if there is a Bh [3]. In this paper we continue the investigation of black hole search problem in the asynchronous case.

The existing investigations on Bhs in asynchronous systems have assumed the presence of a powerful inter-agent communication mechanism, *whiteboards* [3,4], at all nodes. The availability of whiteboards at all nodes is a requirement that is practically expensive to guarantee and theoretically (perhaps) not necessary. In this paper, we consider a less demanding and less expensive inter-communication and synchronization mechanisms that would still empower the team of agents to locate the Bh: the *token model*. In this model, each agent has available a bounded number of tokens that can be carried, placed in a node or/and on a port of the node, or removed from them; all tokens are identical (i.e., indistinguishable) and no other form of communication or coordination is available to the agents.

The problem of Bhs using tokens has been examined in the case of *co-located* agents, that is when all the agents start from the same node. In this case, Bhs is indeed solvable [5]. In [5] it was shown that a team of two or more co-located agents can solve Bhs with $O(n \log n)$ moves and two (2) tokens per agent in a *ring* network.

The problem becomes considerably more difficult if the agents are *scattered*, that is, when they start from many different sites. In particular, with scattered agents, the presence (or lack) of orientation in the ring and knowledge of the team size are important factors. Here, oriented ring means all the agents in this ring are able to agree on a common sense of direction. In the token model, in particular, it is known that in an oriented ring it is possible to locate a Bh with $O(1)$ tokens per agent performing $\Theta(n \log n)$ moves [7].

1.2   Main Results

In this paper we show that, for Bhs in a ring, the token model is computationally and complexity-wise as powerful as the whiteboard model, regardless of the initial position of the agents and of the orientation of the topology.

More precisely, first we prove that in an un-oriented ring, the Bh can be located by a team of three or more scattered agents, each using $O(1)$ tokens; the total amount of moves is $O(n^2)$ in the worst case. We then show that, if there are k ($k \geqslant 4$) scattered agents, the Bh can be located with $O(kn + n \log n)$ moves and $O(1)$ tokens per agent. When k ($k \geqslant 4$) is a constant number, the number of moves used can be reduced to $\Theta(n \log n)$, which is optimal. These results hold even if both agents and nodes are *anonymous*.

Due to space limitations, the proofs of all lemmas are omitted.

2 Model, Observations and Basic Tool

2.1 The Model and Basic Observations

Let $\mathcal{R}$ be a anonymous ring of n nodes (i.e. all the nodes look the same, they do not have distinct identifiers). Operating on $\mathcal{R}$ is a set of k agents $a_1, a_2, ..., a_k$. The agents are *anonymous* (do not have distinct identifiers), *mobile* (can move from a node to a neighboring node) and *autonomous* (each has computing and bounded memory capabilities). All agents have the same behavior, i.e. follow the same protocol, but start at the different nodes (and they may start at different and unpredictable times), each of which is called *homebase* ($\mathcal{H}$ for brevity). The agents can interact with their environment and with each other only through the means of *tokens*.

A token is an atomic entity that the agents can see, place in the middle of a node or/and on a port, or remove. Several tokens can be placed on the same location. The agents can detect the multiplicity, but the tokens themselves are undistinguishable from each other. Initially, there are no tokens placed in the network, and each agent starts with some fixed number of tokens. Most importantly, the tokens are the only means of inter-agent communication we consider. There is no read/write memory (whiteboards) for the agents to access in the nodes, nor is there face-to-face recognition. In fact, an agent notices the presence of another agent by recognizing the token(s) it leaves. When we say two agents *meet*, there are two situations: two agents walking in the same direction *meet*, meaning that one agent catches up with the agent in front of it in the same direction. Here *catch up* means finding the token(s) of the other agent in the same direction. When two agents walking in the opposite direction meet, we mean that both agents find the token(s) of the other agent in the same node.

One of the nodes of the ring $\mathcal{R}$ is a BH. All the agents are aware of the presence of the BH, but at the beginning the location of the BH is unknown. The goal is to locate the BH, i.e. at the end there must be at least one agent that has not entered the BH and knows the location of the BH.

The primary complexity measure is *team size*: the number of agents needed to locate the BH. Other complexity measures we are interested in are *token count*: the number of tokens each agent starts with, and *cost*: the total number of moves executed by the agents (worst case over all possible timings and starting locations).

The computation is asynchronous in the sense that the time an agent sleeps or is on transit is finite but unpredictable. The links obey a FIFO rule, that is, the agents do not overtake each other when traveling over the same link in the same direction. Because of the asynchrony, the agents cannot distinguish between a slow node and the BH. From this we get:

Lemma 1. *[3] It is impossible to find the Black Hole if the size of the ring is not known.*

As the agents are scattered, it could be the case that there is an agent in each neighbor of the BH, and both these agents wake up and make their first move towards the BH. This shows that:

Lemma 2. *[3] Two agents are not sufficient to locate the* Bh *in scattered case without knowing the orientation of the ring.*

3   Algorithm *Shadow Check*

3.1   Basic Ideas, General Description and Communication

We call a node/link *explored* if it is visited by an agent. A *safe (explored) region* consists of contiguous explored nodes and links. We call the last node an agent explored its Last-Safe-Place (LSP for brevity). In the scattered agents case, during the executing of Bhs there are more than one *safe region*s in the ring. Our goal is to merge all the *safe region*s into one, which eventually includes all the nodes and links with the exception of the Bh and the two links leading to the Bh. Let us describe how this goal is going to be achieved.

Upon waking up, an agent becomes a *Junior Explorer (JE)*, exploring the ring to the right (from the viewpoint of the agent) until it meets another agent[1]. When two *JE*s meet, they both become *Senior Explorers (SE)*, and start exploring the ring in opposite directions. We call the explored area between these two *SE*s a *safe region* for them. A *SE* explores the ring, growing its *safe region* and checking after each newly explored node whether the *safe region* contains all the nodes except the Bh. When two *SE*s moving in opposite directions *meet*, the two *safe regions* merge into a bigger *safe region*. The two meeting *SE*s become *Checker*s and check the size of the new *safe region*. There could be more than one such *safe region* in the whole ring. When a *JE* sees a *safe region*(i.e., it encounters a *SE*), it becomes *Passive* (stops being active).

When no unusual event occurs, each *SE* repeats the following cycle: it leaves two (*SE*s use two tokens) tokens on the port (if there is no token on this port) of the unexplored link on which it is going to move next. Once it reaches the node (if it is not the Bh), the *SE* leaves there two tokens on the port from which it did not enter that node. It returns to the previous node, picks up the token(s) on the port it used, then returns to the last explored node. If, between cycles, an agent notices any unusual event (e.g., token situation changes on certain ports of a node), it stops the cycle and acts according to this interruption. The details of possible interruptions are explained later.

The communication and coordination between the agents are described as follows:

- One token on the port means a *JE* is exploring the link via this port.
- Two tokens on the port means a *SE* is exploring the link via this port.
- One token on the port and one token in the middle means this is the node in which two opposite direction *JE*s *meet*.
- One token on each port means this is the node in which one *JE* catches up with another *JE* in the same direction.

---

[1] More precisely, finds a token of another agent.

The details of the algorithm are explained in the next sub-sections. In order to make the algorithm simpler to understand, we describe the procedure "Junior/Senior Explorer" from the viewpoint of the agents, who agree on the same "right" direction. The procedure for all the agents who agree on the same "left" direction can be achieved by changing the word "right" into "left", and "left" into "right".

3.2 Procedure "Initialize" and "Junior Explorer"

Once an agent wakes up, it becomes a JE that will go to the next node to the right immediately. There are 6 possible situations a JE may encounter once in the right neighbor node. A JE will eventually either end up in the Bʜ or become a *Checker* upon *meet*ing a SE or a potential SE, or become a SE upon *meet*ing another JE.

- Case 1
  The agent A puts one token in the middle, then goes back to the left node. If there is a SE caught up with agent A, then A will become a *Checker* to the left. If the agent it just met (let's say B) in the opposite direction also left A a sign (a token in the middle), then A will become a SE to the left. If A is caught up by another JE in the same direction, A will pick up all the tokens, then become a *Checker* to the right.
- Case 2
  The agent A goes back to the left node. If A is caught up by another JE, it will become a *Checker* to the right. If A notices that the SE it just met in the opposite direction left A a sign(a token in the middle), then it will become *Passive* immediately. If the JE A just met left A a sign, then A will become a SE to the left.
- Case 3
  The agent A puts one token on the left port, then goes back to the left node. If A's token is still there, it will move this token to the left port, add one more token on the left port then it becomes a SE to the left. If either A sees the sign a SE it just met left to it, or A is caught up by another JE, it will become *Passive* immediately.
- Case 4
  The agent A goes back to the left node. If A's token is still there, then it will pick the token and then become *Passive*. If A is caught up by a SE, then it will become *Passive*. If A is caught up by another JE, it will pick up the tokens, then become a *Checker* to the right.
- Case 5
  The agent A returns to the left node. If A is caught up by a SE, then it becomes *Passive*. If A is caught up by another JE, it will become a *Checker* to the right. If it notices that the JE it just met left it a sign (a token in the middle), then it will move the two tokens to the left port and become a SE to the left.

– Case 6

The agent A puts a token on the right port then goes back to the left node. If A's token is still there, then it will pick the token and continue as a *JE*. If it is caught up by a *SE*, then it will become *Passive*. If A is caught up by an other *JE*, then it will become a *SE* to the right.

3.3   Procedure "Checker"

A *Checker* is created when an agent realizes it is in the middle of two *SE*s exploring in different directions. The purpose of the *Checker* is to check the distance between the two *SE*s. A *Checker* keeps walking to the right until it either sees the token of a *SE* going to the right, or a node with one token on each port. If the distance is $n - 2$, that means that two agents have died in the Bh, and the only node left is the Bh. Otherwise, it keeps walking to the left until it either sees the token of a *SE* going to the right, or a node with one token on each port. If now the distance is $n - 2$, then it will become *DONE* (the Bh is located), otherwise it becomes *Passive* immediately.

3.4   Procedure "Senior Explorer"

A senior explorer will eventually either end up in the Bh or locate the Bh, or become a *Checker* upon *meet*ing another *SE* or a potential *SE*. A potential *SE* means a status of a *JE* after it either met another *JE* in the same direction or different direction, but before it becomes a *SE*. A *SE* walks to the right node. If it meets another *SE* in the different direction (we say: faces a *SE*), it will pick up all the tokens and become a *Checker* to the right. If it realizes it is the node which two *JE*s in the different directions met, it will then become a *Checker* to the right. If it realizes this node is where two *JE*s in the same direction met, it will then go back to the left port, pick up all the tokens and become a *Checker*. If it meets a *JE* going to the left, then it will pick the token on the left port, put two tokens on the right ports and go back to the left node and pick up the two tokens on the right port. Then the *SE* will execute the *check phase* to the left. If it meets a *JE* going to the right, then it will put one more token on the right port, go back to the left node; pick up the two tokens on the right port, then execute the *check phase* to the left. If the node is empty, the *SE* will then put two tokens on the right ports, go back to the left node; pick up the two tokens on the right port, then execute the *check phase* to the left.

Once a *SE* is in the *check phase*, it walks to the left until it either sees the token of a *SE* going to the right, or a token with one token on each port. If there are $n - 2$ links in the *safe region*, then it will become *DONE*, otherwise it goes back to its LSP. If there is no token on the right port of its LSP, it then will become *Passive*.

3.5   Analysis of Algorithm *Shadow Check*

According to Lemma 2, we assume there are at least three agents in the ring network. The following lemmas and corollary hold.

Lemma 3. *Eventually there is at least one* SE.

Corollary 1. *At most two agents enter the* BH.

Lemma 4. *A* safe region *will be created.*

Lemma 5. *Whenever the length of a* safe region *increases, it will be checked.*

Lemma 6. *The length of a* safe region *keeps increasing until it contains* $n - 2$ *links or* $n - 1$ *nodes.*

Theorem 1. *Algorithm* Shadow Check *correctly locates the* BH *with* k *($k \geqslant 3$) scattered agents in an un-oriented anonymous ring network. The total cost is* $O(n^2)$ *moves and, 5 tokens per agent.*

Proof. According to the above lemmas, three scattered agents are enough to locate the BH.

Now let us analyze the move cost: because there are k scattered agents, there is a maximum of $k/2$ *safe regions* in the ring. In procedure "Senior Explorer", an agent traverses its *safe region* once it explores one more node. There is a maximum $2n$ moves in each such traversal. There are at most n nodes in the ring, which means there are at most n such traversals. So, $O(n^2)$ moves are used. In procedure "Checker", the maximum number of each *check* is $2n$. A *check* can be triggered by either two *safe regions* merging or the *SE* this *Checker* follows exploring one more node. Given, there are no more than $k/2$ merges and n such *check*s, the total number of moves in procedure "Checker" is no more than $2n^2$. Hence, the total move cost is $O(n^2)$.

Now we analyze the token cost: a *JE* uses one token on the port to mark its progress. Once a *JE* meets another *JE*, one extra token is used to mark the node in which the two *JE*s meet and form a pair of *SE*s. This token will stay in the node until the algorithm terminates. A *SE* puts two tokens on the port as soon as it is created. It puts another two tokens on the port of the next node to mark progress. The first two tokens will be picked up and reused when exploring the next node. Hence, at most $1 + 2 + 2 = 5$ tokens are used by each agent.

4 Algorithm *Modified 'Shadow Check'*

4.1 Motivation

In the previous section, we presented algorithm *Shadow Check* that handles the BHs problem in an un-oriented ring with a minimum of 3 scattered agents and 5 tokens per agents. According to Theorem 1, an agent in one of the $k/2$ *safe regions*, traverses its *safe region* every time it explores one more node in order to check the size of this *safe region*. This is due to requiring minimum team size: Since there are only 3 agents in total, and because of the definition of *Checker*, it is obvious that there is at most one *Checker* formed in algorithm *Shadow Check*. So the explorers have to both explore the ring and check the size of the *safe region*. This cost (n^2) moves in the worst case.

After considering what kind of cost would we obtain if we had one more agent, we modify the algorithm slightly. The modified algorithm *Modified 'Shadow Check'* is such that:

- it can handle 4 or more scattered agents instead of 3;
- eventually there will be two *Checker*s formed and $O(kn + n \log n)$ moves are used for an arbitrary k. If k ($k \geqslant 4$) is a constant number, the move cost can be reduced to $\Theta(n \log n)$

4.2 Modification

We can obtain algorithm *Modified 'Shadow Check'* by performing the following modifications on algorithm *Shadow Check*:

1. In procedure "Junior Explorer": change all the action "become *Passive*" of a *JE* in algorithm *Shadow Check*, into "become a *SE* in the same direction reusing the two tokens of the caught up *SE*", whenever this *JE* is caught up by a *SE*.
2. In procedure "Senior Explorer", there are two types of *SE*s: a *SE* with a *Checker* and a *SE* without a *Checker*.
   - a *SE* with a *Checker*: marked as three tokens on the port (the extra token is added by its *Checker*).
   - a *SE* without a *Checker*: marked as two tokens on the port.

   In both procedures, the 'check phase' in algorithm *Shadow Check* is deleted; Instead, as soon as it caught up with a *JE*, it will becomes a *Checker* in the opposite direction;
   - In procedure "a *SE* with a *Checker*": as soon as a *SE* with a *Checker* faces another *SE* with/without a *Checker*, it becomes *Passive*. Otherwise, it continues exploring.
   - In procedure "a *SE* without a *Checker*": as soon as a *SE* without a *Checker* faces another *SE* with/without a *Checker*, it becomes a *Checker*. Otherwise it continues exploring.
3. The procedure "Checker" is modified as follows:
   A *Checker* is created when it realizes it is in the middle of two *SE*s exploring in different directions. Once an agent becomes a *Checker*, it *check*s the size (number of nodes) of the *safe region* once, namely, it walks until the LSP of a *SE* with/without a *Checker*, then changes direction, walks and keeps counting the number of nodes it passes, until it arrives in the LSP of another *SE* without a *Checker*. We call this a *check* and this second *SE* the *SE* of this *Checker*. Let L denote the size of a *safe region*. Now this *Checker* puts an extra token on the port where its *SE* left two tokens. But if what this *Checker* meets is a *SE* with a *Checker*, then this *Checker* leaves a token in the middle of the node and becomes *Passive* immediately. There are two situations that can trigger a *Checker* to *check* again:
   - Merging *check*: there is a token in the middle of the LSP of the *SE* of this *Checker*. This is caused by two *safe region*s merging. This *Checker* then picks up the token in the middle and performs a *check* in order to update L.

– Dividing *check*: this *Checker* followed its *SE* for $\lfloor (n - L)/2 \rfloor$ steps.

If while a *Checker* is following its *SE*, it notices that its *SE* became *Passive* (i.e., no token or not three tokens on the port in the next node), it then keeps walking until it sees the LSP of another *SE*. If it is a *SE* without a *Checker*, then this *Checker* becomes a *Checker* of this *SE*; otherwise, this *Checker* puts a token in the middle of this LSP and becomes *Passive* immediately.

If while a *Checker* is *check*ing the length of its *safe region* L, it notices that the *safe region* contains $n - 1$ links. Then the BH is located: the only node left unexplored is the BH.

4.3   Correctness and Complexity

Given there are at least 4 agents in the ring, we know:

Lemma 7. *At least two* SE*s are formed in algorithm* Modified 'Shadow Check'.

Lemma 8. *Eventually at least two* Checker*s will be formed.*

Theorem 2. *Algorithm* Modified 'Shadow Check' *correctly locates the* BH

Proof. According to Corollary 1 and Lemma 7 and 8, eventually there will be two *Checker*s formed/left that keep checking the size of the *safe region* until the only *safe region* in the ring contains $n - 1$ nodes or $n - 2$ links. Hence the BH is correctly located.

Theorem 3. *Algorithm* Modified 'Shadow Check' *correctly locates the* BH *in an un-oriented ring with* k $(k \geqslant 4)$ *scattered agents, each having 5 tokens. When* k *is arbitrary, the total cost is* $O(kn + n \log n)$. *If* k *is a constant number, then the total move cost is* $\Theta(n \log n)$.

Proof. First we analyze the move cost: a *SE* with/without a *Checker* keeps exploring nodes along the ring without turning back. There are at most n such moves in total. In procedure "Checker", the maximum number of moves in each *check* is $2n$, and there are no more than $\log n$ *Dividing check*s, given a *Checker* does not proceed with the next *Dividing check* until it follows its *SE* for $\lfloor (n - L)/2 \rfloor$ steps. There are no more than $k/2$ *Merging Check*s in total, given k scattered agents can form at most $k/2$ *safe region*s. So the total number of moves in procedure "Checker" is no more than $kn + 2n \log n$. When k $(k \geqslant 4)$ is a constant number, the total number of moves in procedure "Checker" becomes $O(n \log n)$, and the total move cost is $O(n \log n)$. The lower bound follows from the whiteboard model presented in [3]. Hence the total cost of moves is optimal when four or more $(O(1))$ scattered agents searching for a BH in an un-oriented ring.

Now we analyze the token cost: as we know that except for in procedure "Checker", a *Checker* uses one more token compared to a *Checker* in algorithm *Shadow Check*, no other modification affects the number of tokens used by each agent. According to the algorithm, a *Checker* uses a token only once in its lifespan. Also, according to Theorem 1: five tokens per agent are used in algorithm *Shadow Check*. Hence, 5 tokens per agent suffice to locate the BH in algorithm *Modified 'Shadow Check'*.

5   Conclusion

In this paper, we proved that locating the Black Hole in an anonymous ring network using tokens is feasible even if the agents are scattered and the orientation of the ring is unknown. Thus, we proved that, for the black hole search problem, the token model is as powerful as the whiteboard regardless of the initial position of the agents.

From the results we obtained in this paper, we observe that there is a trade-off between the team size (number of agents) and the costs (number of moves and number of tokens used). Since both algorithms we presented require only a constant number of tokens per agent, we are unable to simulate the *distance identity*. (The *distance identity*, presented in [3], is the crucial technique used in order to achieve $\Theta(n \log n)$ moves with optimal team size (3 agents).) But with one more agent, the token model is as powerful as the whiteboard with respect to black hole search in an un-oriented ring. And memorywise our algorithms represent a considerable improvement on the whiteboard model.

References

1. Cooper, C., Klasing, R., Radzik, T.: Searching for black-hole faults in a network using multiple agents. In: Shvartsman, A.A. (ed.) OPODIS 2006. LNCS, vol. 4305, pp. 320–332. Springer, Heidelberg (2006)
2. Czyzowicz, J., Kowalski, D., Markou, E., Pelc, A.: Complexity of searching for a black hole. Fundamenta Informatica 71(2-3), 229–242 (2006)
3. Dobrev, S., Flocchini, P., Prencipe, G., Santoro, N.: Mobile search for a black hole in an anonymous ring. Algorithmica (to appear, 2007)
4. Dobrev, S., Flocchini, P., Prencipe, G., Santoro, N.: Searching for a black hole in arbitrary networks: Optimal mobile agent protocols. Distributed Computing (to appear, 2007)
5. Dobrev, S., Kralovic, R., Santoro, N., Shi, W.: Black hole search in asynchronous rings using tokens. In: Calamoneri, T., Finocchi, I., Italiano, G.F. (eds.) CIAC 2006. LNCS, vol. 3998, pp. 139–150. Springer, Heidelberg (2006)
6. Klasing, R., Markou, E., Radzik, T., Sarracco, F.: Hardness and approximation results for black hole search in arbitrary networks. Structural Information and Communication Complexity 3499, 200–215 (2005)
7. Dobrev, S., Santoro, N., Shi, W.: Scattered black hole search in an oriented ring using tokens. In: Proc. of 9th Workshop on Advances in Parallel and Distributed Computational Models (APDCM'07) (to appear, 2007)

A Decentralized Solution for Locating Mobile Agents

Paola Flocchini and Ming Xie

SITE, University of Ottawa, 800 King Edward, Ottawa, Canada, K1N 6N5
{flocchin,mxie}@site.uottawa.ca

Abstract. In this paper we propose a new strategy for tracking mobile agents in a network. Our proposal is based on a semi-cooperative approach: while performing its own prescribed task, a mobile agent moves keeping in mind that a searching agent might be looking for it. In doing so we want a fully distributed solution that does not rely on a central server, and we also want to avoid the use of long forwarding pointers. Our proposal is based on appropriate delays that the mobile agents must perform while moving on the network so to facilitate its tracking, should it be needed. The searching agent computes a particular searching path that will guarantee the tracking within one traversal of the network. The delays to be computed depend on structural properties of the network. We perform several experiments following different strategies for computing the searching path and we compare our results.

Keywords: Algorithms for mobile agents, agents location, tracking.

1 Introduction

In this paper we consider a classic problem for mobile agents: the *tracking* (or *locating*) problem. An agent, or a group of agents, is sent on a network to locate a particular agent that is instead moving to perform some tasks. Sometimes the tracking is necessary to communicate with the agent or to terminate its task (e.g., see [4,9]).

Other problems involving two mobile agents are somehow related to this one: *pursuit evasion*, and *rendezvous*. In pursuit evasion there is a *competitive* setting, where one agent tries to escape, while the other is chasing it. The problem has been extensively studied in deterministic and especially in randomized environments (e.g., see [1,5,11]). In rendezvous the two agents *cooperate* to find each other; in fact, their goal is to meet somewhere in the network and their actions go towards this common goal. Also rendezvous has been widely investigated, under different scenarios and different assumption (e.g., see [3] and, for a recent survey [10]). In our problem there is no competition, since the moving agent does not try to escape. On the contrary, there is cooperation, since the moving agent is willing to facilitate the task of the locating agent. However, the degree of cooperation is much weaker than in the rendezvous problem. In fact, the moving agent has other tasks to perform, and it does not even know if some agent has been

A.-M. Kermarrec, L. Bougé, and T. Priol (Eds.): Euro-Par 2007, LNCS 4641, pp. 618–628, 2007.
© Springer-Verlag Berlin Heidelberg 2007

sent to locate it. While performing its primary tasks it can also perform some actions in order to help the tracking, in case it has to be performed. The agent is willing to do so at the expenses of its own performances, up to a certain degree.

For this purpose, typical solutions involve: 1) the existence of a central host where the moving agent constantly reports its position 2) having the moving agent leave a trace of its movements on its way (e.g., see [2,7]). Both solutions present obvious disadvantages. In the centralized solution the biggest problems are related to fault tolerance and security (problems that are present every time a centralized solution is employed). In fact, the central database could crash resulting in a complete loss of the information; moreover a third party accessing the central database could immediately determine the positions of all the agents at a given time. The second solution consists of leaving at each node the indication of the node where the agent has moved (*forwarding pointer*). This solution could result in a high space complexity to store all the information (especially considering that several moving agents are usually present in a network).

We propose a totally different approach. The idea is to pre-compute a particular *searching walk* that will be followed by the searching agent whenever the tracking is required. The moving agent moves autonomously and independently to perform its task, without reporting its location and without leaving long traces. The "speed" of its movement, however, is appropriately controlled in such a way that, should the searching agent look for the moving agent, it would locate it within one searching walk. By "controlling the speed" we mean that when the moving agent needs to move over a link, it cumulates a delay (proportional to some network parameter appropriately chosen) before performing the movement. In the paper the network is assumed to be synchronous; preliminary studies suggest that this assumption could be relaxed for a more realistic environment.

We describe how to compute this appropriate delay, we show that the amount is related to a network parameter called MinMax chord. The choice of the optimal search walk is conjectured to be an NP-complete problem. We describe several heuristics to compute various searching walks. We then run some experiments to see what are the performances of the heuristic algorithms in random graphs of various size and degree. The performances measure we consider are the maximum and average delay incurred by the moving, as well as the location time. The results are interesting and motivate further study.

2   Model and Terminology

Although the problem might involve several moving agents and several searching agents, without loss of generality we focus on the behavior of a single pair, so we have a *searching agent* SA and a *moving agent* MA. The algorithm we describe would apply to the case of more moving agents.

We assume the system is synchronous, that is, it takes one unit of time for an agent to traverse a link. The searching agent move at the maximum possible speed (1 time unit per link), while the moving agent "slows down" its movement by waiting an appropriate amount of time at the nodes it encounters on its way.

We assume that the moving agent is followed by a single forwarding pointer; in other words, a trace of the moving agent arrived at node y from link (x, y) is present at node x as long as the moving agent is in y. This fact guarantees that either the moving agent or its trace are always present on a node (even when the moving agent is in transit on a link). The locating problem is considered solved when SA resides on the same node as MA, or when it finds its forwarding pointer.

3 The Searching Walk

The general idea is to determine a traversal of the graph (called *searching walk*) for the searching agent. While the moving agent moves arbitrarily, the searching agent follows this predefined walk. Initially the searching agent is at the "beginning" of the searching walk and the moving agent is obviously "ahead".

The searching walk is a traversal of the graph. Let $T = [y_1, y_2, \ldots, y_k]$ be a traversal of $G = (V, E)$. An extended weighted graph $G' = (V', E')$, based on the traversal T can be constructed by aligning the traversal nodes (each y_i connected to y_{i+1}), and adding the other edges of E as chords, as shown in Figure 1. If a node is visited more than once, there is a chord in G' only between any two consecutive occurrences of the same node in T. Given an edge $(y_i, y_j) \in E'$, let $weight(y_i, y_j)$ denote its weight .

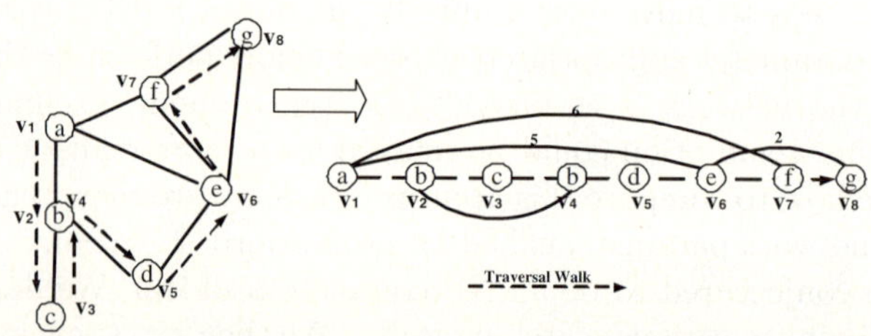

Fig. 1. Traversal Walk with Chords

More precisely, let $G = (V, E)$ be a graph with n nodes, and let $T = [y_1, y_2, \ldots, y_k]$ be a traversal walk of G (with $k \geq n$). Let $f : V' \Rightarrow V$ be a (non injective) function that, for each element y_i of the traversal, returns the corresponding node $f(y_i)$ of V. Let $E(x)$ denote the edges incident to node x in G. We now define a weighted graph $G' = (V', E')$ as follows:

- *Vertices:* $V' = \{y_1, \ldots, y_k\}$ contains one vertex per element of T;
- *Edges :* $(y_i, y_j) \in E'$ if:
  1) $((f(y_i), f(y_j)) \in E)$ AND (there exists no k ($i < k < j$) such that $f(y_i) = f(y_k)$ or $f(y_k) = f(y_j)$).
  2) $f(y_i) = f(y_j)$ with $i \neq j$ AND there exists no k ($i < k < j$) such that $f(y_i) = f(y_k)$.

  Weight: $weight(y_i, y_j) = i - j$

Let us call *physical chord* a chord in G' that corresponds to a link in G, and *virtual chord* a chord in G' that connect two occurrences of the same node of G (thus it does not correspond to any link of G). A virtual chord, in fact, corresponds to a cycle in the traversal walk. For each node $x \in V$, $F^{-1}(x) = \{f^{-1}(x)\}$ contains the corresponding occurrences of x in T.

We now define a node and an edge labeling for graph G. Let $\beta : V \to Z$ be a node labeling function that associate an integer to each node of the network, and $\lambda_x : E(x) \to Z$ be an edge labeling function that associate an integer to each edge incident to node We label the nodes and the edges of G as follows:

- *Vertices:*
  Let $x \in V$. If $|F^{-1}(x)| = 1$, then $\beta(x) = 0$. If $F^{-1}(x) = \{v_{i,1}, v_{i,2}, \ldots, v_{i,m}\}$ $m > 1$, then $\beta(x) = Max_j\{weight(v_{i,j+1}, v_{i,j})\}$ for $j = 1 \ldots m - 1$.
- *Edges:*
  Let $(x, y) \in E$. We define $\lambda_x(x, y) = Max_{a,b,i,j}\{weight(v_{i,a}, v_{j,b})\}$ for $v_{i,a} \in F^{-1}(x)$, $v_{j,d} \in F^{-1}(y))$, and $(v_{i,a}, v_{j,b}) \in E'$.

We compute the delays that the moving agent has to introduce in such a way that the searching agent (SA) is always "behind" the moving agent (MA) along the searching walk, except when it locates it or its forward pointer (FP). With the delays we define below, in fact, it could happen that SA overpasses MA; in this case, however MA is behind the forwarding pointer and it is guaranteed to reach it before it expires.

Suppose the moving agent arrives at node x from link (w, x) and has to move to node y through link (x, y). If $\lambda_x(x, y) < 0$ the agent can move directly without adding any delay because it is moving "away" from the searching agent; if $\lambda_x(x, y) > 0$, it waits the following amount of time.

DELAY
Agent reaching x from (w, x), moving to y through (x, y) in G.

If $\lambda_x(x, y) < 0$
   move-to-y
If $\lambda_x(x, y) > 0$
   WAIT $Max\{\beta(x), \lambda_x(x, y) - 1, \lambda_w(w, x) - 1\}$
   move-to-y

Let $T = [y_1, y_2, \ldots, y_k]$ be the searching walk with SA initially in y_1 when the process starts at time $t = 1$.

Lemma 1. *Let MA arrives at node x at time t from node w. Let del be the time MA has to wait before moving to some node y. If by the time $t + del$ MA has not been located, then after the movement of MA towards y, SA is "before" the first occurrence of $\{f^{-1}(y)\}$ in T (i.e., $t + del + 1 < i$ $\forall v_i \in F^{-1}(y)$).*

Proof. (Sketch) By induction on the movements of MA. It is true at time 1, when MA moves for the first time. Let the lemma be true when MA arrives to node w and let us prove it is true when MA reaches the next node x. The lemma is trivially true if link (x, y) corresponds to a "forward link" in the traversal path, i.e., if $\lambda_x(x, y) < 0$. Let us then consider only "backward" links.

Consider time t when MA arrives in x. At this time, SA is in y_t and, by induction hypothesis, it is either ahead of MA of it will catch it before MA can move. Now MA waits for $Max\{\beta(v), \lambda_w(w, x) - 1, \lambda_x(x, y) - 1\}$ time units. Let m and M be smallest and the largest indices such that $y_m, y_M \in F^{-1}(x)$. We now consider three cases:

- Case $t < m$. If $m - t < \lambda_x(x, y)$, by definition of delay, SA will locate MA before it moves to y (because $del > \lambda_x(x, y)$). Otherwise, SA will be "before" or on y at time $t + del + 1$.
- Case $m < t < M$. Let $y_a, y_b \in F^{-1}(x)$ be the closest occurrences of node x to y_t. SA will reach node x before MA moves to y because, by definition of delay, $del > \beta(x)$ and β is the largest cycle containing x (which is greater than or equal to $b - a$).
- Case $t > M$. SA will reach FA in the next del time units because, by definition of delay, MA stays in x for at least $\lambda_w(w, x) - 1$ time units and thus, FA is in w when SA reaches it at time $t + \lambda_w(w, x) - 1$.

It follows from the previous Lemma that:

Theorem 1. *The searching agent locates the moving agents by the end of its traversal.*

Clearly, one would like to minimize both the location time (which depends on the length of the searching walk) and the delay incurred by the moving agent. In the following we are interested in searching walks of length $O(n)$ and we would like to minimize the maximum delay as well as the average delay incurred by the moving agent. The maximum delay corresponds to the longest (virtual or physical) chord; thus, we would need to find the traversal that minimizes such a chord. More precisely, we define the MinMax Traversal $\mathcal{T}_G$ of G as the traversal that minimizes the maximum weight in G': $\mathcal{T}_G = Min_T\{Max_{i,j}\{w(y_i, y_j)\}\}$ with $(y_i, y_j) \in E'$.

4 Building Good Searching Walks

We conjecture that finding the MinMax traversal of a graph is an NP-complete problem. In the following we propose several heuristic algorithms to construct "good" traversal and we compare them.

General Traversal Algorithm. Consider the following general traversal algorithm that visits the nodes in depth. If the node where the searching agent resides has unvisited neighbours, the agent moves to one of them; if it has no unvisited neighbours, but some nodes have not been visited, it moves through already visited nodes to reach one that has not been visited yet.

In the general traversal algorithm described above, there are two points where we can introduce some variations for obtaining a searching walk that is good for our purposes: 1) How to choose the next node when the current node has several unvisited neighbours. 2) Where to move after visiting a node without unvisited neighbours. In the common *depth-first traversal* (DFT), for example, a random choice is performed for 1) and a backtrack for 2). Obviously an arbitrary DFT does not necessarily result in an efficient searching strategy.

Choosing an Unvisited Neighbouring Node. The following are different strategies to choose one among the unvisited neighboring nodes.

- **Random.** The agent randomly chooses an unvisited neighboring node.
- **Priority Queue.** When the agent moves to a new node, the unvisited neighbouring nodes of that node are stored in a priority queue. When the agent finds more than one unvisited neighboring nodes, it chooses the unvisited neighboring node with highest priority. If none of the unvisited nodes is in the queue:
  - **Basic:** The agent randomly chooses an unvisited neighboring node.
  - **Improved:** The agent checks the unvisited neighboring nodes' unvisited neighbors. If it finds at least one of the unvisited neighboring nodes at distance two in the priority queue, it chooses the one with highest priority and it moves there.
  - **Closest to the queue:** The agent chooses the one that is closest to a node in the priority queue (in case of ties it selects the one closest to a highest priority element in the queue).
- **Closest to start node.** When the agent has more than one unvisited neighboring nodes, it moves to the unvisited neighboring node that is closest to the start node.
- **Neighbour of least visited node.** When the agent has more than one unvisited neighboring nodes, it moves to the unvisited neighboring node whose neighbour has been visited least recently.

Moving to a Non neighbouring Unvisited Node. The following are different strategies to move to an unvisited node when there is no unvisited neighboring node from the current agent's location.

- **DFT.** The agent backtracks to the most recently visited node that has unvisited neighboring node, and then continues the traversal.
- **Greedy.** The agent moves to the nearest unvisited node.
- **BFT.** The agent moves to one of the unvisited node that is closest to the start node.
- **Hybrid.** The agent moves to the nearest unvisited node if there is only one such node; if there are more such nodes, the agent moves to the node among the nearest unvisited nodes that is closest to the starting node of the walk. This strategy combines the *greedy* and *BFT* strategies.

When considering a non-neighbouring unvisited node, we include in the walk the shortest path between the current node and the next. Notice that, in doing so we might visit new nodes.

The strategies described above (except for *Random* and *DFT*) are motivated by the empirical observation that it might be useful to fully visit an area around a visited node before moving to another area of the graph. In fact, visiting a neighbour of a node already visited creates a chord: since we would like to minimize the length of the chords, we would like to visit these nodes as soon as possible. On the other hand, we also want to maintain a walk of size $O(n)$ and to achieve this we cannot revisit nodes already visited too many times.

5   Experimental Results

5.1   Experimental Setup

We now combine the different strategies for choosing the next neighbouring unvisited node and choosing a non-neighbouring unvisited node. For each algorithm, we record the length of the traversal walk, the maximum and average delay for the moving agent. Notice that the maximum delay corresponds to the length of the longest chord.

We run the algorithms on different random topologies of various size and density. For each type of graph, we generated 20 graphs with the same parameters, and we averaged the obtained results. We utilize *Java Universal Network/Graph Framework* [8] library to generate the graphs.

- A1: *random + DFT* .
- A2: *improved priority queue + DFT* .
- A3: *improved priority queue + Greedy*.
- A4: *random + Greedy*.
- A5: *improved priority queue+ BFT*.
- A6: *random + BFT*.
- A7: *improved priority queue + Hybrid*.

We have observed all the results for all the combinations, we however show here only the ones from which we have obtained the most interesting results.

In this set of experiments we are interested in the maximum and average delay incurred by the moving agent, and in the location time, depending on the traversal strategy followed by the searching agent. In all cases, the length of the traversal is only slightly higher than the number of nodes.

In the experiments, we generated 10 random graphs for each type of graph, and obtained the average values for the results. We have run experiments for various graph sizes n ($n = 100, 200, 500, 800, 1000$) and levels of density m (number of edges) ($m = 4n, 5n, 6n, 7n, 8n, 9n, 10n$). For each choice of n and m we generated 10 random graphs to count the average values. In each case we randomly select the starting node.

The graphic in Figure 2 *a*) shows that, among our heuristics, the ones with the best performance in terms of the length of the maximum chord are A_3 and A_7; that is: the *improved priority queue* heuristic as the choice of an unvisited neighbouring node, and either the *Greedy* or the *Hybrid* heuristic for the choice of a non-neighbouring unvisited node. This graph correspond to the case $n = 1000$; the results are quite consistent with different sizes of the graph.

Interestingly, for all heuristics, increasing the density of the network (i.e., its average degree), results in a decrease of the length of the traversal and, most of the times, also of the length of the maximum chord and average delay.

The graphic in Figure 2 b) shows the changes in average delay incurred by the moving agent as the number of edges increases, for the different strategies. Also in this case, the best performances are obtained by A_3 and A_7.

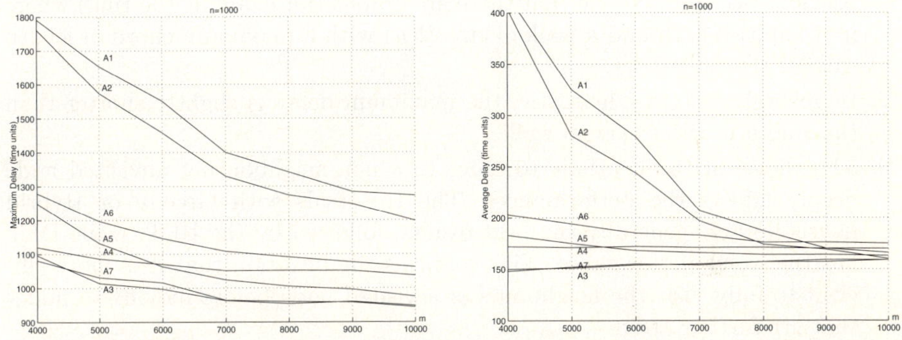

Fig. 2. a) Max delay b) Average delay

The various results for the plain DFT and some combinations of heuristics are reported in the tables below (again for $n = 1000$).

m	A_1: Random + DFT			
	Max. D.	Av. D.	Loc. T.	Length
4000	1791	418	791	1855
5000	1654	325	712	1705
6000	1551	279	669	1617
7000	1400	207	608	1464
8000	1342	187	587	1393
9000	1284	170	554	1319
10000	1273	166	536	1321

m	A_7: Improved Priority+ Hybrid			
	Max. D.	Av. D.	Loc. T.	Length
4000	1081	148	509	1163
5000	1031	152	501	1124
6000	1016	155	510	1098
7000	964	154	498	1075
8000	960	157	479	1067
9000	958	159	502	1060
10000	944	159	491	1049

m	A_4: Random + Greedy			
	Max. D.	Av. D.	Loc. T.	Length
4000	1120	172	506	1149
5000	1086	171	502	1110
6000	1070	172	502	1089
7000	1051	171	499	1076
8000	1052	171	486	1065
9000	1041	170	497	1057
10000	1037	170	492	1051

m	A_5: Improved+BFT			
	Max. D.	Av. D.	Loc. T.	Length
4000	1248	183	621	1329
5000	1132	175	609	1240
6000	1063	168	562	1178
7000	1024	166	556	1144
8000	1005	164	545	1116
9000	996	164	514	1098
10000	976	163	504	1082

Maximum delay, average delay, location time, length of the walk for $n = 1000$.

5.2   Observations

The experiments lead to the following observations.

1. With our best combinations of heuristics (*priority queue and hybrid* and *priority queue and greedy*), the maximum delay is reduced by approximately half compared to a Depth-first traversal; the average delay is reduced of 75 %. The reductions are more evident when the graph is sparse, they become less relevant when the graph is very dense (see Figure 2 and tables in the previous Section). Notice that there are graphs (for example the ring) where it is impossible to find a walk of size $O(n)$ with a maximum chord of length smaller than $O(n)$.

2. In general, with any heuristic, the maximum delay is slightly shorter than the length of the traversal walk.

3. The choice of the heuristic to move to a non-neighbouring unvisited node deeply affects the performances. The traversals with Greedy or Hybrid heuristics give generally the best results, followed by the BFT, while DFT has the worst performance. This confirms the intuition that it is more efficient to fully visit the neighbours of a visited node before moving to nodes that are further apart.

4. While the maximum delay always decreases with the increase of the number of edges, the average delay is more stable and does not display this strong behavior.

5. Combinations A_4 (Random+ Greedy) and A_5 (Improved+ BFT) behaves in a similar way and they both improve the plain depth-first walk. Interestingly, A_4 outperforms A_5 for sparse graphs, while it becomes less efficient then A_5 for more dense graphs. In our experiments this happens roughly when $m = 6n$ Our intuition for this behavior is that the BFT heuristic heavily depends on the density; in fact, when the graph is too sparse BFT gives little chance to find a "good" (close to the start) unvisited node; increasing the number of edges, however, it is more likely that such a "good" node is found thus discovering shorter chords. On the other hand, the Greedy heuristic does not depend on the density since it can always move to the nearest visited node.

6. An interesting observation is that the performances are proportional to the walk's length, in the sense that strategies with longer walks give generally worse performances than strategies with shorter walks. The only exception is A_7 when compared with A_4. The performances of A_7, in fact, are better than those of A_4 although the length of the traversal walk of A_7 is slightly higher (at least for densities $m < 10n$).

7. Fixing the average degree and changing the graph size, the average and maximum delay increases linearly with the graph size. Furthermore, the lower is the lower degree, the more the choice of the heuristic affects the performance (the graphics corresponding to this type of analysis are not shown for lack of space).

6 Conclusion

The contribution of this paper is the proposal of a new approach for locating mobile agents that involves neither forwarding pointers (only a single trace of size one) nor a central server. The approach is based on having the moving agent move at a variable speed, depending on the structural properties of the links it is traversing. At this stage the algorithm could not seem applicable in practice because it is based on very strong assumptions about the environment, which is assumed to be synchronous. Notice however that synchronicity could be relaxed by slightly increasing the length of the trace left by the agents; in fact, preliminary studies suggest that this approach would work also in environments that are not synchronized provided the moving agent leaves a short trace of length 2. Further notice that while computing the delay we have not taken into consideration the time the agent has to actually spend at each node. Depending on its tasks, the agents might have to spend a certain amount of time thus decreasing the forced delay.

We are currently working on several possible improvements that we believe will highly decrease the average delay. The employ of more than a single searching agent would considerably decrease the locating time. For example, static searching agents could be placed at crucial nodes (the ones which have long chords along the searching walk) while a single searching agent could traverse the walk; alternatively, several searching walk could be traversed concurrently thus reducing the locating time. This paper is just a first step towards exploiting the structure of the network for locating purposes, we are now working on improvements like the ones mentioned above to make the technique more applicable and efficient in a practical setting.

References

1. Adler, M., Racke, H., Sivadasan, N., Sohler, C., Vocking, B.: Randomized pursuit-evasion in graphs. In: Int. Colloquium on Automata, Languages and Programming, pp. 901–912 (2002)
2. Alouf, S., Huet, F., Nain, P.: Forwarders vs. centralized server: An evaluation of two approaches for locating mobile agents. Performance Evaluation 49(1-4), 299–319 (2002)
3. Alpern, S., Gal, S.: The theory of search games and rendezvous. Kluwer Academic Publishers, Dordrecht (2003)
4. Baumann, J.: Mobile Agents: Control Algorithms. LNCS, vol. 1658. Springer, Heidelberg (2000)
5. Demirbas, M., Arora, A., Gouda, M.G.: A pursuer-evader game for sensor networks. In: Pro. 6th Symposium on Self-Stabilizing Systems, pp. 1–16 (2003)
6. Diaz, J., Petit, B.J., Serna, M.: A survey of graph layout problems. ACM Computing Surveys 34(3), 313–356 (2002)
7. Fowler, R.J.: The complexity of using forwarding addresses for decentralized object finding. In: 5th ACM Symp. on Principles of Distributed Computing, pp. 108–120. ACM Press, New York (1986)

8. Jung: Java Universal Network/Graph Framework, `http://jung.sourceforge.net/`
9. Lien, Y., Leng, C.W.R.: On the search of mobile agents. In: 7th IEEE Int. Symposium on Personal, Indoor, and Mobile Radio Communications, pp. 703–707. IEEE Computer Society Press, Los Alamitos (1996)
10. Kranakis, E., Krizanc, D., Rajsbaum, S.: Mobile agent rendezvous. In: 13th Int. Coll. on Structural Information and Communication Complexity, pp. 1–9 (2006)
11. Parsons, T.D.: Pursuit-evasion in a graph. In: Jantke, K.P. (ed.) AII 1992. LNCS, vol. 642, pp. 426–441. Springer, Heidelberg (1992)

On Detecting Termination in the Crash-Recovery Model

Felix C. Freiling[1], Matthias Majuntke[2,*], and Neeraj Mittal[3,**]

[1] University of Mannheim, 68131 Mannheim, Germany
[2] Darmstadt University of Technology, 64289 Darmstadt, Germany
[3] The University of Texas at Dallas, Richardson, TX 75083, USA

Abstract. We investigate the problem of detecting termination of a distributed computation in an asynchronous message-passing system where processes may crash and recover. We show that it is impossible to solve the termination detection problem in this model. We identify necessary and sufficient conditions under which it is possible to solve the *stabilizing* version of the problem in which a termination detection algorithm is allowed to make finite number of mistakes. Finally, we present an algorithm to solve the stabilizing termination detection problem under these conditions.

Keywords: asynchronous distributed system, termination detection, crash-recovery failure model, eventual safety, stabilizing algorithm.

1 Introduction

Termination detection involves determining whether a distributed computation has ceased all its activities. It arises when a computation terminates *implicitly* and no process knows about the termination [1]. As a result, a separate algorithm may have to be used to detect termination of the computation. The termination detection problem has been studied quite extensively in fault-free models (*e.g.*, [2,3,4,5,6]). It has also been studied in the crash-stop model (*e.g.*, [7,8,9,10,11,12]). However, little is known about the problem when processes may crash and recover. We believe that the crash-recovery failure model is more realistic than crash-stop failure models because, in practice, to avoid resource exhaustion we must allow crashed processes to recover. However, it is also harder to deal with than the other two failure models as shown by earlier work in this failure model on solving other important distributed computing problems such as consensus [13], reliable broadcast [14] and atomic broadcast [15,14].

In this paper, we investigate the termination detection problem in the *crash-recovery* model. It turns out that it is impossible to solve the problem in this failure model without weakening the problem and/or strengthening the model. The main reason for this impossibility result is that it is not possible to determine the future behavior of a currently down process, that is, whether it will stay down permanently or may recover later.

* Research supported in part by EC IP DECOS, EC NoE ReSIST and DFG Graduiertenkolleg "Cooperative Adaptive and Responsive Monitoring in Mixed Mode Environments" at TU Darmstadt.
** Research supported in part by DFG Graduiertenkolleg "Software for Mobile Communicaton Systems" at RWTH Aachen University.

A.-M. Kermarrec, L. Bougé, and T. Priol (Eds.): Euro-Par 2007, LNCS 4641, pp. 629–638, 2007.

To circumvent the impossibility result, we weaken the problem by allowing a termination detection algorithm to make mistakes, that is, it may falsely announce termination albeit only a finite number of times. We refer to this problem as the *eventually safe termination detection problem*.

Even the eventually safe termination detection problem cannot be solved without strengthening the model. To that end, we make the following two assumptions about the model. First, there are no unstable processes in the system, *i.e.*, there are no processes that crash and recover infinitely often. Second, processes are equipped with an eventually perfect failure detector. We show that both conditions are necessary.

We finally describe an algorithm for solving the stabilizing termination detection problem under the two assumptions. Such an algorithm is very usefull in scenarios where only performance and not correctness of the application is affected by false termination announcements. For example, it is possible to construct an efficient mutual exclusion algorithm in the following way: a first distributed algorithm establishes a spanning tree in the network, while a second algorithm circulates a token in a repeated depth-first traversal of the tree. The second algorithm is only started once the first algorithm has terminated. Suppose several devices use a mutual exclusion algorithm to select a channel for communication. If due to false termination announcement mutual exclusion property is violated, in the worst case two or more devices choose the same channel and their communication will interfere. Eventually, the algorithm satisfies the safety property and will work properly thereafter. Due to lack of space, for the proofs of some lemmas and theorems we refer to [16].

2   Model and Notation

Distributed System.   We assume an asynchronous distributed system consisting of a set of processes, given by $\Pi = \{p_1, p_2, \ldots, p_n\}$, in which processes communicate by exchanging messages with each other over a communication network. A process changes its state by executing an *event*. The system is asynchronous in the sense that there is no bound on the amount of time a process may take to execute an event or a message may take to arrive at its destination.

Failure Model.   We assume that processes may fail by crashing. Further, a crashed process may subsequently recover and resume its operation. While a process is crashed, it does not execute any events. This failure model is referred to as the *crash-recovery model*. In the crash-recovery model, a process may be either *stable* or *unstable*. A process is said to be stable if it crashes (and possibly recovers) only a finite (including zero) number of times; otherwise it is unstable. A stable process can be further classified into two categories: *eventually-up* or *eventually-down* [13]. A process is said to be eventually-up if the process eventually stays up after crashing and recovering a finite number of times; otherwise it is eventually-down. An eventually-up process is said to be *always-up* if it never crashes. Sometimes, eventually-up processes are referred to as *good processes*, and eventually-down and unstable processes are referred to as *bad processes* [13]. A process that is currently operational is called an *up* process, whereas a process that is currently crashed is called a *down* process. We use the phrases "up process" and "live process" interchangeably. Likewise, we use the phrases "down process" and "crashed process" interchangeably.

We assume *eventually-reliable channels* in this paper. Such channels guarantee reliable communication between good processes that do not crash anymore.

Process Incarnations. We assume that each process has access to *volatile storage* and *stable storage*. When a crashed process recovers, we say that the process has a *new incarnation*. At the very least, we use stable storage to distinguish between various incarnations of the same process. Each process maintains an integer in its stable storage that keeps track of its incarnation number, that is, the number of times the process has crashed and recovered. The integer is initially set to 0 for all processes. Whenever a process recovers from a crash, before taking any other action, it reads the value of the integer from its stable storage, increments the value and writes the incremented value back to its stable storage.

Failure Detector for Termination Detection. In this paper, we focus on *realistic failure detectors* which are not capable of predicting the future behavior of a process (*e.g.*, whether a process will stay up forever) [17]. The termination detection algorithm described in this paper needs an *eventually perfect failure detector* [18]. Intuitively, an eventually perfect failure detector is responsible for monitoring the operational state of all processes in the system. It may make mistakes in the beginning. For instance, it may believe that a process is down when, in fact, the process is actually up, and vice versa. However, it should eventually have a correct view of the system.

In this paper, we do not distinguish between the local failure detector at a process and the process itself unless necessary. If process p_i believes process p_j to be down, we say that p_i *suspects* p_j. On the other hand, if p_i believes p_j to be up, we say that p_i *trusts* p_j. We model the output of the failure detector using a list of trusted processes. For a failure detector to be eventually perfect, the list at different processes should satisfy the following properties:

- *Completeness* (two parts): (1) Every bad stable process is eventually permanently suspected by all good processes. (2) Every bad unstable process is either eventually permanently suspected by all good processes, or suspected and trusted infinitely often by all good processes.
- *Accuracy*: Every good process is permanently trusted by all good processes.

Note that if there are no unstable processes in the system, then eventually all good processes agree on which processes are currently up.

3   The Termination Detection Problem

The distributed computation whose termination has to be detected is typically modeled using the following four rules. First, a process is either in *active* state or *passive* state. Second, a process can send a message only if it is active. Third, an active process may become passive at any time. Fourth, a passive process may become active only on receiving a message. Intuitively, an active process is involved in some local activity, whereas a passive process is idle. The termination detection problem involves determining whether the computation has ceased all its activities. Formally,

Definition 1 (termination condition). *A computation is said to have terminated if no process is currently active and, further, no process becomes active in the future.*

We consider a process to be in an active state only when it is up. Further, we conclude that a process, on recovery, restarts in an active state only if it failed in an active state. Any termination detection algorithm should satisfy the following two desirable properties:

- *No false termination announcement (safety):* If the termination detection algorithm announces termination, then the computation has indeed terminated.
- *Eventual termination announcement (liveness):* Once the computation terminates, the termination detection algorithm eventually announces termination.

Note that the termination condition, as formulated in Definition 1, requires reasoning on future states of processes. For failure-free and crash-stop failure models, however, it is possible to reformulate the termination condition so that the condition can be evaluated on a current state of the system and does not require any future knowledge. For instance, for the failure-free model, the termination condition can be redefined as: *all processes are passive and all channels are empty.* Likewise, for the crash-stop model, the termination condition can be redefined as: *all live processes are passive and all channels towards live processes are empty.* However, for the crash-recovery model, in general, it is not possible to formulate the termination condition in a manner that does not require knowledge of future behavior of processes. One of the main reasons is that a termination detection algorithm in general should be able to distinguish between whether a process is only *temporarily down* or is *permanently down.* We now prove this impossibility result:

Theorem 1. *It is impossible to solve the termination detection problem unless it is possible to distinguish between whether a process is temporarily down or permanently down.*

Proof. Consider the initial system state in which only one process, say p_a, is active and all other processes are passive. We construct two possible executions of the system. In the first execution, p_a crashes and never recovers. Clearly, the computation has terminated and any correct termination detection algorithm should eventually announce termination. Assume that some process, say p_c, announces termination at time t. Next, consider the second execution that is identical to the first execution until time t. However, after t, p_a recovers and restarts in an active state. Clearly, the only difference between the two executions is in the behavior of p_a after t. Specifically, in the first execution, p_a is permanently down at t, whereas, in the second execution, it is only temporarily down at t. Since p_c cannot distinguish between the two executions, it announces termination at t in the second execution as well. However, in the second execution, the computation has not terminated at t. $\qquad\qquad\square$

To circumvent the impossibility result, we weaken the termination detection problem by relaxing the safety property. Specifically, a termination detection algorithm is allowed to announce termination *falsely.* However, only a *finite* number of such false announcements are allowed. Formally, a termination detection algorithm should now satisfy the following properties:

- *Finite number of false termination announcements (eventual safety):* Eventually, if a good process announces termination, then the computation has indeed terminated.
- *Eventual termination announcement (liveness):* Once the computation terminates, eventually every good process announces termination.

Note that, even after weakening the safety property, the termination detection problem remains impossible to solve as long as the system contains unstable processes. An unstable process can repeatedly crash and recover in an active state causing the termination detection algorithm to make infinite number of mistakes. Therefore we assume that the system does not contain any unstable process, that is, every process eventually stays either up or down. We denote the resulting model as *crash/finite-recovery model*.

Suppose the underlying distributed system is such that failures are expected to be rare. Further, even the failure detector makes mistakes only rarely. In this case, we would like the termination detection algorithm to also make mistakes only rarely. Formally,

- *No false termination announcements in the absence of failures and false suspicions (zero degradation):* Assume that no process crashes during an execution and no process is falsely suspected of having crashed by the failure detector. Then, a process announces termination only if the computation has terminated.

In the next section, we describe a termination detection algorithm that satisfies eventual safety, liveness and zero degradation properties.

4   An Eventually Safe Termination Detection Algorithm

Our termination detection algorithm uses an eventually perfect failure detector defined in Sect. 2. It turns out that such a failure detector is actually necessary for solving the eventually safe termination detection problem in the stabilizing crash-recovery model. Specifically, we can implement an eventually perfect failure detector using an eventually safe termination detection algorithm as follows. There are n computations in the system, one for each process; the computation for process p_i is denoted by C_i. In C_i, p_i is always active while it is up and all other processes are always passive. Further, no process sends any application message in any computation. Additionally, there are n instance of the termination detection algorithm, and the instance A_i of the algorithm is responsible for observing the computation C_i. When instance A_i announces (respectively, revokes) termination at a process p_j, p_j starts suspecting (respectively, trusting) p_i. It can be easily verified that this transformation correctly implements an eventually perfect failure detector.

We now describe our termination detection algorithm. To avoid confusion, hereafter, we refer to messages exchanged by a distributed computation as *application messages* and those exchanged by a termination detection algorithm as *control messages*.

Each process maintains three different vectors, namely *incarnation vector*, *sent vector* and *received vector*. A process uses the first vector to keep track of the latest incarnation of other processes in the system. It uses the sent vector to keep track of the number of application messages it has sent to the latest incarnation of other processes.

Rules for updating incarnation vector on process p_i:

Variables:

  iv_i: vector $[1..n]$ of incarnation numbers, initially $[0, 0, \ldots, 0]$;

(A1) On sending a message m:
  piggyback iv_i on m;

(A2) On receiving a message m carrying incarnation vector $m.iv$:
  for each j in $[1, n]$ **do**
    $iv_i[j] := \max\{iv_i[j], m.iv[j]\}$;
  endfor;

(A3) On starting new incarnation x after recovery:
  $iv_i[i] := x$;
  // other entries of iv_i may be initialized using stable storage, if applicable

Fig. 1. Rules for maintaining incarnation vector on a process

Finally, it uses the received vector to keep track of the number of application messages it has received from the latest incarnation of other processes *that were sent to its current incarnation*. The incarnation, sent and received vectors for process p_i are denoted by iv_i, $sent_i$ and $received_i$, respectively.

The rules for maintaining the incarnation vector are similar to those for maintaining a Fidge/Mattern vector [19,20]. A process, on recovery, sets its own entry in the incarnation vector to its incarnation number. A process piggybacks the incarnation vector on every message it sends. Further, a process, on receiving a message, updates its incarnation vector by taking the component-wise maximum of each entry in the current vector and the vector received along with the message. Figure 1 describes the actions A1-A3 for maintaining the incarnation vector. Action A1 is executed whenever a process sends a message. Action A2 is executed whenever a process receives a message. Finally, action A3 is executed whenever a process recovers and starts a new incarnation.

We now describe a scheme that a process periodically uses to test if the computation has terminated. As part of the scheme, process requests all processes in the system to send their current local states to it. The local state of a process includes: (1) the incarnation vector, (2) the state with respect to the application, (3) the sent vector, and (4) the received vector. The process waits until it has received a local state from all processes that it currently trusts. It first ascertains that all trusted processes have identical incarnation vectors. If not, the process aborts the current instance of the testing scheme. If yes, the process checks whether the computation has terminated by evaluating the following two conditions:

1. all trusted processes are passive, and
2. sent and received vectors of all trusted processes "match" with each other when restricted to only entries for trusted processes.

If both conditions evaluate to true and the last action by the process was to revoke termination announcement, then the process announces termination. On the other hand, if one of the conditions fails and the last action was to announce termination, then the process revokes that last termination announcement. A process also aborts the current instance of the testing scheme if either its incarnation vector or the output of its failure detector changes. Note that the two conditions for termination are similar to those used in Mattern's channel-counting algorithm [4], which is a fault-intolerant termination detection algorithm.

We say that a system has *stabilized* if (1) each process has stabilized (that is, no more process crashes and recoveries), (2) the failure detector at each process has stabilized (that is, no more changes in the output of the failure detector), and (3) any application message delivered hereafter is sent by the current incarnation of the sender to the current incarnation of the receiver. We show that, once the system has stabilized, the testing scheme satisfies the safety property. Specifically, the two conditions described above evaluate to true only if the computation has terminated.

To ensure liveness, a process *periodically* uses an instance of the above-described scheme to test for termination. Different instances of the scheme are distinguished using an *instance identifier*, which consists of (1) the identifier of the initiating process, (2) its incarnation number and (3) a sequence number. The sequence number helps differentiating between various instances of the scheme initiated by the same incarnation of a process. The sequence number can be stored in the volatile storage.

Actions of the termination detection algorithm TDA-ES for process p_i:

Variables:

> $state_i$: state of p_i with respect to the application (maintained by the application);
> $sent_i$: vector $[1..n]$ of number of application messages send to each process, initially $[0, 0, \ldots, 0]$;
> $received_i$: vector $[1..n]$ of number of application messages received from each process, initially $[0, 0, \ldots, 0]$;
> $announcement_i$: whether termination has occurred, initially false;

(B1) Whenever the jth entry in iv_i advances:
> $sent_i[j] := 0$;
> $received_i[j] := 0$;

(B2) On starting a new incarnation after recovery:
>   **for** $j \in [1..n]$ **do**
> $sent_i[j] := 0$;
> $received_i[j] := 0$;
> $announcement_i := $ false;
>   **endfor**;

(B3) On sending an application message m to process p_j:
> $++ sent_i[j]$;
>   send m to process p_j;

(B4) On receiving an application message m from process p_j:
>   let $m.iv$ denote the incarnation vector piggybacked on m;
>   **if** $(iv_i[i] = m.iv[i])$ **and** $(iv_i[j] = m.iv[j])$ **then**
> $++ received_i[j]$;
>   **endif**;
>   deliver m to the application;

(B5) On invocation of testForTermination( ):
>   send REQUEST message to all processes;

(B6) On receiving REQUEST message from process p_j:
>   send RESPONSE$(iv_i, state_i, sent_i, received_i)$ message to process p_j;

(B7) On receiving RESPONSE$(iv_j, state_j, sent_j, received_j)$ from process p_j:
>   let T_i denote the set of currently trusted processes;
>   **if** (a RESPONSE message has been received from all processes in T_i) **then**
>   **if** $\langle \forall\, p_x, p_y : \{p_x, p_y\} \subseteq T_i : iv_x = iv_y \rangle$ **then**
>   // all processes in T_i have identical incarnation vectors
> $test_i := \langle \forall\, p_x : p_x \in T_i : state_x = $ passive$\rangle \wedge$
> $\langle \forall\, p_x, p_y : \{p_x, p_y\} \subseteq T_i : sent_x[y] = received_y[x]\rangle$;
>   **if** $(announcement_i \wedge \neg test_i)$ **then**
>   // revoke termination announcement
> $announcement_i := $ false;
>   **else if** $(\neg announcement_i \wedge test_i)$ **then**
>   // announce termination
> $announcement := $ true;
>   **endif**;
>   **endif**;
>   **endif**;

Fig. 2. The termination detection algorithm TDA-ES

We refer to our termination detection algorithm as **TDA-ES**. A formal description of the algorithm is given in Fig. 2. It consists of seven actions B1-B7. Action B1 is executed whenever the incarnation vector of a process changes and is invoked from action A2 or action A3. Action B2 is executed when a process recovers and is invoked from action A3. Action B3 (respectively, B4) is executed whenever a process sends (respectively, receives) an application message. Note that action A1 has to be executed after executing action B3. Further, action B4 is invoked after action A2 has been executed. Actions B5, B6 and B7 are executed as part of the testing scheme. Note that **REQUEST** and **RESPONSE** messages carry instance identifier which is not shown in the description.

We now prove that our algorithm satisfies eventual safety, liveness and zero degradation properties. Note that there is no unstable process in the system and the failure detector is eventually perfect. Therefore it follows that:

Proposition 1. *The system eventually becomes stable.*

We say that the incarnation vector at a good process has *stabilized* if its value has stopped changing. Note that the ith entry of the incarnation vector at process p_i is incremented only when p_i recovers. Any other process p_j, with $j \neq i$, simply copies a (new) value into the ith entry of its incarnation vector from the vector it has received along with a message. Therefore, we have:

Proposition 2. *Once the system has become stable, the incarnation vector at a good process eventually becomes stable.*

Every good process periodically uses the testing scheme to test for termination. Once the system becomes stable, by our assumption, all channels between good processes become reliable. Therefore every good process receives infinite number of messages from every other good process. It can be easily verified that:

Proposition 3. *If the incarnation vector at every good process has become stable, then all good processes have identical incarnation vectors.*

We say that a system has become *strongly stable* if the system has become stable and the incarnation vector at every good process has become stable. We refer to the incarnation of a good process that never crashes as the *final incarnation*. Note that an instance of the testing scheme initiated after the system has become strongly stable always completes successfully (that is, is not aborted by its initiator). Also, once the system has become stable, every good process permanently trusts all good processes and permanently suspects all bad processes. This implies that if an instance of the testing scheme is initiated after the system has become strongly stable then its termination conditions are evaluated on local states of all good and only good processes. We now show that our testing scheme is safe and live if it is initiated after the system has become strongly stable.

Lemma 1. *Any instance of the testing scheme initiated after the system has become strongly stable indicates termination only if the computation has terminated.*

Lemma 2. *Any instance of the testing scheme initiated after (1) the system has become strongly stable and (2) the computation has terminated indeed indicates termination.*

The liveness of our algorithm follows from the fact that every good process periodically initiates an instance of the testing scheme to test for termination.

Proposition 4. *If no process crashes during an execution and no process is falsely suspected of having crashed by the failure detector, then the system is strongly stable in the initial state.*

Theorem 2. *TDA-ES satisfies eventual safety, liveness and zero degradation properties.*

Proof. Eventual safety follows from Proposition 1, Proposition 2, Proposition 3 and Lemma 1. Liveness follows from Proposition 1, Proposition 2, Proposition 3 and Lemma 2. Zero degradation follows from Lemma 1 and Proposition 4. □

5 Discussion

In our algorithm, as described above, each message has to carry a vector consisting of n entries. It is possible to optimize our algorithm so that only RESPONSE messages are required to carry a vector. An application message only needs to carry two entries from the incarnation vector of the sender, namely entries corresponding to the sender and the receiver. Specifically, an application message sent by process p_i to process p_j carries entries $iv_i[i]$ and $iv_i[j]$. It can be verified that all propositions and lemmas in the previous section still hold with this modification.

An interesting question to ask is when can the termination detection problem be solved in a *safe* manner under crash-recovery model. We answer this question in [21] where we identify two conditions under which the safe termination detection problem can indeed be solved. These conditions are rather strong compared to the conditions identified in this paper. For example, one of the conditions requires the availability of a *perfect failure detector*, processes to *always restart in passive state after recovery* and processes to *reject old application messages*.

References

1. Tel, G.: Distributed Control for AI. Technical Report UU-CS-1998-17, Information and Computing Sciences, Utrecht University, The Netherlands Technical Report UU-CS–17, Information and Computing Sciences (1998)
2. Dijkstra, E.W., Scholten, C.S.: Termination Detection for Diffusing Computations. Information Processing Letters (IPL) 11(1), 1–4 (1980)
3. Francez, N.: Distributed Termination. ACM Transactions on Programming Languages and Systems (TOPLAS) 2(1), 42–55 (1980)
4. Mattern, F.: Algorithms for Distributed Termination Detection. Distributed Computing (DC) 2(3), 161–175 (1987)
5. Mattern, F.: Global Quiescence Detection based on Credit Distribution and Recovery. Information Processing Letters (IPL) 30(4), 195–200 (1989)
6. Mittal, N., Venkatesan, S., Peri, S.: Message-Optimal and Latency-Optimal Termination Detection Algorithms for Arbitrary Topologies. In: Proceedings of the 18th Symposium on Distributed Computing (DISC), Amsterdam, The Netherlands, pp. 290–304 (October 2004)

7. Venkatesan, S.: Reliable Protocols for Distributed Termination Detection. IEEE Transactions on Reliability 38(1), 103–110 (1989)
8. Lai, T.H., Wu, L.F.: An $(N-1)$-Resilient Algorithm for Distributed Termination Detection. IEEE Transactions on Parallel and Distributed Systems (TPDS) 6(1), 63–78 (1995)
9. Tseng, Y.C.: Detecting Termination by Weight-Throwing in a Faulty Distributed System. Journal of Parallel and Distributed Computing (JPDC) 25(1), 7–15 (1995)
10. Hélary, J.M., Murfin, M., Mostefaoui, A., Raynal, M., Tronel, F.: Computing Global Functions in Asynchronous Distributed Systems with Perfect Failure Detectors. In: IEEE Transactions on Parallel and Distributed Systems (TPDS), September 2000, vol. 11(9), pp. 897–909 (2000)
11. Gärtner, F.C., Pleisch, S. (Im)Possibilities of Predicate Detection in Crash-Affected Systems. In: Proceedings of the 5th Workshop on Self-Stabilizing Systems (WSS),, Lisbon, Portugal, October 2001, pp. 98–113 (2001)
12. Mittal, N., Freiling, F.C., Venkatesan, S., Penso, L.D.: Efficient Reduction for Wait-Free Termination Detection in a Crash-Prone Distributed System. In: Fraigniaud, P. (ed.) DISC 2005. LNCS, vol. 3724, pp. 93–107. Springer, Heidelberg (2005)
13. Aguilera, M.K., Chen, W., Toueg, S.: Failure Detection and Consensus in the Crash Recovery Model. Distributed Computing (DC) 13(2), 99–125 (2000)
14. Boichat, R., Guerraoui, R.: Reliable and Total Order Broadcast in the Crash-Recovery Model. Journal of Parallel and Distributed Computing (JPDC) 65(4), 397–413 (2005)
15. Rodrigues, L., Raynal, M.: Atomic Broadcast in Asynchronous Crash-Recovery Distributed Systems and its Use in Quorum-Based Replication. IEEE Transactions on Knowledge and Data Engineering 15(5), 1206–1217 (2003)
16. Freiling, F., Majuntke, M., Mittal, N.: Termination Detection in an Asynchronous Distributed System with Crash-Recovery Failures. Technical report, TR-2006-008, University of Mannheim (2006)
17. Delporte-Gallet, C., Fauconnier, H., Guerraoui, R.: A Realistic Look At Failure Detectors. In: Proceedings of the IEEE International Conference on Dependable Systems and Networks (DSN), Washington, DC, USA, pp. 345–353 (2002)
18. Chandra, T.D., Toueg, S.: Unreliable Failure Detectors for Reliable Distributed Systems. Journal of the ACM 43(2), 225–267 (1996)
19. Mattern, F.: Virtual Time and Global States of Distributed Systems. In: Parallel and Distributed Algorithms: Proceedings of the Workshop on Distributed Algorithms (WDAG), pp. 215–226 (1989)
20. Fidge, C.J.: Logical Time in Distributed Computing Systems. IEEE Computer 24(8), 28–33 (1991)
21. Mittal, N., Phaneesh, K.L., Freiling, F.C.: Safe Termination Detection in an Asynchronous Distributed System when Processes may Crash and Recover. In: Shvartsman, A.A. (ed.) OPODIS 2006. LNCS, vol. 4305, pp. 126–141. Springer, Heidelberg (2006)

Topic 9
Parallel and Distributed Programming

Luc Moreau, Emmanuel Jeannot, George Bosilca, and Antonio J. Plaza

Topic Chairs

Developing parallel or distributed applications is a hard task and it requires advanced algorithms, realistic modeling, efficient design tools, high performance languages and libraries, and experimental evaluation. This topic provides a forum for presentation of new results and practical experience in this domain. It emphasizes research that facilitates the design and development of correct, high-performance, portable, and scalable parallels program.

We received 24 papers and accepted 7.

"Delayed Side-Effects Ease Multi-Core Programming" and "MCSTL: The Multi-Core Standard Template Library" deal with multi-core programming wich is one of the hot topic in parallel programming. The first paper is focused on language while the other is focused on library.

The library approach is also used in the paper "Library Support for Parallel Sorting in Scientific Computations" while the compilation approach is applied in the paper "Nested Parallelism in the OMPi OpenMP/C Compiler".

Other approaches such as skeleton have been proposed for parallel programming. This approach serves as the basis of two papers in this session: "Domain-Specific Optimization Strategy for Skeleton Programs" and" Management in Distributed Systems: a Semi-Formal Approach".

Finally, such a topic would not be complete without an "application" paper. "Efficient Parallel Simulation of Large-Scale Neuronal Networks on Clusters of Multiprocessor Computers" presents an innovative way to parallelize biological neural network.

A.-M. Kermarrec, L. Bougé, and T. Priol (Eds.): Euro-Par 2007, LNCS 4641, p. 639, 2007.
© Springer-Verlag Berlin Heidelberg 2007

Delayed Side-Effects Ease
Multi-core Programming

Anton Lokhmotov[1,*], Alan Mycroft[1], and Andrew Richards[2]

[1] Computer Laboratory, University of Cambridge
15 JJ Thomson Avenue, Cambridge, CB3 0FD, UK
{anton.lokhmotov,alan.mycroft}@cl.cam.ac.uk
[2] Codeplay Software Ltd, 45 York Place, Edinburgh, EH1 3HP, UK
andrew@codeplay.com

Abstract. Computer systems are increasingly parallel and heterogeneous, while programs are still largely written in sequential languages. The obvious suggestion that the compiler should automatically distribute a sequential program across the system usually fails in practice because of the complexity of dependence analysis in the presence of aliasing.

We introduce the sieve language construct which facilitates dependence analysis by using the programmer's knowledge about data dependences and makes code more amenable to automatic parallelisation.

The behaviour of sieve programs is deterministic, hence predictable and repeatable. Commercial implementations by Codeplay shows that sieve programs can be efficiently mapped onto a range of systems. This suggests that the sieve construct can be used for building reliable, portable and efficient software for multi-core systems.

1 Introduction

The evolution of high-performance single-core processors via increasing architectural complexity and clock frequency has apparently come to an end, as multi-core processors are becoming mainstream in the market. For example, Intel expects [1] that by the end of 2007 most processors it ships will be multi-core.

Homogeneous, shared memory multi-core processors, however, are but a part of the multi-core advent. Another growing trend is to supplement a general-purpose "host" processor with a special-purpose co-processor, which is typically located on a separate plug-in board and connected to large on-board memory. Graphics accelerators have been available since the 1990s and are increasingly used as co-processors for general-purpose computation. AGEIA's PhysX processor [2] is an accelerator for the highly specialized simulation of physical environment. Yet another example is ClearSpeed's SIMD array processor [3] targeted at intensive double-precision floating-point computations. These accelerators containing tens to hundreds of cores can be dubbed deca- and hecto-core to distinguish them from the currently offered dual- and quad-core general-purpose processors.

* This author gratefully acknowledges the financial support by the TNK-BP Cambridge Kapitza Scholarship Scheme and Overseas Research Students Awards Scheme.

A.-M. Kermarrec, L. Bougé, and T. Priol (Eds.): Euro-Par 2007, LNCS 4641, pp. 641–650, 2007.

Computer systems composed of multi-core processors can be fast and efficient in theory but are hard to program in practice. The programmer is confronted with low-level parallel programming, architectural differences between system components, and managing data movement across non-uniform memory spaces— sometimes all at once. Writing parallel programs is hard (as people tend to think sequentially) but testing is even harder (as non-deterministic execution can manifest itself in evasive software errors).

Ideally, the programmer wants to write a high-level program in a familiar sequential language and leave the compiler to manage the complexity of the target hardware. However, modern mainstream programming languages, particularly object-oriented ones, derive from the C/C++ model in which objects are manipulated *by reference, e.g.* by using pointers. While such languages allow for efficient implementation on traditional single-core computers, *aliasing* complicates dependence analysis necessary for sophisticated program transformations including parallelisation.

In this paper we consider an original approach to automatic parallelisation employed in Codeplay's Sieve C++ system [4]. In C99, the programmer can declare a pointer with a `restrict` qualifier to specify that the data pointed to by the pointer cannot be pointed to by any other pointer. In Sieve C++, the programmer makes a stronger statement about code enclosed inside a special block: no memory location defined outside of the block is written to and then read from within the block (we will also say that *the block generates no true dependences* on such memory locations). This guarantee makes code more amenable to auto-parallelisation. Also, the block-based structure maps well to a natural programming style for emerging heterogeneous hierarchical systems.

We describe the basic *sieve* concept in Section 2 and emphasise its importance in Section 3. Section 4 provides experimental evidence. We briefly mention a few recent approaches to parallel programming that are similar to the sieve system in Section 5 and conclude in Section 6.

2   Sieve Concept

We describe the basic sieve concept as an extension to a C-like language. The extension is both syntactic and semantic. Essentially, the programmer encloses a code fragment inside a *sieve block*—by placing it inside a new lexical scope prefixed with the `sieve` keyword. As a semantic consequence, all side-effects on data defined outside the sieve block, are *delayed* until the end of the block.

The name "sieve" has been proposed by Codeplay [4]. We can draw an analogy with a French press, or cafetière. A code fragment inside a sieve block (*cf.* a cylindrical jug with a plunger) is a mix of operations that either have delayed side-effects (*cf.* coffee grounds) or not (*cf.* boiling water). By depressing the plunger, equipped with a sieve or mesh, we collect the side-effects (grounds) at the bottom of the block (jug), leaving operations without side-effects (*cf.* drinkable coffee) that can be freely re-ordered (*cf.* thoroughly enjoyed).

2.1 Preliminaries

We say that a variable is *bound* in a given lexical scope if it is defined within this scope; a variable is *free* in a given lexical scope if it occurs in this scope but is defined outside of this scope.

2.2 Syntax

The programmer opens a new lexical scope and prefixes it with the `sieve` keyword denoting a sieve block:

```
sieve { int b; ... } // sieve block
```

We will call code enclosed in a sieve block as *sieve code*.

2.3 Semantics

Lindley presented the formal semantics of a core imperative language extended with sieves in [5]. We illustrate the sieve semantics by drawing an analogy with calling a function having *call-by-value-delay-result* parameter passing mechanism, which we detail below.

In C-like languages, entrance to a new lexical scope can be seen as equivalent to calling a function (whose body is the scope), arranging that the parameters to this function are the free variables of the scope and passing these by reference. For example, in the following

```
int main() {
    int a; ...
    { ... = a ... a = ... } ...
}
```

the enclosed code fragment accessing the variable `a` can be abstracted as a function call

```
void f(int *ap) { ... = *ap ... *ap = ... }
int main() {
    int a; ...
    f(&a); ...
}
```

By passing `a` by reference we ensure that all modifications to `a` are immediately visible in the program. Note that reads and writes to `a` are treated equally by replacing occurrences of `a` with `*ap`.

The `sieve` keyword changes the semantics of a lexical scope to mean that all modifications to free variables are *delayed* until the end of the scope, whereas all modifications to bound variables remain *immediate*. In accordance with this semantics, we will also refer to free and bound variables as, respectively, delayed and immediate.

Using the function call analogy, we say that

```
int main() {
    int a; ...
    sieve { ... = a ... a = ... } ...
}
```

is equivalent to

```
void f(int *ap) {
    const int ar = *ap;
    int aw = *ap;
    { // sieve block entry
        ... = ar ... aw = ...
    } // sieve block exit
    *ap = aw;
}
int main() {
    int a; ...
    f(&a); ...
}
```

Note the different treatment of reads and writes. On entry to the function the parameter (passed by reference) is copied into local variables `ar` and `aw`. All reads of the parameter are replaced with reads of `ar`, and all writes to the parameter are replaced with writes to `aw`. On exit from the function, `aw` is copied out to the parameter.

We coin the term *call-by-value-delay-result* (CBVDR) for this, as the translation is similar to traditional *call-by-value-result* (used, for example, for `in-out` parameters in Ada), where `ar` and `aw` are coalesced into a single variable.

2.4   Understanding the Change in Semantics

The theory of data dependence [6] helps in understanding how the sieve block semantics departs from that of a standard lexical scope. In the presence of data dependences, the behaviour of sieve code is affected as follows:

1. If a write to a free variable is followed by a read of it (*true dependence*), delaying the write *violates* the dependence.
2. If a read of a free variable is followed by a write to it (*anti-dependence*), delaying the write *preserves* the dependence.
3. If a write to a free variable is followed by another write to it (*output dependence*), delaying the writes *preserves* the dependence *if the order of writes is preserved*.

Since true dependences are violated, it is up to the programmer to ensure that sieve code generates no true dependences on delayed data (this gives the desired equivalence with the conventional semantics). This is hardly restrictive, however, as instead of writing into and subsequently reading from a delayed variable, the programmer can write into and read from a temporary immediate variable, updating the delayed variable on exit from the block.

Anti-dependences present no problem.

Preserving output dependences can be achieved by replacing every write to a delayed variable with a push of *address-value* pair onto a FIFO queue, and applying all queued writes in order on exit from the sieve block. We will refer to such a queue as *side-effect queue*.

Note that the programmer's implicit claim that sieve code generates no true dependences on delayed data, can be easily verified at run-time (and used for debugging) by additionally recording executed reads in the queue and checking that no read from a memory address is followed by a write to the same address.

2.5   Illustrative Example

Consider the following example:

```
int main() {
    int a = 0;
    sieve {
        int b = 0;
        a = a + 1; b = b + 1; print(a, b); // prints 0,1
        a = a + 1; b = b + 1; print(a, b); // prints 0,2
    }
    print(a); // prints 1
}
```

The first two `print` statements behave as expected as writes to the free variable a are delayed until the end of the sieve block, but the result of the third may come as a surprise. This result, however, is easy to explain using the CBVDR analogy, since the sieve block is equivalent to

```
void f(int *ap) {
    const int ar = *ap;
    int aw = *ap;
    {
        int b = 0;
        aw = ar + 1; b = b + 1; print(ar, b); // prints 0,1
        aw = ar + 1; b = b + 1; print(ar, b); // prints 0,2
    }
    *ap = aw; // *ap = 1, since aw == 1
}
int main() {
    int a = 0;
    f(&a);      // passing ap, where *ap == 0
    print(a); // prints 1
}
```

The immediate variable ar is never modified (by construction), hence both assignments to aw inside the sieve block write 1. After the sieve block, the immediate variable aw is copied into the delayed variable a.

This behaviour seems counter-intuitive because the sieve code violates the requirement of the previous section by generating a true dependence on the delayed variable a (the compiler can reasonably warn the programmer).

If the programmer wants to use the updated value of a, he needs to write

```
int main () {
    int a = 0;
    sieve {
        int b = 0, c = a;
        c = c + 1; b = b + 1; print(c, b); // prints 1,1
        c = c + 1; b = b + 1; print(c, b); // prints 2,2
        a = c;
    }
    print(a); // prints 2
}
```

As this code generates no true dependences on delayed variables, the conventional semantics is preserved.

2.6 Function Calls

A common imperative programming style is to use functions for their side-effects. When calling a function inside a sieve block, the programmer can specify whether to execute the function call immediately and have its side-effects delayed (this is natural for functions returning a result) or delay the call itself until the end of the block (this can be useful for I/O).

3 Importance of the Sieve Construct

3.1 Delayed Side-Effects Facilitate Dependence Analysis

Effectively exploiting parallel hardware (whether executing synchronously or asynchronously, in shared or distributed memory environment) often requires the compiler to re-order computations from the order specified by the programmer. However, indirect reads and writes, which are endemic in languages like C/C++, are difficult to re-order, as *alias analysis* is undecidable in theory, and even state-of-the-art implementations often give insufficient information for programs written in mainstream programming languages.

Consider a typical multi-channnel audio processing example

```
for (int i = 0; i < NCHANNELS; i++) process_channel(i);
```

Often, the programmer "knows" that each channel is independent of the others and hence hopes that the code will be parallelised. In practice, this hope is usually misplaced, as somewhere in `process_channel()` there will be indirect memory accesses causing the compiler to preserve the specified (sequential) execution order.

The kernel of the problem is that the programmer writes clear and concise sequential code but has no language-oriented mechanism to express the deep knowledge that sequenced commands can actually be re-ordered.

The sieve construct provides the programmer a way to conclude a treaty with the compiler:

Management in Distributed Systems:
A Semi-formal Approach[*]

Marco Aldinucci[1], Marco Danelutto[1], and Peter Kilpatrick[2]

[1] Department of Computer Science, University of Pisa
{aldinuc,marcod}@di.unipi.it
[2] Department of Computer Science, Queen's University Belfast
p.kilpatrick@qub.ac.uk

Abstract. The reverse engineering of a skeleton based programming environment and redesign to distribute management activities of the system and thereby remove a potential single point of failure is considered. The Orc notation is used to facilitate abstraction of the design and analysis of its properties. It is argued that Orc is particularly suited to this role as this type of management is essentially an orchestration activity. The Orc specification of the original version of the system is modified via a series of semi-formally justified derivation steps to obtain a specification of the decentralized management version which is then used as a basis for its implementation. Analysis of the two specifications allows qualitative prediction of the expected performance of the derived version with respect to the original, and this prediction is borne out in practice.

Keyword: Orchestration, algorithmic skeletons, autonomic computing.

1   Introduction

The muskel system, introduced by Danelutto in [1] and further elaborated in [2], reflects two modern trends in distributed system programming: the use of program skeletons and the provision of means for marshalling resources in the presence of the dynamicity that typifies many current distributed computing environments, e.g. grids. muskel allows the user to describe an application in terms of generic skeleton compositions. The description is then translated to a macro data flow graph [3] and the graph computed by a distributed data flow interpreter [2]. Central to the muskel system is the concept of a *manager* that is responsible for recruiting the computing resources used to implement the distributed data flow interpreter, distributing the fireable data flow instructions (tasks) and monitoring the activity of the computations. The muskel manager is to a certain extent an autonomic manager [4,5]: it adapts the run time behaviour of a muskel program to tolerate faults and maintain a user defined performance contract, much in the sense of what is advocated in [6,7].

---

[*] This research is carried out under the FP6 Network of Excellence CoreGRID funded by the European Commission (Contract IST-2002-004265).

A.-M. Kermarrec, L. Bougé, and T. Priol (Eds.): Euro-Par 2007, LNCS 4641, pp. 651–661, 2007.
© Springer-Verlag Berlin Heidelberg 2007

While the performance results demonstrated the utility of **muskel**, it was noted in [2] that the centralized data flow instruction repository (taskpool) represented a bottleneck and the manager a potential single point of failure. The work reported on here addresses the latter of these issues. The planned reengineering of the **muskel** manager was seen as an opportunity to extend earlier related experiments [8] with the language Orc [9] to investigate if it could usefully be employed in the development of such management software. The intent was not to embark upon a full-blown formal development of a modified **muskel** manager (as was done earlier for the related *Lithium* system [10], or as is normally done when employing other popular formalisms, such as the π-calculus [11]), with attendant formulation and proof of its properties, but rather to discover what return might be obtained from the use of such a formal notation for *modest* effort. In this sense, the aim was in keeping with the lightweight approach to formal methods as advocated by, inter alia, Agerholm and Larsen [12].

Orc was viewed as being apt for two reasons. First, it is an orchestration language, and the job of the **muskel** manager is one of orchestrating computational resources and tasks; and, second, while there are many process calculi which may be used to describe and reason about distributed systems, the syntax of Orc was felt to be more appealing to the distributed system developer whose primary interest lies not in describing and proving formal properties of systems.

The approach taken was to reverse engineer the original **muskel** manager implementation to obtain an Orc description; attempt to derive, in semi-formal fashion, a specification of a modified manager based on decentralized management; and, use this derived specification as a basis for modifying the original code to obtain the decentralized management version of **muskel**. By "semi-formal" we mean that the derivation is presented as a chain of steps in which the terms are described in the (formal) notation of the specification and the steps are justified by rigorous argument of the mathematical textbook variety, but calling also upon domain knowledge and experience when appropriate.

The work described in this paper is the first part of a more complex activity aimed at both removing the single point of failure represented by the **muskel** manager *and* implementing a distributed data flow instruction repository, removing the current related bottleneck. While the second step is still ongoing, the first step provides a suitable vehicle to illustrate the proposed methodology.

Overall, this work is part of a set of articles that are currently being published and that build on the semi-formal framework discussed here. In particular, in [16] the semi-formal approach based on Orc is extended to encompass meta data modeling non-functional aspects related to parallel/distributed program execution, while in [17] the complete approach exploiting Orc to support distributed program development is summarized.

2   muskel: An Overview

muskel is a skeleton based parallel programming environment written in Java. The distinguishing feature of **muskel** with respect to other skeleton environments [13,14] is the presence of an application manager. The muskel user instantiates a manager by providing the skeleton program to be computed, the input and the output streams containing the (independent) tasks to be computed and the results, respectively, and a performance contract modeling user performance expectations (currently, the only contract supported is the `ParDegree` one, requesting the manager to maintain a constant parallelism degree during application computation). The user then requests invocation of the `eval()` method of the manager and the application manager takes care of all the details relating to the parallel computation of the skeleton program.

When the user requires the computation of a skeleton program, the muskel system behaves as follows. The skeleton program is compiled to a macro data flow graph, i.e. a data flow graph of instructions modeled by significant portions of Java code corresponding to user `Sequential` skeletons [3]. A number of remote resources (sufficient to ensure the user performance contract) running an instance of the muskel run time are recruited from the network. The muskel run time on these remote resources provides an RMI object that can be used to compute arbitrary macro data flow instructions, such as those derived from the skeleton program. For each task appearing on the input stream, a copy of the macro data flow graph is instantiated in a centralized `TaskPool`, with a fresh graph id [2]. A `ControlThread` is started for each of the muskel remote resources (`RemoteWorkers`) just discovered. The `ControlThread` repeatedly looks for a fireable instruction in the task pool (the data-flow implementation model ensures that all fireable instructions are independent and can be computed in parallel) and sends it to its associated `RemoteWorker`. That `RemoteWorker` computes the instruction and returns the results. The results are either stored in the appropriate data flow instruction(s) in the task pool or delivered to the output stream, depending on whether they are intermediate results or final ones. In the event of `RemoteWorker` failure, i.e. if either the remote node or the network connecting it to the local machine fails, the `ControlThread` informs the manager and it, in turn, requests the name of another machine running the muskel run time support from a centralized *discovery service* and forks a new `ControlThread` to manage it, while the `ControlThread` managing the failed remote node terminates after reinserting in the `TaskPool` the macro data flow instruction whose computation failed [1]. Note that the failures handled by the muskel manager are *fail-stop* failures, i.e. it is assumed that an unreachable remote worker will not simply restart working again, or, if it restarts, it does so in its initial state. muskel has already been demonstrated to be effective on both clusters and more widely distributed workstation networks and grids [1,2].

3   The Orc Notation

The orchestration language Orc has been introduced by Misra and Cook [9]. Orc is targeted at the description of systems where the challenge lies in organising a set of computations, rather than in the computations themselves. Orc has, as primitive, the notion of a site call, which is intended to represent basic computations. A site, which represents the simplest form of Orc expression, either returns a *single* value or remains silent. Three operators (plus recursion) are provided for the orchestration of site calls:

1. operator > (sequential composition)
   $E_1 > x > E_2(x)$ evaluates E_1, receives a result x, calls E_2 with parameter x. If E_1 produces two results, say x and y, then E_2 is evaluated twice, once with argument x and once with argument y. The abbreviation $E_1 \gg E_2$ is used for $E_1 > x > E_2$ when evaluation of E_2 is independent of x.
2. operator | (parallel composition)
   $(E_1 \mid E_2)$ evaluates E_1 and E_2 in parallel. Both evaluations may produce replies. Evaluation of the expression returns the merged output streams of E_1 and E_2.
3. where (asymmetric parallel composition)
   E_1 where $x :\in E_2$ begins evaluation of both E_1 and $x :\in E_2$ in parallel. Expression E_1 may name x in some of its site calls. Evaluation of E_1 may proceed until a dependency on x is encountered; evaluation is then delayed. The first value delivered by E_2 is returned in x; evaluation of E_1 can proceed and the thread E_2 is halted.

Orc has a number of special sites:

- 0 never responds (0 can be used to terminate execution of threads);
- if b returns a signal if b is true and remains silent otherwise;
- *RTimer(t)*, always responds after t time units (can be used for time-outs);
- *let* always returns (publishes) its argument.

The notation $(\mid i : 1 \leq i \leq 3 : w_i)$ is used as an abbreviation for $(w_1 \mid w_2 \mid w_3)$.

Finally, while Orc does not have an explicit concept of "process", processes may be represented as expressions which, typically, name channels which are shared with other expressions. In Orc a channel is represented by a site [9]. *c.put(m)* adds m to the end of the (FIFO) channel and publishes a signal. If the channel is non-empty *c.get* publishes the value at the head and removes it; otherwise the caller of *c.get* suspends until a value is available.

4   Muskel Manager: An Orc Description

The Orc description presented focuses on the management component of muskel, and in particular on the discovery and recruitment of new remote workers in the event of remote worker failure. The compilation of the skeleton program to a data flow graph is not considered.

The activities of the processes of the muskel system are now described, referring to the Orc specification presented in Fig. 1.

System. The *system* comprises a program, *pgm*, to be executed (for simplicity a single program is considered: in reality a set of programs may be provided here); a set of *tasks* which are initially placed in a *taskpool*; a *discovery* mechanism which

$$system(pgm, tasks, contract, G, t) \triangleq$$
$$taskpool.add(tasks) \mid discovery(G, pgm, t) \mid manager(pgm, contract, t)$$

$$discovery(G, pgm, t) \triangleq (\mid_{g \in G} (\text{ if } remw \neq \text{false } \gg rworkerpool.add(remw)$$
$$\text{where } remw :\in$$
$$(\ g.can\_execute(pgm) \mid Rtimer(t) \gg let(false) \ )$$
$$)$$
$$) \gg discovery(G, pgm, t)$$

$$manager(pgm, contract, t) \triangleq$$
$$\mid i : 1 \leq i \leq contract : (rworkerpool.get > remw > ctrlthread_i(pgm, remw, t))$$
$$\mid monitor$$

$$ctrlthread_i(pgm, remw, t) \triangleq taskpool.get > tk >$$
$$(\quad \text{if } valid \gg resultpool.add(r) \gg ctrlthread_i(pgm, remw, t)$$
$$\mid \text{if } \neg valid \gg (\ taskpool.add(tk)$$
$$\mid alarm.put(i) \gg c_i.get > w > ctrlthread_i(pgm, w, t)$$
$$)$$
$$)$$
$$\text{where } (valid, r) :\in$$
$$(\ remw(pgm, tk) > r > let(true, r) \mid Rtimer(t) \gg let(false, 0) \ )$$

$$monitor \triangleq alarm.get > i > rworkerpool.get > remw > c_i.put(remw)$$
$$\gg monitor$$

Fig. 1. Centralized management: Orc specification

makes available processing engines (remote workers) recruited from a grid, G; and a *manager* which creates control threads and supplies them with remote workers. t is the time interval at which potential remote worker sites are polled; and, for simplicity, also the time allowed for a remote worker to perform its calculation before presumption of failure.

Discovery. It is assumed that the call *g.can_execute(pgm)* to a remote worker site returns its name, g, if it is capable of (in terms of hardware and software resources) and willing to execute the program *pgm*, and remains silent otherwise. The call *rworkerpool.add(g)* adds the remote worker name g to the pool provided it is not already there. The *discovery* mechanism carries on indefinitely to cater for possible communication failure.

Manager. The *manager* creates a number *(contract)* of control threads, supplies them with remote worker handles, monitors the control threads for failed remote workers and, where necessary, supplies a control thread with a new remote worker.

Control thread. A control thread *(ctrlthread)* repeatedly takes a task from the *taskpool* and uses its remote worker to execute the program *pgm* on this task. A result is added to the *resultpool*. A time-out indicates remote worker failure which causes the control thread to execute a call on an *alarm* channel while returning the unprocessed task to the *taskpool*. The replacement remote worker is delivered to the control thread via a channel, c_i.

Monitor. The *monitor* awaits a call on the *alarm* channel and, when received, recruits and supplies the appropriate control thread, i, with a new remote worker via the channel, c_i.

5   Decentralized Management: Derivation

In the muskel system described thus far, the *manager* is responsible for the recruitment and supply of (remote) workers to control threads, both initially and in the event of worker failure. Clearly, if the manager fails, then, depending on the time of failure, the fault recovery mechanism will cease or, at worst, the entire system of control thread recruitment will fail to initiate properly. Thus, the aim is to devolve this management activity to the control threads themselves, making each responsible for its own worker recruitment.

The strategy adopted is to examine the execution of the system in terms of traces of the site calls made by the processes and highlight management related communications. The idea is to use these communications as a means of identifying where/how functionality may be dispersed. In detail, the strategy proceeds as follows:

1. Focus on communication actions concerned with management. Look for patterns based on the following observation. Typically communication occurs when a process, A, generates a value, x, and communicates it to B. Identify occurrences of this pattern and consider if generation of the item could be shifted to B and the communication removed, with the "receive" in B being replaced by the actions leading to x's generation. For example:

   $A : \ldots a1, a2, a3, send(x), a4, a5, \ldots$
   $B : \ldots b1, b2, b3, receive(i), b4, b5, \ldots$

   Assume that $a2$, $a3$ (which, in general, may not be contiguous) are responsible for generation of x, and it is reasonable to transfer this functonality to B. Then the above can be replaced by:

   $A : \ldots a1, a4, a5, \ldots$
   $B : \ldots b1, b2, b3, a2, a3, (b4, b5, \ldots)_{[i/x]}$

2. The following trace subsequences are identified:
   - In control thread: $alarm.put(i) \gg c_i.get > w > ctrlthread_i(pgm, w, t) \ldots$
   - In monitor:

     $alarm.get > i > rworkerpool.get > remw > c_i.put(remw) \gg \ldots$

3. The subsequence $rworkerpool.get > remw > c_i.put(remw)$ of *monitor* actions is responsible for generation of a value (a remote worker) and its forwarding to a *ctrlthread* process. In the *ctrlthread* process the corresponding "receive" is $c_i.get$. So, the two trace subsequences are modified to:
   - In control thread:

     $alarm.put(i) \gg rworkerpool.get > remw > ctrlthread_i(pgm, remw, t) \ldots$
   - In monitor: $alarm.get > i > \ldots$

$systemD(pgm, tasks, contract, G, t) \triangleq$
$\quad taskpool.add(tasks)$
$\quad |i : 1 \leq i \leq contract : ctrlthread_i(pgm, t, G)$

$ctrlthread_i(pgm, t, G) \triangleq discover(G, pgm) > remw > ctrlprocess(pgm, remw, t, G)$

$discover(G, pgm) \triangleq let(remw)$ **where** $remw :\in |_{g \in G}\ g.can\_execute(pgm)$

$ctrlprocess(pgm, remw, t, G) \triangleq taskpool.get > tk >$
$\quad (\ $ if $valid\ \gg\ resultpool.add(r)\ \gg\ ctrlprocess(pgm, remw, t, G)$
$\quad\ |$ if $\neg valid\ \gg\ taskpool.add(tk)$
$\qquad\qquad\qquad | \ discover(G, pgm) > w >\ ctrlprocess(pgm, w, t, G)$
$\quad)$
$\qquad$ **where** $(valid, r) :\in$
$\qquad\qquad (\ remw(pgm, tk) > r > let(true, r)\ |\ Rtimer(t)\ \gg\ let(false, 0)\ )$

Fig. 2. Decentralized management: Orc specification

4. The derived trace subsequences now include the communication of the control thread number, i from $ctrlthread_i$ to the *monitor*, but this is no longer required by *monitor*; so, this communication can be removed.
5. Thus the two trace subsequences become:
   - In control thread:
     $\quad ... \gg rworkerpool.get > remw > ctrlthread_i(pgm, remw, t) ...$
   - In monitor: $... \gg ...$
6. Now the specifications of the processes $ctrlthread_i$ and *monitor* are examined to see how their definition can be changed to achieve the above trace modification, and consideration is given as to whether such modification makes sense and achieves the overall goal.
   (a) In monitor the entire body apart from the recursive call is eliminated thus prompting the removal of the *monitor* process entirely. This is as would be expected: if management is successfuly distributed then there is no need for centralized monitoring of control threads with respect to remote worker failure.
   (b) In control thread the clause:
       $\quad |\ alarm.put(i)\ \gg\ c_i.get > w > ctrlthread_i(pgm, w, t)$
       becomes
       $\quad |\ rworkerpool.get > remw > ctrlthread_i(pgm, remw, t)$
       This now suggests that $ctrlthread_i$ requires access to the *rworkerpool*. But the *rworkerpool* is an artefact of the (centralized) manager and the overall intent is to eliminate this manager. Thus, the action *rworkerpool.get* must be replaced by some action(s), local to $ctrlthread_i$, which has the effect of supplying a new remote worker. Since there is no longer a remote worker pool, on-the-fly recruitment of an remote worker is required. This can be achieved by using a discovery mechanism similar to that of the centralized manager and replacing *rworkerpool.get* by *discover(G, pgm)*:
       $\quad discover(G, pgm) \triangleq let(rw)$ **where** $rw :\in |_{g \in G}\ g.can\_execute(pgm)$

(c) Finally, as there is no longer centralized recruitment of remote workers, the control thread processes are no longer instantiated with their initial remote worker but must recruit it themselves. This requires that

    i. the control thread process be further amended to allow initial recruitment of a remote worker, with the (formerly) recursive body of the process now defined within a subsidiary process, *ctrlprocess*, as shown below.

    ii. the parameter *remw* in *ctrlthread* be replaced by G as the control thread is no longer supplied with an (initial) remote worker, but must handle its own remote worker recruitment by reference to the grid, G.

The result of these modifications is shown in the decentralized manager specification in Fig. 2. Here each control thread is responsible for recruiting its own remote worker (using a discovery mechanism similar to that of the centralized manager specification) and replacing it in the event of failure.

5.1   Analysis

Having derived a decentralized manager specification, the "equivalence" of the two versions must be established. In this context, equivalent means that the same input/output relationship holds, as clearly the two systems are designed to exhibit different non-functional behaviour.

The input/output relationship (i.e. functional semantics) is driven almost entirely by the *taskpool*, whose contents change dynamically to represent the data-flow execution. This execution primarily consists in establishing an on-line partial order among the execution of fireable tasks. All execution traces compliant to this partial order exhibit the same functional semantics by definition of the underlying data-flow execution model. This can be formally proved by showing that all possible execution traces respecting data-dependencies among tasks are functionally confluent (see [10] for the full proof), even if they do not exhibit the same performance.

Informally, one can observe that a global order among the execution of tasks can not be established *ex ante*, since it depends on the program and the execution environment (e.g. task duration, remote workers' availability and their relative speed, network connection speed, etc.). So, different runs of the centralized version will typically generate different orders of task execution. The separation of management issues from core functionality, which is a central plank of the muskel philosophy, allows the functional semantics of the centralized system to carry over intact to the decentralized version as this semantics is clearly independent of the means of recruiting remote workers.

One can also make an observation on how the overall performance of the system might be affected by these changes. In the centralized management system, the discovery activity is composed with the "real work" of the remote workers by the parallel composition operator: discovery can unfold in parallel with computation. In the revised system, the discovery process is composed with core computation using the sequence operator, $\gg$. This suggests a possible price to pay for fault recovery.

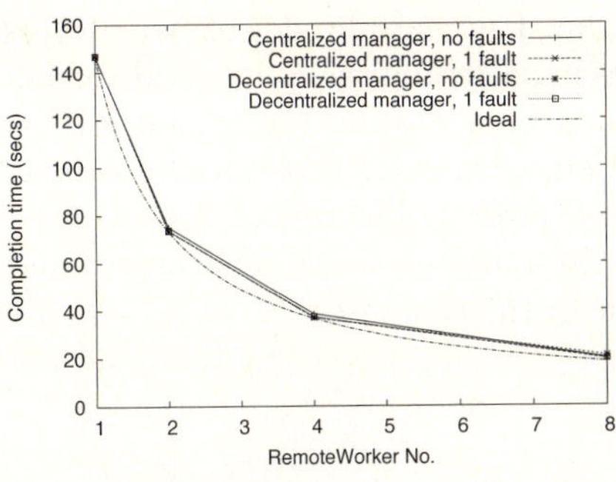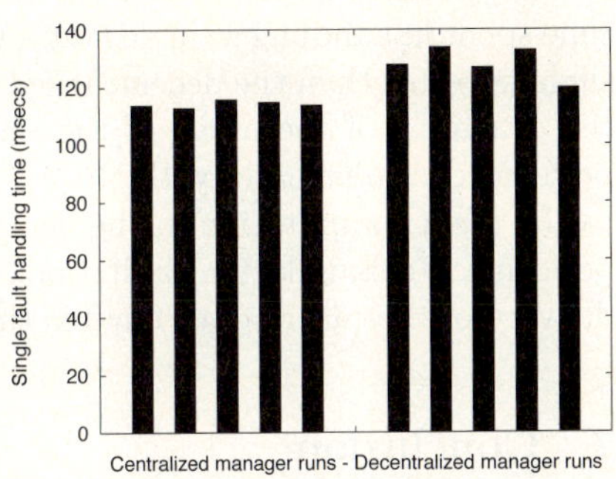

Fig. 3. Scalability (left) and fault handling cost (right) of modified vs. original muskel

6   Decentralized Management: Implementation

Following the derivation of the decentralized manager version outlined above, the existing muskel prototype was modified to introduce distributed fault management and to evaluate the cost, if any, in terms of performance. As shown above, in the decentralized manager, the *discovery(G, pgm, t)* parallel component of the *system(...)* expression become part (the *discover(G, pgm)* expression) of the *ctrlprocess(...)* expression. The *discovery* and *discover* definitions are not exactly the same, but *discover* is easily derived from *discovery*. Thus, the code implementing *discovery(G, pgm, t)* was moved and transformed appropriately to give an implementation of *discover(G, pgm)*. This required the modification of just one of the files in the muskel package (194 lines of code out of a total of 2575, less than 8%), the one implementing the control thread.

Experiments were run using the original and the modified versions to test the functionality and cost of the new implementation. The experiments were run on a Fast Ethernet network of Pentium III machines running Linux and Java 1.5. First the scalability of the decentralized manager version was verified: the scalabilities of the original muskel and of the one with the decentralized version of the manager were measured, with the same skeleton program and input data, to check that no overhead was introduced by the decentralized management, at least in the case where no faults were detected. Figure 3 left shows almost perfect scalability of the decentralized manager version up to 8 nodes, comparable to that achieved when using the original muskel, both in the case of no faults and in the case of a single fault per computation. Then the times spent in managing a node fault in the centralized and decentralized versions were compared. The same skeleton program with the same input data was run using the centralized and decentralized versions of muskel and a number of faults were artificially introduced into the system while the computations were running. In particular, up to 4 faults were introduced per run and the average time taken to handle a fault was measured for each of the two versions. Figure 3 right plots the average time spent in handling a single fault in each run. The centralized version performs slightly better than the decentralized one, as expected: in the centralized version the discovery of the name of the remote machines hosting the muskel RTS is performed *concurrently* with the computation, whereas it is performed *serially* to the main computation in the decentralized version. The rest of the activities performed to handle the fault (lookup of the remote worker RMI object and delivery of the macro data flow) is the same in the two cases.

7   Conclusion

The manager component of the muskel system has been re-engineered to provide distributed remote worker discovery and fault recovery. A formal specification of the component, described in Orc, was developed. The specification provided the developer with a representation of the manager that allowed exploration of its properties and the development of what-if scenarios while hiding the inessential detail. By studying the communication patterns present within the process traces, the developers were able to derive a system exhibiting equivalent core functionality, while having the desired decentralized management properties. The derivation proceeded in a series of semi-formally justified steps, with incorporation of insight and experience as exemplified by the use of expressions such as "reasonable to transfer this functionality" and "such modification makes sense".

The claim is that the creation of such a derivation facilitates exploration (and documentation) of ideas and delivers much return for small investment. Lightweight reasoning about the derived specification gave the developers some insight into the expected performance of the derived implementation relative to its parent. In addition, the authors suggest that Orc is an appropriate vehicle for the description of management systems of the sort described here. Its syntax is small and readable; its constructs allow for easy description of the sorts of activities that typify these systems (in particular the asymmetric parallel composition operator facilitates easy expression of concepts such as time-out and parallel searching); and the *site* abstraction allows clear separation of management activity from core functionality.

The approach has been applied in the context of skeletal systems where the complexity of the orchestration is constrained by the use of skeletons. However, efficient, large distributed systems often have regular structures that can be described using concise parametric definitions. Thus, one may be optimistic that the approach will be feasible for systems of significant size, although, of course, further experimentation is required to confirm this.

Future work will involve tackling the more difficult task of removing the centralized task pool bottleneck, which should provide a stiffer test of the proposed approach. And, the availability of an Orc description makes possible the analysis of system variants with respect to cost and reliability using techniques described in [15].

References

1. Danelutto, M.: QoS in parallel programming through application managers. In: Proc. of Intl. Euromicro PDP: Parallel Distributed and network-based Processing, Lugano, Switzerland, pp. 282–289. IEEE, Los Alamitos (2005)
2. Danelutto, M., Dazzi, P.: Joint structured/non structured parallelism exploitation through data flow. In: Alexandrov, V.N., van Albada, G.D., Sloot, P.M.A., Dongarra, J.J. (eds.) ICCS 2006. LNCS, vol. 3994, Springer, Heidelberg (2006)
3. Danelutto, M.: Dynamic run time support for skeletons. In: Proc. of Intl. PARCO 99: Parallel Computing. Parallel Computing Fundamentals & Applications, pp. 460–467. Imperial College Press, London, UK (EU) (1999)
4. Kephart, J.O., Chess, D.M.: The vision of autonomic computing. IEEE Computer 36, 41–50 (2003)
5. White, S., Hanson, J., Whalley, I., Chess, D., Kephart, J.: An architectural approach to autonomic computing. In: Proc. of the Intl. Conference on Autonomic Computing, pp. 2–9 (2004)
6. Parashar, M., Liu, H., Li, Z., Matossian, V., Schmidt, C., Zhang, G., Hariri, S.: AutoMate: Enabling autonomic applications on the Grid. Cluster Computing 9, 161–174 (2006)
7. Kennedy, K., et al.: Toward a framework for preparing and executing adaptive Grid programs. In: Proc. of NSF Next Generation Systems Program Workshop (IPDPS 2002) (2002)
8. Stewart, A., Gabarró, J., Clint, M., Harmer, T.J., Kilpatrick, P., Perrott, R.: Managing grid computations: An orc-based approach. In: Guo, M., Yang, L.T., Di Martino, B., Zima, H.P., Dongarra, J., Tang, F. (eds.) ISPA 2006. LNCS, vol. 4330, pp. 278–291. Springer, Heidelberg (2006)
9. Misra, J., Cook, W.R.: Computation orchestration: A basis for a wide-area computing. Software and Systems Modeling (2006), doi: 10.1007/s10270-006-0012-1
10. Aldinucci, M., Danelutto, M.: Skeleton based parallel programming: functional and parallel semantic in a single shot. Computer Languages, Systems and Structures 33, 179–192 (2007)
11. Milner, R.: Communicating and Mobile Systems: the π-Calculus. Cambridge University Press, Cambridge (1999)
12. Agerholm, S., Larsen, P.G.: A lightweight approach to formal methods. In: Hutter, D., Traverso, P. (eds.) Applied Formal Methods - FM-Trends 98. LNCS, vol. 1641, pp. 168–183. Springer, Heidelberg (1999)
13. Cole, M.: Bringing skeletons out of the closet: A pragmatic manifesto for skeletal parallel programming. Parallel Computing 30, 389–406 (2004)
14. Kuchen, H.: The Muesli home page (2006), http://www.wi.uni-muenster.de/PI/forschung/Skeletons/
15. Stewart, A., Gabarró, J., Clint, M., Harmer, T.J., Kilpatrick, P., Perrott, R.: Estimating the reliability of web and grid orchestrations. In: Integrated Reserach in Grid Computing, Kraków, Poland, CoreGRID, Academic Computer Centre CYFRONET AGH, pp. 141–152 (2006)
16. Aldinucci, M., Danelutto, M., Kilpatrick, P.: Adding metadata to Orc to support reasoning about grid programs. In: Priol, T., Vanneschi, M. (eds.) Grid and Peer-To-Peer Technologies (Proc. of the CoreGRID Symposium 2007), Springer, Heidelberg (2007)
17. Kilpatrick, P., Danelutto, M., Aldinucci, M.: Deriving Grid Applications from Abstract Models. Technical Report TR-85, CoreGRID (2007), available at http://www.coregrid.net

Nested Parallelism in the OMPi OpenMP/C Compiler

Panagiotis E. Hadjidoukas and Vassilios V. Dimakopoulos

Department of Computer Science
University of Ioannina, Ioannina, Greece
{phadjido,dimako}@cs.uoi.gr

Abstract. This paper presents a new version of the OMPi OpenMP C compiler, enhanced by lightweight runtime support based on user-level multithreading. A large number of threads can be spawned for a parallel region and multiple levels of parallelism are supported efficiently, without introducing additional overheads to the OpenMP library. Management of nested parallelism is based on an adaptive distribution scheme with hierarchical work stealing that not only favors computation and data locality but also maps directly to recent architectural developments in shared memory multiprocessors. A comparative performance evaluation of several OpenMP implementations demonstrates the efficiency of our approach.

1 Introduction

Although nested parallelism was defined from the initial version of OpenMP, several implementation and performance issues still remain open and need to be answered. This necessity stems from the fact that both applications and end-users require such functionality, and is further augmented as multi-core technology tends to increase the number of available processors. Nowadays, several research and commercial OpenMP compilers support more than one levels of parallelism. With a few exceptions, however, most OpenMP implementations have translated "OpenMP threads" to "kernel threads", which are internally mapped to system-scope POSIX threads or native OS threads. These implementations support nested parallelism by extending their highly-optimized and fine-tuned OpenMP runtime library for single-level parallelism.

Despite the fact that the above approach fulfills the requirements for the common case, namely single-level parallelism with dynamic threads (where the runtime system may adjust the number of working threads at will), it can cause serious performance degradation when the number of threads is explicitly requested by the user and exceeds the number of available processors. This excess is quite common in the case of nested parallelism or in multiprogramming (non-dedicated) environments.

This paper presents a new version of the OMPi OpenMP C compiler [1], enhanced by a lightweight thread library that provides efficient runtime support for multiple levels of parallelism. The whole configuration results in significantly

A.-M. Kermarrec, L. Bougé, and T. Priol (Eds.): Euro-Par 2007, LNCS 4641, pp. 662–671, 2007.

low parallelization overheads, especially when the number of OpenMP threads increases, and provides efficient support of nested parallelism. In addition, it simplifies thread management, favors computation and data locality and eliminates the disadvantages of time sharing, including the unavoidable synchronization overheads at the end of inner parallel regions.

The rest of this paper is organized as follows: Section 2 discusses related work. Sections 3 and 4 present the OMPi OpenMP compiler and the lightweight thread library respectively. Section 5 describes the management of nested parallelism. Experimental results are included in Section 6. Finally, Section 7 discusses our ongoing work.

2   Related Work

Several research efforts and compiler vendors support nested parallelism, creating a new team of OpenMP threads to execute nested parallel regions.

The NANOS compiler [2] supports nested regions and OpenMP extensions for processor groups. The runtime support of the NANOS Compiler is provided by an efficient user-level threads library (NthLib), which assumes that dynamic parallelism is always enabled and thus the number of spawned threads never exceeds that of available processors. The Omni compiler [3] supports a limited form of nested parallelism, requiring a user-predefined fixed size for the kernel thread pool, from where threads will be used for the execution of parallel regions. Omni/ST [4], an experimental version of Omni equipped with the Stack-Threads/MP library, provided an efficient though not portable implementation of nested irregular parallelism. The Balder runtime library of OdinMP [5] is capable of fully handing OpenMP 2.0 including nested parallelism. Balder uses POSIX threads as underlying thread library, provides efficient barrier and lock synchronization and uses a pool of threads which is expanded whenever it is necessary.

All vendors that support nested parallelism implement it by maintaining a pool of kernel threads. GOMP [6], the OpenMP implementation for GCC, implements its runtime library (libgomp) as a wrapper around the POSIX threads library, with some target-specific optimizations for systems that support lighter weight implementation of certain primitives. The GOMP runtime library allows the reuse of idle threads from their pool only for non-nested parallel regions, while threads are created dynamically for inner levels.

The OpenMP runtime library of the Sun Studio compiler [7] maintains a pool of threads that can be used as slave threads in parallel regions. The user can control both the number of threads in the pool and the maximum depth of nested parallel regions that require more than one thread. Similarly, the basic mechanism for threading support in the Intel compiler [8] is the thread pool. The threads are not created until the first parallel region is executed, and only as many as needed by that parallel region are created. Further threads are created as needed by subsequent parallel regions. Threads that are created by the OpenMP runtime library are not destroyed but join the thread pool until they are called upon to join a team and are released by the master thread of the subsequent team.

The Fujitsu PRIMEPOWER Fortran compiler [9] also supports nested parallelism. Moreover, if the OpenMP application has only a single level of parallelism then a high performance OpenMP runtime library is used. Finally, the IBM XLC compilers support the execution of nested parallel loops [10].

3   The OMPi Compiler

OMPi is a source-to-source translator that takes as input C source code with OpenMP V.2.0 directives and outputs equivalent multithreaded C code, ready to be built and executed on a multiprocessor. The current version is fully V.2.0 compliant, can target different thread libraries through a unified thread abstraction, and includes many architecture-dependent as well as higher-level optimizations. Finally, the compiler and the runtime libraries include support for the POMP performance monitoring interface. OMPi is implemented entirely in C and has been ported effortlessly on many different platforms, including Intel / Linux, Sun / Solaris and SGI / Irix systems.

OMPi produces multithreaded C code. Its architecture is such that any specific thread library can be supported through a well-defined generic interface, making OMPi quite extensible. It currently supports a number of thread libraries through the generic thread interface. They include a POSIX threads based library (the default thread library target mainly for portability reasons) which is highly tuned for single-level parallelism. Another library can be used in Sun / Solaris machines, where the user has the option of producing code with Solaris threads calls. It should be noted that the generic interface provides for a transparent choice of the underlying thread library. That is, the user source code is not affected in any way—the actual thread library that will be used is only included at linking time.

4   Lightweight Runtime Support

The new internal threading interface in the runtime library of OMPi facilitates the integration of arbitrary thread libraries. In order to efficiently support nested parallelism, a user-level thread library, named `psthreads`, has been developed.

The `psthreads` library implements a two-level thread model, where user-level threads are executed on top of kernel-level threads that act as *virtual processors*. Each virtual processor runs a dispatch loop, selecting the next-to-run user-level thread from a set of ready queues, where threads are submitted for execution. An idle virtual processor extracts threads from the *front* of its local ready queue but steals from the *back* of remote queues. The queue architecture allows the runtime library to represent the layout of physical processors. For instance, a hierarchy can be defined in order to map the coupling of processing elements in current multi-core architectures [11].

Despite the user-level multithreading, the `psthreads` library is fully portable because its implementation is based entirely on the POSIX standard. Its virtual processors are mapped to POSIX threads, permitting the interoperability of OMPi with third-party libraries and the co-existence of OpenMP and POSIX threads in the same program. The primary user-level thread operations are provided by UthLib (Underlying Threads Library), a portable thread package. An underlying thread is actually the stack that a `psthread` uses during its execution. Synchronization is based on the POSIX threads interface. Locks are internally mapped to POSIX mutexes or spinlocks, taking into account the non-preemptive threads of the library. In addition, platform-dependent spin locks are utilized.

The application programming interface of `psthreads` is similar to that of POSIX threads. Its usage simplifies the OpenMP runtime library since spawning of threads is performed explicitly, while thread pooling is provided by the thread library. The thread creation routine of `psthreads` allows the user to specify the queue where the thread will be submitted for execution and whether it will be inserted in the front or in the back of the specified queue. Moreover, there exists a variant of the creation routine that accepts an already allocated thread descriptor. This is useful for cases where the user implements its own management of thread descriptors.

Efficient thread and stack management is essential for nested parallelism because a thread with a private stack should always be created since the runtime library cannot know whether the running application will spawn a new level of parallelism. An important feature of `psthreads` is the utilization of a lazy stack allocation policy. According to this policy, the stack of a user-level thread is allocated just before its execution. This results in minimal memory consumption and simplified thread migrations. Lazy stack allocation is further improved with stack handoff. In peer-to-peer scheduling, a finished thread picks the next descriptor, creates a stack for that thread and switches to it. Usually, this stack is extracted from a reuse queue. Using stack handoff, a finished thread re-initializes its own state, by replacing its descriptor with the following thread's descriptor, and resumes its execution. By allocating the descriptors from the stack of the parent thread (i.e. master), the activation of their recycling mechanism is also avoided.

If native POSIX thread libraries followed a hybrid (two-level or M:N) implementation, the runtime overheads would be reduced, allowing the creation of several threads without additional performance cost, as shown in [12]. However, all vendors have dropped the hybrid model and use a 1:1 mapping of POSIX threads to kernel threads. Fortunately, Marcel [13] is a two-level thread library that provides similar functionality and application programming interface to POSIX threads. Marcel binds one kernel-level thread on each processor and then performs fast user-level context-switches between user-level threads, hence getting complete control of thread scheduling in user-land without any further help from the kernel. We have successfully built and integrated a Marcel-based module into the OMPi compiler, by replacing the corresponding threading calls of the `psthreads` module.

5   Management of Nested Parallelism

According to OpenMP, when an application encounters the first of two nested parallel loops, a new team of threads is created and the loop iterations are distributed among these worker threads. Eventually, each thread will become the master of a new team that will be created for the execution of the inner loop. Assuming that the number of threads spawned in a parallel region is equal to the number (P) of processors, a kernel thread model will result in $P \times P$ threads that compete for hardware resources. Time-sharing can significantly increase implicit synchronization overheads that are related to thread management and dynamic loop scheduling. Even if static loop schedules are used, it is difficult for the runtime library to decide how to bind inner threads to specific processors in order to favor locality. A common approach that handles these problems uses a fixed size pool of threads, limiting thus the number of created threads. Another approach does not create additional threads but partitions the available threads into groups, based on information provided by the programmer [2] or extracted from the loop characteristics [10]. Despite the good locality and the low overheads of grouping, it is hard to determine the number of groups and their size and this can easily cause load imbalance.

We propose a straightforward approach able to handle general unstructured nested parallelism. Due to the lightweight runtime support of `psthreads`, OMPi can support efficiently a large number of threads and multiple levels of parallelism. Moreover, the utilization of non-preemptive threads allows the runtime library to manage parallelism explicitly, which is not possible for the case of kernel threads. Specifically, the OpenMP runtime library utilizes a variant of the all-to-all scheme in order to distribute work across the processors. Threads that are spawned at the first level of parallelism are distributed cyclically and inserted at the back of the ready queues. For inner levels, the threads are inserted in the front of the ready queue that belongs to the virtual processor they were created on. Since an idle virtual processor extracts threads from the front of its local queue and the back of the remote ones, this scheme favors the exploitation of data locality of inner levels of parallelism.

Our approach can be easily generalized to include the latest developments of shared memory architecture, like multi-core and SMT processors. The work stealing mechanism has been designed to work hierarchically, assuming the existence of thread groups, as shown in Fig. 1. Specifically, the virtual processors are organized into hierarchical groups of size that is equal to a power of 2 (i.e. 2, 4, 8, etc), according to the level of hierarchy. Thus, an idle virtual processor first examines the ready queue of its adjacent virtual processor in the two-processor group (level 1, size 2) where it belongs to, then it tests the queues of the rest two processors in the quad-processor group (level 2, size 4), etc. Moreover, due to our two-level thread model, there is a 1:1 mapping between virtual and physical processors and, thus, the queue hierarchy of the runtime library can be mapped directly to the hardware architecture.

The number of groups and their size can be set explicitly using an OpenMP extension (e.g. [2]) or appropriate machine description. In contrast to NANOS

```
        GroupSize = 2;
        while (GroupSize <= Virtual Processors) {
            GroupId = MyID / GroupSize;
            GroupBaseVP = GroupId*GroupSize;
            vp = (MyID + 1) % GroupSize + GroupBaseVP;
            while ((thread == NULL) && (vp != MyID)) {
                if (!visited[vp]) {
                    visited[vp] = 1;
                    thread = DequeueWork(vp);
                }
                vp = (vp + 1) % GroupSize + GroupBaseVP;
            }
            GroupSize *= 2;
        }
        /* execute thread */
```

Fig. 1. Hierarchical work-stealing algorithm for uniform thread groups

groups, our approach supports inter-group stealing and, based on these groups, determines how idle processors access ready queues. Although an idle processor in Omni/ST was able to issue work-stealing requests, the mechanism was heavyweight and lacked portability due to the required intervention of the remote processor. The dependence of Omni/ST on the StackThreads compiler, a patched version of GNU C, did not allow its further usage. On the contrary, work stealing in OMPi/psthreads is performed asynchronously and virtual processors access remote queues in a uniform way. Furthermore, our software configuration allows the integration of any native compiler. Similarly, the Marcel package utilizes the BubbleSched scheduler [14], a framework that allows the distribution of threads over the hierarchy of the computer so as to benefit from cache effects and to avoid NUMA factor penalties as much as possible. The integration of Marcel threads into OMPi will allow the exploitation of the above framework in an OpenMP environment. A further discussion on this issue is beyond the scope of this paper.

6   Experimental Results

We performed our experiments on a Compaq Proliant ML570 server with 4 Intel Xeon III CPUs running Debian Linux (2.6.6). We provide comparative performance results for the Intel C++ compiler (v. 9.1), GNU GCC 4.2, Omni 1.6 and the new version of OMPi. The native compiler for both OMPi and Omni is GNU GCC.

Our first experiment demonstrates our lightweight runtime support as the number of OpenMP threads increases. Specifically, we use the EPCC microbenchmarks [15] to measure the OpenMP runtime overheads using from 2 up to 32 threads on a dedicated machine.

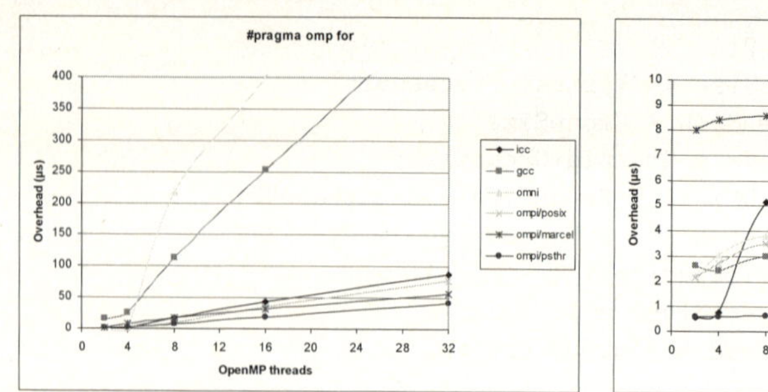
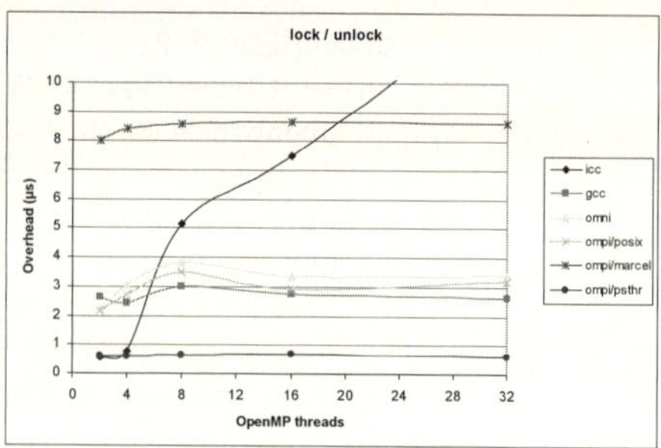

Fig. 2. Runtime overheads for large number of threads

Figures 2 presents the overheads of the `for` and lock/unlock OpenMP constructs. The overhead of `for` is lower if user-level thread libraries (marcel and `psthreads`) are used. The Intel compiler and the OMPi compiler with the native POSIX threads library also achieve good performance. On the contrary, both Omni and GCC exhibit significant runtime overhead as the number of OpenMP threads increases. The overhead of OpenMP locks does not depend on the number of threads for all the cases but the Intel compiler. This is attributed to the spinlock-based synchronization of the Intel compiler, which results in the lowest overhead for 2 and 4 OpenMP threads but fails to maintain stability when the number of threads exceeds the number of physical processors. To avoid this problem for Omni, its mutex-based lock primitives are used in all the experiments.

Our second experiment focuses on the evaluation of OpenMP runtime support for nested parallelism. For this purpose, we have appropriately extended the EPCC microbenchmarks. Specifically, the core benchmark routine for a given construct (e.g. `parallel`) is called several times within a parallel loop. If nested parallelism is tested, the loop is parallelized and these tasks are executed in parallel, otherwise they are executed sequentially as in the original version of the microbenchmarks. We measure the total execution time of the tasks and we compute the mean of the measured runtime overhead for each individual task.

```
omp_set_dynamic(0);
omp_set_nested(1);
t0 = omp_get_wtime();
#if defined(TEST_NESTED_PARALLELISM)
#pragma omp parallel for schedule(dynamic,1)
#endif
for (i = 0; i < ntasks; i++)
    testpr(i);

t1 = omp_get_wtime();
```

Fig. 3. Extended EPCC microbenchmark for nested parallelism

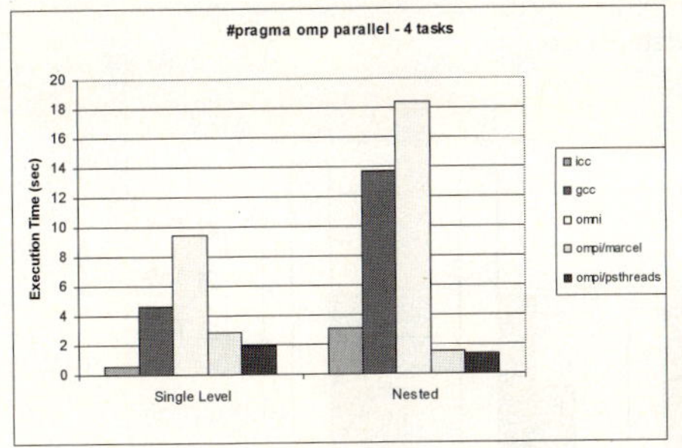
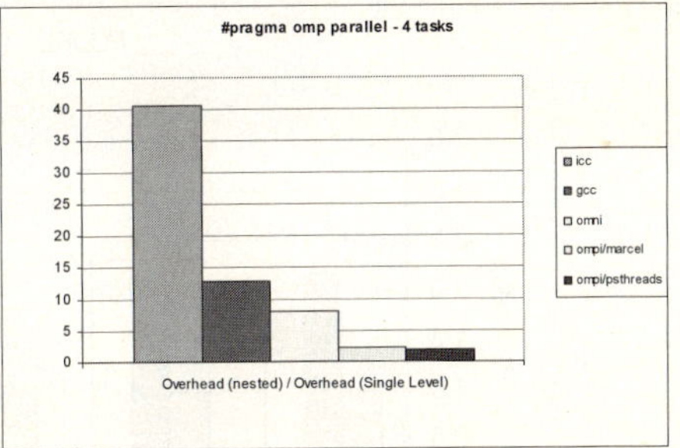

Fig. 4. Runtime behavior of the OpenMP `parallel` construct

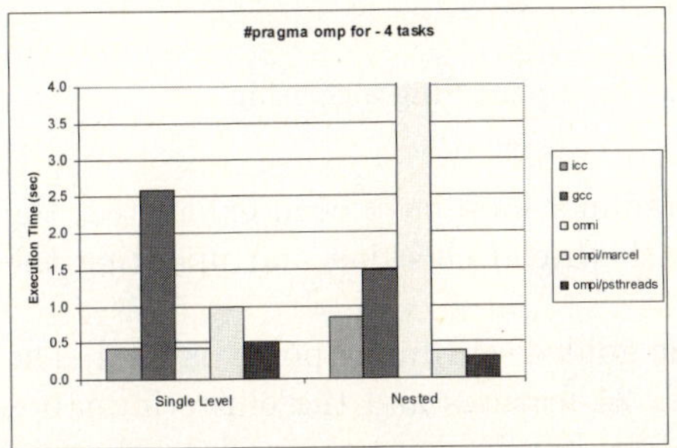
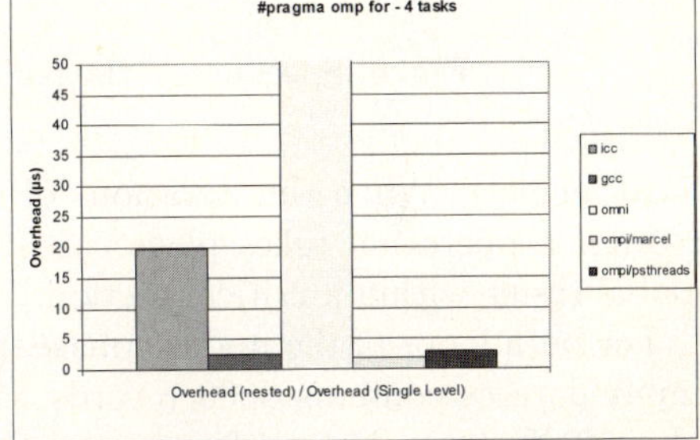

Fig. 5. Runtime behavior of the OpenMP `for` construct

Figures 4 and 5 present the runtime behavior of the `parallel` and `for` constructs respectively. The left figures show the total execution time and the right ones how many times the runtime overhead is increased when nested parallelism is enabled. For the `parallel` construct, both the execution time and the corresponding overhead are significantly increased for the Intel, GCC and Omni compilers. On the other hand, for both configurations of OMPi the total execution time is slightly decreased and the runtime overhead remains almost stable. This is expected because OMPi exploits better the task parallelism of the benchmark and restrains the contention between OpenMP threads. The `for` construct exhibits similar behavior except for the case of GCC, which does not suffer from high contention possibly due to its platform-specific atomic primitives.

The last experiment demonstrates our efficient support of nested parallelism using PCURE [16], an OpenMP implementation of a hierarchical data clustering algorithm. PCURE consists of an initialization phase and the core clustering algorithm, which is repeated until the requested number of clusters has been computed. The asymmetry and non-determinism of the algorithm requires the exploitation of nested parallelism, which is expressed with two parallel nested loops. The inner loop contains a reduction on a pair of data, an operation that is not directly supported by OpenMP. Therefore, a team of threads is spawned and each thread keeps the partial results (minimum distance and index) in its

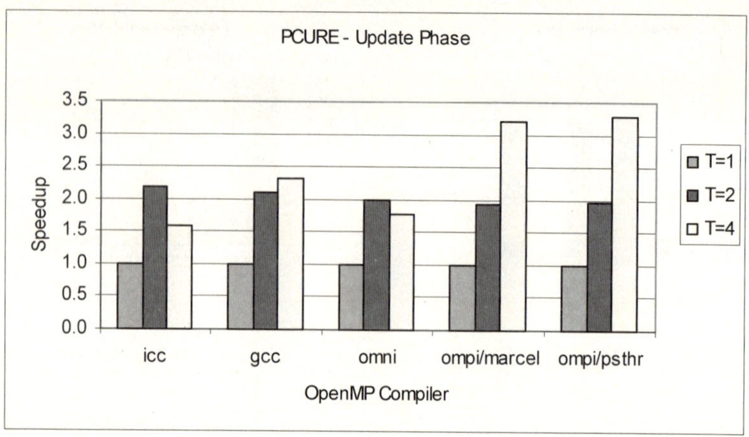

Fig. 6. Speedups of the parallel data clustering algorithm

local memory. When the iterations of the inner loop have been exhausted, the reduction operation takes place, with each thread checking and updating the global result within a critical section.

For both loops of the update phase, the guided scheduling policy is used. The input dataset contains 5000 records with 24 features and the clustering stops after 4000 steps. The performance speedups obtained for the update phase of PCURE are depicted in Figure 6. The Intel and Omni compilers perform best when 2 threads are used because the actual number of kernel threads from both levels of parallelism is equal to the number of physical processors (4). For 4 threads per level, the speedup drops mostly due to the higher contention of threads and the increased overheads of the runtime libraries. For GCC, however, the speedup is improved slightly when 4 threads are used. On the other hand, PCURE scales well for both OMPi configurations (marcel and `psthreads`). Let us note here that PCURE is a data intensive application so its scalability is limited by the low memory bandwidth of the bus-based SMP machine.

7   Conclusions and Future Work

This paper presented an OpenMP implementation based on user-level multi-threading and its advantages on the exploitation of nested parallelism compared to the traditional kernel thread based approaches. Our ongoing work includes the implementation of the workqueuing model in the OMPi compiler and the evaluation of our approach for nested parallelism on large scale multiprocessor systems. The `psthreads` library has also been integrated into the Omni and GCC compilers, while our future plans include open-source OpenMP Fortran compilers. The source code of OMPi is available at `http://www.cs.uoi.gr/~ompi`.

References

1. Dimakopoulos, V.V., Leontiadis, E., Tzoumas, G.: A Portable C Compiler for OpenMP V.2.0. In: Proc. of the 5th European Workshop on OpenMP (EWOMP '03), Aachen, Germany (October 2003)
2. Ayguade, E., Gonzalez, M., Martorell, X., Labarta, J., Navarro, N., Oliver, J.: NanosCompiler: Supporting Flexible Multilevel Parallelism in OpenMP. Concurrency: Practice and Experience 12(12), 1205–1218 (2000)
3. Sato, M., Satoh, S., Kusano, K., Tanaka, Y.: Design of OpenMP Compiler for an SMP Cluster. In: Proc. of the 1st European Workshop on OpenMP (EWOMP '99), Lund, Sweden (September 1999)
4. Tanaka, Y., Taura, K., Sato, M., Yonezawa, A.: Performance Evaluation of OpenMP Applications with Nested Parallelism. In: Dwarkadas, S. (ed.) LCR 2000. LNCS, vol. 1915, Springer, Heidelberg (2000)
5. Karlsson, S.: A Portable and Efficient Thread Library for OpenMP. In: Proc. of the 6th European Workshop on OpenMP (EWOMP '04), Stockholm, Sweden (October 2004)
6. Novillo, D.: OpenMP and automatic parallelization in GCC. In: Proc. of the 2006 GCC Summit, Ottawa, Canada (June 2006)
7. Sun Microsystems.: Sun Studio 10: OpenMP API User's Guide Available at: http://docs.sun.com/app/docs/doc/819-0501
8. Tian, X., Hoeflinger, J.P., Haab, G., Chen, Y.K., Girkar, M., Shah, S.: A compiler for exploiting nested parallelism in OpenMP programs. Parallel Computing 31, 960–983 (2005)
9. Iwashita, H., Kaneko, M., Aoki, M., Hotta, K., van Waveren, M.: On the Implementation of OpenMP 2.0 Extensions in the Fujitsu PRIMEPOWER compiler. In: Proc. of the International Workshop on OpenMP: Experiences and Implementations (WOMPEI '03), Tokyo, Japan (November 2003)
10. Duran, A., Silvera, R., Corbalan, J., Labart, J.: Runtime Adjustment of Parallel Nested Loops. In: Chapman, B.M. (ed.) WOMPAT 2004. LNCS, vol. 3349, Springer, Heidelberg (2004)
11. Nikolopoulos, D.S., Polychronopoulos, E.D., Papatheodorou, T.S.: Efficient Runtime Thread Management for the Nanothreads Programming Model. In: Proc. of the 2nd IEEE IPPS/SPDP Workshop on Runtime Systems for Parallel Programming, Orlando, FL, USA, vol. 1388, pp. 183–194. IEEE Computer Society Press, Los Alamitos (1998)
12. Rufai, R., Bozyigit, M., Alghamdi, J., Ahmed, M.: Multithreaded Parallelism with OpenMP. Parallel Processing Letters 15(4), 367–378 (2005)
13. INRIA, T.R.: Marcel: A POSIX-compliant thread library for hierarchical multiprocessor machines, available at: http://runtime.futurs.inria.fr/marcel
14. Thibault, S.: A Flexible Thread Scheduler for Hierarchical Multiprocessor Machines. In: Proc. of the 2nd International Workshop on Operating Systems, Programming Environments and Management Tools for High-Performance Computing on Clusters (COSET-2), Cambridge, USA (June 2005)
15. Bull, J.M.: Measuring Synchronization and Scheduling Overheads in OpenMP. In: Proc. of the 1st European Workshop on OpenMP (EWOMP '99), Lund, Sweden (September 1999)
16. Hadjidoukas, P.E., Amsaleg, L.: Parallelization of a Hierarchical Data Clustering Algorithm Using OpenMP. In: Proc. the 2nd International Workshop on OpenMP (IWOMP '06), Reims, France (June 2006)

Efficient Parallel Simulation of Large-Scale Neuronal Networks on Clusters of Multiprocessor Computers

Hans E. Plesser[1], Jochen M. Eppler[2,3], Abigail Morrison[4],
Markus Diesmann[3,4], and Marc-Oliver Gewaltig[2,3]

[1] Dept. of Mathematical Sciences and Technology, Norwegian University of Life
Sciences, PO Box 5003, 1432 Ås, Norway
`hans.ekkehard.plesser@umb.no`
[2] Honda Research Institute, Offenbach/Main, Germany
[3] Bernstein Center for Computational Neuroscience, Albert-Ludwigs-University,
Freiburg, Germany
[4] RIKEN Brain Science Institute, Wako-shi, Saitama, Japan

Abstract. To understand the principles of information processing in
the brain, we depend on models with more than 10^5 neurons and 10^9
connections. These networks can be described as graphs of threshold
elements that exchange point events over their connections.

From the computer science perspective, the key challenges are to represent the connections succinctly; to transmit events and update neuron states efficiently; and to provide a comfortable user interface. We
present here the neural simulation tool NEST, a neuronal network simulator which addresses all these requirements. To simulate very large
networks with acceptable time and memory requirements, NEST uses a
hybrid strategy, combining distributed simulation across cluster nodes
(MPI) with thread-based simulation on each computer. Benchmark simulations of a computationally hard biological neuronal network model
demonstrate that hybrid parallelization yields significant performance
benefits on clusters of multi-core computers, compared to purely MPI-
based distributed simulation.

1   Introduction

The neuronal networks in our brains can be described as weighted, directed
graphs, with neurons as nodes and synaptic connections as edges. Neurons communicate by sending and receiving point events (spikes) through their connections (synapses). In the mammalian cortex, each neuron sends connections to
about 10^4 other neurons and receives connections from as many. Just 1 mm^3
cortex contains some 10^5 neurons with 10^9 connections [1]. This represents a
threshold size for simulations, as a realistic number of synapses per neuron can
be combined with realistic sparseness (connection probability ~ 0.1). Brain function emerges from the spatio-temporal patterns of neuronal spike activity, but

A.-M. Kermarrec, L. Bougé, and T. Priol (Eds.): Euro-Par 2007, LNCS 4641, pp. 672–681, 2007.
© Springer-Verlag Berlin Heidelberg 2007

the principles are poorly understood. Progress in understanding brain function therefore depends on simulation studies of large cortical networks.

In large neuronal networks, we can neglect the geometric and biophysical complexity of individual nerve cells and describe neurons as point-like objects with a dynamic state governed by a set of ODEs. The most common state variable is the membrane potential V, which is affected by spikes that arrive at the neuron's synapses. Whenever V crosses a threshold value V_{th}, the neuron produces a spike, which is transmitted to all adjacent neurons with a delay of a few milliseconds. Each connection can have a different delay and weight. Weights may evolve as a result of neuronal activity, a phenomenon known as synaptic plasticity, the biological substrate of learning. The spikes of an individual neuron are rare and occur at rates of 1–50 Hz, whereas the rate of incoming spikes is of the order of 100 kHz due to some 10^4 incoming connections.

Simulating large-scale neuronal networks poses several challenges: (i) 10^9–10^{12} connections must be stored; this requires a distributed representation. (ii) A large number of spikes must be buffered until they are transmitted across the network. (iii) Simulation results must be reproducible down to the level of membrane potentials and spike times. (iv) The object-oriented implementation must be appropriate for the problem domain and allow network and machine level optimizations such as efficient caching.

In this contribution, we describe how the Neural Simulation Tool NEST [2] addresses these issues to efficiently simulate neuronal networks of more than 10^5 neurons and 10^9 synapses. In section 2 we discuss how NEST represents nodes and connections, before describing the update and communication algorithms in section 3. Section 4 demonstrates the performance of our hybrid approach. NEST is available from `www.nest-initiative.org`.

2   Network Representation

A network model consists of nodes, connections, and events, each represented by an abstract base class. Models for neurons and devices inherit from `class Node` and implement the state vector, the internal dynamics, and the responses to different types of events.

NEST distributes a network model over N_{VP} *virtual processes*. A virtual process is a POSIX thread that lives in one of N_{MPI} MPI processes [3,4]. The total number of virtual processes N_{VP} is the number of MPI processes times the number of threads per process: $N_{\text{VP}} = N_{\text{MPI}} \times N_{\text{Thrd}}$. NEST ensures that for a given number of virtual processes N_{VP}, all simulations of a model yield identical results, independent of the combination of N_{MPI} and N_{Thrd}.

Neuron models are often stochastic, consuming many random numbers. To distribute the load of random number generation while obtaining identical simulation results for different combinations of N_{MPI} and N_{Thrd} on $N_{\text{VP}} = \text{const}$ virtual processes, we give each virtual process its own random number generator. By default we use a lagged Fibonacci generator, because it can be seeded to produce non-overlapping sequences of random numbers [5].

2.1 Nodes and Proxies

In a network with n nodes, each node is given a unique integer `gid` in order of creation, and is assigned to a virtual process such that `vid = gid` mod N_{VP}. Each virtual process manages the memory for its nodes. In addition, each virtual process has a vector of size n, called *node list*, with pointers to its own nodes and pointers to proxies for nodes that belong to other virtual process. Within an MPI process, we can collapse the node lists of all its virtual processes into a single list to conserve memory, as illustrated in figure 1.

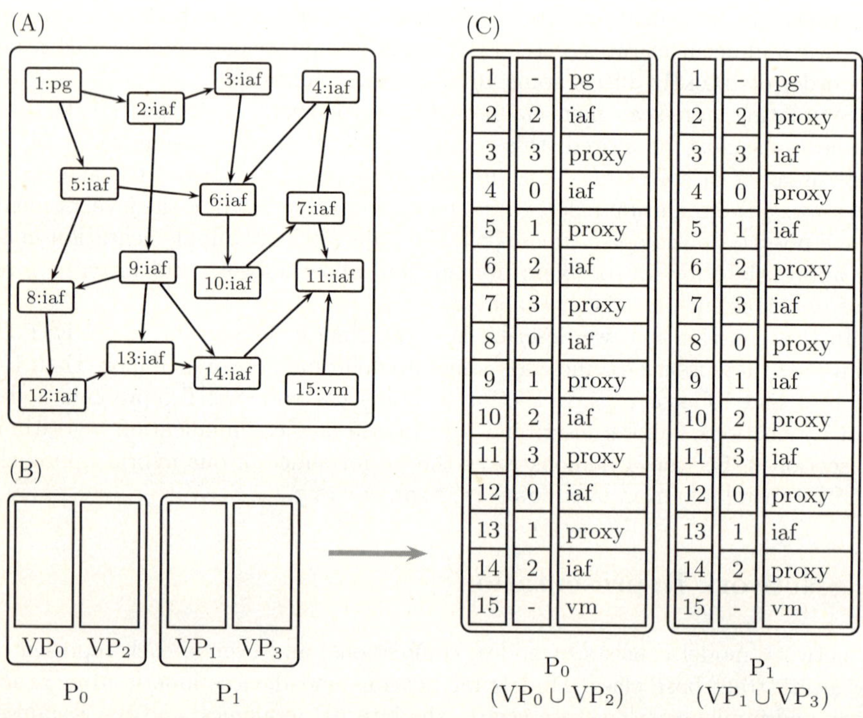

Fig. 1. Distributed and multi-threaded network representation. (A) The network as a directed graph. (B) Sketch of the distribution of four virtual processes onto two MPI processes, P$_0$ and P$_1$. (C) Collapsed node lists of the two MPI processes containing the nodes of two VPs each. The first column shows the node's `gid`, the second contains the VP a node is assigned to; '-' indicates that a node is created for each VP. The third column contains the node type: `pg`, Poisson spike generator device; `vm`, voltmeter device; `iaf`, integrate-and-fire model neurons.

This representation has the following advantages: (i) we can access each node directly with its `gid` as index to the node list on any virtual process; (ii) each virtual process knows the type of each node in the network; (iii) there is no memory overhead for pure multi-threaded simulations.

For a network of n nodes, each requiring S_N bytes of memory, and proxy nodes requiring S_P bytes, the required memory is given by:

$$M_{\text{nodes}} = \frac{n}{N_{\text{MPI}}} S_N + \left(n - \frac{n}{N_{\text{MPI}}} \right) S_P .\tag{1}$$

Typical values are $S_N = 480$ bytes and $S_P = 56$ bytes. Thus, for 10 and more MPI processes, more than half of all memory occupied by objects is occupied by proxy nodes, and more than 90% for more than 80 MPI processes. In absolute numbers, though, M_{nodes} is only around 70 MB for a network of 10^6 nodes distributed across 100 processes, which is negligible compared to the memory required for the connections in the network, cf. sec. 2.2. From a performance aspect, a node list filled with mostly proxies could become suboptimal if the node list became so large that it could no longer be cached efficiently. In this case, a fast hashing lookup may become more efficient.

Devices. So far we assumed that nodes correspond to neurons. We will now discuss nodes representing devices, such as spike injectors and recorders. There are two types of devices: recording devices and stimulation devices. Recording devices measure the state of one or more neurons and write the data to disk. If nodes on different virtual processes are assigned to the same device, each of the virtual processes gets its own instance of the recording device. This has two advantages: (i) the measured data need not be transmitted between virtual processes; (ii) each device instance can write to its own local disk. Stimulation devices supply signals to one or more neurons, thereby manipulating their state. Again, each virtual process has its own instance of a given stimulation device to reduce the amount of data that must be exchanged between virtual processes. For some stimulation devices which produce random signals we must ensure that all instances in different virtual processes produce identical signals. Thus, these devices cannot use the random number generator of the virtual process, but have their own random number generators, initialized with identical seeds.

2.2   Connections

A connection between two nodes is defined by at least four numbers: the `gid` of the sending node, the `gid` of the receiving node, a weight, and a delay. More complicated connections have weights that change, depending on the activity of the connected nodes.

Different types of connections can be implemented by classes that derive from the abstract base class `Connector`. These can implement arbitrarily complex dynamics, provided they only depend on the previous state, the time since the last event, and information available from the target node. The most important applications are synaptic depression [6] and spike-timing dependent plasticity (STDP) [7]. An algorithm for STDP suitable for distributed computing can be found in [8].

Because the connections dominate the memory requirements of large networks, NEST splits them up such that each virtual process only stores the incoming connections to its own nodes. A vector inside each `Connector` contains pointers to all local target nodes, along with the delay (integer), and weight (double). The memory required for connections per MPI process is:

$$M_{\mathrm{conn}} = \frac{n \times c \times S_C}{N_{\mathrm{MPI}}} ,\tag{2}$$

where c is the number of outgoing connections per node and S_C the memory per connection. For connections with constant weight and delay, $S_C = 32$ bytes; plastic synapses require more memory. A network of 10^6 nodes with 10^4 connections each thus requires 32 GB connection memory per process if distributed across 10 MPI processes, but only 3.2 GB per process if distributed across 100 processes.

A `ConnectionManager` stores the connections to all virtual processes of an MPI process in a three-dimensional data-structure. The first dimension is the thread number (virtual process) of the target node. The second dimension is the `gid` of the source node, and the third dimension is the index of the connection type. This memory layout has two advantages: (i) we can construct networks with heterogeneous synaptic dynamics; (ii) it is optimal for multi-threaded event delivery (cf. sec. 3.2) and the efficient implementation of synaptic dynamics [8].

3   Network Update and Event Exchange

Conceptually, NEST evaluates the network model on an evenly spaced time-grid $t_i := i \cdot \Delta$ and at each point, the network is in a well-defined state S_i. Starting at an initial state S_0, a global state transfer function $\mathrm{U}(S)$ propagates the system from one state to the next, such that $S_{t+\Delta} \leftarrow \mathrm{U}(S_t, \Delta)$. As a side effect of $\mathrm{U}(S_t)$, nodes create events that must be delivered to the target nodes after a delay that depends on the connection.

NEST evaluates a network model using the following algorithm:

```
1: T ← 0
2: while  T < T_stop  do
3:     parallel on all vp ∈ N_VP do
4:         deliver all events due
5:         call U(S_T) for all nodes
6:     end parallel
7:     exchange events between VPs
8:     increment network time: T ← T + Δ
9: end while
```

The N_{VP} virtual processes evaluate steps 4 and 5 in parallel, and in step 6 they synchronize to exchange their events in step 7.

Although NEST uses a discrete event framework, it does not use a global event-driven update [9,10]. Event-driven simulation assumes that the communication between nodes is rare and the update of a node is expensive. This does not hold for biological networks, however: If a typical cortical neuron receives spikes from $\sim 10^4$ other neurons at a rate of ~ 10 Hz, the average interval between spike arrivals is ~ 0.01 ms. However, for most neuron models integration steps of $h \sim 0.1$ ms are sufficient. Thus event-driven update would need one order of magnitude *more* updates than time-driven update; see [11] for details.

In the following we describe in more detail (i) how to maximize the time increment Δ and (ii) how to collect, exchange, and deliver events between virtual processes.

3.1 Exploiting Delays for Cache-Efficient Update

Nodes affect each other by exchanging events that arrive at their destination with a delay $d_{ij} > 0$. The time period Δ is the largest permissible temporal desynchronization between any two nodes in the network. Δ may be increased as long as this does not change the order of events. This is equivalent to a system of distributed clocks that synchronize each other with events. Lamport showed that the smallest transmission delay $d_{\min}$ defines the interval at which clocks must be synchronized to maintain the order of events [12]. Accordingly, NEST sets Δ to $d_{\min}$. During this period, all nodes are effectively decoupled.

Most neural simulators use the integration step h of the neuron dynamics as the time increment. Maximizing Δ, typically to ~ 1 ms, i.e. about 10 times larger than h, has two advantages: (i) the virtual processes can run independently for a longer time, thereby reducing the number of synchronizations and thus the communication overhead; (ii) the state-update of each node can run a few tens of integration steps *en bloc*, keeping all required data in the CPU's L1 cache.

3.2 Global Event Exchange

NEST does not transmit individual events between virtual processes, as there are far too many. Instead, for each node that produced an event, the following information is transmitted: the `gid` of the sending node and the time at which the event occurred (address event representation [13]). All other connection parameters, such as the list of weights, delays and targets, are available at each virtual process. With this information, the virtual processes reconstruct the actual events and deliver them to their destinations.

We describe below the buffering and transmission of spike events constrained to a discrete time grid $t_n = nh$. This scheme is easily extended to spikes at arbitrary times [11].

Sender-Side Buffering. Each MPI process has a three-dimensional buffer (*spike register*) to record the nodes that produced a spike-event during the update interval Δ. The first dimension represents the VP, so that they can write without collisions. The second dimension represents the time of the event with one entry per integration step h. The third dimension is a list of `gids`, one for each spike on a given thread at a given time. The total number of spikes per virtual process per update interval is small: even assuming 10^6 neurons firing at 10 Hz and distributed across 20 VPs, only some 500 spikes occur per VP and update interval. With 4 threads per MPI process, the spike register occupies less than 20 kB.

Spike Exchange and Delivery. Before spikes are exchanged between MPI processes, they are copied from the spike register to a communication buffer as follows: their `gids` are written to the buffer, ordered by the integration time step at which the spikes were generated. Sentinels separate spikes generated during different steps. Since the number of integration steps per update cycle is fixed, the receiver can reconstruct the spike time from the sentinels. Each process also maintains buffers to receive the `gids` from other processes. Once all buffers are set up, the spike buffers are exchanged between MPI processes by simultaneous pairwise exchange using CPEX [14,15,16].

Each virtual process delivers the spikes to its nodes in the parallel step 4 of the update algorithm (sec. 3). For each entry of the communication buffer, which now contains both local and remote spikes, it executes the following algorithm.

```
 1: n_sentinels ← 0
 2: if entry is sentinel then
 3:     n_sentinels ← n_sentinels + 1
 4: else
 5:     calculate t_spike from network time and n_sentinels
 6:     for all tgt ∈ local targets do
 7:         send spike time, weight, delay to tgt
 8:         tgt stores spike in its ring buffer according to delay [14].
 9:     end for
10: end if
```

4   Performance

The scaling of large-scale simulations of neural networks depends significantly on the computational load of the individual neuron. The more complex the neuron, the better the scaling, as the ratio of local computation to communication costs increases. We therefore consider the following benchmark to be a hard problem in the field of distributed neural network simulations: the computation load is low, because the neuron and synapse models are simple, but the communication load is high, as the network has a biologically realistic connection density.

Benchmark Simulation. The network consists of 12500 leaky integrate-and-fire neurons (80% excitatory, 20% inhibitory), each receiving input from 10% of all neurons, mediated by alpha-shaped current-injecting synapses with a synaptic delay of 1 ms (total number of synapses: 15.6×10^6). The neurons are initialized with random membrane potentials and receive a constant DC input adjusted to sustain asynchronous irregular firing at 12.7 Hz [17]. For a complete network specification and numerics, see [11].

Simulation times were measured on a cluster of Sun X4100 compute nodes equipped with two dual-core 2.4 GHz AMD Opteron 280 processors, 8GB RAM, and Mellanox MTS2400 Infiniband interconnect under SuSE Linux Enterprise Server 9 using the Scali MPI Connect 4.4 library. Threads were bound to CPU cores using the `taskset` command.

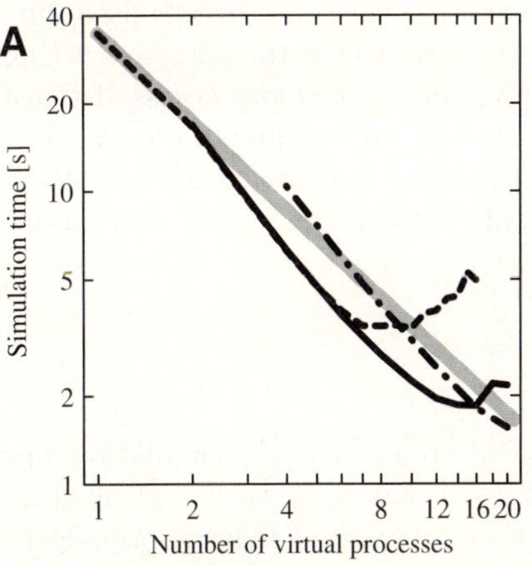
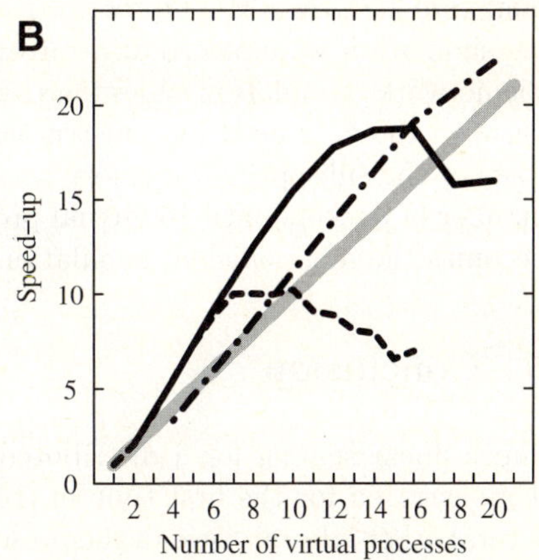

Fig. 2. Performance of different parallelization strategies as a function of the number of virtual processes. Single-thread MPI processes, dashed line; MPI processes with 2 threads, solid line; MPI processes with 4 threads, dash-dotted line. (A) Simulation time for one biological second in double-logarithmic representation. (B) Speed-up. The gray diagonal indicates the slope for a linear speed-up in both cases. Data obtained for simulations of 10 s biological time with a time step of 0.1 ms, averaged over 5 trials.

Results. Figure 2 clearly demonstrates that the parallelization strategy significantly affects the scaling and absolute run-time of the simulation. A purely MPI-parallelized simulation shows supra-linear speed-up up to 8 virtual processes, rapidly saturates, and then undergoes a significant decrease in performance. The supra-linear speed-up is due to increasingly efficient caching [14], and the saturation in performance is due to the communication overhead.

By using a hybrid strategy with two threads per MPI process, such that both threads are bound to the same CPU, the number of MPI processes is halved. This reduces the number of send/receive operations per communication step by a factor of four and results in a performance which is better than the single-threaded case for numbers of virtual processes greater than eight. The performance of this hybrid strategy remains supra-linear up to 16 virtual processes, thus substantially reducing the absolute simulation time.

Reducing the number of MPI processes further by increasing the number of threads per MPI process to four leads to worse performance for small numbers of virtual processors. This is due to the fact that memory allocation is performed by a single thread on each MPI process; as a result of the NUMA architecture, memory access is sub-optimal for the two threads on the non-allocating processor. The role of memory access is corroborated by simulating with two threads per MPI process as above, but binding the threads to different CPUs. This results in a performance which lies between that of the two-thread same-CPU variant discussed above and that of the four-thread variant (data not shown). This analysis is further supported by benchmarks performed on a Sun V40z server with four dual-core 2.2GHz AMD Opteron 875 processors, in which the threads used during simulation were placed at arbitrary cores relative to the thread constructing the network. Simulation times increased with increasing memory-access distance between the core used for construction and those used for simulation [18]. The costs of the sub-optimal memory access outweigh the benefits of decreasing the number of packets until 16 virtual processes, after which the four-thread variant becomes the most efficient simulation strategy.

5   Conclusion

Supra-linear scaling for a distributed biological neural network simulation was demonstrated for the first time in [14]. This result has since been confirmed by several other laboratories. In the present work we show that a hybrid approach to neural network simulation, combining multi-threading and distributed computing techniques, achieves an even better performance than a purely distributed solution. This suggests that the infrastructure of NEST is appropriate for future generations of multiprocessor, multi-core clusters.

The problem studied here was chosen to be particularly hard with respect to communication. In studies with larger neural networks or with more complex dynamics, NEST performance saturates at much larger numbers of processors: Simulation time for a network of 10^5 neurons with 10^9 synapses, driven by Poisson background input, shows supra-linear scaling up to 80 virtual processes on the same hardware. Other laboratories have shown good scaling of large-scale simulations on systems with thousands of processors, albeit on less hard problems [19,20]. The scaling of NEST on such systems remains to be investigated.

The benchmarking results demonstrate the importance of sophisticated memory allocation on modern NUMA machines. Future work on NEST will be concerned with improving memory access times in a hybrid message-passing and multi-threading environment and further optimizing communication with respect to number of packets and latency hiding.

Acknowledgements. HEP acknowledges Anita Woll for the execution of benchmark tests. Benchmarks were performed on supercomputing equipment at the Norwegian University of Life Sciences and the University of Freiburg. Partially funded by DAAD/NFR 313-PPP-N4-lk, DIP F1.2, BMBF Grant 01GQ0420 to the Bernstein Center for Computational Neuroscience Freiburg, and EU Grant 15879 (FACETS).

References

1. Braitenberg, V., Schuz, A.: Cortex: Statistics and Geometry of Neuronal Connectivity, 2nd edn. Springer, Berlin (1998)
2. Gewaltig, M.O., Diesmann, M.: NEST. Scholarpedia (2007)
3. Message Passing Interface Forum: MPI: A message-passing interface standard. Technical Report UT-CS-94-230, University of Tennessee (1994)

4. Lewis, B., Berg, D.J.: Multithreaded programming with pthreads. Sun Microsystems, Mountain View (1998)
5. Knuth, D.E.: The Art of Computer Programming, 3rd edn., vol. 2. Addison-Wesley, Reading, MA (1998)
6. Thomson, A.M., Deuchars, J.: Temporal and spatial properties of local circuits in neocortex. Trends Neurosci. 17, 119–126 (1994)
7. Bi, G.Q., Poo, M.M.: Synaptic modifications in cultured hippocampal neurons: dependence on spike timing, synaptic strength, and postsynaptic cell type. J. Neurosci. 18, 10464–10472 (1998)
8. Morrison, A., Aertsen, A., Diesmann, M.: Spike-time dependent plasticity in balance recurrent networks. Neural Comput. 19, 1437–1467 (2007)
9. Zeigler, B.P., Praehofer, H., Kim, T.G.: Theory of Modeling and Simulation, 2nd edn. Academic Press, Amsterdam (2000)
10. Brette, R., et al.: Simulation of networks of spiking neurons: A review of tools and strategies. J. Comput. Neurosci. (in press, 2007)
11. Morrison, A., Straube, S., Plesser, H.E., Diesmann, M.: Exact subthreshold integration with continuous spike times in discrete time neural network simulations. Neural Comput. 19, 47–79 (2007)
12. Lamport, L.: Time, clocks, and the ordering of events in a distributed system. Communications of the ACM 21, 558–565 (1978)
13. Lazzaro, J.P., Wawrzynek, J., Mahowald, M., Sivilotti, M., Gillespie, D.: Silicon auditory processors as computer peripherals. IEEE Transactions on Neural Networks 4, 523–528 (1993)
14. Morrison, A., Mehring, C., Geisel, T., Aertsen, A., Diesmann, M.: Advancing the boundaries of high connectivity network simulation with distributed computing. Neural Comput. 17, 1776–1801 (2005)
15. Tam, A., Wang, C.: Efficient scheduling of complete exchange on clusters. In: 13th International Conference on Parallel and Distributed Computing Systems (PDCS 2000), Las Vegas (2000)
16. Gross, J., Yellen, J.: Graph Theory and its Applications. CRC Press, Boca Raton, USA (1999)
17. Brunel, N.: Dynamics of sparsely connected networks of excitatory and inhibitory spiking neurons. J. Comp. Neurosci. 8, 183–208 (2000)
18. Woll, A.: Performance analysis of an MPI- and thread-parallel neural network simulator. Master's thesis, Norwegian University of Life Sciences (2007)
19. Djurfeldt, M., Johansson, C., Ekeberg, Ö., Rehn, M., Lundqvist, M., Lansner, A.: Massively parallel simulation of brain-scale neuronal network models. Technical Report Technical Report TRITA-NA-P0513, KTH, School of Computer Science and Communication Stockholm, Stockholm (2005)
20. Migliore, M., Cannia, C., Lytton, W.W., Markram, H., Hines, M.: Parallel network simulations with NEURON. J. Comp. Neurosci. 21, 119–223 (2006)

MCSTL:
The Multi-core Standard Template Library*

Johannes Singler, Peter Sanders, and Felix Putze

Universität Karlsruhe
{singler,sanders,putze}@ira.uka.de

Abstract. Future gain in computing performance will not stem from increased clock rates, but from even more cores in a processor. Since automatic parallelization is still limited to easily parallelizable sections of the code, *most* applications will soon have to support parallelism *explicitly*. The *Multi-Core Standard Template Library (MCSTL)* simplifies parallelization by providing efficient parallel implementations of the algorithms in the C++ Standard Template Library. Thus, simple recompilation will provide partial parallelization of applications that make consistent use of the STL. We present performance measurements on several architectures. For example, our sorter achieves a speedup of 21 on an 8-core 32-thread SUN T1.

1 Introduction

Putting multiple cores into a single processor increases peak performance by exploiting the high transistor budget of modern semiconductor technology. The performance per Watt can also be improved, in particular when used together with reduced clock speeds and moderate instruction level parallelism. Dual-core processors being omnipresent, in the near future, many-cores will be used in virtually all areas of computing, ranging from mobile systems to supercomputers.

To benefit from this increased power, programs have to exploit parallelism. This now becomes mandatory not just for a selected number of specialized programs, but for all nontrivial applications. Because automatic parallelization is still working only for simple programs and explicit parallelization is expensive and outside the qualification of most current programmers, this poses a problem.

This paper addresses a third alternative — easy-to-use libraries of parallel algorithm implementations. While this approach has been successful in numerics for a long time, it has not yet made its way into the mainstream of non-numeric programming. We present our initial work on the **Multi-Core Standard Template Library**. Parallelizing the C++ **Standard Template Library** [2] is a good starting point since it is part of the C++ programming language and offers a widely-known, simple interface to many useful algorithms. Programs that use the STL can thus be partially parallelized by recompilation using the MCSTL.

---

* This is the full and updated version of a Poster Extended Abstract presented at PPoPP 2007 [1].

A.-M. Kermarrec, L. Bougé, and T. Priol (Eds.): Euro-Par 2007, LNCS 4641, pp. 682–694, 2007.
© Springer-Verlag Berlin Heidelberg 2007

Parallelizing the STL is not an new idea. STAPL [3,4] provides parallel container classes that allow writing scalable parallel programs on distributed memory machines. However, judged from publications, only few of the STL algorithms have been implemented, and those that have been implemented sometimes deviate from the STL semantics (e. g. `p_find` in [3]).

The MCSTL limits itself to shared memory systems and thus can offer features that would be difficult to implement efficiently on distributed memory systems, e. g. fine-grained dynamic load-balancing. Except for a few inherently sequential algorithms like binary search, the MCSTL will eventually parallelize all algorithms in the STL following the original semantics and working on ordinary STL random access iterators (e. g., STL vectors or C arrays). This approach brings its own challenges. "Traditional" parallel computing works with many processors, specialized applications, and huge inputs. In contrast, the MCSTL should already yield noticeable speedup for as few as two cores for as many applications as possible. In particular, the amount of work submitted to each call of one of the simple algorithms in STL may be fairly small. In other words, the tight coupling offered by shared memory machines in general and multi-core processors in particular, is not only an opportunity but also an *obligation* to scale *down* to small inputs rather than *up* to many processors. Another issue is that the MCSTL should coexist with other forms of parallelism. The operating system will use some of the computational resources, multiple processes may execute on the same machine and there might be some degree of high-level parallelization using multi-threading within the application. These methods are the easiest way to leverage the power of multi-core processors but might not suffice to fully saturate the machine. In this context, the MCSTL should have dynamic load balancing even when static load balancing would be enough on a dedicated machine. Moreover, parallel algorithms that achieve some limited speedup at the cost of a great increase of total work should be avoided. A better solution are algorithms that use only as much parallelism as can be *efficiently* exploited. Our algorithms use some heuristics to decide on the level of parallelism. This form of efficiency is also important with respect to energy consumption.

More Related Work. Recently, another shared memory STL library has surfaced. MPTL [5] parallelizes many of the simple algorithms in STL using elegant abstractions. However, it does not implement the more complicated parallel algorithms `partition`, `nth_element`, `random_shuffle`, `partial_sum`, `merge`. MPTL has a "naïve" parallelization of quicksort using sequential partitioning. Similarly, there is only a "naïve" implementation of `find` that does not guarantee any speedup even if the position sought is far away from the starting point. MPTL offers a simple dynamic load balancer based on the master worker scheme and fixed chunk sizes.

Most of the algorithms we present here are previously known or can be considered folklore even if we do not cite a specific reference. We view the main contribution of this paper as selecting good starting points and engineering efficient implementations. Interestingly, many of our algorithms were originally developed for distributed memory parallel computing. Very often, such algorithms naturally translate into shared memory algorithms with good cache locality and few costly synchronization operations. The key ideas behind our sorting algorithms are also not new [6,7] but not widely known. It is somewhat astonishing that although there are virtually hundreds of papers on parallel sorting, so few notice that multiway merging can be done with exact splitting, and that the partition in quicksort can be parallelized without changing the inner loop.

Notation. In general, n will refer to the problem size, m to some secondary problem quantity. Often, we are dealing with a sequence $[S[0], \ldots, S[n-1]]$. There are p threads that run in parallel, numbered 0 through $p-1$.

2   Algorithms

We plan to implement all the algorithms provided in the STL for which parallelization looks promising. Figure 1 summarizes the current status of the implementation.

Algorithm Class	Function Call(s)	Status	w/LB	w/oLB
Embarrassingly Parallel	`for_each`, `generate(_n)`, `fill(_n)`, `count(_if)`, `transform`, `unique_copy`, `min/max_element`, `replace`*(_copy)*`(_if)`	impl	yes	yes
Find	`find(_if)`, `mismatch`, `equal`, `adjacent_find`, `lexicographical_compare`	impl	yes	nww
Search	`search(_n)`	impl	yes	nww
Numerical Algorithms	`accumulate`, `partial_sum`, `inner_product`, `adjacent_diff.`	impl	nww	yes
Partition	*(stable_)*`partition`,	impl	yes	nww
Merge	`merge`, `multiway_merge`, *inplace_merge*	impl	pl	yes
Partial Sort	`nth_element`, `partial_sort`*(_copy)*	impl	yes	pl
Sort	`sort, stable_sort`	impl	impl	yes
Shuffle	`random_shuffle`	impl	yes	nww
Set Operations	`set_union`, `set_intersection`, `set_(symmetric_)diff.`	impl	nww	yes
Vector	*valarray operations*	pl		
Containers	*vector, (multi_)map/set, priority_queue operations*	pl		
Heap	*make_heap, sort_heap*	pl		

Fig. 1. Considered STL functions. **impl**emented, (except *italicized*), **pl**anned, **n**ot **w**orth**w**hile, wLB / w/oLB = with / without dynamic load-balancing

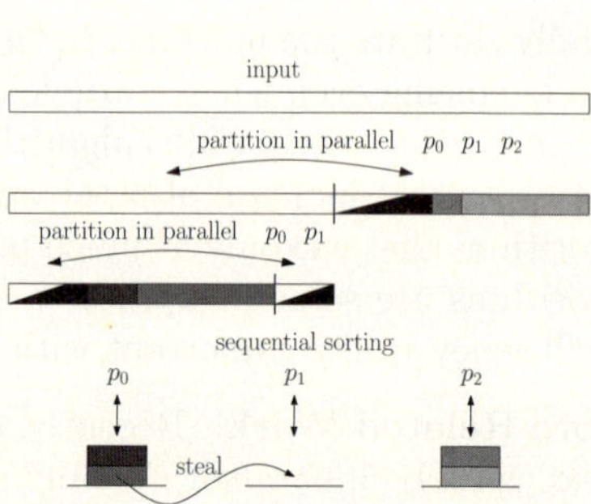

Fig. 2. Schema of parallel balanced quicksort. The ramped parts are already sorted, while the dark grey parts are currently partitioned. The colored parts are remembered on the stack and wait for being (stolen and) processed.

Embarrassingly Parallel Computation. Several STL algorithms may be viewed as the problem of processing n independent atomic jobs in parallel, see the row "Embarrassingly Parallel" in Figure 1 for a list. This looks quite easy on the first glance. For example, we could simply assign $\leq \lceil n/p \rceil$ jobs to each thread. Indeed, we provide such an implementation in order to scale down to very small, fine-grained, and uniform inputs on dedicated machines. However, in general, we cannot assume anything about the availability of cores or the running time of the jobs which might reach from a few machine instructions (quite typical for STL) to complex computations. Hence, we can neither afford dynamic load balancers that schedule each job individually nor should we a priori cluster chunks of jobs together. *Random Polling* or *randomized work stealing* is a way out of this dilemma.[1]

Initially, each thread gets $\leq \lceil n/p \rceil$ consecutive jobs defined by a pair of iterators. A busy thread processes one job after the other. When a thread is done, it looks for more work. If all jobs are already finished, it terminates and so does the algorithms. Otherwise, it tries to steal half of the jobs from a randomly chosen other thread. We implement this without intervention of the victim. Note that this way, the jobs will be dynamically partitioned into a small number of consecutive intervals (this is important for (parallel) cache efficiency), very large jobs will never irrevocably be clustered together, and threads without a processor core assigned to them will lose jobs to stealing (and thus active) threads. To achieve a good compromise between worst-case performance and overhead, the user can optionally devise the algorithm to process the jobs in indivisible chunks of a certain size. Using known techniques (e. g. [10,11]), it can be shown that almost perfect load balancing can be achieved at the cost of only a logarithmic number of work stealing operations per thread.

Find.   Function `find_if` finds the first element in a sequence that satisfies a certain predicate. The functions `find`, `adjacent_find`, `mismatch`, `equal`, and `lexicographical_compare` can be immediately reduced to `find_if`. On the first glance, this looks like just another embarrassingly parallel problem. However, a naïve parallelization may not yield any useful speedup. Assume the first matching element is at position m in a sequence of length n. The sequential algorithm needs time $O(m)$. A naïve parallelization that splits the input into p pieces of size n/p needs time $\Omega(n/p) = \Omega(m)$ if $m = n/p - 1$. In practice, we might even see speedup $\ll 1$ if m is so small that the overhead for coordinating threads becomes overwhelming. Hence, our algorithm starts with a *sequential* search for the first m_0 steps. Only then it starts assigning blocks of consecutive sequence positions to the p threads. Consumption of these blocks is dynamically load-balanced using the fetch-and-add primitive. A thread that finds the element, signals this by grabbing all the remaining work. There is still a difficult trade-off here. Assigning small blocks is good because all threads will learn about termination quickly. On the other hand, small blocks are bad because there is some overhead for a fetch-and-add operation. Therefore, our implementation combines both advantages by starting with a block size of $\underline{m}$ and increasing it by a factor g until a value of $\bar{m}$ is reached. The tuning parameters $\underline{m}$ and $\bar{m}$ allow a flexible compromise between fast termination for small m and low overhead for

---

[1] The method goes back at least to [8], using it for loop scheduling is proposed in [9]. An elegant analysis for shared memory that coined the term work stealing is in [10].

large m. The execution time of our algorithm is independent of n and the term dependent on m is an optimal $O(m/p)$. We need only a single synchronization at the end.

Partial Sum. When computing prefix sums, we synchronize only twice, instead of $\log p$ times, as done by typical textbook algorithms (e. g. [12]). After splitting the sequence into $p + 1$ parts, the partial sums of part 0 and the totals sums of parts $1..p - 1$ are computed in parallel. After processing these intermediate results in a sequential step, the partial sums of parts $1..p$ are computed. To compensate for the first thread possibly taking longer because of having to write back the results in the first step, the user can specify a dilatation factor d. The total running time then is $O(n/p + p)$, the maximum speedup achievable is $(p + d)/(1 + d)$, i. e. $(p + 1)/2$ for $d = 1$.

Sorting and its Kindred. *Partition.* Given a *pivot predicate* P, we are asked to permute $[S[0], \ldots, S[n - 1]]$ such that we have $P(S[i])$ for $i < m$ and $\neg P(S[i])$ for $i \geq m$. This routine is part of the STL and the most important building block for quicksort, selection, etc. We use a parallel algorithm similar to the one by Tsigas and Zhang [7], which has many advantages. Its inner loop is the same as in sequential quicksort, it works in-place, and it is dynamically load-balanced.

The sequential algorithm scans S from both ends until it finds two elements $S[i]$ and $S[j]$ that belong to the "other" side respectively. It swaps $S[i]$ and $S[j]$ and continues scanning. The parallel algorithm works similarly. However, each thread reserves two chunks of a certain size B from each end. It performs the partitioning of those two chunks, until one of them runs empty. If the left chunk runs empty, it reserves a succeeding block using a fetch-and-add primitive. Symmetrically, if the right size runs empty it reserves a preceding block. This process terminates when there are less than B elements left between the left and the right boundary. When all threads have noticed this condition, there is at most one chunk per thread that is partly unprocessed. After calculating various offsets sequentially, each thread swaps its unprocessed part to the "middle" of the sequence. Those remaining elements are treated recursively in this manner, with fewer threads, ending in the sequential call for less than B elements. The running time of this algorithm is bounded by $O(n/p + Bp)$.

m^{th} *Element*[2]. Using the above parallel partitioning algorithm, it is easy to parallelize the well known quickselect algorithm: partition around a pivot chosen as the median of three. If the left side has at least m elements, recurse on the left side. Otherwise recurse on the right side. Switch to a sequential algorithm when the subproblem size becomes smaller than size $2Bp$ where B is the tuning parameter used in `partition`. We get total expected execution time $O(\frac{n}{p} + Bp \log p)$.

Multi-Sequence Partitioning. Given k *sorted* sequences $S_1, \ldots, S_k$ and a *global rank* m, we are asked to find splitting positions $i_1, \ldots, i_k$ such that $i_1 + \cdots + i_k = m$ and $\forall j, j' \in 1..k : S_j[i_j - 1] \leq S_{j'}[i_{j'}]$. The function `multiseq_partition` is

---

[2] For consistency with our notation, where n is the input size, we use m for the requested rank, although the STL function is called `nth_element`.

not part of the STL, but useful for many of the subsequent routines based on merging. Our starting point is an asymptotically optimal algorithm by Varman et al. [6] for selecting the element of global rank m in a set of sorted sequences. It is fairly complicated, and to our knowledge, has been implemented before only for the case that the number of sequences is a power of two and all sequences have the same length $|S_j| = 2^k - 1$ for some integer k.

Explicit care has been taken of the case of many equal elements surrounding the requested rank. To allow stable parallelized merging based on this partitioning, the splitter positions may not be in arbitrary positions in the equal subsequence. In fact, there must not be more than one sequence S_j having a splitter "inside" the equal subsequence. All S_i with $i < j$ must have the splitter at the end of it, all S_i with $i > j$ must have the splitter at its beginning. The running time amounts to $O(k \log k \cdot \log \max_j |S_j|)$.

Merging. Given two sorted sequences S_1 and S_2, STL function `merge` produces a sorted sequence T containing the elements from S_1 and S_2. We generalize this functionality to multiple sorted sequences $S_1, \ldots, S_k$. This is an operation known to be very effective for both cache efficient and parallel algorithms related to sorting (e. g., [13,14,15]).

We can reduce parallel multiway merging to sequential multiway merging using calls to multi-sequence partition with global ranks $\{m/p, 2m/p, \ldots, (p-1)m/p, m\}$, and splitting accordingly. Our implementation of sequential multiway merging is an adaptation of the implementation used for cache-efficient priority queues and external sorting in [13,15]: For $k \leq 4$, we use specialized routines that encode the relative ordering of the next elements of each sequence into the program counter. For $k > 4$, a highly tuned tournament tree data structure keeps the next element of each sequence. The total execution time of our algorithm is $O(\frac{m}{p} \log k + k \log k \cdot \log \max_j |S_j|)$.

Sort. Using the infrastructure presented, we can implement two different parallel sorting algorithms:

(Stable) Parallel Multiway Mergesort: Each thread sorts $\leq \lceil n/p \rceil$ elements sequentially. Then, the resulting sorted sequences are merged using parallel multiway merging. Finally, the result is copied back into the input sequence. The algorithm runs in time $O(\frac{n \log n}{p} + p \log p \cdot \log \frac{n}{p})$. Our implementation of parallel multiway merging allows stable merging — it will always take S_i with the minimum i available when it encounters equal elements. Hence, by using a stable algorithm for sequential sorting we get a stable parallel sorting algorithm.

Load-Balanced Quicksort: Using the parallel partitioning algorithm from Section 2, we can obtain a highly scalable parallel variant of quicksort, as described in [7]. Although this algorithm is likely to be somewhat slower than Parallel Multiway Mergesort, it has the advantage to work in-place and to feature dynamic load balancing.

The sequence is partitioned into two parts recursively. After the sequence has been split to p parts, each thread sorts its subsequence sequentially. However, the length of those subsequences may differ strongly, so normally the overall performance would be poor. To overcome this problem, we implemented the quicksort variant using lock-free double-ended queues to replace the local call stacks. After partitioning a subsequence, the longer part is pushed onto the top end of the local queue, while the shorter part is sorted recursively. When the recursion returns, a subsequence is popped from the top end of the local queue. If there is none available, the thread steals a subsequence from another thread's queue. It pops a block from the bottom end of the victim's queue. Since this part is probably relatively large, the overhead of this operation is compensated for quite well. If the length of the current part is smaller than some threshold, no pushing to the local queue is done any more, so the remaining work is done completely by the owning thread.

The functionality required for the double-ended queue is quite restricted, as is its maximum size. Thus, a circular buffer held in an array and atomic operations like fetch-and-add and compare-and-swap suffice to implement such a data structure, with all operations taking only constant time.

If an attempt to steal a block from another thread is unsuccessful, the thread offers the operating system to yield control using an appropriate call. This is necessary to avoid starvation of threads, in particular if there are less (available) processors than threads. There could still be work available albeit all queues are empty, since all busy threads might be in a high-level partitioning step.

The total running time is $O(\frac{n \log n}{p} + Bp \log p)$, ignoring the load-balancing overhead. Figure 2 shows a possible state of operation.

Random Shuffle. We use a cache-efficient algorithm random permutation algorithm [16] which extends naturally to the parallel setting: In parallel, throw each element into one out of k random bins per thread, pool the corresponding bins, and permute the resulting bins independently in parallel (using the standard algorithm). We use the Mersenne-Twister random number generator [17] which is known to have very good statistical properties even if random numbers are split into their constituent bits, as we do.

3   Software Engineering

Goals. A major design goal of the MCSTL is to provide parallelism with hardly any effort from the user. The interface fully complies to the C++ Standard Template Library in terms of syntax. However, the MCSTL has quite high requirements when it comes to semantics. For example, the operations called in algorithms like `for_each` must be independent and non-interfering with each other. Also, the order of execution cannot be guaranteed any more. The user should call the appropriate algorithms like `accumulate`, if this is undesired. Also, some programs might rely on invariants of algorithms that are not guaranteed by the standard. A common example is `merge`. For the standard merging algorithm, both sequences can overlap without doing any harm. However, this is fatal for the parallel algorithm.

Library Particularities. Many parallel algorithms are published that show excellent scalability results. However, the experiments are usually either limited to a platform, a specific data type as input, and / or assumptions on the input data type. The code is hard to re-use since it is specialized to a specific machine and specific needs.

In contrast to this, a library implementation of a parallel algorithms must be more general. First of all, it need be parameterizable by the data type the input consists of. Secondly, the basic operations are defined by the user, for some algorithms. The data type can carry additional semantics, e. g. the comparison operator might be redefined, or the assignment operator and the copy constructor. The library may not violate any consequences following from this. Also, the running time of the functors may be arbitrarily distributed. For some operations, e. g. prefix sum and accumulation, we assume the execution time to be about the same for each call, while for `for_each`, no such assumption is made.

Implementation. Any parallel library needs a foundation that supports concurrent execution. For our library, this foundation should be both efficient and platform-independent. We have chosen to use OpenMP 2.5 [18]. Additionally, a thin platform-specific layer provides access to a small number of efficiently implemented primitive atomic operations like fetch-and-add and compare-and-swap. Although these primitives are still compiler specific, there is already enough convergence that one can obtain a portable library using this functionality.

Using the MCSTL. Using the MCSTL in a program is extremely simple, just add a few compiler command line options. The library will then use default values for deciding whether to parallelize an algorithm, based on the automatically determined number of cores.

However, if the user wants to customize the calls for maximum performance in any setting, he can do so at little effort. The library allows setting for each algorithm the minimal problem size from which the library should call the parallel algorithm. Also, the number of threads used can be easily specified. The algorithm alternatives and tuning parameters are also accessible to the user. If an algorithm must be executed sequentially by all means, the programmer specifies this by adding `mcstl::sequential_tag()` to the end of the call. The original STL version will then be called without any runtime overhead since the decision is made at compile-time through function overloading.

4   Experimental Results

Testing Procedures. We have tested on four different platforms, namely a Sun T1 (8 cores, 1.0 GHz, 32 threads, 3 MB shared L2 cache), a AMD Opteron 270 (2 cores, 2.0 GHz, 1 MB L2 cache per core), an 2-way Intel Xeon 5140 (Core 2 architecture, 2×2 cores, 2.33 GHz, 2×4 MB shared L2 cache per processor), and a 4-way Opteron 848 (4×1 core, 1.8 GHz, 1 MB L2 cache per processor). For comparability, all programs were compiled with the same GCC 4.2 snapshot[3], which features a reliable implementation of OpenMP.

In the following subsections, we show how the running time relates to the original sequential algorithm provided by the corresponding STL, expressed as speedup. Unless stated explicitly, the Sun T1 is considered. For testing, the parallel execution of all algorithms was forced, to also show the results for small inputs which would not have executed in parallel, normally.

Input Data. Unless stated otherwise, we used uniform random input data for our experiments. All test were run at least 30 times, the running times were averaged. The input data varies with every run, and is generated immediately before executing the algorithm, so it may (partly) reside in cache. This is a realistic setting which in fact favors the sequential version because it can access the data right away, while the many threads of the parallel implementation have to communicate the data via the slow main memory.

Embarrassingly Parallel Computations. We tested the algorithms for performing embarrassingly parallel computations by computing the Mandelbrot fractal, on the 4-way Opteron. For each pixel, for up to i iterations of a computation with complex numbers. Since the computation is interrupted as soon as the point known to be outside the Mandelbrot set, the computation time for the different pixels differs greatly. Speedup of up to 3.4 is achieved for $i = 10000$ iterations per pixel with the dynamical load balancing, whereas static load balancing only gains a factor of at most 2. This shows the superiority of the dynamic load balancing for jobs with highly varying running time.

Find. The naïve parallel implementation of find performs very badly and is far from achieving good speedup, as shown in Figure 3. For the parameters chosen, $m_0 = 1000, \underline{m} = 10000, \bar{m} = 64000, g = 2$, we report an interesting insight. The growing block variant (gb) performs best for most cases. It is only superseded by the fixed size block (fsb) variant in a narrow segment, where the parameters are just right by chance. When switching to the parallel processing there is speedup < 1, for small inputs. Speedup then goes up steeply, too almost full speddup, as long as there are enough cores available.

Partial Sum. The speedups for `partial_sum` (see Figure 5) are only about half of the number of processor, as predicted, because the parallel algorithm performs almost twice the work than the sequential one.

Partition / mth Element. `partition` gains linear speedup for up to eight threads, and benefits from multi-threading up to a factor of 15, as shown in Figure 4. However, there is no speedup at all for small input sizes. This is because there is only little computation for integers, so overhead comes in badly.

Sort. With large elements, the multiway mergesort (mwms) achieves speedup up to 20 on the 8-core T1 (see Figure 7). This is quite impressive, the multi-threading is utilized extensively. With as many threads as cores, speedup 7.5

---

[3] Revision 114849, 2006-06-21.

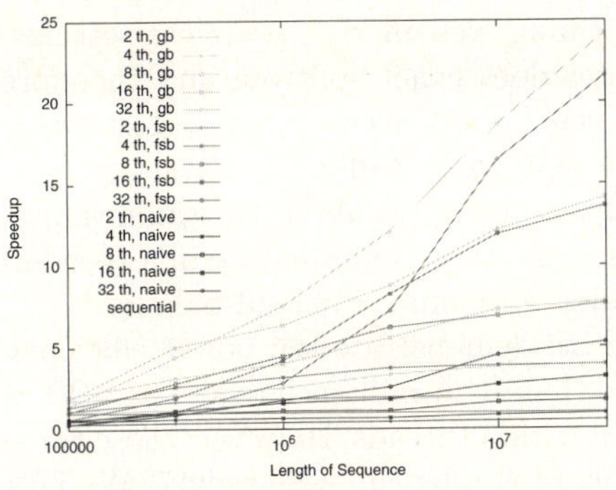

Fig. 3. Finding a 32-bit integer at random position, using different algorithm variants: growing block size (**gb**), fixed-size blocks (**fsb**), and naive

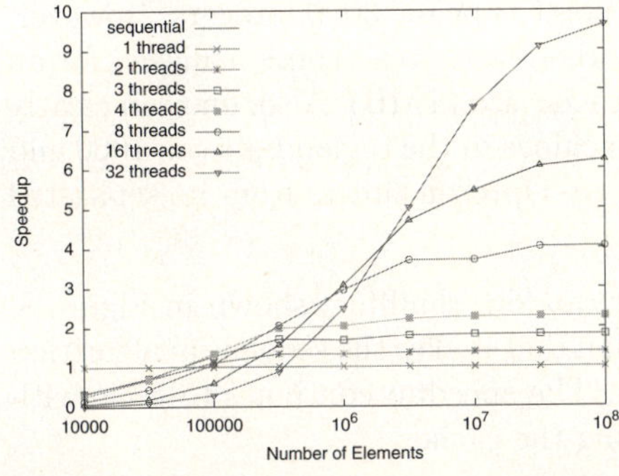

Fig. 4. Partitioning a sequence of 32-bit integers

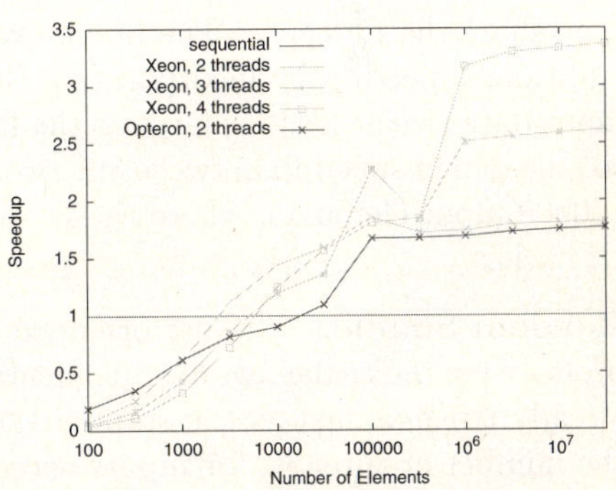

Fig. 5. Computing partial sums of 32-bit integers

Fig. 6. Sorting 32-bit integers on the Xeon and the Dual-Core-Opteron

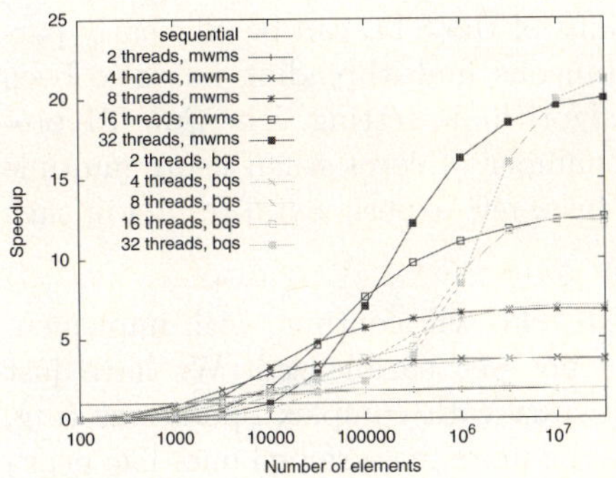

Fig. 7. Sorting pairs of 64-bit integers

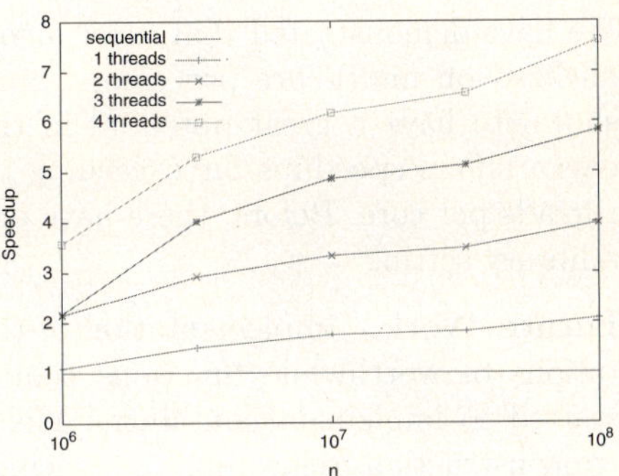

Fig. 8. Random shuffling of 32-bit integers on the 4-way Opteron

is possible. Also, speedup can be achieved with a small number of threads for as few as 3000 elements. Not only performing well in this practical settings, the MCSTL parallel multiway merge sorter does exact splitting and therefore guarantees these good execution times, even for worst-case inputs. Tests have shown that calculating the exact partition only costs negligible time.

The numbers for balanced quicksort (bqs) converge to about the same values, but more slowly, even beating mergesort in some cases, culminating in a speedup of 21. The algorithm works in-place and also is dynamically load-balanced.

With the former example, we also want to demonstrate the power efficiency of multi-core processors. For sorting more than 31.6 million integers, the T1 is about as fast as the 4-way Opteron running with 3 threads. However, those three processors consume about 246 W, while the T1 can be run with only 72 W. This yields a three-times better power-efficiency, although both processors are of the same generation.

The results for the Xeon, presented in Figure 6, show the great influence of the caches on the performances. For one processor, the speedups for two threads are excellent, and also scale down well. The two cores work together very well because they share the L2 cache. The break-even point is below 3000 integers. However, additional speedup by incorporating the other two cores is only achieved for an input data size at least as large as the L2 cache size (4 MB). Also, one can clearly see the gap in speedup between the two machines in the region between 1000 and 100000 input elements, where the Dual-Core-Opteron suffers from its separated L2 caches.

Random Shuffle. The performance for random shuffling, shown in Figure 8, profits from the cache-aware implementation that makes the sequential algorithm already twice as fast as the standard one. The speedup continues to scale with the number of threads, for inputs exceeding the cache.

5   Conclusion

We have demonstrated that most algorithms of the STL can be efficiently parallelized on multi-core processors. Simultaneous multithreading has also been shown to have a great potential in the algorithmic setting. The Sun T1 processor shows speedups far exceeding the number of cores when using multiple threads per core. Before, there have only been few experimental results in such a library setting.

Future Work. Implementation of the MCSTL will continue with implementations of worthwhile functions that are not yet parallelized. We have just started to implement containers. We will start with complex operations (e.g. (re)construction, rehashing), advancing to the more fine grained ones like priority queue updates.

We will add (optional) dynamic load balancing to more functions. In some case, like parallel multiway mergesort or prefix sum, this poses interesting algorithmic problems.

The MCSTL sometimes offers several implementations of functions. Of course, in an easy-to-use library we would like automatic support for selecting an implementation. Some work in this direction has been done in [4]. However, more work is needed because it is not sufficient to *select* an algorithm, we also have to *configure* it to use the right values for tuning parameters such as the value B in `partition`, and, most importantly, the number of threads to be used.

The MCSTL is available freely on our website [19], and can be used by everyone free of charge. We plan to integrate it with the external memory library STXXL [20]. Since in many situations the STXXL is compute bound rather than I/O bound, we expect significantly improved performance for various algorithmic problems on huge data sets.

References

1. Putze, F., Sanders, P., Singler, J.: The multi-core standard template library. In: ACM SIGPLAN Symposium on Principles and Practice of Parallel Programming, pp. 144–145. ACM Press, New York (2007)
2. Plauger, P.J., Stepanov, A.A., Lee, M., Musser, D.R.: The C++ Standard Template Library. Prentice-Hall, Englewood Cliffs (2000)
3. An, P., Jula, A., Rus, S., Saunders, S., Smith, T., Tanase, G., Thomas, N., Amato, N.M., Rauchwerger, L.: STAPL: An Adaptive, Generic Parallel C++ Library. In: Dietz, H.G. (ed.) LCPC 2001. LNCS, vol. 2624, pp. 193–208. Springer, Heidelberg (2003), http://parasol.tamu.edu/groups/rwergergroup/research/stapl/
4. Thomas, N., Tanase, G., Tkachyshyn, O., Perdue, J., Amato, N.M., Rauchwerger, L.: A framework for adaptive algorithm selection in STAPL. In: ACM SIGPLAN Symposium on Principles and Practice of Parallel Programming, pp. 277–288. ACM Press, New York (2005)
5. Baertschiger, D.: Multi-processing template library. Master thesis, Université de Genève (in French) (2006), http://spc.unige.ch/mptl
6. Varman, P.J., Scheufler, S.D., Iyer, B.R., Ricard, G.R.: Merging Multiple Lists on Hierarchical-Memory Multiprocessors. Journal of Parallel and Distributed Computing 12(2), 171–177 (1991)
7. Tsigas, P., Zhang, Y.: A simple, fast parallel implementation of quicksort and its performance evaluation on SUN enterprise 10000. In: 11th Euromicro Conference on Parallel, Distributed and Network-Based Processing, p. 372 (2003)
8. Finkel, R., Manber, U.: DIB – A distributed implementation of backtracking. ACM Transactions on Programming Languages and Systems 9(2), 235–256 (1987)
9. Sanders, P.: Tree shaped computations as a model for parallel applications. In: ALV'98 Workshop on Application Based Load Balancing (1998)
10. Blumofe, R.D., Leiserson, C.E.: Scheduling multithreaded computations by work stealing. Journal of the ACM 46(5), 720–748 (1999)
11. Sanders, P.: Randomized Receiver Initiated Load Balancing Algorithms for Tree Shaped Computations. The Computer Journal 45(5), 561–573 (2002)
12. JáJá, J.: An Introduction to Parallel Algorithms. Addison-Wesley, Reading (1992)
13. Sanders, P.: Fast priority queues for cached memory. ACM Journal of Experimental Algorithmics 5 (2000)
14. Ranade, A., Kothari, S., Udupa, R.: Register Efficient Mergesorting. In: Prasanna, V.K., Vajapeyam, S., Valero, M. (eds.) HiPC 2000. LNCS, vol. 1970, pp. 96–103. Springer, Heidelberg (2000)

15. Dementiev, R., Sanders, P.: Asynchronous parallel disk sorting. In: 15th ACM Symposium on Parallelism in Algorithms and Architectures, pp. 138–148. ACM Press, New York (2003)
16. Sanders, P.: Random permutations on distributed, external and hierarchical memory. Information Processing Letters 67(6), 305–310 (1998)
17. Matsumoto, M., Nishimura, T.: Mersenne twister: A 623-dimensionally equidistributed uniform pseudo-random number generator. ACM Transactions on Modeling and Computer Simulation 8, 3–30 (1998)
18. OpenMP Architecture Review Board: OpenMP Application Program Interface, Version 2.5 (May 2005)
19. Singler, J.: The MCSTL website (June 2006), `http://algo2.iti.uni-karlsruhe.de/singler/mcstl/`
20. Dementiev, R., Kettner, L., Sanders, P.: STXXL: Standard Template Library for XXL data sets. In: Brodal, G.S., Leonardi, S. (eds.) ESA 2005. LNCS, vol. 3669, pp. 640–651. Springer, Heidelberg (2005)

Library Support for Parallel Sorting in Scientific Computations

Holger Dachsel[1], Michael Hofmann[2,*], and Gudula Rünger[2]

[1] John von Neumann Institute for Computing, Central Institute for
Applied Mathematics, Research Centre Jülich, Germany
[2] Department of Computer Science, Chemnitz University of Technology, Germany

Abstract. Sorting is an integral part of numerous algorithms and,
therefore, efficient sorting support is needed by many applications. This
paper presents a parallel sorting library providing efficient implementations of parallel sorting methods that can be easily adapted to a specific
application. A parallel implementation of the Fast Multipole Method
is used to demonstrate the configuration and the usage of the library.
We also describe a parallel sorting method which provides the ability to
adapt to the actual amount of memory available. Performance results for
a BlueGene/L supercomputer[1] are given.

1   Introduction

The task of sorting an arbitrary amount of data according to associated key values is an integral part of various algorithms and applications. As a consequence
there has been active research in the past resulting in numerous contributions in
sequential and parallel sorting [1,2]. Besides computational science in general,
efficient implementations of sorting methods are very important in parallel and
high performance computing. For instance, an integer sort is part of common
benchmarks like NAS Parallel Benchmarks or SPLASH-2. The runtimes spent
for sorting in real-world applications are diverse. For instance, in hierarchical
N-Body methods sorting may require up to 10 percent [3], whereas it is the
major part of the parallel spectral partitioner S-HARP [4]. Thus, the usage
of efficient parallel sorting methods can be essential to obtain good parallel
implementations.

While optimized and ready-to-use libraries exist for many common tasks in
computational science, an appropriate support for parallel sorting is still missing.
Only a few approaches like the POSIX routine `qsort` or the integer sort from
the Zoltan library [5] using quicksort are available. Both of them are comparison
based algorithms and, therefore, they are inappropriate for the common case
of integer or floating point number sorting. Advanced implementations of radix

---

* Supported by Deutsche Forschungsgemeinschaft (DFG).
[1] Measurements are performed on the BlueGene/L system at the John von Neumann Institute for Computing, Jülich, Germany. http://www.fz-juelich.de/zam/
ibm-bgl

sort methods are far more appropriate for that. But due to their fixed working patterns, they also lack the appropriate flexibility to become widely applicable.

This paper presents the main aspects of the configuration and the usage of a parallel sorting library. Primarily intended for applications in parallel scientific computing, this approach tries to combine two major objectives: (1) the generality of a library approach needed to become widely applicable and (2) the adaption to the actual application to obtain good efficiency. The library is written in C, widely configurable, and capable of adapting to various needs of the particular application or hardware environment. It features a separate configuration step to create new versions of the library routines, especially adapted to a certain application (e.g., the elements to be sorted). We also introduce a parallel radix sort method providing resource awareness in terms of memory usage. It is composed of various algorithms implemented in the library and capable of fulfilling the needs of our sample application. A parallel implementation of the Fast Multipole Method (FMM)[6] currently being developed at the John von Neumann Institute for Computing. Performance results are shown for sorting up to 1 billion elements on a BlueGene/L system, using random integer values as well as real data from the sample application.

The rest of this paper is organized as follows. Section 2 introduces the sample application and its need for efficient sorting support, followed by the description of the parallel sorting method in Section 3. Section 4 presents the main aspects of the library approach illustrated by the application. Section 5 presents performance results of the sorting method in a high scaling parallel environment and Section 6 concludes the paper.

2   Sorting Within an FMM Implementation

As a sample application requiring efficient parallel sorting support we consider the Fast Multipole Method, an $\mathcal{O}(n)$ hierarchical N-Body algorithm. The specific parallel implementation used is a three-dimensional FMM for calculating classical coulomb interactions. The main input data is a system of n point charges given by coordinates x_i, y_i, z_i and corresponding charge values q_i, $i = 1, \ldots, n$. The result of the computation is the energy E and the gradient G of the system as well as the potentials p_i. Apart from the energy (which is only a single scalar value), the amount of input and output data depends on the number of particles n of the system.

During the computations the system of particles is hierarchically subdivided into boxes which are enumerated according to a space filling curve scheme. Depending on the positions of the particles, each one is located in a certain box and labeled with the corresponding box number. By sorting the particles according to their box numbers, the locality of the subsequent computations is increased resulting in a more efficient processing of the input data. To preserve the initial ordering of the system, the original indices of the particles (addresses) are stored. These address values can be used to restore the initial ordering of the particles by sorting them according to their addresses. This allows for an integration of

Table 1. Input, output, and administrative data per particle

	particle data	*data types and sizes*	*bytes per particle*
input	positions	$3\times$ `double`	24
	charges	$1\times$ `double`	8
output	gradient	$3\times$ `double`	24
	potentials	$1\times$ `double`	8
administrative	box numbers	$1\times$ `integer`	8
	addresses	$1\times$ `integer`	8

the FMM implementation as a flexible subroutine in various simulations. Table 1 summarizes the data associated with each particle with the particular data types and sizes.

Regarding the number of bytes per particle, one can see that the size of the system is limited by the amount of memory available. For example, a system with about 1 billion particles occupies about 64 GB memory only for input/output data and at least 80 GB including the required administrative data. Even though it is possible to perform these computations with a serial implementation in reasonable time, a parallel shared memory system with about 128 GB main memory has to be used because of the memory requirements. To avoid a further limitation of the size of the system to be computed, the parallel sorting method should be able to work with limited memory usage.

3   A Parallel Radix Sort Algorithm

The choice of a suitable sorting algorithm is strongly influenced by the requirements of the specific application. Because the FMM is an $\mathcal{O}(n)$ algorithm, it is desirable that the sorting algorithm has time complexity $\mathcal{O}(n)$, too. Due to the big amount of data to be sorted, the algorithm should not rely on the availability of a second fully sized output buffer. Moreover, it should be able to operate *in-place* and to adapt to the actual amount of memory available. These requirements have to be met for the sequential as well as for the parallel case and limit the number of methods available. ZZ-sort [7] for example, as a true parallel in-place sorting method, has a larger time complexity than required. In general, comparison based methods are unsuitable, because they provably require $n \log n$ operations in the worst case [1]. Integer sorting in linear time can be achieved using radix sort methods. But recent parallel radix sort methods [8,9] pay no attention to limited memory usage. Due to probabilistic partitioning strategies and all-to-all communication schemes they can hardly be implemented in-place.

A parallel sorting method meeting the requirements described above is a *merge-based* parallel sorting algorithm as described by Tridgell et al. [10]. Figure 1(a) illustrates the two major steps:

1. An arbitrary sequential sorting method executed by all processes in parallel creates locally sorted sequences.
2. The sorted sequences are merged in parallel to form the globally sorted order.

For the in-place sequential sorting method a recursive *most-significant-digit-first* radix sort based on *American Flag* sort [11] is used in this paper. In every recursion step, a set of contiguous keys (starting with all keys) is sorted into bins according to a specific part of the bits of the key values. This is repeated with the keys in the single bins using the radix width r as the number of bits processed in one step. The recursion stops if the number of keys in a bin is below a certain threshold value t. The sorting is finished with an algorithm that is faster for small numbers of keys. The time complexity of this sequential sorting method results from the number of exchange operations for every key. Sorting b-bit integers results in a maximum depth of recursion of $\lceil \frac{b}{r} \rceil$. Each key is exchanged in every recursion step and at most t times according to the fast algorithm finishing the sorting. This results in $\lceil \frac{b}{r} \rceil + t$ exchange operations per key and time complexity $\mathcal{O}(n_s(\lceil \frac{b}{r} \rceil + t))$ for sorting a set of n_s keys. The constants r and t have a strong effect on the performance of the sorting method and their optimal values strongly depend on the hardware system.

The parallel merge step is comprised of several single *merge-exchange* operations with two participating processes at a time. These pairs of processes are determined using classical sorting networks like *batchers-merge-exchange* network shown in Figure 1(b) for 8 processes. The arrows represent single merge-exchange operations executed from left to right. The network consists of 6 consecutive stages denoting the maximum number of merge-exchange operations for every process. For a number of p processes, batchers-merge-exchange network consists of $\frac{1}{2}\lceil \log_2 p \rceil (\lceil \log_2 p \rceil + 1)$ consecutive stages [1].

The merge-exchange operation between two processes is shown in Figure 1(c). The exact number of keys to exchange is determined using a bisection method, followed by the exchange of the keys with point-to-point communication (e.g., in-place with `MPI_Sendrecv_replace`). This reduces the merge-exchange operation

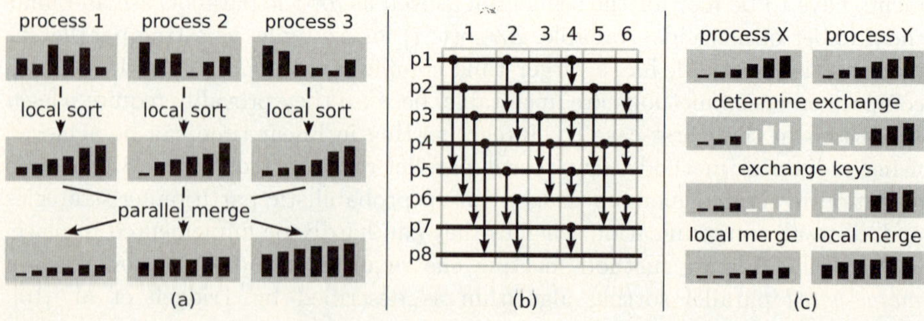

Fig. 1. (a) merge-based parallel sorting with 3 processes, (b) batchers-merge-exchange network for 8 processes, (c) one single merge-exchange operation

Table 2. Space and time complexity of algorithms implemented for the local merge

algorithm	space	time
Two-way Merge [1]	$\mathcal{O}(\min(n_0, n_1))$	$\mathcal{O}(n_0 + n_1)$
Tridgell & Brent [10]	$\mathcal{O}(\sqrt{n_0 + n_1})$	$\mathcal{O}(n_0 + n_1)$
Huang & Langston [12]	$\mathcal{O}(1)$	$\mathcal{O}(n_0 + n_1)$

to two independent local merge operations, one for each process. Table 2 lists several algorithms with varying memory requirements implemented for the local merge. Depending on the size of the subsequences (n_0 and n_1, $n_0 + n_1 = n_s$) to be merged and the amount of memory available the fastest one is chosen.

With each of these algorithms, one merge-exchange operation has time complexity $\mathcal{O}(n_s)$. In every stage of the sorting network, the processes execute their merge-exchange operations in parallel resulting in $\mathcal{O}(n_s \lceil \log_2 p \rceil^2)$ for the parallel merge step. A total number of n keys, equally distributed over p processes results in $n_s = \frac{n}{p}$ keys per process. For the overall parallel sorting method, consisting of local sorting and parallel merging, this results in time complexity $\mathcal{O}(\frac{n}{p} \lceil \log_2 p \rceil^2)$.

4   A Parallel Sorting Library Approach

The previous description shows that efficient and suitable parallel sorting is an expensive and complex task. Instead of a one-fits-all method there is a need for a variety of efficient implementations of different algorithms. For example, the parallel sorting method described in Section 3 consists of (1) a sequential sorting algorithm, (2) a sequential merge algorithm, (3) a sorting network, and (4) functions combining these parts in an appropriate way. The main purpose of the sorting library presented in this paper is to provide implementations of algorithms that can be easily used as or assembled to complete parallel sorting methods.

The sorting library supports sorting of generic elements consisting of a key component and associated data components. Each of these components can be organized in a separate array with a distinct location in memory. A list of elements is fully specified by the number of elements and the base addresses of the single component arrays. The sequential parts of the library rely on simple memory access only, while the parallel routines involve calls to MPI. A distributed list of elements is comprised of the arbitrarily sized local lists of processes participating in a parallel operation, where a global order is determined by their ranks within a given MPI communicator.

Because sorting a list of elements mainly consists of comparing and copying elements, it is necessary to perform these two operations as efficiently as possible. Since this relies on application specific properties (e.g., the type of the elements to be sorted), the sorting library features a separate configuration step before compile time to create new versions of the library especially adapted to the element type of the current application. The library functions can handle the key and data components simultaneously. Besides the instruction to copy a single key value additional instructions are inserted to copy the associated data components as well. While the library is widely configurable, this approach is far more efficient than a configuration at runtime with user-defined copy/comparison functions or many conditional statements. The code generated by the configuration step is less parametrized providing good conditions for compiler optimizations.

4.1   Configuration of the Library

The configuration of the library is made by creating header files with appropriate definitions of preprocessor symbols (macros). At least one main header file is necessary to specify the type of the elements to be sorted. Listing 1.1 shows a small example of a configuration for sorting elements consisting of the input data of the particles from Table 1 according to their associated box numbers.

Listing 1.1. Sample configuration (`input.h`)

```
1    #define SL_USE_MPI

3    /* key section */              /* box numbers */
4    #define sl_key_type_c          long
5    #define sl_key_type_mpi        MPI_LONG
6    #define sl_key_size_mpi        1
7    #define sl_key_integer

9    /* data section */
10   #define SL_DATA0               /* positions */
11   #define sl_data0_type_c        double
12   #define sl_data0_size_c        3
13   #define sl_data0_type_mpi      MPI_DOUBLE
14   #define sl_data0_size_mpi      3

16   ...
```

With `SL_USE_MPI` (line 1) the MPI based parallel parts of the library are enabled. The key component is specified with the `sl_key_...` symbols (lines 4-7) defining the appropriate C and MPI data types and sizes. The symbol `sl_key_integer` is used to signal an integer based key and enables routines like radix sort requiring bitwise operations on key values. Similar symbols are used to specify up to four associated data components that have to be rearranged in the same way as the key value they belong to. For example, the positions of the particles are located in a separate array (three consecutive `double` values per particle) and have to be rearranged together with the box numbers during the sorting. Symbol `SL_DATA0` (line 10) enables an associated data component and `sl_data0_...` (lines 11-14) are used to define the particular data types and sizes. Additional symbols can be used to adapt the comparison operation of key values or the copy operation for a certain component.

A second header file is used to define algorithm specific parameters that may depend on the configuration or the actual hardware. In this header file, performance critical parameters, like the radix width r or the threshold value t for the radix sort method from Section 3, can be adapted. The exact values can be specified manually, for example, derived from runtime observations or specific knowledge about the hardware system.

By calling a separate configuration script, a new instance of the library is created incorporating the given configuration. This new version of the library contains code that corresponds directly to the specific type of the elements to be sorted. To support multiple elements with different types in a single application, it is possible to create multiple configurations in separate header files. For each one (e.g., `input.h` and `output.h`) a separate version of the library is created. Identifiers, like library function names etc., are prefixed with the name of the configuration file (e.g., `input_...` and `output_...`) to distinguish between the different library versions.

4.2   Assembling the Parallel Sorting Method

After the configuration there exist one or several versions of the sorting library, each one especially adapted to a certain kind of elements to be sorted. The main interface for all library functions processing a list of elements is a structure called `elements`. It contains a field (`size`) holding the number of elements in the list and appropriate fields for the memory addresses of the key (`keys`) and the data components (`data0`, `data1`, ...). To distinguish between different kinds of elements, the name of the structure is prefixed too (e.g., `input_elements`, `output_elements`).

Because the FMM application from Section 2 is written in Fortran and, therefore, unable to support call-by-value or the `elements` structure, it is necessary to create appropriate wrapper functions in C. Listing 1.2 shows a sample routine implementing the parallel sorting method from Section 3.

Listing 1.2. Sample routine assembling the parallel sorting method

```
 1    #include "sl_input.h"

 3    void sort_input(long *n, long *box, double *xyz,
 4                    double *q, void *m, long *ms)
 5    {
 6      int size, rank;
 7      input_elements s, sx;

 9      MPI_Comm_size(MPI_COMM_WORLD, &size); MPI_Comm_rank(MPI_COMM_WORLD, &rank);

11      input_elements_alloc_from_block(&sx, m, *ms);

13      s.size = *n;     /* number of elements */
14      s.keys = box;    /* box numbers */
15      s.data0 = xyz;   /* positions */
16      s.data1 = q;     /* charges */

18      input_sort_radix(&s, NULL, -1, -1, -1);

20      input_mpi_mergek(&s, input_sn_batcher, NULL,
21        input_merge2_memory_adaptive, &sx, size, rank, MPI_COMM_WORLD);
22    }
```

In line 1 the automatically generated header file from the sample configuration in Section 4.1 is included. The parameters of the routine (n, box, xyz,

q) are used to initialize the number of elements as well as the key and the data components within an `input_elements` structure named `s` (lines 12-15). The merge-based parallel sorting is done by calling the local radix sort method (line 17) followed by the parallel merge step (lines 19-20) using batchers-merge-exchange network (`input_sn_batcher`). Two more parameters (`m`, `ms`) of the routine are used to allocate a second list of elements (line 10) serving as temporary buffer for the merge algorithms. Depending on the size of the subsequences to be merged and the amount of memory available, the given merge operation (`input_merge2_memory_adaptive`) selects one of the algorithms from Table 2 at runtime.

5   Performance Results

For demonstrating the efficiency of the parallel sorting algorithms implemented in the sorting library, we present performance results on a BlueGene/L system [13]. The BlueGene/L supercomputer is a massively parallel architecture with up to 65,536 dual-processor nodes and a peak performance of 360 teraflops. The system features several networks including a 3D torus interconnect for low-latency (100 ns per hop), high-bandwidth (175 MB/s per link and direction) point-to-point communication. Each node possesses two 700 MHz PowerPC based processors (one dedicated for communication) and 512 MB main memory.

Figure 2 shows the results for sorting equally distributed random integer values. On the left, individual runtimes for the local radix sort and the parallel merge (with the two-way merge algorithm) are shown using 125 million values (maximum problem size on a single node). The runtimes of both parts of the parallel sorting method decrease for an increasing number of processes. For

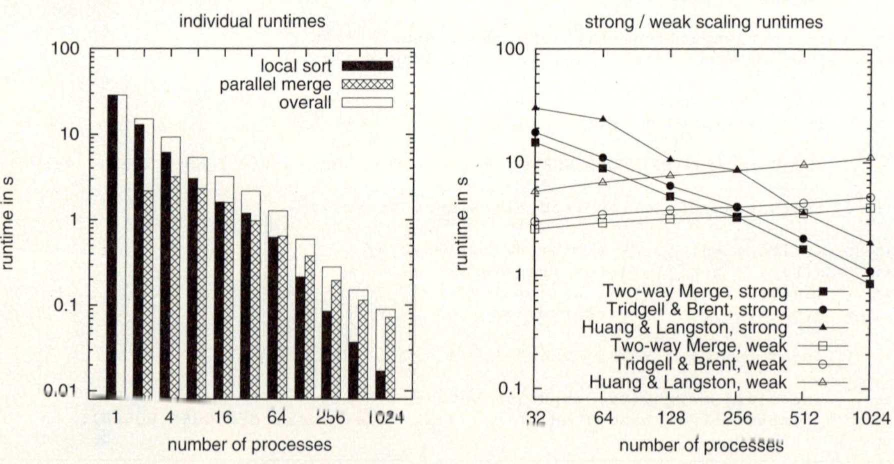

Fig. 2. Runtimes for parallel integer sorting

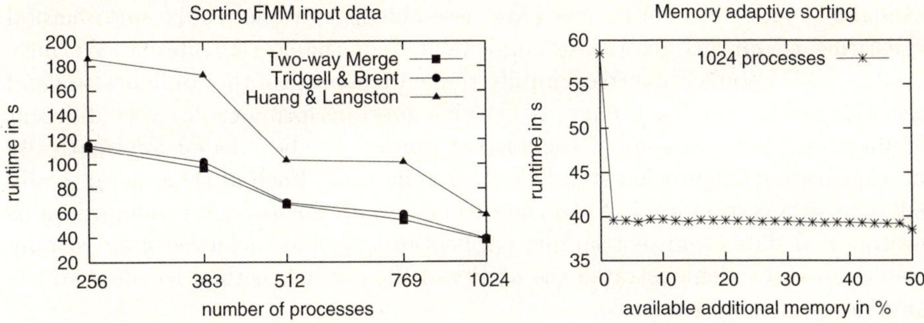

Fig. 3. Runtimes for parallel sorting within the FMM application

smaller numbers of processes the most time consuming part is the local radix sort, while for an increasing number of processes the overall runtime is more and more dominated by the parallel merge part. Scaling this problem size to 1024 processes, we obtain a parallel efficiency of about 32%.

Figure 2 (right) shows runtimes for strong scaling (constant problem size) with a total number of 2^{30} values and weak scaling (constant problem size per process) with 2^{22} values per process. In general, one can see that the parallel sorting method scales well. The strong scaling runtimes constantly decrease while for weak scaling there is a moderate increase resulting from the growing number of stages in the sorting network. Regarding the runtimes of the different merge algorithms used, a clear dependency on the memory usage exists. The fastest parallel sorting is done with the two-way merge algorithm requiring the biggest amount of memory. Using the algorithm of Tridgell et al., the runtimes slightly increase, while with the algorithm of Huang et al., the parallel sorting is at least two times slower. The same applies to the parallel efficiency when going from 32 to 1024 processes. While with the first two algorithms the efficiency remains above 50%, it falls below 40% with the algorithm requiring almost no additional memory.

Figure 3 (left) shows runtime results for sorting the input data of the FMM application, using the configuration from Listing 1.1. Runtimes are added up for 10 subsequent sorting steps within one FMM calculation with 2^{30} particles. Sorting data within this application confirms the good scaling results as well as the dependency on the memory usage. Figure 3 (right) shows runtimes for varying amounts of memory available. Only in the worst case, where almost no additional memory is available, the runtime is very high. With already 2% of additional memory (100% corresponds to a second fully sized output buffer) the runtimes drop to a lower level and another slight decrease occurs only at 50%. Thus, the memory adaptive parallel sorting method provides complete sorting in the worst case while maintaining good performance in all other cases.

6   Summary

In this paper, we have presented a parallel sorting library, which adapts sorting algorithms to the types of the elements to be sorted and the amount of memory available. With this library we have assembled a parallel radix sort method consisting of various algorithms implemented in the sorting library. We have used an FMM application to exemplify the main aspects of the configuration and the usage of the sorting library. To provide good performance for very different applications, new versions of the sorting library can be created automatically in a separate configuration step before compile time. Each of them is especially adapted to a certain kind of elements to be sorted. For sorting random data as well as real data from our sample application, we have achieved good scaling results and have demonstrated the ability of the sorting method to adapt to the amount of memory available.

References

1. Knuth, D.E.: The Art of Computer Programming, vol. III: Sorting and Searching. Addison-Wesley, Reading (1973)
2. Akl, S.G.: Parallel Sorting Algorithms. Academic Press, Inc., London (1990)
3. Hu, Y., Johnsson, S.L.: A data-parallel implementation of O(N) hierarchical N-body methods. In: Supercomputing '96: Proceedings of the 1996 ACM/IEEE conference on Supercomputing, IEEE Computer Society Press, Los Alamitos (1996)
4. Sohn, A., Simon, H.: S-HARP: a scalable parallel dynamic partitioner for adaptive mesh-based computations. In: Supercomputing '98: Proceedings of the 1998 ACM/IEEE conference on Supercomputing (CDROM), IEEE Computer Society Press, Los Alamitos (1998)
5. Devine, K., Boman, E., Heapby, R., Hendrickson, B., Vaughan, C.: Zoltan data management service for parallel dynamic applications. Computing in Science and Engineering 4(2), 90–97 (2002)
6. Greengard, L., Rokhlin, V.: A fast algorithm for particle simulations. Journal of Computational Physics 73, 325–348 (1987)
7. Zheng, S.Q., Calidas, B., Zhang, Y.: An efficient general in-place parallel sorting scheme. The Journal of Supercomputing 14(1), 5–17 (1999)
8. Jiménez-González, D., Navarro, J.J., Larriba-Pey, J.L.: Fast parallel in-memory 64-bit sorting. In: ICS '01: Proceedings of the 15th international conference on Supercomputing, pp. 114–122 (2001)
9. Lee, S.J., Jeon, M., Kim, D., Sohn, A.: Partitioned parallel radix sort. Journal of Parallel and Distributed Computing 62(4), 656–668 (2002)
10. Tridgell, A., Brent, R.P.: A general-purpose parallel sorting algorithm. International Journal of High Speed Computing (IJHSC) 7(2), 285–302 (1995)
11. McIlroy, P.M., Bostic, K., McIlroy, M.D.: Engineering radix sort. Computing Systems 6(1), 5–27 (1993)
12. Huang, B.C., Langston, M.A.: Practical in-place merging. Communications of the ACM 31(3), 348–352 (1988)
13. Adiga, N.R., et al.: An overview of the BlueGene/L supercomputer. In: Supercomputing '02: Proceedings of the 2002 ACM/IEEE conference on Supercomputing, pp. 1–22. IEEE Computer Society Press, Los Alamitos (2002)

Domain-Specific Optimization Strategy
for Skeleton Programs

Kento Emoto, Kiminori Matsuzaki, Zhenjiang Hu, and Masato Takeichi

Graduate School of Information Science and Technology
University of Tokyo
`{emoto,kmatsu}@ipl.t.u-tokyo.ac.jp,`
`{hu,takeichi}@mist.i.u-tokyo.ac.jp`

Abstract. Skeletal parallel programming enables us to develop parallel programs easily by composing ready-made components called *skeletons*. However, a simply-composed skeleton program often lacks efficiency due to overheads of intermediate data structures and communications. Many studies have focused on optimizations by fusing successive skeletons to eliminate the overheads. Existing fusion transformations, however, are too general to achieve adequate efficiency for some classes of problems. Thus, a specific fusion optimization is needed for a specific class. In this paper, we propose a strategy for domain-specific optimization of skeleton programs. In this strategy, one starts with a normal form that abstracts the programs of interest, then develops fusion rules that transform a skeleton program into the normal form, and finally makes efficient parallel implementation of the normal form. We illustrate the strategy with a case study: optimization of skeleton programs involving neighbor elements, which is often seen in scientific computations.

1   Introduction

Recently, the increasing popularity of parallel machines like PC clusters and multi-core CPUs attracts more and more users. However, development of efficient parallel programs is difficult due to synchronization, interprocessor communications, and data distribution that complicate the parallel programs. Many researchers have addressed themselves to developing methodology of parallel programming with ease. As one promising solution, skeletal parallel programming [1, 2] has been proposed.

In skeletal parallel programming users develop parallel programs by composing *skeletons*, which are abstracted basic patterns in parallel programs. Each skeleton is given as a higher order function that takes concrete computations as its parameters, and conceals low-level parallelism from users. Therefore, users can develop parallel programs with the skeletons in a similar way to developing sequential programs.

Efficiency is one of the most important topics in the research of skeletal parallel programming. Since skeleton programs are developed in a compositional style, they often have overheads of redundantly many loops and unnecessary intermediate data. To make skeleton programs efficient, not only each skeleton is implemented efficiently in parallel, but also optimizations over multiple skeletons are necessary.

There have been several studies on the optimizations over multiple skeletons based on *fusion transformations* [3, 4, 5, 6, 7], which were studied in depth in the field of

A.-M. Kermarrec, L. Bougé, and T. Priol (Eds.): Euro-Par 2007, LNCS 4641, pp. 705–714, 2007.

functional programming [8, 9]. In particular, general fusion optimizations [3, 4, 5, 6] have achieved good results both in theory and in practice. For example, Hu et al. [5] proposed a set of fusion rules based on a general form of skeletons named accumulate.

Although the general fusion optimizations so far are reasonably powerful, there is still large room for further optimizations. Due to the generality of their fusion transformations, some overheads in skeleton programs are left through the general fusion optimizations. In many cases such overheads can be removed if we provide a program-specific implementation. Thus, some specific optimizations are important for efficiency of skeleton programs.

In this paper, we study domain-specific optimizations to make skeleton programs more efficient. The target skeleton programs of these optimizations are domain-specific in the sense that the programs are built with a fixed set of skeletons and have some specific way of compositions of the skeletons. With the knowledge of the domain-specific properties, we expect to develop more efficient domain-specific programs.

The main contribution of the paper is a new strategy for domain-specific optimization of skeleton programs. The strategy proposed is as follows. First, we formalize a *normal form* that captures the domain-specific computations. Then, we develop *fusion rules* that transform a skeleton program into the normal form. Finally, we provide an efficient *parallel implementation* of the normal form.

We confirm the usability and effectiveness of the strategy with a case study of optimizing skeleton programs that involve neighbor elements, which is often seen in scientific computations. We formalize a normal form and fusion rules for the class of skeleton programs and developed a small system for fusing skeleton programs into the normal form implemented efficiently in parallel. The experiment results show effectiveness of the domain-specific optimization.

The rest of this paper is organized as follows. Section 2 explains our strategy for domain-specific optimization of skeleton programs. Section 3 gives a case study of optimization for skeleton programs involving neighbor elements. Section 4 discusses the applicability of our strategy and related work. Section 5 concludes this paper.

2   A General Strategy for Domain-Specific Optimization

In skeletal parallel programming, domain-specific programs are often developed with a fixed set of skeletons composed in a specific manner. Based on this observation, we propose the following strategy for developing domain-specific optimizations.

1. Design a normal form that abstracts target computations.
2. Develop fusion rules that transform a skeleton program into the normal form.
3. Implement the normal form efficiently in parallel.

In designing a normal form, we should have the following requirements in mind. A normal form is specified to describe any computation of target programs but should not be too general. A normal form should be specific to the target programs, and should enable us to develop efficient implementation for it. In addition, a normal form should be closed under the fusion rules to maintain the result of optimization in the form. Once we formalize a normal form with fusion rules and efficient implementation, we can perform the optimization easily: we first transform a skeleton program into the normal form with the fusion rules, and then we translate the program in the normal form to an efficient program.

We now demonstrate our strategy with a toy example. We consider programs described by compositions of the following two skeletons map and shift$_\gg$.

$$\begin{aligned}
\mathsf{map}(f, [a_1, \ldots, a_n]) &= [f(a_1), \ldots, f(a_n)] \\
\mathsf{shift}_\gg(e, [a_1, \ldots, a_n]) &= [e, a_1, \ldots, a_{n-1}]
\end{aligned}$$

Here, a list is denoted by lining elements up between '[' and ']' separated by ','. Skeleton map applies given function f to each element of the input list. Skeleton shift$_\gg$ shifts elements of the input list to the right by one, and inserts given value e as the leftmost element. The last element of the input is discarded. An instance of the target skeleton programs is shown below. Since one skeleton has one loop in its implementation, this program has four loops as well as two communication phases in two shift$_\gg$s.

$$ys = \mathsf{shift}_\gg(e_0, \mathsf{map}(f, \mathsf{shift}_\gg(e_1, \mathsf{map}(g, xs)))) \ .$$

First, we design a normal form by abstracting computation of the target programs. Each resulting list consists of two parts in terms of its generation: some left elements are computed from constants introduced by skeleton shift$_\gg$, and the other elements are computed by applying functions of map skeletons to the input list. Thus, we can define a normal form for the programs as a triple $\langle [c_1, \ldots, c_r], [f_1, \ldots, f_m], xs \rangle$: a list of constants, a list of functions, and an input list. For example, the above example program for ys can be described in the normal form as $\langle [e_0, e_1'], [g, f], xs \rangle$ where $e_1' = f(e_1)$.

Then, we define fusion rules to transform any instance of target programs into the normal form where each skeleton is fused with the normal form.

$$\begin{aligned}
xs &\Rightarrow \langle [], [], xs \rangle \\
\mathsf{map}(f, \langle [c_1, \ldots, c_r], [f_1, \ldots, f_m], xs \rangle) &\Rightarrow \langle [f(c_1), \ldots, f(c_r)], [f, f_1, \ldots, f_m], xs \rangle \\
\mathsf{shift}_\gg(e, \langle [c_1, \ldots, c_r], [f_1, \ldots, f_m], xs \rangle) &\Rightarrow \langle [e, c_1, \ldots, c_r], [f_1, \ldots, f_m], xs \rangle
\end{aligned}$$

The rule for map applies the given function f to each constant element, and inserts the function to the list of functions. The rule for shift$_\gg$ inserts the given constant e to the list of constants. It is straightforward to check that the instance above can be transformed into the normal form as shown above. It is worth remarking that any instance of the target programs can be transformed into the normal form using these rules.

Finally, we develop an efficient parallel implementation of the normal form. The programs in the normal form above can be implemented with a single loop and a single communication. For instance, the example program in the normal form is implemented as follows. Here, the input list xs is divided into blocks of the length $bsize$ (>2) and each processor has one of these blocks. Note that indices of arrays start from one.

```
if(proc != last_proc)  send_to_next_proc(xs[bsize-1], xs[bsize]);
if(proc != first_proc) recv_from_prev_proc(v0, v1);

for(i = 3; i <= bsize; i++) { ys[i] = f(g(xs[i-2])); }

if(proc == first_proc) { ys[1] = e0; ys[2] = f(e1); }
else                   { ys[1] = f(g(v0)); ys[2] = f(g(v1)); }
```

In this implementation, skeletons of the original program have been fused into one loop with one communication. As illustrated so far, the example skeleton program has been optimized with the normal form and the fusion rules.

3   A Case Study: Optimizing Skeleton Programs Involving Neighbor Elements

We demonstrate our strategy by a case study of optimizing skeleton programs that involve neighbor elements, which is often seen in scientific computations. Due to space limitation, we omit the details of formal definitions. The details can be found in the technical report [10].

3.1   Target Skeleton Programs

Our target skeleton programs involve neighbor elements using combination of $\text{shift}_\ll$, $\text{shift}_\gg$, zip and map.

$$
\begin{aligned}
Program ::= \; &\mathsf{map}(f, Program) \quad | \quad \mathsf{zip}(Program, Program) \\
| \; &\mathsf{shift}_\ll(e, Program) \; | \; \mathsf{shift}_\gg(e, Program) \\
| \; &x
\end{aligned}
$$

Here, f means a function, e means an element, and x means an input list. We introduce two skeletons zip and $\text{shift}_\ll$ as well as previously defined skeletons map and $\text{shift}_\gg$. Skeleton zip makes a list of pairs of corresponding elements in the given two lists of the same length. Skeleton $\text{shift}_\ll$ shifts elements to the left and inserts the given value as the rightmost element. The first element of the input is discarded.

$$
\begin{aligned}
\mathsf{zip}([a_1, \ldots, a_n], [b_1, \ldots, b_n]) &= [(a_1, b_1), \ldots, (a_n, b_n)] \\
\mathsf{shift}_\ll(e, [a_1, \ldots, a_n]) &= [a_2, \ldots, a_n, e]
\end{aligned}
$$

Note that any instance of $Program$ takes a list and returns a list of the same length, and that each skeleton has parallel implementation [11].

As our running example, consider a simple program for the following recurrence equation obtained by rearranging a difference equation for physical simulation. Here, u_i^n denotes a value of a field u at time n and at location i, and we consider simple boundary conditions: $u_0^n = b_L$ and $u_{N+1}^n = b_R$ for a fixed N.

$$
u_i^{n+1} = c_{-1} u_{i-1}^n + c_0 u_i^n + c_1 u_{i+1}^n
$$

A skeleton program $next$ that computes the values at the next time form the current values of u in parallel is given as follows.

$$
\begin{aligned}
next(u) = \mathbf{let} \; v_{-1}' &= \mathsf{map}(c_{-1}\times, \mathsf{shift}_\gg(b_L, u)) \\
v_0' &= \mathsf{map}(c_0\times, u) \\
v_1' &= \mathsf{map}(c_1\times, \mathsf{shift}_\ll(b_R, u)) \\
\mathbf{in} \; &\mathsf{map}(add, \mathsf{zip}(v_{-1}', \mathsf{map}(add, \mathsf{zip}(v_0', v_1'))))
\end{aligned}
$$

Here, we use intermediate variables v'_{-1}, v'_0, and v'_1 for readability, although the definition of target programs *Program* does not have variables. The correspondence of the program *next* and the above recurrence equation is as follows. First, to generate a list corresponding to the first term $c_{-1}u^n_{i-1}$, we apply shift$_\gg$ to shift the elements, and apply map to multiply the coefficient c_{-1}. Similarly, lists corresponding to the second and the third terms are generated by using shift$_\ll$ and map. Then, zipping these three lists by zip and adding elements by map *add*, we obtain the final result. Since each of the three lists are shifted by shift$_\gg$ and shift$_\ll$, the ith element of the resulting list is $c_{-1}u_{i-1} + c_0u_i + c_1u_{i+1}$.

In the rest of this paper, we explain our idea by using this example *next*.

3.2 Normal Form

The first step of our strategy is to design a normal form that can describe any computation of target programs. In this section, we give a normal form for *Program* (see Section 3.1) that involves neighbor elements by using map, zip, shift$_\gg$ and shift$_\ll$.

Generally, any computation of a target skeleton program is denoted by a triple $[\![\, ls, ce, rs \,]\!]$. Here, ls is a list of computational trees for the left edge of the resulting list, ce is a common computational tree for the center part, and rs is a list of computational trees for the right edge. Thus, we use this triple as our normal form, and denote this triple by using special brackets for readability.

For the example *next*, the leftmost element and the rightmost element of the resulting list are calculated by the following expressions l_1 and r_1. Here, for a fixed index i, $\overrightarrow{u[i]}$ and $\overleftarrow{u[i]}$ denote the ith elements of u from the left and the right respectively.

$$l_1 = add((c_{-1}\times b_L, add((c_0\times \overrightarrow{u[0]}, c_1\times \overrightarrow{u[1]})))) $$
$$r_1 = add((c_{-1}\times \overleftarrow{u[1]}, add((c_0\times \overleftarrow{u[0]}, c_1\times b_R)))) $$

Each computation involves variables ($\overrightarrow{u[0]}$ and $\overrightarrow{u[1]}$, or $\overleftarrow{u[1]}$ and $\overleftarrow{u[0]}$) and a constant (b_L or b_R) introduced by shift$_\gg$ or shift$_\ll$. On the other hand, each element in the center part is calculated by evaluating the following expression ce against each index i. Here, we omit the index i since only the difference from the index is important. Thus, u denotes the ith element, $u_{\ll 1}$ denotes the element on the left of the ith element, and $u_{\gg 1}$ denotes the element on the right.

$$ce = add((c_{-1}\times u_{\ll 1}, add((c_0\times u, c_1\times u_{\gg 1})))) $$

Summarizing these observations, we can denote the whole computation of *next* by a triple $[\![\, [l_1], ce, [r_1] \,]\!]$. Thus, we use this triple as the normal form. These computational trees have the following structures. Here, *add* is denoted by $+$.

$$
\left[\!\left[\;
\left[\begin{array}{c}
+ \\
c_{-1}\times \quad + \\
b_L \quad c_0\times \; c_1\times \\
\overrightarrow{u[0]} \quad \overrightarrow{u[1]}
\end{array}\right],
\begin{array}{c}
+ \\
c_{-1}\times \quad + \\
u_{\ll 1} \quad c_0\times \; c_1\times \\
u \quad u_{\gg 1}
\end{array},
\left[\begin{array}{c}
+ \\
c_{-1}\times \quad + \\
\overleftarrow{u[1]} \quad c_0\times \; c_1\times \\
\overleftarrow{u[0]} \quad b_R
\end{array}\right]
\;\right]\!\right]
\tag{1}
$$

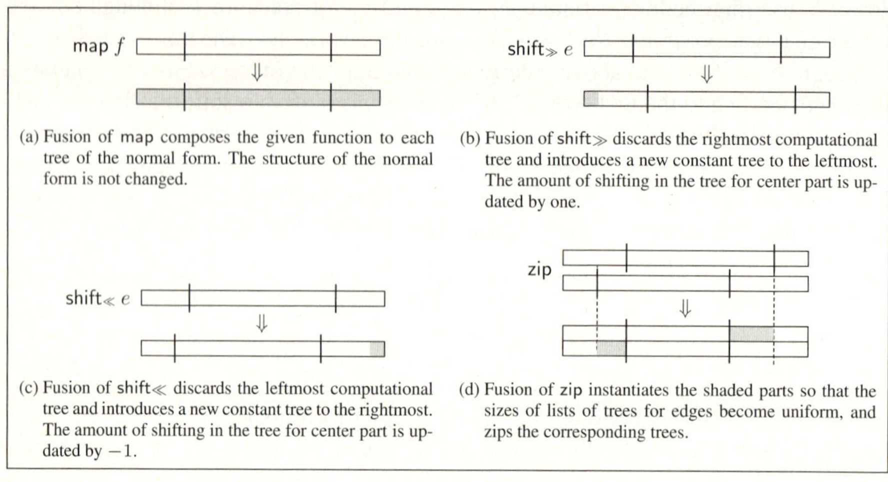

(a) Fusion of map composes the given function to each tree of the normal form. The structure of the normal form is not changed.

(b) Fusion of shift$_\gg$ discards the rightmost computational tree and introduces a new constant tree to the leftmost. The amount of shifting in the tree for center part is updated by one.

(c) Fusion of shift$_\ll$ discards the leftmost computational tree and introduces a new constant tree to the rightmost. The amount of shifting in the tree for center part is updated by -1.

(d) Fusion of zip instantiates the shaded parts so that the sizes of lists of trees for edges become uniform, and zips the corresponding trees.

Fig. 1. An image of fusion rules. Rectangles show the resulting lists. The three parts separated by vertical lines correspond to the triple of a normal form. Changed parts are shaded.

3.3 Fusion Rules for Transformation to a Normal Form

The second step of our strategy is to define fusion rules to transform a skeleton program to a normal form. These rules should be able to transform any of the target skeleton programs to a normal form. In this section, we give fusion rules to transform a skeleton program *Program* (see Section 3.1) into the normal form.

The transformation is done one by one using fusion rules. Figure 1 shows an image of the fusion rules (formal definitions can be found in the technical report [10]). Since our target program involves four kinds of skeletons, there are four fusion rules. As an explanation of these rules, we transform the example *next* into the normal form.

The base case is the transformation of the argument list u. A list u needs only the common computational tree u that is just the element of u. So,

$$u \Rightarrow [\![\, [\,], u, [\,] \,]\!].$$

Next, we fuse shift$_\gg$ to transform shift$_\gg(b_L, u)$. Since shift$_\gg$ introduces the constant b_L to the leftmost element, a new computational tree of the constant b_L is introduced to the new normal form. Also, the amount of shifting in the common tree is updated by one.

$$\text{shift}_\gg(b_L, [\![\, [\,], u, [\,] \,]\!]) \Rightarrow [\![\, [b_L], u_{\gg 1}, [\,] \,]\!]$$

Then, we fuse map to the above result to get the following normal form.

$$\text{map}\,(c_{-1}\times, [\![\, [b_L], u_{\gg 1}, [\,] \,]\!]) \Rightarrow [\![\, [c_{-1}\times b_L], c_{-1}\times u_{\gg 1}, [\,] \,]\!]$$

The constant b_L is replaced by $c_{-1} \times b_L$, and the function $c_{-1}\times$ is composed to the root of the common tree. Similarly, applications of shift$_\ll$ and map to the input list u result in these normal forms:

$$\mathsf{map}\,(c_0\times, u) \Rightarrow [\![\,[\,], c_0 \times u, [\,]\,]\!]\,,$$
$$\mathsf{map}\,(c_1\times, \mathsf{shift}_{\ll}(b_R, u)) \Rightarrow [\![\,[\,], c_1 \times u_{\ll 1}, [c_1 \times b_R]\,]\!]\,.$$

In the last transformation, a new tree is introduced by $\mathsf{shift}_{\ll}$ to the rightmost, and the amount of shifting in the common tree is updated by one to the left.

Next, we perform fusion of zip to transform the following program part.

$$\mathsf{zip}(v_0', v_1') = \mathsf{zip}([\![\,[\,], c_0 \times u, [\,]\,]\!], [\![\,[\,], c_1 \times u_{\ll 1}, [c_1 \times b_R]\,]\!])$$

Since the lengths of edge lists of two normal forms to be zipped are not the same (i.e. $[\,]$ and $[c_1 \times b_R]$ for the right edges), we have to make the lengths uniform by instantiating the common trees for center parts. The instantiation means to fix the indices in the common trees for elements on the edges. The instantiation and the zip of trees result in the following normal form. Here, the instantiation of the common tree of the first normal from $c_0 \times u$ is $c_0 \times u\overset{\leftarrow}{[0]}$, and it is zipped with the rightmost tree of the second normal form to make the new rightmost tree.

$$[\![\,[\,], (c_0 \times u, c_1 \times u_{\ll 1}), [(c_0 \times u\overset{\leftarrow}{[0]}, c_1 \times b_R)]\,]\!]$$

Similarly, we obtain the following normal form for $\mathsf{zip}\,(v_{-1}', \mathsf{map}(add, \mathsf{zip}(v_0', v_1')))$:

$$[\![[(c_{-1} \times b_L, add((c_0 \times u\overset{\rightarrow}{[0]}, c_1 \times u\overset{\rightarrow}{[1]})))]\,,$$
$$(c_{-1} \times u_{\ll 1}, add((c_0 \times u, c_1 \times u_{\gg 1})))\,, [(c_{-1} \times u\overset{\rightarrow}{[1]}, add((c_0 \times u\overset{\leftarrow}{[0]}, c_1 \times b_R)))]\,]\!]$$

Continuing these fusions, we finally obtain the normal form of the example *next*, which is shown in Eq. (1) of Section 3.2.

These four fusion rules and the base case rule can transform any skeleton program defined by *Program* into the normal form. We conclude this fact as a theorem.

Theorem 1. *Any skeleton program defined by* Program *can be transformed into the normal form by using the four fusion rules and the base case rule.*

Proof. This is proven by induction on the structure of *Program*. The base case is shown by the transformation of an input list. Induction cases are shown by the four fusion rules. A formal proof is given in the technical report [10]. □

3.4   Parallel Implementation of Normal Form

The third step of our strategy is to design parallel implementation of the normal form. In this section, we explain it briefly. The details are shown in the technical report [10].

Based on parallel implementation of existing skeletons [11], we design parallel implementation of the normal form with four steps: (1) distribution of input lists, (2) the first local computation, (3) global communication, (4) the second local computation. Figure 2 shows an image of computation of the normal form using two processors.

We will briefly explain the parallel implementation by the example *next*. First, we distribute the input list u by dividing it into two parts: $u = u_1 + u_2$. Each processor has a part of the divided list. Second, in the first local computation, processors calculate partial results in parallel. This is the main part of the whole computation. The computation

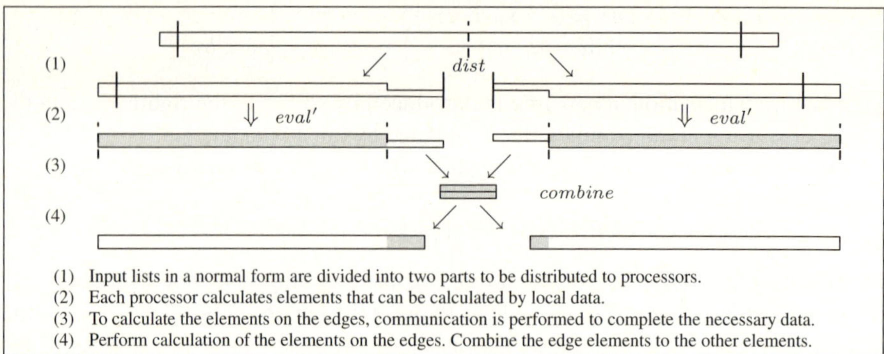

Fig. 2. An image of parallel implementation of the normal form (two processors)

is performed in a single loop in which the common computational tree of the center part is evaluated against each index. Therefore, creating no redundant intermediate data, the main part of the computation is efficient. Since elements on the edges of the distributed results need elements of both u_1 and u_2, these elements are calculated after global communication. Third, in the global communication, neighboring processors communicate incomplete trees to each other to complete trees for their edges. Of course, the first local computation can hide the time of this communication phase. Fourth, in the second local step, each processor calculates the elements on the edges with those completed trees to complete the resulting distributed list. After these four steps, these completed results are gathered to the root processor, or become a new input to another normal form. Distribution of input will be skipped in the latter case.

3.5   Experimental Result

We implemented a small domain-specific optimizer for the case study. The system reads a skeleton program written with our parallel skeleton library SkeTo [11], and generates an optimized C++ code. For the example *next*, we measured running times of a skeleton program written with SkeTo and an optimized program. We used a PC cluster where each node connected with Gigabit Ethernet has a CPU of Intel® Xeon®2.80GHz and 2GB memory, with Linux 2.4.21, GCC 4.1.1, and mpich 1.2.7.

Table 1 shows measured running times and speedups. Running time is of applying *next* 100 times to an input list of 10,000,000 elements. A speedup is a ratio of running time of a sequential program to running time of a parallel program.

The optimized program achieves ten times faster running time than the original skeleton program, and the same running time as a sequential program on one processor. This improvement was gained by elimination of redundant intermediate data and by covering communication time by the computation of center parts in a loop. Also, the optimized program achieves good speedups against the number of processors. These results show effectiveness of the proposed optimization.

Table 1. Running times and speedups of parallel programs against the number of processors

#processors		1	2	4	8	16	24	32	48	64
next	time (s)	210.25	100.84	48.12	24.41	13.31	8.86	6.52	4.70	3.50
	speedup	0.094	0.20	0.41	0.81	1.49	2.24	3.04	4.23	5.67
next_opt	time (s)	19.86	9.64	4.93	2.44	1.26	0.87	0.70	0.54	0.47
	speedup	1.00	2.06	4.03	8.14	15.79	22.76	28.26	36.73	42.22

4   Discussion and Related Work

One of the simplest fusion optimizations so far uses a general form called cataJ [6]. This cataJ has a function applied to each element of the input list, and an associative binary operator used to perform reduction on elements. Thus, cataJ can describe any computation written as composition of any number of map and at most one reduce (a skeleton to perform reduction) at the last. In this sense, cataJ is a normal form of skeleton programs of such compositions.

Hu et al. [5] proposed a general fusion optimization using a general form called accumulate and a set of fusion rules. The accumulate can describe skeleton scan, which calculates an accumulation of the input list with an associative binary operator, as well as map and reduce. So, it can be a normal form of skeleton programs described with compositions of these skeletons. Although accumulate can describe also shift$_\gg$ and shift$_\ll$, it causes some overheads due to lack of consideration of elements on edges. Main overheads are as follows: (1) extensions of elements for uniform manipulation by the associative binary operator, and (2) logarithmic steps of interprocessor communications for general implementation of accumulation. Thus, we need to consider a specific fusion optimization, i.e. a normal form, fusion rules and efficient implementation.

The normal form of the case study extends these fusions optimizations with consideration of elements on edges introduced by shift$_\gg$ and shift$_\ll$. The normal form separates computation of edges from that of center part, so that it does not introduce the overheads accumulate causes. The normal form, instead, cannot deal with scan.

As extension of the normal form, we can add scan to the domain of target programs. The extended normal form has an associative binary operator as well as the triple of the normal form of the case study. Using this normal form, we can successfully optimize, for example, a skeleton program for solving tridiagonal matrix equations. The details are shown in [10].

5   Conclusion

In this paper, we proposed a general strategy for domain-specific optimization of skeleton programs, and showed a case study for programs involving neighbor elements. Our strategy consists of the following three: (1) a normal form that abstracts computation of target skeleton programs, (2) a set of fusion rules to transform a skeleton program into the normal form, and (3) efficient parallel implementation of the normal form. The optimization is performed by transforming skeleton programs into normal forms with efficient implementation. A small system has been implemented and experiment results show effectiveness of proposed optimization. It is our future work to develop support tools for easy development of various domain-specific optimizations.

Acknowledgment

We are grateful to the referees for their detailed and helpful comments. This work was partially supported by Japan Society for the Promotion of Science, Grant-in-Aid for Scientific Research (B) 17300005 , and the Ministry of Education, Culture, Sports, Science and Technology, Grant-in-Aid for Young Scientists (B) 18700021.

References

1. Cole, M.: Algorithmic Skeletons: Structural Management of Parallel Computation. In: Research Monographs in Parallel and Distributed Computing, MIT Press, Cambridge (1989)
2. Rabhi, F.A., Gorlatch, S. (eds.): Patterns and Skeletons for Parallel and Distributed Computing. Springer, Heidelberg (2002)
3. Gorlatch, S., Wedler, C., Lengauer, C.: Optimization rules for programming with collective operations. In: Procedeeings of 13th International Parallel Processing Symposium / 10th Symposium on Parallel and Distributed Processing (IPPS / SPDP '99), San Juan, Puerto Rico, April 12–16, 1999, IEEE Computer Society, Los Alamitos (1999)
4. Wedler, C., Lengauer, C.: On linear list recursion in parallel. Acta Informatica 35(10), Springer (1998)
5. Hu, Z., Iwasaki, H., Takeichi, M.: An accumulative parallel skeleton for all. In: Le Métayer, D. (ed.) ESOP 2002 and ETAPS 2002. LNCS, vol. 2305, Springer, Heidelberg (2002)
6. Matsuzaki, K., Kakehi, K., Iwasaki, H., Hu, Z., Akashi, Y.: A fusion-embedded skeleton library. In: Danelutto, M., Vanneschi, M., Laforenza, D. (eds.) Euro-Par 2004. LNCS, vol. 3149, Springer, Heidelberg (2004)
7. Grelck, C., Scholz, S.B.: Merging compositions of array skeletons in SaC. Parallel Computing 32(7-8) (2006)
8. Wadler, P.: Deforestation: Transforming programs to eliminate trees. In: Ganzinger, H. (ed.) ESOP 1988. LNCS, vol. 300, Springer, Heidelberg (1988)
9. Gill, A.J., Launchbury, J., Jones, S.L.P.: A short cut to deforestation. In: FPCA '93 Conference on Functional Programming Languages and Computer Architecture, Copenhagen, Denmark, June 9–11, 1993, ACM Press, New York (1993)
10. Emoto, K., Matsuzaki, K., Hu, Z., Takeichi, M.: Domain-specific optimization for skeleton programs involving neighbor elements. Technical Report METR2007-05, Department of Mathematical Informatics, University of Tokyo (2007)
11. Matsuzaki, K., Iwasaki, H., Emoto, K., Hu, Z.: A library of constructive skeletons for sequential style of parallel programming. In: InfoScale '06: Proceedings of the 1st international conference on Scalable information systems. ACM International Conference Proceeding Series, vol. 152, ACM Press, New York (2006)

Topic 10
Parallel Numerical Algorithms

Ian Duff, Michel Daydé, Matthias Bollhoefer, and Anne Trefethen

Topic Chairs

Efficient and robust parallel and distributed algorithms with portable and easy-to-use implementations for the solution of fundamental problems in numerical mathematics are essential components of most parallel software systems for scientific and engineering applications.

This topic provides a forum for the presentation and discussion of new developments in the area of parallel and distributed numerical methods. All aspects of the design and implementation of parallel algorithms will be addressed, ranging from discussion of the ideas on which they are based to analyses of their complexities and performance on current parallel and distributed architectures (including clusters and grids). Software design and prototyping in scientific computing or simulation of software environments are also covered.

Methods for the solution of large linear systems are of particular interest because of their widespread occurrence in many fields, particularly in the numerical solution of partial differential equations. However, contributions dealing with new and improved parallel and distributed algorithms for the solution of other problems in numerical linear algebra, linear and non-linear programming, numerical quadrature, differential equations, fast transforms and non-linear systems were also welcome.

Thirteen papers papers were submitted to the track and five were accepted. Three papers consider the parallel implementation of application-related software:

- In "An Efficient Parallel Particle Tracker for Advection-Diffusion Simulations in Heterogeneous Porous Media", the authors consider the contamination of groundwater by migration of pollutants.
- In "A Fully Scalable Parallel Algorithm for Solving Elliptic Partial Differential Equations", the authors compare the probabilistic domain decomposition (DD) method with deterministic methods for solving linear elliptic boundary-value problems.
- In "Locality Optimized Shared-Memory Implementations of Iterated Runge-Kutta Methods", the authors consider the sequential and parallel implementation of Iterated Runge-Kutta methods focusing on the optimization of the locality behaviour. They introduce different implementation variants for sequential and shared-memory computer systems and analyse their performance.

The two remaining papers focus on implementation issues (task scheduling, memory hierarchy, multi-core processor architectures,...) for high performance linear algebra computational kernels:

A.-M. Kermarrec, L. Bougé, and T. Priol (Eds.): Euro-Par 2007, LNCS 4641, pp. 715–716, 2007.
© Springer-Verlag Berlin Heidelberg 2007

- In "Toward Scalable Matrix Multiply on Multithreaded Architectures", the authors analyse the impact of SMP and future multicore architectures on efficient software design.
- In "Task Scheduling for Parallel Multifrontal Methods", the authors present a new scheduling algorithm for task graphs arising from parallel multifrontal methods for sparse linear systems. duling algorithm for task graphs arising from parallel multifrontal methods for sparse linear systems.

An Efficient Parallel Particle Tracker for Advection-Diffusion Simulations in Heterogeneous Porous Media

Anthony Beaudoin[1], Jean-Raynald de Dreuzy[2], and Jocelyne Erhel[3]

[1] University of Le Havre, France
`anthony.beaudoin@univ-lehavre.fr`
[2] Geosciences Rennes, UMR CNRS 6118, France
`Jean-Raynald.de-Dreuzy@univ-rennes1.fr`
[3] Inria Rennes, France
`Jocelyne.Erhel@irisa.fr`
`http://hydrolab.univ-rennes1.fr/`

Abstract. The heterogeneity of natural geological formations has a major impact in the contamination of groundwater by migration of pollutants. In order to get an asymptotic behavior of the solute dispersion, numerical simultations require large scale computations. We have developed a fully parallel software, where the transport model is an original parallel particke tracker. Our performance results on a distributed memory parallel architecture show the efficiency of our algorithm, for the whole range of geological parameters studied.

Keywords: Heterogeneous porous media, advection, diffusion, particle tracker, cluster, distributed memory, speed-up.

1   Introduction

Numerical modeling is an important key for the management and remediation of groundwater resources. Several field experiments have showed that the natural geological formations are highly heterogeneous, leading to preferential flow paths and stagnant regions. The contaminant migration is strongly influenced by these irregular velocity distributions.

In order to account for the limited knowledge of the geological characteristics and for the natural heterogeneity, stochastic approaches have been developed [1]. Here we study the case of a lognormal permeability field with an exponential correlation function. The objective is to compute the dispersion coefficients as functions of time and to find out the asymptotic values of the dispersion coefficients [2]. We use numerical models to compute the velocity field over large spatial domains and to simulate solute transport over large temporal scales [1]. This approach must overcome two main difficulties, memory size and runtime, in order to solve very large linear systems and to simulate over a large number of timesteps. High performance computing is thus necessary to carry out these large scale simulations [1].

A.-M. Kermarrec, L. Bougé, and T. Priol (Eds.): Euro-Par 2007, LNCS 4641, pp. 717–726, 2007.

We have developed a fully parallel software, which is object-oriented, with a modular approach and well-defined interfaces [3]. The permeability field is generated by means of parallel Fast Fourier Transforms and the velocity field is computed by solving a large sparse linear system, using a parallel multigrid solver of the numerical library Hypre [4]. In this paper, we describe a parallel algorithm for advection-diffusion modeling and its performances on clusters of processors. Transport is simulated by a particle tracker algorithm [5], well suited for pure advection and advection-dominated transport processes, because it does not introduce spurious numerical diffusions. Our parallel algorithm takes advantage of a subdomain decomposition and of the non interaction between the particles. This hybrid parallel strategy is quite original, since most of previous work uses either one or the other [6],[7],[8]. We perform many numerical experiments, in order to study the effects on parallel performances of the hydraulic parameters, mainly the size of the computational domain, the heterogeneity of the permeability field and the ratio of advection to diffusion. We also analyze scalability issues by varying the number of processors used. Our results show that our parallel particle tracker is quite robust and efficient in the whole range of interesting values. Using a cluster with 32 processors, we could run simulations with spatial and temporal scales never achieved so far. This allowed us to determine with no ambiguity the asymptotic dispersion coefficients [9].

2   Numerical Model for Advection-Diffusion

2.1   Physical Model

In this paper, we consider a 2D problem where the porous medium defined by a rectangle with dimensions L_x and L_y. As the mean flow direction is the x axis, L_x and L_y are respectively longitudinal and transversal to the mean flow direction. The porous medium, assumed isotropic, is characterized by a random hydraulic conductivity field K, which follows a stationary log-normal probability distribution $Y = ln(K)$, defined by a mean m and a covariance function C given by

$$C(r) = \sigma^2 \exp(-\frac{|r|}{\lambda}),$$

where σ^2 is the variance of the log hydraulic conductivity, $|r|$ represents the separation distance between two points and λ denotes the correlation length scale. The length λ is typically in the range $[0.1m, 100m]$ and the variance σ^2 is in the interval $[0, 7]$. These two ranges encompass most of the generally studied values.

Classical laws governing the steady flow in a porous medium are mass conservation and Darcy law

$$\epsilon V = -K \nabla h, \nabla . V = 0$$

where ϵ is the porosity, V is the Darcy velocity and h is the hydraulic head. Boundary conditions are homogeneous Neumann on upper and lower sides and

Dirichlet $h = 0$ on left side, Dirichlet $h = 1$ on right side. The system should not be too much elongated in order to avoid border effects. We denote by U the mean velocity norm.

An inert solute is injected in the porous medium and transported by advection and dispersion. This type of solute migration is described by the advection - dispersion equation

$$\frac{\partial(\epsilon c)}{\partial t} + \nabla.(\epsilon c V) - \nabla.(\epsilon D \nabla c) = 0$$

where c is the solute concentration and D is the dynamic dispersion tensor. Here, we assume a constant porosity $\epsilon = 1$ and we consider only molecular diffusion, assumed homogeneous and isotropic, with a tensor

$$D = D_m I$$

where D_m is the molecular diffusion coefficient and I denotes the unity tensor.

Boundary conditions and an initial condition complete the transport equation. The initial condition at $t = 0$ is the injection of the solute. In order to overcome the border effects, the inert solute is injected at a given distance of the left side. The ratio of diffusion to advection is measured by the Peclet number defined by

$$Pe = \frac{\lambda U}{D_m}$$

with a typical range of $[100, \infty]$, where ∞ means pure advection. This range covers advection-dominated transport models, whereas diffusion-dominated models require smaller values.

2.2 Numerical Flow

The flow equations are discretized on a regular grid using a classical Finite Volume scheme. Let Δ_x and Δ_y the mesh sizes in each direction. The number of cells in each direction is given by $n_x = L_x/\Delta_x$ and $n_y = L_y/\Delta_y$, with a total number $N = n_x n_y$. The linear system $Ax = b$, obtained from the discretization of flow equations, is characterized by a sparse symmetric positive definite matrix A of order N, with a pentadiagonal structure. The solution x is the discrete hydraulic head and the right-hand side b represents Dirichlet boundary conditions. We solve this sparse linear system by an algebraic multigrid method, with the procedure Boomer-AMG of the Hypre library [4] and we get very good parallel performances.

2.3 Numerical Transport: Particle Tracker

The inert solute transport is simulated by a particle tracker algorithm whose key advantages for this study are the absence of numerical diffusion and the good performances. In this Lagrangian framework, the solution of transport equation consists in evaluating the trajectory of particles [10],[2]. These paths are represented by the Stochastic Differential Equation

$$dX = V dt + \sqrt{2D_m} dW$$

where X is the position of the particle and dW is a Brownian motion process.

Using a first-order explicit scheme for the discretization of transport equation, between a time t and a time $t + \Delta t$, a particle moves at each timestep according to the equation

$$X(t + \Delta t) = X(t) + V \Delta t + \sqrt{2D_m} Z w \sqrt{\Delta t}$$

where Z is a random number drawn from a Gaussian distribution of mean 0 and variance 1 and w is a unitary vector with uniformly distributed orientation. The timestep Δt evolves along the particle path according to the velocity magnitude of the crossed cells. More precisely, the timestep is either proportional to the diffusion time necessary to cross the cell or to the local advection time, which is equal to the cell size divided by the minimum of the velocities computed on the cell borders in both directions. Within each cell, particle velocities V are obtained by a bilinear interpolation of the boundary velocities, because it is the sole interpolation method that respects mass conservation. No-flux boundaries are maintained by bouncing particles off of the boundary. At the free outflow boundary (the right side), particles are allowed to naturally flow out of the domain. The algorithm finishes when all the particles have quit.

2.4   Macro-dispersion Analysis

The main objective of our study is to evaluate the temporal macro-dispersions $D_x(t)$ and $D_y(t)$ in both directions and their asymptotic value when t becomes infinite [1]. Let $(x_m(t), y_m(t))$ the coordinates of the computed trajectory of a particle and let N_p the number of particles. As long as the particles are in the domain, the total mass is constant. The center of mass of the solute distribution in the longitudinal direction is approximated by the first moment

$$\overline{x}(t) = \frac{1}{N_p} \sum_{m=1}^{N_p} x_m(t)$$

and the spread of mass around $\overline{x}$ is approximated by the second moment

$$S_x(t) = \frac{1}{N_p} \sum_{m=1}^{N_p} x_m(t)^2 - \overline{x}(t)^2.$$

Then the temporal longitudinal macro-dispersion is defined by

$$D_x(t) = \frac{dS_r}{dt}(t),$$

with time derivatives approximated by a first-order Taylor expansion. The transversal macro-dipsersion is defined in a similar way.

Asymptotic behavior can be obtained only at very large times. In order to keep all the particles in the domain during long simulated times, the dimensions L_x and L_y must be very large. Therefore, high performance computing is required to deal with these computationally intensive simulations. Also, memory requirements are very high and data must be spread from the beginning in a distributed memory architecture.

3   Parallel Particle Tracker

We have designed a parallel algorithm for parallel distributed memory architectures. We use a SPMD model of programmation, where all processors execute the same program (with conditional statements using the processor number). Communications are based on the message-passing interface provided by the MPI library. In the particle tracker currently used, the particles do not interact so each trajectory can be implemented in a parallel independent process. This is a classical efficient approach, see for example [6]. However, our objective is to run simulations on very large computational domains so we need to distribute data in order to use all memory resources. Therefore, we choose to use a domain decomposition approach, where the velocity field is distributed among processors. Actually, we use the domain decomposition and data distribution arising from permeability generation and flow computation [3]. This methodology is also classical, see for example [7]. Each subdomain is assigned to one processor, which computes the trajectory of the particles inside the subdomain owned by it. Communications with processors owning neighboring subdomains must occur when particles exit or enter the subdomain. This is achieved by means of SEND and RECV operations. We choose non blocking SEND operations so that each processor can first proceed with sending then with receiving. The algorithm must ensure a correct ending, when all particles have left the computational domain [8]. It must also guarantee that the sending buffers are empty when new SEND operations are executed. For these reasons, we choose to add a global synchronisation point after the exchanges of particles. This point allows defining a global state where each processor knows the total number of particles still in the domain. By the way, it guarantees that all RECV operations have been successful and that the algorithm may execute new SEND operations. Thus neither deadlock nor starvation may occur and the algorithm terminates safely.

A difficulty is to balance the computational load between the processors. A specificity of our application is the initial localization of the particles. Indeed, at their initial stage, all particles are on the left part of the domain. If the master processors owning the subdomains with the initial particles launched all particles, then all other processors would be idle, waiting for particles entering their domain. For example, if we divide the domain into slices, most of the processors have two neighbors, on the left and on the right. If the master processor launched all particles together, parallelism would not be efficient at all since only one processor would be active while others would be idle.

Therefore, we have designed a parallel algorithm well-suited for our particular case. We take advantage of the independence of the particles and launch them by bunches. The idea is to initiate a pipeline of particles from the left to the right. At each global synchronisation point, the master processors inject a new bunch of particles, until all particles have been injected into the domain. All processors track the particles in their subdomain, then exchange particles on their boundaries, finally synchronize globally to define the total number of particles in activity. Let us consider as an example the case of a constant velocity field and a transport by pure advection, with a slice decomposition. Then after initiating the pipeline, each processor owns a bunch of particles between two synchronisation points, until draining the pipeline. Thus computations are well-balanced. In the general case, it is not so easy to predict the computational load. Two main factors will impact the parallel algorithm: the heterogeneity and the diffusion. Heterogeneity is measured by the variance of the permeability field, whereas diffusion is measured by the Peclet number.

4 Numerical Results

4.1 Numerical Experiments

We have developed a fully parallel object-oriented software which provides a generic platform to run Monte-Carlo numerical simulations of flow and transport in highly heterogeneous porous media. The software is divided into three main modules, respectively dedicated to the generation of the permeability field, the computation of flow and velocity, the computation of transport and dispersion coeffcents. This modularity allows a great flexibility and portability. The software is written in C++ and can be implemented on machines with Unix, Linux or Windows systems. Graphical functions are written with OpenGL and parallel programming relies on the MPI library. The software integrates open-source components such as sparse linear solvers for flow computation.

In this paper, we show the results of various experiments with the parallel particle tracker. All tests are performed on a SUN cluster composed of two nodes of 32 computers each. Each computer is a 2.2 Ghz AMD Opteron bi-processor with 2 Go of RAM. Inside each node, computers are interconnected by a Gigabit Ethernet Network Interface, and the two nodes are interconnected by a Gigabit Ethernet switch (CISCO 3750). This cluster is a component of the Grid'5000 computing resource installed at INRIA in Rennes. Experiments are done with a variable number of processors P.

The different parameters and their values are summarized below:

$$\begin{cases} \Delta x = \Delta y = 1, \lambda = 10, N_p = 1000, \\ L_x = L_y \in \{512, 1024, 4096, 8192\}, \\ \sigma \in \{0.5, 1, 2, 2.5, 3\}, \\ Pe \in \{10, 100, 1000, \infty\}, \\ P \in \{1, 2, 4, 8, 16, 32, 64\}. \end{cases}$$

We have done other experiments with different values of λ and the choice $\lambda = 10$ is a good tradeoff between the accuracy of the numerical model and the

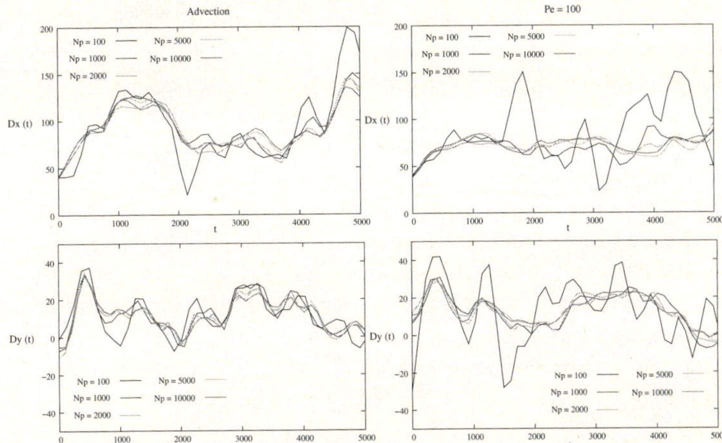

Fig. 1. Convergence analysis: longitudinal and transversal dispersions for varying number of particles N_p. Pure advection $Pe = \infty$ (left) and advection-diffusion $Pe = 100$ (right); heterogeneous case $\sigma = 2.5$. Large domain size $L_x = L_y = 8192$.

computational cost. We analyse and comment below with more details the choice of the number of particles N_p. The values $\sigma = 3$ and $Pe = 10$ are out of scope in geological surveys but we include them in order to check the behavior of our algorithms with extreme values of the parameters. All other values have a strong interest for geological study. The dimension $L_x = 8192$ of the domain is the target value, in order to get with no ambiguity the asymptotic values of the longitudinal and transversal dispersion coefficients. Other dimensions are useful to analyze the complexity and the scalability of our parallel particle tracker.

4.2   Convergence Analysis

Clearly, the CPU time of transport computations is proportional to the number of particles N_p so that this number must be kept low, still ensuring a good convergence of the dispersion coefficients. Therefore, we study the convergence of the dispersion coefficients with the number of particles, in order to choose an optimal number. In Figure 1, we plot the temporal dispersions $D_x(t)$ and $D_y(t)$ for various values of N_p. We find out that the value $N_p = 1000$ is a good tradeoff between convergence and performance. We show only the cases $Pe = \infty$ and $Pe = 100$ with $\sigma = 2.5$ and with $L_x = L_y = 8192$, but other results are similar.

4.3   Performance Analysis

In Figure 2, we plot the CPU time versus the number of processors for two Peclet numbers $Pe = \infty, Pe = 100$ and for two domain sizes $L_x = 1024, L_x = 4096$. In the pure advection case, we can observe good performances for any value of the heterogeneity parameter σ. The CPU time increases slightly with σ and

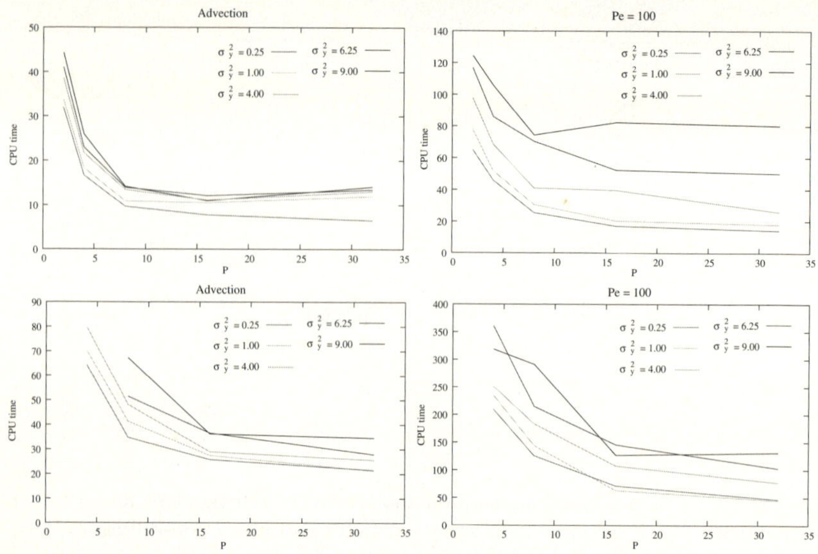

Fig. 2. Performance analysis: CPU time versus the number of processors P. Pure advection $Pe = \infty$ (left) and advection-diffusion $Pe = 100$ (right); small size $L_x = L_y = 1024$ (top) and large size $L_x = L_y = 4096$ (bottom).

performances slightly deteriorate for large σ. From the small to large domain size, the performances improve clearly. The CPU time is significantly reduced from 8 to 16 processors with the large size $L_x = 4096$. In the advection-diffusion case with $Pe = 100$, the trends are similar. However, CPU times are much higher, because the particle tracker includes a random diffusion motion, with generation of random numbers at each timestep. Moreover, the simulated time is larger because the particles stay longer in the domain before reaching the exit at the right side. Performances are not so good with large values of the heterogeneity and the small domain size. A possible reason is some imbalance of work in the parallel algorithm, inducing idle time for some processors. This effect almost disappears for the large domain size and the CPU time decreases from 16 to 32 processors for any kind of heterogeneity.

In order to study scalability issues, in Figure 3, we plot the speed-up versus the number of processors. Because large domain sizes cannot fit in the memory of 2 processors, we compute the speed-up from $P = 4$ with a speed-up arbitrarily equal to 4 for $P = 4$. We conclude that our parallel particle tracker is quite efficient and is reasonably scalable.

4.4   Results for Large Scale Simulations

The dispersion coefficients and their asymptotic values are computed on a large domain size, in order to avoid border effects and to get large simulated times. All the simulations are now done with this large size $L_x = L_y = 8192$ and with

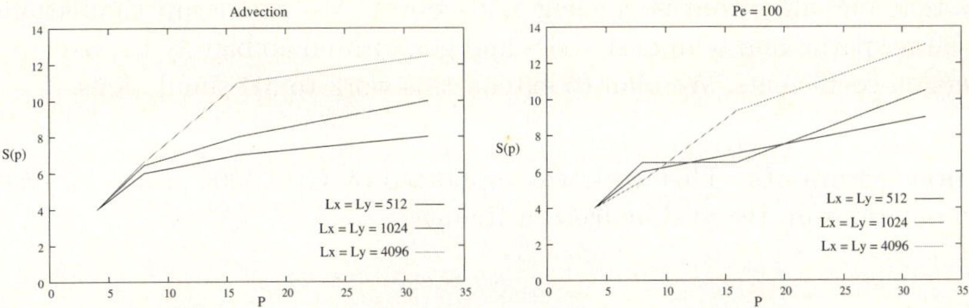

Fig. 3. Scalability analysis: speed-up versus the number of processors P. Pure advection $Pe = \infty$ (left) and advection-diffusion $Pe = 100$ (right); heterogeneous case $\sigma = 2$.

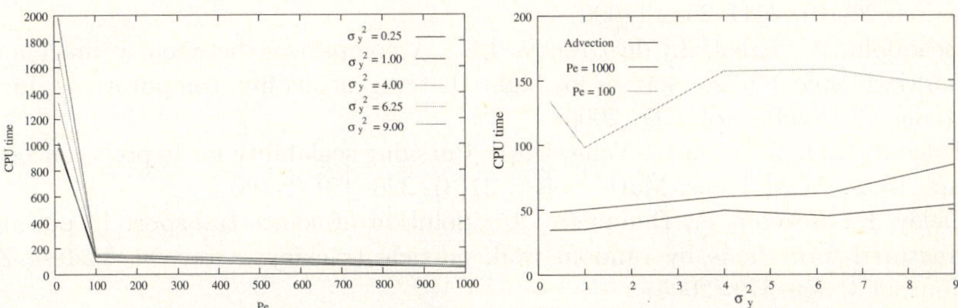

Fig. 4. Impact of diffusion (left) and of heterogeneity (right). Dimension $L_x = L_y = 8192$. Number of processors $P = 32$.

$P = 32$ processors. Figure 4, left, plots the CPU time versus the Peclet number for the different kinds of the heterogeneity. Down to $Pe = 100$, CPU times are not too high. Clearly, the CPU time blows up for the small value $Pe = 10$, indicating an exponential-like behavior. Thus, another numerical model should probably be used for diffusion-dominated transport equations. Indeed, a method such as a Finite Volume method could be relevant since numerical diffusion would not be such an issue. In the range of interest of advection-dominated equations, the particle tracker remains a good choice, efficient for both cost and accuracy. Figure 4, right, plots the CPU time versus the heterogeneity coefficient σ for the different Peclet numbers. The CPU time increases slightly but the largest time is less than twice the smallest time. Therefore, our particle tracker is very well-suited and efficient for all values of interest for the heterogeneity.

5 Conclusion

We have developed a fully parallel software for simulating flow and solute transport in highly heterogeneous porous media. In this paper, we describe an original parallel particle tracker, which relies on both the independence of particles and a subdomain decomposition. The performances for a whole range of parameters show that our algorithm is efficient and robust. We could run simulations at very large spatial and temporal scales and get with no ambiguity the asymptotic dispersion coefficients. We plan to extend this work to 3D simulations.

Acknowledgments. This work was supported by Grid'5000 grants for executing simulations on the grid at Irisa in Rennes.

References

1. Gelhar, L.: Stochastic Subsurface Hydrology, Engelwood Cliffs, New Jersey (1993)
2. Tompson, A., Gelhar, L.: Numerical simulation of solute transport in three-dimensional, randomly heterogeneous porous media. Water Resources Research 26(10), 2541–2562 (1990)
3. Beaudoin, A., Erhel, J., de Dreuzy, J.R.: A comparison between a direct and a multigrid sparse linear solvers for highly heterogeneous flux computations. In: Eccomas CFD 2006, vol. CD (2006)
4. Falgout, R.D., Jones, J.E., Yang, U.M.: Pursuing scalability for hypre's conceptual interfaces. ACM Trans. Math. Softw. 31(3), 326–350 (2005)
5. Delay, F., Ackerer, P., Danquigny, C.: Solution of solute transport in porous or fractured formations by random walk particle tracking: a review. Vadose Zone Journal 4, 360–379 (2005)
6. Huber, E., Spivakovskaya, D., Lin, H., Heemink, A.: The parallel implementation of forward-reverse estimator. In: Wesseling, P., Onate, E., Periaux, J. (eds.) ECCOMAS CFD, TU Delft, The Netherlands (2006)
7. Cheng, J.R., Plassmann, P.: The accuracy and performance of parallel in-element particle tracking methods. In: Proceedings of the Tenth SIAM Conference on Parallel Processing for Scientific Computing, Portsmouth, VA, pp. 252–261 (2001)
8. Kaludercic, B.: Parallelisation of the lagrangian model in a mixed eulerian-lagrangian cfd algorithm. Journal of Parallel and Distributed Computing 64, 277–284 (2004)
9. de Dreuzy, J.R., Beaudoin, A., Erhel, J.: Asymptotic dispersion in 2D heterogeneous porous media determined by parallel numerical simulations. In: INRIA (submitted to Water Resource Research, in revision) (preprint 2007)
10. LaBolle, E., Quastel, J., Fogg, G., Gravner, J.: Diffusion processes in composite porous media and their numerical integration by random walks: generalized stochastic differential equations with discontinuous coefficients. Water Resources Research 36(3), 651–662 (2000)

A Fully Scalable Parallel Algorithm for Solving Elliptic Partial Differential Equations

Juan A. Acebrón[1] and Renato Spigler[2]

[1] Departament d'Enginyeria Informàtica i Matemàtiques,
Universitat Rovira i Virgili, 43007 Tarragona, Spain
`juan.acebron@urv.cat`
[2] Dipartimento di Matematica, Università "Roma Tre", 1, Largo S.L. Murialdo,
00146 Rome, Italy
`spigler@mat.uniroma3.it`

Abstract. A comparison is made between the probabilistic domain decomposition (DD) method and a certain deterministic DD method for solving linear elliptic boundary-value problems. Since in the deterministic approach the CPU time is affected by intercommunications among the processors, it turns out that the probabilistic method performs better, especially when the number of subdomains (hence, of processors) is increased. This fact is clearly illustrated by some examples. The probabilistic DD algorithm has been implemented in an MPI environment, in order to exploit distributed computer architectures. Scalability and fault-tolerance of the probabilistic DD algorithm are emphasized.

1 Introduction and Generalities

Domain decomposition is considered as one of the most natural ways to decouple boundary-value (BV) problems for partial differential equations (PDEs) into subproblems, in order to take advantage of parallel computer architectures, thus allowing for high-performance scientific computing of large-scale problems. The seminal idea of DD can be traced back to the 1870 work of H. A. Schwarz, and consists of splitting the given domain into a number of subdomains, then assigning the task of the numerical solution on each separate subdomain to as many separate processors. A major problem, however, is represented by the need of having the solution on the interfaces, internal to the domain, which divide it into subdomains, while solving BV problems for PDEs is global in character. This means that the solution cannot be obtained even at a single point inside the domain before solving the entire problem. Consequently, some iterative procedures are required, across the chosen (or prescribed) interfaces, in order to determine approximate values of the sought solution inside the original domain. Both, overlapping (Schwarz-type) and not overlapping domains have been considered in the literature so far [13]. In both cases, some additional numerical work is required in advance to decouple the problem into subproblems, and it is unclear whether algorithms based on such strategies might be *scalable* as the number of the subdomains (hence, of the processors) increases unboundedly.

A.-M. Kermarrec, L. Bougé, and T. Priol (Eds.): Euro-Par 2007, LNCS 4641, pp. 727–736, 2007.

In the classical (deterministic) DD method, an important issue is represented by the condition number inherent to the iterative algorithms adopted to precompute interfacial values, see [13]. In fact, the operator governing such an iterative procedure is typically ill-conditioned. Preconditioning is therefore essential, and it is necessary to construct optimal and efficient preconditioners. Optimality means that their spectral condition number (i.e., the ratio between maximum and minimum eigenvalue) is bounded uniformly with respect to the mesh size, h, and to the average diameter, say H, of the subdomains. It is known from the literature that in Schwarz-type domain decomposition methods, that is those with overlapping, the aforementioned condition numbers are actually independent of both, h and H, provided that the overlapping is sufficiently wide. Optimal bounds were indeed found for this case, see [5,7].

A method of completely new type, based on a probabilistically induced domain decomposition (PDD), has been recently proposed which avoids the several problems inherent to the aforementioned traditional approach to DD [1,2,4]. Moreover, the method seems to be specially suited for heterogeneous computing, due to its low communication overhead, and since it is fault-tolerant.

Nowadays, it seems that, using parallel computers with up to a few hundreds of processors in problems with up to few millions of variables, the total computational cost is dominated by that spent by the local solvers. However, machines working in the petaflops regime and endowed with hundreds of thousands or even millions of processors are planned for the near future, and taking full advantage from massive parallel computing would be highly desirable. Indeed, the IBM Blue Gene is expected to break the petaflops barrier within the 2007. With such machines, the issue of *scalability* remains open, at least in some cases. As it was pointed out in [11], Schwarz-type DD methods are not truly scalable, at least in the theoretical sense, since their parallel efficiency in solving elliptic problems is subject to degradation as the number of processors, p, goes to infinity, indeed when p starts being over the thousands. It seems however that things go a little better, in practice, for a number of reasons, as described in [11].

Besides, the possibility of failure of even few processors (even of only one!) is very likely to occur frequently [9]. Therefore, algorithms which are scalable and *fault-tolerant* at the same time would (and will) be extremely important, if not mandatory.

Briefly, the idea of the PDD algorithm consists of generating first the values of the solution at few points inside the domain. Then, an interpolation on such nodes is constructed, and finally the traces of the solution on the interfaces are obtained. This allows to fully decompose the domain into a number of subdomains, and at the same time this procedure seems to be free of the previous drawbacks. In fact, decoupling is complete and no conditioning problem does exist concerning interfacial iteration problems. In [1,2], it was shown that scalability is attained with respect to an arbitrary number of subdomains or processors, and the algorithm is naturally fault-tolerant. The latter property rests on two ingredients, one due to the intrinsic parallelizability of the Monte Carlo methods, the other to the full decoupling into subdomains that can be realized. As for the scalability, it was shown in [1] that the speedup S_p achieved with p processors, assumed to act independently from each other on p subdomains, scales as $S_p \sim c\sqrt{p}$ as $p \to \infty$, c being a constant. The ideal (theoretical) speedup, $S_p \sim cp$, cannot be attained since some (nonnegligeable) time should be spent to compute the values of the solution by Monte Carlo at few points. Here the speedup is defined as the ratio T_1/T_p, where T_1 is the time spent to solve sequentially the given problem on the full domain, while T_p is the time spent for solving the same problem in parallel with p processors.

In the paper [4] the essentials of the PDD method developed in [1,2] have been described, and some additional examples have been presented. In all such papers, however, the code implementing the PDD algorithm was written in OpenMP environment. The numerical results worked out in this paper are based, instead, on a code written in MPI environment. This has been done in order to fully exploit distributed computer architectures. The main purpose of this paper is to compare the performance of the PDD and of the deterministic DD algorithms implemented in MPI environment.

2   The Algorithm

We confine our discussion to the case of the Dirichlet problem for a linear elliptic equation in two dimensions, i.e.,

$$Lu - c(x,y)u = f(x,y), \quad (x,y) \in \Omega \subset \mathbf{R}^2, \qquad u|_{\partial\Omega} = g(x,y), \tag{1}$$

where $L := \sum_{i,j=1}^{2} a_{ij}(x,y)\partial_i\partial_j + \sum_{i=1}^{2} b_i(x,y)\partial_i$ is a linear elliptic operator with smooth coefficients, $c(x,y) \geq 0$, the boundary $\partial\Omega$ of the domain Ω also smooth, as well as the boundary data, g, and the source term, f. The basic idea is to generate only few values of solution, u, by a probabilistic method, all being based on the Monte Carlo method [6].

In some sense, this approach allows to obtain the solution at any point internal to Ω without solving the entire problem in advance. This can be realized by means of the the probabilistic representation of the solution,

$$u(x,y) = E_{x,y}^{L}\left[g(\beta(\tau_{\partial\Omega}))e^{-\int_0^{\tau_{\partial\Omega}} c(\beta(s))\, ds} - \int_0^{\tau_{\partial\Omega}} f(\beta(t))\, e^{-\int_0^{t} c(\beta(s))\, ds}\, dt \right] \tag{2}$$

[8], where $\beta(t)$ is the two-dimensional stochastic process associated to the elliptic operator L, which solves the system of two Ito-type stochastic differential equations (SDEs)

$$d\beta = b(x,y)dt + \sigma(x,y)dW(t). \tag{3}$$

Here $W(t)$ represents the two-dimensional standard Brownian motion (also called Wiener process), and $\tau_{\partial\Omega}$ is the first passage (or hitting) time of the path $\beta(t)$ started at the point (x,y) to $\partial\Omega$. When the operator L is the Laplace operator, Δ, the stochastic process $\beta(t)$ reduces to the standard two dimensional Brownian motion. The drift vector, $b = b(x,y)$ in (3), is the same appearing in

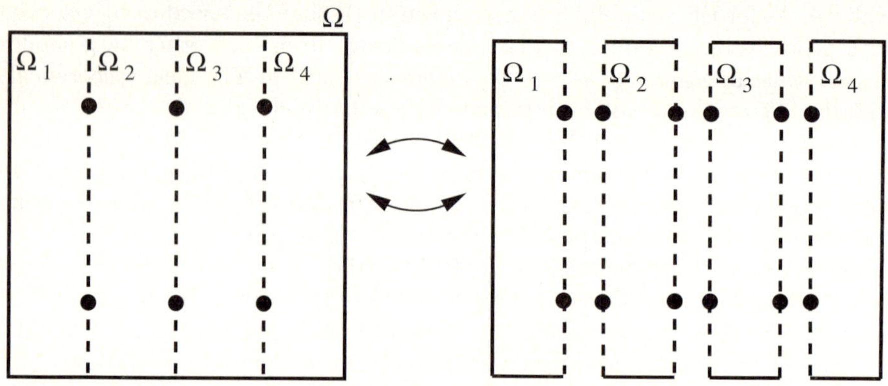

Fig. 1. Sketchy diagram which shows how the PDD algorithm works

the operator L, that is $b = (b_1, b_2)^T$, while the diffusion matrix, $\sigma = \sigma(x, y)$, is related to the coefficients a_{ij} by the relation $\sigma\sigma^T = a \equiv (a_{i,j})_{i,j=1,2}$.

We use the representation formula in (2) to obtain few values of the solution, at some points inside the domain Ω. The expected value is then approximated by an arithmetic mean (known to provide its best estimator) over N realizations of the process β, at the price of a (statistical) error of order of $N^{-1/2}$. The main catch of using a Monte Carlo method rests on this fact, which entails a rather poor accuracy, unless N is taken extremely large. Even for N not very large, the computational cost is high, but in the PDD algorithm we limit its use to very few points. The points where the solution is computed by Monte Carlo simulations are then used as nodes for interpolation. Such interpolation allows to obtain boundary values of solution on certain interfaces internal to the domain, and hence to fully decouple the original problem into subproblems, see Fig. 1. We always use Chebyshev interpolation.

The PDD algorithm can be briefly described as follows:

1. Compute only *few* interfacial values by Monte Carlo simulations.
2. Interpolate on the corresponding nodes to obtain boundary values for the subdomains.
3. Compute the solution to the original problem in each subdomain by standard methods (e.g., finite differences or finite elements).

The Monte Carlo approach is trivially parallelizable, since each of the N problems above can be runned independently of the others. Stated in such terms, clearly, the method appears to be scalable as well. Moreover, it is also naturally fault-tolerant. In fact, assuming that a fraction, n, of the N processors being used fails, meaningful results will be obtained from the remaining $N - n$ processors, and it will suffice just to ignore what could have come from the n processors which have failed. The final error will be slightly worse, of the order of $\mathcal{O}((N - n)^{-1/2})$. Note that

$$(N - n)^{-1/2} \approx N^{-1/2} \left(1 + \frac{1}{2}\frac{n}{N} \right), \quad \text{for } n \ll N, \tag{4}$$

that is $1.005\, N^{-1/2}$ for a small fraction $n/N = 1/100$ (1%) of failing processors.

Even though the probabilistic algorithm we derived and considered here is scalable, naturally fault tolerant, and well suited to grid and heterogeneous computing, it suffers for some weakness, due to the inherently poor accuracy of all Monte Carlo methods. A considerable improvement, however, can be achieved using sequences of "quasi-random numbers" [6,12] instead of sequences of pseudorandom numbers. The pseudorandom numbers are those numbers obtained in practice, when we try to generate truly random numbers, hence are approximately characterized by a statistical distribution. The quasi-random numbers, instead, are deterministic uniformly distributed numbers. Using the latter allows to obtain an error (now deterministic) of order of $\mathcal{O}(N^{-1}\log^{d^*-1} N)$, d^* representing a certain "effective" space dimension. It was shown in [3] that the underlying system of SDEs can indeed be solved numerically in a very efficient way. However, a reordering strategy is required at each step to break somehow the inherent correlations of the sequence of numbers. This makes it difficult to parallelize the algorithm, and raises some doubts on whether using quasi-random numbers can be useful in practice. But if this would be possible, the failure of $n \ll N$ processors would imply the slightly larger error

$$(N - n)^{-1} \log(N - n) = N^{-1} \left(1 - \frac{n}{N} \right)^{-1} \left[\log N + \log \left(1 - \frac{n}{N} \right) \right]$$
$$\approx (1 + \alpha)N^{-1} \left[\log N - \alpha \right]. \tag{5}$$

Here we assumed $d^* = 2$ and set $\alpha := n/N$. With $\alpha = 0.01$, and considering that N will be at least of the order of the thousands, hence $\log N \gg \alpha$ (e. g., $\log N = 3$ against $\alpha = 0.01$), we obtain

$$(N - n)^{-1} \log(N - n) \approx (1 + \alpha)N^{-1} \log N = 1.01\, N^{-1} \log N. \tag{6}$$

Moreover, due to the full uncoupling among the various subdomains, the PDD algorithm exhibits fault-tolerance also when those processors working on small fractions of subdomains fail. In this case, one can fully neglect (in the first instance) the corresponding output, without the need of resorting to restart procedures.

The various sources of numerical error which affect the evaluation of $u(x,y)$ by means of (2) include, besides the finite sample size mentioned above, (i) the truncation error made in the numerical solution of the SDEs in (3), not to mention the round-off errors, (ii) the uncertainty in estimating first exit times (hitting times), and (iii) the numerical quadrature errors in (2). The latter is missing whenever the potential term, $c(x,y)$, and the source, $f(x,y)$, in (1) are identically zero. In particular, estimating accurately first exit times and first exit points (which are also needed when $c(x,y)$ and $f(x,y)$ do not vanish identically), has been often overlooked in the literature. An efficient way to locate accurately first exit times is adopting exponential timesteppings, see [10]. This choice allows in fact to use an explicit analytic form of the hitting probability. All these sources of error have been analyzed in [1,2], and we shall not do it here again.

3   Numerical Examples

Some numerical examples are provided here to compare the performance of the PDD algorithm described in §1 with that obtained by means of certain deterministic DD algorithms. Despite the fact that "pivotal" values generated by the Monte Carlo method are poorly accurate and the Chebyshev interpolation adds some further error, the total error inside of each subdomain is estimated by the boundary errors. This is due to the maximum principle, and the error decays rapidly going inside.

In [1,2,4], a comparison was made only with "parallel finite differences", hence a comparison with a true deterministic DD method was not made there. Moreover, the code was implemented in OpenMP, and runned in a shared memory computer architecture. In addition to the fact that the PDD method outperforms the deterministic DD method, the former wins regarding to the issues of scalability and fault-tolerance. These properties, not easily achieved by the DD deterministic methods, should instead be considered at the present time if one wants to exploit the forthcoming massively parallel supercomputers, equipped with hundreds of thousands or even millions of processors. The numerical examples presented below concern, for the purpose of illustration, partial differential equations on rather elementary domains, indeed, the unit square. Monte Carlo methods, however, are expected to be useful even on domains having highly complex geometries. Here we only stress that our probabilistic DD approach can

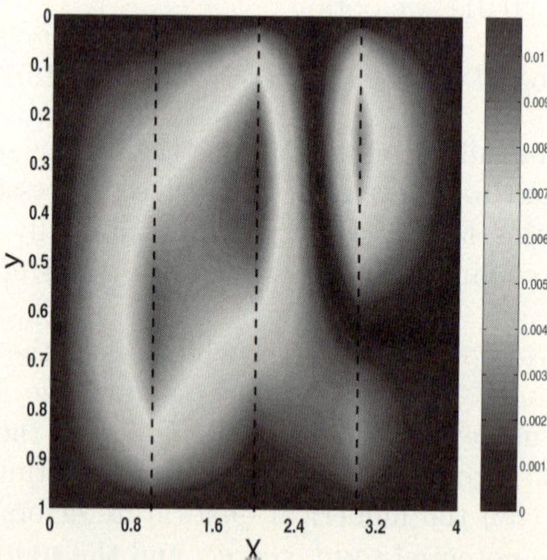

Fig. 2. Example 1. Pointwise numerical error in the PDD algorithm for $p = 4$. Parameters are $N = 10^5$, $\Delta x = \Delta y = 1.25 \times 10^{-3}$, $\lambda = 10^2$.

be easily applied, for instance, to polygonal domains with an arbitrary number of sides. In fact, one of the most delicate points, which contributes significantly to the overall numerical error, is given by the need of evaluating accurately the first exit times and points, see [1,2]. While such a task may be challenging when the boundaries are very irregular, it can be handled easily when the boundary is piecewise linear. We do not enter into details in this paper.

Another related issue is that of having one or more rather general interfaces, inside the domain. Their shape may be prescribed, due to geometrical or physical reasons. The idea is to "approximate" every (sufficiently smooth) interface by a piecewise linear one, i.e., by a polygonal line, evaluating numerically on it few values of the sought solution, to be used for interpolation. This can be accomplished computing the solution by Monte Carlo methods at some more points on such a line. These points should include the endpoints of each segment of the polygonal, say the $k-2$ internal points for a polygonal line of k segments, and perhaps another point on each segment. The overall cost may therefore increase, and load balancing become a little harder.

Finally, we stress that, due to the full decoupling that the PDD method realizes, arbitrary and possible different algorithms can be implemented to solve the problem on each subdomain. In particular, serial solvers can be adopted, taking into account that one will face smaller-size problems on each subdomain. However, the point in serious applications is not that to cope with small-size problems, but, rather, to be able to solve problems that are much bigger.

A code in an MPI environment has been implemented and runned on the MareNostrum supercomputer located at the Barcelona Supercomputing Center (BSC). Below, in order to compare the relative performance of the PDD method against some deterministic method, we solved the same examples in both ways. The chosen deterministic algorithm is extracted from the numerical package pARMS, having chosen the overlapping Schwarz method with a FGMRES iterative method preconditioned with ILUT as local solver, see [14]. We chose the package pARMS for its wide and reliable usage, though better codes may exist for a comparison. The values of the parameters were chosen as follows: For the outer iterations, the Krylov subspace dimension was 20 and the tolerance 10^{-5}, while for the ILUT preconditioner the dropping threshold was 10^{-3} and the amount of fill in any row 20. The inner iterations were set equal to zero. To make the comparison meaningful, we discretized the local problems within the PDD algorithm by finite differences, and solved the ensuing linear algebraic system by the same FGMRES iterative solver preconditioned with ILUT. Here are our numerical examples.

Example 1. Consider the following Dirichlet problem,

$$u_{xx} + u_{yy} = 0 \qquad \text{in } \Omega := (0, p) \times (0, 1), \tag{7}$$

$$u(x,y)|_{\partial\Omega} = g(x,y), \tag{8}$$

where $g(x,y) := \left[\left(x^2 - y^2 \right) / p^2 \right]_{\partial\Omega}$, for the Laplace equation in two dimensions. The solution of such a problem is explicitly known, and is $u(x,y) = (x^2 - y^2)/p^2$

in $\overline{\Omega} = [0,p] \times [0,1]$. Note that the domain was scaled along the x dimension proportionally to the number of processors, p, involved. This has been done in order to keep constant the computational load per processor, being the space discretization fixed to $\Delta x = \Delta y = 1.25 \times 10^{-3}$.

In Fig. 2, the pointwise numerical error made in the PDD method is depicted in a contourplot. Here only two nodes on each interface have been used. Note that the maximum error made in each subdomain is indeed attained on the corresponding boundary. The exponential timestepping used to solve the underlying SDEs (3) was characterized by $\lambda := \langle \Delta t \rangle^{-1}$, Δt being the random exponentially distributed time step used in solving the SDEs, and the bracket denoting its average. Note that maximum error is of order 10^{-2}, which corresponds mainly to the statistical error obtained from the Monte Carlo method at the nodal points. Increasing the accuracy can be attained by increasing the sample size, or resorting to sequences of "quasi-random" numbers keeping fixed the sample size, see [2].

Table 1. CPU time in seconds

Processors	DD	PDD
4	110.49	84.05
8	132.23	84.12
16	163.81	90.84
32	192.22	90.54
64	335.20	88.89
128	14184.35	102.22
256	NA	132.15
512	NA	129.35
1024	NA	148.07

Table 1 shows the CPU time spent in solving the present problem by the two algorithms. Clearly, the PDD outperforms the DD method for any number of processors. This fact becomes more pronounced when the number of processors increases. This behavior can be explained by the high intercommunication overhead existing among processors, which affects strongly the deterministic algorithm. Note that the CPU time for the PDD method remains bounded when the number of processors grows. Testing scalability for larger number of processors (starting for $p = 256$) have been accomplished only for the PDD, because CPU time for DD increases unboundedly. For this reason, in table 1 CPU times are Not Available (NA).

Example 2. The Dirichlet problem

$$u_{xx} + (6x^2 + 1)u_{yy} = 0 \qquad \text{in } \Omega = (0,p) \times (0,1), \tag{9}$$

with the boundary data

$$u(x,y)|_{\partial \Omega} = \left[(x^4 + x^2 - y^2)/2(p^4 + p^2) \right]_{\partial \Omega}, \tag{10}$$

Table 2. CPU time in seconds

$Processors$	DD	PDD
4	115.94	69.38
8	110.80	70.06
16	113.41	70.61
32	133.09	70.12
64	255.07	72.19
128	25827.83	75.73
256	NA	67.08
512	NA	65.67
1024	NA	64.12

has the solution $u(x,y) = (x^4 + x^2 - y^2)/2(p^4 + p^2)$. Again, the CPU times are reported in Table 2. The same comments made in the previous example can be repeated here.

4   Conclusion

A probabilistic method, to accomplish domain decomposition for the numerical solution of linear elliptic boundary-value problems in two dimensions, has been described. The solution is generated by Monte Carlo simulations to solve the associated stochastic differential equations only at very few points inside the domain. A Chebyshev interpolation using such points as nodes is then constructed, and a full splitting into several subdomains, to be handled by separate processors acting concurrently, is made.

A comparison with a deterministic DD algorithm has been made here for the first time. This has been done in an MPI environment. Working in an MPI environment also allows to test the effect of processor intercommunications which beset all deterministic DD algorithms. Besides the competitive results observed in the numerical examples, the PDD method is expected to be competitive concerning scalability and fault-tolerance. These are key issues if one intends to run codes on machines working with hundreds of thousands of processors or more. Finally, we believe the PDD method could be applied, and likely more advantageously, in three or more dimensions. In practice, some more work is needed, especially concerning the important issue of accurately compute first exit times of the trajectories of the underlying stochastic processes from high-dimensional domains.

Acknowledgements

This work was completed during a visit of J.A.A. at the Department of Mathematics, University "Roma Tre", and was supported, in part, by the GNFM of the Italian INdAM. J.A.A. also acknowledges support from the Ministerio de Ciencia y Tecnología (MEC) through the Ramón y Cajal programme. The authors thankfully acknowledge the computer resources, technical expertise and assistance provided by the Barcelona Supercomputing Center-Centro Nacional de Supercomputación.

References

1. Acebrón, J.A., Busico, M.P., Lanucara, P., Spigler, R.: Domain decomposition solution of elliptic boundary-value problems via Monte Carlo and quasi-Monte Carlo methods. SIAM J. Sci. Comput. 27, 440–457 (2005)
2. Acebrón, J.A., Busico, M.P., Lanucara, P., Spigler, R.: Probabilistically induced domain decomposition methods for elliptic boundary-value problems. J. Comput. Phys. 210, 421–438 (2005)
3. Acebrón, J.A., Spigler, R.: Fast simulations of stochastic dynamical systems. J. Comput. Phys. 208, 106–115 (2005)
4. Acebrón, J.A., Spigler, R.: A new probabilistic approach to the domain decomposition method. In: Gerbier, A. (ed.) Mes premieres constructions de programmes. LNCS, vol. 55, pp. 475–480. Springer, Heidelberg (2007)
5. Brenner, S.C.: Lower bounds of two-level additive Schwarz preconditioners with small overlap. SIAM J. Numer. Anal. 21, 1657–1669 (2000)
6. Caflisch, R.E.: Monte Carlo and quasi Monte Carlo methods. In: Acta Numerica, 1–49 (1998)
7. Dryja, M., Widlund, O.B.: Domain decomposition algorithms with small overlap. SIAM J. Sci. Comput. 15, 604–620 (1994)
8. Freidlin, M.: Functional integration and partial differential equations. Annals of Mathematics Studies no. 109, Princeton Univ. Press (1985)
9. Geist, G.A.: Progress towards Petascale Virtual Machines. In: Dongarra, J.J., Laforenza, D., Orlando, S. (eds.) Recent Advances in Parallel Virtual Machine and Message Passing Interface. LNCS, vol. 2840, pp. 10–14. Springer, Heidelberg (2003)
10. Jansons, K.M., Lythe, G.D.: Exponential timestepping with boundary test for stochastic differential equations. SIAM J. Sci. Comput. 24, 1809–1822 (2003)
11. Keyes, D.E.: How scalable is domain decomposition in practice? In: Eleventh International Conference on Domain Decomposition Methods, London, 1998, pp. 286–297 (1998), (DDM.org, Augsburg 1999)
12. Niederreiter, H.: Random number generation and quasi Monte-Carlo methods. SIAM (1992)
13. Quarteroni, A., Valli, A.: Domain decomposition methods for partial differential equations. Oxford Science Publications, Clarendon Press (1999)
14. Li, Z., Saad, Y., Sosonkina, M.: pARMS: a parallel version of the algebraic recursive multilevel solver. Numerical Linear Algebra with Applications 10, 485–509 (2003)

Locality Optimized Shared-Memory Implementations of Iterated Runge-Kutta Methods

Matthias Korch and Thomas Rauber

University of Bayreuth, Department of Computer Science
{matthias.korch,rauber}@uni-bayreuth.de

Abstract. Iterated Runge-Kutta (IRK) methods are a class of explicit solution methods for initial value problems of ordinary differential equations (ODEs) which possess a considerable potential for parallelism across the method and the ODE system. In this paper, we consider the sequential and parallel implementation of IRK methods with the main focus on the optimization of the locality behavior. We introduce different implementation variants for sequential and shared-memory computer systems and analyze their runtime and cache performance on two modern supercomputer systems.

1 Introduction

Owing to their large computational requirements, parallel solution methods for initial value problems (IVPs) of ordinary differential equations (ODEs), defined by

$$\mathbf{y}'(t) = \mathbf{f}(t, \mathbf{y}(t)), \quad \mathbf{y}(t_0) = \mathbf{y}_0, \qquad \mathbf{y} : \mathbb{R} \to \mathbb{R}^n, \quad \mathbf{f} : \mathbb{R} \times \mathbb{R}^n \to \mathbb{R}^n, \quad (1)$$

have been addressed by many authors during the last decades. Among the methods considered are extrapolation methods [1], waveform relaxation techniques [2], and iterated Runge-Kutta (IRK) methods [3,4]. An overview can be found in [2]. Most of these approaches are based on the development of new numerical algorithms with a larger potential for a parallel execution, but with different numerical properties than the classical Runge-Kutta (RK) methods.

In this paper, we consider the parallel and sequential implementation of IRK methods, paying particular attention to the locality of memory references. Based on the classical implicit RK methods

$$\mathbf{Y}_l = \mathbf{y}_\kappa + h_\kappa \sum_{i=1}^{s} a_{li} \mathbf{F}_i, \quad l = 1, \ldots, s; \quad \mathbf{y}_{\kappa+1} = \mathbf{y}_\kappa + h_\kappa \sum_{l=1}^{s} b_l \mathbf{F}_l, \quad (2)$$

with the coefficient matrix $A = (a_{ij}) \in \mathbb{R}^{s,s}$, the weight vector $\mathbf{b} = (b_i) \in \mathbb{R}^s$, the node vector $\mathbf{c} = (c_i) \in \mathbb{R}^s$, and $\mathbf{F}_i = \mathbf{f}(t_\kappa + c_i h_\kappa, \mathbf{Y}_i)$, explicit IRK methods suitable for non-stiff equations introduce the iteration process

$$\mathbf{Y}_l^{(k)} = \mathbf{y}_\kappa + h_\kappa \sum_{i=1}^{s} a_{li} \mathbf{F}_i^{(k-1)}, \qquad l = 1, \ldots, s, \quad k = 1, \ldots, m. \quad (3)$$

A.-M. Kermarrec, L. Bougé, and T. Priol (Eds.): Euro-Par 2007, LNCS 4641, pp. 737–747, 2007.

We choose the 'trivial' predictor

$$\mathbf{Y}_l^{(0)} = \mathbf{y}_\kappa, \qquad l = 1, \ldots, s, \tag{4}$$

to start the iteration process and execute a fixed number of $m = p - 1$ corrector steps (3), where p is the order of the underlying implicit RK method (cf. [5]). Two approximations to $\mathbf{y}(t_{\kappa+1})$ of different order, $\mathbf{y}_{\kappa+1}$ and $\hat{\mathbf{y}}_{\kappa+1}$, are then computed by

$$\mathbf{y}_{\kappa+1} = \mathbf{y}_\kappa + h_\kappa \sum_{l=1}^{s} b_l \mathbf{F}_l^{(m)} \quad \text{and} \quad \hat{\mathbf{y}}_{\kappa+1} = \mathbf{y}_\kappa + h_\kappa \sum_{l=1}^{s} b_l \mathbf{F}_l^{(m-1)}. \tag{5}$$

In comparison to classical RK methods, IRK methods possess a larger potential for parallelism. While, in general, classical explicit RK methods are only suitable for a data-parallel execution exploiting parallelism across the ODE system, IRK methods exhibit an additional degree of method parallelism by enabling an independent computation of the s argument vectors $\mathbf{Y}_l^{(k)}$, $k = 1, \ldots, s$, within each corrector step.

On modern computer systems, which possess deep memory hierarchies, performance and scalability of applications often strongly depend on their locality of memory references. Moreover, large shared-memory systems are often built of physically distributed memory modules, which results in non-uniform memory access times. On such so-called NUMA architectures a good locality behavior is vitally important. Optimizations to increase the locality of memory references have been applied by other authors to many methods from numerical linear algebra (e.g., [6]), but not to ODE solvers. Many popular scientific libraries like LAPACK [7], PHiPAC [8] and ATLAS [9] pay regard to locality and thus can obtain a high efficiency. Modern optimizing compilers [10] try to reorder the instructions of the source program, e.g., of computationally intensive loops [11], to achieve an efficient exploitation of the memory hierarchy. Loop tiling [12] is considered to be one of the most successful techniques. An important analytical model for the effects of loop transformations is described in [13]. Locality optimizations for embedded Runge-Kutta (ERK) methods have been investigated in [14]. [15,16] extend this work by considering parallel implementations and by introducing a pipelining computation scheme which exploits the special access structure of a large class of ODE systems. An investigation of the practical performance of parallel IRK methods for distributed-memory architectures has been presented in [17].

In this paper, we consider parallel IRK implementations for shared-memory architectures. We focus on the optimization of the locality behavior, incorporating our experiences with ERK methods. We present several implementation variants with different loop structures resulting in different memory access patterns. Our focus lies on data-parallel implementations, because they can achieve a higher locality than task-parallel implementations. Moreover, purely task-parallel implementations are only scalable to at most s processors and are therefore not suitable for an execution on large parallel computer systems.

2   Access Distance of a Right-Hand-Side Function

While, in general, the evaluation of one component of the right hand-side-function $\mathbf{f}$, $f_j(t, \mathbf{y})$, may access all components of the argument vector $\mathbf{y}$, many ODE problems only require the use of a limited number of components. In many cases, the components accessed are located nearby the index j. To measure this property of a function $\mathbf{f}$, we make use of the following definitions:

Definition 1. *The access distance of a component function* $f_j(t, \mathbf{y})$, *denoted as* $d(f_j)$, *is the smallest value* b, *such that* f_j *accesses only the subset* $\{y_{j-b}, \ldots, y_{j+b}\}$ *of the components of the argument vector* $\mathbf{y}$.

Definition 2. *The access distance of a vector function* $\mathbf{f}$, *denoted as* $d(\mathbf{f})$, *is the largest access distance of all component functions* f_j *of* $\mathbf{f}$, *i.e.,* $d(\mathbf{f}) = \max_{1 \leq j \leq n} d(f_j)$.

Definition 3. *We say the access distance of* $\mathbf{f}$ *is limited if* $d(\mathbf{f}) \ll n$.

Remark. A function $\mathbf{f}$ with limited access distance $d(\mathbf{f})$ has a banded Jacobian $\frac{\partial \mathbf{f}(t, \mathbf{y})}{\partial \mathbf{y}}$ with bandwidth $2d(\mathbf{f}) + 1$.

3   Equations with Arbitrary Access Structures

The implementation of an IRK method given by Eqs. (3), (4) and (5) leads to a program structure consisting of nested loops. The most time consuming part is the iteration of the corrector steps. Therefore, we will focus our discussion on the loop structure realizing Eq. (3) within one time step. The iteration space of Eq. (3) consists of 4 dimensions:

- $k = 1, \ldots, m$:   Iteration over the corrector steps.
- $l = 1, \ldots, s$:   Iteration over the vectors $\mathbf{Y}_l^{(k)}$.
- $i = 1, \ldots, s$:   Iteration over the summands of $\sum_{i=1}^{s} a_{li} \mathbf{F}_i^{(k-1)}$.
- $j = 1, \ldots, n$:   Iteration over the system dimension, i.e., the elements of a vector of dimension n.

Since, in general, each corrector step k depends on the previous corrector step $k - 1$, the corrector steps must be computed sequentially, and the k-loop must be the outermost loop. But all other loops are independent. Hence, we are free to choose the loop structure inside the k-loop as it is desirable to obtain a high locality. That means we can interchange, merge and split the l-, i- and j-loops, and loop tiling is also possible. One particular consequence of this is that a computation scheme similar to the pipelining technique presented in [15,16], which uses the j-loop as an outer loop surrounding the l- and the i-loops, can be realized for general problems with arbitrary access structures. Moreover, since the iterations of the l-loop are independent, a pipelining of the stages resulting in a diagonal iteration is not necessary.

The realization of the loop structure affects and is affected by the choice of data structures. In general, at each step k of the iteration process (3), only data from the current iteration k and the preceding iteration $k - 1$ is required. Hence, the vectors $\mathbf{Y}_l^{(1)}, \ldots, \mathbf{Y}_l^{(k-2)}$ and $\mathbf{F}_i^{(1)}, \ldots, \mathbf{F}_i^{(k-2)}$ do not have to be kept in memory. Instead, it is sufficient to use the $4s$ vectors $\mathbf{Y}_l^{(\text{cur})}$, $\mathbf{Y}_l^{(\text{prev})}$, $\mathbf{F}_i^{(\text{cur})}$ and $\mathbf{F}_i^{(\text{prev})}$ if we swap the pointers to the respective (prev) and (cur) vectors between the iterations of the k-loop. Further, if we keep the vectors $\mathbf{F}_i^{(\text{cur})}$ and $\mathbf{F}_i^{(\text{prev})}$ in memory, only one additional temporary argument vector $\mathbf{Y} \in \mathbb{R}^n$ is required to compute all $\mathbf{F}_i^{(\text{cur})}$ based on the values of $\mathbf{F}_i^{(\text{prev})}$ if the l-loop is processed sequentially (see implementations (A) and (E) below). However, a task-parallel implementation which processes the iterations of the l-loop simultaneously requires s temporary argument vectors $\mathbf{Y}_1, \ldots, \mathbf{Y}_s \in \mathbb{R}^n$. Another possibility is to store the vectors $\mathbf{Y}_l^{(\text{cur})}$ and $\mathbf{Y}_l^{(\text{prev})}$ instead of the vectors $\mathbf{F}_i^{(\text{cur})}$ and $\mathbf{F}_i^{(\text{prev})}$. Then, a loop structure can be realized which requires only one single scalar variable to handle all results of function evaluations (see implementation (D) and derived implementations below). In addition to the vectors $\mathbf{Y}_l^{(\text{cur})}$ and $\mathbf{Y}_l^{(\text{prev})}$ or $\mathbf{F}_i^{(\text{cur})}$ and $\mathbf{F}_i^{(\text{prev})}$, usually three further vectors of dimension n are required: the two approximation vectors, $\mathbf{y}_{\kappa+1}$ and $\hat{\mathbf{y}}_{\kappa+1}$, and a backup copy of $\mathbf{y}_\kappa$ for the case that the current time step is rejected by the step control.

The choice of data structures is particularly important in parallel implementations. Taking the NUMA character of many modern shared-memory systems into account, distributed data structures corresponding to the physically distributed structure of the memory subsystem are often favorable. Since function evaluations $\mathbf{f}(t, \mathbf{Y})$ may access all components of $\mathbf{Y}$, storing the s argument vectors $\mathbf{Y}_1, \ldots, \mathbf{Y}_s$ separately may increase the working space of a data-parallel implementation and thus lead to a worse locality compared to storing $\mathbf{F}_i^{(\text{cur})}$, $\mathbf{F}_i^{(\text{prev})}$ and only one temporary argument vector $\mathbf{Y}$. But, on the other hand, storing $\mathbf{Y}_1, \ldots, \mathbf{Y}_s$ separately enables alternative loop structures, which can be more efficient for particular problems.

Data-parallel implementations distributing the iterations of the j-loop to different processors can achieve a smaller working space than task-parallel implementations. The reason is that, if we parallelize across the l-loop, the i- and the j-loop of a task-parallel implementation touch all n components of the vectors $\mathbf{F}_i^{(\text{cur})}$ and $\mathbf{F}_i^{(\text{prev})}$. If we parallelize across the i-loop, then the l- and the j-loop access all n components of the vectors $\mathbf{Y}_l^{(\text{cur})}$ and $\mathbf{Y}_l^{(\text{prev})}$. In a data-parallel implementation, on the other hand, which stores the vectors $\mathbf{F}_i^{(\text{cur})}$ and $\mathbf{F}_i^{(\text{prev})}$, a thread uses only n/p components of each vector, where p is the number of processors. But if the argument vectors $\mathbf{Y}_l^{(\text{cur})}$ and $\mathbf{Y}_l^{(\text{prev})}$ are stored, $n/p + 2d(\mathbf{f}) - 1$ components are accessed. Thus, in the latter case, the working space has a similar size as in the task-parallel implementation if the access distance $d(\mathbf{f})$ is large, but it is significantly smaller if the access distance is limited.

Table 1 shows a summary of the implementations we have realized that support equations with arbitrary access structures. All implementation variants have been realized in a sequential version and a data-parallel version. Additionally,

Table 1. Implementations suitable for problems with arbitrary access structures

Implement.	Data structures	Loop struct.	Remarks
(A)	$p \times \mathbf{F}_1^{(\mathrm{cur})}, \ldots, \mathbf{F}_s^{(\mathrm{cur})} \in \mathbb{R}^{n/p}$, $p \times \mathbf{F}_1^{(\mathrm{prev})}, \ldots, \mathbf{F}_s^{(\mathrm{prev})} \in \mathbb{R}^{n/p}$, $\mathbf{Y} \in \mathbb{R}^n$	k–l–i–j	vector oriented: inner loops iterate over system dimension; high spatial locality
(E)	$p \times \mathbf{F}_1^{(\mathrm{cur})}, \ldots, \mathbf{F}_s^{(\mathrm{cur})} \in \mathbb{R}^{n/p}$, $p \times \mathbf{F}_1^{(\mathrm{prev})}, \ldots, \mathbf{F}_s^{(\mathrm{prev})} \in \mathbb{R}^{n/p}$, $\mathbf{Y} \in \mathbb{R}^n$	k–l–j–i	exploits temporal locality of the i-loop, i.e., writes to argument vector components
(D)	$\mathbf{Y}_1^{(\mathrm{cur})}, \ldots, \mathbf{Y}_s^{(\mathrm{cur})} \in \mathbb{R}^n$, $\mathbf{Y}_1^{(\mathrm{prev})}, \ldots, \mathbf{Y}_s^{(\mathrm{prev})} \in \mathbb{R}^n$	k–i–j–l	exploits temporal locality of the l-loop, i.e., reads from results of function evaluations
(Dblock)	$\mathbf{Y}_1^{(\mathrm{cur})}, \ldots, \mathbf{Y}_s^{(\mathrm{cur})} \in \mathbb{R}^n$, $\mathbf{Y}_1^{(\mathrm{prev})}, \ldots, \mathbf{Y}_s^{(\mathrm{prev})} \in \mathbb{R}^n$, $p \times \mathbf{F} \in \mathbb{R}^B$	k–i–j–l–jj	similar to (D), but loop tiling of the j-loop with the l-loop
(PipeDe2m)	$\mathbf{Y}_1^{(\mathrm{cur})}, \ldots, \mathbf{Y}_s^{(\mathrm{cur})} \in \mathbb{R}^n$, $\mathbf{Y}_1^{(\mathrm{prev})}, \ldots, \mathbf{Y}_s^{(\mathrm{prev})} \in \mathbb{R}^n$	k–j–i–l	based on (D); j-loop surrounds l- and i-loop; exploits temporal locality of both loops
(PipeDb2m)	$\mathbf{Y}_1^{(\mathrm{cur})}, \ldots, \mathbf{Y}_s^{(\mathrm{cur})} \in \mathbb{R}^n$, $\mathbf{Y}_1^{(\mathrm{prev})}, \ldots, \mathbf{Y}_s^{(\mathrm{prev})} \in \mathbb{R}^n$	k–j–i–jj–l	similar to (PipeDe2m), but loop tiling of the j-loop with the i-loop
(PipeDb2mt)	$\mathbf{Y}_1^{(\mathrm{cur})}, \ldots, \mathbf{Y}_s^{(\mathrm{cur})} \in \mathbb{R}^n$, $\mathbf{Y}_1^{(\mathrm{prev})}, \ldots, \mathbf{Y}_s^{(\mathrm{prev})} \in \mathbb{R}^n$, $p \times \mathbf{F} \in \mathbb{R}^B$	k–j–i–(jj)–l–jj	similar to (PipeDb2m), but loop tiling expanded to the l-loop

a task-parallel version of implementation (A) has been realized to compare the locality behavior of task- and data-parallel implementations.

4   Exploiting a Limited Access Distance

Similar to ERK methods [15,16], IRK methods can take advantage of a limited access distance of the function $\mathbf{f}$. While the stages of each corrector step are already decoupled, a limited access distance furthermore allows a reduction of the storage space, a pipelining of the corrector steps and a distributed storage of the argument vectors.

A reduction of the storage space can be achieved by using a loop structure in which the l- and the i-loop are inner loops of the j-loop. Then, a limited access distance allows an overlapping of the matrices $Y^{(k)} = \left(y_{l,j}^{(k)} \right) \in \mathbb{R}^{s,n}$, for $k = 1, \ldots, m$, where $Y^{(k)} = \left(\mathbf{Y}_1^{(k)}, \ldots, \mathbf{Y}_s^{(k)} \right)^T$. This can be realized by choosing a blocksize $B \geq d(\mathbf{f})$ and embedding $Y^{(1)}, \ldots, Y^{(m)}$ into a matrix $\bar{Y} = (\bar{y}_{l,j}) \in \mathbb{R}^{s,(m-1)\cdot 2B+n}$ such that $y_{l,j}^{(k)} := \bar{y}_{l,j+(m-k)\cdot 2B}$ and, hence, $y_{l,j}^{(k)}$, $k = 2, \ldots, m$, points to the same memory location as $y_{l,j+2B}^{(k-1)}$. Implementations (PipeDb1m) and (PipeDb1mt) are based on this strategy (see Table 2).

This optimization reduces the working space of one iteration of the outermost loop (the k-loop) nearly by a factor of 2. But this working space still has a size of $\Theta(sn)$ since the j-, the l- and the i-loop are inner loops of the k-loop. In order to obtain an asymptotically smaller working space of the outermost loop,

Table 2. Implementations specialized in problems with a limited access distance

Implement.	Data structures	Loop struct.	Remarks
(PipeDb1m)	$p \times \bar{Y} \in \mathrm{I\!R}^{s,\,m \cdot 2B + n/p}$	$k\!-\!j\!-\!i\!-\!jj\!-\!l$	similar to (PipeDb2m), but the vectors $\mathbf{Y}_l^{(k)}$ are overlapped to reduce space requirements
(PipeDb1mt)	$p \times \bar{Y} \in \mathrm{I\!R}^{s,\,m \cdot 2B + n/p}$, $p \times \mathbf{F} \in \mathrm{I\!R}^{B}$	$k\!-\!j\!-\!i\!-\!(jj)\!-\!l\!-\!jj$	similar to (PipeDb1m), but loop tiling expanded to the l-loop
(ppDb1m)	$p \times \bar{Y} \in \mathrm{I\!R}^{s,\,m \cdot 2B + n/p}$	$j\!-\!k\!-\!i\!-\!jj\!-\!l$	based on (PipeDb1m); j- and k-loop are interchanged using a pipelining approach
(ppDb1mt)	$p \times \bar{Y} \in \mathrm{I\!R}^{s,\,m \cdot 2B + n/p}$, $p \times \mathbf{F} \in \mathrm{I\!R}^{B}$	$j\!-\!k\!-\!i\!-\!(jj)\!-\!l\!-\!jj$	similar to (ppDb1m), but loop tiling expanded to the l-loop

we need to restructure the loops such that the loop with the largest dimension, i.e., the j-loop running from $1, \ldots, n$, becomes the outermost loop. However, if the problem to be integrated has a limited access distance, a pipelining of the corrector steps is possible, which allows an interchange of the k- and the j-loop. This approach is a straight-forward adoption of the pipelining of the stages in ERK methods suggested in [15,16]. It leads to a significant reduction of the working space of the iterations of the outer loop, which now access only $\Theta(sm \cdot 2B)$ distinct vector components. The pipelining of corrector steps has been realized in implementations (ppDb1m) and (ppDb1mt) (see Table 2).

Since many modern shared-memory systems are NUMA systems with a physically distributed memory architecture, a distributed storage of the argument vectors, where each thread keeps a part of each argument vector in its local memory, seems desirable. But in the general case, where the function evaluations $f_j(t, \mathbf{Y})$ may access all components of the argument vector $\mathbf{Y}$, a distributed storage of the argument vectors might not be profitable as it would require an additional level of indirection which maps the indices $j = 1, \ldots, n$ to the associated addresses inside the distributed parts of the vectors. The only alternative would be the replication of the whole argument vector. A limited access distance, however, enables a cost-efficient distributed storage of the argument vectors. Since the function evaluations performed by a processor P on its range of components $J^{(P)} = \left\{ j_{\mathrm{first}}^{(P)}, \ldots, j_{\mathrm{last}}^{(P)} \right\}$ may only access the components $\left\{ j_{\mathrm{first}}^{(P)} - d(\mathbf{f}), \ldots, j_{\mathrm{last}}^{(P)} + d(\mathbf{f}) \right\}$ of the argument vector, the processors can store their own range of components in local vectors, and only $d(\mathbf{f})$ components need to be copied from a processor's predecessor and its successor before a function evaluation can be performed. All of the implementations (PipeDb1m), (PipeDb1mt), (ppDb1m) and (ppDb1mt) make use of a distributed organization of the argument vectors.

5   Experimental Results

Experiments to analyze the scalability and the locality behavior of the implementations have been performed on two of the largest German supercomputer

Table 3. Execution times (in s) of task- and data-parallel versions of (A)

	JUMP				HLRB II			
	Radau I A(5)		Lobatto III C(8)		Radau I A(5)		Lobatto III C(8)	
	3 Threads		5 Threads		3 Threads		5 Threads	
N	Data-Par.	Task-Par.	Data-Par.	Task-Par.	Data-Par.	Task-Par.	Data-Par.	Task-Par.
500	8.5	8.9	11.7	12.9	16.4	33.5	23.2	56.6
750	45.4	47.3	62.7	66.0	103.3	184.2	135.5	299.4
1000	142.2	151.0	200.3	210.9	347.8	591.1	474.9	941.7

Table 4. Comparison of the L2 performance of task- and data-parallel versions of (A) on JUMP. The table shows the number of times a cache line in the L1 data cache was reloaded from the local L2 cache, an L2 cache of another chip on the same MCM or from an L2 cache of another chip on a remote MCM.

| Implementation | Radau I A(5), 3 Threads | | | Lobatto III C(8), 5 Threads | | |
	Local	Same MCM	Other MCM	Local	Same MCM	Other MCM
Data-Parallel	$1.4 \cdot 10^9$	$6.7 \cdot 10^4$	$1.3 \cdot 10^5$	$2.3 \cdot 10^9$	$1.6 \cdot 10^5$	$3.3 \cdot 10^5$
Task-Parallel	$1.4 \cdot 10^9$	$1.0 \cdot 10^7$	$2.3 \cdot 10^7$	$2.3 \cdot 10^9$	$3.0 \cdot 10^7$	$7.2 \cdot 10^7$

systems, the Jülich Multiprocessor (JUMP) at the NIC Jülich and the High End System in Bavaria II (HLRB II) at the LRZ Munich. JUMP is a SMP cluster system consisting of 41 IBM eServer pSeries 690 nodes, each equipped with 32 Power4+ cores running at 1.7 GHz. The nodes are built up of 4 multi-chip modules (MCM), where each MCM carries 4 Power4+ chips, and 2 CPU cores are housed in one chip. The HLRB II is an SGI Altix 4700 system. Currently, the system is equipped with 4096 Intel Itanium 2 Madison 9M processors at a clock rate of 1.6 GHz. As an example problem we use BRUSS2D [2], a typical example of a partial differential equation (PDE) discretized by the method of lines. Using an interleaving of the two dependent variables resulting in a mixed-row oriented ordering of the components (BRUSS2D-MIX, cf. [15,16]), this problem has a limited access distance of $d(\mathbf{f}) = 2N$, where the system size is $n = 2N^2$. Most experiments presented in the following have been performed using the 3-stage method Radau I A(5) as the corrector method.

5.1   Task vs. Data Parallelism

Table 3 shows a comparison of the execution times of a data-parallel and a task-parallel version of implementation (A) using the 3-stage method Radau I A(5) and the 5-stage method Lobatto III C(8). In most experiments performed on JUMP, the data-parallel version is slightly faster than the task-parallel version. On HLRB II, where remote memory accesses are more expensive, the higher locality of the data-parallel version evidently outperforms the task-parallel version. An analysis of the cache performance using performance counter data on JUMP points out that the lower performance of the task-parallel version on this machine can be attributed to an exceedingly higher number of loads from L2 caches of other chips (cf. Table 4).

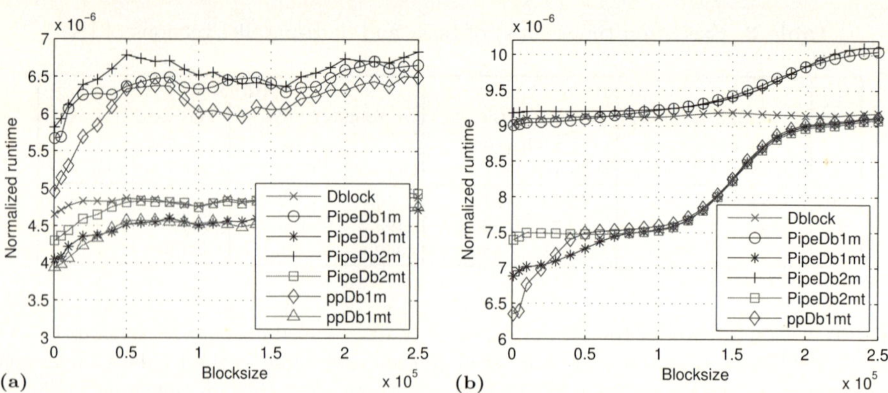

Fig. 1. Normalized sequential runtime as a function of the blocksize measured on (a) JUMP and (b) HLRB II using BRUSS2D-MIX with Radau I A(5) and $N = 1000$

5.2   Choosing a Suitable Blocksize

All implementations which employ loop tiling and all implementations specialized in a limited access distance require the selection of a suitable blocksize B. In case of the specialized implementations, B must at least be as large as the access distance $d(\mathbf{f})$. We choose the blocksize based on an experimental evaluation of the normalized sequential runtime, which is the runtime per time step and per ODE component. A blocksize which leads to a minimum of the normalized runtime would be the optimal choice. Figure 1 shows the normalized runtime as a function of the blocksize measured on JUMP and HLRB II using our example problem. On both systems the optimal blocksize is below 10 000 for all implementations. We choose $B = d(\mathbf{f}) = 2N$, i.e., $B = 2000$ for $N = 1000$, since it is the smallest possible value to be used with the specialized implementations. This value is close enough to the optimal value of all implementations that it enables an accurate comparison.

5.3   Performance Analysis and Comparison of the Implementations

As a first step of the performance analysis, we compare the sequential runtimes of the implementations (Table 5). On both machines we observe that the various implementations obtain significantly differing runtimes. Since all implementations are realizations of Eqs. (3), (4) and (5) only differing in the loop structures and the data structures used, the different runtimes mainly result from the different locality behavior of the implementations. On both systems the best general implementation is (PipeDb2mt), which combines a loop structure where the i- and the l-loop run inside the j-loop with loop tiling of both the i- and the l-loop. The specialization in a limited access distance improves the locality and enables a better runtime. Thus, implementation (PipeDb1mt) runs faster than (PipeDb2mt). The best sequential runtime is obtained by the implementation with the smallest working space, (ppDb1mt).

Table 5. Sequential runtimes measured using BRUSS2D-MIX and Radau I A(5)

Implementation	JUMP			HLRB II		
	$N = 500$	$N = 750$	$N = 1000$	$N = 500$	$N = 750$	$N = 1000$
(A)	26.9	153.6	481.6	61.1	330.9	1071.2
(D)	25.8	142.7	465.2	60.5	322.8	1032.1
(Dblock)	24.8	137.2	454.5	54.0	286.8	907.1
(E)	25.2	140.9	459.8	55.4	294.6	950.9
(PipeDb1m)	25.7	139.9	441.1	53.7	286.5	913.4
(PipeDb1mt)	23.1	126.2	413.0	40.9	218.2	698.5
(PipeDb2m)	25.6	138.0	439.4	54.8	291.9	930.6
(PipeDb2mt)	24.0	135.9	436.5	44.0	234.9	750.3
(PipeDe2m)	31.6	182.6	580.1	55.3	294.2	924.3
(ppDb1m)	25.5	136.1	430.0	n/a[1]	n/a[1]	n/a[1]
(ppDb1mt)	23.0	123.5	395.2	37.7	201.9	644.5

1. Not available due to an incorrect translation by the compiler.

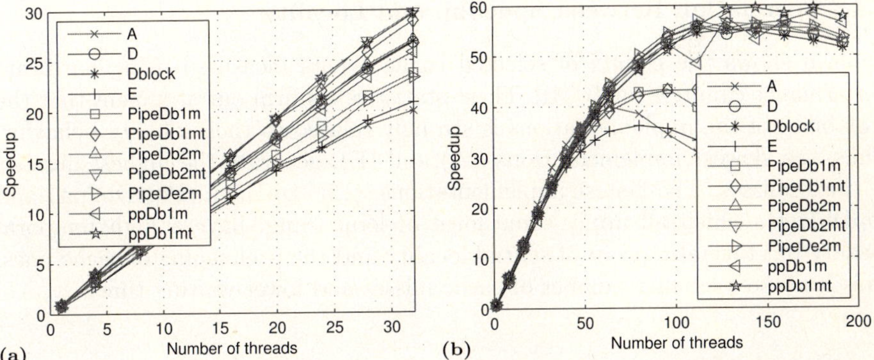

Fig. 2. Speedups measured on **(a)** JUMP and **(b)** HLRB II using BRUSS2D-MIX with Radau I A(5) and $N = 1000$

To analyze the scalability of the implementations, we measured their speedups on JUMP and HLRB II (Fig. 2). On JUMP all implementations show a good scalability. Even the slowest implementation, (A), obtains a speedup of 20.0 on 32 processors. The best implementations are those with the best locality behavior, i.e., (PipeDb2mt), (PipeDb1mt) and (ppDb1mt). They obtain speedups between 28.9 and 29.6. The speedups on HLRB II are satisfactory for small numbers of processors up to 32, because the data-parallel execution leads to a reduction of the working space of each thread and, hence, to a reduction of the number of cache misses. But for larger numbers of processors all data accessed during one time step fits in the cache, and, as a consequence, the scalability no longer profits from these cache effects. Moreover, the communication of more than 32 processors requires the use of one additional level of the hierarchically organized interconnection network. Therefore, the efficiency of the implementations severely degrades if the number of processors is increased further. The highest speedup of 59.5 has been obtained by (ppDb1mt) using 144 processors.

Table 6. Cache access statistics for 32 threads on JUMP measured using BRUSS2D-MIX with Radau I A(5) and $N = 1000$

Implementation	Execution Time (s)	L1 Prefetch Requests	L1 Data Misses		L2 Store Misses	Data from L2	Data from L3	Data from Memory
			Load	Store				
(A)	19.83	$1.3 \cdot 10^8$	$9.3 \cdot 10^7$	$7.4 \cdot 10^8$	$4.4 \cdot 10^7$	$1.3 \cdot 10^8$	$1.1 \cdot 10^7$	$1.9 \cdot 10^6$
(D)	14.78	$7.8 \cdot 10^7$	$2.4 \cdot 10^7$	$2.5 \cdot 10^8$	$1.5 \cdot 10^7$	$8.8 \cdot 10^7$	$4.7 \cdot 10^6$	$4.0 \cdot 10^5$
(Dblock)	14.71	$1.1 \cdot 10^8$	$4.5 \cdot 10^7$	$3.7 \cdot 10^8$	$1.5 \cdot 10^7$	$1.2 \cdot 10^8$	$4.4 \cdot 10^6$	$8.2 \cdot 10^5$
(E)	18.96	$9.8 \cdot 10^7$	$4.1 \cdot 10^8$	$8.5 \cdot 10^8$	$3.6 \cdot 10^7$	$1.0 \cdot 10^8$	$7.6 \cdot 10^6$	$2.8 \cdot 10^5$
(PipeDb1m)	16.66	$8.0 \cdot 10^7$	$3.4 \cdot 10^7$	$2.6 \cdot 10^8$	$6.2 \cdot 10^6$	$9.9 \cdot 10^7$	$2.1 \cdot 10^6$	$7.9 \cdot 10^4$
(PipeDb1mt)	13.66	$1.1 \cdot 10^8$	$3.4 \cdot 10^7$	$3.8 \cdot 10^8$	$6.3 \cdot 10^6$	$1.3 \cdot 10^8$	$2.0 \cdot 10^6$	$1.4 \cdot 10^5$
(PipeDb2m)	14.62	$7.9 \cdot 10^7$	$1.3 \cdot 10^8$	$3.5 \cdot 10^8$	$1.5 \cdot 10^7$	$9.4 \cdot 10^7$	$2.0 \cdot 10^6$	$5.1 \cdot 10^4$
(PipeDb2mt)	13.28	$1.1 \cdot 10^8$	$3.6 \cdot 10^7$	$3.7 \cdot 10^8$	$1.5 \cdot 10^7$	$1.3 \cdot 10^8$	$2.0 \cdot 10^6$	$1.7 \cdot 10^5$
(PipeDe2m)	16.95	$3.5 \cdot 10^7$	$4.7 \cdot 10^7$	$1.4 \cdot 10^7$	$7.6 \cdot 10^6$	$6.4 \cdot 10^7$	$9.7 \cdot 10^6$	$1.1 \cdot 10^5$
(ppDb1m)	16.65	$8.2 \cdot 10^7$	$3.1 \cdot 10^7$	$2.6 \cdot 10^8$	$9.8 \cdot 10^6$	$1.1 \cdot 10^8$	$1.8 \cdot 10^6$	$2.7 \cdot 10^4$
(ppDb1mt)	13.33	$1.2 \cdot 10^8$	$4.7 \cdot 10^7$	$4.2 \cdot 10^8$	$1.0 \cdot 10^7$	$1.4 \cdot 10^8$	$1.8 \cdot 10^6$	$6.8 \cdot 10^4$

5.4   Correlation Between Speedup and Locality

Table 6 shows the counts of selected cache events measured using hardware performance counters on JUMP. These statistics confirm our argument that the scalability of the implementations is strongly related to their locality behavior. Thus, the slowest implementations, (A) and (E), generate the highest number of cache misses. The fastest implementations, (PipeDb1mt), (PipeDb2mt) and (ppDb1mt), which all apply some form of loop tiling, have a high temporal locality but take also profit of spatial locality and the prefetching of cache lines. This leads to a smaller number of cache misses and lower waiting times.

6   Conclusion

In this paper, we have presented several different implementations of iterated Runge-Kutta methods which are suitable for parallel computers with a shared address space. The implementations employ different strategies, i.e., different loop structures and different data structures, to achieve a beneficial utilization of the memory hierarchy. In general, the iteration of the corrector steps must be executed sequentially. But if the problem to be solved has a limited access distance, it can be exploited to reduce the storage space and to reduce the working space of the outer loop by the implementation of a pipelining of the corrector steps. Our runtime experiments performed on two modern supercomputer systems confirm that the locality behavior has a major influence on the scalability of IRK methods. Therefore, data-parallel implementations often obtain a better performance than task-parallel implementations. Loop tiling is one important technique to adjust the size of the working sets to the size of the cache hierarchy. The best scalability, particularly for large numbers of processors up to 144, has been achieved using an implementation which combines loop tiling with a pipelining of the corrector steps.

Acknowledgments. We thank the NIC Jülich and the LRZ Munich for providing access to their supercomputer systems JUMP and HLRB II.

References

1. Ehrig, R., Nowak, U., Deuflhard, P.: Massively parallel linearly-implicit extrapolation algorithms as a powerful tool in process simulation. In: Parallel Computing: Fundamentals, Applications and New Directions, pp. 517–524. Elsevier, Amsterdam (1998)
2. Burrage, K.: Parallel and Sequential Methods for Ordinary Differential Equations. Oxford Science Publications, Oxford (1995)
3. Nørsett, S.P., Simonsen, H.H.: Aspects of parallel Runge-Kutta methods. In: Numerical Methods for Ordinary Differential Equations. LNM, vol. 1386, pp. 103–117 (1989)
4. van der Houwen, P.J., Sommeijer, B.P.: Parallel iteration of high-order Runge-Kutta methods with stepsize control. J. Comput. Appl. Math. 29, 111–127 (1990)
5. Jackson, K.R., Nørsett, S.P.: The potential for parallelism in Runge-Kutta methods. Part 1: RK formulas in standard form. SIAM J. Numer. Anal. 32(1), 49–82 (1995)
6. Choi, J., Dongarra, J.J., Ostrouchov, L.S., Petitet, A.P., Walker, D.W., Whaley, R.C.: Design and implementation of the ScaLAPACK LU, QR and Cholesky factorization routines. Sci. Prog. 5, 173–184 (1996)
7. Anderson, E., Bai, Z., Bischof, C., Blackford, L.S., Demmel, J., Dongarra, J., Croz, J.D., Greenbaum, A., Hammarlin, S., McKenney, A., Sorensen, D.: LAPACK Users' Guide, 3rd edn., SIAM (1999)
8. Bilmes, J., Asanovic, K., Chin, C.W., Demmel, J.: Optimizing matrix multiply using PHiPAC: a portable, high-performance, ANSI C coding methodology. In: 11th ACM Int. Conf. on Supercomputing, ACM Press, New York (1997)
9. Whaley, R.C., Petitet, A., Dongarra, J.J.: Automated empirical optimizations of software and the ATLAS project. Par. Comp. 27(1–2), 3–35 (2001)
10. Allen, R., Kennedy, K.: Optimizing Compilers for Modern Architectures: A Dependence Based Approach. Morgan Kaufmann, San Francisco (2002)
11. McKinley, K.S.: A compiler optimization algorithm for shared-memory multiprocessors. IEEE Trans. Par. Dist. Syst. 9(8), 769–787 (1998)
12. Irigoin, F., Triolet, R.: Supernode partitioning. In: ACM Symposium on Principles of Programming Languages, San Diego, Calif., pp. 319–329. ACM Press, New York (1988)
13. Ghosh, S., Martonosi, M., Malik, S.: Cache miss equations: A compiler framework for analyzing and tuning memory behavior. ACM Trans. Prog. Lang. Syst (TOPLAS) 21(4), 703–746 (1999)
14. Rauber, T., Rünger, G.: Improving locality for ODE solvers by program transformations. Sci. Prog. 12(3), 133–154 (2004)
15. Korch, M.: Effiziente Implementierung eingebetteter Runge-Kutta-Verfahren durch Ausnutzung der Speicherzugriffslokalität. Doctoral thesis, University of Bayreuth, Bayreuth, Germany (December 2006)
16. Korch, M., Rauber, T.: Optimizing locality and scalability of embedded Runge-Kutta solvers using block-based pipelining. J. Par. Distr. Comp. 66(3), 444–468 (2006)
17. Rauber, T., Rünger, G.: Parallel implementations of iterated Runge-Kutta methods. Int. J. Supercomp. App. 10(1), 62–90 (1996)

Toward Scalable Matrix Multiply on Multithreaded Architectures

Bryan Marker[1], Field G. Van Zee[2], Kazushige Goto[2], Gregorio Quintana-Ortí[3], and Robert A. van de Geijn[2]

[1] National Instruments
[2] The University of Texas at Austin
[3] Universidad Jaume I, Spain

Abstract. We show empirically that some of the issues that affected the design of linear algebra libraries for distributed memory architectures will also likely affect such libraries for shared memory architectures with many simultaneous threads of execution, including SMP architectures and future multicore processors. The always-important matrix-matrix multiplication is used to demonstrate that a simple one-dimensional data partitioning is suboptimal in the context of dense linear algebra operations and hinders scalability. In addition we advocate the publishing of low-level interfaces to supporting operations, such as the copying of data to contiguous memory, so that library developers may further optimize parallel linear algebra implementations. Data collected on a 16 CPU Itanium2 server supports these observations.

1 Introduction

The high-performance computing community is obsessed with efficient matrix-matrix multiplication (GEMM). This obsession is well justified, however; it has been shown that fast GEMM is an essential building block of many other linear algebra operations such as those supported by the Level-3 Basic Linear Algebra Subprograms (BLAS) [1,2], the Linear Algebra Package (LAPACK) [3], as well as many other libraries and applications. Moreover, and perhaps more importantly, many of the issues that arise when implementing other dense linear algebra operations can be demonstrated in the simpler setting of GEMM.

The high-performance implementation of GEMM on serial and distributed memory architectures is well understood [4,5,6,7,8]. However, the implementation in the context of multithreaded environments, which includes Symmetric MultiProcessor (SMP), Non-Uniform Memory Access (NUMA), and multicore architectures, is relatively understudied. It is conceivable that architectures with hundreds of concurrent threads will be widespread within a decade, since each processor socket of an SMP or NUMA system will have multiple, possibly many, cores. Thus, issues of *programmability as well as scalability* must be addressed for GEMM and related operations.

A.-M. Kermarrec, L. Bougé, and T. Priol (Eds.): Euro-Par 2007, LNCS 4641, pp. 748–757, 2007.

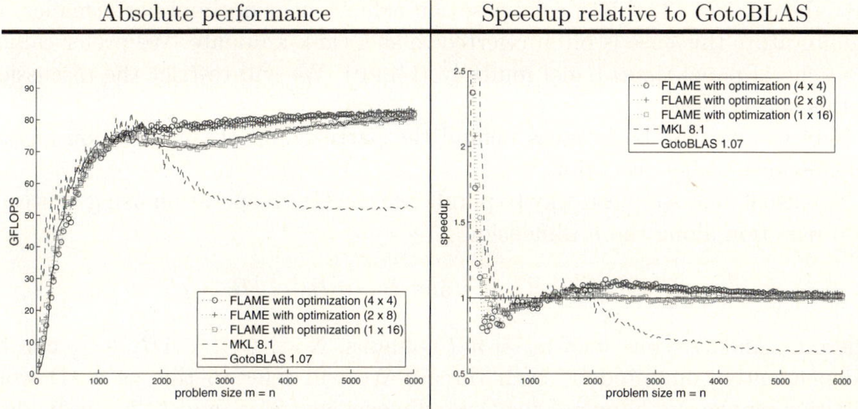

Fig. 1. Performance of our optimized implementations of GEMM when k is fixed at 256 on a 16 CPU Itanium2 server. (For further details, see Section 5).

Our Formal Linear Algebra Methods Environment (FLAME) project studies these programmability[1] and scalability issues. The latter issue is the subject of this paper, which provides empirical evidence for the following observations:

- As was observed for distributed memory architectures in the late 1980s and early 1990s [11,12,13,14], work must be assigned to threads using a two-dimensional partitioning of data [15,16,10].
- In order to achieve near-optimal performance, library developers must be given access to routines or *kernels* that provide computational- and utility-related functionality at a lower level than the customary BLAS interface.

The first issue is of contemporary concern since a number of recent projects still use one-dimensional (1D) partitioning when parallelizing dense linear algebra operations for multithreaded environments [17]. The second is similarly important because it is tempting to simply call sequential BLAS within each thread once the work has been partitioned.

Figure 1 offers a preview of the performance observed when 2D partitioning and low-level optimizations are applied to a special case of parallel GEMM.

2   Partitioning for Parallel Panel-Panel Multiply

Consider the prototypical matrix multiplication $C := AB + C$ with $C \in \mathbb{R}^{m \times n}$, $A \in \mathbb{R}^{m \times k}$, and $B \in \mathbb{R}^{k \times n}$. In a recent paper [5], we argue that the case where k is chosen to be relatively small (in the $k = 256$ range) and m and n relatively large is the most important special case of this operation, for a number of reasons: (1) the general case (m, k, n of any size) can be cast in terms of this special case, (2) most operations supported by LAPACK can (and should) be cast in terms of this

---

[1] How the programmability issue is addressed by FLAME is discussed in [9,10].

special case, and (3) this special case can achieve near-optimal performance. In the literature this case is often referred to as a rank-k update. We prefer calling it a general $\underline{\text{p}}$anel-times-$\underline{\text{p}}$anel multiply (GEPP). We will restrict the discussion to this special case.

Now we describe basic ideas behind the partitioning of computation among threads for parallel execution.

The usual, and simplest, way to parallelize the GEPP operation using t threads is to partition along the n dimension:

$$C = \left(C_0 \big| \cdots \big| C_{t-1} \right) \text{ and } B = \left(B_0 \big| \cdots \big| B_{t-1} \right)$$

where C_j and B_j consist of $n_j \approx n/t$ columns. Now, $C_j := AB_j + C_j$ can be independently computed by each thread. We will refer to this as a 1D work partitioning since only one of the three dimensions (m, n, and k) is subdivided. Clearly, there is a mirror 1D algorithm that partitions along m. In this paper we consider k to be too small to warrant further partitioning.

When distributed memory architectures became popular in the 1980s and 1990s, it was observed, both in theory and in practice, that as the number of processors increased, a 2D work and data partitioning is required to attain scalability[2] within dense linear algebra operations like GEMM. It is for this reason that libraries like ScaLAPACK [18] and PLAPACK [19] use a 2D data and work distribution. While the cost of data movement is much lower on multithreaded architectures, we suggest that the same basic principle applies to exploiting very large numbers of simultaneously executing threads.

To achieve a 2D work partitioning, t threads are logically viewed as forming a $t_r \times t_c$ grid, with $t = t_r \times t_c$. Then,

$$C = \left(\begin{array}{ccc} C_{0,0} & \cdots & C_{0,t_c-1} \\ \hline \vdots & & \vdots \\ \hline C_{t_r-1,0} & \cdots & C_{t_r-1,t_c-1} \end{array} \right), A = \left(\begin{array}{c} A_0 \\ \vdots \\ \hline A_{t_c-1} \end{array} \right), \text{ and } B = \left(B_0 \big| \cdots \big| B_{t_c-1} \right),$$

with $C_{ij} \in \mathbb{R}^{m_i \times n_j}$, $A_i \in \mathbb{R}^{m_i \times k}$, and $B_j \in \mathbb{R}^{k \times n_j}$ ($m_i \approx m/t_r$ and $n_j \approx n/t_c$), so that each task $C_{i,j} := A_i B_j + C_{i,j}$ can be computed by a separate thread.

The purpose of this partitioning is to create parallelism through independent tasks that each perform a sequential GEPP.

3 Anatomy of the Sequential, High-Performance GEPP Algorithm

The following is a minimal discussion of key internal mechanisms of the state-of-the-art GEMM implementation present within the GotoBLAS [5]. This discussion will allow us to explain how to reduce redundant memory-to-memory copies in the multithreaded implementation.

---

[2] Roughly speaking, scalability here should be understood as the ability to retain high performance as the number of processors is increased.

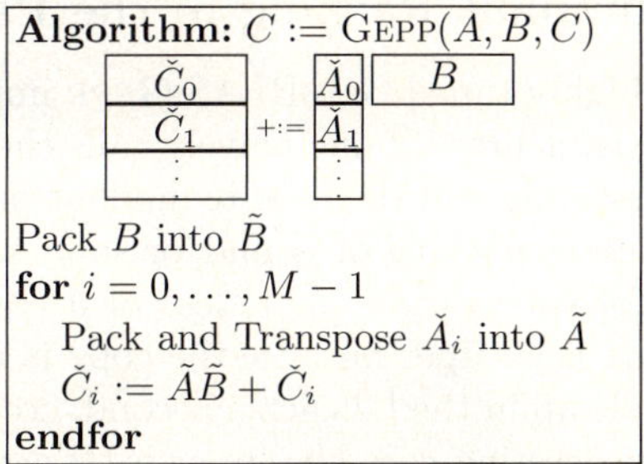

Fig. 2. Outline of optimized implementation of Gepp

Consider again the computation $C := AB + C$ where the k dimension is relatively small. Assume for simplicity that $m = b_m M$ where b_m and M are integers. Partition

$$C = \left(\frac{\check{C}_0}{\frac{\vdots}{\check{C}_{M-1}}} \right) \quad \text{and} \quad A = \left(\frac{\check{A}_0}{\frac{\vdots}{\check{A}_{M-1}}} \right),$$

where $\check{C}_i$ and $\check{A}_i$ have b_m rows. Figure 2 gives a high-performance algorithm for the Gepp operation. This algorithm requires three highly optimized components, or low-level kernels:

- **Pack B:** A routine for packing B into a contiguous buffer[3].
- **Pack and transpose $\check{A}_i$:** A routine for packing $\check{A}_i$ into a contiguous buffer. Often this routine also transposes the matrix to improve the order in which data is accessed by the Gebp kernel routine. This transpose is orchestrated so that the packed and transposed matrix, $\tilde{A}$, is in the L2 cache upon completion.
- Gebp **kernel routine:** This routine computes $\check{C}_i := \tilde{A}\tilde{B} + \check{C}_i$ using the packed buffers. Here, Gebp stands for <u>ge</u>neral <u>b</u>lock-times-<u>p</u>anel multiply.

On current architectures the size of $\check{A}_i$ is chosen to fill about half of the L2 cache (or the memory addressable by the TLB), as explained in [5]. Considerable effort is required to tune each of these kernels, especially Gebp.

Two relevant key insights may be gathered from [5]. First, the cost of packing $\hat{A}_i$ is not much greater than the cost of loading the L2 cache, so that, while significant, it is also an unavoidable cost. This means that the column dimension of B should be large. Second, the cost of packing B is significant and should, therefore, be amortized over as many blocks of A as possible. This calls for a careful look at how packing occurs in a parallel implementation.

---

[3] Note that B is typically embedded in an array with more than k rows and is thus not contiguous.

4   Avoiding Redundant Packing in the Parallel Gepp

The description of the highly tuned GotoBLAS Gepp implementation brings to light the overhead that is incurred if one naively calls the dgemm library routine to perform each task: not only will there be redundant parameter checking, but there will also be redundant packing of B and/or submatrices $\check{A}_i$.

Redundant packing of submatrices $\check{A}_i$ is less of a concern for two reasons. First, redundant packing is cheaper because the copy is not necessarily written back to memory and ends up in the L2 cache. Second, creating redundant copies is unavoidable if each processing core has its own L2 cache; the submatrix has to be loaded into each L2 cache, regardless. Thus, if $t_r \times t_c = 1 \times t$, the naive approach that simply invokes the sequential dgemm routine can be expected to be quite effective.

The packing of matrix B into $\tilde{B}$ is a different matter. It is more expensive since typically B is large enough that this operation requires both reading from and writing to memory. Now, if $t_r > 1$, then the separate calls to dgemm performed by the jth column of threads would each repack B_j. The problem is potentially compounded by the fact that during this redundant packing there is contention for the limited bandwidth to memory. Thus, it can be argued that redundant packing of submatrices of B should be avoided, and, if possible, all available processor-bound threads should be employed during the packing of B.

5   Experiments

In this section we provide early evidence that the issues discussed in this paper are observed in practice. For this we employed a 16 CPU Itanium2 server. While the effects are somewhat limited when there are only 16 simultaneously executing threads, one would expect the effects to become more pronounced as systems utilize larger numbers of CPU cores, as is expected in the near future.

FLAME/C. The experiments were coded using the FLAME/C API, a programming interface that implements common linear algebra operations in an object-based environment [20]. The programmability issues in the multithreaded arena that FLAME/C solves are discussed in a recent paper [9].

Target platform. Experiments were performed on an SGI Altix ccNUMA system containing eight dual-processor Itanium2 nodes. Each pair of CPUs shares 4GB of local memory with the other CPUs to form a logically contiguous address space of 32GB. The Itanium2 microprocessor executes a maximum of four floating-point operations per clock cycle. All CPUs on this system run at 1.5GHz. This allows a peak performance of 6 GFLOPS (10^9 floating-point operations per second) per processor and an aggregate peak attainable performance of 96 GFLOPS, which is represented by the top range of the y-axes in the left column of graphs in Fig. 3. All computation was performed in double precision (64-bit) arithmetic.

Matrix sizes tested. For all experiments $k = 256$, a value for which the GotoBLAS implementation of dgemm is essentially optimal. Dimensions m and n were varied

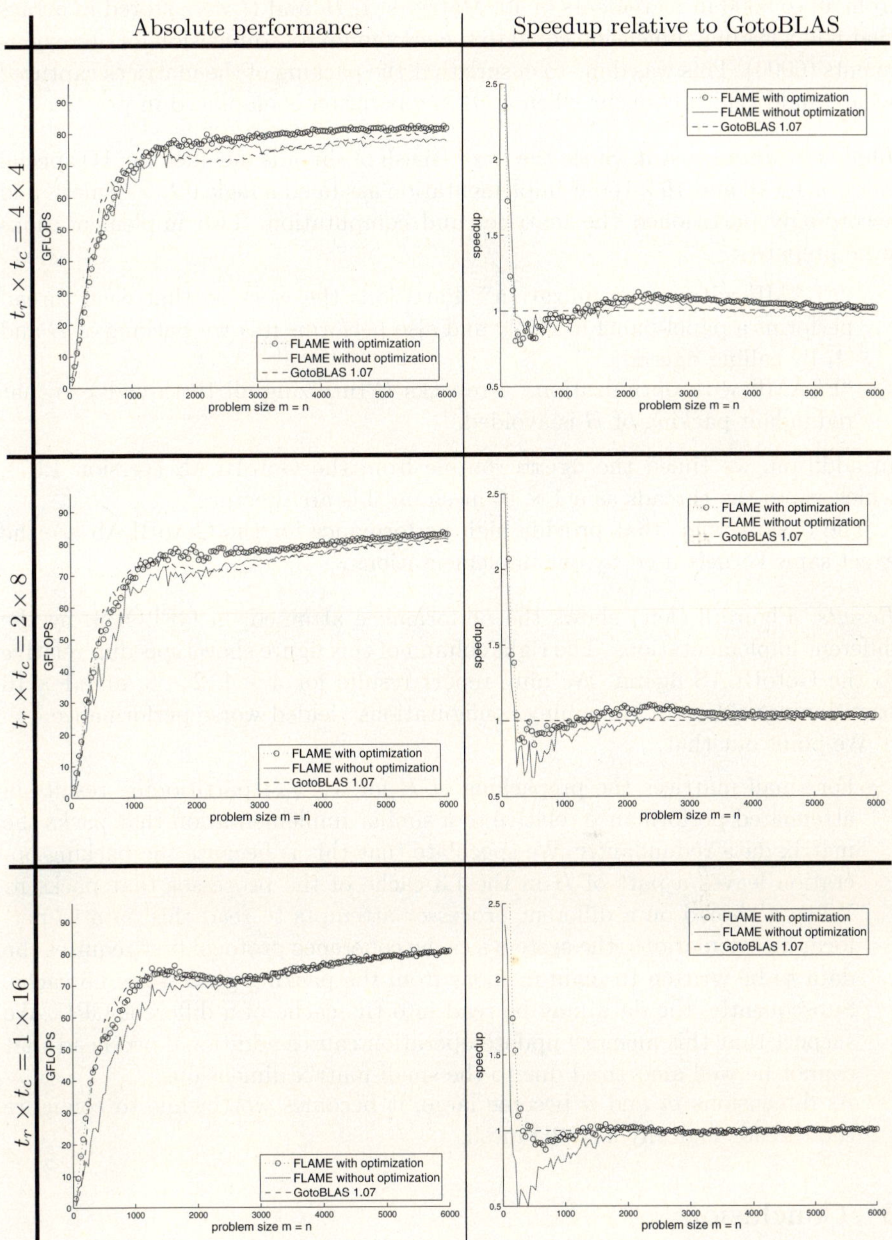

Fig. 3. Performance of conventional and low-level parallelizations (16 threads) of GEPP when m and n are varied and $k = 256$. Absolute performance is shown in the left column for three choices of $t_r \times t_c$ while the corresponding speedup relative to the GotoBLAS dgemm is shown on the right. This figure provides evidence that there is merit to (1) using a 2D partitioning of work and (2) avoiding redundant packing of B.

from 40 to 6000 in increments of 40. Matrices A, B, and C were stored in arrays that had a leading dimension equal to the maximum row dimension for our experiments (6000). This was done to ensure that the packing of the matrices captured what would typically occur when a GEPP operation is employed in practice.

Implementations tested. Since the $t_r \times t_c$ mesh of threads includes the 1D special cases of 1×16 and 16×1, our implementation assumed a logical $t_r \times t_c$ mesh and accordingly partitioned the matrices and computation. Two implementations were prepared:

- "FLAME without optimization." Partitions the work so that each thread performs a panel-panel multiply and also performs its own packing of B and $\check{A}_i$ by calling `dgemm`.
- "FLAME with optimization." Prepacks B (utilizing all 16 threads) so that redundant packing of B is avoided.

In addition, we timed the `dgemm` routine from the GotoBLAS (version 1.07), which views the threads as a 1×16 mesh on this architecture[4].

The basic kernels that provide high performance for the GotoBLAS are the exact same kernels used by our implementations.

Results. Figure 3 (left) shows the performance attained, in GFLOPS, by the different implementations. The right column of this figure shows speedup relative to the GotoBLAS `dgemm`. We only report results for 4×4, 2×8, and 1×16 partitionings since the remaining configurations yielded worse performance.

We point out that,

- For small matrices the prepacking of B for a 4×4 partitioning results in attenuated performance relative to a similar implementation that packs the matrix data redundantly. We speculate that this is because the packing operation leaves a part of $\tilde{B}$ in the L3 cache of the processor that packs it. When a thread on a different processor attempts to read this data to perform its calculations, the system's cache coherence protocol first requires the data to be written to main memory from the packing processor's L3 cache. Subsequently, the data may be read into the cache of a different CPU. We suspect that this memory update operation causes additional overhead that cannot be well amortized due to the small matrix dimensions.
- As dimensions m and n become large, it becomes worthwhile to configure the threads logically as a 2D mesh.

6   Conclusion

We have shown empirically that the parallelization of GEPP on multithreaded architectures with many threads benefits from a 2D decomposition and assignment of work. Moreover, we have demonstrated that it is important to avoid redundant copying of submatrices into contiguous memory.

---

[4] Naturally, one can expect changes in future multithreaded implementations of the GotoBLAS consistent with the insights in this paper.

While we have made similar observations about other linear algebra operations such as symmetric rank-k update [10] and Cholesky factorization [16], the study of GEPP presents the issues in isolation, devoid of tangential complexities such as dependencies (between subpartitions) and load-imbalance (due to matrices with special shape). Moreover, in many ways the Itanium2 server on which the experiments were performed is a much more forgiving architecture than more commonly encountered Pentium- or Opteron-based multithreaded architectures due to its massive memory bandwidth. Finally, given that 16 simultaneously executing threads represents a relatively small level of parallelism, the effects observed here will likely become more dramatic as multicore systems are designed and built with more CPUs.

We envision a range of further studies and development. The GEPP operation is the most important of three frequently used cases of GEMM, which also include GEMP (multiplication of a matrix times a panel of columns) and GEPM (multiplication of a panel of rows times a matrix). Similar experiments on multicore processors are in order given that these systems employ memory architectures different from that of single-core-per-socket SMPs. Interfaces to lower level kernels should be published to help other researchers perform similar experiments more easily. Furthermore, *de facto* standards of such interfaces would allow library developers to realize the performance gains demonstrated in this paper across platforms.

Further Information

For additional information regarding the FLAME project, visit

`http://www.cs.utexas.edu/users/flame/`.

Acknowledgements

This research was partially sponsored by NSF grants CCF-0540926, CCF-0342369, and ACI-0305163. Any opinions, findings and conclusions or recommendations expressed in this material are those of the author(s) and do not necessarily reflect the views of the National Science Foundation (NSF). In addition, Dr. James Truchard (National Instruments) provided an unrestricted grant to our research. Furthermore, the research of Bryan Marker was partially funded by the Undergraduate Research Opportunities Program of the Department of Computer Sciences at The University of Texas at Austin.

Access to the 16 CPU Itanium2 (1.5 GHz) system on which the experiments were performed was provided our collaborators at Universidad Jaume I, Spain. Initial experiments (not reported in this paper) were conducted on a 4 CPU Itanium2 server which was donated to our project by Hewlett-Packard and is administered by UT-Austin's Texas Advanced Computing Center.

As always, we are indebted to our collaborators on the FLAME team.

References

1. Dongarra, J.J., Du Croz, J., Hammarling, S., Duff, I.: A set of level 3 basic linear algebra subprograms. ACM Trans. Math. Soft. 16(1), 1–17 (1990)
2. Kågström, B., Ling, P., Loan, C.V.: GEMM-based level 3 BLAS: High performance model implementations and performance evaluation benchmark. ACM Trans. Math. Soft. 24(3), 268–302 (1998)
3. Anderson, E., Bai, Z., Bischof, C., Demmel, J., Dongarra, J., Croz, J.D., Greenbaum, A., Hammarling, S., McKenney, A., Ostrouchov, S., Sorensen, D.: LAPACK Users' Guide - Release 2.0. SIAM (1994)
4. Gunnels, J.A., Henry, G.M., van de Geijn, R.A.: A family of high-performance matrix multiplication algorithms. In: Alexandrov, V.N., Dongarra, J.J., Juliano, B.A., Renner, R.S., Tan, C.K. (eds.) ICCS 2001. LNCS, vol. 2073, pp. 51–60. Springer, Heidelberg (2001)
5. Goto, K., van de Geijn, R.: Anatomy of high-performance matrix multiplication ACM Trans. Math. Soft. (to appear)
6. Agarwal, R.C., Gustavson, F., Zubair, M.: A high-performance matrix multiplication algorithm on a distributed memory parallel computer using overlapped communication. IBM Journal of Research and Development 38(6) (1994)
7. van de Geijn, R., Watts, J.: SUMMA: Scalable universal matrix multiplication algorithm. Concurrency: Practice and Experience 9(4), 255–274 (1997)
8. Gunnels, J., Lin, C., Morrow, G., van de Geijn, R.: A flexible class of parallel matrix multiplication algorithms. In: Proceedings of First Merged International Parallel Processing Symposium and Symposium on Parallel and Distributed Processing (1998 IPPS/SPDP '98), pp. 110–116 (1998)
9. Low, T.M., Milfeld, K., van de Geijn, R., Zee, F.V.: Parallelizing FLAME code with OpenMP task queues. Technical Report TR-04-50, The University of Texas at Austin, Department of Computer Sciences (December 2004)
10. Van Zee, F.G., Bientinesi, P., Low, T.M., van de Geijn, R.A: Scalable parallelization of FLAME code via the workqueuing model. ACM Trans. Math. Soft. (2007)(submitted)
11. Stewart, G.W.: Communication and matrix computations on large message passing systems. Parallel Computing 16, 27–40 (1990)
12. Lichtenstein, W., Johnsson, S.L.: Block-cyclic dense linear algebra. Technical Report TR-04-92, Harvard University, Center for Research in Computing Technology (January 1992)
13. Hendrickson, B.A., Womble, D.E.: The torus-wrap mapping for dense matrix calculations on massively parallel computers. SIAM J. Sci. Stat. Comput. 15(5), 1201–1226 (1994)
14. Dongarra, J., van de Geijn, R., Walker, D.: Scalability issues affecting the design of a dense linear algebra library. J. Parallel Distrib. Comput. 22(3) (September 1994)
15. Addison, C., Ren, Y.: OpenMP issues arrising in the development of parallel BLAS and LAPACK libraries. In: EWOMP (2001)
16. Chan, E., Ortí, E.S.Q., Ortí, G.Q., van de Geijn, R.: Supermatrix out-of-order scheduling of matrix operations for smp and multi-core architectures. In: SPAA (2007)(submitted)
17. Kurzak, J., Dongarra, J.: Implementing linear algebra routines on multi-core processors with pipelining and a look ahead. LAPACK Working Note 178 UT-CS-06-581, University of Tennessee (September 2006)

18. Blackford, L.S., Choi, J., Cleary, A., D'Azevedo, E., Demmel, J., Dhillon, I., Dongarra, J., Hammarling, S., Henry, G., Petitet, A., Stanley, K., Walker, D., Whaley, R.C.: ScaLAPACK Users' Guide. SIAM (1997)
19. van de Geijn, R.A.: Using PLAPACK: Parallel Linear Algebra Package. MIT Press, Cambridge (1997)
20. Bientinesi, P., Quintana-Ortí, E.S., van de Geijn, R.A.: Representing linear algebra algorithms in code: The FLAME APIs. ACM Trans. Math. Soft. 31(1), 27–59 (2005)

Task Scheduling for Parallel Multifrontal Methods

Olivier Beaumont and Abdou Guermouche

INRIA Futurs LaBRI, UMR CNRS 5800
Bordeaux, France
Olivier.Beaumont@labri.fr, Abdou.Guermouche@labri.fr

Abstract. We present a new scheduling algorithm for task graphs arising from parallel multifrontal methods for sparse linear systems. This algorithm is based on the theorem proved by Prasanna and Musicus [1] for tree-shaped task graphs, when all tasks exhibit the same degree of parallelism. We propose extended versions of this algorithm to take communication between tasks and memory balancing into account. The efficiency of proposed approach is assessed by a set of experiments on a set of large sparse matrices from several libraries.

Keywords: Sparse matrices, multifrontal method, scheduling, memory.

1 Introduction

The solution of sparse systems of linear equations is a central kernel in many simulation applications. Because of their robustness and performance, direct methods can be preferred to iterative methods. In direct methods, the solution of a system of equations $Ax = b$ is generally decomposed into three steps: (i) an analysis step, that considers only the pattern of the matrix, and builds the necessary data structures for numerical computations; (ii) a numerical factorization step, building the sparse factors (e.g., L and U if we consider an unsymmetric LU factorization); and (iii) a solution step, consisting of a forward elimination (solve $Ly = b$ for y) and a backward substitution (solve $Ux = y$ for x).

In this paper, we will work on an existing parallel sparse direct solver, MUMPS [2] (for MUltifrontal Massively Parallel Solver). We will study how to improve the parallel behavior of the solver. The main idea is to use theoretically proved techniques to improve the global behavior of the solver. Thus, Section 2 will be devoted to the presentation of parallel multifrontal method. Then, we will focus in Section 3 on the theoretical model and its application to MUMPS solver. We will present the adaptation of the algorithm for scheduling parallel tasks on homogeneous platforms proposed by Prasanna and Musicus [3,1] in Section 3.2. Finally, we present in Section 4 an experimental comparison between the existing MUMPS scheduling strategies and the techniques inspired from the work of Prasanna and Musicus. We will assess the interest and the limitations of new proposed approaches and then draw conclusions and give some words on future work.

A.-M. Kermarrec, L. Bougé, and T. Priol (Eds.): Euro-Par 2007, LNCS 4641, pp. 758–766, 2007.

2   Parallel Multifrontal Method

We present in this section the parallel multifrontal method as implemented in
the software package MUMPS [2].

2.1   Task Graphs Within MUMPS

MUMPS uses a combination of static and dynamic approaches. The tasks dependency graph is indeed a tree (also called *assembly tree*), that must be processed
from the leaves to the root. Each node of the tree represents the partial factorization of a dense matrix called *frontal matrix* or *front*. Once the partial
factorization is complete, a block of temporary data (i.e. contribution block) is
passed to the parent node. When contributions from all children are available
on the parent, they can be consumed or assembled (*i.e.* summed with the values
contained in the frontal matrix of the parent). Contribution blocks represent
temporary data of the algorithm whereas factors represent final data.

The shape of the tree and costs of the tasks depend on the linear system to
be solved and on the reordering of the unknowns of the problem. Furthermore,
tasks are generally computationally larger near to the root of the tree where the
parallelism of the tree is limited. Figure 1(a) summarizes the different types of
parallelism available in MUMPS:

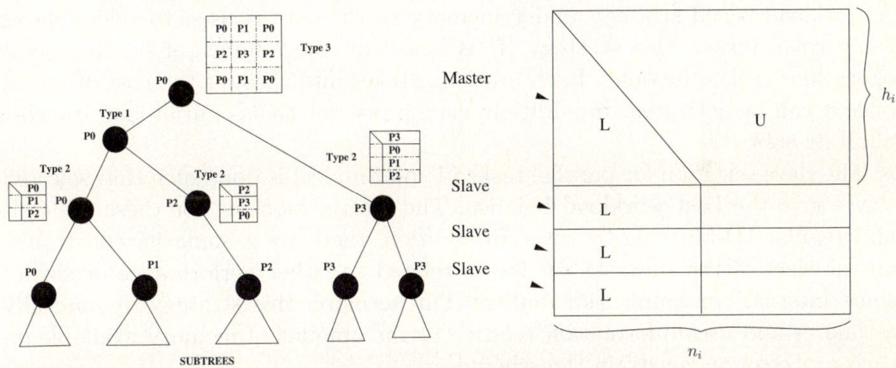

(a) Example of distribution of an assem-   (b) Distribution of slave tasks for unsym-
bly tree over four processors.   metric cases.

Fig. 1. Parallelism management in MUMPS

The first type only uses the intrinsic parallelism induced by the tree (since
branches of the tree can be processed in parallel). A type one node is a sequential
task, that can be activated when results from children nodes have been communicated. Leave subtrees are a set of tasks all assigned to the same processor.
Those are determined using a top-down algorithm [4] and a subtree-to-process

mapping is used to balance the computational work of the subtrees onto the processors. The second type corresponds to parallel tasks; a 1D parallelism of large frontal matrices is applied: the front is distributed by blocks of rows. A *master* processor is chosen statically during the symbolic preprocessing step and all the others (*slaves*) are chosen dynamically by the master from a set of *candidate processors* based on load balance considerations, which can be either the number of floating-point operations still to be done, or the memory usage. The number and the choice of candidate processors is guided by a relaxed *proportional mapping* (see Pothen and Sun [5]) consisting of a recursive assignment of processors to subtrees according to their associated computational work. Note that once the partial factorization is done, the *master* processor eliminates the first block of rows, while slaves perform the updates on the remaining Schur complement (see Figure 1(b)). Finally, the task corresponding to the root of the tree uses a 2D parallelism, and do not require dynamic decisions: ScaLAPACK [6] is applied, with a 2D block cyclic static distribution.

The choice of the type of parallelism is done statically and depends on the height in the tree, and on the size of frontal matrices. The mapping of the masters of parallel tasks is static and only aims at balancing the memory of the corresponding factors. During the execution, several slave selections can be made independently by different master processors.

2.2 Dynamic Scheduling Strategy

A workload-based strategy under memory constraints is used to select slaves for parallel tasks. This strategy [7] is based on the number of floating-point operations still to be done. Each processor takes into account the cost of a task once it can be activated. In addition, each processor has as initial load the cost of all its subtrees.

The slave selection for parallel tasks (Type 2 nodes) is done such that selected slaves give the best workload balance. The matrix blocking for these nodes is an irregular 1D-blocking by rows. In addition, there are granularity constraints on the sizes of the subtasks for issues related to either performance or size of some internal communication buffers. Furthermore, this strategy dynamically estimates and uses information relative to the amount of memory available on each processor to constrain the schedule.

3 Load Balancing and Minimization of Communication Cost

3.1 Related Works

In this section, we consider the problem of finding a schedule that both balances the load throughout the computation and minimizes the overall volume of communications induced by the algorithm. The problem of balancing memory requirements between processors will be addressed in Section 3.2.

Let us recall the algorithm defined in Section 2 from a scheduling point of view. The task graph corresponding to the execution of MUMPS is a tree (see Figure 1(a)) and communications take place along the edges of the tree. Each node of the tree can in turn be executed on several processors, which means that we do not consider tasks at the finest level of granularity: in the context of MUMPS, tasks are associated to partial LU decompositions.

This approach is closely related to malleable tasks scheduling (see [8] for a survey). A malleable task is a computational unit that can be itself processed in parallel. For each possible number of used processor, the time to process the malleable task is given, and communications are taken into account via a penalty factor. In [9], the authors propose a $4(1+\epsilon)$ approximation algorithm for scheduling trees of malleable tasks, if communications between malleable tasks are not taken into account. In the context of MUMPS, these theoretical results can nevertheless be improved, since all malleable tasks correspond to the same routine (partial LU factorization) on different data and for different problem size. In this context, all malleable tasks have the same profile (i.e. the penalty depends only on the number of processors, but not on the specific data). This problem has been addressed by Prasanna and Musicus [3,1]. In their model, it is assumed that the execution time of any malleable task is given by $\frac{L}{p^\alpha}$, where L is the length of the task on one processor, p is the number of processors allocated to this task and α is a penalty factor, that expresses the degree of parallelization of the malleable task (α close to 1 corresponds to ideal parallel task, whereas α close to 0 means that the task is intrinsicly sequential). It is worth noting that α does not depend on L or p (the value of α for our specific application will be discussed in next Section). For trees of such regular malleable tasks, Prasanna and Musicus [1] propose an optimal algorithm (that nevertheless allots rational number of processors to tasks), that will be described in more details in Section 3.2.

3.2   Parallelization of Factorization in MUMPS

Fitting the model. We experimentally measured values of α using frontal matrices having an order (n_i) of 10000 and various sizes of master task (an illustration of h_i is given in Figure 1(b)). We observed that for reasonable ratios of h_i/n_i, α has a constant value of 1.15 (note that if a task has a large master part, we can split it into a chain of tasks that have reasonable size of master tasks). This means that α is constant for all the tasks of the tree. Note that, the α parameter may vary depending on the platform used (network characteristics, processor speed, ...).

However, we also observe super-linear speedups (in the sense that α is larger than 1). This surprising behavior is due to the fact that the processor that handles thes master task does not take part to the processing of the slave part of the task (a processor cannot be a master and a slave in the same partial factorization). Thus, for example, when doubling the number of processors for a given task from 2 to 4, the number of slaves varies from 1 to 3 (inducing a speedup near to 3 in an ideal case). This explains the observed super-linearity.

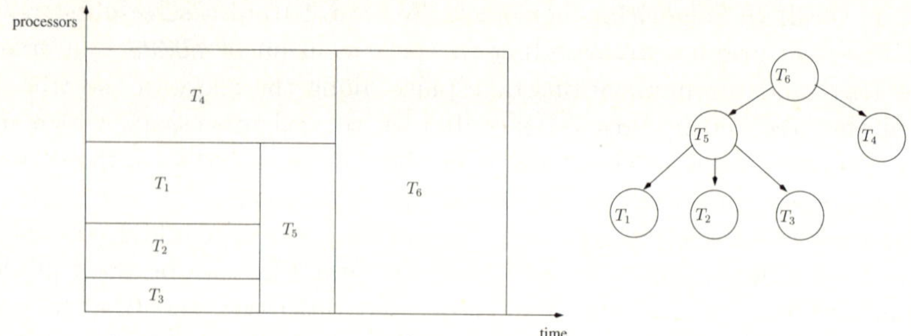

Fig. 2. Description of the optimal schedule

We plan to change the MUMPS management of these tasks to allow a processor to be both a master and a slave in the same task. A more powerful implementation would be to allow the parallelization of the master part of the task, and would remove this super-linear speedup.

Proposed solution. In previous Section, we have shown that current implementation of partial LU factorization induces a super-linear speedup, thus preventing us to use directly optimality results derived by Prasanna and Musicus [3,1]. Nevertheless, we also observed that this super-linear speedup will disappear with the newly version of 1D partial LU factorization. Moreover, the optimal solution for $\alpha > 1$ would lead to execute each task on the whole set of processors, what would induce huge communication costs. Therefore, we propose to use the mapping algorithm proposed by Prasanna and Musicus even if $\alpha > 1$. It has been proved [3] that if $\alpha < 1$, the solution is as depicted in Figure 2. More precisely, the set of processors allocated to a task does not change over time, the set of processors allocated to a given node is the same as the set of processors allocated to its subtree, and all the children of a given task finish their execution at the same time. Given these observations, it is possible to determine the exact number of processors allocated to each task of the tree. For instance, if p_5 denotes the number of processors allocated to task T_5, the number of processors p_i allocated to tasks T_i, $1 \leq i \leq 3$ is given by

$$p_i = p_5 \frac{L_i^{\frac{1}{\alpha}}}{L_1^{\frac{1}{\alpha}} + L_2^{\frac{1}{\alpha}} + L_3^{\frac{1}{\alpha}}},$$

where L_i denotes the execution time of T_i on one processor.

As already noted, the optimal solution associates each task T_i to a fractional number of processors, given as an interval with rational bounds $[l_i, r_i] \subset [0, P]$, where P is the overall number of processors. Therefore, each task T_i is associated to $k_i \geq 0$ processors (denoted as base processors in what follows) that will be completely devoted to the execution of T_i, and possibly $k_i'(k_i' \leq 2)$ extra processors (denoted as candidate processors in what follows), that will be also partially allocated to other tasks during the execution of T_i. Candidate processors for task T_i will be dynamically allocated to the execution of T_i during the execution, given their current load at the beginning of the execution of T_i.

Proposed solution therefore achieves perfect load balancing (all processors work during the whole process) and good locality of communications (since the set of processors allocated to a given task is the union of the set of processors allocated to its children tasks). On the other hand, with the current implementation of partial LU factorization in MUMPS, the length of the schedule is not optimal since $\alpha > 1$. Note that the main difference between this approach and the default MUMPS scheduling strategy (see Section 2), which is based on proportional mapping, is that it cases the communications induced by parallelism into account through the task performance model.

Minimization of memory requirements. In this section, we consider the minimization of memory requirements once the tree of tasks and the set of processors (base processors and candidate processors) allocated to each task have been determined. More specifically, we concentrate on the minimization of the memory needed to store L and U factors and do not consider the memory needed to store intermediate factors that will be sent (and then removed from memory) to the father task. Let us consider the elementary partial factorization depicted in Figure 1(b). The processor responsible for computing the upper part will have to store $n_i h_i$ elements of L and U, whereas the $(p-1)$ processors responsible for the lower part will be in charge of storing $\frac{h_i(n_i-h_i)}{p-1}$ elements of L. In order to decide which processor will be in charge of computing the upper part, we propose a heuristic based on the $\frac{4}{3}$ approximation algorithm Minimum Multiprocessor Scheduling [10] where task lengths are independent of the processor's choice [11]. More precisely, we sort the tree nodes by decreasing values of h_i and we consider tasks in this order, allocating the upper part of task T_i to the less loaded processor, while updating the memory charge for all processors participating to the computation of T_i. We present in next section the results obtained by this simple heuristic.

4   Experimental Results

We should first mention that the algorithms presented in Sections 3.2 and 3.2 have been implemented inside the MUMPS package. In order to compare the proposed algorithm with the default scheduling strategy described in Section 2, we experiment them on several problems (see Table 1) extracted from various sources including Tim Davis's collection at University of Florida[1] or the PARASOL collection[2]. The tests have been performed on the IBM SP system of IDRIS[3] composed of several nodes of either 4 processors at 1.7 GHz or 32 processors at 1.3 GHz. Note that all the experiments are done using unsymmetric matrices. The extension to symmetric ones is natural.

---

[1] http://www.cise.ufl.edu/ davis/sparse/
[2] http://www.parallab.uib.no/parasol
[3] Institut du Dveloppement et des Ressources en Informatique Scientifique.

Table 1. Test problems

Matrix name	Order	nnz	$nnz(L|U) \times 10^6$	Description
CONV3D64	836550	12548250	2693.9	provided by CEA-CESTA; generated using AQUILON (http://www.enscpb.fr/master/aquilon)
GRID	729000	7905960	1430.5	Regular 90-90-90 grid
MHD1	485597	24233141	1250.3	unsymmetric magneto-hydrodynamic 3D problem, provided by Pierre Ramet
ULTRASOUND80	531441	33076161	981.4	Propagation of 3D ultrasound waves, provided by M. Sosonkina

We have tested the algorithms presented in previous Sections on 32 and 64 processors of the above-described platform. By default, we used the METIS package [12] to reorder the variables of the matrices. In the following experimental study we will use `default` to denote the default scheduling strategy of the solver, `P&M` to denote the experiments where we used the Algorithm presented in Section 3.2, and `P&M*` to denote the variant described in Section 3.2. In addition, for each set of experiments, we forced the same tree topology (same amount of splitting etc ...). Finally, in the case of `P&M` approaches, we strictly follow the static schedule, produced during the analysis phase, during factorization.

We report in Table 2 factorization times on the IBM platform using 32 and 64 processors. The factorization time is reduced when using the `P&M` approaches on both 32 and 64 processors. The gains can reach more than 30% (for the conv3d64 on 64 processors) and variants of the Prasanna & Musicus do not strongly differ in terms of factorization time. Note that on 32 processors, the `P&M*` seems to be less efficient than the `P&M` approach. This is principally due to our round-off management when assigning tasks to processors. Indeed, as mentioned in Section 3.2, if a processor is not fully assigned to a task, we choose dynamically if it will take part to the processing of the task or not. These dynamic decisions explain the difference in terms of performances between the two variants. Presented results illustrate the good behavior of both variants and show that they produce a better balanced schedule than the default strategy.

We will now focus on the volume of data exchanged during the factorization. We give in table 3 the volume of communication measured during the factorization. We can see that both `P&M` and `P&M*` give a very reduced amount of communication (up to a factor of 2.5). This reduction is explained by the fact that the algorithms based on Prasanna and Musicus approach have a natural locality in the distribution of tasks over processors (see Section 3.2). In addition,

Table 2. Factorization time (in seconds) on 32 and 64 processors. Not enough memory was available to run the conv3d64 matrix on 32 processors.

Matrix name	Time for facto. (32 procs.)			Time for facto. (64 procs.)		
	default	P&M	P&M*	default	P&M	P&M*
CONV3D64	-	-	-	390	280	274
GRID	237	199	208	169	117	115
MHD1	186	168	175	116	102	99
ULTRASOUND80	89	88	95	82	63	62

Table 3. Volume of communication (in GigaBytes) on 32 and 64 processors. Not enough memory was available to run the conv3d64 matrix on 32 processors.

Matrix name	Vol. of comm. (32 procs.)			Vol. of comm. (64 procs.)		
	default	P&M	P&M*	default	P&M	P&M*
CONV3D64	-	-	-	618	234	229
GRID	184	74	73	308	131	120
MHD1	138	66	61	219	110	103
ULTRASOUND80	79	34	34	167	77	69

Table 4. Peak of memory (in the left) and size of Factors per processors (in the right) (in millions of entries) on 32 and 64 processors. Not enough memory was available to run the conv3d64 matrix on 32 processors.

Matrix name	Mem. peak (32 procs.) / Size of factors (32 procs.)			Mem. peak. (64 procs.) / Size of factors (64 procs.)		
	default	P&M	P&M*	default	P&M	P&M*
CONV3D64	-	-	-	86 / 53	109 / 87	102 / 57
GRID	89 / 55	122 / 77	112 / 68	52 / 27	61 / 36	59 / 35
MHD1	79 / 48	108 / 56	121 / 56	41 / 21	56 / 33	53 / 32
ULTRASOUND80	59 / 37	82 / 49	82 / 49	30 / 19	45 / 25	46 / 25

the number of sequential subtrees assigned to a single processor is smaller in the case of Prasanna and Musicus variants (which means that more work is done without communications).

Finally, we will focus on the memory behavior of the different approaches. We report in Table 4 both the memory peak over the set of processors for performing the factorization and the final size of factors (we give the maximum size of factors over all processors). A first observation is that the memory behavior of the algorithms based on Prasanna & Musicus approach is worst than the `default` one. This is mainly due to the fact that in the `default` strategy, memory constraints are injected in both static and dynamic decisions. This leads to better memory behavior (especially for the management of temporary data (i.e. contribution blocks).

From the factor (terminal data) size point of view, we can see as expected that the `P&M*` approach has a slightly better behavior than the `P&M` one, what illustrates the benefits of the mechanism described in Section 3.2. However, we can also see that `default` strategy gives a more balanced distribution of the size of factors. This is due to the fact that in this approach the layer of sequential subtrees is determined by a combination of workload and memory criteria whereas in the `P&M` and `P&M*` strategies it is built based on workload information only. Thus, the size of sequential subtrees (which are considered as a single task) is bigger in the `P&M` and `P&M*` approaches, what makes the distribution of factors over processors more difficult.

5   Conclusion and Future Work

We presented in this paper a study of scheduling strategies for the parallel multifrontal method implemented in the `MUMPS` software package. We showed that the model of the application fits a state of the art model and how to adapt the scheduling algorithm implemented in the solver. Finally, we presented an experimental study showing the potential of the approach. We observe that new schedules improve the performances of the solver by achieving better load-balancing and a reduce volume of communication. However, we also observe that these techniques induce an increase of memory requirements. This last issue is critical, especially in the area of sparse direct solvers (where memory is often the bottleneck). Thus, we have to work on approaches that will slightly degrade performance to improve the memory behavior, either by injecting memory information during the static allocation or by dynamically relaxing proposed allocation. Another approach could be to study bi-criteria techniques aiming at finding the best tradeoff between these two criteria. Finally, we plan to study how this work can be extended to the context of parallel out-of-core sparse direct solvers (in this new context, I/Os have to be taken into account).

References

1. Prasanna, G.N.S., Musicus, B.R.: Generalized multiprocessor scheduling and applications to matrix computations. IEEE Trans. Parallel Distrib. Syst. 7(6), 650–664 (1996)
2. Amestoy, P.R., Duff, I.S., Koster, J., L'Excellent, J.Y.: A fully asynchronous multifrontal solver using distributed dynamic scheduling. SIMAX 23(1), 15–41 (2001)
3. Prasanna, G.N.S., Musicus, B.R.: Generalised multiprocessor scheduling using optimal control. In: SPAA, pp. 216–228 (1991)
4. Geist, A., Ng, E.: Task scheduling for parallel sparse Cholesky factorization. Int J. Parallel Programming 18, 291–314 (1989)
5. Pothen, A., Sun, C.: A Mapping Algorithm for Parallel Sparse Cholesky Factorization. SISC 14(5), 1253–1257 (1993)
6. Choi, J., Demmel, J., Dhillon, I., Dongarra, J., Ostrouchov, S., Petitet, A., Stanley, K., Walker, D., Whaley, R.C.: ScaLAPACK: A portable linear algebra library for distributed memory computers - design issues and performance. Technical Report LAPACK Working Note 95, CS-95-283, University of Tennessee (1995)
7. Amestoy, P.R., Guermouche, A., L'Excellent, J.Y., Pralet, S.: Hybrid scheduling for the parallel solution of linear systems. Parallel Computing 32(2), 136–156 (2006)
8. Dutot, P.F., Mounie, G., Trystram, D.: Scheduling parallel tasks: Approximation algorithms. In: Leung, J. (ed.) Handbook on Scheduling algorithms: Algorithms, Models and Performance Analysis, CRC press, Boca Raton, USA (2004)
9. Lepere, R., Mounie, G., Trystram, D.: An approximation algorithm for scheduling trees of malleable tasks. European Journal of Operational Research 142, 242–249 (2002)
10. Ausiello, G., Crescenzi, P., Gambosi, G., Kann, V., Marchetti-Spaccamela, A., Protasi, M.: Complexity and Approximation: Combinatorial optimization problems and their approximability properties. Springer, Heidelberg (1999)
11. Hochbaum, D.S., Shmoys, D.B.: Using dual approximation algorithms for scheduling problems: theoretical and practical results. J. ACM **34** (1987) 144–162
12. Karypis, G., Kumar, V.: MeTiS – A Software Package for Partitioning Unstructured Graphs, Partitioning Meshes, and Computing Fill-Reducing Orderings of Sparse Matrices – Version 4.0. University of Minnesota (September 1998)

Topic 11
Distributed and High-Performance Multimedia

Harald Kosh, Laurent Amsaleg, Eric Pauwels, and Björn Jónsson

Topic Chairs

In recent years, the world has seen a tremendous increase in the capability to create, share and store multimedia items, i.e. a combination of pictorial, linguistic, and auditory data. Moreover, in emerging multimedia applications, generation, processing, storage, indexing, querying, retrieval, delivery, shielding, and visualization of multimedia content are integrated issues, all taking place at the same time and - potentially - at different administrative domains. As a result of these trends, a number of novel and hard research questions arise, which can be answered only by applying techniques of parallel, distributed, and Grid computing.

The scope of this topic embraced issues from high-performance processing, coding, indexing, and retrieval of multimedia data over parallel architectures for multimedia servers, databases and information systems, up to highly distributed architectures in heterogeneous, wired and wireless networks.

We received 8 papers this year. We thank all the authors for their submissions The submitted papers mostly dealt with issues in distributed video delivery, parallel coding, storage and transmission of multimedia data. After a rigorous review of these papers by all four committee members, two papers were selected for presentation and publication. The first one, entitled "An Evaluation of Parallelization Concepts for Baseline-Profile Compliant H.264/AVC" is an excellent evaluation of different multithreaded implementations for parallel decoding of H.264/AVC encoded videos. The second one, entitled "DynaPeer: A Dynamic Peer-to-Peer Based Delivery Scheme for VoD Systems" defines and implements an original Virtual Server to enhance the VoD delivery policy in Peer-to-Peer networks.

A.-M. Kermarrec, L. Bougé, and T. Priol (Eds.): Euro-Par 2007, LNCS 4641, p. 767, 2007.
© Springer-Verlag Berlin Heidelberg 2007

DynaPeer: A Dynamic Peer-to-Peer Based Delivery Scheme for VoD Systems*

Leandro Souza[1], Fernando Cores[2], Xiaoyuan Yang[1], and Ana Ripoll[1]

[1] Computer Architecture and Operating System
Universitat Autònoma de Barcelona, 08193 Bellaterra, Spain
[2] Computer Science and Industrial Engineering
Universitat de Lleida, St. Jaume II, 69, 25001 Lleida, Spain
Leandro@aomail.uab.es, Fcores@diei.udl.es, Xiao@aomail.uab.es,
Ana.Ripoll@aub.es

Abstract. Supporting Video-on-Demand (VoD) services in Internet is still a challenging issue due to high bandwidth requirement of multimedia contents and additional constraints imposed by such environment: higher delays and jitter, network congestion, non-symmetrical clients' bandwidth and inadequate support for multicast communications. This paper presents DynaPeer, a peer-to-peer VoD delivery policy designed for Internet environment. Our design defines a Virtual Server, which is responsible for establishing a group of peers, enabling service for new client requests by aggregating the necessary clients' resources. Virtual Server operates in both unicast and multicast environments, thereby improving system performance. To demonstrate the effectiveness of DynaPeer, we have developed an analytical model to evaluate its performance, understood as the server-load reduction due to request service distributed among peers. We conducted a performance comparison study of our proposal with classic unicast, multicast (Patching) and other P2P delivery schemes, such as Pn2Pm, Chaining and Promise, improving their performance by 45%, 59%, 74% respectively, even when taking into account Internet constraints.

Keywords: On-Demand Media Streaming, Peer-to-Peer systems, Internet VoD.

1 Introduction

Advances in network technology will provide the access to new generation, full-interactive and client-oriented services such as Video-on-Demand. Through these services, users will be able to view videos from remote sites at any time. However, serving video files to a large number of clients in an "on demand" and "real time" way imposes a high bandwidth requirement on the underlying network and server.

To spread the deployment of VoD systems, much research effort [4][6][7][11] has been focused on the delivery process of multimedia contents, exploiting both unicast and multicast techniques, trying to reduce the bandwidth consumption and provide better system streaming capacity. In spite of the success of these techniques, their

---

* This research is supported by the MEyC-Spain under contract TIN 2004-03388.

A.-M. Kermarrec, L. Bougé, and T. Priol (Eds.): Euro-Par 2007, LNCS 4641, pp. 769–781, 2007.

scalability requirements to provide service on a large-scale system, such as Internet, is still limited by server and network resources.

Recently research has looked to the peer-to-peer (P2P) paradigm as a solution to decentralize the delivery process among peers, alleviating the server load or avoiding any server at all. P2P systems for streaming video have generated important contributions. In the Chaining delivery policy [7], clients cache the most recently received video information in the buffer and forward it to the next clients using unicast channels. The P2Cast [4] and cache-and-relay [8] allow clients to forward the video data to more than one client, creating a delivery tree or ALM. However, neither chaining or ALM delivery policies consider client output-bandwidth limitation in collaboration process, which limits their usage to dedicated network environments. Other VoD P2P-based architectures such as PROMISE [1], CoopNet [9] or BitVampire [10] assume that a client does not have sufficient output bandwidth to generate the complete information to other clients, using n clients to send the required bandwidth. However, they assume that clients work as proxies storing whole video information. Furthermore, system scalability is compromised due to unicast communication. To solve the scalability problem, in previous works [11] we proposed P^n2P^m architecture that takes advantage of multicast technology on the client side. This architecture works by exploiting the clients non-active resources in two ways: first, it allows clients to collaborate with the server in the delivery of initial portions of video, patches streams; and second, it establishes a group of clients to store the available information of an existent server multicast channel to eliminate it. P^n2P^m also requires that output bandwidth is, at least, the same as video play-rate.

The Internet environment imposes further restrictions to P2P streaming schemes in order to provide VoD service. First, providing service over non-dedicated network environments implies no QoS guaranties, transmission congestion, packet loss and variable point-to-point bandwidth. Second, non-symmetrical clients' bandwidth involves a careful delivery strategy due to clients' output-bandwidth limitation. Third, Internet Service-Provider[1] (ISP) networks differ on supporting (or not) the IP-Multicast delivery technology. Finally, content copyright protection affects content storage limited to non-persistent devices. Thus, content on peers is only available over a limited period of time.

To solve the above challenges, we propose a new delivery scheme called DynaPeer, based on a P2P paradigm for an Internet environment. DynaPeer differs from the previous P2P schemes in certain key aspects. First, DynaPeer works with unicast and multicast communication techniques, depending on the technology available to the ISP network. The combination of unicast and multicast could allow DynaPeer to dynamically exploit the IP multicast mechanism, achieving better network utilization and providing system scalability. Second, this scheme takes into consideration the non-symmetric characteristics of client bandwidth, which is in accordance with current xDSL technology. Third, our delivery scheme assumes the non-homogeneity characteristics founded on a non-dedicated network such as Internet, which allows us to design a realistic delivery scheme for VoD services. To the best of our knowledge, our proposal is the first VoD delivery scheme that

---

[1] Currently, certain ISPs provide multicast technology over the xDSL, through DsLAN technology. Authors in [3], have demonstrated its usage and have proposed a connectivity architecture for multicast ISPs.

combines non-dedicated network environment, asymmetrical connection on the client side and multicast delivery technique for client collaborations.

The remainder of this paper is organized as follows. In section 2 we present DynaPeer design. In section 3, an analytical model to evaluate DynaPeer performance is presented. Section 4 shows the performance evaluation through the analytical model. In Sections 5, we indicate the main conclusions and future works.

2 DynaPeer Design

DynaPeer is not a server-less system; rather, it combines a server-based architecture with a P2P delivery scheme. The server holds the entire system catalogue, acting as seeds for the multimedia content. It is also responsible for establishing every client-collaboration process. DynaPeer takes advantage of client collaboration to decentralize the server-delivery process, eventually shifting streaming load to peers.

The explanation of DynaPeer is divided into 3 parts. Section 2.1 describes the collaboration model of DynaPeer and in sections 2.2. and 2.3, we present P2P delivery schemes over unicast and multicast environments.

2.1 Collaboration Model

The principle of DynaPeer is based on clients' collaborations in which clients (peers) make their idle-resources available so as to generate a complete, or partial, stream for incoming clients. In our system, a peer is an active client who plays a given video and is able to collaborate with the system.

Peers' collaboration capacity is limited by peer resources (bandwidth and storage) and available video data. In our case, we consider that peers have an asymmetrical input/output bandwidth (input bandwidth is, at least, the same as video play-rate and output bandwidth is supposed to be lower than video play-rate) and a limited buffer capacity. Having insufficient output bandwidth to transmit a complete video stream implies that several peers (N_i, defined by the ratio between video i play-rate and peers' output bandwidth) have to collaborate in order to provide service for a complete streaming session. Furthermore, due to copyright protection and peers' limited buffer capacity, peers cannot permanently store a complete video. Therefore, they can only serve, on the fly, video data previously received from an active streaming session and temporally stored on clients' buffer.

All collaborations in DynaPeer are managed by the *Virtual Server* (VS). The objective of a VS is to establish a group of peers, aggregating sufficient resources, enabling the service for new clients' requests. Another important function attributed to the VS is to perform distributed control tasks among peers in a distributed way, minimizing server involvement. A virtual server (Fig. 1), denoted by $VS(j,s,w)$, is a logical entity defined as a set of peers that collaborate in a delivery process to offset s of video j, during a period of time W. The VS's service capacity is achieved by peers' *resource aggregation* and will depend on the number of peers integrating this. The sum of peers' input (I_i) and output-bandwidth (O_i) will determine VS input and output-stream capacity.

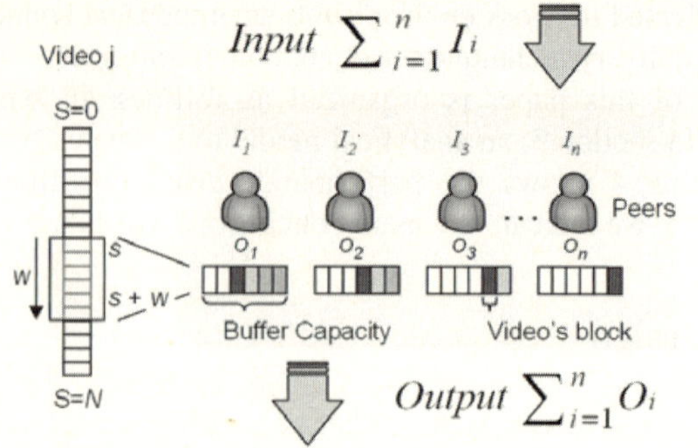

Fig. 1. DynaPeer Virtual Server

Initially, it is assumed that j is the video that all peers forming VS are reproducing. Video data available on VS is defined by s (first video block currently stored on VS) and the *collaboration window W* (period of time that any block remains stored on a VS before it is replaced). Outside $[s, s+W]$, the interval defined by W, the VS is unable to make the collaboration as video data is not available in its buffer. Therefore, to provide full service for a streaming session, DynaPeer policies have to implement a *sliding window* over whole video. In this way, once the collaborative buffer is full, the following blocks received $(s+W, s+W+1,...)$ replace the older blocks $(s, s+1,...)$.

In DynaPeer, each VS is bound to an existent ongoing channel. Thus, the number of peers integrating a VS will depend on the peer's collaboration window and video request arrival-rate. To enlarge the collaboration window, we need to improve peers' buffer capacity (B). DynaPeer manages the peer's buffer by storing only data proportional to the contribution that can be carried out by peers' output-bandwidth (i.e, the video data kept for future collaboration for a video j with a play rate Pr_j, will be determined by Pr_j/O_i relation and buffer capacity). We term this strategy *extended buffer* capacity. The extended buffer allows VS to provide a larger collaboration window, increasing peers' collaboration probability and system efficiency.

The VS manages the collaborations by two different levels: full-stream and partial-stream collaboration. Full-stream collaboration is achieved when the VS has sufficient resources to deliver a full stream to a new client. In this case, the whole video stream will be delivered by the VS. Otherwise, if there are not enough resources, the VS proceeds with the partial stream collaboration. In this case, VS contributes with the new client request proportionally to their service capacity, and the server will be involved in the delivery process, sending data to the client in order to complete the service and to guarantee the QoS. Of course, every VS begins applying partial-stream collaboration and when it has sufficient size and resources, it switches to full-stream.

2.2 DynaPeer Unicast

Assuming that not all ISPs are powered with multicast technology in their access network, our proposal also exploits the delivery scheme by using unicast both from server and client side.

The mandatory condition for the collaboration process is that the requesting peer arrives inside the collaboration window W of the required VS. Following these conditions, if there are no candidate peers available for collaboration in a VS, the server is responsible for opening a channel to serve the incoming request. If the number of peers inside a VS is not enough to take the collaboration, the Vs performs a partial stream collaboration. If there are sufficient candidate peers in the VS (enough resources) to generate a complete stream, a new channel will be opened from the peers to attend incoming request. All requesting peers, automatically, become candidate peers inside a new VS in the system.

Fig. 2a shows a snapshot of Unicast collaboration mechanism in minute six of DynaPeer stream process. In this example, we assume that a video stream must be served by three clients (Ni=3). The first peer (peer 1) is being directly attended by the server and it defines the collaboration window (W_1) for the VS_1. Peers 2 and 3 are also being attended by the server and both are integrated in VS_1. In minute three, the VS_1 has achieved its delivery capacity for one complete stream and when peer 4 makes a request in minute 4 (inside W_1) it is attended by VS_1 (Fig. 2a I), switching to full stream collaboration mode. Automatically, peer 4 starts another Virtual Server VS_2. As peer 5 arrives, the VS_1 does not have available service capacity to serve it. The VS_2 (composed by peer 4) applies partial stream collaboration with the server in the streaming process to peer 5 (fig. 2a II). The same occurs with peer 6 request.

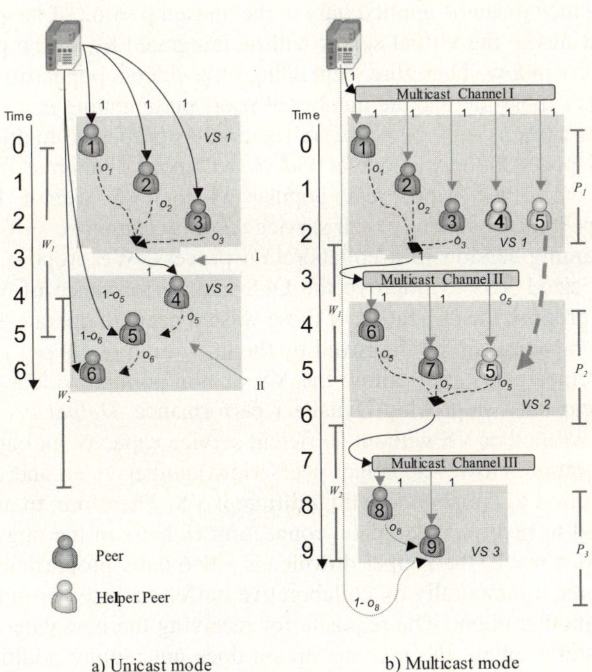

a) Unicast mode   b) Multicast mode

Fig. 2. DynaPeer Snapshot( )

2.3 DynaPeer Multicast

Using the multicast scheme, DynaPeer allows the streaming process for clients in a multi-source/multi-destination way, better exploiting the network capacity of ISPs. The VSs are responsible for creating multicasts channels, from the client's side[2], serving incoming client's requests. In this way, DynaPeer avoids any extra server's resource for serving contents that have already been started by other peers.

The collaboration process for multicast environment works by letting a new peer joining an ongoing multicast channel (*complete stream)* and still receives the entire video data stream. For new requests for the same video, the VS acts in two different ways: First, if an incoming peer can join an ongoing multicast channel, the VS delivers only the missing portion of the requested video in a separate unicast channel, *patch stream*, using the clients' output-bandwidth capacity. The period of time that a peer can join an ongoing multicast channel is called Patching Window (denoted as P time), and it depends on client buffer capacity. Second, if a requesting peer does not have sufficient buffer space for joining the ongoing channel (arrival time $> P$), the VS starts a new multicast channel for the incoming peer. Once patching window finishes, DynaPeer begins the multicast collaborative window, whose size depends on buffer available for collaboration after patching policy. VS only can create a new multicast channel if the next client request arrives inside the collaboration window. Different to unicast delivery, in multicast, the peers need their buffer to store patching information arising from the ongoing channel. Thus, extended buffer capacity for collaboration is more limited, since it can be applied only in the unused portion of the peer's buffer.

In multicast mode, the virtual server will be integrated by all the peers that arrive inside patching window. Therefore, depending on a video's popularity and on clients' requests rate, it is possible that the number of peers participating in a VS can be larger than N_i. In this case, as only N_i peers are required to propagate multicast stream, the remaining VS peers for most popular videos will not collaborate in the streaming process. On the other hand, less popular videos VS cannot have sufficient collaborators peers; consequently their service capacity cannot be sufficient to fulfill a complete streaming session in a collaboration process. We propose to use the idle peers on over-sized VSs to improve the QoS and performance of VoD system. In particular, we propose the utilization of those wasted peers to operate as *Helper peers*.

VS service capacity can be improved by the utilization of *Helper peers*. The main function of Helper peers is to allow the VS of non-popular videos to achieve full collaboration capacity, improving DynaPeer performance. *Helper peers* are allocated to collaborate with other VS without sufficient service capacity for carrying out a full stream collaboration. However, helper peers view another video and do not have the video data required to collaborate with a different VS. Therefore, to assist a VS, they previously need to receive video data, connecting Helpers in the ongoing channel of assisted VS. As a result, the Helper downloads video data, proportional to its output-bandwidth, stores it temporally on collaborative buffer and uses its output-capacity to delivery it to another client. The requisite for receiving the new video before serving it, will be wasteful unless the ingoing stream does not require additional resources. This constraint let this approach feasible only with multicast communications.

---

[2] The mechanism of generating multicast trees from clients to other clients is orthogonal to the analysis presented in this article. For instance, we assume the mechanism proposed in [1].

Fig. 2b shows a snapshot of the system in multicast configuration. Client arrival rates are shown in figures (time bar). Peer 1 has sent a video request to the server that has started a multicast channel to attend it. A few minutes later and inside P_1 time, clients 2, 3, 4 and 5 request video j. Theses clients were joined to multicast channel and they are incorporated to VS_1. In time 2, patching window finishes and DynaPeer begins the multicast collaboration window (W_1). After P_1 time, but also inside W_1 time, peer 6 requests the same content j from the server. DynaPeer selects peers 1, 2 and 3 (N_i=3) to deliver the video and starts a new multicast channel (Channel II) for attending to the client's request. Once channel propagation is made, peer 4 and 5 is set as a helper peer. In time 5, peer 7 request video j. It arrives inside P_2 time and could be joined to multicast channel II. Due to time constraints, peer 8 request was unable to join either multicast channel I or II. The only possible alternative is to create a new channel. The VS_1 is unable to create this channel due to its collaboration window W_1 is surpassed by peer arrival time. Regardless that VS_2 was also incomplete in its total stream capacity to serve the request, it could achieve its completed service capacity by the utilization of VS_1 helper peer 5. At that moment, VS_2 could start the delivery process to the requesting peer, generating multicast channel III. Finally peer 9 arrives in minute 9, and it can join the ongoing multicast channel III.

3 Mathematic Analysis

In this section we present an analytical model for evaluate the DynaPeer performance. The main objective of our model is to evaluate the performance that can be achieved by DynaPeer and related P2P delivery policies. In this case, performance is understood as the server-load reduction (streams) due to request service distributed among peers (S^*).

To perform this analysis some assumptions are made from points of view of architecture, clients and system work-load. The model assumes a VoD system with a single centralized server[3], which stores whole system catalogue. We take in consideration asymmetrical bandwidth behavior for clients. Moreover, this output-bandwidth is not enough to provide video at the required play-rate. Also, the model does not take in consideration clients' or servers' failures during a streaming session.

To undertake the model complexity, we do not handle dynamic behavior of the VoD system (network congestion and jitter, and variable client bandwidth) and we assume average values for clients' output-bandwidths (O) and clients' buffer capacity (B). Furthermore, we suppose that video is encoded with a Constant Bit-Rate and all videos' catalogue has the same length (L) and the same bandwidth requirements (Pr). Symbols used in the analysis are listed in Table 1.

In model development, we have evaluated the streaming capacity required by the server without the utilization of P2P policies as reference, and afterward we have evaluated the server load reduction resulting of the incorporation of DynaPeer schemes. Due to space limitations we only present DynaPeer multicast model.

---

[3] The model can also be directly applied for others architectures composed by multiple servers (Proxy or CDN based architectures).

Table 1. Analytical Model Main Parameters

Symbol	Explanation	Symbol	Explanation
S*	Server Load (streams)	S^i_{C*}	Completed Streams for video i (streams)
M	Video catalog size	S^i_{P*}	Patch Streams for video i (streams)
L	Video Length (min)	W_i	P2P collaboration Window time
B	Peers mean Buffer size in minutes	λ_i	Requests arrival rate video i (req/min)
O	Peer Output Bandwidth (Mbps)	Pr_i	Play rate for Video i
N_i	Number of Peers for serve a stream for video i, $\quad N_i = \dfrac{Pr_i}{O}$	G_i	Number of Collaborative Peers in a Group

3.1 Multicast Performance Analysis

We present two performance models for DynaPeer multicast. First we start by analyzing basic DynaPeer multicast delivery policy. Then we proceed by presenting DynaPeer with Helpers analytical model.

DynaPeer Multicast

In multicast, there are two server costs to evaluate, the full stream cost (complete stream created by the server for incoming clients) and the patch stream cost. Therefore, the total server-load for DynaPeer based policy is given by the sum of total server-load of complete streams and patch streams:

$$S_{DynMul} = \sum_{i=1}^{M} Sc^i_{DynMul} + Sp^i_{DynMul} \tag{1}$$

To achieve DynaPeer functionality, clients able to participate in a collaboration process are grouped inside a collaboration group (G_i). The collaboration group (G_i) is achieved by the total number of candidate peers arriving in the patching window time, which is made up of a VS:

$$G_i = \begin{cases} B * \lambda_i, & \dfrac{1}{\lambda_i} < B \\[3mm] 1, & \dfrac{1}{\lambda_i} \geq B \end{cases} \tag{2}$$

Once patching window has finished, DynaPeer begins the collaborative window (W), whose size depends on the buffer storage available for collaboration after patching policy. Peers' available buffer capacity depends on their relative arrival time inside the patching window. For modeling purposes, we assume the worst collaboration buffer depending on candidate peers inside the collaboration group of video i (G_i), which is multiplied by N_i in order to attain the extended collaboration window:

$$W_i = \begin{cases} B_i - \dfrac{N_i \cdot (N_i - 1)}{\lambda_i}, & G_i \geq N_i \\[3mm] B_i - \dfrac{N_i \cdot (G_i - 1)}{\lambda_i}, & G_i < N_i \end{cases} \tag{3}$$

Then, the full stream cost, for a video i, is achieved by calculating the number of channels that the server must open to serve this video during a period of time (L). In DynaPeer, if there are sufficient peers inside the collaboration group to propagate the requested video i ($G_i \geq N_i$), the server needs only to open first stream. Subsequent streams are managed by peers. Otherwise, peers can collaborate only partially or not at all with the server (requiring the same streams as a central-server using patching policy, Sc^i_{Mul}). Therefore the server-required stream is defines as follows:

$$Sc^i_{DynMul} = \begin{cases} 1, & G_i \geq N_i \\[3mm] Sc^i_{Mul} - \left(Sc^i_{Mul} * \dfrac{G_i}{N_i} \right), & Gi < N_i \;\; and \;\; \dfrac{1}{\lambda_i} \leq W_i \\[3mm] Sc^i_{Mul}, & otherwise \end{cases} \tag{4}$$

P2P patch streams service for video i is managed using the streaming resources from VS_{j-1} and VS_j (under construction) to serve the incoming client's request. If there is at least one peer in the current VS_j, it collaborates proportionally with the main server to send patch streams (Sp^i_{Mu}). The previous VS_{j-1} only helps if it has free streaming resources ($G_i - N_i > 0$):

$$Sp^i_{DynMul} = \begin{cases} Sp^i_{Mul} - \left[\dfrac{Sp^i_{Mul}}{N_i} + \left(|G_i - N_i| * \dfrac{B}{N_i} \right) \right] * \dfrac{Sc^i_{Mul}}{L}, & G_i > 1 \\[3mm] Sp^i_{Mul}, & otherwise \end{cases} \tag{5}$$

DynaPeer Multicast with Helpers

Helpers are only used when there is at least one incoming request arriving inside the collaboration window time. Thus, the incoming request can take advantage of the virtual server created with helpers.

The number of available peers to perform Helper functionality (H_A) is achieved after collaboration is established. This is defined by the total number of peers inside a collaboration group (G_i) and that are not involved in the collaboration process:

$$H_A = \sum_{i=1}^{M} |G_i - N_i| \tag{6}$$

The number of requested Helpers to participate in a collaboration process for a video i is defined by the number of necessary peers to serve a stream, always provided that the collaboration group needs Helpers. Expr. 7 gives the number of requested helpers for video i.

$$H_R^i = \begin{cases} N_i, & (G_i < N_i) \wedge \left(\dfrac{1}{\lambda_i} < W_i\right) \\[2em] 0, & otherwise \end{cases} \tag{7}$$

The number of available helpers are limited. Therefore, we have to decide to which Virtual Server the helpers will be assigned and control when helpers will be exhausted. To resolve the first issue, we assign helpers to those Virtual Servers that have fewer requisites. To control the number of available helpers, we use expr. 6 (available helpers) combined with expr. 8, that evaluates the total number of helpers required by first j more popular videos:

$$H_{TR}(j) = \sum_{i=1}^{j} H_R^i, \qquad \forall j \leq M \tag{8}$$

Using helpers, the server requirements are of one stream only, provided there are sufficient helpers to complete the requisites of video i and also the previous ones ($H_{TR}(i) < H_A$), and that video-request arrival rates support collaboration ($1/\lambda_i < W_i$). If helpers can only partially fulfill the requirements, then only a portion of video streams will be saved by peers. Otherwise, the central server has to manage all streams:

$$Sc_{DynHlpMul}^i = \begin{cases} 1, & \left(H_{TR}(i) < H_A\right) \wedge \left(\dfrac{1}{\lambda_i} < W_i\right) \\[2em] Sc_{DynMul}^i - \dfrac{H_A - H_{TR}(i-1)}{N_i}, & \left(H_{TR}(i-1) < H_A < H_{TR}(i)\right) \wedge \left(\dfrac{1}{\lambda_i} < W_i\right) \\[2em] Sc_{DynMul}^i, & otherwise \end{cases} \tag{9}$$

4 Performance Evaluation

In this section, we show the analytical model results for the DynaPeer delivery scheme, by evaluating the performance contrasted with traditional Unicast and Patching delivery policies, and with other P2P delivery policies such as Promise[5], Chaining[7], and P^{n}2P^m[11].

4.1 Workload and Metrics

In our experiments, we assumed that inter-arrival time of client requests follows a Poisson arrival process with a mean of $1/\lambda$, where λ is the request rate. We used a Zipf-like distribution to model video popularity. The probability of the ith most popular video being chosen is $1/(i^z \cdot \sum_{j=1}^{M} \frac{1}{j^z})$, where M is the catalogue size and z is the skew factor that adjusts the probability function. For the study, the skew factor is fixed to 0.729 (typical video-shop distribution [1]). The time of analysis was 90 minutes, the same as a video length, the output-bandwidth of clients is fixed to 750kbps and video play rate is set to 1500Kbps. The analyze values of the parameters are summarized in table 2.

The comparative evaluation is based on the server load metric that is defined as the mean number of streams required by the server at the end of analysis.

Table 2. Experimentation environment parameters

Parameter	Default Value	Parameter	Default Value
Request Rate	10 requests/minute	Client's Buffer Size	15 minutes
Play rate	1500 Kbps	Client's Output Bandwidth	750 kbps
Video length	90 minutes	Video Catalogue Size	100 videos
Zipf Skew Factor	0.729		

4.2 P2P Delivery Schemes Comparison

Fig. 3 shows the server-load achieved by DynaPeer other P2P approaches. For this test, we have considered the analytical model described in section 3 and the analytical model found in each one of the delivery policies. As we would expect, server load for all P2P architectures is lower than that achieved by pure-Unicast and Patching delivery mechanisms.

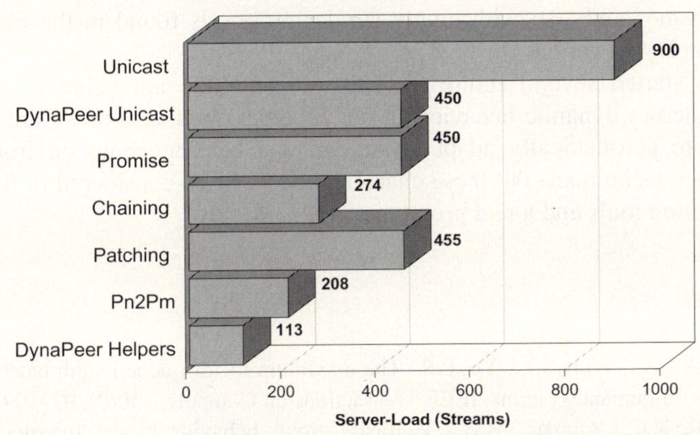

Fig. 3. DynaPeer Comparison

Comparing unicast delivery policies, Promise achieves 50% less server-load while Chaining provides 70% compared with central unicast mode. DynaPeer-Unicast, achieves 50% of server-load reduction, as we find with Promise. The gap performance achieved by DynaPeer in unicast modes occurs because both Promise and Chaining have specific requirements for their functionality. Promise requires peers to have, at least, sufficient buffer capacity to store a whole video (90 minutes), while DynaPeer only assumes a buffer of 15 minutes. On the other hand, Chaining assumes that peer output bandwidth is, at least, the same of a video play-rate, while DynaPeer and Promise assume half the output bandwidth.

Comparing multicast mode, Promise is unable to take advantage of server-load reduction while Chaining keeps reducing server-load in the order of 39% compared with Patching's non-P2P policy. Results show that multicast usage, in conjunction with P2P scheme, provides the best solution for system performance. P^n2P^m (that assumes that peer output bandwidth is the same as video play-rate) and *DynaPeer+Helpers* achieves the best server-load reduction, being 54% and 75% betters than Patching, and 24% and 59% if compared with Chaining. Finally *DynaPeer+Helpers* improve system performance by 45% and 74% if compared with P^n2P^m and Promise, respectively.

5 Conclusion

We have proposed and evaluated a new delivery policy based on a P2P paradigm and multicast communication mechanism for Internet VoD services. Our design defines a Virtual Server, which is responsible for establishing a group of peers, enabling service for new client requests by aggregating the necessary resources. The Virtual Server also performs distributed control tasks among peers by saving server control requirements.

The analytical study shows that DynaPeer policies improve VoD system capacity and decrease the server-load, taking major advantage of client resources to decentralize the delivery process. Compared with traditional unicast and patching delivery schemes and also with highly similar proposals found in the literature, we conclude that DynaPeer has the best performance.

We have started several future research projects. We are going to analyze the impact of Internet dynamic behavior on our schemes. In addition, we are studying a mechanism to automatically adapt DynaPeer to a heterogeneous environment and fault tolerance techniques. All these characteristics will be considered in future work, using simulation tools and a real prototype.

References

1. Aggarwal, C.C., Wolf, J.L., Yu, P.S.: The maximum factor queue length batching scheme for video-on-demand systems. IEEE Transactions on Computers 50(2), 97–109 (2001)
2. Almeroth, K.C., Ammar, M.H.: Multicast group behavior in the Internet's multicast backbone (MBone). IEEE Communications Magazine 35, 124–129 (1997)
3. Cisco Systems. Delivering Multicast Video over Asymmetric Digital Subscriber Line. White Paper, available in 9186a00804f9d11.shtml http://www.cisco.com/en/US/products/ps6598/products_white_paper,(2003)
4. Guo, Y., Suh, K., Kurose, J., Towsley, D.: P2cast: peer-to-peer patching scheme for vod service. In: Proceedings of the 12th Int. Conf. on World Wide Web, pp. 301–309. ACM Press, New York (2003)
5. Hefeeda, M., Habib, A., Botev, B., Xu, D., Bhargava, D.B.: PROMISE: Peer-to-peer media streaming using collectcast. In: Proc. of ACM Multimedia' 03, Berkeley, CA, pp. 45–54. ACM Press, New York (2003)
6. Hua, K.A., Cai, Y., Sheu, S.: Patching: A multicast technique for true video-on-demand services. ACM Multimedia Conf., Bristol, U.K (September 1998)

7. SHua, K.A., Tantaoui, M., Tavanapong, W.: Video delivery technologies for large-scale deployment of multimedia applications. Proc. of the IEEE, 92 (September 2004)
8. Jin, S., Bestavros, A.: Cache-and-relay streaming media delivery for asynchronous clients. In: Proceeding of NGC'02, October 2002, Boston, MA, USA (2002)
9. Padmanabhan, V., Wang, H., Chou, P., Sripanidkulchai, K.: Distributing streaming media content using cooperative networking. In: Proc. NOSSDAV'02, Miami Beach, USA (2002)
10. Liu, X., Vuong, S.T.: A Cost-Effective Peer-to-Peer Architecture for Large-Scale On-Demand Media Streaming. Journal of Multimedia (JMM) 1(2) (May 2006)
11. Yang, X.Y., Hernández, P., Cores, F., Ripoll, A., Suppi, R., Luque, E.: Dynamic Distributed Collaborative Merging Policy to Optimize the Multicasting Delivery Scheme. In: Proc. of 11th Int. Euro-Par 2005 Conference, August 30-September 2005, Lisbon, Portugal (2005)

An Evaluation of Parallelization Concepts for Baseline-Profile Compliant H.264/AVC Decoders

Klaus Schöffmann, Markus Fauster, Oliver Lampl, and Laszlo Böszörmenyi

Department of Information Technology (ITEC), University of Klagenfurt
Universitätsstr. 65-67, 9020 Klagenfurt, Austria
`{ks,mfauster,laszlo}@itec.uni-klu.ac.at`, `olampl@edu.uni-klu.ac.at`

Abstract. Due to the increasing performance requirements of decoding H.264/AVC in HDTV or larger resolutions, new approaches are necessary to enable real-time processing. According to the current trend to parallel computation in all performance classes, decoding of AVC must be mapped to these architectures even though this is complicated by the increased complexity and many data dependencies in the codec. We propose and evaluate different ways of using multithreading to speed-up our .NET implemented decoder. While slice based approaches scale best, this is not a flexible approach because of the reliance on specially encoded streams. Functional partitioning and macroblock pipelining prove to be a good alternative for almost all evaluated videos.

Keywords: AVC, H.264, Parallelization, Decoding, Evaluation, Multi-Threading, Pipelining.

1 Introduction

In recent years, parallelization of computer programs experienced a certain renaissance due to the current architecture of processors used in the mainstream market. CPU manufacturers have changed their CPU architectures from a single-core to a multi-core design. Several independent processors are combined into one package with a shared cache and bus interface. While some current desktop and mobile computers already use a dual-core design, future systems and also entertainment devices will use many cores. In addition, today's multimedia applications require high CPU performance due to advanced video compression standards and the emerging demand for large and detailed digital movies. While current consumer computers still have problems to encode High Definition (HD) videos in real-time, the decoding process is currently fast enough on many modern computers. However, the current trend in the entertainment industry is to go beyond HD and to use Cinema-HD - with a resolution of 4096x2160 pixels - in the entire chain from production to consumption. Future encoders and decoders will have to process approximately four times more data than with current HD videos. Due to the trends mentioned above and since decoding is also part of encoding, future decoders and encoders will have to optimize their processes. They should take advantage of

A.-M. Kermarrec, L. Bougé, and T. Priol (Eds.): Euro-Par 2007, LNCS 4641, pp. 782–791, 2007.

advanced instruction sets (such as MMX/SSE/SSE2 [1]), and possibilities for parallelization provided by the architecture of the CPU. Many multimedia applications are especially well-suited for parallelization because some steps of processing videos can be done independently. However, in order to increase the efficiency of compression, current video coding standards like H.264/AVC [2] have introduced additional dependencies into some parts of the encoding/decoding process. Those dependencies require more advanced parallelization schemes. Manually converting sequential programs into parallel ones is a difficult and error-prone task. In [3] we have already presented a slight language extension to the C# programming language which enables the automatic parallelization of loop operations on independent blocks of data. This extension has been implemented in the Mono [4] compiler and produces multithreaded code, because the run-time systems (Mono, .NET) map multiple threads to the underlying multicore architecture. Other, non thread-based schemes are subject of further study.

In this article, we summarize the results of evaluating several different parallelization schemes for an H.264/AVC baseline-profile decoder. Our decoder has been implemented with the C# programming language using a common style of object-oriented programming. Since the aim of our implementation was to create a portable baseline-profile decoder for the Common-Language-Runtime, we did not use any platform dependent optimizations. We have measured the costs of different parts in the decoding process and used the results as a basis for further parallelization strategies. Although our implementation has high memory requirements and moderate object creation time, the results of our measurements (see figures 1 and 2) show a similar distribution of the workload as in the H.264/AVC reference decoder (as presented in [5]). Although slice level parallelization is an obvious solution achieving good speed-up values (see [6], [7], [8], [9]), it has some disadvantages as well. It can only be applied if frames of a video are separated into slices, which depends on the encoder. Moreover, supporting slice-level parallelization reduces coding efficiency due to the reference limits during the encoding process. Thus, we have evaluated alternative parallelization concepts and analyzed the achievable speed-up. Simple parallelization schemes were generated automatically through our compiler extension as presented in [3]. Some more sophisticated schemes had to be adapted manually (see section 4).

2   Short Overview of H.264/AVC

In H.264/AVC the video bitstream is organized in *Network Abstraction Layer* Units. The most important NAL Unit types are *Sequence Parameter Sets* (SPS), *Picture Parameter Sets* (PPS), and *Slices*. Only slices are relevant for this article, for the other ones please see [2]. A slice consists of a header and a number of macroblocks. A macroblock (MB) represents a 16x16 pixel area of the decoded picture and contains entropy-coded coefficients. Three different types of macroblocks are defined in the baseline-profile: *I-MBs* contain intra-coded coefficients of the corresponding luma and chroma components. An intra-coded macroblock has no reference to other frames but may reference other macroblocks in the current frame. *P-MBs* are inter-coded macroblocks which contain the motion prediction from one frame in the past and the coefficients that encode only the difference to that reference (this is called *residual*). For *Skipped-MBs*, there are no coefficients stored in the bitstream. Such macroblocks use motion-vector-prediction only (see later). A slice is a container for macroblocks and represents either an entire frame or a part of a frame. Each slice of a frame can be decoded independently from other slices in that frame. In general, there are two different types of slices used in H.264/AVC baseline-profile. *I-Slices* which may contain I-MBs and Skipped-MBs, and *P-Slices* which may contain P-MBs, I-MBs, and Skipped-MBs. A macroblock itself is composed of partitions that contain 16x16 (only one partition), 16x8, 8x16, 8x8, 4x8, 8x4 or 4x4 pixels. Since a partition is the elementary unit of motion prediction, encoders typically use large partitions for homogeneous areas and small partitions for areas with varying luma/chroma values. While most of the header fields in an SPS, PPS and a slice are coded using Exp-Golomb Codes, the coefficients of a macroblock are coded using either Context-based Adaptive Binary Arithmetic Coding (CABAC) or Context-based Adaptive Variable Length Coding (CAVLC) techniques, as specified in [2].

The general decoding process for a baseline-profile compliant H.264/AVC decoder consists of the following steps:

Coefficients Parsing: For each macroblock the decoder reads the macroblock-type followed by optional motion vectors, the scaling delta (quantization delta), and the CAVLC coded residual coefficients for luma and chroma. For I-MBs the macroblock-type specifies the chosen intra prediction mode and whether it has been applied to the entire 16x16 block or to 4x4 blocks. The motion vectors of P-MBs specify the position of the referenced data in the reference frame. The process of parsing macroblock coefficients is not parallelizable because at the begining of a slice the start positions of the macroblocks are not known due to variable length coding.

Inverse Transform: After parsing the coefficients, the inverse transformation process is invoked for each macroblock. It is performed on 4x4 blocks of coefficients and consists of two phases: In the first phase the coefficients are scanned in inverse zig-zag scan-order and scaled using a function depending on a context-adaptive scaling value, which is computed from the value of the previous macroblock by adding the scaling delta for the current macroblock. For the first macroblock in a slice the scaling value is initialized to the value specified in the slice header. In the second phase the coefficients are inversely transformed using a DCT-like inverse transformation scheme, which is mainly based on integer additions, subtractions and bit-shifting operations. The inverse transformation process is the only part in the decoding process which is really straightforward to parallelize because it references no other macroblocks. In section 5 we show our results of using parallel inverse transformation.

Inter Prediction: After inverse transformation the inter prediction process can take place for P-MBs. Inter prediction uses motion vector values to add the decoded data from referenced frames to the inverse transformed residual. Motion vectors rely on motion-vector-prediction. Thus, motion vectors of a macroblock specify only the difference to the averaged motion vector values from the macroblock to the left, above, and right-above. Motion-vector-prediction is also applied to Skipped-MBs. The accuracy of motion vectors is 1/4 pixel for luma and 1/8 pixel for chroma. This means that the referenced data has to be interpolated, which is an expensive task in the decoding process. It is possible to parallelize the inter prediction process when operating at macroblock or slice (resp. frame) level. However, as the motion-vector-prediction uses the left/above/right-above macroblock as an input, at least this task has to be done sequentially. In section 5 we show our results of using parallel inter prediction.

Intra Prediction: Intra prediction is applied to I-MBs after inverse transformation, depending on the macroblock-type. Intra prediction may use already predicted surrounding macroblocks (more precisely the left, above, left-above, and right-above MBs) of the current frame as a reference. Similar to P-MBs, the data gathered from the intra prediction is added to the inverse transformed residual. This sum is copied to the decoded buffer which is used as an input for the next step (deblocking filter). It is not a straightforward task to parallelize intra prediction due to the possible existence of referenced macroblocks. A schedule of operations based on references could be calculated and processed in correct order as described in [10].

Deblocking: The deblocking filter (DF) is applied to decoded samples on all 4x4 block edges except to those at frame borders. It uses the already deblocked pair of macroblocks above and to the left as an input for its operation. It is possible to parallelize the deblocking filter when following some processing rules as described in section 4.

More details about H.264/AVC can be found in [2], [5], and [11].

3   Related Work

Research on parallelization of H.264 is primarily done for encoders due to their higher performance requirements in comparison to the decoder. The authors of [6] compare the parallelization of MPEG-2, MPEG-4 and H.264 video encoders for multiprocessor architectures. The parallelization of H.264 is found to be the most difficult due to data dependencies and varying macro block sizes, therefore slice level parallelization is used. Encoding a separate slice per thread leads to a reduction of the workload of up to 96% (for 64 CPUs) but increases bitrate requirements by about 20% for 32 slices. The Intel Hyperthreading architecture is the platform used in [7], where the implementation uses slice level threading as well. The tradeoff between speed-up and compression efficiency is evaluated, leading to the conclusion that it is important to keep the number of slices as low as possible, even though a higher number enables higher parallelism. Therefore, in a later work([12]), a multi-level approach is adopted which processes two frames in a pipeline and uses multiple threads to work on their macroblocks. In [8] a hybrid approach is introduced, which combines the advantages of GoP level (high throughput) with slice level parallelization (low latency) using an encoding cluster. But due to the resulting increase in bitrate, the slice level parallelization is only advantageous for up to 12 slice encoders per GoP. Research and practice indicate a trend to avoid encoding with multiple slices whenever possible, in order to get the best picture quality with regard to the bitrate. While slice level parallelization is an obvious approach to support multithreaded decoding, these results encouraged us to evaluate other approaches to partition the workload.

Most research on H.264 decoders focuses on CPU specific optimizations and hardware/software combinations: For example, [5] discusses optimizations of the reference decoder for media instruction sets. Execution time of the reference decoder is dominated by chrominance and luminance motion compensation and integer transformation. Optimizations achieve a speed up of 10.2x, 2.9x and 4.3x, respectively, for these operations. Entropy decoding and deblocking were barely accelerated at all, and in the optimized decoder they account for half of the decoding time altogether. The authors of [13] provide a detailed analysis of the performance of H.264 decoding when optimized for AltiVec SIMD architectures. While the distribution of time spent at different stages varies with source material and encoding options, the deblocking filter usually dominates because of the limited data level parallelism that could be exploited using AltiVec. In [10] a data-partitioning approach to multithreaded H.264 decoding is described, which does not depend on slices in the encoded video. Macroblock dependencies are resolved by employing a work schedule on the processing order of data partitions. This work targeted a specialized architecture consisting of RISC and media processors and is therefore not directly applicable to general purpose CPUs. An optimized H.264 decoder with block-level pipelining, which is implemented on an ARM platform, is described in [14]. While the implemented optimizations in the software result in a considerable speed-up compared to the reference decoder, the focus is on the integration of a dedicated coprocessor which can process motion compensation and integer IDCT for single macroblocks. Pure software parallelization on shared memory multiprocessors was evaluated for MPEG-2 in [9], where GoP level and slice level parallelism were compared. While both approaches deliver very good speed-ups, parallel processing of entire GoPs requires less synchronization and leads to higher efficiency, but increases delay on start and random access and has extreme memory requirements.

To our best knowledge, at time of writing, no similar evaluation of such a wide variety of parallelization approaches for H.264 decoding has been published.

4   Parallelization Concepts

4.1   Parallelization on Slice-Level

An obvious approach for parallelizing H.264/AVC is to use slices as units of parallelization because slices can be encoded/decoded independently from each other in single a frame. In general, with slice-level parallelism good speed-ups are obtainable; however, it has many disadvantages as well. In order to be useful for a variable number of CPUs, the frame should be partitioned into many slices rather than just a few. The number of slices for a frame depends on the chosen encoder which is usually not suggestible by the decoder. Furthermore, with each additional slice of a frame, the required bitrate increases. As described in [6], using 32 slices increases the bitrate requirements by about 20%. Due to these disadvantages we have analyzed alternative parallelization schemes which are discussed in section 4.2. We have implemented slice-level parallelism using our language extension introduced in [3].

4.2   Parallelization on Macroblock-Level

Due to the results of our measurements of the workload in our H.264/AVC decoder which are illustrated in figures 1 and 2, we have analyzed the individual steps of the decoding process for further parallelization. As shown in the figures the most expensive task is inverse transformation, which has an average share of 42% in the overall decoding process. Another very expensive part is inter prediction with 31% average fraction, although for small video-resolutions (with smaller bitrates) this value decreases. Furthermore, the deblocking filter produces high workload as well, especially for low bitrates.

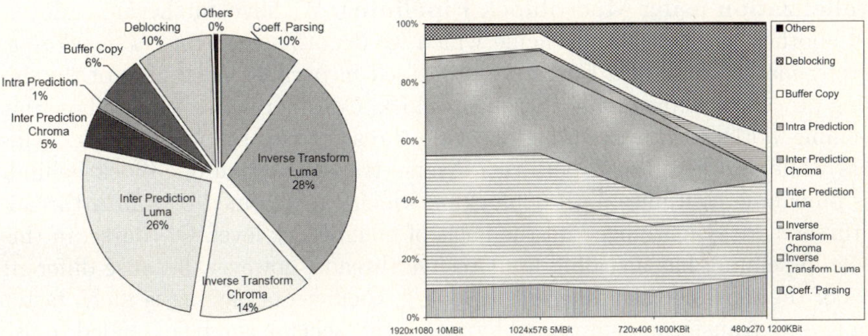

Fig. 1. Load distribution of our H.264/AVC decoder

Fig. 2. Load distribution with different resolutions and bitrates

Parallelization of the Inverse Transformation: Inverse transformation can be easily parallelized because there are no dependencies between the macroblocks. However, using a high number of small processing units[1] results in very poor performance when using the default parallelization behavior of our language extension which interleaves the threads over all blocks of data. Instead, for measuring parallel inverse transformation, we used manually adapted multithreading code which processes evenly sized partitions of consecutive macroblocks. For example, when using four threads for 1000 macroblocks, each thread processes 250 consecutive macroblocks.

[1] As there are e.g. above 8000 macroblocks for a HD resolution video.

Parallelization of the Inter Prediction: Inter prediction can be parallelized when separating motion-vector-prediction, which has to be done sequentially, from the other steps. The remaining steps of inter prediction can be done simultaneously. We used the same strategy for parallelizing inter prediction as with inverse transformation (i.e. uniform areas of consecutive blocks).

Parallelization of the Deblocking Filter: We have developed an approach that supports multiple threads deblocking a single frame simultaneously, while considering that deblocking of a macroblock requires read/write access to the blocks directly above and to the right. As shown in figure 3, it is necessary to synchronize all threads in a staggered pattern[2], where each worker remains at least two macroblocks left of the one above it. This way, a complete parallelization of deblocking is theoretically possible for the most part of an image, excluding 2*(numThreads-1) macroblocks at the top-left and the bottom-right of the frame[3]. This method could be extended by processing vertical ranges starting from the left at same time, but due to the need for more complex synchronization and the good speed-up of the top-down solution, this approach was abandoned.

The same solution seems to be possible for intra prediction and motion-vector-prediction, where similar speedups can be expected.

Parallelization using Macroblock Pipelining: We have, furthermore, developed another parallelization scheme which we denote as *macroblock pipelining*. Thereby macroblocks of a frame are processed in parallel, each part of the decoding process working on another macroblock. Consider figure 4 to see how this pipelining works with multithreading. One thread parses the macroblock coefficients while another thread performs inverse transformation one block behind. This pipelining continues until deblocking which is performed by a fifth thread. All threads are synchronized on the basis of macroblock level. Of course, in this version pipelining has a scaling limit to five threads, however, because different parts of the pipeline take different time (e.g. coefficients parsing is much faster than inverse transform, compare to figure 2), this scheme can be extended to use several threads in the most expensive parts of the pipeline. However, we have not yet performed measurements for this extension.

Fig. 3. Multithreaded deblocking using staggered threads

---

[2] We did not discover other decoders using this technique, but found out that [12] integrates it into the decoder.
[3] Assuming the ideal case that numThreads|imageHeightInMBs.

5 Performance Results and Analysis

For our measurements, we used the *Elephants Dream* [15] video which we encoded with the current release of the x264 encoder [16] using different resolutions (480x270 to 1920x1080) and bitrates (1200KBit/s to 10MBit/s). The videos encoded with x264 do contain only one slice per frame. For comparison we used encodings of QuickTime Pro [17] with default baseline-profile settings. Per default QuickTime Pro uses seven slices per frame. Our measurements have been performed on an SMP system with 4 Intel® Xeon™ CPUs, each 1.5 GHz with 256 KB Cache, and 3.5 GB RAM running Windows Server 2003® with the Microsoft .NET Runtime 2.0. For measuring less than four CPUs, we deactivated the others in the boot.ini file. We did not used the Mono Runtime for our Measurements because in its current version 1.2.2.1, it is much slower than the Microsoft .NET Runtime.

We first measured our parallelization of the elementary processing tasks independently to demonstrate the achievable speed-ups when using multithreaded implementations. The results presented in fig. 5 are calculated by comparing the duration of this elementary task between the sequential and the parallelized (n threads) versions for the input file that provided the best speed-up. Inverse transformation has no data dependencies and is therefore easily parallelizable without much synchronization overhead. It is a processor intensive operation with some data access which explains the good speed-up: Even on a single CPU a speed-up of 1.24 can be observed when using 4 threads, while 8 threads lead to a speed-up of 3.51 on three cores. Inverse transformation is obviously well suited to parallelization, as well as optimization with SIMD instructions or dedicated coprocessors (see related work).

For inter prediction, our results show that only a speed-up of 2.46 is achievable with four CPUs. Because the computational effort of inter prediction is much lower than for inverse transformation but shared memory access is higher, our

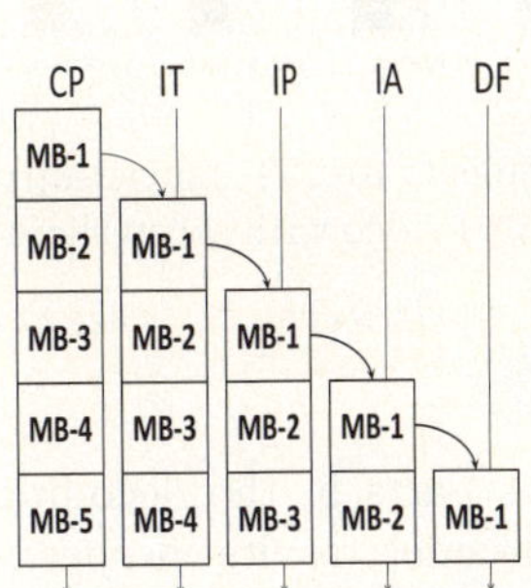

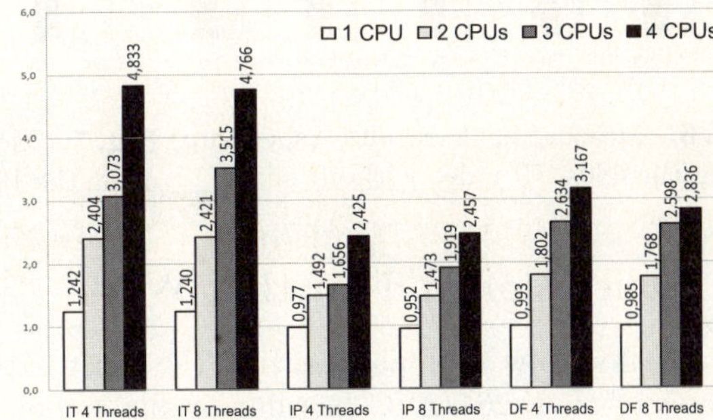

Fig. 4. Parallelization by using macroblock pipelining

Fig. 5. Measurement results showing speed-up when parallelizing different parts of the decoding process (CP=coeff parsing, IT=inverse transformation, IP=inter prediction, IA=intra prediction, DF=deblocking)

measurements show a slow-down if too many threads are used (e.g. eight threads on a single CPU achieve 0.95 speed-up).

As described above, our approach to a multithreaded deblocking filter requires some synchronization between the threads, even though this is a minor influence if all threads work with a similar speed. Therefore the number of threads should be equal to the number of cores available on the system. Using two threads results in a speed-up of 1.86 on two CPUs, but with more threads the speed-up never increases beyond 1.80 on the same configuration due to synchronization overhead.

In figures 6 and 7, the results of combining parallel inverse transformation, inter prediction, and deblocking in direct comparison with macroblock pipelining and slice-level parallelism are given. For a low-resolution video (fig. 6) with only one slice per frame the combination of IT,IP, and DF achieves the best speed-up (1.86) when four threads are in use. Also macroblock pipelining (with five threads) achieves a similar result, while slice-level parallelism causes a slow-down due to the fact that only one slice is used per frame. For a high-resolution video, slice-level parallelism clearly achieves the best speed-up value of 3.2 on four CPUs, because no dependencies between slices of a frame have to be considered. However, it turns out that combining parallel inverse transformation, inter prediction, and deblocking is the best alternative (speed-up of 2.34) to slice-level parallelism.

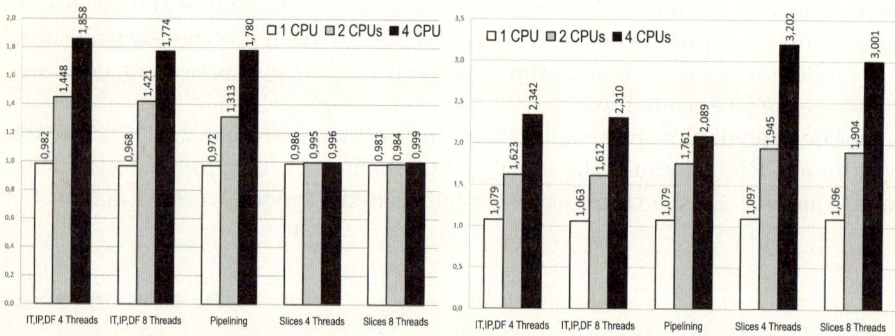

Fig. 6. Measurement results (speed-up) using the 480x270 video without slices

Fig. 7. Measurement results (speed-up) using the 1920x1080 video with seven slices

6   Conclusion and Further Work

Our results show that parallelizing the most expensive parts in the decoding process of an H.264/AVC baseline-profile compliant decoder results in speed-up values which are not far behind those of slice-level parallelization. From our results we can conclude, that parallelizing all three parts, inverse transformation, inter prediction, and deblocking, is the most appropriate alternative to slice-level parallelism. As further work, we continue our measurements with more than four CPUs and an extended version of macroblock-pipelining.

References

1. Zhou, X., Li, E.Q., Chen, Y.K.: Implementation of H.264 Decoder on General-Purpose Processors with Media Instructions. In: SPIE Conf. on Image and Video Communications and Processing (January 2003)
2. ISO/IEC JTC 1/SC 29/WG 11. ISO/IEC FDIS 14496-10: Information Technology - Coding of audio-visual objects - Part 10: Advanced Video Coding (March 2003)
3. Lampl, O., Stellnberger, E., Böszörményi, L.: Programming language concepts for multimedia application development. In: Lightfoot, D.E., Szyperski, C.A. (eds.) JMLC 2006. LNCS, vol. 4228, Springer, Heidelberg (2006)
4. Mono: Open source.net development framework (2007), `http://www.mono-project.com`
5. Zhou, X., Li, E.Q., Chen, Y.K.: Implementation of H.264 decoder on general-purpose processors with media instructions. In: Vasudev, B., Hsing, T.R., Tescher, A.G., Ebrahimi, T., eds.: Image and Video Communications and Processing, 2003. Proceedings of the SPIE, vol. 5022, pp. 224–235 (2003)
6. Jacobs, T., Chouliaras, V., Mulvaney, D.: Thread-parallel MPEG-2, MPEG-4 and H.264 video encoders for SoC multi-processor architectures. In: IEEE Transactions on Consumer Electronics, vol. 52, pp. 269–275. IEEE Computer Society Press, Los Alamitos (2006)
7. Chen, Y.K., Tian, X., Ge, S., Girkar, M.: Towards efficient multi-level threading of H.264 encoder on Intel hyper-threading architectures . In: 18th International Parallel and Distributed Processing Symposium, vol. 1, pp. 63–72 (2004)
8. Rodriguez, A., Gonzalez, A., Malumbres, M.P.: Hierarchical Parallelization of an H.264/AVC Video Encoder. In: International Symposium on Parallel Computing in Electrical Engineering, 2006, pp. 363–368 (September 2006)
9. Bilas, A., Fritts, J., Singh, J.P.: Real-time parallel MPEG-2 decoding in software. Technical Report TR-516-96 (1996)
10. van der Tol, E.B., Jaspers, E.G., Gelderblom, R.H.: Mapping of H.264 decoding on a multiprocessor architecture. In: Vasudev, B., Hsing, T.R., Tescher, A.G., Ebrahimi, T., eds.: Image and Video Communications and Processing 2003. Proceedings of the SPIE, vol. 5022, pp. 707–718 (2003)
11. Richardson, I.: H.264 and MPEG-4 Video Compression: Video Coding for Next Generation Multimedia. John Wiley & Sons Ltd., Chichwster, England (2003)
12. Chen, Y.K., Li, E.Q., Zhou, X., Ge, S.L.: Implementation of H.264 Encoder and Decoder on Personal Computers. Journal of Visual Communications and Image Representations, 509–532 (2005)
13. Alvarez, M., Salami, E., Ramirez, A., Valero, M.: A performance characterization of high definition digital video decoding using H.264/AVC. In: Proceedings of the IEEE International Workload Characterization Symposium, IEEE Computer Society Press, Los Alamitos (2005)
14. Wang, S.H., Peng, W.H., He, Y., Lin, G.Y., Lin, C.Y., Chang, S.C., Wang, C.N., Chiang, T.: A platform-based MPEG-4 advanced video coding (AVC) decoder with block level pipelining. In: Proc. of the 2003 Joint Conference of the Fourth International Conference on Information, Communications and Signal Processing, 2003 and the Fourth Pacific Rim Conference on Multimedia., vol. 1 (December 2003)
15. Orange Open Movie Project: elephants dream (2007), `http://orange.blender.org/`
16. Aimar, L., Merritt, L., et al.: x264 - a free h264/avc encoder (2007), `http://www.videolan.org/developers/x264.html`
17. Apple Inc.: QuickTime Pro (2007), `http://www.apple.com/de/quicktime/pro/`

Topic 12
Theory and Algorithms for Parallel Computation

Nir Shavit, Nicolas Schabanel, Pascal Felber, and Christos Kaklamanis

Topic Chairs

Theory of parallel computation is still a widely unanswered field of research. The lack of truly parallel data structures (such as quantum computing may provide one day) still prevents us from obtaining a unified framework for the design of parallel algorithms, and the design of parallel algorithms remains a case-by-case challenge. As the complexity-related questions have been progressively (and regrettably) abandoned, theoretical research in parallel computing focuses today on two main areas: the classical design of parallel algorithms for specific problems; and very recently, modeling, understanding and formalizing how entities/structures/networks interact in parallel in order to design better parallel algorithms.

Fourteen papers were submitted to this topic and five were accepted. Three of them belong to the first category. In "2D Cutting Stock Problem: a New Parallel Algorithm and Bounds", the authors present new upper and lower bounds for the 2D cutting problem to improve the speed of a parallel implementation of a branch-and-bound strategy that solves the problem exactly. In "Periodic Load Balancing on the N-Cycle: Analytical and Experimental Evaluation", the authors analyze precisely the convergence time of a classical load balancing protocol on a ring by the conductance method. In "Hirschberg's Algorithm on a GCA and its Parallel Hardware Implementation", the authors present a space-efficient implementation on reconfigurable cellular automata of Hirshberg's algorithm for determining connected components in a dense graph.

The last paper belongs to the second category. In "Acyclic Preference Systems in P2p Networks", the authors explore a new formalism based on preference lists to improve the efficiency of P2P networks.

A.-M. Kermarrec, L. Bougé, and T. Priol (Eds.): Euro-Par 2007, LNCS 4641, p. 793, 2007.
© Springer-Verlag Berlin Heidelberg 2007

2D Cutting Stock Problem: A New Parallel Algorithm and Bounds

Coromoto León, Gara Miranda, Casiano Rodríguez, and Carlos Segura

Universidad de La Laguna, Dpto. Estadística, I.O. y Computación
Avda. Astrofísico Fco. Sánchez s/n, 38271 La Laguna, Spain
{cleon,gmiranda,casiano,csegura}@ull.es
http://nereida.deioc.ull.es

Abstract. This work introduces a set of important improvements in the resolution of the Two Dimensional Cutting Stock Problem. It presents a new heuristic enhancing existing ones, an original upper bound that lowers the upper bounds in the literature, and a parallel algorithm for distributed memory machines that achieves linear speedup. Many components of the algorithm are generic and can be ported to parallel branch and bound and A* skeletons. Among the new components there is a comprehensive MPI-compatible synchronization service which facilitates the writing of time-based balancing schemes.

1 Introduction

The Constrained Two Dimensional Cutting Stock Problem (2DCSP) targets the cutting of a large rectangle S of dimensions $L \times W$ in a set of smaller rectangles using orthogonal guillotine cuts: any cut must run from one side of the rectangle to the other end and be parallel to one of the other two edges. The produced rectangles must belong to one of a given set of rectangle types $\mathcal{D} = \{T_1 \ldots T_n\}$ where the i-th type T_i has dimensions $l_i \times w_i$. Associated with each type T_i there is a profit c_i and a demand constraint b_i. The goal is to find a feasible cutting pattern with x_i pieces of type T_i maximizing the total profit:

$$g(x) = \sum_{i=1}^{n} c_i x_i \text{ subject to } x_i \leq b_i \text{ and } x_i \in \mathbb{N}$$

Wang [1] was the first to make the observation that all guillotine cutting patterns can be obtained by means of horizontal and vertical builds of meta-rectangles, Figure 1. Her idea was exploited by Viswanathan and Bagchi [2] to propose a brilliant best first search A* algorithm (VB) which uses Gilmore and Gomory [3] dynamic programming solution to build an upper bound. The VB algorithm uses two lists and, at each step, the best build of pieces (or meta-rectangles) is combined with the already found best meta-rectangles to produce horizontal and vertical builds. Hifi in [4] and later Cung et al. in [5] proposed a modified version of VB algorithm. Niklas et al. in [6] proposed a parallel version of Wang's algorithm. Unfortunately, Wang's method does not always

A.-M. Kermarrec, L. Bougé, and T. Priol (Eds.): Euro-Par 2007, LNCS 4641, pp. 795–804, 2007.

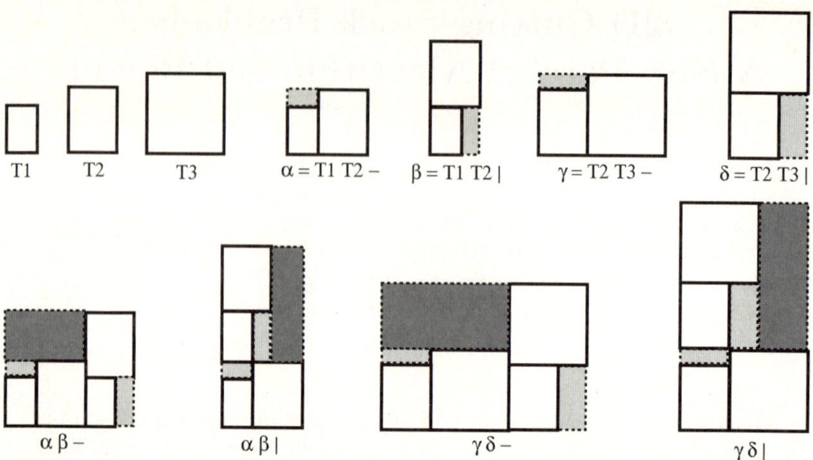

Fig. 1. Examples of Vertical and Horizontal Meta-Rectangles. Postfix notation is used: $\alpha\ \beta\ |$ and $\alpha\ \beta\ -$ denote the vertical and horizontal constructions of rectangles α and β. Shaded areas represent waste.

yield optimal solutions in a single invocation and is slower than VB algorithm. Tschöeke and Holthöfer parallel version [7] starts from the original VB algorithm and uses the Paderborn Parallel Branch and Bound Library, PPBB-LIB [8]. García et al. in [9] presented new data structures and a synchronous parallel algorithm for shared memory architectures. All the approaches to parallelize VB strive against the highly irregular computation structure of the algorithm. Attempts to deal with its intrinsically sequential nature inevitably appears either transformed on an excessively fine granularity or any other source of inefficiency.

This work introduces a set of important improvements when compared with previous work [9]. A new upper bound that lowers the upper bounds in the literature and a new heuristic enhancing existing ones are presented in sections 2.1 and 2.2. Section 3 exposes a parallel algorithm for distributed memory machines achieving linear speedup and good load balancing through the use of a provided timing service. The timing service described in section 4 is portable through platforms and MPI libraries. Section 5 shows the obtained computational results. The conclusions and some lines of future work are given in section 6.

2 Improvements to the Sequential Algorithm

2.1 A New Upper Bound

The new upper bound improves existing upper bounds. It is trivial to prove that is lower than the upper bounds proposed in [2,4,5,7,9]. The calculus of the new upper bound is made in three steps:

1. During the first step, the following bounded knapsack problem is solved using dynamic programming [5,7]:

$$V(\alpha) = \begin{cases} \max \sum_{i=1}^{n} c_i x_i \\ \text{subject to} \qquad \sum_{i=1}^{n}(l_i w_i)x_i \leq \alpha \\ \text{and} \qquad x_i \leq min\{b_i, \lfloor \frac{L}{l_i} \rfloor \times \lfloor \frac{W}{w_i} \rfloor\}, x_i \in \mathbb{N} \end{cases}$$

for all $0 \leq \alpha \leq L \times W$.

2. Then, $F_V(x,y)$ is computed for each rectangle using the equations:

$$\overline{F}(x,y) = \max \begin{cases} \max\{F_V(x,y_1) + F_V(x,y-y_1) & \text{such that } 0 < y_1 \leq \lfloor \frac{y}{2} \rfloor\} \\ \max\{F_V(x_1,y) + F_V(x-x_1,y) & \text{such that } 0 < x_1 \leq \lfloor \frac{x}{2} \rfloor\} \\ \max\{c_i & \text{such that } l_i \leq x \text{ and } w_i \leq y\} \end{cases}$$

where

$$F_V(x,y) = min\{\overline{F}(x,y), V(x \times y)\}$$

3. Finally, substituting the bound of Gilmore and Gomory [3] by F_V in Viswanathan and Bagchi upper bound [2] the new proposed upper bound is obtained:

$$U_V(x,y) = \max \begin{cases} \max\{U_V(x+u,y) + F_V(u,y) & \text{such that } 0 < u \leq L - x\} \\ \max\{U_V(x,y+v) + F_V(x,v) & \text{such that } 0 < v \leq W - y\} \end{cases}$$

2.2   A New Lower Bound

The proposed heuristic mimics Gilmore and Gomory dynamic programming algorithm [3] but substituting unbounded vertical and horizontal combinations by feasible suboptimal ones.

Let be $R = (r_i)_{i=1...n}$ and $S = (s_i)_{i=1...n}$ sets of feasible solutions using $r_i \leq b_i$ and $s_i \leq b_i$ rectangles of type T_i. The cross product $R \otimes S$ of R and S is defined as the set of feasible solutions built from R and S without violating the bounding requirements: i.e. $R \otimes S$ uses $(min\{r_i + s_i, b_i\})_{i=1...n}$ rectangles of type T_i. The lower bound is given by the value $H(L,W)$ computed by the following equations:

$$H(x,y) = \max \begin{cases} \max\{g(S(x,y_1) \otimes S(x,y-y_1)) & \text{such that } 0 < y_1 \leq \lfloor \frac{y}{2} \rfloor\} \\ \max\{g(S(x_1,y) \otimes S(x-x_1,y)) & \text{such that } 0 < x_1 \leq \lfloor \frac{x}{2} \rfloor\} \\ \max\{c_i & \text{such that } l_i \leq x \text{ and } w_i \leq y\} \end{cases}$$

being $S(x,y) = S(\alpha,\beta) \otimes S(\gamma,\delta)$ one of the cross sets, - either a vertical construction $S(x,y_0) \otimes S(x,y-y_0)$ or a horizontal building $S(x_0,y) \otimes S(x-x_0,y)$ - where the former maximum is achieved, i.e. $H(x,y) = g(S(\alpha,\beta) \otimes S(\gamma,\delta))$.

3   Parallel Algorithm

The general operation mode of the parallel scheme follows the structure of the VB algorithm [2]. This exact algorithm uses two lists, OPEN and CLIST. OPEN stores the generated meta-rectangles. At each step, the most promising meta-rectangle from OPEN is moved to CLIST. The main difference between the sequential and parallel scheme lies in the lists management. The parallel scheme replicates CLIST and distributes OPEN among the available processors, p. Figure 2 shows the code for processor k. Initially, the heuristic and the upper bounds are calculated (lines 2-3). CLIST begins empty and the initial builds, $\mathcal{D}$, are distributed among the processors (lines 4-5). At each search step, the meta-rectangle in OPEN with the highest upper bound is chosen (line 9). This build is inserted into CLIST and into the *Pending Combinations* set, PC (lines 10-11). The PC set holds the analyzed meta-rectangles that have not been transferred to other processors. The selected build must be combined with the already found best meta-rectangles to produce new horizontal and vertical builds (lines 12-23). Only the new elements with expectations of success will be inserted into OPEN (lines 16 and 22). Also, OPEN is cleared if necessary (lines 15 and 21).

In order to generate the complete set of feasible solutions, it is necessary to incorporate a synchronization at certain periods of time. That has been implemented using the *synchronization service* explained in section 4. The synchronization subroutine (lines 28-39) is called when a processor has no pending work (line 7) or when an active alarm of the synchronization service goes off. The expiration time of the alarms is fixed by the user, using the *SyncroTime* parameter (line 38). The information given by each processor consists of: its best solution value, the size of its OPEN list and the set of builds that has analyzed since the last synchronization step (line 29). The elements computed by each processor must be inserted into the CLISTs of the other processors (line 32) and also combined among them. Such combinations are uniformly distributed among processors (lines 34 and 35). The current best solution is updated with the best solution found by any of the processors, pruning nodes in OPEN if necessary (line 31). The stop condition is reached when all the OPEN lists are empty (line 6).

The search path followed by the parallel algorithm can be different from the sequential one. In cases where the initial heuristic finds the exact solution, the number of computed nodes is the same. However, in cases where the heuristic does not find the exact solution, changes in the search path may produce modifications on the number of explored nodes.

To have OPEN lists fairly balanced, a parametric method has been designed. This method requires three configuration parameters: *MinBalThreshold*, *MaxBalThreshold* and *MaxBalanceLength*. The method is executed (lines 36-37) after the computation of the pending combinations. It is necessary to sort the set of processors attending to their OPEN size. Processor in position i is associated with a partner located in position $p - i - 1$. That will match the processor with largest OPEN list with the processor with the smallest one, the second largest one with the second smallest and so on. Partners will make an exchange if the one with larger OPEN has more than *MaxBalThreshold* elements and the other has less than *MinBalThreshold*. The number of elements to be exchanged is proportional to the difference of the two OPEN sizes, but it can never be greater than *MaxBalanceLength*.

```
1   search() {
2     BestSol := H(L, W); B: = g(BestSol);
3     h' := U_V;
4     CLIST := ∅;
5     OPEN_k := {T_{k+j*p} / k + j * p < n};
6     while (∃i / OPEN_i ≠ ∅) {
7       if (OPEN_k == ∅) { synchronize(); }
8       else {
9         choose α meta-rectangle from OPEN_k with higher f' = g + h' value;
10        insert α in CLIST;
11        insert α in PC;
12        forall (β ∈ CLIST / x_α + x_β ≤ b, l_β + l_α ≤ L) do {
13          γ = αβ-; l_γ = l_α + l_β; w_γ = max(w_α, w_β); /* horizontal build */
14          g(γ) = g(α) + g(β); x_γ = x_α + x_β;
15          if (g(γ) > B) { clear OPEN_k from B to g(γ); B = g(γ); BestSol = γ; }
16          if (f'(γ) > B) { insert γ in OPEN_k at entry f'(γ); }
17        }
18        forall (β ∈ CLIST / x_α + x_β ≤ b, w_β + w_α ≤ W) do {
19          γ = αβ|; l_γ = max(l_α, l_β); w_γ = w_α + w_β; /* vertical build */
20          g(γ) = g(α) + g(β); x_γ = x_α + x_β;
21          if (g(γ) > B) { clear OPEN_k from B to g(γ); B = g(γ); BestSol = γ; }
22          if (f'(γ) > B) { insert γ in OPEN_k at entry f'(γ); }
23        }
24      }
25    }
26    return(BestSol);
27  }

28  synchronize() {
29    (λ, C, R) = all_to_all (B, |OPEN_k|, PC);
30    if (max(λ) > B)
31      { clear OPEN_k from B to max(λ); B = max(λ); }
32    CLIST = CLIST ∪ (R - PC);
33    PC = ∅;
34    Let be π = {π_0, ..., π_{p-1}} a partition of {R_i ⊗ R'_j / i ≠ j}
35    compute vertical and horizontal combinations of π_k
36    if (balanceRequired(C, minBalThreshold, maxBalThreshold))
37      loadBalance(C, MaxBalanceLength);
38    fixAlarm(SyncroTime);
39  }
```

Fig. 2. Parallel Algorithm Pseudocode for Processor k

4   The Synchronization Service

All synchronizations in the model are done through time alarms (*alarm clocks*). That makes the service independent of the particular algorithm and the MPI implementation. Every process participating in the parallel algorithm fixes the alarm to a certain time value. When the alarm goes off, the corresponding process is informed. If the alarm is fixed with the same time value and then an all-to-all exchange is done when the alarm expires, a synchronous scheme is obtained. The service is initiated on each node by starting a daemon. An *alarm clock manager* is created on each node. This process is in charge of attending all the alarm clocks requests coming from the algorithmic processes. For each received request, the service manager creates a new *alarm clock process* that will communicate to the corresponding requester. Once the communication between the requester and its alarm clock is initiated, their interaction proceeds without any intervention of the manager.

Figure 3 shows the state of one computation node running two MPI processes and the synchronization service. The process at the bottom and its corresponding alarm process have already been initiated. The initialization for the process at the top is presented. First of all, each process in the parallel algorithm must ask for the alarm clock service. The manager process attends each alarm service request creating a new alarm clock process and assigning it to the requester. Then, the algorithmic process can activate the alarm clock specifying a certain amount of time. Once the specified time has passed, the alarm clock will notify the process. In this particular case, after each alarm clock notification, the MPI processes can synchronize their information. If a process finishes its work before the alarm time has expired, it can cancel its alarm and go directly to the synchronization point. If each process cancels the alarm, the synchronization will be reached earlier. This allows the user to better adapt the alarm service behaviour since the alarm can be activated or cancelled at any moment.

The communication between the algorithmic processes and the alarm manager is done through system message queues. The user activation and cancellation of alarms is done through the message queue that was assigned to it. Alarm ex-

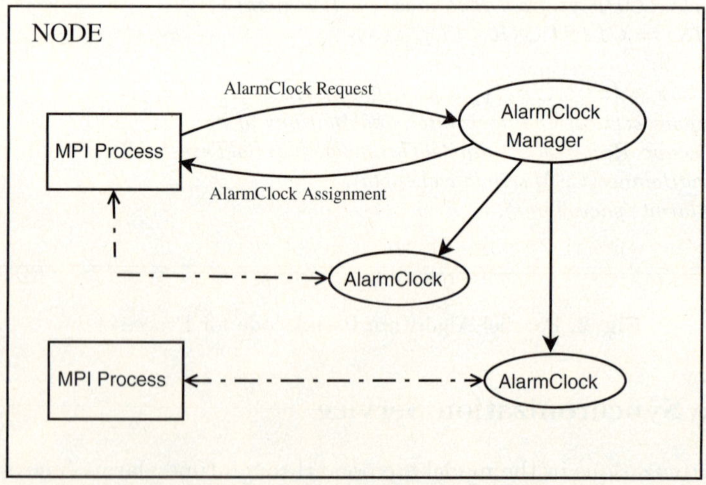

Fig. 3. Synchronization Service Operation

pirations are notified by using a variable allocated in the system shared memory. System signals can be used although it may produce conflicts when combined with some available libraries.

The implemented service can scale to any number of sequential or parallel processes. Users can implement their own time alarms through the system time functions or signals, but then they would have to deal with every implementation detail. Managing multiple alarms inside the same process can be quite complex but doing it with the synchronization service is as easy as for the single alarm case.

5   Computational Results

The instances used in [2,4,5,7,9] are solved by the sequential algorithm in a negligible time. For that reason, the computational study here presented has been performed on some selected instances from the ones available at [10]. Tests have been run on a cluster of 8 HP nodes, each one consisting of two Intel(R) Xeon(TM) at 3.20GHz. The interconnection network is an Infiniband 4X SDR. The compiler and MPI implementation used were *gcc 3.3* and MVAPICH *0.9.7* [11].

Table 1. Lower and Upper Bounds Results

| | Solution | Lower Bound | | Upper Bound | | | | | |
| | | | | V | | | U_V | | |
Problem	Value	Value	Time	Init	Search	Nodes	Init	Search	Nodes
07_25_03	21693	21662	0.442	0.0309	2835.07	179360	0.0312	2308.78	157277
07_25_05	21693	21662	0.436	0.0311	2892.23	183890	0.0301	2304.78	160932
07_25_06	21915	21915	0.449	0.0316	35.55	13713	0.0325	20.83	10310
07_25_08	21915	21915	0.445	0.0318	205.64	33727	0.0284	129.03	25764
07_25_09	21672	21548	0.499	0.0310	37.31	17074	0.0295	25.49	13882
07_25_10	21915	21915	0.510	0.0318	1353.89	86920	0.0327	1107.18	73039
07_50_01	22154	22092	0.725	0.1056	2132.23	126854	0.0454	1551.23	102662
07_50_03	22102	22089	0.793	0.0428	4583.44	189277	0.0450	3046.63	148964
07_50_05	22102	22089	0.782	0.0454	4637.68	189920	0.0451	3027.79	149449
07_50_09	22088	22088	0.795	0.0457	234.42	38777	0.0428	155.35	29124
07_100_08	22443	22443	1.218	0.0769	110.17	25691	0.0760	92.91	22644
07_100_09	22397	22377	1.278	0.0756	75.59	20086	0.0755	61.84	17708

Table 1 presents the results for the sequential runs. The first column shows the exact solution value for each problem instance. The next two columns show the solution value given by the initial lower bound and the time invested in its calculation (all times are in seconds). Note that the designed lower bound highly approximates the final exact solution value. In fact, the exact solution is directly reached in many cases. Last column compares two different upper bounds: the one proposed in [4] and the new upper bound. For each upper bound, the time needed for its initialization, the search time, that is, the time invested in finding the exact solution without including bounds calculations, and the number of computed nodes are presented. Computed nodes are the nodes

Table 2. Parallel Algorithm Results

| | PROCESSORS | | | | | | | | | |
| | 1 | | 2 | | 4 | | 8 | | 16 | | |
PROBLEM	Time	Nodes	Time	Nodes	Time	Nodes	Time	Nodes	Time	Nodes	Sp.
07_25_03	2922.94	157277	1665.26	161200	770.47	157281	384.05	159424	197.82	157603	11.67
07_25_05	3068.19	160932	1738.02	168941	863.23	168867	408.39	165323	206.10	162029	11.18
07_25_06	23.82	10310	11.51	10310	6.36	10310	3.01	10310	1.57	10310	13.26
07_25_08	129.02	25764	61.38	25764	29.98	25764	15.52	25764	8.33	25764	15.48
07_25_09	29.44	13882	13.69	14257	7.02	13916	3.57	13916	2.09	14150	12.44
07_25_10	1140.41	73039	539.89	73039	266.96	73039	132.32	73039	67.94	73039	16.16
07_50_01	1651.51	102662	963.07	102662	598.67	116575	240.93	103545	123.72	102965	12.53
07_50_03	4214.54	148964	2084.77	148964	1057.70	151362	512.12	150644	258.51	149039	11.78
07_50_05	4235.27	149449	2141.41	149449	1077.47	153813	512.43	150937	260.03	149450	11.64
07_50_09	161.38	29124	77.65	29124	40.34	29124	19.45	29124	10.34	29124	14.94
07_100_08	98.96	22644	48.74	22644	25.83	22644	12.60	22644	6.98	22644	13.31
07_100_09	60.05	17708	38.29	19987	18.74	18509	10.59	20584	4.77	18100	12.58

that have been transferred from OPEN to CLIST and combined with all previous
CLIST elements. The new upper bound highly improves the previous bound: the
number of computed nodes decreases, yielding a decrease in the execution time.

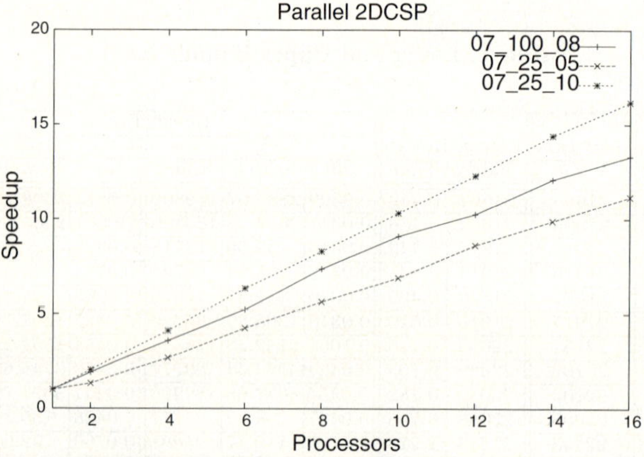

Fig. 4. Parallel 2DCSP: Speedup

Table 2 presents the results obtained for the parallel algorithm. The search
time and the number of computed nodes are shown. For 16-processors, the
speedup in relation to the sequential algorithm is also presented. Both algorithms make use of the improved bounds. Figure 4 represents the speedups for
the three problems with best, worst and intermediate parallel behaviours. Note
that the sequential algorithm and the 1-processor parallel algorithm compute
exactly the same number of nodes, but the parallel implementation introduces
an overhead over the sequential algorithm. When the number of processors increases the parallel algorithm improves its behaviour. In those cases where the
heuristic reaches the exact solution, the parallel and the sequential versions

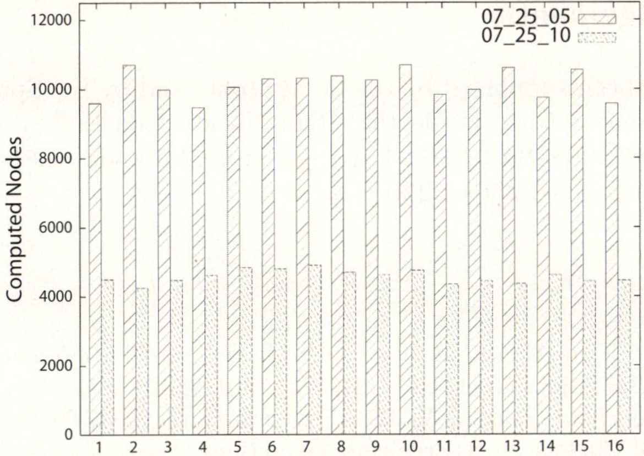

Fig. 5. Parallel 2DCSP: Load Balancing

always compute the same number of nodes and consequently better speedups are achieved. A few superlinear speedups appear due to cache effects.

The number of computed nodes per processor is shown in Figure 5. The chosen problems were those with the best and worst speedups. It clearly indicates that work load is fairly balanced even though the configuration parameters were not tuned in a per-problem basis.

6   Conclusion

This work presents a new lower bound and a new upper bound for the 2DCSP. Computational results prove the quality of such new bounds. A new parallel distributed and synchronous algorithm has been designed from the basis of the inherently sequential VB algorithm. Parallel results demonstrate the almost linear speedups and verify the high scalability of the implementation. Furthermore, a totally application-independent synchronization service has been developed. The service provides an easy way of introducing periodic synchronizations in the user programs. The synchronization service has been decisive for the well operation of the parallel scheme and for the right behaviour of the load balancing model in the presented application.

Some improvements can be added to the current implementation. The first one refers to the load balancing scheme and lies in introducing some method to approximately calculate the work associated to each of the meta-rectangles in OPEN. Instead of considering only the size of the lists, it would be better to consider the work load that they will generate. The other concern is related to the synchronization scheme. At the initial and latest stages of the search, many of the alarms are cancelled because processors do not have enough work. It would be interesting to have an automatic and dynamic way of fixing the time between synchronizations while the search process is progressing.

Acknowledgements

This work has been supported by the EC (FEDER) and by the Spanish Ministry of Education and Science inside the 'Plan Nacional de I+D+i' with contract number TIN2005-08818-C04-04. The work of G. Miranda has been developed under the grant FPU-AP2004-2290.

References

1. Wang, P.Y.: Two Algorithms for Constrained Two-Dimensional Cutting Stock Problems. Operations Research 31(3), 573–586 (1983)
2. Viswanathan, K.V., Bagchi, A.: Best-First Search Methods for Constrained Two-Dimensional Cutting Stock Problems. Operations Research 41(4), 768–776 (1993)
3. Gilmore, P.C., Gomory, R.E.: The Theory and Computation of Knapsack Functions. Operations Research 14, 1045–1074 (1966)
4. Hifi, M.: An Improvement of Viswanathan and Bagchi's Exact Algorithm for Constrained Two-Dimensional Cutting Stock. Computer Operations Research 24(8), 727–736 (1997)
5. Cung, V.D., Hifi, M., Le-Cun, B.: Constrained Two-Dimensional Cutting Stock Problems: A Best-First Branch-and-Bound Algorithm. Technical Report 97/020, Laboratoire PRiSM - CNRS URA 1525. Universitè de Versailles, Saint Quentin en Yvelines. 78035 Versailles Cedex, France (November 1997)
6. Nicklas, L.D., Atkins, R.W., Setia, S.K., Wang, P.Y.: The Design and Implementation of a Parallel Solution to the Cutting Stock Problem. Concurrency - Practice and Experience 10(10), 783–805 (1998)
7. Tschöke, S., Holthöfer, N.: A New Parallel Approach to the Constrained Two-Dimensional Cutting Stock Problem. In: Ferreira, A., Rolim, J. (eds.) Parallel Algorithms for Irregularly Structured Problems, Berlin, Germany, pp. 285–300. Springer, Heidelberg (1995)
8. Tschöke, S., Polzer, T.: Portable Parallel Branch-and-Bound Library - PPBB-LIB, Department of Computer Science, University of Paderborn, D-33095 Paderborn, Germany (December 1996)
9. García, L., León, C., Miranda, G., Rodríguez, C.: A Parallel Algorithm for the Two-Dimensional Cutting Stock Problem. In: Nagel, W.E., Walter, W.V., Lehner, W. (eds.) Euro-Par 2006. LNCS, vol. 4128, pp. 821–830. Springer, Heidelberg (2006)
10. Group, D.O.R.: Library of Instances (Two-Constraint Bin Packing Problem), http://www.or.deis.unibo.it/research_pages/ORinstances/2CBP.html
11. Network-Based Computing Laboratory, Dept. of Computer Science and Eng., The Ohio State University: MPI over InfiniBand Project (2006), http://nowlab.cse.ohio-state.edu/projects/mpi-iba

Periodic Load Balancing on the N-Cycle: Analytical and Experimental Evaluation

Christian Rieß and Rolf Wanka

Computer Science Department, University of Erlangen-Nuremberg, Germany
`sichries@informatik.stud.uni-erlangen.de, rwanka@cs.fau.de`

Abstract. We investigate the following very simple load-balancing algorithm on the N-cycle (N even) which we call *Odd-Even Transposition Balancing* (OETB). The edges of the cycle are partitioned into two matchings canonically. A matching defines a single step, the two matchings form a single round. Processors connected by an edge of the matching perfectly balance their loads, and, if there is an excess token, it is sent to the lower-numbered processor. The difference between the real process where the tokens are assumed integral and the idealized process where the tokens are assumed divisible can be expressed in terms of the local divergence [1]. We show that Odd-Even Transposition Balancing has a local divergence of $N/2 - 1$. Combining this with previous results, this shows that after $O(N^2 \log(KN))$ rounds, any input sequence with initial imbalance K is perfectly balanced. Experiments are presented that show that the number of rounds necessary to perfectly balance a load sequence with imbalance K that has been obtained by pre-balancing a random sequence with much larger imbalance is significantly larger than the average number of rounds necessary for balancing random sequences with imbalance K.

1   Introduction

Background. In the standard abstract formulation of load balancing in a distributed network, processors are modeled as the vertices of a graph and links between them as edges. Each processor initially has a collection (called *load*) of unit-size, integral jobs (called *tokens*). The object is to balance the number of tokens at each processor by transmitting tokens along edges according to some local scheme. This problem has obvious applications to job scheduling and other coordination tasks in parallel and distributed systems. It also arises in the context of finite element computations, and in simulations of physical phenomena.

One load-balancing approach is the *dimension exchange* paradigm [2,3], where the network is decomposed into a sequence $M_1, \ldots, M_d$ of perfect matchings. The edges of the matching are oriented. We write $[i\!:\!j]$ for a single edge connecting processors i and j. Each balancing round consists of d steps, one for each matching. In step k, each pair $[i\!:\!j]$ of processors holding x_i and x_j tokens, resp., that are paired in matching M_k balance their load as closely as possible: their loads

A.-M. Kermarrec, L. Bougé, and T. Priol (Eds.): Euro-Par 2007, LNCS 4641, pp. 805–814, 2007.

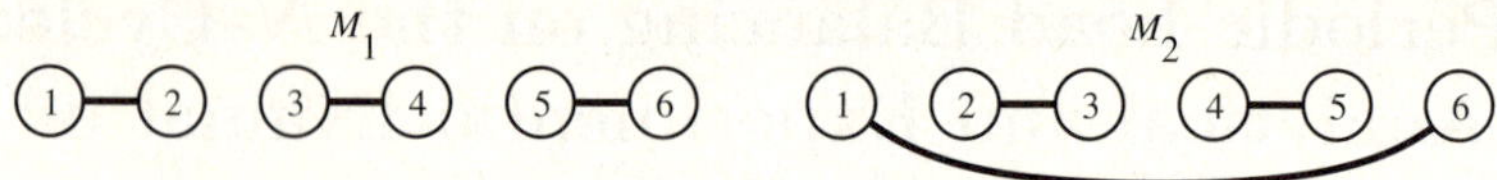

Fig. 1. The two matchings of OETB on the N-cycle for $N = 6$. Note that the excess token goes to the lower-numbered processor.

become $\lceil \frac{x_i + x_j}{2} \rceil$ and $\lfloor \frac{x_i + x_j}{2} \rfloor$, resp., (this means that the excess token, if there is any, is sent to processor i). This model is equivalent to the *(periodic) balancing circuit* paradigm [4]. Such a circuit is composed of a sequence of N wires connected by simple toggling devices called *balancers*. The matchings of the dimension exchange paradigm correspond to the balancers forming a round of d steps. The purpose of such a circuit is to balance the flow of tokens along the wires. Rounds are repeated until the total load is spread among the processors and wires, resp., as evenly as possible.

The Problem. We investigate the following very simple load-balancing algorithm on the N-cycle (N even) which we call *Odd-Even Transposition Balancing*. The edges of the cycle are partitioned into two matchings canonically (e. g., see Fig. 1) that together form a single round. We call this algorithm *Odd-Even Transposition Balancing* (OETB) due to its similarity to Odd-Even Transposition Sort [5, p. 240].

We are interested in upper bounds on the number of rounds necessary to perfectly balance the tokens in terms of the initial imbalance (or discrepancy) K (i. e., $\max_i |x_i - x_j|$) and the cycle's graph theoretical properties. If the tokens are allowed to be subdivided arbitrarily, this *idealized* balancing process can be described in terms of Markov chains (see [6]; for the related *diffusion* paradigm, see [2,7,8]). However, there is a deviation between this idealized process and the actual token process. In [1], the *local divergence* Ψ has been introduced that allows for upper bounding the difference between the two processes. In our setting, Ψ only depends on the matchings of the balancing method. In this paper, we compute the local divergence of OETB exactly.

A further question we address experimentally is the hardness of balancing. We run OETB on random load sequences with certain initial discrepancy K until discrepancy K' is ensured. Then we compare the number of steps necessary to perfectly balance these "pre-balanced" sequences with the number of steps necessary to perfectly balance random load sequences with initial discrepancy K'.

Related Work. The dimension exchange method was introduced by Cybenko [2] and by Hosseini *et al.* [3] in the context of load balancing on the hypercube which explains the name of the paradigm. Balancing circuits were introduced by Aspnes *et al.* [4]. They replace the comparators of some hypercubic sorting circuits on $N = 2^k$ wires by balancers and show that these depth-$O(k^2)$ circuits perfectly balance any input sequence. Interestingly, the number of steps does not depend on the inital discrepancy K of the input sequence. A complementary result was shown by Aharonson and Attiya [9]. They show that, if $N \neq 2^k$, there is no fixed balancing circuit that perfectly balances any input. Therefore, if $N \neq 2^k$, a fixed circuit has to be applied repeatedly to the input sequence, and the number of repetitions depends at least on K.

A similar approach to load balancing is *diffusion* [2,7,8]. Here, for each i, processor i with load x_i and d neighbors shifts about $x_i/(d+1)$ tokens to every neighbor. In [2,7,8], this method is analyzed assuming that the tokens can be subdivided arbitrarily.

In both paradigms, one can identify the actual token process and the idealized process where it is assumed that the load can be arbitrarily subdivided. We refer to $x_i^{(t)}$ as the number of tokens stored in processor i after t rounds, and to $\xi_i^{(t)}$ as the idealized, i. e., fractal load of processor i after t rounds.

Rabani *et al.* [1] introduced the local divergence Ψ of a load-balancing algorithm (regardless whether it is a diffusion or a dimension exchange method) which characterizes the deviation between the idealized and the actual token process. Ψ does not depend on K, only on the algorithm and the used network. It is shown that, for all t, $\max_i |\xi_i^{(t)} - x_i^{(t)}| \leq \Psi$, and that, in general, $\Psi = O(d \cdot (\log N)/(1 - \lambda))$, where d is the degree of the network, N the number of processors, and λ the second eigenvalue of a matrix that describes the algorithm (see Sec. 2). As a consequence, it is shown that any load sequence with initial discrepancy K is transformed into a sequence with imbalance $O(\Psi)$ in $O(\log(K \cdot N)/(1 - \lambda))$ rounds. A further reduction of the imbalance cannot be shown with this approach. For diffusive load balancing on the N-cycle, $\Psi = \frac{3}{4}N$ is proved.

For the dimension exchange model, a sorting-based upper bound of $O(K \cdot N)$ for perfect balancing on networks that have an almost Hamiltonian cycle, is also presented in [1].

Our Contribution. In Sect. 3, we prove $\Psi = N/2 - 1$. In order to compute the exact value we also compute the exact powers of the so called round matrix. As $\lambda = 1 - \Theta(1/N^2)$ for the round matrix, this means that after $O(N^2 \log(KN))$ rounds the discrepancy is at most $N/2 - 1$. After further $\frac{1}{2}N^2$ rounds the load is perfectly distributed among the processors.

In Sect. 4 experiments are presented that show that the number of rounds necessary to perfectly balance a load sequence with imbalance K that has been obtained by pre-balancing a random sequence with much larger imbalance is significantly larger than the average number of rounds necessary for balancing random sequences with imbalance K.

2   Details on the Model and on the Results

Odd-Even Transposition Balancing (OETB) works on the N-cycle, N even, with processors $1, \ldots, N$. The first matching is $M_1 = \{[1{:}2], [3{:}4], \ldots, [N-1{:}N]\}$, the second is $M_2 = \{[2{:}3], [4{:}5], \ldots, [N-2{:}N-1]\} \cup \{[1{:}N]\}$ (e. g., see Fig. 1). The *step matrices* P_1 and P_2 according to M_1 and M_2, resp., and the *round matrix* P are (in Fig. 1, they are presented for $N = 6$):

$$P^{(1)} = \begin{pmatrix} \frac{1}{2} & \frac{1}{2} & 0 & 0 & 0 & 0 \\ \frac{1}{2} & \frac{1}{2} & 0 & 0 & 0 & 0 \\ 0 & 0 & \frac{1}{2} & \frac{1}{2} & 0 & 0 \\ 0 & 0 & \frac{1}{2} & \frac{1}{2} & 0 & 0 \\ 0 & 0 & 0 & 0 & \frac{1}{2} & \frac{1}{2} \\ 0 & 0 & 0 & 0 & \frac{1}{2} & \frac{1}{2} \end{pmatrix}, P^{(2)} = \begin{pmatrix} \frac{1}{2} & 0 & 0 & 0 & 0 & \frac{1}{2} \\ 0 & \frac{1}{2} & \frac{1}{2} & 0 & 0 & 0 \\ 0 & \frac{1}{2} & \frac{1}{2} & 0 & 0 & 0 \\ 0 & 0 & 0 & \frac{1}{2} & \frac{1}{2} & 0 \\ 0 & 0 & 0 & \frac{1}{2} & \frac{1}{2} & 0 \\ \frac{1}{2} & 0 & 0 & 0 & 0 & \frac{1}{2} \end{pmatrix}, P = P^{(1)} \cdot P^{(2)} = \begin{pmatrix} \frac{1}{4} & \frac{1}{4} & \frac{1}{4} & 0 & 0 & \frac{1}{4} \\ \frac{1}{4} & \frac{1}{4} & \frac{1}{4} & 0 & 0 & \frac{1}{4} \\ 0 & \frac{1}{4} & \frac{1}{4} & \frac{1}{4} & \frac{1}{4} & 0 \\ 0 & \frac{1}{4} & \frac{1}{4} & \frac{1}{4} & \frac{1}{4} & 0 \\ \frac{1}{4} & 0 & 0 & \frac{1}{4} & \frac{1}{4} & \frac{1}{4} \\ \frac{1}{4} & 0 & 0 & \frac{1}{4} & \frac{1}{4} & \frac{1}{4} \end{pmatrix}$$

We write $P^{(t,1)} = P^{t-1} \cdot P^{(1)}$, $P^{(t,2)} = P^t$, $P^{(t,1)-1} = P^{(t-1,2)}$, and $P^{(t,2)-1} = P^{(t,1)}$. $P^{(1,1)-1}$ is the identity matrix. E. g., $P^{(2,1)} = P^{(1)} \cdot P^{(2)} \cdot P^{(1)}$. Note that all matrices are block matrices. In the round matrix $P^{(t,k)}$, the denominator is always 2^{2t+k}.

The load sequence after round t, $t \geq 0$, is $x^{(t)} = (x_1^{(t)}, \ldots, x_N^{(t)})$. $x_i^{(t)}$ denotes the number of tokens processor i stores after round t. The *discrepancy* (or imbalance) of a load squence is $\mathcal{D}(x^{(t)}) = \max_{ij} |x_i^{(t)} - x_j^{(t)}|$. $K = \mathcal{D}(x^{(0)})$ is the initial discrepancy. The goal is to determine the number T of rounds required to reduce the discrepancy to some specific value ℓ: we refer to this as ℓ-smoothing. In general, the number of rounds required to ℓ-smooth an initial sequence will depend on both ℓ and the initial discrepancy.

In an idealized setting, single tokens are allowed to be split between the processors involved in a balancing step. In this setting, and with $\xi^{(0)} = x^{(0)}$, it is easy to see that $\xi^{(t)} = \xi^{(0)} \cdot P^t$. The number of rounds t to ℓ-smooth $\xi^{(0)}$ is bounded above by $t \leq 2/(1-\lambda) \cdot \ln(KN^2/\ell)$, where λ is the second eigenvalue of (a symmetrization of) P [1]. For OETB, it is easy to see that $\lambda = 1 - \Theta(1/N^2)$ because the underlying graph is the N-cycle.

In order to relate the deviation between the integral and the idealized process, the local divergence has been introduced [1]. Here, we present it already adapted to OETB on the N-cycle.

Definition 1 ([1]). *The* local divergence *(adapted to OETB and the N-cycle) is*

$$\Psi(P) = \max_l \sum_{t=1}^{\infty} \left(\sum_{[i:j] \in M_1} \left| P_{li}^{(t,1)-1} - P_{lj}^{(t,1)-1} \right| + \sum_{[i:j] \in M_2} \left| P_{li}^{(t,2)-1} - P_{lj}^{(t,2)-1} \right| \right)$$

Theorem 1 ([1]). *The maximum deviation between the idealized process and the integral process satisfies* $\max_i |\xi_i^{(t)} - x_i^{(t)}| \leq \Psi(P^{\mathrm{T}})$ *for all t, where P is the round matrix and P^{T} its transpose.*

Note that the bound depends on the local divergence computed on the transpose of P.

In this paper, we compute the local divergence for OETB exactly:

Theorem 2. *For OETB on the N-cylce, $\Psi(P^{\mathrm{T}}) = N/2 - 1$.*

Corollary 1. *OETB needs $O(N^2 \log(KN))$ rounds to 1-smooth any load sequence with initial discrepancy K.*

3   Computation of the Local Divergence of OETB

This section is devoted to the proof of Theorem 2.

3.1   Obtaining an Expression for $P_{ij}^{(t,k)}$

Simplifying P. Using the block structure of the matrix enables us to work with $n \times n$ matrices, $n = N/2$, by choosing every second row and column from the original matrix. E. g., with $N = 6$ and $n = 3$, we get the reduced $n \times n$ matrix $Q = \begin{pmatrix} \frac{1}{4} & \frac{1}{4} & 0 \\ 0 & \frac{1}{4} & \frac{1}{4} \\ \frac{1}{4} & 0 & \frac{1}{4} \end{pmatrix}$.

No information is lost in the smaller matrix because its entries are exactly the same as the neighboring ones in the full matrix; additionally the calculation gets handier:

$$\Psi(Q^T) = \max_l \sum_{t=1}^{\infty} \sum_{k=1}^{2} \left(|Q_{1l}^{(t,k)-1} - Q_{nl}^{(t,k)-1}| + \sum_{i=1}^{n} |Q_{il}^{(t,k)-1} - Q_{i+1,l}^{(t,k)-1}| \right)$$

Error propagation on the infinite circle. Consider an infinite linear array with the nodes numbered from $-\infty$ to ∞ omitting 0, and let node 1 be the pivot. Fig. 2 shows the error contribution of the surrounding nodes:

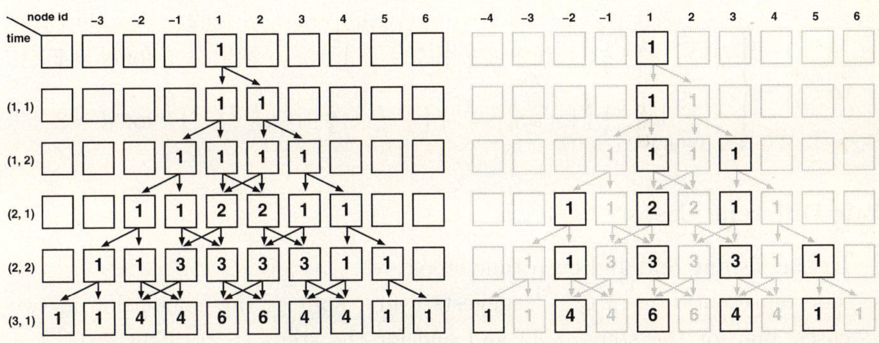

Fig. 2. Token exchange skeleton on an infinite circle – if every second node is left out the binomial coefficients are clearly visible. In order to obtain $P^{(t,k)}$ one has to multiply the shown numbers by $2^{-(2t+k)}$.

We are interested in the contribution of the neighboring nodes to the error in the pivot. Therefore we trace the error contribution of every single node back in time. On the infinite array the binomial coefficients show up, which becomes obvious if every second node is masked out (see Fig. 2 right).

Winding up the infinite array. We will see that winding up the infinite array leads to an explicit expression for the local divergence of OETB.

When winding the series around the cycle with n nodes then every n-th entry is mapped on the same node. The following sum from [5, p. 89] extracts and sums up every n-th entry from a sequence, starting with entry r:

$$\sum_{m \bmod n = r} a_m z^m = \frac{1}{n} \sum_{0 \le k < n} \omega_n^{-kr} G(\omega_n^k z) \ , 0 \le r < m, \tag{1}$$

where $\omega_n = \cos(2\pi/n) + \mathrm{i} \cdot \sin(2\pi/n)$ denotes an n-th primitive root of unity, and $G(z) = \sum_{m=-\infty}^{\infty} a_m z^m$ a generating function for the a_m.

We choose a generating polynomial that fixes the maximum coefficient to z^0, symmetrically surrounded by the other coefficients of $z^{\pm i}$. A double time step from $(t-1, k)$ to (t, k) is obtained by multiplying the existing series of coefficients by

$$\frac{1}{2^2} \left(1 \cdot z^{-1} + 2 + 1 \cdot z^1 \right)$$

This double step avoids the difficulties that arise on handling a single step from (t, k) to $(t, k)+1$, where the existing polynomial must be multiplied alternately by $\frac{1}{2}(z^{-1} + 1)$ and $\frac{1}{2}(1 + z^1)$.

Consequently we distinguish two cases when determining $Q_{ij}^{(t,k)}$, one for the time steps $(t, 1)$ and one for the time steps $(t, 2)$. Note the offset of $\frac{1}{2}$ and that for $d = 2$ one requires an additional factor, which comes from the first single time step:

$$Q_{ij}^{(t,k)} = \begin{cases} \frac{1}{n} \sum_{k=0}^{n} \omega_n^k \cdot \frac{1}{2} \left(\frac{1}{4}\omega_n^k + \frac{2}{4} + \frac{1}{4}\omega_n^{k(n-1)} \right)^t & \text{for } d = 1 \\[2ex] \frac{1}{n} \sum_{k=0}^{n} \omega_n^k \cdot \frac{1}{4} \left(1 + \omega_n^{k(n-1)} \right) \left(\frac{1}{4}\omega_n^k + \frac{2}{4} + \frac{1}{4}\omega_n^{k(n-1)} \right)^t & \text{for } d = 2 \end{cases}$$

Four Different Cases

- Distinguishing odd and even time steps
  The representation of $Q_{ij}^{(t,k)}$ suggests a first split of the calculation in two cases, one for the steps $(t, 1)$ and one for the steps $(t, 2)$. Thus $\Psi(Q^T) = \Psi_1(Q^T) + \Psi_2(Q^T)$ where $\Psi_i(Q^T)$ denotes the sum over all steps (t, i).
- Splitting to $n \equiv 0 \bmod 2$ and $n \equiv 1 \bmod 2$
  Both cases end up in the same result, but the calculation is slightly different.

Lemma 1. $\Psi_1(Q^T) = \frac{n}{2} \ , \Psi_2(Q^T) = \frac{n}{2} - 1 \qquad \text{if } n \equiv 0 \bmod 2$
$\Psi_1(Q^T) = \frac{n}{2} - \frac{1}{2n} \ , \Psi_2(Q^T) = \frac{n}{2} + \frac{1}{2n} - 1 \ \ \text{if } n \equiv 1 \bmod 2$

The calculations for the four cases are very similar. Instead of a full proof of Lemma 1 we present the calculation for $\Psi_2(Q^T)$ in the case that $n \equiv 0 \bmod 2$, which contains all important intermediate steps and is still reasonably short. For a full presentation of all four cases see [10].

3.2   Rewriting the Sum as an Easier Expression

Several simplifications apply in the case of OETB to the original equation:

- block circularity
  Since the round matrix is block circular, it is possible to leave out the maximization over l and put $l = 1$, since all columns are equal up to circular shifts, thus

$$\Psi_2(Q^T) = \sum_{t=1}^{\infty}\left(|Q_{11}^{(t,1)-1} - Q_{n1}^{(t,1)-1}| + \sum_{i=1}^{n} |Q_{i1}^{(t,1)-1} - Q_{i+1,1}^{(t,1)-1}| \right)$$

- symmetry of one column
  Since the entries in one column of Q are symmetrical it is possible to sum over only half of the matrix and multiply it by 2:

$$\Psi_2(Q^T) = \sum_{t=1}^{\infty} 2 \sum_{i=1}^{\lceil n/2 \rceil} |Q_{i1}^{(t,1)-1} - Q_{i+1,1}^{(t,1)-1}|$$

- telescoping the sum
  The sum over the differences $Q_{i1}^{(t,1)-1} - Q_{i+1,1}^{(t,1)-1}$ collapses to a simple difference $Q_{11}^{(t,1)-1} - Q_{\lceil n/2 \rceil,1}^{(t,1)-1}$ due to the unimodality of the winded up binomial coefficients that form the entries of the matrix:

$$\Psi_2(Q^T) = \sum_{t=1}^{\infty} 2 \left(Q_{11}^{(t,1)-1} - Q_{j1}^{(t,1)-1} \right)$$

3.3   Calculating $\Psi(Q^T)$

$$
\begin{aligned}
\Psi_2(Q^T) &= \sum_{t=1}^{\infty} 2 \left(Q_{11}^{(t,1)-1} - Q_{\lceil n/2 \rceil 1}^{(t,1)-1} \right) \\
&= \frac{1}{n} \sum_{t=0}^{\infty} \sum_{k=0}^{n-1} \left(1 - \omega_n^{k\lceil n/2 \rceil}\right)\left(\tfrac{1}{2} + \tfrac{1}{2}\omega_n^{k(n-1)}\right)\left(\tfrac{2}{4} + \tfrac{1}{4}\omega_n^{k} + \tfrac{1}{4}\omega_n^{k(n-1)}\right)^t \\
&= \frac{1}{n} \sum_{k=0}^{n-1} \frac{\left(1 - \omega_n^{k\lceil n/2 \rceil}\right)\left(\tfrac{1}{2} + \tfrac{1}{2}\omega_n^{k(n-1)}\right)}{1 - \left(\tfrac{2}{4} + \tfrac{1}{4}\omega_n^{k} + \tfrac{1}{4}\omega_n^{k(n-1)}\right)} \\
&= \frac{1}{2n} \sum_{k=1}^{n-1} \frac{1 + \cos\left(\tfrac{2\pi k}{n} \cdot (n-1)\right) - \cos\left(\tfrac{2\pi k}{n} \cdot \lceil \tfrac{n}{2} \rceil\right) - \cos\left(\tfrac{2\pi k}{n} \cdot \left(\lceil \tfrac{n}{2} \rceil - 1\right)\right)}{\sin^2\left(\tfrac{\pi k}{n}\right)}
\end{aligned}
$$

The final task is to handle the cases for $\lceil \tfrac{n}{2} \rceil$ properly. As mentioned earlier, we focus on the case $n \equiv 0 \bmod 2$, the other one is substantially the same, but lengthier.

$$\Psi_2(Q^T) = \frac{1}{2n} \sum_{k=1}^{n-1} \frac{1 + \cos\left(\tfrac{2\pi k}{n}\right) - \cos\left(\pi k\right) - \cos\left(\pi k + \tfrac{2\pi k}{n}\right)}{\sin^2\left(\tfrac{\pi k}{n}\right)}$$

remaining balancing time of random load distributions that were balanced down from an initial discrepancy of $k_1 \gg k_0$ to k_0.

Since in the average inputs with larger initial discrepancy need more rounds for 1-smoothing, this leads to the assumption that these additional rounds are "created" by earlier balancing steps: Balancing a randomly chosen input several rounds leaves a load distribution of discrepancy k_i that is very hard to 1-smooth because of the previous balancing steps. See Table 1 and Fig. 3 and 4 for some typical results.

In Figures 3 and 4 is the average progress on a remaining discrepancy of 400 or 100, resp., shown for two cases, for distributions that are

1. pre-balanced from an initial discrepancy of 3200 down to 400 (solid line)
2. randomly chosen with an initial discrepancy of 400 (dashed line)

The two curves show the average number of remaining rounds until the load balancing distributions were 1-smoothed.

References

1. Rabani, Y., Sinclair, A., Wanka, R.: Local divergence of Markov chains and the analysis of iterative load-balancing schemes. In: Proc. 39th IEEE Foundations of Computer Science (FOCS), pp. 694–703 (1998)
2. Cybenko, G.: Dynamic load balancing for distributed memory multiprocessors. Journal of Parallel and Distributed Computing 7, 279–301 (1989)
3. Hosseini, S., Litow, B., Malkawi, M., McPherson, J., Vairavan, K.: Analysis of a graph coloring based distributed load balancing algorithm. Journal of Parallel and Distributed Computing 10, 160–166 (1990)
4. Aspnes, J., Herlihy, M., Shavit, N.: Counting networks. Journal of the ACM 41, 1020–1048 (1994)
5. Knuth, D.E.: The Art of Computer Programming, Sorting and Searching, vol. 3, 2nd edn. Addison-Wesley, Massachusetts (1998)
6. Busch, C., Mavronicolas, M.: A combinatorial treatment of balancing networks. Journal of the ACM 43, 794–983 (1996)
7. Bertsekas, D.P., Tsitsiklis, J.N.: Parallel and Distributed Computations: Numerical Methods. Prentice-Hall, Englewood Cliffs (1989)
8. Boillat, J.E.: Load balancing and Poisson equation in a graph. Concurrency: Practice and Experience 2, 289–313 (1990)
9. Aharonson, E., Attiya, H.: Counting networks with arbitrary fan-out. Distributed Computing 8, 163–169 (1995)
10. Rieß, C.: Load-Balancing auf dem Kreis. Studienarbeit, Department of Computer Science, University of Erlangen-Nuremberg (2006)
11. Bromwich, T.: An Introduction to the Theory of Infinite Series. MacMillan, NYC (1926)

Hirschberg's Algorithm on a GCA and Its Parallel Hardware Implementation

Johannes Jendrsczok[1], Rolf Hoffmann[1], and Jörg Keller[2]

[1] TU Darmstadt, FB Informatik, FG Rechnerarchitektur
Hochschulstraße 10, D-64289 Darmstadt
{jendrsczok,hoffmann}@ra.informatik.tu-darmstadt.de
[2] FernUniversität in Hagen, Fakultät für Mathematik und Informatik
Universitätsstr. 1, D-58084 Hagen
Joerg.Keller@FernUni-Hagen.de

Abstract. We present in detail a GCA (Global Cellular Automaton) algorithm with $3n$ cells for Hirschberg's algorithm which determines the connected components of a n-node undirected graph with time complexity $O(n \log n)$. This algorithm is implemented fully parallel in hardware (FPGA logic). The complexity of the logic and the performance is evaluated and compared to a former implementation using $n(n+1)$ cells with a time complexity of $O(\log^2(n))$. It can be seen from the implementation that the presented algorithm needs significantly fewer resources (logic elements times computation time) compared to the implementation with $n(n+1)$ cells, making it preferable for graphs of reasonable size.

1 Introduction

The GCA (Global Cellular Automata) model [1,2] is an extension of the classical CA (Cellular Automata) model [3]. In the CA model the cells are arranged in a fixed grid with fixed connections to their local neighbors. Each cell computes its next state by the application of a local rule depending on its own state and the states of its neighbors. The data accesses to the neighbor's states are read-only and therefore no write conflicts can occur. The rule can be applied to all cells in parallel and therefore the model is inherently massively parallel.

The GCA model is a generalisation of the CA model which is also massively parallel. It is not restricted to the local communication because any cell can be a neighbor. Furthermore the links to the neighbors are not fixed; they can be changed by the local rule from generation to generation. Thereby the range of parallel applications for the GCA model is much wider than for the CA model.

The CA model suits to all kinds of applications with local communication, like physical fields, lattice-gas models, models of growth, moving particles, fluid flow, routing problems, picture processing, genetic algorithms, and cellular neural networks. Typical applications for the GCA model are – besides all CA applications – graph algorithms, hypercube algorithms, logic simulation, numerical algorithms, communication networks, neuronal networks, games, and graphics.

A.-M. Kermarrec, L. Bougé, and T. Priol (Eds.): Euro-Par 2007, LNCS 4641, pp. 815–824, 2007.

The general aim of our research (supported by Deutsche Forschungsgemeinschaft, project *Massively Parallel Systems for GCA*) is the hardware and software support for this model [4]. Recently we have investigated how graph algorithms can be implemented on the GCA. In [5] we have described the hardware implementation of Hirschberg's algorithm to compute the connected components of an n-node undirected graph [6] on the GCA using $n^2 + n$ cells. The time complexity was $O(log^2(n))$. We use field programmable gate arrays (FPGAs) as hardware platform.

In this paper we present a different hardware implementation of Hirschberg's algorithm on a GCA using only $3n$ cells with two pointers per cell. This implementation will be evaluated with respect to time complexity and hardware complexity. Also this algorithm will be compared to the GCA algorithm with $n^2 + n$ cells in order to find out the advantages. The FPGA reconfigurability allows to implement different GCAs with very low overhead, thus enabling the use of highly efficient co-processors, as argued in a recent journal issue, see [7].

The remainder of the paper is organized as follows. In Section 2 we sketch relations and differences between PRAMs and GCAs. In Sections 3 and 4, we review Hirschberg's algorithm and present how it can be mapped onto a GCA. Section 5 presents results about our hardware realization. Section 6 concludes.

2   GCAs and PRAMs

The state of a GCA cell consists of a data part and an access information part. In a common implementation the access information part contains one or more pointers (Figure 1). The pointers are used to dynamically establish links to global neighbors. We call the GCA model *one handed* if only one neighbor can be addressed, *two handed* if two neighbors can be addressed and so on. Additionally we call the GCA cells *uniform* if all cells have the same transition rule and otherwise *non-uniform*.

As the GCA cells work synchronously and can only read from other cells, the GCA resembles the concurrent-read owner-write (CROW) PRAM model, where each memory location may only be written by a dedicated processor, the

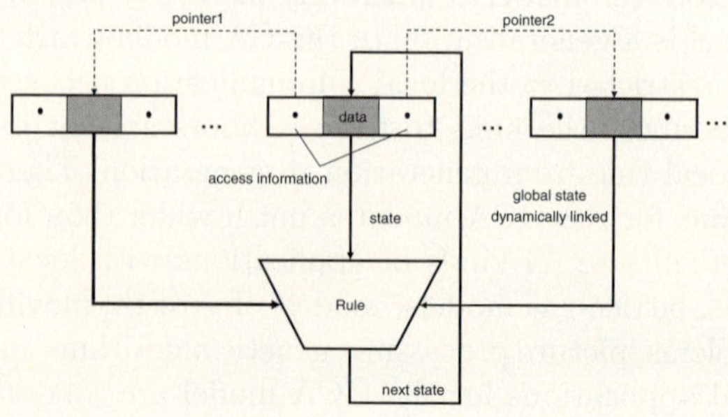

Fig. 1. The operation principle of the GCA

owner. In principle, the GCA is able to implement any PRAM algorithm, as any algorithm consists of a finite number of instructions from a finite instruction set. However, an automaton implementation is particularly advantageous for simple algorithms, which are however available in abundance in the PRAM community. In particular, Hirschberg's algorithm is well-studied on the PRAM model [8].

A general simulation of CRCW or CREW PRAMs onto a GCA can be achieved by sorting the requests according to owners; efficient sorting is available on owner-write PRAMs [9]. Yet, as the PRAM algorithm for a particular problem can be compiled into the GCA rule set when using reconfigurable hardware, i.e. FPGAs, more efficient methods normally are available.

On a PRAM one seeks an algorithm with a short parallel runtime T_p on a number of processors P with $P \cdot T_p = T_s$, where T_s is the sequential complexity of the problem at hand, i.e. a *work-optimal* algorithm. Either, P is driven to its maximum value in the range of the problem size, to explore the limits of parallelism in a problem, or P is chosen so that a practical implementation is available. While the latter case in general can be derived from the first one via Brent's theorem, often direct methods lead to simpler and faster implementations. In a previous paper [5], we have investigated the first case, here we investigate the latter, for the reasons given below.

If the number of processors is taken as a measure of machine *cost* or *price*, and parallel time (for a fixed work) is seen as the inverse of performance, then $P \cdot T_p$ represents a price/performance ratio. While the cost measure may be appropriate for PRAMs because even RISC microprocessors are much more complex than memory cells, in GCAs the memory cost has to be taken into account, because a finite automaton with a few registers is cheap in reconfigurable hardware, while memory cost is comparatively high. This means that the price of a GCA requiring n^2 memory cells does not vary much no matter if one takes n^2 or $n^2 / \log^2(n)$ processors. This enables further simplification of algorithms.

Yet, our feeling was that employing n instead of n^2 processors, and giving each a local memory of size n, not implemented in registers as before but in much denser RAM storage, might still improve the price-performance ratio, and give processor counts that become practical.

Implementing a CROW PRAM algorithm or a GCA requires similar considerations. The memory is mapped to the owners. For non-local read accesses, the *congestion*, i.e. the number of cells reading from one cell, has to be controlled. In the case that we investigate here, where each GCA cell only has a single data value to be accessed, congestion can only occur because of concurrent reading. Yet, this can be ameliorated by appropriate routing networks, such as Ranade's butterfly network. Hence, we will list the congestion numbers in our result table, but not deal further with the congestion and routing problem.

3   Hirschberg's Algorithm

Our example application is Hirschberg's well known algorithm [6] to compute the connected components of an undirected graph on a CREW PRAM. Yet, only a CROW PRAM is really needed. Hirschberg's algorithm was seminal and is work-optimal for dense graphs, i.e. graphs with n nodes and $m = \Theta(n^2)$ edges where the sequential complexity of the problem is $\Theta(m + n) = \Theta(n^2)$. Starting with every single node as a component, the algorithm divides the number of components in every iteration by at least two, so $\log n$ iterations are needed at most. Each iteration needs time $O(\log n)$ on $n^2/\log^2(n)$ processors, therefore the overall time complexity is $O(\log^2(n))$. Each component is represented by its node v_i with the smallest index i. These representing nodes are called super nodes. The index of a component is the index of its super node. Our goal is to show that the algorithm of Hirschberg et al. works efficiently on the GCA with $3n$ cells, for the reasons given in the previous section. Our implementation will need $O(n \log n)$ steps, hence we will see a speedup for graphs with $m = \Omega(n \log n)$ edges, and see maximum speedup for graphs with $m = \Theta(n^2)$ edges. Examples of such graphs appear e.g. when very large graphs are collapsed into smaller ones.

Listing 1.1 shows the original algorithm (reference algorithm) consisting of 6 steps. Each iteration starts with several non connected components. During every iteration, each component searches a connection to another component. First every node of the component searches a connection to a node belonging to another component (step 2). If the node can connect to more than one component, the component with the lowest index is selected. Afterwards the super node picks the component with the lowest index (step 3). The components connect to each other and for each new component a super node is chosen (step 4-6).

```
1. for all i in parallel do C(i) ← i
       do steps 2 through 6 for log n iterations
   2. for all nodes i in parallel do
          T(i) ← min_j{C(j) |A(i,j)=1 AND C(j) != C(i)} if none then C(i)
   3. for all i in parallel do
          T(i) ← min_j{T(j) |C(j)=i AND T(j) != i} if none then C(i)
   4. for all i in parallel do
          C(i) ← T(i)
   5. repeat for log n iterations
          for all i in parallel do T(i) ← T(T(i))
   6. for all i in parallel do
        C(i) ← min{C(T(i)) ,T(i)}
```

Listing 1.1. Pseudo code for the algorithm of Hirschberg et al. on the PRAM (reference algorithm)

The original algorithm was defined for SIMD (single instruction multiple data) parallel processors (e.g. vector machines). Later the algorithm was investigated for the PRAM machines [8]. All these algorithms use a common memory. The algorithm uses the following variables and constants: Input is the adjacency matrix $A = \{A(i,j)|i,j = 1\ldots n\}$. If $A(i,j) = A(j,i) = 1$ then there is a link between node i and node j. $C(i)$ and $T(i)$ are of type integer and hold the number of a node or a super node: $C = \{C(i)|i = 1\ldots n\}$, $T = \{T(i)|i = 1\ldots n\}$. The constant A, the variables C, T and the temporary variables have to be stored in the common memory of the SIMD or PRAM computer.

4 Hirschberg's Algorithm on the GCA

The GCA algorithm uses a cell array Z in order to store the variables and computational rules. The cell array Z consists of three rows with n cells each (Fig. 2):

Row 0 of Z corresponds to the original vector $C(i)$: $Z0 = C = C_0 \ldots C_{n-1}$.

Row 1 of Z corresponds to the original vector $T(i)$: $Z1 = T = T_0 \ldots T_{n-1}$.

Row 2 of Z is used to hold temporary results $Temp(i)$: $Z2 = Temp = Temp_0 \ldots Temp_{n-1}$, as well as the matrix A, one column per cell. The matrix columns are only accessible by the cell holding them[1].

The cells of Z are ordered by a linear index $K = 0 \ldots 3n - 1$. A cell $z = Z(K)$ will be accessed from another cell using the linear index K. For convenience a cell $z = Z(J, I) = Z(K)$ may also be accessed by the row index $J = row(K)$ and the column index $I = column(K)$ using the access functions row and $column$.

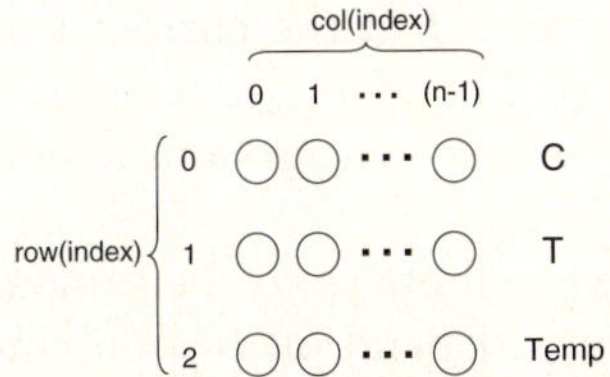

Fig. 2. GCA Field

Each cell $z = (d, p0, p1)$ consists of a data part d and two pointers $p0$ and $p1$. The data part is used for the computation of the connected components storing the node or super node numbers. The pointers dynamically establish links to two other cells (global cells $z^*(p0), z^*(p1)$). The global data is denoted as $d^*(p0)$ and $d^*(p1)$. In general the next cell state $z' = (d', p0', p1')$ depends on its current state z and the states $z^*(p0)$, $z^*(p1)$ of its current neighbors.

A GCA algorithm consists of a sequence of parallel computations. In each computation all cells update their state in parallel in accordance to the local rule. The global state (configuration) of the GCA is given by the cross product of all the local cell states. The configuration changes from time step to time step. In order to emphasize the changing of the configurations with time g the term *generation* is commonly used. The configuration at time g is the gth *generation*.

A GCA algorithm can be clearly represented by a state graph. The state graph consists of states which are reached under certain conditions, e.g. central counter states. In each state two types of operations are performed: *data operations* and *pointer operations*.

The state graph (Fig. 4) shows on the left the computation of the actual pointer p and on the right the data operation of a cell. The pointer p can either be computed in the current generation or one generation in advance. In our

---

[1] If the degree of the graph is known to be low, the matrix columns can be replaced by lists.

Table 1. Generations for each step. Active cells are cells that perform a calculation within a generation. δ is the number of concurrent read accesses to each of the # cells.

Step	State	Active Cells (modifying cell state)	# cells with read access	$\delta = \#$ of concurrent read accesses (congestion)	N algor.		N^2 algor.	
					Gen.	Sub-gen.	Gen.	Sub-gen.
1	0	n	0	0	1		1	
2	1	n	0	0	$2+n$	n	$3 + \log n$	$\log n$
	2	n	$n + 2n$	$2 + 0$				
	3	$2n$	$n + 2n$	$0 + 1$				
3	4	n	$n + 2n$	$0 + 1$	$1 + n$	n	$3 + \log n$	$\log n$
	5	$2n$	$n + 2n$	$0 + 1$				
4	6	n	$n + 2n$	$1 + 0$	1		1	
5	7	n	$n + 2n$	$n + 0$	$\log n$	$\log n$	$\log n$	$\log n$
6	8	n	$n + 2n$	$n + 0$	1		1	

algorithm the pointer is computed in the current generation to be used immediately in the data operation. Therefore the assignment symbol "$=$" is used for pointer operations. In contrast the synchronous assignment symbol "$\leftarrow$" is used for the data operations.

Although in principle each cell obeys to the same uniform algorithm, the operations to be performed may depend on certain conditions. In this algorithm the data operations depend on the positions of a cell in the field, in particular whether a cell is located in C, T or $Temp$. The conditions to distiguish between the three vectors are: $row(index) = 0$ for C, $row(index) = 1$ for T, and $row(index) = 2$ for $Temp$. The GCA algorithm (Fig. 4) consists of 8 states which correspond to the 6 steps of the original algorithm as shown in Table 1.

State 0. The first step of the reference algorithm requires the data of the cell vector C to be set to the corresponding index ($C(i) \leftarrow i$). So the data of the vector C is initialized with the column number of each cell.

State 1. In order to prepare the field for the calculation of the minimum in the next state the vector $Temp$ is set to ∞. Thereby it is possible to identify whether a minimum will have been found in state 2 or not.

State 2. In this state all the $\min_j$ functions of the Hirschberg algorithm are computed in parallel. If the condition $A(i, j) = 1\ AND\ C(j) \neq C(i)$ is fulfilled and $Temp(i)$ is less than $C(i)$, $Temp(i)$ is set to $C(i)$, otherwise $Temp(i)$ remains unchanged. Thus the data of the vector $Temp(i)$ is the minimum of $C(j)$ after n iterations.

In the corresponding GCA algorithm (Fig. 4) each cell is operating on its own and the operations specified in the graph tell each cell what it has to do. In state 2 only the last row ($Temp$) of the cell array with $row(index) = 2$ is activated. Each cell $T(I)$ computes the minimium of the cell $C(I)$ compared to all cells $C(J)$. The pointer $p0$ is used to access $C(I)$ and the pointer $p1$ is successively incremented (using the subgeneration counter) in order to access all J cells (see access pattern Fig. 3).

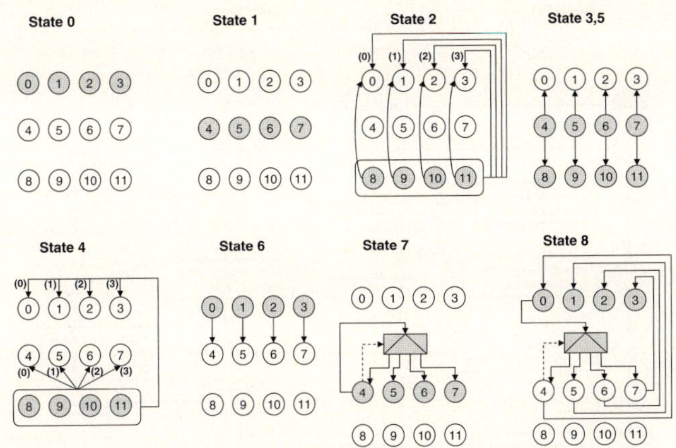

Fig. 3. Access Patterns for $n = 4$. Active cells are shaded.

State 3. After the calculation of the minimum $\min_j$ the value of $C(i)$ is written back in case none of the conditions of generation 2 was true. For this purpose $Temp(i)$ is substituted for the value of $C(i)$ in case $Temp(i)$ equals ∞. Then the value of $Temp(i)$ is set to ∞ in preparation of the next generation.

The cell $T(I)$ uses its pointer $p0$ to access $Temp(I)$ and its pointer $p1$ to access $Temp(I)$. If $p0 = \infty$ then $p1$ is copied to the data part d of $T(i)$, otherwise $p0$ is copied.

State 4. (Similar to state 2). In contrast to State 2 the condition has changed. If the condition $C(j) = i\ AND\ T(j) \neq i$ is fulfilled and $Temp(i)$ is less than $T(j)$, $Temp(i)$ is set to $T(j)$, otherwise $Temp(i)$ remains unchanged. Thus the data of the vector $Temp(i)$ is the minimum of $T(j)$ after n iterations.

State 5. State 5 is identical to state 3.

State 6. Vector T is copied to vector C.

State 7. This state iterates $\log n$ times. Only the vector T (corresponding to Hirschberg's $T(i)$ Vector) is modified. The pointers are data-dependent. The cell $T(i)$ points to the cell $T(T(i))$. Thus the neighbor depends on the value of the cell and it is possible to set the value of $T(i)$ to the value of $T(T(i))$ in one parallel computation.

State 8. State 8 is similar to state 7. In both states the pointers are data dependent. In addition to the previous state the value of $C(T(i))$ is compared to the stored value of $T(i)$. The minimum out of both values is saved as the new value for $C(i)$.

Time complexity. (Fig. 4, Table 1) The steps 1, 4 and 6 can be performed in one generation. Each of the steps 2 and 3 need $1 + n + 1$ respectively $1 + n$ generations, because the computation of the minimum needs n sub generations. Step 5 is repeated $\log n$ times.

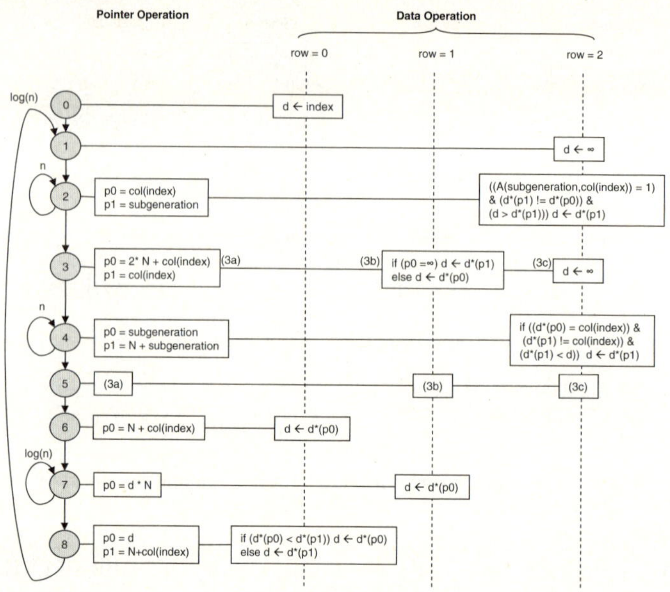

Fig. 4. GCA algorithm with pointer operation (actual access pattern) and data operation

The steps 2 to 6 are executed in $\log n$ iterations. So the total amount of generations is $1 + \log n \cdot (5 + 2n + \log n)$. This corresponds to a time bound of $O(n \log n)$ using $3n$ cells. In a previous GCA implementation [5], $n(n + 1)$ cells were used in order to execute the algorithm as fast as possible. There the minimum function takes only $\log n$ generations instead of n as presented here. Therefore the total amount of generations was $1 + \log n \cdot (3 \log n + 8)$. This corresponds to a time bound of $O(\log^2(n))$ using $n(n + 1)$ cells. In order to distinguish the two algorithms, the algorithm with $3n$ cells is also denoted as "N algorithm" and the algorithm with $n(n + 1)$ cells as "N^2 algorithm".

5   Fully Parallel Hardware Implementation

We have implemented the two GCA algorithms with $3n$ cells (Fig. 6) and with $n(n + 1)$ cells in hardware (FPGA logic) in order to find out the complexity and efficiency. The platform was the ALTERA Quartus synthesis tool and the Stratix II FPGA (EP2S180). Results from the synthesis are shown in Table 2 and Figure 6.

It turned out that the states 7 and 8 are the same in the N and N^2 algorithm. Therefore the synthesis was splitted into three parts: (1) states 0-6 for the N algorithm, (2) the corresponding states for the N^2 algorithm and (3) the states 7-8 for both algorithms (abbreviated N/N^2). If we assume that the register bits have relatively low implementation cost compared to the logic we can focus our comparison on the used logic elements (ALUTs). For the problem size $n = 64$

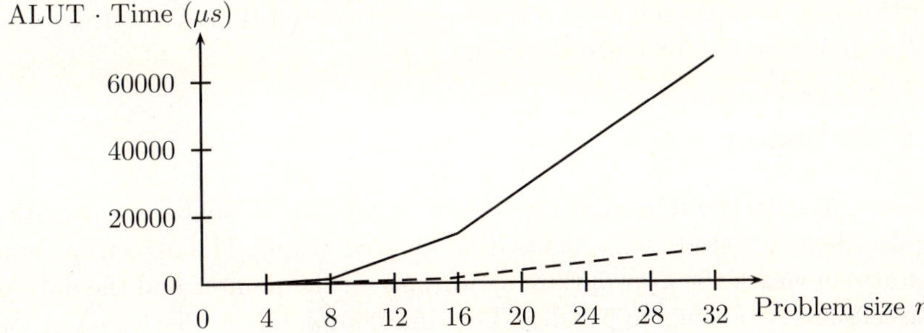

Fig. 5. Resources vs. n (dashed: N algorithm, solid: N^2 algorithm).

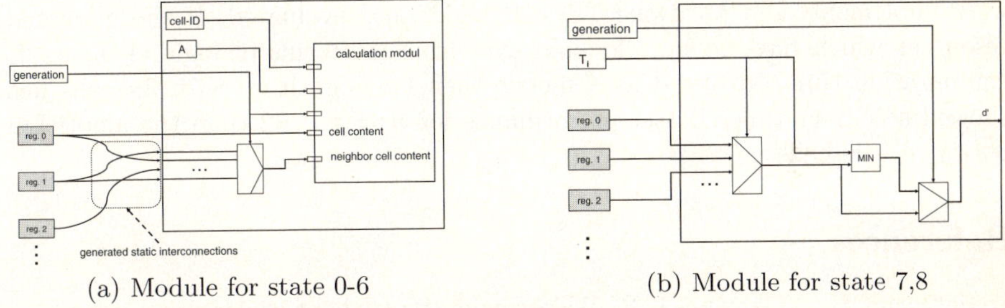

(a) Module for state 0-6   (b) Module for state 7,8

Fig. 6. Hardware modules

Table 2. Synthesis results, case N and N^2 stand for state 0-4, N/N^2 stands for state 5 and 6 and is needed for both algorithms

CASE	PROBLEM SIZE	CELLS	LOGIC ELEMENTS (ALUT)	f_{max} (MHz)	REGISTERS	# CLOCK CYCLES	CALCULATION TIME (ns)
N	4	12	152	126	48	25	197
	8	24	372	88	120	61	692
	16	48	722	84	288	145	1719
	32	96	1853	65	672	341	5176
N^2	4	20	575	82	80	23	279
	8	72	2570	75	360	40	528
	16	272	12328	51	1632	61	1194
	32	1056	56012	71	7392	86	1195
N/N^2	4	4	18	400	8	6	15
	8	8	172	138	32	12	86
	16	16	653	112	80	20	177
	32	32	4888	57	192	30	523

the number of ALUTs needed to implement the states 0-6 are 1,853 for the N algorithm, and 56,012 for the N^2 algorithm whereas the calculation time counterwise is 5.2 μs (N) and 1.2 μs (N^2). Multiplying the number of ALUTs with the calculation time gives us a good measure which corresponds to the resource allocation needed to perform the algorithm. We call that measure *resources* for short. It can be seen from Fig. 5 that the *resources* of the N algorithm (states 0-6) are significantly lower compared to the N^2 algorithm. Therefore the N

algorithm is more economic with respect to the consumption of resources whereas the N^2 algorithm can produce the result faster.

6   Conclusion

We have presented a GCA algorithm with $3n$ cells for Hirschberg's algorithm to compute the connected components of a directed graph. The algorithm consists of 8 states in which the appropriate operations on the pointer and the data parts of the cells are performed in parallel. The time complexity is $O(n \log n)$. A former GCA algorithm with $n(n+1)$ cells can compute the required minimum function, which is the most time consuming part of the whole algorithm in $\log n$ time. Thereby the time complexity can be reduced to $O(\log^2(n))$. Both algorithms were implemented in hardware (FPGA logic) and evaluated. If the allocated resources which have to be allocated over time (in terms of logic elements $\times$ computation time) are used as a metric then the algorithm with $3n$ cells has showed a 5 to 11 times better performance for $n = 4 \ldots 32$ than the algorithm with $n(n+1)$ cells.

References

1. Hoffmann, R., Völkmann, K.P., Waldschmidt, S.: Global Cellular Automata GCA: An Universal Extension of the CA Model. In: Worsch, T. (ed.) ACRI 2000 Conference (2000)
2. Hoffmann, R., Völkmann, K.P., Waldschmidt, S., Heenes, W.: GCA: Global Cellular Automata. A Flexible Parallel Model. In: Malyshkin, V. (ed.) PaCT 2001. LNCS, vol. 2127, pp. 66–73. Springer, Heidelberg (2001)
3. von Neumann, J.: Theory of Self-Reproducing Automata. University of Illinois Press, Urbana and London (1966)
4. Heenes, W., Hoffmann, R., Jendrsczok, J.: A multiprocessor architecture for the massively parallel model GCA. In: International Parallel and Distributed Processing Symposium (IPDPS), Workshop on System Management Tools for Large-Scale Parallel Systems (SMTPS) (2006)
5. Jendrsczok, J., Hoffmann, R., Keller, J.: Implementing Hirschberg's PRAM-Algorithm for Connected Components on a Global Cellular Automaton. In: International Parallel & Distributed Processing Symposium (IPDPS), Workshop on Advances in Parallel and Distributed Computational Models (APDCM) (2007)
6. Hirschberg, D.S.: Parallel algorithms for the transitive closure and the connected component problems. In: STOC '76: Proceedings of the eighth annual ACM symposium on Theory of computing, New York, NY, USA, pp. 55–57. ACM Press, New York (1976)
7. Buell, D., El-Ghazawi, T., Gaj, K., Kindratenko, V.: Guest editors' introduction: High-performance reconfigurable computing. Computer 40(3), 23–27 (2007)
8. Gibbons, A., Ritter, W.: Efficient Parallel Algorithms. Cambridge University Press, Cambridge (1998)
9. Lin, D., Dymond, P.W., Deng, X.: Parallel merge-sort algorithms on owner-write parallel random access machines. In: Lengauer, C., Griebl, M., Gorlatch, S. (eds.) Euro-Par 1997. LNCS, vol. 1300, pp. 379–383. Springer, Heidelberg (1997)

Acyclic Preference Systems in P2P Networks

Anh-Tuan Gai[1], Dmitry Lebedev[2], Fabien Mathieu[2], Fabien de Montgolfier[3], Julien Reynier[4], and Laurent Viennot[1]

[1] INRIA, domaine de Voluceaux, 78153 Le Chesnay cedex, France
[2] FT R&D, 38–40, rue du général Leclerc, 92130 Issy les Moulineaux, France
[3] LIAFA, Université Paris 7, 175 rue du Chevaleret, 75013 Paris, France
[4] Motorola Labs, Parc Algorithmes Saint-Aubin, 91193, Gif sur Yvette Cedex, France

Abstract. In this work we study preference systems suitable for the Peer-to-Peer paradigm. Most of them fall in one of the three following categories: global, symmetric and complementary. All these systems share an acyclicity property. As a consequence, they admit a stable (or Pareto efficient) configuration, where no participant can collaborate with better partners than their current ones.

We analyze the representation of such preference systems and show that any acyclic system can be represented with a symmetric mark matrix. This gives a method to merge acyclic preference systems while retaining the acyclicity property. We also consider properties of the corresponding collaboration graph, such as clustering coefficient and diameter. In particular, the study of the example of preferences based on real latency measurements shows that its stable configuration is a small-world graph.

1 Introduction

Motivation. In most current peer-to-peer (P2P) solutions participants are encouraged to cooperate with each other. Since collaborations may be costly in terms of network resources (connection establishment, resource consumption, maintenance), the number of connections is often bounded by the protocol. This constraint encourages the clients to make a careful choice among other peers to obtain a good performance from the system. The possibility to choose a better partner implies that there exists *a preference system*, which describes the interests of each peer.

The study of such preference systems is the subject of *b*-matching theory. It has started forty-five years ago with the seminal work of Gale and Shapley on *stable marriages* [1]. Although the original paper had a certain *recreational mathematics* flavor, the model turned out to be especially valuable both in theory and practice. Today, *b*-matching's applications are not limited to dating agencies, but include college admissions, roommates attributions, assignment of graduating medical students to their first hospital appointments, or kidney exchanges programs [1,2,3,4]. The goal of the present paper is to expand *b*-matching application domain to P2P networks by using it to model the interactions between the clients of such networks.

A.-M. Kermarrec, L. Bougé, and T. Priol (Eds.): Euro-Par 2007, LNCS 4641, pp. 825–834, 2007.
© Springer-Verlag Berlin Heidelberg 2007

Previous Work. The present work draws on and extends results obtained in [5], where we covered general aspects of the b-matching theory application to the dynamics of the node interactions. We considered preference systems natural for the P2P paradigm, and showed that most of them fall into three categories: global, symmetric, and complementary. We demonstrated that these systems share the same property: acyclicity. We proved existence and uniqueness of a stable configuration for acyclic preference systems.

Contribution. In this article, we analyze the links between properties of local marks and the preference lists that are generated with those marks. We show that all acyclic systems can be created with symmetric marks. We provide a method to merge any two acyclic preference systems and retain the acyclic property. Finally, our simulations show that real latency marks create collaboration graphs with small-worlds properties, in contrast with random symmetric or global marks.

Roadmap. In Section 2 we define the global, symmetric, complementary, and acyclic preference systems, and provide a formal description of our model. In Section 3 we demonstrate that all acyclic preferences can be represented using symmetric preferences. We consider complementary preferences in Section 4, and the results are extended to any linear combination of global or symmetric systems. Section 5 discusses the properties of a stable solution providing an example based on Meridian project measurements [6]. In Section 6 we discuss the impact of our results, and Section 7 concludes the paper.

2   Definition and Applications of P2P Preference Systems

2.1   Definitions and General Modeling Assumptions

We formalize here a b-matching model for common P2P preference systems.

Acceptance Graph. Peers may have a partial knowledge of the network and are not necessarily aware of all other participating nodes. Peers may also want to avoid collaboration with certain others. Such criteria are represented by an *acceptance* graph $G(V, E)$. Neighbors of a peer $p \in V$ are the nodes that may collaborate with p. A configuration C is a subset $\tilde{C} \subset E$ of the existing collaborations at a given time.

Marks. We assume peers use some real marks (like latency, bandwidth,...) to rank their neighbors. This is represented by a valued matrix of marks $m = \{m(i,j)\}$. A peer p uses $m(p,i)$ and $m(p,j)$ to compare i and j. Without loss of generality, we assume that 0 is the best mark and $m(p,i) < m(p,j)$ if and only if p prefers i to j. If p is not a neighbor of q, then $m(p,q) = \infty$. We assume for convenience a peer p has a different mark for each of its neighbors. It implies that a peer can always compare two neighbors and decide which one suits him better.

Preference System. A mark matrix M creates an instance L of a preference system. $L(p)$ is a preference list that indicates how a peer p ranks its neighbors. The relation when p prefers q_1 to q_2 is denoted by $L(p, q_1) < L(p, q_2)$. Note that different mark matrices can produce the same preference system.

Global Preferences. A preference system is global if it can be deduced from global marks $(m(i, p) = m(j, p) = m(p))$.

Symmetric Preferences. A preferences system is symmetric if it can be deduced from symmetric marks $(m(i, j) = m(j, i)$ for all $i, j)$.

Complementary Preferences. A preferences system is complementary if it can be deduced from marks of the form $m(i, j) = v(j) - c(i, j)$, where $v(j)$ values the resources possessed by j and $c(i, j)$ the resources that i and j have in common[1].

Acyclic Preferences. A preferences system is acyclic if it contains no preference cycle. A preference cycle is a cycle of at least three peers such that each peer strictly prefers its successor to its predecessor along the cycle.

Quotas. Each peer p has a quota $b(p)$ (possibly infinite) on the number of links it can support. A *b-matching* is a configuration C that respects the quotas. If the quotas are greater than the number of possible collaborations, then simply $C = E$ would be an optimal solution for all.

Blocking Pairs. We assume that the nodes aim to improve their situation, i.e. to link to most preferred neighbors. A pair of neighbors p and q is *a blocking pair* of a configuration C if $\{p, q\} \in E \setminus C$ and both prefer to change the configuration C and to link with the each other. We assume that system evolves by discrete steps. At any step two nodes can be linked together if and only if they form a blocking pair. Those nodes may drop their worst performing links to stay within their quotas. A configuration C is *stable* if no blocking pairs exist.

Loving Pair. Peers p, q form a loving pair if p prefers q to all its other neighbors and q, in its turn, prefers p to all other neighbors. It implies a strong link which cannot be destroyed in the given preference system.

2.2   Preference Systems and Application Design

Depending on the P2P application, several important criteria can be used by a node to choose its collaborators. We introduce the following three types as representative of most situations:

Proximity: distances in the physical network, in a virtual space or similarities according to some characteristics.
Capacity related: network bandwidth, computing capacity, storage capacity.
Distinction: complementary character of resources owned by different peers.

---

[1] Of course, in this case the preferred neighbor has a larger mark.

Notice that theses types correspond respectively to the definitions of symmetric, global and complementary preference system categories.

Examples of symmetric preferences are P2P applications which optimize latencies. A classical approach for distributed hash-table lists of contacts is selecting the contacts with the smallest round trip time (RTT) in the physical network. In Pastry [7], a node will always prefer contacts with the smallest RTT among all the contacts that can fit into a given routing table entry. More generally, building a low latency overlay network with bounded degree requires to select neighbors with small RTTs. Optimizing latencies between players can also be crucial, for instance, for online real-time gaming applications [8]. Such preferences are symmetric since the mark a peer p gives to some peer q is the same as the mark q gives to p (the RTT between p and q).

Similarly, massively multiplayer online games (MMOG) require connecting players with nearby coordinates in a virtual space [9,10]. Again this can be modeled by symmetric preferences based on the distance in the virtual space. Some authors also propose to connect participants of a file sharing system according to the similarity of their interests [11,12], which is also a symmetric relation.

BitTorrent [13] is an example of a P2P application that uses a capacity related preference system. In brief, a BitTorrent peer uploads to the peers from whom it has the most downloaded during the last ten seconds. This is an implementation of the well known Tit-for-Tat strategy. The mark of a peer can thus be seen as its upload capacity divided by its collaboration quota.

This global preference nature of BitTorrent should be tempered by the fact that only peers with complementary parts of the file are selected. Pushing forward this requirement would lead to another selection criterion for BitTorrent: preference for the peers possessing the most complementary set of file pieces. In other words, each peer should try to exchange with peers possessing a large number of blocks it needs. We call this a complementary preference system. Note, that this kind of preferences changes continuously as new pieces are downloaded. However, the peers with the most complementary set of blocks are those, which enable the longest exchange sessions.

In its more general form, the selection of partners for cooperative file download can be seen as a mix of several global, symmetric, and complementary preference systems.

3 Acyclic Preferences Equivalence

In [5], we have shown that global, symmetric and complementary preferences are acyclic, that any acyclic b-matching preference instance has a unique stable configuration, and that acyclic systems always converge toward their stable configuration. However, since acyclicity is not defined by construction, one may wonder whether other kinds of acyclic preferences exist. This section is devoted to answering this question.

Theorem 1. *Let P be a set of n peers, $\mathcal{A}$ be the set of all possible acyclic preference instances on P, $\mathcal{S}$ be the set of all possible symmetric preference instances on P, $\mathcal{G}$ be the set of all possible global preference instances, then*

$$\mathcal{G} \subsetneq \mathcal{A} = \mathcal{S}$$

This ensures that any acyclic instance can be described by the mean of symmetric marks. As a special case, global mark instances can be emulated by symmetric instances, though the reverse is not true. In the reminder of this section we present the proof of theorem 1: we will first show $\mathcal{S} \subset \mathcal{A}$ and $\mathcal{G} \subset \mathcal{A}$, then $\mathcal{A} \subset \mathcal{S}$, which will be followed by $\mathcal{G} \neq \mathcal{A}$.

Lemma 1. *Global and symmetric preference systems are acyclic.*

Proof. from [5]. Let us assume the contrary, and assume that there is a circular list of peers $p_1, \ldots, p_k$ (with $k \geq 3$), such that each peer of the list strictly prefers its successor to its predecessor. Written in the form of marks it means that $m(p_i, p_{i+1}) < m(p_i, p_{i-1})$ for all i modulo k. Taking the sum for all possible i, we get

$$\sum_{i=1}^{k} m(p_i, p_{i+1}) < \sum_{i=1}^{k} m(p_i, p_{i-1}).$$

If marks are global, this can be rewritten $\sum_{i=1}^{k} m(p_i) < \sum_{i=1}^{k} m(p_i)$, and if they are symmetric, $\sum_{i=1}^{k} m(p_i, p_{i+1}) < \sum_{i=1}^{k} m(p_i, p_{i+1})$. Both are impossible, thus global and symmetric marks create acyclic instances. $\qquad\square$

The next part, $\mathcal{A} \subset \mathcal{S}$, uses the loving pairs described in 2.1. We first prove the existence of loving pairs in Lemma 2.

Lemma 2. *A nontrivial acyclic preference instance always admits at least one loving pair.*

Proof. A formal proof was presented in [5]. In short, if there is no loving pair, one can construct a preference cycle by considering a sequence of first choices of peers. $\qquad\square$

Algorithm 1: Construction of a symmetric note matrix m given an acyclic preferences instance L on n peers

---

$N := 0$

for all p and q, $m(p, q) = +\infty$ (by default, peers do not accept each other)

while *there exists a loving pair $\{i, j\}$* **do**

    $m(i, j) := m(j, i) := N$

    Remove i from the preference list $L(j)$ and j from $L(i)$

    $N := N + 1$

Lemma 3. *Let L be a preference instance. Algorithm 1 constructs a symmetric mark matrix in $O(n^2)$ time that produces L.*

Proof. The matrix output is clearly symmetric. Neighboring peers get finite marks, while others have infinite marks. If an instance contains a loving pair $\{i,j\}$ then $m(i,j) = m(j,i)$ can be the best mark since i and j mutually prefer each other to any other peers. According to Lemma 2 such a loving pair always exists in the acyclic case. By removing the peers i and j from their preference lists, we obtain a smaller acyclic instance with the same preference lists except that i and j are now unacceptable to each other. The process continues until all preference lists are eventually empty. The marks are given in increasing order, therefore when $m(p,q)$ and $m(p,r)$ are finite, $m(p,q) < m(p,r)$ iff the loving pair $\{p,q\}$ is formed before the loving pair $\{p,r\}$, that is iff p prefers q to r.

The algorithm runs in $O(n^2)$ time because an iteration of the **while** loop takes $O(1)$ time. A loving pair can especially be found in constant time by maintaining a list on all loving pairs. The list is updated in constant time since, after i and j becoming mutually unacceptable, each new loving pair contains either i and its new first choice, or j and its new first choice. □

Not All Acyclic Preferences Are Global Preferences. A simple counter-example uses 4 peers p_1, p_2, p_3 and p_4 with the following preference lists:

$L(p_1)$: p_2, p_3, p_4 $L(p_2)$: p_1, p_3, p_4 $L(p_3)$: p_4, p_1, p_2 $L(p_4)$: p_3, p_1, p_2

L is acyclic, but p_1 prefers p_2 to p_3 whereas p_4 prefers p_3 to p_2. p_1 and p_4 rate p_2 and p_3 differently, thus the instance is not global.

4   Complementary and Composite Preference Systems

Complementary preferences appear in systems where peers are equally interested in the resources they do not have yet. As said in Section 2.1, complementary preferences can be deduced from marks of the form $m(p,q) = v(q) - c(p,q)$ (in this case, marks of higher values are preferred).

The expression of a complementary mark matrix m shows that it is a linear combination of previously discussed global and symmetric mark matrices: $m = v - c$, where v defines a global preference system and c defines a symmetric system.

Theorem 2 shows that complementary marks, and more generally any linear combination of global or symmetric marks, produce acyclic preferences.

Theorem 2. *Let m_1 and m_2 be global or symmetric marks. Any linear combination $\lambda m_1 + \mu m_2$ is acyclic.*

Proof. The proof is practically the same as for Lemma 1. Let us suppose that the preference system induced by $m = \lambda m_1 + \mu m_2$ contains a preference cycle $p_1, p_2, \ldots, p_k, p_{k+1} = p_1$, for $k \geq 3$. We assume without loss of generality that m_1 is global, m_2 symmetric and that marks of higher values are preferred for m.

Then $m(p_i, p_{i+1}) > m(p_i, p_{i-1})$ for all i modulo k. Taking a sum over all possible i, we get

$$\sum_{i=1}^{k} m(p_i, p_{i+1}) > \sum_{i=1}^{k} m(p_i, p_{i-1}) = \sum_{i=1}^{k} \left(\lambda m_1(p_i, p_{i-1}) + \mu m_2(p_i, p_{i-1}) \right), \text{ but}$$

$$\sum_{i=1}^{k} \left(\lambda m_1(p_i, p_{i-1}) + \mu m_2(p_i, p_{i-1}) \right) = \lambda \sum_{i=1}^{k} m_1(p_{i-1}) + \mu \sum_{i=1}^{k} m_2(p_{i-1}, p_i)$$

$$= \lambda \sum_{i=1}^{k} m_1(p_{i+1}) + \mu \sum_{i=1}^{k} m_2(p_i, p_{i+1}) = \sum_{i=1}^{k} m(p_i, p_{i+1}).$$

This contradiction proves the Theorem. $\qquad\square$

Theorem 2 leads to the question whether any linear combination of acyclic preferences expressed by any kind of marks is also acyclic. The example bellow illustrates that in general it is not true:

$$M_1 = \begin{pmatrix} 0 & 3 & 1 \\ 2 & 0 & 1 \\ 3 & 1 & 0 \end{pmatrix} , \; M_2 = \begin{pmatrix} 0 & 1 & 2 \\ 1 & 0 & 3 \\ 1 & 2 & 0 \end{pmatrix} , \; M_1 + M_2 = \begin{pmatrix} 0 & 4 & 3 \\ 3 & 0 & 4 \\ 4 & 3 & 0 \end{pmatrix}$$

The preference instance induced by $M_1 + M_2$ has the cycle $1, 2, 3$, while both M_1 and M_2 are acyclic (both produce global preferences).

Note, that a linear combination of two preference system matrices can give duplicates in the marks of a single node, which generates ties in preferences. Ties affect existence and uniqueness of a stable configuration, depending on how they are handled. If a peer prefers a new node to a current collaborator that has the same mark, existence is not guaranteed (but if a stable configuration exists, it is unique). Otherwise, existence stands, but not uniqueness [3].

Application. Theorem 2 provides together with Theorem 1 a way of constructing a tie-less acyclic instance that can take into account several parameters of the network, given that they all correspond to acyclic preferences: the parameters can be first converted into integer symmetric marks using Algorithm 1. Then a linear combination using $\mathbb{Q}$-independent scalars produces distinct acyclic marks.

5   Graph Properties of Stable Configurations

To illustrate the acyclic preferences, in this section, we study the connectivity property of stable configurations corresponding to different systems. In particular, we are using the latency matrix of the Meridian Project [6] as an example of the symmetric marks. The entries of the matrix correspond to the median of the round-trip time and they were measured using King technique [6]. We compare it with random symmetric marks and the global preference system[2].

---

[2] In absence of ties, all global marks are the same up to permutation.

To confront these systems we examine connectivity properties of the corresponding stable configurations. The connectivity was extensively studied since Watts survey [14] on the *small world* graphs. These graphs are known to have good routing and robustness properties. They are characterized by a small (i.e. $O(\log(n))$ mean distances and high (i.e. $O(1)$) clustering. The *clustering coefficient* is the probability for two vertices x and y to be linked, given that x and y have at least one common neighbor.

We consider three networks with different marks and $n = 2500$ peers. Figure 1 shows the properties of the stable configuration for these three marks, as a function of the quota b on the number of links per peer.

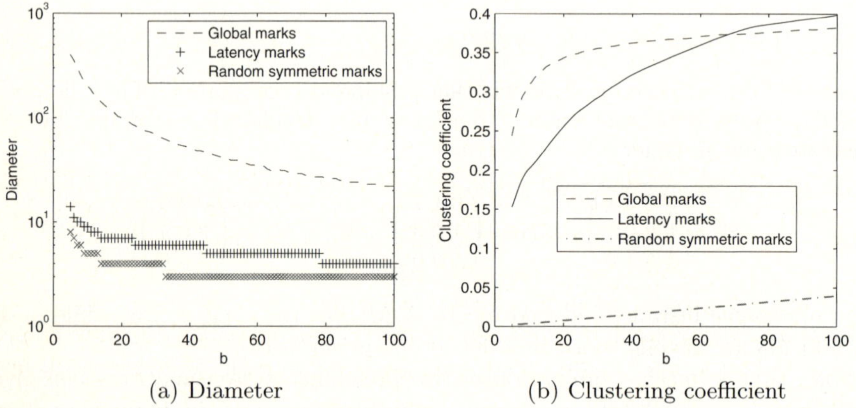

(a) Diameter   (b) Clustering coefficient

Fig. 1. Diameter and clustering coefficient of latency, random symmetric and global marks (2500 nodes) stable configurations. Global marks use an underlying Erdös-Rényi $\mathcal{G}(2500, 0.5)$ acceptance graph.

Global marks produce configuration with disconnected cliques of size $b + 1$ (maximal clustering, and infinite diameter). We have previously observed this *clusterization* effect in [15]. It can be lessened by using an Erdös-Rényi acceptance graph, as it can be seen in Figure 1. Then, the configuration still has a high clustering coefficient, and a high, but finite diameter (of same order of magnitude as $\frac{n}{b}$). This is due to a *stratification* effect: peers only link to peers that have marks similar to them [15].

Random symmetric matrix produces configurations with low diameter and clustering coefficient. There characteristics are similar to those of Erdös-Rényi graphs.

Real latencies from [6] result in both a low diameter and a high clustering coefficient. This indicates that the corresponding stable configuration has *small-world* structure and, therefore, it enjoys nice routing properties. Nevertheless, it is not a *scale-free* network [16], because the degree distribution does not follow a power law (the degrees are bounded by b).

6   Discussion and Future Work

Stability. Decision, whether a stable configuration is a good thing or not, depends on the characteristics and needs of practical applications. If continuous link alteration has a high cost (like in structured P2P networks), or if the stable configuration has appealing properties (like the small-world properties observed for the Meridian latency-based stable configuration), then it is interesting to let the system converge. On the other hand, we have observed that global marks result in a stable configuration with high diameter, which is an undesired feature in most cases. Moreover, some systems like gossip protocols [17] take advantage of constant evolution of the corresponding acceptance graph. In such cases, the eventual convergence would be harmful.

Convergence Speed. The convergence speed is an important characteristic, no matter if the stable solution is desired or not. In the first case, the application is interested in speeding up the process. In the second case, the slower possible speed is preferred instead. Although this question is out of the scope of the present work, our current experiments suggest that the convergence depends on many parameters: the preference system used, the acceptance graph, the activity of peers (details of peers' interaction protocol), the quotas and others. If we use as time unit the mean interval between two attempts of a given peer to change one of its neighbors, then preliminary results show that convergence is logarithmic at best, and polynomial at worst. We plan to provide a complete study on the influence of parameters. This should help understanding existing protocols and making them more efficient.

Dynamics of Preference Systems. We have considered fixed acceptance graph and preference lists. In real applications, arrivals and departures modify the acceptance graph, along with the discovery of new contacts (a toy example is BitTorrent, where a tracker periodically gives new contacts to the clients). The preference system itself can evolve in time. For instance, latency can increase if a corresponding link has a congestion problem. A complementary preference system is dynamic by itself: as a peer gets resources from a complementary peer, the complementarity mark decreases.

All these changes impact the stable configuration of the system. The question is to know whether the convergence speed can sustain the dynamics of preferences or not. Fast convergence and slow changes allow the system to continuously adjust (or stay close) to the current stable configuration. Otherwise, the configurations of the system may always be far from a stable configuration that changes too often. The preferable behavior depends on whether stability is a good feature. This is an interesting direction for future work.

7   Conclusion

In this paper, we gave formal definitions for a b-matching P2P model and analyze the existence of a stable configuration with preference systems natural for P2P environment. The term stability in our case corresponds to Pareto efficiency of the collaboration network, since the participants have no incentives to change such links. We have also showed that in contrast to systems based on intrinsic capacities, a latency-based stable configuration has small-world characteristics.

References

1. Gale, D., Shapley, L.: College admissions and the stability of marriage. American Mathematical Monthly 69, 9–15 (1962)
2. Irving, R.W., Manlove, D., Scott, S.: The hospitals/residents problem with ties. In: Halldórsson, M.M. (ed.) SWAT 2000. LNCS, vol. 1851, pp. 259–271. Springer, Heidelberg (2000)
3. Irving, R.W., Manlove, D.F.: The stable roommates problem with ties. J. Algorithms 43(1), 85–105 (2002)
4. Roth, A.E., Sonmez, T., Utku Unver, M.: Pairwise kidney exchange. Journal of Economic Theory 125(2), 151–188 (2005), available at `http://ideas.repec.org/a/eee/jetheo/v125y2005i2p151-188.html`
5. Lebedev, D., Mathieu, F., Viennot, L., Gai, A.T., Reynier, J., de Montgolfier, F.: On using matching theory to understand P2P network design. In: INOC (2007)
6. Meridian Project: `http://www.cs.cornell.edu/People/egs/meridian/`
7. Rowstron, A., Druschel, P.: Pastry: Scalable, decentralized object location, and routing for large-scale peer-to-peer systems. In: Guerraoui, R. (ed.) Middleware 2001. LNCS, vol. 2218, pp. 329–350. Springer, Heidelberg (2001)
8. Lin, Y.J., Guo, K., Paul, S.: Sync-ms: synchronized messaging service for real-time multi-player distributed games. In: Proc. of the 10th IEEE International Conference on Network Protocols, IEEE Computer Society Press, Los Alamitos (2002)
9. Keller, J., Simon, G.: Solipsis: a massively multi-participant virtual world. In: Intern. Conf. on Parallel and Distributed Techniques and Applications (2003)
10. Kawahara, Y., Aoyama, T., Morikawa, H.: A peer-to-peer message exchange scheme for large-scale networked virtual environments. Telecommunication Systems 25(3) (2004)
11. Le Fessant, F., Handurukande, S., Kermarrec, A.M., Massoulié, L.: Clustering in peer-to-peer file sharing workloads. In: Voelker, G.M., Shenker, S. (eds.) IPTPS 2004. LNCS, vol. 3279, Springer, Heidelberg (2005)
12. Sripanidkulchai, K., Maggs, B., Zhang, H.: Efficient content location using interest-based locality in peer-to-peer systems. In: INFOCOM (2003)
13. Cohen, B.: Incentives build robustness in bittorrent. In: P2PECON (2003)
14. Watts, D.J.: Small Worlds: The Dynamics of Networks between Order and Randomness (Princeton Studies in Complexity). Princeton University Press, Princeton, NJ (2003)
15. Gai, A.T., Mathieu, F., Reynier, J., De Montgolfier, F.: Stratification in P2P networks application to bittorrent. In: ICDCS (2007)
16. Newman, M.E.J.: The structure and function of complex networks. SIAM Review 45(2), 167–256 (2003)
17. Allavena, A., Demers, A., Hopcroft, J.E.: Correctness of a gossip based membership protocol. In: PODC '05: Proceedings of the twenty-fourth annual ACM symposium on Principles of distributed computing, pp. 292–301. ACM Press, New York (2005)

Topic 13
High-Performance Networks

Thilo Kielmann, Pascale Primet, Tomohiro Kudoh, and Bruce Lowekamp

Topic Chairs

Communication networks are likely among the most crucial resources for parallel and distributed computer systems. The last couple of years has seen many significant technical improvements, ranging from advanced on-chip interconnects, via system and storage-area networks, up to dedicated, optical wide-area networks. While this technology push is bringing new opportunities, it also comes with many new research challenges.

This topic is devoted to all kinds of communication issues in scalable compute and storage systems, such as parallel computers, networks of workstations, and clusters. In total, 13 papers were submitted to this topic out of which we have selected the four strongest ones.

Two of the accepted papers are dealing with aspects interconnection networks. In "Integrated QoS Provision and Congestion Management for Interconnection Networks", the authors present a new switch architecture that uses the same resources for both purposes, which is shown to be as effective as previous proposals, but much more cost effective. In "Performance Analysis of an Optical Circuit Switched Network for Peta-Scale Systems", the authors compare the performance of an OCS network, as expected in a peta-scale system, to more traditional networks. They show that an OCS network is comparable with the best networks known so far, however with better cost and flexibility.

The other two accepted papers are dealing with implementing message passing (MPI) libraries on top of modern networks. In "Fast and Efficient Total Exchange on Two Clusters", the authors propose a new algorithm for the collective all-to-all exchange operation across multiple compute clusters, resulting in large performance improvements. Last but not least, "Network Fault Tolerance in Open MPI" describes methods for handling several kinds of network errors while maintaining high-performance communications.

A.-M. Kermarrec, L. Bougé, and T. Priol (Eds.): Euro-Par 2007, LNCS 4641, p. 835, 2007.
© Springer-Verlag Berlin Heidelberg 2007

Integrated QoS Provision and Congestion Management for Interconnection Networks

Alejandro Martínez-Vicente[1], Pedro J. García[1], Francisco J. Alfaro[1],
José-Luis Sánchez[1], Jose Flich[2], Francisco J. Quiles[1], and Jose Duato[2,*]

[1] Departamento de Sistemas Informáticos, Escuela Politécnica Superior
Universidad de Castilla-La Mancha, 02071 - Albacete, Spain
`{alejandro,pgarcia,falfaro,jsanchez,paco}@dsi.uclm.es`
[2] Dept. de Informática de Sistemas y Computadores, Facultad de Informática
Universidad Politécnica de Valencia, 46071 - Valencia, Spain
`{jflich,jduato}@disca.upv.es`

Abstract. Both QoS support and congestion management techniques have become essential for achieving good performance in current high-speed interconnection networks. However, traditional techniques proposed for both issues require too many resources for being implemented. In this paper we propose a new switch architecture that efficiently uses the same resources to offer both congestion management and QoS provision. It is as effective as previous proposals, but much more cost-effective.

1 Introduction

High-speed interconnection networks have become a major issue on the design of computing and communication systems, including systems for parallel computing, since they provide the low-latency and high-throughput demanded by parallel applications. The proliferation of systems based on high-speed networks has increased the researchers' interest on developing techniques for improving the performance of such networks. Moreover, due to the increase in network components' cost and power consumption, it is nowadays very important to propose efficient and cost-effective techniques, trying to use a minimum number of network resources while keeping network performance as high as possible.

For instance, many techniques have been proposed for solving the problem of network performance degradation during congested situations. Congestion is related to the appearance of Head-Of-Line (HOL) blocking, which happens when a packet at the head of a queue blocks[1], preventing other packets in the same queue from advancing, even if they request available resources. This may cause

* This work has been jointly supported by the Spanish MEC and European Comission FEDER funds under grants "Consolider Ingenio-2010 CSD2006-00046" and "TIN2006-15516-C04"; by Junta de Comunidades de Castilla-La Mancha under grant PBC-05-005; and by the Spanish State Secretariat of Education and Universities under FPU grant.
[1] We are considering lossless networks like InfiniBand, Quadrics, or Myrinet.

A.-M. Kermarrec, L. Bougé, and T. Priol (Eds.): Euro-Par 2007, LNCS 4641, pp. 837–847, 2007.

that data flows not contributing to congestion advance at the same speed as congested flows, thereby degrading network performance.

However, although congestion (and HOL bocking) in interconnection networks is a well-known phenomenon, efficient congestion management mechanisms in modern high-speed networks are very rare. On the one hand, traditional, simple solutions are not suitable for modern interconnects. For instance, network overdimensioning is not currently feasible due to cost and power consumption constraints. On the other hand, more elaborated techniques that have been specifically proposed for solving the problems related to congestion have not been really efficient until very recently.

Another way for improving network performance, from an application point of view, is to use techniques for providing Quality of Service (QoS). If communication services must be provided for several types of applications, the QoS consists in guaranteeing a minimum performance to each application, regardless the behavior of the rest of traffic classes. For instance, if we want to guarantee a minimum bandwidth to each traffic class, this minimum must be provided even if there are sources injecting more traffic than they are allowed to.

The usual solution for this problem is to provide a separate virtual channel (VC) for each traffic class. These VCs also provide separate domains of flow control, i.e. there is a separate credit counter for each VC. In this way, two objectives are achieved: firstly, there is no HOL blocking between traffic classes. Secondly, a single traffic class cannot take all the buffer space available, so buffer hogging is avoided. In this way, the performance of different traffic classes only depends on the scheduling and on the amount of injected traffic.

In this paper, we present an efficient and integrated solution for both congestion management and QoS provision problems. We propose a new switch architecture without VCs in which the buffers are managed with a novel congestion management technique: RECN (Regional Explicit Congestion Notification) [1]. Moreover, we add QoS-awareness to this scheme, so as to consider, not only congestion management, but also QoS requirements. The main benefit of this architecture is that it uses the same set of resources for both purposes: congestion management and QoS provision, thus, requiring a lower number of resources than previous proposals.

The rest of the paper is organized as follows. In the following two sections we briefly review proposals for congestion management and QoS. Next, in Section 4, the proposed switch architecture is explained in detail. Section 5 shows an evaluation of the new proposal, based on simulation results. Finally, in Section 6, some conclusions are drawn.

2   Congestion Management

Traditionally, Virtual Output Queues (VOQs) [2] was considered the most effective way to face congestion. In this case, there are at each port as many queues as end-nodes in the network, and any incoming packet is stored in the queue assigned to its destination, thereby avoiding HOL blocking between packets addressed to different destinations. However, although this scheme is very effective, it is not efficient (and even not feasible for medium or large networks) since it requires a considerable number of queues at each switch port, and the silicon area required for implementing such number of buffers strongly increases the cost. A variation of VOQ uses as many queues at each port as output ports in a switch [3], reducing so queue requirements. However, this scheme does not completely eliminate HOL blocking since only switch-level HOL blocking is eliminated.

In contrast to these techniques, RECN [1] completely eliminates HOL blocking while requiring a small number of resources independently of network size, being so a really efficient, scalable and cost-effective technique. Specifically, in order to avoid HOL blocking between congested and non-congested flows, RECN identifies congested flows and puts them in special, dynamically-assigned set aside queues (SAQs).

RECN assumes that packets from non-congested flows can be mixed in the same buffer without producing significant HOL blocking. While standard queues will store non-congested packets, SAQs are dynamically allocated for storing packets passing through a specific congested point. SAQs can be dynamically deallocated when not necessary. Every set of SAQs is controlled by means of a CAM (Content Addressable Memory), in such a way that each CAM line contains information for managing an associated SAQ, including the information required for addressing a congested point.

In this sense, RECN addresses network points by means of the routing information included in packet headers, assuming that source routing is used. For instance, Advanced Switching [4] (AS) packet headers include a turnpool made up of 31 bits, which contains all the turns (offset from the input port to the output port) for every switch in a route. Therefore, CAM lines include turnpools which can be compared with the turnpool of any packet, in order to know if it will pass through the congested point associated to that SAQ. In this way, congested packets can be easily detected. A more detailed description of RECN can be found in [1].

3 QoS Support in Interconnection Networks

In modern interconnection technologies, like InfiniBand or PCI AS, the obvious strategy to provide QoS support consists in providing each traffic class (service level, SL) with a separate VC. This increases switch complexity and required silicon area and, therefore, very few final implementations provide all the VCs proposed in specifications[2].

In order to alleviate this problem some techniques have been proposed that reduce the number of VCs while keeping QoS guarantees. For instance, in [5], a technique for providing full QoS support with only two VCs was proposed. The key idea is that traffic has already been scheduled by the network interfaces.

---

[2] Note that proposals requiring many VCs could be considered if external DRAM is available for implementing the buffers. However, in this case, the low latencies demanded by QoS-requiring traffic could not be provided.

Therefore, there is information regarding the priority of these packets implicit in the order and proportions of packets leaving the end-nodes. For instance, if the end-nodes implement a Weighted Round-Robin [6] policy, packets are injected in the proportions that are configured at the scheduler.

In [7] this technique was combined with RECN by duplicating all the RECN queue structures in two VCs, in order to obtain a switch architecture offering QoS provision and congestion management at the same time.

However, we think that it is possible to achieve full QoS provision in an even more cost-effective, efficient way. Specifically, we propose in this paper a switch architecture that improves the basic RECN mechanism so it can also provide QoS guarantees without introducing additional VCs or queues. The newly proposed architecture takes into account the clear parallelism between the tasks performed by RECN and the use of VCs for QoS provision. Our proposal exploits the RECN queue structure from a new, original approach that efficiently uses these resources for managing congestion while providing QoS at the same time. The benefits of the proposal are obvious since these two important issues on interconnect design would be afforded by a single and very efficient architecture.

4   New Proposal for QoS Provision and Congestion Management

The switch organization that we propose consists of a combination of input and output buffering, which is a usual design for this kind of switches. Note that all the switch components are intended to be implemented in a single chip. This is necessary in order to offer the low cut-through latencies demanded by current applications.

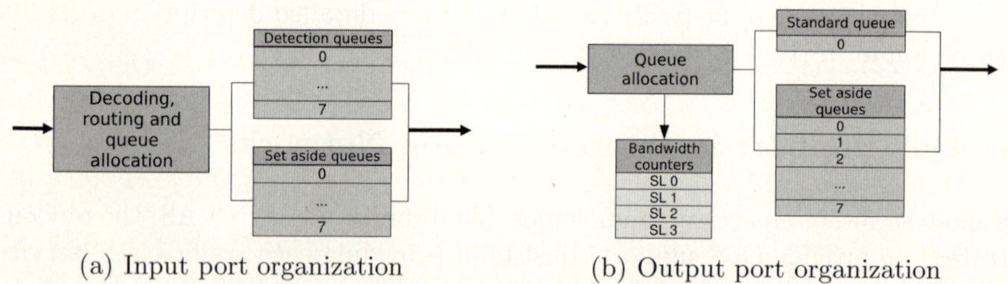

(a) Input port organization   (b) Output port organization

Fig. 1. Input and output ports logical organization

The queue organization of an input port can be seen at Figure 1 (a). This is the standard scheme for a RECN input port, so there are as many detection queues as output ports (8 in this case[3]) for storing non-congested packets, and a group of SAQs (the typical number of SAQs per group is 8 or less) for storing

---

[3] We assume for the sake of simplicity 8 port switches. Anyway, architectures with a different number of ports could be easily deducted.

congested packets. The use of these queues will be discussed later. A CAM is required at each input or output port in order to manage the set of SAQs. The organization of the CAM can be seen in Figure 2 (b). Each CAM line contains all the fields defined by RECN, plus a new field for storing the SL the corresponding SAQ is assigned to.

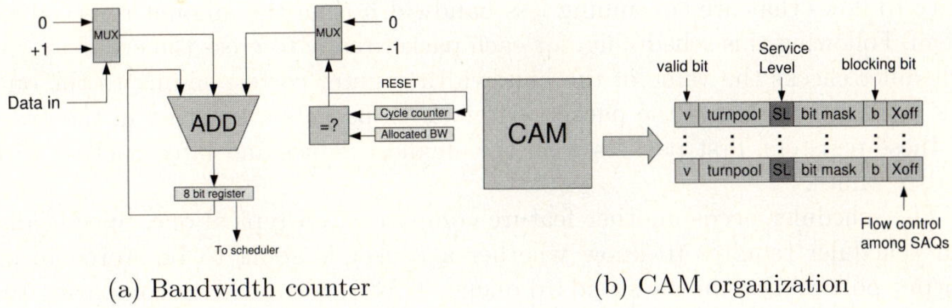

(a) Bandwidth counter   (b) CAM organization

Fig. 2. CAMs and bandwidth counters

Figure 1 (b) shows the organization of output ports. In this case detection queues are not necessary, and a unique standard queue is used for storing non-congested packets. Following also the RECN scheme for output ports, a set of SAQs (8) is also used at the output port. In addition to this RECN queue structure, our proposal introduces at the output ports a set of bandwidth counters, one per SL. These counters must dynamically compute the difference between the reserved bandwidth and the current bandwidth consumption, and they will be used for two purposes: firstly, to perform deficit round-robin scheduling at the switches and secondly, for congestion detection (both issues will be detailed in the following subsections).

The bandwidth counters' structure can be seen at Figure 2 (a). Basically, their behavior is the following: each time a block of 64 bytes from a packet[4] is scheduled to cross towards the output port, the bandwidth counter corresponding to the packet SL is increased by 1. We have assumed a 8-bit register for implementing the bandwidth counter, so its range of values is from -8 kbytes to 8 kbytes[5]. On the other hand, the same register must be automatically decremented at a rate that matches the bandwidth we want to guarantee to the corresponding SL. For instance, if 128 Mbytes/s have been guaranteed to the SL and the internal clock is 100 MHz, the bandwidth counter must be decreased by one every 50 cycles. This decrease is implemented by configuring the "Allocated BW" register with the appropriate number of cycles. When the cycle counter matches the configured cycles, the bandwidth counter register is decreased by one and the cycle counter is reset. All these operations effectively allow to measure the

---

[4] The unit used in our counter is 64 bytes because in PCI AS each credit is 64 bytes and, thus, packets come in 64-byte increments.

[5] These values are adequate to monitor instantaneous traffic behavior.

difference between reserved and consumed bandwidth, while they are simple and do not introduce significant delay or require much silicon area.

4.1   Switch Scheduler

The switch scheduler implements a deficit round-robin (DRR) algorithm based on the bandwidth counters' information. This algorithm consists in giving priority to flows that are consuming less bandwidth than the amount reserved for them. Following this scheduling, for each packet ready to cross the crossbar, the scheduler checks the value of the bandwidth counter corresponding to the output port and the SL of the packet. Afterwards, packets are served in the order of these registers, first packets with the smallest values and later packets with higher values.

The scheduler needs another feature compared to a typical one. Specifically, our scheduler requires to know whether a packet is going to be stored in an output port SAQ or in the standard queue. This is calculated by comparing the turnpool of the packet with the routing and SL information at the CAM. In the case of matching, packets can only be selected if the SAQ is not filled over a certain threshold. By applying this rule, buffer hogging is prevented.

Moreover, the scheduler does not specially treat packets coming from SAQs. However, if such packets are contributing to congestion, the corresponding bandwidth counter will have a high value and they will be penalized by the DRR algorithm. On the other hand, if the packets are no longer contributing to congestion, the corresponding bandwidth counter will have a small value and they will achieve a high priority. This effect contributes to empty unnecessary SAQs as soon as possible, thereby allowing the deallocation of these SAQs (SAQs must be empty for being dellocated).

4.2   Congestion Management

RECN detects congestion both at input or output ports of switches, always by measuring the number of packets stored in the queues. Our new proposal also detects congestion at input and output ports, but in this case these detections do not depend only on queue occupancy. Specifically, in the new proposal, input detections happen when a detection queue at an input port fills (in terms of stored information) over a certain threshold and the value of the bandwidth counter associated to the last received packet SL at the requested output port is over another threshold. The actions to take after a congestion detection at the input port are: first, a SAQ associated to the appropriate SL and with turnpool equal to the output port is allocated at the input port, and, in addition to this, another SAQ, with empty turnpool but associated to the SL is allocated at the corresponding output port. These SAQs with empty turnpool (not considered in original RECN) will store any packet belonging to the associated SL. This is necessary in order to avoid that packets from a single SL completely fill an output buffer.

On the other hand, the conditions for a detection of congestion at an output port are similar. If the arrival of a packet from an input port causes occupancy of the standard queue at this output port to be over a certain threshold and the bandwidth counter of the packet SL is over another threshold, then the detection of congestion takes place. The actions after an output detection are the same as in an input detection: allocation of two SAQs, one at the input port and another with empty turnpool (but associated to the SL) at the output port.

If congestion persists and SAQs start to fill over a certain threshold (known as the propagation level), then information about the corresponding turnpool and SL is propagated backwards the congestion flow, in order to allocate new SAQs for storing packets belonging to this flow wherever these packets are. In the case of propagation from a SAQ at an output port, SAQs are allocated at the input ports that cause the overflow of the output port SAQ. Of course, these input SAQs will have an associated turnpool with one more hop than the output SAQ. SAQs with empty turnpools (only allocated at output ports) may also produce the allocation of SAQs at the input ports, that in this case will have a one-hop turnpool.

If SAQs at input ports fill over the propagation level, a control packet is sent to the preceding switch. This packet includes the turnpool and service level associated to the filled input SAQ, in order to also have an allocated SAQ associated to this information at the receiving output port.

In this way, there will be SAQs at any point where they are necessary in order to store congested packets, thereby eliminating HOL blocking. SAQs can be deallocated when they are not necessary for eliminating HOL blocking at the point where they are allocated. The conditions for SAQ deallocation are exactly the same as in RECN [1].

On the other hand, all the thresholds and propagation levels are constant in the system, and we assume they are properly set up by the network administrator during the configuration phase of the different devices. In our experiments, we have used a set of values obtained after exhaustive tuning which are optimal for a variety of networks designs.

4.3   QoS Provision

In order to provide QoS guarantees, each traffic class is assigned a percentage of link bandwidth. For instance, if there are four traffic classes, each one could be assigned 25%. The total assigned bandwidth must not exceed the bandwidth of any link. At the end-nodes, we assume a traditional DRR implementation with a VC per traffic class, which is something feasible in these devices.

Provided that end-nodes implement this QoS policy, and as long as there is no contention, we have observed that packets pass through the switches in the same proportions as they are injected into the network. The reason is that the switches do not introduce any significant delay when links are not oversubscribed.

However, since there is no admission control, it may happen that any link of the network becomes oversubscribed. In this situation, congestion appears because at this point, one or more traffic classes introduce more traffic than their assignation.

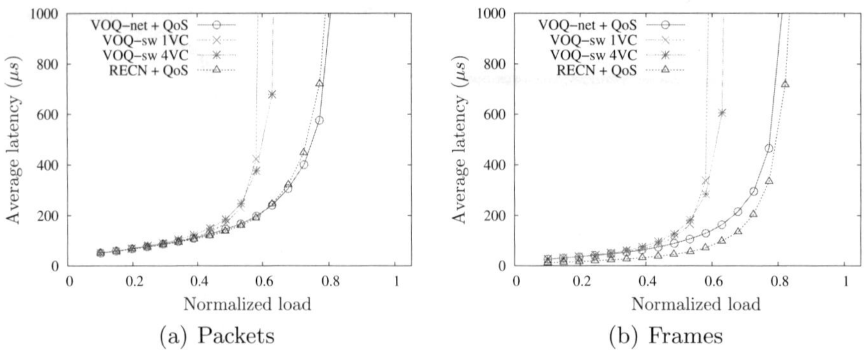

Fig. 4. Performance in video transmission scenario

Regarding individual packets latency, the best results are for the *VOQ-net* architecture, with the *RECN+QoS* architecture very close. The other cases are not able to handle properly the bursts of packets.

If we look at frame's latency, the best results are for our *RECN+QoS* architecture, even better than the *VOQ-net* case. The reason is that our *RECN+QoS* proposal is able to cope better with congestion at any point in the network, while VOQs are designed to handle congested end-nodes. As a consequence, large bursts of packets (like big video frames) progress faster, and hence the better frame-level latency results.

6   Conclusion

Current high-speed interconnection networks demand adequate QoS support and congestion management techniques for achieving good network performance. In this paper we propose a new switch architecture able to face the challenges of congestion management and, at the same time, QoS provision, while being more cost-effective than other proposals, since it uses the same resources for both purposes.

Results have shown that, by means of affordable mechanisms and techniques, we can manage the buffer space and the queues in a very efficient way for addressing our goals. Without VCs and with just some additional queues per port, we can guarantee QoS while eliminating congestion.

References

1. García, P.J., Flich, J., Duato, J., Johnson, I., Quiles, F.J., Naven, F.: Efficient, scalable congestion management for interconnection networks. IEEE Micro 26, 52–66 (2006)
2. Dally, W., Carvey, P., Dennison, L.: Architecture of the Avici terabit switch/router. In: Proceedings of the 6th Symposium on Hot Interconnects, pp. 41–50 (1998)

3. Anderson, T., Owicki, S., Saxe, J., Thacker, C.: High-speed switch scheduling for local-area networks. ACM Transactions on Computer Systems 11, 319–352 (1993)
4. ASI SIG: Advanced switching core architecture specification (2005)
5. Martínez, A., Alfaro, F.J., Sánchez, J.L., Duato, J.: Providing full QoS support in clusters using only two VCs at the switches. In: Bader, D.A., Parashar, M., Sridhar, V., Prasanna, V.K. (eds.) HiPC 2005. LNCS, vol. 3769, pp. 158–169. Springer, Heidelberg (2005), available at http://www.i3a.uclm.es
6. Katevenis, M., Sidiropoulos, S., Courcoubetis, C.: Weighted round-robin cell multiplexing in a general-purpose ATM switch. IEEE J. Select. Areas Commun., 1265–1279 (1991)
7. Martínez, A., García, P.J., Alfaro, F.J., Flich, J., Sánchez, J.L., Quiles, F.J., Duato, J.: A cost-effective interconnection architecture with QoS and congestion management support (2006), available at http://www.i3a.uclm.es

Fast and Efficient Total Exchange
on Two Clusters

Emmanuel Jeannot and Luiz Angelo Steffenel

AlGorille Team, LORIA*
LORIA - Campus Scientifique - BP 239
F-54506 Vandoeuvre-lès-Nancy Cedex, France
Emmanuel.Jeannot@loria.fr, Luiz-Angelo.Steffenel@univ-nancy2.fr

Abstract. Total Exchange is one of the most important collective communication patterns for scientific applications. In this paper we propose an algorithm called $\mathcal{L}G$ for the total exchange redistribution problem between two clusters. In our approach we perform communications in two different phases, aiming to minimize the number of communication steps through the wide-area network. Therefore, we are able to reduce the number of messages exchanged through the backbone to only $2 \times \max(n_1, n_2)$ against $2 \times n_1 \times n_2$ messages with the traditional strategy (where n_1 and n_2 are the number of nodes of each clusters). Experimental results show that we reach over than 50% of performance improvement comparing to the traditional strategies.

1   Introduction

In this paper we address the problem of efficiently perform the *alltoall* (or Total Exchange) communication on two parallel clusters. Total Exchange [1] is one of the most important collective communication patterns for scientific applications, in which each process holds n different data items that should be distributed among the n processes, including itself. An important example of this communication pattern is the `MPI_AlltoAll` operation, where all messages have the same size m.

Although efficient *alltoall* algorithms have been studied for specific networks structures like meshes, hypercubes, tori and circuit-switched butterflies [1,2,3,4], most of the algorithms currently used rely on homogeneous network solutions. These algorithms follow uniform communication patterns between all nodes that prevent an efficient use of both local and distant resources.

In our case, we assume that the parallel application is executed on two different clusters connected by a backbone. This is the case, when for scalability or lack of memory reasons, the application needs to be executed on more than one cluster. In this case, as our experiments will show, standard solutions such as the one implemented in MPI libraries are sub-optimal because they do not use the fact that the performance of the backbone (latency and bandwidth) is lower than the performance of the interconnection network of the cluster.

---

* UMR 7503 CNRS-INPL-INRIA-Nancy2-UHP.

A.-M. Kermarrec, L. Bougé, and T. Priol (Eds.): Euro-Par 2007, LNCS 4641, pp. 848–857, 2007.
© Springer-Verlag Berlin Heidelberg 2007

The main contribution of this paper is to propose an efficient algorithm for executing a Total Exchange communication pattern on two clusters where the latency of the backbone is a performance constraint. Our strategy consists in exploiting the local-area network performance to reduce the number of inter-cluster communication steps. We have compared this algorithm with the standard *alltoall* implementation from the OpenMPI library [5], and the results show that not only we outperform the traditional algorithm in a grid environment but also that our algorithm is more scalable.

This paper is organized as follows: Section 2 introduces the problem of total exchange between nodes in two different clusters. The related works are presented in Section 3. The algorithm we propose is described in Section 4, while the results of the experiments are given in Section 5. Finally, Section 6 presents some conclusions and the future directions of our work.

2 Problem of Total Exchange Between Two Clusters

We consider the following architecture (see Figure 1). Let there be two clusters C_1 and C_2 with respectively n_1 nodes and n_2 nodes. A network, called a backbone, interconnects the two clusters. We assume that a cluster use the same network card to communicate to one of its node or to a node of another cluster. Based on that topology inter cluster communications are never faster than communication within a cluster.

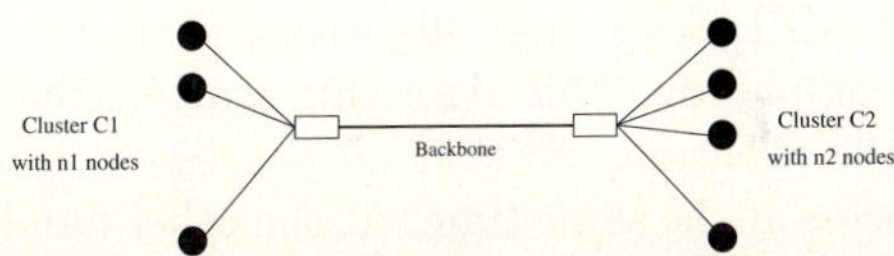

Fig. 1. Architecture for the redistribution problem

Let us suppose that an application is running and using both clusters (for example, a code coupling application). One part of the computation is performed on cluster C_1 and the other part on cluster C_2. During the application, data must be exchanged from C_1 to C_2 using the *alltoall* pattern. *Alltoall* (also called total exchange) is defined in the MPI standard. It means that every node has to send some of its data to all the other nodes. Here we assume that the data to be transfer is different for each receiving node (if the data is the same, the routine is called an *allgather* and is less general that the studied case). Moreover we assume that the size of the data to exchange is the same for every pair of nodes (the case where the size is different is implemented by the *alltoallv* routine: it is more general than our case and will be studied in a future work). Altogether, this means that we will have to transfer $(n_1 + n_2)^2$ messages. The data of all these messages are different but the size of the messages are the same and is given and called m (in bytes).

The question is: how to perform the *alltoall* operation as fast as possible?

Several MPI libraries (OpenMPI, MPICH2, etc.) implement the *allltoall* routine (see Figure 2 for an example). However, these implementations assume that all the nodes are on the same clusters, which means that all the messages have the same importance. However, in our case, some messages are transferred within a cluster (from a node of C_1 to a node of C_1 or form C_2 to C_2) or between the two clusters. In the first case, bandwidth and latency are faster than in the second case. Therefore, there is room for optimizing the transfer time.

```
inbuffer = (void*) calloc (sendcount*size,sizeof(int));
outbuffer = (void*) calloc (recvcount*size,sizeof(int));
(...)
retval = MPI_Alltoall(outbuffer,sendcount,MPI_INT,inbuffer,recvcount,
            MPI_INT,MPI_COMM_WORLD);
(...)
```

Fig. 2. MPI Code extract with an *alltoall* call

3   Related Works

A number of efficient algorithms have been developed for homogeneous clusters with specific network architectures [1,2,3,4]. Generic solutions however rely on direct connections among all nodes, differing only in the communication schedule they use. Indeed, OpenMPI uses a single algorithm that posts all communications and then waits their completion. This algorithm scatters the order of sources and destinations among the processes, so that all processes do not try to send/recv to/from the same process at the same time. At the other hand, MPICH-2 [6] relies on four different algorithms according to the message size and the number of nodes involved in the operation. Indeed, it is worthy of note that the use of the store-and-forward algorithm from Bruck *et al.* [7] for small messages ($m < 256\,bytes$), which behaves well on situations where the latency dominates the bandwidth [8].

Faraj *et al.*[9], on the other side, propose a scheduling algorithm for switched Ethernet clusters. Their approach consists on scheduling communications to maximize simultaneous connections while avoiding collisions at the network bottlenecks. One drawback of this approach, however, is that it supposes uniform communication steps, which does not hold in the case of heterogeneous networks. A similar approach was presented by Sanders *et al.* [10], who proposed a hierarchical factor algorithm to schedule communications in a cluster of multiprocessor machines where each node can only participate in one communication operation with another node at a time.

Another possibility resides on generating communication schedules according to the communication time between each pair of nodes. Different works propose scheduling heuristics for heterogeneous networks [11,12,13], while Goldman *et al.* [8] studies similar heuristics in the context of homogeneous networks with irregular messages (*alltoallv*). While these techniques may provide efficient communication schedules, they suffer from two major drawbacks: first, the complexity of the proposed scheduling heuristics - at least $O(n^3)$ - induces an important extra cost to the operation; second, these heuristics require a relatively high heterogeneity among the communications to improve the performance of the *alltoall* operation. Once the bandwidth dominates the latency (as in the case of large messages), the heterogeneity levels are too small to contribute with the scheduling algorithm.

It is important to note that MagPIe [14] does not implement a grid-aware *alltoall*, preferring the standard approach from MPI. Several reasons contribute to this decision, the first one being that the *alltoall* operation does not fit well MagPIe's hierarchical structure. Indeed, as each process should receive a different set of messages, it becomes too expensive to relay these messages through a single cluster coordinator, a problem similar to that of Bruck's algorithm [7].

Hence, to the best of our knowledge, there is no heuristics that efficiently tackles the problem of the Total Exchange on two clusters.

4 Algorithm

4.1 Design Principles

In order to construct our *alltoall* algorithm we considered the following design principles:

Multi-level Collective Algorithms Are Better Suited for Grids: The recent efforts for grids divide the set of nodes in different clusters organized in a hierarchical structure, following different communication strategies according to the hierarchical level [15]. Indeed, the local area network is usually faster than the wide-area network, and a careful design allow the algorithms to avoid transmitting data through the slow link connecting two clusters. For instance, a hierarchical approach is essential for ensuring wide area optimality for the collective communication algorithms while performing efficiently on the local network [16].

Wide-Area Links Support Simultaneous Transfers Without Performance Degradation: Popular algorithms for collective communications on grids (such as the ones implemented in PACX MPI [17] and MagPIe [16]) define a single coordinator in every cluster, which participates in the inter-cluster data transfers across the wide-area backbone. However, this approach is neither optimal concerning the usage of the wide-area bandwidth, nor well adapted to the *alltoall* problem (gathering data from an entire cluster and sending it through the coordinator becomes too expensive [7] and represents a bottleneck in communications). Actually, simultaneous transfers on these links can help in effectively use the WAN bandwidth [18] while reducing the number of communication steps over the slower link. Moreover as experimentally shown in [19], avoiding contention very seldom improves the transfer time.

Avoiding Centralization Ensures Scalability: If we want to target very large scale environment we have to avoid as much as possible the centralization of information such as message size, schedule pattern, etc. That's why we decided to design a fully distributed algorithm that uses only local information and does not synchronize with other nodes.

Actually, most of the complexity of the All-to-All problem resides on the need to exchange *different* messages with each other process. Indeed, the traditional approach consists in establish connections to each other process in the network (local and distant). However, if we assume that the latency between clusters is higher than intra-clusters ones, it might be useful to send data that has to go from one cluster to the other in one single message. Our propose solution is based on this idea and therefore has two phases. In the first phase only local communications are performed. During this phase the total exchange is performed on local nodes on both cluster and extra buffers are prepared for the second (inter-cluster) phase. During the second phase data are exchanged between the clusters. Buffers that have been prepared during the first phase are sent directly to the corresponding nodes in order to complete the total exchange.

More precisely, our algorithm called *Local Group* or simply $\mathcal{LG}$ works as follow. Without loss of generality, let us assume that cluster $\mathcal{C}_1$ has less nodes than $\mathcal{C}_2$ $(n_1 \leq n_2)$.

Nodes are numbered from 0 to $n_1 + n_2 - 1$, with nodes from 0 to $n_1 - 1$ being on $\mathcal{C}_1$ and nodes from n_1 to $n_1 + n_2 - 1$ being on cluster $\mathcal{C}_2$. We call $\mathcal{M}_{i,j}$ the message (data) that has to be send form node i to node j. The phases are sum-up in Algorithm 1.

First Phase. During the first phase, we perform the local exchange: Process i sends $\mathcal{M}_{i,j}$ to process j, if i and j are on the same cluster. Then it prepares the buffers for the remote communications. On $\mathcal{C}_1$ data that have to be send to node j on $\mathcal{C}_2$ is first stored to node $j \bmod n_1$. Data to be sent from node i on $\mathcal{C}_2$ to node j on $\mathcal{C}_1$ is stored on node $\lfloor i/n_1 \rfloor \times n_1 + j$.

Algorithm 1. The $\mathcal{LG}$ (*Local Group*) algorithm when $n_1 \leq n_2$

```
// Local Phase
   for i = {0, ..., (n₁ + n₂) − 1} do in parallel
      for j = {0, ..., (n₁ + n₂) − 1} do
         if i < n₁ // the sender is on C₁
            send M_{i,j} to j mod n₁
         else // the sender is on C₂
            if j ≥ n₁ // the receiver is on C₂
               send M_{i,j} to j
            else // the receiver is on C₁
               send M_{i,j} to ⌊i/n₁⌋ × n₁ + j
// Inter-cluster Phase
for s = {1, ..., ⌈n₂/n₁⌉}
   for i = {1, ..., n₁ − 1} do in parallel
      if (i + s × n₁ < n₁ + n₂)
         exchange messages between i and j = i + e × n₁
```

Second Phase. During the second phase only n_2 inter-cluster communications occurs. This phase is decomposed in $\lceil n_2/n_1 \rceil$ steps with at most n_1 communications each. Steps are numbered from 1 to $\lceil n_2/n_1 \rceil$ During step s node i of $\mathcal{C}_1$ exchange data stored in its local buffer with node $j = i + n_1 \times s$ on $\mathcal{C}_2$ (if $j < n_1 + n_2$). More precisely i sends $\mathcal{M}_{k,j}$ to j where $k \in [0, n_1]$ and j sends $\mathcal{M}_{k,i}$ to i where $k \in [n_1 \times s, n_1 \times s + n_1 - 1]$.

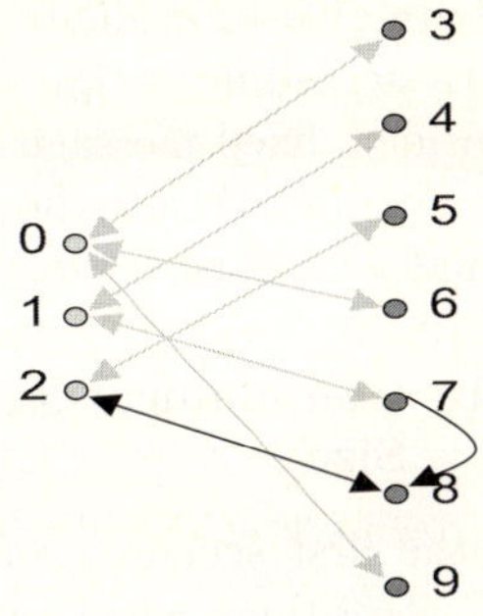
Fig. 3. Example of the 2 phases of the algorithms

Example. Suppose that $n_1 = 3$ and $n_2 = 7$. What happens to the message $\mathcal{M}_{7,2}$ (*i.e* the messages that goes from node 7 on cluster $\mathcal{C}_2$ to node 2 on $\mathcal{C}_1$)? This is illustrated in Figure 3. During the first phase it is stored on node $\lfloor 7/3 \rfloor \times 3 + 2 = 8$ on $\mathcal{C}_2$. Then during the second phase it is sent to node 2 during the step $s = 2$: node 8 sends $\mathcal{M}_{6,2}$, $\mathcal{M}_{7,2}$ and $\mathcal{M}_{8,2}$ to node 2. During this step nodes 1 and 7 and node 0 and 6 exchange data as well, while in the previous step node 0 and 3, 4 and 1 and 5 and 2 exchange data. Finally, only 0 and 9 exchange data in the last step.

4.2   Comparison with the Standard Total Exchange Algorithm

As our algorithm tries to minimize the number of inter-cluster communications between the clusters, we need only $2 \times \max(n_1, n_2)$ messages in both directions against $2 \times n_1 \times n_2$ messages in the traditional algorithm. For instance, the exchange of data between two clusters with the same number of process will proceed in one single communication step of the second phase. At the other hand, if $n_2 \gg n_1$, the total number of communication steps will be similar to the traditional algorithm.

5   Experimental Validation

To validate the algorithm we propose in this paper, this section presents our experiments to evaluate the performance of the `MPI_Alltoall` operation with two clusters connected through a backbone. These experiments were conducted over two clusters of the Grid'5000 platform[1].

We used two clusters, one located in Nancy and one located in Rennes, approximately 1000 Km from each other. Both clusters are composed of HP ProLiant DL145G2 nodes (dual Opteron 246, 2 GHz) and are connected by a private backbone of 10 Gbps. All nodes run Linux, with kernel version 2.6.13.

Two different scenarios have been studied. First, we evaluate the performance of the algorithm in a fixed-size grid for different message sizes. In this approach

---
[1] Experiments presented in this paper were carried out using the Grid'5000 experimental testbed, an initiative from the French Ministry of Research through the ACI GRID incentive action, INRIA, CNRS and RENATER and other contributing partners (see https://www.grid5000.fr).

we are able to evaluate the trade-off between local and wide-area communications according to the amount of data to be exchanged. The second scenario considers fixed message sizes while varying the number of processes. Therefore, we are able to study the scalability aspects of the algorithm. Both scenarios were implemented using Open-MPI 1.1.2 [5] in which we implemented our algorithm.

5.1 Communication Between Two Clusters Varying the Message Size

In the first scenario, we compare the completion time of both standard and $\mathcal{LG}$ algorithms when varying the message size. We conducted experiments for messages of size $m = 2^k$ with $k \leq 24$ (16 MB) in two different environments:

In the first experiment, both clusters C_1 and C_2 have the same number of processes (30 processes each). As this situation corresponds to a 1:1 mapping between processes in both clusters, each process will perform at most one message exchange through the backbone.

In the second experiment we consider C_1 and C_2 having different numbers of processes. Hence, we consider $n_1 = 20$ while $n_2 = 40$. In this scenario, the inter-cluster exchange will proceed in two steps, with processes from C_1 exchanging messages with two processes from C_2.

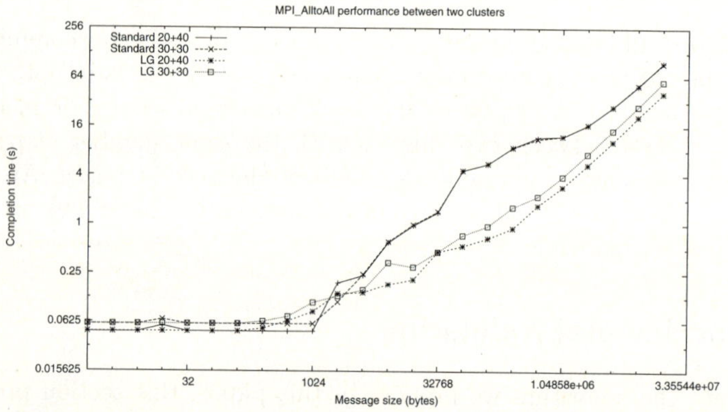

Fig. 4. Performance of the MPI_Alltoall operation in a grid environment

Therefore, in Figure 4 we plot the total communication time obtained for both situations when comparing the performance of the standard `MPI_Alltoall` implementation with the $\mathcal{LG}$ algorithm. Several observations can be made:

- We perform up to 8 times faster than the traditional algorithm according to the message size. Even with large messages we achieve 40% to 60% reduction of communication costs.
- The extra-cost due to message packing observed with small messages is rapidly compensate by the gain on the inter-cluster latency. Indeed, as soon as the traditional algorithm is forced to use more than one single datagram, our strategy presents better results.

– In spite of the different processes distributions, the standard algorithm from OpenMPI does perform almost identically. This evidences the fact that the standard algorithm does not adapts to the network characteristics. At the opposite side we observe that the $\mathcal{LG}$ algorithm performs differently according to the processes distribution, adapting to both scenarios.

5.2 Fixed-Size Messages Varying the Number of Nodes

While the previous experiment demonstrated that our algorithm performs better than the traditional one due to the reduction of inter-cluster communication steps, we are also interested in the scalability behavior of our algorithm. Indeed, the rational usage of both local and remote links should reduce the network contention that usually characterizes an *alltoall* exchange.

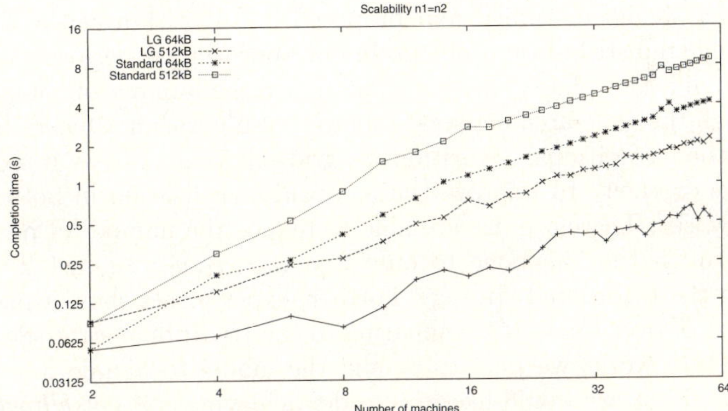

Fig. 5. Performance of the algorithms when varying the number of processes with $n_1 = n_2$

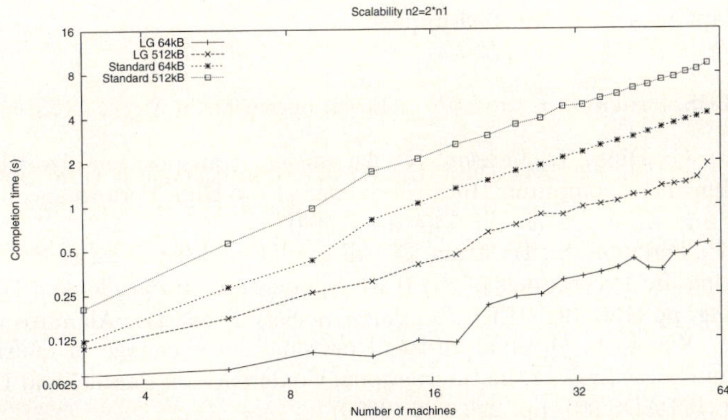

Fig. 6. Performance of the algorithms when varying the number of processes with $n_2 = 2 \times n_1$

Therefore, this experiment compares the performance of both algorithms when we increase the number of interconnected nodes. We evaluate the overall execution time for two different message sizes, 64kB and 512kB under different processes distributions. Therefore, Figure 5 represents a scenario where $n_1 = n_2$, while Figure 6 represents the performance of the algorithms when $n_2 = 2 \times n_1$.

In both experiments we observe that $\mathcal{LG}$ outperforms the standard *alltoall* implemented in Open-MPI. In both graphs we see that the time to perform the *alltoall* with messages of 512kB with $\mathcal{LG}$ is faster than the time to perform the *alltoall* with messages of 64kB with OpenMPI. We see that the slopes of both lines of the $\mathcal{LG}$ algorithm are lower than the slopes of the OpenMPI implementation. This means that, since the y-axis uses a logarithmic scale, the $\mathcal{LG}$ algorithm is more scalable than the OpenMPI implementation.

6   Conclusions and Future Works

In this paper we have studied and proposed an algorithm called $\mathcal{LG}$ for the total exchange redistribution problem. In our approach we perform communications in two different phases, aiming to minimize the number of communication steps through the wide-area network. Indeed, our algorithm achieves better performances than traditional algorithm on grid environments as it exploits the network heterogeneity to improve the bandwidth utilization in both local and remote networks. Therefore, we are able to reduce the number of messages exchanged through the backbone to only $2 \times \max(n_1, n_2)$ against $2 \times n_1 \times n_2$ messages in the traditional strategy. Further, experiments show a performance improvement of over than 50% comparing to the traditional strategies.

In our future works we plan to extend the model to handle more complex distributions. First, we would like to consider achieving efficient *alltoall* communications with more than two clusters. This would allow efficient communications on general grid environments. Second, we would like to explore the problem of total exchange redistribution when messages have different sizes. This problem, represented by the *alltoallv* routine, is more general than our case and does requires adaptive scheduling techniques.

References

1. Christara, C., Ding, X., Jackson, K.: An efficient transposition algorithm for distributed memory computers. In: Proceedings of the High Performance Computing Systems and Applications, pp. 349–368 (1999)
2. Calvin, C., Perennes, S., Trystram, D.: All-to-all broadcast in torus with wormhole-like routing. In: Proceedings of the IEEE Symposium on Parallel and Distributeed Processing, pp. 130–137. IEEE Computer Society Press, Los Alamitos (1995)
3. Yang, Y., Wang, J.: Optimal all-to-all personalized exchange in multistage networks. In: Proceedings of the International Conference on Parallel and Distributed Systems (ICPADS'00), pp. 229–236 (2000)
4. Kalé, L.V., Kumar, S., Varadarajan, K.: A framework for collective personalized communication. In: Proceedings of the International Parallel and Distributed Processing Symposium (IPDPS'03) (2003)

5. Gabriel, E., Fagg, G.E., Bosilca, G., Angskun, T., Dongarra, J., Squyres, J.M., Sahay, V., Kambadur, P., Barrett, B., Lumsdaine, A., Castain, R.H., Daniel, D.J., Graham, R.L., Woodall, T.S.: Open MPI: Goals, concept and design of a next generation MPI implementation. In: Kranzlmüller, D., Kacsuk, P., Dongarra, J.J. (eds.) EuroPVM 2004. LNCS, vol. 3241, pp. 97–104. Springer, Heidelberg (2004)
6. Gropp, W.: Mpich2: A new start for mpi implementations. In: Kranzlmüller, D., Kacsuk, P., Dongarra, J.J., Volkert, J. (eds.) PVM/MPI 2002. LNCS, vol. 2474, p. 7. Springer, Heidelberg (2002)
7. Bruck, J., Ho, C.T., Kipnis, S., Upfal, E., Weathersby, D.: Efficient algorithms for all-to-all communications in multiport message-passing systems. IEEE Transactions on Parallel and Distributed Systems 8(11), 1143–1156 (1997)
8. Goldman, A., Trystram, D., Peters, J.G.: Exchange of messages of different sizes. Journal of Parallel and Distributed Computing 66(1), 1–18 (2006)
9. Faraj, A., Yuan, X.: Message scheduling for all-to-all personalized communication on ethernet switched clusters. In: Proceedings of the IEEE International Parallel and Distributed Processing Symposium (IPDPS), IEEE Computer Society Press, Los Alamitos (2005)
10. Sanders, P., Traff, J.L.: The hierarchical factor algorithm for all-to-all communication. In: Monien, B., Feldmann, R.L. (eds.) Euro-Par 2002. LNCS, vol. 2400, pp. 799–803. Springer, Heidelberg (2002)
11. Bhat, P., Prasanna, V., Raghavendra, C.S.: Adaptive communication algorithms for distributed heterogeneous systems. In: Proceedings of the IEEE International Symposium on High Performance Distributed Computing (HPDC 1998), IEEE Computer Society Press, Los Alamitos (1998)
12. Liu, W., Wang, C.L., Prasanna, V.K.: Portable and scalable algorithms for irregular all-to-all communication. In: Proceedings of the 16th ICDCS, pp. 428–435 (1996)
13. Chun, A.T.T., Wang, C.L.: Contention-aware communication schedule for high-speed communication. Cluster Computing: The Journal of Networks, Software Tools and Application 6(4), 337–351 (2003)
14. Kielmann, T., Bal, H., Gorlatch, S., Verstoep, K., Hofman, R.: Network performance-aware collective communication for clustered wide area systems. Parallel Computing 27(11), 1431–1456 (2001)
15. Steffenel, L.A., Mounie, G.: Scheduling heuristics for efficient broadcast operations on grid environments. In: Proceedings of the Performance Modeling, Evaluation and Optimization of Parallel and Distributed Systems Workshop - PMEO'06 (associated to IPDPS'06), Rhodes Island, Greece, IEEE Computer Society Press, Los Alamitos (2006)
16. Kielmann, T., Hofman, R., Bal, H., Plaat, A., Bhoedjang, R.: Magpie: MPI's collective communication operations for clustered wide area systems. In: Proceedings of the 7th ACM SIGPLAN Symposium on Principles and Practice of Parallel Programming, pp. 131–140. ACM Press, New York (1999)
17. Gabriel, E., Resch, M., Beisel, T., Keller, R.: Distributed computing in a heterogenous computing environment. In: Alexandrov, V.N., Dongarra, J.J. (eds.) PVM/MPI 1998. LNCS, vol. 1497, pp. 180–187. Springer, Heidelberg (1998)
18. Casanova, H.: Network modeling issues for grid application scheduling. International Journal of Foundations of Computer Science 16(2), 145–162 (2005)
19. Jeannot, E., Wagner, F.: Scheduling messages for data redistribution: an experimental study. International Journal of High Performance Computing Applications 20(4), 443–454 (2006)

Performance Analysis of an Optical Circuit Switched Network for Peta-Scale Systems

Kevin J. Barker and Darren J. Kerbyson

Performance and Architecture Laboratory (PAL)
Los Alamos National Laboratory, NM 87545
`{kjbarker,djk}@lanl.gov`

Abstract. Optical Circuit Switching (OCS) is a promising technology for future large-scale high performance computing networks. It currently widely used in telecommunication networks and offers all-optical data paths between nodes in a system. Traffic passing through these paths is subject only to the propagation delay through optical fibers and optical/electrical conversions on the sending and receiving ends. High communication bandwidths within these paths are possible when using multiple wavelengths multiplexed over the same fiber. The set-up time of an OCS circuit is non-negligible but can be amortized over the lifetime of communications between nodes or by the use of multi-hop routing mechanisms. In this work, we compare the expected performance of an OCS network to more traditional networks including meshes and fat-trees. The comparison considers several current large-scale applications. We show that the performance of an OCS network is comparable to the best of the network types examined.

1 Introduction

Recent large-scale procurements indicate that a multi-petaflop system containing tens to hundreds of thousands of processors (or CPU cores) will be built by the end of this decade. The performance of such a system is directly related to the applicability and performance of the interconnection network. An Optical Circuit Switched (OCS) network has recently been proposed that is aimed at solving the low latency and high bandwidth requirements of these systems [1].

Even when considering four or eight cores per processing chip (or socket), the chip-count in such a system will be in the tens of thousands. Given the current trend of increasing chip density and advances in local (near-distance) interconnection technology, we also expect to see an increase in the number of sockets within a compute node. Even with 64 sockets per node, the number of compute nodes will likely be in the thousands. Therefore, the design and implementation of the inter-node communication network is critical in determining both the performance and cost of large HPC systems.

In this work we compare the expected performance of OCS to more traditional networks including meshes and fat-trees. This analysis uses performance models for large-scale applications that have been validated on current large-scale systems.

Traditionally large-scale system designers have had two choices for the high performance network: either topological inflexibility or topological flexibility.

A.-M. Kermarrec, L. Bougé, and T. Priol (Eds.): Euro-Par 2007, LNCS 4641, pp. 858–867, 2007.

Topologically inflexible: networks such as tori and meshes, have the advantage of a cost that scales linearly with machine size. Examples include the IBM BlueGene/L network (current largest is a 32x32x64 3D torus) and the Cray Red Storm network (current largest is a 27x20x24 3D torus/mesh hybrid). A drawback with such networks is that application performance can degrade if the logical communication topology does not map well to the physical network.

Topologically flexible: networks such as multistage Fat-tree networks include Quadrics [2], Myrinet [3], and Infiniband [4]. Such networks, when using adaptive routing, can reduce network contention and enable high performance for a variety of applications that utilize a large range of logical communication topologies. However, cost scales super-linearly with machine size ($N/X \cdot log_x(N)$, switches are required for an X-radix tree containing N nodes) and can become prohibitive in very large systems. In addition, the development cost of a new generation of switches and network adapters can add substantial cost to the system as communication protocols, signaling speeds, and switch radixes scale to keep pace with newer and faster processor chips.

The OCS has the potential to give the best of both worlds. The OCS uses a hybrid network consisting of both Electrical Packet Switching (EPS) and OCS planes constructed using Micro Electro-Mechanical Systems (MEMS) based switches. Although the switching latency associated with MEMS switches is non-negligible, the feasibility of such a hybrid approach was demonstrated for applications with static or slowly changing communication patterns [1].

Like the proposed OCS network analyzed in this work, the recently proposed HFAST (Hybrid Flexibly Assignable Switch Topology) network [5] also makes use of MEMS-based optical circuit switches. The Gemini and Clint projects [6,7] use two types of networks to handle communication traffic in a similar way to OCS. Gemini uses both OCS and Electronic Packet Switched (EPS) switches, while Clint's circuit and packet switched networks are both electronic. In distributed computing where nodes communicate over relatively large distances, there have been several proposed circuit switched networks. Cheetah [8] is aimed at transferring large files between storage centers, while the OptIPuter [9] is aimed at linking multiple resources including visualization and storage.

1.1   Contributions of This Work

A quantitative analysis is presented of the expected performance of a large-scale system that utilizes an OCS network compared to more conventional network architectures. We show that, for a number of applications, the OCS network architecture has excellent performance in both capacity and capability computing modes as well as for weak- and strong-scaling application modes. While the performance of the OCS network is on par with the best traditional network for each application examined, the key to the OCS approach lies in the dynamic allocation of bandwidth to where it is needed by the application, yielding a much more flexible network architecture. Such a broad analysis has not been done before; previous work was constrained to a single application on a single system configuration and focused on the feasibility of the hybrid OCS approach, not on its expected performance [1]. Results from this work indicate the hybrid OCS network architecture is a viable

assumed that a node, which contains 64 quad-core sockets each with 4 inter-node communication links, fits into a single rack and racks are spaced 2m apart in a 2D floor layout. The fat-tree is assumed to use 48-port switches (a 24-ary tree), and the MPI software overhead is 500ns on both the sender and receiver sides.

We assume that messages can be striped across available links. Messages greater than 16 KB are striped across all links within a node, while messages between 2 KB and 16 KB are striped across links available from a single socket. Messages smaller than 2 KB are not striped.

3 Performance Analysis Methodology

Rather than simply considering network latency and bandwidth characteristics, we compare the expected performance of large-scale applications executing on a parallel machine equipped with the networks whose configurations are described in Section 2. Applications considered include a generic one representing a 2D or 3D partitioning of a 3D regular data grid as well as several large-scale scientific codes. Strong and weak scaling modes are used in the generic case while each application is considered in its most appropriate mode of execution.

We utilize detailed performance models for all of the applications. This approach has been used extensively in the past; models for the applications have been published and validated on current systems including 64K nodes of the IBM Blue Gene/L system installed at Lawrence Livermore National Laboratory and 10K nodes of the Cray ASC Red Storm machine at Sandia National Laboratory [10]. On these machines prediction error was less than 10% [11,12,13,14,15]. The use of performance models enables the determination of potential application performance in advance of system implementation.

Our performance models were constructed from a detailed examination of the application both in terms of its static characteristics (as defined by the source code) as well as its dynamic characteristics (the parts of the code that are actually used by an input deck). Through the use of profiling, a structural model is determined which describes the functional "flow" of the application. The structural model includes characteristics such as communication type (e.g., MPI_Send, MPI_Isend, etc.), frequency, and size, as well as computation characteristics such as number of data cells processed on each processor. These characteristics are typically dependent on the particular input deck and the size of the parallel system.

The structural model does not include information related to time, such as computation rate or message latency and bandwidth. Rather, the structural model is combined with hardware performance characteristics, such as message latency, bandwidth, and computation rate obtained from benchmarks or from system specs when the system cannot be benchmarked. System architecture characteristics including network topology are also used. It should be noted that we do not model single processor performance from first principles. Rather we relay on measuring the single processor performance (or using cycle-accurate simulation for a future system) for each of the applications, and concentrate on modeling its parallel behavior.

Once constructed, the model is used to predict performance of current (measurable) systems and hence validated. This is an iterative process which stops when prediction accuracy reaches desired levels – our goal has been to have an error of less than 20%

in this process but we have found that typically the error is considerably less than this. Given a validated model, performance on a non-existent future system can be predicted with some confidence.

4 Performance Analysis – Generic Boundary Exchange

We begin with an analysis of a generic application model that uses a regular data partitioning and boundary exchange communication. This is followed by an analysis of large-scale scientific application performance in Section 5. The generic application partitions a 3D data grid in either two or three dimensions (referred to here as Generic-2D and Generic-3D). Note that in all the following analyses a node is assumed to contain 64 sockets arranged internally in a 6D hypercube.

Performance predictions for a boundary exchange are shown in Figure 2 for strong scaling (for a fixed global grid size of 10^9 cells), while Figure 3 shows predictions for weak scaling (for a fixed problem per processor of 10^7 cells). A single word per boundary cell is communicated in the relative directions. The communication times generally decrease with node count in the strong-scaling case since the boundary surfaces, and hence communication volume decreases. In the weak scaling case, the boundary surface sizes are fixed, resulting in near constant communication times for 2^d or more nodes for a d-dimensional partitioning.

In all cases, fully connected networks with single-hop routing are the worst performers due to lack of available bandwidth between any pair of nodes. The OCS fully-connected network exhibits linearly decreasing bandwidth between node pairs as job size grows; at maximum job size (utilizing the full machine of 256 nodes, 16384 sockets) both fully connected single-hop networks offer the same performance. At smaller configurations, however, the OCS proves to be superior.

The regular communication topologies of the Generic-2D and Generic-3D applications map directly to the 2D and 3D mesh networks, respectively, leading to near optimal performance. However, in cases in which the physical mesh network does not exactly match the application's communication topology, the incurred latency and reduction in bandwidth resulting from multiple hops through the network negatively impact performance, particularly at large scale. In addition, because only a fraction of the inter-node links connect each pair of neighboring nodes, the peak inter-node bandwidth is not as great as with the multi-hop or fat-tree networks. The reconfigurable OCS network is able to match the communication topology of either Generic application, yielding the maximum possible node-to-node bandwidth. However, each packet suffers an SOL latency penalty relative to the mesh networks imposed by having only a single, centrally-located switch component.

In short, if the benefit of reduced latency outweighs the penalty of reduced bandwidth, the 2D or 3D mesh will likely obtain the best performance followed closely by the OCS and Fat-Tree networks (assuming the application's communication topology maps relatively well to the physical machine). The OCS's flexibility puts its performance on par with Fat-Tree networks and far ahead of mesh networks in those cases in which the application's logical and the network's physical topologies do not match. In addition, the OCS network's reduced number of components is attractive in terms of cost and reliability, giving it superior price/performance characteristics.

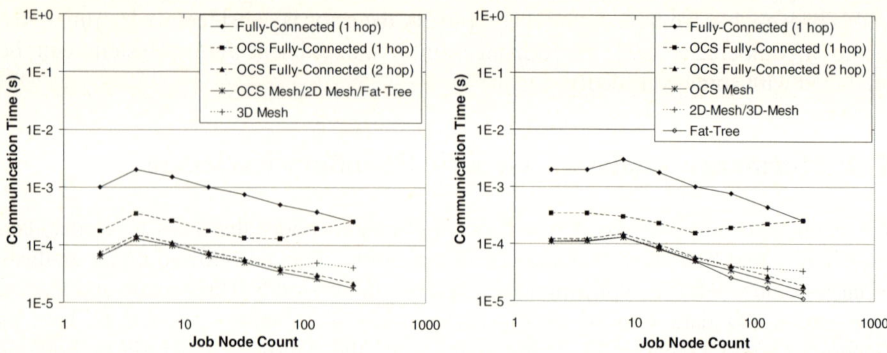

Fig. 2. Boundary exchange times for Generic-2D (left) and Generic-3D (right), strong scaling

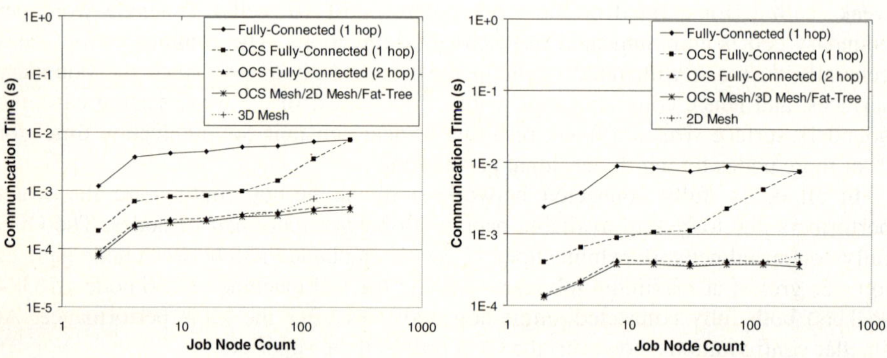

Fig. 3. Boundary exchange times for Generic-2D (left) and Generic-3D (right), weak scaling

5 Application Performance Analysis

Performance models of the applications listed in Table 2 were used to compare the performance of the networks listed in Table 1. We consider two system scenarios here: a 512-socket job executing on a larger capacity system and a 16384-socket capability system in which a single job utilizes the full machine. If we assume each node contains 64 sockets and a socket size of four cores with each core capable of 16 GF/s, then the capability mode provides a peak performance of 1.05 PF/s, while in capacity mode the peak performance available to each job is 32 TF/s.

We calculate the runtime for each application and compare it to the best runtime across networks (Figure 4). The runtime includes both communication and computation performance, allowing us to examine how network performance contributes to overall application performance. Figure 4 shows the relative performance with a value of one indicating the network with the best runtime, and a value greater than one indicating a longer runtime than the best by that factor.

Table 2. Details of the applications used in this analysis

Application	Scaling	Main Communication	Comment and Input
Generic-2D	S or W	2-D regular	2-D partitioning of 3-D data 10^9 cells (s), 10^7 (w)
Generic-3D	S or W	3-D regular	3-D partitioning of 3-D data 10^9 cells (s), 10^7 (w)
HYCOM	Strong	2-D (mod), Row/Col	Hybrid Ocean Model [11]. 1/12 degree (4500x3398x26)
KRAK	Strong	2-D Irregular	Hydrodynamics [12] 4 material cylinder (204,800 cells)
LBMHD	Weak	2-D regular	Magneto-hydrodynamics, 128x128 cells
POP	Strong	2-D partitioning	Parallel Ocean Program [15] 0.1 degree (3600x2400x40 cells)
RF-CTH2	Strong	3-D partitioning	Shock-dynamics (833x191x191 cells)
SAGE	Weak	modified 1-D, reduction	Shock-wave hydrodynamics [14] (100,000 cells)
Sweep3D	Weak	2-D regular with pipeline	Deterministic S_N transport [13] (8x8x12000 cells)

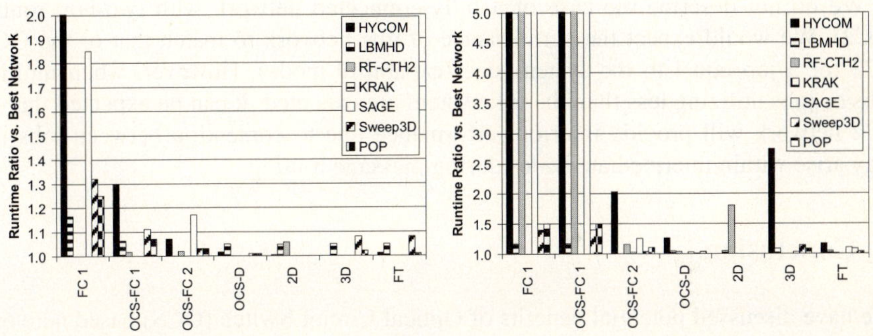

Fig. 4. Application performance on a capacity sized job of 512 sockets (left) and a capability sized job of 16384 sockets (right). In both instances, system size is 16384 sockets.

It can be seen in Figure 4 (left) that there is very little difference in application runtimes on systems equipped with OCS-Dynamic, Fat-Tree, and 2D and 3D mesh networks. Note that the fully-connected network with single-hop routing is a factor of 3.4 and 4.9 times slower than the best network for HYCOM and RF-CTH2 respectively – this is due to the larger bandwidth requirements of these codes.

The situation changes slightly when considering a capability system in Figure 4 (right). Here, the mesh networks start to perform poorly on RF-CTH2 (a 3D code which does not map well to a 2D mesh) and HYCOM (a 2D code which does not map nicely to a 3D mesh). The performance of the OCS-FC1 network and the FC1 network is almost identical at the largest scale. This is expected since these two networks exhibit essentially the same topology at this scale (in a capacity system, the OCS network provides full connectivity for the job only, not the entire system). Note that the ratio of runtimes for HYCOM, RF-CTH2, and SAGE on these network topologies is 6.5, 10.2, and 12.6 respectively. Again the OCS-FC2 and OCS-Dynamic networks result in a performance close to the best performing network in all cases.

A summary of the relative runtimes is included in Table 3. Here, the average runtime ratio is shown across all applications for each network. It can be seen that the OCS-Dynamic network provides the best average runtime, although the average performance of the OCS-Dynamic and Fat-Tree networks are similar. However, recall that we have used an idealized fat-tree network with an assumed minimal congestion (resulting from ideal adaptive routing) within the network; therefore the Fat-Tree result is optimistic. Note that we have not considered network cost in this analysis, but expect that an OCS network will be cheaper and hence a lower price/performance.

Table 3. Average runtime ratio (to the best network) for each network

	FC1	OCS-FC1	OCS-FC2	OCS-D	2D Mesh	3D Mesh	Fat-Tree
Capacity	2.13	1.08	1.05	1.01	1.02	1.02	1.02
Capability	4.87	4.87	1.23	1.06	1.12	1.30	1.06

We do not describe the case of a fully-connected network with two-hop routing (FC2). We would expect the performance of this network to match that of the OCS-FC2 for a job equal to the system size (capability mode). However, when multiple jobs or jobs utilizing less than the full system are executed, it can be expected that the FC2 network will provide inferior performance due to contention between jobs that may arise within intermediate nodes during message routing.

6 Conclusion

We have discussed potential benefits of Optical Circuit Switch (OCS) based networks over traditional direct and indirect networks for large-scale parallel computing systems. Although the number of available OCS planes is limited and optical circuit set-up cost is non-negligible, these factors can be mitigated through architectural decisions such as the number of OCS switch planes and multi-hop routing strategies.

Through the use of detailed and previously validated application performance models, we have been able to study the potential performance of an OCS network relative to several common high performance network types, including mesh networks (2D and 3D), fat-tree networks, and variants of fully connected networks. Such a broad analysis of the potential performance of an OCS network in the realm of high performance computing has not been previously done. The results indicate that the performance of an OCS network should be comparable to the traditional network type that is currently best suited to each of the applications. This results from the flexibility of the OCS design, allowing it to effectively mimic the connectivity of more common direct and indirect network topologies. The true advantage of the OCS network may come from also considering its cost – though has not been quantified.

Acknowledgements. This work was made possible by OCS network concepts initially published in [1] and driven by Eugen Schenfeld of IBM T.J. Watson Research Center. The authors would like to thank Eugen for his enthusiasm and support. This work was funded in part by the DOE Accelerated Strategic Computing (ASC), and the DARPA High Productivity Computing Systems (HPCS) programs. Los Alamos National Laboratory is operated by Los Alamos National Security LLC for the US Department of Energy under contract DE-AC52-06NA25396.

References

1. Barker, K.J., Benner, A., Hoare, R., Hoisie, A., Jones, A.K., Kerbyson, D.J., Li, D., Melhem, R., Rajamony, R., Schenfeld, E., Shao, S., Stunkel, C., Walker, P.: On the Feasibility of Optical Circuit Switching for High Performance Computing Systems. In: Proc. Supercomputing, Seattle (2005)
2. Petrini, F., Feng, W., Hoisie, A., Coll, S., Fractenberg, E.: The Quadrics Network: High-Performance Clustering Technology. IEEE Micro. 22(1), 46–57 (2002)
3. Myricom, http://www.myri.com
4. Infiniband Trade Association, http://www.infinibandta.org/
5. Shalf, J., Kamil, S., Oliker, L., Skinner, D.: Analyzing Ultra-Scale Application Communication Requirements for a Reconfigurable Hybrid Interconnect. In: Proc. Supercomputing, Seattle (2005)
6. Chamberlain, R., Franklin, M., Baw, C.S.: Gemini: An Optical Interconnection Network for Parallel Processing. IEEE Trans. on Parallel and Distributed Processing 13(10), 1038–1055 (2002)
7. Eberle, H., Nilsm Gura, N.: Separated High-bandwidth and Low-latency Communication in the Cluster Interconnect Clint. In: Proc. Supercomputing, Baltimore (2002)
8. Veeraraghavan, M., Zhenga, X., Leeb, H., Gardnerc, M., Fengc, M.: CHEETAH: Circuit-Switched High-Speed End-to-End Transport Architecture. In: SPIE Proc. vol. 5285, pp. 214–225 (2003)
9. Defanti, T., Brown, M., Leigh, J., Yu, O., He, E., Mambretti, J., Lillethun, D., Weinberger, J.: Optical switching middleware for the OptIPuter. IEICE Trans. Commun. E86-B(8), 2263–2272 (2003)
10. Hoisie, A., Johnson, G., Kerbyson, D.J., Lang, M., Pakin, S.: A Performance Comparison Through Benchmarking and Modeling of Three Supercomputers: Blue Gene/L, Read Storm and ASC Purple. In: Proc. SuperComputing, Tampa FL (2006)
11. Barker, K.J., Kerbyson, D.J.: A Performance Model and Scalability Analysis of the HYCOM Ocean Simulation Application. In: Proc. IASTED Int. Conf. on Parallel and Distributed Computing, Las Vegas NV (2005)
12. Berker, K.J., Pakin, S., Kerbyson, D.J.: A Performance Model of the KRAK Hydrodynamics Application. In: Proc. Int. Conf. on Parallel Processing, Columbus OH (2006)
13. Hoisie, A., Lubeck, O., Wasserman, H.J.: Performance and Scalability Analysis of Teraflop-Scale Parallel Architectures using Multidimensional Wavefront Applications. Int. J. of High Performance Computing Applications 14(4), 330–346 (2000)
14. Kerbyson, D.J., Alme, H.J., Hoisie, A., Petrini, F., Wasserman, H.J., Gittings, M.L.: Predictive Performance and Scalability Modeling of a Large-scale Application. In: Proc. Supercomputing, Denver CO (2001)
15. Kerbyson, D.J., Jones, P.W.: A Performance Model of the Parallel Ocean Program. Int. J. of High Performance Computing Applications 19(13) (2005)

Network Fault Tolerance in Open MPI

Galen M. Shipman[1], Richard L. Graham[2], and George Bosilca[3]

[1] Advanced Computing Laboratory, Los Alamos National Laboratory
`gshipman@lanl.gov`
[2] National Center for Computational Sciences, Oak Ridge National Laboratory
`rlgraham@ornl.gov`
[3] University of Tennessee, Dept. of Computer Science
`bosilca@cs.utk.edu`

Abstract. High Performance Computing (HPC) systems are rapidly growing in size and complexity. As a result, transient and persistent network failures can occur on the time scale of application run times, reducing the productive utilization of these systems. The ubiquitous network protocol used to deal with such failures is TCP/IP, however, available implementations of this protocol provide unacceptable performance for HPC system users, and do not provide the high bandwidth, low latency communications of modern interconnects. This paper describes methods used to provide protection against several network errors such as dropped packets, corrupt packets, and loss of network interfaces while maintaining high-performance communications. Micro-benchmark experiments using vendor supplied TCP/IP and O/S bypass low-level communications stacks over InfiniBand and Myrinet are used to demonstrate the high-performance characteristics of our protocol. The NAS Parallel Benchmarks are used to demonstrate the scalability and the minimal performance impact of this protocol. Communication level micro-benchmarks show that providing higher data reliability decreases bandwidth by up to 30% relative to unprotected communications, but provides performance improvements of a factor of four over TCP/IP running over InfiniBand DDR. In addition, application level benchmarks (communication/computation) show virtually no impact of the data reliability protocol on overall run-time.

1   Introduction

The ever increasing complexity and scale of HPC systems increases the likelihood of hardware and software component failure in these systems. The use of commodity (or near commodity) off-the-shelf components to build many such systems further aggravates the problem, as these are often not engineered to provide end-to-end hardware reliability; either ignoring such reliability issues or leaving it to software layers to provide. The ubiquitous software solution for end-to-end reliability is provided by TCP/IP communications stacks. The performance provided by such commonly available stacks does not meet the requirements of the HPC community, providing only a small fraction of the communications performance afforded by the networking hardware.

A.-M. Kermarrec, L. Bougé, and T. Priol (Eds.): Euro-Par 2007, LNCS 4641, pp. 868–878, 2007.
© Springer-Verlag Berlin Heidelberg 2007

While HPC system failures occur in a variety of ways, this paper focuses on a software architecture aimed at detecting and correcting failures in network data transmission. The goal of this design is to provide end-to-end protection against such failures, while providing communication performance commensurate with that of the underlying hardware. These failures may occur in a number of layers including software stacks, firmware or in hardware. They may include transient failures, such as Network Interface Card (NIC) resets, or more permanent failures, such as failed NICs. These failures may result from the normal statistical failure rate associated with large component counts, or may be the result of software or hardware defects, which may be fixed over time. Therefore addressing these issues is required for effective system utilization over the lifetime of these systems.

As costs are incurred when providing these fault tolerant features, the Modular Component Architecture (MCA) [1] of Open MPI is used to provide these as run-time selectable options. When these features are not selected, there is no impact on the the default high-performance configuration of Open MPI.

The remainder of this paper is organized as follows: Section 2 presents a brief overview of previous work. Next, Section 3 discusses the network fault tolerance architecture in Open MPI. Results are discussed in Section 4. Conclusions are discussed in Section 5.

2   Background

There have been several previous efforts to deal with network failure in a manner transparent to the calling application. The TCP/IP [2] stack deals with both transient data corruption, as well as with transient and permanent network failures. However, since TCP/IP is a general purpose network stack, designed to deal with a wide variety of network failures, including lossy data transmissions, flow control, and congestion control issues, these implementations do not provide the level of performance required by HPC applications. Network communication on HPC systems is normally generated only by the applications using these systems, and as such suffer little interference from system services or other applications. In addition, the networks used in these systems often possess a high degree of reliability with low error rates. These operating environments allow for reliability protocols which provide higher overall performance than general purpose reliability protocols which often provide unnecessary (often costly) features.

There have also been several attempts to provide different aspects of network fault tolerance specific to HPC systems. One of the goals of the LA-MPI project [3,4] was to provide reliable network communications. It uses timers on ACKs to detect dropped packets and either the TCP/IP checksum or a Cyclic Redundancy Check (CRC) to detect corrupted packets. The work presented in this paper draws upon prior work in LA-MPI but is entirely new in terms of software and protocol providing better performance while also adding new features such as network fail-over and protocol tuning for specific operating environments. The VMI project [5] also provides a way to deal with network errors, and recently, the MVAPICH project implemented network fail-over using the uDAPL interface [6].

3   Open MPI - Network Fault Tolerance

3.1   Open MPI's Point-To-Point Architecture

Open MPI's point-to-point architecture has been described in great detail elsewhere [7], and will be described very briefly in this section. Figure 1 provides a graphical depiction of this design.

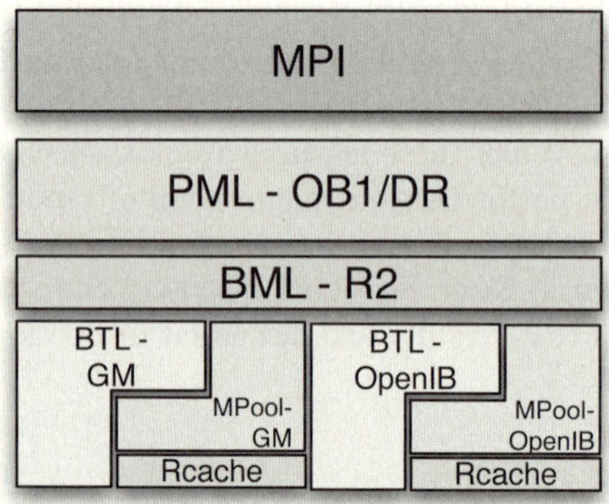

Fig. 1. Open MPI's Layered Architecture

Open MPI uses a layered design, with the aid of the the Modular Component Architecture, to achieve the flexibility in the features this code base offers. The MPI layer interfaces with the point-to-point communications implementation of Open MPI via the Point-To-Point Management Layer (PML). Currently, there are multiple PMLs supported, such as the OB1 PML [8] and the Data Reliability PML (DR), described in this paper.

The OB1 PML is aimed at providing the best point-to-point communications performance possible making use of all available communications resources, and the DR PML provides fault tolerance for this type of communications. Both PML's use the Byte Transfer Layer (BTL), the BTL-Mangement-Layer (BML), the Memory Pool (MPool), and optionally the Registration Cache (RCache) components in implementing these communications protocols.

3.2   Data Reliability DR **Overview**

Open MPI provides transparent user-level reliability over a variety of networks and network APIs. This user-level reliability is encapsulated in a single Modular Component Architecture component, PML DR or just DR. DR implements the point-to-point semantics of MPI while providing several network fault tolerance features. DR provides protection from a number of failure scenarios:

- Dropped Data
- Corrupted Data
- Catastrophic NIC Failure (Fail-over When other Communication Paths Exist)
- Network Agnostic Protection via Local Completion Watchdog Timers and ACK based Timeouts
- Network Specific Protection via Registered Error Handlers

Similar in functionality to PML OB1 [8], DR also makes use of the BTL (byte transfer layer) which abstracts the underlying network API in a uniform manner. This uniform network abstraction API allows DR to provide network fault tolerance in a network agnostic fashion without relying on other network abstraction libraries. The BTL abstraction is high-performance by design and when used with PML OB1 performance is similar to other MPI libraries. In using the high-performance BTLs the additional costs of reliability are isolated to the PML DR component.

3.3   VFRAG Protocol

A unique feature of PML DR is its user-level reliability protocol. A key component of the protocol is the Vector of Fragments (VFRAG). The VFRAG acts as a unit of acknowledgment and allows selective retransmission in a straight forward manner. Each MPI message is divided into N virtual fragments. Each virtual fragment is made up of 64 smaller fragments. The total size of the VFRAG is therefore $64 * S_{max}$ where S_{max} is runtime configurable. For example, given a MPI level message of size $4MB$ and $S_{max} = 16K$ the number of VFRAGS would be $4MB/(16K * 64) = 4$. Figure 2 illustrates the VFRAG structure.

Fig. 2. VFRAG Layout

This reliability protocol begins by allocating (via a free-list) a VFRAG descriptor. The VFRAG descriptor contains two timers, a local completion timer and a remote ACK timer. Upon scheduling fragment 0 of the current VFRAG the local completion timer is initialized and started and the number of pending fragments within the VFRAG is set to 1. Subsequent fragments within the same VFRAG are also scheduled and the number of pending fragments is incremented. As notification is received from the BTL that the fragment was sent and local completion occurred, the number of pending fragments is decremented and the timer is reset with a new timeout (as long as there are pending fragments).

The local completion timer gives some indication of VFRAG progression though it is not definitive. Relying solely on local completion as indication of message progression places reliability of completion semantics with the network interface. In order to protect the VFRAG from unreliable network completion semantics an ACK timer is set when the last fragment of the VFRAG is scheduled for transmission. The ACK message from the receiver contains a bit-mask indicating which of the 64 fragments of the VFRAG were delivered successfully. This two-stage timeout reduces the overhead of reliability by aggregating acknowledgments. In addition, recovery is optimized through local completion timeouts and selective retransmission via the bit-mask ACK.

In addition to the local watchdog timer and the remote ACK timer PML DR can also recover from asynchronous errors from the BTL. During initialization DR registers an asynchronous error handler with the BTL. This error handler can take appropriate action such as failing over traffic to another BTL. Fail-over can also occur based on tunable retransmit levels on a per VFRAG basis. This is allowing fail-overs to occur either as a result of a network reported error or time-outs of the local watchdog timer or remote ACK timer. This is a major difference in contrast of relying solely on network reported errors as described in [6].

This protocol provides several benefits. CRC/Checksum size is configurable based on S_{max}. That is each fragment of size S_{max} carries an associated CRC/Checksum in its fragment header. ACK aggregation is tunable based on the total size of the VFRAG. Selective retransmission is provided by the bit-mask ACK. Along with these benefits there are additional costs including:

1. Latency increased due to specific acknowledgment
2. Bandwidth decreased due to additional protocol overhead
3. CPU Availability decreased due to additional protocol overhead

In recognition of these costs, DR takes advantage of various MPI semantics to limit protocol impact. For example, DR can mark MPI completion of a message as soon as it is buffered instead of waiting for remote ACK of the message thereby hiding some of the additional protocol costs of reliability. In addition, the performance impact can be tuned by selecting the recovery cost/performance ratio appropriate to a given operating environment. For example, if the operating environment includes a highly reliable network with few failures S_{max} can be increased such that each ACK protects a larger amount of data. This increases recovery costs while decreasing the cost of reliability. CSUM/CRCs can be enabled/disabled on a per BTL or global basis thereby allowing the user or system administrator to choose Main-memory to Main-memory protection at a fine or coarse granularity.

While PML DR provides protection from a number of different network failure scenarios, it does not protect against every eventuality. DMA operations initiated by the NIC which corrupt the target address cannot be protected against with DR as the DMA may corrupt random memory locations on the host. Network stack/driver errors can sometimes result in kernel panics or processes hanging in uninterruptible sleep from which the user level process cannot recover.

3.4   Data-Type Engine

The responsibility to pack, unpack and compute the checksum belongs to the Open MPI data-type engine. A full description of the internals of the data-type engine is out of the scope of this paper, thus we will give only a brief description of the data-type engine mechanisms involved in DR.

In order to allow the PML to fragment a message the data-type engine has been modified to work on segments, using another entity called *convertor*. All operations (packing and unpacking) are limited by the number of bytes requested by the PML. In order to satisfy this requirement, the data-type engine keeps track of the last state of the current operation (pack or unpack). Once an operation is completed, the current internal state of the *convertor* is saved. The next operation will continue from this saved position. Therefore, in Open MPI, there is no need to pack the full user data before a send, nor to unpack it in one operation on the receiver side. This approach allows Open MPI to apply different optimization to the packing and unpacking process, as well as to the checksum computation. As an example, we can limit any operation to the amount of data that is cache friendly on the current architecture. Another optimization, is the pipeline that is created using the pack, unpack and the network communication for each of the fragments as illustrated in Figure 3.

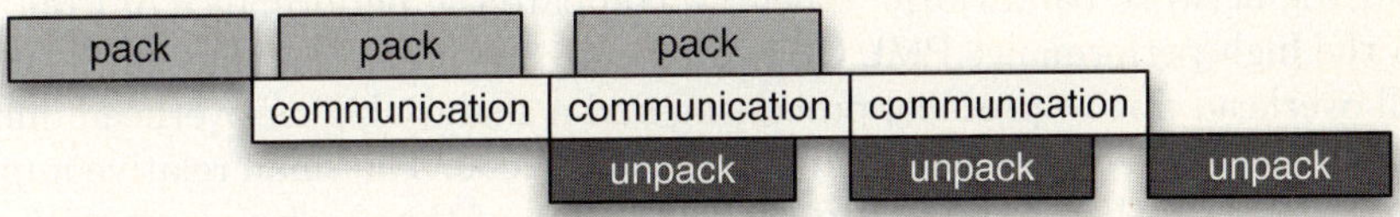

Fig. 3. Overlapping packing/unpacking with the communications using the data-type engine

The default checksum computation is a fast 32 bit algorithm. This algorithm has been modified to work on the same principle as the data-type engine, i.e. in a segmented way. When the checksum is enabled and the network device requires memory copies, the memory copies and the checksum computations are interleaved in order to reduce cache pollution. If there is no need for memory copies the checksum is computed directly on the user buffer.

4   Results

4.1   Experimental Setup

The NAS Parallel benchmark (NPB) [9] were run on a 1290 node cluster (4 segments of 258 nodes). Each node has 2 Single Core AMD Opteron 252 processors, 8 GBytes of memory, 1 Mellanox InfiniBand MT25204 InfiniHost III Ex adaptors connected via a Voltaire SDR switch. All other experiments were performed on a

4 Node test cluster. Each node has 2 Dual Core AMD Opteron 270 processors, 4 GBytes of memory, 2 Mellanox InfiniBand MT25208 InfiniHost III Ex adaptors each on a dedicated PCI-Express 16X bus and 1 Myricom Myrinet 2000 PCI-X "D card" NIC on a 133 MHz PCI-X bus. Myricom adaptors are connected via a Myricom 2000 switch. Mellanox adaptors are connected via a DDR (double data rate) Silverstorm switch. Each node was installed with Fedora Core 5, Open MPI Trunk Revision 12736, OFED 1.1, and GM 2.1.26.

4.2   Results and Analysis

To examine the performance impact of our data reliability protocol we used the NetPipe [10] benchmark as illustrated in Figure 4(a) and Figure 4(b). Four different protocols were examined, Single RDMA GET with registration cache, PML OB1 Copy In/Out using send/recv, PML DR with checksums and PML DR without checksums. On both InfiniBand and Myrinet 2000, the highest performance was obtained using the Single RDMA protocol due to buffer reuse in the NetPipe benchmark and the high performance of RDMA. On InfiniBand PML DR bandwidth performance is substantially lower than the single RDMA protocol although most of this performance difference is a result of using copy in/out protocols, this is not as apparent over Myrinet as the memory bandwidth surpasses the network bandwidth. When we compare the performance of PML DR with the high-performance PML OB1 using copy in/out we see that DR incurs a small overhead due to protocol processing and a slightly higher overhead due to checksum costs on both InfiniBand and Myrinet 2000. The small relative impact of checksums can be attributed to an integration of the checksum/crc with the data-type engine. Of note in the InfiniBand results is the performance degradation at larger message sizes, this is a result of a memory bandwidth bottleneck on this particular architecture/network. When compared to TCP/IP over InfiniBand (using the high-performance IPoIB stack), PML DR provides a substantial performance increase throughout the bandwidth curve. Even in the bandwidth limited Myrinet 2000 the performance of PML DR surpasses TCP/IP over this interconnect.

As discussed earlier PML DR performance can be adapted for a given operating environment. One such adaptation is varying the size of each fragment. Larger fragments provide better performance (to a point) while smaller fragments are better protected by their associated checksum. Figures 5(a) and 5(b) demonstrates the effect of changing the fragment size. While a fragment size of 4096 is better protected by its checksum, the performance degrades over InfiniBand as the upper layer cannot effectively keep the network pipe full. Performance increases as the fragment size increases up to 16K which gives the best performance on InfiniBand in this environment at the expense of a less effective checksum. Recovery costs are also higher at 16K fragment sizes as the retransmission of dropped or corrupted fragments is on entire fragments. Changing the size of the fragments when running over Myrinet 2000 has little impact, again due to the network bottleneck relative to memory bandwidth.

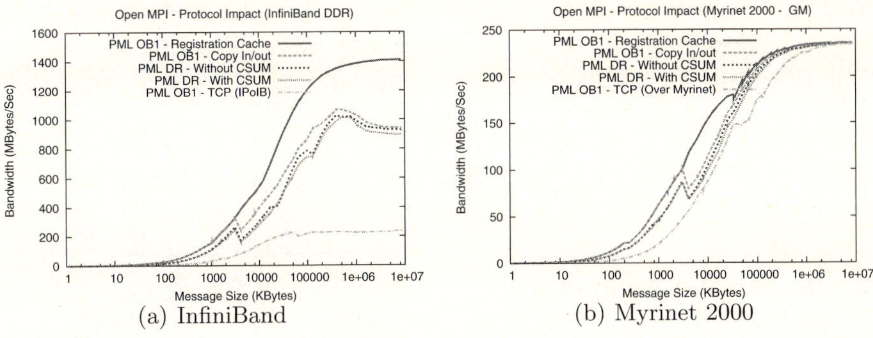

(a) InfiniBand   (b) Myrinet 2000

Fig. 4. PML Protocol Performance

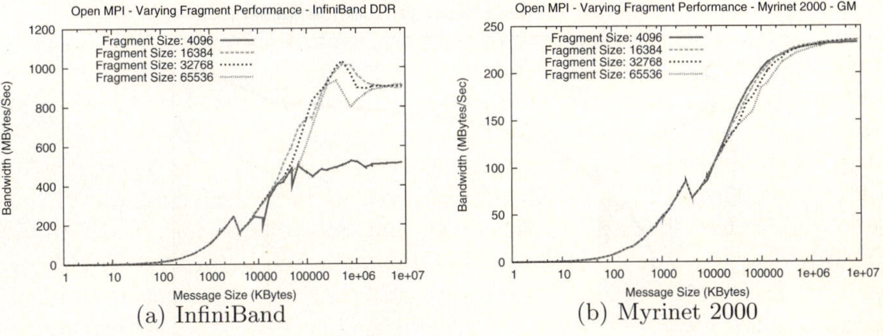

(a) InfiniBand   (b) Myrinet 2000

Fig. 5. Variable Fragment Size - Performance Impact

In order to better assess the performance impact of DR in medium scale clusters the NAS Parallel benchmark (NPB) were used. The BT, CG, MG and SP benchmarks were run with 16 and 64 processors with a problem size of class C (second to largest) and 256 processors with a problem size of class D (the largest). The IS benchmark was only run at 16 and 64 processors with a problem size of class C because class D is not available for IS. Each benchmark was run 3 times and the average runtime is shown as a bar with an additional error bar centered at the top indicating the standard deviation. As illustrated in Figure 6(a) the additional protocol overhead of DR has very little impact on more realistic benchmarks at the class C size with a very small standard deviation (almost not visible in this Figure). Figure 6(b) illustrates the impact on class D size with 256 processors. For each of these runs the performance differences between the 3 protocols is within the standard deviation. The higher standard deviation is expected as the runtime and problem size of class D is more likely to be impacted by memory caching effects and network congestion. These benchmarks indicate that added benefits of DR may come at little or no cost to a wide variety of parallel problems.

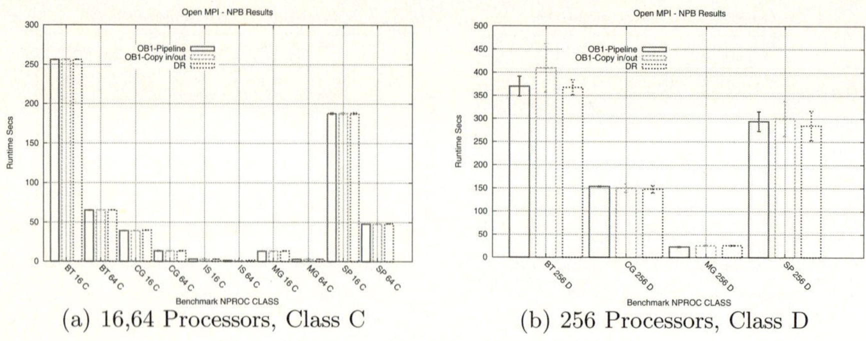

(a) 16,64 Processors, Class C   (b) 256 Processors, Class D

Fig. 6. NPB Benchmarks - InfiniBand

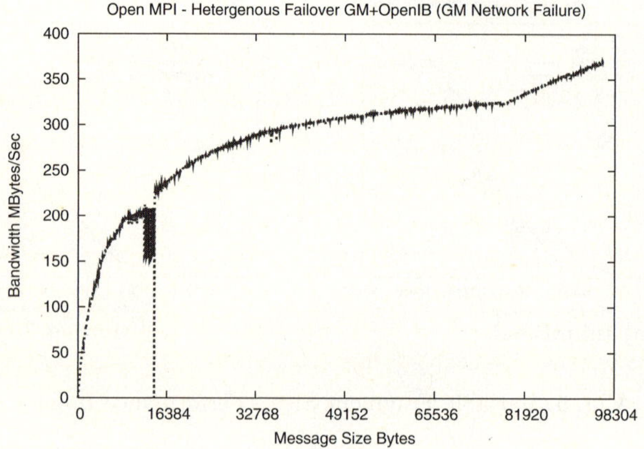

Fig. 7. DR - Failover (Myrinet to InfiniBand)

In addition to network retransmission of dropped and corrupted fragments, PML DR provides network fail-over when more than one path exists to a given peer. In this experiment Myrinet and InfiniBand are both used for message scheduling. Each fragment of the message can be scheduled on either interconnect with the number of fragments scheduled to each based on its relative bandwidth. In Figure 7 we demonstrate the effect of failover on bandwidth. After the Myrinet network on one of the hosts is disconnected from the switch, bandwidth becomes more sporadic as data is rescheduled on timeouts over the InfiniBand interface. Once the number of retransmissions exceeds a configurable threshold, Myrinet BTL is disabled and from that point on all data is scheduled over the InfiniBand interface. Similar results occur when InfiniBand is disconnected. The mechanism is somehow different as InfiniBand will deliver an asynchronous error to PML DR from the OpenIB BTL as a result of a completion queue error. DR responds to the asynchronous error rather than waiting for the exceeding of the retransmission threshold. The ability to detect network failure outside of asynchronous errors allows DR to respond to failure of a variety of networks and network API semantics.

5   Conclusion

As the use of commodity or "near commodity" networks continues to increase in large scale HPC systems, and the size of these systems continues to grow, robust user level libraries can enhance the reliability of network communication. While theoretical error rates remain relatively low for virtually all high-performance networking technologies, some larger scale installations have also shown that hardware, software, or firmware bugs are often long lived and transient. Facilitating useful science throughout the lifetime of these systems is important and may benefit from reliable high-performance network techniques. In this work we have described methods of dealing with network errors that help improve the reliability of network communications, with minimal impact on application performance, thus providing a means to improve the effectiveness of HPC scale simulation clusters.

Acknowledgements. Los Alamos National Laboratory is operated by Los Alamos National Security, LLC for the United States Department of Energy's National Nuclear Security Administration under contract DE-AC52-06NA25396. Project support was provided through ASCI/PSE and the Los Alamos Computer Science Institute, LA-UR-06-8492. This work is funded by subcontract #R7B127 from Rice University under prime contract #12783-001-05 49.

References

1. Squyres, J.M., Lumsdaine, A.: The component architecture of open MPI: Enabling third-party collective algorithms. In: Getov, V., Kielmann, T. (eds.) Proceedings, 18th ACM International Conference on Supercomputing, Workshop on Component Models and Systems for Grid Applications, St. Malo, France, pp. 167–185. Springer, Heidelberg (2004)
2. RFC793: Transmission control protocol. DARPA Internet Program Protocol Specification (September 1981)
3. Aulwes, R.T., Daniel, D.J., Desai, N.N., Graham, R.L., Risinger, L.D., Sukalski, M.W., Taylor, M.A., Woodall, T.S.: Architecture of LA-MPI, a network-fault-tolerant MPI. In: Los Alamos report LA-UR-03-0939, Proceedings of IPDPS (2004)
4. Graham, R.L., Choi, S.E., Daniel, D.J., Desai, N.N., Minnich, R.G., Rasmussen, C.E., Risinger, L.D., Sukalksi, M.W.: A network-failure-tolerant message-passing system for terascale clusters. International Journal of Parallel Programming 31(4) (August 2003)
5. Pakin, S., Pant, A.: VMI 2.0: A dynamically reconfigurable messaging layer for availability, usability, and management. In: Proceedings of The 8th International Symposium on High Performance Computer Architecture (HPCA-8), Cambridge, MA (February 2002)

6. Vishnu, A., Gupta, P., Mamidala, A.R., Panda, D.K.: A software based approach for providing network fault tolerance in clusters with udapl interface: Mpi level design and performance evaluation. In: Proceedings of 2006 International Conference for High Performance Computing, Networking, Storage and Analysis (2006)
7. Graham, R.L., Barrett, B.W., Shipman, G.M., Woodall, T.S., Bosilca, G.: Open mpi: A high performance, flexible implementation of mpi point-to-point communications. Parallel Processing Letters (accepted, January 2007)
8. Shipman, G., Woodall, T., Graham, R., Maccabe, A., Bridges, P.: Infiniband scalability in open mpi. In: Proceedings, 20th IEEE International Parallel & Distributed Processing Symposium, IEEE Computer Society Press, Los Alamitos (2006)
9. Bailey, B., Barton, B.,Carter, D., Fatoohi, F., Frederickson, L., Schreiber, S., Venkatakrishnan, W.: NAS parallel benchmarks (1994)
10. Snell, Q., Mikler, A., Gustafson, J.: NetPIPE: A Network Protocol Independent Performace Evaluator. In: IASTED International Conference on Intelligent Information Management and Systems (June 1996)

Topic 14
Mobile and Ubiquitous Computing

Nuno Preguiça, Éric Fleury, Holger Karl, and Gerd Kortuem

Topic Chairs

Mobile computing has evolved tremendously in the past few years with advances in wireless networks, mobile computing, sensor networks along with the rapid growth of small, portable and powerful computing devices. These advances offer opportunities for the development of new mobile/ubiquitous computing applications and services. Topic 14 of Europar 2007 covers all aspects related with the creation of such systems.

Our topic has attracted sixty six submissions. With the help of external reviewers we ended up accepting nine papers (14% acceptance ratio).

The papers were selected on the basis of their perceived quality, originality and appropriateness to the theme of the topic 14. We thank the authors of all the submitted version for submitting their papers. These papers were organized in three sessions of three papers each: the first one addresses multi-hop wireless networks; the second one presents services for mobile systems; and the last one is devoted to wireless networking connectivity.

In the multi-hop wireless networks session, "An Algorithm for Dissemination and Retrieval of Information in Wireless Ad Hoc Networks" proposes a new algorithm to place data replicas in wireless ad hoc networks that provide good availability and low latency. In "Securing Sensor Reports in Wireless Sensor Networks", the authors propose a security mechanism for early detection of falsely injected data in sensor networks, thus saving unnecessary transmissions and energy. "Surrendering Autonomy: Can Cooperative Mobility Help?" proposes an algorithm that determines how a cooperating node should move to improve communication in battlefield MANETs.

In the services for mobile systems session, "A Context-Dependent XML Compression Approach to Enable Business Applications on Mobile Devices" proposes a context-dependent XML compression approach for helping the deployment of business application on mobile devices. "A Distributed, Leaderless Algorithm for Location Discovery in Specknets" proposes a method for node location discovery in sensor networks (specknets). In "Analysis of a Kalman Approach for a Pedestrian Positioning System in Indoor Environments", the authors propose a positioning system for unprepared indoor environments that uses Kalman filters for providing positioning information in shadow zones.

In the wireless networking session, "Performance of MCS Selection for Collaborative Hybrid-ARQ Protocol" proposes an algorithm to improve throughput by selecting the most suitable MCS level. In "New approaches for Relay selection in

A.-M. Kermarrec, L. Bougé, and T. Priol (Eds.): Euro-Par 2007, LNCS 4641, pp. 879–880, 2007.
© Springer-Verlag Berlin Heidelberg 2007

IEEE 802.16 Mobile Multi-hop Relay Networks", the authors propose a path selection and signaling scheme for relay-enhanced IEEE 802.16 WMANs. presents a theoretical and simulation-based study of the connectivity of bluetooth-based ad hoc networks.

Securing Sensor Reports in Wireless Sensor Networks*

Al-Sakib Khan Pathan and Choong Seon Hong

Department of Computer Engineering, Kyung Hee University
Giheung, Yongin, Gyeonggi, 449-701 Korea
`spathan@networking.khu.ac.kr, cshong@khu.ac.kr`

Abstract. The sensor reports from a wireless sensor network are often used extensively in the decision making process in many systems and applications. Hence, classifying real and false sensor reports is necessary to avoid unwanted results. In this paper, we propose a scheme for securing the sensor reports in a wireless sensor network. We use one-way hash chain and pre-stored shared secret keys to provide data transmission security for the reports that travel from any source node to the base station. To introduce data freshness, our scheme includes an optional key refreshment mechanism which could be applied depending upon the requirement or the application at hand. We present an analysis along with the detailed description of our scheme.

1 Introduction

Wireless sensor networks (WSNs) have promised to provide a great opportunity of gathering specific types of data from specific geographic areas. WSNs can successfully operate even in unattended, hostile or hazardous areas. While this aspect of WSN has made its use very lucrative in many military and public-oriented applications [1], [2], it has also raised a lot of questions and the issue of ensuring security in such types of networks has become a major challenge. It is anticipated that, in most application domains, wireless sensor networks constitute an information source that is a mission critical system component and thus, require commensurate security protection. If an adversary can thwart the work of the network by perturbing the information produced, stopping production, or pilfering information, then the usefulness of sensor networks is drastically curtailed. Thus, it is very crucial in many applications to make sure that, the reports sent from the sensors *in action* are authentic and reach the base station (BS) without any fabrication or modification.

The task of securing wireless sensor networks is however, complicated by the fact that the sensors are mass-produced anonymous devices with severely limited memory, power and communication resources. Also, in most of the cases, the sensors do not have any knowledge of their locations in the deployment environment. Though there are many security issues to address in WSNs, in this paper, we mainly focus on ensuring security for the sensor reports, which the active and legitimate sensors send to the base station (BS). In fact, a single security scheme cannot provide all sorts of security protections in such types of wireless networks.

---

* This work was supported by MIC and ITRC. Dr. C. S. Hong is the corresponding author.

A.-M. Kermarrec, L. Bougé, and T. Priol (Eds.): Euro-Par 2007, LNCS 4641, pp. 881–890, 2007.

Here, we consider a densely deployed scenario of a wireless sensor network. Our main goal is ensuring authenticity and confidentiality of the data that reach from the source sensors to the BS and detecting falsely injected data as early as possible, so that they cannot travel a long way towards the base station, which would save the unnecessary transmissions of the intermediate sensors. Thus it could save the network from wasting its crucial energy resource. We also propose an optional key refreshment mechanism to ensure data freshness in the network which could be employed depending on the requirements or the application at hand.

The rest of the paper is organized as follows: Section 2 states the related works, section 3 presents our assumptions and preliminaries, Section 4 describes our security scheme in detail, performance analysis is presented in section 5, and section 6 concludes the paper with future research directions.

2   Related Works

Ye et al. [3] proposed a statistical en-route filtering (SEF) scheme to detect and drop false reports during the forwarding process. In their scheme, a report is forwarded only if it contains the message authentication codes (MACs) generated by multiple nodes, by using keys from different partitions in a global key pool. Zhu et at. [4] proposed the interleaved hop-by-hop authentication scheme that detects false reports through interleaved authentication. Their scheme guarantees that the base station can detect a false report when no more than t nodes are compromised, where t is a security threshold. In addition, their scheme guarantees that t colluding compromised sensors can deceive at most B noncompromised nodes to forward false data they inject, where B is $O(t^2)$ in the worst case. They also proposed a variant of this scheme which guarantees $B = 0$ and which works for a small t. Motivated by [4], Lee and Cho [5] proposed an enhanced interleaved authentication scheme called the key inheritance-based filtering that prevents forwarding of false reports. In their scheme, the keys of each node used in the message authentication consist of its own key and the keys inherited from its upstream nodes. Every authenticated report contains the combination of the message authentication codes generated by using the keys of the consecutive nodes in a path from the base station to a terminal node.

Our proposed scheme is different from all of the mentioned schemes as we create a logical tree-structure in the network to use OHC for secure data transmission. The OHC ensures the authenticity of the data sent from the sensors to the base station and the confidentiality of the data is ensured with the shared secret keys of the sensors.

3   Network Assumptions, Preliminaries and Threat Model

We consider a wireless sensor network with dense deployment of the sensing devices. In this network, the BS and all the sensors are loosely time synchronized, and each node knows an upper bound on the maximum synchronized error. The sensors deployed in the network have the computational, memory, communication and power resources like the current generation of sensor nodes (e.g., MICA2 motes [6]). Once the sensors are deployed over the target area, they remain relatively static in their respective positions. Each node transmits within its transmission range isotropically (in all directions) so that each message sent is a local broadcast. The link between any pair of nodes in the network is bidirectional, that is, if a node n_i gets a node n_j within its transmission range (i.e. one hop), n_j also gets n_i as its one-hop neighbor. The base station could not be compromised in any way. We assume that, no node could be compromised by any adversary while creating the tree structure in the network (in section 4.1). Initially, each node is equally trusted by the BS. Each sensor in the network has a shared secret key with the BS which is pre-loaded into its memory. The BS keeps an index of the ids of the sensors and the corresponding shared secret keys.

To ensure authenticity of sensor reports, we use one-way hash chain. A one-way hash chain [7] is a sequence of numbers generated by one-way function F that has the property that for a given x, it is easy to compute $y = F(x)$. However, given F and y, it is computationally infeasible to determine x, such that $x = F^{-1}(y)$. An one-way hash chain (OHC) is a sequence of numbers $K_n, K_{n-1}, \ldots, K_0$, such that, $\forall_i : 0 \leq i < n$, $K_i = F(K_{i+1})$. To generate an OHC, first a random number K_r is selected as the seed, and then F is applied successively on K_r to generate other numbers in the sequence.

Due to the use of wireless communications, the nodes in the network are vulnerable to various kinds of attacks. However, dealing with the attack like jamming attack and other attacks [8], [9], [13] is beyond the scope of this paper. We assume that, an adversary could try to eavesdrop on all traffic, inject false packets, and replay older packets. If in any case, a node is compromised, it could be a full compromise where all the information stored in that particular sensor are exposed to the adversary or could be a partial compromise that is, partial information is exposed.

4 Securing Sensor Reports: Our Proposed Scheme

4.1 Initialization of the One-Way Hash Chain Number in the Network

To provide authenticity of the sensor reports, all the intermediate nodes between any particular source node and the base station must be initialized with the basic one-way hash chain number. Let us suppose the initial OHC number is HS_0. To bootstrap the OHC number, the base station first generates a control packet containing HS_0 and a MAC (Message Authentication Code, MAC_{K_i}) for the control packet using a key K_i, where K_i is the number in the key chain number corresponding to time slot t_i. The format of the control packet generated by the base station B is:

$$bcm: \ B|sid|fid|HS_0|MAC_{K_i}(B|sid|fid|HS_0)$$

Here, B is the id (indicates that this message is a control message sent from the base station) of the base station, sid indicates the sender id (required for the subsequent transmissions by the nodes in the network), fid is the id of the selected forwarder node. For base station, sid and fid are set to B.

The initialization message is first received by the one-hop neighbors of the base station. Receiving the message, each node in the one-hop neighborhood stores the value of HS_0 and sets the base station as its forwarder node (in fact, the ultimate destination is the base station). Now, each of these nodes transmits the message again within its own one-hop neighborhood (i.e., local broadcast) with its own id as *sid* and *B* as the *fid*. Any other node that has already got the control message directly from the base station (i.e., any other one-hop neighbor) ignores this packet.

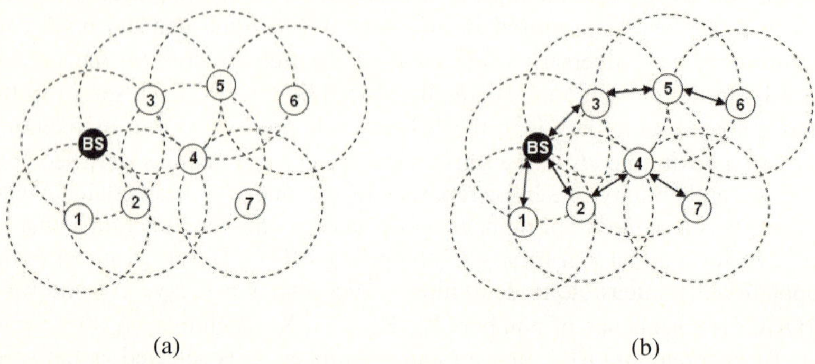

(a)   (b)

Fig. 1. (a) Network in the initial state (b) After initialization phase is over. Here, we have three paths, B → 3 → 5 → 6, B → 2 → 4 → 7 and B → 1 along which the one-way hash chain number is initialized. The dashed circles indicate the transmission ranges of the nodes.

Any node that has not received the message earlier (i.e., two hops away from the base station) receives it and stores the initial OHC number, HS_0. It then sets the id of the sender node as its forwarder node and again locally broadcasts the control message with its own id as *sid*. When it gets the control packet from two or more sender nodes, it picks up the message which it receives first and discards all other messages. However, this node stores the ids of the other senders as neighbor information which are used for later computations. This knowledge is necessary to repair a broken path, which we will discuss later in this paper. When this node does the local broadcast with the modified *sid* and *fid*, the previous sender node eventually knows that it has been selected by its downstream node as a forwarder (as each link is bidirectional). So, the upstream node (in this case, that particular one-hop neighbor of the base station) sets itself as a forwarder for this node. This process continues and eventually a tree-structure is created in the network where, each node has a forwarder node on the way to reach to the base station and a possible downstream node that can send data to it destined to the base station. To authenticate HS_0, *B* releases the key K_i in time slot t_{i+d}. On receiving this key, an intermediate node can verify the integrity and source authentication of HS_0. It is to be noted that, *bcm* won't bring any attack against the network even if the nodes on the other side of the network don't receive K_i at t_{i+d}. Since, the messages that are *MAC*ed by K_i are supposed to be sent out at time slot t, an adversary cannot launch any attacks with K_i when it gets K_i at t_{i+d}. Thus, along each path the initial OHC number is initialized securely.

Let us illustrate the initialization phase with an example. Figure 1 shows the example scenario. At the very beginning, the base station BS transmits the control message *bcm* with initial OHC number and MAC. Nodes 1, 2 and 3 in this case get the initial control message. All of these nodes set base station *B* as their immediate upstream forwarder and set the respective *fids*. Node 4 is within the transmission ranges of both 2 and 3. So, when node 4 gets the message from two different sender nodes, it has to pick up one as the forwarder node. Say, node 4 has chosen 2 as its forwarder. When node 4's turn comes and it transmits the local broadcast message using the control message, node 2 knows that node 4 is its downstream node and sets itself as the forwarder of node 4. Now, when node 3 does the local broadcast, node 5 also gets the message and it could set node 3 as its own forwarder. When node 5 gets the message again from node 4 with node 4 as the *sid*, it simply ignores the message as it has already chosen its forwarder. Also, it could be noticed from the example that, node 1 and node 2 are the one-hop neighbors of the base station and they both get the control message from the same source and both of their *fid*s are B (which is in this case the base station itself). So, when the local broadcasts of node 1 or node 2 reach one another, as previously stated they simply ignore the messages. This process continues until all the nodes in the network are included in the OHC initialization tree. All the nodes get the initial value of OHC number and the network becomes ready for sensor report transmission phase after time slot t_{i+d}. We show the resultant network structure in Figure 1(b) after executing our algorithm on sample network in Figure 1(a).

4.2   Secure Data Transmission

To send the data securely to the sink, each source node n_s maintains a unique one-way hash chain, $HS: <HS_n, HS_{n-1}, \ldots, HS_1, HS_0>$. When a source node, n_s sends a report to the sink using the path created in the sink-rooted tree (for example, a path is $n_s \rightarrow \ldots \rightarrow n_{m-1} \rightarrow n_m \rightarrow B$), it encrypts the packet with its shared secret key with the sink (or base station), includes its own id and an OHC sequence number from HS in the packet. It attaches HS_1 for the first packet, HS_2 for the second packet, and so on. To validate an OHC number, each intermediate node $n_1,\ldots,n_m$ maintains a verifier I_S for each source node, n_s. Initially, I_s for a particular source node is set to HS_0. When n_s sends the ith packet, it includes HS_i with the packet. When any intermediate node n_k receives this packet, it verifies, if $I_s = F(HS_i)$. If so, n_k validates the packet, it forwards it to the next intermediate node, and sets I_s to HS_i. In general, n_k can choose to apply the verification test iteratively up to a fixed number w times, checking at each step whether, $I_s = F(F\ldots(F(HS_i)))$. If the packet is not validated after the verification process has been performed w times, n_k simply drops the packet. By performing the verification process w times, up to a sequence of w packet losses can be tolerated, where the value of w depends on the average packet loss rate of the network. Note that, an intermediate node need not to decrypt the packet rather it checks the authenticity of the packet before forwarding to its immediate forwarder. Figure 2 illustrates the OHC utilization for secure data transmission.

In Figure 2(a), the source node n_s sends the first packet to the base station with the OHC value HS_1. The content of the packet is encrypted with the secret key that it shares with the base station. Getting the packet, the base station performs the authenticity check by verifying the hash chain number and gets the report by decrypting it with the shared key for that particular source node. Figure 2(b) shows a scenario where the packet P_2 could not reach the base station for some reason. In spite of that, the OHC verification is not hampered as for the next packet, the third intermediate node performs the hash verification twice (Figure 2(c)). Here, at the very first attempt it cannot get the value of HS_1 in the verification process but in the second iteration, it verifies it as a valid packet from the source n_s. In fact, in this case, the intermediate node can perform the hash number verification w times where, w is an application dependent parameter. In Figure 2(d) an adversary tries to send a bogus packet with a false hash chain number and it is detected in the next upstream node. Eventually such bogus packet fails to pass the authentication check and is dropped in the very next hop. This feature saves energy of the network as the falsely injected packets cannot travel through the network for more than one hop.

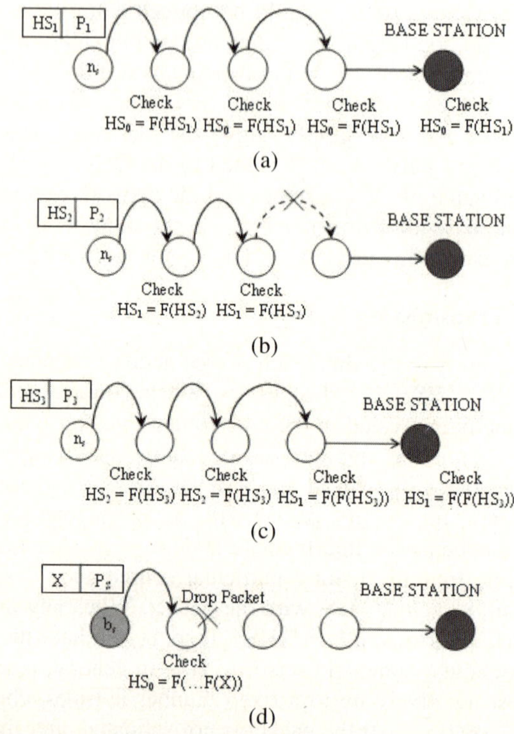

Fig. 2. (a) Authenticated packet delivery to BS (b) A packet could not reach the BS (c) OHC verification method is not hampered (d) Bogus packet detection and dropping

4.3 Optional Key Refreshment of the Sensors

To provide data freshness and to increase the level of security, our scheme has an optional key refreshment mechanism. In this case, the base station periodically broadcasts a new session key to the sensors in the network. The format for this message is:

$$B|K_s| \, MAC_{K_j}(B|K_s))$$

Where, K_j is the number in the key chain number corresponding to time slot t_j. To authenticate K_j, like the OHC initialization phase, B releases the key K_j in time slot t_{j+d}. On receiving this key, the nodes can verify the integrity and source authentication of K_j. Then each node gets the new key by performing an X-OR (exclusive OR) operation with its old key. This method could also be utilized for refreshing the keys of a specific number of nodes. In that case, the base station could simply send the K_s to the specific node by encrypting it with the previous shared secret key. Upon receiving the new key, the node can perform the X-OR operation and could use the newly derived key for subsequent data transmissions.

Changing encryption keys time-to-time has an advantage as it guarantees data freshness in the network. Moreover, it helps to maintain confidentiality of the transmitted data by preventing the use of the same secret key at all the times.

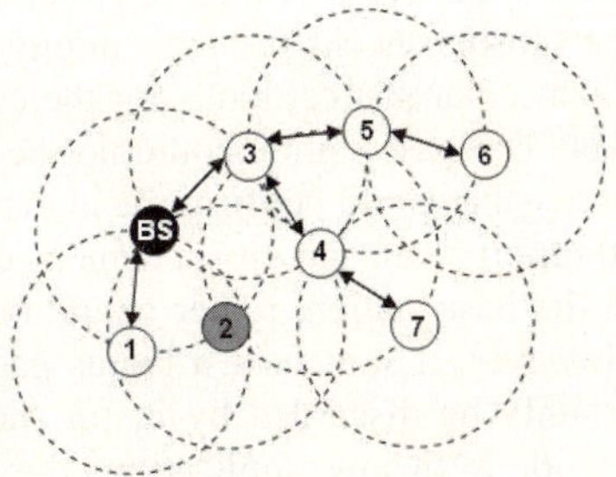

Fig. 3. Broken path recovery using upstream neighbor knowledge

4.4 Repairing a Broken Path and OHC Re-initialization

If in any case, any node between the source node and the base station fails, it could make one or more paths useless. Eventually, in such a case all the downstream nodes along that particular path get disconnected from the base station. To repair such a broken path, we use the stored upstream knowledge of the sensors. We know that, in the first phase each downstream node stores the ids of the one-hop upstream senders of the control message. So, this knowledge could be used for repairing the path.

Let us illustrate it with an example. Say, in Figure 1(b), node 2 is somehow damaged or failed to continue. So, the nodes 4 and 7 get disconnected from the base station. This failure could be detected by node 4. In the first phase, as it stored the id of node 3, it sends a message to node 3 informing that, node 3 would now be its forwarder. When node 4 is a source, for further packet transmission using node 3 it could use the later OHC numbers than that it used for sending the most recent packet to the BS. If there is any node upstream to node 3, through which its (node 4) packets have never been passed, takes the packet with caution and stores the current HS value as the initial HS value for node 4. Based on this HS value of node 4, the subsequent transmissions from node 4 are verified by node 3 and those upstream nodes (if any). The other node, node 7 also follows the same procedure and becomes connected again. As we are considering a highly dense deployment scenario, we think that, in most of the cases, a node might initially get two or more upstream senders who would try to be its forwarder. This procedure works fine as long as no more than w packets are lost on the way, from any source node (after a path is broken due to a node failure). If within the time of repairing the path, more than w packets are lost from a particular source, the OHC chain along that path breaks down. In fact, this is the worst case where all the downstream nodes along the path become invalid to the base station and their sent reports are discarded on the way to reach the base station. To overcome this problem, the entire OHC initialization phase could be made periodic (after certain interval, which is an application dependent parameter). The resultant structure after repairing the broken path of the sample network is shown in figure 3.

5 Analysis of Our Scheme

We analyze the security of our scheme with respect to two design goals; the ability of the base station to detect a false report and the ability of the nodes en-route to detect and filter false reports.

In our scheme, whenever the base station receives a report from any source sensor, it first checks the id of the sensor, checks the authenticity of the report by verifying the OHC number for that particular source, looks for the corresponding shared secret key and decrypts the packet. The base station could not be compromised in any way. So, it is in fact the final entity that could confirm the authenticity, confidentiality and integrity of the transmitted reports. Our security scheme is designed in a way that, any bogus report cannot reach the base station, rather would be detected and dropped by the intermediate nodes. However, if somehow a bogus packet is sent directly to the base station, it would certainly be discarded by it, for the failure of authentication check or because of false id. If in any application, the optional key refreshment mechanism is employed, once the time slot of releasing the new session key is over, the base station first tries to decrypt the incoming packets from any particular source with the X-ORed new key for that node. In case, if it produces garbage result, the BS tries with the previous shared secret key with that node (the previous key could easily be obtained again by X-ORing the most recent session key with the newly computed key for that node). This case might happen when somehow some node cannot get the new session key released by the base station.

We consider two types of attacks that should be detected by the sensors en-route to the base station:

Outsider Attack: In this case, as shown in figure 2(d) that, if an outsider node generates a packet with fake OHC number, the authentication must be failed in the very next node along the path and as a result this packet would never be forwarded. Simple verification of the OHC number prohibits the forwarding of such bogus packets and thus saves crucial energy resource of the network.

Insider Attack: If a legitimate node along any path is compromised, the attacker could grab the OHC sequence and the shared secret key. However, it should be noticed that, to use the OHC numbers successfully, the adversary should also know the last OHC number used by that particular node to send packet to the base station. Otherwise, any arbitrary use of the OHC number from that source might not be forwarded by the next intermediate node because of authentication failure. Now, in case if a node is fully compromised, that is if the adversary obtains all the required information, it actually gets the status of a legitimate node in the network. This fully compromised node could be used to generate false reports with valid authentication numbers. To prevent such type of malicious adversary, there are several factors come into play to detect the abnormal behavior of the node. In our scheme, the BS considers a report legitimate if it is reported by at least δ number of source nodes in the network, where δ is an application dependent parameter. So, the different or modified reports from a single source cannot convince the base station about any event. Also the base station notices the amount of packets generated by a particular source. These are basically the parts of an Intrusion Detection System (IDS) implemented in BS. The detailed description of the IDS is beyond the scope of this paper and will be reported in our future works. The worst case scenario occurs if more than δ number of nodes in the network are somehow compromised. This sort of collaborative and large scale attack could be handled by periodic restructuring of the network. Finding an optimal value of the time interval for periodic restructuring is kept as our future work.

Our scheme ensures replay protection as the OHC numbers are checked as authentic only with later values. If a previous packet is captured by an adversary with the id of a legitimate node, and is sent again later, it would simply be discarded by the intermediate nodes. If an adversary uses a valid OHC number with invalid id, it would be detected by the BS and eventually the adversary would be exposed.

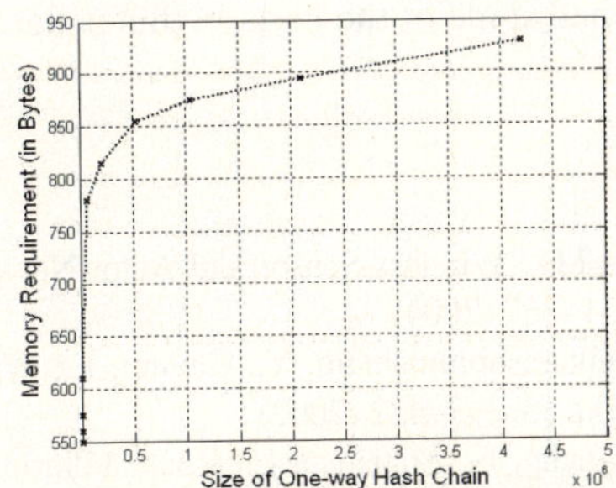

Fig. 4. Memory requirement for One-way Hash Chain generation

The method of generating and storing a long OHC in a sensor node is not straight forward. Naive algorithms require either too much memory to store every OHC number, or too much time to compute the next OHC number. Recently, some efficient OHC generation algorithms for resource-constrained platforms have been proposed [10], [11], [12]. Among these algorithms, the fractal graph traversal algorithm [10] could perform well on the traditional sensor nodes. This algorithm stores only some of the intermediate numbers, called pebbles, of an OHC, and uses them to compute other numbers. If the size of an OHC is n (there are total n numbers in this OHC), the algorithm performs approximately $\frac{1}{2}\log_2 n$ one-way function operations to compute the next OHC number, and requires a little more than $\log_2 n$ units of memory to save pebbles.

The length of an OHC that is needed for a source node is also an important factor. The typical length is between 2^{11} to 2^{22}. If the length of an OHC is 2^{22} and a node uses one OHC number per second, it will take more than a month to exhaust all numbers from this chain. Figure 4 shows the storage requirements for storing pebbles for different lengths of an OHC. This includes a skipjack based one-way function and OHC generation based on [10]. We see that a node needs about 930 bytes to maintain an OHC of length 2^{22}. This includes 256 bytes lookup table for skipjack, which can be shared with other applications. Other than this, each node has to store only a few ids of the upstream sender nodes. Overall, the memory requirement could be well met with today's sensor nodes.

As the re-construction of the network structure and OHC initialization is periodic, new sensors could be added later in the network and they could actively participate in the network after a new tree is created. This feature ensures scalability of our scheme.

6 Conclusions and Future Works

In this paper, we have focused basically on providing security during data transmission from the source sensor nodes to the base station. However, there is a lot of scope for further research in this area. As our future works, we will investigate the energy-efficiency of our scheme in detail and develop an Intrusion Detection System (IDS) to provide supplementary security supports in the network. Also we are currently working on finding out an optimal value of interval for re-initiating the first phase of the scheme so that, the maximum lifetime of the network could be ensured along with data transmission security. Finally, it should be mentioned that, because of the page limitations, we have shortened some of the parts in this paper.

References

1. Akyildiz, I.F., Kasimoglu, I.H.: Wireless Sensor and Actor Networks: Research Challenges. Ad Hoc Networks 2(4), 351–367 (2004)
2. Akyildiz, I.F., Su, W., Sankarasubramaniam, Y., Cayirci, E.: Wireless Sensor Networks: A Survey. Computer Networks 38, 393–422 (2002)
3. Ye, F., Luo, H., Lu, S., Zhang, L.: Statistical En-Route Filtering of Injected False Data in Sensor Networks. IEEE Journal on Sel. Areas in Comm. 23(4), 839–850 (2005)
4. Zhu, S., Setia, S., Jajodia, S., Ning, P.: An Interleaved Hop-by-Hop Authentication Scheme for Filtering of Injected False Data in Sensor Networks. In: Proceedings of S&P, pp. 259–271 (2004)
5. Lee, H.Y., Cho, T.H.: Key Inheritance-Based False Data Filtering Scheme in Wireless Sensor Networks. In: Madria, S.K., Claypool, K.T., Kannan, R., Uppuluri, P., Gore, M.M. (eds.) ICDCIT 2006. LNCS, vol. 4317, pp. 116–127. Springer, Heidelberg (2006)
6. Xbow Sensor Networks, available at: http://www.xbow.com/
7. Lamport, L.: Constructing digital signatures from one-way function. In technical report SRI-CSL-98, SRI International (October 1979)
8. Pathan, A.-S.K., Lee, H.-W., Hong, C.S.: Security in Wireless Sensor Networks: Issues and Challenges. In: Proc. of the 8th IEEE ICACT 2006, vol. II, pp. 1043–1048. IEEE Computer Society Press, Los Alamitos (2006)
9. Wood, A.D., Stankovic, J.A., Son, S.H.: JAM: a jammed-area mapping service for sensor networks. In: 24th IEEE RTSS, pp. 286–297. IEEE Computer Society Press, Los Alamitos (2003)
10. Coppersmith, D., Jakobsson, M.: Almost Optimal Hash Sequence Traversal. In: 6th International Financial Cryptography, Bermuda (March 2002)
11. Jakobsson, M.: Fractal hash sequence representation and traversal. In: 2002 IEEE International Symposium on Information Theory, Switzerland, IEEE Computer Society Press, Los Alamitos (2002)
12. Sella, Y.: On the computation-storage trade-offs of hash chain traversal. In: 7th International Financial Cryptography Conference, Guadeloupe (2003)
13. Karlof, C., Wagner, D.: Secure routing in wireless sensor networks: Attacks and countermeasures. Elsevier's Ad Hoc Network Journal, 293–315 (September 2003)

An Algorithm for Dissemination and Retrieval of Information in Wireless Ad Hoc Networks[*]

Hugo Miranda[1], Simone Leggio[2], Luís Rodrigues[1], and Kimmo Raatikainen[2]

[1] University of Lisbon - Portugal
[2] University of Helsinki - Finland

Abstract. Replication of data items among different nodes of a wireless infrastructureless network may be an efficient technique to increase data availability and improve data access latency. This paper proposes a novel algorithm to distribute data items among nodes in these networks. The goal of the algorithm is to deploy the replicas of the data items in such a way that they are sufficiently distant from each other to prevent excessive redundancy but, simultaneously, they remain close enough to each participant, such that data retrieval can be achieved using a small number of messages. The paper describes the algorithm and provides its performance evaluation for several different network configurations.

1   Introduction

Information management in wireless infrastructureless (ad-hoc) networks is not a straightforward task. The inherently distributed nature of the environment, and the dynamic characteristics of both network topology and medium connectivity, are a challenge for the efficient handling of data. The limited resources and the frequent disconnection of the devices suggest that data should be replicated and distributed over multiple nodes. A data dissemination algorithm for such a decentralised approach should balance the need to provide data replication (to cope with failures) with the need to avoid excessive data redundancy (as nodes may have limited storage capability). Finally, since in wireless networks both bandwidth and battery power are precious resources, the algorithm should also minimise the amount of signalling data.

In this paper, we address the problem of finding adequate locations for the replicas of a data object using a distributed algorithm. The same problem has been addressed before (e.g. [1,2,3,4,5,6]) although with a different set of assumptions.

System Model. We share most of the assumptions described in [5]. In brief, the ad hoc network is composed of cooperative nodes which are producers and consumers of uniquely identifiable data items, composed of a key and a value with

---

[*] This work was partially supported by the MiNEMA programme of the European Science Foundation and by FCT project MICAS, POSC/EIA/60692/2004 through FCT and FEDER.

A.-M. Kermarrec, L. Bougé, and T. Priol (Eds.): Euro-Par 2007, LNCS 4641, pp. 891–900, 2007.

application dependent semantics. Each node has storage space available for storing the items it produces. In addition, the nodes make available limited storage space for keeping replicas of a fraction of all the objects produced by other nodes. The system does not require the space at all nodes to be of the same size.

Replication is used to improve availability and reduce access latency. Also, like in [5], we assume for simplicity that all items are equally sized so that the space made available by each node can be referred in item units instead of bytes.

Contrary to [5], we assume that there is no predictable access pattern to the objects which is known in advance by the nodes and does not change. This access pattern may be used to bias the distribution of the replicas so that the most popular items have more replicas. We are interested in scenarios where these access patterns cannot be derived a priori or even during the lifetime of the system (for instance, short lived objects). Therefore, we aim at distributing data items as evenly as possible among all the nodes that form the network, avoiding clustering of information in sub-areas; an uniform dissemination of data items should leverage lower access latency to any item from any node in the network, i.e, whenever a data item is requested by a node S, the distance to the node that provides the reply should be approximately the same, regardless of the location of S. Naturally, the actual distance depends on multiple parameters, such as the number of nodes in the network, the amount of memory made available at each node, and the number of data items.

There are multiple applications for a distributed storage with these characteristics. Cooperative teams may use it to share photographs, annotations or measurements while on the field [5]. Users on spontaneous networks can use it to advertise SIP records containing their interests to find other users willing to play distributed games or chat [7].

Scope and Contribution of the Paper. Implementing a full system with these characteristics is a complex task that must address multiple challenges and requires several algorithms. The contributions of this paper are the following. Firstly, it proposes an algorithm to perform an initial distribution of the data items that satisfies the requirements above. Additionally, it describes an algorithm for retrieving the information. Problems like updating the data items, shuffling the item distribution to address node movement or disconnection, or tolerating uncooperative nodes are out of the scope of this paper (the interested reader may consult [8]).

2   Overview

An example of the dissemination of an item is depicted in Fig. 1. The dissemination begins with the broadcast of a *registration* message. The item is stored at the producer and included in the message (Fig. 1(a)). The figure depicts in black the nodes that store a replica of the item. *Registration* messages carry a *Time From Storage* (TFS) field which records the distance (in number of hops) from the node sending the message to the known closest copy. The TFS for the message to be forwarded by each node is depicted in the centre of the node.

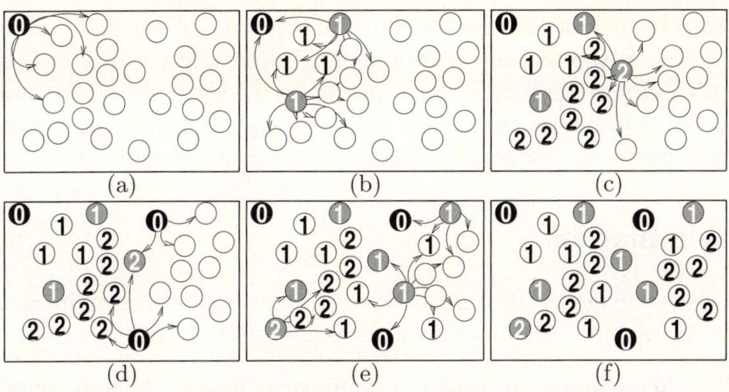

Fig. 1. Example of dissemination of an item

Figs. 1(b) and 1(c) show the progress of the dissemination. Nodes use a message propagation algorithm named Pampa [9] (to be discussed later) to reduce the number of transmissions. Nodes decide whether to forward the message after a small *hold period*, during which they monitor the network, listening for possible retransmissions of the same message. During the hold period, each node computes the lowest value of all the TFS fields it has received in a variable named *mTFS*. In the figure, nodes that forward a registration message but did not store the data item are depicted in gray. When forwarding a registration message, a node sets the TFS field to *mTFS*+1, accounting for the additional hop needed to reach the closest copy of the item.

Central to our algorithm is a constant *Distance Between Copies* (DbC). The DbC dictates the maximum value of the TFS field and, implicitly, the degree of replication of the items. DbC is expected to be small. In this example, we use DbC=2. Fig. 1(d) shows that a node with *mTFS*=DbC at the end of the hold period stores a copy of the item and retransmits the message. The TFS of the message is reset to 0 to let other nodes learn about the newly stored copy and update their *mTFS* variables accordingly (see for example Fig. 1(e)).

The final state of the system after the dissemination of the item is depicted in Fig. 1(f). Although only a small number of nodes have stored the item, a replica is stored at no more than DbC hops away from any of the nodes.

Broadcast Algorithm. We use the Pampa [9] broadcast algorithm to propagate dissemination and query messages. In comparison with a conventional flooding algorithm, Pampa reduces the number of nodes required to transmit a message by having nodes more distant to the previous forwarder to broadcast the message earlier. Nodes closer to the source (i.e., those whose expected additional coverage would be smaller) do not retransmit. Pampa does not require devices to be aware of their location or of the location of their neighbours. Instead, each node uses the Received Signal Strength Indicator (RSSI) of the first retransmission listened to set the hold period. The hold period is set such that nodes with a lower RSSI expire their timers first. During the hold period, nodes count the number of retransmissions listened and, at the end of the hold period, they do not retransmit the message if a predefined threshold was reached. Based on evaluation results presented in [9], in this paper we use a threshold value of 2.

Due to the store-and-forward nature of the algorithm, Pampa is not used as a black-box. Next, we discuss how Pampa was adapted for our purposes.

3   Dissemination

A global overview of the dissemination algorithm was presented in Sec. 2. This section provides additional details on the steps executed by each node.

Forwarding Registration Messages and Storing Items. A node only decides whether to forward (or drop) a *registration* message and whether to store or not the corresponding data item at the end of the hold period. The decision takes as input the following parameters: i) the output of the Pampa's algorithm, that accounts only with the number of retransmissions listened; and ii) the value of $mTFS$. The data item is stored if, at the end of the hold period, $mTFS{=}DbC$. Note that if some other node in the vicinity previously decided to store the data item, it had retransmitted the message with TFS set to zero and, therefore, $mTFS$ would have been reset accordingly. The message is forwarded if the data item was stored or if the output of Pampa's algorithm suggests it.

Computing the Hold Period. The base value for the hold period is given by the underlying Pampa broadcast algorithm. Pampa computes the delay based on the signal strength which, in turn, depends on the relative location of nodes. We have also seen that if a node is the first node in its own vicinity to decide to forward a message and mTFS=DbC, then it stores a copy of the data item. Therefore, depending on the deployment of the nodes, and of the location of the sources of the registration messages, some nodes may end up storing much more items than others. To promote a balanced distribution of items, regardless of the physical location of nodes, our algorithm applies a bias to the base value of the hold period derived by Pampa. The bias is a function of the number of items already stored by the node.

When a node whose storage occupancy ratio is above some threshold receives a *registration* message with TFS=DbC, it multiplies Pampa's hold period by a factor proportional to the occupancy ratio of its storage space. Precisely, the delay is determined by the function $holdPeriod = \text{hP} \times \left(1 + \frac{occup-thresh}{1-thresh} \times \text{bias}\right)$, where hP represents the hold period computed by Pampa, *occup* is the current occupancy ratio of the storage space, and *thresh* and *bias* are configuration parameters indicating respectively the minimal threshold for triggering this function and a weight of this component on the final value of the hold period. In the simulations presented in Sec. 5 *thresh*=0.7 and *bias*=2.0.

Memory Management. We assume that each node has some memory region reserved for storing data items. Nodes keep on adding items to this region until it is completely filled. Only then, nodes are required to drop stored items to make room for new items. Note that our scheme to compute the hold period already attempts to balance the memory occupation among nodes. If a nodes needs to make room for a new item, it will randomly selected one of the previous entries. We note that policies like Least Recently Used (LRU) or other deterministic policies should be avoided in our algorithm. This is because a deterministic criteria applied to different nodes would likely select the same entry for replacement. This undesirable behaviour would eliminate a large number of replicas of the same item resulting in an uneven distribution.

Analytical Properties of the Algorithm. At the end of the dissemination, and assuming a perfect networking environment without message losses and if nodes did not discard any item from their local storage, the following properties can be derived concerning the distance of the nodes to some data item.

All nodes, with the exception of those at the margins of the networked region, should be able to find a copy of the data item at a distance not higher than $\frac{\mathrm{DbC}+1}{2}r$ (where r is the transmission range of the devices) or $\left\lceil \frac{\mathrm{DbC}+1}{2} \right\rceil$ hops. This results from the fact that each node storing a copy of the item will become the "closest copy" to nodes that have served either as predecessors or successors in the dissemination of the registration message. An interesting case happens when DbC is even what may leave some nodes equidistant (in hops) of at least two copies of the item. Fig. 1(f) shows such a node at the centre of the network.

From the previous result it is possible to derive the expected average distance from any node to a data item, assuming an uniform deployment of the nodes. Function τ is given by

$$\tau(\mathrm{DbC}) = \sum_{i=0}^{\mathrm{DbC}} \left(\left\lceil \frac{i+1}{2} \right\rceil \frac{\pi\left(\frac{i+1}{2}r\right)^2 - \pi\left(\frac{i}{2}r\right)^2}{\pi\left(\frac{\mathrm{DbC}+1}{2}r\right)^2} \right) =$$

$\frac{\sum_{i=0}^{\mathrm{DbC}}\left(\left\lceil \frac{i+1}{2} \right\rceil (2i+1)\right)}{(DbC+1)^2}$ and successively partitions a circle with radius corresponding to the DbC in semi-circles centred at the node storing the copy of the item. The function accounts with the proportion of the area contributed by each semi-circle and with the distance in hops of the nodes located in that semi-circle to the node storing the copy. Function τ has the following values for small DbCs: $\tau(2) = 1.55(5), \tau(3) = 1.75$ and $\tau(4) = 2.2$.

4   Data Retrieval

To retrieve some item, a node begins by looking for it in its local storage. If the item is not found locally, the node initiates a search in its vicinity. This is implemented by broadcasting a *query* message with a limited range, given by a variable $qTTL$. This variable is initialised with a small value and is successively adjusted, in order to adapted to the network conditions. Function τ, introduced in the analysis above, is used to set the initial value of $qTTL$. If no answer to the query message is received from the vicinity within a predefined amount of time, the query will be broadcast with a TTL large enough to deliver the message to every node in the network. This broadcast should be avoided as it requires the transmission of as many messages as the data dissemination algorithm.

When a node receives a query message and does not have the item in its local storage, it will have to decide whether to forward or drop the query message. Again this decision is made after an hold period, according to the criteria defined by the underlying Pampa broadcast algorithm. Note that if the TTL field of the message has reached the value of 0 the query is simply dropped. If the query messages is retransmitted, the forwarding node pushes its own address to a *route stack* field of the message, in a route construction process similar to the route discovery algorithm in some source routing protocols for MANETs (e.g. [10]).

If the key is found, the node sends a point to point reply to the source of the query without waiting the delay suggested by Pampa. The reply message follows the path constructed in the *routeStack* field of the query. The *TFS* field of the reply is set to 0 at the origin of the reply and incremented at every intermediate hop to capture the distance at which the data item was found.

The reply message is unreliably forwarded by the intermediate hops, thus there is some probability that the reply is lost. On the other hand, no provision is taken to limit the number of replies sent to the node. Therefore, there is a reasonable probability that at least one of the routes constructed during the query propagation remains valid until the reply is delivered.

When the node that issued the query receives the first reply, it performs corrective measures over the data distribution and the $qTTL$ value. A reply found far away from the source of the query signals an uneven distribution of the item. Therefore, the node that issued the query stores the item if the reply was received from a node located more than DbC hops away. Given that the dissemination algorithm aims at achieving an adequate distribution of the items, the distance (in number of hops) from the source of the query to any item should be approximately the same and will depend mostly of the number of neighbours of the node and their storage space. After each query, $qTTL$ is tuned by weighting its previous value with the distance at which the reply was found (available in the *TFS* field of the reply). The goal is to reduce the number of queries requiring a second broadcast while keeping $qTTL$ as small as possible.

5   Evaluation

We have implemented a prototype of our algorithm in the *ns-2* network simulator v. 2.28. The simulated network is composed of 100 nodes uniformly disposed over a region with 1500mx500m. The simulated network is an IEEE 802.11 at 2Mb/s.

Runs are executed for 900s of simulated time. Each run consisted of 400 queries over a variable number of disseminated data items, as described below. Data items have 300 bytes and are disseminated in time instants selected uniformly between 0 and 400s. Note that the size of the data items is only relevant for estimating the traffic generated at the network; when considering memory availability at each node we have simply taken into account the number of data items stored at each node. Queries start at 200s and are uniformly distributed until the 890s. The nodes performing the queries and the queried items are selected using an uniform distribution. The simulation ensures that only advertised records can be queried.

No warm-up period is defined. All values presented below average 100 independent runs, combining different node deployments, query and dissemination times. The evaluation uses two metrics. The "average distance of the replies" measures the distance (in number of hops) from the querying node to the source of the first reply received. The distance of a reply is 0 if the value is stored in the querying node. The "average number of transmissions per query" measures the total number of query and reply messages (initial transmissions and forwarding) performed by all nodes and divides it by the number of queries.

Theoretic Idealised Model and Saturation Point. The simulation results are compared with an execution of the algorithm, analytically computed for an idealised network where nodes are uniformly distributed and the space made available at the nodes within each circle with radius $\frac{\text{DbC}+1}{2}r$ is sufficient to store all the disseminated items.

The storage capacity in a network region containing n nodes is given by $n \times \left(s + \frac{i}{N}\right)$ where i is the number of items advertised, N the number of nodes in the system (recall that nodes keep the items they advertise in a separate region of the storage space), and s is the storage space made available at each node for data items advertised by other nodes. We define the "saturation points" (SP) of our algorithm as the multiple solutions of the equation $n \times \left(s + \frac{i}{N}\right) = i$. Each solution will correspond to a different configuration that is capable of storing all the data items being advertised and, therefore, that should be able to provide all the replies in the target average distance given by function τ.

In this evaluation we are interested in comparing the implementation of the algorithm with this ideal model, given that it characterises the best results that can be achieved. In particular, to evaluate the performance of the algorithm close to the SP and to compare the performance for different values of DbC.

Sensitivity to Different Network Configurations. The performance of the algorithm is affected by the number of nodes in the neighbourhood of each node, the storage size at every node and the number of items advertised in the network. To evaluate the effect of the variation of each of these parameters individually, we fixed a value for each in a baseline configuration. Each parameter was then individually varied keeping the remaining consistent with the baseline configuration. The number of neighbours was varied by configuring the nodes with transmission ranges between 150 and 325 meters. The number of neighbours was estimated by counting the number of nodes that received each broadcast message on each simulation with the same transmission range. A transmission range of 250m was settled for the baseline configuration. The storage size was varied between 2 and 16 items. In the baseline configuration, each node makes available storage for 10 items. The number of items advertised was varied between 50 and 800. Advertisements were uniformly distributed by the nodes. In the baseline configuration, 200 data items are advertised.

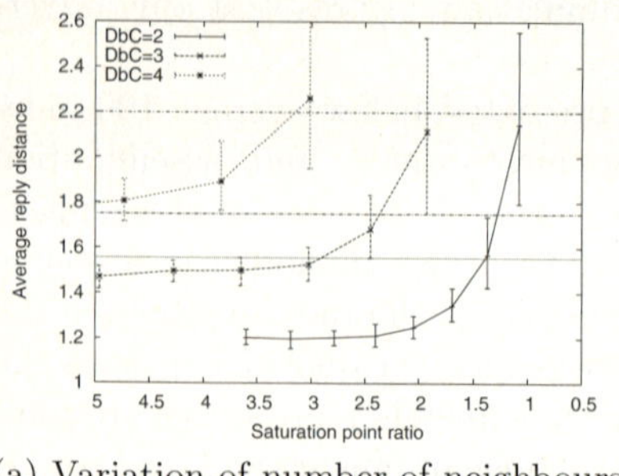

(a) Variation of number of neighbours

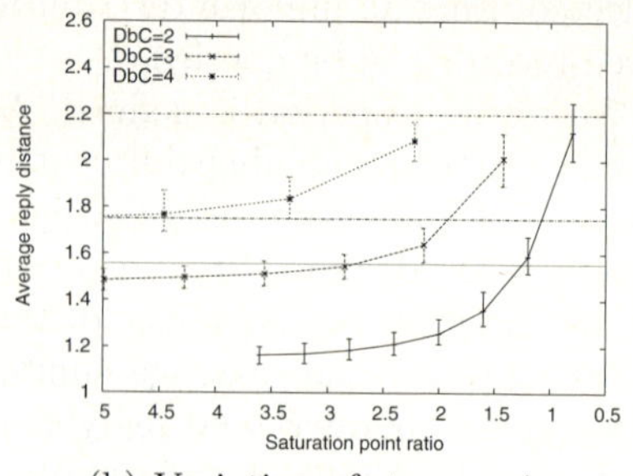

(b) Variation of storage size

Fig. 2. Average distance of the replies

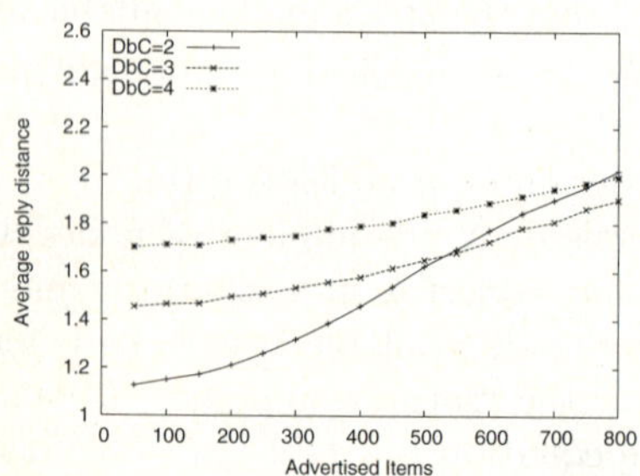

Fig. 3. Variation of number of items

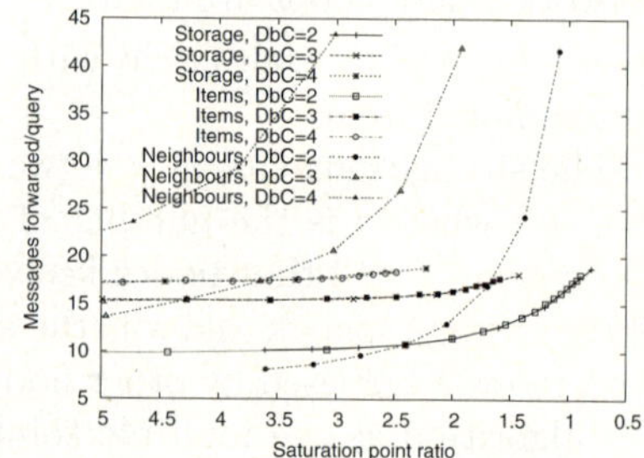

Fig. 4. Frames per query

Note that the baseline configuration is below the SP for all values of DbC. Figure 2 shows the average distance of the replies in the simulations. The x axis harmonises the results by presenting them according to a ratio to the SP given by $\frac{n \times \left(s + \frac{i}{N}\right)}{i}$. Error bars show the highest and lowest average distance of the replies of a subset of the simulations that excluded the 10% with higher and lower values. To facilitate the comparison with the theoretical model, the figures show the values of function τ for every DbC tested.

When the system is below the SP, our algorithm exhibits a smaller average reply distance than the computed for the idealised model. Our approach is creating more replicas than estimated by the idealised model, resulting in an increased proximity of the nodes to the data items. When the system approaches the SP, as expected, the average reply distance increases (as it becomes impossible to store all the items in the target DbC). The target distance is reached with SP > 1. Still, the system continues to provide acceptable results, and in the majority of the cases copies are found within only a few hops in excess of the optimum limit.

We identify two differences between the idealised and the experimental models to justify the discrepancies: *i*) since nodes are randomly deployed, it is unlikely that at every retransmission, there exists one node located precisely over the limit of the transmission radius of the previous source. Smaller distances result in additional hops travelled by the messages, and reduce the effective area (and nodes) that should be accounted in the estimation of the SP; *ii*) concurrent decisions, amplified by the delay in the propagation of the messages may permit to nodes in proximity to simultaneously decide to store the items.

Figure 3 depicts results for different numbers of advertised items (in the x axis). It shows that DbC plays an important role in the performance of the algorithm. It can be seen that the average distance of the replies for the different values of DbC tend to approximate as the storage capacity of the region is reduced. In particular, the lines for DbC=2 and DbC=3 intersect around the SP for DbC=2. This is the expected behaviour of the algorithm, given that, by definition, above the SP, it is not possible to store all information at the target range.

Traffic. The average number of messages per query is presented in Fig. 4. It is interesting to notice the overlap, for each value of DbC, of the lines that capture the behaviour of the system with the size of the storage space and with the number of items. This confirms that when the system is below the SP, none of these factors influences the number of messages transmitted per query. Additionally, we compared the growing ratios of the curves for the average distance of the replies (Fig. 2(b) and 3) and for the number of messages forwarded/query (Fig. 4) in both scenarios. The difference between these ratios is less than 2% when the storage space is changed and less than 7% when the number of items changes. These small values show that the growing of the average distance implies an almost linear grow of the number of messages. This confirms the efficiency of our adaptive mechanism for defining $qTTL$: it prevents the query algorithm from frequently resorting to a full broadcast, even in adverse conditions.

On the other hand, we expect the number of messages to drop significantly when the density increases, because we benefit from the properties of Pampa, which adapts the proportion of nodes retransmitting a message to the network density. Comparing results depicted in Figs. 2(a) and 4, it can be seen that although the distance of the replies tends to stabilise with the grow of the network density, the number of messages continues to diminish. Here, the difference between the ratios is higher than 36%.

6   Related Work

Several papers have addressed the problem of distributing copies of data items in MANETs. However, most of the previous work makes stronger assumptions about the network or the application scenario. Some assume that it is possible to collect statistics about data usage, such as which items are accessed more frequently [5] or obtain similar information from user profiles [4]. Others assume that there is a single data source [6] or that nodes are aware of their location [1,2,3]. In contrast with previous approaches, our work is targeted at spontaneous networks (such as rescue teams) where all nodes need to share many short lived data items. Our algorithm prevents the duplication of data items in neighbouring nodes by counting the number of hops travelled by an item before being stored. Instead of using geographical information, we take advantage of the fine dissemination properties of Pampa to ensure the geographical distribution of the information.

7 Conclusion

This paper has presented an algorithm for retrieving and distributing information in ad-hoc networks. The algorithm is fully distributed. Its main goal is to ensure an even geographical distribution of the data items, so that requests for a given data item are satisfied by some nodes close to the source of the query.

This goal is obtained by combining different techniques. Data items are disseminated with a counter to provide a minimal distance between the copies; a broadcast protocol reduces the number of messages required for propagation and increases the geographical distance between the hops. Finally, an adaptive mechanism allows to limit the propagation of most queries. Simulation results show that the algorithm achieves a fair dissemination of items throughout the network and that a small number of messages is required to retrieve items.

References

1. Ghose, A., Grossklags, J., Chuang, J.: Resilient data-centric storage in wireless sensor networks. IEEE Distributed Systems Online (2003)
2. Liu, D., Stojmenovic, I., Jia, X.: A scalable quorum based location service in ad hoc and sensor networks. In: Proc. of the 3rd IEEE Conf. on Mobile Ad-hoc and Sensor Systems, IEEE Computer Society Press, Los Alamitos (2006)
3. Ratnasamy, S., Karp, B., Shenker, S., Estrin, D., Govindan, R., Yin, L., Yu, F.: Data-centric storage in sensornets with GHT, a geographic hash table. Mobile networks and applications 8(4), 427–442 (2003)
4. Datta, A., Quarteroni, S., Aberer, K.: Autonomous gossiping: A self-organizing epidemic algorithm for selective information dissemination in mobile ad-hoc networks. In: Bouzeghoub, M., Goble, C.A., Kashyap, V., Spaccapietra, S. (eds.) ICSNW 2004. LNCS, vol. 3226, Springer, Heidelberg (2004)
5. Hara, T.: Effective replica allocation in ad hoc networks for improving data accessibility. In: Proc. of the 20th Joint Conf. of the IEEE Computer and Communications Societies (INFOCOM 2001), vol. 3, pp. 1568–1576. IEEE Computer Society Press, Los Alamitos (2001)
6. Yin, L., Cao, G.: Supporting cooperative caching in ad hoc networks. IEEE Transactions on Mobile Computing 5(1), 77–89 (2006)
7. Leggio, S., Miranda, H., Raatikainen, K., Rodrigues, L.: SIPCache: A distributed SIP location service for mobile ad-hoc networks. In: Proc. of the 3rd Conf. on Mobile and Ubiquitous Systems: Networks and Services (MOBIQUITOUS) (2006)
8. Miranda, H., Leggio, S., Rodrigues, L., Raatikainen, K.: A stateless neighbour-aware cooperative caching protocol for ad-hoc networks. DI/FCUL TR 05–23, Department of Informatics, University of Lisbon (2005)
9. Miranda, H., Leggio, S., Rodrigues, L., Raatikainen, K.: A power-aware broadcasting algorithm. In: Proc. of The 17th IEEE Symp. on Personal, Indoor and Mobile Radio Communications (PIMRC'06), IEEE Computer Society Press, Los Alamitos (2006)
10. Johnson, D.B., Maltz, D.A.: Dynamic Source Routing in Ad Hoc Wireless Networks. In: Mobile Computing, pp. 153–181. Kluwer Academic Publishers, Dordrecht (1996)

Surrendering Autonomy:
Can Cooperative Mobility Help?

Ghassen Ben Brahim[1], Bilal Khan[2], Ala Al-Fuqaha[1],
Mohsen Guizani[1], and Dionysios Kountanis[1]

[1] Western Michigan University, Dept. of Computer Science,
Kalamazoo, MI 49009, USA
`http://www.wmich.edu/`
[2] John Jay College of Criminal Justice,
Dept. of Mathematics and Computer Science,
New York, NY 10019, USA
`http://www.cuny.edu/`

Abstract. In this paper, we develop a Cooperative Mobility Model
that captures new salient features of collaborative and mission-oriented
MANETs. In particular, the cost-benefit framework of our model is a
significant advance in modelling heterogenous networks whose nodes
exhibit the complete range of autonomy with respect to mobility. We
then describe the design of CoopSim, a platform for conducting simulation experiments to evaluate the impact of parameter, policy and algorithm choices on any system based on the proposed Cooperative Mobility
Model. We present a small but illustrative case study and use the experimental evidence derived from it to give an initial evaluation of the merits
of the proposed model and the efficacy of the CoopSim software. In our
case study, we propose studying the impact of the proposed model on
improving the end-to-end communication based on the QoS parameter,
namely BER.

Keywords: Cooperative model, mobility, QoS, MANETs.

1 Introduction

The potential applications of MANETs have led, perhaps not surprisingly, to
a surge in research breakthroughs addressing the many technological challenges
which stand in the way of their wide scale adoption. The many challenges include the limitations of wireless RF channels in terms of available bandwidth
and relatively high bit error rates, energy-efficient communication to extend the
network lifetime, QoS aware routing to meet application requirements, and the
design of new protocols to support large networks and handle the limitations of
the underlying wireless RF links.

On the applications side, the demanding requirements of end users in the
military and public-safety sectors have led to the development of a variety of
unmanned platforms [1]. More specifically, end-user demands have driven the

A.-M. Kermarrec, L. Bougé, and T. Priol (Eds.): Euro-Par 2007, LNCS 4641, pp. 901–910, 2007.

development of Unmanned Ground Vehicles (UGVs) and Unmanned Air Vehicles (UAVs) for use within battlefield and public safety missions, e.g. the UAV-Ground Network [2]. These devices are mobile, mission capable, and can be deployed to serve as relay nodes, maintaining mobile communication, serving as mobile power supplies, since they can be easily deployed to travel to remote locations where power is most critically needed, or to support recharging of embedded devices and hardware carried by troops in the field.

The modern battlefield communications network is a MANET comprised of both manned and unmanned elements (e.g. UAVs), the question remains as to the role of cooperation between nodes. Certainly, task-oriented cooperation is to be expected in such a setting, e.g. coordinating the activity of UAVs to achieve a joint objective like radio source localization [3]. Here, however, we pose a more fundamental question: What role can cooperation play in supporting *communication itself?*

Prior work on the question of how cooperation can benefit communication (e.g. See [4,5,6,7,8,9] and others) has approached the issue from the vantage point of a node's willingness to forward messages to the next hop (toward the intended destination) along a multi-hop path. Almost all prior work was colored by the consumer model in which node mobility is considered the sacrosanct domain of the user, autonomously determined and non-negotiable. While this is an appropriate conception of current consumer applications (e.g. cell phone and laptop users) it fails to leverage the unique opportunities present in battlefield MANETs. In the latter setting, mobility is a fundamental resource of every MANET node, and cooperative nodes can potentially contribute their mobility towards the common good vis-a-vis systemic objectives. In this article, we develop a realistic model for cooperation in battlefield MANETs and evaluate the extent to which communications can be improved when constituent nodes are sometimes willing to *be moved.*

2   Budgeted Location-Based Cooperative Model

Our model begins with the model of Basu et al. [10], but extends it by postulating that future MANETs will not be homogeneous in terms of node autonomy. While Basu et al. consider networks consisting of robots and non-robots, we contend that the general setting requires us to consider *heterogenous* networks comprised of nodes which exhibit the entire spectrum of personalities: from defiant autonomy to self-sacrificial cooperativeness. We capture this viewpoint by adopting a cost model for mobility. To wit, every node is willing to move for the sake of the common good, but *for a price*. Each node is assigned a *movement cost* (proportional to distance moved)—this is the price it charges to be moved, say, per meter. Defiant autonomy is exhibited when a node declares this cost to be infinite; self-sacrificial cooperativeness is manifest when this cost is declared to be zero.

3   The CoopSim Platform

We have developed a simulation platform to investigate how parameter, policy and algorithm choices influence the efficacy of systems based on the proposed Cooperative Mobility Model. The CoopSim platform dynamically updates the communication infrastructure by manipulating its heterogenous constituent network elements; network nodes are assumed to have a wide range of characteristics, including mobility costs and available transmission power. CoopSim continuously seeks to fulfill concrete end-to-end QoS requirements for a set of application level (multi-hop) connections between given endpoint pairs. CoopSim achieves this by leveraging cooperative mobility: it determines new locations for cooperative battlefield MANET nodes, while adhering to its mobility budget constraints. In this exposition QoS requirements are stated in terms of maximum acceptable end-to-end connection bit error rates (BER), but we note that CoopSim can seamlessly integrate arbitrary, richer QoS definitions.

The CoopSim platform is implemented as a modular discrete event simulator that is naturally organized in layers. Fig. 1 presents a modular schematic diagram.

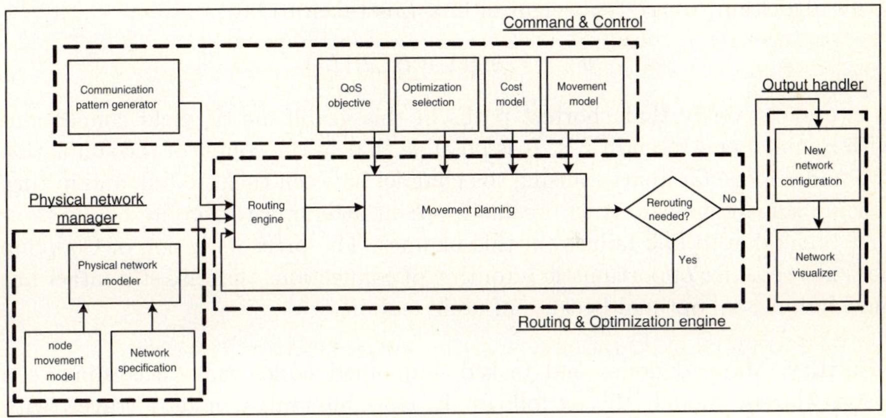

Fig. 1. CoopSim modular architecture

The *lowest layer* of CoopSim represents the **Physical Network Manager**, which consists of a collection of wireless components such as UGVs, manned tanks, etc. Important aspects of this layer include:

Network Discovery. These protocols are used to enable all nodes to discover their neighbors and establish wireless communication channels with them. The design of the network discovery protocol is beyond the scope of this article; a good reference can be found in [11]. For simulation purposes CoopSim assumes that a unidirectional channel connecting a transmitter to a receiver arises whenever the distance separating the two nodes is less than the communication range of the transmitter. A wireless channel forms between two battlefield MANET nodes whenever there is unidirectional channel in both directions.

Channel Characteristics. Suppose we have a pair of nodes at distance D communicating using transmission signal power P over a wireless channel L with noise power P_{noise} through a medium with propagation constant α. The relationship between wireless channel bit error rate (BER) and the received power P_{rcv} is a function of the modulation scheme employed. CoopSim considers non-coherent Binary orthogonal Phase Shift Keying (BPSK) modulation scheme, so $P_{rcv} = P/D^\alpha$, and the instantaneous channel bit error rate is [12,13,14]:

$$BER(L) = \frac{1}{2}\, e^{-\left(\frac{P}{D^\alpha}\right)\frac{1}{P_{noise}}}.$$

The **Routing and Optimization Engine** is the central layer of CoopSim. This layer is responsible for routing the set of connections that need to be maintained and repositioning the cooperative nodes in order to better provide the required QoS. Important aspects of this layer include:

Routing. Connections are routed along shortest paths in the graph using Dijkstra's algorithm, where the weight of link L is taken to be

$$w_L = -\log(1 - BER(L)).$$

It is easy to verify that shortest paths in this graph metric yield connections with minimal end-to-end BER. It is possible that in the course of the simulation two nodes move far apart, causing the channel between them to fail, and in turn causing some connections to break. CoopSim attempts to reroute connections that break due to link failures in this manner. The present version of CoopSim does not consider opportunistic rerouting of connections that are still intact but have become suboptimal because of node mobility.

Mobility. Manned nodes and tasked unmanned nodes move according to a Gauss-Markov model [15], as follows. In time interval n, node i travels with speed $s_{i,n}$ and direction $d_{i,n}$. The mean speed and direction of movement are taken as constants $\bar{s}_i$ and direction $\bar{d}_i$, respectively. Then a node's new speed and direction during the time interval $n + 1$ are given by:

$$s_{i,n+1} = \alpha s_{i,n} + (1 - \alpha)\bar{s}_i + \sqrt{(1 - \alpha^2)s^*_{i,n}}$$

$$d_{i,n+1} = \alpha d_{i,n} + (1 - \alpha)\bar{d}_i + \sqrt{(1 - \alpha^2)d^*_{i,n}}$$

where α represents a continuity-determining constant, and $s^*_{i,n}$ and $d^*_{i,n}$ are random variables with a Gaussian distribution. The coordinates of node i at the end of time interval n are then easily computable as follows:

$$x_{i,n+1} = x_{i,n} + s_{i,n}\cos d_{i,n}$$

$$x_{i,n+1} = x_{i,n} + s_{i,n}\sin d_{i,n}$$

Nodes that are both unmanned and untasked are moved by a mobility planning algorithm. The design and evaluation of such algorithms remains an open area of investigation. Currently, the CoopSim uses our Resultant Algorithm to construct a movement plan; the details of the algorithm are presented in the next section.

The *topmost layer* of CoopSim is called the **Command and Control**. Important aspects of this layer include:

Connections. A connection is defined by a pair of distinct nodes which serve as the source and destination. The Application Layer can generate arbitrary connection topologies based on the structure of the distributed application that is being simulated. In this article, we consider applications in which communication needs are represented by a random set of source-destination pairs.

QoS Requirements. In this exposition, we consider QoS requirements to be defined in terms of maximum acceptable end-to-end BER, but we note that CoopSim can incorporate any computable definition of QoS.

Connection QoS. We compute the BER of multi-hop connections under an end-to-end retransmission scheme. The bit error rate of a connection C which traverses links L_1, L_2, ... L_k can then be computed as follows:

$$BER(C) = 1 - \prod_{i=1}^{k} 1 - BER(L_i).$$

Movement Costs. Command and Control maintains information about each node: whether it is a manned or unmanned asset. Unmanned nodes are further categorized as either tasked or untasked, with tasked nodes having priorities. Every node i declares its movement cost C_i. Manned vehicles and tasked unmanned vehicles are considered quasi-autonomous because they typically declare high movement costs and have their own objective-driven movement; high movement costs make it unlikely they will be moved by the Routing and Optimization Layer. Vehicles that are both unmanned and untasked are considered essentially cooperative; their declared costs reflect the relative logistical expense involved in their deployment.

Mobility Budget. This is the amount of credit to issued by Command and Control to the Routing and Optimization Layer, for funding the movement of cooperative battlefield MANET nodes. The mobility budget is replenished periodically, every T_m time units. In the current simulation, mobility budgets do not accumulate across time intervals.

4   The Resultant Algorithm

Our approach to node mobility planning begins with the following Gedankenexperiment: Consider a single two-hop connection between a source node s and a destination node t, and assume that this connection goes through a cooperative node c. The following two observations are easily be proved by using the well-known Friis' formula [16]:

1. If node c is in line (s, t), then it moves towards s if $BER(c, s) \geq BER(c, t)$, and towards t otherwise; moving node c in a direction that is outside of line (s, t) yields worse connection performance.
2. If node c is not on the line (s, t), then it should move in a direction towards line (s, t).

Making the model more quantitative, we assign weights to the links (c, s) and (c, t); these weights $w(c, s)$ and $w(c, t)$ are taken to be proportional to $BER(c, s)$ and $BER(c, t)$, respectively. The cooperative node c repositions itself by moving in a direction that would improve the total end-to-end connection BER from s to t (see Fig. 2); the direction of movement depends on relative positions of the nodes, as well as the relative magnitudes of $w(c, s)$ and $w(c, t)$.

The previously described Gedankenexperiment suggests a natural analogy between finding the cooperative node movement direction and the problem of resultant forces. Each node c experiences concurrent forces along all its incident links. The magnitude of the force along link L is proportional to $n_L \cdot BER(L)$, where n_L

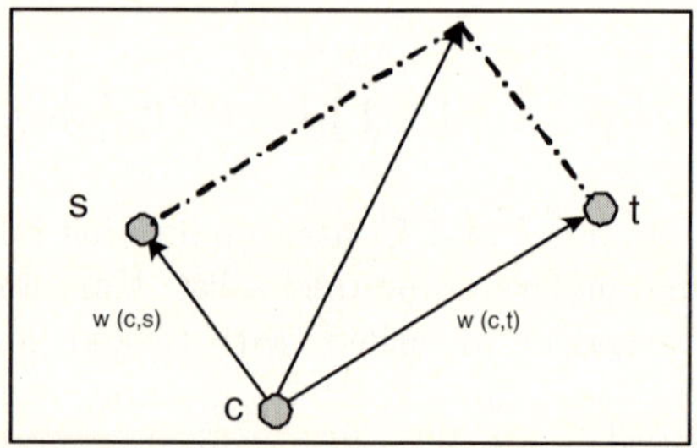

Fig. 2. A Gedankenexperiment on Node Mobility

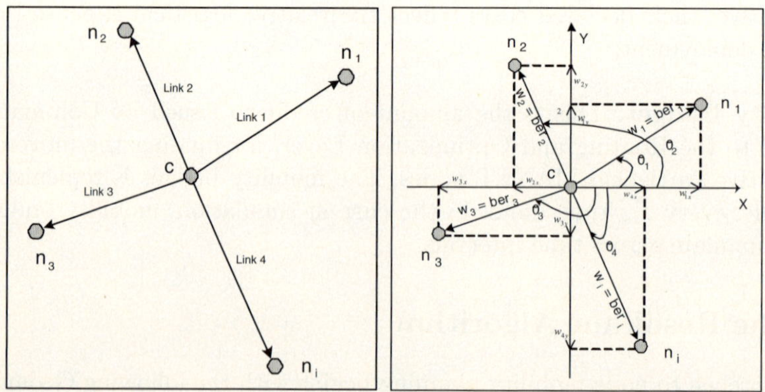

Fig. 3. Resultant Algorithm

is the number of connections which transit over link L. Computing the resultant force can be done in many ways, including standard componentwise analysis by projection onto a set of orthogonal axes (see Fig. 3). After finding the resultant direction, the available movement budget can be used to move the cooperative node. There remains the problem of dividing a global mobility budget among the cooperative nodes. In this preliminary investigation, we consider uniform allocations: each of the N nodes receives $1/N$ fraction of the total mobility budget.

5   Case Study

In this section we give some experimental results to illustrate the types of investigations which can be conducted using the CoopSim platform.

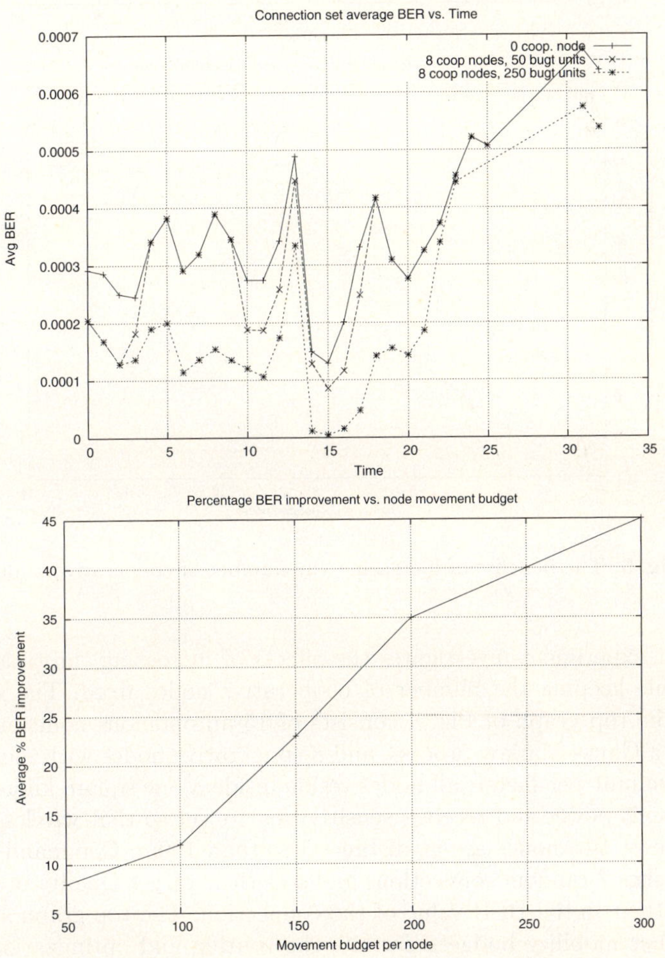

Fig. 4. The benefits of increasing the mobility budget

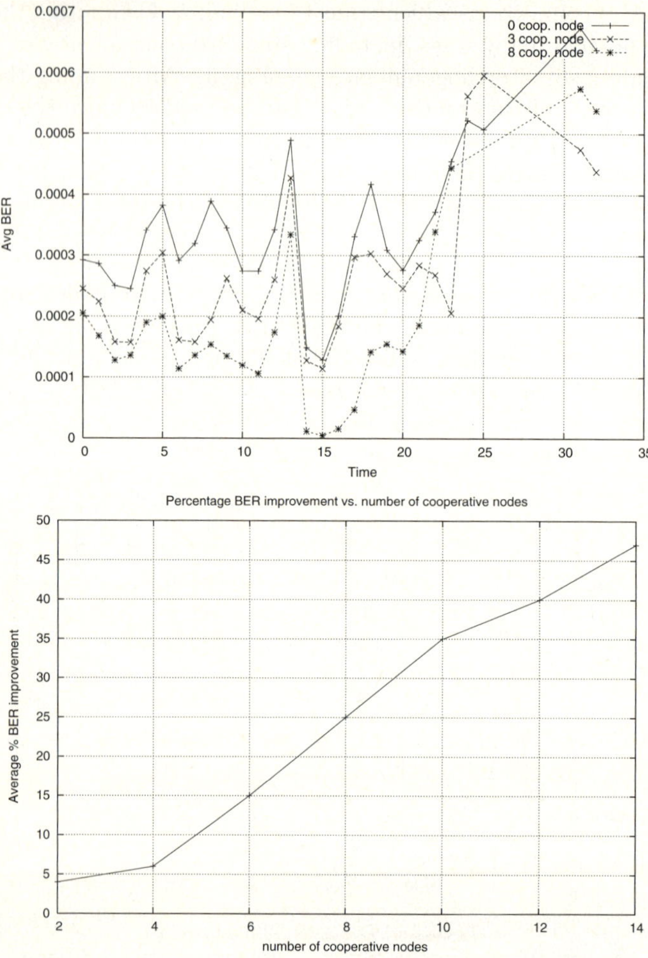

Fig. 5. The benefits of increasing the number of cooperative nodes

The first experiment investigates the effects of increasing the total mobility budget while keeping the number of cooperative nodes fixed. The simulation setup for the top graph of Fig. 4 consists of 15 autonomous nodes moving according to a Gauss-Markov process, and 8 cooperative nodes with mobility cost equal to one unit per meter; all nodes reside inside a one square kilometer grid. Node transmit power and receiver sensitivities are set so that wireless channels arise whenever two nodes are at distance less than 100m. Command and Control establishes 7 random connections and sets their target Quality of Service to be 60% of their initial BER value of the connection. The top graph shows that having higher mobility budgets permits the routing and optimization layer to achieve lower connection BER over time. The bottom chart of Fig. 4 depicts this effect in greater detail by considering the same experimental scenario but with varying mobility budget. The graph shows that a mobility budget of 50 units permits the routing and optimization layer to lower average connection BER by almost 8%, and that increasing the mobility budget to 250 units enables BER reduction of almost 40%. The results indicate that connection BER can be improved almost linearly as the mobility budget increases, even under constant numbers of cooperative nodes.

The second experiment investigates the effects of increasing the number of cooperative nodes while keeping the total mobility budget fixed. The simulation setup for the graph in Fig. 5 consists of 15 autonomous nodes moving according to a Gauss-Markov process, and 0, 3 or 8 cooperative nodes with mobility cost equal to one unit per meter; all nodes reside inside a one square kilometer grid. The mobility budget is fixed at 250 units. Node transmit power and receiver sensitivities are set so that wireless channels arise whenever two nodes are at distance less than 100m. Command and Control establishes 7 random connections and sets their target Quality of Service to be 60% of their initial BER value of the connection. The top graph shows that having more cooperative nodes permits the routing and optimization layer to lower BER more effectively over time, even when the mobility budget is not increased. The bottom chart of Fig. 5 depicts this effect in greater detail by considering the same experimental scenario but with varying numbers of cooperative nodes. The graph shows that with 4 cooperative nodes, the routing and optimization layer can lower average connection BER by almost 8%, and that increasing the number of cooperative units to 12 enables BER reduction of almost 40%. The results indicate that connection BER can be improved almost linearly as the number of cooperative nodes increases, even under constant total mobility budgets.

6   Conclusion and Future Work

The cost-benefit framework of the Cooperative Mobility Model is able to capture MANETs in which nodes exhibit a wide range of autonomy with respect to their mobility. Initial experiments using the CoopSim software demonstrate that with even modest mobility budgets and a few cooperative nodes, it is possible to leverage communication-reactive mobility control in a way that significantly improves MANET communications. The Resultant Algorithm is a promising initial approach towards a distributed mobility planning scheme. Increasing mobility budgets increases the potential benefits of cooperation, while increasing the number of cooperative nodes improves the efficiency with which a mobility budget can be leveraged. Our results are a significant step towards improving MANET operations in battlefield, response & rescue, and contexts involving time-critical mission-oriented deployments of mobile users.

In future work, we will conduct systematic investigations using the CoopSim platform. We will design provably robust and distributed algorithms which leverage mobility in MANETs under the Cooperative Mobility Model, and further evaluate their scalability and performance using both analytic techniques and realistic simulation experiments.

References

1. Lewis, P.J, Torrie, M.R., Omilon, P.M.: Applications Suitable for Unmanned and Autonomous Missions Utilizing the Tactical Amphibious Ground Support (TAGS) Platform. In: Gerhart, G.R., Shoemaker, C.M., Gage, D.W. (eds.) Unmanned Ground Vehicle Technology VI. Proceedings of the SPIE, vol. 5422, pp. 508–519 (2004)
2. Brown, T.X., Argrow, B., Dixon, C., Doshi, S., Thekkekunnel, R.G.: Ad hoc UAV-Ground Network (AUGNet). In: AIAA 3rd Unmanned Unlimited Technical Conference, Chicago, IL, pp. 20–23 (September 2004)
3. Frew, E., Dixon, C., Argrow, B., Brown, T.: Radio source localization by a cooperating uav team, Invited to AIAA Infotech@Aerospace, Arlington, VA (2005)
4. Li, N., Hou, J.C., Sha, L.: Design and Analysis of an MST-Based Topology Control Algorithm. In: IEEE INFOCOM (2003)
5. Khoshnevis, A., Sabharwal, A.: Network channel estimation in cooperative wireless networks. In: Canadian Workshop on Information Theory, Waterloo, Ontario (May 2003)
6. Buchegger, S., Boudec, J.Y.L.: Cooperative routing in mobile ad-hoc networks: Current efforts against malice and selfishness. In: Mobile Internet Workshop. Informatik 2002, Dortmund, Germany (October 2002)
7. Politis, C., Oda, T., Ericsson, N., Dixit, S., Schieder, A., Lach, H., Smirnov, M., Uskela, S., Tafazolli, R.: Cooperative Networks for the Future Wireless World. IEEE Communication Magazine (September 2004)
8. Gerharz, M., de Waal, C., Martini, P., James, P.: A cooperative nearest neighbors topology control algorithm for wireless ad hoc networks. In: Proc. 12th International Conference on Computer Communications and Networks (ICCCN'03), pp. 412–417 (2003)
9. Nosratinia, A., Hunter, T.E., Hedayat, A.: Cooperative Communication in Wireless Networks. IEEE Communication Magazine (October 2004)
10. Basu, P., Redi, J.: Movement Control Algorithms for Realization for Fault-Tolerant Ad-Hoc Robot Networks. IEEE Network (July 2004)
11. Raju, L., Ganu, S., Anepu, B., Seskar, I., Raychaudhuri, D.: Beacon assisted discovery protocol (bead) for self-organizing hierarchical ad hoc networks. In: IEEE Globecom Telecommunications Conference, Dallas, IEEE Computer Society Press, Los Alamitos (2004)
12. Laurer, G.: Packet Radio routing, ch. 11, pp. 351–396. Prentice Hall, Englewood Cliffs (1995)
13. Loyka, S., Gagnon, F.: Performance Analysis of the V-BLAST Algorithm: An Analytical Approach. IEEE Transactions onWireles Communications 3(4) (2004)
14. Proakis, J.G.: Digital Communications. McGraw-Hill, New York (2001)
15. Camp, T., Boleng, J., Davies, V.: A survey of mobility models for ad hoc network research. Wireless Communication & Mobile Computing (WCMC): Special Issue on Mobile Ad Hoc Networking Research, Trends and Applications 2(5), 483–502 (2002)
16. Friis, H.T.: Noise figures of radio receivers. Proc. IEEE 57, 1461–1462 (1969)

A Context-Dependent XML Compression Approach to Enable Business Applications on Mobile Devices

Yuri Natchetoi[1], Huaigu Wu[1], and Gilbert Babin[2,*]

[1] SAP Labs
Montréal, Québec, Canada
Yuri.Natchetoi@sap.com, Huaigu.Wu@sap.com
[2] Information Technologies, HEC Montréal
Montréal, Québec, Canada
Gilbert.Babin@hec.ca

Abstract. As the number of mobile device users increases, the need for mobile business applications development increases as well. However, such development is impeded by the limited resources available on typical mobile phones. This paper presents a context-dependent XML compression approach that enables the deployment of business applications on mobile devices. That is, the compressed XML document is not self-contained and cannot be de-compressed without using information shared between the sender and the recipient. By relying on shared information, we obtain a better compression ratio than existing context-free compression algorithms.

1 Introduction

Nowadays, mobile devices, especially cell phones are everywhere. According to a recent study [1], there are almost 2.5 billions of connected mobile devices in the world. Furthermore, the capabilities of the available phones are also increasing, making it possible to envision complex mobile applications. However, business-related applications running on cell phones are still rare. The major obstacles are the following: 1) limited data storage capability, 2) limited network access capability, 3) limited computation capability, and 4) limited display capability.

Most business applications require the client side to process large amounts of data either locally or through high-speed networks. Currently, most existing cell phones cannot fulfill these requirements. Typical solutions for large data issues in the world of desktop applications are compression and caching. Data are compressed in order to achieve fast transmission, and then de-compressed for access or caching. However, for mobile devices, the compressed data might still be too large for efficient transmission. More importantly, mobile devices lack an efficient way to de-compress the compressed data, and process or store

---

* On sabbatical leave at SAP Labs, Montréal.

A.-M. Kermarrec, L. Bougé, and T. Priol (Eds.): Euro-Par 2007, LNCS 4641, pp. 911–920, 2007.
© Springer-Verlag Berlin Heidelberg 2007

the de-compressed data. Hence, business application development for mobile devices requires an innovative approach to compress and de-compress data. In this paper, we propose to use context-dependent compression on business data. That is, the compressed data does not need to contain a lot of additional data for the de-compressor to interpret. De-compression relies on the presence of shared information between the sender and the recipient of the data. This approach was developed at SAP labs to enable building business applications on regular J2ME-enabled cell phones [2,3]. Data objects are transformed into XML format to be compressed on the server side, and clients de-compress the compressed XML file to restore data objects. The main contributions of our approach are that the context-dependent compression provides a compression rate for business objects exchange higher than existing compression algorithms, and the de-compression algorithm is very simple, which can easily adapt to the capabilities of most cell phones.

XML is frequently used to serialize objects and exchange them through the network. Also, many tools are currently available to compress XML to facilitate transferring and querying. These tools can be categorized as follows:

1. *General compression tools* like Winzip[4], Gzip[5], etc., which compress XML files as regular text files without considering the XML structure,
2. *XML compression tools* such as XMill [6], XMLPPM [7], XAUST [8], which provide better compression ratios than general tools by using the XML structure to optimize the compression results, and
3. *Query-able XML compression tools* including XPress [9], TREECHOP [10], and XGrind [11], which adopt a homomorphic transformation strategy [12] for compressing the XML structure and supporting queries without de-compression at the cost of compression efficiency reduction.

All the aforementioned compression approaches are context-free. The compressed XML document contains all the information required to perform the de-compression, and every transformation is therefore independent. However, in many business applications, object structures are usually predefined, shared on both server and client sides, and rarely changed during run-time. Based on this shared knowledge, the de-compressor can interpret a compressed file even when the XML structure information is not contained by the compressed file. Furthermore, the coding schema for the compression could be optimized to get better compression ratios, because the de-compressor already has some knowledge. The proposed solution combines coding XML file by multiple coding schemata and de-coupling transmission of XML structure information from data transmission. By using multiple coding schemata, we are able to reduce the number of bits required to represent the different symbols transmitted. De-coupling of structure transmission and data transmission drastically reduces the bits transferred, especially for short messages.

This paper is organized as follows. We describe the principles supporting XML compression in Section 2. In Section 3, we describe the proposed compression mechanism. Results from a comparative study with other compression approaches are presented in Section 4. We conclude in Section 5.

```
<PurchaseOrder no="1456">
  <Date>06/05/05</Date>
  <CustomerID>765345</CustomerID>
  <Order>
    <Item>
       <ProductNo>P-4534</ProductNo>
       <Quantity>2</Quantity>
    </Item>
    <Item>
       <ProductNo>P-9182</ProductNo>
       <Quantity>1</Quantity>
    </Item>
  </Order>
</PurchaseOrder>
```

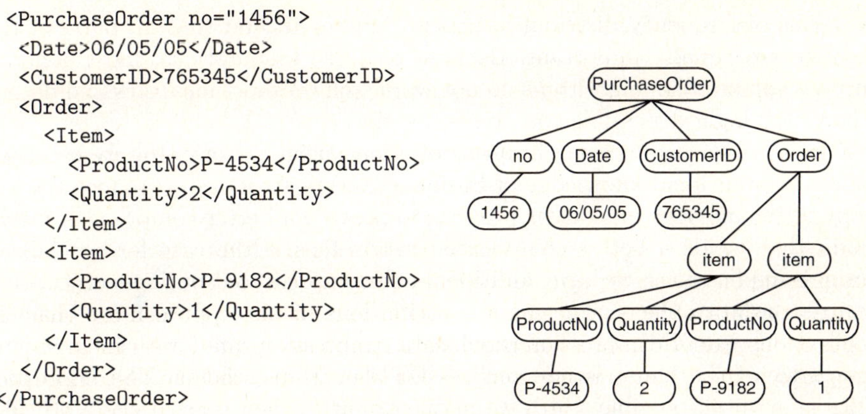

Fig. 1. XML Encoding of a Simple Object **Fig. 2.** Parse Tree for the XML Encoding

2 Principles Supporting XML Compression

Kropf *et al.* [13] have claimed that the information to be transmitted between two systems may be reduced when considering the knowledge (semantic, syntactic, etc.) shared by both systems. Consequently, only the information that is not shared needs to be exchanged, which leads to context-dependent compression.

2.1 Object-Oriented XML

Our compression approach is mostly used to exchange application data in XML format between server and client; messages are relatively small and usually contain a single data object or a collection of data objects. We use a simple algorithme to transform objects to XML files (many tools exist to perform this task, see SOX [14] and XStream [15]). First, the name of the object class becomes the root element of the XML file. Each simple attribute (number or text) of the object is transformed into an attribute of the root element. Each complex attribute (subordinate object) is transformed to a child element. For collection objects (vectors, hash tables, etc.), each contained object is transformed to a child of the wrapping element representing the collection object. This process continues recursively until the whole object has been transformed (see Fig. 1).

2.2 Structured Compression

All general-purpose compression algorithms, such as gzip [5], do not take into account knowledge of internal file structure, business-specific information or data types. They assume that the file to be compressed is composed of plain text with symbol (ASCII characters or their combinations) occurrence frequencies unknown in advance. These algorithms collect their statistics based on a limited data set, build an ad-hoc encoding dictionary, and attach this dictionary to the compressed file. These algorithms assume that every file/message sent is unique (in terms of frequency distribution), is unstructured, and that all parts of the message are equally important. Because of these assumptions, most general-purpose compression algorithms do not work well on short messages and do not achieve the highest possible compression ratios.

If we consider compression in terms of information entropy [16], we can then apply the additional knowledge of business objects structure to reduce the entropy of the message, making it possible to provide a better compression mechanism that yields a better compression ratio. This is the case for specialized business applications; we have additional information about business objects exchanged between the server and the client. For instance, since most business applications use XML as a universal data exchange format, we can use some knowledge of the file structure such as DTD or XML schema. The DTD provides us with all possible states we may encounter when parsing the XML file: tag names, attribute names, and values inside every attribute and inside every tag. The DTD also limits the type of information that may be encountered in each of these states. Consequently, we can determine the probability of meeting certain information in certain states, and can use this knowledge to reduce message entropy. For example, we know that inside a lexical unit delimited by symbols '<' and '>' there is a "very high" probability to encounter one of the tag names listed in the DTD. In fact, most XML compression algorithms use this approach (XMill [6], XGrind [11], Xpress [9]). To further reduce entropy, we build multiple encoding dictionaries, one for each state. This way, we can use longer lexical items (sequences of characters) and their occurrence frequency distribution. Every state has its own set of lexical items and therefore its own dictionary. If more than one state have the same data domain, we can reuse the dictionary. The decision to reuse a dictionary depends on the expected gain in compression.

The advantage of using state-specific dictionaries is that each dictionary is then relatively small, and the resulting encoding requires fewer bits per symbol. Indeed, in many situations, the number of available lexical items per state is finite. We can often list all possible values for these states. In other situations where some items do not exist in a dictionary, we can list the most frequent items, collected by statistical processing. In this latter case, we can introduce "escape" symbols, which are used to inform the parser that the following sequence of symbols is not a part of the dictionary, but a combination of lower-level symbols, like, for example decimal numbers or ASCII characters.

2.3   Combining XML Parser and Compressor

The XML parser implements a finite state machine that navigates the XML document tree (Fig. 2), as determined by an XML schema or DTD. This approach is somewhat similar to XAUST [8], where the compressor uses a recursive finite state machine, one machine per syntactic element. In the approach we propose, we replace the recursive finite state machines by multiple dictionaries with a single finite state machine. Every state corresponds to a specific syntactic unit (XML tag, attribute, value) where symbol combinations have different

Tree Path (key)	Dictionary elements (bits: value)
/	1 : PurchaseOrder
/PurchaseOrder	00: Date 01: CustomerID 10: Order 111: no 1101: ; 1100: \
/PurchaseOrder/Date	1: 06/05/05 0: \
/PurchaseOrder/CustomerID	1: 765345 0: \
/PurchaseOrder/Order/	1: item 0: \
/PurchaseOrder/Order/Item/	0: Quantity 11: ProductNo 101: ; 100: \
/PurchaseOrder/Order/Item/ProductNo	0: P-4534 11: P-9182 10: \
/PurchaseOrder/Order/Item/Quantity	0: 1 11: 1 10: \
/PurchaseOrder/no	1: 1456 0: \

Fig. 3. DTD-Specific Dictionaries of the PurchaseOrder Object

occurrence probabilities. The transitions correspond to the passage from one syntactic unit to the next. The machine can be presented as a graph with nodes corresponding to states and arcs corresponding to transitions. The graph can be automatically created from the DTD or from any formal syntax description such as BNF. Formally, the parser is defined as $F = (A, V, T)$, where A is the set of symbols read by the parser, V are vertices or states and T are transitions $(T : V \times A \to V)$.

Every state $(v_i \in V)$ has its own alphabet of lexical items (i.e., its own dictionary A_i). In fact, the dictionaries are built according to the parse tree following the DTD syntax. Lexical items can be single characters, words, tag names or possible choice options in the data field. Every lexical item $(a \in A)$ has a probability of appearance in the current state $P(a|v_i)$. This probability is determined by collecting statistics on a set of messages, which provides better results than using a single file, as many implementations of gzip and similar compression utilities do. In XML, symbols like: '<', '>', '"', and '&' play the role of delimiters which force a transition in the state machine. To enable the transition from one state to the next, the state-specific alphabet contains not only the set of possible lexical items but also the set of transition delimiters. Formally, $A_i = \{L_i\} \cup \{E_i\} \subset A$, where L_i are possible lexical items for the state v_i and E_i are symbols marking possible transitions leaving state v_i. These state transition delimiters also occur with a certain probability that can be calculated using statistical analysis. Then, every symbol from A_i can be encoded. Figure 3 shows the dictionaries generated based on the tree of Figure 2, where ';' is the end-of-state symbol and '\' is an escape symbol.

The compression module replaces every lexical item in the business object with a bit sequence. Figure 4 shows the resulting bit sequence to encode the XML stream of Figure 1. The length of the sequence depends on the encoding algorithm used (e.g., Huffman encoding [17,18], arithmetic compression [19]). At the end of a state, the encoder adds an end-of-state symbol (';'), which is also encoded in the same dictionary. If it can be determined from the data schema that there is only one lexical item in the state and that the next state

1 111 1 00 1 01 1 10 1 11 0 0 11 101 1 11 11 0 0 101 1101

Fig. 4. Compressed XML Stream for the PurchaseOrder Object

is predetermined, then the end-of-state symbol may be omitted. For example, XML attribute encoding presumes that each attribute name is always followed by an attribute value, and hence the transition system between the attribute name and its value can be omitted.

2.4   De-coupling Dictionaries from Business Messages

One characteristic of business applications is that many messages of the same type will have the same frequency distribution with high probability and therefore can share the same dictionary. This assumption allows the de-coupling of the dictionary from the message. De-coupling helps both with compression ratio and performance as dictionaries can be sent once for many messages, hence making messages shorter. Context-free compression techniques cannot make this assumption, and therefore cannot gain from past messages. Dictionary construction is the most time consuming part of the compression. By using a single dictionary for multiple messages, dictionary construction does not need to be performed for every message. Furthermore, it can be done on the server where the CPU is much more powerful that on the (mobile) client. Once compression dictionaries are compiled, they can be deployed to the client and be ready for a fast streamline compression and de-compression. It is also possible to collect frequency statistics during business operations and periodically optimize and re-compile dictionaries using new statistical data.

When comparing our compression approach to others, we must consider the fact that the dictionary is not sent with every message, but once in a while. Consequently, we must determine the average number of messages that are sent between dictionary updates. This will allow us to compute the *adjusted compression rate* (C_a) as:

$$C_a = \frac{N + pD}{M},$$

where N is the number of bytes sent in the compressed message, D is the number of bytes of the dictionary, p is the probability that the dictionary is sent with the message, and M is the number of bytes of the uncompressed message. Furthermore, if changes in the dictionary are not too significant, i.e. one additional option has been added to the possible attribute selections, then we can send differential updates for the dictionary, including only updated symbols. This is possible since dictionaries are only dependent on states. A change in one dictionary will not change others. Differential updates for business objects will be described in another paper. Every time we change the dictionary, we update its version. When the mobile client initializes the session with the server, it includes the version of local dictionaries into the session handshake. If the version is out of date, the server sends differential dictionary updates first.

3   Implementation

The compression algorithm consists of a compressor and a de-compressor. The compressor is implemented using J2SE and is running on the server. The de-compressor is implemented using J2ME and is running on mobile phones. The compressor's operations can be broken up in two stages. During the first stage, it collects frequency of occurrence statistics and builds a set of dictionaries. This is done by analyzing a sample set of XML files used in the business process. The compressor first determines the frequency of occurrence of all XML elements, attributes, and tag values by state, where every tag, attribute, and value constitutes a separate state. The alphabet for a specific state consists of all the lexical items allowable by the XML schema (L_i) plus two transition symbols (E_i): a generic end-of-state symbol (';') and an escape symbol ('\') for adding new values to the state without rebuilding the whole set of dictionaries. For a given state, we can choose a single generic dictionary to encode symbols not present in the state-specific dictionary, which requires a single escape character. These symbols are also considered in collecting the frequency distribution statistics.

Once the alphabet for each state (A_i) is defined and the frequency of each symbol is determined, the compressor builds a Huffman binary tree [17,18] for each state. These trees are in turn compressed using generic dictionaries also built by using the Huffman compression algorithm.

During the second stage of the compression, actual messages are compressed for transmission to the mobile device. This process requires the compressed dictionaries to be already deployed on mobile phones. The compressed message is wrapped into a standard XML header that has some attributes to identify the XML schema, the compression schema, the encoding, the encryption method, and some other parameters. The attributes can be easily extended because of the flexibility of XML. Clearly, the wrapping schema can also be compressed using the proposed approach, the compression dictionary required being pre-deployed with the compressor.

When de-compressing, the parser is combined with the de-compressor and therefore it has knowledge of the current state. The transitions are determined from the XML schema used. The de-compression algorithm is very simple and allows to uncompress and to de-serialize an XML stream into a business object in a single run, bypassing the conversion phase from binary format to text format. The general algorithm is illustrated as follows:

```
decompress(state,input)
    tree := findDictionaryTree(state)
    do
        token := getToken(tree,input)
        if (token <> end-of-state)
            nextState := getNextState(state,token)
            decompress(nextState,input)
        end-if
    while (token <> end-of-state)
end
```

The de-compression starts at the root element of the XML document (the initial state). Function `findDictionaryTree()` can readily identify the dictionary for this element as dictionaries are organized in a tree structure that follows the DTD syntax. As we move from state to state, we are moving from one dictionary to another. The de-compressor then reads bits from the input stream and immediately converts them into lexical items using the Huffman binary three for the current state (`getToken()`). This is accomplished by reading one byte from the stream and then traversing the Huffman tree one node at a time by shifting single bits. If we reach the end of the current byte, we retrieve the next byte from the input stream. When a leaf node in the Huffman tree is reached, we determine the corresponding lexical item and return it to the parser. The parser changes its state depending on the lexical item identified (`getNextState()`). This method simply implements the transition table T for the current state. We then call the `decompress()` method recursively from this new state. This process is repeated sequentially until the end-of-state symbol for the root element is reached.

This algorithm only requires 1) one encoded tree that reflects the structure of the XML file (based on the DTD or XML schema) and 2) a number of small binary trees that determine Huffman codes for every possible lexical item in each state. The transition from one node to another in the Huffman code tree is similar to the transition in the XML structure and therefore the decoding module is very compact and efficient. It is suitable for the computation capability of mobile phones.

The encoding algorithm is a little bit more complex; however in many business cases the amount of information transferred from server to the mobile device and from the device to the server is highly asymmetric. In this case, instead of encoding the whole XML file, the client can only send back the values modified by the user.

4   Comparison with Other Algorithms

In order to assess the quality of the proposed context-dependent compression algorithm, we compared it with a number of existing context-free compression algorithms, namely: bzip2 [20], gzip, and XMill. We executed these compression algorithms on a number of test cases on a Pentium 4 machine (CPU 3.4 GHz and 2 GB RAM) running Windows XP. The test cases included 7 typical messages used to exchange data between SAP mobile business applications and the SAP mobile infrastructure middle-ware. Each message contains a list objects; the smallest message contains 20 business objects ($3,742$ bytes) and the biggest message contains 436 business objects ($100,029$ bytes). For each test case, we measured the adjusted compression ratio (C_a). Better compression mechanisms will show smaller compression ratios. The results are presented in Table 1, which shows message sizes for the smallest (m_{min}) and largest (m_{max}) messages, and the average (μ) and standard deviation (σ) of C_a over all the test cases.

When comparing gzip and XMill, we see that gzip performed better for smaller messages, but worse for bigger messages. These results are consistent with results from [21]. Table 1 shows that gzip has a better average compression rate than

Table 1. Comparison of compression algorithms; m_{min} is the smallest message, m_{max} is the largest message. μ and σ are the average and standard deviation (σ) of the adjusted compression ratio C_a, p is the probability of sending the dictionary with the message.

	p	Size of m_{min} (Bytes)	Size of m_{max} (Bytes)	μ	σ
Original file		3742	100029	1	0
Context-dependent compression	0.00	127	4108	0.0422	0.0088
	0.10	535	4516	0.0732	0.0333
	0.33	1489	5470	0.1457	0.1260
	0.50	2169	6150	0.1974	0.1923
bzip2		772	8417	0.1520	0.0725
gzip		648	10864	0.1603	0.0617
Xmill		919	9189	0.1690	0.0903

XMill since our test cases tend to be small, which is typical for object exchange in mobile applications. The results also support our claim that context-dependent compression performs better than other algorithms. When all the dictionaries are sent once every 10 messages ($p = 0.1$), the average compression rate is two times better than other compression algorithms. Only when the dictionaries are sent once every 3 messages ($p = 0.33$) is the compression rate close to the other algorithms. Considering the nature of business applications where multiple messages of the same nature are sent between client and server in a short time interval, p is much closer to 0 than to 0.33.

5 Conclusion

The context-dependent compression method provides a high compression ratio and a simple de-compression algorithm that works reasonably on very simple mobile devices. This compression approach performs very well for mobile business applications where an intensive exchange between server and client takes place, using short XML messages, and when the structure of the XML documents is known in advance. This method will probably not prove as efficient for unstructured files or even for XML files having diverse schemata and frequency distributions.

One of the important advantages of this method is a simple de-compression algorithm combined with a simple XML parser. This approach better suits mobile devices, which neither have powerful CPUs to de-compress data nor enough memory to keep all the files uncompressed. Our proposed method requires more resources on the compressing side and a more complex infrastructure for dictionary management, however these requirements are justified for an enterprise environment where the servers normally have high CPU resources. In addition such an environment typically has a business process work-flow that is already in place which makes it simple to manage the synchronization of dictionaries. Overall, there is an increasing demand for the efficient compression methods for mobile client-server applications, which is growing rapidly with the growth of the mobile application market. The context-dependent compression method described in this paper provides good results for this niche and potentially can be used in many applications.

References

1. Taylor, C.: Global mobile phone connections hit 2.5bn (September 2006), `http://www.electricnews.net/frontpage/news-9792607.html`
2. Wu, H., Natchetoi, Y.: Mobile shopping assistant: Integration of mobile applications and web services. In: WWW 2007, Banff, Alberta, Canada, pp. 1259–1260 (May 2007)
3. Dagtas, S., Natchetoi, Y., Wu, H., Shapiro, A.: An integrated wireless sensing and mobile processing architecture for assisted living and healthcare applications. In: HealthNet 2007, Puerto Rico, USA, (June 2007)
4. Winzip (last visit, January 2007), `http://www.winzip.com/`
5. Gailly, J., Adler, M.: gzip 1.2.4 (last visit, January 2007), `http://www.gzip.org/`
6. Liefke, H., Suciu, D.: XMill: An efficient compressor for XML data. In: Proc. of the ACM SIGMOD Int'l Conf. on Management of Data, pp. 153–164. ACM Press, New York (2000)
7. Cheney, J.: Compressing XML with multiplexed hierarchical PPM models. In: Proc. of the IEEE Data Compression Conf., pp. 163–172. IEEE Computer Society Press, Los Alamitos (2000)
8. Harichan, S., Shankar, P.: Evaluating the role of context in syntax directed compression of XML documents. In: IEEE Data Compression Conf., p. 453. IEEE Computer Society Press, Los Alamitos (2006)
9. Min, J.K., Park, M.J., Chung, C.W.: XPRESS: A queriable compression for XML data. In: Proc. of the ACM SIGMOD Int. Conf. on Management of Data, ACM Press, New York (2003)
10. Leighton, G., Müldner, T., Diamond, J.: TREECHOP: A tree-based query-able compressor for XML. In: Proc. 9th Canadian Conf. on Inf. Theory, pp. 115–118 (2005)
11. Tolani, P.M., Haritsa, J.R.: XGRIND: A query-friendly XML compressor. In: IEEE Proc. of the 18th Int'l Conf. on Data Engineering, IEEE Computer Society Press, Los Alamitos (2002)
12. Hopcroft, J.E., Ullman, J.D.: Introduction to Automata Theory, Languages and Computation. Addison-Wesley, Reading (1979)
13. Kropf, P., Babin, G., Hulot, A.: Réduction des besoins en communication de corba. In: Colloque Int'l sur les Nouvelles Technologies de la Répartition (NOTERE'98), Montréal, Canada, pp. 73–84 (October 1998)
14. Sox: Schema for object-oriented xml (last visit, January 2007), `http://www.w3.org/TR/NOTE-SOX/`
15. Xstream (last visit, January 2007), `http://xstream.codehaus.org/`
16. Shannon, C.: A mathematical theory of communication. Bell Syst. Tech. Journal 27, 398–403 (1948)
17. Huffman, D.: A method for the construction of minimum-redundancy codes. In: Proc. of the I.R.E. (1952)
18. Cormen, T.H., Leiserson, C.E., Rivest, R.E., Stein, C.: Introduction to Algorithms, 2nd edn., vol. Section 16.3. MIT Press and McGraw-Hill, Cambridge (2001)
19. MacKay, D.J.: Information theory, inference and learning algorithms. CUP (2003)
20. bzip2 (last visit, January 2007), `http://www.bzip.org/`
21. Ng, W., Yeung, L., Cheng, J.: Comparative analysis of XML compression technologies. World Wide-Web Journal 9(1), 5–33 (2006)

A Distributed, Leaderless Algorithm for Logical Location Discovery in Specknets

Ryan McNally and Damal K. Arvind

Research Consortium in Speckled Computing, School of Informatics,
Edinburgh University, Edinburgh, UK
`{rmcnally,dka}@inf.ed.ac.uk`

Abstract. A speck is intended to be a miniature (5x5x5mm) device that combines sensing, processing, wireless communication and energy storage capabilities [1]. A specknet is an ad-hoc mobile wireless network of specks. The logical location of specks in the network is useful, for reasons ranging from routing data to giving the data sensed a spatial context. This paper presents a novel algorithm for discovering the logical location of specks and updating that information in the face of movement, without recourse to infrastructure support. The proposed algorithm exploits the location constraints implied by the neighbourhood links in order to compute a likely location: one hop neighbours must lie within radio range, two-hop neighbours probably lie outwith radio range. An iterative approach is used to converge on a location estimate that satisfies all constraints. The performance of the location discovery algorithm is evaluated in the SpeckSim simulator for a number of metrics, including location error. The results demonstrate that the quality of the computed locations is within 90% of optimal when used in routing calculations.

1 Introduction

A speck is designed to combine sensing, processing, wireless networking capabilities and a captive power source, all packaged in a volume no larger than a matchstick head. The specknet is a programmable computational network of a large number of specks; in effect, a fine-grained distributed computation platform built on an ad-hoc wireless network substrate consisting of resource-constrained nodes. In addition, the model of distributed computation takes into account some specific features of specknets, such as the unreliability of wireless communication, and a higher than normal failure rate of specks due to the harsh operating conditions, meagre power supply, and less than well tested nodes thanks to large volume manufacturing. This paper addresses the requirement to map the data being sensed, and the information subsequently extracted, to its location within the specknet. In an earlier work [2], we proposed a distributed relaxation-based algorithm for logical location maintenance for resource-constrained mobile wireless ad-hoc networks such as specknets. The algorithm proposed in this paper for logical location discovery of nodes, combined with the earlier published algorithm for location maintenance in mobile ad hoc networks, results in a unique

A.-M. Kermarrec, L. Bougé, and T. Priol (Eds.): Euro-Par 2007, LNCS 4641, pp. 921–930, 2007.

set of methods for logical location in specknets without the need for any special infrastructure support. Some of the challenges of developing location algorithms for specknets include: specks with limited resources, both in terms of energy and memory; unreliability of the nodes and the communication between them; and the requirement for the algorithm to be fully distributed in the interests of scalability and robustness.

2   Related Work

Algorithms for logical location discovery in wireless sensor networks can be classified along the following lines: whether any special infrastructure is required for the functioning of the algorithm; whether the algorithm relies on a central computational resource; what information is available to the nodes, e.g. connectivity, distance between nodes; whether the algorithm running on all the nodes is homogeneous, i.e. whether all the nodes are equal or some are more equal than others.

The algorithm due to Bulusu et al. [3] could be described as using infrastructural support, or as a non-homogeneous network. The approach uses a number of super-nodes with known location (derived from external means or are fixed with location information preprogrammed). These super-nodes periodically broadcast their location, and all the other nodes will set their location as the average of the received locations. This technique is simple and reliable, and can be easily improved by taking received signal strength information into account, or location history in the case of mobile nodes. The drawback is that the number and placement of the super-nodes is critical: ideally, the super-nodes should be deployed in a regular grid, with interlocking areas of radio coverage, which is difficult to guarantee in an ad-hoc network deployment.

Two algorithms that use a central computational resource are described by Doherty et al. [4] and Shang et al. [5]. Both approaches are similar in that they function by gathering the connectivity data of the entire network into the central computer, which performs the necessary computation and relays each node's location back into the network. They differ in the method of performing the computation.

Doherty's technique constructs a set of constraints on each node's location by examining network links. A radio link between two nodes implies that these nodes must be within radio range of each other, which is then used to construct a distance constraint. The set of constraints, in conjunction with the locations of super-nodes, can be used to compute the likely locations of all the nodes in the network.

The approach due to Shang, dubbed MDS-MAP, computes the shortest network path between every pair of nodes in the network. The number of hops in these paths is likely to correspond to the euclidean distance between the two nodes, at least in networks with a uniform distribution of nodes. Once the distances between every pair of nodes is known, multidimensional scaling is used to compute the relative positions of every node in the network. This relative map is useful internally in the network, for tasks such as message routing, and can be combined with knowledge of the absolute locations of a subset of the nodes, in order to reconcile the relative map with the physical locations of the nodes.

The advantages of using a central computing resource for location discovery are considerable. The algorithm can be computationally intensive as this computer need not be resource-constrained, and working on a network-wide view of the data. But the disadvantages are also substantial. The process of routing both incoming connectivity data and outgoing location data will in practice result in communication hotspots in the network around the central computer. As a result the drain on the batteries on the nodes closer to the central computer will be greater, and in due course the central computer may becomes unreachable as the nodes around it are exhausted. Another drawback is the latency between when a node transmits its connectivity data and receives its computed location. The data must make two full traversals of the network, and undergo substantial $(O(n^3)$ for Shang, $O(k^2)$ or $O(k^3)$ for Doherty, for n nodes and k network connections) processing in the central computer. These delays would limit their use in highly mobile networks.

Shang and Ruml have proposed an improvement to their MDS-MAP algorithm [6] that distributes the MDS-MAP process across the network, with every node acting as the central computer for its local neighbourhood, and using the MDS-MAP algorithm to compute a relative location map. These local maps are then stitched together by examining the locations of nodes common to pairs of maps, and computing a transform that reconciles them.

This improvement mitigates the drawbacks of having a central computer, and also improves the performance in cases of non-uniform node distribution. However, the computation and memory requirements are significant, and will scale badly as the density of the network increases. On a resource-constrained platform, as nodes are likely to be, these costs may be prohibitive. Another potential problem is that the relationship between network hops and euclidean distance becomes less accurate as the hop count decreases.

3   Algorithm Description

The algorithm exploits the constraints on possible locations implied by network links but, unlike the algorithm due to Doherty et al. [4], this approach is fully decentralised and runs on a homogeneous network.

It is important to note that the algorithm will generate a coordinate system that is internally consistent to the network, and can be converted with an affine transformation to reflect the nodes' physical locations. Computing the corrective transformation is simply a matter of examining the real and computed locations of a small subset of the nodes.

The following subsections describe the intuition behind the algorithm, with implementation details left to section 4.

Constraints. As illustrated in Fig. 1, a node that knows its location implies inclusive constraints on the possible locations of its one-hop neighbours, i.e.

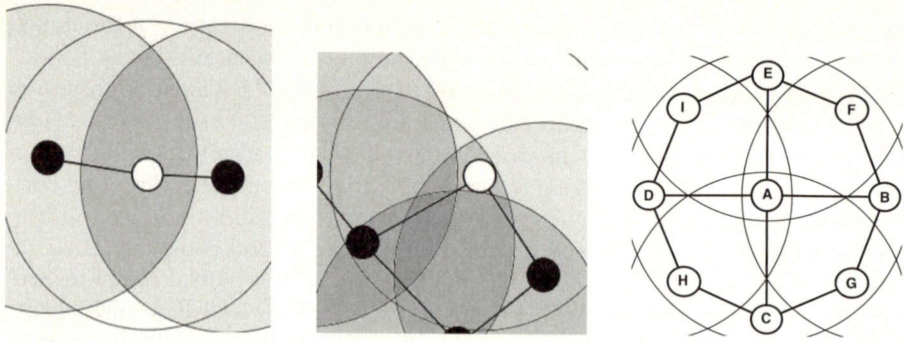

Fig. 1. Loose and tight constraints on node location, and a bridged kernel formation

they must lie within radio range, and exclusive constraints on the locations of its two-hop neighbours, i.e. they probably lie outside the radio range. Once a node detects that its range of possible locations is sufficiently constrained, it commits to a location, and in turn, implies new constraints on its one- and two-hop neighbours. Thus the knowledge of locations will grow from a small kernel of located nodes to encompass the entire network. Even after a node has committed to a location, its estimate is further refined using a numerical relaxation method whenever further constraints are detected.

Kernel Formation. The problem of identifying the initial kernel of located specks is solved by exploiting the idea of mutual ignorance in a neighbourhood. Every node searches for the configuration in which there are four neighbours who do not appear on each others' neighbourhood lists. Once such a configuration is found, it can be asserted that the mutually-ignorant nodes are likely to lie in a 'cross formation', as shown by nodes B, C, D and E in Fig. 1. A further requirement for a valid kernel formation is the presence of bridging nodes F, G, H and I that disambiguate which nodes lie on adjacent arms of the cross formation. Once the formation has been detected by node A, it broadcasts the locations of the mutually-ignorant nodes. The precise values of the computed locations of these nodes are not particularly important, as it will only affect the values of the corrective transform used to reconcile relative and physical locations.

Kernel Incidence. It is probable that more than one kernel will be identified in the network, which will result in a number of competing coordinate systems that must be reconciled. The approach taken here is to have one system take precedence over all the others, based on the identifier (ID) of the original seed node (A in Fig. 1). Normally, nodes will always attempt to locate themselves in the system that has the lowest seed ID, and one system will grow to subsume all the others. However, in practice, it can be the case that the system with the lower seed ID is a worse candidate for growth than that with the higher seed ID, due to factors such as sparseness of the local network. In these cases the growth will stop, and the network will have two distinct coordinate systems. These can be reconciled using coordinate system stitching techniques described by Shang et al. [5], and Moore et al. [7].

If the network is deployed in such a manner as to preclude the possibility of a kernel forming, then an artificial kernel can be introduced into the network to start the growth of a coordinate system. All that is required is a small cluster of nodes that have some idea of their locations relative to each other, and the growth of a coordinate system can be triggered.

Guaranteeing Growth. The growth process might stall as the constraints implied on the unlocated nodes are not tight enough for them to commit to a location. This is more likely to occur in sparse networks. This problem is addressed by lowering the threshold for each node to commit to a location. For example, under ideal circumstances the number of located neighbours that a node needs in order to commit to a location might be set to eight. This value is progressively reduced while that node's neighbours report no changes to their location status. The value is reset to the default value as soon as a neighbour's location status changes. This means that the initial estimate of a node's location is more error-prone, but the estimate is later refined as more neighbours commit to a location and imply further constraints.

Seed Death. A potential problem with algorithms in which unreliable nodes can take on a special role lies in the consequences of that node's demise. Fortunately, the special role of the seed node is vanishingly brief. Once the broadcast that sets the locations of the nodes involved in the kernel formation has been made, the seed node has no further responsibilities and can die without notable adverse effects.

Motion. The data shared by each node with its neighbours is a superset of that used in the Distributed Relaxation algorithm for location maintenance [2]. This allows the nodes to maintain their location information in the face of movement with no additional overhead.

Clique Networks. It is important to note that this algorithm is unable to function in degenerate cases such as a clique network, i.e. a fully-connected network in which every node is able to communicate directly to every other node. The only inference that can be made by examining the connection data of such a network is that all nodes are within radio range of each other, which provides a localisation so coarsely-grained as to be essentially useless.

4   Implementation

A fundamental requirement of the algorithm is that every node share information about its neighbourhood with its one-hop neighbours by making repeated broadcasts. The broadcasts' content and the data stored on each node depends on the progress of the coordinate system growth.

Before a kernel is detected, each node maintains a list of its one- and two-hop neighbours' IDs, and periodically broadcasts a list of its one-hop neighbours'

IDs. This neighbourhood information is all that is required to detect the kernel formation.

Once a kernel has been detected and system growth has begun, nodes discard the two-hop neighbour information and concentrate on their one-hop neighbours. The nodes maintain a list of each neighbour's ID, location, and the ID of the seed node that each neighbour is located with respect to. The broadcasts will contain the same information, along with the details of the sender. On receipt of a broadcast a node will update its records with the senders details. On examination of the received data, if none of the neighbours have committed to a location, the node will attempt to find a kernel formation. If the search is successful, it sets its own location to be at the origin and broadcasts a message ascribing appropriate locations to the mutually-ignorant nodes.

If the node's neighbourhood contains members that have committed to a location, then the node will scrutinise the constraints implied and determine if it can also safely commit to a location. In general, more located neighbours lead to tighter constraints, and so a convenient way to estimate the tightness of constraints is simply to count them. In all cases, preference will be given to locating with lower-ID seeds.

The procedure for committing to a location uses a simple numerical technique for finding suitable coordinates. Initially, the node will set its location to be the mean of the locations of its one-hop neighbours. This initial estimate is then refined by a sequence of relaxation operations. Each constraint on its location is satisfied in turn by applying the least possible change to the location estimate, i.e. if the location estimate is outside the radio range of a one-hop neighbour, the estimated position is moved towards that neighbour's reported location; conversely, if the estimate is within radio range of a two-hop neighbour, it is moved away. These operations are applied iteratively so that the location estimate will converge on a state in which all constraints are satisfied.

Nodes will only update their neighbourhood records with the details of a sender. Second-hand information (about the sender's neighbourhood) is used to refine the node's location estimate through the relaxation procedure, but is not stored for later use.

5   Metrics and Results

SpeckSim is an event-driven behavioural simulator for wireless networks [8]. The simulation results were based on a typical configuration of 200 nodes distributed randomly across a unit square. Radio transmissions are assumed to have a circular propagation with a range of 0.2 units. Each node will attempt a transmission once every second. The MAC protocol used on the nodes is trivially simple: when initiating a broadcast, a carrier sense check is made; if another transmission is detected, the node will wait for a random period before attempting a re-broadcast. Simulations were repeated 100 times for each scenario and the mean and standard deviation of the results were calculated.

It should be noted that the circular radio propagation model is convenient and simple to simulate, but does not reflect the behaviour of a radio broadcast in practice. The effective range of a radio broadcast can vary due to multi-path and fading effects, interference with other broadcasts, and presence of conductive materials. This could result in asymmetric links in the network, i.e., node A can receive broadcasts from node B, but not vice versa. This will cause problems with the relaxation procedure used to refine location estimates, i.e., node A will move its location estimate to within radio range of node B upon receiving B's broadcast, and node B will move its location estimate away from node A when it is informed of A's location by a third node, and the cycle begins again. Thus node A will appear to chase node B around in circles.

This problem can be fixed at the cost of some bandwidth by having each node append the neighbourhood lists of its one-hop neighbours to their broadcasts. Therefore a receiving node can detect when it is part of an asymmetric link and terminate the looping behaviour. The additional bandwidth costs can be assuaged by using Bloom filters [9] to compress the neighbourhood lists at the cost of a negligible margin of error.

Performance is primarily judged by location *Error*, i.e. the distance between a node's actual and computed locations, disregarding the effects of any network-wide skew transformation.

Secondary metrics are based on the impact of location data on location-based routing [10]. A simple greedy algorithm is used to route data between two distant nodes; in the first case, perfect knowledge of the locations of the nodes is used, whereas in the second case location data provided by the algorithm is employed. The two routes are then compared and the following statistcics computed:

Success rate. The number of successful routes found using computed locations, expressed as a fraction of the number of successful routes found using perfect knowledge.

Efficiency. The quotient of the hop-lengths of the two routes. A figure less than one implies that the routes discovered with computed locations were longer than those discovered with perfect knowledge.

The location-based routing metrics highlight that there is much useful information in coordinate systems with large location errors. For example, if the system has a non-uniform scaling factor, i.e., it is stretched in one direction, then the location error will be large, but routing information remains unaffected.

Another metric worth considering is the number of located specks: the proportion of nodes in the network that have been successfully located with respect to a single seed. For example, if the network was split down the middle, and each half had a separate coordinate system, then the metric would be 0.5.

Kernel Incidence. The first graph in Fig. 2 charts the incidence of kernel formations in a random network layout as the network density increases. It shows that there is an abundance of kernel formations for all but the sparsest of networks.

Time. The graph on the right in Fig. 2 shows the performance of the algorithm over time. It can be seen that the network converges into a single coordinate

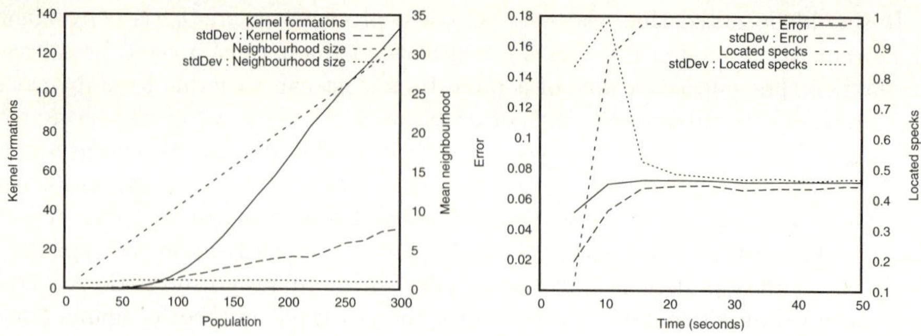

Fig. 2. Kernel incidence and performance with time

system in approximately 15 seconds. The stable error level indicates that the nodes have located themselves to the best of their ability, with all constraints having been satisfied. The routing metrics (not shown) show a steady route success rate of around 0.9, with a route efficiency of around 1.0.

Sparse Networks. Figure 3 investigates the effects of sparseness, which shows that a stable error rate is achieved even in sparse networks, with around 80% of nodes being located with respect to one seed. The location-based routing statistics indicate that when a successful route is found, its efficiency is very close to that of a route found using perfect location information.

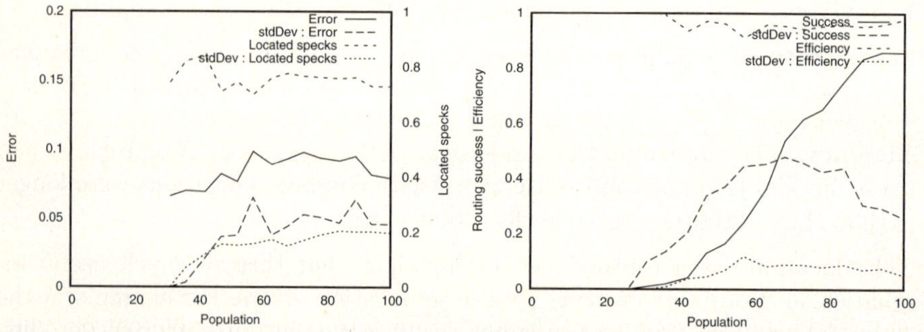

Fig. 3. Sparse networks

Bandwidth Conservation. The previous graphs have been generated without taking the communications bandwidth into account. This is acceptable for static networks as the message frequency can be reduced for low-bandwidth communications. However, communication is one of the most expensive operations for a speck from the point of view of power consumption, and so algorithms must strive to keep messages as terse as possible. The default behaviour of the algorithm is to simply broadcast a node's entire neighbour list. Fortunately, we can place a cap on the number of neighbours for which full details are broadcast without significantly degrading performance. This dramatically reduces the size of each message, and further savings can be made by limiting the precision of location information. Taking these two savings together, Fig. 4 shows that limiting the broadcast neighbourhood details and their precision, will lead to an increase in performance over the default behaviour in low-bandwidth situations, in addition to bandwidth savings of around 60%. This is due to lower channel utilisation and hence a higher message delivery rate. The first graph charts performance and message size as the neighbour limit is increased, while the second graph shows the performance of the algorithm with no limits, and with the number of neighbours limited to 8 and a precision limit of 8 bits.

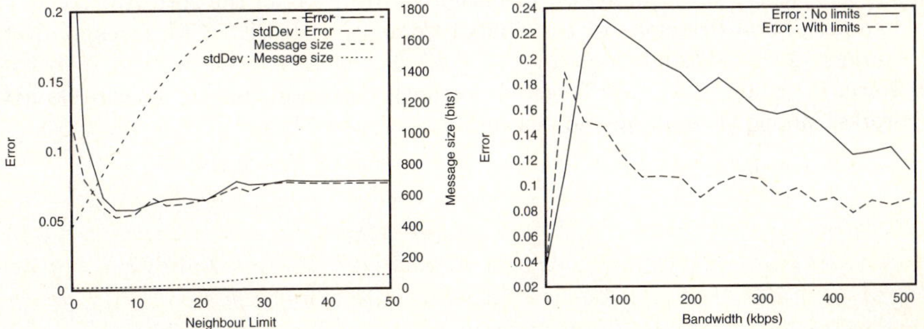

Fig. 4. Bandwidth effects

6   Conclusion and Future Work

We have described a novel method for logical location estimation for mobile networks of resource-constrained devices. This method uses only binary connection information, as opposed to received signal strength or other methods of range estimation. The method does not use any special infrastructure nodes such as beacons with known positions. In addition it is entirely distributed, running on a homogeneous network of nodes. Simulation results have been presented to demonstrate the performance of the algorithm with respect to different parameters. For future work we intend to investigate methods by which disparate coordinate systems may be reconciled, and also the imapct of mobility, unreliable and non-uniform transmissions and obstructions will have on the network formation.

References

1. Arvind, D.: Speckled computing. In: Technical Proceedings of the 2005 NSTI Nanotechnology Conference (May 2005)
2. McNally, R., Wong, K., Arvind, D.: A distributed algorithm for logical location estimation in speckled computing. In: Proceedings of IEEE Wireless Communications and Networking Conference, March 2005, IEEE Computer Society Press, Los Alamitos (March2005)

3. Bulusu, N., Heidemann, J., Estrin, D.: Gps-less low cost outdoor localization for very small devices. IEEE Personal Communications Magazine 7(5), 28–34 (2000)
4. Doherty, L., Pister, K.S.J., Ghaoui, L.E.: Convex position estimation in wireless sensor networks. In: Proc. of the IEEE Infocom, pp. 1655–1663. IEEE Computer Society Press, Los Alamitos (2001)
5. Shang, Y., Ruml, W., Zhang, Y., Fromherz, M.: Localization from mere connectivity (2003)
6. Shang, Y., Ruml, W.: Improved mds-based localization (2004)
7. Moore, D., Leonard, J., Rus, D., Teller, S.: Robust distributed network localization with noisy range measurements. In: Proceedings of Second ACM Conference on Embedded Networked Sensor Systems (SenSys '04), Baltimore, MD, USA, pp. 50–61. ACM Press, New York (2004)
8. McNally, R.: Specksim at www.specknet.org/dev/specksim
9. Mitzenmacher, M.: Compressed bloom filters. In: Proc. of the 20th Annual ACM Symposium on Principles of Distributed Computing. IEEE/ACM Trans. on Networking, 144–150 (2001)
10. Karp, B., Kung, H.T.: GPSR: greedy perimeter stateless routing for wireless networks. Mobile Computing and Networking, 243–254 (2000)

Analysis of a Kalman Approach for a Pedestrian Positioning System in Indoor Environments

Edith Pulido Herrera[1], Ricardo Quirós[1], and Hannes Kaufmann[2]

[1] Universitat Jaume I, Castellón, Spain
pulido@lsi.uji.es, quiros@lsi.uji.es
[2] Vienna University of Technology
Institute of Software Technology and Interactive Systems
Vienna, Austria
kaufmann@ims.tuwien.ac.at

Abstract. In this work we present the design principles of a wearable positioning system for users in unprepared indoor environments. We describe the most suitable technology for our application and we model the dynamics of a walking user. The system uses inertial sensors and a location system based on ultrawideband (UWB). Data fusion is carried out with a Kalman filter. The user position is estimated from data provided by the UWB location system. To update the position and direction of the user we use a dead reckoning algorithm. The use of redundant sensors and the data fusion technique minimises the presence of shadow zones in the environment. We show the advantages of combining different sensors systems.

1 Introduction

The precise determination of the position of users or objects in an environment is a crucial task for many applications. The task was inaccessible a decade ago, for precision location in mobile systems, due to the lack of technology. The development of micro-electromechanical systems (MEMS) has allowed the design of sensors based on the principles of the location systems found in boats or airplanes. We now find on the market several low cost, miniature gyroscopes or accelerometers. This evolution has facilitated the development of applications that use the position of the user, often known as Location Based Services.

There are a great number of applications that need to know the position of a user. Among them we can mention access and security control in enterprises or institutions, civil defense and rescue tasks, or the guidance of users in facilities such as museums or monuments. In the implementation of Location Based Services, the efficiency of the service depends strongly of the location system used.

Present technology allows the design of location systems precise enough for most applications. Systems able to operate in prepared indoor environments have been described. Optical, acoustic and magnetic technologies allow the precise location of a user in indoor environments. Their main disadvantage is that

A.-M. Kermarrec, L. Bougé, and T. Priol (Eds.): Euro-Par 2007, LNCS 4641, pp. 931–940, 2007.

they require complex, expensive infrastructures. On the other side, vision based systems are imprecise, and also require previous preparation of the environment, usually placing markers - fiducials - that assist the recognition process.

In this work we present the design principles of a system for locating users in ad hoc (unprepared) indoor environments. The system uses inertial sensors and specific data fusion techniques. We describe the most suitable technology for our application and we model the dynamics of a walking user. The user's position is estimated from data provided by the system based on UWB. The update of the position and direction of the user is done with dead reckoning algorithm. The use of redundant sensors and the data fusion technique minimises the presence of shadow zones in the environment.

2   Background

Over the past decade research in ubiquitous computing has been impelled by the demand of mobile applications. A fundamental component in mobile systems is the location system. The design of a location system implies the integration of different disciplines and technologies as needed for the application. In this work we propose the design of a location system suitable for unprepared indoor environments. We seek to avoid the use of complex infrastructures while preserving the reliability and robustness of the system.

At the moment, the most widely used technique to determine the position of a user is the combination of different technologies such as optic, acoustic or inertial technologies. There are other technologies, such as those used by cellular telephones (time delay) or the triangulation methods based on wireless networks. We did not consider these technologies because they are not precise enough for most applications. None of them is able to detect with exactitude the position of a user in a room. The main disadvantages of optical or acoustic technologies are that they require complex and expensive infrastructures that usually are not practical. Additionally, the presence of smoke, echoes or other phenomena that affect the transmission or reception of the signal, introduces errors. A direct line-of-sight between the receiver and the transmitter must exist.

Recently the High Sensitivity GPS technology (HSGPS) has been developed for the location of users in indoor environments. The technology solves the problems that conventional GPS reception presents in zones where its signal is obstructed or where there are multi-path effects. Nevertheless, the works of Lachapelle [11], Mezentsev et al, and Mezentsev and Lachapelle, [12], [13], show that HSGPS technology does not work as a stand-alone system, but rather it requires additional technologies for the precise location of users.

2.1   Related Work

The design of location systems based on the fusion of data estimated by different sensors has been described in many investigations. Golding et al.[2] present a system based on machine-learning techniques and multi-sensor data fusion.

They use accelerometers, magnetometers, temperature sensors and light sensors. They show that measuring the characteristics of the environment can help to determine the position of the user, using statistically significant information. The complexity of the method prevents its application in systems that operate in real time.

Gabaglio et al. [4] propose a navigation system based on a Kalman filter. It uses the information provided by magnetic compass, gyroscope, accelerometers and a GPS. Although the system complements GPS when the signal is blocked, the accuracy is lower for particular applications. Kourogi and Kurata [7] propose a location system for Augmented Reality (AR) applications. The user carries a system composed of self-contained sensors (accelerometers, gyroscopes, magnetometers and inclinometers) and a camera. They obtain the data needed to estimate the displacement, register images and fuse the information by means of a Kalman filter. The use of computer vision techniques requires a very large database that is also dependent on the application environment.

Stirling et al. [8] and Foxlin [9] propose location systems based on Inertial Measurement Units (UMI) installed on the feet of the users. The system detects the steps of the user to incrementally calculate its present position. Both works conclude that the system cannot work correctly as a stand-alone system. However, the work is a first step in the search of an integral solution.

3   System Description

In this work we propose a location system based on inertial sensors and an ultra-wideband (UWB) location system. In this section we will describe the hardware used to implement the system, the mechanisms of data gathering and fusion, and the dynamic model proposed for a walking user.

3.1   Hardware

Our goal was to design a system that returns the position (X, Y) and orientation (roll, pitch, azimuth) of human users in indoor environments. In order to select the most suitable sensors for our application we considered characteristics such as the precision, autonomy or the availability of the sensor. Based on these parameters, the following sensors have been selected:

1. **Ubisense:** a location system based on ultrawideband (UWB). It uses a network of sensors installed at well-known positions and a set of tags located in the moving users or objects of the environment. The system uses an Ethernet network for the communication between the different elements. This system can be very useful in large environments. However, it is not suited for environments where larger flat metallic surfaces exist which can cause reflections and therefore affect the transmission and reception of the sensor's signals. We use this system to estimate the positions (X, Y) of the elements to be located.

2. **InertiaCube2:** an Inertial Measurement Unit (IMU), which determines the direction of a user as three Euler angles (*roll, pitch, azimuth*). It has a static accuracy of $1°$ rms and a dynamic accuarcy of $3°$ rms, communication through USB or serial port and its dimensions are suited for mobile applications. In our system, we use this sensor to detect the orientation of the user's head. The sensor rests on a helmet that is worn by the user. See Figure 1.

3. **Xsense:** It is an Inertial Measurement Unit (IMU), which determines the direction like the InertiaCube2. It has a static accuracy less than $0.5°$ for roll and pitch and $1°$ for azimuth, and a dynamic accuracy of $2°$ rms. In our system, we use this sensor to estimate the values of linear acceleration and azimuth of the user. The sensor rests in a belt that is worn by the user. See Figure 1.

The information provided by the sensors is processed in a Tablet PC, as shown in Figure 1.

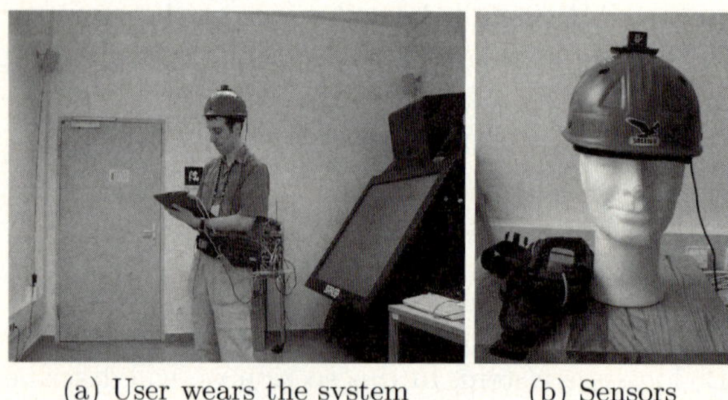

(a) User wears the system   (b) Sensors

Fig. 1. Prototype of wearable System

3.2   Data Reception and Fusion

To gather data from the sensors we use OpenTracker, an open source software package developed at the Vienna University of Technology and the Graz University of Technology. OpenTracker[1] can be used in real time applications and collects data with a timestamp. Opentracker defines the interface with sensors by means of modules implemented in C++. The different modules used are called from parameters defined by an XML configuration file, to gather the data from each sensor. After this data gathering process, the final result is obtained by means of a data fusion process.

Data fusion is a multidisciplinary field that considers all aspects from the modeling of the physical system to the final estimation techniques. Its main goal is to obtain an optimal estimation of a state vector, or vector of variables that allows predicting the behavior of the system. The selected estimation technique is the *Kalman Filter (KF)*. This filter has demonstrated its reliability in navigation

---

[1] More details are in [1].

systems (Brown and Hwang [6]). The Kalman filter inputs are the measured values of a set of parameters. These parameters define the observed state of the dynamic system and are stored in a vector. The filter uses the vector of observed parameters to make a prediction of the state in the next time step . In our work the dynamic system is a user walking around an environment.

3.3 Dynamic Model

The estimation of position is more difficult for a walking user than for vehicles or robots. The added difficulty is due to the unpredictable nature of its trajectory. In order to model the characteristics of the movement of a walking user we must use methods such as *Dead Reckoning* algorithms. In this class of algorithms we consider two fundamental parameters of the walking movement: *the length of the step* of the user and its *azimuth*. The estimation of the length of the step is made from acceleration patterns (Ladetto et al. [5]). Figure 2 shows the behavior of the vertical acceleration, where the peaks of the curve correspond with the foot strike.

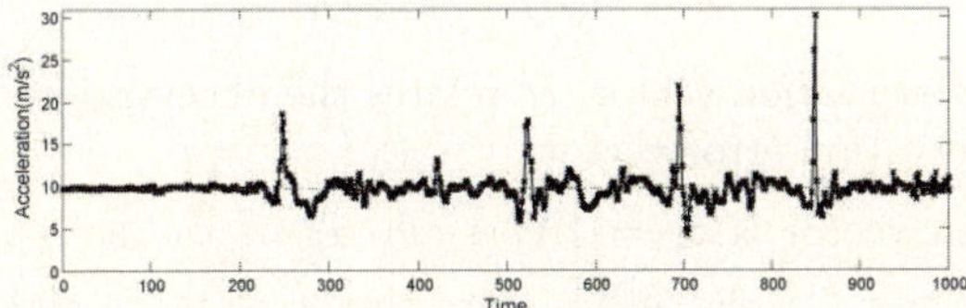

Fig. 2. Acceleration signal from Xsense, when a user walks

The next set of equations model the movement of a walking user. If we use another type of displacement (for example to jump or to slide) it becomes necessary to define another specific dynamic model. Since our system follows a 2D-navigation dynamic scheme, the equations are as follows:

$$X_k = X_{k-1} + s_k Cos\left(\psi_k\right)$$
$$Y_k = Y_{k-1} + s_k Sin\left(\psi_k\right)$$

(1)

where, X_k and Y_k are the information of position provided by the Ubisense, s_k is the length of the step and ψ is the *azimuth* or *heading*.

3.4 Kalman Filter

The data obtained from the Ubisense and the IMUs are integrated with a Kalman filter. The KF requires a model of the system to be defined. In other words, it must implement two models: *state model* and *observation model*.

State Model. The state equation of a Kalman filter is (Brown and Hwang [6]):

$$x_k = f * x_{k-1} + b * u_k + w_{k-1} \tag{2}$$

where, x_k is the state vector, f is the transition matrix that relates the state of a previous time to the current time, b relates the input control u_k and w_k represents the noise state vector.

The state vector in our system is $x_k = [X_k\ Y_k\ \theta_k\ \alpha_k\ \psi_k\ \phi_k\ \beta_k\ \gamma_k]$, the transition matrix $f = I_{8x8}$, $b = [Cos\,(\psi_k)\,Sin\,(\,psi_k)\ 0\ 0\ 0\ 0\ 0\ 0]^T$, the vector u_k represents the length of the step (Kourogi and Kurata [7]). $\theta_k\ \alpha_k\ \psi_k$, are the angles *roll, pitch, heading* from the waist's user respectively; $\phi_k\ \beta_k\ \gamma_k$ are the angles *roll, pitch, heading* from the head's user. Although we don't know the value of the noise at each state, we can approximate the state model as:

$$x_k = f * x_{k-1} + b * u_k \tag{3}$$

Observation Model. This model is defined according to the information provided by the sensors. Its general form is:

$$z_k = H_k * x_k + v_k \tag{4}$$

where, z_k, is the observation vector, H relates the observation and state vector, and v_k is the observation error vector.

The observation vector is $z_k = \left[\tilde{X}_k\ \tilde{Y}_k\ \tilde{\theta}_k\ \tilde{\alpha}_k\ \tilde{\psi}_k\ \tilde{\phi}_k\ \tilde{\beta}_k\ \tilde{\gamma}_k \right]^T$, $H = I_{8x8}$. In the equations (2) and (4), w_k and v_k, are the state noise and the observation noise. These noises are non-correlated, gaussian noises:

$$\begin{aligned} p(w)\ N\,(0,Q) \\ p(v)\ N\,(0,R) \end{aligned} \tag{5}$$

where, Q represents the *state noise covariance* and R represents the *observation noise covariance*.

The KF works in two phases: *the prediction* and *the correction*. In the first phase the filter update the state vector and the error covariance matrix P. The usual equation to calculate P is $P_k = f_k * P_{k-1} f_k^T + Q_{k-1}$. In the correction phase the state is updated by $\hat{x}_k = \hat{x}_k^- + K_k \left(z_k - H * \hat{x}_k^- \right)$, where K is the Kalman gain obtained by: $K = P_k^- H_k^T \left(H P_k^- H^T + R \right)^{-1}$. The update of the error covariances takes place with the equation in the Joseph form $P_k = (I - K_k H_k) P_k^- (I - KH) + KRK^T$.

It is known that the EK provides bad results by factors as an incorrect definition of the system model or the values of the covariances matrices Q or R. We are interested in forcing the filter to converge in spite of having not an exact definition of the model or the right values for Q or R.

We carry out this task keeping in mind the following considerations:

1. In a KF the residual of the measurement or innovation Z, should fulfill:

$$Z_{k+1}^T * P_{k+1}^{-1} * Z_{k+1} \leq K \tag{6}$$

where:

$$Z_{k+1} = z_{k+1} - \hat{z}_{k+1} \tag{7}$$

with

$$\hat{z}_{k+1} = z_k - H * \hat{x}_k^- \tag{8}$$

In the equation (7) z_{k+1}, is the measurement and $\hat{z}_{k+1}$ is the *estimate measurement*. The result of the equation (6) is called *evaluation coefficient* denoted by ec and K is a scalar value defined with the chi-square distribution.

2. We associated an ec for each component of the state vector. This is done with the goal to eliminate only the wrong values. In other words, if a measurement of any of the variables is wrong this is ignored for the fusion, but the rest are fused.
3. According to the observations R is a factor that can be tuned up to force to the convergence of the filter, that is,
   - if $ec \leq K$, R does not change.
   - if $ec \geq K$, R is decremented and the associated measurement is not taken into account for the fusion. K is chosen based on the chi-square distribution with a confidence limit of 95%.

4 Results and Discussion

We made preliminary tests in a small room 4x7 m^2 where we have installed the Ubisense system. We wanted to observe the system in this space because there are more critical situations to localize an user. The data were gathered and post processed with Matlab. To test the system, (1) the user remained still at a position during a short period of time - at different positions throughout the room, (2) walked a predefined path and (3) walked around the room.

In (1) the filter was evaluated without keeping in mind the distances travelled, while in (2) and (3) already existing movement is considered, specifically the length of a step. An approximation of 50 cm has been used. Nevertheless this can be determined in a more sophisticate way that has not been implemented yet.

The functionality of the system can be observed especially in figure 3. The results are more reliable if using criteria evaluation of the residuals than the pure Kalman filter. If the data are processed with the criteria mentioned, wrong data are ignored and the signals are smoothed. On the other hand, the results for a fixed position are very good. After processing the signal filter the algorithm smoothed the signal ignoring the outliers. Sometimes the multi-path phenomenon is responsible for outliers. The results showed the standard deviation in X decreased when using the criteria evaluation from 4 cm (Kalman filter) to 2 cm. In Y direction the decrease of standard deviation was from 6cm (Kalman) to 2 cm (criteria evaluation).

In the experiments, we observed the influence of Q. In the figure 4, we present the results if Q is varied. We decrement Q if the residuals are not between

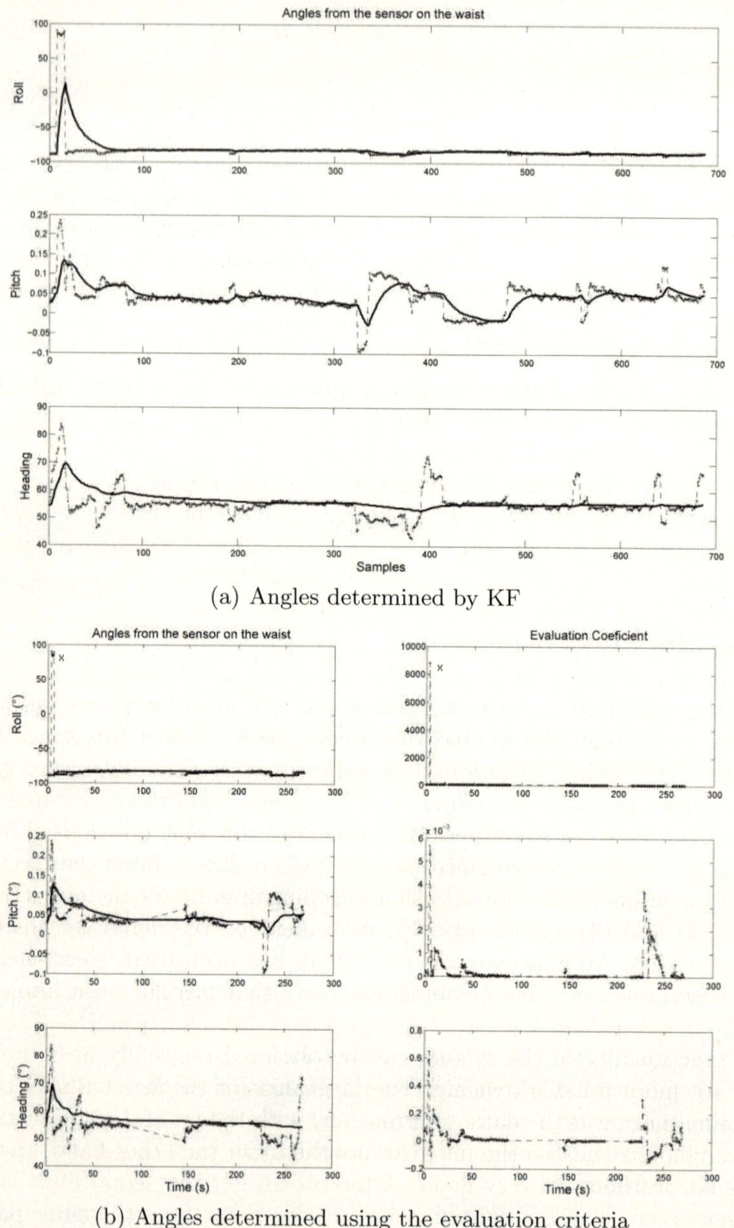

(a) Angles determined by KF

(b) Angles determined using the evaluation criteria

Fig. 3. Example of the elimination of wrong measurements, for the angles provided by Xsense unit

$\pm\sigma$(standard deviation). While the covariance matrix R causes convergence of the KF, Q helps to smooth the signal.

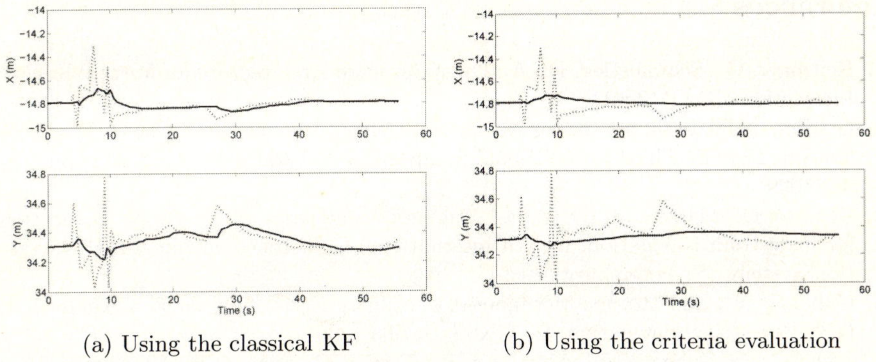

(a) Using the classical KF   (b) Using the criteria evaluation

Fig. 4. Results for position (X, Y)

5   Conclusions and Future Work

We have presented a robust location system for users in indoor environments. The implemented algorithms and the sensors used guarantee an acceptable system performance. The sensors are not affected by typical signals in the environment as magnetism, sounds or changes in the lighting conditions. We can detect if our sensors fail by evaluating the convergence of the Kalman Filter.

With the proposed combination of sensors our system increases the functionality of a portable system. It can work in wide indoor environments and it does not need a previous knowledge of the environment (magnetic disturbances) or big databases as vision based systems.

Although the Ubisense sensor shows problems in environments with metallic surfaces where multi-path phenomena can appear, it is still useful if we do not want to install complex infrastructures. These infrastructures are needed in acoustic or optic technologies. One disadvantage of the Ubisense sytem are multipath effects (reflections) that occur frequently. Nevertheless, we show that our criteria evaluation filter helps to improve this situation considerably. When implementing the filter it is very important to tune the parameters in order to get more suitable results and better performance.

Within the future work we plan the evaluation of the latest HSGPS systems that are on the market, because this may affect the future of the different existing indoor technologies. Also the integration with other kind of sensors as magnetic sensors, for instance.

Acknowledgments

This work has been partially supported by project ALF, grant TIN2005-08863-C03 from Spanish Ministry of Education and Science. We thank Mathis Csisinko for his support with OpenTracker.

References

1. Reitmayr, G., Schmalstieg, D.: An Open Software Architecture for Virtual Reality Interaction VRST (2001)
2. Golding, A.R., Lesh, N.: Indoor Navigation Using a Diverse Set of Cheap, Wearable Sensors. In: The Third International Symposium on Wearable Computers, pp. 29–36 (1999)
3. Ladetto, Q., Merminod, B.: Digital Magnetic Compass and Gyroscope Integration for Pedestrian Navigation. In: 9th International Conference on Integrated Navigation Systems, St-Petersburg (2002)
4. Gabaglio, V., Ladetto, Q., Merminod, B.: Kalman Filter Approach for Augmented GPS Pedestrian Navigation. In: GNSS, Sevilla (2001)
5. Ladetto, Q., Gabaglio, V., Meminod, B., Terrier, P., Schutz, Y.: Human Walking Analysis Assisted by DGPS. In: GNSS, Edinburgh (2000)
6. Brown, R., Hwang, P.Y.C.: Introduction to Random Signals and Applied Kalman Filtering. John Wiley & Sons Inc., New York (1997)
7. Kourogi, M., Kurata, T.: Personal Positioning based on Walking Locomotion Analysis with Self-Contained Sensors and a Wearable Camera. In: Proc. ISMAR, pp. 103–112 (2003)
8. Stirling, R., Collin, J., Fyfe, K., Lachapelle, G.: An Innovative Shoe-Mounted Pedestrian Navigation System. In: CD-ROM Proceedings of GNSS, the European Navigation Conference, pp. 103–112 (2003)
9. Foxlin, E.: Pedestrian Tracking with Shoe-Mounted Interial Sensors. IEEE Computer Graphics and Applications 25-(6), 38–46 (2005)
10. Kim, J.W., Jang, H.J., Hwang, D.-H., Park, C.: A Step, Stride and Heading Determination for the Pedestrian Navigation System. J. of Global Positioning Systems 3(1-2), 273–279 (2004)
11. Lachapelle, G.: GNSS Indoor Location Technologies. J. of Global Positioning Systems 3(1-2), 2–11 (2004)
12. Mezentsev, O., Collin, J., Kuusniem, H., Lachapelle, G.: Accuracy Assessment of a High Sensitivity GPS Based Pedestrian Navigation System Aided by Low-Cost Sensors. In: 11th Saint Petersburg International Conference on Integrated Navigation Systems (2004)
13. Mezentsev, O., Lachapelle, G.: Pedestrian Dead Reckoning-A Solution to Navigation in GPS Signal Degraded Areas? Geomatica 59(2), 175–182 (2005)

Performance of MCS Selection for Collaborative Hybrid-ARQ Protocol

Hanjin Lee, Dongwook Kim, and Hyunsoo Yoon

Korea Advanced Institute of Science and Technology (KAIST)
{hjlee,kimdw,hyoon}@nslab.kaist.ac.kr

Abstract. We propose a Modulation and Coding Scheme (MCS) selection algorithm for the collaborative hybrid Automatic-Repeat-reQuest (ARQ) protocol in order to provide high data rates. The collaborative hybrid ARQ protocol is designed to benefit from diversity gain. It exploits not only the broadcast nature of wireless channels, but also spatial diversity by formation of virtual antenna arrays and collaboration between a base station and relay nodes through Space-Time Block Coding (STBC). The proposed algorithm estimates both the effective Signal-to-Noise-Ratio (SNR) and the average throughput of each MCS level, and then selects the MCS level maximizing the average throughput. Simulation results show that the proposed algorithm outperforms conventional MCS selection algorithms in terms of the total throughput and satisfies packet delay constraints.

1   Introduction

In wireless communication systems, signals experience the shadow fading and multi-path fading, which degrade the performance such as throughput and packet delay. To mitigate these fading effects, spatial diversity techniques using multiple antennas have been thoroughly investigated in many literatures. However, even in single antenna systems, spatial diversity can be exploited through the collaborative hybrid ARQ protocol. In this protocol, a base station and relay nodes form virtual antenna arrays and collaborate through STBC during retransmissions [1][2]. If the base station transmits a packet to a mobile node, not only the mobile node but also the relay nodes near the base station can receive it. When the packet should be retransmitted (the mobile node fails to decode the packet correctly and requests retransmission), the relay nodes which have the successfully decoded packet and the base station transmit a space-time codeword on the same radio channel simultaneously. Therefore, the performance of the mobile node can be improved because it benefits from spatial diversity.

In general, link adaptation, or adaptive modulation and coding (AMC) scheme, is considered indispensable to provide high data rates because of the time varying nature of wireless channels. It denotes the matching of the MCS level to the channel quality, e.g., SNR. To perform link adaptation, a mobile node should measure the channel quality and report it to the base station periodically.

A.-M. Kermarrec, L. Bougé, and T. Priol (Eds.): Euro-Par 2007, LNCS 4641, pp. 941–949, 2007.
© Springer-Verlag Berlin Heidelberg 2007

Then, the base station determines an MCS level according to the information reported by the mobile node and transmits packets in the determined MCS level.

However, the time delay between measurement and transmission may cause packets to be received erroneously at the mobile node because the channel quality may be changed during this time delay. To compensate for this performance degradation of AMC, therefore, most practical systems employ the hybrid ARQ protocol which is combination of two basic error control schemes, ARQ and Forward-Error-Correction (FEC) [3]. The collaborative hybrid ARQ protocol can be regarded as the extension of the conventional hybrid ARQ protocol to benefit from diversity gain. Details of the collaborative hybrid ARQ protocol are presented in Section 2.

Though the hybrid ARQ protocol can compensate for the performance degradation of AMC, determination of MCS levels still has a significant effect on the performance. Too low MCS levels waste radio resources thus degrading throughput. On the other hand, too high MCS levels increase the number of retransmissions, which could violate packet delay constraints. For the conventional hybrid ARQ protocol, an MCS selection algorithm for improving throughput has been proposed in [4]. It estimates the average throughput of each MCS level and selects the MCS level which maximizes the average throughput. However it is not suitable for the collaborative hybrid ARQ protocol because it has no consideration of spatial diversity. In this paper, therefore, we propose an MCS selection algorithm maximizing the average throughput by estimating both the effective SNR and the average throughput of each MCS level for the collaborative hybrid ARQ protocol.

The rest of the paper is organized as follows: Section 2 states the system model of the collaborative hybrid ARQ protocol. Section 3 discusses related works of MCS selection algorithms for the conventional hybrid ARQ protocol. In Section 4, we propose an MCS selection algorithm for the collaborative hybrid ARQ protocol, and its performance is compared with other MCS selection algorithms in Section 5. Finally, we conclude the paper in Section 6.

2   System Model

The system model assumed in this paper is shown in Fig. 1. The channel gain at time t is represented as $h_{i,j}(t)$ where i and j can be a base station, relay node and mobile node. A base station and relay node transmit packets with power P_{BS} and P_{RN}, respectively.

The type of hybrid ARQ used in this paper is synchronous Chase combining. Chase combining implies that erroneous packets are preserved for soft combining with the currently received packet in order to improve SNR thus increasing the probability of successful decoding [5]. On the other hand, synchronous hybrid ARQ implies that retransmissions for a certain hybrid ARQ process are restricted to occur at known time instants [6].

The procedure of packet transmission in the system using the collaborative hybrid ARQ is as follows: A base station transmits a packet to a mobile node.

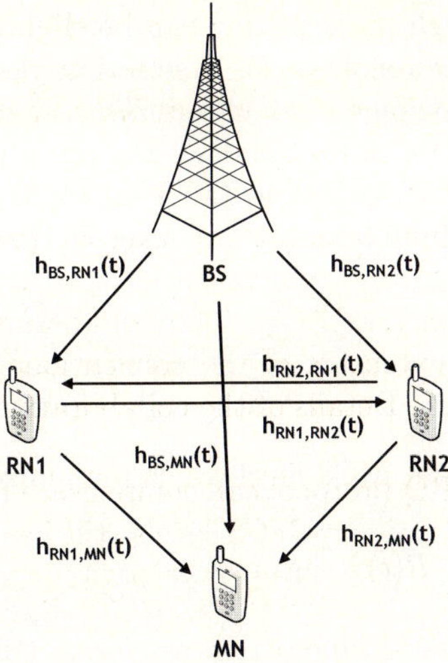

Fig. 1. System model

Because packets are broadcasted in wireless channels, not only the mobile node but also the relay nodes near the base station receive the packet. If the mobile node decodes the packet successfully, it transmits an ACK message to the base station and then transmission of the packet is completed. However, in the case that the mobile node fails to decode the packet, it transmits a NACK message to the base station in order to request retransmission. After the predefined interval, the base station and the relay nodes which have the successfully decoded packet form virtual antenna arrays and transmit a space-time codeword on the same radio channel simultaneously. Therefore, the mobile node receives multiple packets and exploits spatial diversity because each packet is transmitted over a different path thus experiencing independent fading. Prior to decoding the currently received packets at the mobile node, they are combined with previously received packets. After combining them, decoding is performed at the mobile node. Retransmission continues until the base station receives an ACK message from the mobile node or the number of retransmissions exceeds the predefined value T_{max}.

3   Related Works

In most practical systems using the conventional hybrid ARQ protocol, the MCS level is determined in order to satisfy Frame Error Rate (FER) of 10% or 1%; in this case, most packets are successfully transmitted without retransmission or with one retransmission. Such MCS selection algorithms are designed to satisfy packet delay constraints. However, in the case of non-real-time traffic which is insensitive to packet delay, throughput can be improved through fully exploiting time diversity even though packet delay can be slighted increased. For the conventional hybrid ARQ protocol, an MCS selection algorithm has been proposed in [4] to maximize the average throughput. This algorithm estimates the average throughput of each MCS level and selects the MCS level which maximizes the average throughput. If we define MCS set as M, the expected number of transmissions to successfully transmit a packet at time t when the MCS level is $m \in M$ is given by

$$E[Z(t)|m] = \sum_{n=1}^{\infty} n \cdot P\{Z(t) = n|m\} \ . \tag{1}$$

Therefore, we can select the MCS level which maximizes the average throughput at time t as follows:

$$R(t) = \underset{m \in M}{\operatorname{argmax}} \frac{B_m}{\tau \cdot E[Z(t)|m]} \tag{2}$$

where τ denotes the required time per transmission (the frame length multiplied by the retransmission interval), and B_m is the information bits in a frame when the MCS level is m. To calculate $P\{Z(t) = n|m\}$ in (1), we define the effective SNR from the base station to mobile node u at n_{th} transmission as follows (we assume that the first transmission occurs at time t):

$$SB_{MN_u}^{(n)}(t) = \sum_{k=1}^{n} |h_{BS,MN_u}(t + \tau(k-1))|^2 \cdot \frac{P_{BS}}{N} \tag{3}$$

where $h_{BS,MN_u}(t)$ denotes the channel gain from the base station to mobile node u at time t, and N denotes the noise power. In Chase combining, the effective SNR is the sum of SNR of each transmission. With the effective SNR calculated in (3), $P\{Z(t) = n|m\}$ in (1) is then calculated as follows:

$$P\{Z(t) = n|m\} = (1 - F_m(SB_{MN_u}^{(n)}(t))) \cdot \prod_{k=1}^{n-1} F_m(SB_{MN_u}^{(k)}(t)) \tag{4}$$

where $F_m(x)$ represents FER when SNR is x and the MCS level is m. This value can be obtained from the result of link level simulation. Therefore, (2) is represented as follows:

$$R(t) = \underset{m \in M}{\operatorname{argmax}} \frac{B_m}{\tau \cdot \sum_{n=1}^{\infty} n \cdot (1 - F_m(SB_{MN_u}^{(n)}(t))) \cdot \prod_{k=1}^{n-1} F_m(SB_{MN_u}^{(k)}(t))} \tag{5}$$

Moreover, for a given bound on the maximum number of retransmissions T_{max}, (5) is modified as follows [4]:

$$R^{T_{max}}(t) = \underset{m \in M}{\operatorname{argmax}} \frac{X^{T_{max}}(t|m)}{Y^{T_{max}}(t|m)} \tag{6}$$

where $X^{T_{max}}(t|m)$ and $Y^{T_{max}}(t|m)$ are given by

$$X^{T_{max}}(t|m) = B_m \cdot (1 - \prod_{k=1}^{T_{max}+1} F_m(SB_{MN_u}^{(k)}(t))) \tag{7}$$

and

$$Y^{T_{max}}(t|m) = \tau \cdot \sum_{k=1}^{T_{max}+1} k \cdot (1 - F_m(SB_{MN_u}^{(k)}(t))) \cdot \prod_{q=1}^{k-1} F_m(SB_{MN_u}^{(q)}(t))$$
$$+ (T_{max} + 1) \cdot \prod_{k=1}^{T_{max}+1} F_m(SB_{MN_u}^{(k)}(t)) \ .$$

$$(8)$$

4   MCS Selection for the Collaborative Hybrid ARQ Protocol

In this section, we propose an MCS selection algorithm for the collaborative hybrid ARQ protocol. We first estimate the expectation of the effective SNR at n_{th} transmission (we assume that the first transmission occurs at time t as in Section 3). For relay node i, the expectation of the effective SNR at n_{th} transmission when the MCS level is m can be calculated as follows:

$$S_{RN_i}^{(n)}(t|m) = \sum_{k=1}^{n} \{SB_{RN_i}^{(k)}(t) + SR_{RN_i}^{(k)}(t|m)\} \tag{9}$$

where $SB_{RN_i}^{(k)}(t)$ and $SR_{RN_i}^{(k)}(t|m)$ are given by

$$SB_{RN_i}^{(k)}(t) = |h_{BS,RN_i}(t + \tau(k - 1))|^2 \cdot \frac{P_{BS}}{N} \tag{10}$$

and

$$SR_{RN_i}^{(k)}(t|m) = \sum_{j=1,j\neq i}^{L} |h_{RN_j,RN_i}(t + \tau(k - 1))|^2 \cdot (1 - F_m(S_{RN_j}^{(k-1)}(t|m))) \cdot \frac{P_{RN}}{N} \tag{11}$$

where L denotes the total number of relay nodes. The summation notation in (9) reflects the SNR gain of Chase combining. $SB_{RN_i}^{(k)}(t)$ represents the effective SNR from the base station to relay node i at k_{th} transmission, and $SR_{RN_i}^{(k)}(t|m)$ represents the expectation of the effective SNR from the relay nodes which have the successfully decoded packet to relay node i at k_{th} transmission when the MCS level is m. A relay node transmits the packet at n_{th} transmission only if the relay node decoded the packet successfully at $(n-1)_{th}$ transmission. Because there are no relay nodes which can transmit packets at the first transmission, therefore, we have

$$F_m(S_{RN_i}^{(0)}(t|m)) = 1 \qquad \text{for all } t > 0,\ 1 \leq i \leq L,\ m \in M \ . \tag{12}$$

In a similar way, for mobile node u, the expectation of the effective SNR at n_{th} transmission when the MCS level is m can be calculated as follows:

$$S_{MN_u}^{(n)}(t|m) = \sum_{k=1}^{n} \{SB_{MN_u}^{(k)}(t) + SR_{MN_u}^{(k)}(t|m)\} \tag{13}$$

where $SB_{MN_u}^{(k)}(t)$ is defined in (3), and $SR_{MN_u}^{(k)}(t|m)$ is given by

$$SR_{MN_u}^{(k)}(t|m) = \sum_{j=1}^{L} |h_{RN_j,MN_u}(t+\tau(k-1))|^2 \cdot (1 - F_m(S_{RN_j}^{(k-1)}(t|m))) \cdot \frac{P_{RN}}{N} \ . \quad (14)$$

$SB_{MN_u}^{(k)}(t)$ represents the effective SNR from the base station to mobile node u at k_{th} transmission, and $SR_{MN_u}^{(k)}(t|m)$ represents the expectation of the effective SNR from the relay nodes which have the successfully decoded packet to mobile node u at k_{th} transmission when the MCS level is m.

For the collaborative hybrid ARQ protocol, therefore, (4) is modified as follows:

$$P\{Z_P(t) = n|m\} = (1 - F_m(S_{MN_u}^{(n)}(t|m))) \cdot \prod_{k=1}^{n-1} F_m(S_{MN_u}^{(k)}(t|m)) \quad (15)$$

Consequently, we can select the MCS level which maximizes the average throughput by applying (15) to (5) or (6) for the collaborative hybrid ARQ protocol as follows:

$$R_P(t) = \underset{m \in M}{\text{argmax}} \ \frac{B_m}{\tau \cdot \sum_{n=1}^{\infty} n \cdot (1 - F_m(S_{MN_u}^{(n)}(t|m))) \cdot \prod_{k=1}^{n-1} F_m(S_{MN_u}^{(k)}(t|m))} \quad (16)$$

or

$$R_P^{T_{max}}(t) = \underset{m \in M}{\text{argmax}} \ \frac{X_P^{T_{max}}(t|m)}{Y_P^{T_{max}}(t|m)} \quad (17)$$

where $X_P^{T_{max}}(t|m)$ and $Y_P^{T_{max}}(t|m)$ are given by

$$X_P^{T_{max}}(t|m) = B_m \cdot (1 - \prod_{k=1}^{T_{max}+1} F_m(S_{MN_u}^{(k)}(t|m))) \quad (18)$$

and

$$Y_P^{T_{max}}(t|m) = \tau \cdot \sum_{k=1}^{T_{max}+1} k \cdot (1 - F_m(S_{MN_u}^{(k)}(t|m))) \cdot \prod_{q=1}^{k-1} F_m(S_{MN_u}^{(q)}(t|m))$$

$$+ (T_{max} + 1) \cdot \prod_{k=1}^{T_{max}+1} F_m(S_{MN_u}^{(k)}(t|m)) \ . $$

$$(19)$$

5   Simulation Results

We carried out the simulation to compare the performance of various MCS selection algorithms. The simulation environments are summarized in Table 1, and the deployment of the base station and relay nodes is shown in Fig. 1. The number of mobile nodes is set to 10, and their locations are randomly distributed in the shaded area in Fig. 1. The distance between the base station and a mobile node is limited up to 500 m. In each frame, one mobile node is scheduled according to the Proportional Fair (PF) scheduling algorithm [7][8], and the time delay between measurement and transmission is set to 1 frame.

Table 1. Simulation environments

Items	Description
Channel model	Pedestrian B, 1 km/h [9]
Fast fading	Jakes model [10]
Slow fading	Log-normal distribution with standard deviation of 8.9 dB
System bandwidth	10 MHz
Transmission power of base station	43 dBm
Transmission power of relay node	30 dBm
Noise power density	-173 dBm/Hz
Traffic model	Full queue
Distance between base station and relay node	250 m
Distance between base station and mobile node	0 - 500 m
The number of base stations	1
The number of relay nodes	2
The number of mobile nodes	10
Frame length	5 ms
Retransmission interval	3 frame
Maximum number of retransmissions (T_{max})	4
Time delay between measurement and transmission	1 frame
Simulation time	500 frame
Type of hybrid ARQ	Synchronous Chase combining
Scheduling algorithm	Proportional Fair

FERx selects the MCS level as follows:

$$R_{FERx}(t) = \underset{m \in M}{\operatorname{argmax}} \left\{ B_m \middle| F_m(|h_{BS,MN_u}(t)|^2 \cdot \frac{P_{BS}}{N}) < x \right\} \qquad (20)$$

On the other hand, Thrpt selects the MCS level by using (6), and the proposed algorithm selects the MCS level by using (17). In the cases of Thrpt and the proposed algorithm, we calculate the effective SNR under the assumption that the channel gain is unchanged during a hybrid ARQ process because it is difficult to predict the future channel conditions in practice.

Figure 2 shows the cumulative distribution function (CDF) of the average number of transmissions to transmit a packet, which is a measure of packet delay. As shown in Fig. 2, the average number of transmissions of the proposed algorithm is larger than those of any other algorithms. Though the proposed algorithm shows the longest packet delay, we can see that the proposed algorithm satisfies packet delay constraints because there is no case that the average number of transmissions exceeds $T_{max} + 1$.

Figure 3 shows the CDF of the total throughput. The proposed algorithm outperforms FER0.1, FER0.5, FER0.9, and Thrpt algorithms by about 30 %, and FER0.01 algorithm by about 60 % in terms of the total throughput because FERx and Thrpt algorithms are designed without consideration of spatial diversity. Though the proposed algorithm shows longer packet delay than any

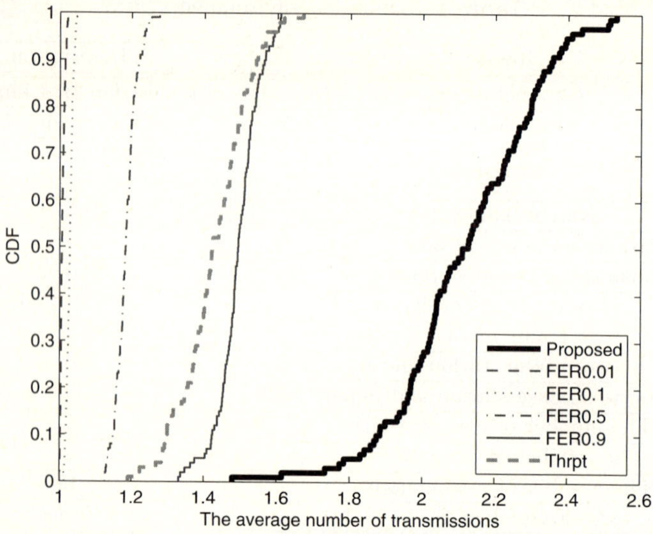

Fig. 2. CDF of the average number of transmissions to transmit a packet

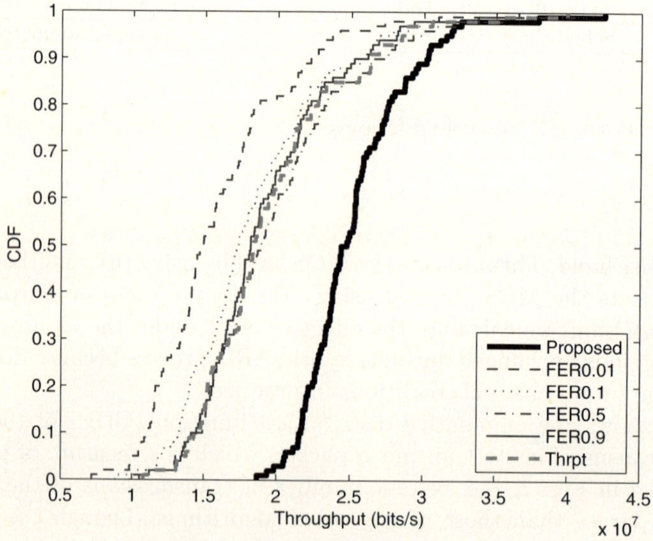

Fig. 3. CDF of the total throughput

other algorithms, selection of higher MCS levels with satisfying packet delay constraints results in higher throughput. Moreover, we can see that FER0.01 algorithm shows the worst throughput because it selects too low MCS levels thus wasting radio resources.

6   Conclusion

In this paper, we have proposed an MCS selection algorithm for the collaborative hybrid ARQ protocol. It selects the MCS level which maximizes the average throughput by estimating the effective SNR and the average throughput of each MCS level. The simulation results have shown that the proposed algorithm outperforms any other MCS selection algorithms by 30 % – 60 % in terms of the total throughput. Though the proposed algorithm shows longer packet delay than other algorithms, it is shown that the proposed algorithm satisfies packet delay constraints.

Acknowledgement

This work was supported by the MOST (Ministry of Science and Technology) / KOSEF (Korea Science and Engineering Foundation) through the AITrc (Advanced Information Technology Research Center) and the MIC (Ministry of Information and Communication), Korea, under the ITRC (Information Technology Research Center) support program supervised by the IITA (Institute of Information Technology Advancement) (IITA-2006-C1090-0603-0015).

References

1. Stanojev, I., Simeone, O., Bar-Ness, Y., You, C.: Performance of Multi-Relay Collaborative Hybrid-ARQ Protocols over Fading Channels. IEEE Communications Letters 10(7) (2006)
2. Zhao, B., Valenti, M.: Practical Relay Networks: A Generalization of Hybrid-ARQ. IEEE Journal on Selected Areas in Communications 23(1), 7–18 (2005)
3. Lin, S., Costello Jr., D.: Error Control Coding, 2nd edn. Pearson Prentice Hall, London (2004)
4. Zheng, H., Viswanathan, H.: Optimizing the ARQ Performance in Downlink Packet Data Systems With Scheduling. IEEE Transactions on Wireless Communications 4(2), 495–506 (2005)
5. Chase, D.: Code Cominng - A Maximum-Likelihood Decoding Approach for Combining an Arbitrary Number of Noisy Packets. IEEE Transactions on Communications 33(5), 385–393 (1985)
6. 3GPP TR 25.814 V7.1.0 (2006-09) Technical Report, 3rd Generation Partnership Project, Technical Specification Group Radio Access Network, Physical layer aspects for evolved Universal Terrestrial Radio Access (UTRA) (Release 7)
7. Bender, P., Black, P., Grob, M., Padovani, R., Sindhushayana, N., Viterbi, A.: CDMA/HDR: A Bandwidth-Efficient High-Speed Wireless Data Service for Nomadic Users. IEEE Communications Magazine 38(7), 70–77 (2000)
8. Jalali, A., Padovani, R., Pankaj, R.: Data Throughput of CDMA-HDR a High Efficiency-High Data Rate Personal Communication Wireless System. In: IEEE Vehicular Technology Conference (VTC), pp. 1854–1858. IEEE Computer Society Press, Los Alamitos (2000)
9. Guidelines for evaluation of radio transmission technologies for IMT-2000. ITU-R M.1225 (1997)
10. Jakes Jr., W.C.: Microwave Mobile Communications. Wiley, Chichester (1974)

New Approaches for Relay Selection in IEEE 802.16 Mobile Multi-hop Relay Networks*

Deepesh Man Shrestha, Sung-Hee Lee, Sung-Chan Kim, and Young-Bae Ko

Graduate School of Information & Communication, Ajou University, South Korea
{deepesh,sunghee,kimsungchan,youngko}@ajou.ac.kr

Abstract. The IEEE 802.16 mobile multi-hop relay (MMR) task group 'j' (TGj) has recently introduced the multi-hop relaying concept in the IEEE 802.16 WirelessMAN, wherein a newly introduced *relay station (RS)* is expected to increase the network performance. In this context, we consider the problem of path selection where several RSs might exist between the base station (so called, MR-BS) and the mobile station (MS). We propose a backward compatible signaling mechanism and a path selection algorithm based on the expected link throughput (ELT). To the best of our knowledge, this is the first contribution that presents path selection algorithm in the IEEE 802.16 networks. Performance of the proposed scheme is compared with shortest hop and no-relay schemes and we show our schemes significantly improves the performance in terms of throughput and latency.

1 Introduction

The existing IEEE 802.16 [1,2] family of standards popularly known as WiMax focuses on providing a broadband wireless access in the metropolitan area network. Although, initially developed for a *point-to-multipoint (PMP)* single hop communication, multi-hop relaying is recently receiving greater attention in the IEEE 802.16 working group (WG), specifically after the commencement of the IEEE 802.16j MMR task group (TG). The main objective of this TG is to introduce a relay station dedicated for relaying data between MR-BS and MS so as to enhance the network coverage and the throughput. Fig. 1 shows the topology of the IEEE 802.16j network. MR-BS is a fixed base station connected to the access network. RSs in general are divided into fixed RS (FRS) installed in a fixed location, nomadic RS (NRS) installed for a temporary duration where events occur and mobile RS (MRS) installed in vehicle such as bus, trains etc. RS can be deployed either in planned or unplanned manner due to which several access links are offered to MSs. Referring to Fig. 1, MS2 can be served either by NRS or FRS through either of three unique paths (MR-BS→FRS1→FRS2, MR-BS→FRS1→NRS and MR-BS→NRS). In such case, at least one *path* from

---

* This work was supported by the MIC (Ministry of Information and Communication), Korea, under the ITRC support program supervised by the IITA, (IITA-2006-C1090-0602-0011) and (IITA-2006-C1090-0603-0015).

A.-M. Kermarrec, L. Bougé, and T. Priol (Eds.): Euro-Par 2007, LNCS 4641, pp. 950–959, 2007.
© Springer-Verlag Berlin Heidelberg 2007

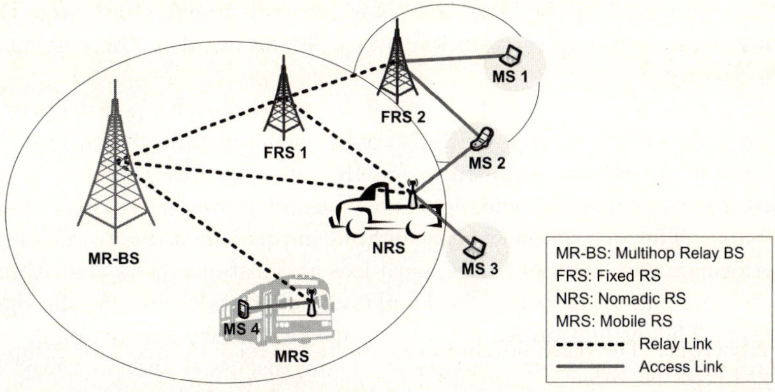

Fig. 1. The IEEE 802.16j Network Topology with several paths and RSs available for relaying between MR-BS and MSs

MR-BS to MS (hereafter just called path) must be selected to forward data packets. For this purpose a naive way would be to let MR-BS randomly choose a path or let all RSs forward the data. However, relaying by all or randomly selected RS might be prohibitive in terms of performance and resource utilization [3]. For example, a relay serving a large number of users might become a bottleneck eventually dropping the packets and enforcing retransmissions. This leads to unnecessary consumption of bandwidth, higher latency and poor QoS. In this paper, we present a centralized path selection algorithm in the medium access control sub-layer (MAC) of the MR-BS. Although there can be distributed approaches, they require RSs of higher complexity with similar capability to that of MR-BS. We assume RSs that are simple and controlled by MR-BS (e.g., for scheduling and resource allocations) such that the installation cost is minimum. We propose a selection metric named expected link throughput (ELT) based on the bandwidth availability and the data-rate to maximize the achievable throughput.

For this purpose, we introduced a new signaling features in the MR-BS and RS during the network entry of the MS and when normal communication is in progress. However, no extra operations are applied on the MS enabled with the legacy protocols [1,2] for backward compatibility [4]. For the performance analysis we extended IEEE 802.16 implementation on the Qualnet simulator version 3.9.5 to support multi-hop relaying. We compare our path selection scheme with single-hop no relays and shortest hop path. We show that the multi-hop schemes and a better path selection protocol enhances throughput with comparable delay.

2   Related Work

We review path selection protocols by segregating (1) RS selection algorithm and 2) related signaling schemes. Since the multi-hop concept is newly established in the IEEE 802.16 networks, we review the closely related RS selection algorithms from the cellular networks. Next, we will briefly summarize the signaling mechanisms for path selection protocols discussed in the current IEEE 802.16j meetings.

[5,6] presents the relay selection in cellular networks with peer-to-peer relaying. They propose selection algorithms based on the distance, path-loss [5] and the signal interference to noise ratio (SINR) [6]. The coverage improvements obtained due to proper RS selection and transmit power selection is presented in this paper. They also show the throughput improvement due to relaying with best performance in order of SINR, path-loss and distance. The algorithms presented in these papers are centralized and does not consider any specific signaling mechanism. The evaluation is performed only for the two-hop relaying.

Several path management schemes are being discussed and presented as the part of contributions in the IEEE 802.16 TGj meetings. Related contributions [7, 8,9,10] propose centralized and distributed signaling mechanisms for the routing path creation, management and data forwarding. [8,9] considers routing path to be established either during the network entry of RS or MS. RS maintains the routing table such that it can forward the packets accordingly. However, these schemes does not consider strategies for selecting appropriate RSs along the path while more than one path exists. In contrary, our proposed scheme is compatible to support such management and forwarding protocols.

Several protocols are proposed in the IEEE 802.16j during the time when this paper is being written. Most of those proposals are not presented with the evaluation and are accepted with the consensus in the IEEE 802.16j meeting for considering it as a standard. While the process of standardization is long, we propose much simpler RS devices that can be implemented with very small changes in the infrastructure already existing for deployment.

3   Preliminaries

In this section we describe MAC management messages and the connection identifiers used in our proposed path selection scheme.

3.1   Connection Identifiers (CID)

In the IEEE 802.16 MAC, CIDs are used to identify the data flow between MR-BS and MS. The 16 bit CID is allocated by the MR-BS and is included in the generic MAC header of both management and data messages. Different classes of CIDs are described in [1]. *Basic CID* is used with control and management messages and to identify MS (instead of 48-bit MAC address). The *Initial ranging CID* is temporarily used by MS during the network entry. The *Broadcast CID* is used with broadcast MAC management messages. *Transport CID* identifies a data flow between MR-BS and MS.

3.2   MAC Management Messages

The description of the following MAC management messages suffices the understanding of our proposed schemes:

- *DCD and UCD* are downlink and uplink channel descriptors generated by the MR-BS that consists of channel information and burst profiles.
- *Downlink MAP (DL-MAP)* and *Uplink MAP (UL-MAP)* describes the bandwidth allocations for downlink and uplink transmissions respectively, with the time offsets for all stations to transmit data. These messages are transmitted periodically by the MR-BS using Broadcast CID.
- *Range Request (RNG-REQ)* and *Range Response (RNG-RSP)* messages are used for ranging operation. Prior is transmitted by MS to request appropriate transmit power corrections, while the later is replied by MR-BS with needed adjustment (e.g. time, power etc).
- The *Report Request (REP-REQ)* is sent by MR-BS to achieve the channel measurement report from MS. In its response, MS replies with the *Report Response (REP-RSP)*.
- The proposed control message *Mobile Station Signal Report MS-SIG-REP* is used for (1) holding the ELT metric of each RS in path and (2) to signal MR-BS for running the path selection alogrithm.

4 Proposed Schemes

In this section, we describe our proposed signaling scheme, our proposed path selection metric and algorithm for path selection. We assume that the network topology is known to the MR-BS prior to running the path selection. All RSs are centrally controlled including their resources such as bandwidth and scheduling etc.

4.1 RS Selection Signaling

The main purpose of the signaling mechanism is to enrich the MR-BS with the path information and their corresponding metrics. The path selection is performed in two situations. First, during initial ranging when the MS joins the network and second, if the MS needs to switch RS (so called intra-BS handover).

In the legacy protocol, the initial ranging is performed for adjusting the transmission parameters of the MS. We embed our signaling scheme during this process to perform path selection as illustrated in Fig. 2. First MS scans the periodically broadcasted MAC management messages such as DCD, UCD, DL-MAP, UL-MAP and achieves synchronization. MS then decodes the initial ranging contention period and transmit the randomly generated ranging code. If the signal quality from MS or forwarding RS is good, MR-BS transmits RNG-RSP with success status. Otherwise MS restarts ranging or scan for the new MR-BS if the status is continue or abort, respectively. In the next step upon success, MR-BS includes the bandwidth allocation and the time offset for the MS to send the RNG-REQ message in the following UL-MAP. In our proposed scheme, RSs learn about this timing from the same UL-MAP and prepare to measure the signal quality. RSs upon hearing RNG-REQ construct a proposed MS-SIG-REP report with the MS MAC address (included in the RNG-REQ) and the corresponding ELT metric. The ELT metric is computed at each upstream RS if

954   D.M. Shrestha et al.

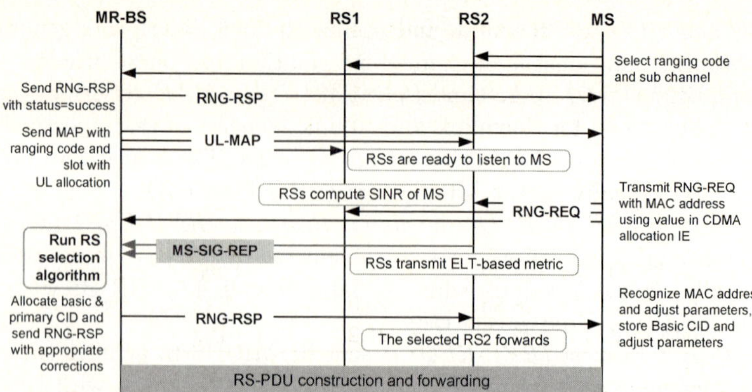

Fig. 2. The example for the proposed CDMA initial ranging during the network entry of MS. RS2 is selected as an access RS. (MS assumed in-coverage of MR-BS and RSs).

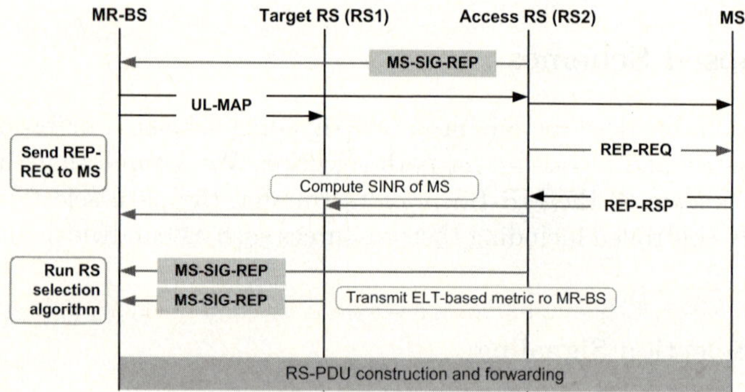

Fig. 3. The proposed intra-BS handover for RS path selection signaling (Target RS1 with better signal is selected in place of access RS2)

there is several RSs in path and piggy-backed with the report until it reaches to the MR-BS. Finally, MR-BS collects the path and metric information to execute path selection algorithm.

The signaling for path selection during the intra-BS handover is shown in Fig. 3. The RS that has been previously selected to serve MS (called access RS) continuously monitors the signal quality. If it degrades below the certain threshold, the access RS sends REP-REQ to the MS and notifies MR-BS about the handover by transmitting MS-SIG-REP. MR-BS includes the time offset for the expected REP-RSP and the Basic CID of MS in the next UL-MAP. As in the previous case, RSs prepare to measure signal quality while receiving REP-RSP from MS. This process is followed by constructing and forwarding MS-SIG-REP to the MR-BS. Note that in the IEEE 802.16j topology it is possible that MS is outside the coverage of MR-BS. To support such scenarios RSs might forward the control and management messages.

Table 1. PHY throughput VS SNR for DL PUSC [12]

Modulation/ code rate	Max. PHY throughput (Mbit/s)	SNR@ 0.01 CBLER (dB)	Cell range(m) with fade margin
QPSK-1/2	7.14	6.0	512
QPSK-3/4	10.71	9.2	421
16-QAM-1/2	14.28	10.3	392
16-QAM-3/4	21.42	14.9	290
64-QAM-2/3	28.56	19.0	222
64-QAM-3/4	32.13	21.2	192

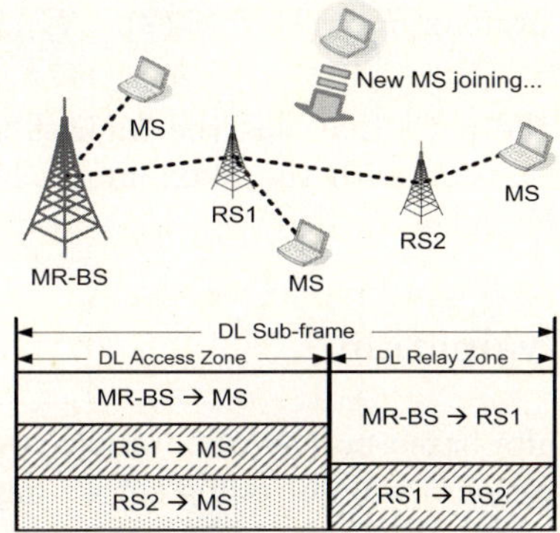

Fig. 4. Example topology and DL sub-frame allocations

4.2 Path Selection Algorithm

From the signaling mechanism, MR-BS receives the MS-SIG-REP from the several link-disjoint paths containing the ELT metrics of both relay and access links. MR-BS collects these messages for a predefined time t seconds after the first message is received. Each RS computes ELT (T_{exp}) of the link between itself and its downstream node as shown in eq. 1.

$$T_{exp}(j) = BW_{avail}(i) \times Data\_rate(j) \tag{1}$$

where, $BW_{avail}(i)$ is the available bandwidth of the ith relay node and the Data_rate(j) is the data-rate on link j supported by the ith RS w.r.t. the received signal to noise ratio (SNR). The available bandwidth is the product of the number of slots and the sub-channel over the downlink subframe. The lowest threshold of the SNR required for a successful communication according to the given modulation scheme is according to the Table 1. The example for the intuition of the path selection based on expected bandwidth is depicted in Fig. 4. We consider a frame structure in [11], in which the access and relay zones are defined as a portion of the frame used for MR-BS/RS to MS and RS transmissions respectively. In this figure, the area filled with diagonal line is the allocated bandwidth for RS1, and the dotted area is for RS2. Rest is allocated to MR-BS itself. Areas represent the maximum bandwidth for each RS which is used for delivering data to MSs. Available bandwidth is amount left after reducing the current used bandwidth to the total allocated bandwidth. Assuming a new MS joins the network, RSs will require additional bandwidth, thus the RS having more available bandwidth might offer better performance.

In the path selection scheme, MR-BS enumerate all the received candidate paths $(p_1..p_n)$, where n is the number of disjoint paths. For each path we find the minimum ELT (T_{min}). The minimum ELT for a ith path p_i is denoted by $T_{min}(i)$. Finally we select a path (p_{sel}) with the maximum among the minimum $T_{min}(i)$ as shown in eq. 2.

$$p_{sel} = \underset{n}{\mathrm{argmax}}\{T_{min}(1), T_{min}(2), ...T_{min}(n)\} \qquad (2)$$

In other words, p_{sel} is the path that has the highest throughput. Thus, our proposed scheme selects a path with the maximum achievable throughput by eliminating the bottleneck links.

5   Performance Evaluation

We now present the results based on the performance evaluation conducted in the Qualnet simulator version 3.9.5. For doing so, we implemented RS features for forwarding the data in the IEEE 802.16 library.

5.1   Simulation Environment

First, we simulated two-hop relaying where only one RS is present between the MR-BS and MS. 3 RSs are placed uniformly about 120 degree apart and 25 MSs are randomly distributed. In the second part, we placed the 6 RSs linearly for observing the performance with more number of hops. In the third part, we randomly distributed 8 RS and placed one MS to compare our scheme with the shortest hop path selection. The CBR traffic is generated in MR-BS and one connection to each MS is established. The transmit power of each node is 15dB and we use 2-ray ground pathloss model. The frame size is fixed at 20ms, divided into 10ms for downlink and uplink. A downlink frame is further divided for MR-BS/RS to transmit towards the downlink. Since the frame size is fixed, downlink duration for each RS is reduced with increasing hops. Following physical parameters as shown in Table 1 are used to calculate the achievable PHY throughput corresponding to the received SNR [12].

Throughput is defined as a ratio of the total number of bytes received to total simulation time (50ms). Average end-to-end latency is the time required for each packet to reach MS from the MR-BS. The delivery ratio is the ratio of received number of packets in all MSs to the total number of packets sent.

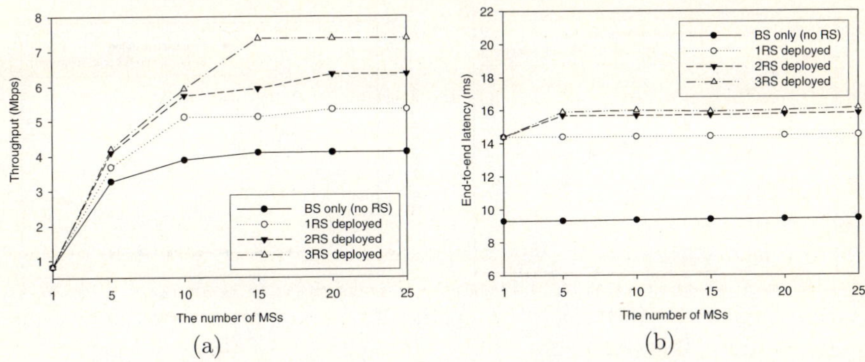

Fig. 5. Average throughput and end-to-end latency in 2-hop topology

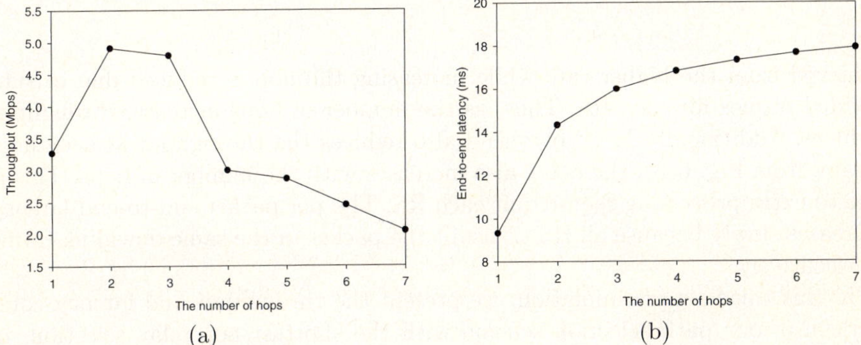

Fig. 6. Average throughput and end-to-end latency with 1 to 5 RSs placed linearly between MR-BS and MS

5.2   Simulation Results

Fig. 5(a) shows the average throughput as the number of MS increases. The no-relaying case shows very low throughput due to low rate than the multi-hop RS topology. As the number of RS is increased, throughput also increases. This means, more RSs participate between two end points. For the fixed number of RS, throughput saturates to the maximum level with the increasing number of connections. Even though the physical data-rate used for one-hop is high (20Mbps), we observe that the maximum application layer throughput is 7Mbps due to the MAC overhead. The end-to-end latency for each packet is slightly affected when the relay is added. The result is shown in Fig. 5(b). The latency is observed even with no RS due to the frame gap between downlink and uplink. In our scheme, latency increases more because of the delayed frame allocation for RS to MS than from BS to MS. The average latency remains almost the same regardless of the number of RSs increased in the deployment region.

Now we present the results for the linear topology. From Fig. 6(a) we see that the throughput nearly doubles from no-relay to one-hop case. As in the previous case, this is due to the use of higher data-rate. However, the benefit

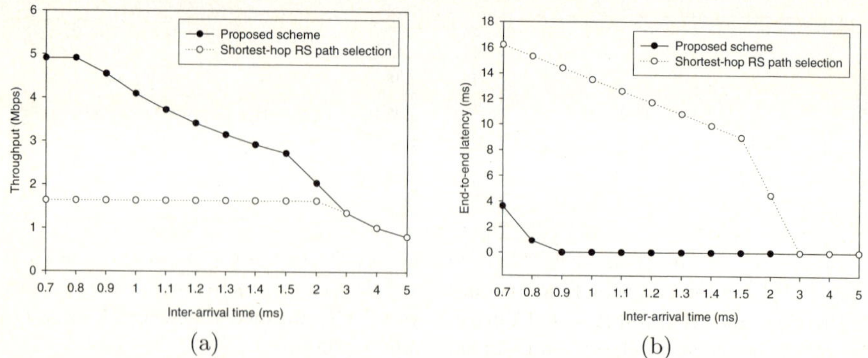

Fig. 7. Average throughput and end-to-end latency vs inter-arrival time with RSs randomly distributed between MR-BS and MS

achieved from the higher rate while increasing the hop is reduced due to subdivided frames for each RS. Thus, as the number of hops increase, throughput reduces. Additionally MAC overhead also reduces the throughput at each hop. Again, from Fig. 6(b), the delay also increases with the number of hops. This is due to extra processing required at each RS. The per packet end-to-end latency increases slowly because all RS transmit the packet in the same downlink frame consecutively.

In this third set of simulation, we present the throughput and latency comparison of our path selection scheme with the shortest hop-relay selection, in which shortest-hop relay path is selected between MR-BS and RS. We decrease the inter-arrival time from 5ms to 0.7ms to increase the traffic in the network. From Fig. 7(a), we observe almost 50% improvement in the throughput of our proposed scheme. The shortest hop relay uses lower data rate due to longer hop distance between the MR-BS to RS and RS to MS. The low average end-to-end latency as shown in Fig. 7(b) is due to packet loss, which shows around 90% improvement by our scheme. Therefore we show that the our path selection strategy shows better performance.

6 Conclusion

We have presented efficient path selection signaling and algorithm for the IEEE 802.16 based multi-hop networks. The proposed algorithm uses unique selection metric that effectively selects high throughput path. We carefully considered reuse of messages and operations already present in the existing standards for backward compatibility. The performance results show the 50% increase in throughput with 90% drop in the average latency with the ELT based path selection compared to that of shortest-hop.

Many issues in IEEE 802.16 based multi-hop networks are open both in PHY and MAC layer. Critical issues such as multi-hop frame structure, scheduling mechanism, support for QoS and many other requirements are under development. In future, we will consider more extensive simulation to compare our scheme with other RS selection schemes.

References

1. IEEE Std 802.16-2004, IEEE Standard for Local and metropolitan area networks, Part 16: Air Interface for Fixed Broadband Wireless Access Systems (2004)
2. IEEE Std 802.16e-2005, IEEE Standard for Local and metropolitan area networks, Part 16: Air Interface for Fixed Broadband Wireless Access Systems Amendment2: Physical and Medium Access Control Layers for Combined Fixed and Mobile Operation in Licensed Bands and Corrigendum 1 (2005)
3. Pabst, R., et al.: Relay-Based Deployment Concepts for Wirelss and Mobile Broadband Radio. IEEE Communications Magazine, 80–89 (2004)
4. Kang, H., et al.: IEEE C802.16j-06/050r4, Proposed Technical Requirements for IEEE 802.16 TGj (2006)
5. Sreng, V., Yanikomeroglu, H., Falconer, D.D.: Relayer Selection Strategies in Cellular Networks with Peer-to-Peer Relaying. IEEE Vehicular Technology Conference 3, 1949–1953 (2003)
6. Hu, H., Yanikomeroglu, H., Falconer D.D., Periyalwar, S.: Range Extension without Capacity Penality in Cellular Networks with Digital Fixed Relays. IEEE Communications Society, Globecom (2004)
7. Lee, H., et al.: Link Adaptive multihop path management for IEEE 802.16j. IEEE C802.16j-06/296 (2006)
8. Wang, H., et al.: Data forwarding and routing path setup for IEEE 802.16j multihop relay networks. IEEE C802.16j-06/212r1 (2006)
9. Saito, K., et al.: Path Selection for RS initial network entry. IEEE C802.16j-06/278 (2006)
10. Zhang, H., et al.: QoS Control Scheme for data forwarding in 802.16j. IEEE C802.16j-07/309r4 (2007)
11. The Relay Task Group of IEEE 802.16: The P802.16j Baseline Document for Draft Standard for Local and Metropolitan Area Networks, IEEE 802.16j-06/026r1 (2006)
12. Eklund, C., et al.: WirelessMAN Inside the IEEE 802.16 Standard for Wireless Metropolitan Networks. IEEE Press, Los Alamitos (2006)

On the Connectivity of Bluetooth-Based
Ad Hoc Networks*

Pilu Crescenzi[1], Carlo Nocentini[1],
Andrea Pietracaprina[2], Geppino Pucci[2], and Carlo Sandri[2]

[1] Dipartimento di Sistemi e Informatica, Università di Firenze, Viale Morgagni 65, 50134,
Firenze, Italy
{piluc,nocentini}@dsi.unifi.it
[2] Dipartimento di Ingegneria dell'Informazione, Università di Padova, Via Gradenigo 6/B,
35131, Padova, Italy
{capri,geppo,sandrica}@dei.unipd.it

Abstract. We study the connectivity properties of a family of random graphs
which closely model the Bluetooth's device discovery process, where each device
tries to connect to other devices within its visibility range in order to establish reliable communication channels yielding a connected topology. Specifically, we
provide both analytical and experimental evidence that when the visibility range
of each node (i.e., device) is limited to a vanishing function of n, the total number
of nodes in the system, full connectivity can still be achieved with high probability by letting each node connect only to a "small" number of visible neighbors.
Our results extend previous studies, where connectivity properties were analyzed
only for the case of a constant visibility range, and provide evidence that Bluetooth can indeed be used for establishing large ad hoc networks.

1 Introduction

A critical problem in setting up mobile multi-hop radio networks, also known as *ad hoc
networks*, is guaranteeing connectivity while minimizing power consumption and, in
some cases, the number of active connections per node. Among others, *Bluetooth* [1]
is a popular enabling technology for ad hoc networks, which was originally introduced
in 1999 by a Special Interest Group formed by more than 1800 manufacturers for the
deployment of Personal Area Networks (PANs), typically consisting of cellular phones,
laptops, wireless peripherals and PDAs. Several arguments have been raised to foster
the use of Bluetooth for the establishment of large ad hoc networks, due to its low
cost, availability, suitability for small devices, and low power consumption (see, for
example, [2]). However, a number of challenges arise in this context, particularly for
what concerns network formation [3,4].

Specifically, Bluetooth features a hierarchical organization where the nodes are
grouped into *piconets*, with each piconet containing one master and multiple slaves.
Piconets are then interconnected through bridge nodes to form a *scatternet*. Scatternet

* This work was supported in part by MIUR of Italy under project *MAINSTREAM*, and by the
EU under the EU/IST Project 15964 *AEOLUS*.

A.-M. Kermarrec, L. Bougé, and T. Priol (Eds.): Euro-Par 2007, LNCS 4641, pp. 960–969, 2007.

formation can be decomposed into three main steps, namely, *device discovery*, *piconet formation*, and *piconet interconnection*. Each of these steps poses interesting algorithmic challenges for which several solutions have been proposed [1]. In particular, during the first step each device attempts at discovering other devices contained within its visibility range and at establishing reliable communication channels with them, in order to form a connected topology, called the *Bluetooth topology*, which underlies the subsequent piconet formation and piconet interconnection steps. Since requiring each device to discover *all* of its neighbors is too time consuming [3], a crucial problem consists of deciding how many neighbors have to be selected in order to guarantee that the resulting Bluetooth topology is connected. Indeed, obtaining connectivity under degree limitations has been indicated in [2] as a major challenge for the adoption of the Bluetooth technology for large ad hoc networks.

In [5] the device discovery step has been effectively modeled as follows. The devices are regarded as a set of n nodes randomly and uniformly distributed in a square of unit side. Each node has a visibility range of $r(n)$, i.e., it can "see" all other nodes within Euclidean distance $r(n)$. Given a function $c(n)$, each node selects as neighbors $c(n)$ visible nodes at random, picking all visible nodes if their number is less than $c(n)$. Observe that the process is unidirectional in nature, however, each link established in this way becomes bidirectional. As a consequence, the final degree of each node may be much higher than $c(n)$, in the case that the node was selected as a neighbor by many other nodes. We refer to $\mathrm{BT}(r(n), c(n))$ as the resulting (undirected) graph.

Previous studies on the connectivity properties of $\mathrm{BT}(r(n), c(n))$ have considered only the case where each node is able to see a constant fraction of all other nodes, that is, the visibility range $r(n)$ is a constant. For this particular case, the experimental analysis conducted in [5] has shown that setting $c(n)$ to a small constant is sufficient to yield connectivity for $\mathrm{BT}(r(n), c(n))$ almost always. The experimental evidence has been later substantiated by the analysis in [6], which shows that, for constant $r(n)$, $c(n) = 2$ is sufficient to achieve connectivity with high probability. Also, in [7] it was proved that constant $c(n)$ (though much larger, in the order of the millions) is also sufficient to guarantee linear expansion of $\mathrm{BT}(r(n), c(n))$. These results suggest that device discovery can be performed efficiently whenever the network is sufficiently small (even though not necessarily a PAN). However, the assumption of constant $r(n)$ becomes quickly unfeasible as the number of devices to be connected increases, which would be the case when adopting Bluetooth for building large ad hoc networks.

In this paper we extend the above studies by providing both analytical and experimental evidence that, when the visibility range is a vanishing function of n, the device discovery step in Bluetooth can still be performed efficiently while guaranteeing connectivity, by letting each device discover only a"small", although non constant, number of neighbors. In particular, we prove that if $r(n) = \Omega(\sqrt{\ln n/n})$, then $\mathrm{BT}(r(n), c(n))$ is connected with high probability as long as $c(n) = \Omega(\ln(1/r(n)))$. We remark that the lower bound on $r(n)$ cannot be improved since it is known that when $r(n) \leq \delta\sqrt{\ln n/n}$, for some constant $0 < \delta < 1$, the *visibility graph* where each node is connected to *all* nodes in its visibility range is disconnected with high probability [8]. A challenging open question is whether the lower bound on the value of $c(n)$ required for connectivity is tight. We give a partial analytical answer to this question by showing that in fact $c(n) = 3$ is sufficient to attain connectivity with high probability, as long as $r(n) \geq n^{-\epsilon}$, for some constant $0 < \epsilon < 1/2$, but each node must choose two of the three neighbors sufficiently close to it.

In the paper we also report on a massive set of experiments conducted in order to assess the real performance of the two previously described protocols. Quite surprisingly, the experiments indicate that, even when the visibility range function is close to the aforementioned lower bound, the number of neighbors needed for connectivity exhibits an extremely weak dependence on $r(n)$: in fact, $c(n) = 3$ suffices almost always, *independently of how the neighbors are chosen*. Moreover, the experiments show that the expected maximum total degree featured by the topologies obtained by choosing three neighbors for each node is much smaller than the one featured by the visibility graph, while the diameter is only slightly larger.

Even though our results were mainly motivated by the question of whether Bluetooth is suitable as a large-scale ad hoc network technology, we believe that they may be of interest for other wireless network scenarios [9].

The rest of the paper is organized as follows. Section 2 analyzes the connectivity of $\mathrm{BT}(r(n), c(n))$ when $c(n)$ is $\Theta(\log(1/r(n)))$. Section 3 analyzes the case of $c(n) = 3$ under further constraints on neighbor selection. Section 4 reports the results of our experiments.

2   Connectivity of $\mathrm{BT}(r(n), c(n))$

Consider a set V of n nodes randomly and uniformly distributed in a unit-side square. Each node $v \in V$ has a visibility range of $r(n)$, i.e., v can "see" all nodes u at Euclidean distance $d(v, u) \leq r(n)$. Let the unit square be tessellated into k^2 square *cells* of side $1/k$, where $k = \lceil \sqrt{5}/r(n) \rceil$. Consequently, any two nodes residing in the same or in adjacent cells are at distance at most $r(n)$: hence, they are visible from one another. Most of our results hold *with high probability* (w.h.p. for short) by which we mean that the probability of the stated event is at least $1 - 1/\mathrm{poly}(n)$, where $\mathrm{poly}(n)$ denotes some polynomial function of n.

The following proposition can be easily proved using Chernoff's bound [10].

Proposition 1. *Let $\alpha = 9/10$ and $\beta = 11/10$. There exists a constant $\gamma_1 > 0$ such that for every $r(n) \geq \gamma_1 \sqrt{\ln n/n}$ the following events occur w.h.p.:*

1. *Every cell contains at least $\alpha n/k^2$ and at most $\beta n/k^2$ nodes.*
2. *Every node has at least $(\alpha/4)\pi n r^2(n)$ and at most $\beta \pi n r^2(n)$ nodes in its visibility range.*

The rest of the section is devoted to the proof of the following theorem.

Theorem 1. *There exist two positive real constants γ_1, γ_2 such that, if $r(n) \geq \gamma_1 \sqrt{\ln n/n}$ and $c(n) = \gamma_2 \ln(1/r(n))$ then $\mathrm{BT}(r(n), c(n))$ is connected w.h.p.*

Let $\epsilon = 1/8$. In the proof of the theorem we distinguish between the case $r(n) \leq n^{-\epsilon}$ and the case $r(n) > n^{-\epsilon}$, which are dealt with separately in the following subsections. Moreover, in both cases we condition on the events expressed by Proposition 1, which occur with high probability.

2.1 Case $\gamma_1 \sqrt{\ln n/n} \leq r(n) \leq n^{-\epsilon}$

We fix the lower bound for $r(n)$ to be the same under which Proposition 1 holds. In the range of $r(n)$ considered in this case, we have that $c(n) = \gamma_2 \ln(1/r(n)) = \Theta(\ln n)$. Let Q be an arbitrary cell and let G_Q denote the subgraph of $\mathrm{BT}(r(n), c(n))$ formed by nodes and edges internal to Q. We first show that every G_Q is connected and then prove that for every pair of adjacent cells there exists an edge in $\mathrm{BT}(r(n), c(n))$ whose endpoints are in the two cells.

Lemma 1. *With high probability, every G_Q is connected.*

Proof. Fix an arbitrary cell Q and let A_Q be the event that, for every partition of the nodes in Q into two nonempty subsets, there is at least an edge with endpoints in distinct subsets. Observe that the subgraph $G_Q \subseteq \mathrm{BT}(r(n), c(n))$ is connected if and only if A_Q occurs. Then:

$$1 - \Pr(A_Q) \leq$$

$$\leq \sum_{s=1}^{\beta n/(2k^2)} \binom{\beta n/k^2}{s} \left(1 - \frac{\alpha n/k^2 - s}{\beta \pi n r^2(n)}\right)^{sc(n)} \left(1 - \frac{s}{\beta \pi n r^2(n)}\right)^{(\alpha n/k^2 - s)c(n)}$$

$$\leq \sum_{s=1}^{\beta n/(2k^2)} \exp\left(s \ln \frac{e\beta n}{sk^2} - \frac{2sc(n)}{\beta \pi n r^2(n)} \left(\frac{\alpha n}{k^2} - s\right)\right)$$

$$\leq \frac{1}{n^2}$$

where the last inequality holds by choosing the constant γ_2 in the expression for $c(n)$ large enough. The lemma follows by applying the union bound over all k^2 cells. $\square$

Lemma 2. *With high probability, for every pair of adjacent cells Q_1 and Q_2 there is an edge $(u, v) \in \mathrm{BT}(r(n), c(n))$ such that u resides in Q_1 and v resides in Q_2.*

Proof. Consider an arbitrary pair of adjacent cells Q_1 and Q_2 and let B_{Q_1,Q_2} denote the event that there is at least one edge in $\mathrm{BT}(r(n), c(n))$ between the two cells. Since we are conditioning on the events described in Proposition 1, we have that

$$1 - \Pr(B_{Q_1,Q_2}) \leq \left(1 - \frac{\alpha n/k^2}{\beta \pi n r^2(n)}\right)^{2c(n)\alpha n/k^2}$$

$$\leq \exp\left(-\frac{\alpha n/k^2}{\beta \pi n r^2(n)}(2c(n)\alpha n/k^2)\right)$$

$$\leq \exp(-\zeta \ln^2 n),$$

where ζ is a positive constant. The lemma follows by applying the union bound over all $O(n)$ pairs of adjacent cells. $\square$

For the case $\gamma_1 \sqrt{\ln n/n} \leq r(n) \leq \delta n^{-\epsilon}$, Theorem 1 follows by combining the results of the above two lemmas.

2.2   Case $r(n) > n^{-\epsilon}$

We generalize and simplify the argument used in [6] for the case $r(n) = \Theta(1)$. Specifically, we first show that $\mathrm{BT}(r(n), c(n))$ contains a large connected component C, and then we show that for every node v there is a path from v to C. We condition on the events that the number of nodes in each cell and in the visibility range of each node are within the bounds stated in Proposition 1, which occur with high probability.

Lemma 3. *For $r(n) > \delta n^{-\epsilon}$ and $c(n) \geq 2$, $\mathrm{BT}(r(n), c(n))$ contains a connected component of size $n/(8k^2)$, w.h.p.*

Proof. The argument is identical to the one used in the proof of Proposition 3 in [6]. In particular, the probability of the existence of the connected component is at least

$$1 - \frac{8k^2 \log_2^2 n}{n} - \frac{1}{9^{\log_2 n}},$$

which is $1 - o(n^{-2/3})$ by our choice of k.   □

Let C be the connected component of size at least $n/(8k^2)$ which, by the above lemma, exists w.h.p. By the pigeonhole principle there must exist a cell Q containing at least $n/(8k^4)$ nodes of C. Let $V(Q, C)$ the set of nodes residing in Q and belonging to C. We have:

Lemma 4. *With high probability, for each node u there exists a path in $\mathrm{BT}(r(n), c(n))$ from u to some node in $V(Q, C)$.*

Proof. Consider a directed version of $\mathrm{BT}(r(n), c(n))$ where an edge (u, v) is directed from u to v if u selected v during the neighbor selection process. Since we are conditioning on the event stated in the second point of Proposition 1, our choice of $c(n)$ implies that the outdegree of each node is *exactly* $c(n)$ w.h.p. Pick an arbitrary node u and run a sequential breadth-first exploration from u in such a directed version of $\mathrm{BT}(r(n), c(n))$. Stop the exploration as soon as m nodes have been discovered but not yet explored (m is a suitable value that will be chosen later). We say that a failure occurs when the edge (v_1, v_2) is considered during the exploration of v_1, and node v_2 had been previously discovered. It is easy to see that at the moment when m nodes are discovered but not yet explored, if at most $c(n) - 1$ failures have occurred, then the total number of nodes discovered up to that moment is at most $2m$. Also, if at most $c(n) - 1$ failures occur before reaching m unexplored nodes, then at most m nodes have been explored. Therefore, from the second point of Proposition 1 it follows that the probability of not reaching m unexplored nodes with less than $c(n) - 1$ failures is at most

$$\binom{m \cdot c(n)}{c(n)} \left(\frac{2m}{(\alpha/4)\pi n r^2(n)} \right)^{c(n)} \leq \left(\frac{8m^2 e}{\alpha \pi n r^2(n)} \right)^{c(n)}. \tag{1}$$

Now suppose that the above event occurs and consider the m unexplored nodes, say $w_1, w_2, \ldots, w_m$, reached via breadth-first exploration from u. We now estimate the probability that $\mathrm{BT}(r(n), c(n))$ contains a path from w_i to a node in $V(Q, C)$. Observe that from the cell containing w_i there is a sequence of at most $2k$ pairwise adjacent cells ending at Q. Specifically, we estimate the probability that $\mathrm{BT}(r(n), c(n))$ contains a path from w_i to $V(Q, C)$ following such a sequence of cells, with the constraint that the path contains one node per cell and these nodes do not belong to the set of at most $2m$ nodes initially discovered from u or to the $m - 1$ paths constructed for any other w_j, with $j \neq i$. This probability is at least $p^{2k}q$, where p is the probability of extending the path one cell further, and q is the probability of ending, in the last step, in a node of $V(Q, C)$. By Proposition 1 we have that

$$p \geq \left(1 - \left(1 - \frac{\alpha n/k^2 - 3m}{\beta \pi n r^2(n)}\right)^{c(n)}\right)$$

$$q \geq \frac{n/(8k^4)}{\beta \pi n r^2(n)} = \frac{1}{8\beta \pi k^4 r^2(n)}.$$

Recall that $c(n) = \gamma_2 \ln(1/r(n)) = \Theta(\ln k)$. If we take $m = o(n/k^2)$ and γ_2 large enough, we have that

$$p^{2k} \geq \tau$$

for some constant $0 < \tau < 1$. It follows that the probability that all of the w_is fail to reach $V(Q, C)$ is at most

$$(1 - \tau q)^m \leq \left(1 - \frac{\tau}{8\beta \pi k^4 r^2(n)}\right)^m = \left(1 - \frac{\tau}{\sigma k^2}\right)^m, \tag{2}$$

for some positive constant σ.

By combining Equations 1 and 2, we get that the probability that u is not connected to $V(Q, C)$ is at most

$$\left(\frac{8m^2 e}{\alpha \pi n r^2(n)}\right)^{c(n)} + \left(1 - \frac{\tau}{\sigma k^2}\right)^m.$$

Now, since $r(n) > n^{-1/8}$, we have that $k = O(n^{1/8})$. If we choose $m = \Theta(n^{1/3})$ we have that $m = o(n/k^2)$, as required above, and $m = \omega(k^2 \ln n)$. This, combined with the choice of $c(n)$, ensures that the above probability is smaller than $1/n^2$. The lemma follows by applying the union bound over all nodes u. $\qquad\square$

For the case $r(n) > n^{-\epsilon}$, Theorem 1 follows by combining the results of the above two lemmas.

3 Achieving $c(n) = 3$ Using a Double Choice Protocol

In the previous section we showed that selecting $c(n) = \Theta(\ln(1/r(n)))$ visible neighbors at random is sufficient to enforce global connectivity for all ranges of $r(n)$ which guarantee connectivity of the visibility graph. Whether these many neighbors are necessary remains a challenging open question. As a step towards this objective, we show that, at least for large enough (yet nonconstant) radii, $c(n) = 3$ always suffices under a slightly different neighbor selection protocol where each node is required to direct the selection of some neighbors within a certain geographical region. More formally, consider again the tessellation of the unit square into k^2 square cells of side $1/k$, with $k = \lceil \sqrt{5}/r(n) \rceil$. Define $\mathrm{BT}(r(n), 2, 1)$ to be the undirected graph resulting by letting each node select two neighbors at random among the nodes residing in its cell, and another neighbor at random among all visible nodes.

Observe that if applied in a practical scenario, the above *double-choice* protocol would require each node to infer geographical information about its location and the location of the nodes in its visibility range. For example, this information could be provided by a GPS device.[1]

Theorem 2. *There exists a constant* ϵ, $0 < \epsilon < 1/2$ *such that if* $r(n) = \Omega(n^{-\epsilon})$, *then* $\mathrm{BT}(r(n), 2, 1)$ *is connected w.h.p.*

Proof. We employ the same approach used in Subsection 2.1. Specifically, we first argue that w.h.p. for all cells Q, the graph G_Q induced by the nodes in Q is connected, and that for every pair of adjacent cells there is an edge with endpoints in the two cells. Since by the first point of Proposition 1, each cell Q contains $\Omega(n^{1-2\epsilon})$ nodes w.h.p., the main result of [6] implies that two neighbors selected by each node in Q suffice to guarantee connectivity of G_Q with probability at least $1 - 1/n^{\delta(1-2\epsilon)}$, for a suitable positive constant $\delta < 1$. Then, choosing ϵ smaller than $\delta/(2(1+\delta))$ and applying the union bound, all cells will be internally connected with high probability. In order to prove connectivity between adjacent cells, we proceed as in the proof of Lemma 2. In particular, consider an arbitrary pair of adjacent cells Q_1 and Q_2, and let B_{Q_1, Q_2} denote the event that there is at least one edge in $\mathrm{BT}(r(n), 2, 1)$ between the two cells. By conditioning on the events described in Proposition 1, we have that

$$1 - \Pr(B_{Q_1, Q_2}) \leq \left(1 - \frac{\alpha n/k^2}{\beta \pi n r^2(n)} \right)^{2\alpha n/k^2}$$

$$\leq \exp\left(-\frac{\alpha n/k^2}{\beta \pi n r^2(n)} (2\alpha n/k^2) \right)$$

$$\leq \exp(-\zeta n^{1-2\epsilon}),$$

where ζ is a positive constant. The theorem follows by applying the union bound over all $O(n)$ pairs of adjacent cells. $\qquad\square$

4   Experiments

We have designed an extensive suite of experiments aimed at comparing the connectivity and other topological properties of the graphs analyzed in the previous sections.[2] In a

---

[1] A full discussion on the feasibility of this approach is outside the scope of this paper, since the analysis of the double-choice protocol is mostly meant to provide evidence that the selection of very few neighbors may suffice in order to build a connected topology.

[2] The implemented code makes use of the Boost Graph Libraries [11] for computing the number of connected components.

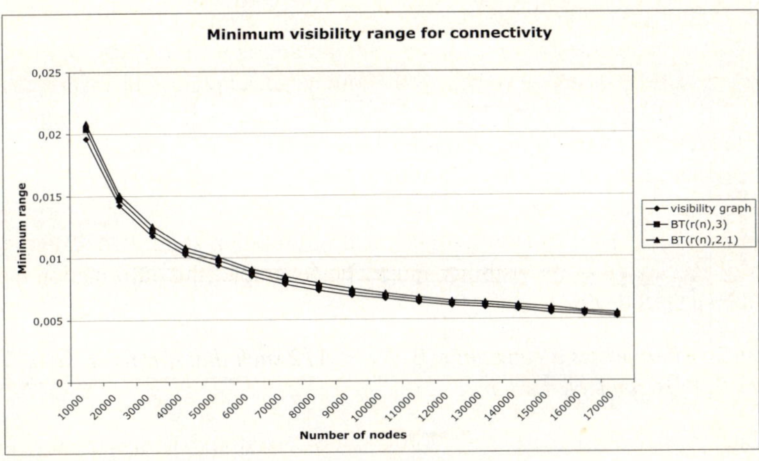

Fig. 1. Comparison of the minimum ranges r_{lb}, r_{sc}, and r_{dc} yielding connectivity in $\mathrm{BT}(r(n), 3)$, $\mathrm{BT}(r(n), 2, 1)$, and in the visibility graph with $r(n) = r_{\mathrm{lb}}$, respectively

first set of experiments, for values of n ranging from 10000 to 170000 with step 10000, we have generated 50 placements of n nodes in the unit square. For each placement, we have determined (through binary search) an approximation to the minimum range r_{lb} which guarantees connectivity of the visibility graph associated with the placement (i.e., the graph where each node connects to all its visible neighbors). Moreover, for the same placement we have determined the minimum range r_{sc} such that the graph $\mathrm{BT}(r_{\mathrm{sc}}, 3)$ turns out to be connected in all of 30 repetitions of the neighbor selection protocol. Finally, for the same placement we have determined the minimum radius r_{dc} such that the graph $\mathrm{BT}(r_{\mathrm{dc}}, 2, 1)$ turns out to be connected in all of 30 repetitions of the neighbor selection protocol. The results of these experiments are depicted in Figure 1 where for every $10000 \le n \le 170000$ the values of r_{lb}, r_{sc}, and r_{dc}, averaged over the 50 placements, are shown. According to the experiments, r_{sc} is very close to r_{lb} (within 5% for all values of n). Moreover, r_{dc} features a similar behavior with a slightly larger value than r_{sc}. Observe that, interestingly, connectivity of $\mathrm{BT}(r(n), 2, 1)$ does not seem to require that $r(n) \in \Omega(1/n^{\epsilon})$ as required by the analysis since it is attained for values of $r(n)$ close to r_{lb}.

In a second set of experiments we measured the maximum degree of the graphs $\mathrm{BT}(r(n), 3)$ and $\mathrm{BT}(r(n), 2, 1)$, and of the visibility graph with visibility range $r(n)$, where $r(n)$ is chosen to be an approximation of the smallest value which guarantees connectivity in all three cases. The results of these experiments are depicted in Figure 2 where, as before, for each value of n the reported values represent the averages over 50 placements. It can be seen that $\mathrm{BT}(r(n), 2, 1)$ exhibits a slightly smaller maximum degree than $\mathrm{BT}(r(n), 3)$, and, clearly, both graphs have a much smaller maximum degree than the visibility graph whose expected maximum degree can be shown to be $\Theta(\ln n)$, when $r(n) \in \Theta(\sqrt{(\ln n)/n})$ is used.

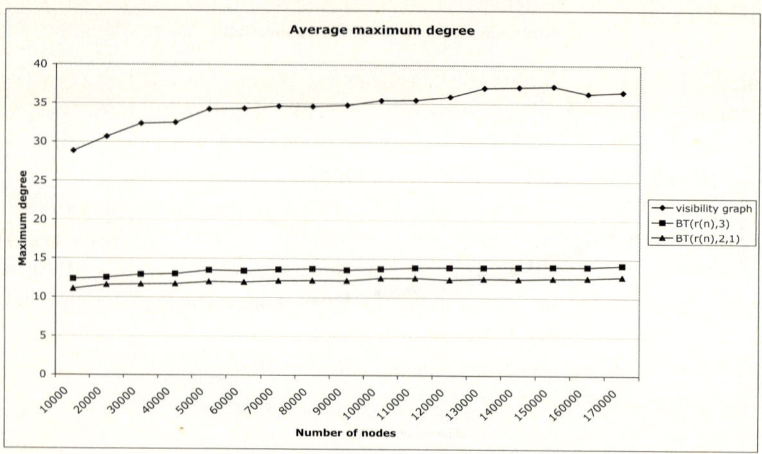

Fig. 2. Comparison of the maximum degree of $\mathrm{BT}(r(n), 3)$, $\mathrm{BT}(r(n), 2, 1)$, and of the visibility graph with range $r(n)$

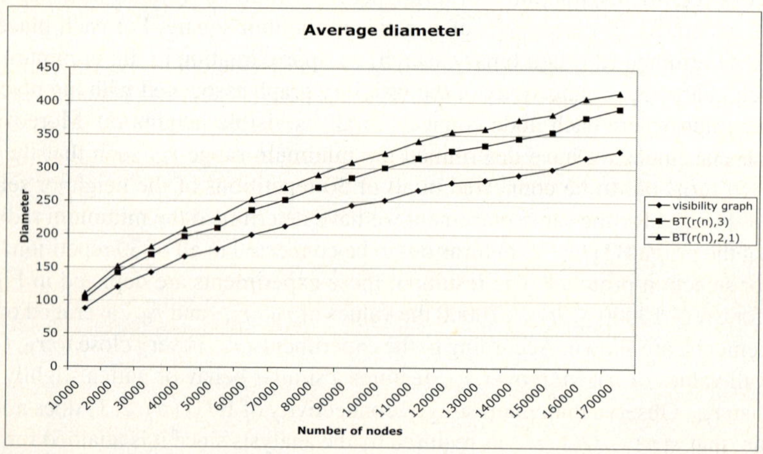

Fig. 3. Comparison of the average diameter of $\mathrm{BT}(r(n), 3)$, $\mathrm{BT}(r(n), 2, 1)$, and of the visibility graph with range $r(n)$

One last set of experiments concerned the estimation of the average diameter of $\mathrm{BT}(r(n), 3)$ and $\mathrm{BT}(r(n), 2, 1)$, and of the visibility graph with visibility range $r(n)$, where $r(n)$ is chosen to be an approximation of the minimum value which guarantees connectivity in all three cases. The results of these experiments are depicted in Figure 3, once again reporting for each n the averages over 50 placements. It can be seen that $\mathrm{BT}(r(n), 3)$ has a smaller diameter than $\mathrm{BT}(r(n), 2, 1)$, and that it has only a slightly larger diameter than the one of the visibility graph.

5   Conclusion

The main theoretical contribution of this paper is a proof of connectivity for the Bluetooth graph when the visibility range $r(n)$ is a vanishing function of the number n of nodes and each node selects only a logarithmic number of neighbors with respect to $1/r(n)$. Also, we introduced a novel neighbor selection protocol based on a double choice mechanism, which ensures connectivity when a total of only three neighbors are selected by each node. In the paper we also report the results of extensive experiments which validate the theoretical findings. In fact, the experiments suggest that the best avenue for future research is to tighten the analytical result on the connectivity yielded by the single choice protocol, while the double choice idea (which could be more complex to implement in practice) seems only needed for the analysis but does not outperform single choice in practice.

References

1. Whitaker, R., Hodge, L., Chlamtac, I.: Bluetooth scatternet formation: a survey. Ad Hoc Networks 3, 403–450 (2005)
2. Stojmenovic, I., Zaguia, N.: Bluetooth scatternet formation in ad hoc wireless networks. In: Misic, J., Misic, V. (eds.) Performance Modeling and Analysis of Bluetooth Networks, pp. 147–171. Auerbach Publications (2006)
3. Basagni, S., Bruno, R., Mambrini, G., Petrioli, C.: Comparative performance evaluation of scatternet formation protocols for networks of Bluetooth devices. Wireless Networks 10(2), 197–213 (2004)
4. Kettimuthu, R., Muthukrishnan, S.: Is Bluetooth suitable for large-scale sensor networks? In: Proc. of the 2005 Intl. Conf. on Wireless Networks, pp. 448–454 (2005)
5. Ferraguto, F., Mambrini, G., Panconesi, A., Petrioli, C.: A new approach to device discovery and scatternet formation in Bluetooth networks. In: Proc. of the 18th International Parallel and Distributed Processing Symposium, pp. 221–228 (2004)
6. Dubhashi, D., Johansson, C., Häggstrom, O., Panconesi, A., Sozio, M.: Irrigating ad hoc networks in constant time. In: Proc. of the 17th ACM Symp. on Parallel Algorithms and Architectures, pp. 106–115 (July 2005)
7. Panconesi, A., Radhakrishnan, J.: Expansion properties of (secure) wireless networks. In: Proc. of the 16th ACM Symp. on Parallel Algorithms and Architectures, pp. 281–285. ACM Press, New York (2004)
8. Ellis, R., Jia, X., Yan, C.: On random points in the unit disk. Random Structures and Algorithms 29(1), 14–25 (2005)
9. Akyildiz, I., Su, W., Sankarasubramaniam, Y., Cayirci, E.: Wireless sensor networks: a survey. Computer Networks 38, 393–422 (2002)
10. Hagerup, T., Rüb, C.: A guided tour of Chernoff bounds. Information Processing Letters 33(6), 305–308 (1990)
11. Siek, J.G., Lee, L., Lumsdaine, A.: Boost Graph Library, The User Guide and Reference Manual. Addison Wesley Professional, Reading MA (December 2001)

Printing: Mercedes-Druck, Berlin
Binding: Stein+Lehmann, Berlin

Lecture Notes in Computer Science

For information about Vols. 1–4572

please contact your bookseller or Springer

Vol. 4708: L. Kučera, A. Kučera (Eds.), Mathematical Foundations of Computer Science 2007. XVIII, 764 pages. 2007.

Vol. 4707: O. Gervasi, M.L. Gavrilova (Eds.), Computational Science and Its Applications – ICCSA 2007, Part III. XXIV, 1205 pages. 2007.

Vol. 4706: O. Gervasi, M.L. Gavrilova (Eds.), Computational Science and Its Applications – ICCSA 2007, Part II. XXIII, 1129 pages. 2007.

Vol. 4705: O. Gervasi, M.L. Gavrilova (Eds.), Computational Science and Its Applications – ICCSA 2007, Part I. XLIV, 1169 pages. 2007.

Vol. 4703: L. Caires, V.T. Vasconcelos (Eds.), CONCUR 2007 – Concurrency Theory. XIII, 507 pages. 2007.

Vol. 4697: L. Choi, Y. Paek, S. Cho (Eds.), Advances in Computer Systems Architecture. XIII, 400 pages. 2007.

Vol. 4685: D.J. Veit, J. Altmann (Eds.), Grid Economics and Business Models. XII, 201 pages. 2007.

Vol. 4683: L. Kang, Y. Liu, S. Zeng (Eds.), Intelligence Computation and Applications. XVII, 663 pages. 2007.

Vol. 4682: D.-S. Huang, L. Heutte, M. Loog (Eds.), Advanced Intelligent Computing Theories and Applications. XXVII, 1373 pages. 2007. (Sublibrary LNAI).

Vol. 4681: D.-S. Huang, L. Heutte, M. Loog (Eds.), Advanced Intelligent Computing Theories and Applications. XXVI, 1379 pages. 2007.

Vol. 4679: A.L. Yuille, S.-C. Zhu, D. Cremers, Y. Wang (Eds.), Energy Minimization Methods in Computer Vision and Pattern Recognition. XII, 494 pages. 2007.

Vol. 4678: J. Blanc-Talon, W. Philips, D. Popescu, P. Scheunders (Eds.), Advanced Concepts for Intelligent Vision Systems. XXIII, 1100 pages. 2007.

Vol. 4673: W.G. Kropatsch, M. Kampel, A. Hanbury (Eds.), Computer Analysis of Images and Patterns. XX, 1006 pages. 2007.

Vol. 4671: V. Malyshkin (Ed.), Parallel Computing Technologies. XIV, 635 pages. 2007.

Vol. 4660: S. Džeroski, J. Todorovski (Eds.), Computational Discovery of Scientific Knowledge. X, 327 pages. 2007. (Sublibrary LNAI).

Vol. 4657: C. Lambrinoudakis, G. Pernul, A.M. Tjoa (Eds.), Trust and Privacy in Digital Business. XIII, 291 pages. 2007.

Vol. 4651: F. Azevedo, P. Barahona, F. Fages, F. Rossi (Eds.), Recent Advances in Constraints. VIII, 185 pages. 2007. (Sublibrary LNAI).

Vol. 4649: V. Diekert, M.V. Volkov, A. Voronkov (Eds.), Computer Science – Theory and Applications. XIII, 420 pages. 2007.

Vol. 4647: R. Martin, M. Sabin, J. Winkler (Eds.), Mathematics of Surfaces XII. IX, 509 pages. 2007.

Vol. 4645: R. Giancarlo, S. Hannenhalli (Eds.), Algorithms in Bioinformatics. XIII, 432 pages. 2007. (Sublibrary LNBI).

Vol. 4644: N. Azemard, L. Svensson (Eds.), Integrated Circuit and System Design. XIV, 583 pages. 2007.

Vol. 4643: M.-F. Sagot, M.E.M.T. Walter (Eds.), Advances in Bioinformatics and Computational Biology. XII, 177 pages. 2007. (Sublibrary LNBI).

Vol. 4642: S.-W. Lee, S.Z. Li (Eds.), Advances in Biometrics. XX, 1216 pages. 2007.

Vol. 4641: A.-M. Kermarrec, L. Bougé, T. Priol (Eds.), Euro-Par 2007 Parallel Processing. XXVII, 974 pages. 2007.

Vol. 4639: E. Csuhaj-Varj'u, Z. Ésik (Eds.), Fundamentals of Computation Theory. XIV, 508 pages. 2007.

Vol. 4638: T. Stützle, M. Birattari, H.H. Hoos (Eds.), Engineering Stochastic Local Search Algorithms. X, 223 pages. 2007.

Vol. 4637: C. Kruegel, R. Lippmann, A. Clark (Eds.), Recent Advances in Intrusion Detection. XII, 337 pages. 2007.

Vol. 4635: B. Kokinov, D.C. Richardson, T.R. Roth-Berghofer, L. Vieu (Eds.), Modeling and Using Context. XIV, 574 pages. 2007. (Sublibrary LNAI).

Vol. 4634: H.R. Nielson, G. Filé (Eds.), Static Analysis. XI, 469 pages. 2007.

Vol. 4633: M. Kamel, A. Campilho (Eds.), Image Analysis and Recognition. XII, 1312 pages. 2007.

Vol. 4632: R. Alhajj, H. Gao, X. Li, J. Li, O.R. Zaïane (Eds.), Advanced Data Mining and Applications. XV, 634 pages. 2007. (Sublibrary LNAI).

Vol. 4628: L.N. de Castro, F.J. Von Zuben, H. Knidel (Eds.), Artificial Immune Systems. XII, 438 pages. 2007.

Vol. 4627: M. Charikar, K. Jansen, O. Reingold, J.D.P. Rolim (Eds.), Approximation, Randomization, and Combinatorial Optimization. XII, 626 pages. 2007.

Vol. 4626: R.O. Weber, M.M. Richter (Eds.), Case-Based Reasoning Research and Development. XIII, 534 pages. 2007. (Sublibrary LNAI).

Vol. 4624: T. Mossakowski, U. Montanari, M. Haveraaen (Eds.), Algebra and Coalgebra in Computer Science. XI, 463 pages. 2007.

Vol. 4622: A. Menezes (Ed.), Advances in Cryptology - CRYPTO 2007. XIV, 631 pages. 2007.

Vol. 4619: F. Dehne, J.-R. Sack, N. Zeh (Eds.), Algorithms and Data Structures. XVI, 662 pages. 2007.

Vol. 4618: S.G. Akl, C.S. Calude, M.J. Dinneen, G. Rozenberg, H.T. Wareham (Eds.), Unconventional Computation. X, 243 pages. 2007.

Vol. 4617: V. Torra, Y. Narukawa, Y. Yoshida (Eds.), Modeling Decisions for Artificial Intelligence. XII, 502 pages. 2007. (Sublibrary LNAI).

Vol. 4616: A. Dress, Y. Xu, B. Zhu (Eds.), Combinatorial Optimization and Applications. XI, 390 pages. 2007.

Vol. 4615: R. de Lemos, C. Gacek, A. Romanovsky (Eds.), Architecting Dependable Systems IV. XIV, 435 pages. 2007.

Vol. 4613: F.P. Preparata, Q. Fang (Eds.), Frontiers in Algorithmics. XI, 348 pages. 2007.

Vol. 4612: I. Miguel, W. Ruml (Eds.), Abstraction, Reformulation, and Approximation. XI, 418 pages. 2007. (Sublibrary LNAI).

Vol. 4611: J. Indulska, J. Ma, L.T. Yang, T. Ungerer, J. Cao (Eds.), Ubiquitous Intelligence and Computing. XXIII, 1257 pages. 2007.

Vol. 4610: B. Xiao, L.T. Yang, J. Ma, C. Muller-Schloer, Y. Hua (Eds.), Autonomic and Trusted Computing. XVIII, 571 pages. 2007.

Vol. 4609: E. Ernst (Ed.), ECOOP 2007 – Object-Oriented Programming. XIII, 625 pages. 2007.

Vol. 4608: H.W. Schmidt, I. Crnkovic, G.T. Heineman, J.A. Stafford (Eds.), Component-Based Software Engineering. XII, 283 pages. 2007.

Vol. 4607: L. Baresi, P. Fraternali, G.-J. Houben (Eds.), Web Engineering. XVI, 576 pages. 2007.

Vol. 4606: A. Pras, M. van Sinderen (Eds.), Dependable and Adaptable Networks and Services. XIV, 149 pages. 2007.

Vol. 4605: D. Papadias, D. Zhang, G. Kollios (Eds.), Advances in Spatial and Temporal Databases. X, 479 pages. 2007.

Vol. 4604: U. Priss, S. Polovina, R. Hill (Eds.), Conceptual Structures: Knowledge Architectures for Smart Applications. XII, 514 pages. 2007. (Sublibrary LNAI).

Vol. 4603: F. Pfenning (Ed.), Automated Deduction – CADE-21. XII, 522 pages. 2007. (Sublibrary LNAI).

Vol. 4602: S. Barker, G.-J. Ahn (Eds.), Data and Applications Security XXI. X, 291 pages. 2007.

Vol. 4600: H. Comon-Lundh, C. Kirchner, H. Kirchner (Eds.), Rewriting, Computation and Proof. XVI, 273 pages. 2007.

Vol. 4599: S. Vassiliadis, M. Berekovic, T.D. Hämäläinen (Eds.), Embedded Computer Systems: Architectures, Modeling, and Simulation. XVIII, 466 pages. 2007.

Vol. 4598: G. Lin (Ed.), Computing and Combinatorics. XII, 570 pages. 2007.

Vol. 4597: P. Perner (Ed.), Advances in Data Mining. XI, 353 pages. 2007. (Sublibrary LNAI).

Vol. 4596: L. Arge, C. Cachin, T. Jurdziński, A. Tarlecki (Eds.), Automata, Languages and Programming. XVII, 953 pages. 2007.

Vol. 4595: D. Bošnački, S. Edelkamp (Eds.), Model Checking Software. X, 285 pages. 2007.

Vol. 4594: R. Bellazzi, A. Abu-Hanna, J. Hunter (Eds.), Artificial Intelligence in Medicine. XVI, 509 pages. 2007. (Sublibrary LNAI).

Vol. 4593: A. Biryukov (Ed.), Fast Software Encryption. XI, 467 pages. 2007.

Vol. 4592: Z. Kedad, N. Lammari, E. Métais, F. Meziane, Y. Rezgui (Eds.), Natural Language Processing and Information Systems. XIV, 442 pages. 2007.

Vol. 4591: J. Davies, J. Gibbons (Eds.), Integrated Formal Methods. IX, 660 pages. 2007.

Vol. 4590: W. Damm, H. Hermanns (Eds.), Computer Aided Verification. XV, 562 pages. 2007.

Vol. 4589: J. Münch, P. Abrahamsson (Eds.), Product-Focused Software Process Improvement. XII, 414 pages. 2007.

Vol. 4588: T. Harju, J. Karhumäki, A. Lepistö (Eds.), Developments in Language Theory. XI, 423 pages. 2007.

Vol. 4587: R. Cooper, J. Kennedy (Eds.), Data Management. XIII, 259 pages. 2007.

Vol. 4586: J. Pieprzyk, H. Ghodosi, E. Dawson (Eds.), Information Security and Privacy. XIV, 476 pages. 2007.

Vol. 4585: M. Kryszkiewicz, J.F. Peters, H. Rybinski, A. Skowron (Eds.), Rough Sets and Intelligent Systems Paradigms. XIX, 836 pages. 2007. (Sublibrary LNAI).

Vol. 4584: N. Karssemeijer, B. Lelieveldt (Eds.), Information Processing in Medical Imaging. XX, 777 pages. 2007.

Vol. 4583: S.R. Della Rocca (Ed.), Typed Lambda Calculi and Applications. X, 397 pages. 2007.

Vol. 4582: J. Lopez, P. Samarati, J.L. Ferrer (Eds.), Public Key Infrastructure. XI, 375 pages. 2007.

Vol. 4581: A. Petrenko, M. Veanes, J. Tretmans, W. Grieskamp (Eds.), Testing of Software and Communicating Systems. XII, 379 pages. 2007.

Vol. 4580: B. Ma, K. Zhang (Eds.), Combinatorial Pattern Matching. XII, 366 pages. 2007.

Vol. 4579: B. M. Hämmerli, R. Sommer (Eds.), Detection of Intrusions and Malware, and Vulnerability Assessment. X, 251 pages. 2007.

Vol. 4578: F. Masulli, S. Mitra, G. Pasi (Eds.), Applications of Fuzzy Sets Theory. XVIII, 693 pages. 2007. (Sublibrary LNAI).

Vol. 4577: N. Sebe, Y. Liu, Y.-t. Zhuang, T.S. Huang (Eds.), Multimedia Content Analysis and Mining. XIII, 513 pages. 2007.

Vol. 4576: D. Leivant, R. de Queiroz (Eds.), Logic, Language, Information and Computation. X, 363 pages. 2007.

Vol. 4575: T. Takagi, T. Okamoto, E. Okamoto, T. Okamoto (Eds.), Pairing-Based Cryptography – Pairing 2007. XI, 408 pages. 2007.

Vol. 4574: J. Derrick, J. Vain (Eds.), Formal Techniques for Networked and Distributed Systems – FORTE 2007. XI, 375 pages. 2007.

Vol. 4573: M. Kauers, M. Kerber, R. Miner, W. Windsteiger (Eds.), Towards Mechanized Mathematical Assistants. XIII, 407 pages. 2007. (Sublibrary LNAI).